Mahale Chimpanzees

50 Years of Research

Long-term ecological research studies are rare and invaluable resources, particularly when they are as thoroughly documented as the Mahale Mountains Chimpanzee Research Project in Tanzania. Directed by the late Toshisada Nishida from 1965 until 2011, the project continues to yield new and fascinating findings about our closest neighbor species.

In a fitting tribute to Nishida's contribution to science, this book brings together 50 years of research into one encyclopedic volume. Alongside previously unpublished data, the editors include new translations of Japanese writings throughout the book to bring previously inaccessible work to non-Japanese speakers. The history and ecology of the site, chimpanzee behavior and biology, and ecological management are all addressed through first-hand accounts by Mahale researchers. The authors highlight long-term changes in behavior, where possible, and draw comparisons with other chimpanzee sites across Africa to provide an integrative view of chimpanzee research today.

This is a major contribution to great ape research, complementing Nishida's last work *Chimpanzees of the Lakeshore* (Cambridge University Press, 2012).

Michio Nakamura is Associate Professor at the Wildlife Research Center, Kyoto University, Japan. He has studied the Mahale chimpanzees since 1994 and is a recipient of the Primate Society of Japan's Takashima Prize.

Kazuhiko Hosaka is Associate Professor at Kamakura Women's University, Japan. His research focuses on the social interactions, hunting, and meat-eating behavior of chimpanzees in relation to human evolution.

Noriko Itoh is a researcher in the Wildlife Research Center, Kyoto University, Japan. She has studied the Mahale chimpanzees since 1995, focusing on grouping behaviors and related environments.

Koichiro Zamma is a researcher in the Graduate School of Asian and African Area Studies and in the Wildlife Research Center, both at Kyoto University, Japan. His research focuses on the impact of ectoparasite infections on grooming and on the sleep-related behaviors of chimpanzees.

Mahale Chimpanzees

50 Years of Research

EDITED BY

MICHIO NAKAMURA
Wildlife Research Center, Kyoto University, Japan

KAZUHIKO HOSAKA
Kamakura Women's University, Japan

NORIKO ITOH
Wildlife Research Center, Kyoto University, Japan

KOICHIRO ZAMMA
Graduate School of Asian and African Area Studies and Wildlife Research Center, Kyoto University, Japan

CAMBRIDGE
UNIVERSITY PRESS

CAMBRIDGE
UNIVERSITY PRESS

University Printing House, Cambridge CB2 8BS, United Kingdom

Cambridge University Press is part of the University of Cambridge.

It furthers the University's mission by disseminating knowledge in the pursuit of education, learning and research at the highest international levels of excellence.

www.cambridge.org
Information on this title: www.cambridge.org/9781107052314

First published 2015

Printed in the United Kingdom by TJ International Ltd. Padstow Cornwall

A catalog record for this publication is available from the British Library

Library of Congress Cataloging in Publication data
Mahale chimpanzees : 50 years of research / edited by Michio Nakamura, Wildlife Research Center, Kyoto University, Japan, Kazuhiko Hosaka, Kamakura Women's University, Japan, Noriko Itoh, Wildlife Research Center, Kyoto University, Japan, Koichiro Zamma, Graduate School of Asian and African Area Studies and Wildlife Research Center, Kyoto University,
pages cm
ISBN 978-1-107-05231-4
1. Chimpanzees – Tanzania – Mahale Mountains National Park. 2. Chimpanzees – Research – Tanzania – Mahale Mountains National Park. I. Nakamura, Michio, 1971–
QL737.P94M34 2015
599.88509678–dc23

2015004541

ISBN 978-1-107-05231-4 Hardback

Additional resources for this publication at www.cambridge.org/9781107052314

CONTENTS

CONTRIBUTORS

SHIHO FUJITA
Joint Faculty of Veterinary Medicine, Kagoshima University, 1-21-24, Korimoto, Kagoshima, 890-0065, Japan

SHUNKICHI HANAMURA
Wildlife Research Center, Kyoto University, 2-24, Tanaka-Sekiden-cho, Sakyo-ku, Kyoto, 606-8203, Japan

HIDEO HASEGAWA
Department of Biology, Faculty of Medicine, Oita University, 1-1, Idaigaoka, Hasama, Oita, 879-5593, Japan

TAKASHI HAYAKAWA
Molecular Biology Section, Department of Cellular and Molecular Biology, Primate Research Institute, Kyoto University, 41-2, Kanrin, Inuyama, Aichi, 484-8506, Japan

HITOSHIGE HAYAKI
Faculty of Humanities and Sciences, Kobe Gakuin University, 518, Arise, Ikawadani-cho, Nishi-ku, Kobe, 651-2113, Japan

KAZUHIKO HOSAKA
Kamakura Women's University, 6-1-3, Ofuna, Kamakura, 247-8512, Japan

MICHAEL A. HUFFMAN
Section of Social Systems Evolution, Primate Research Institute, Kyoto University, 41-2, Kanrin, Inuyama, Aichi, 484-8506, Japan

HIROSHI IHOBE
School of Human Sciences, Sugiyama Jogakuen University, 3-2005, Takenoyama, Nisshin, Aichi, 470-0136, Japan

EIJI INOUE
Graduate School of Science, Kyoto University, Kitashirakawa-Oiwake-cho, Sakyo-ku, Kyoto, 606-8502, Japan

NORIKO ITOH
Wildlife Research Center, Kyoto University, 2-24, Tanaka-Sekiden-cho, Sakyo-ku, Kyoto, 606-8203, Japan

MIEKO KIYONO
Kobe University, 1-1, Rokkodai, Nada-ku, Kobe, 657-8501, Japan

TAKANORI KOORIYAMA
Department of Veterinary Science, School of Veterinary Medicine, Rakuno Gakuen University, 582, Bunkyodai-Midori, Ebetsu, Hokkaido, 069-8501, Japan

NOBUYUKI KUTSUKAKE
Department of Evolutionary Studies of Biosystems, The Graduate University for Advanced Studies, Hayama, Kanagawa, 240-0193, Japan

LINDA F. MARCHANT
Department of Anthropology, Miami University, 124 B Upham Hall, Oxford, OH 45056, USA

TAKUYA MATSUMOTO
Graduate School of Science, Kyoto University, Kitashirakawa-Oiwake-cho, Sakyo-ku, Kyoto, 606-8502, Japan

TAKAHISA MATSUSAKA
Kyoto Seizan College, 26, Saijyo, Ao, Nagaokakyo, Kyoto, 617-0811, Japan

DAISUKE MURAMATSU
Wildlife Research Center, Kyoto University, 2-24, Tanaka-Sekiden-cho, Sakyo-ku, Kyoto, 606-8203, Japan

MICHIO NAKAMURA
Wildlife Research Center, Kyoto University, 2-24, Tanaka-Sekiden-cho, Sakyo-ku, Kyoto, 606-8203, Japan

HITONARU NISHIE
Wildlife Research Center, Kyoto University, 2-24, Tanaka-Sekiden-cho, Sakyo-ku, Kyoto, 606-8203, Japan

TETSUYA SAKAMAKI
Primate Research Institute, Kyoto University, 41-2, Kanrin, Inuyama, Aichi, 484-8506, Japan

MASAKI SHIMADA
Teikyo University of Science, 2525, Yatsusawa, Uenohara, Yamanashi, 409-0193, Japan

DAISUKE SHIMIZU
Laboratory of Physical Anthropology, Department of Zoology, Division of Biological Sciences, Graduate School of Science, Kyoto University, Kitashirakawa-Oiwake-cho, Sakyo-ku, Kyoto, 606-8502, Japan

YUKIO TAKAHATA
Kwansei Gakuin University, 2-1, Gakuen, Sanda, 669-1337, Japan

KOICHIRO ZAMMA
Graduate School of Asian and African Area Studies, Kyoto University, 46, Yoshida-Shimoadachi-cho, Sakyo-ku, Kyoto, 606-8501, Japan; and Wildlife Research Center, Kyoto University, 2-24 Tanaka-Sekiden-cho, Sakyo-ku, Kyoto, 606-8203, Japan

FOREWORD

Frans B. M. de Waal
Living Links, Yerkes National Primate Research Center, Emory University, Atlanta, USA

The Mahale Mountains National Park on the shore of Lake Tanganyika in Tanzania offers a habitat for up to 1000 chimpanzees. This site was among the first chosen for research on the social lives and survival strategies of one of our closest relatives. Study of their behavior began in 1965, when a Japanese scientist, the late Toshisada Nishida of Kyoto University, set up camp there, habituated the apes to his presence, and started following them around on a daily basis.

In contrast to the human anthropologist, who simply asks his subjects about their kinship relations, the primatologist needs many years to collect such critical information. With a species that breeds as slowly and is as long-lived as the chimpanzee, one needs to follow individuals for a long time to know whether or not two adult males are brothers or how many offspring a female rears during her lifetime. Before scientists learned to analyse DNA extracted from fecal or hair samples, the only way to know about genetic relatedness was a long-term project such as the one Nishida set up.

Fifty years later, Nishida is not with us anymore, but his work continues unabated, carried out by former students and young colleagues on a scale that arguably exceeds that of any other chimpanzee field site in Africa. I once spent some time at Mahale, and was impressed by the dedication and sweaty hard fieldwork that is required. In order to follow the chimpanzees around, which is done every day, they obviously first need to be found. Fortunately, chimps are an exceptionally noisy species, often calling from one end of the forest to the other, which helps in locating them. One of Nishida's many discoveries was that these apes form so-called *unit-groups*, whose members rarely travel all together, but disperse over the forest in smaller groups, or "parties," which meet intermittently. Once found, a party is followed through a dense and hilly jungle, full of sections that humans find hard to negotiate. The best way to ensure continued observation is to follow the chimps until they build their beds, or night nests, which they do high up in the trees, and to arrive early next morning at the same tree before they wake up. This method obviously implies that fieldworkers get little rest, and need to travel through dark forest at both dawn and dusk.

I gained deep respect for their endurance, and look at this volume through the same eyes, thinking of the effort that went into collecting all of the precious information it contains. The volume offers an overview of many areas, from the ecology of the park and the food habits of the apes to their health, social organization, occasional violence, endocrinology, internal parasites, and the specific cultural patterns observed at Mahale. Primate culture has become a huge topic focused on habits and behavior patterns that are not genetically, but socially transmitted, producing impressive behavioral diversity in chimpanzee populations across Africa. The Mahale project has been at the forefront of these discoveries, largely because of the very detailed, often videotaped, description of all sorts of behavior patterns. One of this population's many characteristic patterns is the *social scratch* (i.e. one individual deliberately scratching the back of another), which is virtually unknown in other populations in the world.

One significant advance in the study of chimpanzee habits is the discovery that wild chimpanzees consume *Aspilia* leaves, which lack known nutritional value, and are in fact not digested. The apes consume them slowly, mostly in the morning, swallowing the leaves whole, without chewing. The leaves may help them expel parasites. The new field of zoopharmacognosy (i.e. self-medication by animals ingesting plants, insects, or soils) that resulted from this discovery receives attention in this volume, as do many other areas of learned and transmitted behavior at Mahale.

Without making an attempt to review all of the various approaches and topics that are covered, I was struck by the attention to a variety of "political" behaviors, from grooming to coalitions, and to the sharp difference chimpanzees make between in-group and out-group. They are hostile and violent against their neighbors yet show conflict management within their own unit-group. Many developments within and between unit-groups have been followed for years on end, providing a detailed record of shifts in the hierarchy, migrations, and demographic changes. In this context, it is informative to return to notes on historical events, such as one chapter that treats the final days of the last male of K group, in 1982, and the eventual extinction of this unit-group.

The present volume summarizes research projects in first-hand accounts by scientists who have actually worked at Mahale. Some of the material has previously been published, but not always in English. It was never brought together in the coherent whole that we now hold in our hands, however. It offers a treasure trove of information, which is especially useful given our current interest in detailed behavioral comparisons between chimpanzee populations. How wild chimpanzees differ from location to location, either due to genetic factors (e.g. subspecies), ecological differences, or special cultural traditions, helps build an ethnographic record of chimpanzee societies, something thus far not tried for any species other than our own.

Unfortunately, chimpanzees are, like all the apes, under grave existential threat. In the coming 50 years, the species may disappear from all but a few selected sanctuaries and parks. Mahale became Tanzania's 11th national park in 1985, thanks to an initiative by Toshisada Nishida and his teacher, Jun'ichiro Itani, which was followed by the establishment of the Mahale Wildlife Conservation Society in 1994. Not only has Mahale been at the forefront of the accumulation of scientific information on the species, it also remains central for its conservation, setting an inspiring example for other field sites. The combination found at Mahale between detailed record keeping of behavior and ecology, controlled ecotourism, and conservation efforts provides perhaps the best safeguard against extinction of a species that, as no other, sheds light on our own evolutionary background.

PREFACE

Michio Nakamura, Kazuhiko Hosaka, Noriko Itoh, and Koichiro Zamma

Chimpanzee research in the Mahale Mountains, Tanzania, started in 1965 and reaches its 50th year in 2015. Over these 50 years, more than 70 researchers and students have participated in field studies at Mahale. Each of them has focused on particular research topics: from the society, behavior, and ecology of chimpanzees, to the natural environment surrounding them. Such studies have been published in more than 300 academic papers in English and more than 400 articles and books in Japanese. This book aims to commemorate the 50th year of research at Mahale and summarizes the various research outcomes and knowledge gained during this half century. For Western readers, the long-term chimpanzee study at Gombe, started by Jane Goodall, may be more familiar and famous. However, another long-term research project started by the late Toshisada Nishida is equally important and has contributed much in understanding chimpanzees. The Mahale project has been continued seamlessly by Nishida, his colleagues, and students, targeting the wide scope of chimpanzee sociality, behavior, and ecology.

Nishida, as a founder and principal researcher, authored a monograph in 2012 on Mahale's chimpanzees, titled *Chimpanzees of the Lakeshore*, published by Cambridge University Press. Much of the book was his autobiography describing various episodes of his research life in Tanzania. Although the book included several important academic topics to which he dedicated his research career, it was not intended to comprehensively cover the research topics investigated by other researchers at Mahale. Nishida wanted to edit an encyclopedic volume about Mahale, which would complement his personal account. However, his untimely death prevented this. We pursued his goal, making use of the apt opportunity of the 50th year of research at Mahale.

Although this book takes the form of an edited volume, we asked authors to summarize the research outcomes from Mahale for their chapter topics. Thus, the reader will easily learn the key information published from Mahale. We focus on the accumulation of the research effort of 50 years. Thus, we highlight long-term changes in behavior wherever possible. For a wider perspective, we include comparative findings from other chimpanzee research sites across Africa. Finally, we seek to introduce and summarize the publications that were written only in Japanese in English. Most researchers who have worked at Mahale are Japanese and have published many reports about their findings, some of which are important to the scientific community but inaccessible to non-Japanese-speaking researchers.

Although we aimed for this book to be as comprehensive as possible, there must be some topics that remain to be understood. Compared with thousands of years of efforts to understand our own species, humans, from ancient ages, 50 years are insufficient to understand our evolutionary neighbors. We hope that this book will inspire the younger generations to participate in the studies and conservation of this fascinating species and that the research will continue for another 50 years.

ACKNOWLEDGMENTS

First of all, we would like to express our gratitude to William C. McGrew whose encouragement and advice certainly helped us realize our ambitions to publish this book. We also show deepest gratitude to Mahale predecessors and colleagues, Jun'ichiro Itani, Toshisada Nishida, Kenji Kawanaka, Shigeo Uehara, Kohshi Norikoshi, and those listed in Appendix V, who have contributed by the accumulation of common data and maintenance of the research camp. Some of their families also stayed at Mahale and helped the camp management: Haruko Nishida, Hatsuko Kawanaka, Moyo Uehara, Hideko Kakeya, Hitomi Takahata, Kazumi Takasaki, and Kazue Ihobe. We would also like to thank Japanese TV crews, Miho Nakamura, Tamotsu Asou, Masayasu Mori, and Mitsue Matsuya for their various support and collaboration in the field.

The long-term research would not have been possible without help from research assistants at Mahale, Omali Kabule, Mkoli Saidi, Ramadhani Nyundo, Mohamedi Seifu, Sadi Katensi, Jumanne Katensi, Juma Kahaso, Issa Ally Kapama, Haruna Sobongo, Alimasi Kasulamenba, Kabukula Kasulamenba, Athumani Katumba, Haruna Huseni, Ramadhani Kabilambe, Ramadhani Kasakampe, Matumula Nyundo, Rashidi S. Kitopeni, Moshi Bunengwa, Hamishi Bunengwa, Moshi Hamisi (Samola), Rashidi Hawazi (Kijanga), Saidi Musa, Yasini Kiyoya, Hassani Kajugi, Mtunda Mwami, Hamisi Katinkila, Moshi Matumula, Kabumbe Athumani, Bunde Athumani, Luhembe Ramadhani, Abdala Ramadani, Juma Hassani, Makelele Masayuke, Shabani Kabangula, Mwami Rashidi, Bunengwa Hamisi, Deosdery Shija, Hamisi Ramadhani, Shaka Kabugonga, Luhunga Ramadhani, Ramadhani Hamisi, Jafari Zuberi, Maiko Salumu, Moshi Haruna, Baraka Rehani, and Hemedi Rashidi. All have helped us tremendously at different times during the 50 years.

We also owe our gratitude to many predecessors, colleagues, and counterparts who worked or are working at various places in Tanzania, such as Kinji Imanishi, Akira Suzuki, Shigeru Azuma, Kosei Izawa, Takayoshi Kano, Akisato Nishimura, Gen'ichi Idani, Hideshi Ogawa, Midori Yoshikawa, Toshimichi Nemoto, Asami Kanayama, Juichi Itani, Hosea Kayumbo, George Sabuni, Edeus Massawe, Julius Keyyu, Edward Kohi, Pius Kavana, Steven Temu, Tadei Clamsen, Andrei K. Seki, Frederick Malisa, Hassani Mbaga, Crispin Mwinuka, Herman Batiho, Butati Nyundo, Kozo Tomita, Eiko Kimura, and Masaki Nishida.

We thank the Tanzania Commission for Science and Technology, the Tanzania Wildlife Research Institute (formerly Serengeti Wildlife Research Institute), the Tanzania National Parks, the Mahale–Gombe Wildlife Research Centre (formerly Mahale Mountains Wildlife Research Centre), and the Mahale Mountains National Park for their permission to conduct research at Mahale and for logistic support.

Many chapter authors graduated from the Graduate School of Science, Kyoto University. Therefore, we would like to thank its current and former members: Juichi Yamagiwa, Naofumi Nakagawa, Shigeru Suzuki, Chisa Tokimatsu, Ayako Tokunaga, Hidemi Ishida, Kazumichi Katayama, Suehisa Kuroda, Masato Nakatsukasa, Naomichi Ogihara, Yuko Eguchi, Kaoru Imamura, Masato Sawada, Naobi Okayasu, Kaori Kawai, Tomoaki Nishihara, Toru Soga, Daisuke Takekawa, Koichi Kitanishi, Kaoru Adachi, Hiroyuki Takahashi, Yuji Takenoshita,

Nobutaka Kamei, Goro Hanya, Shuhei Hayaishi, Takanori Ohishi, Naoki Matsuura, Mari Nishikawa, Daiji Kimura, Hideaki Terashima, Koji Kitamura, and Kazuyoshi Sugawara. Some of the editors and authors are also indebted to members of the Japan Monkey Centre: Mitsuo Iwamoto, Akira Kato, Naoto Kimura, Yukihisa Mito, Reiko Mizuno, Harumoto Gunji, and Tomo Takano.

We acknowledge *Pan* researchers, such as Yukimaru Sugiyama, Tetsuro Matsuzawa, Takeshi Furuichi, Chie Hashimoto, Gen Yamakoshi, Yasuko Tashiro, Gaku Ohashi, Richard Wrangham, and Vernon Reynolds, for their discussion and/or acceptance to their field sites for comparative research.

We, the editors, would also like to thank colleagues Shiro Kohshima, Miho Murayama, Hideki Sugiura, Masayuki Tanaka, Satoshi Hirata, Naruki Morimura, Michiko Fujisawa, Tadamichi Morisaka, Aya Saito, and Takushi Kishida at the Wildlife Research Center, Kyoto University, and Kazuteru Fukui at the Kamakura Women's University.

The long-term research and conservation activities at Mahale have been financially supported by various funding organizations (see Appendix VII for the list of research grants). We particularly want to thank the Ministry of Education, Culture, Sports, Science and Technology in Japan, the Japan Society for the Promotion of Science, the Japan International Cooperation Agency, the Japan Environment Corporation, the Japan Great Ape Conservation Fund (Tamotsu Aoki and Yuichiro Kitamura), and the Embassy of Japan in Tanzania.

1 Introduction

MICHIO NAKAMURA

1.1 The chimpanzee

The chimpanzee or the common chimpanzee (*Pan troglodytes* Blumenbach 1775, Hominidae) is the closest living species to us, humans (*Homo sapiens*), together with another species of *Pan*, the bonobo (*P. paniscus*). The genome-wide nucleotide divergence between chimpanzees and humans is calculated to be only 1.23% (The Chimpanzee Sequencing and Analysis Consortium, 2005). Two species of *Pan* are more generally called "great apes" together with gorillas (*Gorilla* spp.) and orangutans (*Pongo* spp.). "Great ape," in general terminology, usually excludes humans; however, according to recent taxonomy (Groves, 2001), such a category is no longer applicable. In fact, these apes *and* humans form the family Hominidae and, more specifically, *Homo, Pan*, and *Gorilla* (African linage of Hominidae) form the subfamily Homininae, separated from another subfamily Ponginae (i.e. orangutans; Figure 1.1). There are some arguments that the genus *Pan* should be categorized under *Homo* as a subgenus (Goodman *et al.*, 1998).

The chimpanzee is currently divided into four subspecies (see also Chapter 46): the western chimpanzee (*P. t. verus*), the Nigeria–Cameroon chimpanzee (*P. t. ellioti*), the central chimpanzee (*P. t. troglodytes*), and the eastern chimpanzee (*P. t. schweinfurthii*). The last subspecies is the subject of our research at Mahale. Chimpanzees range in African tropical forests around the equator from Senegal in the northwest to Tanzania in the southeast (Caldecott and Miles, 2005). The species, as a whole, is considered endangered (Oates *et al.*, 2008).

1.2 Research history of chimpanzees

When chimpanzees were first encountered by the western academic world in the seventeenth century (Tyson, 1699), many scholars were fascinated by their striking anatomical similarities to our own species. In the first half of the twentieth century, several studies were conducted in captivity (Köhler, 1917; Yerkes, 1943), which revealed high cognitive abilities of chimpanzees. After the 1960s, when chimpanzee research began in the wild (Goodall, 1965; Chapter 3), scientists discovered even more amazing similarities in terms of behavior. Chimpanzees share many behavioral patterns with humans, such as tool use, hunting, food sharing, medicinal use of plants, and infanticide. In addition, various behavioral patterns differ among populations and are socially inherited from generation to generation within a group, which resembles human culture (Whiten *et al.*, 1999; Chapter 38). Such findings are important for elucidating human evolution and reconstructing the behavior of early hominids, and have captured the interest of the general public. However, we still have much to learn about chimpanzees because they mature slowly and have long lifespans. Thus, further continuing and painstaking research efforts are essential.

1.3 Long-term study sites

Apart from Mahale, there have been dozens of chimpanzee study sites in Africa. Among them, I briefly introduce some long-term study sites (Figure 1.2) because their names frequently appear in

Mahale Chimpanzees: 50 Years of Research, ed. M. Nakamura, K. Hosaka, N. Itoh and K. Zamma. Published by Cambridge University Press.

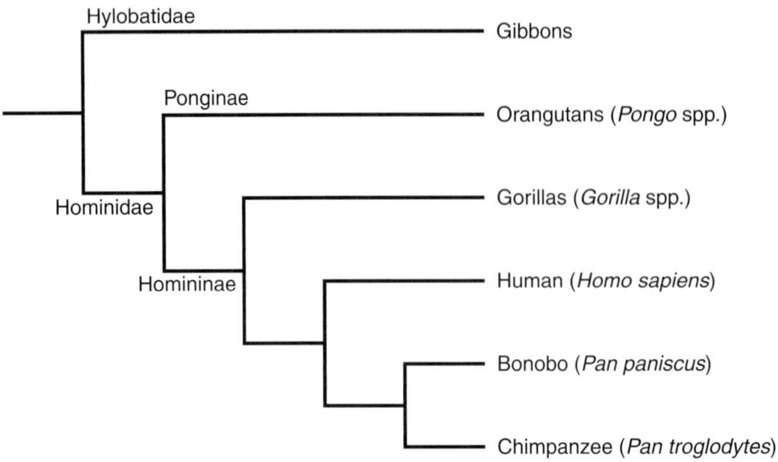

Figure 1.1 Phylogeny of superfamily Hominoidea.

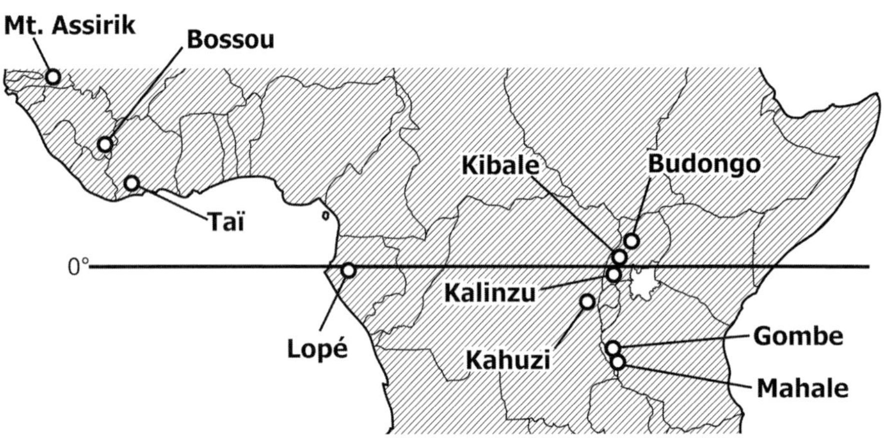

Figure 1.2 Some long-term study sites of chimpanzees.

this volume. Gombe National Park in Tanzania is the oldest and probably most famous chimpanzee research site and is where Jane Goodall started research in 1960 (Wilson, 2012). Several unit-groups have been studied there; among them, the Kasekela group has been the main target of long-term research. Kibale National Park in Uganda is another prominent research site where important findings have emerged (Watts, 2012). There are two study groups at Kibale. One is the Kanyawara group, the studies on which are currently led by Richard Wrangham. Another is the Ngogo group the studies on which are led by David Watts and John Mitani. The Ngogo group is an extraordinarily large unit-group containing more than 150 chimpanzees. The Budongo Forest in Uganda is also one of the oldest study sites (Reynolds, 2005). The Sonso group has been the main target there, and habituation has improved since the 1990s. Bossou in Guinea is another long-term study site led by Japanese researchers (Matsuzawa *et al.*, 2011). Although the Bossou group is an isolated small group, studies focusing on various types of tool use have yielded important findings. The Taï Forest in Côte d'Ivoire is a study site led by Christophe Boesch (Boesch and Boesch-Achermann, 2000). Three groups

(North, South, and Central) have been habituated and inter-unit-group comparisons have produced several outstanding outcomes. Eastern chimpanzees are the target subspecies in Gombe, Kibale, and Budongo, whereas western chimpanzees are the targets in Bossou and Taï. Readers should bear in mind that there are various other chimpanzee study sites (Caldecott and Miles, 2005: 57).

1.4 Introduction to this volume

This volume consists of 50 chapters (or topics). Because each chapter is independent and comprehensive in its topic, readers may begin from any chapter. However, the following chapters are divided into 10 parts with broad affinities among topics. Hereafter, I briefly introduce these parts as a guide map to this volume.

Part I is introductory, giving an overview of the Mahale field site, particularly for readers who are not familiar with it. Chapters in this part provide general information on the field site, research history, chimpanzee distribution at Mahale, and several main individual chimpanzee characters from the study group. Chapter 2 summarizes information such as organizations, abbreviations, and names of places that are used in other chapters.

The initial aim of the Mahale research was to elucidate the social organization of chimpanzees, particularly to look for the origin of the human family. Thus Part II particularly deals with the topics of the chimpanzee society, with chapters introducing the social system, demography, home range, fission–fusion grouping, and intergroup relations. An extraordinary event in the research history, the disappearance of the K-group males and the group extinction in the 1980s, is highlighted in a chapter in this part.

Part III deals with the environment such as climate, flora, and fauna that surrounds Mahale chimpanzees. Long-term monitoring is fundamental in this aspect. Readers can see the decadal changes of climatic parameters, such as temperature and precipitation, vegetation, plant phenology, and relative abundance of large mammals at Mahale. Some rare but interesting observations of chimpanzees encountering other animals, either alive or dead, during the long-term research are also summarized in a chapter.

Part IV basically focuses on what chimpanzees eat. They are mainly ripe-fruit eaters; however, they also eat various parts of plants and meat from mammals, insects, and others. Chapters in this part deal with several aspects such as the diet, the ways chimpanzees eat, and the taste of chimpanzee foods. Many intact seeds are defecated in chimpanzee feces, and a chapter focuses on studies related to such seeds. Unlike many primate species, a chimpanzee frequently shares foods with others, and a chapter in this part discusses this particular topic.

Part V deals with the life history and health of chimpanzees. A general pattern of physical and social development, aging, diseases, and death are included in this part. Because some chimpanzees live long lives, even more than 50 years, many data provided here are the results of painstaking long-term monitoring of the target chimpanzee group. Particular topics of self-medication and conspecific killing, both of which are amazing findings from wild chimpanzee research, are also summarized in this part.

Various social relationships are discussed in Part VI. Male–male relationships are often conspicuous, and various researchers have depicted intriguing details of these relationships. Female–female and male–female relationships, particularly when females are not in estrus, are relatively less well documented. However, improved habituation of female chimpanzees at Mahale enabled us to realize that females are also an important part of chimpanzee society. The topic of orphanhood and allomothering is also included here.

Particular types of social behaviors are the focus of Part VII. As highly social animals, chimpanzees daily direct various behaviors to group mates, such as intimidation displays, aggression and reconciliation, greeting, grooming, sexual behaviors, courtship displays, and social play. Detailed patterns, frequencies, directions, and other observations of such social behaviors are described in each chapter.

Behavioral diversity is another important topic in chimpanzee studies that has attained interest relatively recently, and it is specifically introduced in Part VIII. Several subtle but strange behavioral patterns, repertoire of vocal communication, diversity of play, tool and object use, bed making, nocturnal behaviors, and laterality of hand function are introduced in chapters here. The uniqueness of cultural studies at Mahale is also discussed in a chapter.

Part IX introduces works that connect the field and laboratory. Several noninvasive samples are collected in the field and then analysed in the laboratory. Such samples include hormones from feces or urine, skeletons from dead chimpanzees, DNA from feces, hairs, and others, and parasites found in feces. Specialists in endocrinology, morphology, genetics, and parasitology working on such samples from Mahale contributed to this part, which demonstrates the importance of such samples in addition to direct observations of chimpanzee behaviors.

Finally, in Part X, the relationship between people and chimpanzees is discussed. Particular topics here are tourism at Mahale, the history and culture of the Tongwe people, and conservation efforts by Mahale researchers. A review of anthropological studies on Tongwe people is included here because they are the original residents of Mahale and have supported us as field assistants. Their rich knowledge of nature has contributed to the chimpanzee research at Mahale.

In addition to 50 chapters, the book has several appendices that may be useful for understanding this synthesis of 50 years of research at Mahale.

References

Boesch, C. and Boesch-Achermann, H. (2000). *The Chimpanzees of the Taï Forest: Behavioral Ecology and Evolution*. Oxford: Oxford University Press.

Caldecott, J. and Miles, L., ed. (2005). *World Atlas of Great Apes and Their Conservation*. Berkeley, CA: University of California Press.

The Chimpanzee Sequencing and Analysis Consortium (2005). Initial sequence of the chimpanzee genome and comparison with the human genome. *Nature*, **437**, 69–87.

Goodall, J. (1965). Chimpanzees of the Gombe Stream Reserve. In *Primate Behavior*, ed. I. DeVore. New York: Holt, Rinehart and Winston, pp. 425–73.

Goodman, M., Porter, C. A., Czelusniak, J., *et al.* (1998). Toward a phylogenetic classification of primates based on DNA evidence complemented by fossil evidence. *Molecular Phylogenetics and Evolution*, **9**, 585–98.

Groves, C. (2001). *Primate Taxonomy*. Washington DC: Smithsonian Institution Press.

Köhler, W. (1917 [1962]). [*The Mentality of Apes.*] Trans. K. Miya. Tokyo: Iwanami. Japanese translation.

Matsuzawa, T., Humle, T., and Sugiyama, Y., ed. (2011). *The Chimpanzees of Bossou and Nimba*. Tokyo: Springer.

Oates, J. F., Tutin, C. E. G., Humle, T., *et al.* (2008). *Pan troglodytes. The IUCN Red List of Threatened Species. Version 2014.2*. Available at: www.iucnredlist.org. Accessed August 30, 2014.

Reynolds, V. (2005). *The Chimpanzees of the Budongo Forest: Ecology, Behaviour, and Conservation*. Oxford: Oxford University Press.

Tyson, E. (1699). *Orang-Outang, Sive Homo sylvestris; or the Anatomy of a Pygmie*. London: T. Bennett and D. Brown.

Watts, D. P. (2012). Long-term research on chimpanzee behavioral ecology in Kibale National Park, Uganda. In *Long-Term Field Studies of Primates*, ed. P. M. Kappeler and D. P. Watts. Heidelberg, Germany: Springer, pp. 313–38.

Whiten, A., Goodall, J., McGrew, W. C., *et al.* (1999). Cultures in chimpanzees. *Nature*, **399**, 682–5.

Wilson, M. L. (2012). Long-term studies of the chimpanzees of Gombe National, Park, Tanzania. In *Long-Term Field Studies of Primates*, ed. P. M. Kappeler and D. P. Watts. Heidelberg, Germany: Springer, pp. 357–84.

Yerkes, R. M. (1943). *Chimpanzees: A Laboratory Colony*. New Haven, CT: Yale University Press.

PART I History and overview

2 Overview of the field site: Mahale Mountains and their surroundings

MICHIO NAKAMURA AND NORIKO ITOH

2.1 General introduction to the Mahale Mountains National Park

This chapter introduces the locale of the studies described in this volume, the Mahale Mountains National Park (MMNP), and some other places relevant to Mahale research. The name "Mahali" was used instead of "Mahale" in some early literature. Mahale became a national park on June 14, 1985 (Government Notice No. 262, 1985; Nishida, 1985) after 20 years of research; the main historical introduction is given in Chapter 3. Here, we will first briefly introduce the location, access, geographic features, and the names of some relevant places in and around the park. Then, we will introduce the Kasoje area, which is the main site for our chimpanzee research, in detail. Further information on the Mahale study site can be found in several publications, which give outlines of the site both in English (Nishida, 1990; Itoh *et al.*, 2012a; Nakamura and Nishida, 2012) and Japanese (Nakamura *et al.*, 1999; Itoh *et al.*, 2012b).

MMNP (6°15′S, 29°55′E) is located at the southernmost part of the Kigoma Region, which is at the western end of the United Republic of Tanzania, East Africa (Figure 2.1). Being remote from Dar es Salaam, the economic capital of Tanzania, the population of the Kigoma Region used to be relatively low compared with other regions in Tanzania, at approximately 470 000 based on statistics from 1967 (Anonymous, 1998). However, the population had more than tripled to 1 674 047 by 2002 (Anonymous, 2005a) and had increased to 2 127 930 by 2012 (Anonymous, 2013). Now, the Kigoma Region is the seventh largest region in Tanzania. There are a large number of refugees from neighboring areas such as Burundi and the Democratic Republic of Congo staying in Western Tanzania, mostly in the Kigoma and adjacent Rukuwa Regions. In the year 2000, the number of refugees exceeded 700 000 (Anonymous, 2005b) but recently decreased to approximately 100 000 in 2012 (Anonymous, 2012). Chimpanzees in Tanzania are mostly found in the Kigoma Region, which also includes the world-famous Gombe National Park that was home to Jane Goodall's study of chimpanzees (Goodall, 1986).

Mahale is on the eastern shore of Lake Tanganyika (Figure 2.2), which is the second deepest lake in the world and a part of the Great Rift Valley. Thus, the Mahale Park includes not only inland areas but also a 1.6 km wide strip of coastal waters. The park covers an area of approximately 1600 km^2 (Figure 2.3), which is more than 30 times larger than the Gombe National Park. As indicated by its name, the park is penetrated by the chain of Mahale Mountains that roughly runs from the northwest to the southeast of the park; thus, the park is mostly hilly. The park's highest peak is Mt. Nkungwe (2462 m above sea level; Figure 2.4), which is also the highest peak on the eastern shore of Lake Tanganyika.

2.2 Access to Mahale

Access to Mahale has not been easy compared with other chimpanzee research sites, mainly because Mahale remains unreachable by car. When Toshisada Nishida began research at Mahale in 1965 (see Chapter 3), he had to take a train from Dar es Salaam

Mahale Chimpanzees: 50 Years of Research, ed. M. Nakamura, K. Hosaka, N. Itoh and K. Zamma. Published by Cambridge University Press.

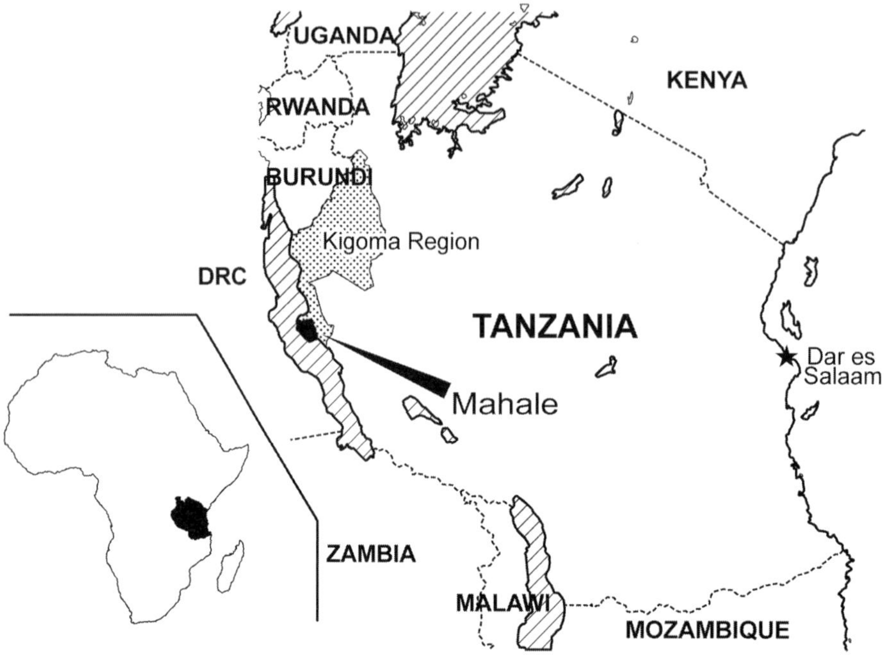

Figure 2.1 Locations of the Kigoma Region and Mahale Mountains National Park in Tanzania.

Figure 2.2 Lake Tanganyika viewed from the Mahale Mountains' main ridge. ©M. Nakamura.

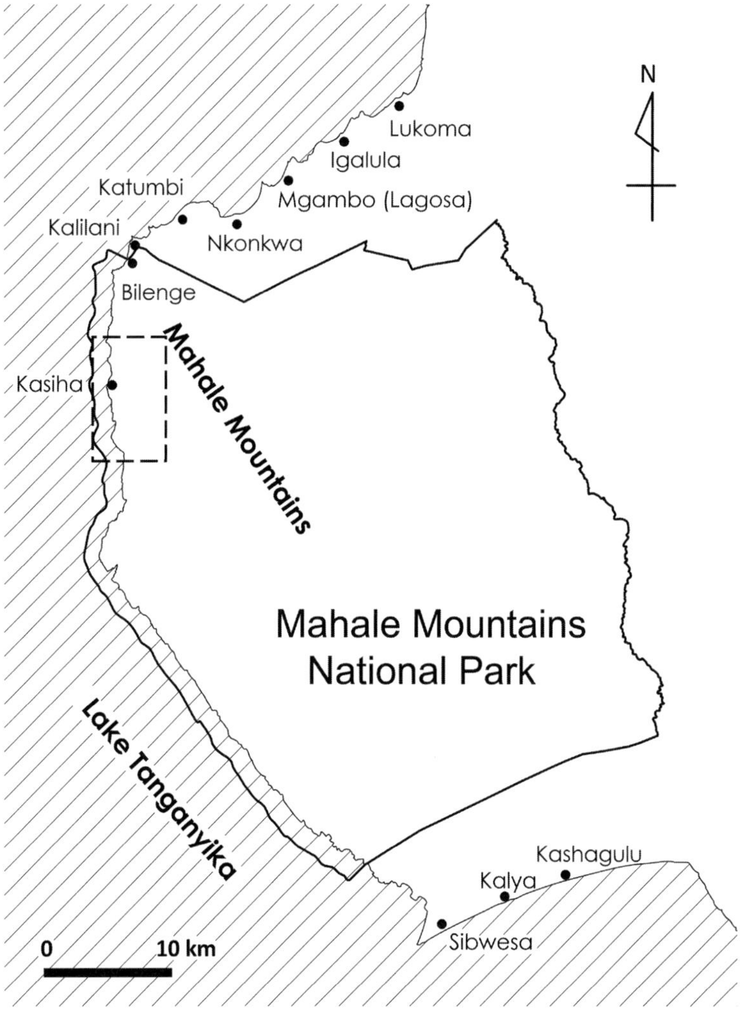

Figure 2.3 Park boundary of the Mahale Mountains National Park. The boundary shown here is tentative because several existing maps show different boundaries. The boundary is currently displayed on-site by signboards and beacons or, in some places, by 20 m wide belts cleared of vegetation (to prevent invasion of fire). However, no existing maps concur with this boundary in practice. In Government Notice No.262 (1985), which should have defined the boundary of the park, a description regarding the southwestern boundary is lacking, and there are obvious errors in the description of the southern boundary. The boundaries shown here are based on Koichiro Zamma's surveys on the locations of signboards, beacons, and the 20 m wide belts devoid of vegetation. The area with a dashed rectangle is shown in Figure 2.7.

to Kigoma, which took 3 days and 2 nights, a steam ship named Liemba from Kigoma to Mgambo for 12 h, and finally a small boat from Mgambo to Mahale for approximately 3 h (for which he had to wait for 4 or 5 days). When he wanted to go out from Mahale, he used a sailboat or a rowboat (Nishida, 1989). Today, regular flights from Dar es Salaam to Kigoma are available (6 days a week in 2013), but the schedule is often subject to change. The road from Dar es Salaam to Kigoma has recently been significantly improved; thus, it is also possible to reach Kigoma by bus or car in 2 to 3 days.

Figure 2.4 Mt. Nkungwe, the highest peak in Mahale. ©M. Nakamura.

All research supplies have to be brought from Dar es Salaam or abroad and sent by parcels to Kigoma, or bought at a local market in Kigoma. These supplies then have to be brought from Kigoma to Mahale via Lake Tanganyika. There are two ways to travel from Kigoma to Mahale. If one uses public transportation, steamships are available once or twice a week from Kigoma to Mgambo (Lagosa), a nearby fishing village to the north of Mahale. It usually takes approximately 8 h to travel from Kigoma to Mgambo, but the schedule is notoriously changeable because of the kinds of cargo carried by the steamships. Boarding a steamship at the Kigoma Port can be a bit chaotic as many porters rush to assist with the baggage, but the trip is usually very comfortable and relaxing once on board in a first-class cabin. After several stops, the steamship reaches Mgambo, where one disembarks to a small boat, often owned by the National Park; this takes tourists to the park station in another 2 to 3 h. Another option for the trip from Kigoma to Mahale is to arrange for a private boat with an outboard engine. The National Park and the research team have their own boats, and several hotels and tour companies at Kigoma own boats that can be chartered by tourists. A boat trip from Kigoma to Mahale takes approximately 10–12 h on a wooden boat and 6–8 h on a fiber boat. Traveling on a smaller boat can sometimes be tough compared with a steamship because the waters can be rough and passengers may get wet and feel cold; thus, a rain suit should be worn.

Most tourists, who do not have to go to Kigoma to buy supplies, now fly directly to the Mahale airstrip (Figure 2.5) on shared chartered flights from Arusha, the second largest city in Tanzania. While this is the easiest way to travel, it is also the most expensive one.

2.3 Organizations and abbreviations

This section introduces some organizations working at or relating to Mahale. Abbreviations mentioned here will be used throughout the book.

2.3.1 Mahale Mountains Chimpanzee Research Project (MMCRP)

This is a project run by a group of researchers conducting research on chimpanzees and other wildlife. The main members originated from KUAPE (see below). They were Jun'ichiro Itani's former students at Kyoto University or Toshisada Nishida's former students at the University of Tokyo or Kyoto University, among others. Nishida had taken a leading role in the project for decades, but the four editors of this volume took on joint executive roles after his death. The members of MMCRP have a project

Figure 2.5 Airstrip located to the north of the Mahale Park. The unpaved surface sometimes prevents take-off and landing during the wet season. ©M. Nakamura.

meeting approximately once a year in Japan. These days, MMCRP arranges researchers' scheduled visits, controls the number of people on-site at one time, and employs research assistants (see Chapter 48).

2.3.2 Kyoto University Africa Primatological Expedition (KUAPE)

This was the first academic expedition to Africa by Japanese academic researchers funded by the Japanese Ministry of Education (now called the Ministry of Education, Culture, Sports, Science and Technology or MEXT). A part of this expedition later became MMCRP. Despite its name, the expedition did not limit its focus to primatology. In addition to a group of primatologists who studied wild chimpanzees, there was a group of anthropologists who studied hunter-gathers, pastoralists, and farmers in Tanzania (Itani, 1977). Nishida joined KUAPE from its fourth expedition (see Chapter 3).

2.3.3 Mahale Wildlife Conservation Society (MWCS)

This is a conservation-focused nongovernmental organization (NGO) that was founded in 1994 (Anonymous, 1995). Hosea Kayumbo of Dar es Salaam University is the chairperson and Nishida was the vice-chairperson. While MMCRP conducts research activity, MWCS conducts wildlife and community conservation activities with its membership dues, donations, and funds from other organizations. Chapter 50 contains more details about the conservation activities. After Nishida's death, Kazuhiko Hosaka became the acting vice-chairperson (chairperson of the Japanese branch), and Eiji Inoue has been an accountant for MWCS.

2.3.4 Tanzania National Parks (TANAPA)

This is a Tanzanian governmental organization that is responsible for all of the national parks in the country. The headquarters are in Arusha. Several world-famous national parks such as Serengeti National Park belong to TANAPA.

2.3.5 Mahale Mountains National Park (MMNP)

This is a national park that belongs to TANAPA. When we write MMNP in this book, we refer to the National Park as an organization. Apart from the chief park warden, there are several park wardens who are specialized as ecologists, community conservationists, accountants, protect wardens, etc. Under these wardens, there are many park staff, such as guides and rangers, some of whom are locally hired, but many are from other parts of Tanzania.

2.3.6 Tanzania Wildlife Research Institute (TAWIRI)

This is a Tanzanian governmental organization that deals with wildlife research in Tanzania. It was formerly called the Serengeti Wildlife Research Institute (SWIRI). The headquarters are in Arusha. All wildlife researchers are supposed to apply to TAWIRI for their research permits and the permits are issued by COSTECH (see below) after their approval. TAWIRI employs several Tanzanian researchers who conduct their research in the country.

2.3.7 Mahale–Gombe Wildlife Research Centre (MGWRC)

This is TAWIRI's research center at Kigoma. It was formerly called the Mahale Mountains Wildlife Research Centre (MMWRC). It consists of a few research staff under a director and helps foreign researchers in various ways, such as with logistics.

2.3.8 Tanzania Commission for Science and Technology (COSTECH)

This is a Tanzanian governmental organization located in Dar es Salaam. It issues research permits for any scientific research conducted in the country. Other organizations such as TAWIRI approve the applications depending on the research theme, and COSTECH issues the permit once a recommendation letter is submitted.

2.3.9 Japan International Cooperation Agency (JICA)

This is a Japanese governmental organization that conducts overseas operations mostly in developing countries. When a survey was conducted to make Mahale a national park in 1975 (Anonymous, 1980, for the outcome of the first JICA project), several Mahale researchers were dispatched to Tanzania as JICA experts. From 1994 to 1997, an additional JICA project was undertaken in order to try to habituate another group of chimpanzees near MMNP headquarters. Chapter 50 contains the names of the JICA experts.

2.3.10 Mahale Ecosystem Management Project (MEMP)

This is a project run by the Frankfurt Zoological Society (FZS) and funded by the European Union (EU). This project compiled a general management plan for Mahale Park in cooperation with MMNP (Anonymous, 2006).

2.4 Villages near Mahale

After the *Ujamaa* settlement policy of the Tanzanian government in 1974, people who had been scattered in small settlements comprising a few households were forced to move to villages called *kijiji cha ujamaa*. The same applied to Mahale (Nishida, 1978). When Nishida started his research at Mahale, there were several small settlements scattered within the area that later became the National Park (Nishida, 1973). The people moved out from the area and gathered in such villages. This section introduces some of these villages near the Mahale Park.

Kalilani: This is a fishing village adjacent to Bilenge headquarters (see below). Villagers are not of the Tongwe (Chapter 49) who used to live in this area, but instead are mostly immigrants from former Zaïre. When Mahale was designated as a national park, Kalilani village was included within the park; thus, most of the villagers had to move out from the place by law. However, because the village became larger before MMNP was in control, there have been several quarrels between the villagers and the park authorities. Villagers are now allowed to live only on the lakeshore.

Figure 2.6 Katumbi village. ©M. Nakamura.

Nkonkwa: This village is second closest to the northern border of the national park after Kalilani. Villagers are mostly the Tongwe who used to live in and around the Mahale area.

Katumbi (Figure 2.6): Formerly, Katumbi was a subvillage of Nkonkwa, but it recently became a formal administrative village. Most research assistants and former assistants live in this village; thus, it is one of the most familiar villages to Mahale researchers. An airstrip owned by MMNP is located at an approximately 20-min walk from this village.

Buhingu: This is the capital of the Buhingu Ward, which includes the villages described in this section. It is also called Lagosa or Mgambo. The steamship from Kigoma stops just offshore from the village (the stop's name is Lagosa).

Lukoma: This is a village northeast of Buhingu. The steamship also stops here. There is a road going inland to Mwese, so this village serves as an intersection of the inland and lakeshore villages.

Kalya: This is a village south of the Mahale Park. The steamship also stops here, and this is the intersection of the inland and lakeshore villages to the south of Mahale.

2.5 Major places in the park

Kasoje area: Kasoje is referred to as "Kasoge" in some publications. Generally, it refers to the northwestern part of the Mahale Park that is relatively flat (Figure 2.7). Nishida (1974) called the low-altitude forest in this area the "Kasoje forest," which was created by the mountain chain that blocks moisture from the lake (Chapter 13). Lowland areas outside Kasoje are basically the Miombo woodland. The long-term chimpanzee research has been mainly conducted within the Kasoje area.

Bilenge: This is the northwestern end of the national park where MMNP headquarters are located. Most of the MMNP staff and their families live here. MEMP is also based here.

Miyako: This is a place around the river mouth of the Miyako Valley. There used to be a researchers' camp (Miyako Camp) in Miyako. When the Kajabala group of chimpanzees (hereafter called the "K group,"

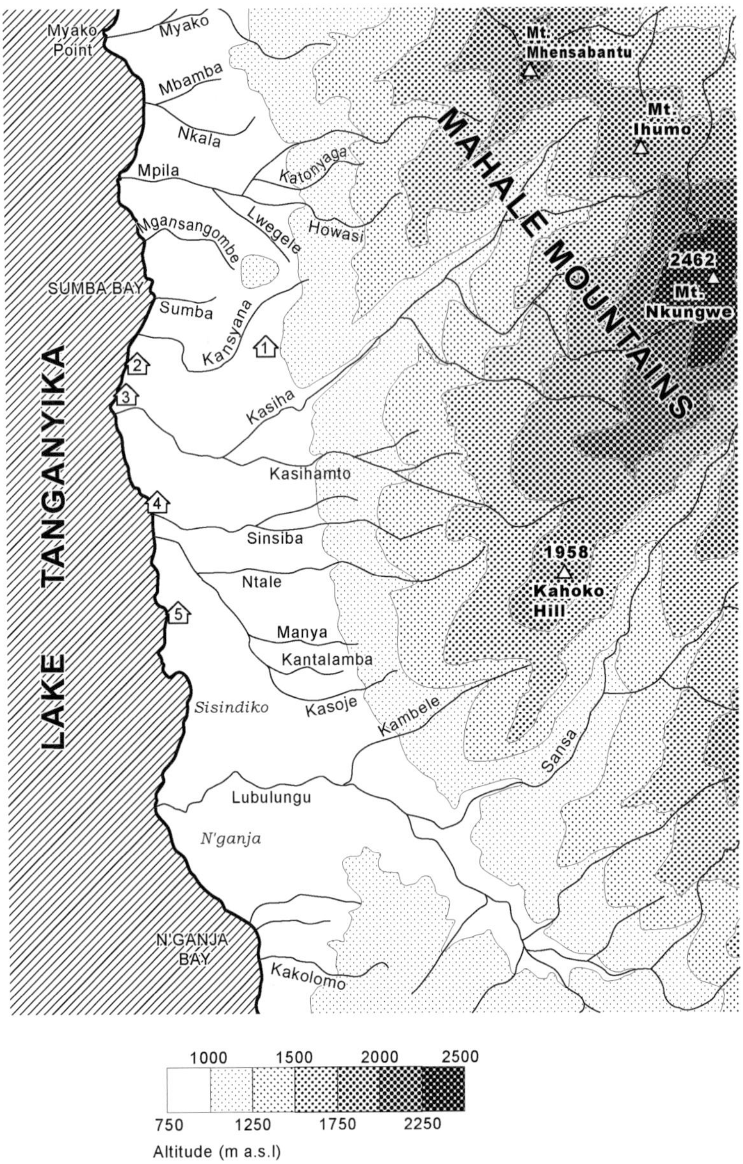

Figure 2.7 Water systems and the locations of research and tourist camps in the Kasoje area. Almost all the small valleys have been named by local Tongwe people. Numbers in house shapes represent (1) Kansyana Research Camp; (2) Flycatcher's Tourist Camp; (3) Kasiha, where MMNP lodging, staff facilities, and research assistants' camps are located; (4) Sinsiba Tourist Camp; and (5) Kangwena Tourist Camp.

see Chapter 3) was the main target of the research, researchers would stay in Miyako Camp. The camp was abandoned after the extinction of the K group (see Chapter 10), and now, only a few remnants of the camp can be seen.

Kasiha: This is an area north of the river mouth of the Kasiha Valley. It was a former settlement of Tongwe people and is the current camp for the staff of MMNP and MMCRP. People use a boat with an outboard engine to travel from Bilenge to Kasiha. It was like a small village until the mid-1990s because the families of MMCRP trackers also lived in Kasiha. Later, with the administration of MMNP, the families of MMCRP trackers moved out of the National Park (mostly to Katumbi). Thc familics of MMNP staff still live in Kasiha. There used to be a campsite where backpackers could pitch tents on the northern beach of Kasiha, but this area is now used by a tour company (Flycatcher).

Kansyana: This has been the MMCRP researchers' camp from the start of their research to date. It is located approximately 1 km inland from Kasiha, adjacent to the Kansyana Valley.

Sinsiba: This is a former settlement and current tourist camp (Nkungwe Camp).

Sisindiko: This is a former fishing camp, now abandoned.

Kangwena: This is a former settlement and current tourist camp (Nomad Camp).

N'ganja: This is a flat and relatively dry habitat south of the Lubulungu River. There was a settlement at the south end of N'ganja.

2.6 Mountains and water systems in the Kasoje area

It rains heavily during the wet season in the Kasoje area because of the mountains to the east and the lake to the west (Chapter 12). The rain water makes several streams that descend the slope of Mahale Mountains from east to west. The original inhabitants of the area, the Tongwe, could name even small streams (Figure 2.7), most of which dry up during the dry season.

The upper streams of the relatively large valleys, such as Miyako, Kasiha, and Ntale, have well-developed V-shaped valleys. Base rocks are often exposed in such V-shaped valleys, and many small and large waterfalls can be seen in these areas

Figure 2.8 A waterfall in the Ntale Valley. ©M. Nakamura.

(Figure 2.8). Because of the steep and complex topography, it is sometimes impossible to follow chimpanzees when they go up the streams and cross the V-shaped valley. The largest valley in the northern Kasoje area is the Kasiha Valley, and there are many large water pools with crystal clear water upstream. Water flows along the riverbed in the lower areas during the dry season, but in the wet season the Kasiha Valley can get flooded after a heavy rain, sometimes preventing people from crossing it for a while.

The Sinsiba Valley is a small stream flowing around the middle of the Kasoje area. In the lowland areas near the lake, its water creates a large swamp known as Sinsiba Swamp (Figure 2.9). There are many water-loving herbaceous plants in the swamp and the Mimikire group (hereafter referred to as the

Figure 2.9 Sinsiba Swamp. ©M. Nakamura.

"M group," see Chapter 3) of chimpanzees often feed on such plants for 1–2 h.

The largest river in this area is called Lubulungu, and it flows with visible surface water even in the dry season. During the wet season, Lubulungu may get flooded to a large degree, flowing around vegetation. Thus, the area is a grassland consisting of elephant grasses, among others.

Mt. Nkungwe (Figure 2.4), the highest peak in the park, is located to the east of Kansyana Camp. There is a path up to Mt. Nkungwe maintained by MMNP, which starts at the northern bank of the Mpila Valley via Mt. Mhensabantu and Mt. Ihumo. One can also climb Mt. Nkungwe from Bilenge via Mt. Pasagulu and Mt. Mhensabantu. The Kahoko Hill, which is located to the southwest of Nkungwe, is another prominent peak that can be observed from the lake, though it is not a part of the Mahale main ridge. The rocky cliff down to the Kambele Valley is very steep but spectacular to see (Figure 2.10).

2.7 Research facilities

In the researchers' camp at Kansyana (Figure 2.11), there are currently two simple buildings with five bedrooms in total, each equipped with a desk and a bed, a library room, and some storage rooms. Water is manually carried to a drum in the dry season, and cooking is done using kerosene stoves and dead wood collected from the forest floor. A minimal solar system is available to charge computers and batteries, and there is a small weather station to monitor daily temperature and precipitation (Itoh *et al.*, 2012a; Chapter 12).

2.8 Tourist facilities

Being isolated from civilization, the only lodging facilities available for tourists are within Mahale Park. MMNP has lodging facilities in Kasiha and Bilenge. There are three tourist camps run by private tour companies: one at Kasiha (Flycatcher Camp), one at

Figure 2.10 Kahoko Hill viewed from the southern side. Base rocks are exposed because of the steep slope. Chimpanzees may climb such slopes with the help of trees and vines but human observers cannot follow without climbing gear and associated technical skills. ©M. Nakamura.

Sinsiba (Nkungwe Camp), and one at Kangwena (Nomad Camp).

2.9 Trail systems

Many areas within Kasoje have very dense undergrowth; thus, observation trails are essential for tracking chimpanzees and making detailed observations (Figure 2.12). Although many other chimpanzee study sites employ a straight grid trail system, it is not possible to do so at Mahale. The dense undergrowth and complex topography of Mahale limit the ability to cut straight trails. Therefore, most of the observation trails at Mahale are former human paths or trails used by chimpanzees or bushpigs that have been widened.

Figure 2.13 shows the main observation trails in the Kasoje area. Trails that run roughly from north to south in the area south of Kansyana Camp are named Routes 1 to 7 (hereafter R-1 to R-7). R-1 starts from Kansyana along the foot of the eastern mountains. The location of the trails and their numbers do not correlate. R-5 runs along the lake, R-2 and R-3 run in the middle, and R-4 runs east of R-1. R-6 and R-7 are relatively new trails that were made in order to be able to walk along the midslope of the steep eastern mountains. In the area north of Kansyana Camp, there is a complex topography with numerous inclines and declines. Thus, there is only one main trail, the Tarimo Road, which runs approximately from north to south.

The trails that run from west to east basically go along the ridges. They are named in alphabetical order from A to P. However, trails A–D were only used when the K group was studied; thus, they are currently almost abandoned and have had no maintenance. Trails E–P run in the area the M group frequently uses. There is a small hill between trails G and J, which makes the topography complex relative to other places; thus, these trails run quite irregularly. There are many other trails that connect these main trails or were made because the chimpanzees frequently used the route. As a consequence, the whole trail system at Mahale is very complicated.

Some of the trails' intersections are named. In principle, these names are a combination of the letter designating the west–east trail and the number in order from the lake. For example, trail F starts at *F0* (at the lakeshore), the next closest intersection from the lake is *F1*, then *F2*, and so on. However, because several additional trails (and therefore also intersections) were made after this general rule was applied, there are some complicated names such as *L1-2*, *L1-3* (both between *L1* and *L2*), *I3N* (meaning north of *I3*), etc. There are also many intersections without names.

There are several good places to listen to chimpanzee vocalizations on the west–east trails

Figure 2.11 Kansyana Camp in 2013. ©M. Nakamura.

Figure 2.12 Chimpanzees walking along an observation trail. ©M. Nakamura.

because they basically run along the ridges. Some of these sites are listed below (see Figure 2.13 for the locations).

E3: An intersection of the E trail and Tarimo Road (①).
Mkokoti Hill: A small hill between the G and F trails (②).
G4-1: A point on the north hillside along the G trail (③).
Kalukula Hill: A small hill at the intersection of R-3 and the K trail (④).
L1-2: An intersection of the L trail and R-3 (⑤).
N1-1: An intersection of the N trail and R-3 (⑥).
Kitanda-cha-simba Hill: A small hill south of the intersection with the P trail and R-2 (⑦).
Kiboko Hill: A small hill south of the P trail between R-3 and R-1 (⑧).

2.10 Outside the Kasoje area

Because long-term research on chimpanzees has been conducted in the Kasoje area and tourists also concentrated in the same area, the vast majority of the park outside the Kasoje area is not utilized for either research or tourism. Such areas include Miombo and bamboo woodlands with relatively sparse vegetation. There are several savanna-dwelling large mammals in these woodlands that cannot usually be observed in the Kasoje area, such as elephants, lions, buffaloes, zebras, wild dogs, and roan antelopes (Appendix III). However, there are also several confirmed groups of chimpanzees in these dry habitats of the Miombo woodland with some riverine forests. Extensive surveys of chimpanzees and other animals in such woodland areas are described in Chapter 4.

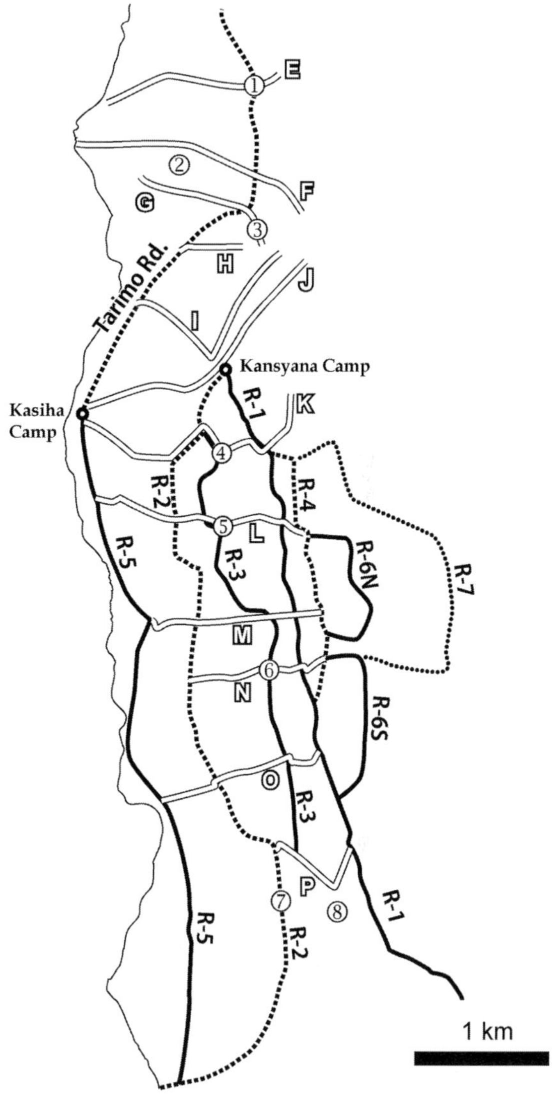

Figure 2.13 Major observation trails in Mahale. Solid or dotted lines show north–south trails and white lines show west–east trails. Good locations for listening to chimpanzee calls are designated by circled numbers: ① E3; ② Mkokoti Hill; ③ G4-1; ④ Kalukula Hill; ⑤ L1-2; ⑥ N1-1; ⑦ Kitanda-cha-simba Hill; and ⑧ Kiboko Hill.

References

Anonymous (1980). *Mahale: Study for the Proposed Mahale Mountains National Park. Final Report, May 1980.* Tokyo: Japan International Cooperation Agency.

Anonymous (1995). Inauguration meeting of Mahale Wildlife Conservation Society held at the University of Dar-es-Salaam on August 12, 1994. *Pan Africa News*, **2**(1), 1.

Anonymous (1998). *Kigoma Region: Socio Economic Profile. Kigoma, Tanzania*. Dar es Salaam: The Planning Commission.

Anonymous (2005a). *The United Republic of Tanzania: 2002 Population and Housing Census: Village and Street Statistics*. Dar es Salaam: National Bureau of Statistics.

Anonymous (2005b). *UNHCR Statistical Yearbook 2004*. Geneva: UNHCR.

Anonymous (2006). *Mahale Mountains National Park: General Management Plan 2006–2016*. Arusha: Tanzania National Parks.

Anonymous (2012). *Population Statistics and Voluntary Repatriation: United Republic of Tanzania*. Available at: http://www.unhcr.org/4f324c069.html.

Anonymous (2013). *The United Republic of Tanzania: 2012 Population and Housing Census: Population Distribution by Administrative Areas*. Dar es Salaam: National Bureau of Statistics.

Goodall, J. (1986). *The Chimpanzees of Gombe: Patterns of Behavior*. Cambridge, MA: Belknap.

Government Notice No. 262 (1985). The national parks ordinance. *Gazette of the United Republic of Tanzania*, **24**, 305–6.

Itani, J., ed. (1977). [*The Chimpanzees*.] Tokyo: Kōdansha. In Japanese.

Itoh, N., Nakamura, M., Ihobe, H., *et al.* (2012a). Long-term changes in the social and natural environments surrounding the chimpanzees of the Mahale Mountains National Park. In *The Ecological Impact of Long-Term Changes in Africa's Rift Valley*, ed. A. J. Plumptre. New York: Nova Science, pp. 249–77.

Itoh, N., Nakamura, M., Zamma, K., Ihobe, H., and Hosaka, K. (2012b). [From field study site (20): Mahale Mountains National Park (Tanzania).] *Japanese Journal of Ecology*, **62**, 83–8. In Japanese.

Nakamura, M. and Nishida, T. (2012). Long-term field studies of chimpanzees at Mahale Mountains National Park, Tanzania. In *Long-Term Studies of Primates*, ed. P. M. Kappeler and D. P. Watts. Heidelberg, Germany: Springer, pp. 339–56.

Nakamura, M., Itoh, N., and Sakamaki, T. (1999). [Site report: The Mahale Mountains National Park, Tanzania.] *Primate Research*, **15**, 93–9. In Japanese with English summary.

Nishida, T. (1973). [*The Children of the Mountain Spirits: Search for the Social Structure of Chimpanzees*.] Tokyo: Chikuma-shobō. In Japanese.

Nishida, T. (1974). [Ecology of wild chimpanzees.] In [*Human Ecology*], ed. R. Ohtsuka, J. Tanaka, and T. Nishida. Tokyo: Kyōritsu-shuppan, pp. 15–60. In Japanese.

Nishida, T. (1978). [Revolution in the Lake Tanganyika Region: A travel to Western Tanzania.] *Tenbō*, **232**, 103–24. In Japanese.

Nishida, T. (1985). The Mahale Mountains declared a national park. *IPPL Newsletter*, **12**, 9.

Nishida, T. (1989). [A journey to the Mahale National Park: its past and present.] *Seien*, **484**, 24–9. In Japanese.

Nishida, T. (1990). A quarter century of research in the Mahale Mountains: an overview. In *The Chimpanzees of the Mahale Mountains: Sexual and Life History Strategies*, ed. T. Nishida. Tokyo: University of Tokyo Press, pp. 3–35.

3 Research history

MICHIO NAKAMURA, KAZUHIKO HOSAKA, AND YUKIO TAKAHATA

3.1 Birth of Japanese primatology and prehistory of chimpanzee studies in Mahale

Why did Japanese researchers start the field study of chimpanzees in Mahale and why have they continued their studies for half a century? In order to answer these questions, we first need to go back to the era when primatology was born in a small country in the Far East and describe the prehistory of Mahale research.

The theoretical background leading to the birth of Japanese primatology is well described in *The World of Living Things* by Kinji Imanishi (Imanishi, 1941). We do not go into the details of his ideas here because the book has already been translated into English (Imanishi, 2002) and there are many English publications that evaluate (support or oppose) and/or introduce Imanishi's ideas and research philosophy (Halstead, 1985; Asquith, 1986, 2007; Sakura *et al.*, 1986; Sano, 1988; Ito, 1991; de Waal, 2001, 2003; Nakamura and Nishida, 2006; Matsuzawa and McGrew, 2008). It is important to mention that Imanishi himself noted that the core idea of his book – and thus, the core idea of his philosophy in general – lies in the fourth chapter, "On Society/Sociality." Society/sociality of nonhuman animals was the main interest of this influential book and the research question on which he based his entire career. His standpoint is concisely given in the sentence below, which he wrote after he and his colleagues began dealing with the question of society/sociality in nonhuman primates:

> What I want to do is research of the history of an animal society that can be understood only by recording and accumulating through descriptions of the animal society instead of the animals therein who do not have language let alone letters to do so. (Imanishi, 1960a, our translation)

With such an interest in mind, Imanishi left for Cape Toi in Miyazaki Prefecture, the southern part of Japan, soon after World War II, together with two undergraduate students, Shunzo Kawamura and Jun'ichiro Itani, in order to study semi-wild horses (*Equus caballus*). They incidentally saw a troop of Japanese monkeys (*Macaca fuscata*) there, which made them decide to study primates instead (Imanishi, 1960a). Soon afterwards, on December 3, 1948, they visited a small island not very far from Cape Toi, called Koshima Islet. This date is said to be the day Japanese primatology was born. Later, Itani proudly wrote about the day in an autobiographical essay:

> We walked around the island, sometimes getting lost in the deep forest. We saw food remains and feces of monkeys but we could not observe the monkeys themselves. However, I would like to say that December 3, 1948, is the beginning of the Japanese monkey studies. (Itani, 1991, our translation)

After this, a group of young researchers led by Imanishi and Denzaburo Miyadi started field studies of Japanese monkeys in several places in Japan, such as Koshima, Takasakiyama, and Arashiyama. These studies led to brilliant findings about the behaviors and society of Japanese monkeys (Itani, 1956, 1959, 1961a; Kawamura, 1956, 1959; Imanishi, 1957a, b, 1960b; Kawai, 1965). With that, the Japan Monkey Centre was established in 1956 in Inuyama, Aichi Prefecture, with the support of the Nagoya Tetsudo Railway company.

Mahale Chimpanzees: 50 Years of Research, ed. M. Nakamura, K. Hosaka, N. Itoh and K. Zamma. Published by Cambridge University Press.

3.2 Off to Africa

In 1958, 2 years after the establishment of the Japan Monkey Centre and 10 years after the start of Japanese primatology, they went to Africa. During the *Japan Monkey Centre's First Gorilla Scientific Expedition*, Imanishi and Itani visited Kenya, Tanganyika (now Tanzania), Uganda, and Belgian Congo (now the Democratic Republic of Congo). Because their main purpose was to clarify the origin of human society by studying nonhuman primates, it was a matter of time before they shifted their target from macaques to apes. They were eager to find the prototype of human society, especially evidence to answer the question about the origin of the human family, in the societies of great apes such as gorillas (*Gorilla* spp.) and chimpanzees (Imanishi, 1977).

Three Gorilla Scientific Expeditions took place (Imanishi, 1960a; Itani, 1961b; Kawai, 1961), but they shifted their target from gorillas to chimpanzees due to the political disorder in Congo. At the end of the third expedition in 1960, Itani visited Uganda and Tanganyika to look for a suitable site for chimpanzee studies. There he met Jane Goodall, who had just started chimpanzee research in Gombe (Itani, 1961b).

A full-scale study of chimpanzees was started in Tanganyika in 1961, under the name KUAPE (see Chapter 2). Imanishi was aware that the research effort might take a long time, so he brought a huge amount of research equipment and built an excellent research station at Kabogo (Nishida, 2012). Following Imanishi's retirement from Kyoto University, Itani led the team from the fourth expedition. In 1963, Itani moved the research station to Kasakati and conducted extensive surveys covering large areas in Western Tanzania from Masito Escarpment to Ugalla (Suzuki, 1969; Izawa, 1970; Kano, 1971, 1972; Itani, 1979).

In 1965, Itani instructed three graduate students to try to habituate chimpanzees in three different ways in three different places, namely habituation to human presence, habituation to a chimpanzee, and habituation with provisioning. Toshisada Nishida was the student who was assigned to habituate with provisioning in Mahale Mountains (Nishida, 1990).

3.3 Research history at Mahale

Nishida (1990) described the academic prehistory of the Mahale area in great detail, including expeditions by explorers and the Oxford team; thus, it is unnecessary to mention it here. The conservation history of Mahale is described in Chapter 50.

The first member of KUAPE to visit Mahale was Shigeru Azuma. He tried to reach Mt. Nkungwe in 1962 (Akisato Nishimura, personal communication). Another member from the fourth expedition, Kosei Izawa, first visited Kasoje in August 1965 for one week, met the chimpanzees four times, and learned that they sometimes raided sugarcane fields (Nishida, 1973a). With this information, Nishida began a full-scale research project in October 1965 and stayed in a humble hut at Kansyana (Figure 3.1). More details on these first days of research at Mahale are available in Nishida's book (2012). MMCRP (see Chapter 2) has continued ever since.

In March 1966, a group of chimpanzees, later named the K group, began to visit the sugarcane fields at Kansyana, which Nishida had developed for provisioning. In 1968, the southern-adjacent M group also began to visit the site. An unintentional advantage of the provisioning site was that it was located in the overlapping area of the two "unit-groups" (more commonly called "communities" by some later primatologists). Nishida clearly recognized the existence of multi-male and multi-female social groups. This differed from Goodall (1965), who first denied the existence of any permanent groups except those of mother–infant/juvenile (Nishida, 1979; see Chapter 6). Nishida and Kenji Kawanaka cooperatively observed the relationships between these two unit-groups throughout the 1970s (Nishida and Kawanaka, 1972; Kawanaka and Nishida, 1975).

Because of the request by Japanese researchers to establish Mahale as a protected area, JICA (see Chapter 2) began to give its support in 1975. The initial JICA project lasted until 1987 (see Chapter 50 for more details), and greatly contributed to learning about chimpanzees and their surrounding natural

Figure 3.1 Kansyana Camp in 1966. © Toshisada Nishida.

environment. After the collection of such basic information, Mahale was finally declared Tanzania's 11th national park in 1985 (Nishida, 1985).

When Nishida started his long-term field study, the local Tongwe people still maintained their traditional lives utilizing slash-and-burn cultivation, fishing, and hunting (see Chapter 49). Small villages of approximately 7–37 persons (Kakeya, 1974), usually composed of patrilineal polygamous families, were scattered within the bush land/wooded savanna of the Tongwe land. Although their population density was very low (approximately 1 person/km^2), their subsistence activity, in particular the slash-and-burn cultivation, may have affected the surrounding vegetation. The Tongwe supported Nishida and colleagues' research as advisers, trading partners, part-time employees, and field assistants, and some of them are co-authors on scientific papers (e.g. Nishida *et al.*, 1979; Uehara and Nyundo, 1983; Huffman and Seifu, 1989).

Following the socialism policy that the Tanzanian government put in place in the 1970s, most villagers were forced to move out of the study area to two villages, Nkonkwa and Kalya (Nishida, 1978; Chapter 2). Most people left the study area in August 1973, and farming, burning, and hunting were subsequently banned (Uehara and Ihobe, 1998). This situation facilitated the foundation of the National Park.

After the prohibition of agriculture, ecological succession took place; the abandoned fields gradually transformed into grasslands (e.g. *Imperata cylindrica* and *Pennisetum purpureum*), shrub land (e.g. *Harungana madagascariensis*), young secondary forests (e.g. *Pycnanthus angolensis*), and semi-deciduous forests (e.g. *Albizia glaberrima* and *Xylopia longipetala*). Along with this transformation, fauna also may have recovered, and the wild animal population appears to have increased (Uehara, 1997; Uehara and Ihobe, 1998; Chapters 15). Regrettably, there is little quantitative data on the changes in the fauna and flora (see Chapter 13). In particular, little is known about the sympatric animal species, except for monkeys, as Uehara (1997) and Uehara and Ihobe (1998) have pointed out.

3.4 Number of researchers who visited Mahale

From the onset of research at Mahale until the 1970s, the number of researchers staying at Mahale at a time was two to three at the most (mean of 2.3 from 1965 to 1978). Itani, the leader of the expedition and Nishida's supervisor, visited Mahale several times but each only for a short duration. Even when two researchers were present, one studied the K group and the other studied the M group.

When the JICA project (see above) began, many researchers joined and stayed at Mahale as JICA experts for longer durations. A total of seven researchers stayed at Mahale in 1979, and thereafter, an average of 7.1 (maximum of 11) researchers have stayed at Mahale every year. Our best estimate is that the total number of researchers from Japan, North America, Europe, and Tanzania who collected some data at Mahale within those 50 years is 82 (see Appendix V).

3.5 Target groups and observation styles

The main target groups and observation styles at Mahale changed during the different stages of research and with varying circumstances. Surveys on unhabituated unit-groups, i.e. groups other than K and M, are described in Chapters 4 and 11.

In 1965, Nishida cleared a bush around Kansyana to make a sugarcane field (Figure 3.2), and the K and M groups started to visit the site. The data obtained at that time were collected from the provisioning site (Figure 3.3). After the chimpanzees became accustomed to the provisioning, sugarcane was imported. In 1966, up to 50 sugarcanes were supplied in a day, but in 1975, the amount was reduced to 10–30 bananas or 5–10 sugarcanes per 10

Figure 3.2 Clearing the fields to set up a provisioning station at Kansyana. ©Toshisada Nishida.

Figure 3.3 Chimpanzees visiting the provisioning site in the early days. ©Toshisada Nishida.

chimpanzees in a day (Nishida, 1979). After 1971, mobile provisioning was introduced, in which researchers carried a small amount of food (e.g. pieces of sugarcane or bananas) to the forest, gave them to the chimpanzees upon encountering them, and then followed the party (Nishida, 1979). In 1971, Miyako Camp (Figure 3.4) was set up to the north of Kansyana (Nishida, 1981) for the observation of the K group; the observation of the M group was based in Kansyana (see Chapter 2 for the names of the places). The K group became smaller in number beginning in the 1970s and went extinct in the early 1980s (see Chapter 10), which resulted in the loss of the opportunity to study intergroup relationships.

At the initial stage of research, *ad libitum* sampling was performed because of the small number of researchers to follow a party; however, the sampling method significantly changed around 1979. Influenced by Altmann's (1974) defining article about observational methods, Nishida (1981) adopted a focal animal sampling technique for the chimpanzees. Younger researchers (Mariko Hiraiwa-Hasegawa, Toshikazu Hasegawa, and Yukio Takahata), who had become accustomed to this sampling method in their studies of Japanese monkeys, also employed it on chimpanzees for different and more detailed topics (see below).

Systematic collaborative data collection at Mahale also started around this time period. Such data enabled the publication of long-term joint works by multiple researchers, including demographic data (e.g. Hiraiwa-Hasegawa *et al.*, 1984; Nishida *et al.*, 1990, 2003), hunting behavior (Nishida *et al.*, 1979), ranging behavior (Nakamura *et al.*, 2013), and ethograms (Nishida *et al.*, 1999). In particular, the demographic data on wild chimpanzees provide valuable hints for speculating on the lifestyle of early hominids (Nishida *et al.*, 1990).

Figure 3.4 Miyako Camp in 1971. ©Toshisada Nishida.

The amount of provisioned foods was further reduced in 1981, with 1–2 (= 150–360 cm) sugarcanes given to a whole group per day during 1979–82, according to Hasegawa (1987). Finally in 1987, provisioning was completely abandoned.

More than 100 members of the M group were identified in 1980 (Hiraiwa-Hasegawa *et al.*, 1984) and a nearly complete record of demographic data has been accumulated to date (see Chapter 7). However, by 1995, there were some old females who were not yet completely habituated to human observers and were therefore witnessed only a few times a year (Nakamura, personal observation). These old females disappeared along with the largest recorded decline in the group size that began from 1994 (Chapter 7); the reason for this is not known. By the 2000s, researchers were able to follow almost every individual, including females and their dependent offspring, except for newly immigrated females (see also Chapter 28).

We began to employ stricter rules for observation (see Chapter 50 for details) after several chimpanzees died of a respiratory disease in 2006 (see Chapter 25).

3.6 Research topics

Research topics over the past 50 years are diverse. There are online records of publications (http://mahale.main.jp/papers/) that list 1023 Japanese publications and 543 English publications up until 2014. It is not possible to summarize all the outcomes in this short chapter, rather the work done will be discussed by topic throughout this book. There seems to be a transition in the types of research topics from year to year, which we will briefly summarize here.

The first Japanese report on Mahale chimpanzees was published in 1967 (Nishida, 1967), followed by an English publication in 1968 (Nishida, 1968). From the onset of the research until the 1970s, social groups were both the main interest in Japanese primatology at that time and the main study topic at Mahale (see Chapter 6). Therefore, the descriptions of general social behaviors enabled by provisioning were important outcomes of that era (Nishida, 1970, 1979). On the other hand, papers on more specific topics were already being published in the 1970s, including topics such as tool use (Nishida, 1973b), social custom/

culture (McGrew and Tutin, 1978), and hunting (Nishida *et al.*, 1979). These topics have since become important when the entirety of the chimpanzee studies is considered.

The research topics diverged in the 1980s, perhaps in connection with the growing number of researchers and the introduction of focal animal sampling. For example, the quantitative analyses of specific types of social interactions such as allomothering (see Chapter 30) and reciprocal grooming between mother and offspring (Nishida, 1988), and basic ecological data on chimpanzee foods (Nishida and Uehara, 1983) and plants in the area (Nishida and Uehara, 1981) became available in the 1980s. The prominent topics in this decade were infanticide (Chapter 26), hunting and meat eating (Chapter 20), tool use (Chapter 42), and the social development of juvenile and adolescent chimpanzees by Hitoshige Hayaki (Chapters 22, 27). The first volume on the Mahale chimpanzees edited in English was published in 1990 and described the outcome of this type of research performed in the 1980s (Nishida, 1990).

The 1990s was a decade of more targeted research. Each researcher or group of researchers was more focused on a specific topic and analysed it in greater detail. For example, there was a series of studies on the medicinal use of plants and pharmacological analyses by Michael A. Huffman and colleagues (Chapter 24), studies on locomotion by Kevin D. Hunt (Chapter 45), and studies on vocalization by John C. Mitani and others (Chapter 39). The research target tended to be adult males (Kawanaka, 1993; Uehara *et al.*, 1994a, b; Nishida and Hosaka, 1996; Nishida, 1997) and estrous females (Matsumoto-Oda and Oda, 1998; Matsumoto-Oda *et al.*, 1998; Matsumoto-Oda, 1999a, b) in the 1990s. This may be due to the difficulty of following some mothers and their offspring.

From 2000 to present, the research topics have further diversified. Comparative studies with other chimpanzee research sites became common, and studies on chimpanzee culture flourished, stimulated by the paper in *Nature* by Whiten *et al.* (1999) (Chapter 38). In addition, some papers on new topics were published, such as DNA analysis (Chapter 46), novel behaviors (Nishida *et al.*, 2009), and diseases (Chapter 25). There continue to be novel findings, including a report that a leopard (*Panthera pardus*) ate a chimpanzee. This is the first report on this type of evidence not only in Mahale but for all the long-term-studied eastern chimpanzees (Nakazawa *et al.*, 2013).

One issue with the chimpanzee studies in Mahale may be that there are fewer reviews and theoretical works (e.g. Nishida and Hiraiwa-Hasegawa, 1987; Hiraiwa-Hasegawa, 1992) compared with a large number of original articles that often favor descriptive styles. As Nakamura (2009a) pointed out, Japanese primatology could not present a theory in Sibatani's (1999) sense, allowing its followers to investigate and mass-produce articles within the framework, without considering the validity of the framework itself. In particular, stemming from the rise of sociobiology/behavioral ecology, Japanese scholars barely followed the transitions in theories and instead were content to continue with field work. Interestingly, although Nishida, Hiraiwa-Hasegawa, and Hasegawa were the earliest Japanese primatologists to accept sociobiological theories, even they did not always develop theoretical works (see Nakamura, 2009a, for the present trend among young Japanese primatologists).

The field of ecology also lacks study; there were only a few studies on other animals or plants conducted at Mahale (see Chapters 13, 14 for plants, Chapter 15 for mammals, Chapter 21 for insects, Chapter 34 for ectoparasites, and Chapter 47 for internal parasites). Except for several secondary works (predation: Nishida *et al.*, 1979; termite fishing: Uehara, 1982; feeding culture/tradition: Nishida *et al.*, 1983; medicinal plants: Huffman and Seifu, 1989), studies on feeding ecology/strategy have scarcely been taken up (food list: Nishida and Uehara, 1983; ranging behavior: Itoh and Nishida, 2007). In such a situation, Takasaki's (1983) study on seed dispersal by chimpanzees should have been a pioneering work concerning the symbiosis between primates and plants. Regrettably, it has not been cited very often.

3.7 Study topics published only in Japanese

We have looked at the transition between research topics above mainly from the superficial perspective of English publications. However, we should be cautious that some research topics are not readily written in the form of journal papers, especially those that are not easily accepted in the Anglophone-dominated academic world (or at least that is what authors may think before they consider submitting an article) and, therefore, have not been written as English articles. Thus, we should introduce some research outcomes that are so far available only in Japanese.

Books are one type of these Japanese publications that have been written for the Japanese public. For example, a series of books by Itani describing the dawn of ape studies in Africa (e.g. Itani, 1961b, 1970, 1977b) have the characteristics of adventure stories, travel stories, and documentaries of yet unknown Africa, and thus attracted a lot of readers in Japan. Some young readers of Itani's books aspired to become primatologists. There are also some books that describe the Mahale chimpanzees' society and behaviors (Nishida, 1973a, 1981, 1994, 2008; Hayaki, 1990; Nakamura, 2009b). They not only educated people about chimpanzees but also contained a lot of interesting episodes and the authors' impressions from the field site that would often be omitted from scientific journal papers. There are two Japanese academic books (Itani, 1977a; Nishida *et al.*, 2002) to which active researchers at that time contributed chapters. These books provided important information in their own language to Japanese researchers in other areas or students who would go on to study primates or human evolution.

Papers written only in Japanese are often those on topics that are difficult, at least for some, to express in the English language because the subtle nuances of epistemological backgrounds can often be adequately expressed only in one's native language (see Nakamura, 2009a). Such topics include communication, social interaction, and society/sociality. For example, Hayaki (1988a, b) and Takahata (1997) began to discuss chimpanzee communication early on.

More recently, some Mahale researchers organized a series of free forums, called *Primatology as Studies of Sociality*, at the 2007–09 academic meetings of the Primate Society of Japan. They tried to discuss primate sociality not only from the perspective of costs and benefits relevant to biological discussions but also by considering monkeys and apes more directly as "social beings." The Mahale researchers contributed papers to each forum (Nishie, 2008; Hanamura, 2010; Nakamura, 2010). An edited volume, titled *Boundary and Conjunction of Social Interaction* (Kimura *et al.*, 2010), was published in 2010. Primatologists, anthropologists, and sociologists gathered together to discuss the theme of social interactions in human and nonhuman primates. Among them, six chapters were written on the Mahale chimpanzees, and two additional chapters on primates were written by the Mahale researchers. Similarly, the academic meeting, *Human Society in Evolutionary Perspective*, was hosted by the anthropologist, Kaori Kawai, who was joined by other anthropologists, philosophers, and primatologists. Here again, the majority of the primatologists are those studying chimpanzees at Mahale (Hanamura, 2013; Itoh, 2013; Nishie, 2013).

In these Japanese writings, the qualitative and minute descriptions of events are often analysed, unlike English journal papers, where quantitative analyses are often predominant. Having actually observed the natural behaviors of chimpanzees in Mahale, the authors analysed the social phenomena, their qualitative aspects, and explored their own philosophies that were, in some part, fertilized at the field site. Readers should bear in mind that, apart from the internationally well-known outcomes based on quantitative data, there are these types of discussions that have emerged from the field studies in Mahale.

References

Altmann, J. (1974). Observational study of behavior: sampling methods. *Behaviour*, **49**, 227–67.

Asquith, P. J. (1986). Imanishi's impact in Japan. *Nature*, **323**, 675–6.

Asquith, P. J. (2007). Source for Imanishi Kinji's views of sociality and evolutionary outcomes. *Journal of Bioscience*, **32**, 635–41.

Goodall, J. (1965). Chimpanzees of the Gombe Stream Reserve. In *Primate Behavior*, ed. I. DeVore. New York: Holt, Rinehart and Winston, pp. 425–73.

Halstead, B. (1985). Anti-Darwinian theory in Japan. *Nature*, **317**, 587–9.

Hanamura, S. (2010). [Long-distance calls by chimpanzees: how they connect actions and organize their social field beyond visual contact.] *Primate Research*, **26**, 159–76. In Japanese with English summary.

Hanamura, S. (2013). [When they keep their ears open for vocalization of others in the distance: chimpanzees' process-oriented conventions and an alternative mode of institution.] In [*Practices, Conventions and Institutions: The Evolution of Human Sociality*], ed. K. Kawai. Kyoto: Kyoto University Press, pp. 167–94. In Japanese.

Hasegawa, T. (1987). [*Sexual Behavior of Wild Chimpanzees: A Comparative Psychological Study*.] Doctoral Dissertation. Tokyo: The University of Tokyo. In Japanese.

Hayaki, H. (1988a). [On chimpanzee communication.] *Seibutsu Kagaku*, **40**, 131–9. In Japanese.

Hayaki, H. (1988b). [How chimpanzees communicate: clues from grooming.] *Kikan Jinruigaku*, **19**, 30–49. In Japanese.

Hayaki, H. (1990). [*The Human within Chimpanzee*.] Tokyo: Shōka-bō. In Japanese.

Hiraiwa-Hasegawa, M. (1992). Cannibalism among non-human primates. In *Cannibalism: Ecology and Evolution among Diverse Taxa*, ed. M. A. Elgar and B. J. Crespi. Oxford: Oxford University Press, pp. 323–38.

Hiraiwa-Hasegawa, M., Hasegawa, T., and Nishida, T. (1984). Demographic study of a large-sized unit-group of chimpanzees in the Mahale Mountains, Tanzania: A preliminary report. *Primates*, **25**, 401–13.

Huffman, M.A. and Seifu M. (1989). Observations on the illness and consumption of a possible medicinal plant *Vernonia amygdalina* (Del), by a wild chimpanzee in the Mahale Mountains National Park, Tanzani. *Primates*, **30**, 51–60.

Imanishi, K. (1941 [2002]). [*The World of Living Things*.] Tokyo: Chūko Classics. In Japanese.

Imanishi, K. (1957a). [Identification: a process of enculturation in the subhuman society of *Macaca fuscata*.] *Primates*, **1**, 1–29. In Japanese.

Imanishi, K. (1957b). Social behavior in Japanese monkeys, *Macaca fuscata. Psychologia*, **1**, 47–54.

Imanishi, K. (1960a). [*The Gorilla*.] Tokyo: Bungei-Shunjū-shinsha. In Japanese.

Imanishi, K. (1960b). Social organization of subhuman primates in their natural habitat. *Current Anthropology*, **1**, 393–407.

Imanishi, K. (1977). [Preface.] In [*The Chimpanzees*], ed. J. Itani. Tokyo: Kōdan-sha, pp. i–xvi. In Japanese.

Imanishi, K. (2002). *A Japanese View of Nature: The World of Living Things*, trans. P. J. Asquith, H. Kawakatsu, S. Yagi, and H. Takasaki. New York: Routledge Curzon.

Itani, J. (1956). [Language before humans: communication of Japanese monkeys, part II.] *Shizen*, **11**(11), 22–7. In Japanese.

Itani, J. (1959). Paternal care in the wild Japanese monkey, *Macaca fuscata fuscata. Primates*, **2**, 61–93.

Itani, J. (1961a). The society of Japanese monkeys. *Japan Quarterly*, **8**, 421–30.

Itani, J. (1961b). [*The Forest of Gorillas and Pigmies*.] Tokyo: Iwanami. In Japanese.

Itani, J. (1970). [*In the Pursuit of Chimpanzees*.] Tokyo: Chikuma-shobō. In Japanese.

Itani, J., ed. (1977a). [*The Chimpanzees*.] Tokyo: Kōdan-sha. In Japanese.

Itani, J. (1977b [1993]). [*The Savanna of Chimpanzees: In Search of Logic of the Wild*.] Tokyo: Heibon-sha Library. In Japanese.

Itani, J. (1979). Distribution and adaptation of chimpanzees in an arid area (Ugalla Area, Western Tanzania). In *The Great Apes*, ed. D. A. Hamburg and E. R. McCown. Menlo Park, CA: Benjamin/Cummings, pp. 55–72.

Itani, J. (1991). [My resumé.] Reprinted in J. Itani (2007). [*Works of Jun'ichiro Itani, Vol. I: Birth of Japanese Primatology*.] Tokyo: Heibon-sha, pp. 7–30. In Japanese.

Ito, Y. (1991). Development of ecology in Japan, with special reference to the role of Kinji Imanishi. *Ecological Research*, **6**, 139–55.

Itoh, N. (2013). [Dual-rings of mode of coexistence and act choices: Chimpanzee group and emergence of institution.] In [*Practices, Conventions and Institutions: The Evolution of Human Sociality*], ed. K. Kawai. Kyoto: Kyoto University Press, pp. 143–66. In Japanese.

Itoh, N. and Nishida, T. (2007). Chimpanzee grouping patterns and food availability in Mahale Mountains National Park, Tanzania. *Primates*, **48**, 87–96.

Izawa, K. (1970). Unit groups of chimpanzees and their nomadism in the savanna woodland. *Primates*, **11**, 1–45.

Kakeya, M. (1974). [Livelihood maintenance system of Tongwe: living environment, subsistence and diet.] *Kikan Jinruigaku*, **5**(3), 3–90. In Japanese.

Kano, T. (1971). Distribution of the primates on the eastern shore of Lake Tanganyika. *Primates*, **12**, 281–304.

Kano, T. (1972). Distribution and adaptation of the chimpanzee on the eastern shore of Lake Tanganyika. *Kyoto University African Studies*, **7**, 37–129.

Kawai, M. (1961). [*The Gorilla Expedition*.] Tokyo: Kappa Books. In Japanese.

Kawai, M. (1965). Newly-acquired pre-cultural behavior of the natural troop of Japanese monkeys of Koshima Islet. *Primates*, **6**, 1–30.

Kawamura, S. (1956). [Culture before humans: with reference to wild Japanese monkeys.] *Shizen*, **11** (11), 28–34. In Japanese.

Kawamura, S. (1959). The process of sub-culture propagation among Japanese macaques. *Primates*, **2**, 43–60.

Kawanaka, K. (1993). Age differences in spatial positioning of males in a chimpanzee unit-group at the Mahale Mountains National Park, Tanzania. *Primates*, **34**, 255–70.

Kawanaka, K. and Nishida, T. (1975). Recent advances in the study of inter-unit relationships and social structure of wild chimpanzees of the Mahali Mountains. In *Proceedings of the Symposium of the Fifth Congress of the International Primatological Society*, ed. S. Kondo, M. Kawai, A. Ehara, and S. Kawamura. Tokyo: Japan Science Press, pp. 173–86.

Kimura, D., Nakamura, M., and Takanashi, K. ed. (2010). [*Boundary and Conjunction of Social Interaction: Studies in Nonhuman Primates, Humans and Conversation*.] Kyoto: Shōwa-dō. In Japanese.

Matsumoto-Oda, A. (1999a). Female choice in the opportunistic mating of wild chimpanzees (*Pan troglodytes schweinfurthii*) at Mahale. *Behavioral Ecology and Sociobiology*, **46**, 258–66.

Matsumoto-Oda, A. (1999b). Mahale chimpanzees: grouping pattern and cycling females. *American Journal of Primatology*, **47**, 197–207.

Matsumoto-Oda, A. and Oda, R. (1998). Changes in the activity budget of cycling female chimpanzees. *American Journal of Primatology*, **46**, 157–66.

Matsumoto-Oda, A., Hosaka, K., Huffman, M. A., and Kawanaka, K. (1998). Factors affecting party size in chimpanzees of Mahale Mountains. *International Journal of Primatology*, **19**, 999–1011.

Matsuzawa, T. and McGrew, W. C. (2008). Kinji Imanishi and 60 years of Japanese primatology. *Current Biology*, **18**, R587–91.

McGrew, W. C. and Tutin, C. E. G. (1978). Evidence for a social custom in wild chimpanzees? *Man*, **13**, 234–51.

Nakamura, M. (2009a). Interaction studies in Japanese primatology: their scope, uniqueness, and the future. *Primates*, **50**, 142–52.

Nakamura, M. (2009b). [*Chimpanzees: What They Have Told Me without Language*.] Tokyo: Chūkō-shinsho. In Japanese.

Nakamura, M. (2010). [What does it mean a society to be complex? Questions about relationship between society and individual.] *Primate Research*, **26**, 131–42. In Japanese with English summary.

Nakamura, M. and Nishida, T. (2006). Subtle behavioral variation in wild chimpanzees, with special reference to Imanishi's concept of *kaluchua*. *Primates*, **47**, 35–42.

Nakamura, M., Corp, N., Fujimoto, M., *et al.* (2013). Ranging behavior of Mahale chimpanzees: a 16 year study. *Primates*, **54**, 171–82.

Nakazawa, N., Hanamura, S., Inoue, E., Nakatsukasa, M., and Nakamura, M. (2013). A

leopard ate a chimpanzee: the first evidence from East Africa. *Journal of Human Evolution*, **65**, 334–7.
Nishida, T. (1967). [The society of wild chimpanzees: success of provisioning at Kasoge.] *Shizen*, **22**(8), 31–41. In Japanese.
Nishida, T. (1968). The social group of wild chimpanzees in the Mahali Mountains. *Primates*, **9**, 167–224.
Nishida, T. (1970). Social behavior and relationships among chimpanzees of Mahali Mountains. *Primates*, **11**, 47–87.
Nishida, T. (1973a). [*The Children of the Mountain Spirits: Search for the Social Structure of Chimpanzees*.] Tokyo: Chikuma-shobō. In Japanese.
Nishida, T. (1973b). The ant-gathering behaviour by the use of tools among wild chimpanzees of the Mahali Mountains. *Journal of Human Evolution*, **2**, 357–70.
Nishida, T. (1978). [Revolution in the Lake Tanganyika Region: a travel to Western Tanzania.] *Tenbō*, **232**, 103–24. In Japanese.
Nishida, T. (1979). The social structure of chimpanzees of the Mahale Mountains. In *The Great Apes*, ed. D. A. Hamburg and E. R. McCown. Menlo Park, CA: Benjamin/Cummings, pp. 73–121.
Nishida, T. (1981). [*The World of Wild Chimpanzees*.] Tokyo: Chūkō-shinsho. In Japanese.
Nishida, T. (1985). The Mahale Mountains declared a national park. *IPPL Newsletter*, **12**, 9.
Nishida, T. (1988). Development of social grooming between mother and offspring in wild chimpanzees. *Folia Primatologica*, **50**, 109–23.
Nishida, T. ed. (1990). *The Chimpanzees of the Mahale Mountains: Sexual and Life History Strategies*. Tokyo: University of Tokyo Press.
Nishida, T. (1994). [*Thirty-six Stories of Wild Chimpanzees*.] Tokyo: Kinokuniya. In Japanese.
Nishida, T. (1997). Sexual behavior of adult male chimpanzees of the Mahale Mountains National Park, Tanzania. *Primates*, **38**, 379–98.
Nishida, T. (2008). [*The Societies of Chimpanzees*.] Tokyo: Tōhō-shuppan. In Japanese.
Nishida, T. (2012). *Chimpanzees of the Lakeshore*. Cambridge: Cambridge University Press.
Nishida, T. and Hiraiwa-Hasegawa, M. (1987). Chimpanzees and bonobos: Cooperative relationships among males. In *Primate Societies*, ed. B. B. Smuts, D. L. Cheney, R. M. Seyfarth, R. W. Wrangham, and T. T. Struhsaker. Chicago, IL: The University of Chicago Press, pp. 165–77.
Nishida, T. and Hosaka, K. (1996). Coalition strategies among adult male chimpanzees of the Mahale Mountains. In *Great Ape Societies*, ed. W. C. McGrew, L. F. Marchant, and T. Nishida. Cambridge: Cambridge University Press, pp. 114–34.
Nishida, T. and Kawanaka, K. (1972). Inter-unit-group relationships among wild chimpanzees of the Mahali Mountains. *Kyoto University African Studies*, **7**, 131–69.
Nishida, T. and Uehara, S. (1981). Kitongwe name of plants: a preliminary listing. *African Study Monographs*, **1**, 109–31.
Nishida, T. and Uehara, S. (1983). Natural diet of chimpanzees (*Pan troglodytes schweinfurthii*): long-term record from the Mahale Mountains, Tanzania. *African Study Monographs*, **3**, 109–30.
Nishida, T., Uehara, S., and Nyundo, R. (1979). Predatory behavior among wild chimpanzees of the Mahale Mountains. *Primates*, **20**, 1–20.
Nishida, T., Wrangham, R., Goodall, J., and Uehara, S. (1983). Local differences in plant-feeding habits of chimpanzees between the Mahale Mountains and Gombe National Park, Tanzania. *Journal of Human Evolution*, **12**, 467–80.
Nishida, T., Takasaki, H., and Takahata, Y. (1990). Demography and reproductive profiles. In *The Chimpanzees of the Mahale Mountains: Sexual and Life History Strategies*, ed. T. Nishida. Tokyo: University of Tokyo Press, pp. 63–97.
Nishida, T., Kano, T., Goodall, J., McGrew, W. C., and Nakamura, M. (1999). Ethogram and ethnography of Mahale chimpanzees. *Anthropological Science*, **107**, 141–88.
Nishida, T., Uehara, S., and Kawanaka, K. ed. (2002). [*The Mahale Chimpanzees: 37 Years of <Panthropology>*.] Kyoto: Kyoto University Press. In Japanese.
Nishida, T., Corp, N., Hamai, M., *et al*. (2003). Demography, female life history, and reproductive profiles among the chimpanzees of Mahale. *American Journal of Primatology*, **59**, 99–121.
Nishida, T., Matsusaka, T., and McGrew, W. C. (2009). Emergence, propagation or disappearance of novel behavioral patterns in the habituated chimpanzees of Mahale: A review. *Primates*, **50**, 23–36.

Nishie, H. (2008). [Chimpanzee culture and sociality: Reconsidering "the metaphor of transmission of knowledge".] *Primate Research*, **24**, 73–90. In Japanese with English summary.

Nishie, H. (2013). ["Who" is the alpha male? Institutional aspects of dominance ranks in chimpanzee societies.] In [*Practices, Conventions and Institutions: The Evolution of Human Sociality*], ed. K. Kawai. Kyoto: Kyoto University Press, pp. 121–42. In Japanese.

Sakura, O., Sawaguchi, T., Kudo, H., and Yoshikubo, S. (1986). Declining support for Imanishi. *Nature*, **323**, 586.

Sano, T. (1988). Halstead's commentary on Imanishi and his theory. *Journal of Social and Biological Structures*, **11**, 251–3.

Sibatani, A. (1999). [*Structuralistic Biology*.] Tokyo: University of Tokyo Press. In Japanese.

Suzuki, A. (1969). An ecological study of chimpanzees in a savanna woodland. *Primates*, **10**, 103–48.

Takahata, Y. (1997). [Monkey talk or what we have learned about "how nonhuman primates communicate".] In [*Natural History of Communication*], ed. Y. Tani. Tokyo: Shin'yō-sha, pp. 61–82. In Japanese.

Takasaki, H. (1983). Seed dispersal by chimpanzees: a preliminary note. *African Study Monographs*, **3**, 105–8.

Uehara, S. (1982). Seasonal change in the techniques employed by wild chimpanzees in the Mahale Mountains, Tanzania, to feed on termites (*Pseudacanthotermes spiniger*). *Folia Primatologica*, **37**, 44–76.

Uehara, S. (1997). Predation on mammals by the chimpanzee (*Pan troglodytes*). *Primates*, **38**, 193–214.

Uehara, S. and Ihobe H. (1998). Distribution and abundance of diurnal mammals, especially monkeys, at Kasoje, Mahale Mountains, Tanzania. *Anthropological Science*, **106**, 349–69.

Uehara, S. and Nyundo, R. (1983). One observed case of temporary adoption of an infant by unrelated nulliparous female among wild chimpanzees in the Mahale Mountains, Tanzania. *Primates*, **24**, 456–66.

Uehara, S., Hiraiwa-Hasegawa, M., Hosaka, K., and Hamai, M. (1994a). The fate of defeated alpha male chimpanzees in relation to their social networks. *Primates*, **35**, 49–55.

Uehara, S., Nishida, T., Takasaki, H., *et al.* (1994b). A lone male chimpanzee in the wild: the survivor of a disintegrated unit-group. *Primates*, **35**, 275–81.

de Waal, F. (2001). *The Ape and the Sushi Master*. New York: Basic Books.

de Waal, F. B. M. (2003). Silent invasion: Imanishi's primatology and cultural bias in science. *Animal Cognition*, **6**, 293–9.

Whiten, A., Goodall, J., McGrew, W. C., *et al.* (1999). Cultures in chimpanzees. *Nature*, **399**, 682–5.

4 Chimpanzee distribution: accumulation of survey reports

KOICHIRO ZAMMA, SHUNKICHI HANAMURA, AND TETSUYA SAKAMAKI

4.1 Introduction

Chimpanzees inhabit a region in equatorial Africa from the coast of Western Africa to the Albertine Rift in East Africa, and four subspecies have been recognized to date (see Chapter 1). They have been shown to live in more than 20 countries, including Tanzania (Kormos *et al.*, 2003; Plumptre *et al.*, 2010).

Tanzania is at the eastern edge of chimpanzee distribution. One of the earliest records on evidence of chimpanzee distribution in Tanzania was by Dollman (1935), who reported that B. W. Savory of the political service observed the footprints and beds of chimpanzees in the Kigoma District and collected a *Pan troglodytes schweinfurthii* skull brought by villagers of Kasoje near the Mahale Mountains. Habituation of chimpanzees started in Gombe (Goodall, 1986), Kabogo (Azuma and Toyoshima, 1965), Kasakati (Izawa and Itani, 1966), Filabanga (Kano, 1971), and Mahale (Nishida, 1990; Chapter 3) in the 1960s. Kano (1972) investigated the distribution of chimpanzees in Tanzania by performing an extensive survey in Western Tanzania, excluding the region to the north of the Malagarasi River, from 1965 to 1967. He estimated a population of about 2000 chimpanzees in 44–51 unit-groups in this area based on information from villagers and direct encounters, vocalizations, and beds of chimpanzees.

As habituation of chimpanzees was successful at two study sites, Gombe and Mahale (Goodall, 1986; Nishida, 1990), many researchers concentrated on observing identified chimpanzees, whereas only few extensive surveys were conducted in Western Tanzania in the 1970s, except for Itani (1990) and Nishida (1989). However, since the 1990s, the chimpanzee populations in Tanzania in areas other than Gombe and Mahale have been surveyed extensively (e.g. Moyer *et al.*, 2006; Yoshikawa *et al.*, 2008; Ogawa *et al.*, 2013), and the conservation status and gene flow of the species in this area have also been discussed (Plumptre *et al.*, 2010; Inoue *et al.*, 2013).

Although the M and K groups in the Kasoje area of Mahale have been studied intensively, the distributions of chimpanzees in other areas of Mahale have not been evaluated systematically. Some surveys have been planned and carried out, but not all were intended primarily to determine the distributions of chimpanzees. Some researchers focused on the vertical distribution of vegetation and/or fauna, and some wished to visit the remains of old villages, but also recorded evidence of chimpanzees in their survey areas. This chapter collates the published and unpublished survey reports and presents a discussion of the distribution of chimpanzees in and around Mahale across the study periods.

4.2 Survey route and data collection

Researchers conducted their surveys according to their specific interests, and we have classified them here into three methods: reconnaissance walks, boat surveys, and chimpanzee group trackings. Reconnaissance walks are exploratory surveys to obtain information about the area and are distinguished from line transect methods by their non-linear routes. Some reconnaissance walk surveys took the old paths of villagers, while others were conducted along rivers and/or mountain ridges.

Mahale Chimpanzees: 50 Years of Research, ed. M. Nakamura, K. Hosaka, N. Itoh and K. Zamma. Published by Cambridge University Press.

Therefore, the information of animal tracks and signs obtained by reconnaissance walks is difficult to consider as indicating precise values of population density of animals, but is useful for verifying the distributions of animal species in the survey area.

Boat surveys are similar to reconnaissance walks except that researchers take a boat and count beds of chimpanzees along the shore (Shimada, 2003). Compared with reconnaissance walks on the ground, a boat on a lake has a number of advantages with regard to looking for chimpanzee beds because of the lack of obstacles, such as leafy trees or woody vines. Researchers are occasionally able to detect a bed on a hill more than 100 m from the boat. Similar to reconnaissance walks, the boat survey method is a valuable means of confirming the presence of animals, but is difficult to use for calculating their density because the detection range changes according to the shape and slope of shore.

Chimpanzee group tracking involves trailing and observing the target unit-group of chimpanzees. This method is useful to distinguish the target unit-group from other groups.

There have been more than 20 such surveys between the 1975 and 2013 (Table 4.1). Five reconnaissance walks and three group tracking surveys were conducted before 1985 in Mahale when the national park was established, and 10 reconnaissance walks, 2 boat surveys, and 2 group tracking surveys have been carried out in this area since 1985. Some other reconnaissance walks have been conducted around Mt. Nkungwe and Mt. Mhensabantu, but the data are not included in Table 4.1 because they were almost identical to those of previous surveys.

The routes of reconnaissance walks and boat surveys were drawn on a map for each period: in the 1960s and before and after establishment of the park (Figure 4.1). We drew the routes based on references or personal communication from the researchers listed in Table 4.1. The routes were retraced on a map following the maps in the reports. In cases where the routes were not shown in the original papers, they were drawn here based on the description in the report.

The actual distance of the survey route will be longer than that on maps because the route on the ground will have a number of obstacles, such as rocks, fallen trees, rivers, cliffs, etc., which the researcher will need to detour around. Researchers now use GPS and can measure the distances of their survey routes with greater precision. However, we do not discuss these data in this chapter for comparison of recent surveys with old data. Thus, all of the survey distances were measured from the maps in Figure 4.1 using the Google Maps™ distance measurement tool (Google, 2014). These are rough measurements but sufficient for comparisons across the study periods.

As mentioned above, although extensive surveys were not always conducted to determine the distribution of chimpanzees, indicators of the presence of chimpanzees, such as direct observations, vocalizations, feces, and beds, were often mentioned in survey reports. The tracks and signs of other animals, such as direct and vocal observations, feces, footprints, fur, spines, skulls, and bones, were also noted in these reports. We compiled the data on maps for each study period to determine the changes in distribution of animals across the study period. The study period was divided into three sections: (1) the first extensive survey conducted by Kano from 1965 to 1967; (2) the period from 1968 to 1985 when the park was established; and (3) after 1985.

4.3 Distribution of chimpanzees

Tracks and signs of chimpanzees were found on all survey routes, and it was confirmed that chimpanzees were distributed widely in and around the national park (Figure 4.2). More chimpanzees were concentrated at the northwest side of the Mahale Mountains where the largest forest in Mahale, the Kasoje forest, is located (Itoh *et al.*, 2012; Chapters 2, 13). The Kasoje forest provides high availability of fruits and other food for chimpanzees (Itoh *et al.*, 2012; Chapters 13, 14, 17), which could explain the concentration of their distribution in this area. The vacancy of indirect information in this area after 1985 in Figure 4.2 may be due to exclusions of those from well-known groups, such as the M and K groups.

Table 4.1 **Records of surveys performed in and around the Mahale Mountains National Park between 1975 and 2013**

	Records of chimpanzee						
Methods[a]	Date	Survey distance (km)[b] Total (in the park)	No of beds (in the park)	Feces	Vocal/ Direct	Survey route in Figure 4.1	Source
W	Aug 14–19, 1975	80 (68)	26 (20)	2	4	a	Nishida, 1975
W	May 16–19, 1976	23 (19)	14 (14)	0	1	b	Nishida, 1976
W	Sept 3–5, 1976	13 (13)	15 (15)	8	2	c	Norikoshi, 1976a
W	Oct 11–13, 1976	16 (0)	34 (0)	–	–	d	Norikoshi, 1976b
W	Sep 16–19, 1977	67 (60)	Present (Present)	–	–	e	Uehara and Nishida, 1978
W	Sep 9–17, 1996	70 (51)	2 (2)	2	1	f	Nakamura and Fukuda, 1999
W	Jun 23–26, 2000	37 (31)	0 (0)	1	2	g	Sakamaki, Mwami, Kasakampe, and Kabangula, unpublished data
W	Oct 12–17, 2003	101 (0)	20 (0)	2	Present	h	Zamma *et al.*, 2004
W	Sep 6–11, 2004	53 (43)	136 (136)	Present	Present	i	Nakamura *et al.*, 2005
W	Aug 13–17, 2006	66 (60)	2 (2)	Present	1	j	Hanamura, Mwami, and Rashidi, unpublished data
W	Nov 23–Dec 1, 2006	164 (52)	11 (11)	0	3	k	Sakamaki, Mwami, and Kabangula, unpublished data
W	Oct 27–28, 2008	43 (4)	8 (8)	0	0	l	Zamma, Bunengwa, and Ramadhani, unpublished data

Table 4.1 (**cont.**)

		Records of chimpanzee					
Methods[a]	Date	Survey distance (km)[b] Total (in the park)	No of beds (in the park)	Feces	Vocal/ Direct	Survey route in Figure 4.1	Source
W	Jan 7–8, 2011	16 (6)	0 (0)	0	0	m	Zamma, Bunengwa, and Rehani, unpublished data
W	Nov 18–19, 2013	31 (31)	5 (5)	0	3	n	Zamma and Kabangula, unpublished data
W	Nov 28, 2013	6 (6)	0 (0)	0	0	o	Zamma, Kabangula, and Ramadhani, unpublished data
B	Mar 20–27, 2002	15 (15)	131 (131)	–	6	p	Shimada, 2003
B	Nov 28, 2013	25 (25)	36 (36)	–	–	q	Zamma, Kabangula, Ramadhani, and Rashidi, unpublished data
T	Oct 16–23, 1975	–	–	30	Present		Nishida and Norikoshi, 1975
T	Aug 26–29, 1980	–	–	Present	Present		Hasegawa and Hasegawa, 1980
T	Oct 9, 1979; Sep 22, Oct 31, Dec 5–10, 1981	–	–	–	Present		Takahata *et al.*, n.d.
T	Jun 18–27, 1998	–	–	–	Present		Itoh *et al.*, 1999
T	Mar 20–27, 2002	–	–	–	Present		Shimada, 2003

[a] W, reconnaissance walk; B, boat census; T, group tracking.
[b] The distances of reconnaissance walks and boat census surveys were calculated using the Google Maps™ distance measurement tool.

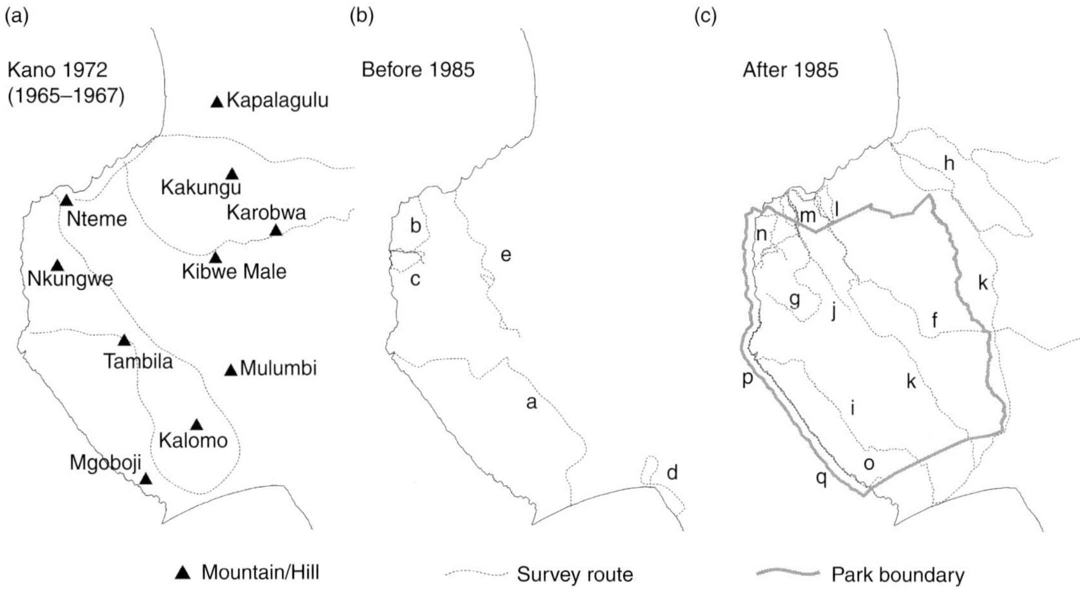

Figure 4.1 Survey routes of reconnaissance walks and boat surveys conducted from 1965 to 1967 (a, modified from Fig. 3 in Kano, 1972), before 1985 (b), and after 1985 (c). Codes indicate the routes as outlined in Table 4.1.

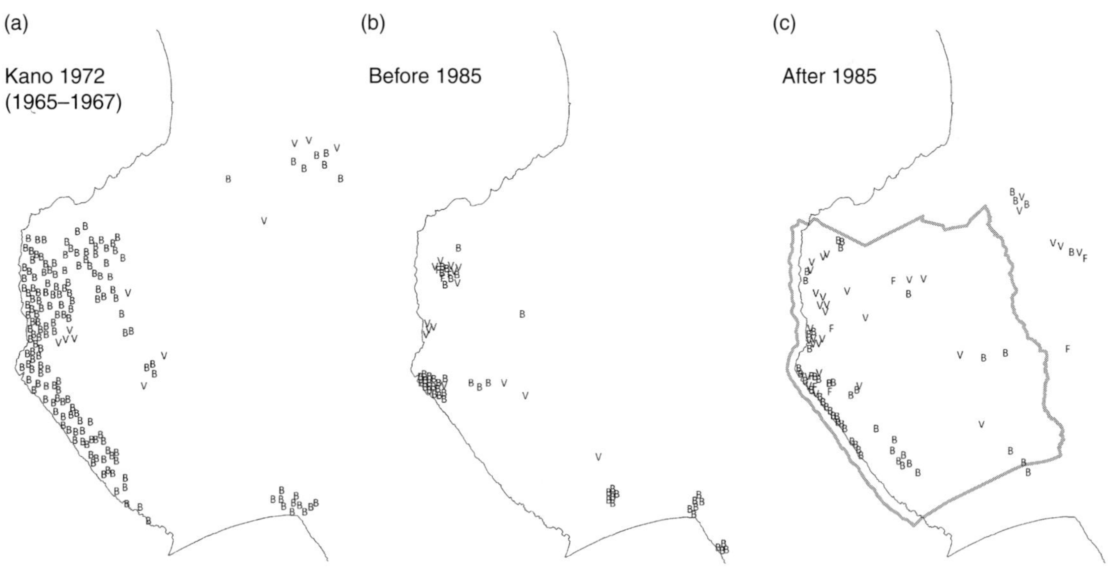

Figure 4.2 Evidence of the presence of chimpanzees obtained from 1965 to 1967 (a, modified from Fig. 6 in Kano, 1972), before 1985 (b), and after 1985 (c). V: vocalization and direct observation; B: bed; F: feces.

It has been suggested that there is a gap in distribution between the Mahale Mountains population and that of the Kakungu–Karobwa Mountains, located about 40 km to the northeast of the Mahale Mountains (Kano, 1972). However, Nakamura and Fukuda (1999) observed beds, feces, and vocalizations of chimpanzees on their survey route along the eastern boundary between the Mahale

Mountains and the Karobwa Hills, suggesting that these two populations may be connected in this area.

There have been no reports of evidence of chimpanzees at the southwest boundary of the park since the 1960s. Human activities in this area have been increasing, with deforestation by fires from village land (Itoh *et al.*, 2012). Although human activity may be one of the reasons for the absence of chimpanzees in this area, vegetation may be another explanation. Kano (1972) suggested that *Brachystegia bussei* was one of the key plant species determining the distribution of chimpanzees in woodland areas. Within the national park, trees of *B. bussei* were found along the eastern shore of Lake Tanganyika, but were not present in the southern part of the Mahale peninsula (Kano, 1972). The lack of this tree species may prevent chimpanzee distribution in this area.

4.4 Unit-group distribution

Kano (1972) estimated the presence of 13–15 unit-groups in the Mahale area (Figure 4.3). Subsequent chimpanzee group tracking surveys identified 12 unit-groups, 8 of which were named B, K, M, N, L, H, S, and U groups (Nishida *et al.*, 1990). The sizes of these named groups were estimated from the 1970s to 1980s as follows: B (60–80), K (10), M (90), N (> 40), L (40–80), H (20–40) S (unknown), and U (unknown) (Nishida *et al.*, 1990). Many of these unit-groups were found in the lower area along the lakeshore, but some were located higher in the mountainous area. One of the groups ranging in the mountainous area was the S group, and feces were found between Mt. Sisaga and Mt. Mfitwa at altitudes over 2100 m a.s.l. (Nishida, 1973). The H group also ranges in the mountainous area. Hasegawa and Hasegawa (1980) observed nine individuals of the H group near Mt. Humo, and two fecal samples were collected on the ridge of Mt. Mhensabantu (1820 and 1860 m a.s.l.).

After the establishment of the national park, some researchers tracked unit-groups other than the M and K groups. Itoh *et al.* (1999) observed a party of chimpanzees belonging to an unknown unit-group that had invaded near the center of the M-group home

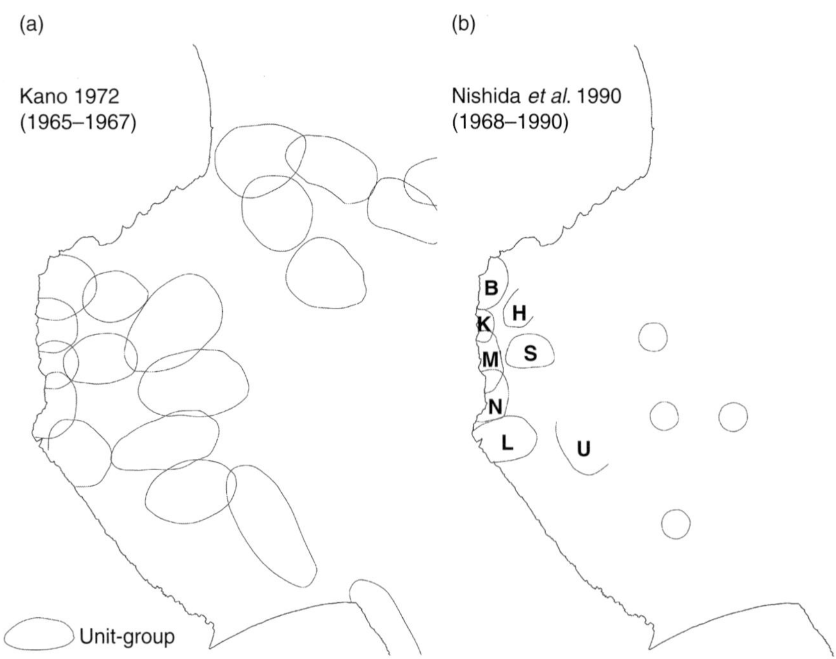

Figure 4.3 Ranges of unit-groups estimated by (a) Kano (1972) and (b) Nishida *et al.* (1990).

range. Shimada (2003) tracked the N and L groups of chimpanzees and confirmed that their ranges were the same as in the 1980s (see also Chapter 11). We have no data regarding unit-groups of the whole area of the park, but comparison of Figures 4.2 and 4.3 shows some evidence of chimpanzees along the southern part of the lakeshore, where no known unit-groups have been identified since the 1960s. Although it is possible that such evidence was left by non-resident parties of chimpanzees occasionally visiting this area, it is more likely that the area is constantly used by the resident members of a group (i.e. the home range of a known group is larger than previously supposed, or there is another unit-group in this area) because the area is covered with *B. bussei.*

4.5 Chimpanzee population density

The population densities of chimpanzees in and out of the park were calculated (Table 4.2). The lifespan of beds, which differs between forest and woodland areas in Mahale (Ihobe, 2005; Zamma and Makelele, 2012; Chapter 43), is one of the key variables for accurate estimation of chimpanzee density (Ghiglieri, 1984; Tutin and Fernandez, 1984; Plumptre and Reynolds, 1996). However, Kano's formula (Kano, 1972) was used to allow comparison with previous estimates of chimpanzee densities:

$$\text{Density } (/\text{km}^2) = [(N \text{ of beds}) \times (1/\text{Detection efficiency}) \times (1/\text{Bed lifespan})]/[(\text{Distance of survey route}) \times (\text{Width of survey route})]$$

We used the same value for each variable as Kano (1972): detection efficiency = 0.7, bed lifespan = 180 days, width of survey route = 0.14 km.

The results shown in Table 4.2 indicate that the population density of chimpanzees in the park was almost identical before and after 1985, when the park was established. However, it was one-tenth of the density in the 1960s calculated by Kano (1972; Table 4.2). It is difficult to accept that the population decreased so rapidly from the 1960s to the 1970s because the distribution of chimpanzees did not change during this period. The low density of chimpanzee population after the 1970s is thought to be due to the bias of survey route and survey effort. As mentioned above, the evidence of the M- and K-group chimpanzees was likely ignored in these surveys, and the data on the area of high population density were not evaluated in this estimation. Even when researchers conducted surveys in the area out of the home ranges of the M and K groups, they investigated a number of topics other than the chimpanzee population density; some were interested in the vertical changes of vegetation and walked on the mountain ridge, while others visited villages and walked along the paths of villagers. Chimpanzee beds were likely built on slopes or hillsides (Hernandez-Aguilar, 2009; Ogawa *et al.*, 2013), but the survey routes were selected to avoid such areas that were unsuitable for walking. In addition, some researchers likely walked quickly to reach their destination, such as campsites with water flow, and could easily miss chimpanzee beds in the trees. They sometimes walked along roads out of the park. These biases may have resulted in the apparent low chimpanzee population density after the 1970s.

For the conservation of chimpanzees, it is important to accurately estimate the population density. However, there are a number of difficulties, including the variability of bed lifespan; it has been reported that the bed lifespan in woodlands is about twice that in forests, although the two study areas were only a few kilometers apart (Ihobe, 2005; Zamma and Makelele, 2012; see Chapter 43). Therefore, it is better to consider that the estimated population density includes wide error ranges and is only a rough indicator of the chimpanzee population in the survey area.

4.6 Distribution of larger mammals

Extensive surveys performed in and around the national park confirmed the presence of 31 species of larger mammals (Table 4.3). The distributions of seven

Table 4.2 **Population density of chimpanzees in and out of the Mahale Mountains National Park**

Survey area	Period	Survey distance (km)	No. of beds	Population density ($/km^2$)[a]	Population density in original article ($/km^2$)	Source
Reconnaissance walk						
in the park	Before 1985	100	49	0.028		Table 4.1
in the park	After 1985	284	164	0.033		Table 4.1
out of the park	Before 1985	32	40	0.071		Table 4.1
out of the park	After 1985	303	20	0.004		Table 4.1
Previous study in the park						
Mahale	1965–1967	–	–	0.350	0.350	Kano, 1972
Kasoje	2005	19.9	69	0.197	1.3	Moyer *et al.*, 2006
Kabesi	2005	26	128	0.279	1.3	Moyer *et al.*, 2006
Mfitwa	2005	?	0	0.000	0	Moyer *et al.*, 2006
Previous study out of the park						
Lilanshimba	1965–1967	–	–	0.290	0.290	Kano, 1972
Lilanshimba	2006	85.3	60	0.040	0.02	Ogawa *et al.*, 2006b; Yoshikawa *et al.*, 2008
Masito	1965–1967	–	–	0.170	0.170	Kano, 1972
Masito	2003–2005	51.3	54	0.060	0.12	Yoshikawa *et al.*, 2008
Filabanga	1965–1966	–	–	–	0.2	Kano, 1971
Filabanga	2001	6	25	0.236	0.16	Moyer *et al.*, 2006
Kasakati	1963–1964	–	–	–	1.10–1.65	Izawa and Itani, 1966
Kasakati, Filabanga, Mukuyu, etc.	1964–1965	–	–	–	0.49–0.71	Suzuki, 1969

Kasakati	2001	3.8	25	0.373	0.26	Moyer *et al.*, 2006
Ugalla	1965–1967	–	–	0.080	0.080	Kano, 1972
Ugalla	1995–2008	332.2	496	0.085	0.1	Yoshikawa *et al.*, 2008
Nguye	1999–2000	98.7	146	0.084	0.06	Moyer *et al.*, 2006
Bhukalai	1999–2000	80.8	134	0.094	0.07	Moyer *et al.*, 2006
Ilumba	2005	82.5	654	0.449	0.06	Moyer *et al.*, 2006
Mttindi	1988	12	8	0.038	0.03	Moyer *et al.*, 2006
Mukuyu	1965–1967	–	–	0.270	0.270	Kano, 1972
Mukuyu	2005	183.7	123	0.038	0.05	Yoshikawa *et al.*, 2008
Karobwa	1965–1967	–	–	0.380	0.380	Kano, 1972
Karobwa	2003	183.7	123	0.038	0.03	Ogawa *et al.*, 2006a; Yoshikawa *et al.*, 2008
Kapalagulu	2005	18.9	65	0.195	0.6	Moyer *et al.*, 2006
Ntakata (west)	2005	11.6	18	0.088	0.5	Moyer *et al.*, 2006
Ntakata (east)	2005	11.5	41	0.202	0.8	Moyer *et al.*, 2006
Lubalisi Hills	2005	12.3	129	0.595	3.2	Moyer *et al.*, 2006
Mwese Hills	2005	20	70	0.198	0.7	Moyer *et al.*, 2006
Lugalla Hills	2005	15.5	420	1.536	4.8	Moyer *et al.*, 2006
Wansisi	1965–1967	–	–	0.120	0.120	Kano, 1972
Wansisi	2001	19.2	29	0.086	0.02	Ogawa *et al.*, 2004; Yoshikawa *et al.*, 2008
Wansisi Hills	2005	12.3	60	0.277	0.4	Moyer *et al.*, 2006
Lwazi	2003–2005	140.8	40	0.016	0.01	Yoshikawa *et al.*, 2008
Southern Tanganyika Lakeshore	?	71.4	28	0.022	0.2	Moyer *et al.*, 2006

[a] Population density was calculated using the formula published by Kano in 1972. See also text.

Table 4.3 **Evidence of animals, such as direct observation, vocal observation, footprints, and feces, observed in each survey**

			Survey route[b,c]														
English name	Latin name	Code[a]	a	b	c	d	e	f	g	h	i	j	k	l	m	n	o
Primates																	
Galago	Galagonidae	G											y				
Vervet monkey	*Chlorocebus pygerythrus*	V				o								o			
Red-tailed monkey	*Cercopithecus ascanius*	R	y		y	o			y	o						y	
Blue monkey	*Cercopithecus mitis*	B							y								
Yellow baboon	*Papio cynocephalus*	Y	y			o	y		y	o		y	o		y	y	
Eastern red colobus	*Procolobus rufomitratus*	C	y	y	y		y		y		y	y				y	
Angolan black-and-white colobus	*Colobus angolensis*	W		y													
Ungulates																	
Roan antelope	*Hippotragus equinus*	R	y			o		y			y	y	y				
Hartebeest	*Alcelaphus lichtensteini*	H				o											
Waterbuck	*Kobus ellipsiprymnus*	T															y
Klipspringer	*Oreotragus oreotragus*	K		y		o											
Sharpe's grysbok	*Raphicerus sharpei*	S		y													
Bush duiker	*Sylvicapra grimmia*	U				o	y		y	o	y				y		
Eland	*Taurotragus oryx*	E				o											
Bushbuck	*Tragelaphus scriptus*	B	y	y		o	y		y	o		y	y	y	y	y	y
African buffalo	*Syncerus caffer*	F	y			o		y	y		y	y	y	y	y	y	y
Giraffe	*Giraffa camelopardalis*	G				o											
Common warthog	*Phacochoerus africanus*	W				o	y				y	y	y				
Bushpig	*Potamochoerus larvatus*	I	y			o	y		y	o	y	y	y	y	y	y	
Hippopotamus	*Hippopotamus amphibius*	P				o											

Common zebra	*Equus quagga*	Z				o											
Others																	
Cape crested porcupine	*Hystrix africaeaustralis*	P	y			o	y				y	y	y			y	
African civet	*Civettictis civetta*	C							y								
Common dwarf mongoose	*Helogale parvula*	M					y										
Mongoose	Herpestidae	M									y		o				
Spotted hyena	*Crocuta crocuta*	H	o				y	y	y		y						
Leopard	*Panthera pardus*	L	y			o			y		y	y		y			
Lion	*Panthera leo*	O				o		y									
Aardvark	*Orycteropus afer*	A				o			y				o	y		y	
African elephant	*Loxodanta africana*	E				o		y	y		y	y	y				

[a] The letter codes of the species are shown in Figures 4.4 (primates), 4.5 (ungulates), and 4.6 (others).

[b] Survey routes a–o correspond to those in Table 4.1 and Figure 4.1.

[c] The symbol "y" indicates that evidence of animals was observed inside or inside and outside of the park. The symbol "o" indicates that evidence of animals was observed only outside of the park.

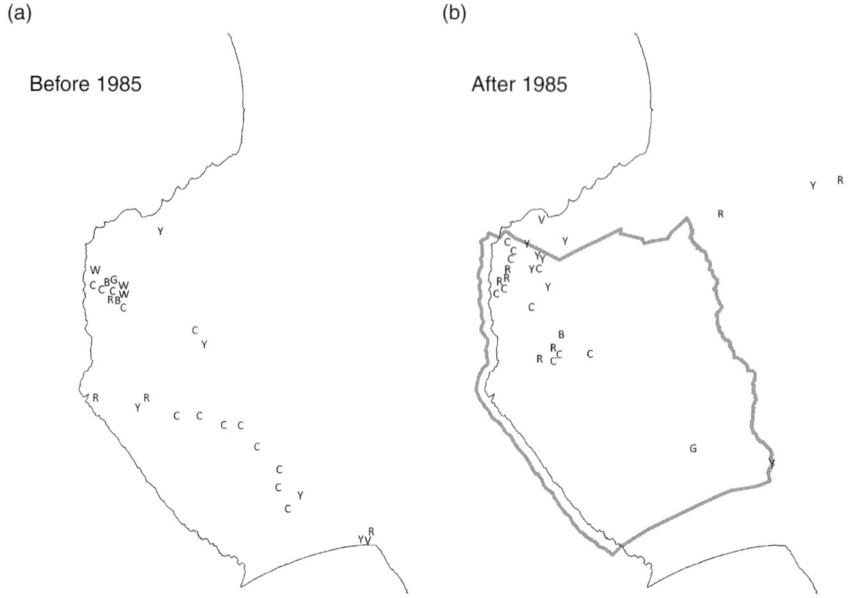

Figure 4.4 Evidence of the presence of primates (a) before and (b) after 1985. B: blue monkey. C: red colobus; G: thick-tailed galago; R: red-tailed monkey; V: vervet monkey; W: Angolan black-and-white colobus; Y: yellow baboon (see Table 4.3).

primate species were similar to that of chimpanzees: they were observed in the Kasoje area (Figure 4.4). Yellow baboon was more likely to be found at the periphery of the park with dry woodland vegetation. Angolan black-and-white colobus was found only in the high montane forest of the northwest part of the park (Nishida *et al.*, 1981), but information from villagers suggested that they might also be distributed in the hills of the southeast boundary (Juma Hassani, personal communication).

In contrast to the distribution of primates, 15 species of ungulates seemed to be spread widely throughout the area (Figure 4.5). The dominant species within the park were bushbucks, bushpigs, and warthogs. Other large ungulates, such as roan antelopes and waterbucks, were found in the southern part of the park. Elephants and African buffaloes also traveled in the park (Figures 4.5 and 4.6) where vegetation is covered with woodland with more grasses. On the other hand, leopards tended to be found in the northwestern part in the Kasoje area.

4.7 Conclusion

There were no marked changes in the distributions of chimpanzees and other larger mammals in Mahale across study periods. On the other hand, according to villagers' impressions, many mammals escaped into the park from outside after its establishment. This may be based on their experience of decreasing encounters with animals outside the park. Further surveys are required both within and outside the park to confirm these reports.

The distributions of chimpanzees and the other mammals are related to those of vegetation. The vegetation in the area has been changing due to the direct and indirect effects of human activities, such as deforestation and climate change (Itoh *et al.*, 2012; Chapter 13). However, the distributions of animals did not change markedly. In this study, we could not accurately clarify the population densities of chimpanzees and other mammals, and further monitoring in more appropriate ways is required over the next 50 years.

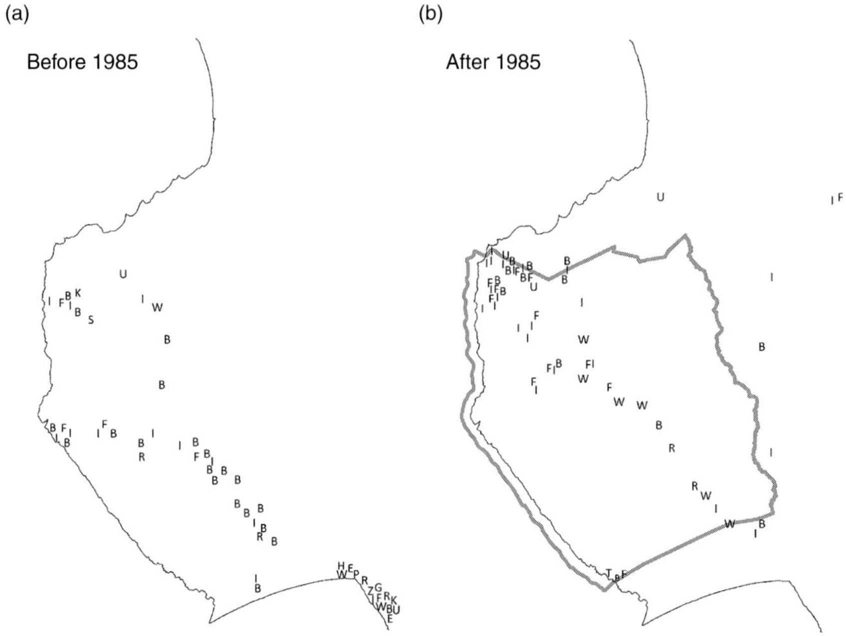

Figure 4.5 Evidence of the presence of ungulates (a) before and (b) after 1985. B: bushbuck; D: blue duiker; E: eland; F: buffalo; G: giraffe; H: hartebeest; I: bushpig; K: klipspringer; P: hippopotamus; R: roan antelope; S: Sharpe's grysbok; T: waterbuck; U: bush duiker; W: warthog; Z: zebra (see Table 4.3).

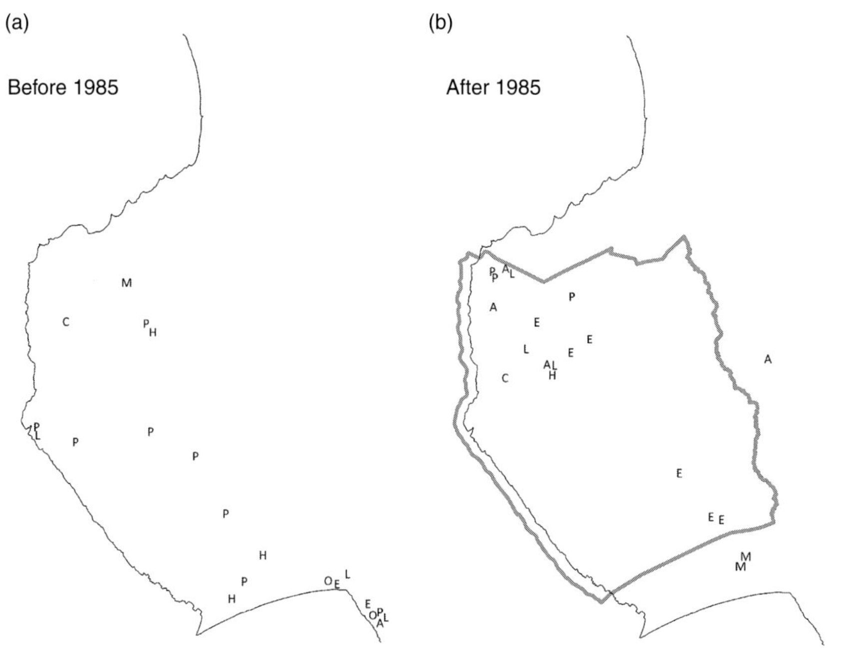

Figure 4.6 Evidence of the presence of other animals (a) before and (b) after 1985. A: aardvark; C: civet; E: elephant; H: spotted hyena; L: leopard; M: dwarf mongoose and other mongoose species; O: lion; P: crested porcupine (see Table 4.3).

References

Azuma, S. and Toyoshima, A. (1965). [Chimpanzees in Kabogo Point Area, Tanganyika.] In [*Monkeys and Apes*], ed. S. Kawamura and J. Itani. Tokyo: Chūō-kōron-sha, pp. 127–63. In Japanese.

Dollman, G. (1935). The occurrence of the chimpanzee in Tanganyika Territory. *Proceedings of the Linnean Society of London*, **148**(1), 15–16.

Ghiglieri, M. P. (1984). *The Chimpanzees of Kibale Forest*. New York: Columbia University Press.

Goodall, J. (1986). *The Chimpanzees of Gombe: Patterns of Behavior*. Cambridge, MA: Belknap.

Google (2014). *Distance Measurement Tool of Google Maps*. https://support.google.com/maps/answer/1628031?hl=en.

Hasegawa, T. and Hasegawa, M. (1980). The eighth ascent of Mt. Nkungwe: chimpanzees of the Humo-group and other larger mammals of the main ridge of Mahale Massif. *Mahale Mountains Chimpanzee Research Project Survey Report*, **10**.

Hernandez-Aguilar, R. A. (2009). Chimpanzee nest distribution and site reuse in a dry habitat: Implications for early hominin ranging. *Journal of Human Evolution*, **57**, 350–64.

Ihobe, H. (2005). Life span of chimpanzee beds at the Mahale Mountains National Park, Tanzania. *Pan Africa News*, **12**, 10–12.

Inoue, E., Tashiro, Y., Ogawa, H., *et al.* (2013). Gene flow and genetic diversity of chimpanzee populations in East Africa. *Primate Conservation*, **26**, 67–74.

Itani, J. (1990). Safari surveys of the vegetation and the chimpanzee groups in the northern half of Mahale Mountains. In *The Chimpanzees of the Mahale Mountains: Sexual and Life History Strategies*, ed. T. Nishida. Tokyo: University of Tokyo Press, pp. 37–61.

Itoh, N., Sakamaki, T., Hamisi, M., *et al.* (1999). A new record of invasion by an unknown unit-group into the center of M group territory. *Pan Africa News*, **6**, 8–10.

Itoh, N., Nakamura, M., Ihobe, H., *et al.* (2012). Long-term changes in the social and natural environments surrounding the chimpanzees of the Mahale Mountains National Park. In *The Ecological Impact of Long-Term Changes in Africa's Rift Valley*. ed. A. J. Plumptre. New York: Nova Science, pp. 249–77.

Izawa, K. and Itani, J. (1966). Chimpanzees in Kasakati Basin, Tanganyika (1): Ecological study in the rainy season 1963–1964. *Kyoto University African Studies*, **1**, 73–156.

Kano, T. (1971). The chimpanzees of Filabanga, western Tanzania. *Primates*, **12**, 229–46.

Kano, T. (1972). Distribution and adaptation of the chimpanzee on the eastern shore of Lake Tanganyika. *Kyoto University African Studies*, **7**, 37–129.

Kormos, R., Boesch, C., Bakarr, M. I., and Butynski, T., ed. (2003). *West African Chimpanzees. Status Survey and Conservation Action Plan*. Gland: IUCN.

Moyer, D., Plumptre, A. J., Pintea, L., *et al.* (2006). *Surveys of Chimpanzees and Other Biodiversity in Western Tanzania*. Unpublished Report to United States Fish and Wildlife Service.

Nakamura, M. and Fukuda, F. (1999). Chimpanzees to the east of the Mahale Mountains. *Pan Africa News*, **6**, 5–7.

Nakamura, M., Nishie, H., and Mwinuka, C. (2005). Survey of the southern part of the Mahale Mountains. *Pan Africa News*, **12**, 5–8.

Nishida, T. (1973). [*The Children of the Mountain Spirits: Search for the Social Structure of Chimpanzees*.] Tokyo: Chikuma-shobō. In Japanese.

Nishida, T. (1975). A general survey of the southern part of the Mahale Mountains. Preliminary draft. *Mahale Mountains Chimpanzee Research Project Survey Report*, **1**.

Nishida, T. (1976). The sixth ascent of Mt. Nkungwe, the paramount peak of the Mahale Massif. *Mahale Mountains Chimpanzee Research Project Survey Report*, **4**.

Nishida, T. (1989). A note on the chimpanzee ecology of the Ugalla Area, Tanzania. *Primates*, **30**, 129–38.

Nishida, T. (1990). A quarter century of research in the Mahale Mountains: an overview. In *The Chimpanzees of the Mahale Mountains: Sexual and Life History Strategies*, ed. T. Nishida. Tokyo: University of Tokyo Press, pp. 3–35.

Nishida, T. and Norikoshi, K. (1975). Distribution of the unit-groups of chimpanzees in the Kasoje area in the *Garcinia huillensis* season of 1975. Preliminary draft. *Mahale Mountains Chimpanzee Research Project Survey Report*, **3**.

Nishida, T., Itani, J., Hasegawa, M., and Hasegawa, T. (1981). A newly-discovered population of *Colobus angolensis* in East Africa. *Primates*, **22**, 557–63.

Nishida, T., Takasaki, H., and Takahata, Y. (1990). Demography and reproductive profiles. In *The Chimpanzees of the Mahale Mountains: Sexual and Life History Strategies*, ed. T. Nishida. Tokyo: University of Tokyo Press, pp. 63–97.

Norikoshi, K. (1976a). Investigation of the Kasiha Valley. *Mahale Mountains Chimpanzee Research Project Survey Report*, **5**.

Norikoshi, K. (1976b). Investigation in the Lubangagulu Mountain. *Mahale Mountains Chimpanzee Research Project Survey Report*, **6**.

Ogawa, H., Moore, J., Kanamori, M., and Kamenya, S. (2004). Report on the chimpanzees of the Wansisi and Makomayo areas, Tanzania. *Pan Africa News*, **11**, 3–5.

Ogawa, H., Moore, J., and Kamenya, S. (2006a). Chimpanzees in the Ntakata and Kakungu areas, Tanzania. *Primate Conservation*, **21**, 97–101.

Ogawa, H., Sakamaki, T., and Idani, G. (2006b). The influence of Congolese refugees on chimpanzees in the Lilanshimba area, Tanzania. *Pan Africa News*, **13**, 21–2.

Ogawa, H., Yoshikawa, M., and Idani, G. (2013). The population and habitat preferences of chimpanzees in non-protected areas of Tanzania. *Pan Africa News*, **20**, 1–5.

Plumptre, A. J. and Reynolds, V. (1996). Censusing chimpanzees in the Budongo Forest, Uganda. *International Journal of Primatology*, **17**, 85–99.

Plumptre, A. J., Rose, R., Nangendo, G., *et al.* (2010). *Eastern Chimpanzee* (Pan troglodytes schweinfurthii)*: Status Survey and Conservation Action Plan 2010–2020*. Gland: IUCN.

Shimada, M. (2003). A note on the southern neighboring groups of M group. *Pan Africa News*, **10**, 11–14.

Suzuki, A. (1996). An ecological study of chimpanzees in a savanna woodland. *Primates*, **10**, 103–48.

Takahata, Y., Hasegawa, T., Hayaki H., and Takasaki H. (n.d.). A brief survey on the home range of the N-Group chimpanzees. *Mahale Mountains Chimpanzee Research Project Survey Report*, 11.

Tutin, C. E. G. and Fernandez, M. (1984). Nationwide census of gorilla (*Gorilla g. gorilla*) and chimpanzee (*Pan t. troglodytes*) populations in Gabon. *American Journal of Primatology*, **6**, 313–36.

Uehara, S. and Nishida, T. (1978). A general survey to Bugansa. *Mahale Mountains Chimpanzee Research Project Survey Report*, 7.

Yoshikawa, M., Ogawa, H., Sakamaki, T. and Idani, G. (2008). Population density of chimpanzees in Tanzania. *Pan Africa News*, **15**, 17–20.

Zamma, K. and Makelele, M. (2012). Comparison of the longevity of chimpanzee beds between two areas in the Mahale Mountains National Park, Tanzania. *Pan Africa News*, **19**, 25–8.

Zamma, K., Inoue, E., Mwami, M., *et al.* (2004). On the chimpanzees of Kakungu, Karobwa and Ntakata. *Pan Africa News*, **11**(1), 8–10.

5 Who's who

KAZUHIKO HOSAKA

5.1 Introduction

Long-term research, spanning 50 years, on the wild chimpanzees at Mahale has revealed that they live in unit-groups consisting of individuals with abundant personality. This chapter provides biographical descriptions of the Mahale chimpanzees whose names have appeared repeatedly in publications. What we have observed of their societies may well be a "history" resulting from interactions between chronological social changes and chimpanzee personalities. For example, some of the alpha males may have attained their status almost entirely by their own endeavors, while others may have received their status solely because of interactions between other individuals. In another instance, one male that had survived an attempted infanticide grew up to be a low-ranking adult male. There are several examples of females with diverse histories of reproductive success and relations with adult males.

The two four-digit numbers in parentheses after an individual name, e.g. "(1940–1978)," indicate the birth and the death years for that chimpanzee. A question mark denotes an estimated year. When a natal female was assumed to have emigrated, the year of emigration is shown instead of the death year by attaching an "e" to the year, e.g. "e2010." An asterisk (*) after an individual's name indicates that the individual is introduced elsewhere in this chapter.

5.2 Males born in K group

5.2.1 Kasonta (1940?–1978)

Figure 5.1. Mother unknown. The alpha male of the K group from the onset of research to 1976. Nishida (2012) recollected that, when he was attempting to habituate chimpanzees of the K group by provisioning, Kasonta was so shy that Nishida did not notice him and mistook another adult male, Kajabala, for the alpha male. However, soon afterwards, it became apparent that Kasonta was a tyrannical alpha male who would monopolize estrous females and bananas and sugarcane (Nishida, 2012). His last years as the alpha male were well documented in terms of dominance-related interactions with Kamemanfu* and Sobongo*, marking the first study of chimpanzee political tactics and ostracism of the alpha male in the wild (Chapter 27).

Figure 5.1 Kasonta in 1976. ©Toshisada Nishida.

Mahale Chimpanzees: 50 Years of Research, ed. M. Nakamura, K. Hosaka, N. Itoh and K. Zamma. Published by Cambridge University Press.

5.2.2 Kamemanfu (1940?–1982)

Figure 5.2. Mother unknown. He is famous for having shown an ambivalent attitude in his alliance policy when two higher-ranking males, Kasonta* and Sobongo*, were competing for the alpha status. Nishida (1983) called it "allegiance fickleness" and suggested that it is a strategy exploited by post-prime males who can no longer be alphas, but can influence the outcome of conflicts between higher-ranking males. After the disappearance of all the other K-group males in 1979, he maintained his unit-group as the alpha until its extinction (Chapter 10).

Figure 5.2 Kamemanfu. ©Toshisada Nishida.

5.2.3 Sobongo (1958?–1979)

Figure 5.3. He was presumed to be Wabwema's son when he was identified as an adolescent. He vigorously challenged and finally dethroned Kasonta* in May 1976. He continued to be the alpha male of the K group for 2.5 years until 1979, when he abruptly disappeared for unknown reasons (Chapter 10).

Figure 5.3 Sobongo. ©Toshisada Nishida.

5.2.4 Limongo (1972–1993?)

Figure 5.4. Wamikambi's son. He survived the extinction of the K group (Uehara *et al.*, 1994; Uehara, 2003). Unlike another ex-K-group male, Masudi*, who transferred to the M group with his mother, Limongo appeared to have no opportunities for such transfer for two reasons. First, Limongo's mother was never observed to associate with the M group up to her disappearance in 1981. Second, when the K group lost all its adult males in 1982, he had reached the age of 10, which might have been too old to transfer to another unit-group. Even after all the other individuals of the K group transferred to the M group or disappeared, Limongo kept ranging in the ex-K group's home range and was sometimes observed together with ex-K-group chimpanzees, such as Masudi* and some females. Interestingly, Limongo and Masudi* had similar faces. Some researchers speculated that they were paternal half-siblings sired by Sobongo*. On February 7, 1994, Mtunda Mwami observed Limongo traveling alone on the trail to the south of Nkala Valley and emitting pant-hoots. Another sighting (Chapter 11) suggested that he might have lived even after that.

Figure 5.4 Limongo in 1992. ©Shigeo Uehara.

5.2.5 Masudi (1977–2006)

Figure 5.5. As Wantendele's son, he was born in the K group but transferred to the M group with his mother in the early 1980s during the K group's extinction (Chapter 10). This was an exceptional case of male transfer to a non-natal unit-group. It took almost 2 years for the mother–son pair to establish themselves in the M group. In 1981, they were observed being severely attacked twice by the M-group members. In both cases, most of the attacks were directed at the mother, but researchers feared that both mother and the son would be killed if left alone and intervened to protect them (Nishida and Hiraiwa-Hasegawa, 1985). In October 1982, when Wantendele was in estrus, the mother–son pair began to associate with the M-group members without being attacked (Takahata, 1985). Takahata and Takahata (1989) documented how Masudi formed relationships with M-group chimpanzees during that time. They implied that he might have eventually depended on two adult males, Bakali* and Musa*, who had showed tolerance during the course of transfer. Even after he was accepted as a member of M group, his mother's relationship with the group remained unstable because she preferred to stay with the K group when she was not in estrus. In July 1983, Masudi lost his newborn brother due to infanticide (Chapter 26). Takahata (1985) described a scene where Masudi approached and stared at Ntologi*, who was eating his brother, for several minutes. When Masudi lost his mother at the age of 13 to lion predation (Nishida, 2012), he had a 3-year-old sister, Maggie, who survived with the assistance of a non-kin adolescent female, Tula (Chapter 30).

Figure 5.5 Masudi in 2005. ©K. Hosaka.

When he grew up, he became a low-ranking male who was not motivated to rise up through the dominance hierarchy, but managed to attain middle ranking in his later years. As for copulations with estrous females, he often employed the "ambush" strategy, in contrast to the possessiveness or consortship of other males.

Masudi did not forget the taste of bananas for a long time after researchers abandoned provisioning. Nishida (2010a) reported that he was forced to hide bananas inside the locked store because Masudi once stole a bunch from a table in the outdoor room of Kansyana Camp. On February 3, 2001, however, he walked past Kansyana Camp and found the store open because a researcher was preparing his breakfast. Masudi had no hesitation in entering the store and quickly dashing out with a bunch of green plantain bananas. He ran off surprisingly fast, climbed the tree to taste them, and discarded them because they were not sweet. This implies that chimpanzees may have a sense of guilt when stealing hidden foods. Masudi disappeared during the epidemic of infectious respiratory disease in 2006 (Chapter 25), and is assumed to have died.

5.3 Males born in M group

5.3.1 Kajugi (1955?–1982)

Figure 5.6. Mother unknown. M group's alpha male from 1976 to 1979. He formed an intimate relationship with a prime adult male, Lubulungu, but was rather estranged from the then beta male, Ntologi* (Kawanaka, 1990), who later became the alpha. He was observed sharing the meat of an infant chimpanzee with other individuals in a case of cannibalism that occurred in June 1979, only about a month before his loss of alpha status (Kawanaka, 1981). He was another victim of ostracism after dethronement (Chapter 27).

Figure 5.6 Kajugi in 1982. ©Shigeo Uehara.

5.3.2 Ntologi (1955?–1995)

Figure 5.7. Assumed to be Wanaguma's son. He has the longest record of being an alpha male in the M group, holding the role during 1979–91 and 1992–5. In March 1991, he was ostracized by an alliance of high-ranking males that were against him. Therefore, he ranged separately from the other M-group members until returning as the alpha male on January 20, 1992 (Chapter 27). He maintained his second stint until 1995 owing to his new alliance with Kalunde*,

Figure 5.7 Ntologi in 1992. ©K. Hosaka.

the very male who primarily formed the old alliance to ostracize him. In April 1995, he finally lost the alpha status to a challenge from the beta male Nsaba*, and again experienced ostracism. About 4 months later, he came back as a low-ranking male who pant-grunted at every other adult male he met. In November 1995, he was unexpectedly found dying, and his life ended in a room at Kansyana Camp (Chapter 26). It is plausible that our knowledge of the Mahale chimpanzees, obtained from the beginning of the 1980s to the first half of 1990s, might have been biased by Ntologi's strong personality.

5.3.3 Kalunde (1963?–present)

Figure 5.8. Identified as an orphaned adolescent, his mother was unknown. At that time, he would follow Kagimimi (Nishida, 1994). Beginning in 1987, when he became the beta male, he was a long-term rival of the alpha male Ntologi*. In March 1991, Kalunde usurped the alpha status, but lost it less than 10 months later via a challenge from Nsaba*. However, Kalunde did not let the role go to Nsaba*, but restored it to the then ostracized Ntologi*, thus forming a formidable alliance between the two consecutive former alpha males. His second alpha stint came in January 1996 when the then alpha male Nsaba* disappeared abruptly, and this lasted for about 2 years. Since then, he has served as an important ally for several successive alpha males, performing quite well politically.

Figure 5.8 Kalunde in 2009. ©K. Hosaka.

Additionally, he has intrigued researchers with his use of nasal probes (Chapter 42), and continues to be the oldest known Mahale male (Chapter 23).

5.3.4 Nsaba (1973–1997)

Figure 5.9. Sada's son. In 1980, he had a younger sister, Tula, but did not develop a close relationship with her, since he already preferred associating with adult males rather than with his mother (Hayaki, 1988). By 1985, Nsaba had dominated his adolescent peer males (Hayaki *et al.*, 1989). Even after Nsaba grew up to be a high-ranking male, he was fond of playing with his favorite infants, in particular Darwin* in 1991 and Brutus, son of Wakampompo, in 1993. Once, when he performed a charging display, he hung from woody vines and swung back and forth with piloerection while carrying Brutus on his belly. He became the alpha in 1995 but disappeared for unknown reasons on the last day of 1996.

Figure 5.9 Nsaba (left) and Fanana* (right) in 1993. ©K. Hosaka.

5.3.5 Fanana (1978?–present)

Figures 5.9 and 5.10. Probably a son of Fahari, who was identified by only a few researchers because of her shyness. In 1993, when he was on the cusp of adolescence and adulthood, he outranked the lowest-ranking adult male, Masudi*, but was often attacked by another low-ranking male, Bembe*. Fanana was the alpha male in 1997–2003, when he often walked together with Kalunde*. In 2003, he started to move away from other group members for unknown reasons (but was sporadically observed by researchers) (Nishie, 2004a). After a while, he met other M-group males and pant-grunted at Alofu* (Nishie, 2004b), then disappeared from the group again. After Alofu* lost his alpha position in 2007, Fanana finally returned to the M group and is now an ordinary mid- to low-ranking older male.

Figure 5.10 Fanana in 2009. ©K. Hosaka.

5.3.6 Alofu (1982–present)

Figure 5.11. Wakusi's* son. The alpha male in 2003–7. He might be the initiator of a peculiar behavior, termed "nipple press" (Chapter 37), typically accompanied by submissive greeting behavior. It was not certain when he began this habit, but he displayed it at least by the age of 7 in 1989 (Toshisada Nishida, personal communication). Afterwards, several other individuals, including his mother Wakusi*, were observed performing this behavior. This habit continued even after he became the alpha male. He was often observed turning his face away from pant-grunting individuals with a painful expression as if being the alpha felt like a burden to him (Nakamura, 2006). After losing his alpha status, he remained a beta male for over a year.

Figure 5.11 Alofu in 2008. ©K. Hosaka.

5.3.7 Pimu (1988–2011)

Figure 5.12. Fatuma's* son. At the age of 13, he already had confirmed dominance over all the adult females (Shimada, 2013a). Around 2005, he began to use human researchers or tourists like tools for intimidation displays, e.g. by slapping, throwing a bough, etc. (Chapter 31). In the dry season of 2007, he outranked Alofu* and became the alpha. He was a formidable alpha male who managed to dominate high-ranking adult males such as Alofu* and Primus* without frequent assistance from his coalition partners. When he was an alpha, he showed three idiosyncratic behaviors. First, he had a habit of slapping the back of a female with his hand once just before copulation. Second, he was occasionally observed copulating with his estrous mother, which shocked Tongwe research assistants, who believed that it was a deviation from their ethical norms. Third, when sharing meat with other chimpanzees after successfully hunting a red colobus, he often left the kill on the ground, started charging displays, and then retrieved the carcass as if to pronounce that the meat was inaccessible even when he was not holding it. Some researchers speculated that his stint as the alpha would last as long as Ntologi*. However, in October 2011, he fell victim to a prolonged gang attack by several M-group males and died, which is quite unusual for an alpha male (Kaburu *et al.*, 2013; see also Chapter 26).

Figure 5.12 Pimu in 2008. ©K. Hosaka.

5.3.8 Primus (1991–present)

Figure 5.13. Pinky's* son. Some researchers speculated that his father may be Toshibo* based on their facial resemblance. During the outbreak of an infectious respiratory disease in 1993 (Chapter 25), he almost died. At the age of 9, he was observed boldly fighting adult male colobus monkeys in the canopy before finally retreating. At the age of 17, being the gamma male, he was often observed screaming after receiving directed charging displays by the beta male, Alofu*, who had lost the alpha position to Pimu* the year before. One year later, he began to perform intimidation displays in front of Alofu*, indicating that he had almost become dominant over Alofu*, even though neither of them pant-grunted at each other. He has been the alpha of the M group since the death of Pimu* in 2011.

Figure 5.13 Primus in 2008. ©K. Hosaka.

5.3.9 Bakali (1954?–1993)

Figure 5.14. Mother unknown. He was the beta male in the early years of Ntologi's* first stint as the alpha male. In his later years, Bakali became an important ally for the alpha males Ntologi* and Kalunde*. Additionally, he became a "grandpa-like" male who liked to play with infants and juveniles (Chapter 23).

Figure 5.14 Bakali in 1993. ©Toshisada Nishida.

5.3.10 Musa (1957?–1993)

Figure 5.15. Mother unknown. He was known as a particularly small adult male, with limbs shorter than those of even some adult females (Gunji *et al.*, 1998). Musa seldom played with infants or juveniles (Chapter 23), and he never displayed a "relaxed face with droop lip" (Goodall, 1986). He had a tendency to "discipline" adolescent males, while being tolerant to and often being followed by orphans (Chapter 23; also see Masudi's biography above). When the low-ranking Masudi* slapped the metal-wall house, Musa charged him angrily (Chapter 31). While never highly ranked, he was an important ally to Kalunde* and Ntologi* in his later years. When Musa died of old age and illness, his decomposed body was found by the M-group chimpanzees. Ntologi* observed the body for 13 min, while Kalunde* just ran past it in the course of a charging display (Hosaka *et al.*, 2000).

Figure 5.15 Musa in 1993. ©K. Hosaka.

5.3.11 Shike (1970?–1992)

Figure 5.16. Gwepala's son. Shike was known for his frequent, impressive charging displays, which started in adolescence, using his long limbs (Nishida, 2012; Chapters 27, 31). In the 1980s, some researchers thought that he would become the alpha male. However, at the end of November 1991, while he was still the beta male, he suffered a serious illness and was presumed dead about 2 months later (Chapter 25). His disappearance may have greatly influenced the power balance among higher-ranking M-group males at that time.

Figure 5.16 Shike in 1991. ©K. Hosaka.

5.3.12 Jilba (1975?–1996)

Figure 5.17. Tango's son. When he was a young adult male, he was insubordinate to the alpha male Kalunde* for unknown reasons. He did not pant-grunt at Kalunde*, and sometimes even attacked him. He was attacked by a gang of M-group chimpanzees in October 1991 (Nishida *et al.*, 1995) and June 1992 (Shigeo Uehara, personal communication). After the first attack, he was ostracized and joined Ntologi*, who was ranging alone at that point, but after the second attack he kept ranging with the M group. On another occasion in 1992, the palm side of his left forearm was so severely injured, possibly by a bushpig bite (Toshisada Nishida, personal communication), that his left hand had impaired function thereafter. Additionally, in 1993, the whole middle finger of his left hand was bitten off in a coalitional attack by Gwekulo* and Wakasunga when he was trying to initiate consortship with estrous Wakasunga. In December 1995, his skeleton was found (see Chapter 26).

Figure 5.17 Jilba (right) with Ntologi* (left) when both were ostracized in 1991. ©K. Hosaka.

5.3.13 Aji (1973?–1996)

Figure 5.18. Mother unknown, and identified as an orphan. He survived by following several adult males, particularly Rashidi (Chapter 30). Because he was extremely tame and unaffected by human presence, he was favored by tourists and TV crew cameras. In 1991, in spite of not playing with them, he was often followed by juvenile and adolescent males, including some orphans.

Figure 5.18 Aji in 1993. ©K. Hosaka.

5.3.14 Bembe (1972?–1995)

Figure 5.19. Wansombo's* son. He was known for his frequent association with his mother even in adulthood. Some researchers hypothesized that the remaining white tufted hair on his rump was related to his extended dependence on his mother. Bembe was a typical low-ranking male who apparently showed little interest in raising his dominance rank. In 1993, however, he performed directed charging displays whenever he encountered two of the other three lower-ranking adult males, Fanana* (young adult) and Musa* (old adult). He was a very skilled pant-hooter, and often called in chorus and stamped with his mother. His closest friend was Toshibo*, whom he frequently played with even in adulthood.

Figure 5.19 Bembe in 1993. ©K. Hosaka.

5.3.15 Toshibo (1976–1995)

Figure 5.20. Wasobongo's son. Named after the son of Toshisada Nishida. He was a "too-habituated" chimpanzee that had tasted bananas and sugarcane provisioned for behavioral observation (Chapter 3). In 1991, I observed that he crawled closer to a researcher while lying on his side on the ground when he became tense due to an approach by the then alpha male Kalunde*. In 1992, he began to challenge the then gamma male Kalunde* (Toshisada Nishida, personal communication). Whenever Toshibo noticed Kalunde* walking without his ally Ntologi*, he charged him, but was unsuccessful in eliciting a submissive response. In 1993, the challenges continued. One day, when Toshibo charged Kalunde*, the two stood upright facing each other and leaped bipedally ("bipedal contest"), as was often the pattern of their duel. However, in this case, an adult female, Ikocha*, intervened in the duel in favor of Kalunde*. She came up screaming and rammed into Toshibo, who fell rolling down the slope and ran away screaming loudly. He became the beta male in 1995 beneath the then alpha male Nsaba*, but later disappeared for unknown reasons.

Figure 5.20 Toshibo in 1995. ©K. Hosaka.

5.3.16 Dogura (1981?–2004)

Figure 5.21. Presumed to be Dodoma's son. Although he had a relatively small body, he had upward mobility in dominance rank. In 1991, even as an early adolescent, he showed excellent skill in hunting monkeys. He sometimes joined Kalunde's* meat-eating cluster possibly because, at least in some cases, he may have actually killed the monkey, and was being robbed of his prey. Meat possessors may be tolerant to the captors of prey (Chapter 20). In 1993, while still an adolescent, he started to challenge an older male, Masudi*. Later, Dogura became the beta male, but died because of disease and emaciation (Chapter 25).

Figure 5.21 Dogura. ©Michio Nakamura.

5.3.17 Bonobo (1981–present)

Figure 5.22. A son of Wakampompo, who was a large-bodied and apparently dominant female in the M group (Nishida, 1989). He grew up to be one of the

tamest chimpanzees. At first, he was only habituated to relatively short people, like most Japanese researchers and Tongwe assistants. In 1991, when he encountered Craig B. Stanford, a tall guest who visited Mahale, on the trail, he looked uneasy on the tree and threw a branch towards him. An incident where he carried a dead infant is described in Chapter 26. He appeared to lack motivation to raise his dominance rank, but he was once the beta male, probably due to his large body size, which he inherited from his mother.

Figure 5.22 Bonobo eating a lemon in 2005. ©K. Hosaka.

5.3.18 Darwin (1988–present)

Figure 5.23. Dar's son. At the age of 3, he was a very calm infant whose face was spotlessly white and a favorite "baby" of the young, high-ranking adult male Nsaba*, who approached and played with Darwin whenever they encountered each other. When Darwin was orphaned in 1994, he began routinely ranging with other M-group members (Nakamura, 2011), and he grew up to be an easy-to-observe, low-ranking male. Pimu*, Darwin's* peer and a contrasting "ambitious" male, was relatively tolerant of Darwin. When Pimu* became the alpha male, Darwin became a grooming partner of the alpha male for the first time, since he had been excluded from the political network of adult males. In 2008, when Darwin captured a juvenile red colobus, he solicited aid from the then alpha male Pimu* to chase off some scrounging individuals, such as Bonobo*, and they shared the small meal together in a relatively small cluster (Chapter 20).

Figure 5.23 Darwin in 2005. ©K. Hosaka.

5.4 Females

5.4.1 Chausiku (1958?–1990)

Figure 5.24. Wabunengwa's daughter. Born in the K group, she was Nishida's most important research collaborator. In the acknowledgements of his book (Nishida, 1994), he expressed his special gratitude to "the late Ms. Chausiku Binti Kajabala for her openhearted personality, by which my research advanced a great deal." Later, he noted that this acknowledgement was not directed to a local woman, but to this female chimpanzee (Nishida, 2003). She adopted a "let-alone" principle for child rearing, thus offering nulliparous females many opportunities for babysitting. Katabi (1974–1980; Figure 42.2 in Chapter 42) was her first offspring (male). Nishida (1981) meticulously described how Katabi developed as an infant and died tragically at the age of 6 following separation from his mother, who transferred to the M group in 1979 while he remained in the K group. She gave birth to four additional offspring. The first three died in infancy, one by infanticide (Chapter 26). Chopin (1985–1993) was her last son. When Huffman and Seifu (1989) reported their first observations of chimpanzee consumption of a medicinal plant, *Vernonia amygdalina*, they described in detail how Chausiku showed this rare behavior and recovered from illness, and how Chopin observed this. Chopin was orphaned when Chausiku was among the victims of the first reported case of lion predation on chimpanzees (Chapter 16).

Figure 5.24 Chausiku nursing Katabi. ©Toshisada Nishida.

5.4.2 Gwekulo (1962?–present)

Figure 5.25. Mother unknown. Gwekulo immigrated into the K group in 1972 and transferred to the M group in 1979. She is presumably sterile since she has never borne offspring. Instead, she takes care of the infants of her female friends (Chapter 30). In 1993, when Gwamwami* was carrying her mummified infant (Chapter 25), Gwekulo followed her while carrying a colobus arm (Hosaka *et al.*, 2000), as if she imagined that she was also carrying a mummy. She sometimes shows "masculine" behavior, such as intimidation displays (Nishida, 2006). Occasionally, she has formed a coalition with another female, particularly a friend (Chapters 28, 29, 31). In one case she formed a formidable coalition with Pinky* and Nkombo* to support the alpha male Kalunde* in his fight against the challenging Fanana* (Nishida, 2012). She stopped sexual cycling for more than 11 years (Chapter 23).

Figure 5.25 Gwekulo in 2009. ©K. Hosaka.

5.4.3 Nkombo (1970–present)

Figure 5.26. Wangala's daughter. Born in the K group and transferred to the M group in 1981. She is an aunt of Nsaba* (her elder sister is Nsaba's* mother), but they may not understand their own kinship because her M-group-born nephew was already preferentially associating with adult males. In the early 1990s, she was observed daily following high-ranking adult males like Kalunde* and Nsaba* (Hosaka, 2012). Around 1996, her relation with Kalunde* became intimate (see Chapter 29). She gave birth in 1993 and possibly in 1999, but never afterwards.

Figure 5.26 Nkombo in 2009. ©K. Hosaka.

5.4.4 Gwamwami (1966?–1996)

Figure 5.27. Mother unknown. Immigrated to the K group in 1975 and transferred to the M group in 1979. Tragically, she gave birth to nine offspring but none survived. In 1987, she suddenly lost her 5-month-old daughter, and researchers suspected an occurrence of infanticide although the evidence was circumstantial (Hamai *et al.*, 1992). In 1992, she lost her daughter Garbo (1989–1992) to an unknown injury, and the carcass was snatched by the beta male Nsaba* and eaten. Her next baby died of infectious respiratory disease (Chapter 25). Gwamwami carried its mummified body for at least 111 days, the longest record of dead infant carrying by a wild animal (Hosaka *et al.*, 2000).

Figure 5.27 Gwamwami in 1993. ©Toshisada Nishida.

5.4.5 Abi (1982–present) and Ako (1981?–present)

Abi (Figure 5.28) is Wasobongo's daughter. After she lost her mother to lion predation in 1989, she was observed to follow and groom her elder brother Toshibo* (Nishida, 2009). In 1991, Abi was often observed skipping playfully towards her brother, who was busy interacting with other adult males but quite tolerant to his sister. Ako (Figure 5.29), a twin-like

Figure 5.28 Abi in 2008. ©K. Hosaka.

friend of Abi since childhood, is Imelda's daughter and an orphan. Both remained in the natal M group and have successfully reared several children, despite their relatively slender bodies.

Figure 5.29 Ako (middle) with her 8-month-old infant son, Agano (left), and 7-year-old daughter, Acadia (right), in 2005. ©K. Hosaka.

5.4.6 Wansombo (1952?–1994)

Figure 5.30. Identified as an adult in the M group. She once had a big family with children and grandchildren, including a probable daughter, Katyentye (1967?–1996), and a son, Bembe*. Katyentye stayed in the M group and had two daughters, Totzy (1982–2012; see Nishida, 2008a) and Tamara (1988–1996). In the first half of 1991, Wansombo's family comprised as many as six members, including Bunde (1980?–1991), her last daughter, who was ill at that time and disappeared. Interestingly, Totzy also remained in the M group and bore several offspring. Totzy's son Teddy (2001–present) is the only remnant of M group's second biggest family of all time.

Figure 5.30 Wansombo in 1986. ©Toshisada Nishida.

5.4.7 Opal (1971?–2006)

Figure 5.31. Immigrated to the M group in 1983. She also had a big family in the M group, since her first daughter, Ruby (1986–2009), stayed in the group and had children. In 2004–6, Opal's family consisted of seven: herself, Ruby, Opal's three sons (Orion, 1991–present; Oscar, 1998–2006; and an unnamed infant), and Ruby's two daughters (Rubicon, 1998–e2010, and Rhea, 2004–6). Rubicon was the victim of an attempted infanticide but survived with support from Opal and another female (Chapters 26, 28, 29). Orion is currently a high-ranking adult male.

Figure 5.31 Opal in 2005. ©K. Hosaka.

5.4.8 Pinky (1972?–2006)

Figure 5.32. Immigrated to the M group in 1983. Her son Primus* is the current alpha male. Her daughter Puffy (2000–present: Shimada, 2013b) was orphaned but adopted by Gwekulo* and recently gave birth in the natal M group. Pinky and Opal* immigrated to M group in the same year and became stable association partners ranging in the northern area of M group's home range with their dependent offspring (Nishida 2007a).

Figure 5.32 Pinky in 2004. ©Michio Nakamura.

5.4.9 Fatuma (1963?–present)

Immigrated to the M group in 1974. She has been a skillful hunter since her twenties (Hosaka *et al.*, 2001; Chapter 20). She used to share meat with her son Pimu*, but after he became the alpha, she received meat from him. Her pant-grunts are so noisy that many researchers can detect her presence by her sounds even when she is not visible. Besides Pimu*, she has had several children (Figure 5.33). Three of her daughters have already emigrated to other groups. She is currently rearing a male infant, Figaro (2010–present). Fatuma is an old female who has been carefully observed for the longest time in the M-group research history (Chapter 23), but she remains wary of humans (Nishida, 2005a), a trait which Pimu* seemed to have inherited (see above).

Figure 5.33 Fatuma (left) and her 1-year-old daughter, Fimbi (right), in 2008. ©K. Hosaka.

5.4.10 Wakusi (1961?–present)

Figure 5.34. Identified as a young adult in the M group. Alofu's* mother. One of her daughters, Ai (1988–2002), was a lively infant but became emaciated due to an unknown disease in her

adolescence (Shimada, 2011) and died. Wakusi began performing nipple presses, the strange habit started by her son Alofu* (Chapter 37). She is the oldest female in the history of Mahale research, and surprised researchers when she bore two offspring in what was estimated to be her forties, although they both died of illness (Nishida, 2007b; Chapter 23).

Figure 5.34 Wakusi (left) grooming Masudi* (right) in 2005. She has shown the symptom of presbyopia since her late thirties. ©K. Hosaka.

5.4.11 Calliope (1960?–2012)

Figure 5.35. Identified as an adult in the M group. In the 1970s and 1980s, Calliope was one of the shyest M-group females, but she gradually became observable to humans in the 1990s (Nishida, 2010b). She lived to an estimated age of 52 years (Chapter 23). Her reproductive profile is available elsewhere (Nakamura and Nishida, 2012). Her oldest son, Carter (1985–present), is a low-ranking adult male in the M group. Another son, Cadmus (1991–2010), appeared highly motivated to raise his rank, but disappeared at the age of 18 for unknown reasons.

Figure 5.35 Calliope (right) grooming her son, Carter (left), in 2008. ©K. Hosaka.

5.4.12 Ikocha (1965?–present)

Figure 5.36. Identified as a young adult in the M group in 1979. Since her younger days, she has been wary of humans but has employed a strong attitude against chimpanzees (see the above biography of Toshibo*). In 1992, when she held a colobus monkey carcass and the then alpha male Ntologi* tried to snatch it, she never let it go and screamed loudly, and the "tug-of-war" looked so funny that some Tongwe assistants laughed.

She bore many children, such as Iwan (male, 1982–1997), Isisu (female, 1988–1994), Ivana (female, 1993–e2005), and Imani (female, 1998–e2009), who have already died or disappeared from the M group. What impressed me was that she left her seriously sick daughter Isisu to her fate in a determined manner. At

Figure 5.36 Ikocha with her newborn daughter, Iris, in 2009. ©K. Hosaka.

first, Isisu tried to follow old Wansombo*, who walked more slowly than Ikocha. After fighting her sickness for 3 months and receiving temporary "piggyback" assistance from several non-kin (Tula, Nick, Linda, and Totzy) at the final stage, she disappeared. Currently, as one of the oldest females (Chapter 23), Ikocha still has two children, a son, Ichiro (2002–present), and a daughter, Iris (2009–present).

5.4.13 Christina (1975?–present)

Figure 5.37. Immigrated to the M group in 1987. She is a female, greedy for meat (Nishida, 2008b). Once, she even abandoned carrying her dead infant when rushing to scrounge for meat (Hosaka *et al.*, 2000). She often threw noisy temper tantrums whenever she could not obtain meat from the sharers. In one case, Ikocha*, who appeared irritated by her temper tantrums, punched her face, and her eyelid was swollen for several days. This behavior declined once she bore children, and she became a good mother, very fond of playing with her children (Nishida, 2008b). She has two living children, an adult son, Christmas (1995–present), and a young adult daughter, Xantip (2000–present; see Shimada, 2014), who stayed in the M group and gave birth in 2012. Christina's recent infant XT11 (2011–2012) was unable to walk on her own (Matsumoto *et al.*, in prep.).

Figure 5.37 Christina (left) with her 1-year-old son (middle front), Alofu* (middle rear), and Gwekulo* (right) in 2009. ©K. Hosaka.

5.4.14 Miya (1980?–2006)

Immigrated to the M group in 1991. When she appeared as a stranger, several adult males and females intimidated her by hanging from woody vines and swinging, but Musa* approached and groomed her. She quickly became accustomed to humans (Nishida, 2005b). She had a habit of always shaking her head sideways slightly. She bore three children (Figure 5.38). Michio (1996–present) experienced severe weight loss but recovered. Now he is a large-bodied young adult male. His younger sister Mitsue (2001–e2012) was orphaned by their mother's death in 2006.

Figure 5.38 Miya (middle front) grooming her son Michio while being groomed by Kalunde* with her daughter Mitsue standing next to her in 2005. ©K. Hosaka.

5.4.15 Cynthia (1982?–present)

After immigration to the M group in 1993, she quickly became accustomed to humans. At that time, her face and swollen sexual skin were spotless white. After losing several infants, she finally succeeded in rearing a son, Caesar (1998–present; Shimada, 2012), and a daughter, Cissy (2008–2013). Caesar was attracted to his younger sister and often cared for her by grooming and playing with her (Figure 5.39; Chapter 30).

Figure 5.39 Cynthia (middle) is groomed by Orion (left) while grooming her son Caesar (right rear), who is fond of caring for his sister Cissy (right front), in 2009. ©K. Hosaka.

5.4.16 Pulin (1970?–1993)

Figure 5.40. Immigrated to the M group in 1982. She was among the tamest females and easy for researchers to follow. Of the victims of the outbreak of infectious respiratory disease in October 1993 (Chapter 25), her body was found the earliest. Her daughter Pipi (1990–e2002) was orphaned at the age of 3 years and 4 months, but survived by following and pant-grunting at adult males, and with encouragement from senior orphans Tula and Maggie. The following year, Pipi formed strong allomother–offspring relationships with Gwekulo* (Chapter 30).

Figure 5.40 Pulin. ©Toshisada Nishida.

References

Goodall, J. (1986). *The Chimpanzees of Gombe: Patterns of Behavior*. Cambridge, MA: Belknap.

Gunji, H., Hosaka, K., Huffman, M. A., *et al.* (1998). Four free-ranging chimpanzee skeletons from the Mahale Mountains, National Park, Tanzania. *Anthropological Science*, **106**, 185.

Hamai, M., Nishida, T., Takasaki, H., and Turner, L. A. (1992). New records of within-group infanticide and cannibalism in wild chimpanzees, *Primates*, **33**, 151–62.

Hayaki, H. (1988). Association partners of young chimpanzees in the Mahale Mountains National Park, Tanzania. *Primates*, **29**, 147–61.

Hayaki, H., Huffman, M. A., and Nishida, T. (1989). Dominance among male chimpanzees in the Mahale Mountains National Park, Tanzania: a preliminary study. *Primates*, **30**, 187–97.

Hosaka, K. (2012). [Introduction to Mahale chimps 19: Nkombo.] *Mahale Chimpun*, **19**, 6. In Japanese.

Hosaka, K., Matsumoto-Oda, A., Huffman, M. A., and Kawanaka, K. (2000). [Reactions to dead bodies of conspecifics by wild chimpanzees in the Mahale Mountains, Tanzania.] *Primate Research*, **16**, 1–15. In Japanese with English summary.

Hosaka, K., Nishida, T., Hamai, M., Matsumoto-Oda, A., and Uehara, S. (2001). Predation of mammals by the chimpanzees of the Mahale Mountains, Tanzania. In *All Apes Great and Small, Vol. I: African Apes*, ed. B. M. F. Galdikas, N. E. Briggs, L. K. Sheeran, G. L. Shapiro, and J. Goodall. New York: Kluwer Academic/Plenum, pp. 107–30.

Huffman, M.A. and Seifu, M. (1989). Observations on the illness and consumption of a possibly medicinal plant *Vernonia amygdalina* (Del.), by a wild chimpanzee in the Mahale Mountains National Park, Tanzania. *Primates*, **30**, 51–63.

Kaburu, S. S. K., Inoue, S., and Newton-Fisher, N. E. (2013). Death of the alpha: within-community lethal violence among chimpanzees of the Mahale Mountains National Park. *American Journal of Primatology*, **75**, 789–97.

Kawanaka, K. (1981). Infanticide and cannibalism in chimpanzees, with special reference to the newly observed case in the Mahale Mountains. *African Study Monographs*, **1**, 69–99.

Kawanaka, K. (1990). Alpha male's interactions and social skills. In *The Chimpanzees of the Mahale Mountains: Sexual and Life History Strategies*, ed. T. Nishida. Tokyo: University of Tokyo Press, pp. 171–87.

Nakamura, M. (2006). [The alpha male who does not like to receive pant-grunts?] *Mahale Chimpun*, **7**, 4. In Japanese.

Nakamura, M. (2011). [Introduction to Mahale chimps 17: Darwin.] *Mahale Chimpun*, **17**, 9. In Japanese.

Nakamura, M. and Nishida, T. (2012). Long-term field studies of chimpanzees at Mahale Mountains National Park, Tanzania. In *Long-Term Studies of Primates*, ed. P. M. Kappeler and D. P. Watts. Heidelberg: Springer, pp. 339–56.

Nishida, T. (1981). [*The World of Wild Chimpanzees*.] Tokyo: Chūkō-shinsho. In Japanese.

Nishida, T. (1983). Alpha status and agonistic alliance in wild chimpanzees (*Pan troglodytes schweinfurthii*). *Primates*, **24**, 318–36.

Nishida, T. (1989). Social interactions between resident and immigrant female chimpanzees. In *Understanding Chimpanzees*, ed. P. Heltne and L.A. Marquardt. Cambridge, MA: Harvard University Press, pp. 68–89.

Nishida, T. (1994). [*Thirty-six Stories of Wild Chimpanzees*.] Tokyo: Kinokuniya. In Japanese.

Nishida, T. (2003). [Introduction to Mahale chimps 1: Chausiku.] *Mahale Chimpun*, **1**, 4. In Japanese.

Nishida, T. (2005a). [Introduction to Mahale chimps 6: Fatuma.] *Mahale Chimpun*, **6**, 7. In Japanese.

Nishida, T. (2005b). [Introduction to Mahale chimps 5: Miya.] *Mahale Chimpun*, **5**, 7. In Japanese.

Nishida, T. (2006). [Introduction to Mahale chimps 7: Gwekulo.] *Mahale Chimpun*, **7**, 7. In Japanese.

Nishida, T. (2007a). [Introduction to Mahale chimps 10: Pinky.] *Mahale Chimpun*, **10**, 4. In Japanese.

Nishida, T. (2007b). [Introduction to Mahale chimps 9: Wakusi.] *Mahale Chimpun*, **9**, 5. In Japanese.

Nishida, T. (2008a). [Introduction to Mahale chimps 12: Totzy.] *Mahale Chimpun*, **12**, 5. In Japanese.

Nishida, T. (2008b). [Introduction to Mahale chimps 11: Christina.] *Mahale Chimpun*, **11**, 7. In Japanese.

Nishida, T. (2009). [Introduction to Mahale chimps 13: Abi.] *Mahale Chimpun*, **13**, 7. In Japanese.

Nishida, T. (2010a). [Introduction to Mahale chimps 15: Masudi.] *Mahale Chimpun*, **15**, 5–6. In Japanese.

Nishida, T. (2010b). [Introduction to Mahale chimps 16: Calliope.] *Mahale Chimpun*, **16**, 6. In Japanese.

Nishida, T. (2012). *Chimpanzees of the Lakeshore: Natural History and Culture at Mahale*. Cambridge: Cambridge University Press.

Nishida, T. and Hiraiwa-Hasegawa, M. (1985). Responses to a stranger mother–son pair in the wild chimpanzee: a case report. *Primates*, **26**, 1–13.

Nishida, T., Nakamura, M., and Hamai, M. (1995). A within-group gang attack on young adult male

chimpanzee: ostracism of an ill-mannered member? *Primates*, **36**, 207–11.
Nishie, H. (2004a). [Kansyana report: disappearance of the alpha male Fanana and subsequent stories.] *Mahale Chimpun*, **3**, 5–6. In Japanese.
Nishie, H. (2004b). [Kansyana report: a battle at Kasiha Valley. The fall of the alpha male Fanana and enthronement of Alofu.] *Mahale Chimpun*, **4**, 6–7. In Japanese.
Shimada, M. (2011). [Introduction to Mahale chimps 18: Ai.] *Mahale Chimpun*, **18**, 8. In Japanese.
Shimada, M. (2012). [Introduction to Mahale chimps 20: Caesar.] *Mahale Chimpun*, **20**, 7. In Japanese.
Shimada, M. (2013a). [Introduction to Mahale chimps 22: Pimu.] *Mahale Chimpun*, **22**, 6. In Japanese.
Shimada, M. (2013b). [Introduction to Mahale chimps 21: Puffy.] *Mahale Chimpun*, **21**, 4. In Japanese.
Shimada, M. (2014). Homosexual interactions among young female wild chimpanzees: An example of social pretend play? *Pan Africa News*, **21**, 6–8.
Takahata, H. and Takahata, Y. (1989). Inter-unit group transfer of an immature male of the common chimpanzee and his social interactions in the non-natal group. *African Study Monographs*, **9**, 209–20.
Takahata, Y. (1985). Adult male chimpanzees kill and eat a male newborn infant: newly observed intragroup infanticide and cannibalism in Mahale National Park, Tanzania. *Folia Primatologica*, **44**, 161–70.
Uehara, S. (2003). [Introduction to Mahale chimps 2: Limongo.] *Mahale Chimpun*, **2**, 6. In Japanese.
Uehara, S., Nishida, T., Takasaki, H., *et al.* (1994). A lone male chimpanzee in the wild: the survivor of a disintegrated unit-group. *Primates*, **35**, 275–81.

PART II Social organization

6 Social system: features and variations

NORIKO ITOH AND MICHIO NAKAMURA

6.1 Introduction

Primate social systems have drawn the attention of anthropologists interested in reconstructing the social lives of early humans, irrespective of the actual aims of each primatologist who studies them (Fedigan, 1982). The term "social system" covers a wide range of topics such as a mating and reproductive systems, spatiotemporal cohesiveness, age–sex composition, and social interactions and intra-relationships. In this chapter, the findings on the social system of chimpanzees are reviewed. First, the relevant aspects of the features of the Mahale M group are briefly summarized. Then we proceed to the topic of how a chimpanzee "social unit" was identified in early studies. We further compare some aspects of social system among several long-term chimpanzee study sites to investigate their similarities and variations.

6.2 Social system of the Mahale M group

Because the relevant information of the social system at Mahale is scattered throughout this book, a guide to the chapters is provided in this section.

Mahale chimpanzees form bisexual social units called unit-groups, each comprising multiple males and females of various ages (Nishida, 1967, 1968). Unit-groups have semi-closed features, with females transferring between unit-groups at sexual maturity and males remaining in their natal group (a male philopatry system; Chapters 7, 10, 46). This practice functions as incest avoidance (Chapter 35), given that mating promiscuously occurs mainly within the unit-group (Chapters 35, 46).

Chapter 7 presents demographic information of the M group. Its size ranged from 45 to 100 during the last 50 years. The largest size was observed after it experienced a "large-scale" transfer of females from the K group (Chapter 10) and may represent an overcrowded condition. Afterward, the size decreased to the minimum, slowly recovered to around 60, and has been stable for 10 years. Although the overall sex ratio has been constant at 0.5, the ratio of adult males to adult females shows an increasing trend. Immigrations and emigrations of adolescent females regularly occur.

All unit-group members do not range together in a tight, cohesive manner. The members are dispersed and frequently meet and part with one another over a day to a week, month, or more. The features of the M-group fission–fusion pattern are described in Chapter 9 and partly in Chapter 28. Interaction patterns and social relationships within the M group are described in detail mainly in Parts VI and VII.

6.3 On the social unit of chimpanzees

6.3.1 Basic social unit

From the start of Japanese primatology, the social unit was regarded as important (e.g. Itani, 1977, 1980, 1985), and still is (e.g. Nakagawa, 2013), for understanding the evolution of social structure in primates, including humans. The social unit of primates is a sociologically, ecologically, and biologically important entity. However, it took a long time to identify the social unit in chimpanzees, largely because of their fluid fission–fusion grouping pattern of forming temporal parties of various compositions.

Mahale Chimpanzees: 50 Years of Research, ed. M. Nakamura, K. Hosaka, N. Itoh and K. Zamma. Published by Cambridge University Press.

Figure 6.1 The K group was named after the male at the front, Kajabala. ©Toshisada Nishida.

As described later, the identification of the social unit of chimpanzees at Mahale was possible because of the long-term observations and habituation of two neighboring unit-groups (K and M; Figures 6.1, 6.2). However, it should also be noted that underpinning the finding was Itani's (1977, 1980, 1985) conceptual framework of the basic social unit (BSU) of primates in hypothesizing the evolution of primate social structure (see also Adachi, 2013). With several revisions resulting from additional findings, Itani's framework was largely completed by 1984 (Itani, 1985). A BSU conforms to the following characteristics or principles: (1) a BSU has and maintains bisexual structure through both the sexual and nonsexual periods; (2) a BSU is a semi-closed system toward lone individuals and other BSUs; (3) a BSU has temporal continuity that possesses a mechanism for regulating structural (size and composition) consistency; and (4) each *specia* (a sociological entity corresponding to a biological species [Itani, 1985: 596]) has only one type of BSU. When multiple levels of aggregations exist, only one level is considered as the BSU, and other (higher or lower) units are considered as more recently developed structures. The features of Mahale chimpanzees summarized earlier are generally consistent with this concept of BSU.

Itani's concept of animal society inherited the insights of Imanishi and Carpenter, who acknowledged the possibilities of culture and cultural behaviors in nonhumans (Itani, 1985). In other words, they presumed that, even in nonhuman societies, individuals not only form a society but are also influenced *by* their society. Thus, the identification of a social unit is indispensable to the full understanding of behaviors.

6.3.2 Elucidating the chimpanzee social unit at Mahale

Before the identification of the social unit at Mahale, Itani and his colleagues had already conducted studies at Budongo, Uganda (Itani, 1961), and Kabogo, Tanzania (Azuma and Toyoshima, 1961; Itani, 1964). From these pilot studies, they already had an idea that chimpanzees form a large social unit and that what researchers could actually observe at one time was only a part (currently known as a "party"; Chapter 9)

Figure 6.2 The M group was named after the male on the left, Mimikire. ©Toshisada Nishida.

of the social unit. Succeeding studies at Kabogo (Azuma and Toyoshima, 1965), Filabanga, Tanzania (Itani and Suzuki, 1967), and Kasakati, Tanzania (Izawa and Itani, 1966; Suzuki, 1969; Izawa, 1970) in the 1960s also supported the idea. For example, Itani and Suzuki (1967) confirmed a large group of chimpanzees (43 individuals) in progression, although it may not have been the entire social unit. Izawa (1970) succeeded in the individual identification of some of possible unit-groups. At this point, all studies assumed that fission–fusion occurs between members of a larger social unit in relation to seasonal variation in the food environment. They considered the further possibilities of whether such a large-sized group has subunits (e.g. Itani, 1966) or not (e.g. Nishida, 1967). Itani (1966) also implied some kind of suprastructure, which he called a "regional community," composed of several large-sized groups (note that this usage of "community" completely differs from that later applied to the social unit of chimpanzees by Western researchers).

Following these studies, Nishida (1967, 1968) identified all individuals coming to the feeding site in small parties and accumulated the records of their interactions at Mahale. Thus, he finally confirmed that there is a social unit in chimpanzees with stable membership and denied the existence of a subunit. He called this social unit a "unit-group" and explained that the chaotic fission–fusion was to be understood as a phenomenon occurring within the unit-group. The finding was further strengthened with the confirmation of female transfer between unit-groups (Nishida, 1968; Nishida and Kawanaka, 1972) and inter-unit-group avoidance relationships (Nishida and Kawanaka, 1972), the latter of which had been previously suggested by Izawa (1970).

6.3.3 Controversies over social unit

At the time, a few other studies were ongoing at different sites. Goodall assumed that chimpanzees, which she individually identified, of the entire area of the Gombe National Park, Tanzania (35.4 km^2 [Wilson, 2012]), formed a "community" (van Lawick-Goodall, 1968; see also Bygott, 1974). This whole community was considered indivisible into any permanent social structures. Similarly, at Beni (Congo), Kortlandt (1962) observed two different types of parties (nursery and sexual groups) that sometimes intermingled but did not assume sociological links between these parties.

Reynolds (1963), who observed Budongo chimpanzees in 1962, presumed that parties (termed as "groups") were loosely connected and termed this higher level of aggregation an "open group" (Reynolds, 1966). Subsequently, Sugiyama (1968) conducted an intensive survey there in 1966–7 by individual identification and concluded that there are several "regional populations." These may be similar to Nishida's unit-groups in that their members utilize almost the same area, know each other, and form parties easily. At the same time, they seem to differ in that their relationship with other regional populations was nonantagonistic (Sugiyama, 1968). The Budongo researchers' early acknowledgment of the social units may be because of several favorable conditions. First, there were multiple neighboring groups. Second, group members frequently contacted each other by vocalizations (Reynolds, 1963; Sugiyama, 1968). Third, Budongo chimpanzees, at least occasionally, formed large parties: Itani (1961) saw a party of around 17 chimpanzees during his short stay in 1960, and Reynolds (1963) observed up to 50 individuals at a time aggregating to eat *Maesopsis* fruits.

At Gombe, it took another 5 years from validation of the unit-group at Mahale for researchers to accept the existence of social units (van Lawick-Goodall, 1973) within the whole "community" (*sensu* van Lawick-Goodall, 1968). However, to complicate things, the very same term "community" was somehow applied to a social unit this time. We will return to this terminological issue later. In any case, since then, it has become common to assume a social unit for chimpanzees.

Sugiyama started to investigate chimpanzees at Bossou (Guinea) in 1976 (Sugiyama and Koman, 1979). The Bossou group lacks neighboring groups, but similar to the Budongo groups, its members are always in an auditorily contactable area of 6 km^2 encircled by human settlements and activities (see Sugiyama and Fujita, 2011). Sugiyama considered this group to be comparable with the social units in other sites, but he simply called it a "group" or "population" (e.g. Sugiyama and Koman, 1979; Sugiyama, 1981, 2004). His students and colleagues at Bossou use the term "community," in the sense now synonymously used with "unit-group" by Western researchers. Ghiglieri (1984), who started research in 1976 at Kibale (Uganda), discriminated two groups (Kanyawara and Ngogo groups) as different social units, which were referred to as "communities." Similarly, Boesch and Boesch-Achermann (2000), who have been studying at Taï (Côte d'Ivoire) since 1976, referred to a social unit as a "community." Yamagiwa *et al.* (1992), who started chimpanzee research at Kahuzi-Biega (Democratic Republic of Congo) in 1991, identified three "unit-groups."

Social units found in these studies, whatever they are called, are similar (fission–fusion social system, semi-closed nature, succeeding over generations, bisexual unit) and generally fit all criteria of a BSU. However, as seen earlier, most researchers now use the term "community" for a chimpanzee social unit. Some acknowledge the original terminology as "community (unit-group)," and only researchers at Mahale and Kahuzi commonly use the term "unit-group." Apart from the neglect of the priority of the term "unit-group" (see de Waal, 2003), the term "community" for the social unit is problematic for several reasons. First, the most common usage of the term in biology is for an assemblage of populations of two or more different species in the same habitat. Even when we restrict its usage in primatology, the term had already been used for hypothetical higher structures of primate society, including several BSUs (Imanishi, 1961; Itani, 1966; Nishida, 1968; note that this terminology follows its usage in anthropology or sociology rather than ecology). Furthermore, even if we neglect those studies by the Japanese, this term had originally been used in the chimpanzee research by a leading researcher to mean a clearly different concept (van Lawick-Goodall, 1968), as described earlier. Thus to avoid confusion, we use the term "unit-group" precisely meaning a BSU of chimpanzees.

We may take into account that there are more local variations in structure, organization, history, and so on of unit-groups than have previously been expected. This consideration may partly clarify the reason why different terms have been employed for unit-groups of different sites. In addition to local variations after multi-decadal monitoring, we are now aware that there are chronological variations within a unit-group.

6.4 Variation in demography

Because of the extreme longevity and slow reproduction of chimpanzees, our knowledge of their demography is still limited. Although variation in living habitats makes comparison across sites difficult, accumulated knowledge about chronological changes and inter-unit-group variances of demography may be much larger than expected in the early days.

6.4.1 Variation in unit-group sizes

Presently known unit-group sizes of chimpanzees vary from 12 for the Bossou group (Sugiyama and Fujita, 2011) to 148 for the Ngogo group (Mitani, 2006). Where decadal-scale demography is available, most unit-groups have experienced large fluctuations in size, including Mahale unit-groups. During 1966–83, the K group increased to 33 but subsequently decreased and eventually went extinct (Chapter 10). The reason for this decline is difficult to identify, but the demographic structure may have been unhealthy from the beginning of the observation (Chapter 10). As summarized earlier, the M group also experienced a large fluctuation of the unit-group size.

The initial targeted Gombe unit-group split into Kasekela and Kahama groups in 1971–2 (Goodall, 1986). Goodall considered that this split probably started in the early 1960s but was interrupted by provisioning. After this drastic change, the demography of the Kasekela group seems to have been relatively stable until 1983, except for a decline in the percentage of immatures (infants, juveniles, and early adolescents) from 51% in 1978 to 43% in 1983.

Sugiyama and Fujita (2011) reported that the Bossou group size has been stable at around 20 during 1976–2002, with several small fluctuations. However, after an epidemic of a flu-like disease in 2003, the size once shrank to a minimum of 12. As the size had been stable for decades at around 20, the authors suspect that the carrying capacity of the environment is the limiting factor for this unit-group.

The Taï unit-group also experienced two large declines in 1982–95. The initial size 74–82 (1982–7) declined to 50 by 1991 and to 29 by 1995 because of leopard predation and Ebola epidemics, respectively (Boesch and Boesch-Achermann, 2000). The Taï North group also declined from 42 to 21 in 1992–2001 (Lehmann and Boesch, 2008).

As we can see, the known major causes of large demographic changes vary between and within unit-groups. Similar events may have different impacts on different unit-groups with different initial sizes and/or compositions. Evaluation of demographic changes and causes awaits a more detailed comparison of demography and possible influential events between sites.

6.4.2 Variations in births, survival, and sex ratio

Births and deaths are the major parameters determining the size of a unit-group. The mean annual birth rate is 1.1 (37 births/34 years) at Bossou (Sugiyama and Fujita, 2011), 3.11 (59/19) at Gombe (Goodall, 1986), 5.00 (75/15) at Taï (Boesch and Boesch-Achermann, 2000), and 5.48 (181/33) for Mahale M group (Chapter 7). Mortality is the highest in infancy and decreases with age at all studied sites. Inter-unit-group variation in infant mortality rate is also known, but the underlying factors are variable (see Emery Thompson, 2013; Chapter 7).

Apart from ecological constraints, birth and infant mortality rates are influenced by the number of females and are regulated by dispersal regime and demographic histories. At Taï, a combination of the dominant–subordinate rank of the mother and the infant's sex influences the length of interbirth interval (Boesch and Boesch-Achermann, 2000). Female–

female competition also results in different effects on reproductive profiles (birth rate, interbirth interval, and infant survival) at Gombe (Pusey *et al.*, 1997) and Kanyawara (Emery Thompson *et al.*, 2007), where habitat and habitat use among females are heterogeneous (see Chapter 28 for a review). The effect of female–female relationships on overall demography is still under evaluation.

Although sex-biased births have not been found in any unit-groups, adult sex ratios show variation (see Boesch and Boesch-Achermann, 2000). They also chronologically change; the outstanding case is at Gombe, where an almost 1:1 (males:females) ratio in 1968 changed to 1:3 after 10 years (Goodall, 1986). In the Mahale M group, it changed from 1:3 to 1:2 over 30 years (Chapter 7). Several mechanisms in combination could produce a female-biased sex ratio. First is the imbalance between female immigration and emigration (Boesch and Boesch-Achermann, 2000). Second is the inter-sex difference in survival rate. Although there is some variation by age class, males' survivorship is generally lower than that of females. For the Mahale M group, survival rates of females were higher than those of males at up to 9 years (Chapter 7). Taï females show higher survival rates than males in the period from 15 to 38 years of age (Boesch and Boesch-Achermann, 2000). Males' survivorship at Mahale is also reduced by the absence of their mothers even after weaning (Nakamura *et al.*, 2014). Such an effect may eventually cause a skewed adult sex ratio, given that as many as 37 males have experienced orphanhood in the Mahale group to date (Nakamura *et al.*, 2014; Chapter 30). Single events such as large-scale female immigration and unit-group fission may also affect overall sex ratios.

6.4.3 Variation in dispersal regime

For all known studied unit-groups, adolescent females usually disperse and males stay in their natal unit-group. Because most female immigrants stay in the unit-group after giving birth (Chapter 7), female immigration/emigration is also an important parameter determining unit-group size and composition. It is known that inter-unit-group variation in the rates of female immigration is large (Gombe: 0.6 immigrants/year [11/19; Goodall, 1986]; Taï: 0.8 [13/15; Boesch and Boesch-Achermann, 2000]; Budongo: 1.2 [11/9; Reynolds, 2005]; Mahale M: 1.2 [41/34], excluding immigrants from the K group [Chapter 7]). Decisions on emigration are not unveiled (Chapter 11); thus, known variations in emigration rate (Taï: 88% [7/8; Boesch and Boesch-Achermann, 2000]; Mahale M: 75% [24/32; Chapter 7]; Gombe: 36–57% [5–8/14; Pusey *et al.*, 1997]) are still under evaluation (see also Chapter 11).

These variables may also chronologically change, so that short-term monitoring can yield underestimates/overestimates. Immigration occurred almost constantly for the M group after 1980 (Chapter 7). However, both Gombe (Goodall, 1986) and Taï (Boesch and Boesch-Achermann, 2000) data demonstrate that female immigration did not occur in some consecutive years.

With respect to immigration, the Bossou group has distinct features in several aspects (Sugiyama and Fujita, 2011). In 33 years, no female immigrations occurred, a circumstance assumed to be "a by-product of human habituation and the close proximity of the chimpanzees' core area to the village" (Sugiyama and Fujita, 2011: 30). Moreover, all natal females disappeared by the age of 15 years. The striking difference with other unit-groups is that all but two of the natal males also disappeared. Among these, some are considered to have emigrated (Sugiyama, 2004). In other sites, male transfers were exceptional. Two cases are of Mahale K-group juveniles, who transferred to the M group with their mothers in the course of the K-group extinction (Chapters 5, 10). Another is of an orphaned Gombe juvenile who emigrated to another unit-group following his presumed sister (Pusey, 1979). Typically, dispersal regimes are highly phylogenetically conservative traits in primates (see Strier, 2000). If so, male transfers at Bossou, if there are really such, may be explained by its isolated habitats having reached the upper limit of their carrying capacity and by the absence of immigrant females for decades.

Two observations of unfamiliar male visitors in Bossou were regarded as support for male emigration (Sugiyama and Fujita, 2011). Temporary disappearance of males for up to a year has also been reported from Budongo; these males might have visited another group (Reynolds, 2005). The possibility of temporary visits may not necessarily mean male emigration but need further attention.

6.4.4 Variation in inter-unit-group relationships

Inter-unit-group relations (see Chapter 11) may also be important for understanding the social system of chimpanzees. They too seem variable, from aggressive to avoidance to affiliative relationships both at population level and chronologically. However, detailed and quantitative information of inter-unit-group relationships is still scarce. To validate and reveal underlying factors, we need to wait for further multi-decadal-scale studies from different populations.

6.5 Social system and behaviors

A social system is viable when multiple individuals interact with each other, directly or indirectly. Frequencies of certain type of social behaviors are typically used to detect certain social relationships. Social behaviors (and consequently relationships) are also extremely variable both within and between unit-groups and are nested in ecology and demography (including slow individual development) regimes (Mitani and Watts, 1999; Mitani *et al.*, 2002; Mitani, 2006). On the other hand, the fission–fusion system is the sole common feature among chimpanzee unit-groups. In this section, we focus on some different perspectives to understand a social system through the social behaviors.

6.5.1 Bond/core models

Early studies expected some kind of strong bond or core for unity of a fluid social unit of chimpanzees. For example, Nishida (1968, 1979) emphasized "highly sociable males," who frequently, directly, and explicitly interacted with each other, as being the core, and some researchers followed this line (e.g. Kawanaka, 1984; Sakamaki, 2009). To an extreme, Wrangham (1979) considered that social units be restricted only to males. However, after some debate (Itani, 1980; Pusey, 1980; Uehara, 1981; Kawanaka, 1984), this "male-only community" model seems to have been moderated to a "male-bonded" model (see Williams *et al.*, 2002), which is similar to that of Nishida's.

Based on the findings from Taï, Boesch (1991) proposed a "bisexually bonded model," where both sexes equally contribute to the formation of the social unit in terms of sociality and range use. Because Bossou chimpanzees (e.g. Sugiyama, 1988) were also considered bisexually bonded, an eastern/western subspecies difference was emphasized with respect to the "bonds." Such simple differentiation is now in question based on recent studies (see Chapter 28). We should also note that the early image of male centrality at Mahale may have been formed under an unusual social circumstance by "large-scale female immigration" (Chapters 7, 28).

Another supportive phenomenon for the male-centralized view is the highly organized hierarchical/linear dominant–subordinate relationship among males. When a dominant–subordinate relationship is central to the integration of a unit-group, females are regarded as less significant because they are subordinate to all males (e.g. Bygott, 1974). However, no studies have empirically validated the importance of inter-sex dominant–subordinate relationships to unit-group unity. Furthermore, dominant–subordinate relationships between sexes may not be as simple as currently assumed. For example, Nishida (Nishida, 1970; Nishida and Kawanaka, 1972) reported that a K-group female who tended to travel alone was often one-sidedly greeted by two adult males. Such overt expression of "reversed" male–female dominant–subordinate relationships may be rare but warrants further attention.

6.5.2 Nonbond/noncore model

Mori (1977) demonstrated that a Japanese monkey (*Macaca fuscata*) troop maintained its unity not through frequent interactions but through avoidances. In a study on patas monkeys (*Erythrocebus patas*), Rowell and Olson asserted that "[d]isplays should draw the attention of the observer to the dynamic aspects of social organization, whereas distance regulation is perhaps a more useful indicator of its conservative aspects." (1983: 53). If we take these studies seriously, there is no basis to the premise that a social "bond/core," represented by frequent explicit social behaviors within a social unit, contributes to the unity or viability of a social unit. Adachi (2013) demonstrated the importance of such antistructural aspects for assemblages of various species, including mixed-species assemblages.

In this sense, chimpanzees' fission–fusion system may need further attention. Although chimpanzees stay away from each other, unit-group members are not *always* apart but meet each other at various intervals and for various durations throughout their lives. Thus, their fission–fusion itself could be considered as an interaction, acting as a system of unit-group viability. Here, intra- and inter-unit-group variations in fission–fusion phenomena, including party size, party formation, duration of each party, and party compositions may be considered as different conventional processes where each unit-group maintains its social system *in situ* (see also Chapter 38). For evaluation, sufficient long-term multiple studies to assess socioecological variables such as home-range size and limiting factors, demographic histories, and unit-group size and composition, as well as application of universal methods to assess ecological conditions are desirable. However variable they may be, all chimpanzee societies commonly retain a fission–fusion system, which can thus be considered a conservative trait.

6.5.3 Social behavior and the social system

Social behaviors are commonly counted and calculated to investigate gradients of dyadic social relationships, which eventually enable us to describe social structure or organization (these terms being differently or synonymously used depending on the study). This hierarchical concept of behavior/relationship/structure developed by Hinde (1976) implies that a social system is decomposable into individual behaviors.

Itani (1987), however, regarded a social system as undecomposable and discussed social behaviors as a means of coexistence of social-unit members. Here, social behaviors are recursively assessed for the meaning or quality of behavior in relation to the actually existing social system. Apart from sexual and mother–offspring interactions, he considered two principles governing primate societies: *inequality* (e.g. dominant–subordinate relationships) and *equality* (e.g. food sharing).

He considered that the principle of inequality in chimpanzee society has become somewhat collapsed or at least unreliable for their coexistence in the evolutionary process. He drew this conclusion because, compared with the more rigid dominant–subordinate relationships in macaques such as Japanese monkeys, those in chimpanzees are more variable and unstable (see Chapters 27, 28). Under the "collapse" of the inequality principle, pluralistic social norms for coexistence were assumed to be required. He argued that chimpanzee society is in some situations governed by the principle of equality, as evidenced by food sharing (Chapter 20). Behaviors such as variable greeting behaviors accompanied by facial, vocal, and gestural expressions and appeasement behaviors were exemplified as adjustments to the "collapse" of the inequality principle. Although Itani's arguments attracted less attention in subsequent studies, we believe that they still have validity for understanding chimpanzee sociality and reconsidering human evolution.

6.6 Possible means of chimpanzee expansion

Group fission occurs in some primate species as a means by which they regulate and expand their distribution and population size (Schaller, 1976; Izawa, 1999; Fedigan and Jack, 2001). However, there has been only one case of unit-group fission observed for chimpanzees (Bygott, 1974), and it resulted in the extinction of one of the fission groups (Goodall,

1986). Thus, we currently do not know whether group fission in chimpanzees commonly results in emergence of a new unit-group.

There are, however, a few suggestive cases of formation of new groups. Izawa (1970) witnessed some small-sized groups in the vast Kasakati area and suspected that they would grow into a large-sized group. The Mahale Y group may be another example of a small splinter group, given that it somehow suddenly appeared in the 1990s in the vacant area formally used by the K group (Chapter 11). If a group split is possible only when such a vacant area is available, a chance to observe a unit-group fission in chimpanzees may be rare because they *cannot* split even in need. Such vacancies are becoming fewer and/or reduced in size these days partly because of human activities, including deforestation.

Furthermore, we are ignorant of whether there are any social or structural constraints on chimpanzees' unit-group size. What is currently known is that it can reach up to around 150. If group fission is constrained by difficulty in obtaining additional space, unit-group size may increase but then crash at the limit of, at least, carrying capacity of the home range. Such a situation is not favorable for this endangered species (Chapters 1, 50).

It is overwhelmingly difficult to validate whether the social system we are observing now is a result of adaptation, random drift, or human pressure (see Rowell, 1979). We may be looking at chimpanzees who have already lost enough extra room for unit-group fission to occur. If the Kasakati and the Mahale Y-group cases were really splinter groups, it may have been the very rare cases of newly formed unit-groups.

References

Adachi, K. (2013). The sociology of anti-structure: toward a climax of groups. In *Groups: The Evolution of Human Sociality*, ed. K. Kawai. Kyoto: Kyoto University Press and Trans Pacific Press, pp. 21–41.

Azuma, S. and Toyoshima, A. (1961). Progress report of the survey of chimpanzees in their natural habitat, Kabogo Point area, Tanganyika. *Primates*, **3**, 61–70.

Azuma, S. and Toyoshima, A. (1965). [Chimpanzees in Kabogo Point Area, Tanganyika.] In [*Monkeys and Apes*], ed. S. Kawamura and J. Itani. Tokyo: Chūō-kōron-sha, pp. 127–63. In Japanese.

Boesch, C. (1991). The effect of leopard predation on grouping patterns in forest chimpanzees. *Behaviour*, **117**, 220–41.

Boesch, C. and Boesch-Achermann, H. (2000). *The Chimpanzees of the Taï Forest: Behavioural Ecology and Evolution*. New York: Oxford University Press.

Bygott, J. D. (1974). *Agonistic Behaviour and Dominance in Wild Chimpanzees*. Ph.D. thesis. Cambridge: University of Cambridge.

Emery Thompson, M. (2013). Reproductive ecology of female chimpanzees. *American Journal of Primatology*, **75**, 222–37.

Emery Thompson, M., Kahlenberg, S. M., Gilby, I. C., and Wrangham, R. W. (2007). Core area quality is associated with variance in reproductive success among female chimpanzees at Kibale National Park. *Animal Behaviour*, **73**, 501–12.

Fedigan, L. M. (1982). *Primate Paradigms: Sex Roles and Social Bonds*. Chicago, IL: University of Chicago Press.

Fedigan, L. M. and Jack, K. (2001). Neotropical primates in a regenerating Costa Rican dry forest: a comparison of howler and capuchin population patterns. *International Journal of Primatology*, **22**, 689–713.

Ghiglieri, M. P. (1984). *The Chimpanzees of Kibale Forest: A Field Study of Ecology and Social Structure*. New York: Columbia University Press.

Goodall, J. (1986). *The Chimpanzees of Gombe: Patterns of Behavior*. Cambridge, MA: Belknap.

Hinde, R. A. (1976). Interactions, relationships and social structure. *Man*, **11**, 1–17.

Imanishi, K. (1961). [The origin of human family: a primatological approach.] *The Japanese Journal of Ethnology*, **25**, 119–38. In Japanese with English abstract.

Itani, J. (1961). [*The Forest of Gorillas and Pigmies*.] Tokyo: Iwanami. In Japanese.

Itani, J. (1964). [*Animals of Africa*.] Tokyo: Kawade-shobō. In Japanese.

Itani, J. (1966). [Social organization of chimpanzees.] *Shizen*, **8**, 17–30. In Japanese.

Itani, J. (1977). Evolution of primate social structure. *Journal of Human Evolution*, **6**, 235–43.

Itani, J. (1980). Social structures of African great apes. *Journal of Reproduction and Fertility*, **Suppl. 28**, 33–41.

Itani, J. (1985). The evolution of primate social structures. *Man*, **20**, 593–611.

Itani, J. (1987). *[The Evolution of Primate Society*.] Tokyo: Heibon-sha. In Japanese.

Itani, J. and Suzuki, A. (1967). The social unit of chimpanzees. *Primates*, **8**, 355–81.

Izawa, K. (1970). Unit groups of chimpanzees and their nomadism in the savanna woodland. *Primates*, **11**, 1–45.

Izawa, K. (1999). [Comparison on six troops of Japanese monkey in Kinkazan.] *Japanese Monkeys of Miyagi Prefecture*, **10**, 1–11. In Japanese.

Izawa, K. and Itani, J. (1966). Chimpanzees in Kasakati Basin, Tanganyika (1). Ecological study in the rainy season, 1963–1964. *Kyoto University African Studies*, **1**, 73–156.

Kawanaka, K. (1984). Association, ranging, and the social unit in chimpanzees of the Mahale Mountains, Tanzania. *International Journal of Primatology*, **5**, 411–34.

Kortlandt, A. (1962). Chimpanzees in the wild. *Scientific American*, **20**, 349–67.

van Lawick-Goodall, J. (1968). The behaviour of free-living chimpanzees in the Gombe Stream Reserve. *Animal Behaviour Monographs*, **1**, 161–311.

van Lawick-Goodall, J. (1973). Cultural elements in a chimpanzee community. In *Precultural Primate Behaviour*, ed. W. W. Menzel. Basel, Switzerland: S. Karger, pp. 144–84.

Lehmann, J. and Boesch, C. (2008). Sexual differences in chimpanzee sociality. *International Journal of Primatology*, **29**, 65–81.

Mitani, J. C. (2006). Demographic influences on the behavior of chimpanzees. *Primates*, **47**, 6–13.

Mitani, J. C. and Watts, D. P. (1999). Demographic influences on the hunting behavior of chimpanzees. *American Journal of Physical Anthropology*, **109**, 439–54.

Mitani, J. C., Watts, D. P., Pepper, J. W., and Merriwether, D. A. (2002). Demographic and social constraints on male chimpanzee behaviour. *Animal Behaviour*, **64**, 727–37.

Mori, A. (1977). Intra-troop spacing mechanism of the wild Japanese monkeys of the Koshima troop. *Primates*, **18**, 331–57.

Nakagawa, N. (2013). The function and evolutionary history of primate groups: Focusing on sex differences in locational dispersal. In *Groups: The Evolution of Human Sociality*, ed. K. Kawai. Kyoto: Kyoto University Press and Trans Pacific Press, pp. 79–110.

Nakamura, M., Hayaki, H., Hosaka, K., Itoh, N., and Zamma, K. (2014). Orphaned male chimpanzees die young even after weaning. *American Journal of Physical Anthropology*, **153**, 139–43.

Nishida, T. (1967). [The society of wild chimpanzees: success of provisioning at Kasoge.] *Shizen*, **22**(8), 31–41. In Japanese.

Nishida, T. (1968). The social group of wild chimpanzees in the Mahali Mountains. *Primates*, **9**, 167–224.

Nishida, T. (1970). Social behavior and relationship among wild chimpanzees of the Mahali Mountains. *Primates*, **11**, 47–87.

Nishida, T. (1979). The social structure of chimpanzees of the Mahale Mountains. In *The Great Apes*, ed. D. A. Hamburg and E. R. McCown. Menlo Park, CA: Benjamin/Cummings, pp. 73–121.

Nishida, T. and Kawanaka, K. (1972). Inter-unit-group relationships among wild chimpanzees of the Mahali Mountains. *Kyoto University African Studies*, **7**, 131–69.

Pusey, A. E. (1979). Intercommunity transfer of chimpanzees in Gombe National Park. In *The Great Apes*, ed. D. A. Hamburg and E. R. McCown. Menlo Park, CA: Benjamin/Cummings, pp. 465–79.

Pusey, A. E. (1980). Inbreeding avoidance in chimpanzees. *Animal Behaviour*, **28**, 543–52.

Pusey, A. E., Williams, J., and Goodall, J. (1997). The influence of dominance rank on the reproductive success of female chimpanzees. *Science*, **277**, 828–31.

Reynolds, V. (1963). An outline of the behaviour and social organisation of forest-living chimpanzees. *Folia Primatologica*, **1**, 95–102.

Reynolds, V. (1966). Open groups in hominid evolution. *Man*, **1**, 441–52.
Reynolds, V. (2005). *The Chimpanzees of the Budongo Forest: Ecology, Behaviour, and Conservation*. Oxford: Oxford University Press.
Rowell, T. E. (1979). How would we know if social organization were not adaptive? In *Primate Ecology and Human Origins: Ecological Influences on Social Organization*, ed. I. S. Bernstein and E. O. Smith. New York: Garland STPM Press, pp. 1–22.
Rowell, T. E. and Olson, D. K. (1983). Alternative mechanisms of social organization in monkeys. *Behaviour*, **86**, 31–54.
Sakamaki, T. (2009). Group unity of chimpanzees elucidated by comparison of sex differences in short-range interactions in Mahale Mountains National Park, Tanzania. *Primates*, **50**, 321–32.
Schaller, G. B. (1976). *The Mountain Gorilla: Ecology and Behavior*. Chicago: University of Chicago Press.
Strier, K. B. (2000). From binding brotherhoods to short-term sovereignty: the dilemma of male Cebidae. In *Primate Males: Causes and Consequences of Variation in Group Composition*, ed. P. M. Kappeler. Cambridge: Cambridge University Press, pp. 72–83.
Sugiyama, Y. (1968). Social organization of chimpanzees in the Budongo Forest, Uganda. *Primates*, **9**, 225–58.
Sugiyama, Y. (1981). Observations on the population dynamics and behavior of wild chimpanzees at Bossou, Guinea, in 1979–1980. *Primates*, **22**, 435–44.
Sugiyama, Y. (1988). Grooming interactions among adult chimpanzees at Bossou, Guinea, with special reference to social structure. *International Journal of Primatology*, **9**, 393–407.
Sugiyama, Y. (2004). Demographic parameters and life history of chimpanzees at Bossou, Guinea. *American Journal of Physical Anthropology*, **124**, 154–65.
Sugiyama, Y. and Fujita, S. (2011). The demography and reproductive parameters of Bossou chimpanzees. In *The Chimpanzees of Bossou and Nimba*, ed. T. Matsuzawa, T. Humle, and Y. Sugiyama. Tokyo: Springer, pp. 23–34.
Sugiyama, Y. and Koman, J. (1979). Social structure and dynamics of wild chimpanzees at Bossou, Guinea. *Primates*, **20**, 323–39.
Suzuki, A. (1969). An ecological study of chimpanzees in a savanna woodland. *Primates*, **10**, 103–48.
Uehara, S. (1981). [The social unit of wild chimpanzees: a reconsideration based on the diachronic data accumulated at Kasoje in the Mahale Mountains, Tanzania.] *Journal of African Studies*, **20**, 15–32. In Japanese with English summary.
de Waal, F. M. (2003). Silent invasion: Imanishi's primatology and cultural bias in science. *Animal Cognition*, **6**, 293–9.
Williams, J. M., Pusey, A. E., Carlis, J. V., Farm, B. P., and Goodall, J. (2002). Female competition and male territorial behaviour influence female chimpanzees' ranging patterns. *Animal Behaviour*, **63**, 347–60.
Wilson, M. (2012). Long-term studies of the chimpanzees of Gombe National Park, Tanzania. In *Long-Term Field Studies of Primates*, ed. P. M. Kappeler and D. P. Watts. Heidelberg, Germany: Springer, pp. 357–84.
Wrangham, R. W. (1979). Sex differences in chimpanzee dispersion. In *The Great Apes*, ed. D. A. Hamburg and E. R. McCown. Menlo Park, CA: Benjamin/Cummings, pp. 481–9.
Yamagiwa, J., Mwanza, N., Spangenberg, A., *et al.* (1992). Population density and ranging pattern of chimpanzees in Kahuzi-Biega National Park, Zaire: a comparison with a sympatric population of gorillas. *African Study Monographs*, **13**, 217–30.

7 Demography of the M group

MICHIO NAKAMURA

7.1 Introduction

Demographic information is very important and fundamental to many aspects of animal studies, including those on humans. However, it is not easy to collect and accumulate sufficient demographic data in wild mammals, especially when they have a long lifespan. Several reports have already described the demographic characteristics of chimpanzees at Mahale (Hiraiwa-Hasegawa *et al.*, 1984; Nishida *et al.*, 1990, 2003). Demographic data have also been published from other long-term study sites such as Gombe, Tanzania (Teleki *et al.*, 1976; Goodall, 1983, 1986); Bossou, Guinea (Sugiyama, 2004; Sugiyama and Fujita, 2011); Taï, Côte d'Ivoire (Boesch and Boesch-Achermann, 2000); and Budongo, Uganda (Reynolds, 2005). Such data are useful for further investigation of more specific topics, often in comparison with humans, such as menopause and post-reproductive lifespan (Emery Thompson *et al.*, 2007; Hawkes and Smith, 2010), age-specific mortality rates (Hill *et al.*, 2001; Hawkes *et al.*, 2009; Bronikowski *et al.*, 2011), and causes of deaths (Williams *et al.*, 2008) that may have important implications for the studies of human evolution.

Nishida *et al.* (2003) performed detailed analyses on a large demographic dataset for the Mahale chimpanzees, covering the time period from the onset of research to 1999. Thus, the aim of this chapter is to present additional demographic data of the Mahale M group from 2000 onward and to analyse the data focusing on the long-term changes of some demographic characteristics. The demographic data of the extinct K group is presented in Chapter 10.

7.2 Dataset

7.2.1 Overall data

The members of MMCRP (see Chapter 2) have been recording the presence/absence of chimpanzees on a daily basis at Mahale from the early days of research. With this accumulation of daily attendance records, we know the first observed days of newborn or new immigrant chimpanzees and the final observation days of individuals. We can only retrospectively know the final days from long-term monitoring because it is not uncommon for researchers to miss a chimpanzee for weeks or sometimes even months because of their fission–fusion sociality. As the death of an individual is not usually confirmed, an individual is assumed to have died when he/she has disappeared from the group, except for females who may have transferred to unknown groups (Chapter 6; see below for the basis to assume female emigrations).

For the number of individuals in the unit-group each year, the number of individuals on December 31 is used to represent the number for the year. Data analyses were conducted in 2013; thus, group size data shown here are until 2012. For those whose birth years were estimated, such as immigrant females and old individuals, it is assumed that the individuals advanced 1 year in age on January 1 each year.

Although the demographic dataset of the M group until 1999 is basically the same as that presented by Nishida *et al.* (2003), there are some small inconsistencies in the numbers. This is due to the partial failure to reconstruct the data extraction

Mahale Chimpanzees: 50 Years of Research, ed. M. Nakamura, K. Hosaka, N. Itoh and K. Zamma. Published by Cambridge University Press.

method used by Nishida, who, as the first author, was responsible for the data analyses of the paper (Nishida *et al.*, 2003). The inconsistencies are mostly for those individuals observed in the early 1980s; although all the M-group members were identified by 1980 (Hiraiwa-Hasegawa *et al.*, 1984), there seemed to be some observers who were not perfectly clear on some chimpanzees in the >100 unit-group. Also, the age estimations for some individuals were not consistent during that time, and it is not very clear how Nishida treated some individuals referred to as "temporal visitors" in Hiraiwa-Hasegawa *et al.* (1984). For the dataset in this chapter, the original attendance sheets were re-evaluated when anything was unclear. However, it was difficult to reconstruct the dataset completely because the recording protocol was not consistent at that time, and tentative names were sometimes used for newcomers or temporal visitors. Thus, the group size may be slightly underestimated for the 1980s. However, because there were not many inconsistencies, it is believed there is no large effect on overall trends. No such ambiguity is present after 1995.

7.2.2 Data on female reproductive life history

For immigrant females, data from 41 females who had immigrated into the M group between 1980 and 2013 as adolescents were used. Females who had immigrated from the K group to the M group in the 1980s were not included because these transfers might have occurred in strong connection with the extinction of the former group (Chapter 10). Furthermore, some older parous females, who are usually not supposed to emigrate to a different group, were actually observed to do so during this unusual event. The ages of the new immigrants were estimated when they were first observed in comparison with females of known ages based on body size, face colors, genital swellings, etc. Some immigrants did not stay in the M group and were assumed to transfer again to a different group. At Mahale, we usually assume a second transfer has occurred when an immigrant female disappears from the M group before giving birth and without an apparent health problem.

The data on 32 natal females who had been born to the M group between 1980 and 2003 and had reached at least 9 years of age, the possible age for emigration, by the time of data analyses in 2013 were used. Although Nishida *et al.* (2003) did not include the female, Ako (AK), in their discussion of natal females (probably because her mother's identity was uncertain), she was included as a natal female following a more recent paper (Inoue *et al.*, 2008). As it was estimated that she was born in 1981, she is considered here to have been born on January 1, 1981, by the calculations used in this study. For natal females who were observed soon after their births, usually within a month, the first observed dates were used as their birthdays. We usually assume that a natal female has emigrated to another group when she disappears from the M group after experiencing an estrous cycle but before giving birth and without an apparent health problem.

7.3 Overall trends in demography

7.3.1 Changes in the age–sex structure

Figure 7.1 shows the age–sex structure changes in the M group during 1980–2012. The size of the M group was always much larger than that of the K group (see Chapter 10 for the demographic data of the K group), and it once had more than 100 members. The large increase in its number in the early 1980s corresponds to the immigration of several females from the K group during the course of its extinction and the accompanying births. From the latter half of the 1980s, the M group experienced several declines caused by diseases (Chapter 25), lion predations (Tsukahara, 1992), etc., and counter-increases. The largest decline was observed during 1994–6 for unknown reasons; the smallest group size of 45 was seen in 1996, which was less than half of the group size when it was at its largest. After the smallest recorded group size in 1996, the M group gradually increased its number to approximately 60, and there have been relatively few fluctuations in the past 10 or so years.

Compared with the reversed age structure of the K group (see Chapter 10), the M group has a relatively

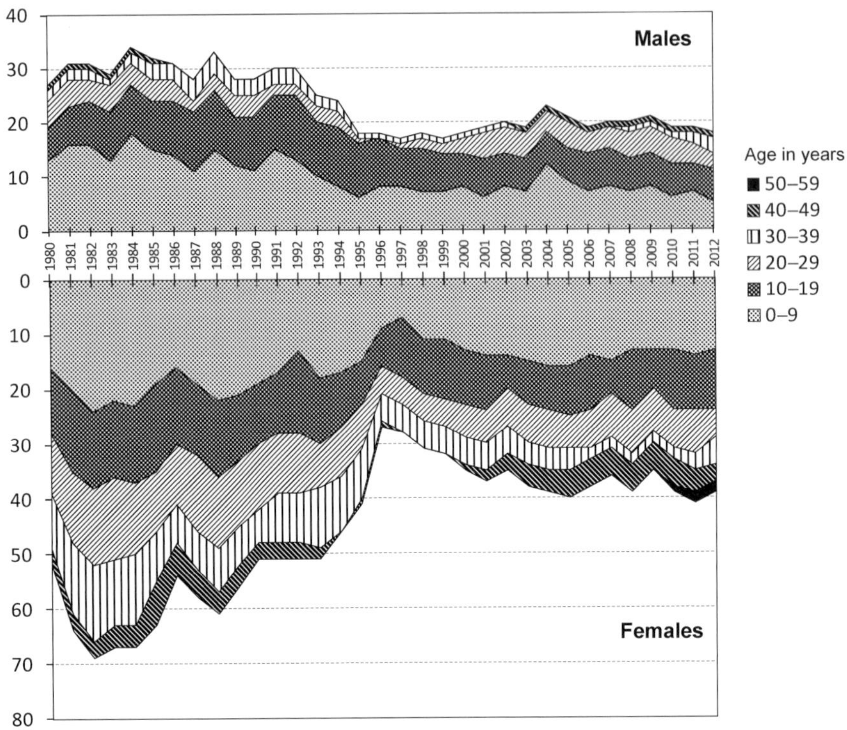

Figure 7.1 Change in the M group's age–sex structure during 1980–2012.

healthier age–sex pyramid structure on average (Figure 7.2), with younger generations (under 19 years) comprising more than half the group. However, the proportion of older generations (30 years or older) has been gradually increasing in recent years (Figure 7.1). The median age of males per year is increasing with each year (Kendall's rank correlation, $\tau = 0.63$, $p = 8.49 \times 10^{-7}$). However, it is possible that the males' ages were underestimated in the early period (see Chapter 23). On the other hand, the median age of females is decreasing with each year ($\tau = -0.63$, $p = 6.90 \times 10^{-7}$). This may be largely due to a large number of middle-aged (twenties and thirties) females in the 1980s, some of whom were immigrants from the K group. In recent years, there were several females 50 years or older. Although there were no chimpanzees this old until recently, this may also be due to an underestimation of ages for older individuals in the early period (see Chapter 23 for more discussion about old-aged individuals).

7.3.2 Changes in the sex ratio

The number of adult males per adult female is increasing each year (Figure 7.3, Kendall's rank correlation test: $N = 34$, $\tau = 0.55$, $p < 0.000001$). There were only an average of 0.26 adult males per adult female in the 1980s, but there were an average of 0.32, 0.42, and 0.50 adult males per adult female in the 1990s, the 2000s, and 2012, respectively. A large change in the adult sex ratio also occurred at Gombe, where there was almost an equivalent number of adult males and females in 1968, but the ratio of the females to males tripled 10 years later (Goodall, 1986). Contrary to the change in the adult sex ratio, the overall sex ratio has remained quite consistent in the Mahale M group throughout the research period (approximately 0.5 males per female). The latter figure is also similar in the K group, despite a large difference in the group size. When the number of males and females in these two groups are plotted (Figure 7.4), the data points are almost on one straight line with few outliers.

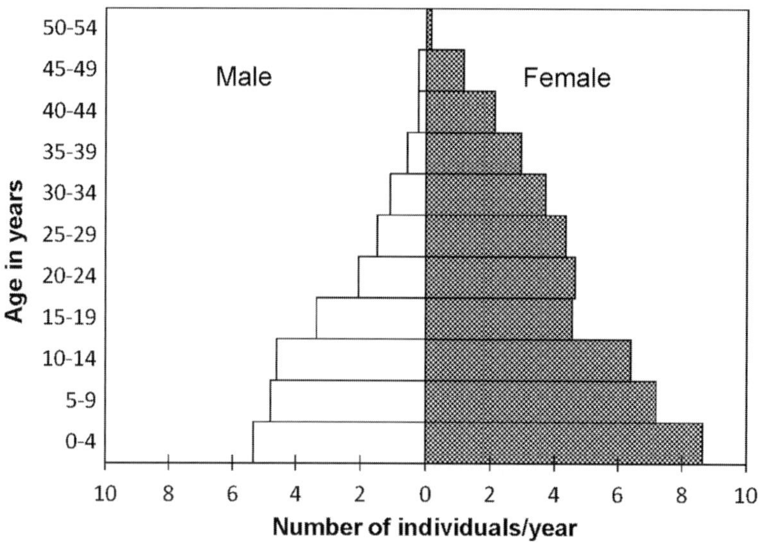

Figure 7.2 Average age–sex pyramid of the M group during 1980–2012. The number of individuals in each age–sex category in different years was averaged throughout the period.

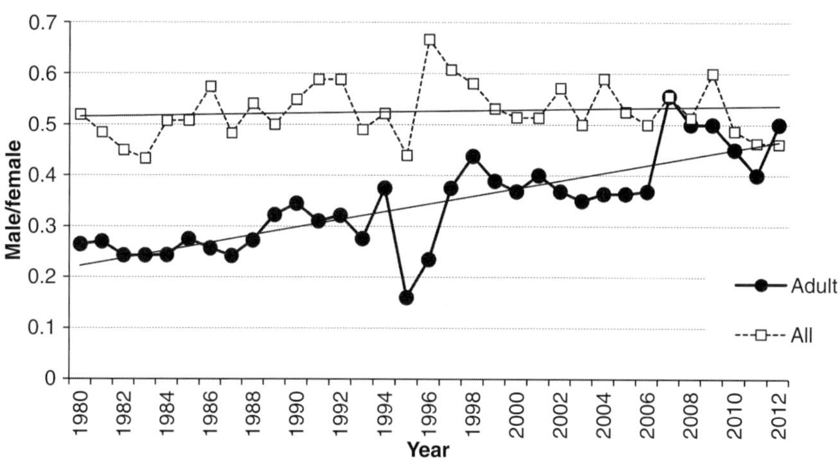

Figure 7.3 Changes in the sex ratio (male/female) of the M-group chimpanzees.

7.3.3 Changes in the number of births per year

There were 181 births (including some stillborns) confirmed during 1980–2012 in the M group (mean = 5.48 ± 2.58 SD per year, median = 5, range = 2–12), including 74 male infants, 85 female infants, and 22 infants with unknown sex (those who died before researchers could confirm their sexes). There were more female births, but there was no significant difference if we compared the sexes by year (Wilcoxon signed-rank test, $T = 174$, $p = 0.27$). The yearly birth number fluctuates (Figure 7.5) with an approximately 2-year cycle of increases and

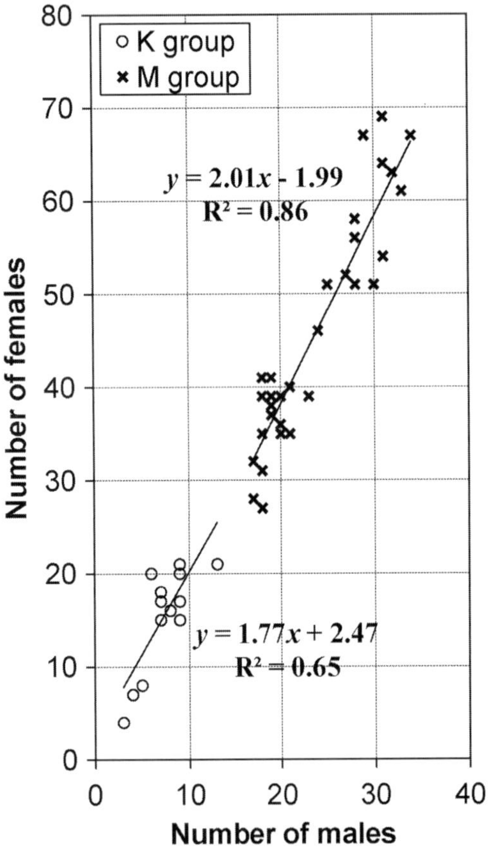

Figure 7.4 Number of females plotted against number of males for the K and M groups. Each point represents the data for a year.

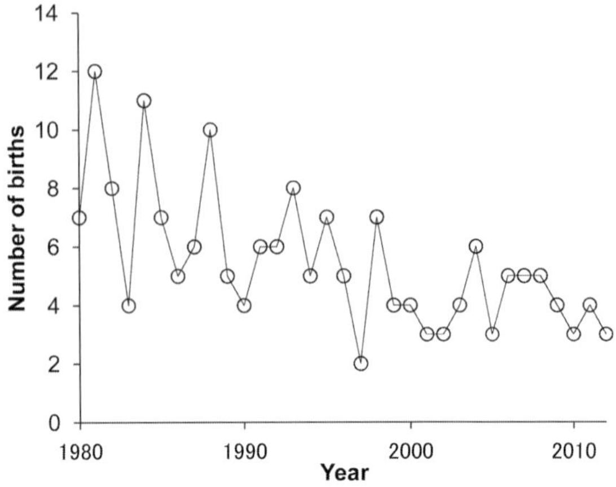

Figure 7.5 Number of births in the M group in each year.

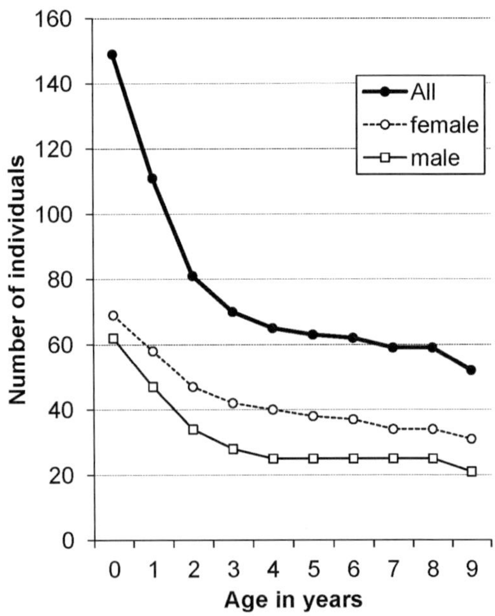

Figure 7.6 Survival up to 9 years in the M group. Data from individuals born between 1980 and 2004 were used here. The sum of male and female numbers does not equal the total infant number because there are some unsexed infants.

decreases. This is understandable because when many females give birth in a year, there will be few estrous females in the following years, thereby reducing the potential number of births that year. There has been a declining trend from 1980 to 2012 (Kendall's rank correlation, $\tau = -0.48$, $p = 0.00019$), which roughly corresponds with the decline in the number of adult females (see Figure 7.1).

7.3.4 Survivorship

Figure 7.6 shows the survival curve of the 149 chimpanzees (62 males, 69 females, and 18 unsexed) born between 1980 and 2004 in the M group until they were 9 years old. Overall, 74.5% of the newborns lived until the age of 1 year, and 42.3% survived until the age of 5 years. The risk of death declines after 5 years of age and 34.9% survived until the age of 9 years. The males have similar proportions of survival compared with all the chimpanzees, namely 75.8%, 40.3%, and 33.9%, respectively. On the other hand, the females' survival rates are higher at 84.2%, 55.1%, and 44.9%, respectively.

Corresponding figures from other chimpanzee study sites reported 88% survival until the age of 5 years at Taï (Boesch and Boesch-Achermann, 2000), 85% until the age of 5 years at Kanyawara, Uganda (Emery Thompson *et al.*, 2007), 79% until the age of 4 years at Bossou (Sugiyama and Fujita, 2011), and 77.6% until the age of 1 year, and 71.8% until the age of 5 years at Ngogo, Uganda (Watts, 2012). Combined survival rates from multiple field sites until the ages of 1 and 5 years are 78% and 52% for males and 82% and 60% for females, respectively (Hill *et al.*, 2001). Thus, the infant chimpanzees' survival rate at Mahale is much lower than any of these figures.

7.3.5 Causes of death

Until 1999, the major cause of death was disease (48%), followed by senescence (24%), and within-species aggression (16%) (Nishida *et al.*, 2003). The situation may not largely differ after 2000 because the M group experienced 12 losses in 2006 due to a disease outbreak (Hanamura *et al.*, 2008). The largest cause of deaths at Gombe was also disease (Williams *et al.*, 2008). Deaths due to diseases are described in Chapter 25 and deaths caused by within-species aggression are described in Chapter 26.

7.4 Female reproductive life history

Figures 7.7 and 7.8 show female reproductive life histories of immigrant females and natal females, respectively. In general, immigrations and births occurred at a relatively constant rate throughout the research period.

7.4.1 Emigration and immigration

The median of the estimated ages at the time of immigration to the M group after 1980 (excluding ex-K-group immigrants during the course of their group's extinction) was 11.00 years (mean = 11.24 ± 1.30 SD, N = 41). Nine out of 41 immigrant females were assumed to have transferred again to different groups within one to a few years without giving birth in the M group. Twenty-four out of 32 natal females transferred from the M group, and the median age of emigration was 11.00 years (mean = 11.38 ± 1.42, N = 24), which is the same as the 11.00 years reported by Nishida *et al.* (2003) because of overlapping data. Because the estimated age at immigration is almost identical to the known age at emigration, it can be said that the age estimation at immigration seems to be adequate.

There was not a single case of the immigration of a parous female or even a suspected case (such as an unidentified mother suddenly appearing in the group) after 1990. However, there were several cases of parous females' immigration during the course of extinction of the K group (cases not shown in Figure 7.7).

7.4.2 Year of first delivery

Among the 41 immigrant females in the M group, 28 gave birth and the median duration from the time of immigration to the first delivery was 3.14 years (mean = 3.11 ± 1.44, N = 28). This is longer than the 32 months (= 2.67 years) reported by Nishida *et al.* (2003), probably because their analysis included some immigrants from the K group to the M group, some of whom were older and might not have transferred if the K group continued to exist. The median estimated age at the time of first delivery for the 28 immigrant females was 14.28 years (mean = 14.50 ± 1.69, N = 28).

Among the 32 natal females in the M group, six gave birth and the median age at the time of first delivery was 13.08 years (mean = 13.57 ± 1.81, N = 6), which is the same as the 13.1 years reported by Nishida *et al.* (2003) because of the previously mentioned data overlap. There was a period when no natal females remained in the M group after a few natal females gave birth in 1998. However, in 2012 and 2014, two natal females in the M group,

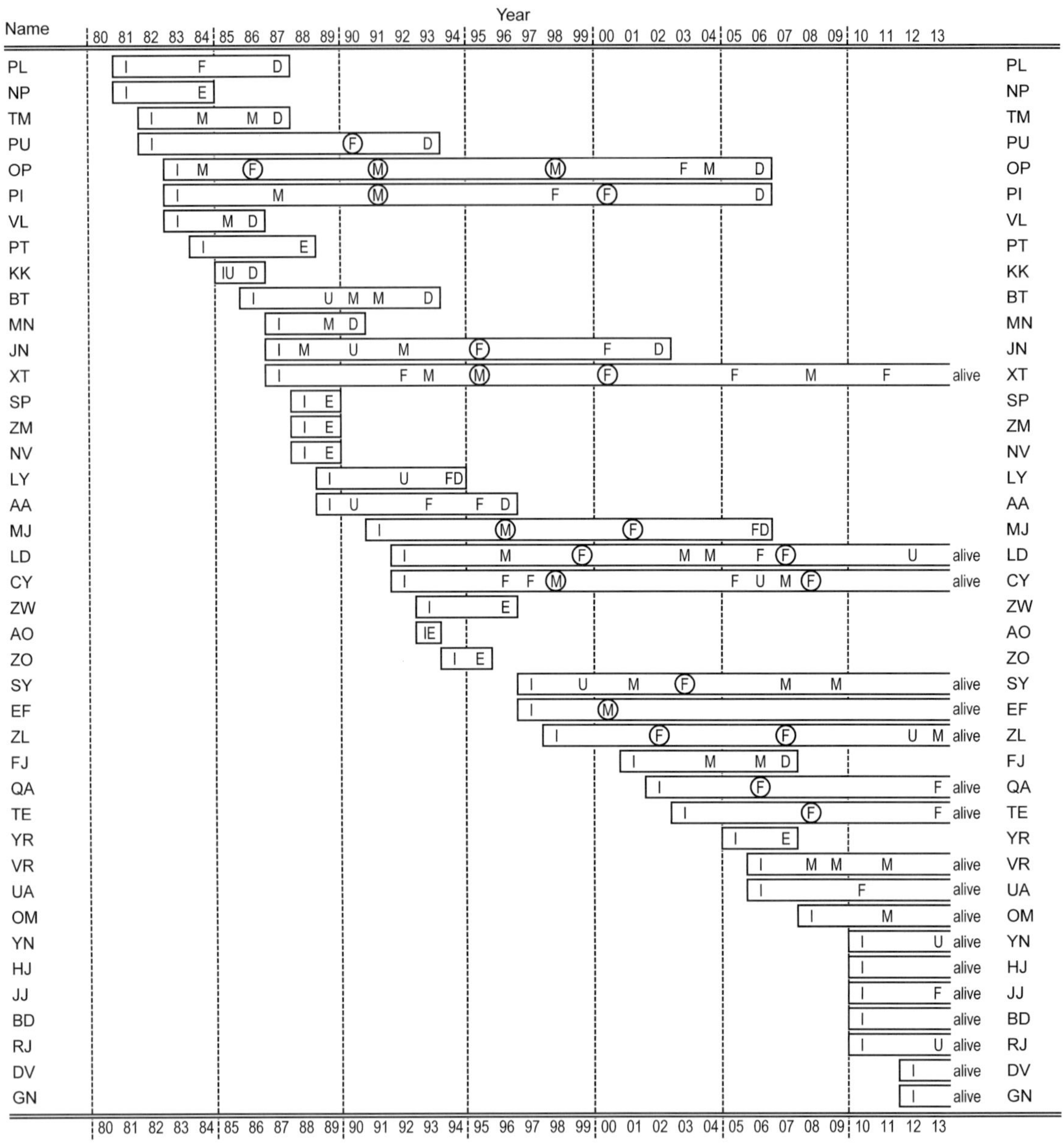

Figure 7.7 Reproductive life histories of females immigrating to the M group between 1980 and 2013. Each horizontal bar indicates the presence of a female in the group. Capital letters in the bars indicate that the following events occurred during the year. I, immigration; E, emigration; D, death; M, birth of a male; F, birth of a female: U, birth of an unsexed infant. Circled M and F represent successful reproduction. The "F" in LY's row represents a twin birth. See Appendix VI for abbreviations of individual names.

Xantip (XP) and Puffy (PF), gave birth at the ages of 12 and 14, respectively (Puffy gave birth in 2014, thus the case was not included in the analyses in this chapter).

Although the median age at the time of first delivery was approximately 1 year older in the immigrant females, there was no significant difference in age between the immigrants and natal

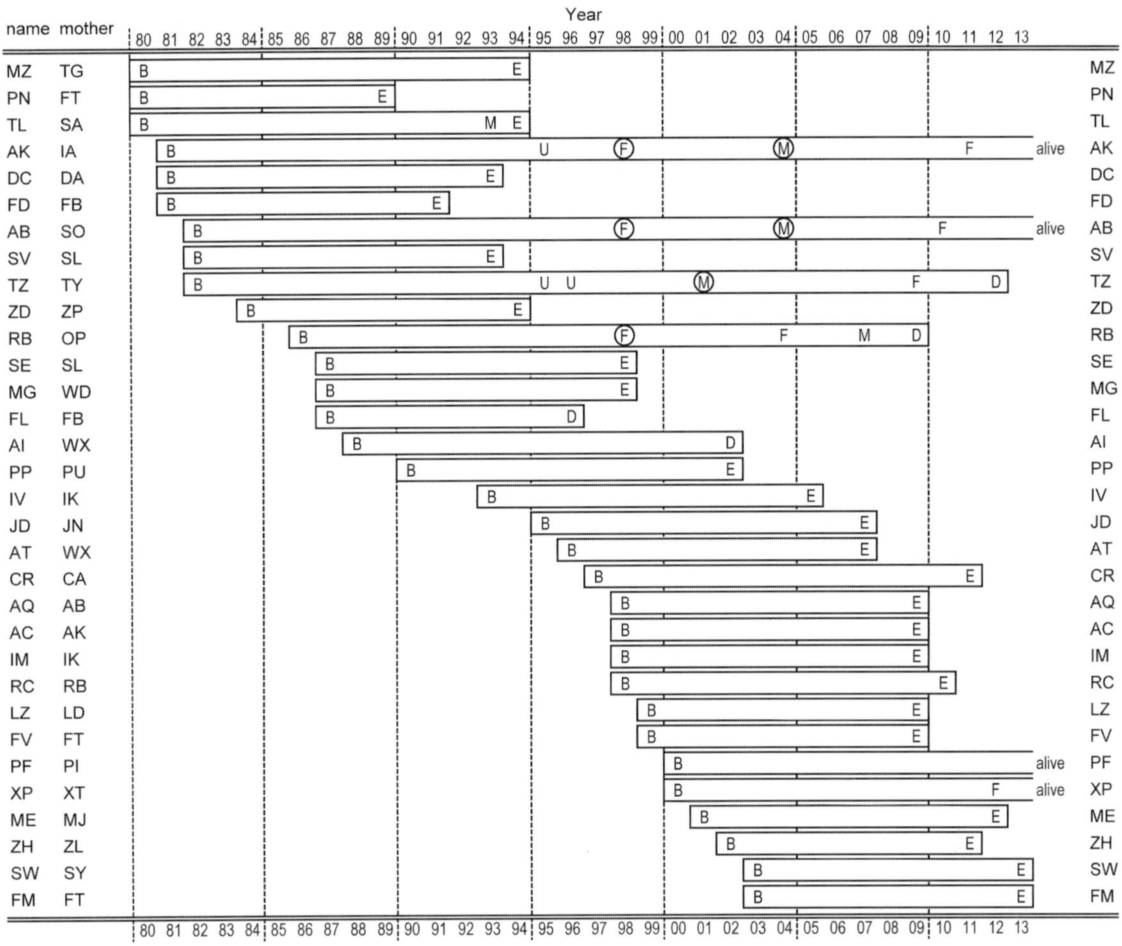

Figure 7.8 Reproductive life histories of natal females born into the M group between 1980 and 2003. Each horizontal bar indicates the presence of a female. Capital letters in the figure indicate that the following events occurred during the year. B, birth of a female; E, emigration; D, death; M, birth of a male; F, birth of a female: U, birth of an unsexed infant. Circled M and F represent successful reproduction. FL was assumed to have died during the largest reported decline of the group because she was not in estrus and was still ranging with her mother before she disappeared. AI was also assumed to have died because she seriously weakened before she disappeared (Shimada and Matumula, 2004). See Appendix VI for abbreviations of individual names.

females (Wilcoxon rank sum test, $W = 117$, $p = 0.15$). There has been only one twin birth, from the female Lynso (LY in Figure 7.7) (Matsumoto-Oda, 1995).

7.4.3 Year of the first successful reproduction

As we have previously discussed, many chimpanzee infants die before weaning, and this is typically more common when the mothers are young adults. Thus, giving birth does not necessarily mean that the female has reproduced successfully. Here, a *successful reproduction* is defined as the delivery of a child who survived at least until weaning (usually 4–5 years old). The median duration from immigration to successful reproduction was 5.91 years (mean = 5.61 ± 1.82 SD, $N = 13$) and the median estimated age was 16.23 years (mean = 16.84 ± 2.45 SD, $N = 13$). For natal females, the median age for successful reproduction was 17.07 years (mean = 16.51 ± 3.29 SD, N = 4). Although the latter sample size was small, there was no significant difference between immigrant and natal females (Wilcoxon rank sum test, W = 24, p = 0.87).

7.4.4 Interbirth interval after successful reproduction of both sexes

Both immigrant and natal females were included in this analysis. After the successful reproductions of male and female offspring, the median interbirth interval was 6.36 years (mean = 6.19 ± 1.03 SD, N = 9) and 5.14 years (mean = 5.18 ± 0.65 SD, N = 14), respectively. Thus, it takes longer for the female chimpanzees in Mahale to produce the next offspring when they are raising male offspring than when raising female offspring, and this difference was significant (Wilcoxon rank sum test, W = 100, p = 0.021). The tendency of a longer interbirth interval for male offspring is consistent with the results reported by Nishida *et al.* (2003).

7.5 Discussion

7.5.1 Demographic changes and chimpanzee society

As Mitani and colleagues (Mitani *et al.*, 2002; Mitani, 2006) emphasized, demography can affect several aspects of chimpanzees' behaviors and society. The Mahale M group, as shown in Figure 7.1, experienced very large changes in its size during the several decades studied. A part of these changes may have been related to the extinction of the adjacent K group and the associated mass immigration of females. Compared with the relatively stable group size in the 2000s–10s (approximately 60 members), the size in the 1980s was very large (over 100 members). On the other hand, the home-range size in the early 1980s (Hasegawa, 1987), when the group size was the largest, was comparable to the home-range sizes in later years (Nakamura *et al.*, 2013). This may imply that the M group had more individuals in the 1980s than the area could afford or it was beyond the carrying capacity. Provisioning was still ongoing at that time, even though it was reduced (see Chapter 3), and it might have provided additional food sources and/or attracted more chimpanzees than the group could contain. An alternate, unlikely theory would be that the M-group chimpanzees currently utilize more area than is actually required for the current number of chimpanzees. It is important to remember that the data on the M group's demography reported by Hiraiwa-Hasegawa *et al.* (1984) (data from 1979 to 1982) and Nishida *et al.* (1990) (data from 1980 to 1988) were somehow influenced by the unusual event of extinction of the neighboring group, which happened in the late 1970s to the early 1980s.

We still do not know the definitive cause of the largest recorded decline in the group size that started in 1994. However, we hypothesize that the group size in the 1980s was too large for the area; if the group size was larger than the carrying capacity, it was only a matter of time before the group size would decline. Therefore, the dataset reported by Nishida *et al.* (2003) from 1980 to 1999 covered the probable overcrowding period (1980s), the following probable reactive declining period (until 1996), and the following few years.

Several studies on various topics using quantitative methods flourished at Mahale in the 1980s (see Chapter 3). However, some characteristics of the Mahale chimpanzees described during that period might have been partially influenced by demographic characteristics (i.e. the likelihood of too many individuals or resource competition).

There are some observations that support the idea that the 1980s' M group was overcrowded. For example, most of the infanticides at Mahale occurred before 1990 and there was not a single confirmed case of infanticide after 2000 (see Chapter 26). Unlike other study sites where intergroup infanticides were more common, infanticides at Mahale were mostly intragroup. Although this is merely speculative, a group size that was larger than its carrying capacity could have been the cause of these repeated intragroup infanticides. An idea of a "male-centric society" was chiefly drawn from the studies in the

1980s at Mahale but no longer seems applicable. For example, when I first visited Mahale in 1994, there were still more than 70 chimpanzees in the M group, and there was no doubt that it was a "male-centric society." The adult males formed a large "core" party with adolescent males and estrous females following them, and several mothers could be identified as "peripheral" because they were seen only few times a year when the "core" party was followed. However, the situation had changed when I revisited Mahale in 1996 after the largest recorded decline in the group size had brought the group size down to less than 50. I sometimes saw almost all the parous females ranging in one large party, even without any adult males, and I could no longer assign the "peripheral" status to any females because almost everybody was ranging together in the high fruiting season (see Figure 8.5 in Chapter 8; Chapter 9; discussions in Nakamura *et al.*, 2013). Thus, the image of the "male-centric society" with frequent infanticides that was appropriate for the M group in the 1980s was no longer accurate. We need more detailed analyses with quantitative data to address this issue.

Along with the group size, the sex ratio can also probably cause changes in the chimpanzees' society and behaviors. The ratio of females in the Mahale M group declined from the 1980s to the present. Thus, male competition over mates may increase and female competition over food may decrease. Such parameters may be important in the discussions of reproductive strategies and some of the behavioral patterns of each sex. Further research is required on this topic as well.

7.5.2 High mortality of infant chimpanzees at Mahale

It is obvious that infant mortality is much higher in Mahale than other chimpanzee study sites. Compared with other sites, it is not especially high in chimpanzees less than 1-year-old, but is higher in chimpanzees between 1 and 5 years old. We still do not know why this occurred at Mahale, but it could again be explained by the assumption that the group size was too large in the 1980s. There were more females in the 1980s than in the later decades, so the number of births during that period was also larger. Repeated infanticides (Chapter 26) and a potentially high level of resource competition because of the large group size in the 1980s may explain the overall high rate of infant mortality. However, this may not be a complete explanation because the survivorship of even recent infants remains lower than at other sites. This may be related to the fact that infants are victims of respiratory diseases more frequently than other age classes at Mahale (Chapter 25). We still need a comparative research on various demographic parameters between sites.

7.5.3 Female life history

Nishida *et al.* (2003) provided a detailed discussion on the female reproductive life history and most of the figures presented in this chapter are comparable to theirs. Females typically transfer when they are approximately 11 years old and first give birth when approximately 13–14 years old. However, because they often fail to wean their first offspring, females give birth to their first surviving child when they are approximately 17 years old.

Although Nishida *et al.* (2003) emphasized the cost of transfer because immigrant females give birth later than natal females, I could not find a significant difference in the dataset used in this chapter. There are not many females at Mahale who stay and give birth in their natal group, so even a slight difference in the datasets may affect the statistical significance. Looking at individual cases, among six natal females who gave birth in the M group, two, Ruby and Xantip (RB and XP in Figure 7.8), did so when they were 12 years old, and their mothers were still alive. Ruby's first delivery and child rearing was successful, and she is the youngest on record to successfully reproduce in the M group. Xantip gave birth in 2012; thus, more time is needed to know whether she will successfully wean her infant. For Ruby, the support of her mother, Opal, might have been important for her success. Although Opal herself had an infant son, Ruby's infant daughter was often observed following her grandmother, Opal, instead of her mother, Ruby, from a pretty early age (Chapter 30). Another natal female,

Totzy, first gave birth at the age of 13 when her mother was still alive, but she lost the infant very young. Totzy's mother died the next year, and Totzy did not have her first successful reproduction until she was 19 years old. This is the second-oldest age for the first successful reproduction among this dataset, including immigrants.

There were four immigrant females who gave birth at an estimated age of 12, of which one, Zola, successfully weaned her first child (as Zola's first birth was in 2002, it was not included in the dataset by Nishida *et al.*, 2003). Thus, it is currently difficult to make the generalization that immigrant females give birth later than natal females. Although there must be some costs for transferring to different groups, we need some more information to conclude whether the transfer itself is costly for successful reproduction.

References

Boesch, C. and Boesch-Achermann, H. (2000). *The Chimpanzees of the Taï Forest: Behavioral Ecology and Evolution*. Oxford: Oxford University Press.

Bronikowski, A. M., Altmann, J., Brockman, D. K., *et al.* (2011). Aging in the natural world: comparative data reveal similar mortality patterns across primates. *Science*, **331**, 1325–8.

Emery Thompson, M., Jones, J. H., Pusey, A. E., *et al.* (2007). Aging and fertility patterns in wild chimpanzees provide insights into the evolution of menopause. *Current Biology*, **17**, 2150–6.

Goodall, J. (1983). Population dynamics during a 15 year period in one community of free-living chimpanzees in the Gombe National Park, Tanzania. *Zeitschrift für Tierpsychologie*, **61**, 1–60.

Goodall, J. (1986). *The Chimpanzees of Gombe: Patterns of Behavior*. Cambridge, MA: Belknap.

Hasegawa, T. (1987). [*Sexual Behavior of Wild Chimpanzees: A Comparative Psychological Study*.] Doctoral Dissertation. Tokyo: The University of Tokyo. In Japanese.

Hanamura, S., Kiyono, M., Lukasik-Braum, M., *et al.* (2008). Chimpanzee deaths at Mahale caused by a flu-like disease. *Primates*, **49**, 77–80.

Hawkes, K. and Smith, K. R. (2010). Do women stop early? Similarities in fertility decline in humans and chimpanzees. *Annals of the New York Academy of Sciences*, **1204**, 43–53.

Hawkes, K., Smith, K. R., and Robson, S. L. (2009). Mortality and fertility rates in humans and chimpanzees: how within-species variation complicates cross-species comparisons. *American Journal of Human Biology*, **21**, 578–86.

Hill, K., Boesch, C., Goodall, J., *et al.* (2001). Mortality rates among wild chimpanzees. *Journal of Human Evolution*, **40**, 437–50.

Hiraiwa-Hasegawa, M., Hasegawa, T., and Nishida, T. (1984). Demographic study of a large-sized unit-group of chimpanzees in the Mahale Mountains, Tanzania: a preliminary report. *Primates*, **25**, 401–13.

Inoue, E., Inoue-Murayama, M., Vigilant, L., Takenaka, O., and Nishida, T. (2008). Relatedness in wild chimpanzees: Influence of paternity, male philopatry, and demographic factors. *American Journal of Physical Anthropology*, **137**, 256–62.

Matsumoto-Oda, A. (1995). First record of a twin birth in chimpanzees of the Mahale Mountains National Park, Tanzania. *African Study Monographs*, **16**, 159–64.

Mitani, J. C. (2006). Demographic influences on the behavior of chimpanzees. *Primates*, **47**, 6–13.

Mitani, J. C., Watts, D. P., Pepper, J. W., and Merriwether, D. A. (2002). Demographic and social constraints on male chimpanzee behavior. *Animal Behaviour*, **64**, 727–37.

Nakamura, M., Corp, N., Fujimoto, M., *et al.* (2013). Ranging behavior of Mahale chimpanzees: a 16 year study. *Primates*, **54**, 171–82.

Nishida, T., Takasaki, H., and Takahata, Y. (1990). Demography and reproductive profiles. In *The Chimpanzees of the Mahale Mountains: Sexual and Life History Strategies*, ed. T. Nishida. Tokyo: University of Tokyo Press, pp. 63–97.

Nishida, T., Corp, N., Hamai, M., *et al.* (2003). Demography, female life history and reproductive profiles among the chimpanzees of Mahale. *American Journal of Primatology*, **59**, 99–121.

Reynolds, V. (2005). *The Chimpanzees of the Budongo Forest: Ecology, Behaviour, and Conservation*. Oxford: Oxford University Press.

Shimada, M. and Matumula M. (2004). Chimpanzee attitude toward a seriously weakened adolescent female at Mahale. *Pan Africa News*, **11**(1), 6–8.

Sugiyama, Y. (2004). Demographic parameters and life history of chimpanzees at Bossou, Guinea. *American Journal of Physical Anthropology*, **124**, 154–65.

Sugiyama, Y. and Fujita, S. (2011). The demography and reproductive parameters of Bossou chimpanzees. In *The Chimpanzees of Bossou and Nimba*, ed. T. Matsuzawa, T. Humle, and Y. Sugiyama. Tokyo: Springer, pp. 23–34.

Teleki, G., Hunt Jr., E. E., and Pfifferling, J. H. (1976). Demographic observations (1963–1973) on the chimpanzees of Gombe National Park, Tanzania. *Journal of Human Evolution*, **5**, 559–98.

Tsukahara, T. (1992). Lion eat chimpanzees: the first evidence of predation by lions on wild chimpanzees. *American Journal of Primatology*, **29**, 1–11.

Watts, D. P. (2012). Long-term research on chimpanzee behavioral ecology in Kibale National Park, Uganda. In *Long-Term Studies of Primates*, ed. P. M. Kappeler and D. P. Watts. Heidelberg: Springer, pp. 313–38.

Williams, J. M., Lonsdorf, E. V., Wilson, M. L., *et al.* (2008). Cause of death in the Kasekela chimpanzees of Gombe National Park, Tanzania. *American Journal of Primatology*, **70**, 766–77.

8 Home range

MICHIO NAKAMURA

8.1 Introduction

This chapter describes the home range of the Mahale M-group chimpanzees. The home range is the area where an animal lives, ranges, and utilizes resources (Burt, 1943). The size of a home range may depend on the availability of food and the relationship with the home ranges of adjacent conspecific individuals or groups. The term "territory" is sometimes used to describe the home range of chimpanzees (Boesch and Boesch-Achermann, 2000; Herbinger *et al.*, 2001; Mitani *et al.*, 2010) primarily because males may display defensive behaviors against neighboring unit-groups, such as boundary patrolling (Watts and Mitani, 2001) and lethal inter-unit-group aggression (Manson and Wrangham, 1991; Mitani *et al.*, 2010). However, here I employ the neutral term "home range," because such defensive behaviors have rarely been observed at Mahale, at least in recent times.

In group-living mammals, including many species of primates, the home range of a group member is usually equivalent to the home range of the group as a whole. Therefore, group members can potentially utilize the entire home range. In some chimpanzee populations, particularly, at Gombe, Tanzania (Wrangham and Smuts, 1980) and Kibale, Uganda (Chapman and Wrangham, 1993), females stay in small core areas, whereas males have a larger "male community range" (Wrangham, 1979) or a "male-defended boundary" (Williams *et al.*, 2002) covering the core areas of several females. However, in other places such as Taï, Côte d'Ivoire (Lehmann and Boesch, 2005) and Budongo, Uganda (Fawcett, 2000), the male and female ranging patterns are more similar. As discussed later, in some seasons, Mahale females also range widely within the home range, and there is an implication that females are largely responsible for the long-term changes in home-range size (Nakamura *et al.*, 2013). For this reason, here I assume that the home range is, at least potentially, utilized equally by males and females.

8.2 Overview of the home-range size in 1994–2009

Long-term data on the ranging of M group from 1994 to 2009 have been published (Nakamura *et al.*, 2013); thus, only the main results are summarized here (check the original article for some possible sampling biases). The area utilized by the M group during the 16 years from 1994 to 2009 was 25.2 km^2 based on a grid-cell method and 27.4 km^2 based on a minimum convex polygon (MCP) method (Figure 8.1). These values are well within the reported sizes of chimpanzee home ranges. The chimpanzees used an average of 70% of the total area per year. After 5 years, they had utilized only 80% of the total area, and after 11 years of research, 95% of the area had been used. The M-group chimpanzees spent more than half of the time in 15% of their total home range, and the size of this high-use area fluctuated little year by year. This suggests that the M-group chimpanzees consistently utilize a limited area within the home range but sometimes visit peripheral areas. The monthly mean home-range size was 7.6 km^2, but the

Mahale Chimpanzees: 50 Years of Research, ed. M. Nakamura, K. Hosaka, N. Itoh and K. Zamma. Published by Cambridge University Press.

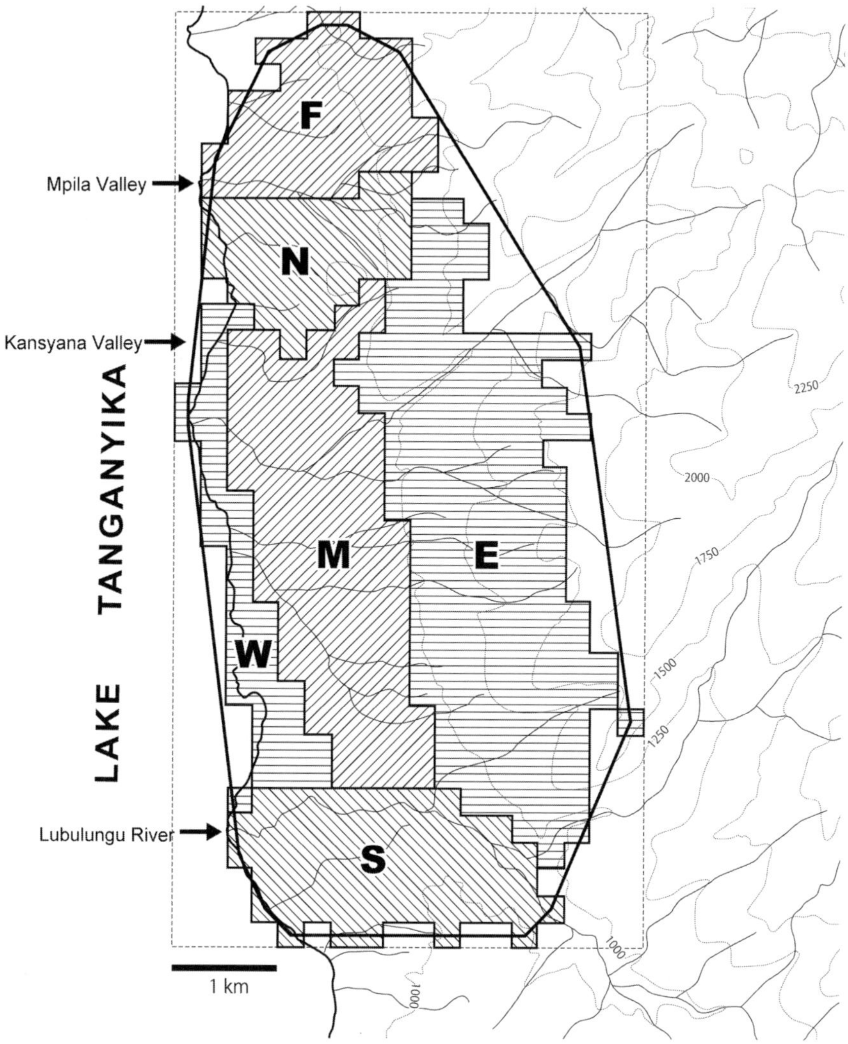

Figure 8.1 Areas used by M-group chimpanzees during 1994–2009, which are divided into six blocks. The solid polygon shows the MCP. The size of the outermost rectangle (dotted line) is 8.75 km × 4.5 km.

size differed considerably from month to month. The highest monthly mean size was comparable with a yearly home-range size. On average, the monthly home-range size was smallest in April (the end of the wet season) and largest in August (the dry season), but the pattern differed from year to year. In recent years, it appears that the M group made less use of the northern part of the home range but increased visits to the eastern part of the range. The yearly home-range size was not correlated with the number of adult males but was correlated with the number of adult females. This is in contrast to other East African populations in which males protect the home range and the number of males determines the size of the home range.

8.3 Characteristics of the areas in the home range

In this section, the characteristics of the different areas within the M group's home range are introduced. For the readers' convenience, I have divided the home range into six blocks (Figure 8.1). The blocks were basically divided according to geographic characteristics such as large rivers, valleys, and hills and by the ranging characteristics of the chimpanzees (see Chapter 2 for the names of the valleys and observation trails).

8.3.1 Far north block (F)

This block includes the north of the Mpila Valley and the F trail. Chimpanzees often visit this area during the *Saba comorensis* fruiting season, but do not stay for a long time in one visit. However, they did stay longer in the area one June when *Mimusops bagshawei* bore a large crop of fruit (Itoh *et al.*, 1999). During 1994–2009, *C4* was the northernmost point that the M group reached, but they very rarely visited that part of the block. The M group tended to return southward once it reached the Nkala Valley (this tendency has been observed since the 1970s: Kawanaka and Nishida, 1977). From the 1960s to the 1980s, this area overlapped with the range of K group, and currently, it overlaps with the range of Y group (Sakamaki and Nakamura, 2007; Chapter 11). The Y group has not yet been habituated, but most recent observations of the group have been around the Nkala Valley. Because the M group has not visited the area regularly, particularly in the last 10 years some observation trails are disappearing. Because the undergrowth is thick over most of the area, following chimpanzees in this block is difficult.

8.3.2 North block (N)

This block includes the south of the Mpila Valley and the north of the Kansyana Valley and F–J observation trails. There is a small hill in the center of the area, with small undulations in the topography. Although there are many observation trails, because of the topography and thick undergrowth, it is not easy to follow chimpanzees in this block. There is a patch of *Garcinia huillensis* on the northern slope of the hill. Because this species is an important food for Mahale chimpanzees, in most years, they visit and remain in this area from September to November. Chimpanzees also visit the western part of the area in May–June to feed on *Psydrax parviflora*.

8.3.3 Middle block (M)

This block includes the relatively flat area between the Kansyana Valley to the north and the Lubulungu River to the south. K–P observation trails run in this block. The area is covered with evergreen Kasoje forest that includes several seasonally fruiting species such as *Saba*, *Myrianthus arboreus*, *Ficus vallis-choudae*, and *Pseudospondias microcarpa*, and the M group utilizes this area throughout the year. In particular, *Saba*, a large liana species, is present throughout the region (Chapter 13). In the *Saba* fruiting season, approximately from August to January, the M group spends a lot of time in the area. There are many observation trails, and the off-trail undergrowth is not thick, therefore following chimpanzees in this block is relatively easy. As a result, observations may be biased to this area. For example, even when many chimpanzees are calling in the higher, less accessible area to the east (the east block), we may encounter and observe a smaller party of chimpanzees in this middle block.

8.3.4 West block (W)

This block includes the lakeshore and the area west of the R-5 trail. The topography is flat and its southern part is mostly covered by Miombo woodland. There used to be several Tongwe settlements and crop fields in the area, but these have been replaced with tourist camps. Typically, the undergrowth is not thick but some areas, such as former crop fields under succession, have become densely covered with thorny bushes. Chimpanzees sometimes visit there to feed on plants such as *Pterocarpus tinctorius* and *Parinari curatellifolia*. Until the 1990s, *Harungana madagascariensis* in this area was a very important food resource for the M group during June, but as a result of the succession of vegetation, this species has almost disappeared

(Chapter 13). Accordingly, the relative importance of the lakeshore area has decreased, at least in the *Harungana* season.

8.3.5 East block (E)

This block includes the steep eastern slopes that climb up to the main ridge of the Mahale Mountains, east of the R-1 trail. This area may overlap with the range of another unit-group to the east (Itoh *et al.*, 1999). Although there are several observation trails (the highest, R-7, was established in 1994), all are very steep, and it is often impossible to follow chimpanzees off trail. We have been able to confirm that chimpanzees reach 1200–1400 m a.s.l., but it is possible that they go even higher. We sometimes hear calls from above the highest point we have been able to reach. From recent observations, chimpanzees most often use this area in June–July to feed on *Psydrax parviflora, Synsepalum brevipes*, and *Toddalia asiatica*. There are also several large trees of *Pseudospondias* and *Ficus* spp. along the deep valleys in the area. When the fruits of these trees are plentiful, chimpanzees may stay in the higher areas and may not descend to middle block for more than a week.

8.3.6 South block (S)

This block includes the area south of the Lubulungu River, the largest river in the Kasoje area. The river always contains plenty of water even during the dry season. After heavy rain in the wet season, neither chimpanzees nor humans can cross the river. The lowland area is mostly flat, and the trees are relatively sparse. Therefore, it is not difficult to follow chimpanzees, although there are few observation trails. However, the observations are not always enjoyable because there are many vines of *Mucuna* spp. in the area. The beans of these vines have tiny needle-like hairs that cause itchy blisters on contact with the skin. The area overlaps with the range of N group to the south (Shimada, 2003). The M-group chimpanzees frequently visit the area in June to feed on *Azanza garckeana* and *Sterculia quinqueloba*. As mentioned in Section 8.4.4, there is a *Garcinia* patch further to the south of the area but, according to the 1994–2009 data, the use of the patch has not been confirmed.

8.4 Longer-term changes of the home range

Several reports described or estimated the home range of the M group before Nakamura *et al.* (2013). However, most did not calculate the estimated home-range size but only presented maps of the home

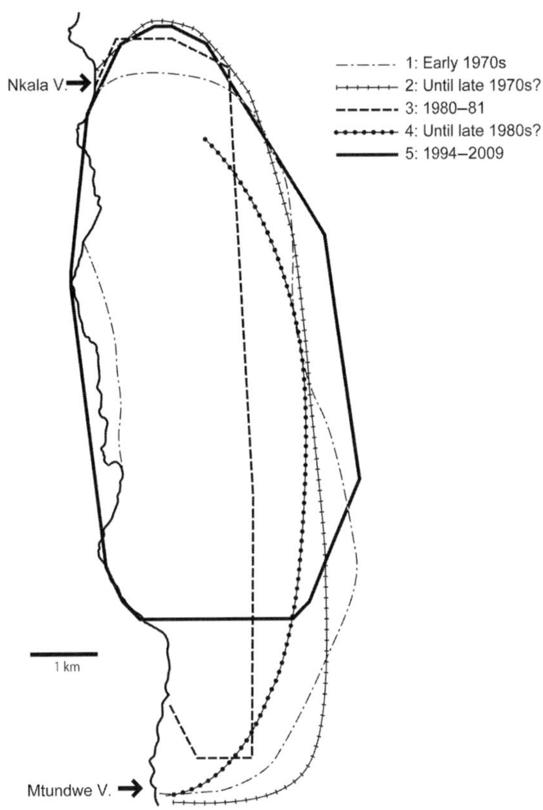

Figure 8.2 Comparison of the home range of the M group reported in the literature from different periods. Outlines of the home range redrawn from the following papers: (1) Kawanaka and Nishida (1974); (2) Anonymous (1980); (3) Hasegawa (1987); (4) Nishida *et al.* (1990); and (5) Nakamura *et al.* (2013).

range. Figure 8.2 shows outlines of the home range according to previous studies (Kawanaka and Nishida, 1974; Anonymous, 1980; Hasegawa, 1987; Nishida *et al.*, 1990: outlines 1–4, respectively) and the MCP method for 1994–2009 (Nakamura *et al.*, 2013: outline 5). Several other authors have published maps of the M-group home range in Japanese books or book chapters (e.g. Nishida 1981; Takahata, 1986; Hayaki, 1990; Norikoshi, 2002). However, I have not included these maps here because most resemble those in Figure 8.2 and are probably drawn from previous studies. Except for the home range described by Hasegawa (1987), who used an MCP method, the papers did not explain the methods used to estimate the size of the home range. The maps were probably drawn to include areas known to be used by chimpanzees for certain periods (Nishida, 1974). Therefore, it is not very useful to calculate the areas within each outline in Figure 8.2, but the maps do show some general trends of changes in the home range.

8.4.1 Data periods

First, I summarize the periods of data used to produce the home ranges shown in Figure 8.2 and describe what kind of data was used for the estimation. In the figure from Kawanaka and Nishida (1974) (outline 1 in Figure 8.2), there is a caption stating "after Nishida (1974)" although the two figures are not identical. In the original description by Nishida (1974), there is no statement about the data period used to generate the home-range outline. However, there is another map in Nishida (1974) showing the travel routes of chimpanzees during October–November 1966, which covers a smaller area than the area included in outline 1 in Figure 8.2. This suggests that outline 1 was probably drawn based on data from the late 1960s to the early 1970s, integrating the available ranging data. Nishida (1974) not only estimated the area to be 13.4 km^2 but also mentioned that the southern boundary was still uncertain. In another paper that presented an almost identical figure, Kawanaka and Nishida (1977) clearly stated that the southern boundary was Mtundwe Valley and that the M group never crossed Nkala Valley to the north. They estimated the area to be 17 km^2 on the map and 34 km^2 taking into account the undulations of the rugged topography. Outline 1 has a vacant space on the lakeshore from Kasiha to Kangwena, although the authors do not mention why the space exists. It is likely that because human settlements were still present along the lakeshore (Chapter 2) and because chimpanzees were not fully habituated to human presence, the M group did not use the lakeshore to avoid contact with humans. There are no descriptions about the data periods or methods used to produce outline 2 (Anonymous, 1980) or outline 4 (Nishida *et al.*, 1990), which was also adopted in Nishida's later book (2012). I assume outlines 2 and 4 were drawn based on accumulated ranging information obtained up until the papers were written, i.e. until the late 1970s and 1980s, respectively. Outline 3 by Hasegawa (1987), also presented in his English language paper (Hasegawa, 1990), was drawn from his observation data for the periods July 1980 to June 1981 and August 1981 to May 1982. He used 400 m × 400 m grids and estimated the range size to be 19.4 km^2. Although Hasegawa's method is different from ours (Nakamura *et al.*, 2013), the home-range size in 1981–2 is well within the range of annual home-range sizes in the later years (13.8–22.2 km^2).

8.4.2 Change in the northern boundary

Outlines 2, 3, and 5 in Figure 8.2 show almost identical northern boundaries but the northern boundary of outline 1 lies a bit to the south, at the Nkala Valley. Until the early 1970s, before the K group began to decrease in size (see Chapter 10), the northern boundary might have been around the valley. However, Kawanaka and Nishida (1974) observed that the M group did go beyond the Nkala Valley to the north, invading the home range of the K group. In the late 1970s, when the K group began to decrease in size, the area to the north of the valley might have gradually been included in the M group's home range. It remains unclear why Nishida *et al.* (1990) (outline 4) show the boundary far to the south and why there is an incomplete line to the lakeshore. No explanation was given for this statement.

8.4.3 Change in the eastern boundary

All earlier eastern boundaries were drawn to the west of the eastern boundary of the most recent home range, outline 5 in Figure 8.2. This is most evident in outline 3 and may be explained by the steep mountain slope that rises to the east, which might have limited observations in the earlier years. It was not until 1994 that an additional observation trail (the highest observation trail at Kasoje) opened up access to this eastern part of the home range (Nakamura *et al.*, 2013). Therefore, few observations were made in that area until the mid-1990s, and the estimation of the eastern boundary was made parsimoniously. Alternatively, because the M group increased the frequency of visits to the eastern mountainous areas during 1994–2009, it is possible that the M group is gradually extending its range eastward, even in the longer term.

8.4.4 Change in the southern boundary

The most apparent difference between earlier home ranges and the recent home range is the southern boundary. In outlines 1–4, the boundary is drawn far to the south of outline 5. Most former maps set the Mtundwe Valley as the southern boundary of the M group. This information may be based on a survey conducted in 1975. For example, Nishida (1981) and Norikoshi (2002) mentioned the 1975 survey when they addressed the issue of the southern border of the M group's home range.

In this survey conducted on 8 days (October 16–23, 1975), five teams of researchers aimed to observe chimpanzees in different places in the Kasoje area to determine the distribution of different unit-groups in the region (Nishida and Norikoshi, 1975). As mentioned in the title of the report, the survey was conducted in the *Garcinia* fruiting season. According to the original description by Nishida and Norikoshi (1975), the M group utilized the upper streams of the Kakolomo and Ilanga valleys just north of the Mtundwe. Norikoshi (2002: 102) later wrote that "it was revealed by the survey that the M group does utilize Ilanga but does not go beyond Mtundwe to the south" (my translation). It seems most papers published after 1975 defined Mtundwe as the southern border because of the results of this survey.

Hasegawa (1987) drew the southern border based on his own data. In 1980, he noted that it was only in October–November, again *Garcinia* season, that chimpanzees utilized the southernmost area. In his figure, he indicates that chimpanzees utilized a few blocks, including the upper stream of the Mtundwe Valley (corresponding to the right bottom corner of outline 3 in Figure 8.2). These blocks were detached from the major blocks in the home range.

The southern borders in outlines 2 and 4 appear to be based on observations in the *Garcinia* season in 1975, whereas the southern border in outline 3 was based on observations in the *Garcinia* season in 1980. Therefore, when chimpanzees visited the *Garcinia* patch in the fruiting season (at least in 1975 and 1980), the southern boundary extended considerably. Why did the 1994–2009 data not include visits to that patch? One possibility is that the M group reduced the southernmost boundary of the home range. This needs further verification, but it is plausible that the M group did not have to use the southern patch because of frequent visits to the eastern area and because they could also stay in the northern area, formerly shared with the K group, after the extinction of that group. Another possibility is that visits to the *Garcinia* patch in Mtundwe occur very infrequently. There is another large patch of *Garcinia* in the northern part of the home range (see Section 8.3); therefore, recently the M group has stayed in the northern patch in the *Garcinia* fruiting season. It is possible that 1975 and 1980 were lean years for the northern patch, and the group had to travel to the southern patch. It should also be noted that the northern patch is much closer to the researchers' camp (a 30-min walk), but to reach the southern patch, we have to walk for more than 2 h and then climb to a mountain distant from the observation trail. Therefore, researchers could have missed some visits by the M group to the southern patch, and the complete lack of observations in the southernmost areas could be an observational bias. Another problem is that the Mtundwe Valley is not included in

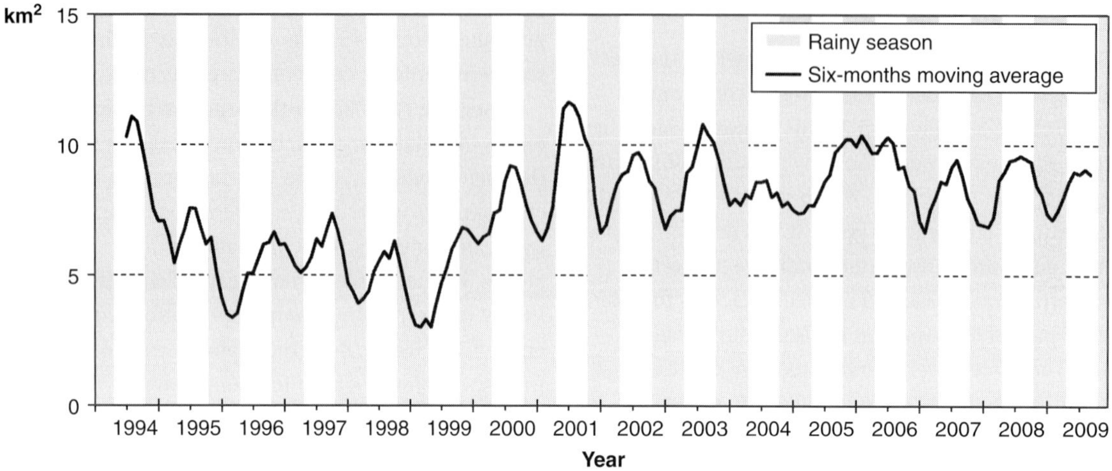

Figure 8.3 Six-month moving average of the size of the home range of the M-group chimpanzees from April 1994 to December 2009.

the blank maps recently used for recording the ranging routes of chimpanzees.

8.5 Seasonal change in ranging

8.5.1 Dispersing season and gathering season

There is an apparent seasonal difference in the gregariousness of chimpanzees at Mahale that strongly affects the ranging pattern. In the *dispersing* season, roughly from February to July, the chimpanzees split into several small parties, and each party ranges in a small area. In the *gathering* season, roughly from August to January, the chimpanzees form one large nomadic party (*sensu* Itoh and Nishida, 2007), sometimes including all group members, ranging in large areas throughout the home range, usually searching for *Saba* fruits. The dispersing season and gathering season are not dichotomously divided. Typically, the nomadic party size is smallest in March–April, increases gradually to a maximum in September–November, and then decreases again. An obvious increase and decrease is observed every year, but the timing and the extent may change from year to year. Probably according to such changes in the party size, the moving averages of the monthly home range used by the M group (Figure 8.3) also show a periodicity in which the area is smallest around the end of the wet season (smallest in April on average) and largest around the mid to late dry season (largest in August on average).

Figure 8.4 shows the overlap in the monthly home range in different years. The extent of the overlap has been shown as overlap indices in Nakamura *et al.* (2013). In the gathering season from August to January, the ranging pattern is relatively stable over the years. In each year, the range is oval shaped with the major axis from north to south, and, with some exceptions in August, M-group chimpanzees seldom go to the eastern part of their home range. The barycenters are located close to each other during the gathering season, particularly in the west–east axis.

Conversely, in the dispersing season, there is less overlap between years. In particular, in June, the locations of the barycenters differ considerably from year to year. This is because different species bear fruits in June of different years. This was discussed in detail in Nakamura *et al.* (2013).

8.5.2 Sex differences in ranging

The difference in ranging areas between male and female chimpanzees at Mahale (Hasegawa, 1990) is not as large as at Gombe and Kibale (Wrangham and

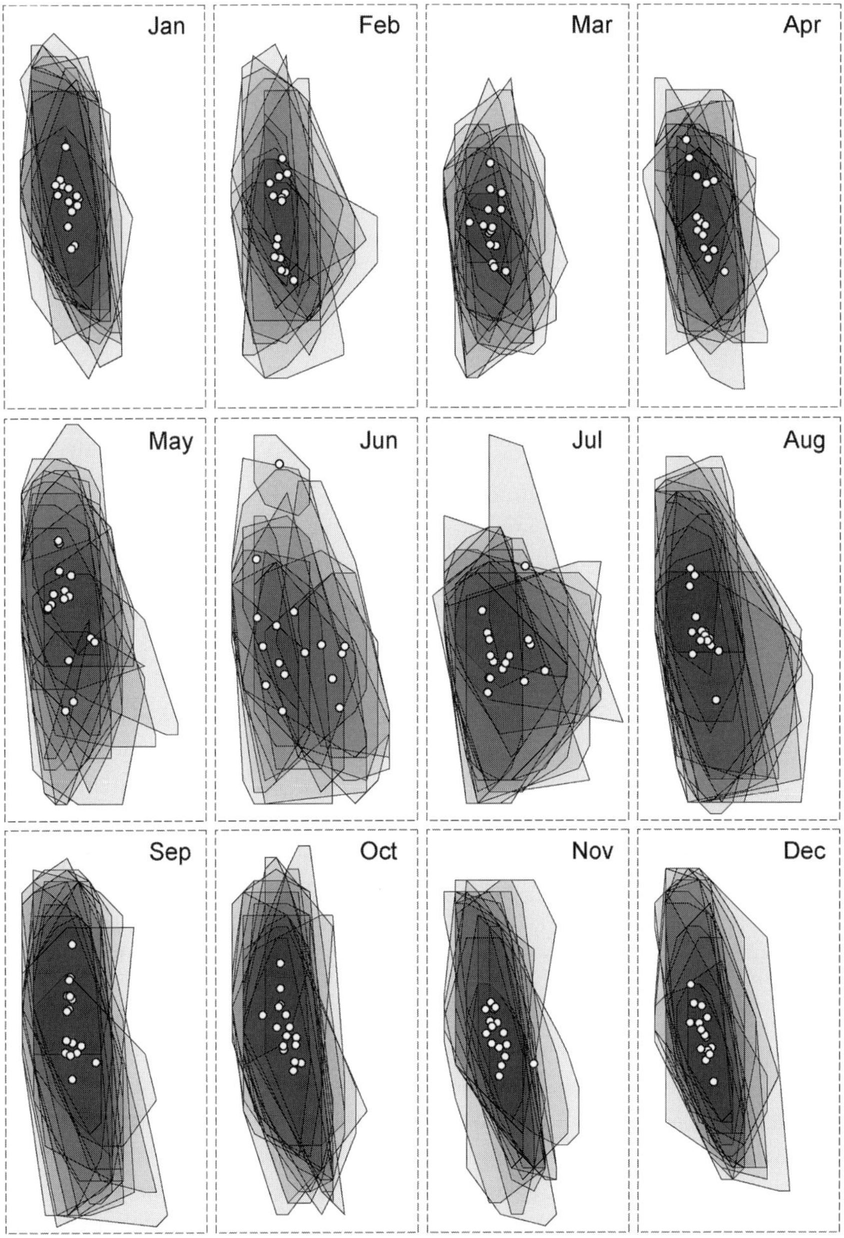

Figure 8.4 Overlap of the monthly home range of the M-group chimpanzees from 1994 to 2009. The polygons show the monthly home range, and the white circles show the barycenter in a year. The darker the gray color, the more overlap there is between years. Dotted rectangles correspond to those in Figure 8.1.

Smuts, 1980; Chapman and Wrangham, 1993). Particularly in the gathering season, or in the *Saba* season, most of the M-group chimpanzees, males and females, form a large nomadic party and range widely within the home range. Consequently, at this time, all members of the M group have an almost identical ranging area. Later, I present an example from the ranging data for September 2004.

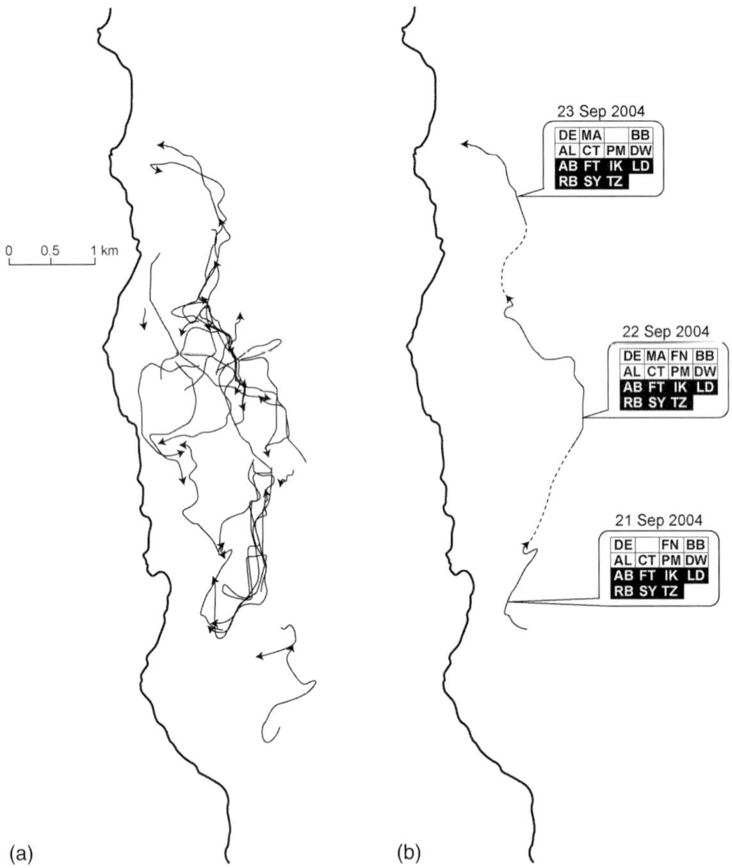

Figure 8.5 (a) Daily travel routes of the M-group chimpanzees in September 2004. One line corresponds to one day, and the arrow indicates the direction of travel. (b) The observed (solid line) and supplemented (dashed line) routes of September 21–23, 2004. The abbreviated names of adult males (black letters) and mothers (white letters) are shown in the balloons.

Ranging in a typical *Saba* season

Figure 8.5a shows all the daily travel routes of chimpanzees in September 2004. The overall ranging pattern is typical of the season, with the chimpanzees moving back and forth from north to south. Three days (September 21–23) are shown in Figure 8.5b. It is not always possible to follow the chimpanzees from bed to bed, so the gaps between the observed routes are marked with dashed lines. A balloon shows the names of adult males (16 years or older) and mothers (those who have dependent offspring and are not showing estrous cycle) observed for each day. In summary, all the adult males and mothers, except for Masudi (MA) on the 21st and Fanana (FN) on the 23rd (both males), were observed on these 3 days. Similarly, on the 20th, only Fanana was not observed, and on the 24th, all chimpanzees were observed. The names in a balloon represent all the chimpanzees observed in a day, but this does not necessarily mean they were together in a face-to-face party (*sensu* Itoh and Nishida, 2007) for the entire day. However, because all the mothers were observed consistently on these 3 days in very different locations in the home range, it is natural to assume that, in a broad sense, they were ranging together.

It is important to note that these 3 days were not special but typical of the *Saba* season (see more examples in Chapter 9). For example, for all the 28 days in this month with available data, the sizes of nomadic parties exceeded more than half of the unit-group size (hereafter, I will refer to such nomadic parties as "large parties"). The median attendance rate of adult males in the large parties (observed days for a particular individual divided by 28) was 1.00 (range: 0.21–1.00), and for mothers, the rate was 0.93 (range: 0.86–1.00). This means almost all of the adult males were observed on all 28 days of the month. An exception was Fanana whose attendance rate was very low at 0.21. He had fallen from the alpha status in 2003 (Nishie, 2004a, b; Chapters 5, 27) and had often been observed walking alone. It is important to emphasize the high attendance rate for mothers. Although females with dependent offspring are thought to be "asocial" and tend to range alone only with their offspring (see Chapter 28), these data do not support such a tendency. Every mother was present in large parties for more than 80% of the 28 days, and some females were present every day, similar to adult males. Even the female with lowest attendance rate, Totzy (TZ), was only absent twice for 2 days each time. Such high attendance of females was observed almost every year although the extent did vary from year to year. The average attendance rate of females in large parties relative to adult males in September during 1994–2009 was 0.81 (±0.14, $N = 16$) (MMCRP, unpublished data).

Therefore, in *Saba* season, similar to adult males, the majority of mothers join large parties. Because such large parties usually range long distances from north to south during this season (Figure 8.5), the mothers do not stay in small core areas but range widely within the home range. As a result, there is little difference in the ranging areas for males and females in the *Saba* season. Most of the M-group chimpanzees range together in a large mixed-sex party and only some individuals sometimes break away. This may be similar to the ranging pattern at Taï described by Boesch and Boesch-Achermann (2000).

Ranging in dispersal season

It is not easy to determine the ranging pattern in the dispersal season because the party size is usually smaller and the smaller parties tend to emit less frequent calls. Consequently, the discovery rate is biased to parties with relatively more males. However, it appears likely that the sex differences in ranging, as often emphasized in some other East African populations, are more obvious during this season at Mahale.

Even in the dispersing season, several males, often including the alpha, tended to gather together in a party and were often joined by adolescent males and nulliparous young females. However, some males were observed walking alone or in groups of two or three. Some females were not seen for long periods and were found walking alone with their dependent offspring. It is not easy to observe such females on consecutive days because, once they are lost, it is almost impossible to find them again. Although Mahale females do not stay in small core areas during the gathering season (*Saba* season), individual females may range in relatively small areas in the dispersing season because each one tends to be located in a similar area when observed. These small areas may correspond to the core areas of other East African populations.

8.6 Characteristics of the home range of Mahale chimpanzees compared with other chimpanzee study sites

In some chimpanzee study sites in East Africa, particularly in Gombe and Kibale, females typically range in small core areas, whereas males range in larger areas, including the core areas of the females (Wrangham and Smutts, 1980). Therefore, the groups' ranging area is often referred to as a "male community range" (Wrangham, 1979) or a "male-defended boundary" (Williams *et al.*, 2002). This reminds me of a strong impression that I had when I visited Kibale in 2000 to observe the Kanyawara chimpanzees. In the list of Kanyawara chimpanzees, females were divided into three categories: northern females, central females, and southern females. To my surprise, when I went to the northern part of the area I observed northern females, and when I went south I observed southern females. This is not likely to happen at

Mahale, particularly in the gathering season when all the females can be observed throughout the wide area from the north to the south of the group's home range.

West African chimpanzees show a different ranging pattern compared with the chimpanzees of Gombe and Kibale. At Taï (Boesch and Boesch-Achermann, 2000) and Bossou (Sugiyama and Koman, 1979), most individuals range widely within the home range. When I visited Bossou in 2003, I could observe almost all individuals whenever I encountered chimpanzees in different places in their home range. The home-range use and ranging pattern of the Mahale chimpanzees may be unique in that it is more like West African chimpanzees in the gathering season and more like other East African chimpanzees in the dispersing season.

References

Anonymous (1980). *Mahale: Study for the Proposed Mahale Mountains National Park. Final Report, May 1980*. Tokyo: Japan International Cooperation Agency.

Boesch, C. and Boesch-Achermann, H. (2000). *The Chimpanzees of the Taï Forest: Behavioral Ecology and Evolution*. Oxford: Oxford University Press.

Burt, W. H. (1943). Territoriality and home range concepts as applied to mammals. *Journal of Mammalogy*, **24**, 346–52.

Chapman, C. A. and Wrangham, R. W. (1993). Range use of the forest chimpanzees of Kibale: Implication for the understanding of chimpanzee social organization. *American Journal of Primatology*, **31**, 263–73.

Fawcett,K. A. (2000). *Female Relationships and Food Availability in a Forest Community of Chimpanzees*. Ph.D.thesis. Edinburgh: University of Edinburgh.

Hasegawa, T. (1987). [*Sexual Behavior of Wild Chimpanzees: A Comparative Psychological Study*.] Doctoral Dissertation. Tokyo: The University of Tokyo. In Japanese.

Hasegawa, T. (1990). Sex differences in ranging patterns. In *The Chimpanzees of the Mahale Mountains: Sexual and Life History Strategies*, ed. T. Nishida. Tokyo: University of Tokyo Press, pp. 99–114.

Hayaki, H. (1990). [*The Human within Chimpanzee*.] Tokyo: Shōka-bō. In Japanese.

Herbinger, I., Boesch, C., and Rothe, H. (2001). Territory characteristics among three neighboring chimpanzee communities in the Taï National Park, Côte d'Ivoire. *International Journal of Primatology*, **22**, 143–67.

Itoh, N. and Nishida, T. (2007). Chimpanzee grouping patterns and food availability in Mahale Mountains National Park, Tanzania. *Primates*, **48**, 87–96.

Itoh, N., Sakamaki, T., Hamisi, M., *et al.* (1999). A new record of invasion by an unknown unit-group into the center of M group territory. *Pan Africa News*, **6**, 8–10.

Kawanaka, K. and Nishida, T. (1974). Recent advances in the study of inter-unit-group relationships and social structure of wild chimpanzees of the Mahali Mountains. In *Proceedings from the Symposia of the Fifth Congress of the International Primatological Society, Nagoya, Japan, August 1974*, ed. S. Kondo, M. Kawai, A. Ehara, and S. Kawamura. Tokyo: Japan Science Press, pp. 173–86.

Kawanaka, K. and Nishida, T. (1977). [Chimpanzees of the Mahale Mountains (II): Intergroup relationships between unit groups.] In [*The Chimpanzees*], ed. J. Itani. Tokyo: Kōdan-sha, pp. 369–694. In Japanese.

Lehmann, J. and Boesch, C. (2005). Bisexually bonded ranging in chimpanzees (*Pan troglodytes verus*). *Behavioral Ecology and Sociobiology*, **57**, 525–35.

Manson, J. H. and Wrangham, R. W. (1991). Intergroup aggression in chimpanzees and humans. *Current Anthropology*, **32**, 369–90.

Mitani, J. C., Watts, D. P., and Amsler, S. J. (2010). Lethal intergroup aggression leads to territorial expansion in wild chimpanzees. *Current Biology*, **20**, R507–8.

Nakamura, M., Corp, N., Fujimoto, M., *et al.* (2013). Ranging behavior of Mahale chimpanzees: a 16 year study. *Primates*, **54**, 171–82.

Nishida, T. (1974). [Ecology of wild chimpanzees.] In [*Human Ecology*], ed. R. Ohtsuka, J. Tanaka, and T. Nishida. Tokyo: Kyōritsu-shuppan, pp. 15–60. In Japanese.

Nishida, T. (1981). [*The World of Wild Chimpanzees.*] Tokyo: Chūkō-shinsho. In Japanese.

Nishida, T. (2012). *Chimpanzees of the Lakeshore: Natural History and Culture at Mahale*. Cambridge: Cambridge University Press.

Nishida, T. and Norikoshi, K. (1975). Distribution of the unit-groups of chimpanzees in the Kasoje area in the *Garcinia huillensis* season of 1975 (preliminary draft). *Mahale Mountains Chimpanzee Research Project Survey Report*, **3**.

Nishida, T., Takasaki, H., and Takahata, Y. (1990). Demography and reproductive profiles. In *The Chimpanzees of the Mahale Mountains: Sexual and Life History Strategies*, ed. T. Nishida. Tokyo: University of Tokyo Press, pp. 63–97.

Nishie, H. (2004a). [Disappearance of the alpha male Fanana and its consequences.] *Mahale Chimpun*, **3**, 5–6. In Japanese.

Nishie, H. (2004b). [The fight at Kasiha Valley: the fall of Fanana from the alpha status and Alofu's takeover.] *Mahale Chimpun*, **4**, 6–7. In Japanese.

Norikoshi, K. (2002). [Vegetation in the Kasoje area and the ranging pattern: changes during 20 years around the establishment of the national park.] In [*The Mahale Chimpanzees: 37 Years of <Panthropology>*], ed. T. Nishida, S. Uehara, and K. Kawanaka. Kyoto: Kyoto University Press, pp. 101–28. In Japanese.

Sakamaki, T. and Nakamura, M. (2007). Preliminary survey of unhabituated chimpanzees in the Mahale Mountains National Park, Tanzania: Behavioral diversity across neighboring unit-groups and intergroup relationships. In *Formation of a Strategic Base for Biodiversity Studies*. Kyoto: The 21st Century COE Program of Kyoto University, pp.278–80.

Shimada, M. (2003). A note on the southern neighboring groups of M group in the Mahale Mountains National Park. *Pan Africa News*, **10**,11–14.

Sugiyama, Y. and Koman, J. (1979). Social structure and dynamics of wild chimpanzees at Bossou, Guinea. *Primates*, **20**, 323–39.

Takahata, Y. (1986). [Chimpanzees of the Mahale Mountains National Park, Tanzania: their ecology and society.] In [*Anthropology of Natural Societies: Living in Africa*], ed. J. Itani and J. Tanaka. Tokyo: Academia Shuppan-kai, pp. 7–41. In Japanese.

Watts, D. P. and Mitani, J. C. (2001). Boundary patrols and intergroup encounters in wild chimpanzees. *Behaviour*, **138**, 299–327.

Williams, J. M., Pusey, A. E., Carlis, J. V., Farm, B. P., and Goodall, J. (2002). Female competition and male territorial behaviour influence female chimpanzees' ranging patterns. *Animal Behaviour*, **63**, 347–60.

Wrangham, R. W. (1979). Sex differences in chimpanzee dispersion. In *The Great Apes*, ed. D. A. Hamburg and E. R. McCown. Menlo Park, CA: Benjamin/Cummings, pp. 481–9.

Wrangham, R. W. and Smuts, B. B. (1980). Sex differences in the behavioural ecology of chimpanzees in the Gombe National Park, Tanzania. *Journal of Reproduction and Fertility, Supplement*, **28**, 13–31.

9 Fission–fusion grouping

SHUNKICHI HANAMURA

9.1 What is "fission–fusion"?

Wild chimpanzees form multi-male, multi-female unit-groups among which females transfer upon reaching sexual maturity (Chapter 6). Chimpanzee unit-groups are similar to social groups of many other primates in that group membership does not change except for births, deaths, or transfers of individuals of one or both sexes. However, there is a remarkable difference in the degree of spatial and temporal cohesiveness. In many group-living primates, all or most of the group members basically range in one cohesive troop by keeping relatively close, often visible, interindividual distances. In contrast, the members of a chimpanzee unit-group, except for pairs of mothers and their dependent offspring, repeatedly meet and part at various time intervals; consequently, they split into "parties" of various sizes, compositions, and durations while ranging.

In the case of the M-group chimpanzees at Mahale, about 50–80 group members (Chapter 7) repeatedly meet and part in their home range (Chapter 8). Any two members may keep several kilometers apart from each other, and even if they do meet or hear each other, they do not always subsequently range together. Two particular individuals may not meet with each other for days, weeks, or even months. On the other hand, over the course of one day, the same pair may repeatedly meet at different places and spend some time together after short separations. Because all the members of this unit-group repeatedly meet and part over various time intervals, a particular chimpanzee may meet with no, or few, other members in a day; conversely, he/she may meet almost all members of the group in a day. This fluid grouping pattern, which occurs on various spatial and temporal scales, has been represented by the term "fission–fusion" in chimpanzee studies.

The term was first used in primatology by Kummer (1971: 45, 50) for expressing grouping patterns of hamadryas (*Papio hamadryas*) and gelada (*Theropithecus gelada*) baboons living in multilevel societies: these species commonly form bands consisting of a number of one-male, multi-female units (OMUs) that can range separately, but share a home range, although the latter species form looser bands with a higher level of social units (Snyder-Mackler *et al.*, 2012).

However, the fission–fusion grouping patterns observed in these baboons differ from that of chimpanzees because the membership of each OMU is basically steady, unlike that of chimpanzee parties (for a review, see Grueter *et al.*, 2012). Moreover, in the case of hamadryas baboons, all the OMUs in a band travel in the same direction – with distances among the furthest OMUs not exceeding 1 km – and gather at a fixed sleeping cliff every night (Schreier and Swedell, 2012). There are several cases of separations along clan lines, but each clan also consists of stable members (OMUs) (Schreier and Swedell, 2012). Compared to such baboon fission–fusion, the characteristics of chimpanzee fission–fusion are: that it occurs at the level of individuals, not stable subgroups such as OMUs and clans; that it can occur in much more varied spatial and time scales; and that it continues without periodical assemblage of

Mahale Chimpanzees: 50 Years of Research, ed. M. Nakamura, K. Hosaka, N. Itoh and K. Zamma. Published by Cambridge University Press.

all the unit-group members in one place. Because of this endless, individual-based fission–fusion, many pioneers have proposed various terms and hypotheses to describe the social unit of chimpanzees. Although some researchers predicted that chimpanzees would form a stable group, a unit-group, despite repeated meeting and parting (e.g. Itani and Suzuki, 1967), this was not ascertained until Nishida's (1968) study at Mahale (see Chapter 6). The method he used to reveal this embodies the characteristics of chimpanzee fission–fusion: he continued to identify chimpanzees who appeared at the provisioning site – alone or in various social combinations – every day for around 2 years; he then confirmed that the accumulated number of individuals did not increase beyond a certain point.

Other primate species known to form groups with individual-based fission–fusion similar to chimpanzees are spider monkeys (*Ateles belzebuth*: Izawa *et al.*, 1979; *A. geoffroyi*: Chapman, 1990), southern muriquis (*Brachyteles arachnoides*: Coles *et al.*, 2013), and the congeneric species of chimpanzees, the bonobo (*Pan paniscus*: Kuroda, 1979). However, fission–fusion dynamics seen in these species may not be identical to that of chimpanzees because spider monkeys and muriquis have much smaller home ranges (1.7–3.4 km^2: Shimooka, 2005; Coles *et al.*, 2013), and bonobos usually form larger and more cohesive parties than chimpanzees (Furuichi, 2009).

Recently, it has been suggested that some non-primate group-living mammals show various degrees of fission–fusion dynamics at the level of individuals or stable subgroups, such as hyenas (*Crocuta crocuta*: Smith *et al.*, 2008), elephants (*Loxodonta africana*: Wittemyer *et al.*, 2005; Fishlock *et al.*, 2008; *Elephas maximus*: de Silva *et al.*, 2011), dolphins (*Tursiops aduncus*: Smolker *et al.*, 1992; *Lagenorhynchus obscurus*: Pearson, 2009), and bats (*Myotis bechsteinii*: Kerth *et al.*, 2011). Accordingly, Aureli *et al.* (2008) have offered a framework for evaluating the differences in fission–fusion dynamics among various aggregations of mammals, including primates, to replace the dichotomy between fission–fusion and non-fission–fusion societies. In this framework, chimpanzees are considered to show a higher degree of fission–fusion dynamics.

Numerous studies have been conducted to investigate the phenomenon of chimpanzee fission–fusion, and have focused on party size and the contributing ecological and social factors. In this chapter, I first introduce two levels of parties observed in Mahale with possible inter-site differences in the size of these parties, and the relationships between them. I then review some studies exploring factors affecting party size. Although many studies have also focused on different types of party compositions, mainly to examine sex differences and social structure, that issue is beyond the scope of this chapter.

9.2 Two levels of parties

A party is a partial aggregation of unit-group members that is temporarily formed in the course of fission–fusion. Definition of party varies among studies, which affects the results of party size estimations to some extent, together with other factors, such as degree of habituation and method of locating parties (Chapman *et al.*, 1993). However, it may be useful to distinguish two party levels that are observed in Mahale: face-to-face and nomadic parties (Itoh and Nishida, 2007).

9.2.1 The face-to-face party

Because of their individual-based fission–fusion, chimpanzees may remain alone, or associate with other individuals. Many studies have focused on the number of individuals comprising temporary gatherings, in which chimpanzees can engage in visual or tactile interactions, such as grooming and mating, and can potentially share a food patch. This kind of party is considered to be "the minimum unit in which members remain in visible contact with each other," and I refer to such a party as a "face-to-face party" (Itoh and Nishida, 2007: 88; for examples, see Figure 9.1). However, visibility changes depending largely on the environment *in situ*; thus, identifying

Figure 9.1 Examples of face-to-face parties. (a) Mixed-sex party resting on the ground; (b) all-female party moving along a trail; (c) all-female party feeding in a tree; (d) all-female party resting in a tree. ©S. Hanamura.

an identical party at different situations is often difficult. It is also difficult for observers to accurately determine party membership, because any chimpanzee can leave or join the party at any time.

In such situations, many researchers who conduct focal animal observations define the members of such a party as all individuals (either including or excluding dependent individuals) present within the visual range of the researcher, which is assumed to be similar to the visual range of the focal animal (Boesch, 1991, 1996; Fawcett, 2000; Itoh and Nishida, 2007; Wakefield, 2008). Some researchers may confine their visual range to within approximately 25 m of the focal animal (Wittiger and Boesch, 2013) or define party members as all individuals who are visible or known to be present within a 35 m radius of the approximate party center (Newton-Fisher *et al.*, 2000). Since visibility at Mahale is heterogeneous (see Figure 9.1), Itoh and Nishida (2007) continuously recorded spatial positions of all members to minimize observation bias based on the degree of visibility. The authors also defined members who had left the party as individuals being out of sight for more than 1 min, to exclude cases in which individuals were temporarily absent due to poor visibility. For studies estimating average party size within a given period, dependency between data points matters. Thus, to avoid interdependence of the data, appropriate time intervals between sampling points may be set (Itoh and Nishida, 2007; Wakefield, 2008). These data can be obtained via scan sampling or continuous recording. Researchers interested in the process of membership changes often employ the latter sampling method. In such cases, the party is considered to have

Figure 9.1 (cont.)

changed whenever its membership changes (Boesch, 1991, 1996; Wittiger and Boesch, 2013).

If the study chimpanzees are not well habituated, researchers may estimate the party size from the number of individuals observed on initial daily sighting, since chimpanzees can easily flee from humans (Tutin *et al.*, 1983; Mitani *et al.*, 2002). Some may estimate party size from the number of chimpanzees present when they come to a particular place with good visibility, such as a food tree (Basabose, 2004) or a provisioning site (Nishida, 1968).

Despite such methodological differences, it can be considered that all of these studies define a face-to-face party as those members who are assumed to remain in visible contact with or ready to contact each other at a certain time and place. The mean face-to-face party size (excluding dependent individuals) for the Mahale M group was four; the party size for K group and for some well-habituated unit-groups in other study sites was also around four to six (reviewed in Itoh and Nishida, 2007). Thus, there seems to be little difference in mean face-to-face party size among study sites. However, maximum face-to-face party size varies by study site: it reaches 76% of the unit-group size in the Mahale M group, and such large face-to-face parties have also been observed in the Mahale K group (97%), Budongo, Uganda (78%), and Bossou, Guinea (100%), but not in Gombe, Tanzania (58%) and Taï, Côte d'Ivoire (54%) (reviewed in Itoh and Nishida, 2007).

9.2.2 The nomadic party

Since the early days of research, researchers studying the Mahale chimpanzees have recognized much looser gatherings, in which all or most unit-group members may be involved during consecutive periods (Nishida, 1968, 1979, 1990, Nishida and Kawanaka, 1972). When "members are not always in visible contact but stay within acoustic range and roughly travel in the same direction on a given day," they are considered a "nomadic party" (Itoh and Nishida, 2007: 89). Some Mahale researchers (e.g. Matsumoto-Oda *et al.*, 1998), previously called a similar gathering merely a "party," but this may fail to distinguish these looser gatherings from face-to-face parties. The nomadic party size at Mahale has been estimated from the number of individuals observed by several researchers on a daily basis. Those individuals observed on the same day but who ranged in different parts of the home range without meeting each other during that day, are regarded as forming different nomadic parties (Matsumoto-Oda *et al.*, 1998; Matsumoto-Oda, 2002; Itoh and Nishida, 2007).

Although little attention has been paid to this kind of party, several researchers in other study sites have described somewhat similar loose gatherings beyond visual range, and their association with loud vocalizations and drumming (Budongo: Reynolds and Reynolds, 1965; Kasakati, Tanzania: Izawa, 1970; Bossou: Sugiyama and Koman, 1979; Taï: Boesch, 1991; Kanyawara, Uganda: Chapman *et al.*, 1995; see Chapter 39 for vocalizations). Some of these researchers refer to such a loose gathering as a "party" (Sugiyama and Koman, 1979), a "group" (Boesch, 1991), or an "acoustic subgroup" (Chapman *et al.*, 1995) to distinguish it from a gathering that roughly corresponds to our face-to-face party; i.e. "compact association," "party," or "subgroup," respectively.

If we regard these loose gatherings as synonymous, the mean nomadic party size relative to the unit-group size varies by study site (Bossou, 72% [8.6/12]; Kanyawara, 19% [7.1/37, including dependents]; Mahale M, 49% [15.5/31.5]); the maximum percentages reach 100% in Bossou and Mahale, but only 57% in Kanyawara (reviewed in Itoh and Nishida, 2007). Although more comparable data from other study sites are needed, these results suggest that there are large differences in unit-group cohesiveness among study sites. Given that smaller unit-groups tend to show higher cohesiveness (Boesch, 1996; Lehmann and Boesch, 2004), the data for the small-sized Bossou group (cited from Sugiyama and Koman, 1979; see also Sugiyama, 1984) may be biased by the effect of demography on unit-group cohesiveness.

Moreover, Mahale researchers have experienced two seasonal phases regarding chimpanzee cohesiveness (e.g. Nishida, 2012: 34) that coincide

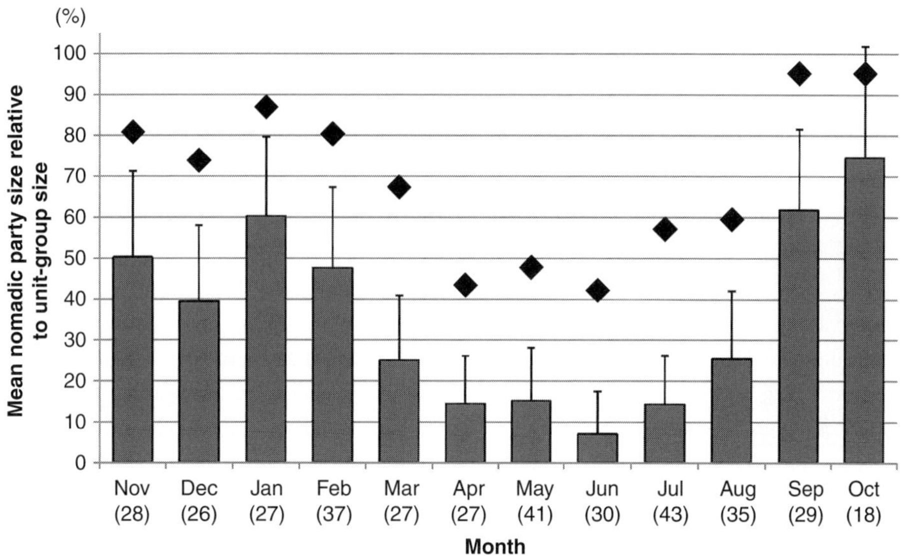

Figure 9.2 Seasonal change in nomadic party size from November 2005 to October 2006 (error bars, SD; diamonds, maximum; numbers in parentheses, monthly number of nomadic parties used in the analysis). The size was estimated by a method similar to that of Itoh and Nishida (2007), excluding individuals less than 7 years old (the corresponding unit-group size: 41–47).

with seasonal fluctuations in nomadic party size; i.e. the congregating season, when chimpanzees tend to form a larger nomadic party, and the dispersing season, when they tend to split into smaller parties (see Itoh and Nishida [2007: Figure 2] for the 1997–2001 period and Figure 9.2 for the 2005–6 period). Such drastic seasonal changes in unit-group cohesiveness and formation of a large nomadic party have seldom been reported from other study sites. Therefore, Itoh and Nishida (2007) concluded that the phenomenon might be unique to unit-groups in Mahale, and possibly to some other unit-groups in areas surrounding Mahale (e.g. Kasakati: Izawa, 1970). This phenomenon is also closely linked to the chimpanzees' unique ranging patterns (Chapter 8) and their auditory interactions (Figure 9.3).

9.2.3 Relationships between nomadic party size and face-to-face party

The seasonal change in nomadic party size of the Mahale chimpanzees is also related to the size variation and the membership fluidity of face-to-face parties (Itoh and Nishida, 2007). In the congregating season, face-to-face party size fluctuates frequently and drastically throughout the day with large maximum size, and the cumulative number of individuals met by a chimpanzee increases with time, suggesting fluidity of face-to-face party membership (see Figure 9.4a). In the dispersing season, the size and membership of a face-to-face party are more stable, and often equal to those of the small nomadic party of the day (see Figure 9.4b). Thus, repeated meeting and parting (or repetition of lags in movements) among various unit-group members at the level of the face-to-face party can be considered an important mechanism for generating a large nomadic party (Itoh, 2003).

The emitting and hearing of various loud vocalizations and sounds would play an important role in these fission–fusion dynamics (Itoh, 2003; also Figure 9.3). For example, there is a long-distance call, the pant-hoot, which is emitted by both sexes in various contexts, including when two face-to-face parties meet (Goodall, 1986; Hanamura, 2013; Chapter 39). Face-to-face party members often utter pant-hoots together, and such a chorus is occasionally "exchanged" between different parties

Figure 9.3 Hearing an out-of-sight vocalization (an adult female, Ako, and her two children). ©S. Hanamura.

(Goodall, 1986; Newton-Fisher, 1999; Hanamura, 2010b, 2013). In the Mahale M group, the frequency of pant-hoots seems to change drastically depending on the seasons: in the congregating season, it is relatively easy for researchers to find chimpanzees because of their frequent pant-hoots, whereas it is just the contrary in the dispersing season (Hanamura, 2010a). During the focal sessions shown in Figure 9.4, I heard 35 pant-hoots outside and 18 pant-hoots within the face-to-face party of my focal animal in the congregating season, and heard only 2 and 0, respectively, in the dispersing season. Further investigations should be conducted to obtain more details about the relationship between interactions through pant-hoots and the fission–fusion dynamics at Mahale.

9.3 Factors affecting party size and grouping patterns

9.3.1 Estrous females

Many studies at various sites have suggested that, to some extent, a greater number of estrous females increases the size of a face-to-face party and/or the number of sexually mature males in the party (Mahale K: Nishida, 1979; Gombe: Goodall, 1986; Bossou: Sakura, 1994; Taï North: Boesch, 1996; Anderson *et al.*, 2002; Budongo: Newton-Fisher, 1999; Fawcett, 2000; Ngogo, Uganda,: Mitani *et al.*, 2002; Kanyawara and Ngogo: Emery Thompson and Wrangham, 2006; Taï South: Wittiger and Boesch, 2013). Similar results from Kalinzu, Uganda, have been reported for 1-h, and 1-day party sizes, which represent the cumulative number of individuals observed during 1 h, and the total observation hours in a day (Hashimoto *et al.*, 2001).

In the Mahale M group, the relationship between the number of estrous females and nomadic party size has been investigated; the results are similar to those described above (Matsumoto-Oda *et al.*, 1998; Matsumoto-Oda, 1999). Moreover, Matsumoto-Oda (1999) revealed that cycling females, even when not in estrus, were observed in large mixed-sex nomadic parties more often than lactating females, and hence proposed the following dual mechanism for the construction of such a large mixed-sex party. First, cycling females are attracted to the party that contains the alpha male, and other males, in order to evaluate future mates. Second, when cycling females are in estrus, males who were not initially associated with such a party are attracted to join.

9.3.2 Place, time, activity, and social relationships

Some studies have suggested that particular dangerous places affect face-to-face party size. Chimpanzees may form a large party when they range in open land, where they are easily detected by large carnivores who are potential predators (Filabanga,

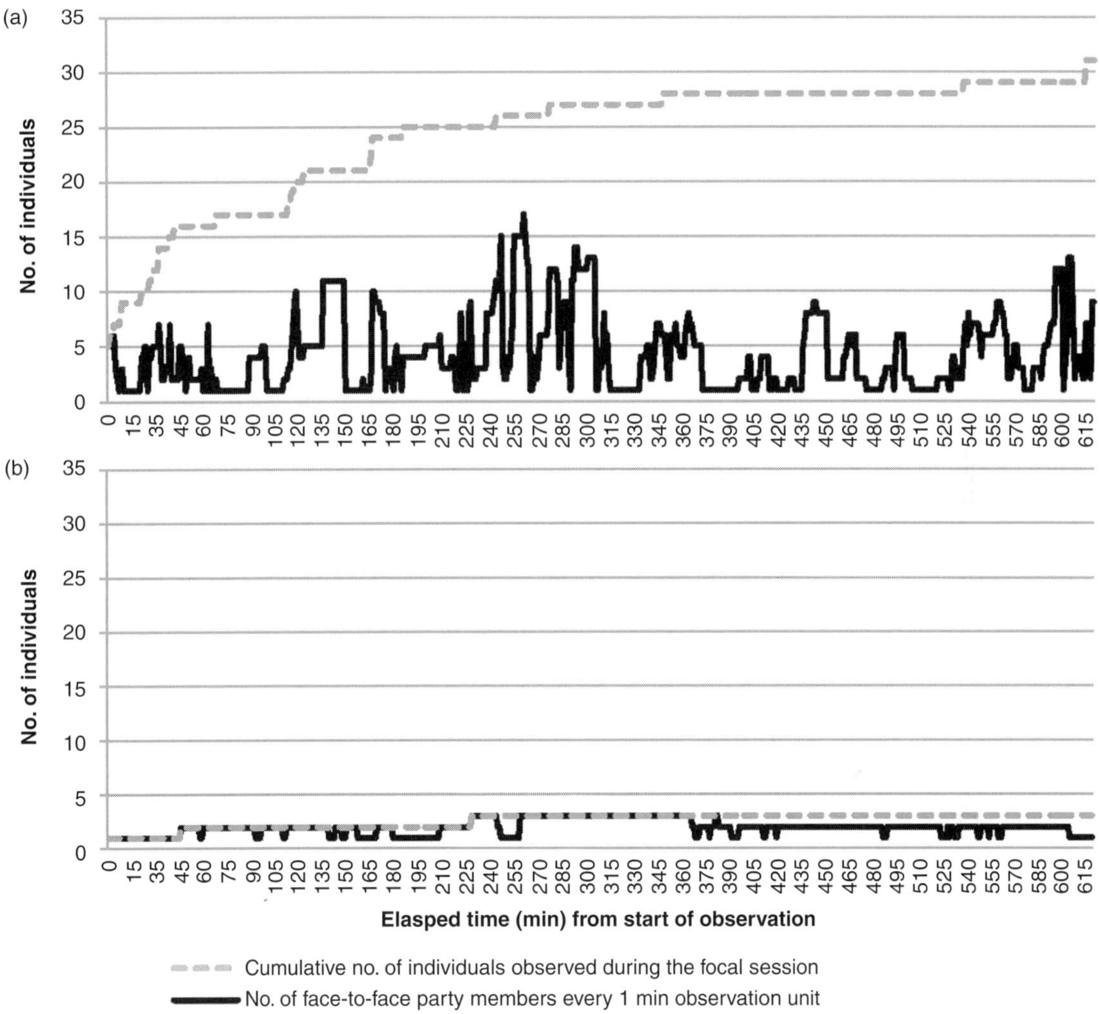

Figure 9.4 Size variation and membership fluidity of face-to-face party of a focal animal on a day when nomadic party size was large or small. (**a**) An example in the congregating season during the focal observation of an adolescent female, Vera, on September 28, 2006. (**b**) An example in the dispersing season during the focal observation of a lactating adult female, Totzy, on May 5, 2006. The *y*-axis shows the number of individuals, excluding individuals under 9 years old.

Tanzania: Itani and Suzuki, 1967), and when crossing human roads (Bossou: Sakura, 1994). When chimpanzees range in peripheral areas of the home range, in which they might potentially encounter individuals of adjacent unit-groups, the number of males may increase, at least in the cumulative number of individuals observed during 1 hour (Kanyawara: Wilson *et al.*, 2007; see also Chapter 11).

It has also been suggested that time of day and type of activity affect face-to-face party size: chimpanzees tend to form larger parties in the morning and the evening, and when hunting and meat sharing (Taï: Boesch, 1996; Anderson *et al.*, 2002). Chimpanzees living in more arid habitats have increased party size when they travel, especially when they undertake long-distance travel from one part of their broad home range to another in order to search for food or water (Assirik, Senegal: Tutin *et al.*, 1983).

Although not systematically investigated, at least some activities in the Mahale M group seem to correlate with face-to-face party sizes. For example, the chimpanzees sometimes form a huge gathering in one place when they hunt and share meat (see Chapters 20), groom (see Chapter 34), and travel together in a line (Nishida, 2012: 39–41), although of shorter distance than at Assirik.

Several studies have suggested the influence of social relationships on face-to-face party size. The coalition strategy of males affects their party size: males in a strong coalition may form smaller parties because they prefer to range by themselves away from opponents (Budongo: Newton-Fisher, 1999). Similarly, association cliques of females observed at the unusually large unit-group of Ngogo may restrict their party size (Wakefield, 2008).

Such male–male coalitions (see Chapter 27) and female–female friendships (see Chapter 28) may likely affect the party formation at Mahale. It is also possible that affiliative relationships between particular male–female pairs (see Chapter 29) influence party size although systematic studies have not been conducted.

9.3.3 Fruit availability and seasonal changes in grouping patterns

One of the most-investigated factors affecting party size is fruit availability; the relationship between this and party size has been analyzed in terms of the scramble competition model. This model predicts that when more fruit is available, chimpanzees *can* form larger parties because of the lower degree of competition (cf. Wrangham *et al.*, 1996).

Estimating fruit abundance by density and/or patch size of fruiting trees, some studies have supported this model at the face-to-face party level (Kanyawara: Isabirye-Basuta, 1988; Taï North: Doran, 1997; Budongo: Fawcett, 2000; Ngogo: Mitani *et al.*, 2002), while other studies do not, or only partially, support it (Budongo: Newton-Fisher *et al.*, 2000; Taï North: Anderson *et al.*, 2002; Kahuzi, Democratic Republic of Congo: Basabose, 2004; Ngogo: Wakefield, 2008; Taï South: Wittiger and Boesch, 2013). Such discrepancies might reflect interannual variation within study sites. At the level of looser gatherings, somewhat similar to nomadic parties, the model was supported at least for the number of males in Kanyawara (acoustic subgroups in Chapman *et al.*, 1995) but not in Kalinzu (1-h and 1-day parties in Hashimoto *et al.*, 2001).

Such between- and within-site differences can occur for the following reasons: (1) relative fruit abundance does not affect party size where or when the absolute fruit abundance is high (Newton-Fisher *et al.*, 2000; Hashimoto *et al.*, 2001); (2) the distribution patterns of plants and length of the fruiting season, rather than overall fruit abundance, affect party size (Newton-Fisher *et al.*, 2000; Basabose, 2004; Wittiger and Boesch, 2013); (3) since plant phenology changes from year to year, it may not affect party size in the same way in different years (Doran, 1997; Fawcett, 2000; Wakefield, 2008); and (4) fruit availability does not equally influence all age-sex classes (Chapman *et al.*, 1995; Fawcett, 2000; Wakefield, 2008; Wittiger and Boesch, 2013). In addition, the aforementioned non-food factors, such as the number of estrous females, can obscure the effect of fruit availability (Doran, 1997; Anderson *et al.*, 2002). As such, the relationship between party size and fruit availability requires a multifaceted understanding.

In the Mahale M group, monthly mean nomadic party size was positively correlated with fruiting plant density (Itoh and Nishida, 2007). However, monthly mean face-to-face party size had no correlation with either fruiting plant density or potential patch size (including liana species), and was roughly stable (between three and five) throughout the year (Itoh and Nishida, 2007).

These results were considered to be related to the phenological and vegetation characteristics of *Saba comorensis*, the fruit species most frequently consumed by the M group chimpanzees (Chapter 17). Indeed, the community-wide fluctuation pattern of fruit availability is basically shaped by the fruiting pattern of *Saba*, because it has a longer and more stable fruiting season than other species, and is densely distributed throughout the M group's home range (Itoh, 2004; Chapters 13, 14). In terms of the scramble competition model, these characteristics of *Saba* largely contribute to creating conditions where the Mahale chimpanzees can gather in the high fruiting season. However, because *Saba* is a liana species, each liana provides only a limited physical space for co-feeding (Itoh and Nishida, 2007). This patch feature of *Saba* may function to limit face-to-face party size in the high fruiting season, and may countervail the possible relationship between increased patch size of fruiting trees and greater face-to-face party size (Itoh and Nishida, 2007). Thus, in the high fruiting season, the M-group chimpanzees can travel with many unit-group members, but they often split into small face-to-face parties while feeding on the hyper-dispersed but small-patched liana species, *Saba*. After feeding, they may meet again in various compositions for grooming, resting, traveling together, and so on; consequently, they form a large nomadic party in this season.

In sum, although on average the M-group chimpanzees remain in small face-to-face parties in different seasons, they form a large nomadic party in the *Saba* season. Although this seasonality of nomadic party size does not contradict the scramble competition model, the model does not readily explain why M-group chimpanzees actually form large nomadic parties in the *Saba* season. Nishida (2012: 39–41) proposed several explanations in terms of possible benefits; for example, that such congregations are effective in hunting, searching for mating partners, and preventing potential attacks from predators. Although such benefits are not readily observable, given that similar seasonal changes in nomadic party size have been observed for years (quantitative data available for 1993–4: Matsumoto-Oda *et al.*, 1998; Matsumoto-Oda, 2002; 1997–2001: Itoh and Nishida, 2007; 2005–6: Figure 9.2), this is a *constant* as well as unique grouping pattern of the M-group chimpanzees that is closely linked to their preference for *Saba* and its phenological and vegetation characteristics. Therefore, it is possible that the formation of a large nomadic party in the *Saba* season has become a kind of "convention" for these chimpanzees (Itoh, 2013; see also Chapter 38).

9.4 Conclusion

In the relatively high fruiting season, chimpanzees of the Mahale M group form a large but loose gathering in which most of the unit-group members repeatedly meet and part. In the low fruiting season, on the other hand, they split into small but cohesive gatherings. Because the formation of such huge nomadic parties, traveling together beyond visual range, and drastic seasonal changes in the size have seldom been reported in other study sites, this is one of the unique characteristics of the grouping patterns of the M-group chimpanzees. In contrast, within visual range, these chimpanzees associate with only 3–5 unit-group members on average in any season, and such face-to-face party size is the same as that in other study sites. However, at Mahale, size variation and membership fluidity of face-to-face parties change depending on the nomadic party size. This seasonal change in the dynamics of face-to-face parties may also be one of the unique characteristics of their grouping patterns. Further understanding may be provided through investigation of the relationships between fission–fusion dynamics and auditory interactions, such as interactions among some temporary face-to-face parties constituting a nomadic party and those among some small nomadic parties in the dispersing season.

References

Anderson, D. P., Nordheim, E. V., Boesch, C., and Moermond, T. C. (2002). Factors influencing fission–fusion grouping in chimpanzees in the Taï National Park, Côte d'Ivoire. In *Behavioural Diversity in Chimpanzees and Bonobos*, ed. C. Boesch, G. Hohmann, and L. F. Marchant. Cambridge: Cambridge University Press, pp. 90–101.

Aureli, F., Schaffner, C. M., Boesch, C., *et al.* (2008). Fission–fusion dynamics. New research frameworks. *Current Anthropology*, **49**, 627–54.

Basabose, A. K. (2004). Fruit availability and chimpanzee party size at Kahuzi montane forest, Democratic Republic of Congo. *Primates*, **45**, 211–19.

Boesch, C. (1991). The effects of leopard predation on grouping patterns in forest chimpanzees. *Behaviour*, **117**, 220–42.

Boesch, C. (1996). Social grouping in Taï chimpanzees. In *Great Ape Societies*, ed. W. C. McGrew, L. F. Marchant, and T. Nishida. Cambridge: Cambridge University Press, pp. 101–13.

Chapman, C. A. (1990). Association patterns of spider monkeys: the influence of ecology and sex on social organization. *Behavioral Ecology and Sociobiology*, **26**, 409–14.

Chapman, C. A., White, F. J., and Wrangham, R. W. (1993). Defining subgroup size in fission–fusion societies. *Folia Primatologica*, **61**, 31–4.

Chapman, C. A., Wrangham, R. W., and Chapman, L. J. (1995). Ecological constraints on group size: an analysis of spider monkey and chimpanzee subgroups. *Behavioral Ecology and Sociobiology*, **36**, 59–70.

Coles, R. C., Lee, P. C., and Telebi, M. (2013). Fission–fusion dynamics in southern muriquis (*Branchyteles arachnoides*) in continuous Brazilian Atlantic forest. *International Journal of Primatology*, **33**, 93–114.

Doran, D. (1997). Influence of seasonality on activity patterns, feeding behavior, ranging, and grouping patterns in Taï chimpanzees. *International Journal of Primatology*, **18**, 183–206.

Emery Thompson, M. and Wrangham, R. W. (2006). Comparison of sex differences in gregariousness in fission–fusion species: reducing bias by standardizing for party size. In *Primates of Western Uganda*, ed. N. E. Newton-Fisher, H. Notman, J. D. Paterson, and V. Reynolds. New York: Springer, pp. 209–26.

Fawcett, K. A. (2000). *Female Relationships and Food Availability in a Forest Community of Chimpanzees*. Ph.D. thesis. Edinburgh: University of Edinburgh.

Fishlock, V., Lee, P. C., and Breuer, T. (2008). Quantifying forest elephant social structure in Central African bai environments. *Pachyderm*, **44**, 19–28.

Furuichi, T. (2009). Factors underlying party size differences between chimpanzees and bonobos: a review and hypotheses for future study. *Primates*, **50**, 197–209.

Goodall, J. (1986). *The Chimpanzees of Gombe: Patterns of Behavior*. Cambridge, MA: Belknap.

Grueter, C. C., Chapais, B., and Zinner, D. (2012). Evolution of multilevel social systems in nonhuman primates and humans. *International Journal of Primatology*, **33**, 1002–37.

Hanamura, S. (2010a). [Chimpanzee interactions through long-distance call and auditory co-presence: the contexts and contingency of hearers' actions.] In [*Boundary and Conjunction of Social Interaction: Studies in Nonhuman Primates, Humans and Conversation*], ed. D. Kimura, M. Nakamura, and K. Takanashi. Kyoto: Shōwa-dō, pp. 185–204. In Japanese.

Hanamura, S. (2010b). [Long-distance calls by chimpanzees: how they connect actions and organize their social field beyond visual contact.] *Primate Research*, **26**, 159–76. In Japanese with English summary.

Hanamura, S. (2013). [When they keep their ears open for vocalization of others in the distance: chimpanzees' process-oriented conventions and an alternative mode of institution.] In [*Practices, Conventions and Institutions: The Evolution of Human Sociality*], ed. K. Kawai. Kyoto: Kyoto University Press, pp. 167–94. In Japanese.

Hashimoto, C., Furuichi, T., and Tashiro, Y. (2001). What factors affect the size of chimpanzee parties in the Kalinzu Forest, Uganda? Examination of fruit abundance and number of estrous females. *International Journal of Primatology*, **22**, 947–59.

Isabirye-Basuta, G. (1988). Food competition among individuals in a free-ranging chimpanzee community in Kibale Forest, Uganda. *Behaviour*, **105**, 135–47.

Itani, J. and Suzuki, A. (1967). The social unit of chimpanzees. *Primates*, **8**, 355–81.

Itoh, N. (2003). [Dynamics of gathering.] In [*Origin and Evolution of Humanity*], ed. M. Nishida, K. Kitamura, and J. Yamagiwa. Kyoto: Shōwa-dō, pp. 233–62. In Japanese.

Itoh, N. (2004). [*The Plant Phenology and Fission–Fusion Social System in the Mahale Mountains National Park*.] Doctoral Dissertation. Kyoto: Kyoto University. In Japanese.

Itoh, N. (2013). [Dual-rings of mode of coexistence and act choices: chimpanzee group and emergence of institution.] In [*Practices, Conventions and Institutions: The Evolution of Human Sociality*], ed. K. Kawai. Kyoto: Kyoto University Press, pp. 143–66. In Japanese.

Itoh, N. and Nishida, T. (2007). Chimpanzee grouping patterns and food availability in Mahale Mountains National Park, Tanzania. *Primates*, **48**, 87–96.

Izawa, K. (1970). Unit groups of chimpanzees and their nomadism in the savanna woodland. *Primates*, **11**, 1–46.

Izawa, K., Kimura, K., and Nieto, A. S. (1979). Grouping of the wild spider monkey. *Primates*, **20**, 503–12.

Kerth, G., Perony, N., and Schweitzer, F. (2011). Bats are able to maintain long-term social relationships despite the high fission–fusion dynamics of their groups. *Proceeding of the Royal Society B*, **279**, 2761–7.

Kummer, H. (1971). *Primate Societies: Group Techniques of Ecological Adaptation*. Chicago, IL: Aldine Atherton.

Kuroda, S. (1979). Grouping of the pygmy chimpanzees. *Primates*, **20**, 161–83.

Lehmann, J. and Boesch, C. (2004). To fission or to fusion: effects of community size on wild chimpanzee (*Pan troglodytes verus*) social organisation. *Behavioral Ecology and Sociobiology*, **56**, 207–16.

Matsumoto-Oda, A. (1999). Mahale chimpanzees: grouping patterns and cycling females. *American Journal of Primatology*, **47**, 197–207.

Matsumoto-Oda, A. (2002). Behavioral seasonality in Mahale chimpanzees. *Primates*, **43**, 103–17.

Matsumoto-Oda, A., Hosaka, K., Huffman, M. A., and Kawanaka, K. (1998). Factors affecting party size in chimpanzees of the Mahale mountains. *International Journal of Primatology*, **19**, 999–1011.

Mitani, J. C., Watts, D. P., and Lwanga, J. S. (2002). Ecological and social correlates of chimpanzee party size and composition. In *Behavioural Diversity in Chimpanzees and Bonobos*, ed. C. Boesch, G. Hohmann, and L. F. Marchant. Cambridge: Cambridge University Press, pp. 102–11.

Newton-Fisher, N. E. (1999). Association by male chimpanzees: a social tactic? *Behaviour*, **136**, 705–30.

Newton-Fisher, N. E., Reynolds, V., and Plumptre, A. J. (2000). Food supply and chimpanzee (*Pan troglodytes schweinfurthii*) party size in the Budongo Forest Reserve, Uganda. *International Journal of Primatology*, **21**, 613–28.

Nishida, T. (1968). The social group of wild chimpanzees in the Mahali Mountains. *Primates*, **9**, 167–224.

Nishida, T. (1979). The social structure of chimpanzees of the Mahale Mountains. In *The Great Apes*, ed. D. A. Hamburg and E. R. McCown. Menlo Park, CA: Benjamin/Cummings, pp. 73–121.

Nishida, T. (1990). A quarter century of research in the Mahale Mountains: an overview. In *The Chimpanzees of the Mahale Mountains: Sexual and Life History Strategies*, ed. T. Nishida. Tokyo: University of Tokyo Press, pp. 3–35.

Nishida, T. (2012). *Chimpanzees of the Lakeshore: Natural History and Culture at Mahale*. Cambridge: Cambridge University Press.

Nishida, T. and Kawanaka, K. (1972). Inter-unit-group relationships among wild chimpanzees of the Mahali Mountains. *Kyoto University African Studies*, **7**, 131–69.

Pearson, H. C. (2009). Influences on dusky dolphin (*Lagenorhynchus obscurus*) fission–fusion dynamics in Admiralty Bay, New Zealand. *Behavioral Ecology and Sociobiology*, **63**, 1437–46.

Reynolds, V. and Reynolds, F. (1965). Chimpanzees of the Budongo Forest. In *Primate Behavior: Field Studies of Monkeys and Apes*, ed. I. DeVore. New York: Holt, Rinehart and Winston, pp. 368–424.

Sakura, O. (1994). Factors affecting party size and composition of chimpanzees (*Pan troglodytes verus*) at Bossou, Guinea. *International Journal of Primatology*, **15**, 167–83.

Schreier, A. L. and Swedell, L. (2012). Ecology and sociality in a multilevel society: ecological determinants of spatial cohesion in hamadryas baboons. *American Journal of Physical Anthropology*, **148**, 580–8.

Shimooka, Y. (2005). Sexual differences in ranging of *Ateles belzebuth belzebuth* at la Macarena, Colombia. *International Journal of Primatology*, **26**, 385–406.

de Silva, S., Ranjeewa, A. D. G., and Kryazhimskiy, S. (2011). The dynamics of social networks among female Asian elephants. *BMC Ecology*, **11**, 17.

Smith J. E., Kolowski, J. M., Graham, K. E., Dawes, S. E., and Holekamp, K. E. (2008). Social and ecological determinants of fission–fusion dynamics in the spotted hyena. *Animal Behaviour*, **76**, 619–36.

Smolker, R. A., Richards, A. F., Connor, R. C., and Pepper, J. W. (1992). Sex differences in patterns of association among Indian ocean bottlenose dolphins. *Behaviour*, **123**, 38–69.

Snyder-Mackler, N., Beehner, J. C., and Bergman, T. J. (2012). Defining higher levels in the multilevel societies of geladas (*Theropithecus gelada*). *International Journal of Primatology*, **33**, 1054–68.

Sugiyama, Y. (1984). Population dynamics of wild chimpanzees at Bossou, Guinea, between 1976 and 1983. *Primates*, **25**, 391–400.

Sugiyama, Y. and Koman, J. (1979). Social structure and dynamics of wild chimpanzees at Bossou, Guinea. *Primates*, **20**, 323–39.

Tutin, C. E. G., McGrew, W. C., and Baldwin, P. J. (1983). Social organization of savanna-dwelling chimpanzees, *Pan troglodytes verus*, at Mt. Assirik, Senegal. *Primates*, **24**, 154–73.

Wakefield, M. L. (2008). Grouping patterns and competition among female *Pan troglodytes schweinfurthii* at Ngogo, Kibale National Park, Uganda. *International Journal of Primatology*, **29**, 907–29.

Wilson, M. L., Hauser, M. D., and Wrangham, R. W. (2007). Chimpanzees (*Pan troglodytes*) modify grouping and vocal behaviour in response to location-specific risk. *Behaviour*, **144**, 1621–53.

Wittemyer, G., Douglas-Hamilton, I., and Getz, W. M. (2005). The socioecology of elephants: analysis of the processes creating multitiered social structures. *Animal Behaviour*, **69**, 1357–71.

Wittiger, L. and Boesch, C. (2013). Female gregariousness in western chimpanzees (*Pan troglodytes verus*) is influenced by resource aggregation and the number of females in estrus. *Behavioral Ecology and Sociobiology*, **67**, 1097–111.

Wrangham, R. W., Chapman, C. A., Clark-Arcadi, A. P., and Isabirye-Basuta, G. (1996). Social ecology of Kanyawara chimpanzees: implications for understanding the costs of great ape groups. In *Great Ape Societies*, ed. W. C. McGrew, L. F. Marchant, and T. Nishida. Cambridge: Cambridge University Press, pp. 45–57.

10 Disappearance of K-group male chimpanzees: re-examination of group extinction

YUKIO TAKAHATA

10.1 Introduction

Since the pioneering work of Toshisada Nishida (1968), the male-bonded/patrilineal social structure of chimpanzees has received much attention from anthropologists and primatologists (Itani, 1972; Wrangham, 1979). Close-kin-related males form a male bond, occupy a territorial range (food resources), and monopolize immigrated and non-kin-related females (reproductive resources). In contrast, females leave their natal groups and transfer to other groups (Chapter 6).

Adult male chimpanzees should experience within- and inter-unit-group (community) competitions in the male-bonded social structure, similar to adult female primates that compete with one another in the female-bonded social structure (van Schaik, 1983; Chapter 27). Adult males cooperatively defend resources against males of neighboring groups, but occasionally exhibit severe aggression toward the same-group males (a field study by Nishida, 2012; a zoo study by de Waal, 1986). Such a complex relationship should provide a basis for evolutionary selection of social intelligence, i.e. the "political" skills described by de Waal (1982).

Intergroup antagonism occasionally escalates into lethal aggression (see Chapter 11). The most famous case was reported in Gombe National Park, Tanzania, where five adult males and one adult female of the Kahama group were observed being injured or were thought to have been injured by adult males of the neighboring Kasekela group. Consequently, the Kahama group became extinct (Goodall, 1986). Such lethal intergroup hostility has been primarily discussed on the basis of a male reproductive strategy model, in which males benefit their group by attacking males from neighboring groups, only in carefully selected contexts, imposing little risk of harm to the aggressors (Wrangham and Glowacki, 2012).

Wrangham and colleagues further discussed these cases with reference to the evolution of violence and warfare among humans (Wrangham, 1999; Wrangham and Wilson, 2004; Crofoot and Wrangham, 2009). For example, Wilson and Wrangham (2003) concluded that human males similarly compete for territory, food, and females, showing a preference for low-risk attacks and intergroup dominance and that chimpanzee studies illustrate the promise of behavioral biology.

Sussman (1999) criticized the discussion of Wrangham and colleagues as a variation of the "man-the-hunter/man-the-killer myth" popularized by Ardrey (1961). Sussman stressed that few direct observations of lethal aggression have been reported and that such aggression is an incidental by-product of human intervention. In addition, Sussman pointed out that the bonobo (*Pan paniscus*), a closely related ape, frequently shows peaceful encounters between neighboring unit-groups (e.g. Idani, 1991).

On the other hand, Mitani *et al.* (2010) reported that similar lethal intergroup aggression leads to territorial expansion in the wild chimpanzee population at Ngogo, Kibale, Uganda. Wrangham and Glowacki (2012) reviewed five of the longest term studies and concluded that peaceful encounters between unit-groups have not yet been reported (Watts and Mitani, 2001; Muller, 2002; Wrangham *et al.*, 2006).

Mahale Chimpanzees: 50 Years of Research, ed. M. Nakamura, K. Hosaka, N. Itoh and K. Zamma. Published by Cambridge University Press.

Nishida *et al.* (1985) reported that a unit-group (K group) in Mahale drastically declined in number and became extinct as a reproductive social unit in 1982. Nishida proposed that the loss of adult males was partly attributed to inter-unit-group aggression, although no direct observations were made. Sussman criticized this speculation, and suggested that chimpanzees were killed by predators such as lions (Hart and Sussman, 2005).

In this chapter, the extinction of the K group is re-examined, with particular focus on the last days of the "male left behind" and "the decisions of adult females and their offspring."

10.2 Background of the K and M groups

In the early stage of his study, Nishida (1968) found that chimpanzees formed multi-male and multi-female groups, with a repeating "fission and fusion" of members (see Chapters 6, 9). He called such a group a "unit-group," which is a synonym for "community," as defined by Goodall (1986; Chapter 6). Among 5–6 unit-groups, the K group was habituated in 1966, and the M group was partly habituated in 1968 (Nishida, 1979).

The K group consisted of 20–30 chimpanzees (Figure 10.1). This group was continually observed until 1983, when all the mature males disappeared and most female members immigrated to the M group (Nishida *et al.*, 2003).

Some members of the M group were not identified until the 1980s, and it was thought to have 91 members in 1980 (Nishida *et al.*, 1990). The group size ranged from 85 to 102 in the 1980s and gradually decreased in the 1990s, with a minimum of 46 in 1997 (Nishida *et al.*, 2003; Chapter 7). This decline in the group size may have been partly because of disease and predation by lions (Tsukahara, 1993).

Nishida (1979) observed that the large-sized M group was dominant over the small-sized K group in the overlapping area of their ranges. When the K group was aware of the M group's approach by their vocal outbursts, the K group retreated into the northern part of their range. This displacement generally occurred during the fruiting season from September to May, probably because the staple food species (e.g. *Saba comorensis*) were abundant in the overlapping area. Young females transferred between these groups. For example, five adolescent females born in the K group transferred to the M group from 1966 to 1983.

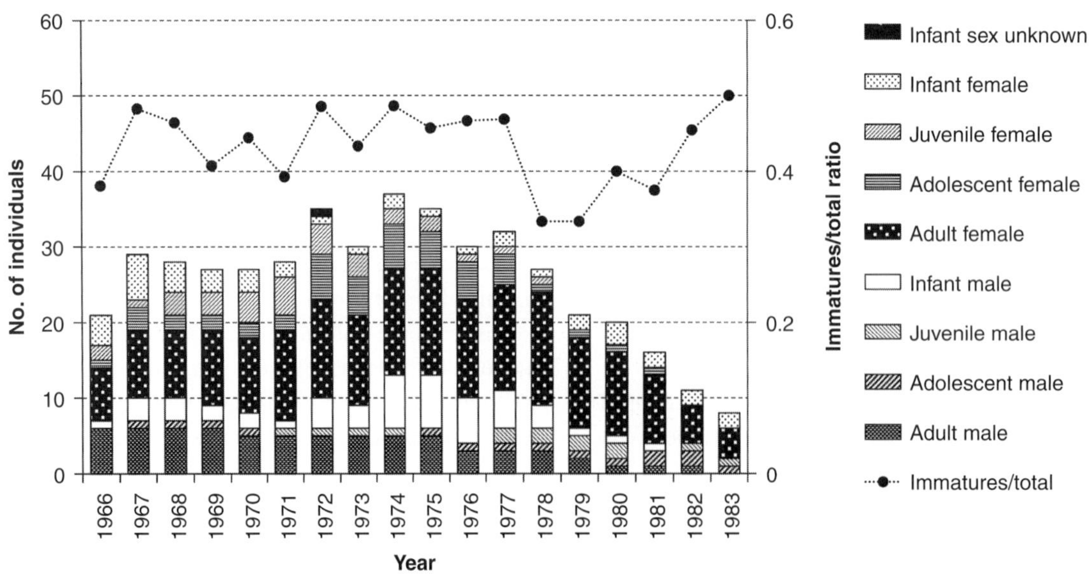

Figure 10.1 Age–sex composition of the K group (1966–83).

10.3 The demographic change in the K group from 1966 to 1983

Figure 10.1 shows the change in the age–sex structure of the K group during 1966–83. The K group reached 34 in members at one stage, but it was generally a small group with a mean of 23.5 (± 7.6 SD) members throughout the period. As previously reported (Nishida *et al.*, 1985), the number of K-group members gradually declined from the latter half of the1970s, and they finally became extinct. It is obvious from the data that, from the onset of the research, the group had very few young males (10–19 years old) compared with older adult males. Thus, the 18-year-study period can be divided into three parts: 1966–74 (the developing period), 1975–8 (the stationary period), and 1979–83 (the declining period).

10.3.1 1966–1974: The developing period

The size of the K group increased during 1966–74. Twenty-eight chimpanzees were observed in the K group's range in 1967 and 33 in 1974. The number of adult males ranged from five to six, and they formed a strong social bond, i.e. a "male bond." The approximate mortality (disappearance rate) of adult males (number of deaths [disappearance]/number of adult males/year) was 0.041. This number is similar to the calculated approximate mortality of 0.042 for 11 adult M-group males during 1979–88.

In contrast, Figure 10.1 shows that few males survived to adulthood, although the reason is unknown. Only one male (Sobongo) grew to adulthood from 1965 to 1983, which may have arrested the alternation of adult male generations. The age–sex pyramid of the average age–sex structure throughout this period (Figure 10.2) shows an unhealthy group structure, at least for males, where the young age class and the mid-to-old age classes are reversed in number. In retrospect, it appears that the K group never had a demographic structure to maintain itself for a long period of time.

Fourteen infants were born during 1966–74. The approximate birth rate (number of births/number of adult females per year) was 0.16, which is slightly higher than that recorded for the M group (0.13) (Nishida *et al.*, 2003). The approximate infant mortality (number of infant deaths/number of infants) was 0.16, which was slightly lower than that recorded for the M group (0.22) (calculated from Table 3 in Nishida *et al.*, 2003).

Adolescent females generally left their natal groups. Seven females disappeared from the K group, whereas seven immigrated into the K group (Nishida

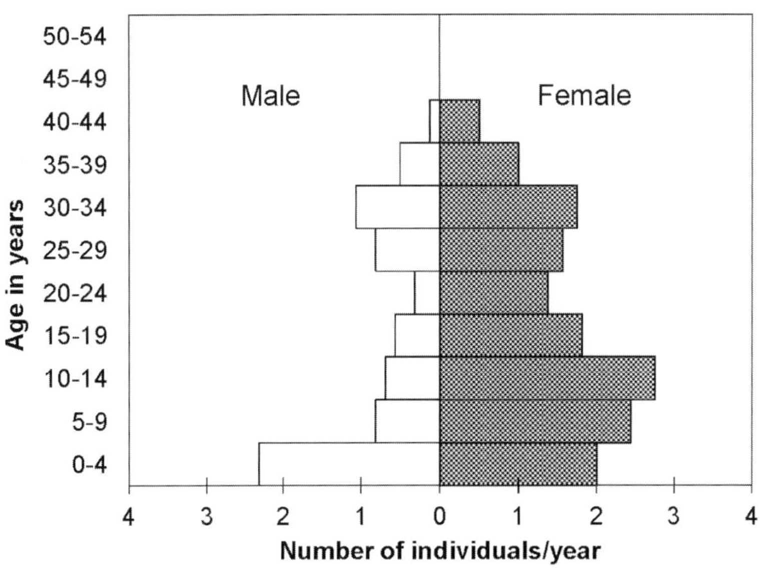

Figure 10.2 The average age–sex pyramid of the K group during 1967–82 (illustrated by M. Nakamura). Numbers of individuals in each age–sex category in different years were averaged throughout the period.

et al., 1985). Thus, female emigration and immigration were well balanced.

Direct group encounters were seldom observed. Nishida (1979) reported that intergroup contacts were observed on 37 "contact days" in 1974. However, 29 of these were acoustic ones. Of the remaining eight visual contacts, in five cases, the K-group chimpanzees encountered ex-K-group members (i.e. the chimpanzees born in the K group or those that belonged to the K group and subsequently transferred to the M group). Thus, stranger-to-stranger visual contacts were only observed in three cases. The following cases suggest a severe hostility between male chimpanzees and an acceptance of female chimpanzees of different groups.

Case 10.1

On January 21, 1974, a large party from the M group had penetrated the boundary of their range northward, and approached the Miyako Valley, the core area of the K group's range. On that day, the K group appeared at the temporary feeding site at 12:32 h. The K-group males pant-hooted at 12:45 h, resulting in vocal outbursts from the M group. Kasonta (the alpha male of the K group) and other chimpanzees slapped the ground, emitting barking calls. At 12:46 h, several adult M-group males (Kidole and Ntologi) appeared at the fringe of the feeding site, slapping the ground and tree trunks, and vocalizing loudly. The chimpanzees in both groups repeated the same actions around the feeding site. At 12:50 h, the K-group members retired to an area further upstream, and at 12:53 h, the M-group members returned southward or eastward. The M-group chimpanzees called out intermittently until 14:07 h.

This was one of the only "true" intergroup contacts, in which large parties of both groups encountered each other.

Case 10.2

Toward the end of January 1974, the majority of the M-group chimpanzees had migrated southward away from the overlapping area of the two groups. However, several M-group members were still ranging around the Kansyana feeding site. On January 31, an adult male from the M group (Lubulungu) appeared at the feeding site at 16:49 h. Then, the core party of the K group suddenly approached the feeding site at 16:53 h, after the lapse of 5 months. On hearing the foliage rustling, Lubulungu fled. The K-group members then occupied the feeding site. At 17:27 h, several K-group chimpanzees stood bipedally and moved westward. At 17:34 h, Ndilo (an ex-K-group female) was seen in the woods, and Sobongo (an adult K-group male) rushed toward her. Sobongo returned to the feeding site at 17:37 h, and Ndilo appeared again at 17:44 h. At 18:01 h, Kasonta and other adult males rushed toward a strange chimpanzee (probably Musa, an adult M-group male), and the stranger fled. At 18:08 h, another adolescent M-group female (Fatuma) appeared and pant-barked at the K-group chimpanzees, together with Ndilo. At 18:45 h, Fatuma finally joined the K-group members to feed on elephant grass.

Case 10.3

Violent aggression was observed in this case. When the core party of the K group visited the feeding site on February 1, 1974, an M-group male (Mimikire) appeared alone. Three K-group males (Kasonta, Kasanga, and Kajabala) rushed towards him. Kasonta chased Mimikire for 200 m and caught him. The two males wrestled, and Kasonta forced Mimikire to the ground and bit him on the right thigh. Finally, Mimikire managed to break free and flee.

10.3.2 1975–8: The stationary period

Although the K group maintained a similar group size (approximately 30) during this period, there may have been a large rift in its continuance. First, two adult males (Kasanga and Kajabala) disappeared in April and October 1975, leaving three adult males behind. The approximate mortality of adult males during this period was 0.27, which was considerably higher than that during 1966–74.

Subsequently, the male bond collapsed. From March to May 1976, the remaining three males entered into severe and complex competition for the alpha status (Nishida, 1983). The alpha male (Kasonta) was challenged by a young beta male (Sobongo), and the old gamma male (Kamemanfu) behaved as if he intended to outmaneuver the other two males. Kasonta was eventually dethroned by Sobongo, but Kasonta reassumed the alpha male status in September 1977. However, Kasonta permanently disappeared in January 1978.

The demographic parameters also exhibited several changes. Only one female immigrated into the K group, but 11 (7 females and 4 males) left the K group. Consequently, the number of chimpanzees in the K group decreased from 29 in 1975 to 19 in 1978.

The approximate birth rate decreased to 0.09, partly because several cycling females transferred to the M group. In contrast, the approximate infant mortality increased to 0.23. Therefore, in 1978, the proportion of immature chimpanzees relative to the group size decreased to 0.33 (Figure 10.1), which was considerably smaller than that recorded for the M group (0.53) (Nishida *et al.*, 2003).

10.3.3 1979–1983: The declining period

Sobongo (the K-group alpha male) disappeared in June 1979, leaving one old male (Kamemanfu) and one adolescent male (Masisa) behind. Kamemanfu and Masisa survived until 1982; however, they could neither reproduce nor prevent cycling females from transferring into the M group. Thus, reproductive activity in the K group largely declined. Only one infant was born during this period, and the approximate birth rate was 0.024.

Kamemanfu eventually disappeared in early 1982. According to Mohamedi Seifu, a field assistant, Kamemanfu was last observed on January 9, 1982, with Masisa, Wamikambi (an adult female), and Limongo (an adolescent male). The approximate adult male mortality rate was 0.4.

After Sobongo disappeared, resident females began to transfer to the M group; i.e. a "large-scale transfer," as described by Nishida *et al.* (1985). The timing of their transfer was correlated with the resumption of sexual swelling (estrus or ovulation). In contrast, no females immigrated to the K group.

Of the five immature sons of adult females, two (Lulemyo and Masudi) transferred to the M group with their mothers and were eventually accepted by the M-group males. One (Katabi) was deserted by his mother (Chausiku) and remained in the K-group range. Chausiku returned to the K group in 1980, but Katabi became seriously ill and probably died. Another son (Milembe) disappeared with his mother (Wakasila). The last son (Limongo) moved with his mother (Wamikambi) within the K-group range.

The M group began to frequently invade the K-group range (Nishida *et al.*, 1985). For example, in December 1979, an M-group party invaded Miyako Valley, the core area of the K-group range. Another neighboring group, the B group, also began to invade the K-group range.

10.3.4 The last days of a male left behind (Masisa) from November to December 1982

In October 1982, I was sent to Mahale by JICA (see Chapter 2) as a scientific expert. My main study subject was the M group, and I had little time to search for Masisa.

On November 26, I walked northward from Kansyana Camp to search for the M group in the overlapping area of the two group ranges. At 09:53–10:14 h, I observed several unhabituated females and juveniles from the M group, but was unable to catch up to the core party.

Near the Miyako Camp at 13:29 h, I saw Masisa unexpectedly appear from the bush along the lakeshore. He seemed to have lost weight. He ate a piece of sugar cane that I gave him, but at 13:38 h, he stopped eating and became tense, when he heard long calls from the M group. From 13:42 to 14:24 h, calls were heard intermittently from a close range. Masisa continued to eat the sugar cane but looked in the direction of the calls. I stopped observations at 14:25 h, and returned to the Kansyana Camp.

From December 13, 1982, I stayed in the Miyako Camp to inquire about Masisa. On December 16 at 07:52 h, I heard the long calls from the Mpila Valley, probably from the members of the M group. At 08:16 h, Masisa suddenly appeared at the Miyako Camp. I threw two pieces of sugar cane toward Masisa, and he rushed towards them. However, he stumbled on the rough ground and fell. He then picked up the sugar cane and crept into the bush. His hind legs had become very thin, and he looked unhealthy compared with the way he looked in November.

Masisa changed postures continually (e.g. lying down or sitting up straight) and chewed the sugar cane. Subsequently, he fed on *Pterocarpus tinctorius* leaves. At 09:52 h, he stopped eating, lay down, and intermittently scratched himself. At 10:56 h, Masisa crept back into the bush. He moved again at 11:10 h, but then lay down on the research trail. At 11:22 h, he tore the bark of a vine (*Annonaceae* sp.) and began to fish for arboreal ants (*Camponotus* sp.) until 12:16 h. At 12:20 h, I stopped observations because of the heavy rain. This was the last sighting of Masisa.

Only lactating and post-reproductive females and their dependent offspring remained within the K-group range after Masisa had disappeared. Thus, the K group had become extinct as a reproductive social unit.

10.4 Discussion

10.4.1 How and why did the K group become extinct?

The extinction of the K group was probably caused by the following multiple factors, which include demographic and social factors, as well as male and female reproductive strategies: (1) Few male infants survived to adulthood in the K group, which could be a fundamental factor in its extinction. (2) Around 1975, the number of adult males decreased. Such a situation may have expanded the "imbalance of power" (Wrangham, 1999) between the K and M groups. (3) Subsequently, male bonding collapsed, although it is uncertain whether the decrease in adult males resulted in the escalation of within-group conflicts. (4) No new females immigrated into the K group after 1976. Female chimpanzees may have avoided immigration based on the principle of "female choice." (5) After 1979, when only Kamemanfu remained, resident females began to leave the K group (large-scale transfer). It appeared that they chose the M-group males as mating partners, likely based on the principle of "female choice." (6) In 1983, Kamemanfu and Masisa disappeared. The M-group chimpanzees accepted the cycling females and several dependent sons from the K group and expanded their range. Eventually, only lactating or post-reproductive females remained with their dependent offspring within the former K-group range.

Two basic questions remain unanswered. Why did few immature males survive to adulthood in the K group, and why did the K-group adult males disappear one after another?

Other than the stochastic results of accidental death, infanticide may be a reason why few immature males survived to adulthood. For example, six infanticide episodes were recorded in Mahale from 1974 to 1984, and all of the victims were male infants (Takahata, 1985). Incidentally, whether male infants tended to be selected as victims is open to debate. Hamai *et al.* (1992) pointed out that, in Mahale, all the mothers of victims were immigrant females, and they tended to give birth to sons significantly more often than daughters (17 males, 6 females, 3 unsexed). If this birth sex ratio is applied, the probability of being killed between male and female infants is not statistically significant.

The reason for the disappearance of the K-group adult males has been controversial. Nishida *et al.* (1985) speculated that at least some of the adult K-group males were killed by the M-group chimpanzees. Hart and Sussman (2005) severely criticized this conclusion and speculated that the K-group males were killed by large-sized carnivores such as lions.

I believe that the two old-aged males (Kasagula and Kamemanfu) probably died of senility and that Masisa died from psychological stress, as described above. There should have been sufficient food resources for Masisa to survive. However, he could not tolerate the stress after the disappearance of Kamemanfu.

Several scenarios are possible for the other four males (Kajabala, Kasanga, Kasonta, and Kaguba), including emigration, diseases, killed by conspecifics (K or M groups), or killed by large carnivores such as lions.

Of these possibilities, "voluntary emigration" is unlikely. In Mahale, there are no observations suggesting that adult male chimpanzees immigrated to non-natal groups. Moreover, a solitary male cannot secure a range or mating partners. For example, in July 1979, the alpha male of the M group (Kajugi) was probably defeated and banished by a new alpha male (Ntologi). Kajugi traveled alone for several years, and he eventually disappeared. This suggests that emigration is quite risky for adult male chimpanzees.

"Killed by lions" is also unlikely because lions had not settled in the study area. If lions visited temporarily, researchers or villagers would have noticed their presence. On the other hand, the possibility of "killed by leopards" should be considered. Although there had been no previous evidence of leopard predation on chimpanzees in Mahale (Nishida, 2012), recently, Nakazawa *et al.* (2013) reported that, of the 142 leopard scats collected within the range of the M group, one contained several bones of a female chimpanzee.

Further discussion on whether the K-group males were banished by other males, died of "diseases," or were "killed by other chimpanzees" is rather pointless because there are no convincing data. However, considering the high mortality of adult males from 1975 to 1979, as Nishida (2012) speculated, the possibility that some of them, particularly Sobongo, were killed by M-group males, cannot be denied (also see Chapter 26 for a relevant discussion).

10.4.2 Intergroup antagonistic relationships among males and female choice

Regardless of whether K-group males were killed by M-group males, the present study indicates several remarkable features of the chimpanzee society: (1) antagonism between the adult males of different groups, which should have evolved from a male reproductive strategy (Nishida, 2012); (2) fragility of the male bond; (3) female choice and their "passport" to transfer to other groups (i.e. estrus/ovulation); and (4) the extremely delicate position of the immature offspring of such females.

The present study shows unexpected fragility of the male bond. Based on the data observed during the developing period, Nishida (1979) described K-group males forming a strong social bond with one another. However, potential competition among adult males may have lurked under such a male bond. Furthermore, some unexpected changes in the social relationships (e.g. sudden death or the disappearance of adult male members) resulted in excess within-group competition, which brought about the collapse of the K group.

Adult females did not hesitate to emigrate when the K group became unstable. This "large-scale transfer" should be regarded as female choice because such migration could lead to a greater number of available mates and decrease the possibility of attacks from large neighboring groups (Nishida *et al.*, 1985).

Of course, females are restrained in several ways. One is sexual cycling. With regard to this, Nishida (2012) stressed that "sexual receptivity" should be a "passport" to transfer between groups. Lactating females remained within the K group's range, but they left K group after they resumed cycling (Nishida *et al.*, 1985).

Another restraint is offspring. If a female had offspring, particularly sons, those male offspring face the risk of infanticide, and they are occasionally abandoned by their mothers, as reported by Hiraiwa-Hasegawa and Hasegawa (1988). In other cases, they can transfer to neighboring groups together with their mothers, as reported by Takahata and Takahata (1989).

10.4.3 Comparison with female-bonded primate societies

Group extinction and/or fusion have been observed in several species of female-bonded primates. The group extinction process in Japanese macaques (*Macaca fuscata yakui*) may be a "mirror image" of what occurred in the K group (Takahata *et al.*, 1994). The birth rate initially decreased and infant mortality increased in a group of wild Japanese macaques, which was probably due to the severe intergroup competition for food resources. When the group size decreased, males left the group one after another, probably based on the principle of "male reproductive strategy." When only one or two females remained with dependent offspring, the last male deserted them before the mating season, and the group ceased to be a reproductive social unit. Then, the remaining females finally transferred to the neighboring groups with their offspring, which was different from the case of K-group males. Similar female transfer or group fusions have been observed in vervet monkeys (Isbell *et al.*, 1991). Compared with such cases observed in cercopithecoid species, the present study showed a severe hostility between adult male chimpanzees of different groups. This observation also exemplifies the flexibility of the choice of a female chimpanzee to belong to a unit-group.

On finishing writing this chapter, I heartily thank the chimpanzees of the K and M groups for tolerating me.

References

Ardrey, R. (1961). *African Genesis: A Personal Investigation into Animal Origins and Nature of Man*. New York: Atheneum.

Crofoot, M. C. and Wrangham, R. W. (2009). Intergroup aggression in primates and humans: The case for a unified theory. In *Mind the Gap: Tracing the Origins of Human Universals*, ed. P. M. Kappeler and J. B. Silk. London: Springer, pp. 171–95.

Goodall, J. (1986). *The Chimpanzees of Gombe: Patterns of Behavior*. Cambridge, MA: Belknap.

Hamai, M., Nishida T., Takasaki, H., and Turner, A. (1992). New records of within-group infanticide and cannibalism in wild chimpanzees. *Primates*, **33**, 151–62.

Hart, D. and Sussman, R. W. (2005). *Man the Hunted: Primates, Predators, and Human Evolution*. Cambridge, MA: Westview Press.

Hiraiwa-Hasegawa, M. and Hasegawa, T. (1988). A case of offspring desertion by a female chimpanzee and the behavioral changes of the abandoned offspring. *Primates*, **29**, 319–30.

Idani, G. (1991). Relations between unit-groups of bonobos at Wamba, Zaire: encounters and temporary fusions. *African Study Monographs*, **11**, 153–86.

Isbell, L. A., Cheney, D. L., and Seyfarth, R. M. (1991). Group fusions and minimum group sizes in vervet monkeys (*Cercopithecus aethiops*). *American Journal of Primatology*, **25**, 57–65.

Itani, J. (1972). A preliminary essay on the relationship between social organization and incest avoidance in nonhuman primates. In *Primate Socialization*, ed. F. E. Poirier. New York: Random House, pp. 165–71.

Mitani, J. C., Watts, D. P., and Amsler, S. J. (2010). Lethal intergroup aggression leads to territorial expansion in wild chimpanzees. *Current Biology*, **20**, 507–8.

Muller, M. N. (2002). Agonistic relations among Kanyawara chimpanzees. In *Behavioural Diversity in Chimpanzees and Bonobos*, ed. C. Boesch, G. Hohmann, and L. Marchant. Cambridge: Cambridge University Press, pp. 112–24.

Nakazawa, N., Hanamura, S., Inoue E., Nakatsukasa, M., and Nakamura, M. (2013). A leopard ate a chimpanzee: first evidence from East Africa. *Journal of Human Evolution*, **65**, 334–7.

Nishida, T. (1968). The social group of wild chimpanzees in the Mahale Mountains. *Primates*, **9**, 167–224.

Nishida, T. (1979). The social structure of chimpanzees of the Mahale Mountains. In *The Great Apes*, ed. D. A. Hamburg and E. R. McCown. Menlo Park, CA: Benjamin/Cummings, pp. 73–121.

Nishida, T. (1983). Alpha status and agonistic alliance in wild chimpanzees (*Pan troglodytes schweinfurthii*). *Primates*, **24**, 318–36.

Nishida, T. (2012). *Chimpanzees of the Lakeshore: Natural History and Culture at Mahale*. Cambridge: Cambridge University Press.

Nishida, T., Hiraiwa-Hasegawa, M., Hasegawa T., and Takahata, Y. (1985). Group extinction and female transfer in wild chimpanzees in the Mahale National Park, Tanzania. *Zeitschrift für Tierpsychologie*, **67**, 284–301.

Nishida, T., Takasaki, H., and Takahata Y. (1990). Demography and reproductive profiles. In *The Chimpanzees of the Mahale Mountains*, ed. T. Nishida. Tokyo: University of Tokyo Press, pp. 63–97.

Nishida, T., Corp, N., Hamai, M., *et al.* (2003). Demography, female life history, and reproductive profiles among the chimpanzees of Mahale. *American Journal of Primatology*, **59**, 99–121.

van Schaik, C. P. (1983). Why are diurnal primates living in groups? *Behaviour*, **87**, 120–44.

Sussman, R. W. (1999). The myth of man the hunter, man the killer and the evolution of human morality. *Zygon*, **34**, 453–71.

Takahata, H. and Takahata Y. (1989). Inter-unit group transfer of an immature male of the common chimpanzees and his social interactions in the non-natal group. *African Study Monographs*, **9**, 209–20.

Takahata, Y. (1985). Adult male chimpanzees kill and eat a male newborn infant: newly observed intragroup infanticide and cannibalism in Mahale National Park, Tanzania. *Folia Primatologica*, **44**, 161–70.

Takahata, Y., Suzuki, S., Okayasu, N., and Hill, D. (1994). Troop extinction and fusion in wild Japanese macaques of Yakushima Island, Japan. *American Journal of Primatology*, **33**, 317–22.

Tsukahara, T. (1993). Lions eat chimpanzees: the first evidence of predation by lions on wild chimpanzees. *American Journal of Primatology*, **29**, 1–11.

de Waal, F. (1982). *Chimpanzee Politics: Power and Sex among Apes*. New York: Harper and Row.

de Waal, F. (1986). The brutal elimination of a rival among captive male chimpanzees. *Ethology and Sociobiology*, **7**, 237–51.

Watts, D. P. and Mitani, J. C. (2001). Boundary patrols and intergroup encounters in wild chimpanzees. *Behaviour*, **138**, 299–327.

Wilson, M. L. and Wrangham, R. W. (2003). Intergroup relations in chimpanzees. *Annual Review of Anthropology*, **32**, 363–92.

Wrangham, R. W. (1979). Sex differences in chimpanzee dispersion. In *The Great Apes*, ed. D. A. Hamburg and E. R. McCown. Menlo Park, CA: Benjamin/Cummings, pp. 481–9.

Wrangham, R. W. (1999). Evolution of coalitionary killing. *Yearbook of Physical Anthropology*, **42**, 1–30.

Wrangham, R. W. and Glowacki, L. (2012). Chimpanzee violence is a serious topic: a response to Sussman and Marshak's critique of "Demonic Males: Apes and the Origins of Human Violence". *Global Nonkilling Working Papers*, **1**, 29–50.

Wrangham, R. W. and Wilson, M. (2004). Collective violence: comparisons between youths and chimpanzees. *Annual New York Academy of Science*, **1036**, 233–56.

Wrangham, R. W., Wilson, M. L., and Muller, M. N. (2006). Comparative rates of violence in chimpanzees and humans. *Primates*, **47**, 14–26.

11 Intergroup relationships

TETSUYA SAKAMAKI AND MICHIO NAKAMURA

11.1 Introduction

Chimpanzees live in multi-male and multi-female unit-groups that exhibit fission–fusion grouping patterns (Chapters 6, 9). Long-term studies at Mahale have revealed two contradictory characteristics of the unit-group: on the one hand, it is rigid and closed, and on the other, it is open and fluid (Nishida, 1979, 1990). The former refers to the stable membership and ranging area of a unit-group, the social bonding within a group, and the antagonistic relationships between groups. By contrast, the transfer of individuals between unit-groups and the transmission of behaviors and genes, some of which may be shared in regional populations, reflect the open, fluid quality of the unit-group. Although group fission is an open and fluid aspect of the unit-group, no cases have been observed at Mahale. One case was observed at Gombe (Tanzania) where the Kahama group split from the main Kasekela group (Goodall *et al.*, 1979).

The fundamental rigid and closed characteristics of the unit-group, such as membership and bonding mechanisms, are introduced in Chapters 6, 8, 9, 27–30. The relationship between the K and M groups and the extinction of the former are described in detail in Chapter 10. In this chapter, we first review the intergroup dynamics at Mahale with a focus on the M group and its neighbors, and we then describe antagonistic intergroup relationships, which are characterized by avoidance or displacement behavior. In the latter half of the chapter, we focus on the open and fluid aspect of unit-groups, that is, nonantagonistic intergroup relationships, which include female transfers between unit-groups and the associated transmission of behaviors and genes.

11.2 Intergroup dynamics at Mahale

11.2.1 Displacement relationship between unit-groups

Unlike studies of social interactions within a group, the investigation of intergroup relationships is limited by the methodological difficulties involved in simultaneously surveying multiple groups (Cheney, 1987). Field surveys conducted prior to those at Mahale revealed agonistic relationships between chimpanzee unit-groups. Early Japanese primatologists seeking to identify the social unit of chimpanzee society (Nishida, 1990: 10–11) conducted studies at Kabogo, Filabanga, and Kasakati in the savanna woodland of Western Tanzania and found that two or more large groups ranged continuously, each with a nomadic range of 100 km^2 or more, some of which overlapped extensively (Azuma and Toyoshima, 1962; Itani and Suzuki, 1967; Suzuki, 1969; Izawa, 1970; Kano, 1971). These nomadic groups had seasonal migration patterns and avoided encounters with one another.

The artificial feeding site at Mahale was inadvertently established in the overlapping area of two unit-groups; thus, the researchers were able to observe the displacement relationship between the two unit-groups. In this dominance–subordinance

Mahale Chimpanzees: 50 Years of Research, ed. M. Nakamura, K. Hosaka, N. Itoh and K. Zamma. Published by Cambridge University Press.

relationship, the subordinate K group avoided the dominant M group (Nishida, 1968, 1979). It is likely that the size and composition of the M group, particularly the number of adult males, was the basis of the group's dominant status (Nishida, 1968, 1979; Nishida and Kawanaka, 1972).

11.2.2 Groups neighboring the K and M groups

At Mahale, the K group was intensively studied from the beginning of 1965; however, the neighboring M group to the south also visited the artificial feeding site (Nishida, 1979, 1990). Six unit-groups were tentatively identified in the study area (Nishida, 1968); however, these early observations have since been revised (see also Chapter 4). In 1975, five teams simultaneously followed different chimpanzee groups for a week and concluded that the home range of the B group, situated north and adjacent to the K group, was considerably larger than initially supposed (Nishida and Norikoshi, 1975; Figure 4.3 in Chapter 4). Evidence collected throughout the 1970s supported the existence of a fourth unit-group (H group) that ranged in the mountain area to the east of the K and M groups (Nishida, 1976; Norikoshi, 1976; Hasegawa and Hasegawa, 1980). A fifth unit-group (N group) was identified south of the M group's home range; researchers occasionally heard their calls while following the M group in the southern peripheral area (Nishida, 1990; see below for further details). Furthermore, a sixth unit-group (S group) may exist in the southeast region of the M-group home range, although the ranging area of this group has not been clearly established (Nishida, 1990; Itoh *et al.*, 1999; Chapter 4).

11.2.3 Extinction of the K group

In the late 1970s and 1980s, researchers observed the extinction of the K group (see Chapter 10). When the number of adult males decreased to one in 1979, all of the parous and nulliparous cycling females permanently emigrated to the M group (Nishida *et al.*, 1985). Direct interactions between members of the K and M groups were observed in this period, during which both group members showed unusual strain; however, their behaviors were not agonistic (Kawanaka, 1982).

Observations of the remaining K-group males during the group extinction suggested that it is difficult for males to transfer between unit-groups. Early in 1983, at most, seven individuals remained in the K group: one adolescent male (Limongo), three adult females, and their offspring (Uehara *et al.*, 1994). In 1987, Limongo was left alone at the age of 15 after the others emigrated permanently to the M group or disappeared. In the 1990s, Limongo was observed only sporadically within the former range of the K group, and no contact with members of the M group was observed (Uehara *et al.*, 1994). In November 1998, Mtunda Mwami (personal communication) observed an adult male with two females along Lake Tanganyika in the former K group's range. The two females fled into the bush when they saw the human; however, the male remained to lick rocks along the lake. This male may have been Limongo, as he appeared to be habituated to humans. This case demonstrates the strong philopatric tendency of adult males.

Although rare, two cases of male transfer between unit-groups were observed at Mahale. Both cases were sons (Lulemyo and Masudi) who transferred from the K group with their mothers and were eventually accepted by the M group (Takahata and Takahata, 1989; Chapter 10).

11.2.4 Invasion of an unknown group

Since the extinction of the K group, as Nishida (2012: 187) wrote, “No one who visits Mahale today can imagine that at one point, in their heyday, K group had 30 members whose unit-group ranged between B and M groups.” M-group chimpanzees occasionally heard the calls of other unit-groups in the southern peripheral area, whereas the northern adjacent area remained vacant after the extinction of the K group. In the late 1990s, evidence suggested the presence of unknown groups in the eastern and northern peripheral areas of the M group's range.

In June 1998, an unknown group of chimpanzees invaded the M group's range from the east (Itoh *et al.*, 1999). This incident was the first observation of a clear intrusion into the M-group home range following the extinction of the K group. The incursion may have been related to the loss of relative fighting power, as the M group's size had decreased markedly between 1995 and 1996, particularly with respect to the number of adult males (Nishida *et al.*, 2003; Chapter 7). Furthermore, the fact that the M group remained in the far northern area of their range during the incursion, which was unusual, may have contributed to the incident. This unknown group has not been seen since the incursion.

In the late 1990s, another unknown group appeared to have moved into the northern area adjacent to part of the M group's range that was left vacant following the disappearance of the K group. Researchers heard the vocalizations of unknown chimpanzees when following the M group in the north and northeastern peripheral areas of their range. Moreover, M-group chimpanzees were observed to attack an unknown female with an infant in the northern area of their range in December 2000 (Kutsukake and Matsusaka, 2002). This female may have been from the unknown northern group. Sporadic evidence confirming the existence of this northern group has accumulated throughout the 2000s (e.g. Uehara, 2002).

A trial habituation of this northern group, led by the director of The East Entertainment, Inc. (Chapter 50), has been ongoing since 2002, and we conducted intensive surveys of this group between 2005 and 2007 (Sakamaki and Nakamura, 2007; Sakamaki *et al.*, 2007).

11.2.5 Appearance of the Y group

The only remaining member of the K group, Limongo, was not observed during intensive surveys conducted between 2005 and 2007 in the northern group; thus, we concluded that this group was different from the former K group, and named it the Miyako group (Y group; Figure 11.1). The group size and the age–sex composition of the Y group have not been confirmed because habituation has not yet been achieved. Judging from the frequency of vocalizations and ranging pattern, it is likely that the M group is larger than and dominant over the Y group (Nakamura *et al.*, 2013).

We do not know the origin of the Y group; however, four possible scenarios exist. First, a remnant of the former K group, i.e. Limongo, may have acquired new members and formed the Y group. Second, the H group from the eastern mountain area or a splinter of

Figure 11.1 A male chimpanzee of the Y group. ©M. Nakamura.

this group may have moved down into the vacant area. Third, the S group from the southeast mountain area or a splinter of this group may have moved to the vacant area. Fourth, the B group in the farther northern areas or a splinter of this group may have moved to the vacant area. Although it is difficult to confirm the true scenario, we believe it is most likely that the H group moved into the vacant area because the Y group ranged farther east or higher in the mountains than did the former K group (Sakamaki and Nakamura, 2007).

11.3 Antagonistic relationships between unit-groups

11.3.1 Territoriality and lethal aggression

In the 1970s, researchers at Gombe were shocked when they observed an incident of intergroup aggression, including coalitionary killings, which led to the extermination of a unit-group (Goodall *et al.*, 1979; Goodall, 1986). About the same time, the Mahale K group was losing adult males. This may have been caused, at least in part, by intergroup aggression, although we have no direct evidence of this (Nishida *et al.*, 1985; Chapters 10, 26). Following several subsequent observations of fatal attacks during intergroup encounters at other study sites, it became apparent that lethal aggression is likely to occur under certain conditions. In general, the attackers significantly outnumbered the victims, indicating that males are unlikely to attack neighbors without overwhelming numeral superiority, which greatly reduces their own risk. These observations provided the basis for the "imbalance of power" hypothesis (Manson and Wrangham, 1991; Wrangham, 1999; Wilson and Wrangham, 2003).

In contrast to other study sites, territoriality and lethal aggression between unit-groups is rare at Mahale. Although within-group killings of infants and adult males have been reported (Chapter 26), lethal raids of neighboring unit-groups have not been observed at Mahale.

11.3.2 Boundary patrol

Boundary patrols, in which males travel silently around the periphery of their territory checking for evidence of other chimpanzees groups, were first reported at Gombe (Goodall, 1986). Male chimpanzees at Gombe (Goodall, 1986), Taï, Côte d'Ivoire (Boesch *et al.*, 2008), and Kibale, Uganda (Wrangham, 1999; Watts and Mitani, 2001; Mitani and Watts, 2005; Watts *et al.*, 2006) routinely patrol their territories, occasionally making deep incursions into those of their neighbors. However, Mahale chimpanzees rarely conduct boundary patrols in a similar fashion; they generally avoid encounters, or their intergroup interactions are in the form of displacement.

In 1999–2000 (1261 h of observation over 178 days; Sakamaki, unpublished data), Sakamaki observed that the M-group chimpanzees occasionally conducted mass expeditions to peripheral areas of their range, but did not conduct the typical boundary patrols observed at Gombe. In a notable case, on May 8, 2000, six adult males and one estous adult female traveled silently to the northern boundary. The other M-group chimpanzees remained behind. While traveling, the seven occasionally sought physical contact such as extending a hand or embracing. The volume of pant-grunts directed to the alpha male was low and apparent only as panting. This behavior was suggestive of a boundary patrol. Subsequently, however, other members of the M group, including many female and immature individuals, followed the leading party, and the researchers heard a loud burst of vocalizations. We could not confirm whether the calls were made by the M group alone or by the M group and a neighboring group.

Mass excursions to the peripheral range areas are occasionally observed at Mahale. For instance, during the *Saba comorensis* fruiting season, chimpanzees form large nomadic parties including males and females and may make mass expeditions to the peripheral areas (Nakamura *et al.*, 2013; Chapters 8, 9).

11.3.3 Exchange of calls and unit-group avoidance

The intergroup interactions observed at Mahale are primarily exchanges of loud calls, including counter-calls, and avoidance (see also Chapter 39). For example, during the 1999–2000 study of the M group, the calls of an unidentified unit-group in the northern area (later designated as the Y group) were heard on 11 days. In response, the M-group chimpanzees emitted counter-calls on five occasions. Counter-calls are a loud chorus of calls, usually pant-hoots, made by adult males who then move rapidly one to a few hundred meters toward the unfamiliar vocalizations. They may make loud calls again after moving forward. On two other occasions, the chimpanzees responded with a chorus of loud calls but did not run toward the unfamiliar vocalizations, and on four occasions, we were unable to observe the behaviors of the M-group chimpanzees directly.

The calls of unfamiliar chimpanzees have been heard occasionally in the eastern region of the M group's range. On September 2, 2008, when the M group was in the lowland area near the Kasiha Valley (see Chapter 2), the calls of unknown chimpanzees were heard upstream, and the M-group males made counter-calls showing excitement (Kazuhiko Hosaka, personal communication). The next day, Nakamura confirmed vocalizations higher in the valley and found a new well-trod chimpanzee trail leading to the northern area, where the Y group often ranged, suggesting that the Y-group chimpanzees occasionally visited the higher areas of the valley.

Furthermore, interactions with the southern N group have been observed. On December 20, 2001, Nakamura heard loud calls from a hill in the southern area of the M-group range. M-group members, including all of the adult males, were at the ridge on the other side of the river, which was more than 1 km away from the vocalizations. On hearing the calls of this neighboring group, the M-group males emitted counter-calls and hugged each other. The exchange of calls continued intermittently for about 2 h, and then the M group returned to the north without a physical encounter with the N group.

11.3.4 Functional significance of intergroup aggression

Hypotheses concerning the functional significance of intergroup aggression have been investigated at various study sites. One study suggests that male chimpanzees may benefit from intergroup aggression by excluding rival males. A lethal attacks on males of a neighboring group reduces the strength of the group, which may be an ultimate explanation for cooperative male aggression against neighbors (Wrangham, 1999). Several lines of evidence support this notion. At Gombe, after the Kahama group was annihilated, the neighboring Kasekela group expanded into the vacant territory (Goodall, 1986; see also Wilson *et al.*, 2004). Lethal attacks by the Ngogo chimpanzees at Kibale were concentrated on their northeast boundary, and then the group expanded its range to the northeast in 2009 (Mitani *et al.*, 2010). At Taï, the number of adult males was the best predictor of home-range size, suggesting a significant influence of relative fighting power; however, the effect of food availability on home-range size is unclear because of changes in various factors such as overall chimpanzee density at the study site (Lehmann and Boesch, 2003).

A second hypothesis proposes that the defense and expansion of their territories benefit males by increasing the amount of food available for themselves, their mates, and offspring, and/or by attracting potential immigrant females. After the Ngogo group expanded its range to the northeast in 2009, they began to routinely travel and feed in the expanded territory, which they had rarely entered previously, except during boundary or hunting patrols (Mitani *et al.*, 2010; Watts, 2012). Large-scale transfers of parous and nulliparous females between unit-groups were recorded at Mahale, Gombe, Taï, and Budongo (Uganda), as described below. Such large-scale transfers may occur when a neighboring unit-group loses most of its adult males.

Furthermore, females may benefit from the defense and expansion of territory through increased safety and food resources. Data from the Kasekela group at Gombe collected over 18 years (1975–92) revealed that females reproduced more quickly, individuals traveled in larger parties, and males encountered receptive females more often when the territory size was larger (Williams *et al.*, 2004). Furthermore, the analysis of 34 years (1967–2000) of body mass data revealed that the Kasekela chimpanzees were heavier when territory size was larger and population density was lower (Pusey *et al.*, 2005). However, the risks to females during the unstable periods when males are challenging or being challenged in the service of territorial expansion, such as injury or infanticide by neighboring males, are not known.

11.3.5 Mortality through intergroup aggression and extra-group paternity

Mortality resulting from intergroup aggression may be extremely high during particular periods. The highest reported number of chimpanzee mortalities resulting from intergroup aggression was at Ngogo (Mitani *et al.*, 2010). Between 1999 and 2009, four adult males and nine immature chimpanzees of an unhabituated group were killed by Ngogo chimpanzees. The impact of intergroup aggression on population structure is an important subject for future investigation (Wrangham *et al.*, 2006; Williams *et al.*, 2008).

The increasing number of genetic studies of paternity may reveal whether males of a certain group have been successful in preventing outside males from mating with their females. Recent studies have shown that the rate of extra-group paternity varies among populations. No extra-group paternity has been reported at Mahale (Inoue *et al.*, 2008) or Budongo (Reynolds, 2005), and only one infant in the Kalande group was fathered by a member of a different group, a Kasekela male, at Gombe (Rudicell *et al.*, 2010). Extra-group paternity accounted for 7–11% of the offspring born into groups at Taï (Vigilant *et al.*, 2001; Boesch *et al.*, 2006), and of the four infants investigated at Bossou (Guinea), one case of extra-group paternity was found (Sugiyama *et al.*, 1993).

11.4 Nonantagonistic relationships between unit-groups

11.4.1 Female transfer

Male intergroup interactions are most often agonistic, whereas females peacefully transfer between unit-groups. Males typically remain in the natal group for their entire lives, whereas females generally migrate among unit-groups in their adolescence (Nishida and Kawanaka, 1972; Nishida, 1979; Chapter 6). Because temporary visitors and permanent immigrants have been observed in unit-groups (Nishida, 1979; Nakamura and Nishida, 2012; Chapter 7), it appears that some adolescent females move among two or more unit-groups. However, females generally settle in the unit-group in which they first give birth (Nishida *et al.*, 2003; Chapter 7). When the K group existed, temporary visits and permanent transfers of K- and M-group females in the overlapping range area were often observed (Nishida, 1979). Researchers are not familiar enough with other unit-groups to identify the origin of immigrant females into the M group, after the K group's extinction.

Female immigration into the M group was steady between 1980 and 2009 (0–2 females per year; Nakamura and Nishida, 2012). In 2010, five adolescent females immigrated to (Hayakawa *et al.*, 2011) and became permanent members of the M group. This case differs from the large-scale transfer from the K group (described below) in that no parous females with offspring immigrated, and no evidence was found to suggest the extinction of a neighboring group. The immigration of five females in 2010 may have been a mere coincidence. Recently, some immigrant females were found to habituate very quickly. It typically takes a few days or weeks of habituation before a chimpanzee will allow observers to approach within 10 m (Nishida *et al.*, 2009); thus, it was surprising that an adolescent immigrant, Yuri, tolerated an observer within close proximity the second day after her immigration in 2005 (Nakamura and Itoh, 2005). The boldness displayed by this chimpanzee suggested that she had emigrated from the Y group, which was undergoing habituation;

however, we had no direct evidence to support this notion.

11.4.2 Large-scale transfer

A large-scale transfer from the K to the M group occurred when most of the adult males disappeared from the K group (Nishida *et al.*, 1985; Chapter 10). During the process, nulliparous and multiparous females with offspring transferred between unit-groups. Similar large-scale transfers have been reported at other study sites. All cases occurred when an adjacent unit-group lost or appeared to have lost most of the males. At Gombe, when the Kahama males disappeared, all of the Kahama females had transferred to the Kasekela or other groups by the end of 1977 (Goodall, 1986). When the size of the Kalande group decreased in the 2000s, most likely due to simian immunodeficiency virus (SIV) infections, and the number of adult males was reduced to one, parous females began to visit the Kasekela group (Rudicell *et al.*, 2010). At Budongo, at least 13 female immigrants, several with dependent offspring, transferred to the Sonso group between 2001 and 2006. The migration was most likely the result of the breakup of a neighboring group or a major habitat disturbance (Emery Thompson *et al.*, 2006; Townsend *et al.*, 2007). Parous females have been observed to make voluntary visits to neighboring groups at Taï (Boesch *et al.*, 2008). Thus, a drastic decrease of the number of adult males in a unit-group appears to be a trigger for females to emigrate from the group.

11.4.3 Factors determining emigration and immigration

Further study is needed to fully understand the factors that lead to adolescent female emigration. At Mahale, four females born in the M group in the 1980s were exceptional in that they gave birth and stayed in their native group. This may be explained by the fact that the size of the M group decreased when these females were adolescents (Nishida *et al.*, 2003; Nakamura and Nishida, 2012), thus reducing the competition for local resources. Alternatively, a lowered risk of incest may explain the females' behavior because several males disappeared during the period (Nishida, 2012). Recently, however, two M-group-born females stayed in the group and gave birth. Although no obvious change in the group size was apparent in this case, the effect of competition for local resource on emigration remains unclear.

The factors that determine immigrant females' choice to settle in a new group are not known. Both adolescent and large-scale female transfer between unit-groups appear to be voluntary. No clear evidence has been found to suggest that male aggression directly forced females to emigrate or immigrate. Female recruitment by males would not be necessary for their immigration. Group size and composition, or local resource and/or mate competition, may affect female transfers. Moreover, the attractiveness of males in a new group and the social interactions between the immigrant and resident females at the initial stage of immigration (see Chapters 28, 29) may influence the choice of new immigrants to settle.

11.4.4 Cultural transmission

Female immigration may contribute to genetic and behavioral transmission among groups (see Chapter 46 for genetic variations among unit-groups at Mahale). Our focus here is the behavioral variation and diversity within and between unit-groups at Mahale, some of which may be transmitted between unit-groups.

Several unit-groups share local cultures (Chapters 38, 42). For example, leaf clipping is a custom observed in the context of courtship in the K and M groups. Circumstantial evidence of leaf clipping was found in the Y-group range (Uehara, 2002). The M-group custom of leaf grooming was performed by a new immigrant female, Yuri, on the first day of her immigration, suggesting that it was shared by her natal group (Nakamura and Itoh, 2005; see also Chapters 34, 42). The grooming hand-clasp is a classic example of cultural variation (Chapters 34, 38). Yuri performed the grooming hand-clasp the first day of her immigration (Nakamura and Itoh, 2005). This grooming pattern has been customarily performed by the M and K groups; however, one adult female who transferred from the K to the M group, persisted in using K-group customs years after her immigration to the M group (see Chapter 38).

It may be difficult to transmit some potentially group-specific behaviors between unit-groups. Immigrant females may display novel behaviors in the new unit-group. One example is algae eating in a stream (Chapter 38). An adolescent female, Sally, who immigrated into the M group in January 1997, was seen to enter a stream and eat algae using her hands in September and October 1997 (Sakamaki, 1998). However, no other chimpanzees were observed imitating her behavior, and Sally was never again observed eating algae (Nishida *et al.*, 2009).

During the intensive habituation of the Y group, we found circumstantial evidence of feeding items unique to this group (Sakamaki *et al.*, 2007). For example, wadges of young *Cissus oliveri* stems were frequently found in the Y group's range but never in that of the M group. However, in 2013, a wadge of *Cissus* was found in the middle of the M group's range, suggesting that it may have been eaten by a new immigrant female from the Y group.

It may be difficult for a new behavioral pattern such as algae eating to propagate from a single newcomer to several members of a unit-group. In contrast, it is likely that newcomers readily acquire established group behavioral patterns. In fact, new immigrant females quickly become habituated to human observers after immigration (Nishida *et al.*, 2009). Given the difficulty of transmitting new behavioral patterns to other unit-groups, the question remains as to why and how local cultures become common across several unit-groups. Such cultural transmission may occur if multiple females simultaneously transfer from the same natal group to the same new group (Nishida *et al.*, 2009). Moreover, group fission and large-scale transfers may provide opportunities for the propagation of new behavioral patterns to several unit-groups; however, such cultural transmission is difficult to verify because group fission and large-scale transfers are rare.

11.5 Toward future studies

11.5.1 Variations in intergroup relationships

The long-term study at Mahale, which has been ongoing for five decades, has revealed that intergroup dynamics are changeable, from the displacement relationship between the K and M groups in 1960s and extinction of the K group around 1980 to the appearance of the new Y group in the former K group's range in the 1990s. The frequency of intergroup lethal aggression may vary across populations (Boesch *et al.*, 2008). Thus, chronological changes in particular populations must be considered in addition to local variations. Observation over a relatively short period, such as a decade, may not be long enough to reveal all aspects of intergroup relationships. For example, mortality resulting from intergroup aggression may change every few years, as noted above. Furthermore, the constraints of environmental, demographic, and sociological factors may be related to local variations and chronological changes in intergroup relationships. Further studies are needed to clarify the relationships among these factors according to variations and changes in intergroup relationships.

11.5.2 Influence of human activities

Regrettably, human activities may influence intergroup dynamics. At Gombe, the habitat loss outside the national park and poaching near the park's boundary may have influenced intergroup dynamics (Pusey *et al.*, 2007; Wilson, 2012). Furthermore, SIV infection may have an impact on population dynamics, although the effect of human activity in this regard is not clear (Rudicell *et al.*, 2010). At Budongo, deforestation in the northern area adjacent to the range of the Sonso group may have triggered a large-scale transfer of females into the group (Reynolds, 2005; Emery Thompson *et al.*, 2006). A zoonosis caused a drastic decrease in some of the groups at Taï (Boesch and Boesch-Achermann, 2000; Lehmann and Boesch, 2004). For a discussion of the problems surrounding zoonoses and human activity at Mahale, see Chapters 25 and 50, respectively. Furthermore, we need to consider the effects of human observers on intergroup interactions; researchers following chimpanzees in a habituated group may inadvertently frighten away unhabituated neighbors during intergroup encounters (Wrangham *et al.*, 2006). The way in which we evaluate and control the influence of human activities on chimpanzee populations is a critical issue for now and into the future.

11.5.3 Female perspective

The role of males in intergroup interactions is primarily antagonistic in nature, whereas females are vital for the transmission of local behavioral patterns and genes, which may be shared by several unit-groups. Compared with our knowledge of strong male bonds in chimpanzee unit-groups (Chapter 27), our understanding of the role of females in influencing population structure is less clear. For example, females may have close relatives and familiar individuals in neighboring groups because most females transfer between unit-groups as adolescents. While hearing and exchanging loud calls with a neighboring group, females may recognize the calls of ex-familiars. However, it is not known whether relationships with ex-familiars in neighboring groups influence intergroup relationships.

The relative position of females in relation to males may be linked to intergroup interactions. The females at Taï participate in intergroup aggression more often than do females at other sites (Boesch *et al.*, 2008). Moreover, support for the victims of intergroup aggression at Taï is remarkable; thus, several individuals were rescued before they got injured (Boesch *et al.*, 2008). Taï females are more gregarious than is typical for eastern chimpanzees (Boesch and Boesch-Achermann, 2000), and this gregariousness may explain, at least in part, the frequency of females participation in intergroup aggression (see Chapters 8, 28 about female gregariousness at Mahale). In bonobos (*Pan paniscus*), the closest relatives of chimpanzees, females establish intimate bonds with one another, and their status is equal or equivocal to that of males. Moreover, they engage in non-agonistic interactions with members of neighboring groups (Idani, 1990). In contrast, female chimpanzees in East Africa have recently been observed engaging in severe aggression (Muller, 2007; Pusey *et al.*, 2008). The complexities of females' involvement in intergroup interactions are considerable.

In this chapter, we reviewed observations of chimpanzee intergroup dynamics made over five decades at Mahale. However, five decades is not long enough to fully understand intergroup dynamics. The Mahale Mountains National Park contains multiple unit-groups (Chapter 4) that experience minimal disturbance from humans (Chapter 50), making the area suitable for the study of intergroup relationships. Furthermore, genetic studies are beginning to reveal the gene flow in the Mahale population (Chapter 46). Unraveling the complexities of how intergroup relationships affect gene flow in the chimpanzee population undoubtedly makes this one of the most intriguing topics in primatology.

References

Azuma, S. and Toyoshima, A. (1962). Progress report of the survey of chimpanzees in their natural habitat, Kabogo point area, Tanganyika. *Primates*, **3**, 61–70.

Boesch, C. and Boesch-Achermann, H. (2000). *The Chimpanzees of the Taï Forest: Behavioural Ecology and Evolution*. New York: Oxford University Press.

Boesch, C., Kohou, G., Néné, H., and Vigilant, L. (2006). Male competition and paternity in wild chimpanzees of the Taï Forest. *American Journal of Physical Anthropology*, **130**, 103–15.

Boesch, C., Crockford, C., Herbinger, I., *et al.* (2008). Intergroup conflicts among chimpanzees in Taï National Park: lethal violence and the female perspective. *American Journal of Primatology*, **70**, 519–32.

Cheney, D. L. (1987). Interactions and relationships between groups. In *Primate Societies*, ed. B. B. Smuts, D. L. Cheney, R. M. Seyfarth, R. W. Wrangham, and T. T. Struhsaker, Chicago: University of Chicago Press, pp. 267–81.

Emery Thompson, M., Newton-Fisher, N. E., and Reynolds, V. (2006). Probable community transfer of parous adult female chimpanzees in the Budongo Forest, Uganda. *International Journal of Primatology*, **27**, 1601–17.

Goodall, J. (1986). *The Chimpanzees of Gombe: Patterns of Behavior*. Cambridge, MA: Belknap.

Goodall, J., Bandora, A., Bergman, E., *et al*. (1979). Intercommunity interactions in the chimpanzee population of the Gombe National Park. In *The Great Apes*, ed. D. A. Hamburg and E. R. McCown. Menlo Park, CA: Benjamin/Cummings, pp. 13–53.

Hasegawa, T. and Hasegawa, M. (1980). The eighth ascent of Mt. Nkungwe: chimpanzees of the Humo-group and other larger mammals of the main ridge of Mahale Massif. *Mahale Mountains Chimpanzee Research Project Survey Report*, **10**.

Hayakawa, T., Nakashima, M., and Nakamura, M. (2011). Immigration of a large number of adolescent female chimpanzees into the Mahale M group. *Pan Africa News*, **18**, 8–10.

Idani, G. (1990). Relations between unit-groups of bonobos at Wamba, Zaire: encounters and temporary fusions. *African Study Monographs*, **11**, 153–86.

Inoue, E., Inoue-Murayama, M., Vigilant, L., Takenaka, O., and Nishida, T. (2008). Relatedness in wild chimpanzees: influence of paternity, male philopatry, and demographic factors. *American Journal of Physical Anthropology*, **137**, 256–62.

Itani, J. and Suzuki, A. (1967). The social unit of chimpanzees. *Primates*, **8**, 355–81.

Itoh, N., Sakamaki, T., Hamisi, M., *et al*. (1999). A new record of invasion by an unknown unit group into the center of M group territory. *Pan Africa News*, **6**, 8–10.

Izawa, K. (1970). Unit groups of chimpanzees and their nomadism in the savanna woodland. *Primates*, **11**, 1–46.

Kano, T. (1971). The chimpanzee of Filabanga, western Tanzania. *Primates*, **12**, 229–46.

Kawanaka, K. (1982). A case of inter-unit-group encounter in chimpanzees of the Mahale Mountains. *Primates*, **23**, 558–62.

Kutsukake, N. and Matsusaka, T. (2002). Incident of intense aggression by chimpanzees against an infant from another group in Mahale Mountains National Park, Tanzania. *American Journal of Primatology*, **58**, 175–80.

Lehmann, J. and Boesch, C. (2003). Social influences on ranging patterns among chimpanzees (*Pan troglodytes verus*) in the Taï National Park, Côte d'Ivoire. *Behavioral Ecology*, **14**, 642–9.

Lehmann, J. and Boesch, C. (2004). To fission or to fusion: effects of community size on wild chimpanzee (*Pan troglodytes verus*) social organization. *Behavioral Ecology and Sociobiology*, **56**, 207–16.

Manson, J. H. and Wrangham, R. W. (1991). Intergroup aggression in chimpanzees and humans. *Current Anthropology*, **32**, 369–90.

Mitani, J. C. and Watts, D. P. (2005). Correlates of territorial boundary patrol behaviour in wild chimpanzees. *Animal Behaviour*, **70**, 1079–86.

Mitani, J. C., Watts, D. P., and Amsler, S. J. (2010). Lethal intergroup aggression leads to territorial expansion in wild chimpanzees. *Current Biology*, **20**, R507–8.

Muller, M. N. (2007). Chimpanzee violence: femmes fatales. *Current Biology*, **17**, R365–6.

Nakamura, M. and Itoh, N. (2005). Notes on the behavior of a newly immigrated female chimpanzee to the Mahale M group. *Pan Africa News*, **12**, 20–2.

Nakamura, M. and Nishida, T. (2012). Long-term field studies of chimpanzees at Mahale Mountains National Park, Tanzania. In *Long-Term Field Studies of Primates*, ed. P. M. Kappeler and D. P. Watts. Heidelberg, Germany: Springer, pp. 339–56.

Nakamura, M., Corp, N., Fujimoto, M. *et al*. (2013). Ranging behavior of Mahale chimpanzees: a 16 year study. *Primates*, **54**, 171–82.

Nishida, T. (1968). The social group of wild chimpanzees in the Mahali Mountains. *Primates*, **9**, 167–224.

Nishida, T. (1976). The sixth ascent of Mt. Nkungwe, the paramount peak of the Mahale Massif. *Mahale Mountains Chimpanzee Research Project Survey Report*, **4**.

Nishida, T. (1979). The social structure of chimpanzees of the Mahale Mountains. In *The Great Apes*, ed. D.

A. Hamburg and E. R. McCown. Menlo Park, CA: Benjamin/Cummings, pp. 73–121.
Nishida, T. (1990). A quarter century of research in the Mahale Mountains: an overview. In *The Chimpanzees of the Mahale Mountains: Sexual and Life History Strategies*, ed. T. Nishida. Tokyo: University of Tokyo Press, pp. 3–35.
Nishida, T. (2012). *Chimpanzees of the Lakeshore: Natural History and Culture at Mahale*. Cambridge: Cambridge University Press.
Nishida, T. and Kawanaka, K. (1972). Inter-unit-group relationships among wild chimpanzees of the Mahali Mountains. *Kyoto University African Studies*, **7**, 131–69.
Nishida, T. and Norikoshi, K. (1975). Distribution of the unit-groups of chimpanzees in the Kasoje area in the *Garcinia huillensis* season of 1975. *Mahale Mountains Chimpanzee Research Project Survey Report*, **3**.
Nishida, T., Hiraiwa-Hasegawa, M., Hasegawa, T., and Takahata, Y. (1985). Group extinction and female transfer in wild chimpanzees in the Mahale National Park, Tanzania. *Zeitschrift für Tierpsychologie*, **67**, 284–301.
Nishida, T., Corp, N., Hamai, M., *et al.* (2003). Demography, female life history, and reproductive profiles among the chimpanzees of Mahale. *American Journal of Primatology*, **59**, 99–121.
Nishida, T., Matsusaka, T., and McGrew, W. C. (2009). Emergence, propagation or disappearance of novel behavioral patterns in the habituated chimpanzees of Mahale: a review. *Primates*, **50**, 23–36.
Norikoshi, K. (1976). Investigation of the Kashiha Valley. *Mahale Mountains Chimpanzee Research Project Survey Report*, **5**.
Pusey, A. E., Oehlert, G. W., Williams, J. M., and Goodall, J. (2005). Influence of ecological and social factors on body mass of wild chimpanzees. *International Journal of Primatology*, **26**, 3–31.
Pusey, A. E., Pintea, L., Wilson, M. L., *et al.* (2007). The contribution of long-term research at Gombe National Park to chimpanzee conservation. *Conservation Biology*, **21**, 623–34.
Pusey, A., Murray, C., Wallauer, W., *et al.* (2008). Severe aggression among female *Pan troglodytes schweinfurthii* at Gombe National Park, Tanzania. *International Journal of Primatology*, **29**, 949–73.
Reynolds, V. (2005). *The Chimpanzees of the Budongo Forest: Ecology, Behaviour, and Conservation*. New York: Oxford University Press.
Rudicell, R. S., Jones, J. H., Wroblewski, E. E., *et al.* (2010). Impact of simian immunodeficiency virus infection on chimpanzee population dynamics. *PLoS Pathogens*, **6**(9), e1001116.
Sakamaki, T. (1998). First record of algae-feeding by a female chimpanzee at Mahale. *Pan Africa News*, **5**, 1–3.
Sakamaki, T. and Nakamura, M. (2007). Preliminary survey of unhabituated chimpanzees in the Mahale Mountains National Park, Tanzania: Behavioral diversity across neighboring unit-groups and intergroup relationships. In *Formation of a Strategic Base for Biodiversity Studies*. Kyoto: The 21st Century COE Program of Kyoto University, pp. 278–80.
Sakamaki, T., Nakamura, M., and Nishida, T. (2007). Evidence for cultural differences in diet between two neighboring unit groups of chimpanzees in the Mahale Mountains National Park, Tanzania. *Pan Africa News*, **14**, 3–5.
Sugiyama, Y., Kawamoto, S., Takenaka, O., Kumazaki, K., and Miwa, N. (1993). Paternity discrimination and inter-group relationships of chimpanzees at Bossou. *Primates*, **34**, 545–52.
Suzuki, A. (1969). An ecological study of chimpanzees in a savanna woodland. *Primates*, **10**, 103–48.
Takahata, H. and Takahata, Y. (1989). Inter-unit group transfer of an immature male of the common chimpanzee and his social interactions in the non-natal group. *African Study Monographs*, **9**, 209–20.
Townsend, S. W., Slocombe, K. E., Emery Thompson, M., and Zuberbühler, K. (2007). Female-led infanticide in wild chimpanzees. *Current Biology*, **17**, R355–6.
Uehara, S. (2002). Evidence of the leaf-clipping behavior by a chimpanzee of an unhabituated group at Mahale. *Pan Africa News*, **9**, 3–4.
Uehara, S., Nishida, T., Takasaki, H., *et al.* (1994). A lone male chimpanzee in the wild: the survivor of a disintegrated unit-group. *Primates*, **35**, 275–81.
Vigilant, L., Hofreiter, M., Siedel, H., and Boesch, C. (2001). Paternity and relatedness in wild chimpanzee communities. *Proceedings of the National Academy of Sciences of the USA*, **98**, 12890–5.

Watts, D. P. (2012). Long-term research on chimpanzee behavioral ecology in Kibale National Park, Uganda. In *Long-Term Field Studies of Primates*, ed. P. M. Kappeler and D. P. Watts. Heidelberg, Germany: Springer, pp. 313–56.

Watts, D. P. and Mitani, J. C. (2001). Boundary patrols and intergroup encounters in wild chimpanzees. *Behaviour*, **138**, 299–327.

Watts, D. P., Muller, M. N., Amsler, S. J., Mbabazi, G., and Mitani, J. C. (2006). Lethal intergroup aggression by chimpanzees in the Kibale National Park, Uganda. *American Journal of Primatology*, **68**, 161–80.

Williams, J. M., Oehlert, G. W., Carlis, J. V., and Pusey, A. E. (2004). Why do male chimpanzees defend a group range? *Animal Behaviour*, **68**, 523–32.

Williams, J. M., Lonsdorf, E. V., Wilson, M. L., *et al.* (2008). Causes of death in the Kasekela chimpanzees of Gombe National Park, Tanzania. *American Journal of Primatology*, **70**, 766–77.

Wilson, M. L. (2012). Long-term studies of the chimpanzees of Gombe National Park, Tanzania. In *Long-Term Field Studies of Primates*, ed. P. M. Kappeler and D. P. Watts. Heidelberg, Germany: Springer, pp. 357–84.

Wilson, M. L. and Wrangham, R. W. (2003). Intergroup relations in chimpanzees. *Annual Review of Anthropology*, **32**, 363–92.

Wilson, M. L., Wallauer, W. R., and Pusey, A. E. (2004). New cases of intergroup violence among chimpanzees in Gombe National Park, Tanzania. *International Journal of Primatology*, **25**, 523–49.

Wrangham, R. W. (1999). Evolution of coalitionary killing. *Yearbook of Physical Anthropology*, **42**, 1–30.

Wrangham, R. W., Wilson, M. L., and Muller, M. N. (2006). Comparative rates of violence in chimpanzees and humans. *Primates*, **47**, 14–26.

PART III Ecology

12 Climate and climatological trends in the Kasoje forest

NORIKO ITOH

12.1 Introduction

All living and non-living things are influenced in many ways by climate. Every species of plant is greatly influenced by climate, and climate seasonality influences plant phenology, which in turn influences the lives of animals. Animal lives may also be directly influenced by climate, and humans are no exception.

Climate is itself influenced by the various activities of living and non-living things. Latitude and altitude largely determine climatic variations, and local topography and vegetation shape local climatic behaviors. Both plants and animals can influence climatic variables, the most critical of which may be the human activities inducing global climatic change.

In this chapter, the general features and the long-term changes of Mahale's climate are summarized. An understanding of the local climate of Mahale is important for understanding its basic ecology. In addition, knowing the location and specific topography of Mahale (Chapter 2) can provide significant insights into wider areas such as Tanzania, East Africa, the Albertine Rift, and finally the entire earth (Seimon and Picton Phillips, 2012).

With this perspective, under the supervision of an expert in the field, Dr. Anton Seimon, some Mahale climatic data (1989–2008) were analysed (Itoh *et al.*, 2012). From this 20-year dataset, signs of global climatic change were detected. I then added data collected 6 years prior to and 5 years after this 20-year dataset, and reanalysed the data from January 1, 1983, through December 31, 2013, using the same procedure.

All data were collected at the weather station near the Kansyana Camp (Chapter 2; Figure 12.1). See Appendix IV for details of the dataset used in the analysis. I excluded data obtained on days dated February 29 to standardize the data across years. For daily periodicity, temperature and humidity data that were collected using a data logger (Ondotori RTR-57U, T&D Corporation) from June 17, 2010, were used. The data were collected on an hourly basis, which is too long an interval to obtain accurate minimum and maximum temperatures for each day, but which enables us to observe the general pattern of daily temperature fluctuations (Section 12.3).

12.2 Precipitation

Mahale is very moist, with an annual mean precipitation over the 31-year period of 1751 mm. The driest year was 2000, but the total precipitation was still more than 1000 mm (1357 mm).

The hydrological year begins on July 1 (middle of the dry season) followed by a long wet season that begins in early October (Figure 12.2, consistent with Itoh *et al.*, 2012). The wet season is interrupted by a relatively short dry season starting around the middle of May. Analysis results show that the precipitation rate peaks (above 10 mm/day) during three multi-day periods between November and April. The early wet season receives the highest levels of precipitation, more than 12 mm/day (7-day running mean), around December 1. The highest daily mean record shifted in time and decreased in amount from 19.2 mm on

Mahale Chimpanzees: 50 Years of Research, ed. M. Nakamura, K. Hosaka, N. Itoh and K. Zamma. Published by Cambridge University Press.

Figure 12.1 Weather station at Kansyana. ©N. Itoh.

December 20 (Itoh *et al.*, 2012) to 18.6 mm on November 30 with the 31-year dataset. However, the maximum precipitation over the 31-year period occurred on December 20, 2006, totaling 220.0 mm. This rainfall caused a large landslide near our camp (Zamma, 2007).

The Mahale precipitation pattern of interannual and decade-scale variability is known to be high (Itoh *et al.*, 2012). Departures from the normal mean precipitation, as a 365-day running mean, are shown in Figure 12.3 (comparable with Figure 12.3 in Itoh *et al.*, 2012). The annual mean is 4.8 mm/day. A drying trend is evident (dashed line) at a rate of 7.7 mm/year, which is more moderate in comparison with the 20-year period analysis (11.7 mm/year; Itoh *et al.*, 2012). The range of short period anomalies in the 31-year dataset does not differ from the 20-year analysis: precipitation was 50% above the mean in 1997–8 and 33% below the mean in 1993. High-amplitude year-to-year variations between 1993 and 2001 are still evident but could be traced back at least from the late 1980s. The moderated variability, observed since 2001, continued to the end of the record. El Niño-Southern Oscillation (ENSO)-related patterns were not observed in Mahale except for the extremely high precipitation period in 1997–8 (Itoh *et al.*, 2012).

Another feature which was evident over the 20-year dataset was a temporal shift in the annual precipitation distribution during the hydrological year. The "rainy season has on average decreased in length by several weeks, and has significantly more rainfall occurring in the first peak of the wet season in November–December, and significantly less in the third peak from mid-February through mid-April" (Seimon and Picton Phillips, 2012: 23). This trend is still evident, or at times even more pronounced in the current dataset, which suggests that this shift may be due to climatic change and not the cyclical behavior of decadal climatic variability.

Seimon and Picton Phillips (2012) predicted that changes in the annual precipitation patterns would occur mid-century in the Albertine Rift region, in which Mahale is located in the far south. One change was a shift in the distribution of annual precipitation for which they predicted a large increase in the

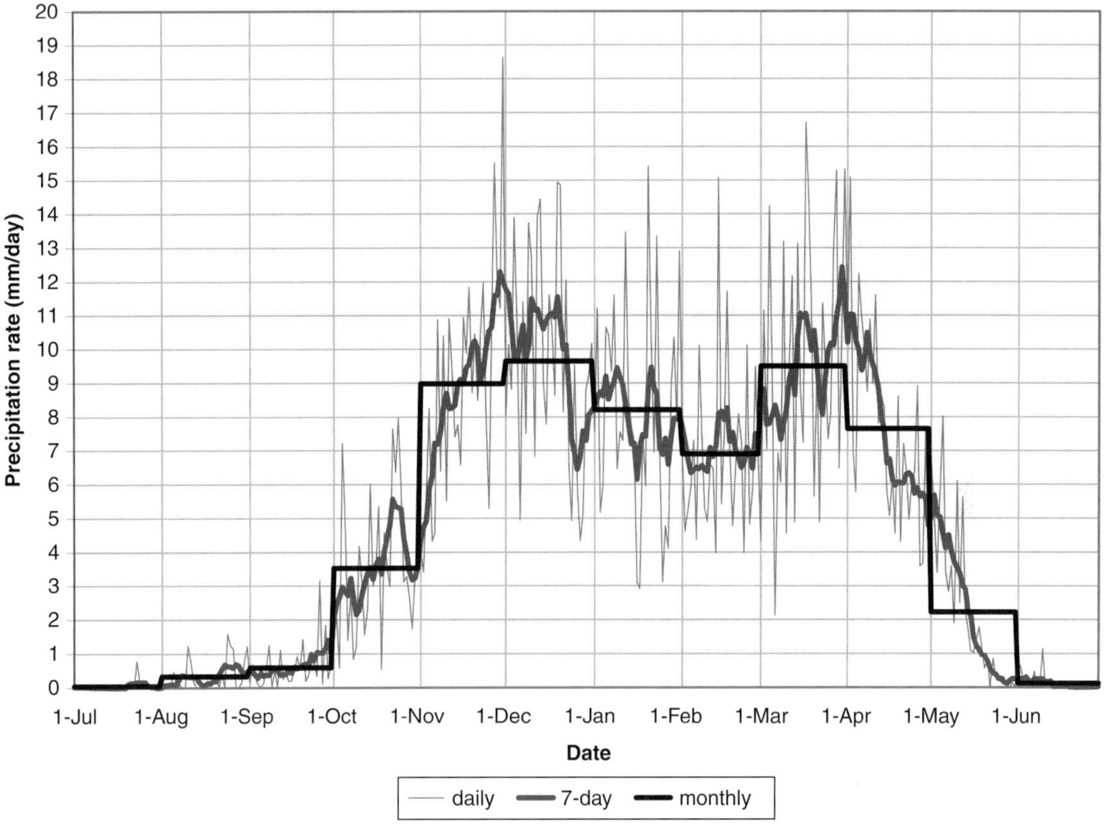

Figure 12.2 Hydrological year precipitation rate: pluviogram of daily mean, 7-day running mean, and monthly mean precipitation.

November–December period, and which was observed for Mahale only. Therefore, Mahale might be the first to experience the change within the Albertine region. The second change was a predicted increase in annual precipitation of 3%, 7%, and 17% in 2030, 2060, and 2090, respectively (Seimon and Picton Phillips, 2012). Although most of the region, including the sites inhabited by chimpanzees, showed this increasing trend, the Mahale trend was contrary to the predicted trends for unknown reasons (see also Plumptre, 2012). However, the Mahale trend does match with future projections in southern Tanzania (Seimon and Picton Phillips, 2010). Further careful monitoring is needed to validate the drying trend precisely and to determine whether the trend is consistent with global climatic change or is within natural decadal-scale variability (Itoh *et al.*, 2012).

12.3 Daily fluctuation pattern of temperature and humidity

Figure 12.4 shows the hourly mean temperatures. Temperature starts to rise around 08:00 h and continues until 14:00–15:00 h, exceeding 26.5°C. Then it begins to decline but the decline slows from 21:00 h. The coldest hour is 06:00 h (19.12°C). This pattern does not vary between months or years. June, July, and August show the coldest mornings and nights, and August, October, and September show the hottest daytimes. All these months fall within the dry season, with the exception of October, which is the start of the wet season.

As Figure 12.5 shows, humidity is high (>70%) throughout the day. It is fairly stable during the night but starts to decline at 10:00 h, and reaches its

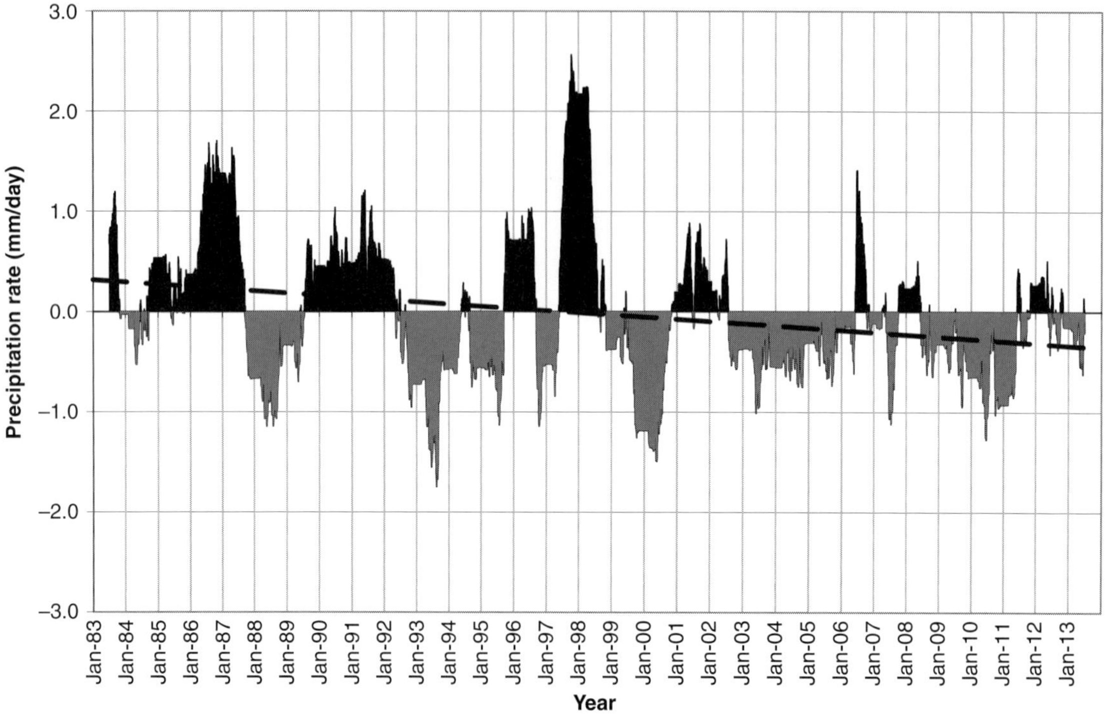

Figure 12.3 Time-series anomalies (365-day running mean) in the precipitation rate. Black and gray lines indicate values above and below normal, respectively.

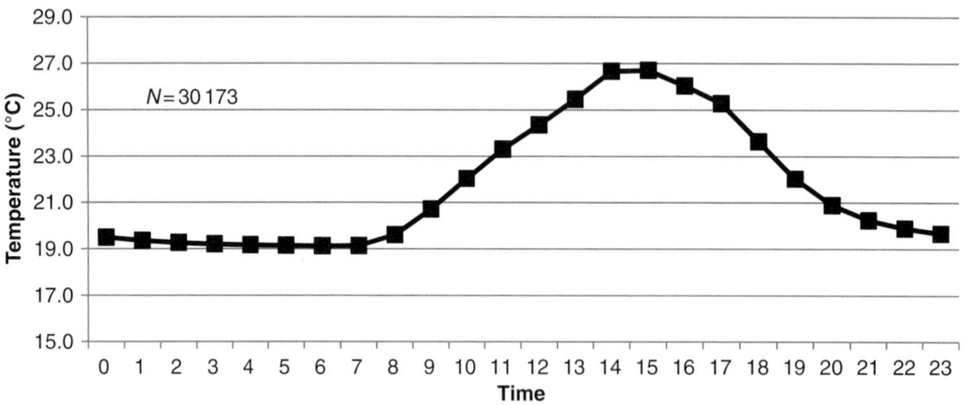

Figure 12.4 Daily fluctuation pattern of mean temperature.

minimum at 14:00–15:00 h. These hours coincide to within a few hours after temperature begins to rise and the hours with maximum temperatures, respectively. The monthly variation also corresponds to the variations in temperature and precipitation, showing lowest mean humidity in daytime from June through October.

12.4 Temperature

Figure 12.6 shows the 5-day running mean and monthly mean of the minimum and maximum temperatures. The mean maximum temperature is 28.9°C (N = 10 337, range 19.1–38.6°C). The hottest months are mid-August–October, with a mean

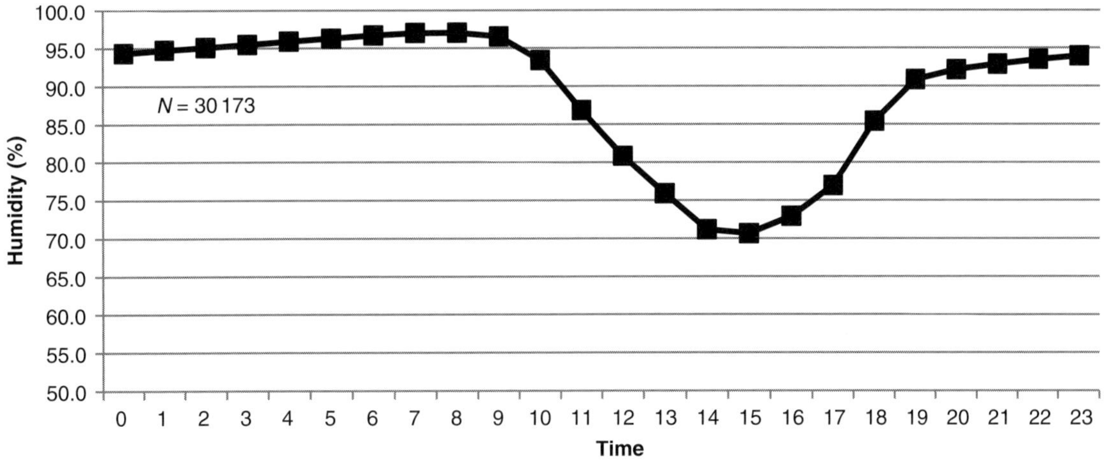

Figure 12.5 Daily fluctuation pattern of mean humidity.

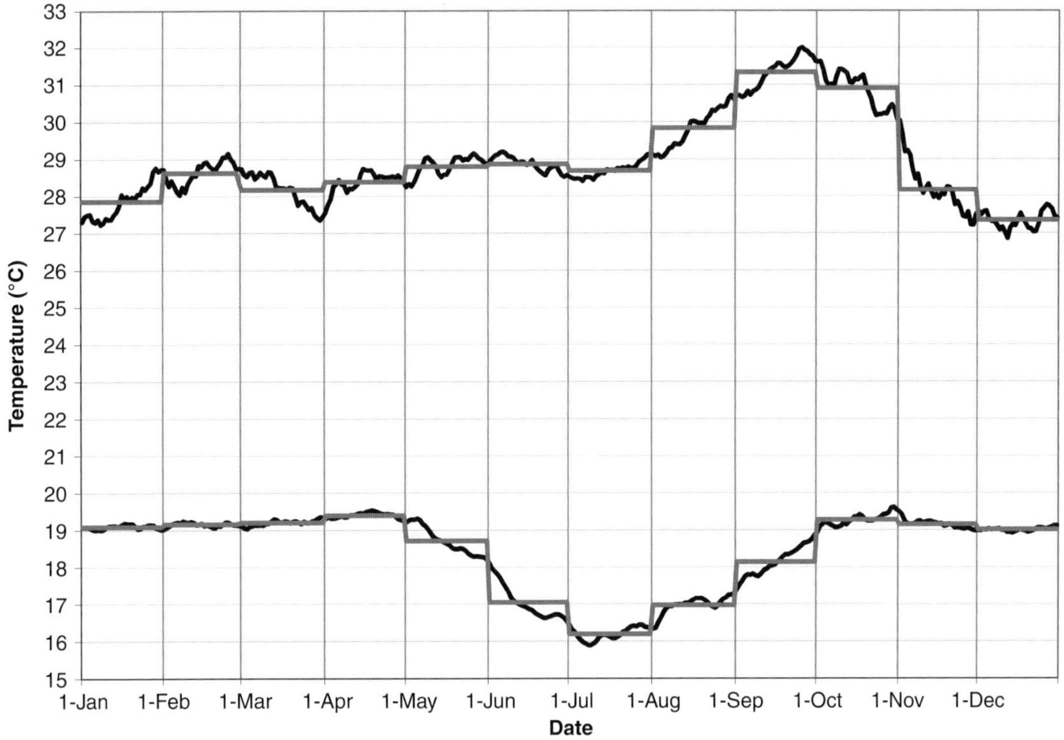

Figure 12.6 Annual fluctuations in minimum and maximum temperatures. Black and gray lines indicate 5-day running mean and monthly mean, respectively.

monthly maximum temperature above 30°C. The mean minimum temperature is 18.4°C (N = 10 476, range 13.2–23.0°C). The coldest months are June–September, with a mean monthly minimum temperature below 17.0°C. The diurnal range of the mean temperature is large, more than 10°C, from May to October. This period corresponds with the dry season. The largest diurnal range is in

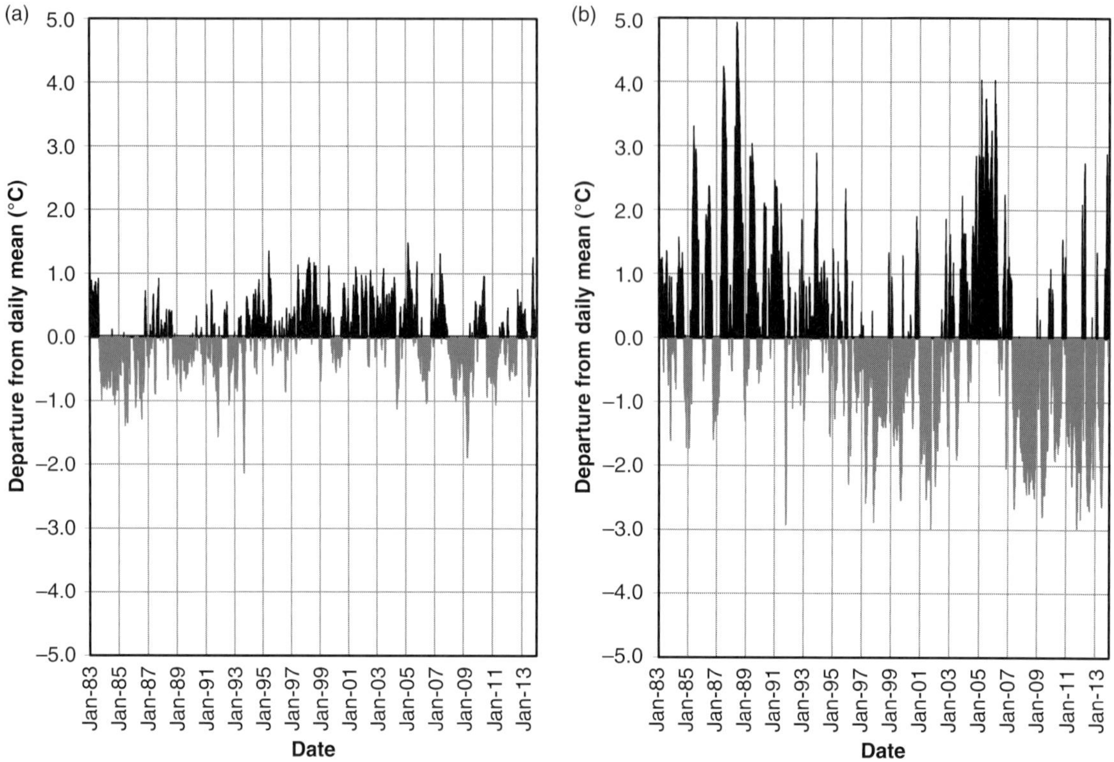

Figure 12.7 Time-series anomalies (31-day running mean) in daily temperature. (a) Minimum temperature anomaly relative to the normal mean. (b) Maximum temperature anomaly relative to the normal mean. Scales are the same.

September (13.2°C) and the lowest is in December (8.3°C).

Mahale's mean minimum and maximum temperatures are 1–10°C higher than those of most chimpanzees study sites, for which long-term records (over 10 years) have recently been published (Budongo, Uganda: 16.4 and 27.7°C [Babweteera *et al.*, 2012], Kibale, Uganda: 14.9 and 20.2°C [Chapman *et al.*, 2005], Bwindi, Uganda: 13.4–14.4 and 18.0–19.1°C [Kasangaki *et al.*, 2012], Kahuzi-Biega, DRC: 13.8 and 26.5°C [Yamagiwa *et al.*, 2012]), but the mean temperatures are lower than one unusually dry and hot chimpanzee habitat, Toro-Semliki, Uganda (19.1 and 33.7°C [Patrick *et al.*, 2012]).

Time-series anomalies in the daily temperature were smoothed by applying a 31-day running mean, and are shown in Figure 12.7 (comparable with Figure 12.4 in Itoh *et al.*, 2012). Variations in the daily minimum temperature anomalies are relatively stable, rarely exceeding ±1°C (Figure 12.7a). Colder years at the end of the record in the previous analysis (2006 and 2008) continue. However, viewing the trend from 2010, there seems to be annual-scale variability, which will require further monitoring to confirm. Looking over the record, the cooler periods (1984–93 and 2006–13) and the warmer period (in between) may be due to decadal-scale variability.

Variation in the daily maximum temperature anomalies from the mean values is far greater than that in the minimum anomalies (Figure 12.7b). This may reflect a dependence on the cloud cover variability, which is a future task for analysis (Itoh *et al.*, 2012). There has been a tendency to alternate between positive and negative anomaly periods for several consecutive years. In addition, a great amplitude of the maximum temperature variability's (8°C, from +5 to −3°C) departure from the mean has also been confirmed in both the 20-year and 31-year datasets.

The small amplitude of the daily minimum temperature variability and the great amplitude of the maximum temperature variability observed for Mahale suggest a warming trend (Itoh *et al.*, 2012). The latter exceeds the most severe projections for temperature change in the global climate warming models and needs further testing for verification (Itoh *et al.*, 2012).

12.5 Future interests

The results provided in this chapter come from data from the Kasoje Forest, where the weather station is located in the middle of the vast Mahale area. Experience tells us that rainfall patterns could differ in the far northern and far southern areas, and also in areas with higher altitudes. Average temperatures also differ with altitude. To grasp the overall climatological trends in the Mahale area and to verify long-term changes in climate, continued monitoring and monitoring in the higher altitude areas is essential. Since our results suggest some influences from climate change on the vegetation in Mahale (Chapter 13), and some direct and indirect (due to changes in vegetation) influences on plant phenology are also expected (Chapter 14), more caution may be needed with respect to any conclusions about animal responses in the long term. In addition, an assessment of the effects of changes in vegetation and other related changes in climate is required.

References

Babweteera, F., Sheil, D., Reynolds, V., *et al.* (2012). Environmental and anthropogenic changes in and around Budongo Forest Reserve. In *The Ecological Impact of Long-Term Changes in Africa's Rift Valley*, ed. A. J. Plumptre. New York: Nova Science, pp. 31–53.

Chapman, C. A., Chapman, L. J., Struhsaker, T. T., *et al.* (2005). A long-term evaluation of fruiting phenology: importance of climate change. *Journal of Tropical Ecology*, **21**, 31–45.

Itoh, N., Nakamura, M., Ihobe, H., *et al.* (2012). Long-term changes in the social and natural environments surrounding the chimpanzees of the Mahale Mountains National Park. In *The Ecological Impact of Long-Term Changes in Africa's Rift Valley*, ed. A. J. Plumptre. New York: Nova Science, pp.249–77.

Kasangaki, A., Bitariho, R., Shaw, P., Robbins, M., and McNeilage, A. (2012). Long-term ecological and socio-economic changes in and around Bwindi Impenetrable National Park, south-western Uganda. In *The Ecological Impact of Long-Term Changes in Africa's Rift Valley*, ed. A. J. Plumptre. New York: Nova Science, pp. 117–39.

Patrick, R., Patrick, D., and Hunt, K. D. (2012). Long-term changes at Toro-Semliki Wildlife Reserve. In *The Ecological Impact of Long-Term Changes in Africa's Rift Valley*, ed. A. J. Plumptre. New York: Nova Science, pp. 55–71.

Plumptre, A. J. (2012). Long term changes in Africa's western rift valley: synthesis of main findings. In *The Ecological Impact of Long-Term Changes in Africa's Rift Valley*, ed. A. J. Plumptre. New York: Nova Science, pp. 279–93.

Seimon, A. and Picton Phillips, G. (2010). *Climatological Assessment of the Albertine Rift for Conservation Applications.* WCS Albertine Rift Climate Assessment Whitepaper Report No. 1.

Seimon, A. and Picton Phillips, G. (2012). Regional climatology of the Albertine Rift. In *The Ecological Impact of Long-Term Changes in Africa's Rift Valley*, ed. A. J. Plumptre. New York: Nova Science, pp. 9–30.

Yamagiwa, J., Basabose, A. K., Kahekwa, J. *et al.* (2012). Long-term changes in habitats and ecology of African apes in Kahuzi-Biega National Park, Democratic Republic of Congo. In *The Ecological Impact of Long-Term Changes in Africa's Rift Valley*, ed. A. J. Plumptre. New York: Nova Science, pp. 203–25.

Zamma, K. (2007). [Landslide at Mahale.] *Mahale Chimpun*, **9**. In Japanese.

13 Mahale flora: its historical background and long-term changes

NORIKO ITOH AND MICHIO NAKAMURA

13.1 Introduction

The number of studies on tropical forests has been increasing for some time. While the interpretation of available data is often contested, there is a consensus on the need for more information (Sheil *et al.*, 2000). Our understanding of tropical forest succession is still not well known (Richards, 1952). Global climate change is an additional factor that makes the task even more challenging. These climatic changes make the study and long-term monitoring of tropical forests even more urgent. In this chapter, we present general information on the floral and vegetational diversity at Mahale and the historical background of the forest. Some recent changes in the lowland forest are also discussed.

13.2 Mahale floral diversity

The level of floral diversity in the modern East African forest is much lower than in South America or Malaysia. However, the East African forest was thought to be richer during the Miocene (Lind and Morrison, 1974). As a contributor to Lind and Morrison (1974), A. C. Hamilton assumed that this change was caused by the past extreme climatic vicissitudes during either the Pleistocene or Pliocene (Lind and Morrison, 1974).

Apart from such mass changes, floral diversity *in situ* is modified by climate, topography, and the history of the region, including activities of humans and animals. Assembling and analysing floral lists is the first and most important step toward understanding and predicting the future dynamics of the ecosystem *in situ*. Moreover, this basic work is indispensable at Mahale because of its location and unique topography (Chapter 2), the continuing succession (see 13.5), and the climate changes likely to take place (Chapter 12).

Over 600 species of plants from Mahale have already been listed (Nishida and Uehara, 1981). However, the overall number of vascular plant species in the Mahale national park is predicted to exceed 2000; it is likely to include numerous rare and endemic plants (Lovett, 1994, cited in Turner, 2006). As the revised list (Itoh, in prep.; Appendix I) comprises only 882 species and 26 additional suspected species, it is still far from complete.

13.3 Historical background of the forest

On the basis of fossil records, the African forest may be dated back to the late Cretaceous and early Tertiary period when Gondwanaland began to drift apart from Laurasia (65 myr BP) (Lovett, 1993). When the major rifts at the end of Miocene (*c.* 10 myr BP) divided east–west continuum, Mahale acted as a transactional zone for western (Guineo-Congolian) plants migrating to Eastern Africa (Lovett, 1993).

During the Pleistocene, climatic fluctuations are assumed to have caused major extinctions and reduced the area of tropical forest in Africa (Wasser and Lovett, 1993). The composition of Mahale fauna suggests that the area has also experienced several

Mahale Chimpanzees: 50 Years of Research, ed. M. Nakamura, K. Hosaka, N. Itoh and K. Zamma. Published by Cambridge University Press.

expansions of tropical rain forest from the west during the Pleistocene (Anonymous, 1980). The effects of these periods of expansion remain visible as the elements common with the western forest. The major vegetation type in Mahale, "Miombo woodland," shares some features with woodlands in Zambia, Angola, and Western Tanzania. Many types of the highly diversified vegetation characteristic to Mahale (Anonymous, 1980; Nishida and Uehara, 1981; Nishida, 1990) originate from Guineo-Congolian forest (reviewed by Turner, 2006).

Apart from geological and climatological factors, human activities have also profoundly modified the vegetation of East Africa. In Lind and Morrison (1974: 205), A. C. Hamilton stated, "[I]t is often extremely difficult to envisage what the natural vegetation would be like." Mahale and especially the Kasoje area have been affected by various human activities for a long time. After the establishment of the national park (Chapter 2; Part X), such activities have been restricted, and the forest is considered to be in the succession stage. However, the process has not been closely studied at Mahale or many other tropical forests (Richards, 1952). Some studies have tried to predict the succession path of the Kasoje forest (see Section 13.5). However, succession processes are complex and greatly vary depending on the macro- and micro-environments and histories. Long-term quantitative observation is indispensable to fully elucidate these phenomena (Richards, 1952).

Lilian Pintea analysed the satellite images of Mahale between 2001 and 2007 and found a substantial forest loss along and even inside the northern and southern peripheries of the park (Itoh *et al.*, 2012; see also Figure 2.3, Chapter 2). The homogeneous pattern of deforestation in the north suggests that deforestation in this area (especially north of the beacons, which are locally recognized as the boundary) is caused by conversion of the forest to farmlands. Deforestation along and inside the southern boundary was more heterogeneous than in the north, indicating other activities (i.e. firewood and timber collection or fire spreading from the adjacent village lands).

13.4 Vegetation

Vegetation of Mahale is divided into three major types: woodland, montane forest, and Kasoje forest (lowland forest) (Anonymous, 1980).

The woodland, generally referred to as the Miombo woodland, covers three-fourths of the whole park (Anonymous, 1980). This open-land vegetation area is covered with dry woodland or lowland bamboo bush, sometimes with small patches of forest. There are several sub types of woodland (some types may be called "savanna") whose classification differs in various reports (Anonymous, 1980; Nishida and Uehara, 1981; Nishida, 1990; Turner, 2006). They are *Brachystegia* woodland, the dominant type at Mahale, *Combretum* woodland, *Uapaca* woodland, *Diplorhynchus–Pterocarpus* woodland, *Acacia* woodland, and *Oxyanthera* bamboo woodland.

Highland vegetation above 1500 m a.s.l. might be one of the least studied areas (Turner, 2006). Besides the forest, there are two distinct vegetation belts in the area (Anonymous, 1980). One is *Yushania alpina* bamboo bush, covering ravines near the mountain ridge. There is some grassland on the long main ridge of Mahale, where many wild flowers can be seen toward the end of the wet season and at the beginning of the dry season.

The Kasoje forest is the most complex vegetation type with various small vegetation patches distributed in a mosaic pattern. This lowland forest has developed in the western side of the Mahale Mountains chain. The altitude of the area ranges from 780 m (lakeshore) to around 1300 m a.s.l. This "forest island" (Nishida, 1990) in the middle of the Miombo woodland contains a number of moist-forest species as the topography favors high humidity (Anonymous, 1980; Turner, 2006; Itoh *et al.*, 2012).

Linda A. Turner performed a quantitative analysis of the vegetation in this region in the mid-1990s. She classified the Kasoje forest into five types: lowland forest (61.5%) comprising *Erythrophleum* forest and *Xylopia–Pycnanthus* forest, colonizing forest (11.5%) dominated by species like *Senna spectabilis* and *Croton sylvaticus*, woodland (*Combretum* and other species, 19.2%), Miombo woodland (*Brachystegia bussei*, 3.8%), and swamp (3.8%) (Turner, 2006). Itoh also conducted a quantitative survey in the late 1990s and obtained similar results for the lowland forest (Itoh, 2002).

Forest patches dominated by *Senna* are now limited thanks to the prolonged efforts to eradicate them. This exotic tree was introduced to the Kasoje area in 1967, naturalized, and by the mid-1990s, became a threat to the regeneration of local plant communities (Turner, 1996). Thus, Toshisada Nishida initiated the eradication program, which was later carried out by TANAPA on a larger scale. It was funded by ANC Production and MWCS (Nishida, 1996; Turner, 1996; Lukosi, 1997; Wakibara, 1998; Wakibara and Mnaya, 2002) (for details of institutions, see Chapter 2).

13.5 Recent changes in the "lowland forest" in the Kasoje area

Many ecologists have reported drastic worldwide changes in tropical forests (e.g. increase in tree mortality rate, turnover rate, and liana density) since the twentieth century. These tendencies may be related to the global climate change and/or other anthropogenic changes. In this section, we summarize the results of the preliminary analyses to show the changes in the lowland forest community at Mahale over approximately 16 years.

In our study, we set up a transect in the lowland forest, from the mouth of the Mpila Valley to Lubulungu River. An observational trail (Chapter 2) was used for convenience (see Williamson, 1988). Woody plants within 5 m from the center of the transect on both sides were monitored. The transect was divided into 10 m × 10 m plots, which were used as plant location units.

The initial length of the transect was 8.38 km. However, because of flooding, turnover of trees and lianas, and other micro-disturbances, the transect route gradually changed, turning and swerving. In the end, its length was 8.69 km. We found that 206 plots had been disturbed to various degrees since the initial survey, as judged by the location of trees in each plot (we did not take the location of lianas into consideration as they are more mobile than trees). The details of the inventories are summarized in Table 13.1.

For analyses, the whole data were divided into the following datasets:

A. Data restricted to the initial transect/plots.
B. Data selected from plots that did not change in datasets A and C.
C. Data from all plots in the second inventory.

All lianas (or woody vines) with diameter at breast height (DBH) ≥5 cm and trees with DBH ≥10 cm within the transect were included in the study. *S. spectabilis* was excluded because this species was being eradicated (see Section 13.4). As it came to our attention that the palms, *Elaeis guineensis* and *Phoenix reclinata*, may also be eradicated by the park authorities, alternatives were not checked for these species (our impression was that there were few individuals left).

Trees were divided into five life-form categories at the species level: T (species excluding all other four categories), BT (trees that tend to grow by branching stems under breast height or from the ground), SH (shrubs), SG (stranglers), and P (palms). Note that a life-form category of a given species defined here may not be universal. For example, *Myrianthus arboreus* grows as a large tree in other areas, but at Mahale, it is usually a small tree with numerous small stems (Figure 13.1). Lianas were categorized at the species level, although some may grow as shrubs under some conditions.

Identification was first made in the field by local assistants and Itoh, using local names (see Table 13.1) and verified by at least two staff members. Local and Latin names listed in Nishida and Uehara (1981) were matched, and the features were re-checked from the literature. Some plants were identified by specialists at a herbarium.

The survival status of all the individuals registered in the initial survey was marked in the final inventory as follows: alive, dead, absent (those that could not be found because of death or a change in the transect routes), fallen (those that had fallen down but were

Table 13.1 **Study contents**

	Initial inventory	Final inventory		
Observation year	1996–1998	2006, 2007	2011, 2012	
Censused individuals	Original	Original	Original	Recruited
DBH measure[a] – *Life-form* – *Tree* – T	✓	✓	▲	✓
DBH measure[a] – *Life-form* – *Tree* – BT	✓	✓	▲	✓
DBH measure[a] – *Life-form* – *Tree* – SH	✓	✓	▲	✓
DBH measure[a] – *Life-form* – *Tree* – SG	✓	✓	*	✓
DBH measure[a] – *Life-form* – *Tree* – P	✓	✓	*	excluded
DBH measure[a] – *Life-form* – *Liana*	✓	▲	*	*
Location (mapping)	✓	✓	✓	
ID tagged	✓	✓	✓	
Survival status	–	✓	✓	
Transect length	8.38 km	8.69 km (overlap: 6.63 km)		
Transect width	10 m	10 m		
Collaborators	HBK, KA, MBK, MN, TS	HBK, MBK, RSK, MR, SK		

✓ fully checked, ▲ partially measured, * checked threshold only.

[a] Threshold of diameter at breast height (DBH) size: tree ≥10 cm, liana ≥5 cm.

Tree: T = species excluding BT, SH, SG, and P; BT = trees that tend to grow by branching stems under breast height or from the ground; SH = shrubs; SG = stranglers; P = palms.
Collaborators: HBK = Mr. Hamisi Bunengwa Kasagula; MBK = Mr. Moshi Bunengwa Kasagula; KA = Mr. Kabumbe Athumani; MN = Dr. Michio Nakamura; TS = Dr. Tetsuya Sakamaki; RSK = Mr. Rashidi S. Kitopeni; MR = Mr. Mwami Rashidi; SK = Mr. Shaka Kabugonga.
Plants were identified by HBK, KA, MBK, and Itoh.

still alive), and shrunk (to DBH under the threshold size). Phenological data (Chapter 14) were also used whenever possible to decide whether the absent individuals had died.

As the sizes of life-form categories other than T did not change much, the analysis of the sizes was done for T only. All DBH data were calculated in units of 5 cm. For individuals with multiple stems below the breast height, DBH sizes of all stems were added.

13.5.1 Forest size

Basal area (BA) of trees (calculated using DBH) is one of the measures for forest maturity (Williamson, 1988). Mean BA in the two lowland forest types was 32.5 m^2/ha in Turner's (2006) random transect, which is close to the pantropical mean for mature forest (32 m^2/ha). In our study, the value calculated for T category in the initial survey was similar (31.6 m^2/ha, $N = 2049$, dataset A). In the restricted dataset B, the original mean BA was slightly smaller (30.3 m^2/ha, $N = 1504$).

Final BA was calculated using only alive and newly recruited individuals (i.e. final individuals). DBH size for 11 individuals could not be re-measured because of their dangerous location or obstruction by many climbers and shrubs. For these individuals, the values from the initial survey were used. The final BA value (dataset B) was 33.4 m^2/ha ($N = 1470$).

To standardize the time lag (5–6 years) of measurements for the newly recruited individuals and original trees, the values of original individuals were

Figure 13.1 *Myrianthus* with numerous stems. ©*N. Itoh.*

weighted using growth rate. The growth rate was obtained from the rates calculated using two DBH measures (sum of DBHs for all stems of the individuals and the maximum DBH size for each individual) for original trees. These individuals were selected from every 50 m plot and measured both in 2005–6 and 2011–12 ($N = 269$). The corrected approximate BA was 33.1–33.7 m²/ha.

Overall, BA/ha increased slightly, although the density in dataset B declined from 226.9 to 221.7 individuals/ha. This result may indicate an increase in tree size; to evaluate it, the original and final DBH sizes of T-category members in dataset B were compared.

Figure 13.2 displays $\log_{10}$ of number of individuals in each DBH size class in the original (left) and final (right) inventory. There were no differences between the distribution patterns in the two inventories. However, the proportion of large individuals (≥100 cm) increased from 2% to 3%, of medium-sized individuals (30–100 cm) decreased from 50% to 45%, and of smaller individuals (<30 cm) increased from 50% to 55%. The maximum DBH size observed in the initial survey was 200 cm, and in the final survey was

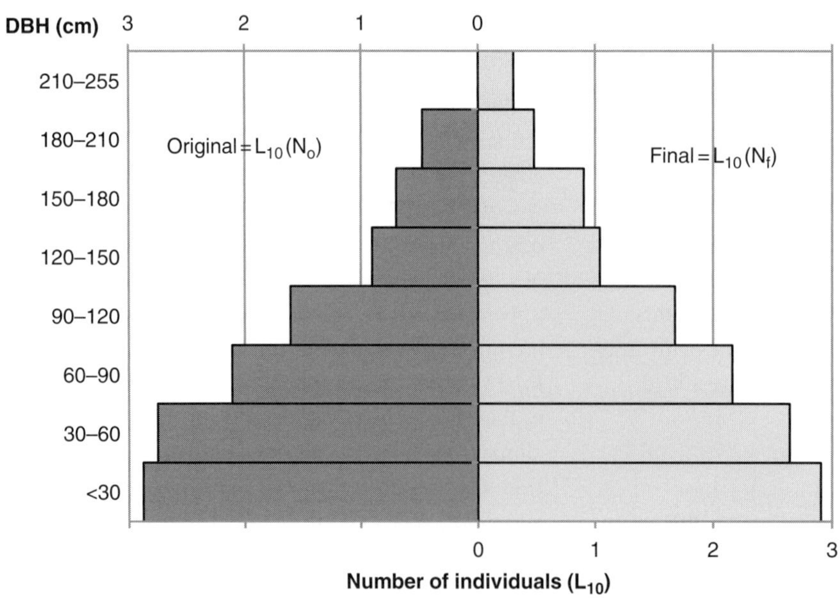

Figure 13.2 Distribution of individuals (T) in each DBH class.

255 cm. There was only one individual with DBH larger than 200 cm in the initial survey, but in the final survey, there are three.

Many disturbed parts of the transect were excluded from dataset B, which could lead to overestimates. However, the BA/ha of the final individuals in dataset C was also slightly larger (31.7 m^2/ha) than the original BA/ha (datasets A and B).

In summary, the forest size slightly increased, and the increase seems to be due to the rise in the DBH and number of large trees (≥100 cm). More detailed analysis and successive measurements are needed to understand the underlying factors of such a heterogeneous growth pattern. As the profiles of DBH classes of trees are affected by the history of forest clearing, the patterns observed in Kasoje do not strictly follow those observed in other tropical forests (Turner, 2006). This phenomenon may include the inhibition of tree growth by lianas (see 13.5.5).

13.5.2 Species records

Species-area curve (Figure 13.3) shows that the number of new species reached a plateau at around 5 km, the distance similar to that in the initial survey (Itoh, 2002). Thus, the collected samples probably captured the majority of species in the community.

Table 13.2 shows the list of all species observed (N = 4671 individuals). Species composition did not substantially differ between the original and final inventories and the absent and newly recorded species did not affect the density (≤4 individuals). The number of species also did not significantly change. The original number was 115 and 101 in dataset A and B, respectively. The final numbers were 102 in dataset B and 112 in dataset C.

The diversity (number of species/ha) of this lowland forest in Turner's random transect falls within the estimates of other African forests where the wild chimpanzees have been studied (35–41 in Gombe, Tanzania; Lopé, Gabon; Budongo and Kibale, Uganda; Kahuzi-Biega, and DRC) (Turner, 2006). However, the diversity measured in our transect was far lower than that in Turner's transect in both the original and final studies (15 species/ha).

13.5.3 Mortality dynamics of trees

Tree deaths create gaps that play an important role in the process of regeneration and maintenance of

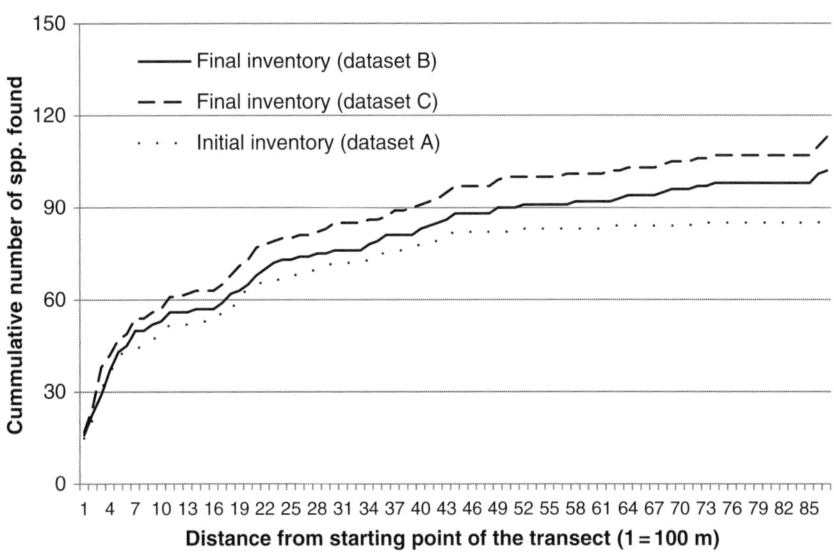

Figure 13.3 The number of species as a function of distance from the starting point of the transect.

Table 13.2 **Plant species observed in the transect in 16 years**

					Original		Final	
Latin names	Abbreviation	MLID	Local names	LF	Dataset A	Dataset B	Dataset B	Dataset C
Albizia adianthifolia	Albadi	470	Mutanga type A	T	✓	✓		✓
Albizia glaberrima	Albgla	1427	Kafunampasa type A	T	✓	✓	✓	✓
Annona senegalensis	Annsen	39	Bhufila	T	✓	✓	✓	✓
Anthocleista schweinfurthii	Antsch	411	Mughongoghongo	T	✓	✓	✓	✓
Antidesma venosum	Antven	299	Mutimpu	T	✓	✓	✓	✓
Aporrhiza paniculata	Apopan	699	Muhoko type A	T	✓	✓	✓	✓
Artabotrys monteiroae	Artmon	40	Lujongololo bhusie	LS	✓	✓	✓	✓
Azanza garckeana	Azagar	421	Mutobho	T	✓	✓	✓	✓
Baissea major	Baimaj	52	Ikonjogholo bhuhele	LS	✓	✓	✓	✓
Baphia capparidifolia	Bapcap	553	Ijubhila	LS	✓	✓	✓	✓
Bauhinia petersiana	Baupet	125	Iketya	LS	✓	✓	✓	✓
Bauhinia thonningii	Bautho	126	Musakansaka	T	✓			✓
Bersama abyssinica	Beraby	454	Sighonfi type B	T	✓	✓	✓	✓
Blighia unijugata	Bliuni	701	Kasindabhilangulubhe no.1	T	✓	✓	✓	✓
Bridelia atroviridis	Briatr	300	Kamilaninga	BT	✓	✓	✓	✓
Bridelia micrantha	Brimic	302	Munyonyi type A	T	✓	✓		
Celtis africana	Celafr	748	Kahefu	T	✓	✓	✓	✓
Chaetachme aristata	Chaari	749	Lisumata	T	✓	✓	✓	✓
Chassalia cristata	Chacri	644	Lisisusanandi	LS	✓		✓	✓
Chionanthus niloticus	Chinil	532	Kasindabhilangulubhe no.2	T	✓	✓	✓	✓
Cissus petiolata	Cispet	778	Lighaghaja Lamba	LS	✓	✓	✓	✓
Cissus sp.	CissSp	1410	zunk6300	LS				✓

Citrus limon	Citlim	690	Limauwe	T	✓	✓	✓	✓
Clausena anisata	Claani	691	Sibhingabhasimu type A	T	✓	✓	✓	✓
Combretum collinum	Comcol	170	Sifunfwe	T	✓	✓	✓	✓
Combretum molle	Commol	172	Mulama	T	✓	✓	✓	✓
Cordia africana	Corafr	114	Mukibhu ghwakakamba	T	✓	✓	✓	✓
Cordia millenii	Cormil	115	Mukibhu ghwesimbwa	T	✓	✓	✓	✓
Croton sylvaticus	Crosyl	956	Ibhonobhono	T	✓	✓	✓	✓
Culcasia falcifolia	Culfal	73	Nandi	LS	✓	✓	✓	✓
Dalbergia malangensis	Dalmal	566	Itambuka	LS	✓	✓	✓	✓
Deinbollia fulvotomentella	Deiful	703	Mpongolwela	SH			✓	✓
Diplorhynchus condylocarpon	Dipcon	54	Musongati	T	✓			
Dombeya kirkii	Domkir	728	Katobho	T	✓	✓	✓	✓
Dombeya sp.	Domsp2	1482	Katobhotobho		✓	✓	✓	✓
Dracaena mannii	Draman	22	Bhulonje	BT	✓	✓	✓	✓
Elaeis guineensis	Elagui	544	Sighasi	P	✓	✓	✓	✓
Erythrina abyssinica	Eryaby	574	Muko	T	✓	✓	✓	✓
Faidherbia albida	Faialb	1278	Ikese	T			✓	✓
Ficus artocarpoides	Ficart	488	Mululu	SG	✓			✓
Ficus asperifolia	Ficasp	489	Kankolonkombe	LS	✓	✓	✓	✓
Ficus cyathistipula	Ficcya	492	Kajimonsole type D	SG	✓	✓		
Ficus exasperata	Ficexa	493	Lwagho	T	✓	✓	✓	✓
Ficus sur	Ficsur	500	Ikubhila	T	✓	✓	✓	✓
Ficus sycomorus	Ficsyc	501	Ikuku type B	T	✓			
Ficus thonningii	Fictho	502	Kajimonsole type A	SG	✓	✓	✓	✓
Ficus trichopoda	Fictri	994	Muhololo	T	✓	✓	✓	✓
Ficus vallis-choudae	Ficval	504	Ihambwa	T	✓	✓	✓	✓
Ficus sp.	Ficsp2	1419	Kajimonsole type G	SG			✓	✓
Ficus sp.	Ficsp1	1406	Kajimonsole type undet.	SG	✓	✓		

Table 13.2 (**cont.**)

Latin names	Abbreviation	MLID	Local names	LF	Original Dataset A	Original Dataset B	Final Dataset B	Final Dataset C
Funtumia africana	Funafr	55	Mutwentwe no.2	T	✓	✓	✓	✓
Garcinia huillensis	Garhui	383	Kasolyo	T	✓	✓	✓	✓
Grewia flavescens	Grefla	742	Bhunkukuma	LS	✓	✓	✓	✓
Grewia mollis	Gremol	744	Mukole	T	✓			✓
Harungana madagascariensis	Harmad	385	Mutunu	T	✓	✓	✓	✓
Illigera pentaphylla	Illpen	387	Ilyungulyungu type C	LS			✓	✓
Jasminum dichotomum	Jasdic	533	Kapotwabhakulu	LS	✓	✓	✓	✓
Jatropha curcas	Jatcur	312	Libhutubhutu	T	✓	✓	✓	✓
Keetia gueinzii	Keegue	649	Bhuntafwanengwa bhosa	LS	✓	✓	✓	✓
Keetia venosa	Keeven	650	Bhuntafwanengwa bhusie	LS	✓	✓	✓	✓
Landolphia owariensis	Lanowa	57	Mpila	LS	✓	✓	✓	✓
Lannea schimperi	Lansch	36	Kabhumbu	T	✓	✓		
Lecaniodiscus fraxinifolius	Lecfra	706	Kafulujeghe	T	✓	✓	✓	✓
Markhamia lutea	Marlut	106	Kabhulampako type A	T	✓	✓	✓	✓
Memecylon sp.	MemeSp	443	Kamungwe	T	✓	✓	✓	✓
Milicia excelsa	Milexc	508	Mukamba	T	✓	✓	✓	✓
Mimusops bagshawei	Mimbag	718	Mulonje type C	T	✓	✓	✓	✓
Mussaenda arcuata	Musarc	654	Mumpindu	LS	✓	✓		
Myrianthus arboreus	Myrarb	509	Isakama type A	BT	✓	✓	✓	✓
Ochna mossambicensis	Ochmoss	526	Kabhukobhuko type B	T	✓	✓	✓	✓
Oncoba spinosa	Oncspi	337	Kaposo	BT	✓	✓	✓	✓
Oxyanthus speciosus	Oxyspe	658	Lintinfu	T	✓	✓	✓	✓
Pandanus sp.	Pansp	547	Ikofu	T	✓			✓
Parinari curatellifolia	Parcur	629	Mubhula type A	T	✓	✓	✓	✓

Parkia filicoidea	Parfil	484	Iseha	T	✓			
Pericopsis angolensis	Perang	588	Mubhanga	T	✓	✓	✓	✓
Phoenix reclinata	Phorec	546	Lusanda	P	✓	✓	✓	✓
Phytolacca dodecandra	Phydod	608	Ntembotelemiya	LS	✓	✓	✓	✓
Premna sp.	Presp	768	Lulindi type A	T	✓	✓	✓	✓
Pseudospondias microcarpa	Psemic	37	Bhuhono	T	✓	✓	✓	✓
Psidium guajava	Psigua	517	Mpela	T	✓	✓	✓	✓
Psychotria capensis	Psycap	665	Lyakusona	SH	✓	✓	✓	✓
Psydrax parviflora	Psypar	1420	Kamwibhi	T	✓	✓	✓	✓
Pterocarpus tinctorius	Ptetin	1488	Mwenje	T	✓			✓
Pterygota macrocarpa	Ptemac	732	Mukungu	T	✓	✓	✓	✓
Pycnanthus angolensis	Pycang	513	Lulumasya	T	✓	✓	✓	✓
Rauvolfia caffra	Raucaf	61	Mutwentwe no.1	T	✓			✓
Rothmannia longiflora	Rotlon	678	Nyundon ghosi	T	✓	✓	✓	✓
Rytigynia acuminatissima	Rytacu	680	Lisisusalintinfu	T	✓			
Saba comorensis	Sabcom	62	Ilombo	LS	✓	✓	✓	✓
Salacia sp.	Salsp	165	Ighandamakungu	LS	✓	✓	✓	✓
Sapindus saponaria	Sapsap	708	Bhusolo	T	✓		✓	✓
Schrebera alata	Schala	535	Mubhasa	T	✓	✓	✓	✓
Schrebera trichoclada	Schtri	536	Mukutuka	T	✓	✓	✓	✓
Scolopia sp.	Scosp	339	Musilantumbalo	T	✓	✓	✓	✓
Spathodea campanulata	Spacam	108	Lipulu	T	✓	✓	✓	✓
Sterculia tragacantha	Stetra	735	Kakubhabholo	T	✓	✓	✓	✓
Stereospermum kunthianum	Stekun	110	Mutelele	T	✓	✓	✓	✓
Strychnos cocculoides	Strcoc	415	Libhwaje	T			✓	✓

Table 13.2 (**cont.**)

					Original		Final	
Latin names	Abbreviation	MLID	Local names	LF	Dataset A	Dataset B	Dataset B	Dataset C
Synsepalum brevipes	Synbre	722	Mulyansekesi	T	✓	✓	✓	✓
Syzygium guineense	Syzgui	520	Musabhasabha type A	T	✓	✓	✓	✓
Tabernaemontana pachysiphon	Tabpac	65	Ikolyoko no.2	BT	✓	✓	✓	✓
Tarenna pavettoides	Tarpav	684	Mukombelonda type B	T	✓	✓	✓	✓
Terminalia kaiseriana	Terkai	176	Kaselenje	T	✓	✓	✓	✓
Tetracera potatoria	Tetpot	280	Katwa	LS	✓	✓	✓	✓
Toddalia asiatica	Todasi	693	Katapansima	LS	✓	✓	✓	✓
Trema orientalis	Treori	750	Muhehu	T	✓	✓	✓	✓
Trichilia sp.	Trisp	451	Kamoko	T	✓	✓	✓	✓
Uapaca nitida	Uapnit	329	Lulobhe	T	✓	✓	✓	✓
Uvaria angolensis	Uvaang	45	Lujongololo bhosa	LS	✓	✓	✓	✓
Vepris nobilis	Vepnob	1355	Mupemba	T			✓	✓
Vernonia amygdalina	Veramy	231	Mujonso	SH	✓	✓		
Vitex doniana	Vitdon	769	Bhufulu	T	✓			✓
Vitex fischeri	Vitfis	771	Lisisusakafulu	T	✓	✓	✓	✓
Voacanga africana	Voaafr	67	Ikolyoko no.1	BT	✓	✓	✓	✓
Xylopia longipetala	Xyllon	1287	Kahwibhili	T	✓	✓	✓	✓
Zanha golungensis	Zangol	709	Sitalya	T	✓	✓	✓	✓
Ziziphus abyssinica	Zizaby	626	Kaghobhole	T	✓	✓	✓	✓
Species not identified but belongs to different species listed above								
Sp. Kakambakamba	SpKaka	1415	Kakambakamba	T			✓	✓
Sp. Ng'unga	SpNgun	1418	Ng'unga	T	✓			✓

Sp. Nr1121	Sp1121	1411	zunkNr1121	T	✓	✓		
Sp. Nr1271	Sp1271	1412	zunkNr1271	T	✓	✓		
Sp. Nr6202	Sp6202	1409	zunk6202	LS			✓	✓
Sp. Siponda no.1	SpSipo	1407	Siponda no.1	T	✓	✓	✓	✓
Sp. Siponda no.2	SpSip2	1408	Siponda no.2	LS				✓
Species that possibly belongs to species listed above								
Keetia sp.	Keetia	1417	Bhuntafwanengwa type undet.	LS			✓	✓
Sp. Ikonjogholo	SpIkon	1414	Ikonjogholo type undet.	LS			✓	✓
Sp. Lujongololo type undet.	SpLujo	1416	Lujongololo type undet.	LS			✓	✓

Abbreviation: Six characters from genus and specific epithet (three from each).

MLID: reference number in Appendix I.

LF: life-form.

Trees: T, BT, SH, SG, P.

Lianas: LS (some individuals in this category could be shrubs as they can change their life-form depending on the habitat).

species diversity. Gap sizes vary depending on how the tree died (i.e. gaps are small if it dies slowly limb by limb but large if a large tree falls over) (Richards, 1952).

Of the 2470 trees in dataset A, 65.0% survived, 24.4% died, and 8.4% were missing. We assume that the majority of absent individuals died long before the second inventory. The remaining 2.3% of trees are the targets that we had in mind (shrunk 0.7%, problem 0.1%, fall 1.5%). The annual rate of survival was calculated using the formula below:

$$p = 1 - (1 - N_q/N_o)^{1/t}, \qquad (13.1)$$

where N_0 and N_q are the numbers of individuals at the beginning and of individuals with survival status in question at the end of the census interval, *t*. This formula is equivalent to the formula commonly used to calculate mortality rate.

Trees shrunk to below threshold value (<10 cm DBH) at 0.05% per year. Most of these trees had several stems and lacked major stem(s) in the second inventory or might have been suffering from some type of disorder (i.e. dying).

There were two trees that were categorized as "problem" (annual rate 0.01%). One was *Dracaena* with all three stems cut off at about 3 m height. The other was *Pycnanthus*, broken at approximately 2 m height, thus losing all branches. Although these trees were still alive in 2012, no leaves were observed.

Trees fell at an annual rate of 0.09%. The majority of these were *Aporrhiza, Spathodea*, and *Pseudospondias* (36%). Although trees of the former two species were still alive in 2012, they are likely to die within several years. In contrast, *Pseudospondias* trees are long lived, and some even grow vertical stems from the mother trunk. The maximum observation period of such a mother trunk was 13 years, and the tree is still alive. A similar life cycle was also observed for other species, including *Spathodea* individuals under some condition.

13.5.4 Tree density dynamics

To see the effect of the initial tree density on the current density distribution pattern across species, mortality and recruitment rates in dataset B were compared ($N = 2338$, 86 species including unidentified ones). Palms were excluded from this analysis (the contribution of palms to the overall density was 2%). For the analysis, death and survival statuses described above were added; they were not likely to contribute to the forest structure or function, at least not in the ordinal way.

Annual mortality and recruitment rates were calculated using Equation (13.2) (Sheil *et al.*, 1995) and Equation (13.3) (Sheil *et al.*, 2000), respectively.

$$m = 1-(1-[N_o-N_s]/N_o)^{1/t} \qquad (13.2)$$

$$r = 1-(1-N_r/N_f)^{1/t} \qquad (13.3)$$

where N_0 and N_s are the number of original and surviving individuals, N_r is the number of individuals recruited, and N_f is the final number of individuals observed at the end of the census interval in years, *t*. Equations (13.2) and (13.3) are symmetrical in statistical terms (Sheil *et al.*, 2000).

Overall change in tree density

Tree stand annual mortality and recruitment rates were 2.54% and 2.46%, respectively. The overall density of trees decreased from 266.21 trees/ha to 262.75 trees/ha.

There has been a significant upward trend in the turnover (mortality and recruitment) rates for tropical forest trees with DBH ≥10 cm since the 1950s because of direct and/or indirect effects of climate change (Phillips and Gentry, 1994). Although *m* and *r* are the true annual rates (Sheil *et al.*, 1995), the formulae shown below are used in studies comparing turnover rates across sites to calculate the "exponential mortality (recruitment, turnover) coefficients" (Sheil *et al.*, 1995).

$$\lambda = \ln(N_o/N_f)/t \qquad (13.4)$$

for mortality (Sheil *et al.*, 1995) and

$$\lambda = \ln(N_f/N_s)/t \qquad (13.5)$$

for recruitment (Phillips *et al.*, 1994).

As these measures are known to be affected by the time interval between censuses (Sheil and

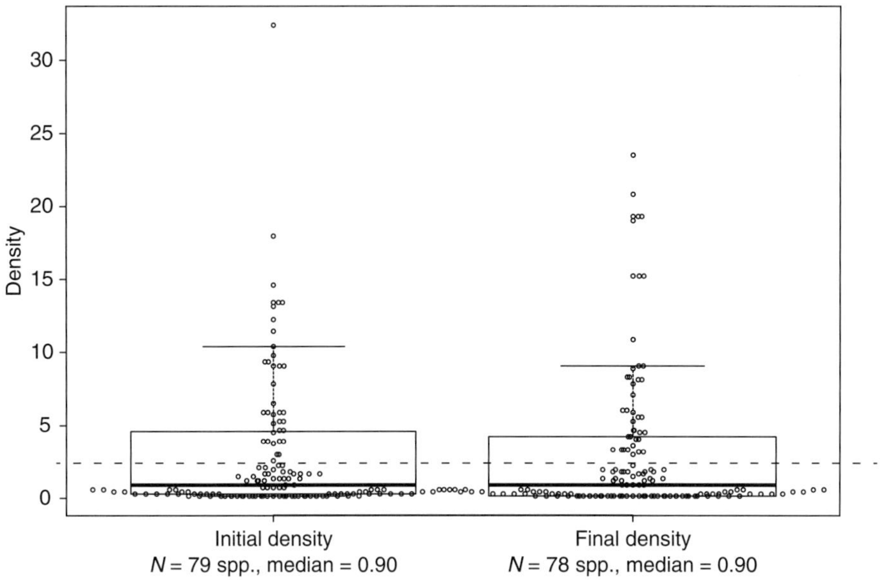

Figure 13.4 Density distribution pattern across tree species in the initial and final inventories. Open circles = species. Dashed line = 2.5 individuals/ha line.

May, 1996), the standardizing formula below (Lewis *et al.*, 2004) was used for comparisons with other tropical forests.

$$\lambda_{corr} = \lambda * t^{0.08} \quad (13.6)$$

The census-corrected turnover estimate (average of mortality and recruitment) of the Kasoje forest was 1.61%, and it was similar to the Old and New World averages (1.81 ± 0.19% and 1.80 ± 0.15%, respectively; Lewis *et al.*, 2004). The nearest estimate was that of a mature forest in Mishana, Peru, 1.68% (λ_{corr} from Phillips *et al.*, 1994). Thus, the Kasoje forest might also be influenced by climate change. However, as this forest is also under succession, we need further monitoring at the species level.

The overall density distribution patterns across species were similar in the initial and final inventories (Figure 13.4). Here, species with more than 2.5 trees/ha (above the dotted line in Figure 13.4) were considered "common" in the forest and were used for further analysis. One non-common species in the initial inventory, *Trema orientalis*, was included in the analysis as it became common in the final inventory (1.81 to 3.02). The contribution of non-common species densities to the overall tree stand was limited (only around 0.85% at most, both in initial and final inventories).

The change in density was first examined by calculating the ratio of the final number of trees to the original number of trees (PF hereafter). Although this value does not take into account the density or the contribution to the overall tree stand, it can be an indicator of the magnitude of density change per species.

The boxplot on the left-hand side of Figure 13.5 shows PF values for initially common species. Species with PF values below the lower hinge were assumed to be decreasing more rapidly than those above. The boxplot on the right-hand side shows PF values for the common species in the final inventory. The species with PF values above the upper hinge were increasing more vigorously than those below the hinge. Three other species (names shown in italics) just below the upper hinge were also considered for the next analysis, although the trend for these species was more moderate. Density changes in relation to mortality and recruitment rates of these species are examined below.

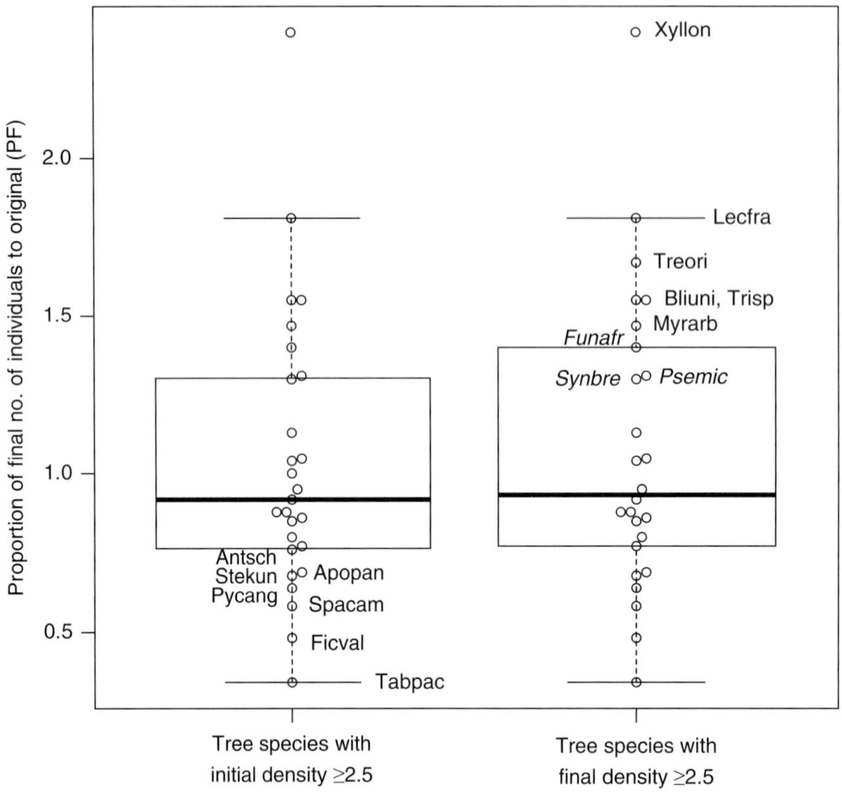

Figure 13.5 PF distribution pattern among common tree species. Abbreviated species names are listed in Table 13.2.

Decreasing-density species

Figure 13.6 shows the mortality and recruitment rates calculated for the species in Figure 13.5. Species whose density was decreasing rapidly had a mortality rate not lower than the median (0.02) and a recruitment rate not higher than the median (0.02), and the former was always higher than the latter. The trees of these species represented 31.2% (83.0 trees/ha) of all trees in the initial survey, but decreased to 18.7% in the final inventory (49.0 trees/ha).

Aporrhiza is unusual in comparison with other species because its decline (2.21% to 1.55% occupation or 15th to 21st density rank) is more due to the failure in recruitment than the high mortality rate. This tree was first recognized at Mahale in the initial inventory, and its density should have increased after a sufficient succession period. However, it is also possible that it was confused with other species by the earlier researchers.

Tabernaemontana, which showed the lowest PF, was initially the 10th most common tree species (3.51% occupation). In the final survey, it fell to 25th position (1.21%). Its mortality rate was also the highest. In 2006 and 2007, Itoh observed many trees with most or all leaves atrophied; many of these were dead in 2011 and 2012.

Ficus vallis-choudae had the second-lowest PF and the second-highest mortality rate. This species originally ranked 6th in density, representing 4.59% of tree stand, which then decreased to 2.24% (ranked 15th). This tree has a patchy distribution pattern occurring along rivers or in moist valley bottoms in Kasoje (Turner, 2006). This type of location means that many individuals are directly affected by floods caused by heavy rains. The frequency of floods might have increased in recent years (see Chapter 12). Some individuals were found lying in the riverbed. The recruitment rate for this species was the second lowest.

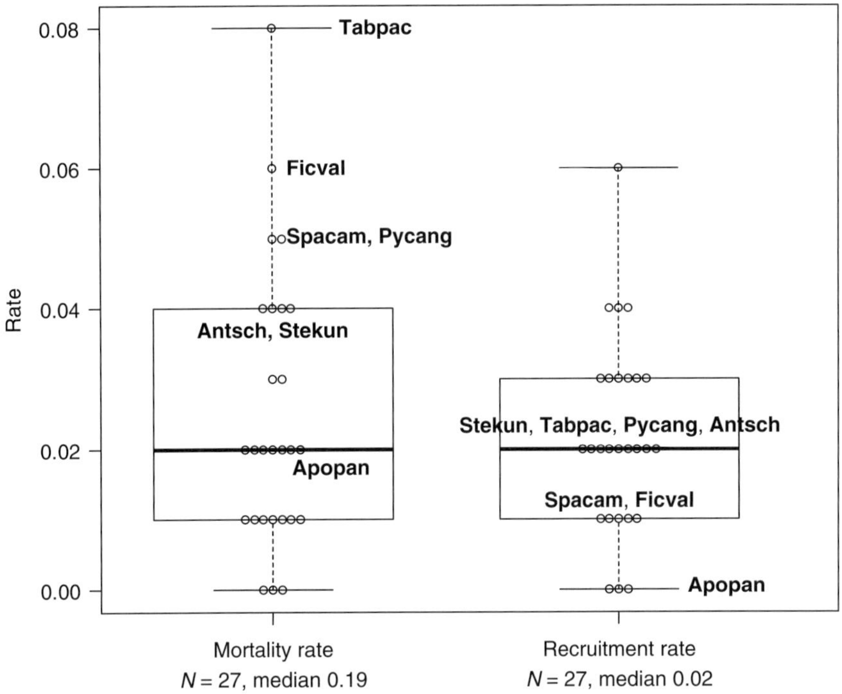

Figure 13.6 Mortality and recruitment rates among tree species with decreasing density (common species in the initial inventory). Abbreviated species names are listed in Table 13.2.

In terms of the impact of the mortality rate on the forest community, *Pycnanthus* might be the most important species. Turner (2006) called it a "hyper-dispersion" species because of its high density and continuous spatial distribution pattern in Kasoje. In our transect, this species also had an extremely high density (originally 12.18% of all trees, whereas the second-ranking species constituted only 6.74%). Although this species still has a high density, its contribution has become much more moderate (7.92%), and it fell to second place in the ranking.

The mortality rate for *Spathodea* was the same as that for *Pycnanthus*, but the recruitment rate was lower. It also used to have a high density (ranked 12th, 3.40% occupation), which decreased in the final inventory (17th, 2.01%). This species, known commonly as African tulip, occurs in colonizing and secondary forests in Kasoje (Turner, 2006). A reduction in the species density may be a natural phenomenon due to forest succession.

Stereospermum is a conspicuous tree because of its scaly bark. Originally, this tree had a high density (ranked 7th, 4.31% occupation), which decreased in the final survey (12th, 2.99%). These changes in density tend to occur in woodland and wooded grassland in Kasoje (Turner, 2006), but in our transect, they happened on the slopes and mountain ridges. In Kenya and South Africa, this species is found in both environments (Coates Palgrave *et al.*, 1993; Beentje *et al.*, 1994).

Considering its contribution rate, *Anthocleista* is the least important among the decreasing species (ranked 26th to 28th, 0.96% to 0.75% occupation). The magnitude of the decline for this species was the smallest, predicted by both PF and a small difference between the mortality and recruitment rates.

Increasing-density species

Here, we discuss the species that are increasing, those which became common in the final datasets

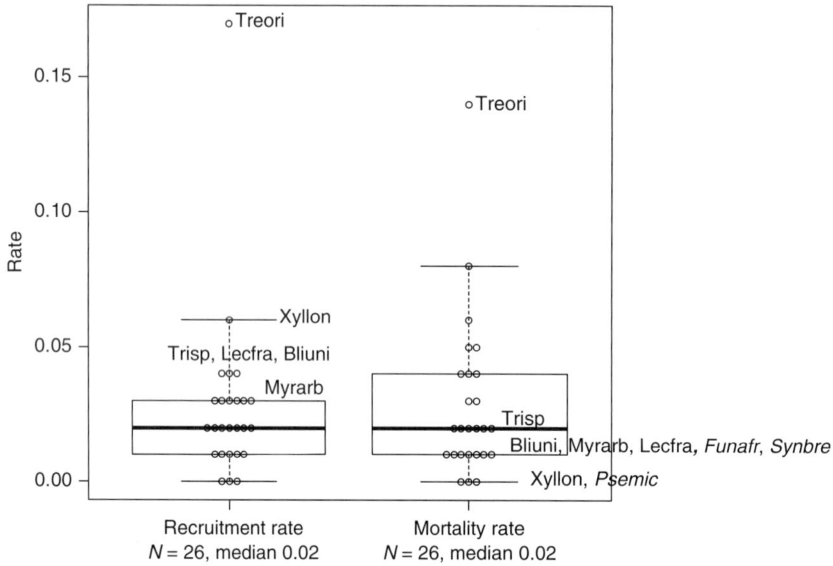

Figure 13.7 Mortality and recruitment rates among tree species with increasing density (common species in the final inventory). Abbreviated species names are listed in Table 13.2.

(Figure 13.7). These species showed a high recruitment rate and a low mortality rate, except for *Trema*, which had a high turnover rate. Overall, the density of these species increased from 36.2 trees/ha to 60.2 trees/ha, representing 13.8% and 22.6% of all trees, respectively.

Before looking into details, some explanation is needed for *Trema*. Its extremely high recruitment and mortality rates may be understandable as this tree often occurs as a pioneer in clearings (Hubbard *et al.*, 1952; Turner, 2006). However, as it can be found in a variety of habitats (Coates Palgrave *et al.*, 1993), further habitat analysis is needed to understand the turnover occurring at Mahale. Nevertheless, the contribution of *Trema* to the overall tree number is still limited (ranked 30th to 26th, 0.68% of initial and 1.15% of final stands).

Xylopia stands out because of its low mortality (0.2%) and high recruitment (5.5%), and PF rates. These properties are reflected in the changes in its ranking (a rise from 21st to 7th place) and increased contribution to the tree stand (1.70% to 4.13%). This tree, in combination with *Pycnanthus*, is characteristic for a typical lowland forest in Kasoje, *Pycnanthus–Xylopia* forest, and tends to be found in closed forests (Turner, 2006).

Lecaniodiscus had a high PF, next to the *Xylopia*. This plant also rose in its density ranking (from 22nd to 13th) and occupation (1.47% to 2.70%). *Lecaniodiscus* tends to be present in the secondary forest and woodland (Turner, 2006). In this transect, it was found in a relatively closed forest. This is also a fast-growing species as there were many small individuals (<10 cm DBH) that we did not observe for other species.

Blighia is a tall tree in Kasoje. Since the 1990s, it showed a relatively high occupation level (ranked 9th, 3.68%), which then substantially increased (5th, 5.80%). Typically, this tree grows in the secondary forest and woodland (Turner, 2006).

Trichilia is also a tall tree found in Kasoje. It was identified only to the genus level, but we assumed that it was the species identified by Turner (2006) as *Trichilia emetica*, which grows in a riverine forest. This tree also increased its density, but it represented a small proportion of the forest stand (25th to 18th rank, 1.13% to 1.78%).

Myrianthus is a small- to medium-size tree. The low mortality rate of this tree may reflect its life strategy – branching into numerous stems. Even if most of the stems die, the individual survives and grows new stems. Similar to *Blighia*, *Myrianthus* represents a large proportion of the tree stand in both inventories (5th to 3rd, 4.93% to 7.35%).

The densities of two of the remaining three species retained their top 10 ranks and one joined the top 10 in the final inventory; *Funtumia* increased in rank from 14th to 8th place (2.44% to 3.44% occupation). This tree usually grows in the moist forests of Kenya and South Africa (Coates Palgrave *et al.*, 1993; Beentje *et al.*, 1994). Although the species was newly recognized (or identified) in this transect, it is common throughout the Kasoje forest, typically in moist areas.

The density of *Pseudospondias* originally ranked second, but it became the top-ranking tree (6.74% to 8.96% occupation). Its increase, unlike in other cases, was affected more by the low mortality rate (minimum, close to that of *Xylopia*) than the recruitment rate (around the median). This is a forest tree usually seen in wet areas (Turner, 2006) with one of the largest DBHs (maximum 200 cm).

Synsepalum is a tree found in the closed forests (Turner, 2006) that also retained its high density (rank 3rd to 4th, 5.50% to 7.23% occupation) and had moderate recruitment and mortality rates.

Observed changes in relation to succession

From the observation of small old forest patches called "*isigo lyetabami*" (traditional burial ground for village headmen) that escaped from slash-and-burn agriculture, Nishida (1990) assumed that *Albizia glaberrima, Xylopia longipetala, Pseudospondias microcarpa, Parkia filicoidea, Julbernardia seretii*, and *Pycnanthus angolensis* represent the climax stage of the lowland forest. He predicted that ecological succession will proceed from an abandoned field to elephant grass bush and *Harungana–Ficus* forest, and then through *Pycnanthus* forest to the climax *Albizia–Xylopia–Julbernardia* forest.

This succession scenario is supported by the following observations. Many elephant grass patches along the transect found during Itoh's first visit in 1995 disappeared (although some dense patches of *Marantochloa leucantha* were observed in the new gaps). *Harungana* was rare in the transect at the beginning of the survey; since then, most of the individuals frequently observed near the lakeside have disappeared. This pioneer species grows densely after the field is abandoned, forming a pure forest within 20 years, but soon gives way to the next succession stage (Norikoshi, 2002). In 1998, Norikoshi also observed the disappearance of many *Harungana* trees and their replacement by *Pycnanthus, Pseudospondias*, and *Saba comorensis*, after a 20-year interval. In our initial survey, *Pycnanthus* was the densest species. An increase in the density of *Xylopia* and *Pseudospondias* supports the next stage predictions.

However, if we take other candidate species into consideration, the scenario might be more complex than assumed. *Pycnanthus* is now declining. The only two large *Parkia* trees in the transect died and no recruitments were observed, even under the threshold size. No *Julbernardia* trees have been found in the transect during the16-year period, and the changes for *Albizia* are ambiguous.

In general, the Kasoje forest seems to be in the process of reforestation or succession. However, colonizing species such as *Croton sylvaticus* do not seem to be decreasing, as the recruitment rate (2%) counterbalances its mortality rate (3%), thus maintaining a relatively high occupation (3.40% to 3.16%, ranked 11th to 10th). The high turnover rate found for *Trema*, typical in forest clearing, is another controversial observation.

Turner (2006) mentioned that micro-environmental factors (climate, soil, drainage, and/or differences in the history of human land use) may influence the succession process by manifold routes from the pioneer phase to maturity. She also mentioned Uehara's personal communication that the predicted pattern described above is observed in some areas and not in others. There is even the possibility that the forest may not reach the climax because of the complex topography impeding succession or intermittently causing various disturbances. A high liana density may also contribute to this phenomenon, as described below. We need a further decadal-scale monitoring of the forest with detailed analyses to understand the succession path(s) and the influence of climate change.

13.5.5 Lianas "in the shadow of trees"

Lianas reach their greatest diversity and biomass in tropical forests (Grubb *et al.*, 1963; Gentry and Dodson, 1987; Gentry, 1991, 1992; Balfour and Bond, 1993; Dewalt *et al.*, 2000). African tropical forests have a higher density of lianas than Asian and neotropical forests (Gentry, 1991).

Apart from the structural importance of lianas, their productivity and exploitation by animals are also important. Lianas support more leaves per unit BAs than trees (Putz, 1983; Gerwing and Farias, 2000). Lianas also provide a valuable food source for animals (Emmons and Gentry, 1983), such as chimpanzees (Nishida and Uehara, 1983; Turner, 2006) and others at Mahale (personal observation). These plants have played an important role in the locomotory adaptation of arboreal vertebrates by providing a different canopy structure at the continental level (Emmons and Gentry, 1983). In spite of their obvious importance, our understanding of their ecology is more limited than for other vascular plants (Jacobs, 1976; Gentry and Dodson, 1987; Gentry, 1991; Nabe-Nielsen, 2001; Schnitzer and Bongers, 2002).

Recent studies have revealed that lianas play vital roles in forest dynamics, such as maintenance of species diversity, transpiration, and carbon sequestration of the forest (reviewed in Schnitzer and Bongers, 2002). For example, micro-disturbances such as tree-fall gaps have long been hypothesized as a part of the general mechanism maintaining overall plant diversity. Although there is little evidence of this mechanism affecting the trees, several studies have revealed that liana species diversity is maintained by gaps (Balee and Campbell, 1990; Pérez-Salicrup and Barker, 2000). In contrast with the commonly accepted succession paths, lianas can dominate (Pérez-Salicrup and Barker, 2000) and remain in a low-canopy state for over 13 years (Augusto *et al.*, 2000; Schnitzer *et al.*, 2000).

Lianas affect trees in various ways thereby accelerating the forest dynamics, although studies of long-term interactions are still rare (Ingwell *et al.*, 2010). Lianas may slow down tree growth (Clark and Clark, 1990; Ingwell *et al.*, 2010) and contribute to the host-tree mortality risk (Putz, 1984a, b; Laurance *et al.*, 2001; Phillips *et al.*, 2005; Ingwell *et al.*, 2010). If the "collateral" death (i.e. pulling down neighboring trees when the host tree falls) is taken into account, the contribution of lianas to tree mortality may be even larger than commonly accepted (Putz, 1984a, b; Phillips and Gentry, 1994; Phillips *et al.*, 2005). Such disturbances favor lianas (Putz, 1984a, b; Laurance *et al.*, 2001; Schnitzer and Carson, 2001), thus possibly forming a positive feedback loop for these species.

There is some evidence for an increasing abundance of lianas in paleotropical and neotropical forests after 1983 (Phillips and Gentry, 1994), and the trend has been continuing (Phillips *et al.*, 2002; Swaine and Grace, 2007; Chave *et al.*, 2008; Ingwell *et al.*, 2010). Both the abundance of this species and the leaf-litter production has been increasing, including the areas of old-growth forests (Wright *et al.*, 2004). The rising concentration of atmospheric CO_2, which is advantageous for lianas but not trees, is suspected to be one important factor (Sasek and Strain, 1988; Phillips *et al.*, 1994; Granados and Körner, 2002).

Thus, the effects of lianas on forest dynamics are substantial, and their significance is still growing, considering the changes in climate. However, most studies of lianas come from tropical forests in the New World. Lianas are important in Kasoje because of their high density and usage by chimpanzees and other animals (Itoh, 2002, 2004; Turner, 2006; Itoh and Nishida, 2007). As both climate change and forest succession are continuing, the accurate evaluation of density and changes in the composition of lianas may be difficult. However, we hope that the basic information we supply below is an important first step in the further evaluation of such changes.

Overall features and changes

Estimates and methods used here are the same as above. However, as it was difficult to identify the individual lianas that had lost their ID numbers (the stems died, the labels dropped, or were torn off), the only reliable data that we could supply was the number of lianas in the initial and final inventories.

In the initial inventory (dataset B), 20 liana species formed 26.4% of the forest stand. Let us first compare these values with those of Turner (2006), who used a random transect with DBH ≥3 cm (instead of ≥5 cm used in this study), also in the 1990s. The reported percentage is close to 26%, the highest liana density, in her *Erythrophleum* forest. However, in her *Xylopia–Pycnanthus* forest, which is much more similar to our studied forest, only 11 species formed 7% of the forest stand; our results differed substantially. This discrepancy may be partially due to the differences between the transects used in the two studies (random *versus* observation trail) because lianas are generally denser in disturbed areas. However, this does not fully explain the differences as we found that the density of lianas was increasing.

Among the initial 20 species, only the one *Mussaenda arcuata* individual disappeared, and three new species ($N = 8$) appeared. Thus, we had a total of 22 species in the final inventory. Three more unidentified species ($N = 9$) possibly belong to one of these 22 species.

The proportion of lianas in the forest stand increased to 32.1% because of the increase in liana density (95.5 to 124.0 lianas/ha) and decrease in tree density. Considering that the forest size fell within the range of a mature tropical forest (see Section 13.5.1), the worldwide trend of increase in liana abundance may also contribute to the phenomena observed in this forest.

Species-specific changes

Initially, most (around 70%) species had densities of less than 5 lianas/ha, and this did not change in the final liana stand. Figure 13.8 shows that *Saba* had originally been super abundant and its numbers increased after 16 years. This single species constituted 37% and 41% of all lianas in the initial and final inventories, respectively. It represented 10% of initial and 13% of the final forest stand, which

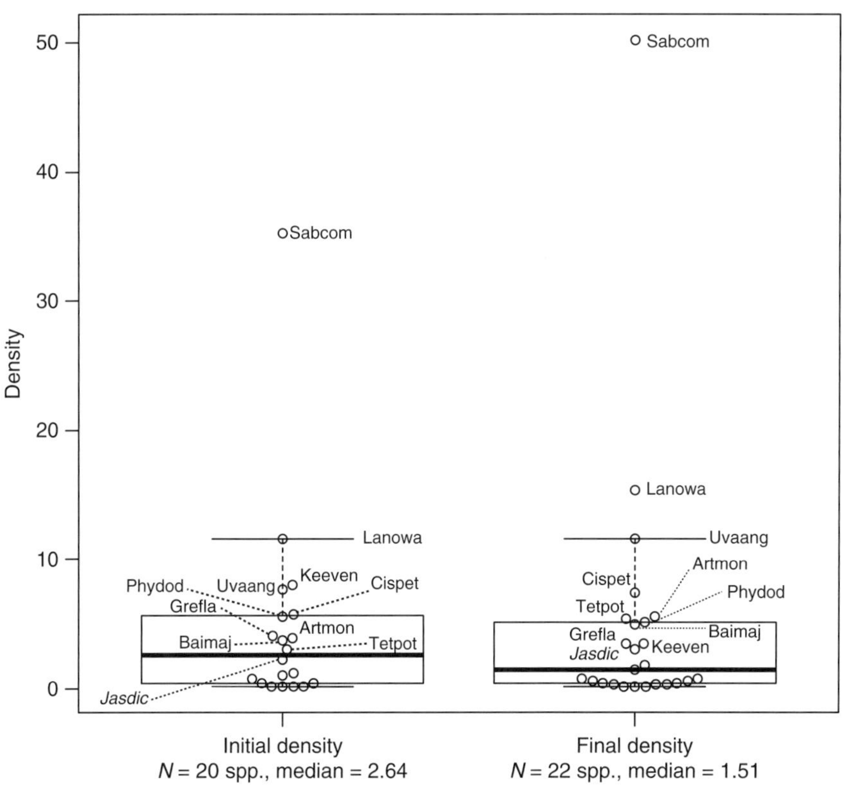

Figure 13.8 Density distribution pattern for liana species. Abbreviated species names are listed in Table 13.2.

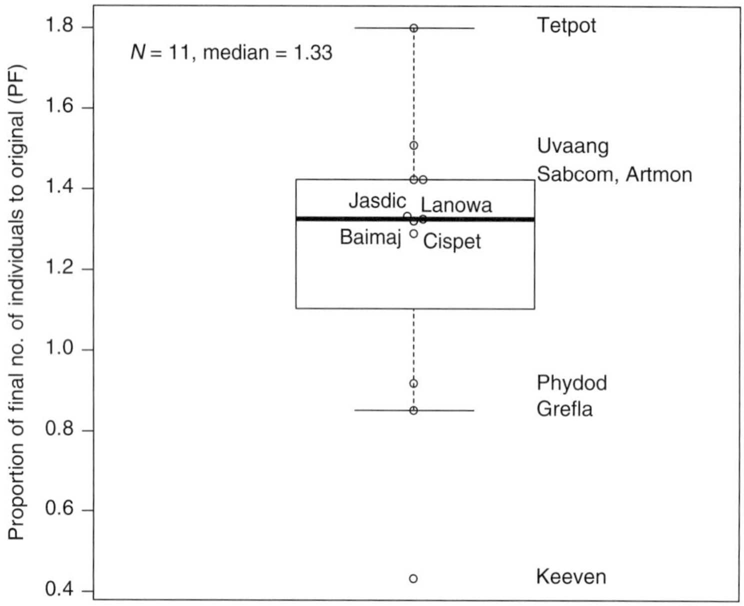

Figure 13.9 Pattern of PF distribution among common liana species. Abbreviated species names are listed in Table 13.2.

made it the top-ranking species in abundance among all species. This species has been referred to as a "hyper-dispersion" species, as has *Pycnanthus* (Turner, 2006). The fruit of this liana is heavily consumed by the chimpanzees at Mahale and strongly affects chimpanzee behavior (feeding: Chapter 17; home-range utility: Chapter 8; grouping pattern: Itoh and Nishida, 2007; Itoh, 2013). *Landolphia* density also increased, but its proportion remained the same (12%) during these 16 years.

Figure 13.9 shows PF values for 11 common liana species whose densities were greater than 2.5 lianas/ha in both inventories. We included *Jasminum* for further analysis, although its density was 2.3 lianas/ha in the initial inventory because it was clustered near the other 10 species.

The decrease in *Keetia venosa* density was substantial (7.99 to 3.47 lianas/ha). *Grewia* and *Phytolacca* also showed a decreasing trend. Many individuals of *Grewia* were initially observed in the form of shrubs rather than climbers. Almost half of them were found dead in the final inventory. On the other hand, *Tetracera* (3.02 to 5.43 lianas/ha), *Uvaria* (7.69 to 11.61 lianas/ha), *Saba*, and *Artabotrys* (3.92–5.58 lianas/ha) showed increasing trends.

Most species whose density increased were forest species. *Tetracera*, which typically appears in the colonizing forest, was an exception (Turner, 2006). The decreasing species were also forest species except for *Grewia*, which grows in the drier forest areas at the forest edges.

This vigorous increase in liana species may partly explain why our initial inventory (using trail transect) showed a higher proportion of lianas than Turner's *Xylopia–Pycnanthus* forest (using random transect), even though both studies were conducted in the 1990s. A combination of several factors may make lianas more abundant on and around the trail rather than away from the trails. Fruits of increasing liana species, *Uvaria*, *Saba*, and *Artabotrys*, are frequently eaten and the seeds are dispersed by the chimpanzees (Chapter 19). As chimpanzees frequent the trails, the seeds are more likely to be deposited there, thus making lianas more abundant on and around trails rather than in other areas.

Future studies should explore the liana–tree relationships in more detail, including the tree-climbing patterns of lianas, the degree of liana infestation, and the correlation between tree and liana densities.

References

Anonymous (1980). *Mahale: Study for the Proposed Mahale Mountains National Park. Final Report, May 1980*. Tokyo: Japan International Cooperation Agency.

Augusto, A., Tabanez, J., and Viana, V. M. (2000). Patch structure within Brazilian Atlantic forest fragments and implications for conservation 1. *Biotropica*, **32**, 925–33.

Balee, W. and Campbell, D. G. (1990). Evidence for the successional status of liana forest (Xingu River basin, Amazonian Brazil). *Biotropica*, **22**, 36–47.

Balfour, D. and Bond, W. (1993). Factors limiting climber distribution and abundance in a southern African forest. *Journal of Ecology*, **81**, 93–100.

Beentje, H., Adamson, J., and Bhanderi, D. (1994). *Kenya Trees, Shrubs, and Lianas*. Nairobi: National Museums of Kenya.

Chave, J., Olivier, J., Bongers, F., *et al.* (2008). Above-ground biomass and productivity in a rain forest of eastern South America. *Journal of Tropical Ecology*, **24**, 355.

Clark, D. B. and Clark, D. A. (1990). Distribution and effects on tree growth of lianas and woody hemiepiphytes in a Costa Rican tropical wet forest. *Journal of Tropical Ecology*, **6**, 321–31.

Coates Palgrave, K., Drummond, R. B., Moll, E. J., *et al.* (1993). *Trees of Southern Africa*. Cape Town, South Africa: Struik.

Dewalt, S. J., Schnitzer, S. A., and Denslow, J. S. (2000). Density and diversity of lianas along a chronosequence in a central Panamanian lowland forest. *Journal of Tropical Ecology*, **16**, 1–19.

Emmons, L. and Gentry, A. H. (1983). Tropical forest structure and the distribution of gliding and prehensile-tailed vertebrates. *American Naturalist*, **121**, 513–24.

Gentry, A. H. (1991). The distribution and evolution of climbing plants. In *The Biology of Vines*, ed. F. E. Putz and H. A. Mooney. New York: Cambridge University Press, pp. 3–49.

Gentry, A. H. (1992). Tropical forest biodiversity: distributional patterns and their conservational significance. *Oikos*, **63**, 19–28.

Gentry, A. H. and Dodson, C. (1987). Contribution of nontrees to species richness of a tropical rain forest. *Biotropica*, **19**, 149–56.

Gerwing, J. J. and Farias, D. L. (2000). Integrating liana abundance and forest stature into an estimate of total aboveground biomass for an eastern Amazonian forest. *Journal of Tropical Ecology*, **16**, 327–35.

Granados, J. and Körner, C. (2002). In deep shade, elevated CO_2 increases the vigor of tropical climbing plants. *Global Change Biology*, **8**, 1109–17.

Grubb, P., Lloyd, J., Pennington, T., and Whitmore, T. (1963). A comparison of montane and lowland rain forest in Ecuador I. The forest structure, physiognomy, and floristics. *Journal of Ecology*, **51**, 567–601.

Hubbard, C. E., Turrill, W. B. M.-R., and Polhill, R. M., ed. (1952 in continuation). *Flora of Tropical East Africa*. London: Crown Agents.

Ingwell, L. L., Joseph Wright, S., Becklund, K. K., Hubbell, S. P., and Schnitzer, S. A. (2010). The impact of lianas on 10 years of tree growth and mortality on Barro Colorado Island, Panama. *Journal of Ecology*, **98**, 879–87.

Itoh, N. (2002). [Foods in the forest: chimpanzee food density and spatial distribution.] In [*The Mahale Chimpanzees: 37 Years of <Panthropology>*], ed. T. Nishida, S. Uehara, and K. Kawanaka. Kyoto: Kyoto University Press, pp. 77–100. In Japanese.

Itoh, N. (2004). [*The Plant Phenology and Fission–Fusion Social System in the Mahale Mountains National Park*.] Doctoral Dissertation. Kyoto: Kyoto University. In Japanese.

Itoh, N. (2013). [Dual-rings of mode of coexistence and act choices: Chimpanzee group and emergence of institution.] In [*Institutions: Evolution of Human Sociality*], ed. K. Kawai. Kyoto: Kyoto University Press, pp. 141–66. In Japanese.

Itoh, N. and Nishida, T. (2007). Chimpanzee grouping patterns and food availability in Mahale Mountains National Park, Tanzania. *Primates*, **48**, 87–96.

Itoh, N., Nakamura, M., Ihobe, H., *et al.* (2012). Long-term changes in the social and natural environments surrounding the chimpanzees of the Mahale Mountains National Park. In *The Ecological Impact of Long-Term Changes in Africa's Rift Valley*, ed. A. J. Plumptre. New York: Nova Science, pp. 249–77.

Jacobs, M. (1976). The study of lianas. *Flora Malesiana Bulletin*, **29**, 2610–17.

Laurance, W. F., Pérez-Salicrup, D., Delamônica, P., *et al.* (2001). Rain forest fragmentation and the structure of Amazonian liana communities. *Ecology*, **82**, 105–16.

Lewis, S. L., Phillips, O. L., Sheil, D., *et al.* (2004). Tropical forest tree mortality, recruitment and turnover rates: calculation, interpretation and comparison when census intervals vary. *Journal of Ecology*, **92**, 929–44.

Lind, E. M. and Morrison, M. E. S. (1974). *East African Vegetation*. London: Longman.

Lovett, J. C. (1993). Climatic history and forest distribution in eastern Africa. In *Biogeography and Ecology of the Rain Forests of Eastern Africa*, ed. J. C. Lovett and S. K. Wasser. Cambridge: Cambridge University Press, pp. 23–9.

Lovett, J. C. (1994). Zambezian regional centre of endemism: CPD site AF33 Mahale-Karobwa Hills, Tanzania. In *Centres of Plant Diversity*, ed. S. D. Davis, V. H. Heywood, and A. C. Hamilton. Cambridge: WWF-IUCN-ODA, pp. 184–7.

Lukosi, N. (1997). A brief note on possible control of *Senna spectabilis*, an invasive exotic tree at Mahale. *Pan Africa News*, **4**, 18.

Nabe-Nielsen, J. (2001). Diversity and distribution of lianas in a neotropical rain forest, Yasuní National Park, Ecuador. *Journal of Tropical Ecology*, **17**, 1–19.

Nishida, T. (1990). A quarter century of research in the Mahale Mountains: An overview. In *The Chimpanzees of the Mahale Mountains: Sexual and Life History Strategies*, ed. T. Nishida. Tokyo: University of Tokyo Press, pp. 3–35.

Nishida, T. (1996). Eradication of the invasive, exotic tree *Senna spectabilis* in the Mahale Mountains. *Pan Africa News*, **3**(2), 6–7.

Nishida, T. and Uehara, S. (1981). Kitongwe name of plants: a preliminary listing. *African Study Monographs*, **1**, 109–31.

Nishida, T. and Uehara, S. (1983). Natural diet of chimpanzees (*Pan troglodytes schweinfurthii*): Long-term record from the Mahale Mountains, Tanzania. *African Study Monographs*, **3**, 109–30.

Norikoshi, K. (2002). [Vegetation of Kasoje research area and ranging pattern: Changes during 20 years before and after the establishment of national park.] In [*The Mahale Chimpanzees: 37 Years of <Panthropology>*], ed. T. Nishida, S. Uehara, and K. Kawanaka. Kyoto: Kyoto University Press, pp. 101–288. In Japanese.

Pérez-Salicrup, D. R. and Barker, M. G. (2000). Effect of liana cutting on water potential and growth of adult *Senna multijuga* (Caesalpinioideae) trees in a Bolivian tropical forest. *Oecologia*, **124**, 469–75.

Phillips, O. L. and Gentry, A. H. (1994). Increasing turnover through time in tropical forests. *Science*, **263**, 954–8.

Phillips, O. L., Hall, P., Gentry, A., Sawyer, S., and Vasquez, R. (1994). Dynamics and species richness of tropical rain forests. *Proceedings of the National Academy of Sciences of the USA*, **91**, 2805–9.

Phillips, O. L., Martínez, R. V., Arroyo, L., *et al.* (2002). Increasing dominance of large lianas in Amazonian forests. *Nature*, **418**, 770–4.

Phillips, O. L., Martínez, R. V., Mendoza, A. M., Baker, T. R., and Vargas, P. N. (2005). Large lianas as hyperdynamic elements of the tropical forest canopy. *Ecology*, **86**, 1250–8.

Putz, F. E. (1983). Liana biomass and leaf area of a "tierra firme" forest in the Rio Negro Basin, Venezuela. *Biotropica*, **15**, 185–9.

Putz, F. E. (1984a). How trees avoid and shed lianas. *Biotropica*, **16**, 19–23.

Putz, F. E. (1984b). The natural history of lianas on Barro Colorado Island, Panama. *Ecology*, **65**, 1713–24.

Richards, P. W. (1952). *The Tropical Rain Forest: An Ecological Study*. London: Cambridge University Press.

Sasek, T. W. and Strain, B. R. (1988). Effects of carbon dioxide enrichment on the growth and morphology of kudzu (*Pueraria lobata*). *Weed Science*, **36**, 28–36.

Schnitzer, S. A. and Bongers, F. (2002). The ecology of lianas and their role in forests. *Trends in Ecology & Evolution*, **17**, 223–30.

Schnitzer, S. A. and Carson, W. P. (2001). Treefall gaps and the maintenance of species diversity in a tropical forest. *Ecology*, **82**, 913–19.

Schnitzer, S. A., Dalling, J. W., and Carson, W. P. (2000). The impact of lianas on tree regeneration in tropical forest canopy gaps: evidence for an alternative pathway of gap-phase regeneration. *Journal of Ecology*, **88**, 655–66.

Sheil, D. and May, R. M. (1996). Mortality and recruitment rate evaluations in heterogeneous tropical forests. *Journal of Ecology*, **84**, 91–100.

Sheil, D., Burslem, D. F., and Alder, D. (1995). The interpretation and misinterpretation of mortality rate measures. *Journal of Ecology*, **83**, 331–3.

Sheil, D., Jennings, S., and Savill, P. (2000). Long-term permanent plot observations of vegetation dynamics in Budongo, a Ugandan rain forest. *Journal of Tropical Ecology*, **16**, 865–82.

Swaine, M. D. and Grace, J. (2007). Lianas may be favoured by low rainfall: evidence from Ghana. *Plant Ecology*, **192**, 271–6.

Turner, L. A. (1996). Invasive plant in chimpanzee habitat at Mahale. *Pan Africa News*, **3**(1), 5.

Turner, L. A. (2006). Vegetation and chimpanzee ranging in the Mahale Mountains National Park, Tanzania. *Memoirs of the Faculty of Science Kyoto University (Series of Biology)*, **18**, 45–82.

Wakibara, J. V. (1998). Observations on the pilot control of *Senna spectabilis*, an invasive exotic tree in the Mahale Mountains National Park, Western Tanzania. *Pan Africa News*, **5**, 4–6.

Wakibara, J. V. and Mnaya, B. J. (2002). Possible control of *Senna spectabilis* (Caesalpiniaceae), an invasive tree in Mahale Mountains National Park, Tanzania. *Oryx*, **36**, 357–63.

Wasser, S. K. and Lovett, J. C. (1993). Introduction to the biogeography and ecology of the rain forests of eastern Africa. In *Biogeography and Ecology of the Rain Forests of Eastern Africa*, ed. J. C. Lovett and S. K. Wasser. Cambridge: Cambridge University Press, pp. 3–7.

Williamson, E. A. (1988). *Behavioural Ecology of Western Lowland Gorillas in Gabon*. Ph.D. thesis. Stirling, UK: University of Stirling.

Wright, S. J., Calderón, O., Hernández, A., and Paton, S. (2004). Are lianas increasing in importance in tropical forests? A 17-year record from Panama. *Ecology*, **85**, 484–9.

14 Patterns and trends in fruiting phenology: some important implications for chimpanzee diet*

NORIKO ITOH AND DAISUKE MURAMATSU

14.1 Introduction

Community-level plant phenology in the Kasoje forest assessed in the transect (Chapter 13) over 4 years (May 1997–June 2001) showed a roughly 1-year cycle (Itoh, 2004; Itoh *et al.*, 2012). It demonstrated leaf-flushing activity peaking in the early wet season (November–January), flowering in the early dry season (around July), and fruiting from the late dry to early wet season (September–December) (Itoh, 2004; Itoh *et al.*, 2012). Thus, the fluctuation pattern does not strictly follow climatic seasons, which has one long wet season starting in early October interrupted by a relatively short dry season starting in mid-May (Chapter 12).

Most plants in the transect were food plants for chimpanzees (Itoh, 2002, 2004). The availability of *Saba comorensis* fruit strongly affects the grouping patterns (Itoh and Nishida, 2007) and the ranging patterns of chimpanzees (Nakamura *et al.*, 2013; Chapter 8). This large liana species is the most abundant among all species in the transect and increases its density in the long term (Chapter 13). Thus, this species represents the overall phenological pattern of the Kasoje forest (Itoh, 2004).

Short-term phenological trends often do not persist (Chapman *et al.*, 1999); and our 4-year record was definitely too short to examine those tendencies. The importance of long-term phenological studies has been brought to the attention of the scientific community by the recent global climatic changes. The number of such studies has been increasing; some of them have detected changes in plant phenology in response to climate change (e.g. Chapman *et al.*, 2005). Some recent changes in fruit production have been reported for communities with relatively long-term records: a decreasing trend in Budongo, Uganda (Babweteera *et al.*, 2012), an increasing trend in Kibale, Uganda (Chapman *et al.*, 2012), and a decreasing trend in peak fruit periods in Nyungwe, Rwanda (Chao *et al.*, 2012) (see Plumptre, 2012 for a summary). However, the exact causes have not been identified and long-term records are scarce in tropical regions (Chapman *et al.*, 2005).

Climate, vegetation, and plant phenology are not only related to each other but also constitute a basic structure for animal existence. Considering the long-term changes in climate (Chapter 12) and vegetation (Chapter 13) at Mahale, a continuous monitoring of plant phenology, which directly affects the health and survival of chimpanzees, is essential. Although interrupted for some months, phenological monitoring has continued up to now in the Kasoje forest, the home range of the M-group chimpanzees (Chapters 4, 8). In this chapter, we made a preliminary analysis of 16 years (191 months) of phenological

* We thank Dr. Linda Turner, who initiated the phenological survey in Mahale and supplied important advice for constructing the monitoring framework. Without support from Drs. Koichiro Zamma, Hiroshi Ihobe, and Michio Nakamura, this long-term monitoring project would not have continued. Wildlife Conservation Society (WCS) project on long-term phenology in Albertine Rift led by Dr. Andrew Plumtre encouraged us to continue monitoring and also financially supported our project between 2010 and 2012. We also thank the local assistants for their monitoring and other researchers for their support.

Mahale Chimpanzees: 50 Years of Research, ed. M. Nakamura, K. Hosaka, N. Itoh and K. Zamma. Published by Cambridge University Press.

data to examine the temporal patterns and long-term changes, focusing on fruit food species important for chimpanzees (Chapter 17).

14.2 Dataset

New leaves, flowers, and fruits were monitored monthly for the individuals selected among the ID-tagged woody plants in the transect (see Chapter 13 for details). Flowering of fig species (*Ficus exasperata*, *F. sur*, and *F. vallis-choudae*) was not monitored as the inflorescence is enclosed inside the fruit. Sixteen years (May 1997–August 2013) of data, collected mainly by local assistants, is analysed here. Data for even months from December 2007 to August 2008 are unavailable.

For species and/or individuals with short phenophases, we might have missed the timing of peaks and/or troughs. A study comparing different sampling methods for plant phenological monitoring recommended a fortnightly observation (Morellato *et al.*, 2010). In our experience, monthly monitoring was sufficient to observe flushing episodes. However, we might have missed some flowering episodes, thus obtaining a low flowering rate throughout the study period. Monthly intervals between the monitoring of fruiting events might be too long to discriminate maturation stages (i.e. unripe fruits may ripen within a month). Thus, we do not discriminate between ripe and unripe fruits. This is also effective for some species where a visual assessment of the maturation stages is difficult.

Species and individuals that met the following criteria within the transect were used for the analyses (17 species, listed in Table 14.1).

1. Mature individuals at the onset of the vegetational survey (in dataset B, Chapter 13). Our observations showed that lianas with DBH ≥5 cm (per stem) were mature enough to produce fruit. For trees, a 30 cm threshold of total DBH of each individual (see Morellato *et al.*, 2010) was used following Williamson (1988), with some exceptions where species-specific threshold level were confirmed (see Table 14.1). Non-alive (dead, absent, fallen, etc.) individuals (see Section 13.5.3 in Chapter 13) were excluded from the date when they lost the "alive" status, if known, or from January 2000 if the date was unknown.
2. Species that are important fruit foods for the M-group chimpanzees (Table 17.2 in Chapter 17).
3. Different sample sizes affect phenological patterns, and at least 15 individuals should be monitored (Morellato *et al.*, 2010). However, it has been also recommended that small sample-sized species be included (with caution) in tropical forests where many species have low densities (Morellato *et al.*, 2010). Moreover, sample sizes may diminish because of death or other problems during the long-term monitoring (see also Chapter 13). Thus, records for the species with at least four individuals monitored in a given month were used to reduce sampling errors (Sun *et al.*, 1996).

14.3 Definitions for analysis

Illustrating phenological data is difficult and relies mainly on graphical representations (Newstrom *et al.*, 1994a, b; Chapman *et al.*, 2005). We used multiple measurements to illustrate and evaluate fruiting patterns. First, we summarized the definitions of the terms. Newstrom *et al.* (1994a) proposed a conceptual framework to resolve difficulties in describing phenological diversity and terminological confusion (also reviewed in Sakai, 2001). We follow their terminology with some qualifications and some additional terms (terms with additional qualifications are shown with "*" and additional terms with "+") (see also Figure 14.1).

Cycle	A period that consists of a consecutive episode and interval.
Episode	A portion of the cycle that is in the phenophase.
Interval	A portion of the cycle that is not in the phenophase.
Duration	A period for which an episode/ interval remains in a given

Table 14.1 **Analysis of 17 important chimpanzee fruit foods**

IC[a]	Species	MLID[b]	Density (ind./ha)[c]	DBH (cm)[d]	*N* (ind.)[e]	*N* (mon.)[f]	Comments
Consistently staple (regularly eaten frequently in most years)							
1	*Saba comorensis*	62	33.25 (29.86–35.29)	5	165.83 (43–234)	191	
2	*Ficus sur*	500	2.84 (2.56–3.32)	20	15.92 (10–20)	191	Maturity determined from the dataset
3	*Pseudospondias microcarpa*	37	10.99 (10.71–11.31)	30	55.06 (35–74)	191	Dioecious. Males not excluded from analysis
4	*Ficus exasperata*	493	6.52 (4.98–7.84)	30	38.18 (18–52)	191	
5	*Cordia millenii*	115	5.28	30	29.06 (23–35)	191	
6	*Citrus limon*	690	0.60	15	4	121–123	Maturity determined from the dataset which also is the observed minimum size for this species in the dataset
7	*Pycnanthus angolensis*	513	11.68 (7.84–16.89)	30	55.77 (14–88)	191	Two months (May, June 2012) having only 14 individuals monitored
Staple (regularly eaten frequently in many years)							
8	*Ficus vallis-choudae*	504	4.86 (2.71–7.09)	30	25.3 (12–47)	191	
9	*Myrianthus arboreus*	509	6.90 (6.64–7.84)	30	26.29 (10–40)	191	Dioecious. Males not excluded from analysis
11	*Tabernaemontana pachysiphon*	65	1.85 (1.21–2.71)	20	10.25 (4–18)	188–190	Maturity determined from the dataset
12	*Psydrax parviflora*	1420	0.93 (0.75–1.06)	30	5.53 (4–7)	189–190	

Other						
13	*Garcinia huillensis*	383	1.32 (1.06–1.51)	30	8.60 (5–10)	191
15	*Azanza garckeana*	421	0.72 (0.60–0.90)	30	4.72 (4–6)	163–167
17	*Cordia africana*	114	3.00 (2.71–3.47)	30	17.72 (13–23)	191
19	*Landolphia owariensis*	57	10.57 (9.05–11.61)	5	56.32 (25–77)	191
20	*Synsepalum brevipes*	722	6.74 (6.64–7.09)	30	33.96 (22–47)	191
23	*Uvaria angolensis*	45	6.88 (5.58–7.99)	5	37.79 (11–51)	191

[a] Importance categories and ranks (Table 17.2, Chapter 17).

[b] Species-specific ID used for Mahale flora (Itoh, in prep.). Full names are given in Appendix I.

[c] Density of mature individuals (see note *d*) in the initial vegetation inventory. Recruited individuals and those that reached maturity in the later inventories are excluded from all analyses.

[d] Threshold of diameter at breast height (DBH) of a mature individual. For trees, total DBH size was used if there were multiple stems. As we have just started to assess species-specific threshold levels, not all species are strictly defined.

[e] Mean number of individuals monitored per month and the range in parentheses. Small sample size (<15) might have introduced a bias (see text for detail).

[f] Total number of months monitored (range means that the number of monitored months differed for different phenological events).

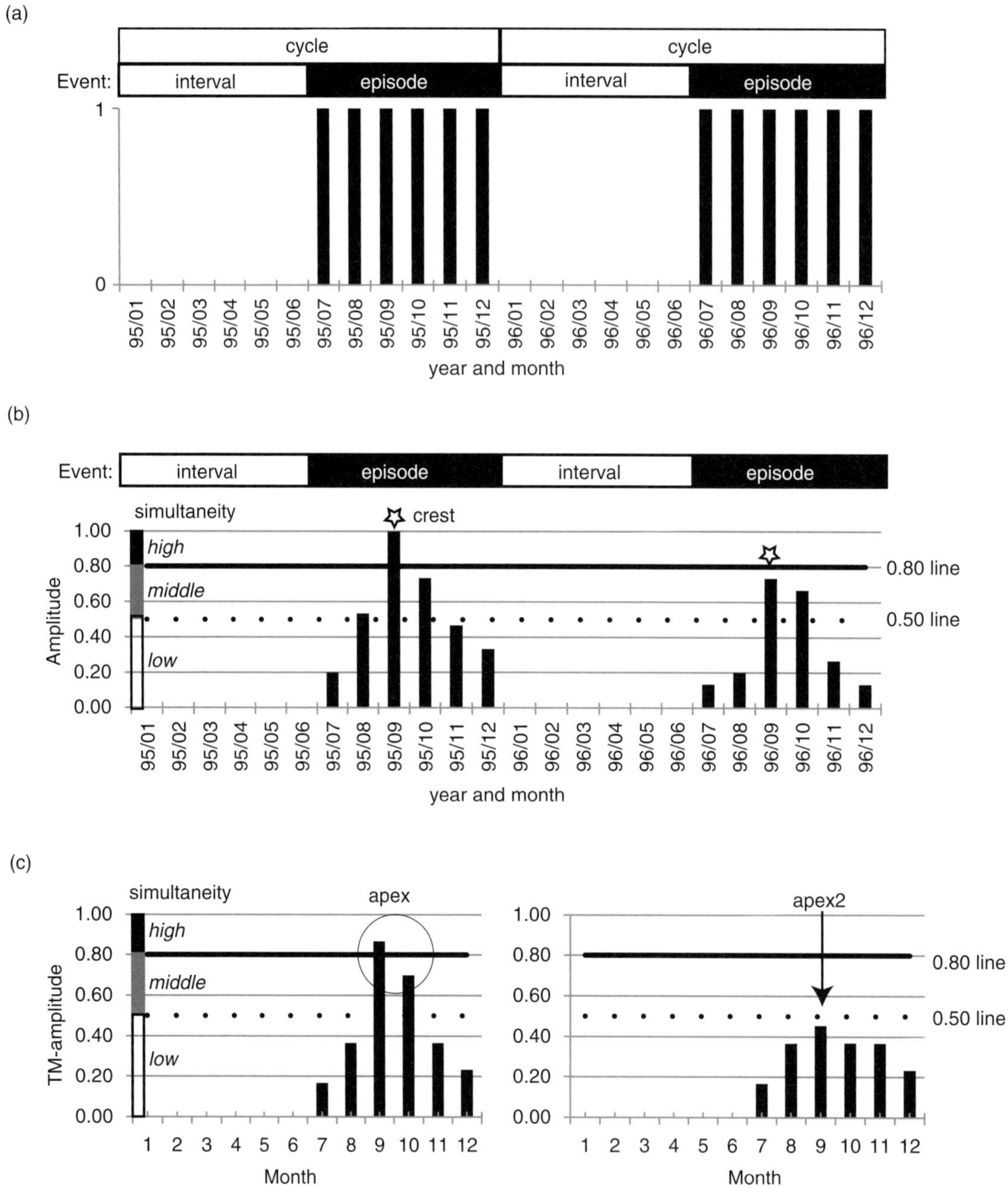

Figure 14.1 Diagram illustrating the terms used in the study.

a. An example showing individuals in a phenophase each month. Cycle frequency in this idealized figure is once per year; it represents an annual pattern.
b. An example showing changes in amplitude for each month. Periodicity is assessed using this dataset in autocorrelogram analysis.
c. Examples of TM-amplitude calculated as mean of monthly amplitudes throughout the record. In the left-hand-side figure, the species shows apex months. In the figure on the right, the amplitudes are too low, and the month with maximum amplitude is referred to as apex2 instead.

cycle; in our dataset, one month is the minimum time unit.

Frequency* — Number of cycles per unit time. Four arbitrary classes based on an annual timescale are defined as follows.

Continual* — Always in phenophase having none or only brief breaks ≤3 months; it may be interrupted for more than 3 months in some years. We restrict such longer interruptions to less than twice per decade.

Sub-annual — More than one cycle per year.

Annual — One major cycle per year.

Supra-annual — One cycle over more than a year.

Amplitude* — Activity calculated here as monthly proportion of individuals in the phenophase among the individuals monitored in each species. An average of amplitudes in the same calendar month in different years is referred to as **total monthly amplitude** (**TM-amplitude**, hereafter).

Crest⁺ — The month with the maximum amplitude within a cycle. If there is another month separated by an interval or low amplitude month (s), it is identified as a second crest.

Apex⁺ — Months with TM-amplitude ≥50%. If the overall amplitudes remain low, the month with the maximal amplitude is identified as an apex (**Apex2**). An apex could be clear or vague depending on the surrounding amplitudes.

Vertex⁺ — Significant lag location in autocorrelogram.

Periodicity⁺ — Recurrence of an event, with amplitude taken into account, throughout the observation period; see the explanation of autocorrelogram below.

Regularity* — Variability in lengths of cycles. Positive and negative vertices in autocorrelogram are used for regularity estimation.

Simultaneity⁺ — Occurrence of the same episode for all or most individuals in the same month in each species. We define the intensity of simultaneity with amplitudes or TM-amplitudes ≥0.8 as high, 0.5–0.8 as middle, and <0.5 as low.

Trough⁺ — Month with the minimum amplitude within an event.

Seasonal — The term is used only when a phenological event is associated with a climatic season.

14.4 Fruiting phenology

When considering frugivore behaviors, a variation in the temporal availability of fruits (timing, duration, and intensity) and rare events (severe scarcity and bumper crops) are important elements. Van Schaik *et al.* (1993) reviewed the evidence of occasional famine leading to mass mortality, annual-scale scarcity leading to dietary switch, and adaptation of seasonal breeding. As we lack such behavioral data and mass population decrease (1994–6; Chapter 7) occurred before our phenological records, such analysis is out of our scope here. However, annual variability in fruit availability is still important (see also Chapters 8, 9, 17). Thus, general fruiting phenology of important food species was studied. Spatial variation is also important, but we concentrated our assessment on temporal variation because of limited space.

For chimpanzees, the presence or absence of specific food items might be important (e.g. for the fruit eaten by chance and/or the fruit highly preferred but rarely eaten because of time and/or space limitations or low availability). However, for the frequently eaten foods, fluctuations in amplitude may be more important. Taking into account the significance of these different factors for chimpanzees, several issues were assessed for each of the three important food categories (consistently staple = CF, staple = SF, and other = OF; Chapter 17).

First, frequency pattern in terms of presence/absence of 17 plant species were examined (Figure 14.2). Most species had different frequency patterns in some parts of the record; however, the patterns seem to have changed after around the mid-2000s. Thus, we classified the frequency patterns of a species on the basis of the first half of the record. Species are arranged from top to bottom in the relative order of typical continual pattern to typical annual pattern. No supra-annual species were found. Intermediate species tended to have irregular or complicated fruiting patterns, which made the classification difficult (*Tabernaemontana pachysiphon, Synsepalum brevipes, Landolphia owariensis, Psydrax parviflora, Cordia africana, Myrianthus arboreus*, and *Azanza garckeana*). Thus, all 17 species can be placed in the continual–annual continuum, including the intermediate or complicated patterns in-between.

Next, to examine the accessibility of each food item within each episode, the amplitude was analysed. We first assessed the periodicity and its regularity using the acf function implemented in the statistical package R (R Core Team, 2014) (Figure 14.3, right). This function plots an autocorrelogram, which represents the autocorrelation coefficients, with a lag at 1-month intervals. Linear interpolation was employed to handle missing values. The highest vertex in the autocorrelogram is at the 0-lag (correlation coefficient is 1), and the following highest positive vertex gives an estimate of the periodicity. A biannual periodicity, for example, has the second-highest vertex at 6-month lag, followed by peaks at around 12-, 18-, 24-... month lag.

Then, the apex and simultaneity levels of monthly average were assessed to examine the seasonal accessibility. These values were identical to those in the TM-amplitude figure (Figure 14.3, left).

Lastly, more detailed characteristics were presented by plotting the monthly amplitude as a function of time for the whole study period (Figure 14.4).

14.4.1 General patterns

Below, we describe the results for each food category. The results are presented in the following subgroups (R1–R5): frequency (R1, Figure 14.2), periodicity and regularity (R2, Figure 14.3), apex and simultaneity (R3, Figure 14.3), other detailed information (R4: crests, intensity, and duration of each episode; Figure 14.4), and long-term changes (R5, Figures 14.2, 14.4). See Table 14.2 for a summary.

Consistently staple food species

R1 Out of the seven CF species, five exhibited the annual pattern and two (*F. sur* and *Citrus limon)* exhibited the continual pattern. *Saba*, a liana, had extremely long fruiting episodes with 1–3-month intervals in January–March and/or May–June. There was an exceptionally long interval in 2001 (January–May), and intervals occurred twice in 2003 and 2005. Besides these unusual periods, the pattern of *Saba* may be categorized as a subcategory of annual, a "pulsed annual," pattern (Newstrom *et al.*, 1994a). Moscovice *et al.* (2007) found a fruiting episode of *Saba* during all 11 months of the monitoring period in 2003 in Rubondo Island (Tanzania). *Pycnanthus angolensis* had a pattern similar to that of *Saba* but with intervals that were more regular.

R2 All species exhibited a clear and regular 12-month periodicity. There were three exceptions: *F. sur* and *Citrus* lacking periodicity, and irregular pattern of *Pycnanthus.*

R3 All five cyclical species exhibited clear apexes, most of which (80%) peaked at the end of the dry season (September) and/or at the onset of the wet season (October). *Saba* had a prolonged and pronounced fruiting apex in comparison with other species, with the exception of *F. sur* and *Citrus*, which lacked clear apexes. *Cordia millenii* peaked in the mid-wet season (January).

R4 *Saba* was a species with stable high crests (>0.8) in most events (except for the two events at the beginning) and lengthy duration with relatively high amplitude (>0.5) during each episode. *Citrus* was remarkable for remaining in high simultaneity

Figure 14.2 Fruiting phenology throughout the study (fruiting periods shown in black).

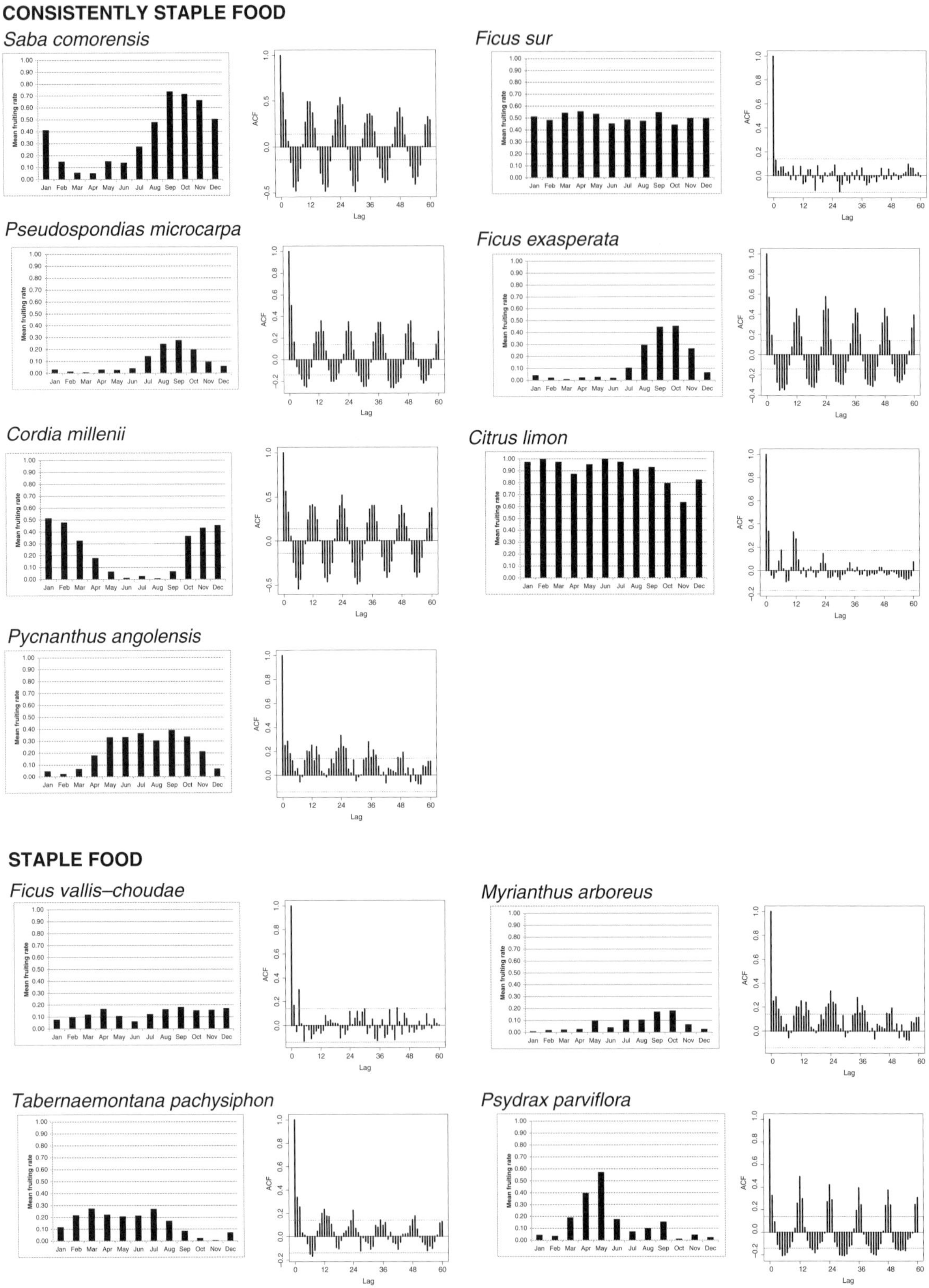

Figure 14.3 Fruiting peaks, simultaneity, and periodicity.

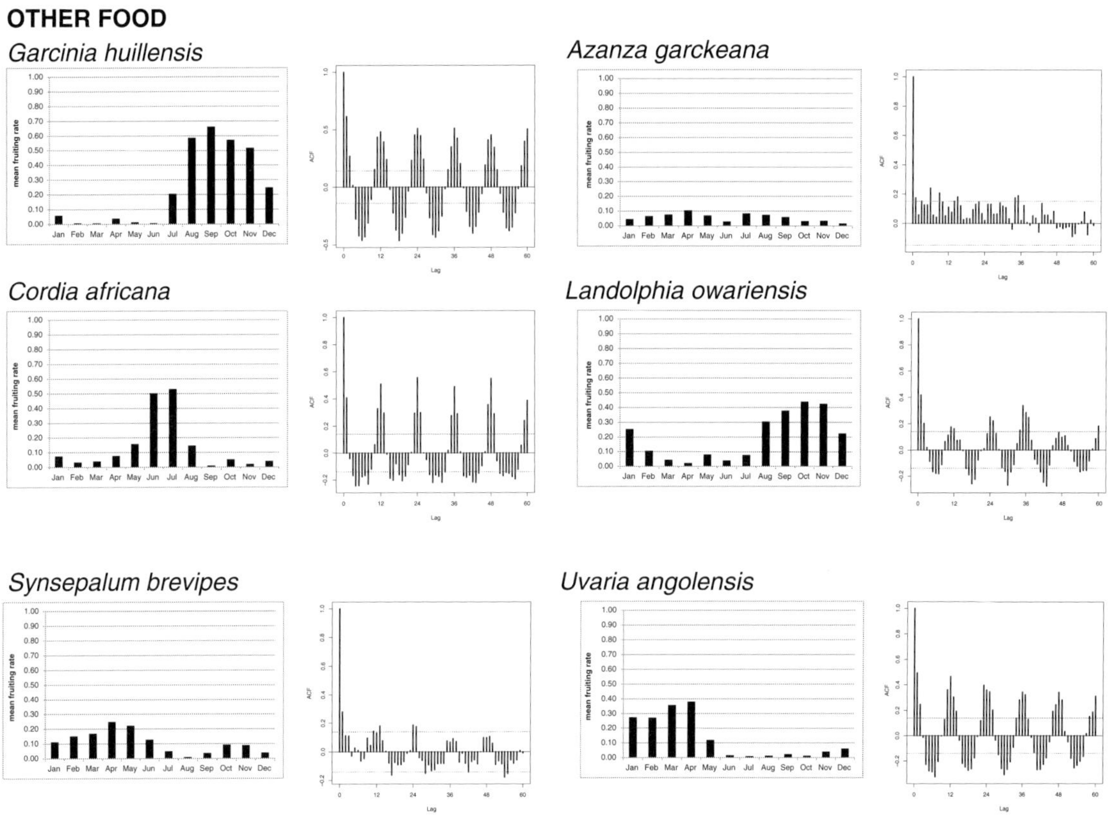

Figure 14.3 (cont.)

throughout the record (81%). The amplitude for *F. sur* remained relatively high (>0.5) for more months than for *Saba* (55% vs. 37% of all monitored months), but the intensity for *F. sur* rarely exceeded 0.8 (4% vs. 18% for *Saba*). In contrast, *F. exasperata* and *Pseudospondias microcarpa* generally had short fruiting episodes, low crests, and low intensity in each episode. *Pseudospondias* showed a relatively high annual variation in intensity and duration of fruiting episodes, even if we omit the complex records of 2009. However, this might have been partly caused by the fact that the males were neither specified nor excluded. *C. millenii* had the intermediate pattern of these two species and *Saba* in terms of duration and simultaneity (0.5–0.8) with occasional high simultaneity events (the exact crests fall around the beginning of these years or at the end of the year before). *Pycnanthus* fruited for long periods although the simultaneity remained at low–middle level with a large annual variation, which might have resulted in reduced regularity.

R5 *Saba* extended its fruiting episodes to the beginning of the consecutive year since 2006 (the trend that had been observed occasionally before). From 2005, the variation of amplitude of *F. sur* seems to have increased. Irregular small events at the beginning of each calendar year observed for *F. exasperata* and *Pseudospondias* have become regular since 2009 (or 2008). Intervals and troughs for *C. millenii* have shortened since 2009 (or 2008). All these results indicate that the accessibility to these foods has increased in terms of duration, timing, and/or amount (if the density of the plant was not decreasing). However, as the fruiting of *Citrus* after 2009 (or 2008) had frequent troughs (<0.5), it may indicate an overall decreasing trend. For *Pycnanthus*, the intervals shortened at the middle of the record (2004–8 or 2009) and became more irregular at the end, but the annual variation might have decreased.

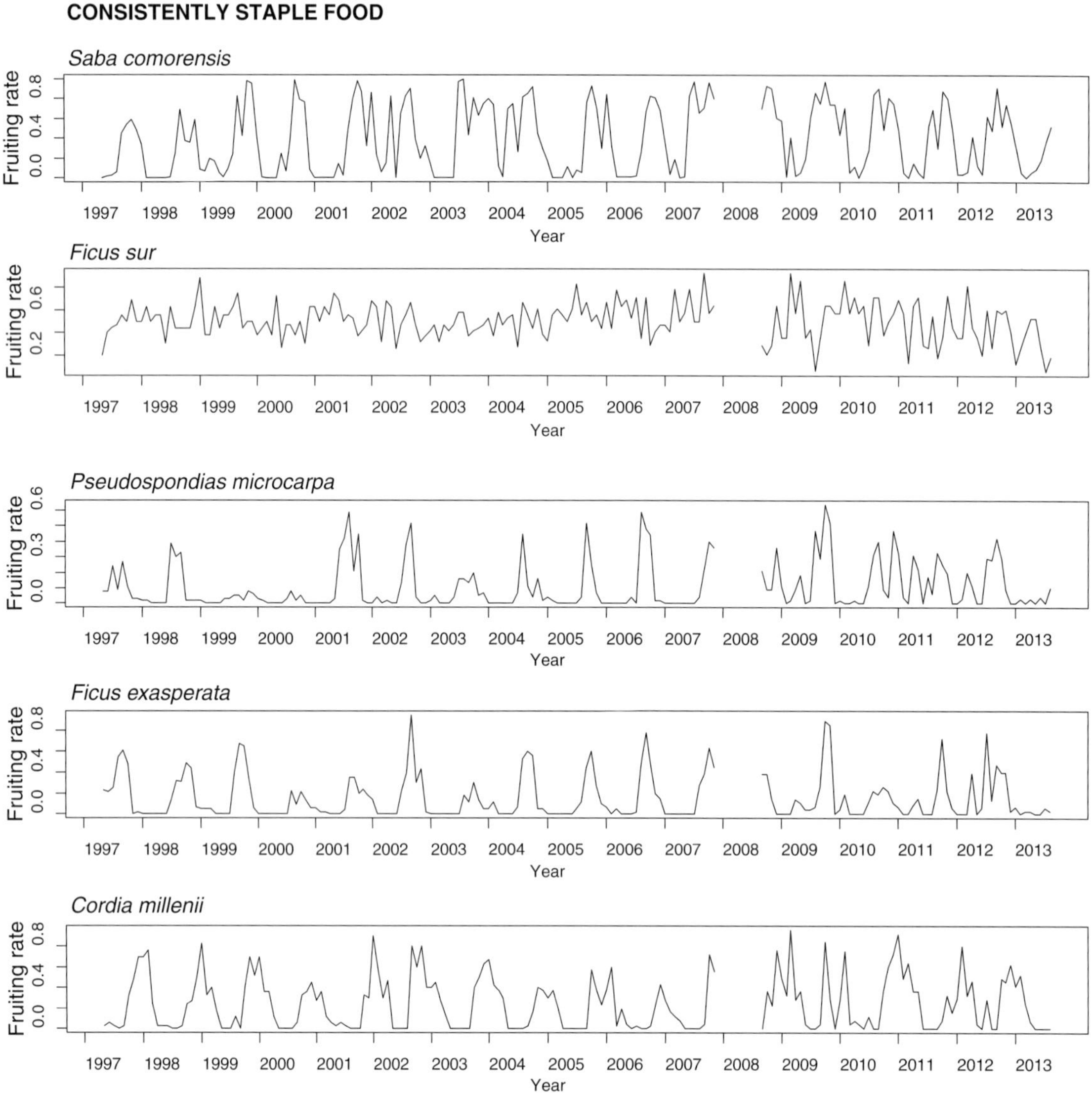

Figure 14.4 Amplitude time-course of fruiting phenology.

Staple foods

R1 Three of the four SF species exhibited the intermediate pattern (*Tabernaemontana, Psydrax*, and *Myrianthus arboreus*). The patterns observed in each species varied in different years. *F. vallis-choudae* showed the continual pattern with longer durations in the middle of the record (2001–6) and shorter durations in other periods.

R2 In contrast to R1 (with the exception of *F. vallis-choudae*), all three SF species had a 12-month periodicity, although only *Psydrax* exhibited a clear and regular periodicity. The inconsistencies might be due to the amplitude of fruiting.

R3 *Psydrax* had a clear apex at the end of the wet season (May). Apexes of remaining species were weak and vague, which might have been caused by prolonged and/or asynchronous fruiting. All patterns seem to have a slight bimodality. Note that *Myrianthus* males might have been included.

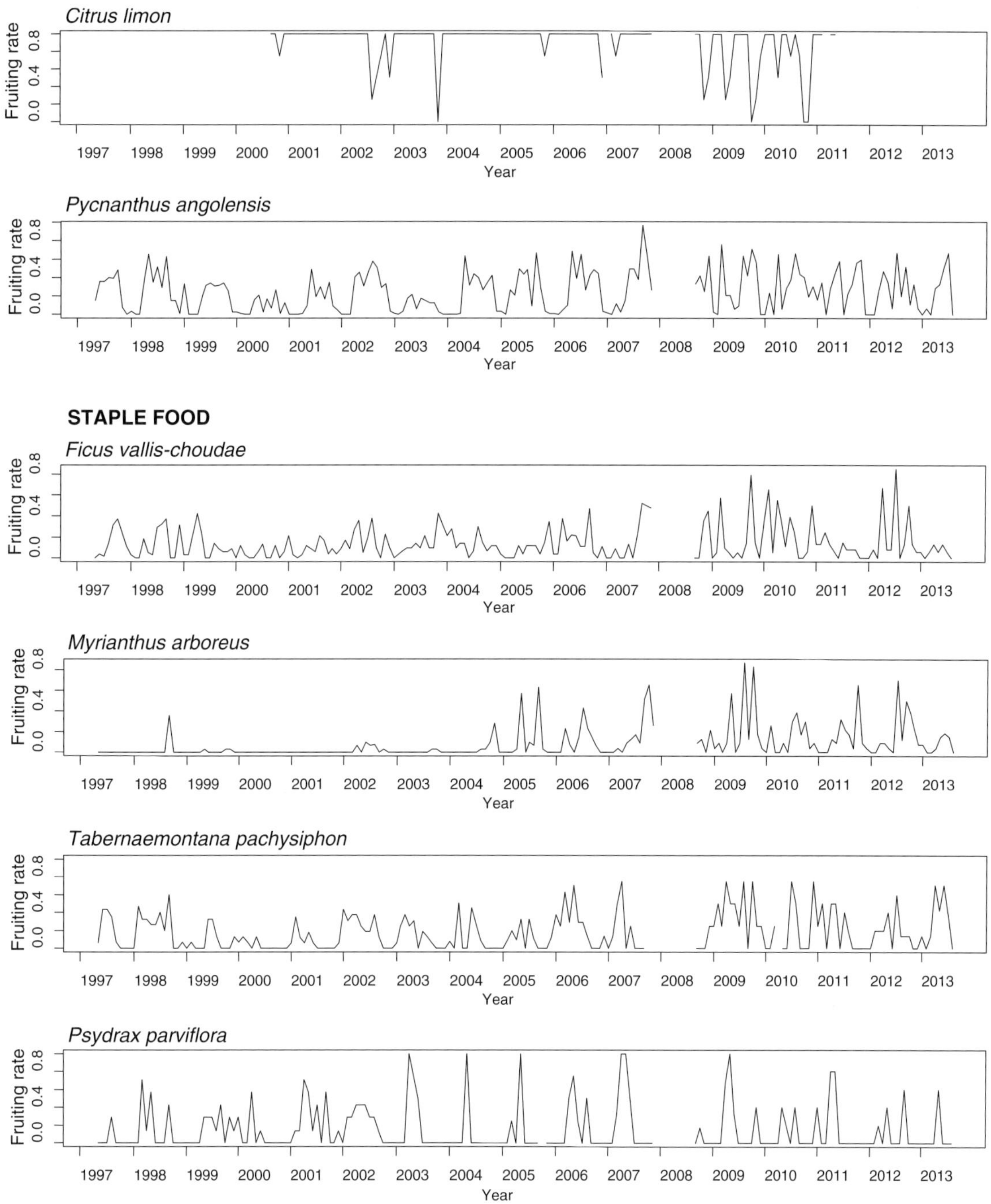

Figure 14.4 (cont.)

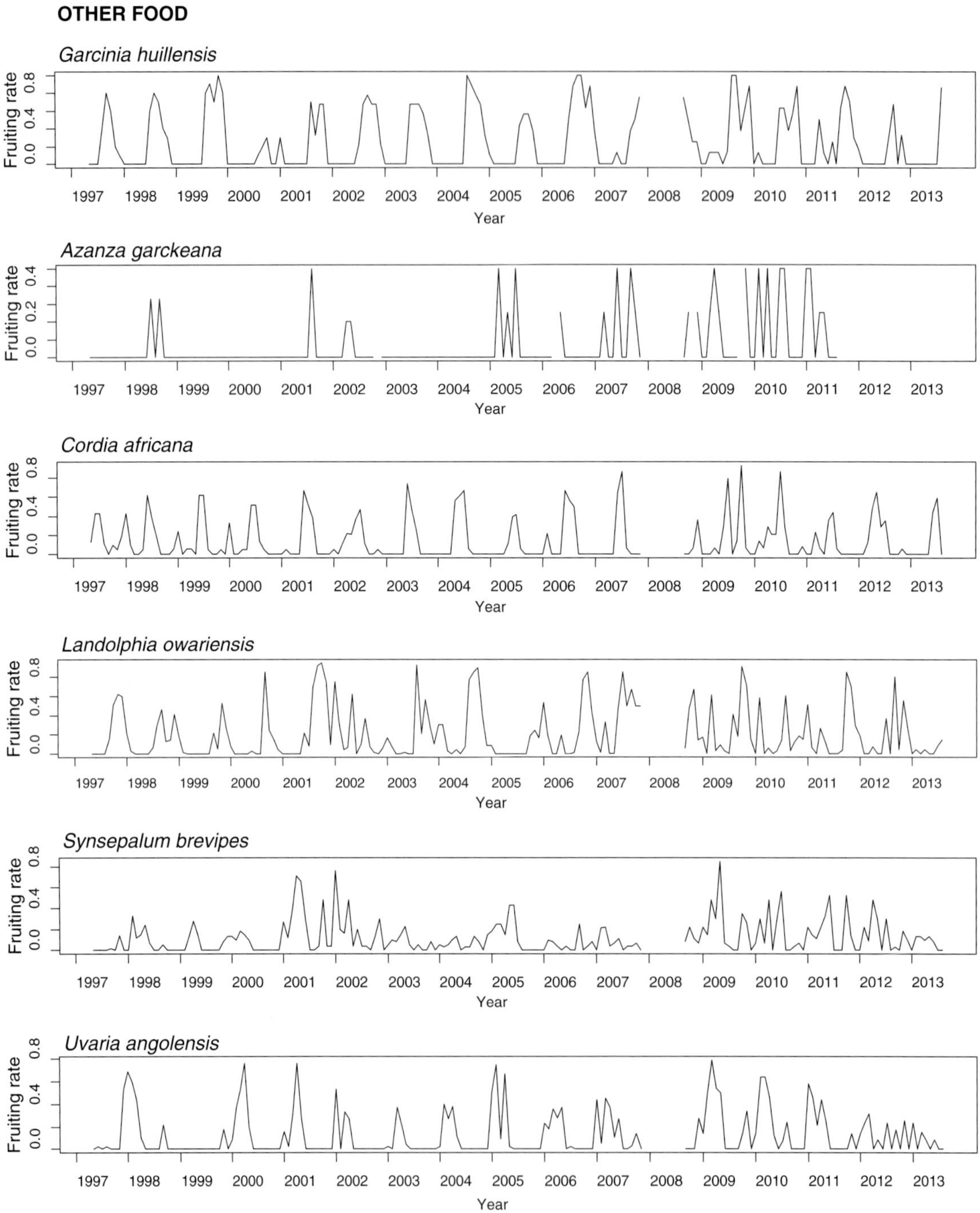

Figure 14.4 (cont.)

Table 14.2 **Summary of fruiting phenology**

		Frequency				Apex[a]		Interval[b]	
	Species	Class[c]	Simultaneity[d]	Periodicity[e]	Fruit	Flower	Flush	Flower–flush	Fruit–flower
CONSISTENTLY STAPLE	*Saba comorensis*	A	M	12 reg.	Sep–Dec (Sep)	May–Jul (Jun)	Mar–Dec (Jul)[f]	0	5
	Ficus sur	C	M	none	Nov–Jan, Mar–May, Sep (Apr)[f]	NA	A2: Nov[f]	NA	NA
	Pseudospondias microcarpa	A	L	12 reg.	A2: Sep	A2: Jul	Dec	0	2
	Ficus exasperata	A	L	12 reg.	A2: Oct	NA	Oct–Dec (Nov)	NA	NA
	Cordia millenii	A	M	12 reg.	Jan	A2: Sep	Sep–Nov (Sep)	0	3
	Citrus limon	C	H	none	all months (Feb, Jun)[f]	Nov	Nov	0	none
	Pycnanthus angolensis	A	L	12 irr.	A2: Sep	Feb–Mar (Feb)	A2: Mar	0	5
STAPLE	*Ficus vallis-choudae*	C	L	none	A2: Sep[f]	NA	A2: Jun–Jul[g]	NA	NA
	Myrianthus arboreus	I	L	12 irr.	A2: Oct	A2: Apr	Nov–Sep (Nov)[f]	0	3
	Tabernaemontana pachysiphon	I	L	12 weak but reg.	A2: Mar	Oct–Nov (Oct)	Nov–Dec (Nov)	0	0
	Psydrax parviflora	I	M	12 reg.	May	A2: Feb	Mar, Nov–Dec (Nov)	0	2

Table 14.2 (**cont.**)

		Frequency			Apex[a]			Interval[b]	
	Species	Class[c]	Simultaneity[d]	Periodicity[e]	Fruit	Flower	Flush	Flower–flush	Fruit–flower
OTHER	*Garcinia huillensis*	A	M	12 reg.	Aug–Nov (Sep)	A2: Jul	Aug	0	2
	Azanza garckeana	I	L	none	A2: Apr	A2: Mar	Mar–Apr, Nov–Jan (Dec)	2	14
	Cordia africana	I	M	12 reg.	Jun–Jul (Jul)	Apr–May (May)	Apr	5	2
	Landolphia owariensis	I	L	12 weak but reg.	A2: Oct	Jul	all except Feb (Jul)[f]	0	3
	Synsepalum brevipes	I	L	12 weak & irr.	A2: Apr	A2: Nov	A2: Nov	0	too weak
	Uvaria angolensis	A	L	12 reg.	A2: Apr	Dec	Nov–Dec (Dec)	0	4

[a] Apex months (≥50% of individuals fruiting) shown in Figure 14.3. When there were multiple apexes, the maximum month is shown in parentheses. A2 = Apex2, indicates the maximum month but below the 50%-line.

[b] The numbers indicate the lag point with the maximum positive and significant cross-correlation values. In X–Y relationship (flower–flush, fruit–flower), Y was assumed to be the predictor. Only positive lag points were assessed.

[c] Each letter indicates frequency class category shown in Figure 14.2. C: continual; A: annual; I: intermediate. *Saba* has pulsed annual pattern (subcategory of the annual pattern).

[d] Mean level of simultaneity assessed in Figure 14.3. H: high (≥80%); M: middle (50–80%); L: low (<50%).

[e] Periodicity and regularity shown in autocorrelogram (Figure 14.3). 12: 12-month periodicity; reg.: regular periodicity; irr.: irregular periodicity; weak: weak correlation.

[f] Although the apexes are identical by definition, clear trough periods were lacking, thus having similar intensity throughout year.

[g] Both months had equal intensity.

R4 *F. vallis-choudae* fruited for most months (78%) but at low amplitudes. *Tabernaemontana* fruited for longer periods although the simultaneity remained low in most months. These two species had large annual variations. *Myrianthus* also exhibited a large annual variation in all measured variables (crest location, intensity, and duration); this may reflect long (over 7–8 years) lean and bumper periods.

R5 *F. vallis-choudae* exhibited an increased simultaneity from 2009 (or 2008) to 2012. *Myrianthus* sporadically fruited up to 2002; this was followed by annual (2003–4) and sub-annual frequency patterns (2005–6) with increased intensity, and continual or pulsed annual patterns since 2007 with an even higher intensity for some episodes. *Psydrax* had short episodes but its simultaneity was augmented for most crests after 2004. *Tabernaemontana* showed increased intensity of episodes and crests from 2006 (except for 2012), but they were interrupted by increasingly deeper and more frequent troughs. Overall changes for *F. vallis-choudae* and *Myrianthus* indicate an increased accessibility in terms of duration, timing, and/or amount in the latter half of the record.

Other foods

R1 Among six OF species, four (67%) followed the intermediate pattern (*Azanza, C. africana, Landolphia,* and *Synsepalum*). All these species changed their fruiting pattern in the early- or mid-2000s. Up to 2004, fruiting of *Azanza* was occasional, followed by several short intervals and episodes per year. The immaturity of some individuals in the earlier period does not explain the low fruiting level because all individuals had fruited even before 2004. The fruiting pattern of *C. africana* in 2003–7 should be regarded as annual rather than intermediate. Intervals for *Landolphia* and *Synsepalum* have shortened since 2001, although this trend was interrupted in 2005. The remaining two species (*Uvaria angolensis* and *Garcinia huillensis*) had annual patterns, although episodes for *Uvaria* were frequently interrupted with brief intervals.

R2 In contrast to R1, most species (67%) showed a regular 12-month periodicity (most notable in the case of *C. africana*) although *Landolphia*'s periodicity was weak. For *Synsepalum*, the inconsistency was not strong as it had a weak and irregular annual periodicity. *Azanza* showed a trend similar to the results shown in section R1.

R3 The fruiting peaks for 50% of OF species (*Azanza, Synsepalum, Uvaria*) were in April (end of the wet season), with a low simultaneity. *Azana* and *Synsepalum* had small secondary apexes in mid-dry and early wet seasons, respectively, which may correspond to the irregularity or absence of periodicity described in R2. The fruiting of the remaining 50% of OF species (*C. africana, Garcinia,* and *Landolphia*) peaked during the dry and early wet (July–October) seasons.

R4 *Azanza* and *Synsepalum* showed a high year-to-year variability in the intensity of crests and events, duration of intervals, and the timing/frequency of crests and episodes. Other species also showed some variability but were stable in comparison to these species. The variability of amplitude for *Garcinia* and *Uvaria* episodes may have some long-term (several-year) periodicity, which was reported by the local assistants.

R5 Small events at the beginning of the calendar year became regular for *Garcinia* during 2009–12 (might be dated back to 2007), and the major episodes tended to be interrupted by intervals or troughs since 2009 (or 2008). The frequency of episodes of *Azanza* has increased since 2005. In 2005, it had three episodes, but we did not determine whether it followed a sub-annual pattern. The intervals *C. africana* were lengthened in the middle of the record (2003–7). The number of crests for *Landolphia* has increased from 1–2 to 2–3 per year since 2009 (could be dated back to 2008). *Synsepalum* and *Uvaria* alternated between low and high intensities for several consecutive years. The highly variable pattern at the end of *Uvaria*'s record might obscure a long-term trend. The accessibility of *Azanza* might be increasing in terms of amount and duration. However, a longer period of monitoring is needed to decide whether such results indicate a natural short-term variability or whether we are witnessing a turning point in a long-term change.

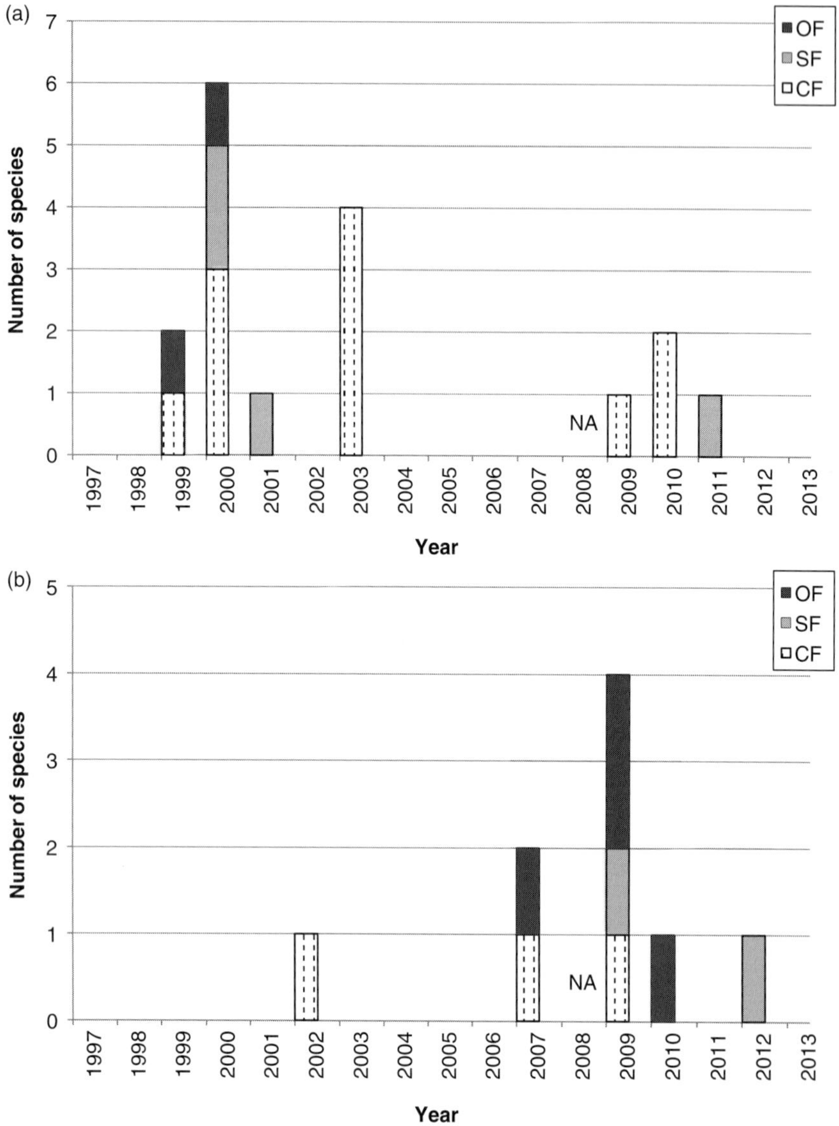

Figure 14.5 Number of species involved in rare events per year. (a) Lean events, (b) bumper events. *For species with cycles that span two calendar years, the event is shown between those years.

14.4.2 Rare events

Here, we examine the possible lean and bumper periods for each species in Figure 14.4. These periods contain rare events with large departures from the norm. For the species with low values in most episodes, only the bumper periods may be of interest, and for the species with high values, only the lean periods may be of interest.

Relatively lean events were defined as episodes with intensity smaller than one-third the maximum amplitude for the species, and bumper events were defined as episodes with the maximum intensity exceeding 0.8. Among these, we considered events rare only if they occurred, on average, once per 5 years.

Species that experienced lean events were *Pseudospondias, F. exasperata, Pycnanthus*, and *Citrus* among CF, *F. vallis-choudae* and

Tabernaemontana among SF, and *Garcinia* and *Uvaria* among OF. Figure 14.5a shows the number of species with lean events in each year. Most such events occurred before 2004.

Bumper events were detected for *F. exasperata, Pycnanthus* (CF), *F. vallis-choudae, Myrianthus* (SF), *C. africana*, and *Synsepalum* (OF). Figure 14.5b shows that such bumper events were more frequent in the second half of the record.

A comparison of these two figures demonstrates that lean events occurred more frequently in the first half, and bumper events occurred more in the second half of the record. Among climatological variables, rainfall and maximum temperature showed large departures from normal means during this monitoring period (Chapter 12). There were severe decreases in rainfall between 1999 and 2000, and from 2007, some cooler periods in both minimum and maximum temperature intervals. Although the changes in climatological variables might have affected the frequency of lean and bumper events, we found no straightforward correlations throughout the record. However, to understand such complex phenomena, we need further investigation of relationships between phenological events and abiotic and biotic factors.

14.5 Other phenological events

Van Schaik *et al.* (1993) proposed a basic framework to understand adaptive significance of the phenological-event timing. They examined possible ultimate factors by comparing records from various tropical forest sites and concluded that abiotic factors play a major role.

They found that many tropical forest communities (or species, in some cases), flushing and flowering episodes coincide. Two underlying mechanisms were proposed. The first mechanism was the water-limitation hypothesis. Water stress has a negative effect on growth, cell expansion, and production of new organs (van Schaik *et al.*, 1993). Among 53 seasonally dry tropical sites, the authors found flushing coincidence in 67% of sites and flowering in 60% of sites. Both coincidences occurred at the onset of the wet season (only 33% in both flushing and flowering for weak seasonal forest). In those seasonally dry forests, fruiting also tended to occur simultaneously; here a different underlying mechanism was suspected (i.e. minimizing seedling mortality in the coming wet season).

The second mechanism postulated was the insolation-limitation hypothesis, only applicable in the absence of water limitation. Here, tropical plants are predicted to produce young leaves during the period of high assimilation potential (i.e. high insolation); young leaves photosynthesize and control transpirational water loss most efficiently. Flowering is predicted to occur close to the flushing episode as transferring directly assimilates into growing organs. This is assumed to be energetically more efficient than storing them for later events. Among the 32 tropical sites, 31% had flushing and flowering peaks in the same month and 59% had in two consecutive months. The hypothesis does not predict the timing of fruiting in relation to these phenological trends.

Here, we briefly examine whether important chimpanzee food plants at Mahale show trends similar to those observed in other tropical forests. To establish intervals between different phenological events, lag points (month) in maximum positive and significant cross-correlation function values were used (summarized in Table 14.2; figs were excluded as flowering data was unavailable). The results show that the majority of species synchronized flushing and flowering in the same month (86%, $N = 14$), although the intervals between these events and fruiting were variable.

Among these synchronous species, no species peaked at the onset of the wet season (October). The nearest species may be *C. millenii* with maximum flushing and flowering peaks in September and the flushing peak continuing to November and *Tabernaemontana* with a flowering peak in October–November and a flushing peak in November–December. These species might be under water stress. Intervals between fruiting and any other phenological episodes for these species had no tendency for their peaks to coincide at or near the onset of the wet season.

As we do not have insolation records, the figures provided by van Schaik *et al.* (1993: Figure 1) tracing solar maxima along the latitude were used for estimates. According to the data, the sun passes directly over Mahale around October. The flushing and flowering of 64% and 29% of the species clustered around this period, respectively (species having apex month[s] falling during September–November).

Thus, our brief analysis shows that insolation, rather than water, may be one of the key factors at Mahale. The results also indicate that the majority of important food species are not under water stress.

14.6 Summary and future concerns

Among the frequency classes, annual and continual patterns were consistent with periodicity values. Intermediate patterns were more variable, but generally, the species exhibited weak and/or irregular 12-month periodicity and had weak apexes because of asynchronous fruiting activities.

Among CF and SF species, *F. sur, Citrus*, and *F. vallis-choudae* were always accessible. The differing importance of these species for chimpanzees cannot be explained only by the temporal patterns for these species; however, a high simultaneity level in the fruiting activity of the CF species may be one important feature. Investigations of spatial distribution patterns, nutrition, tastes, and behavioral data on food selection and/or preference may be also needed. Apart from these always-accessible food plants, all species exhibited periodicity with varying durations of accessibility.

CF species showed a high predictability with regular periodicity. Most species peaked from the end of the dry season to the onset of the wet season (80% of five periodical species). The most important food, in terms of frequent feeding and its lengthy duration, *Saba*, has a stable fruiting pattern and long accessibility periods. Among CF species, only *C. millenii* peaks in the wet season, when it is frequently consumed by chimpanzees (Chapter 17). CF species are generally accessible during long periods, except for *F. exasperata* and *Pseudospondias*. Fruits of *Pycnanthus* may be utilized because of its irreplaceable nutritional properties. Chimpanzees eat the pericarp of this fruit, which contains higher levels of crude lipids (77.2%) but less carbohydrate (16.9%) than other fleshy fruit at Mahale (4% and 80%, respectively, $N = 7$ species) (Itoh, 2004).

Among the periodical SF species, only *Myrianthus* had a fruiting peak at the onset of the wet season. Interestingly, although its predictability in terms of frequency and periodicity was generally low, chimpanzees frequently eat this fruit (over 50% of observation days per month) during the apex month, October (see Table 17.3), as was estimated from the means of all datasets. In comparison with other SFs, the apex for *Psydrax* was clearly at the end of the wet season and its accessibility period was more limited. The fruit of this species is extensively eaten (but during a short period) almost every year, making it stand out among the SF species (see Table 17.3).

The OF species were relatively predictable in terms of frequency and generally peaked from end of the wet season to the early dry season; in this, they differed from CF species. However, peaks tended to be weak with a large annual variability in amplitude. *Garcinia* and *C. africana* were the exceptions, having a clear apex in the dry season. Among the OF species, *Landolphia* exhibited the most stable pattern of interannual variations in intensity, but the timing of crests was variable. The fruiting pattern of *Garcinia* was comparable to that of CFs, with regular periodicity, long-lasting episodes, and high crests in most events. Thus, there might be some factors affecting the way chimpanzees consume this fruit, other than phenological patterns. The distribution pattern might be such a factor as this species forms a single-species forest stand in only two areas within the habitat (see Chapter 8). However, a longer-term periodicity with 1–3 highly productive years may also affect the observed feeding pattern.

To summarize, the predictability of fruiting phenology, in terms of periodicity and intensity, may be one of the key factors in the selection of fruit foods by chimpanzees. The effect of continuing changes in climate and vegetation will be comprehensively evaluated in the future as it affects other frugivores and the whole ecosystem of

Mahale. Below, we briefly summarize some possible effects.

Some caution is needed in the interpretation of the actual availability of foods as changes in vegetation are detected (Chapter 13). Most plant species did not show long-term decreasing trends in accessibility. Among the 17 species, *Tabernaemontana* and *F. vallis-choudae* showed the most substantial decrease in density (Chapter 13). The density of *Pycnanthus* also decreased, but as this species was superabundant at the beginning, it was still abundant at the end of the study. Thus, the response to a decrease in the density of this species, if continued, might be delayed. However, *Saba, Pseudospondias* (CFs), *Myrianthus* (SF), *Landolphia, Synsepalum*, and *Uvaria* (OFs) showed increasing-density trends. These trends in the food categories other than CF may affect the feeding frequency in the future. *Saba, Landolphia*, and *Uvaria* are lianas; this life-form competes over resources with trees (Chapter 13). Thus, an increase in lianas at Mahale (Chapter 13) may also affect the phenology of trees.

Apart from its cyclical behavior, the climate shows both short-term and long-term changes (Chapter 12). Severe decrease in rainfall may cause a shortage of fruit as demonstrated by the lean-event results. More specifically, *C. millenii* (CF) and *Tabernaemontana* (SF) with their synchronous flowering and flushing at the onset of the wet season might be negatively affected by a decrease in rainfall. As *Tabernaemontana* showed a decreasing-density trend, further investigation is needed to see whether this species responds to the decreasing rainfall at Mahale (Chapter 12).

Monitoring of plant phenology, climate, and vegetation is the most basic activity necessary for understanding animal lives and establishing conservation management plans. However, research projects and conservation programs tend to rely on short-term ecological data for a variety of reasons (insufficient funds and/or workforce, difficulties in obtaining clear-cut results, etc.). In the short run (2–5 years), ecological conditions may seem stable or follow their own cyclical behavior. Even our 16-year phenological study seems too short to draw a firm conclusion. Continuous, multi-decadal-scale monitoring is needed. Even short interruptions in monitoring could limit the understanding of long-term changes, as was the case in our study. To communicate the importance of long-term research, we should improve collaboration with governmental and funding organizations, as well as other researchers (see also Plumptre, 2012, for the importance of long-term monitoring).

References

Babweteera, F., Sheil, D., Reynolds, V., *et al.* (2012). Environmental and anthropogenic changes in and around Budongo Forest Reserve. In *The Ecological Impact of Long-Term Changes in Africa's Rift Valley*, ed. A. J. Plumptre. New York: Nova Science, pp. 31–53.

Chao, N., Mulindahabi, F., Easton, J., *et al.* (2012). Long-term changes in a montane forest in a region of high human population density. In *The Ecological Impact of Long-Term Changes in Africa's Rift Valley*, ed. A. J. Plumptre. New York: Nova Science, pp. 167–202.

Chapman, C. A., Wrangham, R. W., Chapman, L. J., Kennard, D. K., and Zanne, A. E. (1999). Fruit and flower phenology at two sites in Kibale National Park, Uganda. *Journal of Tropical Ecology*, **15**, 189–211.

Chapman, C. A., Chapman, L. J., Struhsaker, T. T., *et al.* (2005). A long-term evaluation of fruiting phenology: importance of climate change. *Journal of Tropical Ecology*, **21**, 31–45.

Chapman, C. A., Chapman, L. J., Ghai, R., *et al.* (2012). Complex responses to climate and anthropogenic changes: an evaluation based on long-term data from Kibale National Park, Uganda. In *The Ecological Impact of Long-Term Changes in Africa's Rift Valley*, ed. A. J. Plumptre. New York: Nova Science, pp. 73–94.

Itoh, N. (2002). [Foods in the forest: chimpanzee food density and spatial distribution.] In [*The Mahale Chimpanzees: 37 Years of <Panthropology>*], ed. T. Nishida, S. Uehara, and K. Kawanaka. Kyoto: Kyoto University Press, pp. 77–100. In Japanese.

Itoh, N. (2004). [*The Plant Phenology and Fission–Fusion Social System in the Mahale Mountains National Park*.] Doctoral Dissertation. Kyoto: Kyoto University. In Japanese.

Itoh, N. and Nishida, T. (2007). Chimpanzee grouping patterns and food availability in Mahale Mountains National Park, Tanzania. *Primates*, **48**, 87–96.

Itoh, N., Nakamura, M., Ihobe, H., *et al.* (2012). Long-term changes in the social and natural environments surrounding the chimpanzees of the Mahale Mountains National Park. In *The Ecological Impact of Long-Term Changes in Africa's Rift Valley*, ed. A. J. Plumptre. New York: Nova Science, pp. 249–77.

Morellato, L. P. C., Camargo, M. G., D'Eça Neves, F. F., *et al.* (2010). The influence of sampling method, sample size, and frequency of observations on plant phenological patterns and interpretation in tropical forest trees. In *Phenological Research*, ed. I. L. Hudson and M. R. Keatley. Amsterdam: Springer, pp. 99–121.

Moscovice, L. R., Issa, M. H., Petrzelkova, K. J., *et al.* (2007). Fruit availability, chimpanzee diet, and grouping patterns on Rubondo Island, Tanzania. *American Journal of Primatology*, **69**, 487–502.

Nakamura, M., Corp, N., Fujimoto, M., *et al.* (2013). Ranging behavior of Mahale chimpanzees: a 16 year study. *Primates*, **54**, 171–82.

Newstrom, L. E., Frankie, G. W., and Baker, H. G. (1994a). A new classification for plant phenology based on flowering patterns in lowland tropical rain forest trees at La Selva, Cost Rica. *Biotropica*, **26**(2), 141–59.

Newstrom, L., Frankie, G., Baker, H., and Colwell, R. (1994b). Diversity of long-term flowering patterns at La Selva. In *La Selva: Ecology and Natural History of a Neotropical Rain Forest*, ed. L. A. McDade, K. S. Bawa, H. A. Hespenheide, and G. S. Hartshorn. Chicago, IL: University of Chicago Press, pp. 142–60.

Plumptre, A. J. (2012). Long term changes in Africa's western rift valley: synthesis of main findings. In *The Ecological Impact of Long-Term Changes in Africa's Rift Valley*, ed. A. J. Plumptre. New York: Nova Science, pp. 279–93.

R Core Team (2014). *R: A Language and Environment for Statistical Computing, Vienna, Austria*: R Foundation for Statistical Computing. Available at: http://www.R-project.org.

Sakai, S. (2001). Phenological diversity in tropical forests. *Population Ecology*, **43**(1), 77–86.

van Schaik, C. P., Terborgh, J. W., and Wright, S. J. (1993). The phenology of tropical forests: adaptive significance and consequences for primary consumers. *Annual Review of Ecology and Systematics*, **24**(1), 353–77.

Sun, C., Kaplin, B. A., Kristensen, K. A., *et al.* (1996). Tree phenology in a tropical montane forest in Rwanda. *Biotropica*, **28**(4b), 668–81.

Williamson, E. A. (1988). *Behavioural Ecology of Western Lowland Gorillas in Gabon*. Ph.D. thesis. Stirling, UK: University of Stirling.

15 Mammalian fauna

HIROSHI IHOBE

15.1 Overview

Mahale is a biogeographically interesting region; its fauna includes both savanna-adapted species of eastern and southern Africa and forest-adapted species of western and central Africa and the Congo Basin (Carcasson, 1966; Itani, 1977b; Kielland, 1978). Appendix III shows the list of mammals inhabiting the Mahale Mountains National Park. At least 70 mammal species of 11 orders have been found in the park. Of particular interest are the African giant squirrel (*Protoxerus stangeri*), red-legged sun squirrel (*Heliosciurus rufobrachium*), brush-tailed porcupine (*Atherurus africanus*), bushy-tailed mongoose (*Bdeogale crassicauda*), banded mongoose (*Mungos mungo*), blue duiker (*Cephalophus monticola*), and Sharpe's grysbok (*Raphicerus sharpei*). Some of these sighting records represent considerable extensions of the known ranges of these species (Nishida, 1990). The list of animals from mammals to insects and their vernacular names of the Tongwe land, including Mahale, were described by Itani (1977a). Fukuda (2001) also reviewed the fauna of the park.

15.1.1 Primates

Seven species of diurnal primates, chimpanzees (*Pan troglodytes*), yellow baboons (*Papio cynocephalus*), red colobus (*Procolobus rufomitratus*), Angolan black-and-white colobus (*Colobus angolensis*), red-tailed monkeys (*Cercopithecus ascanius*), blue monkeys (*C. mitis*), and vervet monkeys (*Chlorocebus pygerythrus*), and at least three species of nocturnal primates, greater galagos (*Otolemur crassicaudatus*), South African galagos (*Galago moholi*), and dwarf galagos (*Galagoides* sp.), inhabit the national park. Detailed descriptions of the chimpanzees are provided in other chapters (see Chapters 4–8). In this chapter, some information about primates other than chimpanzees is presented.

Yellow baboons (Figure 15.1) inhabit the park, whereas olive baboons (*Papio anubis*) inhabit Gombe National Park, Tanzania, located about 150 km away from Mahale. The border between the two species of baboons is the Malagarasi River. Yellow baboons mainly inhabit the woodland along the lakeshore in the park. They form multi-male and multi-female groups of 40 to 60 animals. Chimpanzees were first observed eating yellow baboons in 1997 (Nakamura, 1997). Since then, eating of yellow baboon meat has occurred sporadically at relatively low frequencies (see Chapter 20). This is partly because yellow baboons have invaded the forest away from the lakeshore and competition between the two species has increased (Nishida, 1997, 2002; Matsumoto-Oda and Kasagula, 2000; see also Chapter 16).

In Table 1.1 of Nishida (1990), the scientific name of the red colobus at Mahale (Figure 15.2) was given as *Colobus badius*. In recent studies, however, red colobus have been divided into several species (Kingdon *et al.*, 2013) and the species inhabiting Mahale is now classified as the eastern red colobus (*Procolobus rufomitratus*). Red colobus digest their food in much the same way as do ruminant animals. They have a sacculated forestomach housing a multitude of microbes that ferment the ingested food,

Mahale Chimpanzees: 50 Years of Research, ed. M. Nakamura, K. Hosaka, N. Itoh and K. Zamma. Published by Cambridge University Press.

Figure 15.1 A yellow baboon. ©Michio Nakamura.

Figure 15.2 A red colobus. ©Michio Nakamura.

allowing the colobus to feed on large amounts of leaves. They form multi-male and multi-female groups of about 30 animals at Mahale (Ihobe, 2002). The most interesting aspect of red colobus is that they are the most frequent target of chimpanzee hunting (Chapter 20). The proportion of red colobus among the targets of chimpanzee hunting is more than 80% (Hosaka *et al.*, 2001). This is almost the same as at other chimpanzee research sites like Gombe (Stanford *et al.*, 1994), Kibale Forest in Uganda (Watts and Mitani, 2002), and Taï Forest in Côte d'Ivoire (Boesch and Boesch, 1989). It is unclear why chimpanzees prefer red colobus (see Chapters 16, 20).

The Angolan black-and-white colobus in the park are probably a new subspecies (Nishida *et al.*, 1981). Detailed information of their ecology and behavior has not been collected because they inhabit only higher altitudes near the peak of Mt. Nkungwe.

Red-tailed monkeys (Figure 15.3) are the most abundant primates in the park (Uehara and Ihobe, 1998). They form one-male groups of about 20 animals. Chimpanzees seldom hunt this species, despite their abundance, because its agility appears to hinder

Figure 15.3 A red-tailed monkey. ©Michio Nakamura.

predation by chimpanzees (see Chapters 16, 20). Ihobe (2000) suggested that polyspecific associations between this species and red colobus occur less often than can be expected by chance at Mahale. Both species may not benefit from polyspecific associations, considering that they employ different anti-predator strategies against chimpanzees (see Chapter 16). In contrast, red-tailed monkeys from Kibale seem to initiate, maintain, and terminate associations with red colobus, probably because the associations may reduce the high risk of eagle predation on red-tailed monkeys (Teelen, 2007).

Blue monkeys form similar one-male groups to red-tailed monkeys. Their population densities at Mahale are very low, about one-tenth that of red-tailed monkeys. The reason for this low density is unclear.

Nishida (1990) used the classic scientific name of vervet monkeys (Figure 15.4), *Cercopithecus aethiops*. Recent studies, however, have classified vervet monkeys into several species (Kingdon *et al.*, 2013), and the scientific name of the species at Mahale is *Chlorocebus pygerythrus*. Vervet monkeys form multi-male and multi-female groups containing 20 to 30 animals at Mahale. Only one or two groups inhabit the area around the Kasiha camp (Chapter 2) in the Kasoje area of the park (Uehara and Ihobe, 1998).

In Nishida (1990), the scientific name and English name of "greater galagos" were given as *Galago crassicaudatus* and thick-tailed galagos, respectively. However, a group of greater galago species was recently separated from other galagos to form another genus, *Otolemur*. Greater galagos (Figure 15.5) are the biggest species among galagos, and they weigh about 1 kg. Another species of galagos was described as Senegal galagos (*Galago senegalensis*) in Nishida (1990). South African galagos, however, is the correct species, reconsidering the taxonomy and distribution of galagos. Their distribution pattern corresponds to the distribution of Miombo forest. Moyer (2006) reported the presence of another species of galagos, dwarf galagos, but identification of this species has not been conducted. Each species has loud calls for defending their territories that sound like human babies, so they are sometimes called "bush babies."

Figure 15.4 A vervet monkey. ©Michio Nakamura.

Figure 15.5 A greater galago. ©Michio Nakamura.

Population densities of diurnal primates have been periodically surveyed since 1995 in the Kasoje area. The changes in their population densities are described later.

15.1.2 Chiroptera

At least five species of Chiroptera, bats, belonging to three families have been confirmed to inhabit the park (Stanley, 2004; Moyer, 2006). Among them, two species are fruit bats and three are insect-eating bats. Singing fruit bats (*Epomops franqueti*) are medium-sized and inhabit forests and forest–woodland mosaics. Epauletted fruit bats (*Epomophorus* sp.) are suspected to inhabit the park (Stanley, 2004), but identification of this species has not been completed.

Tomb bats (*Taphozous perforates*) have a wide distribution and mainly hunt moths. Evening bats (*Scotoecus albofuscus*) are distributed in woodland savannas and dry acacia country and eat insects. One species of pipistrelle bat (*Pipistrellus* sp.) is suspected to inhabit the park (Stanley, 2004), but identification of this species has not been completed.

15.1.3 Soricomorpha and Macroscelidea

Order Insectivora is divided into several orders by the recent classification and Soricomorpha and Macroscelidea were formally grouped in Insectivora. At least two species of Soricomorpha and one species of Macroscelidea have been confirmed to inhabit the park. From Soricomorpha, one species of climbing shrew (*Sylvisorex megalura*) and at least two species of white-toothed shrew (*Crocidura* spp.) of Soricidae seem to inhabit the park (Stanley, 2004; Moyer, 2006). Table 1.1 of Nishida (1990) names one species of white-toothed shrew (*Crocidura monax*) inhabiting the park. However, because the identification source is unclear and because more than 100 species are known in this genus, we have to be careful about its species identity. There is one species of chequered elephant shrew (*Rhynchocyon*

Figure 15.6 A chequered elephant shrew. ©Michio Nakamura.

cirnei) (Figure 15.6) of Macroscelididae in Macroscelidea in the park.

15.1.4 Rodents

At least 18 species of rodents of seven families have been confirmed in the park.

Two species of Sciuridae, squirrels, have been confirmed to inhabit the park, the red-legged sun squirrel (*Heliosciurus rufobrachium*) and the African giant squirrel (*Protoxerus stangeri*). Moyer (2006), however, pointed out that other squirrels are often misidentified as red-legged sun squirrels and it is possible that another type of sun squirrel inhabits the park. Population densities of squirrels have been investigated since 1996 (Uehara and Ihobe, 1998) and their densities were almost equal until 2011 (Ihobe, 2012).

One species of African dormouse (*Graphiurus* sp.) of Gliridae is suspected to inhabit the park (Stanley, 2004). One species of silky blesmol (*Heliophobius* sp.) of Bathyergidae is also suspected to inhabit the park. This species was inaccurately referred to as a "mole rat" in Table 1.1 of Nishida (1990).

Two species of Hystricidae, porcupines, inhabit the park, the cape crested porcupine (*Hystrix africaeaustralis*) (Figure 15.7) and the brush-tailed porcupine (*Atherurus africanus*). Nishida (1990) classified the former as the crested porcupine (*Hystrix galeata*), but according to a more recent investigation of mammals in Mahale (Moyer, 2006), it is likely that cape crested porcupines actually inhabit the park.

One species of cane rat (greater cane rat: *Thryonomys swinderianus*) of Thryonomyidae and at least two species of Nesomyidae (forest pouched rat: *Cricetomys gambianus* and climbing mouse: *Dendromus* sp.) inhabit the park. Nishida (1990) referred to the former as "giant rats" (*Cricetomys emini*). However, according to a more recent investigation of small mammals in Mahale (Stanley, 2004), forest pouched rats may be more appropriate.

At least nine species of Muridae inhabit the park. Among them, information on two species, *Mus tenellus* and *Praomys jacksoni* that appeared in Table 1.1 of Nishida (1990) may need reconsideration; a recent investigation of small mammals in Mahale by experts (Stanley, 2004) could not identify the animals

Figure 15.7 A cape crested porcupine. ©Koichiro Zamma.

Figure 15.8 An African clawless otter. ©Michio Nakamura.

of these genera to the species level. Thus, it may be safer to say that *Mus* sp. and *Praomys* sp. exist at Mahale, but species level identifications should be done in the future. Seven other species (groove-toothed rat, *Otomys* sp.; brush-furred rat, *Lophuromys* sp.; montane wood mouse, *Hylomyscus* sp.; thicket rat, *Grammomys* sp.; rusty-nosed rat, *Oenomys hypoxanthus*; single-striped grass mouse, *Lemniscomys rosalia*; and common striped grass mouse, *L. striatus*) were identified in recent studies on small mammals (Stanley, 2004; Moyer, 2006).

15.1.5 Carnivores

Seventeen species from six families (Canidae, Mustelidae, Herpestidae, Hyaenidae, Viverridae, and Felidae) of this order have been confirmed to inhabit the park.

Two species of Canidae inhabit the park, side-striped jackals (*Canis adustus*) and wild dogs (*Lycaon pictus*). Both species are rarely observed in the Kasoje area, but there have been some sightings. Side-striped jackals have a wide distribution in savanna and thickets in sub-Saharan Africa, and there is some informal information from local residents (Michio Nakamura, personal communication) that these jackals are commonly observed near Katumbi village (see Chapter 2) outside the park. Wild dogs also had a very wide distribution in Africa, but it is now fragmented by human activities, such as livestock grazing, direct persecution, and diseases from domestic dogs.

Three species of Mustelidae, ratels (*Mellivora capensis*), African clawless otters (*Aonyx capensis*) (Figure 15.8), and spotted-necked otters (*Lutra maculicollis*), have been confirmed to inhabit the park.

Ratels, also known as honey badgers, inhabit not only sub-Saharan Africa, but also the Middle East and India. Their main food is social insects such as bees, but they also feed on small mammals and birds. The English name of *Aonyx capensis* was given as "cape clawless otter" in Table 1.1 of Nishida (1990), but another name, African clawless otter, is now more commonly used. The two species of otters have wide distributions. We can occasionally observe them around Kasiha River near Kansyana Camp (Chapter 2).

In Nishida (1990), mongooses were included in Viverridae, but they have been placed in Herpestidae

Figure 15.9 A genet with relatively large blotches. ©Michio Nakamura.

according to the recent classification. Three species of mongooses, banded mongooses (*Mungos mungo*), white-tailed mongooses (*Ichneumia albicauda*), and bushy-tailed mongooses (*Bdeogale crassicauda*), were listed in the table, and two more species, slender mongooses (*Herpestes sanguineus*) and common dwarf mongooses (*Helogale parvula*) were confirmed to inhabit the park in a recent study (Moyer, 2006). Among them, banded mongooses form large groups containing around several dozen animals, and they can sometimes be observed during studies on chimpanzees or monkeys at the Kasoje area.

One species of Hyaenidae, spotted hyenas (*Crocuta crocuta*), inhabits the park. They are rarely observed in the Kasoje area, but there has been some evidence of them, such as their white feces and footprints.

At least two species of Viverridae inhabit the park, genets (*Genetta* sp. or spp.) (Figure 15.9) and African civets (*Civettictis civetta*). In Table 1.1 of Nishida (1990), two species of genets, common genets (*Genetta genetta*) and large-spotted genets (*Genetta tigrina*), are listed. In fact, two or more species of genets may inhabit the park since genets with different sizes of spots have been observed (Michio Nakamura, personal communication). Kingdon *et al.* (2013), however, indicated that large-spotted genets are only found in South Africa, thus they are possibly absent around Mahale. Because the revision of classifications of *Genetta* is underway, exact identifications of genets in the park by specialists are awaited. African civets are found in sub-Saharan Africa in forested or partly forested mosaics.

Four species of Felidae, African wild cats (*Felis silvestris*), serval cats (*Felis serval*), leopards (*Panthera pardus*), and lions (*Panthera leo*), inhabit the park. Among them, serval cats and lions are rarely observed in the Kasoje area, but there have been some sightings.

African wild cats are found in most of the African continent except for deserts and rainforests. They are known to feed on small mammals up to the size of a hare or small antelope. Serval cats are distributed in savanna areas and they also feed on small mammals.

Leopards (Figure 15.10) have a wide distribution from the Kamchatka Peninsula of eastern Eurasia to Africa, and the populations inhabiting Africa seem to be one subspecies. We sometimes hear vocalizations

Figure 15.10 A leopard. ©Michio Nakamura.

of leopards in the Kasoje area, even near the research camp, and evidence of their presence, such as feces, is found everywhere in the park. Leopards eat antelopes, small mammals like rodents, birds, and arthropods; red colobus fur and bones have been found in their feces at Mahale (Fukuda, 2001). Their feeding habits may strongly affect mammal populations at Mahale. A recent study showed that they also ate chimpanzees at Mahale, and this was the first record of leopard consumption of chimpanzees in East Africa (Nakazawa *et al.*, 2013). There is some possibility that predation pressure on chimpanzees by leopards affects the population dynamics of chimpanzees since such a phenomenon has been observed at Taï (Boesch and Boesch-Achermann, 2000). Meanwhile, chimpanzees kill infant leopards at Mahale (Hiraiwa-Hasegawa *et al.*, 1986; for details see Chapter 16) and are one of the natural enemies of leopards.

Lions had a wide distribution from Africa to India, but nowadays, outside Africa they inhabit only the Gir Forest National Park. They are one of the biggest "cats" in the world. They do not inhabit the Kasoje area, but there were occasional sightings of lions in the home range of the M group at least until 1989 (Hiraiwa-Hasegawa *et al.*, 1986; Tsukahara, 1993). Two lions that visited in 1989 were suspected to have killed four chimpanzees of the M group (Tsukahara, 1993).

15.1.6 Pholidota

One species of Pholidota, the giant pangolin (*Smutsia gigantea*) (Figure 15.11), inhabits the park. Early Mahale researchers (e.g. Itani, 1977; Nishida, 1990), listed cape pangolins (*Manis temminckii*) as present at Mahale, but a recent study on mammals in the park confirmed the existence of giant pangolins (Moyer, 2006). Although Moyer (2006) noted that Mahale may be a rare location where two pangolin species are sympatric, a reassessment of an old photograph of a pangolin (in Itani *et al.*, 1973) at Mahale seems to

Figure 15.11 A giant pangolin. ©Michio Nakamura.

show that the species is actually a giant pangolin. Thus, it is likely that former descriptions of cape pangolins at Mahale are due to misidentification of giant pangolins. Chimpanzees at Bossou, Guinea, were observed to eat tree pangolins (*Phataginus tricuspis*) (Sugiyama and Koman, 1987), but chimpanzees at Mahale have not been observed to hunt pangolins.

15.1.7 Tubulidentata

One species of Orycteropodidae, aardvarks (*Orycteropus afer*) (Figure 15.12), inhabits the park. Aardvarks are distributed in sub-Saharan Africa. They are nocturnal and it is difficult to observe them directly, but holes that they are suspected to have dug are found in the Kasoje area.

Figure 15.12 An aardvark. ©Michio Nakamura.

15.1.8 Hyracoidea

Two species of this order in the Procaviidae family, hyraxes, have been confirmed to inhabit the park, the yellow-spotted hyrax (*Heterohyrax brucei*) and the southern tree hyrax (*Dendrohyrax arboreus*). The latter is not found in the Kasoje area. Iida (2010) briefly reported on the ecology of yellow-spotted hyraxes from observations on the lakeshore of Lake Tanganyika in the Kasoje area. They make colonies of several dozen animals in rock outcrops.

15.1.9 Proboscidea

One species of Elephantidae, African elephants (*Loxodonta africana*), inhabits the park. They are not found in the Kasoje area; instead, they occur in drier habitat on the eastern side of the Mahale Mountains of the park.

15.1.10 Perrissodactyla

One species of Equidae, common zebras (*Equus quagga*), inhabits the park. They are listed as Grant's zebras (*Equus bruchelli*) in Table 1.1 of Nishida (1990), but a recent classification of zebras adopted this English name and Latin name. They are not found in the Kasoje area; instead, they inhabit the same drier part of the park as elephants.

15.1.11 Cetartiodactyla

Eleven species of three families (Hippopotamidae, Suidae, and Bovidae) of this order have been confirmed to inhabit the park.

Hippopotamuses (*Hippopotamus amphibius*) are often seen around the river mouth of the Lubulungu River in the Kasoje area (Chapter 2) because this river consists of a large amount of water even in the dry season, whereas other rivers in the area dry up in this season. Many paths that hippopotamuses use for feeding on grasses at night are observed around the river and around Sinsiba Swamp (Chapter 2).

Two species of Suidae inhabit the park, bushpigs (*Potamochoerus larvatus*) (Figure 15.13) and common warthogs (*Phacochoerus africanus*) (Figure 15.14). The scientific names of these two species were given as *Phacochoerus aethiopicus* and *Potamochoerus porcus*, respectively, in Table 1.1 of Nishida (1990), but these were incorrect and the correct names are as above.

Bushpigs occupy a wide range of habitats and are highly omnivorous. They are solitary except during the mating season or in mother-and-infants groups. Young animals or infants of bushpigs have been observed to be hunted by chimpanzees at Mahale (Uehara *et al.*, 1992) and Gombe (Goodall, 1986). Population densities of this species are unclear

Figure 15.13 A bushpig. ©Michio Nakamura.

Figure 15.14 A common warthog. ©Michio Nakamura.

because of their nocturnal habit (Uehara and Ihobe, 1998).

Warthogs are widely distributed and they typically live in arid and open areas. They have "warts" on their faces, and their very narrow tail is held vertically when they run. Mothers and their offspring form groups while adult males usually live solitarily. Warthogs were not observed in the Kasoje area until the 1980s, but they can be easily observed today (Uehara and Ihobe, 1998). Moreover, chimpanzees were observed to start to hunt warthogs after the 1980s (Uehara *et al.*, 1992).

Eight species of Bovidae inhabit the park. African buffaloes (*Syncerus caffer*) range across all but the driest parts of sub-Saharan Africa and also inhabit the park. Evidence of this species, however, is rarely observed in the Kasoje area.

Bushbucks (*Tragelaphus scriptus*) (Figure 15.15) are distributed across much of sub-Saharan Africa. They are water-dependent but can subsist on dew. Adult males weigh about 80 kg, and adult females weigh about 60 kg. Only males have horns. They are ruminants and largely depend on leguminous herbs and growing grasses. Males live solitarily and females and their offspring form small groups. Young and infant bushbucks have been observed to be hunted by chimpanzees at Mahale (Uehara *et al.*, 1992) and Gombe (Goodall, 1986), and scavenging behavior of chimpanzees on this species was observed at Mahale (Hasegawa *et al.*, 1983; see Chapters 16, 20). Bushbuck population densities have decreased from the late 1990s (Ihobe and Uehara, 1999). The reasons for this decrease seem to be hunting by chimpanzees and leopards and changes in vegetation from woodland to forest (Uehara, 2003).

Two species of duikers inhabit the park, bush duikers (*Sylvicapra grimmia*) and blue duikers (*Cephalophus monticola*). Bush duikers are distributed in the savannas and woodlands of sub-Saharan Africa and are longer legged and larger eared than forest duikers like blue duikers. Bush duikers are probably absent around Kasoje.

Blue duikers (Figure 15.16) are small duikers; adults weigh 4–10 kg. They are widely distributed from the Niger River to Western Kenya in forests and forest

Figure 15.15 A bushbuck. ©Kazuhiko Hosaka.

Figure 15.16 A blue duiker. ©Michio Nakamura.

mosaics. They prefer wet environments, because their population densities are higher in the forest than in the woodland at Mahale (Uehara and Ihobe, 1998). They form bonded pairs or live solitarily. They are ruminants but heavily depend on fruits. Chimpanzees of K group were more likely to hunt blue duikers (Uehara *et al.*, 1992) than chimpanzees of M group (Hosaka *et al.*, 2001). However, this lower proportion in the latter may be due to increased red colobus hunting, or the selectivity of prey species of chimpanzees may have changed because the population densities of blue duikers have increased in recent years as described below. Chimpanzees at Taï ignore blue duikers, and this species does not belong to chimpanzees' prey images (Boesch and Boesch, 1989).

Sharpe's grysboks (*Raphicerus sharpei*) are small antelopes; adults weigh about 10 kg. They prefer low thickets and secondary growth. They are predominantly nocturnal and it is very difficult to observe them directly. Klipspringers (*Oreotragus oreotragus*) are also small antelopes. They are found from Ethiopia to South Africa and favor steep, rocky hillsides, escarpments, and valleys. Waterbucks (*Kobus ellipsiprymnus*) have been recently confirmed to inhabit the park (Koichiro Zamma, personal communication). They were observed in the northern limit of the park. Roan antelopes (*Hippotragus equinus*) are tall, horse-like antelopes with massive, arched horns. These four species of antelopes are not found in the Kasoje area.

15.2 Population changes of mammals

The population densities of some medium- and large-sized mammals have been periodically surveyed since 1995 in the Kasoje area. Censuses were done in 1995, 1996, 2000, 2002, and 2011 (Uehara and Ihobe, 1998; Uehara 2003; Ihobe, 2012; Itoh *et al.*, 2012). Uehara and Ihobe (1998) described the details of the methods of the census. Subject species of the census were red colobus, red-tailed monkeys, blue monkeys, yellow baboons, forest squirrels (two species of squirrels combined), warthogs, blue duikers, and bushbucks (see Table 15.1 for the summary).

Red colobus do not seem to have decreased since 1995 (Itoh *et al.*, 2012), although red colobus hunting by chimpanzees frequently occurred in this period. This agrees with the previous estimation that chimpanzee hunting did not seem to affect red colobus populations and annual predation rates for red colobus (1.1–3.8%) were supposed to be less than their population growth rates at Mahale (Boesch *et al.*, 2002).

Chimpanzee hunting, however, may cause changes in population densities of red colobus in specific areas (Itoh *et al.*, 2012). The frequency of red colobus hunting in the northern part of the home range of M-group chimpanzees has decreased in recent years (MMCRP, unpublished data). This decrease in hunting in the northern part may explain the increase in group densities of red colobus there. Meanwhile, group densities of red colobus in the southern part of the home range of the chimpanzees were highest in 1996 and decreased after that. This decrease in group densities may be caused by frequent hunting by chimpanzees in this area in recent years (MMCRP, unpublished data).

Population densities of red-tailed monkeys and blue monkeys do not seem to have changed in this period (Itoh *et al.*, 2012).

Although group densities of yellow baboons may not have changed since 1995, individual densities or numbers of this species may have increased because of increases in group size (Itoh *et al.*, 2012). In the previous study, the group size of this species was estimated to be about 40 (Uehara and Ihobe, 1998), but more than 60 baboons in a group were observed in 2002 (Ihobe, 2012). Such increases in individual densities of baboons may have caused their invasion into chimpanzee habitat since the late 1990s (Nishida, 1997, 2002; Matsumoto-Oda and Kasagula, 2000).

There are no apparent trends in individual densities of forest squirrels (Itoh *et al.*, 2012) or population densities of warthogs during the period (Ihobe, 2012).

The density of blue duikers has increased. Decreased blue duiker hunting by chimpanzees in recent decades (Hosaka *et al.*, 2001) may explain this increase (Itoh *et al.*, 2012).

The density of bushbucks has decreased and chimpanzee hunting may also have caused this change (Itoh *et al.*, 2012). Estimated chimpanzee predation rates for bushbucks before 1996 might have been higher than their population growth rates (Ihobe and Uehara, 1999). The same situation

Table 15.1 **Summary of group or individual densities of some mammals at Kasoje based on Uehara and Ihobe (1998), Uehara (2003), Ihobe (2012), and Itoh et al. (2012). Numbers represent groups/km^2 for primates and individuals/km^2 for other species**

	Red-tailed monkey		Red colobus		Yellow baboon		Blue monkey		Blue duiker		Forest squirrels		Warthog		Bushbuck
Year	F	W	F	W	F	W	F	W	F	W	F	W	F	W	W
1995	4.2	2.8	3.3	0.9	–	1.8	–	–	n.i.	n.i.	n.i.	n.i.	n.i.	n.i.	n.i.
1996	6.9	3.8	4.2	1.0	0.2	1.5	0.9	–	19.5	7.8	4.5	0.8	0.7	4.6	7.2
2000	7.7	4.2	3.5	0.9	0.2	1.6	0.4	–	36.9	16.6	10.2	3.0	0.8	4.5	1.5
2002	6.4	3.5	4.1	0.4	0.3	1.9	0.2	–	27.9	15.7	5.1	–	0.7	–	1.3
2011	5.8	5.1	2.1	1.1	–	1.1	0.4	0.4	37.6	28.7	8.8	6.1	–	–	–

F: forest; W: woodland

n.i.: not investigated.

–: frequencies of encounters were fewer than three times for each vegetation type during the study periods and densities were not indicated.

was observed at Gombe. Predation rates for bushbucks by chimpanzees and olive baboons (*Papio anubis*) might exceed their population growth rate (Wrangham and van Zinnicq Bergmann Riss, 1990).

Although apparent changes in vegetation have not been detected since 1996, changes in the population densities of blue duikers and bushbucks may be related to secondary succession (Itoh *et al.*, 2012). Blue duikers were often found in the forest, but bushbucks were often found in the woodland. In addition, such a change in vegetation type may explain the increase in group densities of red-tailed monkeys and red colobus in some areas, because these monkeys are often found in the forest (Itoh *et al.*, 2012).

Leopard predation may affect individual densities for some species. Unlike other eastern chimpanzee study sites, where leopards seem to be extinct, leopards are commonly seen at Mahale. Nakazawa *et al.* (2013) further suggested that there are comparable, or even more, sightings of leopards at Mahale than at Taï, where predation pressure by leopards is known to be high. Bones and hairs of red colobus were sometimes found in feces of leopards (Fukuda, 2001), and leopard predation may also explain the decrease in bushbucks (Uehara, 2003). It has been confirmed that leopards at Mahale prey upon animals as large as an adult chimpanzee (Nakazawa *et al.*, 2013). Crowned hawk eagles inhabit Mahale (Itani, 1977a; Nishida, 1990) and they may prey on small- or medium-sized mammals. A raptor (species not confirmed) was observed preying on an adult blue duiker (Michio Nakamura, personal communication). We have to conduct studies on these animals to understand the predator–prey relationships more precisely.

15.3 Conclusion

Since Nishida (1990) reported on the mammal fauna of the park, more than 20 further species inhabiting the park have been identified and information on several species has been modified. Some of the progress has come from identifying several species of small mammals, such as bats and rodents, (Stanley, 2004; Moyer, 2006), since Nishida (1990) noted that "bats, shrews, and many rodents still remain to be identified." However, classifications of several species have not been completed to date; further studies on mammals other than chimpanzees are required to elucidate the fauna of mammals in the park. In addition, studies on mammals other than chimpanzees will be needed to understand the ecosystem in the park. For example, Ihobe (2002) described anti-chimpanzee hunting strategies of red colobus. Nobuko Nakazawa and Mia Otani recently started studies of leopards (e.g. Nakazawa *et al.*, 2013), and Eriko Iida studied hyraxes (Iida, 2010). Such studies will contribute to understanding not only the ecology and behavior of the target species but also interspecific relationships among mammals of the park (see also Chapter 16).

Further studies on population densities will be needed from several viewpoints. At first, such studies will contribute to conservation activities in the park. Data on changes in the populations of mammals are necessary for determining the order of priority of conservation of mammals. In addition, such data will be useful for evaluating conservation strategies. Second, these studies will provide information on relationships among mammals. In particular, information on changes in potential prey are useful for understanding the nature of chimpanzee hunting because the availability of potential prey in at least some cases at Mahale seems to have affected predation frequency by the chimpanzees (Nishida and Uehara, 1983; Uehara *et al.*, 1992). Chimpanzees at Mahale are rather euryphagous in terms of their selectivity towards prey species (Nishida, 1974; Boesch and Boesch, 1989), and the opportunistic nature of their hunting behavior appears to have been reconfirmed on the basis of the abundance of potential prey (Uehara and Ihobe, 1998; Chapter 20). Further studies on changes in the population densities of such mammals will also reveal some of the characteristics of hunting by leopards and raptors because target mammals of population surveys are potential prey of these animals.

References

Boesch, C. and Boesch, H. (1989). Hunting behavior of wild chimpanzees in the Taï National Park. *American Journal of Physical Anthropology*, **78**, 547–73.

Boesch, C. and Boesch-Achermann, H. (2000). *The Chimpanzees of the Taï Forest: Behavioural Ecology and Evolution*. Oxford: Oxford University Press.

Boesch, C., Uehara, S., and Ihobe, H. (2002). Variations in chimpanzee-red colobus interactions. In *Behavioural Diversity in Chimpanzees and Bonobos*, ed. C. Boesch, G. Hohmann, and L. F. Marchant. Cambridge: Cambridge University Press, pp. 221–30.

Carcasson, R. H. (1966). Lepidoptera Rhopalocera collected by the Kyoto University African Anthropoid Expedition in the Kigoma area of western Tanganyika. *Kyoto University African Studies*, **1**, 11–72.

Fukuda, F. (2001). [*The Animals in the African Forest*.] Tokyo: Jinruibunka-sha. In Japanese.

Goodall, J. (1986). *The Chimpanzees of Gombe: Patterns of Behavior*. Cambridge, MA: Belknap.

Hasegawa, T., Hiraiwa, M., Nishida, T., and Takasaki, H. (1983). New evidence on scavenging behavior in wild chimpanzees. *Current Anthropology*, **24**, 231–2.

Hiraiwa-Hasegawa, M., Byrne, R. W., Takasaki, H., and Byrne, J. M. E. (1986). Aggression toward large carnivores by wild chimpanzees of Mahale Mountains National Park, Tanzania. *Folia Primatologica*, **47**, 8–13.

Hosaka, K., Nishida, T., Hamai, M., Matsumoto-Oda, A., and Uehara, S. (2001). Predation of mammals by the chimpanzees of the Mahale Mountains, Tanzania. In *All Apes Great and Small, Vol. 1: African Apes*, ed. B. M. F. Galdikas, N. E. Briggs, L. K. Sheeran, G. L. Shapiro, and J. Goodall. New York: Kluwer Academic Publishers, pp. 107–30.

Ihobe, H. (2000). [Red colobus vs. chimpanzees: predator–prey relationships among primates.] In [*Primate Ecology: Dynamism between the Environment and Behavior*], ed. Y. Sugiyama. Kyoto: Kyoto University Press, pp. 61–84. In Japanese.

Ihobe, H. (2002). [Anti-chimpanzee hunting strategies of red colobus.] In [*The Mahale Chimpanzees: 37 Years of <Panthropology>*], ed. T. Nishida, S. Uehara, and K. Kawanaka. Kyoto: Kyoto University Press, pp. 245–60. In Japanese.

Ihobe, H. (2012). A further study of the impacts of chimpanzee hunting on the population of several mammal species at Mahale, Tanzania. Paper presented at the 24th Congress of International Primatological Society, Cancun, Mexico.

Ihobe, H. and Uehara, S. (1999). [A preliminary report on the impact of chimpanzee hunting on mammal populations at Mahale, Tanzania.] *Primate Research*, **15**, 163–9. In Japanese with English summary.

Iida, E. (2010). [Strange animals inhabiting rock outcrops: hyraxes.] *Mahale Chimpun*, **16**, 1–2. In Japanese.

Itani, J. (1977a). [Ethnozoology of the Batongwe.] In [*The Natural History of Man*], ed. J. Itani and R. Harako. Tokyo: Yūzan-kaku, pp. 441–537. In Japanese.

Itani, J. (1977b). [Introduction.] In [*The Chimpanzees*], ed. J. Itani. Tokyo: Kōdan-sha, pp. 3–56. In Japanese.

Itani, J., Nishida, T., and Kakeya, M. (1973). [*The Eastern Hinterland of Lake Tanganyika: Its Nature and People*.] Tokyo: Chikuma-shobō. In Japanese.

Itoh, N., Nakamura, M., Ihobe, H., *et al*. (2012). Long-term changes in the social and natural environments surrounding the chimpanzees of the Mahale Mountains National Park. In *The Ecological Impact of Long-Term Changes in Africa's Rift Valley*, ed. A. J. Plumptre. New York: Nova Science, pp. 249–77.

Kielland, J. (1978). A provisional checklist of the Rhopalocera of the eastern side of Lake Tanganyika. *Tijdschrift voor Entomologie*, **121**, 147–237.

Kingdon, J., Happold, D., Butynski, T., *et al*. (2013). *Mammals of Africa Vol. I–VI*. London: Bloomsbury Natural History.

Matsumoto-Oda, A. and Kasagula, M. B. (2000). Preliminary study of feeding competition between baboons and chimpanzees in the Mahale Mountains National Park, Tanzania. *African Study Monographs*, **21**, 147–57.

Moyer, D. C. (2006). *Biodiversity of Mahale Mountains National Park, Tanzania*. New York: Wildlife Conservation Society.

Nakamura, M. (1997). First observed case of chimpanzee predation on yellow baboons (*Papio cynocephalus*) at the Mahale Mountains National Park. *Pan Africa News*, **4**, 9–11.

Nakazawa, N., Hanamura, S., Inoue, E., Nakatsukasa, M., and Nakamura, M. (2013). A leopard ate a chimpanzee: the first evidence from East Africa. *Journal of Human Evolution*, **65**, 334–7.
Nishida, T. (1974). [Ecology of wild chimpanzees.] In [*Human Ecology*], ed. R. Ohtsuka, J. Tanaka, and T. Nishida. Tokyo: Kyōritsu-shuppan, pp. 15–60. In Japanese.
Nishida, T. (1990). A quarter century of research in the Mahale Mountains: an overview. In *The Chimpanzees of the Mahale Mountains: Sexual and Life History Strategies*, ed. T. Nishida. Tokyo: University of Tokyo Press, pp. 3–35.
Nishida, T. (1997). Baboon invasion into chimpanzee habitat. *Pan Africa News*, **4**, 11–12.
Nishida, T. (2002). Competition between baboons and chimpanzees at Mahale. *Pan Africa News*, **9**, 23–6.
Nishida, T. and Uehara, S. (1983). Natural diet of chimpanzees (*Pan troglodytes schweinfurthii*): Long-term record from the Mahale Mountains, Tanzania. *African Study Monographs*, **3**, 109–30.
Nishida, T., Itani, J., Hiraiwa, M., and Hasegawa, T. (1981). A newly-discovered population of *Colobus angolensis* in East Africa. *Primates*, **22**, 557–63.
Stanford, C. B., Wallis, J., Matama, H., and Goodall, J. (1994). Patterns of predation by chimpanzees on red colobus monkeys in Gombe National Park, 1982–1991. *American Journal of Physical Anthropology*, **94**, 213–28.
Stanley, W. T. (2004). *The Small Mammals of Mahale Mountains National Park*. Chicago, IL: The Field Museum.
Sugiyama, Y. and Koman, J. (1987). A preliminary list of chimpanzees' alimentation at Bossou, Guinea. *Primates*, **28**, 133–47.
Teelen, S. (2007). Influence of chimpanzee predation on associations between red colobus and red-tailed monkeys at Ngogo, Kibale National Park, Uganda. *International Journal of Primatology*, **28**, 593–606.
Tsukahara, T. (1993). Lions eat chimpanzees: the first evidence of predation by lions on wild chimpanzees. *American Journal of Primatology*, **29**, 1–11.
Uehara, S. (2003). Population densities of diurnal mammals sympatric with the chimpanzees of the Mahale Mountains, Tanzania: comparison between the census data of 1996 and 2000. *African Study Monographs*, **24**, 169–79.
Uehara, S. and Ihobe, H. (1998). Distribution and abundance of diurnal mammals, especially monkeys, at Kasoje, Mahale Mountains, Tanzania. *Anthropological Science*, **106**, 349–69.
Uehara, S., Nishida, T., Hamai, M., *et al.* (1992). Characteristics of predation by the chimpanzees in the Mahale Mountains National Park, Tanzania. In *Topics in Primatology, Vol. 1: Human Origins*, ed. T. Nishida, W. C. McGrew, P. Marler, M. Pickford, and F. B. M. de Waal. Tokyo: University of Tokyo Press, pp. 143–58.
Watts, D. and Mitani, J. C. (2002). Hunting behavior of chimpanzees at Ngogo, Kibale National Park, Uganda. *International Journal of Primatology*, **23**, 1–28.
Wrangham, R. W. and van Zinnicq Bergmann Riss, E. (1990). Rates of predation on mammals by Gombe chimpanzees, 1972–1975. *Primates*, **31**, 157–70.

16 Interspecific relationships

KAZUHIKO HOSAKA AND HIROSHI IHOBE

16.1 Introduction

The chimpanzee in the wild lives in a community of multiple living organisms that interact with each other in various ways. From an ecological aspect, chimpanzees are first known as predators of numerous plants (Chapter 17), insects (Chapter 21), birds, and mammals (Chapter 20). Second, they are the potential prey of some predators such as humans (Chapter 50) and large carnivores. Third, they are often in competition with other animals for food sources. From a behavioral aspect, chimpanzees may interact with other animals in a social manner, e.g. agonistically, peacefully, and playfully.

The purpose of this chapter is to review interactions between chimpanzees and various sympatric vertebrates in Mahale (Chapter 15; Appendix III). Furthermore, we discuss how a better understanding of their interspecific relationships could be useful in further studies of behavioral ecology and human evolution.

Most of the interspecific interactions reviewed in this chapter were observed when tracking chimpanzees. Therefore, a bidirectional perspective, such as predator–prey ecology, is clearly insufficient. One exception is a study comparing the anti-predator strategies of two arboreal monkey species against chimpanzees.

Responses of chimpanzees upon encounter with sympatric animals may provide insight into how chimpanzees view the species. They may have acquired particular images of the animal as a result of evolution or individual development. Because chimpanzees are vigilant, demonstrate avoidance, and are aggressive against potential predators, it may be meaningful to explore how they categorize other animals into "danger vs. non-danger." Second, because chimpanzees tend to consume only a portion of rich fauna (Chapter 20), they may stick to their "prey image" (Boesch and Boesch, 1989; Ihobe, 1993) when determining which animal to eat. Thus, it may also be of interest to explore how chimpanzees categorize other animals into "food vs. non-food." Third, chimpanzees may see some other animals as competitors, when the proportion of overlapping foods is large.

However, their view of other animals as danger, food, or competitor, may vary. For example, animals of the same species may be eaten in some cases, but not in others. Such ambiguity may reflect flexibility in the recognition of other species by chimpanzees, which may explain why the prey profile of chimpanzees varies between sites (Chapter 20).

16.2 Relationships with sympatric carnivores

16.2.1 Leopards

For a long time, large felines have been thought to prey upon chimpanzees in some situations. Studies of savanna-living chimpanzees in Western Tanzania reported encounters between chimpanzees and a leopard (Izawa and Itani, 1966) or a young adult lion (Kano, 1972). They suggested that large felines could

Mahale Chimpanzees: 50 Years of Research, ed. M. Nakamura, K. Hosaka, N. Itoh and K. Zamma. Published by Cambridge University Press.

be dangerous because some chimpanzees, if not all, showed excitement associated with fear and threat. Nishida (1968) insisted that Mahale chimpanzees do not see leopards as predators, based on his observation that an adult male chimpanzee successfully threatened a distant leopard by shaking vegetation without showing fear or alarming others. Van Lawick-Goodall (1968) and Pierce (2009) reported four cases of chimpanzee encounters with leopards from early studies at Gombe, Tanzania. In these, chimpanzees showed emotional responses filled with screams and wraas, together with threatening behavior such as shaking vegetation and investigative behavior toward leopards, while they sometimes ignored leopards that were not nearby.

Do leopards prey upon chimpanzees?

The first evidence of leopard predation on chimpanzees came from Taï, Côte d'Ivoire, the study site of western chimpanzees, where six chimpanzees died and four were injured probably by leopards (Boesch, 1991). Boesch concluded that 39% of chimpanzee deaths over 5 years could be ascribed to leopard predation. Although he insisted that regular leopard predation occurs there, he also proposed that one "chimp-killer" might have been responsible for all or most predations. Fecal analyses of Taï leopards conducted before and after those 5 years support the latter, because it was rare to find chimpanzee remains in leopard scats (0/215 feces in Hoppe-Dominik, 1984; 1/200 feces in Zuberbühler and Jenny, 2002), suggesting that chimpanzees are not usually an important prey of leopards. Jenny and Zuberbühler (2005) updated the Taï data by adding two separate cases of a dead juvenile chimpanzee being dragged and scavenged by a leopard.

Leopard predation on chimpanzees was suggested in two other subspecies of chimpanzees. Furuichi (2000) found a dismembered carcass of a central chimpanzee in Petit Loango, Gabon, suggesting that it was eaten by a leopard based on a leopard scat and footprints near the carcass. Nakazawa *et al.* (2013) first reported evidence of an adult female eastern chimpanzee of Mahale being eaten by a leopard, based on morphological investigations and DNA analyses of chimpanzee bones found in a leopard scat. Hart *et al.* (1996) found chimpanzee remains in leopard scats collected in the Ituri Forest, one of the habitats of eastern chimpanzees, based on microscopic examinations of the sample. However, since there was no direct observations of killing, scavenging remains a possibility.

Agonistic relationships between chimpanzees and leopards

Relationships between chimpanzees and leopards are generally hostile. At Mahale in 1984, Hiraiwa-Hasegawa *et al.* (1986) observed 33 chimpanzees surrounding a leopard den containing a mother and at least one cub; the latter was dragged out and killed. Although 10 chimpanzees were observed holding the kill by turns, it was not eaten. Boesch (1991) observed that Taï chimpanzees mobbed a leopard, forcing it into a narrow hole using a weapon (hitting and stabbing with a fallen branch) to attack the leopard. Jenny and Zuberbühler (2005) showed that one adult female leopard avoided chimpanzees: in 6 out of 15 cases when the leopard detected a group of chimpanzees 100–250 m away by their drumming and/or screaming, she started moving or changed her route to avoid them.

During 18 months between 1991 and 1994, Hosaka recorded five cases of M-group chimpanzee responses to leopards. These suggest that a chimpanzee reacts to a leopard in diverse ways depending on the situation. They may show intense reactions such as wraas and flight, mounting others from behind, and taking shelter on a tree. They may also show moderate reactions such as vigilance without particular emotions, and even approach to scavenge fresh kills left by leopards.

On the morning of December 7, 1991, researchers noticed that chimpanzees had encountered a leopard because at least two females emitted wraas and fresh leopard footprints were found on the ground nearby. The alpha male, Kalunde, walked close behind his aged ally, Bakali, and the two sat together on a fallen tree. When an adult female, Wakilufya, and a juvenile male then sat close behind Kalunde, and another adult, female perched on a nearby tree emitted a wraa although the leopard was gone. All of the chimpanzees except for Bakali that were sitting on the same fallen tree rushed to climb nearby trees or vines. About 1 min later, Kalunde climbed down from the tree slowly and started to depart to the north followed by Bakali and Wakilufya.

On March 31, 1993, Hosaka observed a young adult male, Toshibo, emitting wraas and sheltering in a tree for at least 5 h, presumably after detecting a leopard because a leopard footprint was spotted beneath the tree on which Toshibo was sitting. There were no other chimpanzees nearby. Toshibo had been eating *Cordia millenii* fruits alone and was excited, looking intently down from a tree and calling wraas for about 12 min. He then calmed down and emitted sporadic calls such as huus and extended grunts. Throughout, he did not descend from the tree but spent most of his time resting and sometimes moved branch by branch to adjacent trees to find fruits. On another day in the same month, a Tanzanian research assistant observed a very similar case of prolonged sheltering by two adult females Wakasunga and Pulin with an infant, in a presumed encounter with a leopard.

On January 15, 1994, when six chimpanzees, including the alpha male, Ntologi, an adult male, Jilba, and an adult female, Fatuma, walked along the trail to Kansyana Camp (Chapter 2), they burst out chorusing waa-barks in response to a leopard roar from a nearby thicket. Twenty minutes later, the third-ranking male, Kalunde, joined them and initiated a pant-hoot in chorus. Then Ntologi departed for the camp, followed by Kalunde, Jilba, and Fatuma in single file. Forty seconds later, Jilba stood bipedally and pricked up his ears toward the eastern thicket. Thirty seconds later, a leopard's rasping or sawing call came from the east nearby. Ntologi sat, turning back to Kalunde. Jilba grinned, approached bipedally, and mounted Kalunde, who mounted Ntologi having just stood up quadrupedally. The three males, joined in a row by mounting, emitted shrill barks (Figure 16.1a–c). Several waa-barks broke out in response. Ntologi quickly went ahead. Jilba and Kalunde followed him, but stopped one by one to sniff the same wayside leaves as if to check whether the leopard had passed by (Figure 16.1d; see Matsumoto-Oda *et al.* [2007] for "sniffing behaviors" in chimpanzees). Two minutes after the leopard call, the three chimpanzees reached the camp and chorused a pant-hoot with Kalunde drumming a metal house. Within 7 min, five additional drumming displays were performed by the males.

In contrast, chimpanzees react less intensely to distant leopards. On August 21, 2008, on the riverbed of the Kasiha (Chapter 2), several chimpanzees including the alpha male, Pimu, heard a leopard rasping call from more than 300 m upstream. They had no specific response, except that a female turned her head to gaze intently upstream.

Twice in the late evening of July 9, 1993, a party of chimpanzees including the alpha male Ntologi heard a leopard rasping call from a 100-m distance. However, the chimpanzees had no particular reactions and continued to feed on leaves. They even moved closer to the leopard call and made their beds for sleeping. The next morning, six chimpanzees including the beta male Nsaba found and scavenged a fresh kill bushbuck that

Figure 16.1 Responses of chimpanzees to a leopard roar. (a) The three adult males (from front, Jilba, Kalunde, and Ntologi) have formed a row by mounting. (b) After mounting, Jilba (left) and Kalunde (right) remain standing bipedally behind Ntologi (middle). (c) Ntologi started to walk. Jilba was grinning and still touching Kalunde's back. (d) Jilba and Kalunde sniffing the same wayside leaves as if to check whether the leopard had passed by. ©K. Hosaka.

had been left on the ground by the leopard (Hosaka *et al.*, 2001). Interestingly, all animals remained silent while eating and three males left at an extremely slow pace as if not to be detected by the leopard.

16.2.2 Lions

Lions prefer to inhabit savanna or grassland and are sympatric with savanna-living chimpanzees, such as in Ugalla, Tanzania (Kano, 1972), and Mt. Assirik, Senegal (Tutin *et al.*, 1981). In September 1966, at a small forest along the Ugalla River, Kano (1972) observed about 15 chimpanzees being driven into a tree after spotting a male lion, followed by loud calls made with confusion and fear.

At Mahale in the early days of research, there were sporadic visits by a lone lion passing through the study site once every 2–3 years (Tsukahara, 1993). In the 1980s, lions visited Kasoje at least twice. First, in October 1984, at least two lions visited Kansyana Camp (Hiraiwa-Hasegawa *et al.*, 1986). At the artificial feeding site, four adult females noticed lions resting in the thickets and began emitting loud vocalizations such as wraas for 1.5 h, followed by hoo calls (probably "huu" in Goodall [1986]; see also Chapter 39), single wraas, and screams. Other notable responses were scratching, defecation, and branch shaking. The lions did not move from the ground throughout the episode. Second, at least two lions visited Mahale and stayed for most of 1989 (Tsukahara, 1993). Analyses of lion scats and daily records of chimpanzee presence suggested that at least four chimpanzees were eaten by lions between June and September 1989. This was the first evidence of lion predation on wild chimpanzees. Responses of chimpanzees to lions included alarm calls, whimpers, tree climbing, and silence (Tsukahara, 1993).

16.2.3 Other carnivores

In Mahale, there are at least 15 other carnivores sympatric with chimpanzees (Appendix III). Some, such as spotted hyena, wild dogs, side-striped jackals, could be dangerous to chimpanzees, but encounters between chimpanzees and these carnivores have not been observed in Kasoje. In August–September 2005, Hosaka identified one cluster of footprints and two hyena scats on a trail utilized by chimpanzees. Thus, chimpanzees might have been aware of the hyena roaming in their habitat.

Genets, civets, and mongooses are abundant in Kasoje, but encounters between these and chimpanzees have not been reported. From Gombe, Teleki (1973a) reported a case where an infant showed excitement in encountering a genet while its mother ignored it. At present, there is no evidence that chimpanzees see smaller carnivores as potential predators against conspecific infants. However, although rare, Mahale chimpanzees may compete with other carnivores for meat. On January 17, 2001, the 23-year-old male Masudi appropriated a blue duiker kill from another predator, perhaps an African civet. When Masudi was alone eating fruits on the tree, a loud, prolonged scream of a duiker burst out from the ground. Human observers instantly realized that the duiker must have been captured by a small ambush predator. Masudi quickly descended from the tree and rushed into the bush, obtaining a fresh kill of an adult duiker. Although the predator was not visible because of the dense undergrowth, it was presumed to be either a civet or a ratel considering its hunting method and prey size. Masudi climbed the tree with the kill, scavenged meat around the head, and discarded the whole body.

Nishida *et al.* (1979) recorded a rare case in which K-group chimpanzees hunted a white-tailed mongoose. In contrast, Wrangham (1975) described an anecdote about a special relationship between Gombe chimpanzees and a male banded mongoose at the artificial feeding site. Adult chimpanzees ignored the mongoose, while juveniles and adolescents attempted to play with him. Interestingly, frequent encounters between the two species generated a "troop member" feeling among the chimpanzees. When Wrangham caught the mongoose within sight of chimpanzees, female and juvenile chimpanzees ran toward him in a threatening manner and slapped him.

16.3 Relationships with major sympatric primates

16.3.1 Red colobus

The red colobus is the second most abundant monkey in the Kasoje forest (Chapter 15) and an important prey species for sympatric predators such as leopards and chimpanzees, as well as for Tongwe inhabitants who use the colobus coat as part of their traditional bellows (Nishida, 1972). Interestingly, in Mahale and in every other long-term study site where chimpanzees and red colobus are sympatric, red colobus is eaten by chimpanzees more frequently than other forest monkeys (Chapters 15, 20).

On the basis of field data collected in 1995 and 1998, Ihobe (2002) investigated specific predator–prey relationships from the prey perspective at Mahale, and suggested that red colobus at Mahale adopt a "stand and defend strategy," as does the red colobus at Gombe (Stanford, 1998). When red colobus monkeys detect chimpanzees, they emit alarm calls to elicit anti-predator responses of troop members (Ihobe, 2002). An adult female colobus moved lower in the trees to hide silently in the foliage of vine leaves, while an adult male colobus moved further down to defend the females and offspring (Ihobe, 2002). Ironically, this anti-predator strategy may be the very reason chimpanzees can catch and hunt red colobus frequently.

This anti-chimpanzee strategy of red colobus at Mahale and Gombe is different from that of counterparts at Taï (Bshary and Noë, 1997). In this region, red colobus individuals are more elusive and move silently in the opposite direction through the canopy when they hear chimpanzees at distance. Such a strategy may have developed due to the closed canopy structure there (Bshary and Noë, 1997). Alternatively, a large-bodied chimpanzee at Taï could easily kill a small-bodied red colobus of any age or sex, while this would not be the case for small-bodied chimpanzees at Gombe and Mahale.

The chimpanzee–red colobus relationship has changed due to long-term changes in the flora and fauna at Mahale (Chapters 13, 15). In the early 1990s, adult male colobus were occasionally successful at deterring chimpanzees, and sometimes attacked chimpanzees directly in the trees. At that time, no colobus at Mahale had been observed to launch a surprise attack on chimpanzees when they were not attempting to hunt (Hosaka, personal observations), whereas this occasionally occurred in the early years at Gombe (Busse, 1977). However, in the early 2000s, some researchers noticed that the colobus living southwest of Kansyana Camp to the lower reaches of the Kasiha Valley had started to act aggressively toward chimpanzees and humans. For example, on September 8, 2005 when the aged adult male Kalunde was walking alone on the trail near the Kasiha, a large adult male colobus charged Kalunde, vigorously descending and barking from a tree. Kalunde ran away rapidly westward. The colobus stopped and sat on a lower branch. There have been similar observations to this, and in some cases, adult male colobus have descended to the ground to threaten and chase adult chimpanzees away.

16.3.2 Red-tailed monkeys

Red-tailed monkeys are the most abundant forest monkey in Kasoje (Chapter 15), but are much less frequently hunted by the M-group chimpanzees than are red colobus (Chapter 20). Ihobe (2002) noted that red-tailed monkeys adopt an anti-predator strategy that contrasts with that of red colobus. Red-tailed monkeys silently and quickly move in the opposite directions once they detect chimpanzees, while red colobus remain in the trees and hide behind foliage or defend themselves. It is plausible that chimpanzees have difficulty hunting elusive red-tailed monkeys.

Interestingly, however, Mahale chimpanzees sometimes show indifference to red-tailed monkeys even when they are in the same tree. On November 16, 1991, Hosaka observed such an instance when nine chimpanzees shared a fruiting tree of *Pseudospondias microcarpa* beside Kansyana Camp with an adult red-tailed monkey. There were three adult male chimpanzees in the tree but they were engaged in grooming behavior. An adult female, Silafu, an eager meat eater and skillful hunter, was also there but continued to feed on fruits, and never showed an interest in the red-tailed monkey. After sharing the tree for 84 min, the red-tailed monkey moved to an adjacent tree, avoiding Serena, the 4-year-old daughter of Silafu, who was approaching and charging it playfully while on the same branch.

Tolerance of chimpanzees toward guenons was reported from other study sites of eastern chimpanzees. At Budongo, Uganda, red-tailed and blue monkeys are sometimes found nearby, and occasionally feed in the same tree (Reynolds and Reynolds, 1965). John and Reynolds (1997) observed an adolescent female chimpanzee grooming a red-tailed monkey at Budongo. At Gombe, an adolescent chimpanzee was observed in relaxed contact play with an adult red-tailed monkey for 30 min (Teleki, 1973b).

16.3.3 Yellow baboons

The interspecific relationship between chimpanzees and yellow baboons has changed during long-term research at Mahale. Although baboons are abundant at Kasoje, they prefer woodland to forest and so they encounter chimpanzees less frequently than do forest monkeys (Chapter 15).

According to Nishida (2002), in 1973–4, baboon troops at Kasoje inhabited only two dispersed open areas, because villagers drove them off. However, since villagers moved out of Kasoje in 1975, baboons have expanded their ranges into the lakeshore area between the Kansyana Valley and Lubulungu River (Chapter 2). In the early 1990s, their ranges extended to cover more forested areas inland.

Nishida (1997) observed the first encounter between chimpanzees and baboons at Kasoje in 1991. In January 1992, Hosaka noticed that a troop of baboons occasionally reached the Kansyana Camp to feed on oil palm mesocarp when the chimpanzees were far away. Three months later, those baboons stole dried fish from the camp kitchen (Shigeo Uehara, personal communication).

In the early 1990s when Mahale chimpanzees encountered baboons, they never attempted to hunt, but rather avoided contact or showed aggression to drive them away. Once in 1993, on the riverbed of the Kasiha Valley, an adult male, Aji, was observed to charge, with piloerection, toward baboons feeding on fruits on a fig tree, and successfully took over the food patch. Thus, chimpanzees appeared to be dominant over baboons in direct competition for food.

However, as baboons expanded deep into the home range of the M group, the diet overlap grew and chimpanzees began to face more severe feeding competition with baboons (Nishida, 1997, 2002; Matsumoto-Oda and Kasagula, 2000). Baboons have an advantage in that they harvest unripe fruits earlier than do chimpanzees, who prefer ripe fruits. Nishida (2002) speculated that such a disadvantage forced chimpanzees to use eastern and highland parts of their home range more frequently around 1994 than they had previously.

The interspecific relationship changed when Mahale chimpanzees began to sporadically prey upon baboons in 1997 (Chapter 20).

16.4 Responses to encounters with other sympatric animals

16.4.1 Bovids and suids

Blue duiker is the second most frequently hunted prey animal of the Mahale chimpanzees (Chapter 20). Although the density of blue duiker has increased over the past two decades (Chapter 15), which is possibly due to decreased predation by chimpanzees (Chapter 20), chimpanzees seem to have retained their image of this bovid as food. Bushbuck fawn and bushpig piglets are also eaten by Mahale chimpanzees, although this has been rare since the 1990s (Chapter 20).

The warthog is a diurnal animal that is abundant in the woodland of Kasoje, and is a prey species (but rarely killed) for Mahale chimpanzees (see Chapter 20 for details). In usual encounters, chimpanzees ignore warthogs. In some cases, chimpanzees show fearful and threatening responses such as wraas, barks, swinging hands downward, and charging toward

Figure 16.2 Pimu charging and chasing a warthog. ©K. Hosaka.

warthogs. The following describes an instance with intense responses.

On August 28, 2008, when chimpanzees were moving along the trail, one adult female, Effie, encountered two warthogs and emitted a wraa. Three adult males nearby, Pimu (alpha), Fanana, and Darwin, hastened to climb a tree, but descended on seeing the warthogs coming into sight. Pimu and Fanana resumed grooming without vigilance against the warthogs. However, a young adult male, Orion, and an adolescent male, Christmas, watched the warthogs intently. One warthog abruptly ran away passing the chimpanzees, which frightened Orion into wraas. The other warthog followed, which Christmas then threatened by shaking vines. Chimpanzees in the way stepped aside, and Pimu charged and chased the warthog (Figure 16.2).

16.4.2 Rodents

Mahale chimpanzees, especially females or juveniles, sometimes prey on squirrels (Chapter 20); however, Zamma (2002) reported that an adult female, Nkombo, captured a red-legged sun squirrel without eating it, instead playing with it until it died. Among seven other chimpanzees around her, only one juvenile male, Christmas, was attracted by the event and showed playful and exploratory behaviors toward the squirrel without eating it.

Matsusaka (2007) reported a rare encounter between seven chimpanzees and a porcupine at Mahale. No chimpanzee made a hunting attempt. The porcupine assumed a defensive posture throughout the event. Four chimpanzees (adult female Fatuma, 9-year-old male Primus, 8-year-old Cadmus, 2-year-old female Imani) threatened the porcupine by shaking the thicket nearby and hitting the ground with a stick. Chimpanzees seemed to explore reactions from the porcupine cautiously: they maintained a distance, and used sticks or vines instead of touching or biting. Thus, those behaviors could be regarded as "exploratory-threat behaviors" (Matsusaka, 2007).

16.4.3 Aardvark

No encounters between chimpanzees and living aardvarks have been observed. Hosaka *et al.* (2014) observed two cases where wild chimpanzees encountered aardvark carcasses.

First, on the morning of August 17, 2005, chimpanzees encountered a fresh aardvark carcass that is presumed to have been killed by a leopard a few hours previously (Figure 16.3). Nineteen chimpanzees, some of whom even turned back the way they had gone, gathered at the spot where an adult male, Orion, and an adult female, Sally, emitted

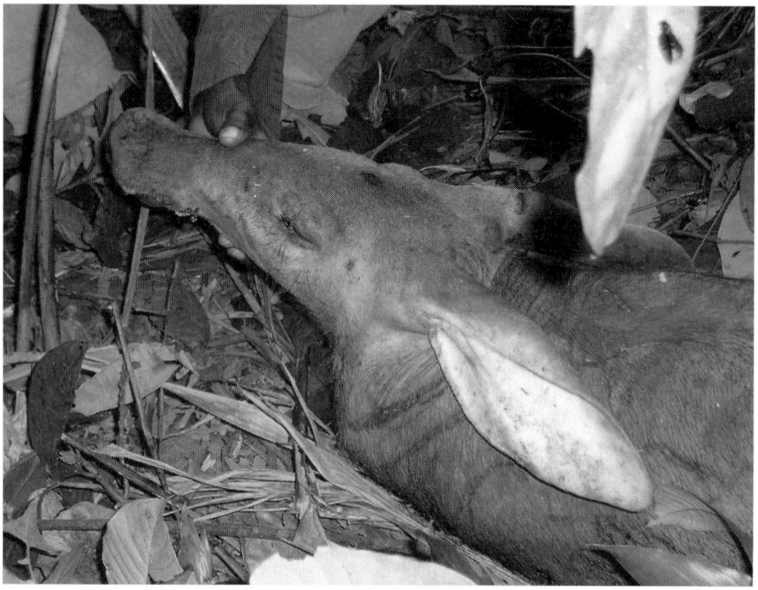

Figure 16.3 An aardvark carcass, presumably killed by a leopard, with many fresh claw scratches on its surface. ©K. Hosaka.

wraas. The two never approached the carcass but continued to stare at it from a tree at a distance of 20 m. However, 16 other chimpanzees approached within 10 m of the carcass and showed various responses such as huus, intense staring, sniffing, touching, leaping back, and self-scratching. None tried its meat. After 75 min, they left the carcass and continued on their way. It is likely that the leopard had been hiding or resting in the vicinity because the carcass had been moved when a researcher revisited the spot later that afternoon.

Next, on September 4, 2005, 26 chimpanzees encountered a putrid aardvark carcass. Only three chimpanzees (13-year-old male Cadmus, 7-year-old male Oscar, 6-year-old female Liz) took a detour to inspect the carcass. Oscar and Liz showed intense, ambiguous reactions reflecting curiosity and fear toward the carcass, repeatedly approaching and retreating. Cadmus was the only individual who touched the carcass in each case.

These observations have three explanations. First, Mahale chimpanzees are not interested in aardvark as food, because they did not eat aardvark even when they came by its newly killed body. Second, it seems probable that their intense responses to the fresh carcass reflected their awareness that a leopard had just killed it. Third, exploratory behaviors of the three immatures might have reflected their curiosity toward a nocturnal and unfamiliar animal, while this could also be interpreted parsimoniously as an example of general responses to dead bodies of any species.

16.4.4 Snakes

Although pythons have been considered a potential predator of chimpanzees (Izawa and Itani, 1966), no encounters between living pythons and chimpanzees had been observed in the wild until Zamma (2011) videotaped the first interaction at Mahale. In this episode, continuous wraas emitted by one chimpanzee appeared to attract the other 13 chimpanzees, including the alpha male, Pimu. At first, the entangled vines made it difficult to see the python, but when it came into sight, wraas stopped, and huus were uttered by one or two chimpanzees over the course of 2 min. Other notable reactions included leaping back, shaking vines, staring intently, which seemed to reflect mixed emotions of fear, avoidance, threat, and curiosity.

Chimpanzee encounters with various kinds of smaller snakes are likely to be more frequent considering how often they are detected by human observers. Although van Lawick-Goodall (1968) reported six such encounters, no published reports are available from Mahale. In 1993, Hosaka observed an encounter between several chimpanzees and a small snake (not identified), during which the snake moved along a vine at a height of 2–3 m. An adult male, Aji, stared at it intently and shook the vine, emitting huus sporadically.

16.4.5 Birds

Mahale chimpanzees sometimes eat birds (Chapter 20). However, in comparison with some major prey mammals, it seems that Mahale chimpanzees have no established "prey image" of specific birds. Chimpanzees may target eggs and nestlings rather than specific birds (Fujimoto and Shimada, 2008). In addition, there are individual differences in their image of birds as food. For example, an adult female, Totzy, was the bird-eater in two out of three reported cases (Fujimoto and Shimada, 2008). When she captured a trumpeter hornbill nestling, an older female, Nkombo, appropriated the bird but only carried it for a few minutes before abandoning it without eating.

It is likely that some individuals may acquire a habit of eating specific birds. From 1985 to 1993, keeping free-range chickens to a minimum in the courtyards of lodgings inside the park was allowed; this gave chimpanzees opportunities to hunt chickens. Masudi was observed hunting chickens three times; once in 1985 (Michael A. Huffman, personal communication), and twice in June–July 1991 (Miya Hamai, personal communication). On July 6, 1993, Hosaka observed two early adolescent males, Chopin and Nick, running out of a courtyard with their mouths full of chicks. At that time, Nick had another habit of eating sunbird eggs. Whenever he passed Kansyana Camp, he would climb a small citrus tree in which a sunbird had built a nest. If he found eggs, he would throw them into his mouth.

Fujimoto and Shimada (2008) reported a rare case of a juvenile male, Oscar, eating a black kite with his two female kin, although they failed to describe the background to this behavior. At the beginning of the wet season, from October to November, black kites migrate from the north to Mahale and are often found exhausted on the riverbed. It is easy for chimpanzees to capture them, because they rest on the rocks and are unable to fly away. Chimpanzees usually flail or sway black kites to enhance their charging display, but rarely eat them (Nishida, 2012).

16.5 Discussion and future perspectives

16.5.1 Leopard, colobus, and baboon

To discuss interspecific relationships between Mahale chimpanzees and sympatric animals, it would be interesting to begin with the three mammals: leopard as a potential predator, red colobus as the main prey, and yellow baboon as a competitor for foods. For the past 25 years, these relationships have been noticeable and are changing on a longitudinal basis. This is why we emphasize the need for long-term observations of these mammals.

It seems probable that leopards are recognized by Mahale chimpanzees as a dangerous animal, and that this influences their daily behaviors such as vigilance (Kutsukake, 2006), bed making (Chapter 43), and grouping. In addition, the frequency of encounters between chimpanzees and leopards may increase and the relationship between the two species may change accordingly. Kutsukake (2006) recorded 10 auditory encounters between chimpanzees and leopards during his 1100-h observation time spanning about a year. This frequency seemed higher than the five auditory or visual encounters recorded over the course of 18 months in 1991–4.

Nobuko Nakazawa began her research of leopards at Mahale in 2012. If it is possible to observe leopards sometime in the future, it would be interesting to simultaneously monitor chimpanzees and leopards to see how the two species adjust their distance from each other as Jenny and Zuberbühler (2005) did at Taï. In this chapter, we suggest that the red colobus of Kasiha began a "preventive attack" against chimpanzees in the 2000s. Two explanations seem possible. First, the attack might have developed to counter the high predation pressure. As the area is located at the center of the M-group home range (see Chapter 8), chimpanzees pass there very often. Thus, it is reasonable to assume that the colobus of that area have faced high predation pressure from chimpanzees. Second, the change of vegetation might have caused the new tactic to emerge. In the late 1990s, the invasive alien plant tree *Senna spectabilis* was eradicated from Kasiha (Chapter 13). Accordingly, in the 2000s, the vegetation in the area changed into a more open forest consisting of low trees and bushes. Such a change may have hindered colobus from hiding and moving to adjacent trees when detecting approaching chimpanzees. Thus, the surprise attack and ground charging of adult male colobus might have developed as a desperate tactic to prevent chimpanzee hunting.

Unfortunately, however, only limited information is available concerning anti-predator behaviors of colobus to compare with chimpanzee hunting behaviors on a longitudinal basis. Even when there is no researcher in the field who studies colobus, some basic information about colobus troops, including anti-predator behaviors, should be gathered at regular intervals. In addition, to understand the conservation status of red colobus at Mahale, it is necessary to regularly collect demographic information because there have been some reports that chimpanzee hunting may have impacted on the population growth at other study sites (Chapters 15, 20).

Long-term research on the yellow baboons is necessary considering that there are limited data on them at Mahale. Although the frequency of chimpanzee predation on baboons has remained low since the first record in 1997 (Chapter 20), baboons might have inherited their experience of chimpanzee predation because there are only four troops of baboons inside the M-group home range (Nishida, 2002). Basic and regular information regarding sympatric baboons would benefit chimpanzee researchers greatly in their studies of feeding competition, and predator–prey interaction between the two species.

16.5.2 Conservatism in "prey image" of *Pan* species

Finally, we discuss conservatism in the "prey image" of chimpanzees in relation to human evolution. Ihobe (1993) suggested that chimpanzees have evolved their meat-eating habit independently of humans, referring to the low frequency of this behavior in their kin species, bonobos (*Pan paniscus*). Furthermore, because chimpanzees consume only a portion of sympatric animals, unlike African hunter-gatherers who consume and hunt almost all animals, he proposed that the early humans might have extended their prey image substantially to consider "all meat is food" at some point of their evolution. This shift may have enabled humans to adapt themselves to a wider range of ecological environments.

Although red-tailed monkeys are abundant in Kasoje, they are much less frequently hunted by Mahale chimpanzees than are red colobus. In this chapter, we have suggested reasons to explain the elusive nature of red-tailed monkeys (Ihobe, 2002). In fact, chimpanzees do not hunt red-tailed monkeys in groups as they do red colobus, although a cohort of males has been observed to hunt red-tailed monkeys by silently stalking them on rare occasions (Chapter 20). Thus, chimpanzees have multiple daily encounters with red-tailed monkeys without hunting, which may keep their prey image of this monkey at a low level. At the same time, red-tailed monkeys may also keep their "predator image" of chimpanzee at a low level. Without these assumptions, it would be difficult to understand why chimpanzees and red-tailed monkeys coexisted commensally in the same tree at Mahale and other study sites.

It seems plausible that Mahale chimpanzees have no prey image of aardvark and leopard, since they did not eat any meat even though they encountered a fresh carcass. Aardvarks are nocturnal animals that chimpanzees scarcely encounter in the daytime. Leopards are a potential predator of chimpanzees that may be well avoided in the forest. In addition, there are individual differences in reactions to captured squirrels or birds. Some eat them, while others toy with them. Furthermore, similar differences are found among chimpanzees of different populations. A Taï chimpanzee did not eat the captured duiker but toyed with it (Boesch and Boesch, 1989; also see Chapter 20). Thus, conservatism of chimpanzee prey image is reconfirmed by anecdotes reviewed in this chapter. Such conservatism may be related to the stenophagous nature of chimpanzees, that they hardly exploit fish, amphibians, or reptiles as food sources.

References

Boesch, C. (1991). The effects of leopard predation on grouping patterns in forest chimpanzees. *Behaviour*, **117**, 220–41.

Boesch, C. and Boesch, H. (1989). Hunting behavior of wild chimpanzees in the Taï National Park. *American Journal of Physical Anthropology*, **78**, 547–73.

Bshary, R. and Noë, R. (1997). Anti-predation behaviour of red colobus monkeys in the presence of chimpanzees. *Behavioral Ecology and Sociobiology*, **41**, 321–33.

Busse, C. D. (1977). Chimpanzee predation as a possible factor in the evolution of red colobus monkey social organization. *Evolution*, **31**, 907–11.

Fujimoto, M. and Shimada, M. (2008). Newly observed predation of wild birds by M-group chimpanzees (*Pan troglodytes schweinfurthii*) at Mahale, Tanzania. *Pan Africa News*, **15**, 23–6.

Furuichi, T. (2000). Possible case of predation on a chimpanzee by a leopard in the Petit Loango Reserve, Gabon. *Pan Africa News*, **7**, 21–3.

Goodall, J. (1986). *The Chimpanzees of Gombe: Patterns of Behavior*. Cambridge, MA: Belknap.

Hart, J. A., Katembo, M., and Punga, K. (1996). Diet, prey selection and ecological relations of leopard and golden cat in the Ituri Forest, Zaire. *African Journal of Ecology*, **34**, 364–79.

Hiraiwa-Hasegawa, M., Byrne, R. W., Takasaki, H., and Byrne, J. M. (1986). Aggression toward large carnivores by wild chimpanzees of Mahale Mountains National Park, Tanzania. *Folia Primatologica*, **47**, 8–13.

Hoppe-Dominik, B. (1984). Etude du spectre des proies de la panthère, *Panthera pardus*, dans le Parc National de Taï en Côte d'Ivoire. *Mammalia*, **48**, 477–88.

Hosaka, K., Nishida, T., Hamai, M., Matsumoto-Oda, A., and Uehara, S. (2001). Predation of mammals by the chimpanzees of the Mahale Mountains, Tanzania. In *All Apes Great and Small Vol. 1: African Apes*, ed. B. M. F. Galdikas, N. Erickson Briggs, L. K. Sheeran, G. L. Shapiro, and J. Goodall. New York: Kluwer Academic Publishers, pp. 107–30.

Hosaka, K., Inoue, E., and Fujimoto, M. (2014). Responses of wild chimpanzees to fresh carcasses of aardvark (*Orycteropus afer*) in Mahale. *Pan Africa News*, **21**, 19–22.

Ihobe, H. (1993). [Meat and the beast: prey image of bonobos, chimpanzees and humans.] *Journal of African Studies*, **42**, 61–8. In Japanese.

Ihobe, H. (2002). [Anti-chimpanzee hunting strategies of red colobus.] In [*The Chimpanzees at Mahale: 37 Years Study of <Panthropology>*], ed. T. Nishida, S. Uehara, and K. Kawanaka. Kyoto: Kyoto University Press, pp. 245–60. In Japanese.

Izawa, K. and Itani, J. (1966). Chimpanzees in Kasakati Basin, Tanganyika: I. Ecological study in the rainy season, 1963–1964. *Kyoto University African Studies*, **1**, 73–156.

Jenny, D. and Zuberbühler, K. (2005). Hunting behaviour in West African forest leopards. *African Journal of Ecology*, **43**, 197–200.

John, T. and Reynolds, V. (1997). Budongo Forest chimpanzee grooms a redtailed monkey. *Pan Africa News*, **4**, 6.

Kano, T. (1972). Distribution and adaptation of the chimpanzee on the eastern shore of Lake Tanganyika. *Kyoto University African Studies*, **7**, 37–129.

Kutsukake, N. (2006). The context and quality of social relationships affect vigilance behaviour in wild chimpanzees. *Ethology*, **112**, 581–91.

van Lawick-Goodall, J. (1968). The behaviour of free-living chimpanzees in the Gombe Stream Reserve. *Animal Behaviour Monographs*, **1**, 161–311.

Matsumoto-oda, A. and Kasagula, M. B. (2000). Preliminary study of feeding competition between baboons and chimpanzees in the Mahale Mountains National Park, Tanzania. *African Study Monographs*, **21**, 147–57.

Matsumoto-Oda, A., Kutsukake, N., Hosaka, K., and Matsusaka, T. (2007). Sniffing behaviors in Mahale chimpanzees. *Primates*, **48**, 81–5.

Matsusaka, T. (2007). Exploratory–threat behaviors in wild chimpanzees encountering a porcupine. *Pan Africa News*, **14**, 29–31.

Nakazawa, N., Hanamura, S., Inoue, E., Nakatsukasa, M., and Nakamura, M. (2013). A leopard ate a chimpanzee: first evidence from East Africa. *Journal of Human Evolution*, **65**, 334–7.

Nishida, T. (1968). The social group of wild chimpanzees in the Mahali Mountains. *Primates*, **9**, 167–224.

Nishida, T. (1972). A note on the ecology of the red-colobus monkeys (*Colobus badius tephrosceles*) living in the Mahali Mountains. *Primates*, **13**, 57–64.

Nishida, T. (1997). Baboon invasion into chimpanzee habitat. *Pan Africa News*, **4**, 11–12.

Nishida, T. (2002). Competition between baboons and chimpanzees at Mahale. *Pan Africa News*, **9**, 23–6.

Nishida, T. (2012). *Chimpanzees of the Lakeshore: Natural History and Culture at Mahale*. Cambridge: Cambridge University Press.

Nishida, T., Uehara, S., and Nyundo, R. (1979). Predatory behavior among wild chimpanzees of the Mahale Mountains. *Primates*, **20**, 1–20.

Pierce, A. H. (2009). An encounter between a leopard and a group of chimpanzees at Gombe National Park. *Pan Africa News*, **16**, 22–4.

Reynolds, V. and Reynolds, F. (1965). Chimpanzees of the Budongo Forest. *In Primate Behavior: Field Studies of Monkeys and Apes*, ed. I. DeVore. New York: Holt, Rinehart, and Winston, pp. 368–424.

Stanford, C. B. (1998). *Chimpanzees and Red Colobus*. Cambridge, MA: Harvard University Press.

Teleki, G. (1973a). Notes on chimpanzee interactions with small carnivores in Gombe National Park, Tanzania. *Primates*, **14**, 407–11.

Teleki, G. (1973b). *The Predatory Behavior of Wild Chimpanzees*. Lewisburg, PA: Bucknell University Press.

Tsukahara, T. (1993). Lions eat chimpanzees: the first evidence of predation by lions on wild chimpanzees. *American Journal of Primatology*, **29**, 1–11.

Tutin, C. E. G., McGrew, W. C., and Baldwin, P. J. (1981). Responses of wild chimpanzees to potential predators. In *Primate Behavior and Sociobiology*, ed. A. B. Chiarelli and R. S. Corruccini. Berlin: Springer, pp. 136–41.

Wrangham, R. W. (1975). *The Behavioral Ecology of Chimpanzees in Gombe National Park, Tanzania*. Ph.D. thesis. Cambridge: Cambridge University.

Zamma, K. (2002). A chimpanzee trifling with a squirrel: pleasure derived from teasing? *Pan Africa News*, **9**, 9–11.

Zamma, K. (2011). Responses of chimpanzees to a python. *Pan Africa News*, **18**, 13–15.

Zuberbühler, K. and Jenny, D. (2002). Leopard predation and primate evolution. *Journal of Human Evolution*, **43**, 873–86.

PART IV Feeding

17 Diet and feeding behavior

NORIKO ITOH AND MICHIO NAKAMURA

17.1 Introduction

Observing the daily feeding behavior of chimpanzees can offer a fascinating perspective to better understand their ever-changing environment, society or sociality, behavior, and bodies, and how these factors are connected as a nexus. This nexus, where so many factors exert an influence and interact, also underlies the difficulty experienced by observers. Many of these factors are optional depending on chimpanzees' decisions *in situ*, and, therefore, their quantitative evaluation and determination is challenging.

Studies detailing chimpanzee feeding behavior are relatively scarce. The early work by Wrangham (1975) at Gombe, Tanzania, may be the most detailed study on feeding within the nexus. However, there are some recent works that offer a different perspective (e.g. Corp and Byrne, 2002a, b, 2004). Environmental assessments have also progressed (e.g. Turner, 2006a, b). Here, we review such works to provide information on the complex feeding nexus of the Mahale M-group chimpanzees. For some of the additional results, data from Itoh (2004) were used (referred to as NI1997 hereafter, see note at the end). Food availability was also reanalysed according to the re-census of the monitoring transect (see Chapters 13, 14). Species names are revised following Itoh (in prep., corrected names are available in Appendices I and II). For ambiguous species names, ID numbers from the Mahale floral list are added in brackets.

17.2 Dietary repertoire

Chimpanzees are omnivorous, but their diet mainly consists of fruits (van Lawick-Goodall, 1968; Nishida, 1968). Here, we briefly mention non-plant food types that have variable nutritional values, as they are summarized in other chapters.

17.2.1 Non-plant foods

Chimpanzees consume animal matter. Mammals and birds consumed by Mahale chimpanzees are outlined in Chapter 20 and Chapter 21 lists the insect types eaten.

We never observed Mahale chimpanzees to eat mushrooms. However, chimpanzees at some other study sites eat mushrooms (e.g. Yamakoshi, 1998).

Mahale chimpanzees consume (eat, lick) various inorganic substances such as rock, stone, streambed sandy soil, termite-mound soil, and drink water (Nishida and Uehara, 1983). Licking of stones and rocks usually occurs at lakeshores and riversides. However, a past observation noted a huge precipitous rock cliff that was repeatedly visited and licked by several M-group chimpanzees (Nishida and Uehara, 1983; Chapter 18).

In a review of primate geophagy, Krishnamani and Mahaney (2000) stated that various reasons may exist for geophagy; mineral supplementation, toxin adsorption, diarrhea control, and gut pH adjustment. In the case of Mahale, termite-mound soil ingestion is suggested to function as a temporary relief from gastrointestinal ailments (Mahaney *et al.*, 1996).

Mahale Chimpanzees: 50 Years of Research, ed. M. Nakamura, K. Hosaka, N. Itoh and K. Zamma. Published by Cambridge University Press.

Some plant species are also ingested as self-medication (Chapter 24).

While direct drinking of water is not frequent at Mahale, the chimpanzees sometimes drink from rivers and the Lake Tanganyika (Nishida, 1980) or lick water from tree hollows using their fingers. Occasionally, they use tools to obtain drinking water (Chapter 42).

17.2.2 Plant-food repertoire

Chimpanzees consume almost all parts of the plants they forage for food (Figure 17.1). Below, the predominant plant parts (hereafter referred to as food type) eaten by chimpanzees are summarized from Nishida and Uehara (1983) with some additional information (see also Appendix II).

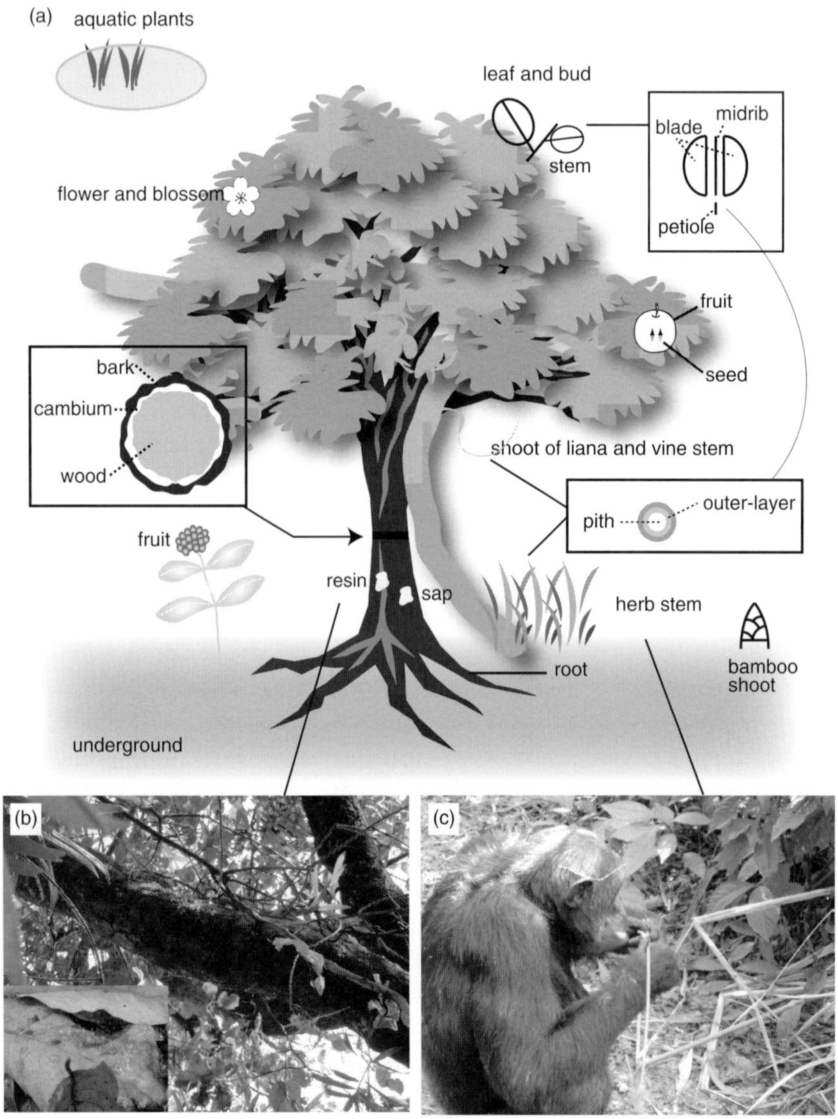

Figure 17.1 (a) A schematic diagram of plant parts eaten by M-group chimpanzees. Chimpanzees eat most parts and minutely discriminate parts wanted and unwanted. (b) Unusually abundant resin from *Combretum molle* tree. It not only seeped from trunks and branches, but dropped on the ground (chimpanzees ate both arboreally and on the ground, picking the dropped resin). (c) An adult male eating the pith of *Marantochloa leucantha*. ©N. Itoh.

Dominant in their diet are fruits, leaves (mainly young leaves), buds, young stems of vines and lianas, seeds, and flowers. Fruits, leaves, and young stems may be partially eaten after some processing. This processing involves pulp extraction from a fruit and discarding midribs and petioles to consume only the leaf blade. Seeds are swallowed or spat out (Chapter 19). The seeds of unripe fruits are sometimes extracted and consumed. Seeds of ripe fruits such as *Pterocarpus tinctorius* are also eaten. Piths of young terrestrial herb stalks (Gramineae and Zingiberaceae) are also commonly eaten. Piths of the lower parts of immature stems are eaten while upper parts of mature grasses are occasionally consumed (e.g. *Pennisetum purpureum*). Bamboo shoots are rarely eaten by the M group.

Older shoot systems, the tree trunk itself, or any organic compound seeping from the bark is also consumed (see also Chapter 24). Licking the sap of trees is commonly observed. While the specific substance has not been identified, chimpanzees sometimes repeatedly visit specific trees to lick the sap. Resins are occasionally eaten, possibly because of the limited amounts available. For some particular species (e.g. *Brachystegia bussei* and *Sterculia tragacantha*), the chimpanzees eat only the inner bark or cambium, whereas for other species (e.g. *Pycnanthus angolensis* and *Vernonia amygdalina*), the outer and inner barks are eaten. Bark-eating behavior is not common for the M group, but it was common for the K group (see Chapter 10) as an "emergency food" during the lean period (Nishida, 1976).

In the M group, a few unidentified species of woody roots emerging from the ground were bitten off and chewed. Root-digging behavior has not been observed (Nishida and Uehara, 1983), but Nakamura once observed an adult female pull and chew a root of *Pseudospondias microcarpa*.

The plant parts eaten also depend on specific plant species. Many parts of some plants are eaten, whereas for others, only specific parts are eaten. Hereafter, we term this combination of a food plant species and a specific plant part as an "item."

Differentiation of plant parts eaten among plant species, maturity stages, etc., suggest that chimpanzees are very specific in what they eat. On the other hand, they may simultaneously eat several plant parts such as flowers and young leaves of *Baphia capparidifolia* [MLID:553] or fruits and leaves of *Ficus asperifolia*. However, they may choose to eat any one part of the aforementioned plants, even though both parts are available.

17.3 Feeding behavior

17.3.1 How food items are acquired

Acquiring a food item on-site requires variable tasks at different stages. Here, the strategies chosen by Mahale chimpanzees to acquire a variety of plant-food items are reviewed.

Figure 17.2 is a rough sketch of foraging tasks required at different stages. For most items, search and selection can be visually accomplished. However, for fruits like *Tabernaemontana pachysiphon* that do not change in color with maturity, ripeness is determined by sniffing.

The desired item can be acquired in different ways. Although the main method is detachment from the plant organ, some items may be recovered from the ground (e.g. pods of *Bauhinia thonningii*, fruits of *Pseudospondias*, *Myrianthus arboreus*, and *Citrus limon*). Chimpanzees may obtain items from other individuals through food sharing (Chapter 20).

Most leaves, blossoms, and drupe- and berry-type fruits are detached by picking with fingers or directly biting off the plant. In the case of fruits, leaves, and blossoms, chimpanzees may break off or pull the flowering or fruiting branch, hold it with one hand, and then pick or bite off the items. Some fruits may be collected in their hands, feet, and/or mouth before eating (Figure 17.3).

For some items, several further tasks are required. These tasks may be combined, repeated, or skipped during the course of repetition. Leaves may be piled in one hand and then completely put into the mouth. The

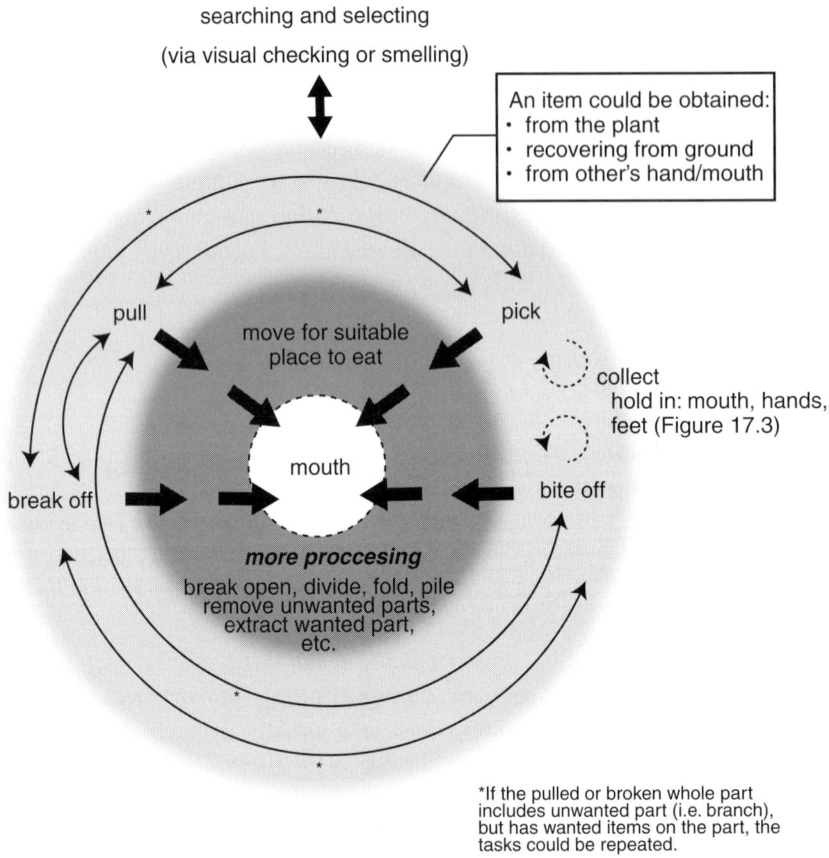

Figure 17.2 A schematic diagram of tasks required for eating *in situ* at various stages. Some tasks could be repeated, iterated, or optional, depending on the item, its condition, physical and social situation at the site, and/or physical capacity or age of the individual chimpanzee. For eating fruits like *Tabernaemontana pachysiphon* and *Myrianthus arboreus*, as most trees are small, chimpanzees may move to other places (e.g. a nearby tree with large branches or even to the ground) to settle and eat. These fruits are the largest of all plant-food items available in Mahale. *Tabernaemontana* has an unusually hard and thick shell. *Myrianthus* fruit forms a drupaceous whole, and each drupe-like fruit is tightly adnated to the perianth. Thus, these fruits need additional processing before consumption, which may be a difficult task for immature chimpanzees.

petiole of picked leaves may be bitten off or snapped off by fingers.

More complex tasks are required for some items. Corp and Byrne (Corp and Byrne, 2002a, b, 2004; Byrne and Corp, 2003) conducted detailed research on processing of two types of difficult-to-process plant foods: (1) embedded foods such as pulp inside tough-shelled large fruits and (2) physically defended leaves.

Corp and Byrne (2002b) found that the processing of *Saba* fruits (Figure 17.4) requires considerable manual dexterity at several stages such as acquiring, opening, dividing into smaller sections, removing unwanted parts, and lastly extracting and eating the pulp. All stages have alternative procedures such as opening with teeth or hands, which could be iterated or entirely optional. These procedures could be performed unimanually or bimanually (see Chapter 41 for laterality). The choice of procedures may be dependent on individual preferences and also depends on a variety of circumstances (e.g. in a tree, or unusually large or small item). Developmental change in these procedures was also investigated in the paper.

Figure 17.3 Collecting before eating. (a–c) Collecting *Citrus limon* fruits. (d–f) *Myrianthus arboreus* fruits and a male eating it after collection. ©N. Itoh.

Figure 17.4 Fruit of *Saba comorensis* is a difficult-to-process large-sized food item. It has a tough outer skin, 0.5–1 cm thick, with numerous tightly packed seeds each coated with pulp. (a) The liana in fruit. (b) Immature fruits on the liana. Mature fruits reach around 10 cm in diameter. (c) Seeds coated with yellow-orange fresh pulp inside (leftover from chimpanzee probably a few days before). (d) Leftover from chimpanzee with pulp and seed already extracted. ©N. Itoh.

Similar techniques may be applicable to *Citrus* fruits. Discarded parts also differ (Figure 17.5). Although we never observed chimpanzees eating pericarps, some individuals spit out seeds and/or segments while others swallow the whole fruit. Some carefully peel the segments with fingers and/or lips, whereas others spit out the juice sack after squeezing juice. We are continuing to profile each individual's techniques.

Corp and Byrne (2002a) investigated the manual processing of several leaves. They identified 96 distinct techniques, such as folding a leaf with the lip before putting it into the mouth, in total for processing leaves. Physically defended leaves, such as those with rough hairs, required 2.5 times more techniques than undefended leaves. Here, flexible and hierarchical organizations of the processes were found.

Processing of pith may not be so variable. The chimpanzees carefully select stems of terrestrial herbs and arboreal vines. What they are discriminating against is yet to be identified. Chimpanzees first bite off stems of terrestrial herbs near the ground, hold the stem with both hands, bite the middle to open up the inner part, and then start eating. Piths of vines seem much softer and easier to process. They may start eating without detaching the long stringy stem.

Other techniques

Insect eating may be the best studied feeding behavior among non-plant foods of Mahale chimpanzees. In

Figure 17.5 Variation in leftovers of *Citrus limon* fruits, indicating individual differences or preferences. (a) A beautiful leftover. (b) Similar to A, but this individual spits out seeds. (c) This individual eats all parts including seeds and segments and only pericarps are left. (d) A not so good-looking leftover. ©N. Itoh.

particular, insect eating with tools is famous (Chapter 42), but eating of ants, larvae, and eggs from within dead branches is also common (Chapter 21). The events of meat consumption (Chapter 20) are usually protracted, partly due to the difficulty chimpanzees encounter in processing (e.g. difficulty in biting off parts), but it is also due to various social interactions, including sharing that occurs during these events (Chapter 20).

Some plants are ingested in different ways from usual food. For example, chimpanzees sometimes put leaves into the mouth when they make a wadge of fruits (Chapter 19) or meat (Chapter 20). Because such leaves are not usually eaten as foods and even dead leaves are eaten, acquiring nutrients from these leaves may not be the goal. Possibly, chewing fibrous leaves may protract the taste of certain fruits or meat; however, further studies are required. In addition, some medicinal plants are eaten in unusual ways (e.g. leaf swallowing; Chapter 24).

17.3.2 Where feeding occurs

Besides on-site movements in search for a stable place to process and eat foods, feeding behavior depends on the scale of food availability. Some food items and types are ubiquitously available, whereas others are available within limited spatial areas. Examples of a widely available item are fruits of *Saba* and *Pycnanthus*, which are hyper-dispersion species (Chapter 13) and have high–middle simultaneity in fruiting (Chapter 14). Most other fruits are usually more limited in their spatiotemporal distribution (Turner, 2006b; Itoh, 2004; Chapters 13, 14).

Some leaf and pith foods are limited in spatial distribution, especially those growing in swamps. Swamps and seasonally flooded areas occupy only 3.8% of the chimpanzees' home range and are patchy in distribution (Turner, 2006b), and among them, Sinsiba and Nganja (Chapter 2) are the places chimpanzees frequently visit.

The structure of a food site (food patch) also varies. A feeding tree separated from other trees with some gaps can be easily recognized as a patch. A vine entangled in a single tree may also be recognizable in a similar way. Foods from terrestrial vegetation are usually distributed in clumps of various sizes. Lianas, especially those of the vigorous liana species *Saba*, are difficult to define as a patch because they grow over several tree canopies (Itoh and Nishida, 2007). Defining a patch is also difficult when multiple items are available (see Isabirye-Basuta, 1988).

17.3.3 When feeding occurs

Feeding can occur at any time during daylight, although Wrangham (1977) once observed a Gombe chimpanzee eating at midnight. There are two feeding peaks (in the morning and evening) observed at several study sites (Kasakati, Tanzania: Izawa and Itani, 1966; Mahale: Nishida, 1974; Gombe: Wrangham, 1977; Taï, Côte d'Ivoire: Doran, 1997; Budongo, Uganda: Newton-Fisher, 1999b).

17.4 Fluctuation in dietary repertoires of plant foods

To date, 407 items of 224 species have been confirmed to be eaten by Mahale chimpanzees (Appendix II). The repertoire is not a simple reflection of the presence of a specific plant but also reflects culture (see Chapter 38). We know little about what this large catalog means for chimpanzees, but several authors (e.g. Wrangham, 1975; Nishida and Uehara, 1983) speculated that it is important for adaptation to a wide geographical range.

Plants are characterized by their reproductive cycle in relation to abiotic and biotic fluctuations (Chapter 14). Thus, the maximum breadth of the dietary repertoire during a given period could depend on plant phenology. At the same time, chimpanzees do not fully consume all items available but select some, which could also shape the repertoire breadth during a specific period.

17.4.1 Dietary breadth

Mahale chimpanzees eat an average of 12.4 items per day (Matsumoto-Oda *et al.*, 1998), and around 20–40 items per month (Turner, 2006b). The annual (or equivalent lengths) dietary breadth at different study sites is shown in Table 17.1. Although the number of items consumed may depend on the total observation length and could vary across years, it is apparent that chimpanzees of all sites consume a variety of items.

Wrangham (1975) estimated that chimpanzees in general may eat 200–300 different items over a few years. He also suggested that chimpanzees maintain some variety in their dietary choice within a given period because the dietary breadth of Gombe chimpanzees did not vary between seasons and because chimpanzees commonly do not deplete food patches. Turner (2006b) found that the number of available food species did not correlate with the number of species eaten at Mahale. The number of species consumed monthly and the proportion of fruits eaten at Mahale showed a positive (although not significant) correlation (Turner, 2006b; NI1997, Kendall's rank correlation, $\tau = 0.30$, $p = 0.2$, $N = 8$). Although Matsumoto-Oda *et al.* (1998) assumed low food availability when chimpanzees ate more food items based on such tendencies found in two different primate species, the premise does not seem to hold, at least for Mahale chimpanzees. Rather, maintaining a certain level of food variety regardless of food availability seems important. Nutritional requirements might be related to this tendency, which requires further investigation.

17.4.2 Heterogeneous consumption of plant-food types and items

General

Despite a wide dietary repertoire, chimpanzees concentrate on a few key items on a monthly and/or annual basis at Mahale (Nishida, 1974; Itoh, 2004; Turner, 2006b) and other sites (Gombe: Wrangham, 1975; Kibale, Uganda: Chemurot *et al.*, 2012; Budongo: Newton-Fisher, 1999b; Fawcett, 2000).

Table 17.1 **Number of food items eaten at Mahale and other study sites**

Study sites	Country	Study period	Duration (month)	Number of food items	Notes	Source
Mahale	Tanzania	1994–5	10	70	Belonging to 59 plant and 6 animal species, termite-mound soil, and water	Turner, 2006b
Mahale	Tanzania	1997–8	12	83	76 were plant items belonging to 60 species, including 29 fruit food species. Insects and vertebrates were lumped as one item each and 18 unconfirmed items were excluded	This study (NI1997)
Kasakati	Tanzania	1964–5	17	78	Plant items only	Suzuki, 1969
Gombe	Tanzania	1972–3	12	77	Plant items only	Wrangham, 1975
Ipassa	Gabon	1971–2	NA	141	Plant items only. 144 unidentified items were excluded	Hladik, 1977
Bossou	Guinea	1995	12	160	Belonging to 110 plant species	Yamakoshi, 1998
Kibale	Uganda	1976–8, 1981	25	50	Belonging to 21 plant species excluding unidentified ones	Ghiglieri, 1984
Budongo	Uganda	1994–5	15	118	Belonging to 58 plant species	Newton-Fisher, 1999b
Budongo	Uganda	1997–8	19	91	Belonging to 49 plant species	Fawcett, 2000

Mahale chimpanzees predominantly rely on fruits in terms of time spent feeding, accounting for more than 50% of their monthly feeding time (Itoh, 2004; Turner, 2006b), although the monthly fruit availability fluctuated (Itoh, 2004; Itoh and Nishida, 2007).

Table 17.2 lists frequently eaten fruit species over a 6-year period (2008–13). Some fruit species were frequently eaten almost every year (e.g. *Saba*), while others (e.g. *Uvaria*) were frequently eaten only in particular year(s). This indicates that the relative importance of a fruit species as a food item for chimpanzees must be assessed over the longer term. To evaluate such long-term importance of food species, Nishida (1974, 1991) proposed the classification of items in terms of their relative importance to chimpanzees. Revised relative importance categories are shown in Table 17.2. CF are those consumed most regularly during most years. SF are also regularly eaten in many years but not to the extent of CF. While the CF and SF roughly overlap with those classified as major and important fruits during 1965–72 (Nishida, 1974), some species such as *Antidesma membranaceum* [MFID:298], *Julbernardia* sp., *Grewia forbesii, Parkia filicoidea*, and *Uapaca nitida* do not achieve a ranking in our list. On the other hand, the most important species ("major food" in Nishida, 1974 and CF species here) are generally stable, except for *Garcinia huillensis*, which does not rank within the current results.

The extent of consumption of particular fruits shows year-to-year variation. The months that each fruit items are heavily consumed also vary across years (Table 17.3). *Saba* stands out as a fruit species that shows consistently high and lengthy consumption across all years. This result, even in rough estimation, is consistent with that of Itoh (2004). Feeding patterns shown here seem to be generally compatible as phenological features, as CF species often show a high predictability of fruiting time and have a relatively lengthy fruiting duration (Chapter 14). The general correspondence of food item selection with their phenological characteristics does not mean that chimpanzees are selecting items according to their relative availability (see below). In addition, the climatic season cannot be used as an indicator of food availability at Mahale (Chapter 14).

17.4.3 Case study: 1997–98

To determine how Mahale chimpanzees decide which fruit items to eat in relation to availability, data from Itoh (2004) were used with some statistical revisions. Seeds, which comprised a small proportion within the dataset, were lumped with fruits. For relative estimates of feeding efforts, percentages of time spent eating different food types or items per time spent on all plant foods were calculated (referred hereafter as feeding rate or FR). Monthly availability of each food species (shown as D_m in figures) was estimated by multiplying density (numbers of individuals per ha) with fruiting rate (Chapter 14) (modified from Sun *et al.*, 1996).

Monthly mean FRs showed that the most feeding time was spent on fruits (68.7%), followed by leaves (16.2%), piths (11%), flowers (3.9%), and resin and sap (0.1% for each) (N = 9 months). The monthly availability of 15 fruit species where phenological records were available (see Table 17.4; marked with asterisks) did not correlate with calculated FRs, except for *Psydrax parviflora*, which showed a positive significant correlation ($\tau = 0.73$, $p < 0.01$, $N = 9$). These results suggest the chimpanzees heavily relied on fruit items; however, the decision of which items to eat was not solely made upon their availability.

To determine feeding patterns for each fruit item, FRs of all 27 fruit items eaten were plotted against frequencies (i.e. proportion of months chimpanzees ate the species; Figure 17.6). Four general groups can be distinguished as indicated by the dotted circles. Three groups can be distinguished from the frequencies: low (<0.25: Group 1), high (>0.5: Group 2) and intermediate. The intermediate group was further divided into two on the basis of differing FRs (Groups 3 and 4).

As FR significantly correlates with frequency (τ = 0.52, $p < 0.01$, $N = 55$), Group 1 species tend to be associated with low FRs, but this does not necessarily mean that these are non-important or non-preferred species. Within Group 1, there are species that

Table 17.2 **Importance index category 1 (IIC-1) for frequently eaten fruit species during 2008–2013**[a]

	Latin name	Local name	MLID[b]	Importance category[c]	Rank in each observed year 2008	2009	2010	2011	2012	2013	Rank over 6 years	Total days eaten
*	*Saba comorensis*	Ilombo	62	CF	1	1	1	1	1	1	1	683
*	*Ficus sur*	Ikubhila	500	CF	2	2	2	2	2	3	2	321
*	*Pseudospondias microcarpa*	Bhuhono	37	CF	6	3	4	4	4	5	3	213
*	*Ficus exasperata*	Lwago	493	CF	3	5	3	5	3	6	4	209
*	*Cordia millenii*	Mukibhu ghwesimbwa	115	CF	5	9	13	3	5	4	5	178
*	*Citrus limon*	Limauwe	690	CF	8	4	8	6	8	2	6	160
*	*Pycnanthus angolensis*	Lulumasya	513	CF	4	8	5	14	6	14	7	124
*	*Ficus vallis-choudae*	Ihambwa	504	SF	9	6	8	9	12	8	8	111
*	*Myrianthus arboreus*	Isakama type A	509	SF	14	14	6	7	7	10	9	99
	Ficus asperifolia	Kankolonkombe	489	SF	12	10	10	9	8	12	10	89
*	*Tabernaemontana pachysiphon*	Ikolyoko no. 2	65	SF	10	7	11	12	15	12	11	73
*	*Psydrax parviflora*	Kamwibhi	1420	SF	20	12	15	7	14	8	12	67
*	*Garcinia huillensis*	Kasolyo	383	OF	7	16	15	14	15	15	13	65
	Dioscorea sp.	Sihama type undet. CF	1443	SF	10	45	39	14	28	8	14	47
*	*Azanza garckeana*	Mutobho	421	OF	14	16	13	30	13	29	15	44
	Psychotria peduncularis	Lulyolwakape	824	OF	27	29	18	12	10	17	15	44
*	*Cordia africana*	Mukibhu ghwakakamba	114	OF	16	11	15	25	22	17	17	40

Table 17.2 (**cont.**)

	Latin name	Local name	MLID[b]	Importance category[c]	Rank in each observed year						Rank over 6 years	Total days eaten
					2008	2009	2010	2011	2012	2013		
	Aframomum sp. (possibly)[d]	Itungulu type undet. CF	1444	OF	18	14	25	20	19	25	18	37
*	*Landolphia owariensis*	Mpila	57	OF	24	25	21	11	40	10	19	36
	Diplorhynchus condylocarpon	Musongati	54	OF	13	21	31	24	28	17	20	33
*	*Synsepalum brevipes*	Mulyansekesi	722	OF	27	12	39	34	11	29	20	33
*	*Uvaria angolensis*[e]	Lujonghololo bhosa	45	OF	45	36	**6**	34	28	15	23	28
	Mimusops bagshawei	Mulonje type C	718	OF	45	21	39	14	40	42	36	13
	Total observation days				233	193	188	266	244	222		1346

[a] Only the number of days that fruits of a species were eaten was used for analysis irrespective of amount eaten or time spent for feeding in a day (from the MMCRP [Chapter 2] records). The ranking of each fruit species was given for each year and the whole 6-year period as descending order of the number of days for respective periods. Species included in this list are those ranked in the top 20 in any of the whole 6 years or the top 10 in any single year (shown in bold). As the food item was only identified on-site for local names, matching of Latin names was conducted later only by the authors (names obtained from Itoh, in prep.). Thus, local names are also left here. Names with * are those showing phenological patterns and are assessed in Chapter 14. Family names are omitted here to save space but are available in Appendix I. Data for May 2008, March and April 2009, February 2010, September and October 2013 are excluded because of too few observation days (<5 days).

[b] Unique ID number for Mahale flora (Itoh, in prep.; Appendix I).

[c] CF: consistently staple food that is included in the top five over multiple years; SF: staple food that is included in the top 10 for multiple years; OF: other food, the remaining species.

[d] Probably only one species included here.

[e] There is a possibility of the other species under the local supra-name, Lujonghololo, being included. We assigned this species as it is most commonly eaten among Lujonghololo.

Table 17.3 **Importance index category 2 (IIC-2) based on heavy monthly consumption**[a]

Latin name	Importance category[b]	2008	2009	2010	2011	2012	2013
Saba comorensis	Major	Jan–Mar, Oct–Dec	Jan, Aug–Dec	Jan, Aug–Dec	Jan, Oct–Dec	Jan, Jul–Dec	Jan, Sep–Dec
Ficus sur	Major	Apr	May			Feb, Apr	
Pseudospondias microcarpa	Major	Sep	Aug–Sep		Aug–Sep	Oct	Sep
Ficus exasperata	Major	Jan		Oct	Oct	Oct–Nov	
Cordia millenii	Major	Mar–Apr	Feb		Feb–Apr	Mar–Apr	Feb–Mar
Citrus limon	Important						May
Pycnanthus angolensis	Important	Sep					
Myrianthus arboreus	Major			Oct		Oct	Oct
Psydrax parviflora	Major		Jun	Jun	May		May
Dioscorea sp.	Important	Mar–Apr					
Psychotria peduncularis	Important					Apr	
Synsepalum brevipes	Important		May				
Uvaria angolensis	Important			May–Jun			
Mimusops bagshawei	Important		May				

[a] The dataset used here is the same as that of Table 17.2. We assumed a heavy monthly consumption of an item when chimpanzees ate the item for more than 50% of observation days in the month. For example, *Saba* was eaten on 18 out of 24 observation days (75%) in January 2008. Thus, January is listed here.

[b] Major: species with heavy monthly consumption in multiple years; Important: species with a heavy monthly consumption in a single year.

chimpanzees extensively ate within a particular period. For example, *Psydrax* and *Mimusops* occupied 22.2% (May) and 41.2% (June) monthly FR, respectively. The species in this group display various fruiting patterns, including timing, duration, periodicity, and intensity for those known species (Chapter 14), and feeding pattern.

Group 2 contains only *Saba*, which stands out for its prolonged usage and high concentration. This fruit has always been a preferred item in Mahale occupying 2–3 times the feeding time (NI1997) and frequency (Nishida, 1974) compared with the second-ranking species. Thus, *Saba* can be called "super-important" item. This liana is characterized by the most stable periodicity, intensity, and lengthy duration of fruiting across years (Chapter 14).

Group 3 contains two species that fulfill different requirements for chimpanzees. *Psychotria peduncularis* is a very short shrub and chimpanzees eat its fruit on the ground. The fruits are available only during the non-fruiting season of *Saba*, and occupied 22.2% of the FR in May. While the fruiting season of *F. exasperata* corresponds with the high fruiting season of *Saba* (Chapter 14), chimpanzees extensively consumed the fruit of *F. exasperata* in October (50.4% FR), which corresponds to the decrease in FR of *Saba* during this period to 8.3%. Thus, it seems that chimpanzees chose to extensively eat this item over *Saba* during a short period.

Group 4 contains relatively long fruiting species (Chapter 14). *Pycnanthus* were eaten over a short period in a high concentration (26.2% FR in November). The three fig species never became dominant in FRs (<15%): *F. sur* and *F. asperifolia* are eaten relatively frequently but in small amounts estimated by the time spent eating.

Non-fruit food types are also important for Mahale chimpanzees, sometimes even influencing fruit-eating patterns. Flower eating was highly limited in time and FR of *Cordia millenii* peaked to as much as 21.4% in August (rank 5 in overall FR). Leaves and piths were consumed throughout the year. The pith of *Aframomum* sp. ranked fourth in overall FR and was eaten in all months except October and peaked in December (20.1% FR). The leaves of *F. exasperata* and *F. asperifolia* are also in the top 10 items in the overall FRs, but monthly FRs never exceeded 20%. Leaf eating (including young stem) on *B. capparidifolia* (rank 13) showed a clear peak in June (23.7% FR). As these food types are characterized by different nutritional values compared with fruits and were regularly eaten throughout the year, for 3–10 species and 14.1%–50.3% monthly FRs, the chimpanzees were possibly obtaining different nutritional requirements from various food types and items. It is also known that Budongo chimpanzees spend time eating young leaves at the expense of eating fruits during some periods (Fawcett, 2000).

Apart from plants, meat eating strongly influences feeding activities or overall activities of the day. When chimpanzees catch several colobus monkeys in a day, they may continue eating them for almost the entire day. As hunting has some seasonality (Chapter 20), the influence of meat eating may differ depending on the seasons.

Long-term studies of different unit-groups from a variety of aspects have taught us that "chimpanzees do not exist within a habitat where the amount of resource at any given time can be simply characterized as 'abundant' or 'scarce'" (Turner 2006b: 68). The concept of food abundance or availability is quite complex as the value of the environment, including nutritional quality, may change with time (Wrangham, 1975). Although nutrition of some fruits were analysed at Mahale (Itoh, 2004), further research is required. We are beginning to understand some of the multiple intertwined factors that may affect their diet or food selection. Further long-term studies, including those from different habitats, may help obtain a better picture of chimpanzees' food environment, and in turn, their feeding behaviors.

17.5 Feeding with others

For animals living in social groups, feeding also involves coping with other individuals. Few detailed

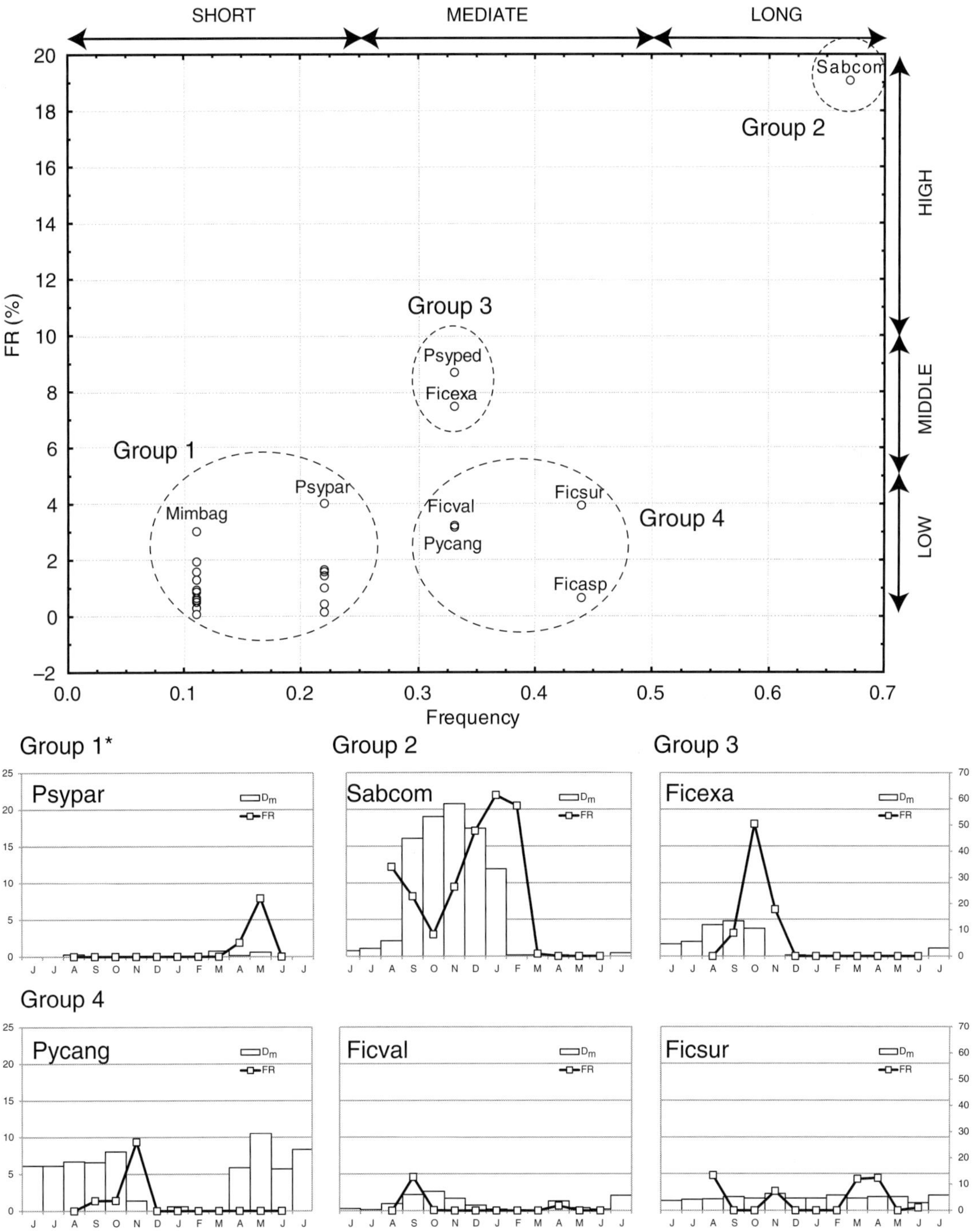

Figure 17.6 Scatterplot of FR against frequency for each eaten fruit item (top). Species names are abbreviated (see Table 17.4). Some examples of FR (second y-axis) in relation to fruiting phenology (D_m, first y-axis) on monthly bases are shown for each clustered group (bottom). The scales are the same for all figures (max FR = 70%, D_m = 25). Phenological record: June 1997–July 1998. Feeding record: August 1997–June 1998 (includes small sample sized months, February and March, for reference). *Group 1 includes various phenological and feeding patterns.

Table 17.4 **List of top 25 items eaten by the M-group chimpanzees during August 1997–January 1998 and April–June 1998**

Species (ID for Mahale flora)[a]	Abbreviation	Local name[b]	Part eaten[c]	FR (%)	Frequency
**Saba comorensis* (62)	Sabcom	Ilombo	F	17.28	0.67
Psychotria peduncularis (824)	Psyped	Lulyolwakape	F	7.84	0.33
**Ficus exasperata* (493)	Ficexa	Lwagho	F	6.74	0.33
Aframomum sp. (possibly) (1444)	AfrZINb	Itungulu undet. CF	P	4.75	0.89
Cordia millenii (115)	Cormil	Mukibhu ghwesimbwa	Bl	3.83	0.22
**Psydrax parviflora* (1420)	Psypar	Kamwibhi	F	3.67	0.22
**Ficus sur* (500)	Ficsur	Ikubhila	F	3.55	0.44
Ficus exasperata (493)	Ficexa	Lwagho	L	3.42	0.44
Ficus asperifolia (489)	Ficasp	Kankolonkombe	L	3.40	0.67
**Ficus vallis-choudae* (504)	Ficval	Ihambwa	F	2.92	0.33
**Pycnanthus angolensis* (513)	Pycang	Lulumasya	F	2.87	0.33
Mimusops bagshawei (718)	Mimbag	Mulonje type C	F	2.75	0.11
Baphia capparidifolia (553)	Bapcap	Ijubhila	L	2.63	0.44
Pennisetum purpureum (369)	Penpur	Iswe	P	1.81	0.44
**Uvaria angolensis* (45)	Uvaang	Lujongololo bhosa	F	1.78	0.11
**Synsepalum brevipes* (722)	Synbre	Mulyansekesi	F	1.51	0.22
Psidium guajava (517)	Psigua	Mpela	F	1.46	0.22
**Tabernaemontana pachysiphon* (65)	Tabpac	Ikolyoko no.2	F	1.45	0.11
**Pseudospondias microcarpa* (37)	Psemic	Bhuhono	F	1.45	0.22
Marantochloa leucantha (438)	Marleu	Kalubhwabhakulu	P	1.36	0.44
Uapaca nitida (329)	Uapnit	Lulobhe	F	1.34	0.22
Pterocarpus tinctorius (592)	Ptetin	Mwenje	L	1.28	0.33
Sp. "Lumpululu"	"Lumpululu"	Lumpululu type undet. CF	L	1.20	0.44
Ficus thonningii (502)	Fictho	Mujimo	F	1.15	0.11
Diplorhynchus condylocarpon (54)	Dipcon	Musongati	F	0.93	0.22

[a] See Itoh (in prep.) for details on identification. Names with * are those for which phenological relations have been checked (Chapter 14).

[b] Local names are provided here as most plants were identified by local name with the aid of local assistants on-site.

[c] Bl: flowers and blossoms; F: fruits and seeds; P: piths of herbs, lianas stems, and leaf petioles; L: leaves.

studies on the social aspects of feeding have been conducted. However, tolerance and patience towards others are observed even in some of the interactions considered to be competitive.

Direct aggressive confrontations or quarrels over a food item are rare at Gombe (Wrangham, 1975) and Mahale. Even when these do occur, it is difficult to determine whether it is food related. For example, on October 11, 1997, an adolescent female, Maggie, was eating *F. exasperata* fruits on a tree with two adult males. Another adult male, Fanana, arrived and ate for 3 min, after which he directed a charging display toward her. She screamed and fled down the tree. As Maggie had to cease eating, this event disturbed her feeding activity. However, as Fanana also left the tree just after, it may be difficult to say that this event occurred over food. Although such interactions are easily interpreted in terms of competition over food, one needs careful observation in consecutive time series and the consequence (i.e. the gained benefit) needs to be carefully assessed.

"Competitive" interactions over food seem to occur in more implicit ways, rather than snatching food directly from others. Wrangham (1975, 1977) classified four types of competitive interaction. The most frequently observed type is type I, "supplanting" (0.19/min). Type II is "following into site," where an individual waits nearby and occupies the place as soon as the other individual finishes eating (0.08/min). This is observed when the feeding site is small and occupying another site nearby is difficult. Type III is "protecting the site," where an individual interrupts another's feeding by pulling the whole branch with attached food items out of reach (0.16/min). Type IV is "occupying the preferred site," which is similar to type II but differs in that there are other individuals eating nearby (0.16/min).

Such interactions can occur in a multilayered way within a consecutive time series. For example, in one occurrence within the Mahale chimpanzees, an adult female, Totzy, approached an adult male, Bonobo, who was ant fishing alone. Eventually, she sat between Bonobo and the fishing tree. He soon gave up fishing. After this relatively apparent supplanting (type I), Bonobo waited nearby and resumed fishing after she left (type II) (Itoh, 2002).

Although there are "non-competitive" interactions during eating, such as food sharing (Chapter 20), the majority of interactions including just staying nearby may be interpreted as "competitive." However, Nishie (2008, 2010, 2012) revealed the interactions during ant fishing were not simply competitive or affiliative, but rather, the chimpanzees continuously and redundantly responded to each other's actions.

Another important aspect on "feeding with others" is the fission–fusion grouping pattern characteristic to chimpanzees (Chapters 6, 9). Individuals are relatively free to decide when to leave from a specific location, where to go, and what to do next. Wrangham (1975, 1977) investigated why an individual parts from a party and revealed that if both chimpanzees in a pair are satiated, they tend to leave the feeding site together. However, if not, the possibility of leaving alone or together is similar. This indicates that the experience of eating together promotes individuals to resume eating and subsequently stay together, at least for a short period.

Reasons for individuals leaving food sites were difficult to investigate (Wrangham, 1975, 1977). Studies on party size in relation to activities may be suggestive on this issue. At all studied sites, traveling parties tend to be smaller than feeding parties (Gombe: Wrangham and Smuts, 1980; Taï: Boesch, 1996; Budongo: Newton-Fisher, 1999a). This may be a consequence of differing decisions in timing and reason to leave, irrespective of the party size from which they are to leave. Not only on-site food-related reasons but also off-site social reasons can promote an individual to leave a particular place (Wrangham, 1975). Chimpanzees may gather in response to another's food calls or find foods by following others (Wrangham, 1975, 1977). Party formation and feeding are intermingled, in that individuals temporarily eating alone are also involved in a longer time scale of interaction.

Study of the various social interactions occurring within the feeding nexus should shed more light on the broader context of their sociality for which there still is a vast unexplored area.

Note

Focal observation data on feeding were collected during August 1997–July 1998 for 10 adult females, 5 adult males, 3 adolescent females, and 2 adolescent males (see Itoh and Nishida, 2007, for more details). All focal data were analysed to select those that lasted at least 60 min without intervals. Because of small sample sizes, July, February, and March were excluded from statistical tests. The observation time unit is 1 min (total OU = 14 590, including February–March OU = 1126). Non-plant foods and leaves consumed while wadging fruits or meat were excluded from the analysis.

References

Boesch, C. (1996). Social grouping in Taï chimpanzees. In *Great Ape Societies*, ed. W. McGrew, L. Marchant, and T. Nishida. Cambridge: Cambridge University Press, pp. 101–13.

Byrne, R. W. and Corp, N. (2003). Acquisition of skilled gathering techniques in Mahale chimpanzees. *Pan Africa News*, **10**, 4–7.

Chemurot, M., Isabirye-Basuta, G., and Sande, E. (2012). Amount of plant foods eaten and sexual differences in feeding among wild chimpanzees (*Pan troglodytes*) of Kanyawara Community. *ISRN Zoology*, **2012**, 1–5.

Corp, N. and Byrne, R. W. (2002a). Leaf processing by wild chimpanzees: Physically defended leaves reveal complex manual skills. *Ethology*, **108**, 673–96.

Corp, N. and Byrne, R. W. (2002b). The ontogeny of manual skill in wild chimpanzees: Evidence from feeding on the fruit of *Saba florida*. *Behaviour*, **139**, 137–68.

Corp, N. and Byrne, R. W. (2004). Sex difference in chimpanzee handedness. *American Journal of Physical Anthropology*, **123**, 62–8.

Doran, D. (1997). Influence of seasonality on activity patterns, feeding behavior, ranging, and grouping patterns in Taï chimpanzees. *International Journal of Primatology*, **18**, 183–206.

Fawcett, K. A. (2000). *Female Relationships and Food Availability in a Forest Community of Chimpanzees*. Ph.D. thesis. Edinburgh, UK: University of Edinburgh.

Ghiglieri, M. P. (1984). *The Chimpanzees of Kibale Forest: A Field Study of Ecology and Social Structure*. New York: Columbia University Press.

Hladik, C. M. (1977). Chimpanzees of Gabon and chimpanzees of Gombe: some comparative data on the diet. In *Primate Ecology: Studies of Feeding and Ranging Behaviour in Lemurs, Monkeys, and Apes*, ed. T. H. Clutton-Brock. New York: Academic Press, pp. 481–501.

Isabirye-Basuta, G. (1988). Food competition among individuals in a free-ranging chimpanzee community in Kibale Forest, Uganda. *Behaviour*, **105**, 135–47.

Itoh, N. (2002). Feeding act of wild chimpanzees: To eat as a social event. *Anthropological Science*, **110**, 48.

Itoh, N. (2004). [*The Plant Phenology and Fission–Fusion Social System in the Mahale Mountains National Park*.] Doctoral Dissertation. Kyoto: Kyoto University. In Japanese.

Itoh, N. and Nishida, T. (2007). Chimpanzee grouping patterns and food availability in Mahale Mountains National Park, Tanzania. *Primates*, **48**, 87–96.

Izawa, K. and Itani, J. (1966). Chimpanzees in Kasakati Basin, Tanganyika. (1) Ecological study in the rainy season, 1963–1964. *Kyoto University African Studies*, **1**, 73–156.

Krishnamani, R. and Mahaney, W. C. (2000). Geophagy among primates: Adaptive significance and ecological consequences. *Animal Behaviour*, **59**, 899–915.

van Lawick-Goodall, J. (1968). The behaviour of free-living chimpanzees in the Gombe Stream Reserve. *Animal Behaviour Monographs*, **1**, 161–311.

Mahaney, W. C., Hancock, R. G. V., Aufreiter, S., and Huffman, M. (1996). Geochemistry and clay mineralogy of termite mound soil and the role of geophagy in chimpanzees of the Mahale Mountains, Tanzania. *Primates*, **37**, 121–34.

Matsumoto-Oda, A., Hosaka, K., Huffman, M., and Kawanaka, K. (1998). Factors affecting party size in chimpanzees of the Mahale Mountains. *International Journal of Primatology*, **19**, 999–1011.

Newton-Fisher, N. E. (1999a). Association by male chimpanzees: a social tactic? *Behaviour*, **136**, 705–30.

Newton-Fisher, N. E. (1999b). The diet of chimpanzees in the Budongo Forest Reserve, Uganda. *African Journal of Ecology*, **37**, 344–54.

Nishida, T. (1968). The social group of wild chimpanzees in the Mahali Mountains. *Primates*, **9**, 167–224.

Nishida, T. (1974). [The ecology of wild chimpanzees.] In [*The Human Ecology*], ed. R. Ohtsuka, J. Tanaka, and T. Nishida. Tokyo: Kyōritsu Shuppan, pp. 15–60. In Japanese.

Nishida, T. (1976). The bark-eating habits in primates, with special reference to their status in the diet of wild chimpanzees. *Folia Primatologica*, **25**, 277–87.

Nishida, T. (1980). Local differences in responses to water among wild chimpanzees. *Folia Primatologica*, **33**, 189–209.

Nishida, T. (1991). Primate gastronomy: cultural food preferences in nonhuman primates and origins of cuisine. In *Chemical Senses, Vol. 4: Appetite and Nutrition*, ed. M. I. Friedman, M. G. Tordoff, and M. R. Kare. New York: Marcel Dekker, pp. 195–209.

Nishida, T. and Uehara, S. (1983). Natural diet of chimpanzees (*Pan troglodytes schweinfurthii*): long-term record from the Mahale Mountains, Tanzania. *African Study Monographs*, **3**, 109–30.

Nishie, H. (2008). [Chimpanzee culture and sociality: reconsidering "the metaphor of transmission of knowledge".] *Primate Research*, **24**, 73–90. In Japanese with English abstract.

Nishie, H. (2010). [Never-ending interaction: "redundant" interaction among wild chimpanzees.] In [*Boundary and Conjunction of Social Interaction: Studies in Nonhuman Primates, Humans and Conversation*], ed. D. Kimura, M. Nakamura, and K. Takanashi. Kyoto: Shōwadō, pp. 387–96. In Japanese.

Nishie, H. (2012). [*Chimpanzee Culture and Sociality.*] Doctoral Dissertation. Kyoto: Kyoto University. In Japanese.

Sun, C., Kaplin, B. A., Kristensen, K. A., *et al.* (1996). Tree phenology in a tropical montane forest in Rwanda. *Biotropica*, **28**(4b), 668–81.

Suzuki, A. (1969). An ecological study of chimpanzees in a savanna woodland. *Primates*, **10**, 103–48.

Turner, L. A. (2006a). Seed dispersal of a tropical forest tree, *Pycnanthus angolensis* (Myristricaceae) by chimpanzees in the Mahale Mountains National Park, Tanzania. *Memoirs of the Faculty of Science Kyoto University (Series of Biology)*, **18**, 35–43.

Turner, L. A. (2006b). Vegetation and chimpanzee ranging in the Mahale Mountains National Park, Tanzania. *Memoirs of the Faculty of Science Kyoto University (Series of Biology)*, **18**, 45–82.

Wrangham, R. W. (1975). *The Behavioural Ecology of Chimpanzees in Gombe National Park, Tanzania*. Ph.D. thesis. Cambridge: University of Cambridge.

Wrangham, R. W. (1977). Feeding behaviour of chimpanzees in Gombe National Park, Tanzania. In *Primate Ecology*, ed. T. H. Clutton-Brock. New York: Academic Press, pp. 504–38.

Wrangham, R. W. and Smuts, B. B. (1980). Sex differences in the behavioural ecology of chimpanzees in the Gombe National Park, Tanzania. *Journal of Reproduction and Fertility*, **Suppl. 28**, 13–31.

Yamakoshi, G. (1998). Dietary responses to fruit scarcity of wild chimpanzees at Bossou, Guinea: possible implications for ecological importance of tool use. *American Journal of Physical Anthropology*, **106**, 283–95.

18 Taste of chimpanzee foods*

TAKASHI HAYAKAWA

18.1 Food repertoires of chimpanzees

Long-term studies have revealed patterns of food choice by wild chimpanzees (Chapter 17). Chimpanzees consume a variety of plants and animals as foods and sometimes consume abiotic substances such as minerals. However, they do not consume all the available animals and plants in their environment. They also show geographical and individual variation in what and how they eat. Therefore, the diversity and variability of food repertoires of chimpanzees have been topics of interest in the feeding ecology of chimpanzees. Here, I review chimpanzee foods in terms of taste physiology with the aim of clarifying mechanisms of food choice by chimpanzees.

A primary list of food repertoires of chimpanzees at Mahale has been reported (Nishida and Uehara, 1983). Based mainly on this report and information of existing plant species within the Tongwe territory (Nishida and Uehara, 1981), Nishida *et al.* (1983) showed that there were >472 plant species at Mahale but chimpanzees there consumed only 198 species and 328 parts of them as foods, indicating that chimpanzees selectively consume plants to be ingested, although more systematic analysis is required because the food list has been updated (Appendix II). The food repertoire of chimpanzees is not geographically or chronologically constant. This inconsistency is particularly significant in plant species introduced by sympatric living humans (Chapter 38). No parts of oil palm (*Elaeis guineensis*) are commonly eaten by chimpanzees at Mahale despite its presence in the vegetation, but fruits and the pith of leaf fronds of oil palms are eaten by chimpanzees at Gombe, Tanzania (Nishida *et al.*, 1983). Moreover, chimpanzees at Bossou, Guinea, eat kernels of oil palms using hammer and anvil stones to crack their shell, as well as their fruits, bases of young leaves, and woody tissues (Sugiyama and Koman, 1992). It was reported that M-group chimpanzees at Mahale started to eat fruits of mangoes, lemons, and guavas in the 1980s (Takasaki, 1983; Takahata *et al.*, 1986). These novel fruits are now eaten by almost all the individuals of the M group, suggesting that social learning has altered the food repertoire within a group. Rare feeding behaviors by particular individuals are sometimes reported, such as algae feeding by a female immigrant (Sakamaki, 1998) and chewing of oil palm fronds by an adolescent female (Zamma *et al.*, 2011). Although these two behaviors have not spread to other chimpanzees, such rare feeding behaviors also may represent the flexibility of the food repertoire of chimpanzees.

18.2 Function of taste in food choices of chimpanzees

What factors determine the diversity and variability of food choices of chimpanzees? The existence of group-specific food repertoires and feeding techniques indicates that social learning greatly

* I thank Dr. Hirohisa Hirai, Dr. Hiroo Imai, Dr. Yasuhiro Go, Dr. Toshifumi Udono, Dr. Tohru Sugawara, Dr. Atsushi Matsui, and Ms. Nami Suzuki-Hashido for supporting molecular studies of bitter taste receptors and for valuable discussion.

Mahale Chimpanzees: 50 Years of Research, ed. M. Nakamura, K. Hosaka, N. Itoh and K. Zamma. Published by Cambridge University Press.

influences food choice in chimpanzees. On the other hand, the observation that chimpanzees sometimes display seemingly capricious food choice at the individual level indicates that chimpanzees can modify their own food repertoire by individual learning. Senses such as vision, olfaction, and gustation can possibly play important roles in learning of foods to be eaten because developing individuals need to recognize and memorize the appearance, texture, and chemical characteristics of food items. Taste may be particularly important because it is directly linked to quality of food items.

Mammals generally have the following five basic tastes: sweetness, umami, saltiness, bitterness, and sourness (Chandrashekar *et al.*, 2006; Yarmolinsky *et al.*, 2009). The first three tastes detect beneficial compounds (nutrition and mineral) to be ingested. For example, sweetness is due to sugar, umami due to amino acids, and saltiness due to sodium chloride. The remaining two tastes indicate harmful compounds not to be ingested. Plant secondary metabolites, such as alkaloids, are usually poisonous and are detected by bitterness. The ingestion of decomposed foods can be rejected using the sour taste to detect oxidative products derived from organic materials (rancidification). The sensing of highly acidic foods by sourness may also be important for reducing erosive tooth wear. Although bitterness and sourness are generally unpleasant tastes, "learned" humans know that all bitter and sour foods are not always physiologically harmful, but are sometimes beneficial, and often acquire a preference for their ingestion. Some chimpanzee foods at Mahale also include compounds that are bitter or sour for humans, such as leaves and pith of *Vernonia amygdalina* (bitter), aril of *Pycnanthus angolensis* (bitter), and fruits of *Citrus limon* (lemon; sour but pleasant, at least for humans) (Nishida and Uehara, 1983; Nishida *et al.*, 2000).

Therefore, I have proposed two alternative reasons why chimpanzees sometimes ingest seemingly unpleasant-tasting foods. First, chimpanzees may be able to compensate such "unpleasant" tastes of food items with the knowledge, acquired by social or individual learning, that they are edible. They may even find some of such food items highly pleasant by mixture with other senses, as humans do. Second, chimpanzees may have different physiology from humans with respect to such food items. Recent molecular studies have revealed similarities and differences in taste physiology between humans and chimpanzees and among chimpanzee individuals. Such physiological differences can deceive researchers investigating the mechanism of ingestion of seemingly unpleasant foods by chimpanzees.

18.3 Tasting studies of plant foods of chimpanzees

The easiest way to understand the taste of wild chimpanzee foods is for researchers themselves to taste them. Many field researchers have often evaluated foods of wild animals with their own sense of taste and realized that such foods include both pleasant and unpleasant tastes (Nissen, 1931; Izawa and Itani, 1966; Sabater Pi, 1979; Nishida, 1991). For example, Nishida noted his terrible experience at Mahale when trying to eat the aril of *Pycnanthus angolensis*, which is frequently eaten by the Mahale chimpanzees but seldom by local people (Nishida, 2012: 48–9) (Figures 18.1, 18.2a). He was so

Figure 18.1 A female chimpanzee at Mahale eating fruits of *Pycnanthus angolensis*. Their arils are bitter for humans. ©T. Hayakawa.

Figure 18.2 Bitter (at least for humans) plant foods of chimpanzees at Mahale. (a) A fruit of *Pycnanthus angolensis*. (b) Leaves of *Vernonia amygdalina*. (c) Leaves of *Trema orientalis*. (d) Leaves of *Tinospora caffra*. ©T. Hayakawa.

overwhelmed by the bitter and unpleasant taste of the aril that he immediately spat it out.

Based on such experiences, he performed a systematic evaluation of the taste of chimpanzee foods at Mahale (Nishida *et al.*, 2000). He recorded the taste of almost all plant-food items eaten by focal adult males for a total of 13 months, comprising 2–3 months each in 1991, 1992, 1994, 1995, and 1997. He succeeded in recording the taste of 114 plant-food items (34% of all the chimpanzee plant-food items recorded at that time). His taste categories were sweetness, saltiness, bitterness, sourness, and astringency with three levels (slightly, normally, and very) in addition to insipidity (no taste). A combination of multiple taste categories was allowed for a food item, such as both sweet and bitter taste. He also provided descriptive information for each food item, such as pleasantness, umami, fattiness, pungency, and specific flavor. The plant-food items were most frequently characterized as insipid, followed by sweet and then bitter taste, and pleasant and palatable items made up 69% of all the items. None of the items were salty. Only the ripe fruit of *Uvaria angolensis* was regarded as being umami. He also analysed the length of time that chimpanzees spent eating each food item from June to August 1994 and from September to November 1995. Because the predominant fruits, as measured by feeding duration, in each period had different types of taste categories, the tastes of the temporal food repertoires were different over the periods. His conclusion was that chimpanzee plant foods were acceptable to him most of the time in both periods.

Researchers who have had a chance to attempt to taste foods of both wild apes and wild monkeys have often reported that the food repertoires of chimpanzees are more pleasant and palatable than those of monkeys (Nishida *et al.*, 2000; Wrangham, 2009). This distinction may reflect the phylogenetic proximity of humans and apes in physiology

Table 18.1 **Bitter plant-food items of chimpanzees at Mahale**

Plant	Part	Bitterness for TN[a]	Bitterness for TH[b]	Used as foods by local people[b]	Medicinally used by local people[b]
Parkia filicoidea	Unripe seed	Very bitter	Bitter and astringent	No	No
Vernonia amygdalina	Pith of stalk/ branch	Very bitter	Very bitter	No	Gastrointestinal
Tinospora caffra	Leaf, shoot, and bud	Bitter	Bitter	No	Gastrointestinal
Trema orientalis	Leaf, shoot, and bud	Bitter	Slightly bitter/ Insipid[c]	No	Gastrointestinal
Baphia capparidifolia	Blossom	Bitter and sweet	Bitter and sweet	No	No
Pycnanthus angolensis	Fruit	Bitter/Bitter and sweet	Very bitter	No	Cathartic
Diplorhynchus condylocarpon	Seed	Slightly bitter	Slightly bitter	No	Cathartic
Pteridium aquilinum	Pith of shoot	Slightly bitter	Slightly bitter	No	No
Pterocarpus tinctorius	Seed	Slightly bitter	Bitter	No	No

[a] Nishida *et al.*, 2000.
[b] The investigation was performed from October to December, 2010. I tasted some of the bitter plant foods of chimpanzees previously reported (Nishida *et al.*, 2000). My tasting was not necessarily performed during focal sampling of chimpanzees. I also interviewed several Tongwe people about whether they use the plant part as foods and/or medicines.
[c] The leaves of *Trema orientalis* tasted insipid to me in October, but bitter in November.

involving taste and food metabolism. Genetic studies have shown distinct differences between humans and monkeys in the bitter taste receptor gene repertoire and ligand specificity of the umami taste receptor (Toda *et al.*, 2013; Hayakawa *et al.*, 2014).

18.4 Bitter plant foods of chimpanzees

The food repertoires of chimpanzees included several food items with unpleasant taste (Nishida *et al.*, 2000). They may use such unpleasant food items for medicinal purposes (Chapter 24). Leaves and pith of *V. amygdalina* (Figure 18.2b) are among the very bitter food items that may be used as medicines by chimpanzees. Based on an observation of a female chimpanzee who recovered from her poor health after chewing pith of *V. amygdalina*, Huffman and Seifu (1989) suggested that chimpanzees use it for self-medication. Subsequently, antischistosomal substances (sesquiterpene lactones and steroid glucosides) were extracted from *V. amygdalina* as the bitter components (Jisaka *et al.*, 1992). The observation that sympatric local people also use *V.*

amygdalina as a medicine supports the possibility of self-medication by chimpanzees. Therefore, screening bitter plant foods of chimpanzees and asking local people about their use of such plants may help to identify potentially medicinal plants.

I performed tasting of the bitter food items of chimpanzees and interviewed local people about their use (Table 18.1). The levels of bitterness differed from those reported by Nishida (Nishida *et al.*, 2000). This difference may be because of seasonal differences in capacity to produce bitter compounds of each plant and individual differences in sense of taste between the tasters themselves. The leaves of *Trema orientalis* (Figure 18.2c) did not taste bitter to me in October, but were bitter in November, suggesting a seasonal change in bitterness. I also attempted to taste the pith of *Tinospora caffra* (Figure 18.2d), which is ingested by chimpanzees at Mahale as well as the leaves, and found it very bitter. At that time, I asked three Japanese researchers and five Tongwe assistants in our camp to taste it. Three of them (two Tongwe assistants and I) said "very bitter," one of the Japanese said "normally bitter," and the others said "slightly bitter." Although more controlled experiments are required, this case indicated that individual differences in taste exist. Approximately half of the bitter chimpanzee foods I tested were medicinal for local people, mainly for gastrointestinal diseases (Table 18.1). Pharmacological analysis would be useful in the future to identify the reasons for the ingestion of bitter foods by chimpanzees at Mahale.

18.5 Tastes of non-plant foods of chimpanzees

None of the chimpanzee plant-food items reported is salty, and only one item (the ripe fruit of *U. angolensis*) is umami (Nishida *et al.*, 2000). Salty taste indicates sodium chloride, which is contained in animal rather than plant foods. Umami taste indicates free amino acids, which are contained in both plant and animal foods. However, plant foods often contain a large amount of sweet sugar and may rarely show a dominant umami taste. Chimpanzees at Mahale also feed on mammals, birds, insects, and abiotic substances (Nishida and Uehara, 1983). Chimpanzees may use their senses of salty and umami taste for these non-plant foods. Some psychological and electrophysiological studies have shown taste responses to salt and umami taste in chimpanzees (Steiner and Glaser, 1984; Hellekant and Ninomiya, 1991; Hladik and Simmen, 1996; Hellekant *et al.*, 1997; Ueno *et al.*, 2004).

Figure 18.3 Chimpanzees at Mahale licking a rock along the shore of Lake Tanganyika (©K. Hosaka).

Chimpanzees at Mahale may also use their sense of salty and umami taste for "miscellaneous" foods such as breast milk for infants, their own or others' mucus, semen, vaginal secretions, and blood (Nishida and Uehara, 1983). Although it is more difficult to taste these foods than plant foods, some researchers have attempted to do this. Chimpanzees at Mahale sometimes display rock-licking behavior, not observed at Gombe (Nishida, 1980). They lick rocks along the shore of Lake Tanganyika, in riverbeds of large rivers, or in rocky outcrops in the forest (Figure 18.3). According to researchers who ventured to lick the same or surrounding surfaces of rocks licked by chimpanzees, the taste was insipid; therefore, the nutritional significance of rock-licking behavior is unknown (Nishida, 2008: 135–6; Nishida *et al.*, 2010; Takuya Matsumoto, personal communication).

18.6 Genetic variation in bitter taste receptors in humans and chimpanzees

To characterize the tastes of chimpanzee foods, researchers have attempted to taste them, asked local people about their use, and extracted their chemical components. However, we should also consider physiological differences between chimpanzees and humans because there is a possibility that researchers themselves misinterpret taste in chimpanzees by relying on human-specific taste physiology. Although direct comparison of psychological experiments between humans and chimpanzees is difficult, the bitter taste threshold for quinine hydrochloride is seemingly higher in chimpanzees than in humans (Glaser, 1986; Hladik and Simmen, 1996). Moreover, the threshold for quinine hydrochloride is variable among human populations (Hladik and Simmen, 1996). Threshold variation within and between species is particularly prominent in the bitter taste of an artificial compound, phenylthiocarbamide (PTC). In 1931, it was found that nontasters of PTC bitterness account for several dozen percent of human populations and their phenotype is inherited as a recessive trait (Snyder, 1931). It was also reported that chimpanzees and some other nonhuman primates have intraspecific (within species) nontaster variation for PTC bitterness (Fisher *et al.*, 1939; Chiarelli, 1963). For this reason, studies involving tasting chimpanzee foods should be reconsidered, in view of intra- and interspecific differences in taste sensitivity in humans and chimpanzees.

Taste sensitivity variation is, of course, affected by environmental factors such as food habits and taste disorders. However, recent molecular studies have revealed that variation in bitter taste sensitivity is also linked to genetic variation in bitter taste receptor proteins called TAS2R proteins. In 2000, TAS2R proteins were first identified as membrane proteins of taste-receptor cells of the mammalian tongue and surrounding taste organs, expressed from *TAS2R* genes (Adler *et al.*, 2000; Chandrashekar *et al.*, 2000). TAS2 R proteins recognize bitter compounds in food and send electrophysiological signals to sensory neurons. PTC bitterness is recognized by one of the TAS2R proteins, TAS2R38, and the nontaster phenotype for PTC is caused in both humans and chimpanzees by genetic variation in TAS2R38 (Kim *et al.*, 2003; Bufe *et al.*, 2005; Wooding *et al.*, 2006) (Figure 18.4). The cause of the nontaster haplotype is different between humans and chimpanzees; three amino acid substitutions in the TAS2R38 protein result in the inability to recognize PTC in nontaster humans (PAV type, taster haplotype; AVI type, nontaster haplotype), whereas the loss of the start codon in the *TAS2R38* gene causes the truncation of the amino acid sequence of the TAS2R38 protein in nontaster chimpanzees.

The best-known compounds analogous to PTC are glucosinolates, which are contained in cruciferous vegetables; therefore, it would appear that glucosinolates and their derivatives stimulate TAS2R38 causing variation in sensitivity to the bitterness of cruciferous vegetables in humans (Sandell and Breslin, 2006; Meyerhof *et al.*, 2010; Wooding *et al.*, 2010). However, the habitats of wild chimpanzees do not appear to contain cruciferous vegetables (Nishida and Uehara, 1983; Sugiyama and Koman, 1992). Genetic variation in chimpanzee TAS2R38 may also be associated with other compounds such as the nortriterpene limonin contained in citruses, to which human TAS2R38 responds (Meyerhof *et al.*, 2010). Citrus fruits, such as lemon, are ingested by wild chimpanzees; however, a functional assay will be required to elucidate the relationship between chimpanzee TAS2R38 and limonin.

18.7 Geographical variation and natural selection of *TAS2R38* in chimpanzees

Intraspecific variation is also interesting in terms of geographical variation. This is because the population ratio of AVI (nontaster) haplotypes of *TAS2R38* is variable among human populations (44–9% in Asians and Europeans; 5% in Southwest Native Americans), although this geographical difference may result from population history rather than natural selection

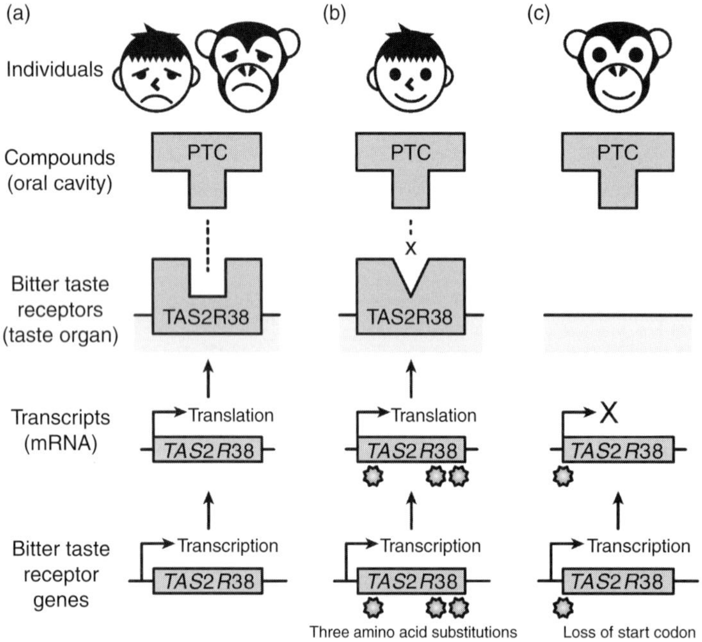

Figure 18.4 Molecular mechanism of bitter taste perception and its variation in humans and chimpanzees. This figure shows a relationship between one of the bitter taste receptors, TAS2R38, and the ligand, phenylthiocarbamide (PTC). (a) The *TAS2R38* gene encodes a bitter taste receptor protein, TAS2R38. Its transcription and translation is mediated by messenger RNA (mRNA). TAS2R38, which is expressed on the taste organ, specifically recognizes PTC in the oral cavity and evokes a bitter taste in the individual via sensory neurons (Kim *et al.*, 2003). (b) Nontaster-type *TAS2R38* in humans. The haplotype has three amino acid substitutions (from PAV to AVI), which reduce PTC specificity (Kim *et al.*, 2003; Bufe *et al.*, 2005). (c) Nontaster-type *TAS2R38* in chimpanzees. The haplotype has lost its start codon (undergone pseudogenization), and therefore the gene cannot be translated into the normal bitter taste receptor (Wooding *et al.*, 2006).

(Wooding *et al.*, 2004). Does such a geographical difference exist in chimpanzee *TAS2R38*? Sugawara *et al.* (2011) found that the nontaster haplotype of *TAS2R38* accounts for 76% of wild-born western chimpanzees (*Pan troglodytes verus*) in Japan. However, chimpanzees are distributed across equatorial Africa. My collaborators and I also identified *TAS2R38* variations in other subspecies of chimpanzees including eastern chimpanzees (*P. t. schweinfurthii*) in the Mahale M group (Hayakawa *et al.*, 2012; Hayakawa, in prep.). The DNA of chimpanzees in the Mahale M group is extracted from noninvasive samples, such as feces and urine, collected from identified individuals during focal sampling (Chapter 46). None of the chimpanzees, except the western subspecies, were found to have the nontaster haplotype (Figure 18.5). We showed that this geographical difference is inconsistent with selective neutrality ($F_{ST} = 0.614$, $p < 0.05$). In other words, some kind of selective pressure, such as food environmental differences, region-specific food habits, probably led to the geographical variation in chimpanzee *TAS2R38*.

18.8 Region-specific bitter taste as a possible origin of the regional food repertoire

Humans and chimpanzees have 26 and 28 functional *TAS2R* genes, respectively, including *TAS2R38* (Go *et al.*, 2005; Sugawara *et al.*, 2011; Hayakawa *et al.*, 2014). Each of the *TAS2R* genes encodes a receptor that recognizes different bitter ligand repertoires that

Table 18.2 **Human TAS2R proteins and their ligands. This table shows some of the already known receptor–ligand relationships (Meyerhof *et al.*, 2010)**

Right: TAS2R Bottom: compound	4	7	10	14	16	31	38	39	40	43	46
PTC	–	–	–	–	–	–	+	–	–	–	–
D-salicin	–	–	–	–	+	–	–	–	–	–	–
Cnicin	–	–	–	–	–	–	–	–	–	–	+
Aloin	–	–	–	–	–	–	–	–	–	+	–
Caffeine	–	+	+	+	–	–	–	–	–	+	+
Quinine	+	+	+	+	–	+	–	+	+	+	+

"+" indicates the ligand of the corresponding TAS2 R.

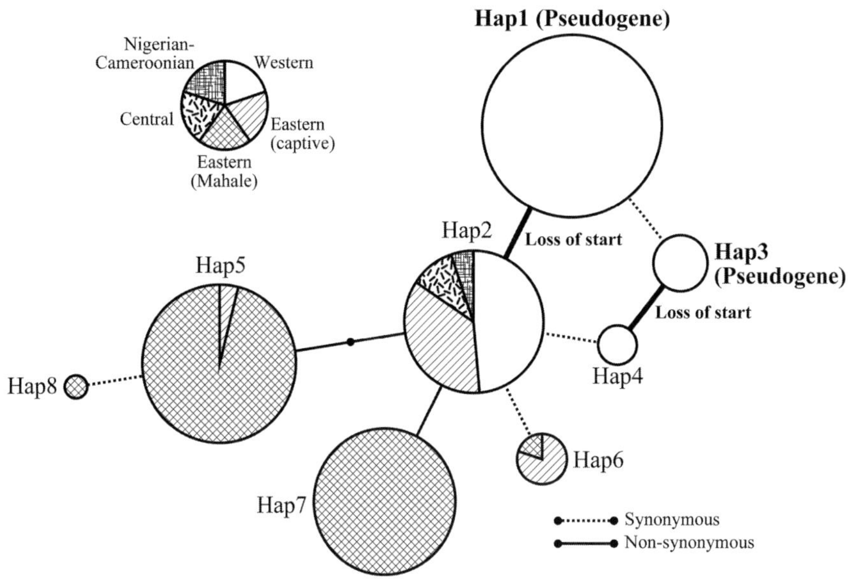

Figure 18.5 The median-joining network for *TAS2R38* haplotypes in chimpanzees. The network was calculated using NETWORK v4.6.1.1 (Bandelt *et al.*, 1999). Five groups were sampled: four groups of wild-born chimpanzees kept in Japan from four geographical subspecies (46 western, 10 eastern, 2 central, and 1 Nigerian–Cameroonian chimpanzees) and 45 wild eastern chimpanzees at Mahale (Hayakawa *et al.*, 2012; Hayakawa, in prep.). The geographical origins of the four groups in Japan are unknown; therefore, information beyond the level of subspecies is unavailable. Circles represent haplotypes. Pattern within the circle indicates each group. Areas within pattern-coded parts of the circles indicate numbers of sampled chromosomes. Pseudogenized *TAS2R38* (nontaster type) are found only in western chimpanzees, indicating that this is a difference at the subspecies level. Eastern chimpanzees in Japan and at Mahale share several haplotypes, but chimpanzees at Mahale have two unique haplotypes (Hap7 and Hap8), indicating that this is a difference at a more local level.

overlap with one another (Meyerhof *et al.*, 2010) (Table 18.2). For example, in humans, TAS2R16 specifically recognizes β-glucopyranosides such as D-salicin, whereas caffeine is recognized by the following multiple receptors: TAS2R7, TAS2R10, TAS2R14, TAS2R43, and TAS2R46. Not only *TAS2R38* but also other *TAS2R*s have functional variation for ligand sensitivity in humans (Soranzo *et al.*, 2005; Pronin *et* *al.*, 2007; Dotson *et al.*, 2008; Thalmann *et al.*, 2013). A genome-wide association study found that sensitivity to quinine hydrochloride is associated with variation in a chromosomal region containing *TAS2R* genes in humans (Reed *et al.*, 2010).

These studies help us to understand interspecific differences of taste between humans and chimpanzees and intraspecific variations of taste within chimpanzees. For example, sesquiterpene lactones are recognized by TAS2R46 in humans (Brockhoff *et al.*, 2007; Meyerhof *et al.*, 2010). As we have seen above, the bitter pith of *V. amygdalina* that is ingested by chimpanzees contains bitter sesquiterpene lactones. If chimpanzee TAS2R46 also recognizes sesquiterpene lactones of *V. amygdalina*, such specificity of chimpanzee TAS2R46 may contribute to the intriguing feeding behavior on *V. amygdalina*.

My collaborators and I identified genetic variations in all 28 *TAS2R* genes in three subspecies of chimpanzees and compared them with those in western chimpanzees, previously determined by Sugawara *et al.* (2011), to measure region-specific selective pressures on chimpanzee *TAS2R* genes (Hayakawa *et al.*, 2012). We identified 172 protein haplotypes, including 16 nonfunctional haplotypes (pseudogenes and whole gene deletions), in the four subspecies of chimpanzees we sampled. Approximately two-thirds of them were unique to each subspecies. One of the haplotypes specific to western chimpanzees (TAS2R16 haplotype 2) is known as that of a sensitivity-changing receptor (Imai *et al.*, 2012). Although it has not been observed that other species-specific haplotypes also change their sensitivity to bitter compounds, this result suggested the marked diversification of chimpanzee *TAS2R* genes at the subspecies level.

Geographical differences in the *TAS2R* gene repertoire among subspecies of chimpanzees are also supported from the perspective of natural selection. We performed the Tajima's D and π_N/π_S tests, which can reject selective neutrality as the null hypothesis, for the nucleotide diversity of *TAS2R* genes in each subspecies of chimpanzee (Hayakawa *et al.*, 2012). Based on summary statistics such as haplotype frequencies, the number of variable sites, and amino acid substitutions, these tests can provide presence of natural selection as alternative hypotheses (Nei and Gojobori, 1986; Tajima, 1989). As a result, statistical values of Tajima's D and π_N/π_S suggest that the diversification of *TAS2R* genes in western chimpanzees resulted from balancing selection, whereas in eastern chimpanzees, purifying selection dominated as the evolutionary form of the diversification of *TAS2R* genes. Such a distinction was particularly strong in *TAS2R* genes that were specifically acquired by the anthropoid lineage (Hayakawa *et al.*, 2014). Because the anthropoid-specific TAS2R proteins (at least in humans) recognize various kinds of plant-derived toxins such as camphor, aristolochic acid, aloin, and sesquiterpene lactones (Kuhn *et al.*, 2004; Behrens *et al.*, 2009; Meyerhof *et al.*, 2010), these *TAS2R* genes may have undergone stronger selective pressure from the plant-feeding repertoires of chimpanzees. Therefore, we clearly showed that the genetic basis of the sense of bitter taste of chimpanzees has been guided by natural selection. This selection may account for region-specific sense of bitter taste in chimpanzees.

18.9 Heritable taste as a determinative factor of food choice

Studies of taste in chimpanzees have been performed in interdisciplinary science fields. Field studies have described food repertoires of chimpanzees associated with learning and assessed taste using human researchers' own sense of taste. Comparison between multiple field sites has identified region-specific food repertoires of chimpanzees resulting from both environmental and learning factors. Chemistry and ethnopharmacology have provided a physiological function for each taste compound contained in chimpanzee foods. Psychological and electrophysiological studies on captive chimpanzees have revealed taste sensitivities of chimpanzees themselves. Molecular studies have revealed the presence of heritable taste and history of natural selection.

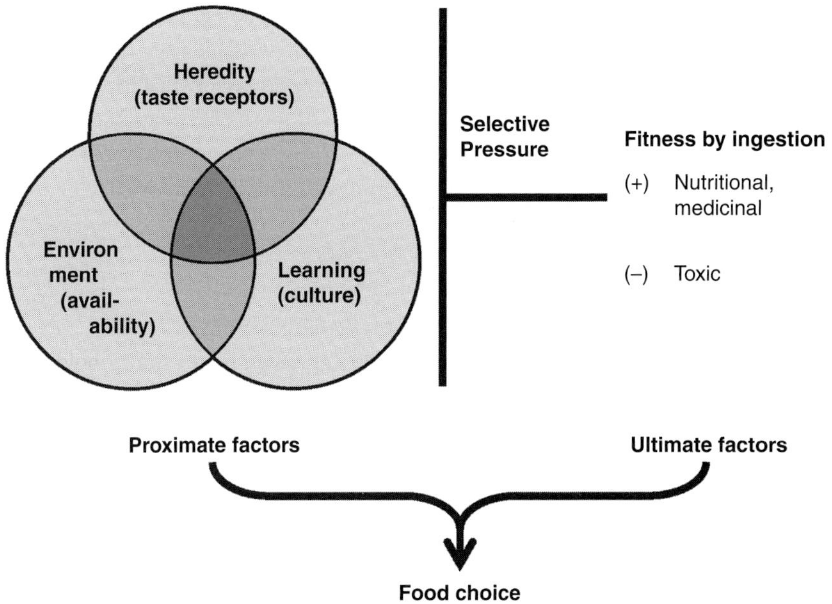

Figure 18.6 The three determinative factors of food choice: heredity, environment, and learning. Interaction among these proximate factors determines the food choice of chimpanzees. As ultimate factors, nutritional, medicinal, and toxic components of foods are expected to exert selective pressure that influences these proximate factors.

Therefore, I propose that it will be sufficient to integrate the complex results from interdisciplinary fields of taste studies if feeding ecology studies of chimpanzees in future are able to account for the three proximate factors, i.e. heredity, environment, and learning, eliciting food choices (Figure 18.6). For example, we will be able to investigate *V. amygdalina* ingestion by chimpanzees at Mahale in terms of these three factors. The main bitter components of *V. amygdalina* are sesquiterpene lactones, which may be recognized by TAS2R46 in humans; therefore, to understanding the ingestion behavior, we must confirm whether TAS2R46 haplotypes of chimpanzees also recognize sesquiterpene lactones. If sensitivity differences from humans and/or sensitivity variations among chimpanzees exist in the relationship between TAS2R46 and sesquiterpene lactones, we should reconsider our interpretations of the taste of *V. amygdalina*. Therefore, we will be able to determine whether chimpanzees at Mahale decide to eat the pith of *V. amygdalina* as bitter or nonbitter food.

With respect to the environmental factor, food availability should also be considered. *Vernonia* species are broadly distributed over equatorial Africa. However, *Vernonia* eating has been observed only in eastern chimpanzees (Huffman, 2003). For example, western chimpanzees at Bossou do not ingest *Vernonia* species, despite its presence in the vegetation (Sugiyama and Koman, 1992). Five *Vernonia* species were identified in the flora of Bossou, but only one species, *V. conferta*, was present in secondary and deciduous forest, and the others in scrub forest or savanna. We will need to survey the availability of *Vernonia* species in habitats of western chimpanzees.

Finally, if the use of *V. amygdalina* in chimpanzees is indeed for self-medication, the behavior may be inseparable from learning (Huffman, 1997). Identifying the thresholds in chimpanzees themselves for bitterness of *V. amygdalina* may yield clues to the ways in which self-medicating chimpanzees learn sensory signals including bitterness.

The discovery of the molecular mechanism of inherited taste has greatly advanced our understanding of the sensory world of chimpanzees. The molecular bases not only of bitter taste but also of other tastes have been elucidated (Chandrashekar *et al.*, 2006; Yarmolinsky *et al.*, 2009). Environment and learning have often been proposed as determinative factors of food choice, but heritable taste has received little consideration as such a factor (Nishida *et al.*, 1983). The integration of these proximate factors in feeding behavior will lead to our understanding the ultimate basis of food choice of chimpanzees.

References

Adler, E., Hoon, M. A., Mueller, K. L., *et al.* (2000). A novel family of mammalian taste receptors. *Cell*, **100**, 693–702.

Bandelt, H. J., Forster, P., and Röhl, A. (1999). Median-joining networks for inferring intraspecific phylogenies. *Molecular Biology and Evolution*, **16**, 37–48.

Behrens, M., Brockhoff, A., Batram, C., *et al.* (2009). The human bitter taste receptor hTAS2R50 is activated by the two natural bitter terpenoids andrographolide and amarogentin. *Journal of Agricultural and Food Chemistry*, **57**, 9860–6.

Brockhoff, A., Behrens, M., Massarotti, A., Appendino, G., and Meyerhof, W. (2007). Broad tuning of the human bitter taste receptor hTAS2R46 to various sesquiterpene lactones, clerodane and labdane diterpenoids, strychnine, and denatonium. *Journal of Agricultural and Food Chemistry*, **55**, 6236–43.

Bufe, B., Breslin, P. A., Kuhn, C., *et al.* (2005). The molecular basis of individual differences in phenylthiocarbamide and propylthiouracil bitterness perception. *Current Biology*, **15**, 322–7.

Chandrashekar, J., Mueller, K. L., Hoon, M. A., *et al.* (2000). T2Rs function as bitter taste receptors. *Cell*, **100**, 703–11.

Chandrashekar, J., Hoon, M. A., Ryba, N. J., and Zuker, C. S. (2006). The receptors and cells for mammalian taste. *Nature*. **444**, 288–94.

Chiarelli, B. (1963). Sensitivity to PTC (phenyl-thio-carbamide) in primates. *Folia Primatologica*, **1**, 88–94.

Dotson, C. D., Zhang, L., Xu, H., *et al.* (2008). Bitter taste receptors influence glucose homeostasis. *PLoS ONE*, **3**, e3974.

Fisher, R. A., Ford, E. B., and Huxley, J. (1939). Taste-testing the anthropoid apes. *Nature*, **144**, 750.

Glaser, D. (1986). Geschmacksforschung bei Primaten. *Vierteljahrsschrift der Naturforschenden Gesellschaft in Zürich*, **131**, 92–110.

Go, Y., Satta, Y., Takenaka, O., and Takahata, N. (2005). Lineage-specific loss of function of bitter taste receptor genes in humans and nonhuman primates. *Genetics*, **170**, 313–26.

Hayakawa, T., Sugawara, T., Go, Y., *et al.* (2012). Eco-geographical diversification of bitter taste receptor genes (*TAS2Rs*) among subspecies of chimpanzees (*Pan troglodytes*). *PLoS ONE*, **7**, e43277.

Hayakawa, T., Suzuki-Hashido, N., Matsui, A., and Go, Y. (2014). Frequent expansions of the bitter taste receptor gene repertoire during evolution of mammals in the Euarchontoglires clade. *Molecular Biology and Evolution*, **31**, 2018–31.

Hellekant, G. and Ninomiya, Y. (1991). On the taste of umami in chimpanzee. *Physiology and Behavior*, **49**, 927–34.

Hellekant, G., Ninomiya, Y., and Danilova, V. (1997). Taste in chimpanzees II: single chorda tympani fibers. *Physiology and Behavior*, **61**, 829–41.

Hladik, C. M. and Simmen, B. (1996). Taste perception and feeding behavior in nonhuman primates and human populations. *Evolutionary Anthropology*, **5**, 58–71.

Huffman, M. A. (1997). Current evidence for self-medication in primates: a multidisciplinary perspective. *Yearbook of Physical Anthropology*, **104**(S25), 171–200.

Huffman, M. A. (2003). Animal self-medication and ethno-medicine: exploration and exploitation of the medicinal properties of plants. *The Proceedings of the Nutrition Society*, **62**, 371–81.

Huffman, M. A. and Seifu, M. (1989). Observations on the illness and consumption of a possibly medicinal plant *Vernonia amygdalina* (Del.), by a wild chimpanzee in the Mahale Mountains National Park, Tanzania. *Primates*, **30**, 51–63.

Imai, H., Suzuki, N., Ishimaru, Y., *et al.* (2012). Functional diversity of bitter taste receptor TAS2R16 in primates. *Biology Letters*, **8**, 652–6.

Izawa, K. and Itani, J. (1966). Chimpanzees in the Kasakati Basin, Tanganyika. (1) Ecological study in the rainy season, 1963–1964. *Kyoto University African Studies*, **1**, 73–156.

Jisaka, M., Kawanaka, M., Sugiyama, H., *et al.* (1992). Antischistosomal activities of sesquiterpene lactones and steroid glucosides from *Vernonia amygdalina*, possibly used by wild chimpanzees against parasite-related diseases. *Bioscience, Biotechnology, and Biochemistry*, **56**, 845–6.

Kim, U. K., Jorgenson, E., Coon, H., *et al.* (2003). Positional cloning of the human quantitative trait locus underlying taste sensitivity to phenylthiocarbamide. *Science*, **299**, 1221–5.

Kuhn, C., Bufe, B., Winnig, M., *et al.* (2004). Bitter taste receptors for saccharin and acesulfame K. *The Journal of Neuroscience*, **24**, 10260–5.

Meyerhof, W., Batram, C., Kuhn, C., *et al.* (2010). The molecular receptive ranges of human TAS2 R bitter taste receptors. *Chemical Senses*, **35**, 157–70.

Nei, M. and Gojobori, T. (1986). Simple methods for estimating the numbers of synonymous and nonsynonymous nucleotide substitutions. *Molecular Biology and Evolution*, **3**, 418–26.

Nishida, T. (1980). Local differences in responses to water among wild chimpanzees. *Folia Primatologica*, **33**, 189–209.

Nishida, T. (1991). Primate gastronomy: Cultural food preferences in nonhuman primates and origins of cuisine. In *Chemical Senses, Vol. 4: Appetite and Nutrition*. ed. M. L. Friedman, M. G. Tordoff, and M. R. Kare. New York: Marcel Dekker, pp. 195– 209.

Nishida, T. (2008). [*Learning from Animals' "Feeding": New Edition*.] Kyoto: Kyoto University Press. In Japanese.

Nishida, T. (2012). *Chimpanzees of the Lakeshore: Natural History and Culture at Mahale*. Cambridge: Cambridge University Press.

Nishida, T. and Uehara, S. (1981). Kitongwe name of plants: A preliminary listing. *African Study Monographs*, **1**, 109–31.

Nishida, T. and Uehara, S. (1983). Natural diet of chimpanzees (*Pan troglodytes schweinfurthii*): Long-term record from the Mahale Mountains, Tanzania. *African Study Monographs*, **3**, 109–30.

Nishida, T., Wrangham, R. W., Goodall, J., and Uehara, S. (1983). Local differences in plant-feeding habits of chimpanzees between the Mahale Mountains and Gombe National Park, Tanzania. *Journal of Human Evolution*, **12**, 467–80.

Nishida, T., Ohigashi, H., and Koshimizu, K. (2000). Tastes of chimpanzee plant foods. *Current Anthropology*, **41**, 431–8.

Nishida, T., Zamma, K., Matsusaka, T., Inaba, A., and McGrew, M. C. (2010). *Chimpanzee Behavior in the Wild: An Audio-Visual Encyclopedia*. Tokyo: Springer.

Nissen, H. W. (1931). A field study of the chimpanzee: observations of chimpanzee behavior and environment in western French Guinea. *Comparative Psychology Monographs*, **8**, 1–122.

Pronin, A. N., Xu, H., Tang, H., *et al.* (2007), Specific alleles of bitter receptor genes influence human sensitivity to the bitterness of aloin and saccharin. *Current Biology*, **17**, 1403–8.

Reed, D. R., Zhu, G., Breslin, P. A., *et al.* (2010). The perception of quinine taste intensity is associated with common genetic variants in a bitter receptor cluster on chromosome 12. *Human Molecular Genetics*, **19**, 4278–85.

Sabater Pi, J. (1979). Feeding behaviour and diet of chimpanzees (*Pan troglodytes troglodytes*) in the Okorobiko Mountains in Rio Muni (West Africa). *Zeitschrift für Tierpsychologie*, **50**, 265–81.

Sakamaki, T. (1998). First record of algae-feeding by a female chimpanzee at Mahale. *Pan Africa News*, **5**, 1–3.

Sandell, M. A. and Breslin, P. A. (2006). Variability in a taste-receptor gene determines whether we taste toxins in food. *Current Biology*, **16**, R792–4.

Snyder, L. H. (1931). Inherited taste deficiency. *Science*, **74**, 151–2.

Soranzo, N., Bufe, B., Sabeti, P. C., *et al.* (2005). Positive selection on a high-sensitivity allele of the human bitter-taste receptor TAS2R16. *Current Biology*, **15**, 1257–65.

Steiner, J. E. and Glaser, D. (1984). Differential behavioral responses to taste stimuli in non-human primates. *Journal of Human Evolution*, **13**, 709–23.

Sugawara, T., Go, Y., Udono, T., *et al.* (2011). Diversification of bitter taste receptor gene family in western chimpanzees. *Molecular Biology and Evolution*, **28**, 921–31.

Sugiyama, Y. and Koman, J. (1992). The flora of Bossou: its utilization by chimpanzees and humans. *African Study Monographs*, **13**, 127–69.

Tajima, F. (1989). Statistical method for testing the neutral mutation hypothesis by DNA polymorphism. *Genetics*, **123**, 585–95.

Takahata, Y., Hiraiwa-Hasegawa, M., Takasaki, H. and Nyundo, R. (1986). Newly acquired feeding habits among the chimpanzees of the Mahale Mountains National Park, Tanzania. *Human Evolution*, **1**, 277–84.

Takasaki, H. (1983). Mahale chimpanzees taste mangoes: towards acquisition of a new food item? *Primates*, **24**, 273–5.

Thalmann, S., Behrens, M., and Meyerhof, W. (2013). Major haplotypes of the human bitter taste receptor TAS2R41 encode functional receptors for chloramphenicol. *Biochemical and Biophysical Research Communications*, **435**, 267–73.

Toda, Y., Nakagita, T., Hayakawa, T., *et al.* (2013). Two distinct determinants of ligand specificity in T1R1/T1R3 (the umami taste receptor). *The Journal of Biological Chemistry*, **288**, 36863–77.

Ueno, A., Ueno, Y., and Tomonaga, M. (2004). Facial responses to four basic tastes in newborn rhesus macaques (*Macaca mulatta*) and chimpanzees (*Pan troglodytes*). *Behavioural Brain Research*, **154**, 261–71.

Yarmolinsky, D. A., Zuker, C. S., and Ryba, N. J. (2009). Common sense about taste: from mammals to insects. *Cell*, **139**, 234–44.

Wooding, S., Kim, U. K., Bamshad, M. J., *et al.* (2004). Natural selection and molecular evolution in PTC, a bitter-taste receptor gene. *American Journal of Human Genetics*, **74**, 637–46.

Wooding, S., Bufe, B., Grassi, C., *et al.* (2006). Independent evolution of bitter-taste sensitivity in humans and chimpanzees. *Nature*, **440**, 930–4.

Wooding, S., Gunn, H., Ramos, P., *et al.* (2010). Genetics and bitter taste responses to goitrin, a plant toxin found in vegetables. *Chemical Senses*, **35**, 685–92.

Wrangham, R. (2009). *Catching Fire: How Cooking Made Us Human*. London: Profile books.

Zamma, K., Nakashima, M., and Ramadhani, A. (2011). Mahale chimpanzees start to eat oil palm. *Pan Africa News*, **18**, 6–8.

19 Seeds from feces: implications for seed dispersal and fecal analyses

MICHIO NAKAMURA AND NORIKO ITOH

19.1 Introduction

Chimpanzees are mainly fruit eaters (Chapter 17), and being one of the largest primate species, they can swallow relatively large seeds together with the fruit pulp. Swallowed seeds are then often defecated intact in their feces (Figure 19.1). This is of special importance for understanding seed dispersal and feeding habits. Seeds in chimpanzee feces retain their ability to germinate and can be transported long distances. It is considered to be beneficial for a plant seed be carried away from its parent to avoid competition with other seeds (but this assumption may not be as universally applicable, as previously thought; see Herrera, 2002, for a review). If seed dispersal by chimpanzees is significant for particular plant species, it is possible that chimpanzees are contributing to the reproduction and maintenance of the forests they inhabit (for a review of the interaction between seed dispersal and forest regeneration, see Chapman, 1995). In addition, by identifying seeds from feces, researchers can study the diet of chimpanzees even when direct feeding behaviors are unobservable. Interpreting chimpanzees' diets by fecal analyses is often less precise than direct observations of feeding. However, this method still has several advantages over, and can be used as a supplement to, direct observations.

In this chapter, we first introduce some common plant species whose seeds are often recovered from Mahale chimpanzees' feces and describe their characteristics. Second, we review studies of seed dispersal at Mahale. Finally, we provide examples of fecal analyses. It should be noted that some of the information has not been systematically collected at Mahale (fecal analyses were conducted systematically during the initial period of research, but the results have not yet been published formally), and, therefore, the results we present here describe preliminary findings. Throughout this chapter, we have used updated Latin names (following Appendix I of this volume) for plants, although some different names (e.g. synonyms) have been used in previous studies.

19.2 Seeds defecated in feces and those that are not

Seeds from numerous plant species are found in Mahale chimpanzees' feces. Some commonly seen species are *Saba comorensis, Landolphia owariensis, Pseudospondias microcarpa, Pycnanthus angolensis, Cordia millenii, Cordia africana, Synsepalum brevipes, Myrianthus arboreus, Garcinia huillensis*, and several species of *Ficus* (Figure 19.2). Most of these fruit species (with the exception of *Ficus* that contain thousands of tiny seeds in fruits) have relatively large seeds (see below for seed sizes) and the seeds are often tightly adhered to fruit pulp, and cannot be easily detached (Nishida, 2001). Chimpanzees may eat these fruits by nibbling in the mouth and then spitting out the seeds after consuming only the pulp. However, more often they swallow the fruit pulp together with seeds, especially when many fruits are eaten at the same time. While there have been no systematic investigations at Mahale on whether they swallow or spit out seeds, according to a study at Kibale, Uganda, chimpanzees swallowed seeds in 77% of fruit-feeding bouts, and this

Mahale Chimpanzees: 50 Years of Research, ed. M. Nakamura, K. Hosaka, N. Itoh and K. Zamma. Published by Cambridge University Press.

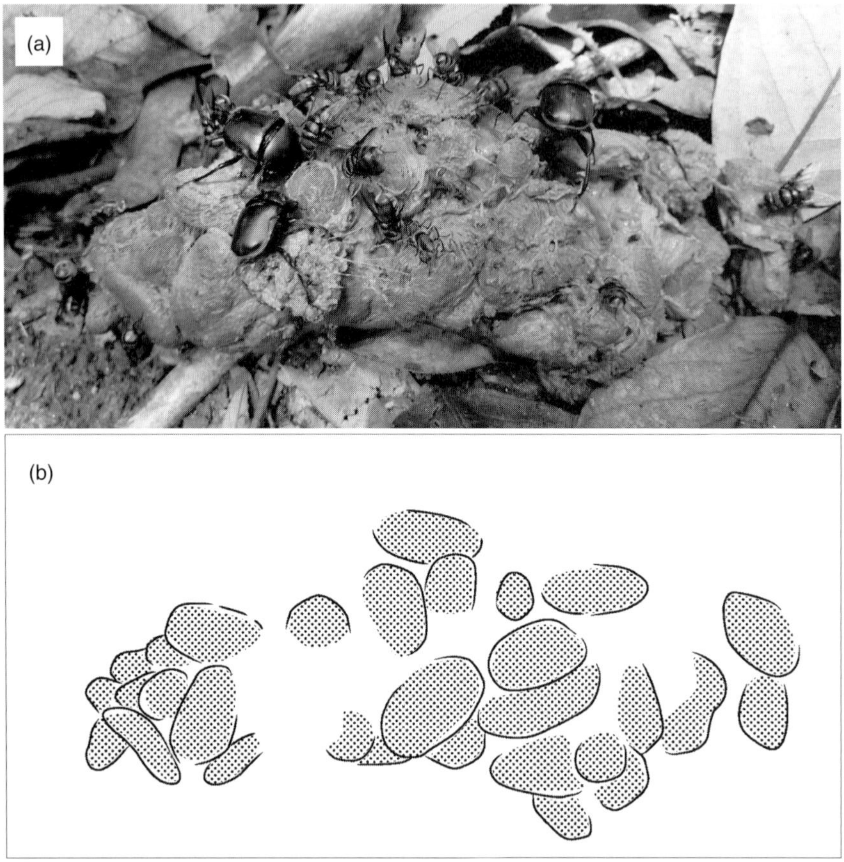

Figure 19.1 (a) A chimpanzee fecal clump at Mahale containing many *Saba* seeds. (b) At least 30 seeds are visible from outside. ©M. Nakamura.

was a much larger proportion compared to red-tailed monkeys (*Cercopithecus ascanius*) in the same forest (Lambert, 1999). When seeds are defecated in feces, fruit pulp is usually digested completely. However, *Cordia* spp. have very sticky fruit pulp, and a proportion of the pulp often remains adhered to seeds, even after defecation (Nakamura, personal observation). Chimpanzees at Mahale eat several species of *Ficus* (see Appendix II) but it is often difficult to macroscopically identify the exact species of seeds in the feces.

There are several fruit species at Mahale that are not swallowed immediately but are, instead, eaten by "wadging" in the mouth (Figure 19.3). Typically, when wadging, a chimpanzee puts mouthfuls of fruit between the lips and incisors, and, without biting, squeezes out the juice (Goodall, 1986). The fruit parts (i.e. skin, seeds, and fibers) that remain after extracting the juice are called a wadge, and this is often discarded. However, occasionally, chimpanzees will eat and swallow the wadge. When a wadge is discarded, many undestroyed seeds are also included. The plant species at Mahale whose fruits are often eaten by wadging include *Uvaria angolensis, Tabernaemontana pachysiphon, Psydrax parviflora, Syzygium guineense, Keetia venosa, Psychotria peduncularis*, and *Harungana madagascariensis* (Figure 19.4). *Pseudospondias* fruits are often swallowed whole, but occasionally wadged and discarded. For certain *Ficus* spp., chimpanzees tend to bite and ingest unripe green fruits and, consequently, swallow the seeds. However, when many ripe fruits are available they are often eaten by wadging. This has also been observed at Kibale (Lambert, 1999).

Figure 19.2 Seeds dispersed by Mahale chimpanzees. (a) *Cordia millenii*, (b) *Synsepalum brevipes*, (c) *Myrianthus arboreus*, (d) *Pseudospondias microcarpa*; (e) *Mimusops bagshawei*, (f) *Saba comorensis*, (g) *Toddalia asiatica*, (h) *Psychotria peduncularis*, (i) *Landolphia owariensis*, (j) *Psydrax parviflora*, (k) *Tabernaemontana pachysiphon*, (l) *Uvaria angolensis*, (m) *Pycnanthus angolensis*, (n) *Garcinia huillensis*. ©M. Nakamura.

Even the seeds of plant species whose fruits are often eaten by wadging can be found in feces. Therefore, for these species, it is accurate to say that a portion of seeds is swallowed during the process of wadging, while some are also discarded in wadges. At Nyungwe, Rwanda, *Syzygium guineense* seeds are also found in both wadges and chimpanzee feces, and it is the only plant species whose seeds are found in

Figure 19.3 A chimpanzee feeding on *Syzygium* fruits by wadging. ©M. Nakamura.

wadges (Gross-Camp *et al.*, 2009). It is, therefore, obvious that fecal analysis alone cannot be used to evaluate the relative importance of the fruit species in chimpanzees' diets which are also consumed by wadging.

Seeds of plant species that are bitten and eaten by chimpanzees cannot, of course, be detected in their feces. At Mahale, examples of such plant species are *Pterocarpus tinctorius, Diplorhynchus condylocarpon*, and *Sterculia quinqueloba* whose endosperms are often consumed (Figure 19.5). Moreover, especially when fruits are unripe and scarce, Mahale chimpanzees occasionally bite and eat *Saba* seeds, which are typically swallowed with fruit pulp.

19.3 Traits of seeds from feces

According to data from other study sites, average chimpanzee fecal clumps weigh 65.7 g at Dja, Cameroon (Poulsen *et al.*, 2001), 81.1 g at Kibale (Wrangham *et al.*, 1994), and 106.1 g at Nyungwe (Gross-Camp *et al.*, 2009). Fecal clumps at Mahale may lie within the typical weight range of 50 and 120 g, as recorded by Wrangham *et al.* (1994). Mean chimpanzee digestive retention time was 31.5 h at Kibale (Lambert, 1997, cited in Lambert and Garber, 1998) and 23.6 h (range = 19–27 h) in an experiment on captive chimpanzees (Idani, 1986). This shows that seeds are defecated approximately 1 day after being swallowed by a chimpanzee. Furthermore, based on observations of adult males at Gombe, Tanzania, Wrangham *et al.* (1994) estimated that a chimpanzee defecates 6.7 times per day on average.

The number of seeds contained in a fecal clump may differ depending on the size of the seeds and the season. In a study of the Mahale Y group between September and November 2006 (Nakamura and Sakamaki, unpublished data), all 41 fecal clumps collected contained seeds, and the mean number of seeds (excluding *Ficus* seeds) per clump was 43.1 (maximum 108). In support of this, a larger dataset from Nyungwe showed a similar figure of 40.6 seeds (excluding *Ficus* and *Urera*) (Gross-Camp *et al.*, 2009). At Dja, however, only 19 seeds on average were recovered from chimpanzee feces (Poulsen *et al.*,

Figure 19.4 Wadges discarded by Mahale chimpanzees. (a) *Pseudospondias microcarpa*, (b) *Ficus* sp., (c) *Psychotria peduncularis*, (d) *Psydrax parviflora*, (e) *Syzygium guineense*, (f) *Uvaria angolensis*. ©M. Nakamura.

2001). Such local differences may partially represent different plant community compositions and phenology, and/or abundances of fruit-eating competitors. At Mahale, on average, seeds from 3.2 different plant species (±1.2 SD, range = 1–6) were observed. Similarly, at Kibale, seeds from 2.8 species were recorded (Wrangham *et al.*, 1994). However, data from Dja (Poulsen *et al.*, 2001) and Nyungwe (Gross-Camp *et al.*, 2009) showed lower averages (both had a mean species number of 1.7).

In our Mahale dataset, the largest numbers of seeds for each plant species (or type) were: 92 for *Cordia africana*; 77, *Pseudospondias*; 73, *Saba/Landolphia*; 68, *Garcinia*; and 35, *Pycnanthus*. However, the largest number of seeds recorded for *Pycnanthus* was 63 in another study (Turner, 2006). According to Turner's (2006) estimation, M-group chimpanzees, as a whole, can disperse more than 1 million *Pycnanthus* seeds during its peak fruiting season. Chimpanzees may possibly disperse more seeds for certain other

Figure 19.5 A leftover of *Pterocarpus tinctorius* with endosperm eaten by a chimpanzee. ©M. Nakamura.

plant species, such as *Saba*, whose seeds are contained more in feces and whose fruiting season lasts longer.

Mahale chimpanzees can swallow and defecate a seed up to 30 mm in length and 15 mm in width. Table 19.1 shows the sizes of particular seeds from fruits commonly eaten at Mahale. The largest seeds found in chimpanzee feces at Mahale are those of *C. millenii*, averaging 28.76 mm in length. Seeds of this species are the largest among chimpanzee-dispersed seeds at Kibale (Wrangham *et al.*, 1994) and are also dispersed by Budongo chimpanzees in Uganda (Bakuneeta *et al.*, 1995). Among the fruits at Mahale that were categorized as "major foods" by Nishida (1991), three species of *Ficus* (*F. sur, F. exasperata*, and *F. vallis-choudae*; all have seeds smaller than 1 mm) and *Harungana* (small fruits approximately 2–3 mm in size) have not been listed in Table 19.1. Since the late 1990s, *Harungana* has not been frequently utilized as a food source because of declining tree numbers in lowland areas, probably due to vegetation succession (Chapter 13). Thus, *Harungana* may no longer be a "major food" for Mahale M-group chimpanzees (Chapter 17). This means that, currently, except for *Ficus*, many of the major fruit foods for the M-group chimpanzees have relatively large seeds.

Many fruits frequently consumed by M-group chimpanzees have seeds that are large in size but are still swallowable by chimpanzees. This implies that such plants are adapted for seed dispersal by large-bodied apes. African apes are often sympatric with smaller-bodied frugivorous primates such as guenons, but these monkeys do not usually swallow seeds larger than 5 mm. For example, red-tailed monkeys at Kibale only swallow seeds up to 4 mm in size (Lambert, 1999). Red-tails at Mahale also do not swallow but, instead, drop large seeds (Turner, 2006). Cercopithecoid monkeys are basically seed spitters (Lambert, 1999), who put fruits in their cheek pouches and spit out the seeds after detaching the pulp. Consequently, they do not usually disperse seeds by swallowing and defecating in a remote place.

The characteristics of fruits eaten frequently at Mahale vary greatly. For example, the fruits of *C. millenii, Pycnanthus*, and *Pseudospondias* contain only one seed, while *Saba* and *Tabernaemontana* fruits contain multiple seeds. Contrastingly, *Myrianthus* fruit forms a syncarp containing many seeds. Fruit sizes range from approximately 2 cm (*Pseudospondias*) to 15–25 cm (*Tabernaemontana*). Interestingly, however, there is much less variation in

Table 19.1 **Sizes of various plant seeds observed in chimpanzee feces and/or wadges**

Plant species	Family	Mean length[a] (mm ± SD)	Mean width[b] (mm ± SD)	Mean ellipticity	*N*	Food category[c]	Photo in Figure 19.2
Cordia millenii	Boraginaceae	28.76 ± 5.22	14.43 ± 0.99	0.52	9	Major	a
Pycnanthus angolensis	Myristicaceae	24.00 ± 2.56	14.52 ± 1.50	0.61	25	Major	m
Myrianthus arboreus	Cecropiaceae	19.83 ± 1.69	11.12 ± 1.46	0.56	22	Important	c
Saba comorensis	Apocynaceae	19.06 ± 1.81	9.97 ± 1.05	0.53	25	Major	f
Mimusops bagshawei	Sapotaceae	18.06 ± 1.40	9.62 ± 0.94	0.54	13	N/A	e
Synsepalum brevipes	Sapotaceae	17.61 ± 3.05	10.93 ± 1.35	0.63	6	N/A	b
Pseudospondias microcarpa	Anacardiaceae	15.67 ± 2.06	8.00 ± 0.84	0.52	20	Major	d
Landolphia owariensis	Apocynaceae	14.64 ± 1.92	8.49 ± 1.05	0.59	28	Important	i
Garcinia huillensis	Guttiferae	12.48 ± 1.05	8.58 ± 0.99	0.69	19	Major	n
Tabernaemontana pachysiphon	Apocynaceae	10.98 ± 1.49	5.67 ± 0.63	0.52	9	N/A	k
Uvaria angolensis	Annonaceae	10.87 ± 1.26	6.69 ± 1.03	0.62	13	Important	l
Psydrax parviflora	Rubiaceae	5.40 ± 0.43	4.58 ± 0.40	0.85	25	Important	j
Psychotria peduncularis	Rubiaceae	5.35 ± 0.22	3.90 ± 0.33	0.73	22	Important	h
Toddalia asiatica	Rutaceae	3.83 ± 0.23	2.62 ± 0.16	0.69	8	N/A	g

[a] The longest axis of a seed.

[b] The second longest axis of a seed orthographical to the longest axis.

[c] Based on Nishida's (1991) distinction. Although Nishida did not define these distinctions, "major foods" refer to foods that mainly influence the daily ranging of chimpanzees at least in some months almost every year. "Important foods" refer to those which partially influence ranging patterns (Itoh, 2002). Nishida (1991) did not list species shown as N/A; however, all of these species have recently strongly influenced the ranging of chimpanzees in June and July in some years (see Chapter 8). Therefore, these are considered as important foods.

the seed sizes of these fruits. Importantly, the width of many of these seeds is in the range 8–15 mm, which may limit swallowing. Most of the seeds described here are ovoid and elongate in shape, and the average ellipticity of seeds (i.e. width/length) tends to be smaller when the seed length is larger (Nakamura and Itoh, 2015). This means that larger seeds have relatively narrower widths, which may permit them to be swallowed more easily by frugivores. For example, although the largest seed of *C. millenii* has an average length of 28.76 mm, its width is only 14.43 mm. The longer seeds of this species also tend to be relatively narrower, i.e. even when a seed becomes longer, it does not become wider. This characteristic may allow chimpanzees to swallow seeds lengthwise. If any seed was spherical, with a diameter of 28.76 mm, chimpanzees would possibly be unable to swallow them and would therefore not utilize the fruit.

At Kibale, Wrangham *et al.* (1994) measured the lengths of 24 seed species that were found in chimpanzee feces. The largest one was 27.4 mm (*C. millenii*), while among the most commonly found species the largest was 20 mm (*Monodora myristica*). Seed lengths of six common species (only those with available size data) at Kibale averaged 13.58 ± 5.29 mm. Whereas, at Mahale, mean seed lengths of common species (Table 19.1) are larger than those measured at Kibale. However, we cannot make a direct comparison here as Mahale samples were not collected as systematically as those from Kibale. Most chimpanzee seed-dispersal studies have assessed seed size solely on length (Wrangham *et al.*, 1994; Lambert and Garber, 1998; Poulsen *et al.*, 2001; Gross-Camp *et al.*, 2009), although Lambert (1999) provided some width data. However, as discussed above, when we consider how easily a seed can be swallowed, the second largest axis (which is likely to be orthogonal to the direction of the digestive tract) may be more critical (Nakamura and Itoh, 2015). It is assumed that larger animals usually have larger gape widths and can thus swallow larger seeds (however, New World monkeys tend to swallow relatively larger seeds compared with Old World monkeys: Lambert and Garber, 1998). Therefore, it is often emphasized that chimpanzees are one of the most important dispersers of large seeds in African forests (e.g. Wrangham *et al.*, 1994). Still, there is no clear single criterion for "large" seeds, as sizes of ≥2 mm, ≥5 mm, and ≥2 cm have all been used in previous studies (Wrangham *et al.* 1994; Lambert and Garber, 1998; Poulsen *et al.*, 2001; Gross-Camp and Kaplin, 2005). If the first two criteria are applied to the Mahale dataset, almost all fruits frequently eaten by chimpanzees contain large seeds.

19.4 Chimpanzees as seed dispersers

As mentioned above, chimpanzees can swallow relatively large seeds. A larger size may be advantageous for a seed as it can contain more nutrition for germination and subsequent growth of the seedling. Moreover, because chimpanzees can travel long distances within the length of time of their digestive retention (i.e. about a day), seeds can be dispersed far from the parent tree in a variety of vegetation types, conferring an advantage to plants that have their seeds consumed by chimpanzees

19.4.1 Germinability of seeds from feces

Seeds from chimpanzee feces germinate under natural conditions (Figure 19.6), and it is known that they have higher germinability than those that did not pass through the digestive tracts of chimpanzees. In a simple experiment performed at Mahale, seeds of *Saba/Landolphia, Myrianthus, Pycnanthus*, and *Pseudospondias* were collected from chimpanzee feces and fallen fruits, and then planted in a camp garden (Takasaki, 1983). Although all *Saba/Landolphia* seeds decayed, the other three species germinated, with increased germination rates observed for seeds recovered from chimpanzee feces compared with seeds from fallen fruit. This was the first study to show that seeds that pass through chimpanzee digestive tracts have high germinability. Takasaki and Uehara (1984) carried out an additional experiment for *Saba/Landolphia, Garcinia, Azanza garckeana, Parkia filicoidea, Zanha golungensis*, and

Figure 19.6 *Pycnanthus* seeds, possibly from a chimpanzee deposit, germinating under natural conditions. ©M. Nakamura.

Strychnos innocua seeds from chimpanzee feces, and confirmed that all species of seeds germinated. Subsequently, in a similar experiment performed at Kibale, Wrangham *et al.* (1994) found that, for 10 species of plants, a higher proportion of chimpanzee-dispersed seeds germinated than non-dispersed seeds. Moreover, chimpanzee-dispersed seeds were also found to germinate earlier.

These experiments were performed at research camps, and not where the seeds were actually deposited. Therefore, it should be considered that the deposited location (i.e. the site of defecation) may also affect germinability and seedling survival. To address this, Turner (2006) monitored 564 *Pycnanthus* seeds at 30 different locations, and found that seedling survival was greater at canopy-covered forest sites (69.9%) compared with forest gaps (36.1%). Furthermore, 4 months after deposition, 143 (44.7%) of the original 320 seedlings (25.4% of the original number of seeds) were still alive at deposition sites.

Although no quantitative investigations on germinability of other plant species have been conducted at Mahale, there have been some descriptive accounts. Turner (2006) observed germination of *Synsepalum* seeds from chimpanzee feces deposited on rocks, and Nishida (2008) includes a photo of *C. millenii* seeds germinating from chimpanzee feces. Germinating *Mimusops* seeds from feces have also been observed (Nakamura, personal observation). These casual observations emphasize the need for a systematic investigation of other species whose seeds are dispersed by chimpanzees at Mahale.

At Mahale, Turner (2006) also noted that *Psychotria* and *Syzygium* seeds in chimpanzee wadges germinated. According to Gross-Camp and Kaplin (2005), the survival rate of seeds from *Syzygium* wadges was higher than that of other seeds from feces at Nyungwe. Although *Syzygium* is the only plant whose seeds are contained in wadges at Nyungwe, there are more fruit species whose unharmed seeds are included in wadges at Mahale. Therefore, it would be prudent to also investigate the fate of seeds in wadges with regard to seed dispersal by chimpanzees.

19.4.2 Plant–chimpanzee interaction

Generally, seedlings and saplings of a plant are observed more frequently near the parent tree. If a chimpanzee-dispersed plant species does not fit this general pattern, then it may be possible that chimpanzees are capable of influencing the distribution of particular plant species. In this regard, by counting the number of individual plants at different growth stages, Wakibara (2005) investigated the distributions of chimpanzee-dispersed plant species: *Pseudospondias, Pycnanthus, Syzygium, C. millenii*, and *Saba*. He concluded that the hyper-dispersion of *Saba* (see Chapter 13) appears to mirror the profound intensive and extensive dispersal of its seeds by chimpanzees. As repeatedly emphasized (Chapters 8, 17), *Saba* is a very important, preferred species for Mahale chimpanzees; its fruits can be eaten throughout almost half the year. *Saba* is a liana species with woody vines that reach down to the soil surface and have the potential to form new roots. It is also often seen to extend small vines into forest gaps during the early stages of succession. As a result, *Saba* can easily and efficiently increase its biomass through vegetative reproduction, which may determine its hyper-dispersion. However, intensive feeding and mass seed dispersal by chimpanzees may also contribute to the unique distribution of this species (see Chapter 13 for relevant discussion). *Saba* fruits are also eaten by guenons, such as red-tailed monkeys, but because the fruits have a relatively thick and hard cover, guenons require a lot of time and effort to pierce the cover before they can access the flesh. On the other hand, because adult chimpanzees can easily split the cover into two halves with their teeth and hands, then swallow the flesh whole together with the seeds, they can consume a lot of fruits in a much shorter time and with less effort. Such fruit traits also suggest that *Saba* is better adapted to being eaten by larger frugivores, such as chimpanzees.

Large fruit seeds (such as *Saba*) often have tightly adhering pulp. Nishida (2001) speculated that this may be an adaptation of plants to ensure that seeds are swallowed whole by large frugivores. He also proposed that thick-shelled fruits (an extreme case is *Tabernaemontana*; see Nishida, 2012: 45, for a photograph) are especially adapted for large-bodied frugivores, as almost no other frugivores but chimpanzees can break open the shell of *Tabernaemontana* at Mahale.

Through studies of seed dispersal by various animal species, coevolution of one-to-one relationships between plant and animal dispersers has been doubted (Chapman, 1995; Lambert and Garber, 1998). The only possible exception to this is the coevolution between gorillas (*Gorilla gorilla*) and *Cola lizae* (Tutin *et al.*, 1991). Chapman (1995) reported that there are no such highly interdependent relationships between chimpanzees and plants. However, more ovoid and elongate shapes of larger seeds swallowed by chimpanzees at Mahale imply that these plants may have evolved so that the seeds can be swallowed by large frugivores such as great apes (Nakamura and Itoh, 2015). In addition, primates other than apes are not known to disperse seeds frequently by swallowing in African forests (Lambert, 1999); apes disperse a large number of seeds despite their low density (Poulsen *et al.*, 2001); and endozoochorous plants tend to have larger fruits in the paleotropics than the neotropics (Mack, 1993). Taking all of these into account, further considerations may be needed on the interdependence between African tropical plants with large fruits, and large-bodied frugivores capable of swallowing seeds whole.

19.5 Feces as an informative source of diet

Seeds from chimpanzee feces are also important for identifying the types of fruit eaten during a particular period (see McGrew *et al.*, 2009, for a standard fecal analysis protocol). Along with fruit seeds, fibers from leaves, mammal bones or hairs, and insect chitin can also be found in feces. In order to understand how chimpanzees and others animals feed it is best to make direct observations. However, when this is not possible, for example, during the early stages of habituation or when there has been no habituation, fecal analysis is a very informative method (e.g. Kasakati, Tanzania: Izawa and Itani, 1966; Suzuki, 1969; Assirik, Senegal: McGrew *et al.*, 1988; Lopé, Gabon: Tutin and Fernandez, 1993; Kahuzi, DRC: Basabose, 2002; Nyungwe, Rwanada: Gross-Camp *et al.*, 2009).

Also at Mahale, quantitative fecal analyses were conducted at an early stage of the research when direct observations of feeding in the forest were not easy to perform. For example, between June 1975 and May 1976, Nishida and Norikoshi (n.d.a, n.d.b) collected and analysed 989 fecal clumps both from K- and M-group chimpanzees. During each month, the percentage of fecal clumps containing each fruit species (or type) was used as a simple parameter to estimate the relative importance of fruit species. Figure 19.7 highlights the species (types) with percentages greater than 50% in any one month. Nishida and Norikoshi discussed the results mainly in relation to between-group differences. However, here we will mostly focus on fecal clump data for the M group (Figure 19.7, solid lines) (although some results from the K group [dashed lines] will also be discussed) and compare these to their recent feeding habits that have been studied through long-term direct observations.

First, the fecal analysis revealed that *Saba/Landolphia* is an obvious, important food source. In this analysis, the two species were grouped together because their seeds were indistinguishable from each other. However, from the direct observations of M-group feeding, they were found to feed more often on *Saba* than *Landolphia*. We, therefore, assume that the seeds found in the feces mostly represent *Saba* (hereafter we just refer to *Saba* for *Saba/Landolphia*). In general, 27.9% of M-group feces (38.6% for K group) contained *Saba* seeds (second only to *Ficus*, in terms of proportion of feces containing seeds), and the seeds were found over half a year between September and February. Most notably, from December through to January, 100% of M-group fecal clumps (November–December for the K group) contained *Saba* seeds. This high reliance on *Saba* is consistent with recent feeding tendencies of the M-group chimpanzees that have been directly observed eating many *Saba* fruits almost every day from August to January (Chapter 17), and whose feces almost always contain *Saba* seeds. Although *Ficus* is also an important food at Mahale (the fruit commonly consumed by both K and M groups), its seeds were found in only 32.9% of M-group feces. This is a smaller percentage compared to recordings at other study sites. For example, *Ficus* seeds were found in 89.9%, 83.5%, 91.9%, and 67.1% of chimpanzee fecal clumps at Kanyawara, Kibale (Wrangham *et al.*, 1994), Kahuzi (Basabose, 2002), and Nyungwe (Gross-Camp *et al.*, 2009), respectively. These are all more than double the proportions found in the M group. Particularly, at Kahuzi, *Ficus* was among the top three species in 76 out of 78 months studied, and was, thus, assumed to be the preferred food rather than a fallback food (Basabose, 2002). On the other hand, in the Mahale M group, *Ficus* was found most frequently in April when the frequency of leaves found in feces was also highest (implying relatively less fruit in the diet). The proportion of *Saba*-containing and *Ficus*-containing fecal clumps were inversely correlated (Kendall's rank correlation, $\tau = -0.65$, $p = 0.006$), suggesting that the M-group chimpanzees utilize *Ficus* when the preferred *Saba* is not available (except for *F. exasperata*; Chapter 17).

Next, we focus on the uniqueness of feeding habits in June (and sometimes July) at Mahale. The frequency of leaves in chimpanzee feces was low in June, suggesting that they rely more on fruits, but instead of relying on a single or a few fruit species they consume multiple fruit species, including *Azanza, Psydrax, Toddalia*, and *Harungana*. Although *Psydrax* was not a particularly important food source for the M group in this dataset, in certain years chimpanzees have been known to eat *Psydrax* intensively (Chapter 17). As mentioned above, the importance of *Harungana* has declined in recent years. Although it has not been shown in this 1-year dataset, in different years, other fruit species are intensively eaten in June, including *Cordia africana, Synsepalum*, and *Uvaria*. Consequently, the frequency of seeds from these plants found in chimpanzee feces is also expected to be high during these years.

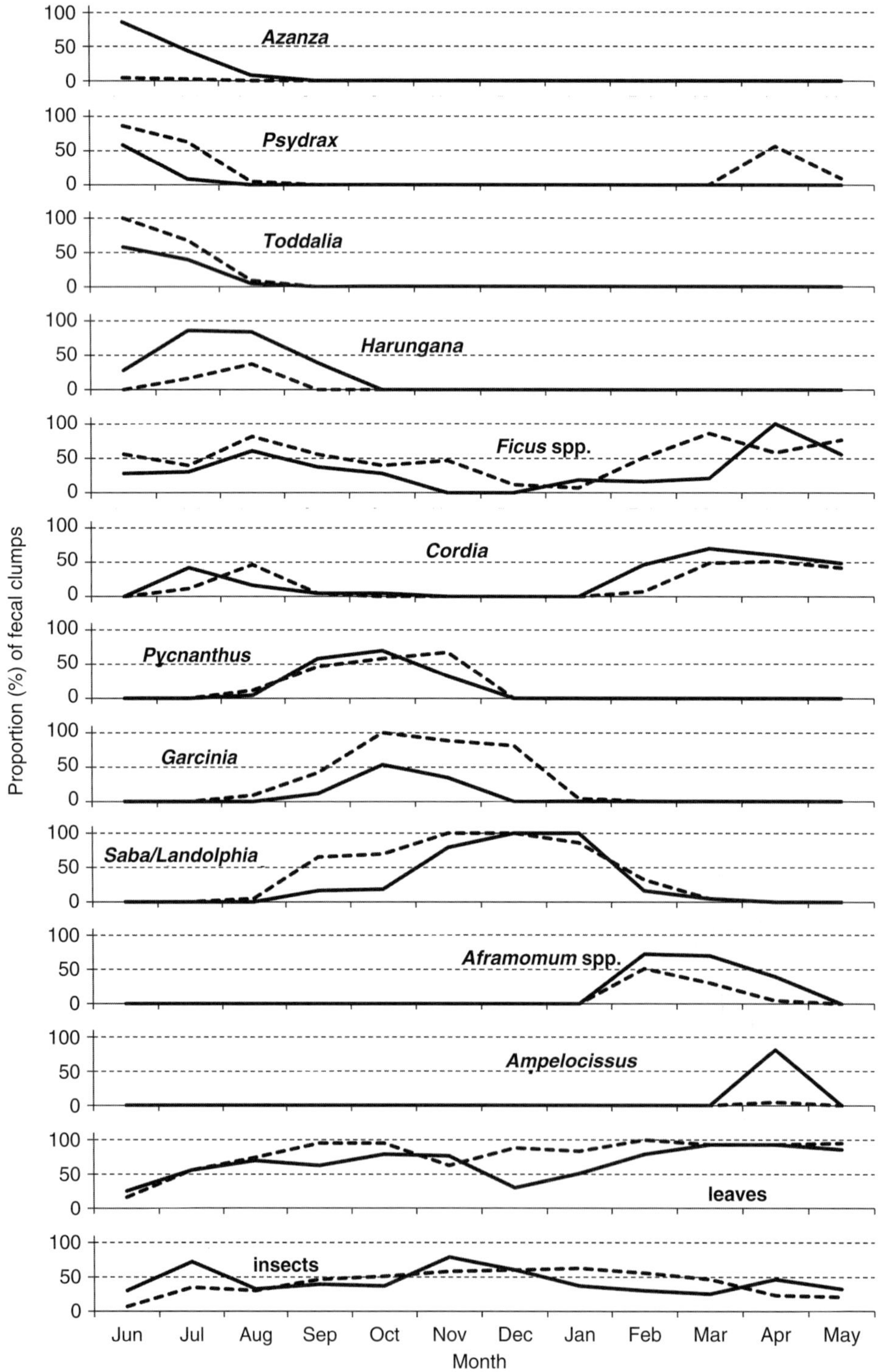

Figure 19.7 Fecal analyses from June 1975 to May 1976 (data manually read from hand-written figures of Nishida and Norikoshi, n.d.a, n.d.b). Vertical axes show the proportion (%) of fecal clumps containing each plant species (or type) per

Pseudospondias is another important example of a plant species not represented in Figure 19.7. In the 1976–7 fecal analyses (Norikoshi, 2002), this species exceeded 50% of seeds found in feces for certain months and was utilized from September to November. Based on our direct feeding observations, *Pseudospondias* is frequently and intensively eaten in particular years. *Zanha* provides an extreme example of interannual variation. A single large tree of this species bore many fruits in November 1996 and the M-group chimpanzees visited the tree and ate fruit almost every day. Subsequently, however, the fruits have very rarely been eaten. Over the last 5 years, the M-group chimpanzees ate *Zanha* fruits only on a single day in December 2008, December 2010, and November 2013 (Koichiro Zamma, personal communication). Therefore, fecal analysis in a single year may not represent feeding behavior or diet of chimpanzees in the long term. However, the overall patterns derived from this simple measure (i.e. percentage of fecal samples containing a seed species) seem broadly consistent with actual seasonal feeding patterns of the M-group chimpanzees.

Finally, we should mention that the information from feces is still important, even today, for well-habituated M-group chimpanzees. Mahale is mountainous, and chimpanzees sometimes climb steep mountains that cannot be reached by human observers. On such occasions, examining feces would tell us what they had been eating in these areas. For example, throughout 10 days in June 2012, Nakamura observed that 38 food items were eaten by the M group. Twenty items were fruits and five items could only be confirmed from indirect evidence, such as seeds from feces. After chimpanzees come down from a more elevated part of their range in June, we often observe many *Toddalia* seeds in their feces. However, as *Toddalia* have a decreased abundance in lowland areas (Itoh, personal observation), they are seldom eaten there; direct observation alone may underestimate the importance of this species.

19.6 Discriminating *Saba* and *Landolphia* seeds from feces

Fruits of *Saba* and *Landolphia* are easily distinguishable from each other by color and size. However, the shape and color of their seeds resemble each other (f and i of Figure 19.2), and are not easily distinguishable. Because these fruits are available during similar seasons, *Saba* and *Landolphia* have been clumped together in most previous seed dispersal and fecal analyses studies at Mahale (Nishida and Norikoshi, n.d.a, n.d.b; Takasaki, 1983; Takasaki and Uehara, 1984). However, because *Saba* is the most important fruit species, it may be useful to have some criteria to distinguish its seeds from those of *Landolphia*.

Thus, Nakamura (2014) conducted linear discriminant analysis on 25 *Saba* and 28 *Landolphia* seeds (Table 19.1), using length and width measures. The following formula yielded the discriminating function of the two species: $D_{sp}(L, W) = -0.455 \times L - 0.367 \times W + 11.06$ where L is the length and W is the width of a seed in millimeters. When D_{sp} value is negative, the seed is judged to be that of *Saba*, and when it is positive, it is judged to be that of *Landolphia*. By reevaluating the original dataset with this formula, three seeds of each species were misidentified as the opposite species. Therefore, discrimination accuracy was 88.68% (47/53). When data for 24 unidentified *Saba/Landolphia*-shaped seeds from a chimpanzee fecal clump were trialed using this formula, all but one were judged to be *Saba* (95.83% = 23/24).

Caption for Figure 19.7 (cont.)

analysed clump in a month. Solid and dashed lines show the M and K groups, respectively. Species (or type) data are shown for *Azanza garckeana, Psydrax parviflora, Toddalia asiatica, Harungana madagascariensis, Ficus* spp., *Cordia millenii, Pycnanthus angolensis, Garcinia huillensis, Saba comorensis/Landolphia owariensis, Aframomum* spp., leaves, and insects whose proportion exceeded 50% in any one month. *Saba* and *Landolphia*, multiple *Ficus* species, and multiple *Aframomum* species are combined respectively because of difficulties identifying seeds.

19.7 Conclusion and future perspectives

Chimpanzee feces contain important information reflecting their feeding habits. Researchers may tend to think (at least it seems to be the case at Mahale) that research using fecal samples is a preliminary measure, and is, therefore, only necessary when habituation has not been completed. However, seeds (and other materials) from feces can potentially tell us a variety of things even after chimpanzees have been habituated completely. We are yet to have a complete understanding on what and how much chimpanzees eat and defecate, the effects of these fundamental activities on forest reproduction and maintenance through seed dispersal, and the possible interrelationships between plants and apes. As such, there are many studies yet to be conducted at Mahale to address these issues, and we look forward to carrying out more systematic investigations in the near future.

References

Bakuneeta, C., Johnson, K., Plumptre, R., and Reynolds, V. (1995). Human uses of tree species whose seeds are dispersed by chimpanzees in the Budongo Forest, Uganda. *African Journal of Ecology*, **33**, 276–8.

Basabose, A. K. (2002). Diet composition of chimpanzees inhabiting the montane forest of Kahuzi, Democratic Republic of Congo. *American Journal of Primatology*, **58**, 1–21.

Chapman, C. A. (1995). Primate seed dispersal: Coevolution and conservation implications. *Evolutionary Anthropology*, **4**, 74–82.

Goodall, J. (1986). *The Chimpanzees of Gombe: Patterns of Behavior*. Cambridge, MA: Belknap.

Gross-Camp, N. and Kaplin, B. A. (2005). Chimpanzee (*Pan troglodytes*) seed dispersal in an afromontane forest: microhabitat influences on the postdispersal fate of large seeds. *Biotropica*, **37**, 641–9.

Gross-Camp, N. D., Masozera, M., and Kaplin, B. A. (2009). Chimpanzee seed dispersal quantity in a tropical montane forest of Rwanda. *American Journal of Primatology*, **71**, 901–11.

Herrera, C. M. (2002). Seed dispersal by vertebrates. In *Plant–Animal Interactions: An Evolutionary Approach*, ed. C. M. Herrera and O. Pellmyr. Oxford: Blackwell, pp. 185–208.

Idani, G. (1986). Seed dispersal by pygmy chimpanzees (*Pan paniscus*): a preliminary report. *Primates*, **27**, 441–7.

Itoh, N. (2002). [Food in the forest: Chimpanzees' food density and spatial distribution.] In [*The Mahale Chimpanzees: 37 Years of <Panthropology>*], ed. T. Nishida, S. Uehara, and K. Kawanaka. Kyoto: Kyoto University Press, pp. 77–100. In Japanese.

Izawa, K. and Itani, J. (1966). Chimpanzees in Kasakati Basin, Tanganyika: (I) Ecological study in the rainy season 1963–1964. *Kyoto University African Studies*, **1**, 73–156.

Lambert, J. E. (1999). Seed handling in chimpanzees (*Pan troglodytes*) and redtail monkeys (*Cercopithecus ascanius*): implications for understanding hominoid and cercopithecine fruit-processing strategies and seed dispersal. *American Journal of Physical Anthropology*, **109**, 365–86.

Lambert, J. E. and Garber, P. A. (1998). Evolutionary and ecological implications of primate seed dispersal. *American Journal of Primatology*, **45**, 9–28.

Mack, A. L. (1993). The sizes of vertebrate-dispersed fruits: A Neotropical–Paleotropical comparison. *The American Naturalist*, **142**, 840–56.

McGrew, W. C., Baldwin P. J., and Tutin C. E. G. (1988). Diet of wild chimpanzees (*Pan troglodytes verus*) at Mt. Assirik, Senegal: I. Composition. *American Journal of Primatology*, **16**, 213–26.

McGrew, W. C., Marchant, L. F., and Phillips, C. A. (2009). Standardised protocol for primate faecal analysis. *Primates*, **50**, 363–6.

Nakamura, M. (2014). Discriminating *Saba* and *Landolphia* seeds from chimpanzee feces at Mahale. *Pan Africa News*, **21**, 3–6.

Nakamura, M. and Itoh, N. (2015). Larger chimpanzee-dispersed seeds are elongated at Mahale, Tanzania: possible consequence of plant–disperser interaction? *Journal of Tropical Ecology*, **31**, 183–6.

Nishida, T. (1991). Primate gastronomy: cultural food preferences in nonhuman primate and origins of cuisine. In *Chemical Senses. Vol. 4. Appetite and Nutrition*, ed. M. I. Friedman, M. G. Tordoff, and M. R. Kare. New York: Marcel Dekker, pp. 195–209.

Nishida, T. (2001). [*Learning from Animals' "Feeding"*.] Tokyo: Kagawa Nutrition Publishing Division. In Japanese.

Nishida, T. (2008). [*The Societies of Chimpanzees*.] Tokyo: Tōhō-shuppan. In Japanese.

Nishida, T. (2012). *Chimpanzees of the Lakeshore: Natural History and Culture at Mahale*. Cambridge: Cambridge University Press.

Nishida, T. and Norikoshi, K. (n.d.a). The study of food on a basis of fecal analysis of the chimpanzees of the K-group and the M-group, June–November, 1975. *Mahale Mountains Chimpanzee Research Project Ecological Report*, **1**.

Nishida, T. and Norikoshi, K. (n.d.b). The study of food on fecal analysis of the chimpanzees of the K-group and the M-group, December 1975–May 1976. *Mahale Mountains Chimpanzee Research Project Ecological Report*, **2**.

Norikoshi, K. (2002). [Vegetation in the Kasoje area and the ranging pattern: changes during 20 years around the establishment of the national park.] In [*The Mahale Chimpanzees: 37 Years of <Panthropology>*], ed. T. Nishida, S. Uehara, and K. Kawanaka. Kyoto: Kyoto University Press, pp. 101–28. In Japanese.

Poulsen, J. R., Clark, C. J., and Smith, T. B. (2001). Seed dispersal by a diurnal primate community in the Dja Reserve, Cameroon. *Journal of Tropical Ecology*, **17**, 787–808.

Suzuki, A. (1969). An ecological study of chimpanzees in a savanna woodland. *Primates*, **10**, 103–48.

Takasaki, H. (1983). Seed dispersal by chimpanzees: a preliminary note. *African Study Monographs*, **3**, 105–8.

Takasaki, H. and Uehara, S. (1984). Seed dispersal by chimpanzees: supplementary note 1. *African Study Monographs*, **5**, 91–2.

Turner, L. A. (2006). Seed dispersal of a tropical forest tree, *Pycnanthus angolensis* (Myristricaceae) by chimpanzees in the Mahale Mountains National Park, Tanzania. *Memoirs of the Faculty of Science Kyoto University (Series of Biology)*, **18**, 35–43.

Tutin, C. E. G. and Fernandez, M. (1993). Composition of the diet of chimpanzees and comparisons with that of sympatric lowland gorillas in the Lopé Reserve, Gabon. *American Journal of Primatology*, **30**, 195–211.

Tutin, C. E. G., Williamson, E. A., Rogers, M. E., and Fernandez, M. (1991). A case study of a plant–animal relationship: *Cola lizae* and lowland gorillas in the Lopé Reserve, Gabon. *Journal of Tropical Ecology*, **7**, 181–99.

Wakibara, J. V. (2005). Abundance and dispersion of some chimpanzee-dispersed fruiting plants at Mahale, Tanzania. *African Journal of Ecology*, **43**, 107–13.

Wrangham, R. W., Chapman, C. A., and Chapman, L. J. (1994). Seed dispersal by forest chimpanzees in Uganda. *Journal of Tropical Ecology*, **10**, 355–68.

20 Hunting and food sharing

KAZUHIKO HOSAKA

20.1 Introduction

20.1.1 Hunting behavior

Since Nishida *et al.* (1979) first documented the hunting and meat-eating behavior of Mahale chimpanzees, several publications have appeared in the course of the long-term research (Kawanaka, 1982; Norikoshi, 1983; Takahata *et al.*, 1984; Uehara *et al.*, 1992; Hosaka *et al.*, 2001). Although the picture of hunting at Mahale has changed a great deal, the report by Nishida *et al.* (1979) is often still cited as the data from Mahale. Therefore, it is reasonable to start this chapter by describing the chronological change and the current status of hunting at Mahale, which has been fairly stable since the 1990s.

Comparison of hunting and meat eating with other study sites is also an important topic. Until the end of the 1980s, Gombe, Tanzania, from where the first evidence of chimpanzee hunting was documented (Goodall, 1963), was the only source of information for comparative studies and many investigators have reported on hunting behavior there (e.g. Teleki, 1973; Wrangham, 1975; Busse, 1978; Goodall, 1986; Gilby *et al.*, 2006).

Since 1990, a lot of publications about chimpanzee hunting and meat-sharing behavior have appeared (e.g. Wrangham and van Zinnicq Bergmann Riss, 1990; Uehara *et al.*, 1992; Stanford *et al.*, 1994b; Hosaka *et al.*, 2001), possibly as a response to a stimulating paper by Boesch and Boesch (1989), which compiled the first detailed data about the hunting behavior of western chimpanzees from Taï Forest, Côte d'Ivoire. They argued that Taï chimpanzees employ "cooperative hunting," which they maintain is different from hunting at Gombe and Mahale (see below).

After scientists grasped the interpopulational similarities in the behavior of chimpanzees from all three sites hunting for red colobus, more attention has been paid to why chimpanzees hunt in groups and what factors affect their motivation. An explanation for this trend was suggested by the "meat-for-sex hypothesis" (Stanford *et al.*, 1994a), that the presence of an estrous female in a foraging party may influence the decision of male chimpanzees in the party to hunt red colobus. Although this gained popularity beyond the scientific community, it was opposed by researchers from Mahale, Gombe, and Kibale (Ngogo and Kanyawara), Uganda, on the basis of analyses of group hunting and meat sharing (Hosaka, 1998, 2002; Mitani and Watts, 1999, 2001; Watts and Mitani, 2002a, b; Gilby *et al.*, 2006).

Evidence of chimpanzees hunting or meat eating has been reported from many other sites, such as Bossou, Guinea (Hockings *et al.*, 2012), and Fongoli, Senegal (e.g. Gašperšič and Pruetz, 2004; Bogart *et al.*, 2008; Pruetz and Marshack, 2009). See Chapter 42 for details about hunting with tools in chimpanzees.

20.1.2 Food sharing

Food sharing has been one of the central tenets of human evolution studies (e.g. Isaac, 1978). Thus, food sharing by primates, chimpanzees in particular, has intrigued many scholars.

McGrew and Feistner (1992) mentioned that chimpanzees are characterized by *passive* sharing

Mahale Chimpanzees: 50 Years of Research, ed. M. Nakamura, K. Hosaka, N. Itoh and K. Zamma. Published by Cambridge University Press.

("tolerated scrounging") of meat among group members, including non-kin adults, while humans and callitrichids engage in *active* sharing of plant foods among family members.

In other words, chimpanzees share food typically in response to begging behavior. Such passiveness has been noted since the publications from Mahale about food sharing between chimpanzees of both mother–infant and non-kin pairs (e.g. Nishida, 1973; Nishida *et al.*, 1979; Nishida and Turner, 1996). Begging behavior is also observed as a submissive gesture from subordinates to dominants (Chapter 33), irrespective of the age or sex of individuals. Thus, food sharing is an interesting topic in terms of evolution of gestures indicating social relationships and social cognition of chimpanzees.

The food most frequently shared among chimpanzees, particularly non-kin adults, is meat. Nishida *et al.* (1992) hypothesized that the alpha male tended to share meat with his mother, former and current female mating partners, and male coalition partners. They also showed that he never shared meat with the second-ranking males and rank-ascending young adult and adolescent males. This "political tactic hypothesis" has not yet been examined at Mahale, but in combination of the "meat-for-sex hypothesis," it has stimulated studies of the meat-sharing behavior of chimpanzees (e.g. Gilby *et al.*, 2008; Gomes and Boesch, 2009).

20.2 Animals that wild chimpanzees eat

20.2.1 Chronological changes in prey profile at Mahale

At Mahale, the proportion of red colobus among all eaten prey drastically increased from 14% (N = 50) in 1966–81 (Nishida *et al.*, 1979; Kawanaka, 1982; Norikoshi, 1983; Takahata *et al.*, 1984) and 56.3% (N = 103) in 1983–9 (Uehara *et al.*, 1992) to 83.3% (N = 295) in 1990–5 (Hosaka *et al.*, 2001). From the 1990s on, this tendency to prey upon red colobus has remained constant; 84.3% (N = 214) in 1996–2009 (Hosaka and Nishida, 2010; Table 20.1). Thus, the current Mahale chimpanzees have the same level of selectivity for red colobus as their counterparts in other populations (Gombe, 82%, Stanford *et al.*, 1994b; Taï, 78%, Boesch and Boesch, 1989; Ngogo, 91%, Mitani and Watts, 1999).

In contrast, the proportion of artiodactyl species among prey decreased from 1960–89 to 1990–5, but this

Table 20.1 **Prey species of Mahale chimpanzees from 1996 to 2009**

English name	Latin name	No. kills	%
Eastern red colobus	*Procolobus rufomitratus*	214	84.3
Red-tailed monkey	*Cercopithecus ascanius*	2	0.8
Yellow baboon	*Papio cynocephalus*	8	3.1
Blue duiker	*Cephalophus monticola*	17	6.7
Bushbuck	*Tragelaphus scriptus*	1	0.4
Bushpig	*Potamochoerus larvatus*	3	1.2
Squirrel	–	1	0.4
Other rodents	–	3	1.2
Unidentified	–	5	2.0
Total		254	100.0

Source: Hosaka and Nishida (2010). Note that this table is derived from pooled data collected by Toshisada Nishida, Kazuhiko Hosaka, and Michio Nakamura and is not conclusive in that it does not reflect all the MMCRP data collected during this period.

is most probably the result of numerical artifacts due to the increase of predation on red colobus (Hosaka *et al.*, 2001). Compared with the data from 1990–5, both blue duikers and bushpigs showed a slight decrease in percentages of all eaten prey in the updated data from 1996–2009, but this is not statistically significant (blue duikers: from 7.8 to 6.7%, bushpigs: 1.7 to 1.2%) (Hosaka and Nishida, 2010).

Mahale chimpanzees have never been observed to eat warthogs since one piglet was eaten in 1988. From 2000 to 2009, I observed six encounters between chimpanzees and warthogs during a total 643.4 h of observation, but no hunting attempt was recorded. However, Mahale chimpanzees may retain a "prey image" for warthogs, because there were two possible hunting attempts on September 18, 2008, and December 9, 2011, in which adult males chased warthogs with piglets (Michio Nakamura, personal communication).

On January 5, 1997, the first observation of chimpanzee predation on yellow baboon was recorded (Nakamura, 1997). Up to 2009, at least eight baboons had been killed (Nishie, 2004; Hosaka and Nishida, 2010). Formerly, Mahale chimpanzees had virtually no opportunity to hunt baboons, simply because no encounter between them was witnessed before 1991 (Chapter 16). It is plausible that increased encounters with baboons facilitated the addition of this species to the chimpanzee prey repertoire. However, baboon hunting at Mahale is still rare, which could be explained by the low frequency of encounters with this animal. According to my data from 2000 to 2009 (observation time = 643.3 h), chimpanzees' encounter rate with yellow baboons per hour was 0.02, compared with 0.16 for red colobus and 0.06 for red-tailed monkeys (Hosaka and Nishida, 2010).

To date, Mahale chimpanzees have been observed killing only immature baboons (one subadult, three juveniles, two infants, two unknown) (Hosaka and Nishida, 2010). In all except two cases, adult males ate baboon meat. However, in at least three cases, adult males did not show any persistence to obtain meat (Nishie, 2004; Hosaka and Nishida, 2010). Furthermore, the excitement level of beggars seemed lower for eating the meat of baboons than for eating the meat of red colobus, although it reflected only two observations (Nishie, 2004; Hosaka and Nishida, 2010). From my own observation, there was no excitement characteristic to group hunting on red colobus prior to the capture of a baboon. Chimpanzees may not have fully acquired a "prey image" for baboons, though more data are necessary to make such a statement conclusive.

Despite some changes of prey profile, it seems plausible that Mahale chimpanzees prey upon red colobus most frequently and most selectively in comparison to other prey species. The proportion of red colobus in all prey species has been stable at >80% since the 1990s, which is very similar to the status in other long-term study sites in which red colobus are sympatric.

Reasons why predation on red colobus was not frequent in the first two decades of Mahale research can be explained out in two ways. First, the encounter rate between chimpanzees and red colobus must have increased since the environment of Kasoje has undergone a transition to become more forested after Mahale became a protected area. Second, 20 of 50 observed cases of meat eating in 1966–81 at Mahale were derived from the K group, which was much smaller with fewer males than the M group. Considering the effect of number of skillful hunters on group hunting on red colobus, it is reasonable that chimpanzees of a small unit-group can hardly kill red colobus. There are some other explanations such as observational bias (see Hosaka *et al.*, 2001, for details).

20.2.2 Diversity of prey species across different populations

Uehara (1997) maintained that chimpanzees predominantly prey on red colobus across all habitats in which the two species are sympatric. At the Budongo Forest, Uganda, one of the long-term study sites of eastern chimpanzees is not inhabited by red colobus. Instead, Budongo chimpanzees prey upon Abyssinian colobus, *Colobus guereza*, which is a minor prey species in Kibale (Newton-Fisher *et al.*, 2002). At Kalinzu, another long-term site in Uganda not inhabited by red colobus, two successful hunts (Hashimoto *et al.*, 2000) of a red-tailed monkey and a blue monkey, and one unsuccessful hunt (Ihobe, 2001) of Abyssinian colobus by chimpanzees were recorded.

At Bossou, Sugiyama (1981) reported the first direct observation of hunting and meat eating by western chimpanzees. The tree pangolin (*Manis tricuspis*) eaten in that episode has been confirmed captured and eaten by Bossou chimpanzees in six cases (Hockings *et al.*, 2012). Haplorhine primates other than humans and chimpanzees are absent at Bossou. It seems interesting to note that Bossou chimpanzees do not appear to regard some other sympatric mammals (squirrels, duikers, and hyraxes) as their prey. In encountering these animals, they did not consume but just played with them even when they eventually killed the animals.

20.2.3 Chimpanzee and red colobus: predator–prey relationships

Ubiquitous high predation of chimpanzee on red colobus has drawn attention from behavioral ecologists interested in predator–prey relationships (Chapters 15, 16). If such relationships between the two species have persisted since long ago, it is possible that anti-predator behavior has evolved for red colobus, although Mahale red colobus seemed to have no effective strategy against chimpanzee hunting as compared with their counterpart in Taï (see Chapter 16 for details).

Busse (1977) first explored whether chimpanzee predation might have an impact on red colobus social organization. Based on 39 kills of red colobus at Gombe during 1973 and 1974, combined with the assumption of red colobus population there at 300–500 individuals, he estimated that chimpanzees killed about 8–13% of the sympatric red colobus population. Wrangham and van Zinnicq Bergmann Riss (1990) also calculated rates of predation on mammals using data on 75 kills recorded during focal observation at Gombe in 1972–5. Thus, they concluded that Gombe chimpanzees killed 204 prey per year, varying between 99 and 420 prey per year in periods of low and high predation rate. The authors claimed that red colobus in Gombe sustain a remarkably high rate of predation (from 20 to >40%).

According to the data in Uehara *et al.* (1992), chimpanzees of Mahale M group were estimated to have killed 33.6 red colobus per year from September 1983 to January 1990. Using this estimate, Ihobe (2000) calculated that the annual kill rate of red colobus by Mahale chimpanzees was 33.6, corresponding to 1.3–2.0% of the estimated red colobus population (Uehara and Ihobe, 1998). Applying the same estimation method to the data collected from February 1990 to November 1995 (Hosaka *et al.*, 2001) yields an annual rate of 97.3 red colobus kills; 3.6–5.8% of the prey population (Ihobe, 2000).

On the basis of the number of kills I observed in two study periods during 1991–4, Hosaka *et al.* (2001) estimated that Mahale chimpanzees killed 153 red colobus per year. Restricting the data to focal sampling yielded the estimate of 146 colobus killed per year. Combined with the estimated red colobus population (Uehara and Ihobe, 1998), the impact of chimpanzee predation was calculated to be 5.5–9.0% of the prey population (Ihobe, 2000).

Based on these data, Ihobe (2000) insisted that the impact of chimpanzee predation on the red colobus population was not catastrophic at Mahale, because it was no larger than the assumed rate of the population's natural increase, while admitting that chimpanzee predation is a limiting factor for growth of the population. This agrees with a result of red colobus density which did not seem to have decreased since 1995 (Chapter 15).

A similar trend seems applicable to Taï: Boesch (1994a) gave an annual kill rate of 125 colobus. Combined with the estimated number of the monkeys (Noë and Bshary, 1997), the annual predation rate of chimpanzee is calculated at 2.5–2.8% of the red colobus population.

In contrast, recent studies using demographic data of red colobus from Gombe (Fourrier *et al.*, 2008) and Ngogo (Teelen, 2008) suggested that the red colobus population could go extinct in the near future if chimpanzee hunting remains as high as current levels.

Further studies at Mahale will be necessary to evaluate which view is more accurate, as this is valuable information for devising a strategy for wildlife conservation at Mahale (Chapter 50).

20.2.4 Scavenging

Scavenging is another way of procuring meat in natural habitats. For instance, leopards, another forest predator species, are known to scavenge carrion (Bailey, 1993). However, wild chimpanzees rarely eat meat of carrion they encounter in the forest. Van Lawick-Goodall (1968) described some cases in which Gombe chimpanzees encountered dead bodies of other mammals such as rats, baboons, or bushpigs without eating the meat, even when some of them were fresh.

At Mahale, Hasegawa *et al.* (1983) reported four cases in which chimpanzees scavenged meat from mammals not killed by themselves. Strictly, caution is needed in interpreting two cases as scavenging, because dead bodies of blue duikers were collected by human observers and provisioned chimpanzees that excitedly devoured the meat even though one of the bodies was not fresh. In the other two cases, chimpanzees scavenged meat from dead bodies of bushbuck that were freshly killed and cached (or left over) by leopards.

On July 10, 1993, I observed another case of scavenging in which six chimpanzees found a freshly killed subadult bushbuck cached by a leopard and scavenged it (Hosaka *et al.*, 2001; see also Chapter 16). Three males spent at least 50 min eating the meat and then began to move to the south, while the females remained at the site to continue to eat the meat for an additional 3 h.

In contrast, Gombe chimpanzees were observed to show more ambivalent behavior than their counterparts at Mahale (Muller *et al.*, 1995). In this case, although chimpanzees encountered a freshly killed bushbuck carcass, most of them, except an infant orphan, did not eat its meat but expressed curiosity as if they were inspecting the carcass of their own species. There are no observations of chimpanzees scavenging leopard kills from Taï, either (Boesch and Boesch-Achermann, 2000).

In the early days at Gombe, chimpanzees were observed to appropriate freshly killed mammals directly from olive baboons (Morris and Goodall, 1977; Goodall, 1986). At Mahale, there was an observation that an adult male chimpanzee appropriated a freshly killed blue duiker from an unknown mid-sized predator (Chapter 16).

Watts (2008) added new information that scavenging is also rare at Ngogo and argued that chimpanzees avoid scavenging because fresh carcasses are rare in the forest and because chimpanzees may not recognize certain species as potential prey. The latter reasoning seems to be supported by a case in which a freshly killed aardvark was not scavenged at Mahale (Chapter 16).

20.2.5 Cannibalism motivated by meat-eating inclination?

As reviewed in Chapter 26, despite occurring at a low rate, infanticidal cannibalism has been reported from several wild chimpanzee research sites (also see Chapter 27). At Mahale, eight intragroup and two intergroup infanticides were confirmed or suspected (Chapter 26).

Although most considerations have focused on infanticide rather than cannibalism, Hiraiwa-Hasegawa (1992) doubted that this behavior could be unequivocally explained by adaptive significance. Instead, she presented another argument that chimpanzees may exploit infant flesh as food. The protein taken from infant flesh is not insignificant, given that the average size of infants eaten could be around 3 kg and that most mammal prey eaten by Gombe chimpanzees weighed less than 6 kg (Teleki, 1973). She further speculated that "the differences in the occurrence of cannibalism between the populations may be related to differences in the available protein in [the] diet."

20.2.6 Non-mammal prey

It is relatively rare that chimpanzee prey upon non-mammal vertebrates, but they sometimes eat birds (see also Chapter 16). Fujimoto and Shimada (2008) reviewed and updated information about bird eating by wild chimpanzees; chimpanzees appear to have no favorite avian prey and they opportunistically eat avian eggs, nestlings, fledglings, chicks, or injured/fatigued adults.

Mahale chimpanzees showed individual differences in their response to captured birds; sometimes kill and play, sometimes consume after capture (see Chapter 16). This seems a little similar to the individually different responses of Bossou chimpanzees who reportedly consumed birds in two out of six captures (Hockings *et al.*, 2012; Yu *et al.*, 2013). Data from Mahale and Gombe appeared to indicate that Bossou counterparts consume birds less frequently than Tanzanian chimpanzees. However, paucity of information made it difficult to examine between-site differences in chimpanzee tendency to consume captured birds.

20.3 Food sharing in wild chimpanzees

20.3.1 Parental food sharing

Based on observations at Gombe, Silk (1978) found that food sharing among mother and infant chimpanzees offers infants opportunities to learn foods they cannot find and process independently, arguing that food sharing may facilitate an infant's transition from dependent suckling to independent foraging.

At Mahale, Hiraiwa-Hasegawa (1990) also studied the development of infant solicitation for food from mothers and drew a conclusion opposite to Silk's (1978) in that infants were not successful in obtaining the difficult foods more often than easy ones, although difficult foods were solicited more often. On the basis of more observational data, Nishida and Turner (1996) concluded that infant chimpanzees solicited more often for and were successful in obtaining difficult foods more often. They also revealed that the rate of infant solicitation for food drastically increased at the age of 2 years, then gradually decreased, and virtually disappeared at the age of 7 years.

Nishida and Hosaka (2001) argued that mothers provision such kind of foods that infants cannot procure or process by themselves; thereby mothers give infants opportunities to learn the taste of foods. Thus, parental food sharing could be viewed as "indirect teaching" because it is useful in acquiring a mature food repertoire.

20.3.2 Plant-food sharing

The sharing of plant foods occurs primarily between kin. In pioneering days at Gombe and Mahale, provisioned foods such as bananas and sugarcane were frequently shared among chimpanzees. McGrew (1975) observed 457 transfers of bananas at Gombe, among which 391 (86%) occurred between mother–offspring pairs, and 47 (10%) from adult males to unrelated adult females. At Mahale, Nishida (1973) described how chimpanzees shared sugarcane with each other.

At Bossou, where western chimpanzees dwelled near to a human village, chimpanzees sometimes intruded into papaya fields to share the fruits (Hockings *et al.*, 2007; Ohashi, 2007).

Nakamura and Itoh (2001) reported two cases of plant-food sharing among mature chimpanzees at Mahale. In one case, when a young adult male shared pieces of a lemon fruit with another male peer and an estrous female. In the other case, an adult male begged an older but lower-ranking male for the fruit pulp of *Tabernaemontana pachysiphon*, which was being chewed in the latter's mouth, and succeeded in obtaining pulp portions 15 times. In September 2008, I observed another case of plant-food sharing, in which an old male, Kalunde, approached and peered at the former alpha male, Alofu, who was feeding on a bunch of fruits of *Pseudospondias microcarpa* in his hand. Kalunde succeeded in co-feeding on the fruits with Alofu, sitting side by side (Figure 20.1). The two adult males had been association partners since Alofu was an alpha male.

Slocombe and Newton-Fisher (2005) described details of how an adult male who was holding a ripe fruit of *Monodora myristica* responded to the begging of an adult female. The fruit, which was large and

Figure 20.1 Alofu peers at Kalunde who is eating a bunch of fruits of *Pseudospondias microcarpa*. ©K. Hosaka.

heavy (>1 kg) as a chimpanzee food item, was shared between the two chimpanzees. The authors suggested that nutritional gain was not the dominant factor driving the begging behavior of the adult female, because she invested too much energy and took an unnecessary risk of injury for only a small portion of food she obtained.

Pruetz and Lindshield (2012) reported relatively high rates of plant-food sharing for non-kin chimpanzees in savanna environments at Fongoli.

20.3.3 Kind of foods shared among non-kin

Nishida and Hosaka (2001) indicated non-kin adult chimpanzees tend to share (1) food which is too large for a single individual to consume (e.g. sugarcane, meat, large fruit); and (2) food which requires skills for procurement and processing (e.g. sugarcane, meat, nuts). At Mahale, sugarcane and meat fit both types well, although sharing of sugarcane was observed only in the times of artificial provisioning. Bonobos (*Pan paniscus*) are known to share large-sized (>10 kg) fruits such as *Treculia africana* and *Anonidium mannii* (Kuroda, 1984; White, 1994; Hirata *et al.*, 2010). In other words, it is plausible that some foods are acquired excessively and/or limitedly, which makes room for other individuals to scrounge for such valuable foods. To discuss what is special about food sharing in *Pan* species, it seems necessary to consider such "economic value" of particular foods (Nishida and Hosaka, 2001).

20.4 How do chimpanzees procure meat?

20.4.1 Meat sharing

The basic pattern of meat sharing in chimpanzees is the same as sharing of other foods in that it occurs only as a response to begging behavior (Figure 20.2). "Meat possessor" is defined as an individual holding meat to which others are prevented from having access (Figure 20.3). More simply, "meat possessor" could be defined as an individual to whom begging behavior is directed.

Even in mother–infant pairs, mothers do not share meat unless infants beg them for it. Usually, mothers relinquish a smaller portion of meat to infants as a result of infant solicitation. Zamma (2005) reported a rare observation that an infant female rejected a small meat portion provided by her mother and continued to beg for the bigger chunk of meat that her mother was holding.

When meat sharing occurs, some individuals eager to eat meat scrounge around a meat possessor, forming a "meat-eating cluster" (Teleki, 1973;

Kawanaka, 1982; Nishida *et al.*, 1992). During a brief period just after prey capture until the meat-eating cluster forms, it seems that meat possessor status is insecure, thus allowing "dismemberment" (Teleki, 1973) or "division" (Boesch and Boesch, 1989), and "snatching" (Nishida *et al.*, 1979) or "stealing" (Uehara *et al.*, 1992). An adolescent male who has captured a prey is most likely to run away with the kill, because he will lose it to adults if he stays. Once the meat-eating cluster is structured, scroungers turn into beggars, who begin to show begging behavior to solicit the possessor for meat (Figure 20.4). The possessor sometimes averts their eyes from beggars or begins an intimidation display, but remains in the cluster.

Figure 20.2 Masudi (left) begs Alofu (right) for meat by stretching hand. ©K. Hosaka.

Figure 20.3 The meat possessor (an alpha male Alofu, right) shares a colobus kill with his old mother (left). Alofu's important ally (Kalunde, center) is appoaching to have access to the meat. ©K. Hosaka.

20.4.2 Are males the sole hunters?

Predatory behavior has been regarded as mainly a male activity and the amount of meat intake by males considerably exceeds that by females (McGrew, 1979). Such a view was supported by Uehara (1986), who, based on fecal analyses, found that prime adult or old males of the large-sized M group (about 100 individuals at that time) fed more on vertebrates, while adult females fed more on *Camponotus* ants (Chapter 21). In contrast, feces of the small-sized K-group chimpanzees (about 20–30 individuals) did not yield similar results. He argued such a group difference resulted in a larger per capita intake of meat among adult males of the M group. Further, he suggested another explanation that frequent interactions among many adult males of the M group may reduce the time allotted for ant fishing.

Even though sexual differences in predatory tendency may exist, we should not neglect the roles of individuals other than adult males, especially adult females, in large-group hunting. To what extent do

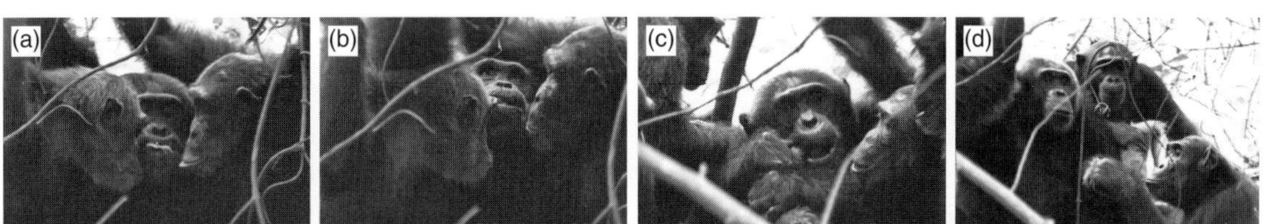

Figure 20.4 Ntologi's meat sharing. (a) Two adult females, Wakilufya (left) and Dar (right), peer into Ntologi's mouth, which contains meat. (b) Ntologi averts his eyes from beggars. (c) A young adult female, Amisa, joins as a beggar. (d) Dar begins to eat meat Ntologi is holding. Amisa is also grasping meat. ©K. Hosaka.

females contribute to prey capture? How much meat do they consume compared to males? These questions should be examined, as large-group hunting is a mixture of the actions of active and passive hunters, rather than the actions of active hunters alone.

The contribution of adult females as active hunters in group hunts appears to be greater than previously thought. At Mahale, adult females were responsible for 22.2% (1990–5, $N = 117$) and 18.9% (1996–2009, $N = 95$) of the captures of colobus during the group hunts in which the captors were confirmed. At Taï, the contribution of adult females in prey capture (18.4%) seems similar to that at Mahale. Goodall (1986) stated that the corresponding figure (10.6%) from Gombe might have been underestimated, as most researchers observed adult males when hunts occurred. At Ngogo, the corresponding figure (1.1%) might also have been underestimated, since the adult females there were not as well habituated as adult males; however, the possibility of interpopulation variation in the active involvement of adult females in group hunting remains.

20.4.3 Female contribution to group hunting

For females, access to colobus meat is chiefly confined to group hunting because females struggle to kill colobus alone, unlike males (Section 20.5.2). Boesch (1994a) stressed that chimpanzees "fear" red colobus that are aggressive towards predators. In order for females and immatures to pursue colobus, they have to overcome this fear. The presence of other individuals may help them challenge aggressive prey. Even if they do not capture prey themselves, they still have opportunities to gain meat by scrounging. In conclusion, group hunting benefits females and immatures when they are involved in the hunt themselves or when they scrounge for meat.

Further research should reveal how individuals behave in group hunting, and how meat is distributed among individuals, more precisely. For example, adult females may initiate group hunting by calls signaling colobus groups, because the more often group hunts occur, the more opportunities they will have to obtain meat. If such involvement were valued more highly, then it would confirm the idea that group hunts are not the sum of solitary hunts (Section 20.5.1). Finally, it is important to determine whether variations in individual behavior during group hunting reflect meat distribution patterns. It may be important to judge whether active hunters reciprocate any indirect contribution of other individuals to their captures, or whether they simply tolerate scrounging on the part of specific individuals, irrespective of their contribution to the hunting process.

20.4.4 Estimates of meat consumption

Interestingly, Mahale adult females obtain as much meat as adult males (adult males 46.0 g/day or 16.8 kg/year, adult females 44.8 g/day or 16.4 kg/year) (Hosaka, 2002). This is a rough estimate, as it is based on the estimated weight of a colobus by age, and on two assumptions: (1) an individual eats the same amount of meat when participating in a meat-eating session, and (2) an individual of the same age and sex has access to meat at the same frequency. Although the estimation should be improved by using more accurate measures, differences in the meat consumption rate among age–sex classes may reflect two important aspects of meat eating after large-group hunting. First, there is a marked difference between the meat consumption rate of the alpha male (233.7 g/day or 85.3 kg/year) and other adult males (46.0 g/day). Second, active hunting is not the only way to access meat in group hunting, since adult females may rival adult males in this respect.

At Gombe, based on assumptions of body weights for each prey species, Wrangham and van Zinnicq Bergmann Riss (1990) estimated that per capita meat intake was 25 kg/year for adult males, 8 kg/year for adult females, and 3 kg/year for offspring.

At Taï, Boesch and Boesch-Achermann (2000) calculated per capita meat intake to be 186 g/day (= 67.9 kg/year) for adult males and 25 g/day (= 9.1 kg/year) for adult females.

At Ngogo, based on their estimate of annual prey availability (about 6.3 kg of red colobus; about 7.7 kg of all prey species per individual in the unit-group), Watts and Mitani (2002a) concluded that per capita availability of red colobus meat was 15–20 kg/year for adult males (though it varied from 2 to 40 kg depending on individuals); about 3.7 kg for adult females (though it might be closer to 10 kg for those who were present during red colobus hunts).

20.5 Why do chimpanzees hunt in groups?

20.5.1 Is group hunting just summation of solitary hunts?

Many scientists are intrigued by the fact that chimpanzees engage in "group hunts," especially in hunts for forest monkeys. Initial discussions on this topic centered on whether chimpanzee group hunts should be called "cooperative hunts," as described by Teleki (1973). Busse (1978) criticized Teleki's view; according to his analyses, lone male chimpanzees at Gombe are capable of killing red colobus monkeys, and group hunting does not favor "a single individual" on the basis of either hunting success or meat consumption. He also cast doubt on the interpretation of hunting behaviors as group behaviors, noting that each chimpanzee hunter chases red colobus prey separately in space or time. Nishida (1981) held the same view, based on his observation of Mahale chimpanzees, and suggested that group hunts should be called "simultaneous solitary hunts."

Scientists resumed this argument about chimpanzee group hunting, stimulated by the first detailed reports from Taï. Boesch and Boesch (1989) stressed the population differences between Taï, Gombe, and Mahale, insisting that only Taï chimpanzees engage in "cooperative hunting" on a regular basis. They maintained that Taï chimpanzee hunters focus on the same prey and, in most cases, reach the highest level of organization with respect to their hunting roles, i.e. chimpanzees perform complementary actions. Boesch (1994b) adopted the term "simultaneous solitary hunts" to describe group hunts at Gombe and Mahale in order to emphasize the dichotomy between "cooperative hunting" and "(virtually) solitary hunting."

However, group hunts without clear-cut cooperation are not the same as summations of solitary hunts. The presence of other individuals in hunts can influence the motivations and outcomes of the hunts, either technically or psychologically. For example, numerous analyses from Gombe (Stanford *et al.*, 1994a), Mahale (Hosaka, 1998), and Ngogo (Mitani and Watts, 1999, 2001; Watts and Mitani, 2002a) have suggested that chimpanzees undertake hunts more frequently and more successfully when there are more estrous females (Gombe) or adult males (Mahale and Ngogo) present.

Hosaka *et al.* (2001) and Watts and Mitani (2002a, b) stressed the importance of "bystanders" in group hunts, while Boesch and Boesch (1989) counted only "active hunters" for their analyses. Bystanders may be considered a pool of potential active hunters, in that they usually monitor active hunters and prey in the trees from the ground; in addition, they often switch their behaviors when they detect a good chance to make a capture. Furthermore, they may remain below and adopt other low-cost tactics, such as a "wait-for-monkey-to-fall" strategy. Besides these technical effects created by the presence of passive hunters, even pure bystanders can influence group hunts psychologically (Hosaka *et al.*, 2001). For example, the excitement produced by individuals barking at prey may help potential hunters initiate and accomplish hunts.

20.5.2 Collaboration in small-group male hunting parties

Boesch (1994b) maintained that "cooperative hunting" in Taï chimpanzees is an evolutionarily stable strategy for two reasons. First, a hunting success rate is low for single hunters, due to the high canopy of the forest, and increases with group size if the definition of hunters is confined to active hunters only. Second, a social mechanism exists to limit access to meat by non-hunters ("cheaters").

Hosaka *et al.* (2001) described the 18 episodes of solitary and small-group (two or three individuals) encounters of Mahale chimpanzees with three species (red colobus, red-tailed monkey, and blue duiker). Interestingly, several distinct patterns characterize such episodes, as compared to normal group encounters with red colobus.

1. The red-tailed monkey, which was not hunted by groups of more than three individuals, was a target of chimpanzee hunting.
2. Monkey targets, either red colobus or red-tailed, were either solitary individuals or individuals that were temporarily isolated from troops.
3. The participants in all episodes were exclusively males (mostly adult males).
4. During the encounters with monkeys, males sometimes walked in single file and stopped to watch the prey intently.
5. In two hunts for monkeys, males performed a "collaboration" in terms of the criteria of Boesch and Boesch (1989).
6. Before, during, and after the episodes, males behaved *silently*, which was in striking contrast to large-group hunts, because the latter were *noisy*, especially after prey capture.

In small-group hunting, male hunters may refrain from vocalizing when a kill is made, possibly to limit access by scroungers and to eat meat by themselves. Otherwise, vocalizing would invite females to scrounge for meat. This is in contrast to most group hunts, in which males hunt colobus in the presence of females and tolerate scrounging.

20.5.3 Hunting seasonality

Hunting at Mahale peaked in the dry season around August and September (Hosaka *et al.*, 2001), which was similar at Gombe (Stanford *et al.*, 1994b). It seems reasonable to stress the importance of the similarity in seasonality of nomadic party size between Mahale and Gombe. According to Itoh and Nishida (2007), August and September at Mahale roughly correspond to one of the peaks of large nomadic parties. Since several studies have showed that chimpanzee party size is positively correlated with hunting fluctuation (Stanford *et al.*, 1994a; Hosaka, 1998; Mitani and Watts, 1999), hunting seasonality is explained well by seasonality of nomadic party size which is most probably related to food availability.

Takahata *et al.* (1984) implied that a greater salt requirement may increase chimpanzee hunting in the dry season, because chimpanzees lose salt through sweating in that season. In one hunting season (August and September) in 2008, there was a binge of rock-licking behavior at a rocky area of the lakeshore during a 2-week period. Interestingly, a hunting binge in that season occurred slightly later.

Recently, Gilby and Wrangham (2007) postulated an interesting idea suggesting that "risk-prone hunting" by chimpanzees increases during periods of high diet quality, based on the assumption that large-bodied omnivores are most likely to adopt "abundance/risk" strategy in contrast to the "scarcity/risk" strategy adopted by small animals with high metabolic rates susceptible to the threat of starvation.

20.5.4 Hunting binges caused by social changes?

Several authors mentioned that chimpanzees engaged in hunting "crazes" or "binges" in which they were more highly motivated to hunt than at other periods of time (Norikoshi, 1983; Goodall, 1986; Stanford *et al.*, 1994b; Mitani and Watts, 1999).

In Mahale, statistical analysis of systematic sampling data collected in two study periods revealed that the red colobus hunting fluctuated more irregularly in the study period of 1991–2 than that of 1993–4 (Hosaka, 1998; Figure 20.5). More specifically, Figure 20.5 indicates that there were two hunting binges in which Mahale chimpanzees engaged in more frequent predation on red colobus: August to mid-September 1991 and late January to mid-February. Hosaka (2002) pointed out that the first binge corresponded to the period in which Kalunde established his alpha status after ostracizing the former alpha male Ntologi; the second binge corresponded to the period in which Ntologi retrieved his alpha status (Chapter 27). At Mahale, alpha males share meat disproportionately with their allies as a political tool. Hosaka (2002) claimed that alpha males may have facilitated group hunting either by increasing their own hunting frequency or generating the psychological

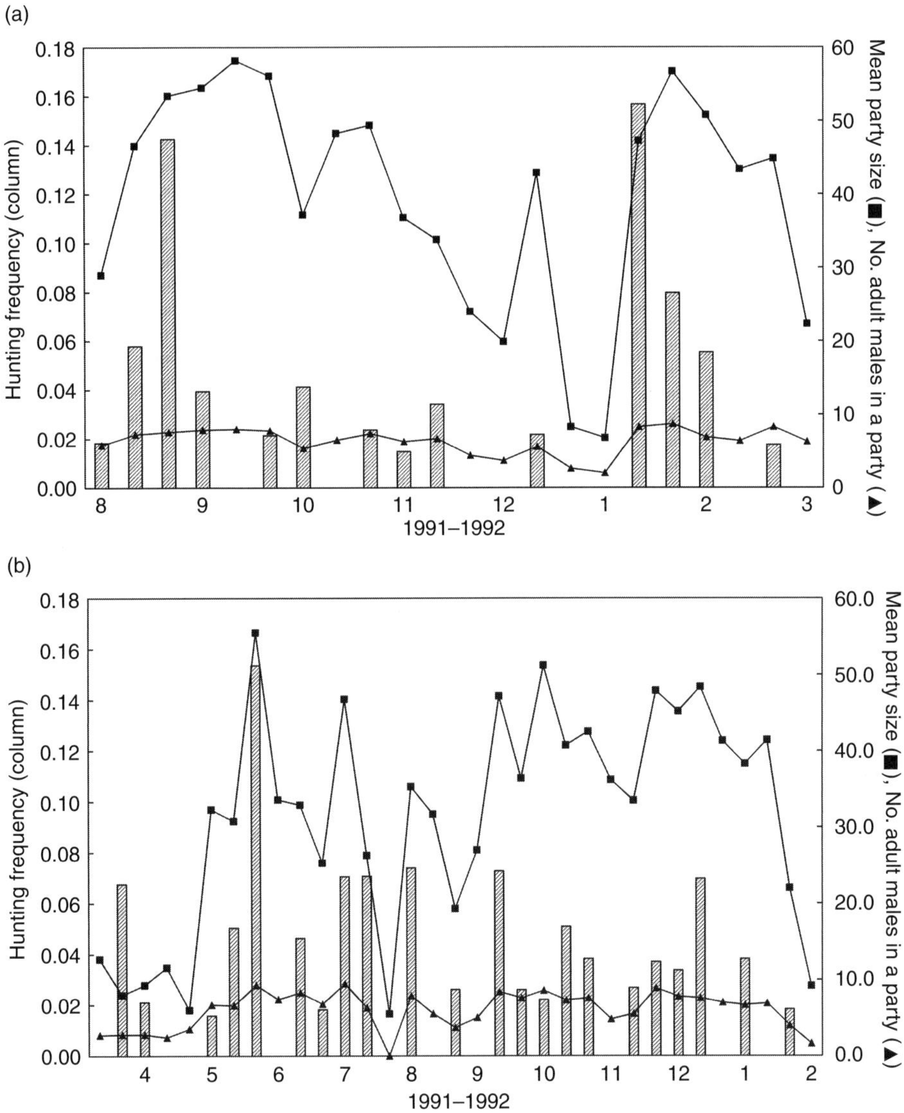

Figure 20.5 The relationship between hunting frequency and mean party size (or number of males in a party). (a) August 1991–March 1992. (b) March 1993–February, 1994. ©K. Hosaka.

atmosphere within a party to obtain opportunities for meat sharing.

20.5.5 Skillful hunters as catalysts in group hunting

Boesch and Boesch (1989) and Stanford *et al.* (1994b) suggested that there are persistent hunters or good hunters and that their presence in a party may influence hunting decisions or hunting success.

At Mahale, both alpha males and some skillful hunters seemed to act as catalysts of group hunting (Hosaka *et al.*, 2001). As for alpha males, it seems reasonable that they benefit politically from meat sharing in that they could draw others' attention and demonstrate their power to stabilize social status. In

1991–5, among 11 individuals whose numbers of captures of red colobus ranked 1–10, there were seven adult males, two adolescent males, and two females (Fatuma and Linda). First, all adult males that experienced alpha status during this study tended to rank high in number of captures (Kalunde, Ntologi, and Nsaba). Second, two adolescent males (Alofu and Dogura) ranked high, but also often lost carcasses (6/8 for Alofu and 5/7 for Dogura) due to snatching by adult males.

Gilby *et al.* (2008) termed such highly motivated hunters "impact males," and, on the basis of their analysis on the hunting data at Kanyawara from 1996 to 2004, revealed that two impact hunters had persistently high individual hunting rates and acted as hunting "catalysts." That is, the presence of impact hunters makes hunting more likely both by their own efforts and by increasing the likelihood that others hunt.

20.5.6 Why chimpanzees share meat with non-kin?

The harassment hypothesis

Also called "sharing-under-pressure" (de Waal, 1989). The idea of a "harassment" hypothesis (Gilby, 2006) has persisted as the most parsimonious explanation, implying a "passive" benefit. That is, when the nutritional cost of holding food is higher than the nutritional benefit of doing so due to the persistent, sometimes aggressive, behavior of a beggar, a portion may be relinquished to the beggar, namely, food sharing occurs. This was originally proposed by Wrangham (1975) for meat sharing among adult male chimpanzees. Nishida *et al.* (1979) also indicated that chimpanzees share food reluctantly only in response to the persistent demands of beggars. The behavioral pattern of sharing in response to begging resembles that of human food sharing referred to as "tolerated theft" (Blurton Jones, 1987), which emphasizes the passive attitude of a sharer, or "demand sharing" (Peterson, 1993), which emphasizes social and symbolic significance of the practice.

The social display hypothesis

The first version was advocated by Teleki (1973) as what we might call a "dominance display" hypothesis. Although he mentioned it as an alternative hypothesis to explain why Gombe chimpanzees practice predatory behavior, he stated that an individual, and especially a low-ranking adolescent or young adult male, may in the long run benefit by raising their social status from repeatedly achieving a temporary control role. Since he did not mention reciprocity on the part of beneficiaries, but emphasized the "control role" of a meat possessor, the focal point of his assertion may lie in the social display.

The reciprocity hypothesis

The reciprocity hypothesis proposes that food is shared in exchange for something beneficial to food possessors. The origin of this hypothesis is very old, dating back to more than 60 years ago when Nissen and Crawford (1936) reported food sharing in captive immature chimpanzees. De Waal (1989) presented the first evidence of reciprocity in food sharing among the captive chimpanzees of Arnhem Colony, the Netherlands. He showed that chimpanzees exchange food (branches and leaves) for food. Moreover, grooming between two individuals during the morning affected the probability of food sharing later in the day (de Waal, 1997). In wild populations, Mitani and Watts (1999) and Watts and Mitani (2002b) showed evidence for the first time at Ngogo that adult males shared meat reciprocally.

Political tactic hypothesis

Nishida *et al.* (1992) proposed that meat sharing by an alpha male, Ntologi, may be used as a coalition tactic, a "political tool," as it were. This hypothesis could be explained as a variant of either the social display hypothesis or reciprocity hypothesis. Their argument was based on the fact that individuals receiving meat from Ntologi corresponded well to Ntologi's grooming and association partners. This hypothesis would be convincing if repeated observations were obtained showing the role of meat control which is switched to new alpha males whenever alpha rank reversal takes place.

Meat-for-sex hypothesis

The meat-for-sex hypothesis itself originated when Tutin (1979) pointed out that females showed a preference for males who spent most of the time with them, and who showed the highest frequencies of affiliative behaviors, such as grooming and food sharing. Stanford *et al.* (1994a) proposed that males may hunt red colobus at least partially to obtain meat for use as a tool to attract estrous females. Their idea was based on their finding that, among social variables such as foraging party size, number of males in the foraging party, etc., the presence of one or more swollen females was the best predictor for chimpanzees to decide on monkey hunting. Despite lack of meat-sharing evidence, this idea gained popularity under the name of "meat-for-sex" hypothesis.

Hosaka (1998), Mitani and Watts (2001), and Gilby *et al.* (2010) opposed the meat-for-sex hypothesis in terms of short-term interactions after meat transfer based on data of group hunting and meat sharing. However, Gomes and Boesch (2009) supported the hypothesis on a long-term basis; using 22-month-data from Taï, they showed that female chimpanzees copulate more frequently with males who share meat with them.

20.6 For future studies

Since Hosaka *et al.* (2001), no comprehensive report on hunting behaviors by Mahale chimpanzees has been published. Since 1994, a project of fruiting phenology has been ongoing (Chapter 14). Thus, it is now possible to examine hunting seasonality in terms of plant phenology at Mahale, which will further assess several other factors such as nutrition, ecology of prey (mammals, insects), party size and composition, and social relationships among individuals.

Similarly, there has been no follow-up study to test Nishida's "political tactic hypothesis," although MMCRP has accumulated data of meat sharing for five additional reigns of alpha males (see Chapter 5) after Nsaba disappeared at the end of 1995. If this hypothesis holds true only for Mahale M group, it would be interesting to explore differences in meat-sharing policy among chimpanzees of different study sites.

References

Bailey, T. N. (1993). *The African Leopard: Ecology and Behavior of Solitary Felid*. New York: Columbia University Press.

Blurton Jones, N. G. (1987). Tolerated theft, suggestions about the ecology and evolution of sharing, hoarding and scrounging. *Social Science Information*, **26**, 31–54.

Boesch, C. (1994a). Chimpanzees-red colobus monkeys: a predator-prey system. *Animal Behaviour*, **47**, 1135–48.

Boesch, C. (1994b). Cooperative hunting in wild chimpanzees. *Animal Behaviour*, **48**, 653–67.

Boesch, C. and Boesch, H. (1989). Hunting behavior of wild chimpanzees in the Taï National Park. *American Journal of Physical Anthropology*, **78**, 547–73.

Boesch, C. and Boesch-Achermann, H. (2000). *The Chimpanzees of the Taï Forest: Behavioural Ecology and Evolution*. Oxford: Oxford University Press.

Bogart, S. L., Pruetz, J. D., and Kante, D. (2008). Fongoli chimpanzee (*Pan troglodytes verus*) eats banded mongoose (*Mungos mungo*). *Pan Africa News*, **15**, 15–17.

Busse, C. D. (1977). Chimpanzee predation as a possible factor in the evolution of red colobus monkey social organization. *Evolution*, **31**, 907–11.

Busse, C. D. (1978). Do chimpanzees hunt cooperatively? *The American Naturalist*, **112**, 767–70.

Fourrier, M., Sussman, R.W., Kippen, R., and Childs, G. (2008). Demographic modeling of a predator–prey system and its implication for the Gombe population of *Procolobus rufomitratus tephrosceles*. *International Journal of Primatology*, **29**, 497–508.

Fujimoto, M. and Shimada, M. (2008). Newly observed predation of wild birds by M-group chimpanzees (*Pan troglodytes schweinfurthii*) at Mahale, Tanzania. *Pan Africa News*, **15**, 23–6.

Gašperšič, M. and Pruetz, J. D. (2004). Predation on a monkey by savanna chimpanzees at Fongoli, Senegal. *Pan Africa News*, **11**(2), 8–10.

Gilby, I. C. (2006). Meat sharing among the Gombe chimpanzees: harassment and reciprocal exchange. *Animal Behaviour*, **71**, 953–63.

Gilby, I. C. and Wrangham, R. W. (2007). Risk-prone hunting by chimpanzees (*Pan troglodytes schweinfurthii*) increases during periods of high diet quality. *Behavioral Ecology and Sociobiology*, **61**, 1771–9.

Gilby, I. C., Eberly, L. E., Pintea, L., and Pusey, A. E. (2006). Ecological and social influences on the hunting behaviour of wild chimpanzees, *Pan troglodytes schweinfurthii. Animal Behaviour*, **72**, 169–80.

Gilby, I. C., Eberly, L. E., and Wrangham, R. W. (2008). Economic profitability of social predation among wild chimpanzees: Individual variation promotes cooperation. *Animal Behaviour*, **75**, 351–60.

Gilby, I. C., Emery Thompson, M., Ruane, J. D., and Wrangham, R.W. (2010). No evidence of short-term exchange of meat for sex among chimpanzees. *Journal of Human Evolution*, **59**, 44–53.

Gomes, C. M. and Boesch, C. (2009). Wild chimpanzees exchange meat for sex on a long-term basis. *PLoS ONE*, **4**, e5116.

Goodall, J. (1963). Feeding behaviour of wild chimpanzees: a preliminary report. *Symposium of the Zoological Society of London*, **10**, 39–47.

Goodall, J. (1986). *The Chimpanzees of Gombe: Patterns of Behavior*. Cambridge, MA: Belknap.

Hasegawa, T., Hiraiwa, M., Nishida, T., and Takasaki, H. (1983). New evidence on scavenging behavior in wild chimpanzees. *Current Anthropology*, **24**, 231–2.

Hashimoto, C., Furuichi, T., and Tashiro, Y. (2000). Ant dipping and meat eating by wild chimpanzees in the Kalinzu Forest, Uganda. *Primates*, **41**, 103–8.

Hiraiwa-Hasegawa, M. (1990). Role of food sharing between mother and infant in the ontogeny of feeding behavior. In *The Chimpanzees of the Mahale Mountains. Sexual and Life History Strategies*, ed. T. Nishida. Tokyo: University of Tokyo Press, pp. 265–75.

Hiraiwa-Hasegawa, M. (1992). Cannibalism among non-human primates. In *Cannibalism: Ecology and Evolution among Diverse Taxa*, ed. M. A. Elgar and B. J. Crespi. Oxford: Oxford University Press, pp. 323–38.

Hirata, S., Yamamoto, S., Takemoto, H., and Matsuzawa, T. (2010). A case report of meat and fruit sharing in a pair of wild bonobos. *Pan Africa News*, **17**, 21–3.

Hockings, K. J., Humle, T., Anderson, J. R., *et al.* (2007). Chimpanzees share forbidden fruit. *PLoS ONE*, **2**, e886.

Hockings, K. J., Humle, T., Carvalho, S., and Matsuzawa, T. (2012). Chimpanzee interactions with nonhuman species in an anthropogenic habitat. *Behaviour*, **149**, 299–324.

Hosaka, K. (1998). [*Predatory Behavior of Chimpanzees in the Mahale Mountains.*] Doctoral Dissertation. Kyoto: Kyoto University. In Japanese.

Hosaka, K. (2002). [Hunting and meat-eating behavior.] In [*The Mahale Chimpanzees: 37 Years of <Panthropology>*], ed. T. Nishida, S. Uehara, and K. Kawanaka, Kyoto: Kyoto University Press, pp. 219–44. In Japanese.

Hosaka, K. and Nishida, T. (2010). Long-term changes in hunting behavior of the Mahale chimpanzees. *International Primatological Society XXIII Congress Kyoto 2010, Abstracts*, p. 337.

Hosaka, K., Nishida, T., Hamai, M., Matsumoto-Oda, A., and Uehara, S. (2001). Predation of mammals by the chimpanzees of the Mahale Mountains, Tanzania. In *All Apes Great and Small Vol. I: African Apes*, ed. B. M. F. Galdikas, N. E. Briggs, L. K. Sheeran, G. L. Shapiro, and J. Goodall. New York: Kluwer Academic/Plenum Publishers, pp. 107–30.

Ihobe, H. (2000). [Red colobus vs. chimpanzees: predator-prey relationships among primates.] In [*Primate Ecology: Dynamism between the Environment and Behavior*], ed. Y. Sugiyama. Kyoto: Kyoto University Press, pp. 61–84. In Japanese.

Ihobe, H. (2001). Hunting attempt by chimpanzees on Abyssinian colobus at the Kalinzu Forest, Uganda. *Pan Africa News*, **8**, 31–2.

Isaac, G. (1978). The food-sharing behavior of protohuman hominids. *Scientific American*, **238**, 90–108.

Itoh, N. and Nishida, T. (2007). Chimpanzee grouping patterns and food availability in Mahale Mountains National Park, Tanzania. *Primates*, **48**, 87–96.

Kawanaka, K. (1982). Further studies on predation by chimpanzees of the Mahale Mountains. *Primates*, **23**, 364–84.

Kuroda, S. (1984). Interaction over food among pygmy chimpanzees. In *The Pygmy Chimpanzee: Evolutionary Biology and Behavior*, ed. R. L. Sussman. New York: Plenum, pp. 301–24.

van Lawick-Goodall, J. (1968). The behaviour of free-living chimpanzees in the Gombe Stream Reserve. *Animal Behaviour Monographs*, **1**, 161–312.

McGrew, W. C. (1975). Patterns of plant food sharing by wild chimpanzees. In *Contemporary Biology, ed.* S. Kondo, M. Kawai, and A. Ehara. Basel, Switzerland: Karger, pp. 304–9.

McGrew, W. C. (1979). Evolutionary implications of sex differences in chimpanzee predations and tool use. In *The Great Apes*, ed. D. A. Hamburg and E. R. McCown. Menlo Park, CA: Benjamin/Cummings, pp. 441–63.

McGrew, W. C. and Feistner, A. T. C. (1992). Two nonhuman primate models for the evolution of human food sharing: chimpanzees and callitrichids. In *The Adapted Mind: Evolutionary Psychology and the Generation of Culture*, ed. J. H. Barkow, L. Cosmides, and J. Tooby. Oxford: Oxford University Press, pp. 229–43.

Mitani, J. C. and Watts, D. P. (1999). Demographic influences on the hunting behavior of chimpanzees. *American Journal of Physical Anthropology*, **109**, 439–54.

Mitani, J. C. and Watts, D. P. (2001). Why do chimpanzees hunt and share meat? *Animal Behaviour*, **61**, 915–24.

Morris, K. and Goodall, J. (1977). Competition for meat between chimpanzees and baboons of the Gombe National Park. *Folia Primatologica*, **28**, 109–21.

Muller, M. N., Mpongo, E., Stanford, C. B., and Boehm, C. (1995). A note on scavenging by wild chimpanzees, *Folia Primatologica*, **65**, 43–7.

Nakamura, M. (1997). First observed case of chimpanzee. predation on yellow baboons (*Papio cynocephalus*) at the Mahale Mountains National Park. *Pan Africa News*, **4**, 9–11.

Nakamura, M. and Itoh, N. (2001). Sharing of wild fruits among male chimpanzees: Two cases from Mahale, Tanzania. *Pan Africa News*, **8**, 28–31.

Newton-Fisher, N. E., Notman, H., and Reynolds, V. (2002). Hunting of mammalian prey by Budongo Forest chimpanzees. *Folia Primatologica*, **73**, 281–3.

Nishida, T. (1973). [*The Children of the Mountain Spirits*.] Tokyo: Chikuma-shobō. In Japanese.

Nishida, T. (1981). [*The World of Wild Chimpanzees*.] Tokyo: Chūkō-shinsho. In Japanese.

Nishida, T. and Hosaka, K. (2001). [Food-sharing in primates.] In [*Hominization*], ed. T. Nishida, Kyoto: Kyoto University Press, pp. 255–302. In Japanese.

Nishida, T. and Turner, L. A. (1996). Food transfer between mother and infant chimpanzees of the Mahale Mountains National Park, Tanzania. *International Journal of Primatology*, **17**, 947–68.

Nishida, T., Uehara, S., and Nyundo, R. (1979). Predatory behavior among wild chimpanzees of the Mahale Mountains. *Primates*, **20**, 1–20.

Nishida, T., Hasegawa, T., Hayaki, H., Takahata, Y., and Uehara, S. (1992). Meat-sharing as a coalition strategy by an alpha male chimpanzee? In *Topics in Primatology, Vol. 1: Human Origins*, ed. T. Nishida, W. C. McGrew, P. Marler, M. Pickford, and F. B. M. de Waal. Tokyo: University of Tokyo Press, pp. 159–74.

Nishie, H. (2004). Increased hunting of yellow baboons (*Papio cynocephalus*) by M group chimpanzees at Mahale. *Pan Africa News*, **11**(2), 10–12.

Nissen, H. W. and Crawford, M. P. (1936). A preliminary study of food-sharing behavior in young chimpanzees. *Journal of Comparative Psychology*, **22**, 383–419.

Noë, R. and Bshary, R. (1997). The formation of red colobus–diana monkey associations under predation pressure from chimpanzees. *Proceedings of the Royal Society B*, **264**, 253–9.

Norikoshi, K. (1983). Prevalent phenomenon of predation observed among wild chimpanzees of the Mahale Mountains. *Journal of Anthropological Society, Nippon*, **91**, 475–9.

Ohashi, G. (2007). Papaya fruit sharing in wild chimpanzees at Bossou, Guinea. *Pan Africa News*, **14**, 14–16.

Peterson, N. (1993). Demand sharing: reciprocity and the pressure for generosity among foragers. *American Anthropologist*, **95**, 860–74.

Pruetz, J. D. and Lindshield, S. (2012). Plant-food and tool transfer among savanna chimpanzees at Fongoli, Senegal. *Primates*, **53**, 133–45.

Pruetz, J. D. and Marshack, J. L. (2009). Savanna chimpanzees (*Pan troglodytes verus*) prey on patas monkeys (*Erythrocebus patas*) at Fongoli, Senegal. *Pan Africa News*, **16**, 15–17.

Silk, J. B. (1978). Patterns of food sharing among mother and infant chimpanzees at Gombe National Park, Tanzania. *Folia Primatologica*, **29**, 129–41.

Slocombe, K. E. and Newton-Fisher, N. E. (2005). Fruit sharing between wild adult chimpanzees (*Pan troglodytes schweinfurthii*): a socially significant event? *American Journal of Primatology*, **65**, 385–91.

Stanford, C. B., Mpongo, E., Wallis, J., and Goodall, J. (1994a). Hunting decisions in wild chimpanzees. *Behaviour*, **131**, 1–18.

Stanford, C. B., Wallis, J., Matama, H., and Goodall, J. (1994b). Patterns of predation by chimpanzees on red colobus monkeys in Gombe National Park, 1982–1991. *American Journal of Physical Anthropology*, **94**, 213–28.

Sugiyama, Y. (1981). Observations on the population dynamics and behavior of wild chimpanzees at Bossou, Guinea, in 1979–1980. *Primates*, **22**, 435–44.

Takahata, Y., Hasegawa, T., and Nishida, T. (1984). Chimpanzee predation in the Mahale Mountains from August 1979 to May 1982. *International Journal of Primatology*, **5**, 213–33.

Teelen, S. (2008). Influence of chimpanzee predation on the red colobus population at Ngogo, Kibale National Park, Uganda. *Primates*, **49**, 41–9.

Teleki, G. (1973). *The Predatory Behavior of Wild Chimpanzees*. Lewisburg, PA: Bucknell University Press.

Tutin, C. E. G. (1979). Mating patterns and reproductive strategies in a community of wild chimpanzees (*Pan troglodytes schweinfurthii*). *Behavioral Ecology and Sociobiology*, **6**, 29–38.

Uehara, S. (1986). Sex and group differences in feeding on animals by wild chimpanzees in the Mahale Mountains National Park, Tanzania. *Primates*, **27**, 1–13.

Uehara, S. (1997). Predation on mammals by the chimpanzee (*Pan troglodytes*). *Primates*, **38**, 193–214.

Uehara, S. and Ihobe, H. (1998). Distribution and abundance of diurnal mammals, especially monkeys, at Kasoje, Mahale Mountains, Tanzania. *Anthropological Science*, **106**, 349–69.

Uehara, S., Nishida, T., Hamai, M., *et al.* (1992). Characteristics of predation by the chimpanzees in the Mahale Mountains National Park, Tanzania. In *Topics in Primatology, Vol. 1: Human Origins*, ed. T. Nishida, W. C. McGrew, P. Marler, M. Pickford, and F. B. M. de Waal. Tokyo: University of Tokyo Press, pp. 143–58.

de Waal, F. B. M. (1989). Food sharing and reciprocal obligations among chimpanzees. *Journal of Human Evolution*, **18**, 433–59.

de Waal, F. B. M. (1997). The chimpanzee's service economy: food for grooming. *Evolution and Human Behavior*, **18**, 375–86.

Watts, D. P. (2008). Scavenging by chimpanzees at Ngogo and the relevance of chimpanzee scavenging to early hominin behavioral ecology. *Journal of Human Evolution*, **54**, 125–33.

Watts, D. P. and Mitani, J. C. (2002a). Hunting behavior of chimpanzees at Ngogo, Kibale National Park, Uganda. *International Journal of Primatology*, **23**, 1–28.

Watts, D. P. and Mitani, J. C. (2002b). Hunting and meat sharing by chimpanzees at Ngogo, Kibale National Park, Uganda. In *Behavioural Diversity in Chimpanzees and Bonobos*, ed. C. Boesch, G. Hohmann, and L. F. Marchant. Cambridge: Cambridge University Press, pp. 244–55.

White, F. J. (1994). Food sharing in wild pygmy chimpanzees, *Pan paniscus*. In *Current Primatology Vol. II: Social Development, Learning and Behavior*, ed. J. J. Roeder, B. Thierry, J. R. Anderson, and N. Herrenschmidt. Strasbourg, France: Universite Louis Pasteur, pp. 1–10.

Wrangham, R.W. (1975). *The Behavioral Ecology of Chimpanzees in Gombe National Park, Tanzania*. Ph.D. thesis. Cambridge: Cambridge University.

Wrangham, R.W. and van Zinnicq Bergmann Riss, E. (1990). Rates of predation on mammals by Gombe chimpanzees, 1972–1975. *Primates*, **31**, 157–70.

Yu, L., Zogbila, B., and Matsuzawa, T. (2013). Newly observed bird consumption by a chimpanzee (*Pan troglodytes*) at Bossou, Guinea, West Africa. *Pan Africa News*, **20**, 8–10.

Zamma, K. (2005). Rejecting a bit of meat to get more. *Pan Africa News*, **12**, 8–10.

21 Insect-feeding behavior and insect fauna: with special reference to plant–insect relationships

MIEKO KIYONO

21.1 Introduction

Recently, the importance of insect feeding by apes has been re-evaluated based on nutritional analyses of various insects. For example, Deblauwe and Janssens (2008) suggested that chimpanzees and gorillas (*Gorilla gorilla*) in Dja, Cameroon, select some insect foods based on their nutritional contents. O'Malley and Power (2012) found that chimpanzees of the Kasekela group in Gombe, Tanzania, preferred to feed on insects that are high in energy, fat, and protein, and that have high potential macronutrient yields. Their analyses also showed that insects are nutritionally rich foods and are comparable to vertebrate meat on a per-gram basis. Such studies have highlighted the importance of insects as prey, and have helped in our understanding of *why* chimpanzees feed on insects. Although these studies provide a good insight into insectivory by chimpanzees, another question, i.e. *how* chimpanzees feed on insects, remains unanswered. Because many insects are typically hidden in plants or underground, it may not be easy for an animal to locate them. Different insect species have also acquired defensive strategies to deter predators. Chimpanzees need to understand the characteristics of insect species to overcome these defensive strategies if they are to find and feed on them. In other words, detailed observation of insect-feeding behaviors may provide insight into the knowledge that chimpanzees have about the relationship between insects and the environment.

At Mahale, a list of insects fed on by chimpanzees was published based mainly on observations of the K-group chimpanzees (Nishida and Uehara, 1983). The current main study group at Mahale is the M group, and although many researchers have accumulated knowledge about the foods eaten by the group (Chapter 17; Appendix II), there has as yet been no systematic attempt to compile a list of insect foods. Therefore, in this chapter, after briefly summarizing the insect fauna at Mahale, a list of insects that are known to be eaten by M-group chimpanzees is described including feeding behavior. Then, more detailed descriptions of insects that are eaten by chimpanzees are provided. Finally, I discuss some of the observations that appear to show chimpanzees selecting insect-dwelling plants based on their characteristics. In this chapter, insect-feeding behaviors *without* the use of tools are considered. More details regarding tool use are available in Chapter 42.

The term "insect feeding" in this chapter includes catching, processing, and consumption of insects. When chimpanzees seemed to search for insects and/or actively seek out prey, the term "insect forage" is used.

21.2 Insect fauna of Mahale

Although no systematic attempts have been made to completely characterize the insect fauna on Mahale, there have been a few studies on some taxonomic groups of insects. Itani (1964, 1977a) collected butterflies from the Kabogo area, which is about 50 km north of Mahale and shares similar vegetation, and identified 187 species. According to Itani (1977a), the butterfly fauna in Kabogo was composed of seven groups representing different paleobiogeographic

Mahale Chimpanzees: 50 Years of Research, ed. M. Nakamura, K. Hosaka, N. Itoh and K. Zamma. Published by Cambridge University Press.

distributions. Itani (1977b) also interviewed the local Tongwe people (Chapter 49) about the terminology they used and their knowledge about animals, including insects, in the Tongwe land, including Mahale. Sasaki and Nishida (1999) investigated fly fauna at Mahale using traps, insect nets, and traps baited with chimpanzee excrement. Sasaki (2005) more specifically reported 17 species (four genera) of Tabanid flies. Collins and McGrew (1987) surveyed termite fauna in several areas within Mahale by scan sampling. Fuse (2013) identified ant species living in dead branches in the Kasoje area, Mahale. A summary of the insects that are present in Mahale is provided in Table 21.1. The list is far from complete, and the number of species will increase as investigations proceed.

In this paragraph, examples of insects that I often saw in Mahale are introduced. One of the cicadas stridulating in the dry season is probably *Ugada*

Table 21.1 **List of insects identified or observed in Mahale**

ORDER Family Species	Common name	Eaten by chimpanzees[a]	Source[b]
COLEOPTERA		K, M	(4)
Buprestidae	jewel beetle		(1), (8)
Cerambycidae	longhorn beetle		(1), (3), (8)
Scarabaeidae	dung beetle		
Kheper sp.			(8)
Scarabaeus spp.			(1)
Gymnopleurus spp.			(1)
Copris spp.			(1)
Lampyridae			(1)
Dytiscidae			(1)
Cybister sp.	water beetle		(8)
DIPTERA			
Muscidae	house fly		
Musca spp.			(1)
Glossina spp.			(1)
Calliphoridae			
Lucilia spp.			(1)
Tabanidae			
Chrysops longicornis			(10)
Ancala africana			(10)
Tabanus biguttatus			(10)
Tabanus gratus			(10)
Tabanus insignis			(10)
Tabanus taeniola			(10)
Tabanus thoracinus			(10)

Table 21.1 **(cont.)**

ORDER Family Species	Common name	Eaten by chimpanzees[a]	Source[b]
Haematopota albihirta			(10)
Haematopota divisapex			(10)
Haematopota edax			(10)
Haematopota fasciatapex			(10)
Haematopota hirta			(10)
Haematopota insidiatrix			(10)
Haematopota lamborni			(10)
Haematopota nigripennis			(10)
Haematopota theobaldi			(10)
Haematopota unicolor			(10)
ORTHOPTERA			
Anostostomatidae			(1)
Gryllidae			
Brachytrupes membranaceus	sand cricket	K?	(1), (2), (3), (8)
Phalangopsidae			(1)
Gryllotalpidae			
Gryllotalpa africana	mole cricket		(8)
Tettigoniidae	katydid		(1), (8)
Bradyporidae			
Enyaliopsis sp.?	corn cricket		(8)
Pyrgomorphoidae			
Taphronota sp.?	grasshopper		(1)
Acrididae			
Nomadacris septemfasciata	red locust		(1)
PHASMATODEA	stick insect		(1)
PHTHIRAPTERA	louse		(1)
HOMOPTERA			
Cicadidae	cicada		
Ugada limbalis?			(8)
Flatidae			(8)
STERNORRHYNCHA			
Psyllidae	aphid		
Phytolima lata		K, M	(3)
Paracopium sp.?		M	(8)

Table 21.1 **(cont.)**

ORDER Family Species	Common name	Eaten by chimpanzees[a]	Source[b]
HEMIPTERA			
Coreidae			(1)
Cimicidae			
Cimex sp.			(1)
Reduviidae	assassin bug		(1), (8)
Coreidae			(1)
Belostomatidae			
Limnigeton sp.?	giant water bug		(8)
Nepidae			
Laccotrephes sp.?	common water scorpion		(8)
NEUROPTERA			
Myrmeleontidae	antlion		(1), (8)
HYMENOPTERA			
Formicidae	ant		
Aenictus sp.			(7)
Camponotus vividus		K	(3)
Camponotus brutus		K, M, Y	(3), (7), (11), (12)
Camponotus maculatus		K	(3)
Camponotus (Myrmotrema) sp.		K	(3)
Camponotus sp. (*chrysurus*-complex)		M	(12)
Camponotus (Tanaemyrmex) sp.			(7)
Cataulacus egenus		M	(7)
Cataulacus brevisetosus		M	(7)
Cataulacus mocquerysi			(7)
Crematogaster (Atopogyne) *clariventris*		K	(13)
Crematogaster (Atopogyne) sp. 1			(7)
Crematogaster (Sphaerocrema) sp. 3		M	(6)
Crematogaster (Sphaerocrema) sp. 11		M	(6)
Crematogaster (Sphaerocrema) sp. 24		M	(6)
Crematogaster (Crematogaster) sp. 18		M	(6)
Crematogaster (Crematogaster) sp. 23			(7)

Table 21.1 **(cont.)**

ORDER Family Species	Common name	Eaten by chimpanzees[a]	Source[b]
Crematogaster (Orthocrema) sp. 25		M	(6)
Crematogaster (Decacrema) sp. 12			(7)
Crematogaster (Decacrema) sp. 21			(6)
Crematogaster (Decacrema) sp. 26			(7)
Dorylus nigricans-complex (=*molestus*)			(7)
Dorylus gerstaeckeri			(7)
Dorylus kohli			(7)
Lepisota sp.		M	(7)
Monomorium trake		K, M	(3), (7)
Monomorium affabile			(7)
Oecophylla longinoda		K, M	(3)
Plagiolepis sp.			(7)
Pachycondyla analis			(7)
Pachycondyla sp.			(7)
Pheidole spp. (7 species)			(7)
Polyrhachis cf. *phidias*		M	(7)
Polyrhachis decemdentata		M	(7)
Technomyrmex spp. (2 species)			(7)
Tetramorium aculeatum		K	(3)
Tetramorium decem		M	(7)
Tetraponera sp.			(7)
Vespidae			
Belanogaster sp.			(1), (7)
Polistes spp.			(1)
Aspidae			
Apis mellifera (*adansonii?*)	honey bee	K, M	(1), (3)
Trigona spp.		K, M	(1), (3)
Xylocopa (Koptortosoma) sp.	carpenter bee	K, M	(3), (7)
Blastophaga sp.			(7)
Sphecidae			
Sceliphron spp.			(7)
ISOPTERA			
Termitidae	termite		
Coptotermes sp.			(5)

Table 21.1 **(cont.)**

ORDER Family Species	Common name	Eaten by chimpanzees[a]	Source[b]
Microcerotermes sp.			(5)
Termitinae sp.			(5)
Basidentitermes sp.			(5)
Megagnathatermes sp.			(5)
Macrotermes ?herus		B	(5)
Odontotermes spp.		M	(3), (5)
Pseudacanthotermes spiniger		K, M	(3), (7)
Pseudacanthotermes militaris		K, M	(3), (7)
Rhinotermitidae			(1)
LEPIDOPTERA	moth and butterfly		
Crambidae			
Cirrhochrista sp. or *Conogethes* sp.		M	(9)
Cossidae			
Zeuzera sp.		M	(9)

[a] The letters represent chimpanzee unit-groups confirmed to eat the insects. The data of B group are from Uehara (1982). The data of K group are from Nishida and Hiraiwa (1982), Nishida and Uehara (1983), and Uehara (1986). The data of M group are from Uehara (1986) and observations from 2006 to 2011 by Takahisa Matsusaka, Koichiro Zamma, and Mieko Kiyono. Data of Y group are from Sakamaki *et al.* (2007).

[b] (1) Itani (1977b); (2) Itani *et al.* (1973); (3) Nishida and Hiraiwa (1982); (4) Uehara (1986); (5) Collins and McGrew (1987); (6) Fuse (2013); (7) Identified by Seiki Yamane; (8) Kiyono, personal observation (identified based on The Species File Software, online); Picker *et al.* (2004); Schabel (2006); (9) Identified by Yoshitaka Sakamaki; (10) Sasaki (2005); (11) Sakamaki *et al.* (2007); (12) Nishie (2011); (13) Goodall (1986).

limbalis (Figure 21.1). Many sand pit traps made by larvae of Myrmeleontidae (antlions) are commonly observed in the sand. Brightly colored locusts were seen occasionally (Figure 21.2). Sand crickets (*Brachytrupes membranaceus*) (Figure 21.3) are common in Mahale. Local people and chimpanzees are known to eat the crickets (Itani *et al.*, 1973), although I have not seen them doing so. There are at least five species of dung beetles. The size of the largest dung beetle (species unidentified) is about 50 mm in length, (Figure 21.4, left). A commonly observed dung beetle (*Kheper* sp.) is about 35 mm, and is metallic dark red in color (Figure 21.4, right). A group of homopteran insects (Flatidae) that resembled a moth were also observed (Figure 21.5). Therefore, Mahale is home to a wide variety of insects.

21.3 Studies of insect-feeding behaviors of chimpanzees in Mahale

At Mahale, the first detailed study on insect-feeding was on the ant-gathering behavior of the K-group chimpanzees (Nishida, 1973). Following this initial report, Nishida and Hiraiwa (1982) reported that chimpanzees fed on 25 species of insects and described the natural history of ant-feeding behaviors

Figure 21.1 A common cicada (probably *Ugada limbalis*). ©M. Kiyono.

Figure 21.2 A brightly colored locust observed in the forest. ©M. Kiyono.

Figure 21.3 A sand cricket (*Brachytrypes membranaceus*). ©M. Kiyono.

Figure 21.4 Dung beetles. The right one appears to be *Kheper* sp. ©M. Kiyono.

in chimpanzees, mainly from the K group. As already mentioned, Nishida and Uehara (1983) compiled the list of chimpanzee food in Mahale, which included several insect foods. Kawanaka (1990) observed the M-group chimpanzees feeding on the eggs, pupae, larvae, and imagoes (alates) of *Crematogaster* ants living in dead branches and stems.

Figure 21.5 A homopteran insect in the forest. ©M. Kiyono.

Although ants are the most frequent prey species, termites are also eaten by Mahale chimpanzees (e.g. Nishida and Uehara, 1980). Uehara (1982) clearly showed that seasonality in the feeding behavior of K-group chimpanzees on termites (*Pseudacanthotermes spiniger*) was affected by the natural history of termites. Kiyono-Fuse (2008) reported that an M-group chimpanzee captured swarming termites (*P. spiniger*) using wet hairs, which may be the same technique as was observed in the K-group chimpanzees. Both K- and M-group chimpanzees were observed to fish for *P. spiniger*, although this occurred only rarely (Nishida and Uehara, 1980; Takahata, 1982).

Group and/or sex differences in insect feeding were also elucidated: Nishida and Uehara (1980) first indicated a between-group difference in termite feeding at Mahale, where termites were habitually fished by chimpanzees in the B group, whereas this behavior was rare in the K group. The authors suggested that this difference may be due to local heterogeneity of the home-range habitat that influences the termite fauna. In an attempt to identify sex differences in required subsistence, Uehara (1986) discovered that female chimpanzees were more insectivorous than males in both the M and K groups.

Uehara (1986) suggested that the functional objective of feeding on animals, including insects, is obtaining nutrients such as minerals but the importance may differ among species and/or feeding technique (i.e. fishing) with an acknowledgment of previous studies. Nishida (1973) suggested that ant fishing using tools is performed in leisure time. Nishida and Hiraiwa (1982) also suggested that chimpanzees are seeking formic acid from soldiers' venom as a spice or snack rather than as nutrients in *Camponotus* ant fishing. For further support for this hypothesis, Uehara (1986: 11) mentioned that ant fishing, the common feeding behavior for *Camponotus*, "appears too inefficient to provide significant nutrition." Nishie (2011) noted that the estimated macro- and micro-nutritional intake per day from *Camponotus* ant fishing by M-group chimpanzees was markedly lower than the estimated recommended intake, thus he regarded ant fishing to be a probable "leisure" activity rather than an important activity for obtaining subsistence. Uehara (1986) also suggested that this alternative hypothesis may be applicable to feeding on the workers of weaver ants (*Oecophylla longinoda*) as they also have venomous glands. At the same time, he considered, together with other small ants, *Camponotus* ant eating without tools and weaver ant eating may also have nutritional importance as chimpanzees sometimes eat eggs, larvae, and pupae of these ants. Termite feeding has been commonly recognized as nutritionally oriented behavior from early on (Nishida and Hiraiwa, 1982; Uehara, 1986). Kawanaka (1990) discussed that ants living in dead branches and stems (mainly *Crematogaster* spp.) might be nutritionally important for the growth of male chimpanzees in Mahale.

Recently, a study on the nutritional function of insect feeding was conducted in Kasekela chimpanzees from Gombe, with respect to termite fishing (*Macrotermes*), ant dipping (*Dorylus*), ant fishing (*Camponotus*), weaver ant feeding (*Oecophylla*), and termite alate feeding (*Macrotermes* and *Pseudacanthotermes*) (O'Malley and Power, 2014). The authors concluded that all forms of insectivory provided small but probably non-trivial amounts of fat and protein; however, they also found that *Camponotus* ant fishing provides significantly lower mass intake, thus generally supporting studies at Mahale reviewed above.

According to O'Malley and Power (2012), gross energy (GE) of *Crematogaster* sp. was 227.5 kcal/100 g, and the value was the highest among ants and bees (*Apis, Camponotus, Crematogaster, Dorylus, Oecophylla,* and *Pachycondyla*), except for termites. Thus, the recent study may support Kawanaka's idea that *Crematogaster* feeding is nutritionally important. When GE values were compared on the basis of foraging units ("per-insects," "per-dip," or "per-nest"), the results were different from those mentioned above (O'Malley and Power, 2012). Among ants, bees, and termites (*Apis, Camponotus, Dorylus, Oecophylla, Pachycondyla, Cubitermes, Macrotermes,* and *Pseudacanthotermes*), *Oecophylla longinoda* showed the highest GE values. Unfortunately, *Crematogaster* ants were excluded in this per foraging unit analysis; however, because chimpanzees usually feed on a whole nest of *Crematogaster* ants, namely, the eggs, pupae, larvae, and imagoes (alates), their per foraging unit GE value should also be high.

21.4 Characteristics of insect prey and related species

21.4.1 *Camponotus* ants

The M-group chimpanzees frequently feed on *Camponotus* ants by ant fishing and by breaking dead branches. Six species of *Camponotus* ants were recognized at Mahale among which *Camponotus vividus, C. brutus, C. maculatus, C.* (Myrmotrema) sp., and *C.* sp. (*chrysurus*-complex) are preyed on via ant fishing (see Chapter 42).

The quantity of *Camponotus* ants consumed differs among the different unit-groups. Uehara (1986) revealed that fecal samples from the M-group females contained more *Camponotus* ants than those from the K-group females, which may represent different distributions of ants over the range of the two chimpanzee groups. Furthermore, based on fecal analysis, the Y-group chimpanzees (see Chapter 11) seem to consume more *C. brutus* than do the M-group chimpanzees (Sakamaki *et al.*, 2007). *Camponotus* ant fishing has also been observed in other chimpanzee populations (see Chapter 42 for more details).

Although some *Camponotus* ants are known to build nests in the tree canopy (Hölldobler and Wilson, 1990), it is not known whether they prefer particular plant species for nesting. Various host-tree species are used in *Camponotus* ant fishing by chimpanzees in both the K and M groups (Nishida and Hiraiwa, 1982; Nishie, 2011). Host trees, *Spathodea nilotica* (*S. campanulata* in Fuse, 2013), *Cordia millenii,* and *Cordia africana,* which were used during *Camponotus* ant fishing by the M group (Nishie, 2011), were also used for feeding on ants in dead branches and stems (Fuse, 2013). In particular, the second most frequently used plant species for *Camponotus* ant fishing, *Cordia millenii,* was the fifth most common dead tree species used for *Crematogaster* ant feeding (Fuse, 2013). The fruits of this plant species are also one of the five main staple fruit foods (Chapter 17); thus this species is an

Figure 21.6 A big carton-like nest on a tree nested by *Crematogaster* (*Atopogyne*) sp. ants. ©M. Kiyono.

important plant species for chimpanzees of the M group, and supplies both ants and fruits.

21.4.2 *Crematogaster* ants

Crematogaster ants are more often eaten by Mahale chimpanzees than are *Camponotus* ants (Nishida and Hiraiwa, 1982; Nishida and Uehara, 1983; Uehara, 1986). Eleven species of *Crematogaster* ants were identified in Mahale; however, not all species were eaten by chimpanzees. Ants of the subgenus *Atopogyne* have an aggressive nature, and are not preyed upon by the M-group chimpanzees (Fuse, 2013), but are preyed upon by the K-group chimpanzees (Goodall, 1986). They build big carton-like nests on trees (Figure 21.6) and leave conspicuous trails that continue to the nest. The abdomens of these ants are triangular and orange in color. Subgenera *Sphaerocrema* and *Crematogaster* (*Crematogaster* sp. 3 and *C.* sp. 18), which live in dead branches, are more frequently eaten by chimpanzees than are other *Crematogaster* ants. *C.* sp. 3 and *C.* sp. 18 resemble each other in appearance, as their abdomens are triangular and black in color; however, the latter adults' size are larger (4.84 mm, $N = 22$) than the former (3.18 mm, $N = 22$). Subgenus *Decacrema* is distributed in the stems of herbs such as Zingiberaceae (local name, Itungulu) and *Pennisetum purpureum* (Iswe). The abdomens of these ants are triangular and orange in color, and these are the smallest ants (2.94 mm, $N = 9$) among the *Crematogaster* found in Mahale.

The genus *Crematogaster* tends to have strong relationships with certain plant species. *Crematogaster* ants play roles in the pollination and seed dispersal of some plant species, and appropriate plants as domiciles (Hölldobler and Wilson, 1990). These ants are known to use four types of nest sites: (1) cavities in live branches and stems excavated by wood-boring beetles; (2) cavities in stems and branches that are naturally hollow or contain piths soft enough to be easily excavated by ants; (3) cavities in barks; and (4) roots of epiphytes (Hölldobler and Wilson, 1990). The ants that are frequently preyed upon by Mahale chimpanzees, use type-2 nest sites, as described above.

21.4.3 Weaver ants (*Oecophylla longinoda*)

This species is relatively large (8 mm in body length) and exclusively arboreal (Hölldobler and Wilson, 1990). This ant is reported to be eaten by various populations of chimpanzees (McGrew, 1992). The workers create natural enclosures for their nests by first pulling leaves together. These ants build nests with leaves of plants such as *Mangifera indica, Brachystegia spiciformis*, and *Pseudospondias microcarpa* (Olotu *et al.*, 2013). Nishida and Hiraiwa (1982) reported that these ants live in the *Brachystegia* woodland (often in *Brachystegia spiciformis*) and in gallery forests (often in *Pseudospondias microcarpa*).

The M-group chimpanzees feed on these ants by taking the whole leaf nest in their hand and quickly placing their mouth over it, probably because this ant species is very aggressive. When an adult male, Bonobo, fed on the nest of *O. longinoda* on August 4, 2006, he moved away from the place where he grabbed a part of the nest. Shunkichi Hanamura (personal communication) observed an adult female, Tarnie, eating *O. longinoda* on August 8, 2006. She then picked away at the nest of ants and continued to feed on ants walking on her body, but she soon discarded the nest. This may be because many ants came out of the nest. The very aggressive nature of these ants makes it difficult for chimpanzees to feed on them. However, there are no records of the ants of this species being eaten with the aid of tools.

21.4.4 Other ants

Mahale chimpanzees also feed on *Cataulacus egenus, C. brevisetosus, Polyrhachis* cf. *phidias, P. decemdentata, Tetramorium aculeatum, T. decem*, and *Monomorium trake* that live in dead branches and stems. As *T. aculeatum* has a long, slim, stick-like body, and a sting, it is easy to distinguish it from *Crematogaster* and *Camponotus* ants. *Lepisota* sp., *Technomyrmex* spp., *Tetraponera* sp., and *Plagiolepis* sp. are also present in dead branches and stems; however, these species are not preyed upon by chimpanzees.

Pachycondyla analis, known as the Matabele ant (Figure 21.7), is a widespread and relatively common ponerine species found in most parts of Africa (Corbara and Dejean, 2000), as well as in Mahale. This ant species feeds specifically on termites, and is therefore called a termitophage (Corbara and Dejean, 2000). Their nests often extend down to 0.7 m below the surface. Each colony uses a single nest where eggs and larvae are kept together with the queen in one chamber, while the cocoons are often placed in the sun outside the nest entrance (Yusuf, 2010). Chimpanzees in Mahale have never been observed feeding on this species. These ants are reported to be present in Dja, Cameroon, but are not known to be eaten by chimpanzees (Deblauwe and Janssens, 2008).

Dorylus spp., also known as driver ants, are preyed upon by chimpanzees in Assirik (Senegal), Fongoli (Senegal), Tenkere (Sierra Leone), Bossou (Guinea), Seringbara (Guinea), Taï (Côte d'Ivoire), Gashaka (Nigeria), Gombe, Goualougo (Republic of Congo), Ngogo (Uganda), and Kalinzu (Uganda) (Schöning *et al.*, 2008), but never by Mahale chimpanzees. Schöning *et al.* (2005) categorized *Dorylus* species into three groups, namely, epigaeic, intermediate, and hypogaeic, based on their lifestyles. Epigaeic species hunt on the ground surface and in the vegetation, and have earth nests that are often large and conspicuous. Intermediate species hunt in the leaf litter but not in the vegetation. Their swarm raids can be conspicuous but their nests are usually not detectable. Hypogaeic species usually hunt in the soil and their nests are also undetectable. Thus, they are rarely seen in the leaf litter or on the ground. At Mahale, there are three species of *Dorylus: D. molestus* (*Dorylus nigricans*-complex, Seiki Yamane's identification), *D. kohli* (Schöning *et al.*, 2008), and *D. gerstaeckeri* (Seiki Yamane's identification). According to the above categorization, *D. molestus* is classified as epigaeic, and *D. kohli* and *D. gerstaeckeri* are classified as intermediate. *Aenictus* sp. similar to *Dorylus* spp. also occur at Mahale, but are never fed on by the M-group chimpanzees.

21.4.5 Termites

There are eight genera of termites living in the Kasoje area (Collins and McGrew, 1987), among which *Pseudacanthotermes spiniger, P. militaris*, and

Figure 21.7 Worker ants of *Pachycondyla analis* with pupae. ©M. Kiyono.

Odontotermes spp. are abundant (Uehara, 1982; Collins and McGrew, 1987). *Macrotermes* spp., the main termite targets for tool use in other chimpanzee study sites (McGrew, 1992), have not been reported in the Kasoje area where the K and the M groups range (Chapters 4, 8), but their presence was confirmed at Bilenge where the unhabituated B group ranges (Collins and McGrew, 1987). *P. spiniger* seems to be more abundant in Mahale, and these termites were observed in gallery forest rather than in the woodland (Uehara, 1982). They built a cone-like dome (up to 2 m in height and 1 m in diameter) in which winged reproductives gather every year before the dispersal night. Swarming of *P. spiniger* takes place during the daytime at the end of the wet season and reproductives leave the mother colony at dusk between 18:00 and 18:30 h (Connétable *et al.*, 2012). On the other hand, *P. militaris* generally make subterranean nests, and are therefore difficult to find (Uehara, 1982); however, they can sometimes be covered by a flat dome 10–20 cm in height (Connétable *et al.*, 2012). Their dispersal flights occur during the day (between 11:00 and 18:30 h), which is earlier than those of *P. spiniger* (Connétable *et al.*, 2012).

Uehara (1982) summarized the utilization of these *Pseudacanthotermes* termites by the K-group chimpanzees as follows: (1) eating soil from a termite mound; (2) termite fishing; (3) destroying the tower of a mound with their hands and capturing them directly from it; (4) feeding on termites living inside sugarcane and dead trees; and (5) feeding on winged reproductive termites that swarm in the air. Currently, chimpanzees of the M group have also been found to utilize these termites in three out of the five ways (1, 3, and 5). Although a few cases of termite fishing were observed among the M-group chimpanzees in 1981, these were by immigrants from the K group who were observed to fish for termites in 1977 when they still belonged to the K group (Takahata, 1982). Thereafter, none of the M-group chimpanzees have been observed to do so.

As Uehara (1982) noted for the K group, chimpanzees of the M group also fed on winged reproductives of *Pseudacanthotermes* termites on April 13, 2007 (Kiyono-Fuse, 2008). This was in the same month as the K-group chimpanzees showed this behavior in 1977 (Uehara, 1982).

It is quite rare for the M-group chimpanzees to find termites when they break open dead branches or stems, even though I systematically investigated such behaviors (see below). *Odontotermes* spp. occur at Mahale, but their ecology has not yet been investigated.

21.4.6 Other insects

The K-group chimpanzees sometimes fed on moth larvae (Nishida and Hiraiwa, 1982); however, the details of this behavior have not been described. I observed two such cases during 703.2 h spent observing the M-group chimpanzees (Table 21.2). They broke the stems of live plants (species unidentified) to obtain larvae, mostly from the end of March to the beginning of April. According to Yoshitaka Sakamaki (personal communication), these larvae resembled *Zeuzera* sp., Cossidae, which often inhabit trees of tea (*Camellia sinensis*) and coffee (*Coffea* spp.) in Asia. The existence of a larva in a stem is often revealed by wood debris at the exit of the larval nest (Figure 21.8). Chimpanzees may also exploit this cue to detect a stem inhabited by this larva. I sometimes observed chimpanzees breaking stems open after staring at them. Although it is difficult to prove, chimpanzees might do so after noticing the wood debris. On December 10, 2010, an adolescent male, Caesar, ate larvae of Crambidae (Yoshitaka Sakamaki's identification) living inside a nest made with a rolled leaf of *Baphia capparidifolia* (Koichiro Zamma, personal observation). On December 15, 2010, Koichiro Zamma observed that an adult male, Pimu, broke fruits of *Ficus sur* in half with his teeth and selectively ate larvae living inside the fruits. The larvae were probably of *Cirrhochrista* sp. or *Conogethes* sp., according to Yoshitaka Sakamaki (personal communication). The M-group chimpanzees may like to feed on moth larvae inhabiting several parts of a plant.

On the other hand, chimpanzees may not aim to feed on fig wasp larvae which live inside figs. Fig wasps (*Blastophaga* sp.) have previously been observed in the fecal samples of the K- or M-group chimpanzees (Uehara, 1986). Nishida (2012) also noted that the M-group chimpanzees sometimes consume fig wasps when eating figs, although this could be accidental. I agree with Nishida's conclusion because I sometimes observed that some of the fig fruits discarded on the ground by M-group chimpanzees were inhabited by wasps.

I once observed a few M-group chimpanzees inserting tools into a hole that resembled the nest of carpenter bee, *Xylocopa* (*Koptortosoma*) sp. (Seiki Yamane's identification). Although I could not confirm what they were trying to eat in this case, K-group chimpanzees are known to eat the larvae (Nishida and Uehara, 1983) and honey (Nishida and Hiraiwa, 1982) of this bee. This species excavates nest burrows in tree branches and structural timbers, constructing a series of cells separated by partitions of wood scrapings glued with saliva (Picker *et al.*, 2004). Takahisa Matsusaka (personal communication) also observed that an adult male, Cadmus, tried to feed on the contents of a probable carpenter bee nest on November 22, 2000.

The M-group chimpanzees eat aphids (*Phytolima lata*, according to Nishida and Uehara, 1983) that inhabit the leaves of *Chlorophora excelsa* as did the K-group chimpanzees (Nishida and Uehara, 1983). Chimpanzees seem to feed on leaves or galls with aphids. More observations are needed to conclude whether they aim to feed on plants infested by insects or not. Moreover, chimpanzees of the M group fed on aphids (species unidentified but probably of *Paracopium*) in the insect galls of *Clerodendrum* sp. (Linkumbwe). This feeding behavior was also reported in Gombe (Goodall, 1986).

21.4.7 Searching behaviors for insects

Nishida and Hiraiwa (1982) noted that chimpanzees generally make use of established knowledge regarding the distribution of host trees for ant fishing, because chimpanzees generally make tools before

Table 21.2 **Number of observed bouts per insect-feeding behavior per month (no data for September 2006)**

Observed feeding techniques for each insect type	2006						2007					
	Jun	Jul	Aug	Oct	Nov	Dec	Jan	Feb	Mar	Apr	May	Total
Ants												
breaking dead branches	2	4	5	1	10	15	26	11	5	0	1	75
fishing		1	2		2	3	5	8		1		22
Larvae of moth												
breaking live stems									1	1		2
Carpenter bee												
breaking dead branches							1					
Aphids in galls		3	2									5
Termite												
winged reproductives									2	1		3
mound feeding		1						2	1			4
Focal observation time (h)	49.3	130.3	81.6	5.0	34.7	19.6	60.3	66.6	51.6	27.1	28.3	554.6
Nonfocal observation time (h)	0	0	0	8.5	22.4	55.7	28.5	15.9	0	4.6	13.0	148.6

Figure 21.8 Wood debris at the exit of the nest of a larva (probably of *Zeuzera* sp.). ©M. Kiyono.

moving to the ant-fishing site. They also reported that the chimpanzees seem to "discover" nest-entrance holes after visually inspecting branches and inspecting nests through smelling.

Uehara (1982) established the way to observe "searching" or "finding" behaviors for termites. He described that the feeding techniques for termites are related to the prey's phenology. Eating soil from termite mounds supplies minerals, and helps in the absorption of certain components from their vegetable diets, such as tannins (but see also Chapter 17). In addition, the consequence of eating soil from termites makes them visually and tactually aware of the phenological changes of the termites' reproductive cycle. Uehara (1982) suggested that chimpanzees have advanced cognitive ability for recognizing temporal changes in environmental factors, which enables them to choose the technique most suitable for their purpose.

21.5 Ant-foraging behavior: utilizing plant–ant relationships

Chimpanzees can feed on concealed ant colonies (eggs, pupae, larvae, and reproductives) by locating them in dead branches and stems. This behavior has been described in several papers (Nishida, 1973; Nishida and Hiraiwa, 1982; Uehara, 1982; Kawanaka, 1990; Nishida, 2012). Table 21.2 shows that M-group chimpanzees foraged on ants in dead branches throughout the year except April (and no data on September). A detailed investigation into how chimpanzees forage on ants in dead branches revealed that chimpanzees can efficiently forage on *Crematogaster* ants because they might utilize the loose relationship between several plant species and ants (Fuse, 2013).

The ant-inhabited ratio in chimpanzee-foraged dead branches and stems was higher than that in the branches available in the home range. Furthermore, chimpanzees more often foraged on *Crematogaster* spp. than was expected from their occurrence in the litter units. Although the majority of prey in such foraging was *Crematogaster* ants, it was found that *Camponotus* and *Cataulacus* ants were also foraged. Among *Crematogaster* spp., chimpanzees mostly ate *C.* sp. 3 and *C.* sp. 18, as described above. The mean frequency of this foraging behavior was 0.18 bouts per hour. This ant-foraging behavior seemed to increase from November to March during the wet season. The reason for the increase in the ant-foraging behavior during the wet season is not known. Turner (2006) noted that chimpanzees fed on ants of *Crematogaster* spp., particularly in September and October. Because my observation time was limited in September and October, it is difficult to discuss whether there is an interannual difference of seasonality of ant-foraging behavior.

Ants were found to be associated with several plant species and they often make nests in spaces within the dead parts of the plants. When the occurrence of ants in dead branches and stems in quadrats was investigated, I found that ants do prefer particular plant species for their nests. Three plant species (*Vernonia subligera, Dracaena usambarensis,* and *Senna spectabilis*) were more frequently broken off by chimpanzees. The part of dead branches and stems which was broken off by chimpanzees was defined as "litter unit" (Fuse, 2013). The number of litter units broken off by chimpanzees was 129 (*S. spectabilis*), 117 (*V. subligera*), and 64 (*D. usambarensis*). These three plant species were more often occupied by *Crematogaster* ants. Chimpanzees of the K group also seemed to prefer ants living in the dead branches of trees, often *Dr. usambarensis*, and in the dry hollow grasses (Nishida and Hiraiwa, 1982). Chimpanzees may know which plant species are frequently associated with ants. Further study is necessary to identify how chimpanzees recognize relationships between plant species and *Crematogaster* ants.

21.6 The importance of observation on insect-feeding behaviors with plants

Long-term studies have revealed that Mahale chimpanzees feed on a variety of insects, particularly social insects. Although social insects are obtained in large yields, it is not always easy to locate them. We still lack empirical data on how they locate insects and precise analysis on plant–insect relations at species level, which should be investigated to fully understand the insect-feeding behavior of chimpanzees. Moreover, we must bear in mind that most insect–plant relationships are not as close as those observed in symbiosis. An ant species does not usually have a relationship with only one plant species, but with several plant species (Seiki Yamane, personal communication). To note, it may be a characteristic of Mahale chimpanzees' insect feeding that targeted insects, such as *Camponotus* and *Crematogaster* ants, typically inhabit plants. This is in contrast to most of the other chimpanzee study sites where main insect prey, such as *Dorylus* driver ants and *Macrotermes* termites, live underground. We also need to consider the ecology of non-social insects in relation to chimpanzee feeding. For example, certain moth species seem to prefer one or a few plant species. Although the frequency of moth feeding is low at Mahale, there is a possibility that chimpanzees utilize such insect–plant relationships to search for moths.

The climate and vegetation of Mahale is changing (Chapters 12, 13), which means that the environment for insects is also changing accordingly. We may need to see how these changes influence insect fauna and ecology as well as how chimpanzees cope with such changes in order to obtain insect foods. Continuing efforts to maintain long-term investigations at Mahale are essential in order to detect such decadal changes in the interactions among chimpanzees, insects, and plants.

References

Collins, D. A. and McGrew, W. C. (1987). Termite fauna related to differences in tool-use between groups of chimpanzees (*Pan troglodytes*). *Primates*, **28**, 457–71.

Connétable, S., Robert, A., and Bodereau, C. (2012). Dispersal flight and colony development in the fungus-growing termites *Pseudacanthotermes spiniger* and *P. militaris*. *Insect Sociaux*, **59**, 269–77.

Corbara, B. and Dejean, A. (2000). Adaptive behavioral flexibility of the ant *Pachycondyla analis* (=*Megaponera foetens*) (Formicidae: Ponerinae) during prey capture. *Sociobiology*, **36**, 465–83.

Deblauwe, I. and Janssens, P. J. (2008). New insight in insect prey choice by chimpanzees and gorillas in southeast Cameroon: the role of nutritional value. *American Journal of Physical Anthropology*, **135**, 42–55.

Fuse, M. (2013). Chimpanzees detect ant-inhabited dead branches and stems: a study of the utilization of plant–ant relationship in the Mahale Mountains, Tanzania. *Primates*, **54**, 385–97.

Goodall, J. (1986). *The Chimpanzees of Gombe: Patterns of Behavior*. Cambridge, MA: Belknap.

Hölldobler, B. and Wilson, E. O. (1990). Symbioses between ants and plants. In *The Ants*, ed. B. Hölldobler and E. O. Wilson. Cambridge, MA: Harvard University Press, pp. 530–56.

Itani, J. (1964). [*Animals of Africa*.] Tokyo: Kawade-shobō. In Japanese.

Itani, J. (1977a). [*The Chimpanzees*.] Tokyo: Kōdan-sha. In Japanese.

Itani, J. (1977b). [Ethnozoology of the Batongwe.] In [*The Natural History of Man*], ed. J. Itani and R. Harako. Tokyo: Yūzan-kaku, pp. 441–537. In Japanese.

Itani, J., Nishida, T., and Kakeya, M. (1973). [*The Eastern Hinterland of Lake Tanganyika: Its Nature and People*.] Tokyo: Chikuma-shobō. In Japanese.

Kawanaka, K. (1990). Age differences in ant-eating by adult and adolescent males. In *The Chimpanzees of the Mahale Mountains: Sexual and Life Strategies*, ed. T. Nishida. Tokyo: University of Tokyo Press, pp. 207–22.

Kiyono-Fuse, M. (2008). Use of wet hair to capture swarming termites by a chimpanzee in Mahale, Tanzania. *Pan Africa News*, **15**, 8–12.

McGrew, W. C. (1992). *Chimpanzee Material Culture: Implications for Human Evolution*. Cambridge: Cambridge University Press.

Nishida, T. (1973). The ant-gathering behavior by the use of tools among wild chimpanzees of the Mahali Mountains. *Journal of Human Evolution*, **2**, 357–70.

Nishida, T. (2012). *Chimpanzees of the Lakeshore: Natural History and Culture at Mahale*. Cambridge: Cambridge University Press.

Nishida, T. and Hiraiwa, M. (1982). Natural history of a tool-using behavior by wild chimpanzees in feeding upon wood-boring ants. *Journal of Human Evolution*, **11**, 73–99.

Nishida, T. and Uehara, S. (1980). Chimpanzees, tools, and termites: another example from Tanzania. *Current Anthropology*, **21**, 671–2.

Nishida, T. and Uehara, S. (1983). Natural diet of chimpanzees (*Pan troglodytes schweinfurthii*): long-term record from the Mahale Mountains, Tanzania. *African Study Monographs*, **3**, 109–30.

Nishie, H. (2011). Natural history of *Camponotus* ant-fishing by the M group chimpanzees at the Mahale Mountains National Park, Tanzania. *Primates*, **52**, 329–42.

Olotu, M. I., du Plessis, H., Segunic, Z. S., and Manianiaa, N. K. (2013). Efficacy of the African weaver ant *Oecophylla longinoda* (Hymenoptera: Formicidae) in the control of *Helopeltis* spp. (Hemiptera: Miridae) and *Pseudotheraptus wayi* (Hemiptera: Coreidae) in cashew crop in Tanzania. *Pest Management Science*, **69**, 911–18.

O'Malley, R. C. and Power, M. L. (2012). Nutritional composition of actual and potential insect prey for the Kasekela chimpanzees of Gombe National Park, Tanzania. *American Journal of Physical Anthropology*, **149**, 493–503.

O'Malley, R. C. and Power, M. L. (2014). The energetic and nutritional yields from insectivory for Kasekela chimpanzees. *Journal of Human Evolution*, **71**, 46–58.

Picker, M., Griffiths, C., and Weaving, A. (2004). *Field Guide to Insects of South Africa*. Cape Town: Struik.

Sakamaki, T., Nakamura, M., and Nishida, T. (2007). Evidence of cultural differences in diet between two neighboring unit groups of chimpanzees in Mahale Mountains National Park, Tanzania. *Pan Africa News*, **14**, 3–5.

Sasaki, H. (2005). Tabanid flies (Dipterra: Tabanidae) of the Mahale Mountains National Park, Tanzania, East Africa. *Journal of Rakuno Gakuen University*, **30**, 93–8.

Sasaki, H. and Nishida, T. (1999). Notes on the flies associated with wild chimpanzees at Mahale Mountains National Park, Tanzania, East Africa. *Medical Entomology and Zoology*, **50**, 151–5.

Schabel H. G. (2006). *Forest Entomology in East Africa: Forest Insects of Tanzania*. Dordrecht, the Netherlands: Springer.

Schöning, C., Kinuthia W., and Franks, N. R. (2005). Evolution of allometries in the worker caste of *Dorylus* army ants. *OIKOS*, **110**, 231–40.

Schöning, C., Humle, T., Möbius, Y., and McGrew, W. C. (2008). The nature of culture: technological variation in chimpanzee predation on army ants revisited. *Journal of Human Evolution*, **55**, 48–59.

Species File Software (online). http://help.speciesfile.org/index.php/Main_Page, accessed April 17, 2014.

Takahata, Y. (1982). Termite-fishing observed in the M group chimpanzees. *Mahale Mountains Chimpanzee Research Project Ecological Report*, **18**.

Turner, L. A. (2006). Vegetation and chimpanzee ranging in the Mahale Mountains National Park, Tanzania. *Memoirs of the Faculty of Science Kyoto University (Series of Biology)*, **18**, 45–82.

Uehara, S. (1982). Seasonal changes in the techniques employed by wild chimpanzees in the Mahale Mountains, Tanzania, to feed on termites (*Pseudacanthotermes spiniger*). *Folia Primatologica*, **37**, 44–76.

Uehara, S. (1986). Sex and group differences in feeding on animals by wild chimpanzees in the Mahale Mountains National Park, Tanzania. *Primates*, **27**, 1–13.

Yusuf, A. A. (2010). *Termite Raiding by the Ponerine Ant Pachycondyla analis (Hymenoptera: Formicidae): Behavioral and Chemical Ecology*. Ph.D. thesis. Pretoria, South Africa: University of Pretoria.

PART V Life history and health

22 Development and growth: with special reference to mother–infant relationships

TAKUYA MATSUMOTO AND HITOSHIGE HAYAKI

22.1 Introduction

A notable characteristic of primate life history is the very long period of immaturity (Ross, 1998); compared with other primate taxa, great apes have a more prolonged period of immaturity. Long-term research on wild chimpanzee populations across different study sites has revealed a great deal about the parameters of life history and development (Goodall, 1983, 1986; Hiraiwa-Hasegawa *et al.*, 1984; Sugiyama, 1984; Nishida *et al.*, 1990; Boesch and Boesch-Achermann, 2000). At the same time, we are beginning to recognize that there are large differences among individuals and study populations.

The continual research on Mahale chimpanzees has accumulated demographic and behavioral data, and revealed the general developmental patterns in this wild chimpanzee group. Females with infants are usually more afraid than males of human observers when habituation is ongoing, but a large advantage of the Mahale site is that both males and nursing females are so habituated we can observe their interactions in calm situations.

This chapter introduces the growth and development of wild chimpanzees in Mahale. We first describe the four developmental stages often used in chimpanzee studies, focusing on physical and social aspects. We then focus on the earliest stage of development, i.e. infancy. By reviewing topics about the mother–infant relationship and interactions with other group members, we elaborate on the process of social maturity in wild chimpanzees.

22.2 Developmental stages

The developmental stages of chimpanzee growth are typically divided into four phases: infancy, juvenility, adolescence, and adulthood, based on the degree of physical, behavioral, and social maturity (Goodall, 1983, 1986; Hiraiwa-Hasegawa *et al.*, 1984; Nishida *et al.*, 1990; Nishida, 2002, 2012). Although Gombe (Tanzania) and Mahale researchers employ almost identical definitions for these developmental stages, some researchers at other study sites use different definitions (e.g. Bossou, Guinea; Sugiyama, 1981). In other words, there are some local and/or subspecific differences in the rate of development even among wild populations. Table 22.1 shows the general life history pattern of Mahale chimpanzees. Below, we describe these four developmental stages (Figure 22.1) mainly according to the general patterns of eastern chimpanzees (*Pan troglodytes schweinfurthii*) in Mahale and Gombe.

22.2.1 Infancy

Infancy (Table 22.1) is characterized by breastfeeding and sharing of an overnight bed (or nest) with the mother, and terminated by weaning and resumption of the mother's estrus. The mother usually carries the infant when she is traveling for a long time, and hugs him/her when it is cold or raining, or when a disturbance occurs. The average birth weight in captivity is 1.80 kg, with no significant sex difference (Gavan, 1953).

Mahale Chimpanzees: 50 Years of Research, ed. M. Nakamura, K. Hosaka, N. Itoh and K. Zamma. Published by Cambridge University Press.

Table 22.1 **General life history pattern of Mahale chimpanzees**[a]

Earliest observed age[b]	Average age	Event[b]	Relevant chapters	References
–	0 yr	Birth, start of infancy	22	Hiraiwa-Hasegawa *et al.*, 1984; Nishida *et al.*, 1990; Nishida, 2002, 2012
–	3 wk	Look around	22	Hiraiwa-Hasegawa, 1986
2 mo	4 mo	Ride on mother's back	22	Hiraiwa-Hasegawa, 1986
–	3 mo	Play with mother	22, 36, 40	Hiraiwa-Hasegawa, 1986
–	3.5 mo	Mouth plant-food items	22	Hiraiwa-Hasegawa, 1986
3 mo	5 mo	Emit play panting	39	Hiraiwa-Hasegawa, 1986
–	5 mo	Ingest plant-food items	22	Hiraiwa-Hasegawa, 1986
–	5 mo	Play with non-mother individuals	36, 40	Hiraiwa-Hasegawa, 1986
–	6 mo	Move away from mother	22	Hiraiwa-Hasegawa, 1986
10 mo	2 yr	Groom mother	34	Nishida, 1988
1 yr	–	Leaf clipping	35, 42	Nishida, 1980
1 yr	1.1 yr	Make a day bed	43	Hiraiwa-Hasegawa, 1989
1.2 yr	1.5–2 yr	Perform pant-grunts as greeting	33	Hiraiwa-Hasegawa, 1986, 1989
–	2 yr	Walk unaided in long-distance travel	22	Nishida, 2012
–	2–2.5 yr	Make pant-hoot-like vocalizations	39	Hiraiwa-Hasegawa, 1986
2 yr	–	Leaf-pile pulling	42	Nishida and Wallauer, 2003
2.2 yr	–	First copulation with adult females (M)	35	Hasegawa, 1987
2.3 yr	–	Drink water with tool	42	Matsusaka *et al.*, 2006
2.2 yr	3 yr	Ant fishing	42	Nishida and Hiraiwa, 1982; Nishie, 2011, unpublished data
3.1 yr	–	An orphan may survive at this age	30	Nishida, 2012
4.8 yr (M), 3.7 yr (F)	4– yr	Mutual grooming	34	Nishida, 1988
4.3 yr (M), 5.3 yr (F)	4–6 yr	Grooming hand-clasp with mother	34, 38	Nakamura and Nishida, 2013
–	4 yr	Drum natural objects	42	Nishida, 1994a, 2003b
–	4–5 yr	Weaning period	7, 22, 30	Hiraiwa-Hasegawa, 1986, 1989
–	5 yr	Start of juvenility	22	Hiraiwa-Hasegawa *et al.*, 1984; Nishida *et al.*, 1990; Nishida, 2002, 2012

Table 22.1 (**cont.**)

Earliest observed age[b]	Average age	Event[b]	Relevant chapters	References
5.3 yr	5–6 yr	Sleep alone on a night bed	43	Hiraiwa-Hasegawa, 1989
–	5 yr	No longer carried by mother	22, 30	Hiraiwa-Hasegawa, 1986
–	5 yr	Playful harassment of adult females (M)	22, 29, 32, 40, 42	Nishida, 2003a
–	5–13 yr	Most intense period of allomothering	30	Nishida, 1983
–	8–9 yr	**Start of adolescence**	22	Hiraiwa-Hasegawa *et al.*, 1984; Nishida *et al.*, 1990; Nishida, 2002, 2012
–	8–9 yr	Show irregular genital swelling (F)	22, 35	Nishida, 2012
–	8 yr	Travel away from mother and with adult males (M)	22, 43	Hayaki, 1988
–	8 yr	Harass adult females (M)	29, 33	Hayaki *et al.*, 1989
–	8–9 yr	Treated harshly by adult males	22	Nishida, 2012
–	9 yr	Ejaculate (M)	35	Hasegawa, 1991
–	10–11 yr	Catch colobus monkeys	20	Nishida, 2012
9.5 yr	10.2 yr	First maximal genital swelling (F)	35, 44	Nishida, 2002; Nishida *et al.*, 2003
–	10–11 yr	First copulation with adult males (F)	35	Hasegawa, 1991
9.3 yr	11.2 yr	Emigration (F)	7, 11, 35	MMCRP record; Chapter 7
–	10–13 yr	Adolescent infertility (F)	22	Nishida *et al.*, 2003
12.0 yr	13.5–14.5 yr	First delivery (F)	7, 35, 44	MMCRP record; Chapter 7; Nishida *et al.*, 2003
–	13–15 yr 15–16 yr	**Start of adulthood (F)** **Start of adulthood (M)**	22	Hiraiwa-Hasegawa *et al.*, 1984; Nishida *et al.*, 1990; Nishida, 2002, 2012
–	16– yr	Drum metal walls (M)	42	Nishida, 1994a, 2003b
–	16– yr	Selected as a mating partner (F)	35	Hasegawa and Hiraiwa-Hasegawa, 1983; Takasaki, 1985
–	16 yr	Become dominant over any adult female (M)	29	Nishida, 2003a
–	16.5–17 yr	Successful reproduction (i.e. wean the first infant) (F)	7	Chapter 7
–	18–30 yr	Reach the highest dominance rank in life (M)	27	MMCRP record; Chapter 27

Table 22.1 (**cont.**)

Earliest observed age[b]	Average age	Event[b]	Relevant chapters	References
–	40– yr	Some show estrus without delivering (F)	23, 35	Nishida, 2002
26 yr	40 yr	Cataracts observed	23, 25	Chapter 23; Nishida, 1994b
–	40 yr	Presbyopia observed	23, 25	Chapter 23; Nakamura and Nishie, 2012
52 yr	–	Current record for the oldest age	23	Nakamura and Nishie, 2012

[a] The ages shown here are typical or earliest-known ones and there may be some individual deviations. There have also been some studies employing age categories different from those shown here.
[b] (F) and (M) indicate events that are applicable only to females and males, respectively.

Figure 22.1 (a) A 17-year-old adult female (left) and her 1-year-old infant male (right). (b) A 6-year-old juvenile female grooming another individual. (c) A 10-year-old adolescent female with genital swelling. (d) A 12-year-old adolescent male. (e) A 22-year-old adult male (left) is grooming the leg of a 32-year-old male (right). ©T. Matsumoto.

When the mother is grooming others, resting, or feeding, the infant stays close to her. Around the age of 1.5 years, the infant leaves the mother and increases its distance away to over 1.5 m (Hiraiwa-Hasegawa, 1986; van de Rijt-Plooij and Plooij, 1988). After about 2 years of age, the infant begins walking on its own, either in front of or behind the mother in long-distance travel, though an infant still stays within the sight of the mother. The infant will rush to the mother's breast or return to her back in a dangerous situation, or if it feels restless. This close mother–infant relationship continues throughout infancy. If an infant does not accompany the mother, the mother may make soft "hoo hoo hoo" sounds, similar to the whimpers of an infant, to attract its attention. If she fails to get the infant's attention, she may go ahead of the infant, or drag the infant directly with a hand (Nishida, 2012).

Male infants will begin to mate playfully with adult or adolescent females, including the mother, without ejaculation (Hiraiwa-Hasegawa, 1987), while female infants do not engage in such sexual activities. More details of infant development and the relationship with the mother will be described in the latter half of this chapter.

22.2.2 Juvenility

Juvenility (Table 22.1) comes after the weaning period. A juvenile can walk when ranging and climb up a tree by itself. Therefore, a juvenile is physically independent of the mother to some extent. However, we should note that some juveniles in Mahale are continually carried by the mother, especially in cases where there are no younger siblings.

A juvenile can make its own overnight bed and sleep in it by itself; its body size is now big enough to regulate body temperature without the mother (Bloomsmith *et al.*, 2003). However, juveniles need to follow and feed with mothers for several years even after weaning (Goodall, 1986). A juvenile associates frequently with its mother, although the mother–offspring association gradually decreases as the offspring matures (Hayaki, 1988).

If a juvenile loses sight of the mother, he/she will start to whimper, and if the mother cannot be found, will climb up a tree to look for her and will scream even if other members of the group are there (e.g. Matsusaka, 2003). Juvenile males with adolescent or adult older brothers tend to become independent from their mothers earlier (Nishida, 2012). Juvenile females more often follow their mothers than do male juveniles (Hayaki, 1988). This is partly because juvenile females are more attracted to their younger siblings than are juvenile males, while males are more attracted to adult males and playmates than are females.

Juveniles frequently play solitarily and socially with various age/sex partners. Males especially will sometimes approach other individuals and may throw sticks at them, or slap them and then run away. Juveniles seem to use such "harassment" to test the reactions of the target individuals to confirm their own social status. A juvenile male sometimes exhibits playful harassment of adult females (Nishida, 2003a), who rarely make counterattacks on the juvenile (Nishida, 2012).

22.2.3 Adolescence

Adolescence (Table 22.1) is characterized by sexual maturity and a spurt in weight gain (Pusey, 1990; Hamada *et al.*, 1996). Adolescents are independent of the mother for subsistence and protection. They can handle and process any kind of food. Typically, they can catch colobus monkeys by 10–11 years old (see Chapter 20 for more details).

Male adolescence can be defined as the period by which males can ejaculate (Nishida *et al.*, 1990; Chapter 35) until they reach full body size and join male–male social relationships (Chapter 27). They start traveling away from the mother, and often travel with adult males by following middle-aged adult males of relatively low rank (Kawanaka, 1993).

Adolescent males, unlike juveniles, are tolerated less and are treated harshly by adult males. When an adolescent male approaches an adult male to greet him, the adult male sometimes chases, slaps, or stamps on him, or even bites him. This is partly because adolescent males have started to become sexual rivals of adult males. All adolescent males perform pant-grunts to all adult males and they seem sensitive to the presence of adult males (Hayaki, 1985; Chapter 33). Late adolescent males may groom with adult males, but they tend to receive less grooming than they give (Pusey, 1978; Hayaki, 1992).

Adolescent males also often harass adult or adolescent females, as do juveniles. However, their harassment is now more aggressive and serious than playful. A harassed individual often displays counter-behavior such as barking or, together with other females, chasing the harasser. This can often provoke the male to escalate the harassment into a charging display. Therefore, the harassment of adult females by adolescent males is thought to be a normal male developmental process, through which males finally become dominant over all adult females (Nishida, 2003a; Chapter 29).

Adolescence in females starts when they begin to show irregular genital swellings at about 8 years. Although females start to mate frequently with males when they are about 10 years old (Pusey, 1990), there is a time lag before they become pregnant for the first time. Most adolescent female copulations do not result in pregnancy for physiological reasons; this is called "adolescent infertility." Adolescent females also often copulate almost playfully with sexually immature males, including infants. Most females emigrate from their natal group to other groups when they are in estrus but before they get pregnant; this probably functions as a mechanism to avoid inbreeding (Hiraiwa-Hasegawa, 1987; Pusey, 1990). It is quite difficult to determine when female adolescence ends, but it is reasonable to assume that it continues until the age of 13 years, since females have their first offspring at about that age (Nishida *et al.*, 2003).

A typical behavior of adolescent females is "alloparental" or babysitting behavior (Nishida, 1983; Chapter 30). An adolescent female may be allowed to hold and take care of infants of more tolerant mothers (Uehara and Nyundo, 1983).

22.2.4 Adulthood

Adulthood (Table 22.1) is characterized by physical and social maturity. According to the measurement of chimpanzee weights at Mahale, adult males average 42.0 kg and adult females 35.2 kg (Uehara and Nishida, 1987; Chapter 45).

An adult male increases his associations with older males. Prime and old adult males are most frequently in proximity with each other, and young adult males (usually late teens) consistently attempt to approach them. They, even though physically full-grown, tend to have less proximity partners than older adult males. Thus, it seems that they are not yet sufficiently qualified to join the coalitions formed by their seniors.

Most adult females are immigrants from other groups who remain permanently in the group after they give birth. However, there are some females who do not leave, but stay and give birth in their natal group (Nishida *et al.*, 1990; Chapter 7). Adult females, when they are not in estrus, sometimes travel only with their dependent offspring, which is in contrast to adolescent females, especially immigrants, who often travel with adult males. Although parous adult females are often referred to as relatively "asocial," they also form various associations and social relationships with females and males in the group after immigration (Chapters 28, 29).

22.3 Mother–infant relationships and infant development

We introduce here some characteristics about the mother–offspring relationship and infant development to elaborate on the process of social maturity of wild chimpanzees.

22.3.1 Birth

During the final 2 months of pregnancy, captive females behave with extreme caution and try to avoid strenuous exercise (Yerkes and Tomilin, 1935), but such self-protective behaviors are not usually obvious in the wild. It is possible, however, that most females travel alone, or only with their dependents, during the final days of pregnancy, thus avoiding strenuous group activities (van Lawick-Goodall, 1967; van de Rijt-Plooij and Plooij, 1987). Nishida (2012) has called this period of isolation "maternity leave."

We can rarely observe births in the wild. Rare observations reveal that mothers consume the placenta and umbilical cord. The mother holds her newborn baby with her hand and walks tripedally (Goodall and Athumani, 1980; Zamma *et al.*, 2012). A newborn baby on the day of birth can suckle the mother's nipple and let out staccato vocalizations (Nishida, 2012; Zamma *et al.*, 2012). When a mother with a newborn meets other group members for the first time, many of them show interest in the mother and her baby, and seek proximity to them. The mother's "maternity leave" can last a day to several weeks after birth, depending on the individual.

Chimpanzee mothers usually give birth to only one infant, and only one case of a twin birth has been observed at Mahale, in 1994 (Matsumoto-Oda, 1995). Each of the twin infants was estimated to be only two-thirds the normal size. The mother often stopped to rest every 10–20 m while traveling, and often walked bipedally for 2–3 m at a time, gripping the twins by the nape of their necks. Neither infant was seen from 1 week after the birth.

22.3.2 Locomotion and transport

For the first few days after the birth, a mother supports her infant with her hand and thighs while she is walking (van Lawick-Goodall, 1967; Zamma *et al.*, 2012) and holds the baby constantly while she is sitting or lying. After several days, the infant can cling to its mother's belly on its own by grasping the mother's hair. For the first 6–9 months, an infant is often transported by the mother in the ventro-ventral position (Nicolson, 1977) with the infant clinging to the mother's belly, grasping her sides with both hands. Infants more than 3 months old sometimes try to ride on the mother's back when she is walking, but dorsal transporting is not common until the infant is a year old. After 1 year, an infant often tries to ride on its mother's back, and the mother sometimes places the infant up on her back. At first, the infant lies prone on the mother's back with its head near the mother's neck or shoulder, and grips her hair. Later, the infant sits or stands with all four limbs on the mother's back. The former transport posture is called "ride quadrupedal" (Nishida, 2012), and the latter posture "jockey position" (van Lawick-Goodall, 1967) or "ride jockey" (Nishida, 2012).

During the period when dorsal transporting is observed, the infant starts to walk by itself near the mother. If the mother moves off a short way without the infant, the infant usually hurries after her, grabs on to her rump, and pulls itself up onto her back while she continues to walk. An infant during the weaning period is more likely to ride on the mother's back than it was several months before (Nishida *et al.*, 1990). The mother may sometimes remove an older infant from her back and encourage him or her to walk independently. Typically, a juvenile no longer rides on the mother's back when they travel, especially if the mother also has a younger infant. However, some observations have shown that a mother will carry two children at once – an infant on the belly and a juvenile on the back (e.g. Nishida *et al.*, 2010) – especially in dangerous situations.

22.3.3 Breastfeeding

Breastfeeding has an obvious adaptive value for an infant (Lee, 1996). Chimpanzee infants depend on their mother's milk for a longer period after birth, compared to other primates. Although suckling is usually thought to be a response to the infant's rooting, a more proximate explanation of suckling may be that it reduces the mother's breast pain. One female, for example, was observed to self-suckle after losing her infant (Matsumoto-Oda, 1997). The typical suckling rate is about 1.5 min per hour and an infant sucks both of the mother's nipples during a suckling bout (Matsumoto, unpublished data). However, older infants tend to begin with the left nipple and spend more time suckling this nipple; left nipple preference may be facilitated by mothers' tendency to support their infants with their left arm (Nishida, 1993; Chapter 41). Suckling time decreases by 3 years of age, but increases again at 4 or 5 years (see below for further details).

Suckling bouts are usually terminated by the infant. Aggressive maternal restriction of suckling is rare in wild chimpanzees (Nicolson, 1977). However, some bouts are interrupted or terminated by the mother jerking the nipple out of the infant's mouth.

22.3.4 Weaning conflict

In the weaning period, the conflict between mother and infant becomes intense (Hiraiwa-Hasegawa, 1990b). There are large differences among mother–infant pairs concerning the age at which the infant is weaned. In the weaning period, the mother tries to force the infant to cease nursing, while the infant tries to continue. From an evolutionary perspective, this weaning conflict may be considered the result of a difference in interests of the mother and the infant. It is beneficial for the infant to retain the mother's help as long as possible, but it is costly for the mother to continue nursing the infant if it is large enough to eat by itself – prolonged weaning may decrease or delay the potential benefits of giving birth to the next offspring.

Nishida (1990) observed occurrences of deceptive behaviors in infants being weaned, suggesting that infants may behave deceptively to induce maternal investment. For example, infants during weaning may suck the mother's nipple for a relatively longer amount of time, and try to ride on the mother's back or cling to her belly even when they travel shorter distances. Weaning conflict usually ends when the mother finally gives birth to the next baby (Goodall, 1986).

22.3.5 Food sharing

The main subject of food sharing among adults is meat (Chapter 20). Plant-food items are rarely shared among adults, but are often shared between a mother and her infant offspring (Silk, 1978). Food sharing (Figure 22.2) occurs only when the infant chimpanzee begs for food. It may be that food sharing helps infants to acquire information about food items, helping them to master the food repertoire of the natal group (Hiraiwa-Hasegawa, 1990c; Nishida and Turner, 1996).

Figure 22.2 Food sharing between a 2-year-old infant male and his mother. ©T. Matsumoto.

22.3.6 Feeding

Before being weaned, infant primates need to learn how to select suitable food from the variety of potential food items present in their environment, because of the immaturity of their digestive organs and morphology (Aiello and Wheeler, 1995). Chimpanzees are ideal subjects with which to investigate the acquisition process of food selection. Their relatively long infancy compared with other animals facilitates monitoring of the relation between food selection and developmental events (Hiraiwa-Hasegawa, 1990b).

Infants start to supplement their diet of milk with substantial amounts of solid food during their second year. By the age of 3 years, they are eating most of the food items in the adult food repertoire of the natal group. However, infants sometimes eat food items that are not typically eaten by adults, especially when their mothers are not feeding, but resting or grooming. In addition, they eat smaller amounts of mature leaves than do mothers (Matsumoto, in prep.). It is possible that development plays a part in the original selection of an infant's food items.

22.3.7 Play

Play is a regular mother–offspring interaction, with the mother initiating play in most cases (Pusey, 1978). According to Nicolson (1977) and Hiraiwa-Hasegawa (1986), the first mother–infant play is seen as early as several months after birth; the infant actively participates in play interactions and shows a play face. The frequency of play tends to increase as the infant grows older. The usual playmate of an infant less than 1 year is the mother. They play with each other even on a night bed. However, an infant less than 1 year can play with non-mother individuals (i.e. allomothers) if the mother allows. An infant over 3 years of age more often plays with other individuals than with their mother (Nishida, 2012). See

Chapters 36 and 40 for more details about social play and diversity of play.

22.4 Discussion and future perspectives

22.4.1 Developmental stages

We described four developmental stages in eastern chimpanzees, based mainly on long-term research at Mahale and Gombe. Primate social development is often divided into various stages that can be defined in terms of particular significant events in a life history, such as weaning and pregnancy (Mason, 1965). Although there are typical ages for these developmental stages, there is also a great deal of local and individual variation in chimpanzee development. For example, female age at first pregnancy is about 2 years younger at Bossou compared with that of females in other study groups (Goodall, 1983, 1986; Hiraiwa-Hasegawa *et al.*, 1984; Sugiyama, 1984; Nishida *et al.*, 1990; Boesch and Boesch-Achermann, 2000). These variations among both individuals and populations could be a result of several factors, such as the individual's social status in the group, the demographic structure of the group, or ecological differences between sites. Therefore, future studies of chimpanzee development may need to refer to wider social and ecological contexts that are interwoven with each other.

Average life history parameters of chimpanzees can often prove useful in understanding the typical pattern of development of the species, and in making comparisons with other species, especially with humans, from an evolutionary perspective. However, we should also be aware that there are few strict boundaries, or shifts, between developmental stages and there are large differences in the rate of physical and social maturity among individuals, sexes, and mother–offspring pairs. Although typical classification of developmental stages (such as the four stages described in this chapter) may give us the impression that these stages are clearly separable, this is not, of coursc, true. We should emphasize here that development is a process. The weaning period can provide a good example: as we described in this chapter, infants are confronted with a physiologically and socially complicated situation. They must keep growing, but the amount of breastfeeding gradually decreases. The mothers try to wean them, while they try to retain the mothers' protection. There is no fixed solution for this kind of conflict, and each mother–offspring pair copes with this in its own way. In addition to collecting average life history parameters using larger datasets, we may also need to focus more on the developmental process of each individual chimpanzee in order to have a complete understanding of the life history of chimpanzees.

22.4.2 Flexibility of maternal investment

In this chapter, we also described several features of the mother–offspring relationship and infant development. These are often discussed according to maternal investment theory (Trivers, 1972), which explains the mother–offspring relationship in terms of trade-offs between maternal efforts towards each offspring, and the costs of this investment in terms of subsequent reproduction. In this section, we raise two issues concerning this theory that may need to be studied in more detail in the future.

First, it is easy to assume that maternal investment contributes to the offspring mainly in infancy. One major component of maternal effort is the provision of milk to ensure offspring growth. Transport is also an important maternal investment. While it typically seems that mothers easily carry infants, a case of a physically disabled infant observed recently at Mahale (Matsumoto *et al.*, in prep.) implies that transport of an infant is quite a costly behavior, especially when the infant cannot coordinate itself to being carried properly. Although such major maternal investments are usually given only before weaning, Nakamura *et al.* (2014) showed that orphaned chimpanzee sons at Mahale died younger than expected even if they lost their mothers after weaning. Because a mother and her son live for a long time in the same group, it is possible that a chimpanzee son continues to receive some maternal investment even after weaning, in terms of sharing food patches, support in conflict with other individuals, and so on. In other words, maternal investments may not be restricted to those given to infants, such as breastfeeding and transportation. Although maternal investments in infancy are often the focus of study, we may also have to pay attention to investment after infancy, especially in long-lived mammals like chimpanzees.

Second, the flexibility of maternal care should be taken into consideration. The theory of parental investment predicts that, in dimorphic species, the offspring of the larger sex will receive greater maternal investment than the other sex, because increased investment before weaning has a greater effect on the overall fitness of individuals of one sex. Chimpanzees are sexually dimorphic, with males larger than females. Boesch (1997) reported the evidence for dominant females in Taï, Côte d'Ivoire, investing more in sons; however, in Mahale, sex-biased maternal investments before weaning have not been confirmed (Hiraiwa-Hasegawa, 1990a). There may be local and individual differences in ways of rearing infants. The case of the disabled infant at Mahale mentioned above also demonstrates the flexibility of maternal behavior. Sometimes, the mother coped with the disabled infant in unusual ways, such as laying down the infant on the ground while she groomed with others. Such flexibility, and a strong mother–infant bond, may have helped the infant to survive for up to 23 months in the wild. Although this case may be a very unusual and conspicuous example of maternal flexibility, other mothers may also change their maternal behaviors in response to the differing needs and conditions of their infants. Thus, in future studies, we may have to consider such flexibility more fully when we discuss local and individual differences in maternal care.

22.4.3 Life history of chimpanzees and insight into human evolution

Studies of primates both in captivity and in the wild have accumulated knowledge about the life history traits of various species. Comparing life history characteristics of different species living in different environments, or in different social structures, is important in understanding the evolutionary background of how they were shaped with natural selection. As the closest living species to humans, chimpanzee life history research provides important insights into human evolution. Some scholars believe that human ancestors were able to change their living habitat from that typically used by other apes, partly because humans acquired an additional developmental stage: "childhood." In humans, childhood falls between the infancy and juvenility periods of typical nonhuman primates, and is defined as the period when the child is already weaned but still strongly depends on caretakers for food and protection (Bogin and Smith, 1996). It probably first appeared in *Homo erectus* (Thompson and Nelson, 2011). In addition to childhood, humans also have a greatly elongated adolescence accompanying the conspicuous adolescent growth spurt (Bogin and Smith, 1996; Bogin, 1997). Some have claimed that such special features of human life history were crucial for the success of the repeated migrations to new habitats by our human ancestors, and for their increasing brain size.

Comparison of life history characteristics with extant apes can be a key to understanding human evolution. However, we should note that most arguments on this account often focus on the adaptation of adults. For example, they say that human ancestors "stopped" nursing earlier than other apes and "gave" weaning food to infants instead of mother's milk. In other words, immature individuals are often assumed to be passive agents accepting the changes made by adults. However, as we described in this chapter, chimpanzee infants are not always passive, but often actively ask their mothers to share plant foods, and try to eat food items that are not typically eaten by adults. In this regard, we also need to think of the possibility that when human ancestors had to make large changes in their life history variables, immature individuals, as well as adults, must have played an active role. To our knowledge, no detailed studies on wild chimpanzees have examined the mother–infant relationship from the viewpoint that the infant is also an active participant in maintaining the relationship. Further research needs to consider the weaning period – which is one of the most drastic turning points in a mother–offspring relationship – from the perspective of infants, who are also active agents in coping with various environmental factors. In order to do this, we need to monitor the developmental process of infant chimpanzees in relation to the wild environment. Although we have learned a lot in these 50 years, we definitely have more to do in order to understand the entire life history of chimpanzees.

References

Aiello, L. C. and Wheeler, P. (1995). The expensive tissue hypothesis. *Current Anthropology*, **36**, 199–221.

Bloomsmith, M. A., Kuhar, C., Baker, K., *et al*. (2003). Primiparous chimpanzee mothers: behavior and success in a short-term assessment of infant rearing. *Applied Animal Behaviour Science*, **84**, 235–50.

Boesch, C. (1997). Evidence for dominant wild female chimpanzees investing more in sons. *Animal Behaviour*, **54**, 811–15.

Boesch, C. and Boesch-Achermann, H. (2000). *The Chimpanzees of the Taï Forest: Behavioural Ecology and Evolution*. Oxford: Oxford University Press.

Bogin, B. (1997). Evolutionary hypotheses for human childhood. *Yearbook of Physical Anthropology*, **40**, 63–89.

Bogin, B. and Smith, B. H. (1996). Evolution of the human life cycle. *American Journal of Human Biology*, **8**, 703–16.

Gavan, J. A. (1953). Growth and development of the chimpanzee: a longitudinal and comparative study. *Human Biology*, **25**, 93–143.

Goodall, J. (1983). Population dynamics during a 15 year period in one community of free-living chimpanzees in the Gombe National Park, Tanzania. *Zeitschrift für Tierpsychologie*, **61**, 1–60.

Goodall, J. (1986). *The Chimpanzees of Gombe: Patterns of Behavior*. Cambridge, MA: Belknap.

Goodall, J. and Athumani, J. (1980). An observed birth in a free-living chimpanzee (*Pan troglodytes schweinfurthii*) in Gombe National Park, Tanzania. *Primates*, **21**, 545–9.

Hamada, Y., Udono, T., Teramoto, M., and Sugawara, T. (1996). The growth pattern of chimpanzees: somatic growth and reproductive maturation in *Pan troglodytes*. *Primates*, **37**, 279–95.

Hasegawa, T. (1987). [*Sexual Behavior of Wild Chimpanzees: A Comparative Psychological Study*.] Doctoral Dissertation. Tokyo: The University of Tokyo. In Japanese.

Hasegawa, T. (1991). [Sexual behavior of wild chimpanzees.] In [*The Cultural History of Primates*], ed. T. Nishida, K. Izawa, and T. Kano. Tokyo: Heibonsha, pp. 371–80. In Japanese.

Hasegawa, T. and Hiraiwa-Hasegawa, M. (1983). Opportunistic and restrictive matings among wild chimpanzees in the Mahale Mountains, Tanzania. *Journal of Ethology*, **1**, 75–85.

Hayaki, H. (1985). Copulation of adolescent male chimpanzees, with special reference to the influence of adult males, in the Mahale National Park, Tanzania. *Folia Primatologica*, **44**, 148–60.

Hayaki, H. (1988). Association partners of young chimpanzees in the Mahale Mountains National Park, Tanzania. *Primates*, **29**, 147–61.

Hayaki, H. (1992). [Grooming partners of young males in a chimpanzee unit-group at the Mahale Mountains National Park, Tanzania.] *Bulletin of the Faculty of Humanities and Sciences, Kobe Gakuin University*, **4**, 83–105. In Japanese.

Hayaki, H., Huffman, M. A., and Nishida, T. (1989). Dominance among male chimpanzees in the Mahale Mountains National Park, Tanzania: a preliminary study. *Primates*, **30**, 187–97.

Hiraiwa-Hasegawa, M. (1986). [*Development and Mother–Infant Relationship of Wild Chimpanzees in Mahale National Park*.] Doctoral Dissertation. Tokyo: The University of Tokyo. In Japanese.

Hiraiwa-Hasegawa, M. (1987). Infanticide in primates and a possible case of male-biased infanticide in chimpanzees. In *Animal Societies: Theories and Facts*, ed. Y. Ito, J. L. Brown, and J. Kikkawa. Tokyo: Japan Scientific Societies Press, pp.125–39.

Hiraiwa-Hasegawa, M. (1989). Sex differences in the behavioral development of chimpanzees at Mahale.

In *Understanding Chimpanzees*, ed. P. G. Heltne and L. A. Marquardt. Cambridge, MA: Harvard University Press, pp. 104–15.

Hiraiwa-Hasegawa, M. (1990a). Maternal investment before weaning. In *The Chimpanzees of the Mahale Mountains: Sexual and Life History Strategies*, ed. T. Nishida. Tokyo: University of Tokyo Press, pp. 257–66.

Hiraiwa-Hasegawa, M. (1990b). A note on the ontogeny of feeding. In *The Chimpanzees of the Mahale Mountains: Sexual and Life History Strategies*, ed. T. Nishida. Tokyo: University of Tokyo Press, pp. 277–83.

Hiraiwa-Hasegawa, M. (1990c). The role of food sharing between mother and infant in the ontogeny of feeding behavior. In *The Chimpanzees of the Mahale Mountains: Sexual and Life History Strategies*, ed. T. Nishida. Tokyo: University of Tokyo Press, pp. 267–75.

Hiraiwa-Hasegawa, M., Hasegawa, T., and Nishida, T. (1984). Demographic study of a large-sized unit-group of chimpanzees in the Mahale Mountains, Tanzania: a preliminary report. *Primates*, **25**, 401–13.

Kawanaka, K. (1993). Age differences in spatial positioning of males in a chimpanzee unit-group at the Mahale Mountains National Park, Tanzania. *Primates*, **34**, 255–70.

van Lawick-Goodall, J. (1967). Mother–offspring relationships in free-ranging chimpanzees. In *Primate Ethology*, ed. D. Morris. London: Weidenfeld & Nicolson, pp. 287–346.

Lee, P. C. (1996). The meanings of weaning: growth, lactation, and life history. *Evolutionary Anthropology*, **5**, 87–98.

Mason, W. A. (1965). The social development of monkeys and apes. In *Primate Behavior: Field Studies of Monkeys and Apes*, ed. I. DeVore. New York: Holt, Rinehart and Winston, pp. 514–43.

Matsumoto-Oda, A. (1995). First record of a twin birth in chimpanzees of the Mahale Mountains National Park, Tanzania. *African Study Monographs*, **16**, 159–64.

Matsumoto-Oda, A. (1997). Self-suckling behaviour by a wild chimpanzee. *Folia Primatologica*, **68**, 342–3.

Matsusaka, T. (2003). [Where is my mother? A story of a chimpanzee who got lost.] *Mahale Chimpun*, **2**, 5–6. In Japanese.

Matsusaka, T., Nishie, H., Shimada, M., *et al.* (2006). Tool-use for drinking water by immature chimpanzees of Mahale: prevalence of an unessential behavior. *Primates*, **47**, 113–22.

Nakamura, M. and Nishida, T. (2013). Ontogeny of a social custom in wild chimpanzees: age changes in grooming hand-clasp at Mahale. *American Journal of Primatology*, **75**, 186–96.

Nakamura, M. and Nishie, H. (2012). Death of the oldest female at Mahale and some notes about longevity of wild chimpanzees. *Pan Africa News*, **19**, 19–21.

Nakamura, M., Hayaki, H., Hosaka, K., Itoh, N., and Zamma, K. (2014). Orphaned male chimpanzees die young even after weaning. *American Journal of Physical Anthropology*, **153**, 139–43.

Nicolson, N. A. (1977). A comparison of early behavioral development in wild and captive chimpanzees. In *Primate Bio-Social Development*, ed. S. Chevalier-Skolnikoff and F. E. Poirier. New York: Garland, pp. 529–60.

Nishida, T. (1980). The leaf-clipping display: a newly-discovered expressive gesture in wild chimpanzees. *Journal of Human Evolution*, **9**, 117–28.

Nishida, T. (1983). Alloparental behavior in wild chimpanzees of the Mahale Mountains, Tanzania. *Folia Primatologica*, **41**, 1–33.

Nishida, T. (1988). Development of social grooming between mother and offspring in wild chimpanzees. *Folia Primatologica*, **50**, 109–23.

Nishida, T. (1990). Deceptive behavior in young chimpanzees: An essay. In *The Chimpanzees of the Mahale Mountains: Sexual and Life History Strategies*, ed. T. Nishida. Tokyo: University of Tokyo Press, pp. 285–90.

Nishida, T. (1993). Left nipple suckling preference in wild chimpanzees. *Ethology and Sociobiology*, **14**, 45–51.

Nishida, T. (1994a). Review of recent findings on Mahale chimpanzees: implications and future research directions. In *Chimpanzee Cultures*, ed. R. W. Wrangham, W. C. McGrew, F. B. M. de Waal and P. G. Heltne. Cambridge MA: Harvard University Press, pp. 373–96.

Nishida, T. (1994b). [*Thirty-six Stories of Wild Chimpanzees*.] Tokyo: Kinokuniya. In Japanese.

Nishida, T. (2002). [Demography.] In [*The Mahale Chimpanzees: 37 Years of <Panthropology>*], ed.

T. Nishida, S. Uehara, and K. Kawanaka. Kyoto: Kyoto University Press, pp. 171–202. In Japanese.

Nishida, T. (2003a). Harassment of mature female chimpanzees by young males in the Mahale Mountains. *International Journal of Primatology*, **24**, 503–14.

Nishida, T. (2003b). Individuality and flexibility of cultural behavior patterns in chimpanzees. In *Animal Social Complexity and Intelligence*, ed. F. B. M. de Waal and P. Tyack. Cambridge MA: Harvard University Press, pp. 392–413.

Nishida, T. (2012). *Chimpanzees of the Lakeshore: Natural History and Culture at Mahale*. Cambridge: Cambridge University Press.

Nishida, T. and Hiraiwa, M. (1982). Natural history of a tool-using behavior by wild chimpanzees in feeding upon wood-boring ants. *Journal of Human Evolution*, **11**, 73–99.

Nishida, T. and Turner, L.A. (1996). Food transfer between mother and infant chimpanzees of the Mahale Mountains National Park, Tanzania. *International Journal of Primatology*, **17**, 947–68.

Nishida, T. and Wallauer, W. (2003). Leaf-pile pulling: an unusual play pattern in wild chimpanzees. *American Journal of Primatology*, **60**, 167–73.

Nishida, T., Takasaki, H., and Takahata, Y. (1990). Demography and reproductive profiles. In *The Chimpanzees of the Mahale Mountains: Sexual and Life History Strategies*, ed. T. Nishida. Tokyo: University of Tokyo Press, pp. 63–97.

Nishida, T., Corp, N., Hamai, M., *et al.* (2003). Demography, female life history, and reproductive profiles among the chimpanzees of Mahale. *American Journal of Primatology*, **59**, 99–121.

Nishida, T., Zamma, K., Matsusaka, T., Inaba, A., and McGrew, W. C. (2010). *Chimpanzee Behavior in the Wild: An Audio-Visual Encyclopedia*. Tokyo: Springer.

Nishie, H. (2011). Natural history of *Camponotus* ant fishing by the M group chimpanzees at the Mahale Mountains National Park, Tanzania. *Primates*, **52**, 329–42.

Pusey, A. E. (1978). *The Physical and Social Development of Wild Adolescent Chimpanzees (Pan troglodytes schweinfurthii)*. Ph.D. thesis. Stanford, CA: Stanford University.

Pusey, A. E. (1990). Behavioural changes at adolescence in chimpanzees. *Behaviour*, **115**, 203–46.

van de Rijt-Plooij, H. H. C. and Plooij, F. X. (1987). Growing independence, conflict and learning in mother–infant relations in free-ranging chimpanzees. *Behaviour*, **101**, 1–86.

van de Rijt-Plooij, H. H. C. and Plooij, F. X. (1988). Mother–infant relations, conflict, stress and illness among free-ranging chimpanzees. *Developmental Medicine and Child Neurology*, **30**, 306–15.

Ross, C. (1998). Primate life histories. *Evolutionary Anthropology*, **6**, 54–63.

Silk, J. B. (1978). Patterns of food sharing among mother and infant chimpanzees at Gombe National Park, Tanzania. *Folia Primatologica*, **29**, 129–41.

Sugiyama, Y. (1981). Observations on the population dynamics and behavior of wild chimpanzees at Bossou, Guinea, in 1979–1980. *Primates*, **22**, 435–44.

Sugiyama, Y. (1984). Population dynamics of wild chimpanzees at Bossou, Guinea, between 1976 and 1983. *Primates*, **25**, 391–400.

Takasaki, H. (1985). Female life history and mating patterns among the M group chimpanzees of the Mahale National Park, Tanzania. *Primates*, **26**, 121–9.

Thompson, J. L. and Nelson, A. J. (2011). Middle childhood and modern human origins. *Human Nature*, **22**, 249–80.

Trivers, R. (1972). Parental investment and sexual selection. In *Sexual Selection and the Descent of Man 1871–1971*, ed. B. Campbell. Chicago, IL: Aldine, pp. 136–79.

Uehara, S. and Nishida, T. (1987). Body weights of wild chimpanzees (*Pan troglodytes schweinfurthii*) of the Mahale Mountains National Park, Tanzania. *American Journal of Physical Anthropology*, **72**, 315–21.

Uehara, S. and Nyundo, R. (1983). One observed case of temporary adoption of an infant by unrelated nulliparous females among wild chimpanzees in the Mahale Mountains, Tanzania. *Primates*, **24**, 456–66.

Yerkes, R. M. and Tomilin, M. I. (1935). Mother–infant relations in chimpanzee. *Journal of Comparative Psychology*, **20**, 321–59.

Zamma, K., Sakamaki, T., and Kitopeni, R. S. (2012). A wild chimpanzee birth at Mahale. *Pan Africa News*, **19**, 3–5.

23 Gerontology

KAZUHIKO HOSAKA AND MICHAEL A. HUFFMAN

23.1 Introduction

Nonhuman primates, captive macaques in particular, have been widely studied as models of human aging with regard to reproductive physiology and cognition, while studies in apes, our closest relatives, are lacking in the wild due to their long lifespan (Erwin and Hof, 2002; Lacreuse and Herndon, 2009). Because long-term studies of apes, such as those at Gombe, Tanzania, and Mahale, have reached their 50th year, it is now possible to analyse the lifelong history of individual chimpanzees in terms of activities, reproduction, and social behaviors.

First, the difference between "aging" and "old age" should be clarified. Every animal experiences the multifaceted development of reproductive, physical, and cognitive functions, and upon reaching maturity, biological aging begins with the gradual deterioration of cells, organs, bones, neurons, and the brain. That is, if any of these functions peak after maturity, it is reasonable to state that aging has started even before the peak of that function. In the case of chimpanzees, which like ourselves are highly social animals, we may introduce the concept of "social aging," whereby physical aging is associated with a reduction in social activities.

On the other hand, old age is one stage of an animal's life history. Therefore, one can ask the question "Is there a certain age beyond which the animal is classified as being old aged?" The debate as to whether or not female chimpanzees experience menopause, as we will see below, is an example of one such question. If there is a certain biological age around which reproduction, including the rearing of offspring, ends, it may be appropriate to define the period after that age until death as old age.

However, because aging can occur in reproduction and in several other physical and social aspects, it is reasonable to look at old age multi-dimensionally. For example, a male can begin to decline in the dominance hierarchy after spending years in a higher rank due to some social or physical reason (Chapters 5, 27). Thus, old age in the social dimension ("socially aged") could be determined by each individual's history and not by species-specific age. It is interesting to see if there is any social status specific to such socially aged chimpanzees, as well as "elders" or "seniors" among humans.

Considering that aging is a gradual process, it may be more productive to view the beginning of old age as a transitional shift rather than a discontinuous switch from prime age. This may be important when discussing the relationship between "aging" and "old age." Thus, we will discuss a broad concept for old age in addition to a traditional, narrower concept.

23.2 Longevity of chimpanzees in the wild

How long do chimpanzees live? To answer this question as accurately as possible it is necessary to discuss reproduction, sexuality, and life after the cessation of reproduction of old female chimpanzees in comparison to the exceptional postmenopausal longevity of humans.

The oldest chimpanzee ever reliably recorded was Gamma, a female in the Yerkes Regional Primate

Mahale Chimpanzees: 50 Years of Research, ed. M. Nakamura, K. Hosaka, N. Itoh and K. Zamma. Published by Cambridge University Press.

Research Center, who died at the age of 59.4 years (Hakeem *et al.*, 1996). Her menstrual bleeding was last confirmed at the age of 53 years, while she showed anogenital swelling until the age of 57. Furthermore, her overall health condition had been good in her final 2 years (Lacreuse *et al.*, 2008). The oldest chimpanzee in Japan is Johnny of Kobe Oji Zoo, a male who is presumed to be 64 years old, according to a press release of the Kobe city government dated July 16, 2014. He came to the zoo in 1955 at the estimated age of 5 years. The press release reported that Cherry, Johnny's first child who was a 51-year-old female, had died at the zoo 2 days previously.

When field research began at Gombe, scientists assumed that chimpanzees cannot live as long in the wild as in captivity, and Teleki *et al.* (1976) reported that the maximum age at death was estimated to be 33 years for males and 37 years for females based on 10-year data. Such underestimation of longevity was probably caused by an underestimation of age at the initial stage of long-term research at Gombe. It seems likely that researchers later noticed this discrepancy and Williams *et al.* (2008) made an upward revision of the age at death of a famous old female, Flo, from 43 (Goodall, 1986) to 53 years.

This was also true for the first 25 years of research undertaken in Mahale. It seems that the age of chimpanzees estimated at that time was biased toward them being younger (Nakamura and Nishie, 2012), presumably because researchers referred to the early demography data of Gombe chimpanzees. Thus, it is possible that Mahale chimpanzees who were identified in the 1970s and 1980s who looked "old" were estimated to be in their thirties. Eventually, at the pioneering stage, the ages of the oldest males and females were estimated to be 30 and 31, respectively, in the K group, and 28 and 36, respectively, in the M group.

Over time researchers have realized that wild chimpanzees can live to be over 50 years old (Emery Thompson *et al.*, 2007; Nakamura and Nishie, 2012; Nishida, 2012). In the Mahale K group, there were one male and two females whose estimated ages reached 40; the oldest male was 42 and the oldest female was 46 (Tables 23.1, 23.2). In the M group, 4 males and 20 females reached the estimated age of 40 years (Tables 23.1, 23.2). Kalunde is the oldest male, who reached the age of 51. Wakusi was the oldest female who reached the age of 53. However, taking into account estimation bias at earlier stages of research, there should have been more chimpanzees that reached their forties (see Chapter 7). Furthermore, according to our own observations, several individuals, such as Wanaguma, Happy, Wansombo, and Kalindimya looked very old and they most likely reached their fifties.

When we studied Mahale chimpanzees intensively from 1991 to 1993, there were three senior males, Bakali, Ntologi, and Musa (Chapters 5, 20, 27, 31). They were identified as peer young adult males when M group was habituated at the beginning of the 1970s. Among these, only Ntologi lived to be 40, according to the estimated ages. However, Musa clearly looked old, and researchers had no doubt of this when they described him as an "old male" (e.g. Nishida, 1994, 2012; Hosaka *et al.*, 2000; Shimizu *et al.*, 2002; Gunji *et al.*, 2003). Therefore, in this chapter, we describe all three of these chimpanzees as old males.

23.3 Physical aging

23.3.1 Pathology

Considerable research has been carried out on the aging process in nonhuman primates in captivity as a model of pathological aging in humans (Erwin and Hof, 2002). However, for wild chimpanzees, knowledge about physical changes or functional decline with aging remains limited.

At Mahale, the dead bodies of two old chimpanzees, Wansombo and Musa, were collected in 1993 and osteological analyses were conducted (see Chapter 45). Of note, Wansombo was special in that the whole skeleton was obtained because she was found weak and unable to move along the trail. She died about 10 h after being discovered (Hosaka *et al.*, 2000). When physical anthropologists examined her skeleton, it was found that her bone mineral density was abnormally low (Gunji *et al.*, 2003), and that traces of periodontal disease were conspicuous (Shimizu *et al.*, 2002).

Table 23.1 **List of female chimpanzees at Mahale that lived to the age of 40 and their reproductive profile**

Name	Unit-group	Year of birth	Year of first observation	Estimated age at first observation	Age at death/ disappearance	Age at last birth	Age at last estrus	Post-reproductive lifespan (PRL)[a]	Post-cycling lifespan (PCL)[b]
Wakusi (WX)	M	1961[c]	1979?	18	53[d]	44[e]	52[e]	8	–[f]
Calliope (CA)	M	1960[c]	1980?	20	52	37	51	10	1
Gwekulo (GW)	K→M	1962[c]	1972	10	52[d]	–[g]	40	–[g]	11[d]
Fatuma (FT)	M	1963[c]	1974	11	51[d]	47[e]	46[e]	–[h]	4[f]
Wanaguma (NG)	M	1940[c]	1973	33	50	Unknown	46	?	4
Dar (DA)	M	1945[c]	1980?	35	49	43	49	1	0
Ikocha (IK)	M	1965[c]	1979?	14	49[d]	44[e]	48[e]	0	–[f]
Ganmo (GA)	M	1940[c]	1974	34	48	39	48	5	0
Eagnes (EA)	M	1945[c]	1979?	34	48	47	45	0	3
Happy (HA)	M	1945[c]	1981?	36	47	35	47	7	0
Wamikambi (WM)	K	1936[c]	1966	30	46	36	44	5	2
Gwepala (GP)	M	1940[c]	1973	33	45	38	44	2	1
Nkombo (NK)	K→M	1970	1970	0[i]	44[d]	23[e]	43[e]	21	–[f]
Wabwema (WW)	K	1936[c]	1967	31	43	41	?	1	?
Wabunengwa (WB)	K→M	1943[c]	1966	23	43	31	40	9	3
Wangala (WN)	K→M	1945[c]	1966	21	43	33	39	5	4
Wankuma (UM)	M	1945[c]	1979?	34	42	35	41	6	1

Wansombo (BO)	M	1952[c]	1973	21	42	36	41	4	1
Silafu (SL)	M	1955[c]	1973	18	42	39	39	1	3
Mpoko (MP)	M	1945[c]	1979?	34	40	31	40	?	0
Tango (TG)	M	1950[c]	1973	23	40	20	?	10	?
Wakasunga (WA)	K→M	1954[c]	1966	12	40	30	39	1	1

[a] Defined as the number of years from reproductive cessation to the female's death. Reproductive cessation comes two ways: one is when the last offspring reached age 5, the other is when the last offspring died.

[b] Defined as the number of years from cycling cessation to the female's death.

[c] Presumed on the basis of estimated age at first observation. Some data are different from those published by Nishida *et al.* (2003), in which ages were estimated by assumption of female age at immigration and first birth.

[d] Alive and to be updated.

[e] Nurturing or cycling.

[f] Cycling.

[g] Presumed to be sterile.

[h] Nurturing.

[i] Nkombo is the only female whose birth year and mother were known in this list.

Table 23.2 **List of male chimpanzees at Mahale that lived to the age of 40**

Name	Unit-group	Year of birth[a]	Year of first observation	Estimated age at first observation	Age at death/ disappearance
Kalunde (DE)	M	1963	1973?	10	51[b]
Kalindimya (KL)	M	1938	1966?	28	48
Kamemanfu (KM)	K	1940	1966	26	42
Mimikire (MK)	M	1938	1966?	28	41
Ntologi (NT)	M	1955	1972	17	40

[a] Presumed.
[b] Alive in 2014.

Presbyopia and cataracts are age-related symptoms that can be observed in the wild. Wakusi, around her forties, started to stretch her hands to have a comfortable distance between eyes and fingers when she groomed others (Toshisada Nishida, personal communication; see also Figure 5.34). Similar behavior was also observed for Calliope (Nakamura and Nishie, 2012). As video data of grooming behavior have accumulated for years, it is possible to investigate when presbyopia began for those females. In their later years, Wanaguma and Wakasunga had cloudiness in their eyes that appeared to be cataracts (our observations; see also Nishida, 2012). However, Nakamura (personal communication) observed that Sally was beginning to show the same symptom in one of her eyes at the estimated age of 26 years in 2012. It is uncertain whether her estimated age is too young or her symptom was unrelated to aging.

23.3.2 Brain science

Can age-related structural and functional changes in brains be observed in chimpanzees as well as in humans? Brain science may help us to understand the behavior and sociality of old chimpanzees.

Herndon *et al.* (1999) investigated the brain weight of 76 chimpanzees who died at the Yerkes Center. Their data covered individuals from infants to the aged, the oldest of which were 59.4 years (females) and 45 years (males). The results show that the weight of chimpanzee brains decreases with age at a rate of about 0.9 g/year, which is comparable to that observed in rhesus macaques but is much slower than the rapid decrease in brain weight observed in aged humans. This means that although there is a gradual decrease of brain weight with advancing age, there is no evidence of a significant reduction in brain weight that is typical of old age in chimpanzees. In support of this view, studies examining age effects on cerebral cortex shrinkage using MRI (Sherwood *et al.*, 2011) and on gene expression in the cerebral cortex (Fraser *et al.*, 2005) have concluded that the patterns of age-related changes in the brain differ significantly between humans and chimpanzees. Therefore, from the perspective of brain science, human old age is relatively new in evolutionary history.

23.4 Reproduction and sexuality of aged chimpanzees

23.4.1 Post-reproductive lifespan

The fact that humans spend many years in their old age even after reproduction has ceased has attracted the interest of evolutionary biologists. For example, Hamilton (1966) noted the long post-reproductive lifespan of women and explained that it may evolve if older women still have beneficial effects on the survival and reproduction of their descendants. Under such a theoretical framework, the "grandmother hypothesis" was proposed by anthropological ecologists studying hunter-gatherers (Hawkes *et al.*, 1997), which has gained considerable interests among scholars. Hawkes *et al.* (1997) showed that the foraging time of Hadza women was positively correlated with body weight gain of the weaned grandchild. They argued that the "grandmother effect" helps to shorten the interbirth interval of the daughter, which favors the evolution of an extended postmenopausal lifespan. Such an explanation presupposes that apes have neither grandmother support for child rearing nor an extended postmenopausal lifespan. Therefore, researchers have awaited the answer to this question in chimpanzees, which are the closest living relatives to humans.

By analysing long-term data from Mahale, Nishida *et al.* (1990) concluded that chimpanzees experience a non-negligible post-reproductive lifespan. They implied that chimpanzees have a menopause, but noted that female chimpanzees did not share parenting with daughters, arguing that the evolution of menopause is not unique to humans but is an outcome of sexual selection in mammals. A similar argument was made by Nishida *et al.* (2003) when the demographic data were updated. Among 25 females observed from middle age (18–33 years) to death in older age (31–48 years), eight females (23.5%) lived for more than 5 years, a period sufficiently long for maternal care, after bearing their last offspring. Nishida *et al.* (2003) argued that these females experienced post-reproductive lives, while Nishida (2012) admitted that these data clearly show that the post-reproductive lifespan of chimpanzees is much shorter than that of humans.

Takahata *et al.* (1995) showed that Japanese macaques (*Macaca fuscata*) in the Arashiyama B troop also had a post-reproductive lifespan similar to that observed in Mahale chimpanzees. There was an average time lag of 6.0 years from the last delivery of offspring to death in females over the age of 20 years. Subtracting 1.5 years that is required for parenting, those macaques had a 4.5-year post-reproductive lifespan, which was equivalent to about 16% of their average lifespan of 27.3 years. Excluding the oldest female, female macaques in the post-reproductive lifespan continued to copulate with males, suggesting the existence of excessive sexual behavior in Japanese macaques. Thus, the authors claimed that there is a need to distinguish between the end of reproduction (the cessation of parturition) and the end of sexuality (the cessation of estrus). The former is related to the question of whether menopause exists in nonhuman primates and the latter deals with a perspective on sexuality in nonhuman primates that may continue after the loss of reproductive ability due to aging, whether it can be called menopause or not. Such a distinction may be useful to reevaluate non-reproductive sexuality among nonhuman primates that was believed to be unique to humans (Section 23.4.3).

23.4.2 Do chimpanzees experience menopause?

Although Gould *et al.* (1981) concluded that chimpanzees do not experience menopause based on analyses of menstrual cycles and hormone dynamics, the controversy has persisted as these results were based on small sample sizes.

Videan *et al.* (2006) studied menstrual cycles and hormone dynamics of 14 captive female chimpanzees and argued that chimpanzees experience menopause from a physiological standpoint. They showed that the ratio of days with maximal tumescence in a menstrual cycle decreased as chimpanzees became older, the cycle was prolonged, and days with detumescence in a cycle increased. Furthermore, based on age-related changes in the levels of luteinizing hormone (LH) and follicle-stimulating hormone (FSH), they argued that chimpanzees reach perimenopause at the age of 30–35 years and menopause at the age of 35–40.

On the other hand, Emery Thompson *et al.* (2007) provided contrasting evidence by analysing the long-term data regarding fertility and mortality in wild female chimpanzees from six study sites, and argued for the grandmother hypothesis. The authors concluded that menopause was not a characteristic stage of chimpanzee life history. Admitting that chimpanzee birth rates follow a similar pattern of decline to that of humans until the beginning of their forties, the authors emphasized that female chimpanzees continue to give birth throughout their entire lifespans, providing they are in good health. Even though chimpanzees live for more than 50 years, they live for shorter periods than do humans. Thus, the life of most females of this species is focused on reproduction. Considering that it is common for human females to live a healthy life for several decades after menopause, it is plausible that human and chimpanzee females employ different life history strategies.

23.4.3 When does sexuality end?

Even if healthy chimpanzees in old age continue to reproduce, is it possible that chimpanzees maintain post-reproductive sexuality as do Japanese macaques? The gap between the post-reproductive lifespan (PRL) and the post-cycling lifespan (PCL) may provide some clarification (Table 23.1). PRL is defined as the period between reproductive cessation and death. Two types of reproductive cessation are defined according to Nishida *et al.* (2003): one is when the last offspring survives for 5 years, the assumed period of maternal care, and the other is when the last offspring dies. PCL is defined as the period between the last estrus to death.

In 1992, Happy (Table 23.1), who looked very old, showed estrus and copulated only in February and September (Nishida, 1997) but disappeared after the last estrus. Thus, her PCL was 0 years. She did not give birth after her son, Hanby, who was 12 years old when she died. Therefore, her PRL was 7 years. It seems possible that Happy maintained post-reproductive sexuality for 7 years despite the difficulty in confirming when she actually lost her reproductive ability.

Conversely, some females at Mahale ceased estrus and copulation well before death. For example, Gwekulo showed her last estrus in December 2002 when her estimated age was 40 years. Interestingly, she did not show any signs of estrus over the following 11 years (Table 23.1) although she was healthy enough to coordinate social and foraging activities with the core males of M group. Researchers have always believed that she is sterile (Chapters 5, 30), although the reason for her presumed sterility is unknown. Whether it is an endocrine, traumatic, or pathological problem responsible for inhibiting her ovulation or fertilization, it is possible that the same factor caused her early cessation of estrus. She was, nevertheless, receptive until her thirties. It seems unusual for a female showing no apparent decline in health to cease sexual behaviors. Hasegawa (1989) and Takahata *et al.* (1995) showed that old females, Wanaguma and Wabunengwa, showed lower frequencies of copulation than other females. Wabunengwa did not show copulation at all even when she was in estrus (Takahata *et al.*, 1995).

23.4.4 Reproductive activities in aged chimpanzees

As discussed, it is evident that female chimpanzees over 40 years of age are active in their sexual behaviors as long as they are healthy, and it is possible that some females can reproduce even after 50 years of age. Furthermore, older females seem to have high sexual attractiveness in that males prefer mating with older, not younger, females (Muller *et al.*, 2006). Such a tendency may evolve if the older females are more capable of child rearing than their younger counterparts, since the survival of offspring depends on maternal care in chimpanzees.

As of February 2014, there are five females over 40 years of age in Mahale (Table 23.1). Among these, Fatuma, gave birth in January 2010 at age 47, the oldest ever recorded at Mahale. It will be interesting to observe what happens in terms of her reproductive life after her son is weaned. Similarly, Ikocha, who had given birth in August 2009 at the estimated age of 44, resumed estrus in September 2013 and if she succeeds to give the next birth, she will break the oldest birth record at Mahale.

Compared to females, male sexuality is relatively simple. For example in September 2014, the current oldest male Kalunde, whose estimated age is 51, followed some estrous females and even succeeded in a copulation with a newly immigrated female (Masaki Shimada, personal communication). In 1991–3, old males Bakali and Musa showed similarly high sexual activities despite their low dominance ranks (Nishida, 1997; Hosaka, unpublished data).

However, estrous females tend to avoid copulation with old males (Nishida, 1997). Pusey (1980) assumed that this is a mechanism of incest avoidance between a female remaining in her natal group and her father, but Nishida (1997) opposed this idea based on his finding that all females avoiding copulation with old males were non-natal. Instead, he speculated that females avoid it because the sperm quality of the older male may be degraded due to aging.

23.5 Social status of aged chimpanzees

23.5.1 Behavior and sociality of aged chimpanzees

Studies on macaques have shown that in comparison to younger individuals, old animals increase the distance and decrease interactions with other individuals (Hauser and Tyrrell, 1984; Nakamichi, 1984). This pattern also holds true for chimpanzees according to Huffman (1990) who studied behavioral manifestations of old age in Mahale chimpanzees. He was, however, cautious in seeing relative age as an absolute indicator of behavioral or social performance, because variations in life history such as the presence or absence of offspring, health, non-kin social relationships, and reproductive state are also important factors.

At the same time, Huffman (1990) found that old females tended to join a core nomadic party more than younger adult females do. This leads to a narrowing of their network to grooming mainly with a few prime adult males. Furthermore, he showed that old males, despite their declining rank, remained associated with males in the core nomadic party.

Old individuals of each sex are not reciprocal groomers (Huffman, 1990). They receive grooming one-sidedly and perform less mutual (simultaneous bidirectional) grooming than younger animals. Kawanaka (1989) showed that adolescent or young adult males unidirectionally groom elders while the alpha and other prime adult males often groom reciprocally with each other (Chapter 27). If we regard such reciprocity in grooming as part of social maturity, passive grooming of old individuals is in good contrast to that of younger animals. Further research is necessary to confirm whether this pattern can be generalized, but it is tempting to investigate the social meaning of grooming because there seems to be some particular social role or status of old age as we see in the following section.

23.5.2 Does any social role emerge as a function of old age in nonhuman primates?

McDonald Pavelka (1990) pointed out that many studies on nonhuman primates have presumed the social roles and status of old individuals and questioned the existence of a behaviorally distinct subgroup of "old females" in Japanese macaques. She concluded that there were no clear differences in behavior and social interactions of old females with other adult females. Thus, she argued that biological aging does not create any distinct social situation in which any role specific to old females emerges.

Even if social class is not associated with old age in macaques, one may expect it in chimpanzees. Although such social status or role could easily be explained by personality or long tenure, rather than old age, it may be useful to see similar cases from observations in the wild to examine whether old age is a meaningful category or not.

23.5.3 Tolerance of alpha males toward old allies in agonistic interactions

In 1989–93, two old males, Bakali and Musa, were important allies of the two alpha males, Ntologi and Kalunde (Chapter 27), both of whom preferred constant proximity with these two "socially aged" males although their physical strength and dominance rank were declining. It is evident that these alphas prevented these old allies from forming affiliative relations with rivals. For instance, Ntologi showed repeated separating interventions when Bakali was with his rival, Nsaba (Nishida and Hosaka, 1996).

There were cases in which adult males tended to tolerate such old males (e.g. Kalindimya and Kagimimi; Huffman, 1990). For example, Hayaki *et al.* (1989) reported that an old male, Kagimimi, even attacked or threatened higher-ranking young males without suffering any counterattacks. Similarly, Nishida (1994) noted that when the alpha male, Kalunde, interrupted an old ally, Musa, from snatching meat from a female, Kalunde was chased away by Musa and then screamed in a tree. However, alpha males are not always tolerant of old allies. In the same period, Hosaka (personal observation) observed that Musa was counter-charged by Kalunde after Musa charged Kalunde due to conflict over access to an estrous female. In February 1992, 1 month after Ntologi came back to the M group as the alpha male, Ntologi and Bakali sat side by side staring at an estrous female in a tree. Bakali slowly approached the female ahead of Ntologi. Next, Ntologi swiftly caught up with and kicked Bakali down from the tree. As such, alpha males are not always tolerant of old allies in terms of access to estrous females.

A similar tolerance toward old individuals is common when chimpanzees share meat (see Chapter 20 for details; Huffman, 1990). In Section 23.6, we will consider issues regarding agonistic interactions and meat sharing together and discuss why socially aged individuals are tolerated in chimpanzee society.

23.5.4 Relationships between socially aged and immature chimpanzees

Hayaki (1988) reported that orphan chimpanzees often follow particular mid- to low-ranking males (Chapter 30). Some, if not all, males followed by the orphans are old males. Takahata (1990) pointed out that males followed by immature males (including non-orphans) are calm natured and frequently interact with young males. Thus, it is possible that former high-ranking males will begin to act like "easy-to-follow" males once they become socially aged.

Bakali, the beta male in the early 1980s (Takahata, 1990), became a "good old" playmate of youngsters when he became low-ranking in his later years. Table 23.3 describes Bakali's play in 1991 (total observation time during focal animal sampling: 34.4 h). In addition to some "political" plays with Kalunde or Ntologi, it is evident that he played with a wide range of youngsters, infants, and adolescent males in particular.

Wakasunga's son, Linta, was Bakali's favorite playmate and they played together in eight sessions during this period (Table 23.3; Figure 23.1). The frequency of play sessions with orphans (7-year-old Chopin, 9-year-old Nick, and 14-year-old Masudi) was significantly higher than with young males with live mothers ($\chi^2 = 12.3$, $p < 0.005$, $df = 1$), although this may be due to better association with the former. Whether such relationships between old males and youngsters contribute to the survival of the orphans needs to be examined in future studies.

In contrast, another old male, Musa, played less frequently with any partners, including young males. In 1991, he played with only three young males during 97.3 h of focal follows. However, he exhibited some behaviors that Bakali did not, namely "directing-to-stay behavior" and "punishment."

"Directing-to-stay behavior" was observed once in 1991 when Musa was resting with an orphan male, Nick, who was a frequent follower of Musa at that time. When other adult males approached Musa and started to groom him, Nick looked nervous and was about to leave the place slowly, as if not to be noticed by the males. Musa, while being groomed by others, quickly extended his hand toward Nick and grabbed him by the back. Nick collapsed on his belly for a moment then sat up. Next, Musa started to groom him on the back. As a result, Nick stayed inside the adult males' grooming cluster. It is uncertain whether Musa intentionally invited Nick to the cluster, but this anecdote is worth noting, because in this manner orphan males may benefit from following old males so that they can assimilate themselves into male bonding groups.

"Punishment" was observed in October 1991. When Musa and the alpha male Kalunde finished feeding

Table 23.3 **Play partners of Bakali in August 1991–March 1992 (total observation time during focal animal sampling: 34.4 h)**

Name	Sex[a]	Age	No. sessions[b]	Total duration (min)	Mean duration[c] (min)	Remark
Ntologi	M	36[d]	1	4.5	4.5	Just after Ntologi regained alpha status
Kalunde	M	28[d]	5	14.6	3.7	All sessions occurred only when Kalunde maintained his alpha status
Masudi	M	14	3	6.1	3.1	Orphaned a year before
Bonobo	M	10	1	–	–	
Alofu	M	9	3	10.2	10.2	
Nick	M	9	2	1.1	1.1	Orphaned two years before
Linta	M	7	8	6.7	2.2	
Chopin	M	7	3	7.3	2.4	Orphaned a year before
Carter	M	6	1	10.2	10.2	
Argo	M	2	8	25.6	3.2	
Silafu	F	36[d]	2	2.2	2.2	
Ai	F	3	1	–	–	
Maggie	F	4	1	–	–	Orphaned a year before
Serena	F	4	5	21.5	4.3	Daughter of Silafu
Tamara	F	3	1	3.3	3.3	
All			45	113.3	3.7	

[a] M, Male; F, Female.
[b] A session of play was defined as a set of bouts in maximum intervals of 3 min (Hayaki, 1985).
[c] Defined as the mean value that divides total duration by number of sessions. The calculation was based exclusively on sessions where the times of both the start and end of play interactions were recorded.
[d] Presumed.

and were descending from the tree, the 10-year-old Bonobo defecated above them and a dropping hit Kalunde's right shoulder. After reaching the ground, Kalunde wiped the feces with a leaf and emitted a cough bark at Bonobo 1 min later. Musa, who was sitting next to Kalunde, then charged Bonobo, pushed him down, and bit him on the back. Bonobo screamed. Musa returned to Kalunde and started to groom his shoulder. Twenty-eight minutes later, when Bonobo was about to walk away from the two males resting in 3 m proximity, Musa got up, chased and barked at Bonobo, who then screamed. This anecdote may be cautiously interpreted as a charging instigated by Kalunde's mild threat against Bonobo. However, it could be interpreted as punishment for bad behavior, whether or not Musa had such an intention.

Relationships between old males and youngsters, as described above, have never been elaborated to include the adoption of infants by males at Mahale. On the other hand, some childless adult females formed stable relationships, and acted like "foster mothers" toward orphans (Chapter 30).

Figure 23.1 Bakali (front center) playing finger wrestling with his favorite playmate, Linta. ©K. Hosaka.

23.6 Conclusion and future perspectives

Interdisciplinary comparisons of life history between great apes and humans suggest that chimpanzees have no "old age" in the same sense as humans do. That is, there seems to be no discrete age after which chimpanzees lose their own reproductive ability and assume a new role to assist reproduction of their offspring. Rather, chimpanzee aging may mirror standard aging patterns observed in nonhuman mammals whose reproductive functions decline gradually in the same way as do somatic cells and the brain.

However, it would be too hasty to conclude that chimpanzees have no old age. Aging is a multi-dimensional phenomenon. It is still possible to explore social status or the role of the aged in the society of wild chimpanzees, where individuals form complex but flexible relationships based on personal histories.

As regards to "social status," we considered that old chimpanzees are groomed in a one-sided manner and are tolerated by younger individuals in agonistic interactions and meat sharing. Two concepts, "respect" and "restraint," may be useful in explaining the emergence of social status.

Respect may arise in several ways. First, given that chimpanzees have a good memory, other individuals remember the prime of an old individual. Bakali, a former high-ranking male, was an example of this in his later years. It is possible that memories such as these might be transformed into respect. The old individual gathering respect may retain some social influence depending on personal history (see Chapter 5 for Kalunde's case). Second, old individuals of both sexes may be respected because of practical benefits. Their skills and knowledge are based on rich experience that could be useful in either social or foraging situations. For example, foraging with old individuals may be beneficial for younger individuals to learn about potential dangers or rare fruit trees of which only a few individuals are aware. Kummer (1995) described how hamadryas baboons benefited from following an old male in their clan who had rich knowledge about easy routes and water sources.

Restraint may resemble self-handicapping shown by adults toward infants or juveniles in social play (Hayaki, 1985). Similarly, younger and stronger individuals might empathize with the physical frailty of old individuals, which leads to tolerance toward these feeble individuals in the context of aggressive interactions or meat sharing. In fact, although the skeleton of Wansombo was categorized as osteoporotic, her bones were free from fracture (Gunji *et al.*, 2003; Chapter 45). Considering that a hard blow can cause severe injury to such a feeble body, it is possible that chimpanzees will control their aggression toward an old individual.

The "social role" of old individuals has been mostly discussed in terms of caring for non-kin immature animals, including orphans. However, if we consider the current old individuals over 40 years of age in the M group, among five old females, only Gwekulo, presumably sterile, was observed to care for non-kin infants and youngsters, although she started this behavior during her adolescence. Nkombo, who is cycling but has not had her own offspring for the past 20 years, prefers to walk with adult males, Kalunde, in particular (Chapters 5, 29). Three other old females are active in reproduction, and so they have never taken the role of caring for non-kin.

Kalunde, the only living old male, seldom played with youngsters when he became the alpha male for the first time in 1991 (Chapter 27), yet he came to play with infants and juveniles more frequently as he became older; which resembled Bakali's pattern. There is, however, no observation that he ever cared for youngsters in his more recent declining years.

Gerontology is a promising area for future research in M-group chimpanzees since more males and females will reach the age of 40, and the accuracy of age estimation has been greatly improved. Given that there is great variation among individuals, the longitudinal study of each individual is valuable to observe all age-related changes and to examine if there is respect and restraint in behavior of other individuals. Next, we should explore the emergence of social status or social roles in chimpanzee society through a cross-sectional study of old individuals.

References

Emery Thompson, M., Jones, J. H., Pusey, A. E., *et al.* (2007). Aging and fertility patterns in wild chimpanzees provide insights into the evolution of menopause. *Current Biology*, **17**, 2150–6.

Erwin, J. M. and Hof, P. R., ed. (2002). *Aging in Nonhuman Primates*. New York: Karger.

Fraser, H. B., Khaitovich, P., Plotkin, J. B., *et al.* (2005). Aging and gene expression in the primate brain. *PLoS Biology*, **3**, 1653–61.

Goodall, J. (1986). *The Chimpanzees of Gombe: Patterns of Behavior*. Cambridge, MA: Belknap.

Gould, K. G., Flint, M., and Graham, C. E. (1981). Chimpanzee reproductive senescence: a possible model for evolution of the menopause. *Maturitas*, **3**, 157–66.

Gunji, H., Hosaka, K., Huffman, M. A., *et al.* (2003). Extraordinarily low bone mineral density in an old female chimpanzee (*Pan troglodytes schweinfurthii*) from the Mahale Mountains National Park. *Primates*, **44**, 145–9.

Hakeem, A., Sandoval, G. R., Jones, M., *et al.* (1996). Brain and life span in primates. In *Handbook of the Psychology of Aging, 4e*, ed. J. E. Birren and K. W. Schaie. San Diego, CA: Academic Press, pp. 78–104.

Hamilton, W. D. (1966). The moulding of senescence by natural selection. *Journal of Theoretical Biology*, **12**, 12–45.

Hasegawa, T. (1989). Sexual behavior of immigrant and resident female chimpanzees at Mahale. In *Understanding Chimpanzees*, ed. P. G. Heltne and L. A. Marquardt. Cambridge, MA: Harvard University Press, pp. 90–103.

Hauser, M. D. and Tyrrell, G. (1984). Old age and its behavioral manifestations: A study on two species of macaque. *Folia Primatologica*, **43**, 24–35.

Hawkes, K., O'Connell, J. F., and Blurton Jones, N. G. (1997). Hadza women's time allocation, offspring provisioning, and the evolution of long postmenopausal life spans. *Current Anthropology*, **38**, 551–77.

Hayaki, H. (1985). Social play of juvenile and adolescent chimpanzees in the Mahale Mountains National Park, Tanzania. *Primates*, **26**, 343–60.

Hayaki, H. (1988). Association partners of young chimpanzees in the Mahale Mountains National Park, Tanzania. *Primates*, **29**, 147–61.

Hayaki, H., Huffman, M. A., and Nishida, T. (1989). Dominance among male chimpanzees in the Mahale Mountains National Park, Tanzania: a preliminary study. *Primates*, **30**, 187–97.

Herndon, J. G., Tigges, J., Anderson, D. C., *et al.* (1999). Brain weight throughout the life span of the chimpanzee. *The Journal of Comparative Neurology*, **409**, 567–72.

Hosaka, K., Matsumoto-Oda, A., Huffman, M. A., *et al.* (2000). [Reactions to dead bodies of conspecifics by wild chimpanzees in the Mahale Mountains, Tanzania.] *Primate Research*, **16**, 1–15. In Japanese with English summary.

Huffman, M. A. (1990). Some socio-behavioral manifestations of old age. In *The Chimpanzees of the Mahale Mountains: Sexual and Life History Strategies*, ed. T. Nishida. Tokyo: University of Tokyo Press, pp. 237–55.

Kawanaka, K. (1989). Age differences in social interactions of young males in a chimpanzee unit-group at the Mahale Mountains National Park, Tanzania. *Primates*, **30**, 285–305.

Kummer, H. (1995). *In Quest of the Sacred Baboons: A Scientist's Journey*. Princeton, NJ: Princeton University Press.

Lacreuse, A. and Herndon, J. G. (2009). Nonhuman primate models of cognitive aging. In *Animal Models of Human Cognitive Aging*, ed. J. L. Bizon and A. Woods. Totowa, NJ: Humana Press, pp. 29–59.

Lacreuse, A., Chennareddi, L., Gould, K. G., *et al.* (2008). Menstrual cycles continue into advanced old age in the common chimpanzee (*Pan troglodytes*). *Biology of Reproduction*, **79**, 407–12.

McDonald Pavelka, M. S. (1990). Do old female monkeys have a specific social role? *Primates*, **31**, 363–73.

Muller, M. N., Emery Thompson, M., and Wrangham, R. W. (2006). Male chimpanzees prefer mating with old females. *Current Biology*, **16**, 2234–8.

Nakamichi, M. (1984). Behavioral characteristics of old female Japanese monkeys in a free-ranging group. *Primates*, **25**, 192–203.

Nakamura, M. and Nishie, H. (2012). Death of the oldest female at Mahale and some notes about longevity of wild chimpanzees. *Pan Africa News*, **19**, 19–21.

Nishida, T. (1994). Review of recent findings on Mahale chimpanzees implications. In *Chimpanzee Cultures*, ed. R. W. Wrangham, W. C. McGrew, F. B. M. de Waal, and P. Heltne. Cambridge, MA: Harvard University Press, pp. 373–96.

Nishida, T. (1997). Sexual behavior of adult male chimpanzees of the Mahale Mountains National Park, Tanzania. *Primates*, **38**, 379–98.

Nishida, T. (2012). *Chimpanzees of the Lakeshore: Natural History and Culture at Mahale*. Cambridge: Cambridge University Press.

Nishida, T. and Hosaka, K. (1996). Coalition strategies among adult male chimpanzees of the Mahale Mountains, Tanzania. In *Great Ape Societies*, ed. W. C. McGrew, L. F. Marchant, and T. Nishida. Cambridge: Cambridge University Press, pp. 114–34.

Nishida, T., Takasaki, H., and Takahata, Y. (1990). Demography and reproductive profiles. In *The Chimpanzees of the Mahale Mountains: Sexual and Life History Strategies*, ed. T. Nishida. Tokyo: University of Tokyo Press, pp. 63–97.

Nishida, T., Corp, N., Hamai, M., *et al.* (2003). Demography, female life history, and reproductive profiles among the chimpanzees of Mahale. *American Journal of Primatology*, **59**, 99–121.

Pusey, A. E. (1980). Inbreeding avoidance in chimpanzees. *Animal Behaviour*, **28**, 543–52.

Sherwood, C. C., Gordon, A. D., Allen, J. S., *et al.* (2011). Aging of the cerebral cortex differs between humans and chimpanzees. *Proceedings of the National Academy of Sciences of the USA*, **108**, 13029–34.

Shimizu, D., Gunji, H., Hashimoto, H., *et al.* (2002). The four chimpanzee skulls collected in the Mahale Mountains, Tanzania. *Anthropological Science*, **110**, 251–66.

Takahata, Y. (1990). Social relationships among adult males. In *The Chimpanzees of the Mahale Mountains: Sexual and Life History Strategies*, ed. T. Nishida. Tokyo: University of Tokyo Press, pp. 149–70.

Takahata, Y., Koyama, N., and Suzuki, S. (1995). Do the old aged females experience a long post-reproductive life span? The cases of Japanese macaques and chimpanzees. *Primates*, **36**, 169–80.

Teleki, G., Hunt, E. E., and Pfifferling, J. H. (1976). Demographic observations (1963–1973) on the chimpanzees of Gombe National Park, Tanzania. *Journal of Human Evolution*, **5**, 559–98.

Videan, E. N., Fritz, J., Heward, C. B., *et al.* (2006). The effects of aging on hormone and reproductive cycles in female chimpanzees (*Pan troglodytes*). *Comparative Medicine*, **56**, 291–9.

Williams, J. M., Lonsdorf, E. V., Wilson, M. L., *et al.* (2008). Causes of death in the Kasekela chimpanzees of Gombe National Park, Tanzania. *American Journal of Primatology*, **70**, 766–77.

24 Chimpanzee self-medication: a historical perspective of the key findings

MICHAEL A. HUFFMAN

24.1 Introduction

The adaptive value of warding off predators, whether larger and faster, or smaller and stealthier than oneself should not be surprising, yet science has been slow to recognize the ability of animals to avoid, suppress, or eliminate parasites and other agents of disease that equally compromise their health and well-being. While it appears that an awareness of, and even practical applications based on, the behavior of animals as a source of medicinal wisdom has been around for much of our own species history (e.g. Huffman, 2001, 2002, 2007), science has been slow to appreciate this. The first scientist to propose that self-medication may exist in primates, and other animals, was the neotropical ecologist, Daniel Janzen. He based his assertion on field observation and the ethnographic literature of animals using plants for, among other ailments, parasite infection (Janzen, 1978).

In recent years a growing body of evidence, both anecdotal and empirical, has shown the ability of a diverse array of animal species to engage in self-medication (e.g. Huffman, 1997; Engel, 2002; Villalba and Landau, 2012; Su *et al.*, 2013; Suárez-Rodríguez *et al.*, 2013). Janzen's (1978) paper was the starting point for scientific debate on this topic.

The goal of this chapter is to provide a descriptive chronology of the key findings on chimpanzee self-medication at Mahale in order to place the pioneering contribution of work at this site into the larger context of the development of the study of self-medication in animals.

24.2 Self-medicating apes?

The first example of self-medication in chimpanzees is a behavior called leaf swallowing. Jane Goodall first noticed whole, unchewed, leaves of *Aspilia pluriseta* in the dung of chimpanzees in the 1960s, but she was not sure of their significance at the time (Goodall, 2013). The published record of whole leaves in chimpanzee dung first appears in Richard Wrangham's 1975 Ph.D. thesis of the feeding ecology of Gombe (Tanzania) chimpanzees. When Wrangham visited Mahale for the first time in 1971, he told Toshisada Nishida about this behavior, remarking that "chimpanzees swallowing *Aspilia* leaves might be similar to humans drinking coffee," since they do so early in the morning (Nishida, 2012: 58). Nishida first recognized *Aspilia* leaves in chimpanzee dung at Mahale in 1975 (Nishida, 2012). Here he found that the folded, undigested, leaves of *Aspilia mossambicensis* and *A. rudis* were in the dung. Nishida also confirmed that they were being swallowed mostly in the early morning hours after chimpanzees left their sleeping nests (Figure 24.1).

Wrangham and Nishida (1983) later described the leaf-swallowing behavior at Gombe and Mahale and discussed it in terms of a curious non-nutritional feeding habit with potential, but unknown, medicinal benefits. This paper was based on information taken from the records of chimpanzee fecal contents at both sites collected originally for the study of their feeding ecology.

Mahale Chimpanzees: 50 Years of Research, ed. M. Nakamura, K. Hosaka, N. Itoh and K. Zamma. Published by Cambridge University Press.

Figure 24.1 (a) *Aspilia mossambicensis* and (b) the adult female, Linda, of M group swallowing leaves of that species. ©M. A. Huffman.

Wrangham, Nishida, and their colleague Shigeo Uehara later provided *Aspilia* spp. samples collected in the field to phytochemist and ethnobotanist Eloy Rodriquez, at the University of California Irvine. Rodriguez analysed the leaves for their potential bioactive properties. In 1985, they published a second short paper proposing that chimpanzees use *Aspilia* spp. for the powerful antibiotic compound thiarubrine A, which was said to occur at identical quantities in the leaves of both species (Rodriguez *et al.*, 1985).

The thiarubrine A hypothesis was first promoted as a "silver bullet" cure, capable of killing 70–80% of a chimpanzee's parasites in a single dose (Rodriguez *et al.*, 1985; Rodriguez and Wrangham, 1993). Due to the fact that the thiarubrines are highly susceptible to degradation by sunlight and stomach acid (Page *et al.*, 1992), the authors proposed several ad hoc interpretations for how thiarubrine A could be delivered to the source of parasite infection unaltered. First, they reasoned that swallowing in the early morning hours was done to avoid sun-induced degradation of its chemical activity. It was later noted, however, that leaf swallowing also occurs at midday, on an empty stomach (Huffman *et al.*, 1996b; Huffman and Caton, 2001). Rodriguez and colleagues also proposed that by not chewing, the folded leaves served as a protective "time-release capsule," delivering the active ingredients past the stomach to the intestines, the location of infection. Citing the early report of Janzen (1978), they proposed that *Aspilia* leaves were used to treat parasite infection, but no systematic studies of self-medication had yet been done. They had no direct evidence to demonstrate either a cause or effect for leaf swallowing. In 1987, Takasaki and Hunt (1987) reported yet another species (*Lippia plicata* Baker) being swallowed by chimpanzees at Mahale. Newton and Nishida (1990) later proposed yet another ad hoc explanation for the function of folding and swallowing, speculating that this was a form of sub-lingual medication, i.e. absorbing the active compounds via the underside of the tongue. However, observations of this behavior in both the wild and captivity (e.g. Huffman and Caton, 2001; Huffman and Hirata, 2004; Huffman *et al.*, 2010) do not support their argument that leaves are temporarily held under the tongue before swallowing. Scientists and the public met these early reports of leaf swallowing with both fascination and skepticism (e.g. Page *et al.*, 1992; Sapolsky, 1998).

24.3 Bitter pith chewing: a new self-medicative behavior from Mahale

The next investigation of self-medication in chimpanzees started out quite serendipitously in 1987 with the chance observation by myself and Mohamedi Seifu Kalunde (Tanzania National Parks) of a patently sick adult female, Chausiku, ingesting the bitter pith of *Vernonia amygdalina* (Huffman and Seifu, 1989). These initial descriptive observations formed the basis of the first long-term field study of animal self-medication (1989–2000) initiated by us in close collaboration with Koichi Koshimizu, Hajime Ohigashi (Kyoto University), their graduate students, Mitsuo Jisaka, Daisuke Izutsu, and international collaborators. Together, we formed a consortium called the "CHIMPP Group," an acronym for Chemo-ethology of Hominoid Interactions with Medicinal Plants and Parasites (Huffman, 1994). This work led to the first direct evidence for self-medication in an animal species.

Based on the initial observations of bitter pith chewing in 1987, parasitological surveys (August 1989–February 1998) and unpublished feeding records (from fecal remains) of K-group (January 1976–December 1977), I tested the hypothesis for a possible antiparasitic function for bitter pith chewing. Data collected on the 30 K-group members continuously over a 24-month period between 1976 and 1978 by the Mahale staff showed a statistically significant tendency for the more frequent occurrence of bitter pith chewing during the rainy season ($N = 36/38$ days occurred in rainy season, $z = 67.41$, $p < 0.0001$; weighted for differential length in seasons; in Huffman *et al.*, 1990). A peak in the frequency of bitter pith chewing was noted during December–January just after the onset of the rainy season (Huffman *et al.*, 1996b). This preliminary analysis, *albeit* of two unconnected datasets, provided a plausible link between parasite infection and self-medication in chimpanzees.

A co-author on this first report, Nishida was quick to support these conclusions in a short communication co-authored with Kawabata describing the parasites of K and M groups, in which he noted a possible link between a February peak in *Oesophagostomum* and *Strongyloides* parasite infections and leaf swallowing from fecal samples collected at Mahale between September 1975 and February 1976 (Kawabata and Nishida, 1991). While neither preliminary study clearly demonstrated an antiparasitic effect nor provided direct evidence for a mechanism responsible for either bitter pith chewing or leaf swallowing, Nishida again suggested the possibility of the pharmacological activity of thiarubrine A (Kawabata and Nishida, 1991). The definitive link between a parasite species and self-medication was only to come later from research on *V. amygdalina* bitter pith chewing at Mahale.

24.3.1 Behavioral ecology of bitter pith chewing

Within the home range of the M group at Mahale, *V. amygdalina* is neither abundant nor evenly distributed. This tree species prefers open sunlit locations, and typically occurs singly along or near water. Chimpanzees remove the bark and leaves from soft terminal branch segments of the pith (Figure 24.2). This inner portion is chewed, sucking out the bitter juices within, and spitting out the juice-depleted fibers. The length of pith ingested in a single bout is relatively small, ranging from portions of 5 to 120 cm × 1 cm (mean 54.57 cm, SD 41.01, $N = 7$; Huffman, 1997). The entire process can last from less than 1 up to 8 min (mean 2.9 min, SD 2.59, $N = 16$; Huffman, 1997).

Use is rare and seasonal, occurring most frequently during the rainy season months (Huffman *et al.*, 1997). Between 1985 and 1997, bitter pith chewing was observed by Huffman seven times in M group during a total of 728 h of focal observation (one feeding bout per 104 h; Huffman *et al.*, 1996b; Huffman, 1997). The peak in frequency of bitter pith chewing occurred in the

Figure 24.2 (a) Adult male, Jilba, chewing on bitter pith of *Vernonia amygdalina* and (b) a fresh pith with leaves of the same species. ©M. A. Huffman.

months of December and January, just after the onset of the rainy season, the peak period of *Oesophagostomum stephanostomum* reinfections (Fig. 2 in Huffman *et al.*, 1996b). All but two of the eight cases of bitter pith chewing recorded between 1971 and 1974 in K group also occurred between November and February (Huffman *et al.*, 1990).

24.3.2 Parasitic infection and general health of the individual at the time of bitter pith chewing

The first longitudinal study of the intestinal parasite fauna of chimpanzees with the explicit goal of verifying the possible causes of illness associated with self-medication was conducted at Mahale from 1989 by Huffman *et al.* (1996b). M-group individuals were infected by parasites from three genera of nematode: *Strongyloides fuelleborni, Trichuris trichiura*, and *O. stephanostomum*; one genus of cestode: *Bertiella studeri*; one genus of trematode: *Dicrocelium lanceatum* and four genera of protozoa, *Entamoeba coli, Endolimax nana, Iodamoeba buetschlii*, and *Giardia lamblia* (Huffman *et al.*, 1997, 2009; see Chapter 47 for a comprehensive list).

For the first time, individually identified chimpanzees were systematically monitored with the goal to detect weekly, monthly, and yearly changes in infection levels. For M-group chimpanzees monitored in 1991–2 and 1993–4, among all parasite species detected, a statistically significant increase during the rainy season was detected only for the incidence of *O. stephanostomum* (Fisher's exact test, $p = 0.0001$, $p = 0.01$, respectively in Huffman *et al.*, 1997). *O. stephanostomum* infections (93%, 14/15) were also associated significantly more frequently with bitter pith chewing ($N = 3$) and leaf swallowing ($N = 17$) during the same two study periods than either *Trichuris* (50%, $N = 7$) or *Strongyloides* (40%, $N = 6$) infections (Fisher's exact test for *O.s.* – *T.t.*, $p = 0.0142$; O.s. – *S.f.*, 0.005 in Huffman *et al.*, 1997).

At Mahale, mature conspecifics in proximity to a patently ill individual chewing *V. amygdalina* bitter pith typically show little or no interest in the plant (e.g. Huffman and Seifu, 1989; Huffman *et al.*, 1997), though some cases of several individuals pith chewing together from the same tree have been observed (Kazuhiko Hosaka, personal communication). Infants of ill mothers have also been observed to taste the pith discarded by their mothers, characteristic of mother–infant co-feeding on food items. Interestingly, in one instance, Fatuma, prevented her adolescent young daughter, Penelope, from picking up a piece of *V. amygdalina* pith and bark discarded by the patently ill adult male Musa, who had dropped it with leaf remains (Huffman, 1997). When Penelope moved over to investigate, Fatuma put her foot on the remains, preventing her from picking them up, and swiftly carried her away. While actively attempting to give medicinal plants to ill individuals has never been observed, this case of preventing an infant from ingesting a plant part normally avoided by adults is suggestive of one mechanism for instructive teaching.

Only one case of ingesting leaves of *V. amygdalina* has been reported in M group, and that was in 1993 for the same ill male mentioned above. After inspecting the small tree lacking any usable young pith, Musa removed and ingested a leaf from it (Huffman *et al.*, 1996b, 1997). He died a few days later of undetermined causes, but his *O. stephanostomum* infection had progressed to the point that he was passing adult worms through his urethra the last day he was seen alive. The worms had spread from their normal habitat in the intestines to neighboring vital organs. *O. stephanostomum* is the most hazardous species of nodular worm found in the great apes (Brack, 1987). Repeated infection, expected to occur in the wild, causes significant complications including diarrhea, severe abdominal pain, secondary bacterial infection, weight loss, and weakness resulting in high mortality (Brack, 1987).

Observations on the general state of health at the time *V. amygdalina* bitter pith is chewed were first available from cases recorded between 1987 and 1996 in M group (Huffman *et al.*, 1997). In 4 out of 10 such cases, ill health was evidenced by the presence of diarrhea, malaise, and nematode infection. In the two cases based on focal observations over multiple days, recovery within 20–24 h from a lack of appetite, malaise, and constipation or diarrhea was demonstrated (Huffman and Seifu, 1989; Huffman *et al.*, 1993). In the later case, the egg per gram feces (EPG) level of the *O. stephanostomum* infection was measured before and in days following ingestion of *Vernonia* bitter pith. The individual's EPG dropped from 130 to 15 within 20 h. No change occurred in this individual's concurrent *Trichuris* infection (Huffman *et al.*, 1993). When compared with seven other individuals with *O. stephanostomum* infections monitored over the same period, this was the only case in which such a dramatic drop in EPG was noted. A decrease was detected in only two of seven other individuals monitored over the same period (Huffman *et al.*, 1993). In one case, the drop was minimal (−15 EPGs). In the other case, a total drop of 510 EPGs over 18 days was noted. Unfortunately, the patently ill male, Aji, was absent from the group for most of this time, making it impossible to speculate about a cause for this dramatic change. In the remaining control cases, *O. stephanostomum* EPG levels increased over time. The average increase was 69.9 EPG (SD = 84, range 5–236; Huffman *et al.*, 1993), representative of the overall trend for increased reinfection by nodular worms at the beginning of the rainy season (Huffman *et al.*, 1997).

This was the first evidence to suggest that the serious effects of *O. stephanostomum* on chimpanzees was both an important stimulus of and target for both bitter pith chewing (Huffman *et al.*, 1996b) and leaf swallowing (Huffman *et al.*, 1996b, 1997; Huffman and Caton, 2001).

24.3.3 Pharmacological and ethnomedicinal evidence for the therapeutic value of *Vernonia amygdalina*

The above noted recovery time of 20–24 h after bitter pith chewing in two M-group chimpanzees is comparable to that of local human inhabitants, the Tongwe people (Chapter 49), who use cold concoctions of this plant as a treatment for malaria, intestinal parasites, diarrhea, and stomach upset among other ailments (Huffman *et al.*, 1993, 1996a). These observations led to further investigation of the possible pharmacological properties of *V. amygdalina* against parasite infection. The phytochemical analysis of *V. amygdalina* collected from Mahale conducted by Ohigashi, Koshimizu, and colleagues revealed the presence of two major classes of bioactive compounds, the sesquiterpene lactones and

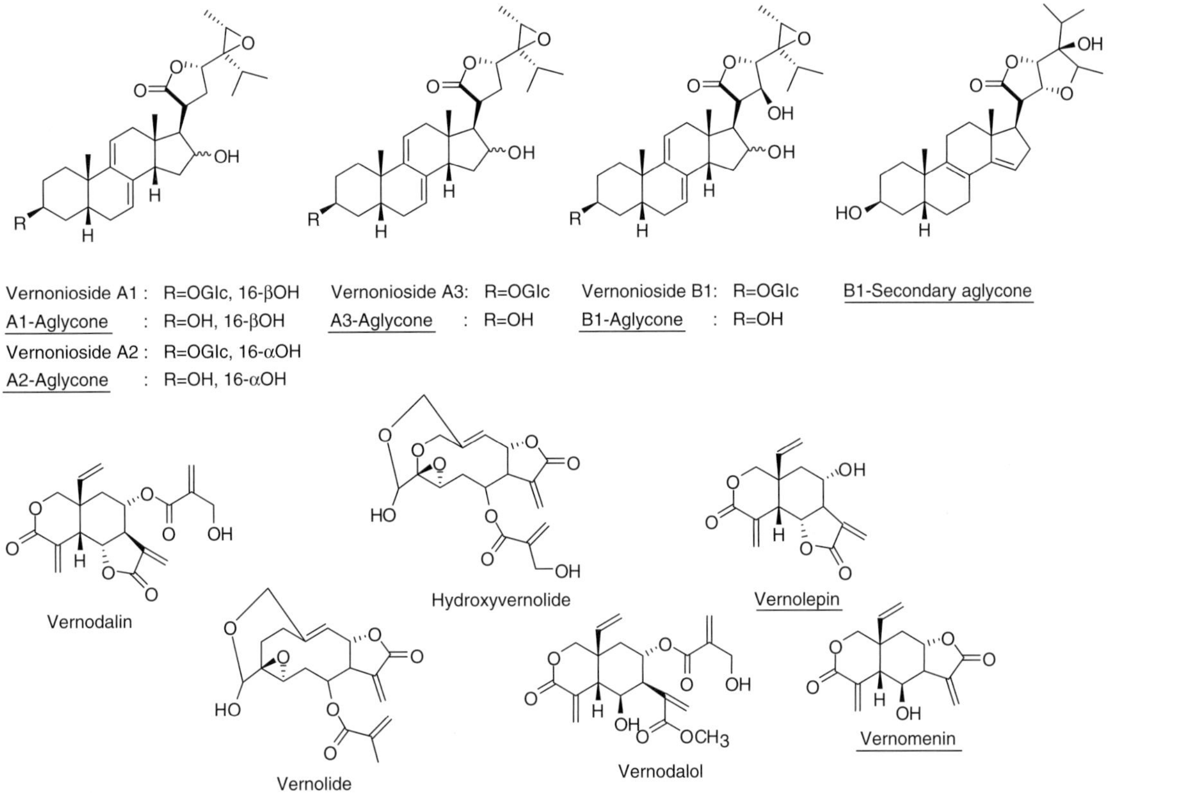

Figure 24.3 The chemical structures of compounds isolated from *Vernonia amygdalina.*

a newly discovered group, the steroid glucosides. A total of four known sesquiterpene lactones (vernodalin, vernolide, hydroxyvernolide, and vernodalol), seven new stigmastane-type steroid glucosides (vernonioside A1–A4, B1–B3), and two freely occurring aglycones of these glucosides (vernoniol A1, B1) were isolated (Ohigashi *et al.*, 1991; Jisaka *et al.*, 1992a, b, 1993a, b). Figure 24.3 shows the chemical structures of the newly described steroid glucosides and the widely reported sesquiterpene lactones found in *V. amygdalina*. The compounds from both groups were present at varying concentrations in the leaf, bark, root, and pith (Jisaka *et al.*, 1992b).

The sesquiterpene lactones in *V. amygdalina* are also found in *V. colorata* and a number of other *Vernonia* spp. These compounds are all well known for their anthelmintic, anti-amoebic, anti-tumor, and antibiotic properties (e.g. Toubiana and Gaudemer, 1967; Kupchan *et al.*, 1969; Jisaka *et al.*, 1992a, b).

In vitro tests demonstrated anti-schistosomal activity of the pith's most abundant steroid glucoside (vernonioside B1) and sesquiterpene lactone (vernodalin). At 200 ppm, vernonioside B1 significantly inhibited both movement and egg laying. At 20 ppm, however, the effects of B1's primary (vernoniol B1) and secondary aglycones were even more active than vernonioside B1 itself, especially against the movement of schistosomes. Similarly, the inhibition of movement activity of vernonioside A4's aglycone, vernoniol A4, was remarkably higher than that of vernonioside itself (Ohigashi *et al.*, 1994). While these aglycones are naturally present within the pith, they are not as abundant as either vernonioside B1and A4 (Jisaka *et al.*, 1992a). However, in vitro their concentrations increased as B1or A4 were metabolized into these more effective forms after digestion (hydrolysis) was simulated in the laboratory (Ohigashi *et al.*, 1994). From this, it was interpreted that the bioactivity of the pith would actually be enhanced by ingestion, not reduced. These findings are consistent with the observed decline in *O. stephanostomum* EPG level 20 h after the adult female Fatuma ingested *V. amygdalina* pith (Huffman *et al.*, 1993).

European members of the CHIMPP Group analysed the antiparasitic activity of the sesquiterpene lactones provided by Ohigashi and Koshimizu. P. Timon-David (Universite D'Aix Marseille, France) used *Leishmania donovani infantum*, and G. C. Kirby, D. Allen, Colin W. Wright, David C. Warhurst, and J. D. Phillipson (University of London, UK) used a K1 multi-drug resistant strain of *Plasmodium falciparum* to demonstrate significant pharmacological activity (Ohigashi *et al.*, 1991, 1994). The sesquiterpene lactones showed significant plasmodicidal activity although the IC_{50} values were more than 20 times higher than that of the common commercial agent, chloroquine diphosphate (Ohigashi *et al.*, 1994). The steroid glucosides showed weaker plasmodicidal activity. However, the activity of their aglycones was significantly higher in both plasmodicidal and amoebicidal activities. In particular, a significant increase in the plasmodicidal activity of vernoniol A4 was noted (Ohigashi *et al.*, 1994).

Nishida later speculated that chimpanzees might be seeking the high protein contents from *Vernonia*'s pith (Nishida, 2012). Perhaps because of the difficulty of analysing nutritional value of these compounds, his analysis apparently included pith and bark too. The nutritional argument is weak as the analysis ignored all the behavioral, pharmacological, and parasitological evidence for the use of this plant by chimpanzees. Given the fact that the major secondary compounds of the ingested pith are toxic, even if there was some nutritional value in the bits of pith inadvertently ingested in the process of sucking out the bitter juices, it is unlikely they provide any meaningful source of nutrition since the plant is so rarely used and the pith is regularly spat out.

24.4 A reassessment of leaf-swallowing behavior

Despite the initial report of thiarubrine A from *Aspilia* spp. leaves (Rodriguez *et al.*, 1985), Neil Towers (University of British Colombia), a leading expert on these compounds and their biological properties, was unsuccessful in replicating these findings in *Aspilia* plants raised in the UBC Botany Department's greenhouse (Towers *et al.*, 1985; Constable and Towers, 1989). In a further attempt to resolve the issue, Jon Page (University of British Columbia), a student of Tower's, went to Mahale in 1991 and collected his own fresh *A. mossambicensis* plant samples for analysis. Page too was unable to find thiarubrine A in the leaves, only a small amount appeared in the roots. Based on his overall analysis of the leaves, he proposed an alternative hypothesis for ingestion of *Aspilia* (Page *et al.*, 1992), suggesting that the uterostimulatory properties of kaurene diterpenes that he isolated from *A. mossambicensis* might have some role in fertility regulation (Page *et al.*, 1992). This was an interesting idea, but no effort was made to investigate its possible effect on chimpanzee reproduction.

In the meantime, more observations and reports of leaf swallowing by chimpanzees, bonobos, and lowland gorillas at sites across Africa began to appear. By 1992 Huffman had compiled a list of 22 plant species whose leaves were folded and swallowed whole at eight field sites (Huffman, 1993). Six of these species (six different genera) were used for leaf swallowing by chimpanzees at Mahale.

Additional tests to replicate the thiarubrine research of Rodriguez and colleagues were conducted together by Page and myself, but again not even the minutest traces of thiarubrine A, B, D, E, or their degraded traces in the leaves of three key *Aspilia* species (*A. mossambicensis, A. rudis*, and *A. africana*) were found after over 100 tests. I collected these leaves directly from plants used by chimpanzees at Mahale and Gombe, as well as from specimens collected at two other locations in central Kenya and eastern Tanzania (Huffman *et al.*, 1996b; Page *et al.*, 1997).

The chief phytochemist responsible for overseeing the original research, Eloy Rodriguez continues to defend his results, while Nishida, a co-author of these papers but not involved in the analysis himself, later wrote that "the experimental results proved to be based on a contamination of leaves and roots!" (Nishida, 2012: 58). Exactly what happened in the original analyses remains controversial.

Three other independent lines of evidence made any unified phytochemical mode of action for leaf swallowing unlikely: (1) leaf swallowing is known to have an effect only on two of the many different parasites (*O. stephanostomum, B. studeri*) carried by chimpanzees at the time of leaf swallowing (Wrangham, 1995; Huffman *et al.*, 1996b; 1997); (2) regardless of the species of leaf swallowed, at the time of expulsion, these worms are alive and can remain so up to several days afterward (Huffman *et al.*, 1997); and (3) indirect evidence from field bioassays on baboon *Strongyloides* larvae, using leaves swallowed by chimpanzees at Kibale, Uganda, found no antiparasitic activity (Messer and Wrangham, 1996). The "thiarubrine A hypothesis" for leaf-swallowing behavior further came into doubt.

The methodology and experience we developed from the study of bitter pith chewing was next applied at Mahale (1992–14) to directly investigate the mode and function of leaf-swallowing behavior (Huffman *et al.*, 1996b; Page *et al.*, 1997). I also conducted comparative research at Gombe (1995, 1996) and at Budongo, Uganda (1995–8) in collaboration with Vernon Reynolds (Oxford University) and independent researcher Paula Pebsworth. Modifying the Mahale research paradigm for chimpanzee self-medication research Sabrina Krief (Museum National d'Histoire Naturelle, Paris) and colleagues began behavioral, parasitological, and phytochemical studies at Kibale (2000–) (e.g. Krief *et al.*, 2005, 2006).

A series of field and laboratory studies we conducted at Mahale provided the first direct evidence demonstrating signs of ill health, parasite infection at the time of leaf swallowing, and physical expulsion of *O. stephanostomum* adult parasite worms. The one common trait among all the leaves selected for use is their characteristic hispid surface, created by the dense mass of tiny silicate trichomes (hairs). Currently, African great apes at 18 sites swallow the leaves of over 40 species of rough hispid leaves (Huffman, 2010; McLennan and Huffman, 2012; Fruth *et al.*, 2014). Other animal species now known to display leaf-swallowing behavior use the same criteria in their leaf selection (Huffman, 1997; Su *et al.*, 2013).

Based on the presence of adult *O. stephanostomum* worms found attached to the surface by trichomes or captured within the folds of these undigested *Aspilia* leaves defecated by chimpanzees, Huffman and colleagues (1996b) proposed the "Velcro effect" to describe the observed mode of parasite capture and expulsion. This and the fact that the adult worms came out alive, and remained so even when left for several days together with *Aspilia* leaves in sealed plastic bags, led us to conclude that unlike bitter pith, leaf swallowing could not easily be explained by pharmacological properties of the plant.

In a re-evaluation of this proposed mechanism, based on further observations, Huffman and Judith Caton (Australian National University) concluded that folding and swallowing whole unchewed leaves acted to stimulate the gastrointestinal tract into rapid movement and physically purges the parasites of an ill chimpanzee 6 h after ingestion (Huffman and Caton, 2001).

24.5 Medicinal value of the diet

The gap between medicine and food may at times be blurred. Much of what is consumed by wild animals contains an array of secondary compound(s). However, Johns (1990) argued that the non-nutritive chemicals commonly present in animal diets and those of our early hominid ancestors have been replaced by herbal medicines and pharmaceuticals in modern humans. The possible medicinal contribution in the primate diet received little attention as an aspect of self-medication until it was first described for chimpanzees at Mahale (Huffman, 1997). I

Table 24.1 **Some "medicinal foods" in the diet of Mahale chimpanzees, with special reference to those ingested items with known ethnomedicinal treatment of parasite infection and gastrointestinal upset**

Plant taxa	Prescribed treatment [part used]
Acanthaceae	
Asystasia gangetica (L.) T. And.	Nematodes, vermifuge, colic [lf]
Lannea schimperi (A.Rich.) Engl.	Diarrhea, dysentery [lf]
Annonaceae	
Annona senegalensis Pers.	Dysentery, diarrhea, intestinal upset [bk]
Combretaceae	
Combretum molle G. Don	Hookworms, stomach ache [lf]
Euphorbiaceae	
Croton sylvaticus Hochst. ex. Krause	Purgative [lf]
Malvaceae	
Hibiscus cannabinus L.	Stomach ache, dysentery [lf]
Menispermaceae	
Stephania abyssinica (Dillon and A. Rich.) Walp.	Round worms, purgative [lf]
Mimosaceae	
Parkia filicoidea Oliv.	Purgative, diarrhea [sd]
Moraceae	
Ficus exasperata Vahl.	Roundworms, anthelmintic [lf]
Myristicaceae	
Pycnanthus angolensis (Welw.) Warb.	Purgative, laxative [bk]
Olacaceae	
Ximenia americana L.	Vermifuge [fr]
Papilionaceae	
Erythrina abyssinica DC.	Antiseptic, stomach ache, parasites [bk]
Tiliaceae	
Grewia platyclada K. Schum	Stomach ache [bk]
Vitaceae	
Cissus oliveri (Engl.) Gilg.	Diarrhea [lf]
Leea quineensis G. Don	Colic [lf]

Plant part consumed: lf, leaf; fr, fruit; sd, seed; bk, bark.
Sources of ethnomedicinal use: Githen (1949); Watt and Breyer-Brandwinjk (1962); Kokwaro (1976); Abbiw (1990); Neuwinger (1996).

classified those plant foods with demonstrated biological activity as "medicinal foods" (Huffman, 1997). Table 24.1 lists some potential medicinal foods in the diet of M-group chimpanzees with known ethnomedicinal treatments for parasites or related gastrointestinal upsets. Colleagues and I later looked at this in detail for chimpanzees at Budongo (Pebsworth *et al.*, 2006), gorillas (Cousins and

Huffman, 2002), and Japanese macaques (MacIntosh and Huffman, 2010; Huffman and MacIntosh, 2012).

24.5.1 Fruits and leaves of potential medicinal value

Secondary compounds as well as inorganic materials are considered to be a plant's front-line defense against most non-specialized herbivores (Howe and Westley, 1988). These items are toxic to the consumer or deter consumption by reducing palatability or digestibility (Howe and Westley, 1988, Wink *et al.*, 1993), yet a number of them can be found in the diet.

24.5.2 Bark and wood of potential medicinal value

Bark and wood are characteristically highly fibrous, heavily lignified, sometimes toxic, relatively indigestible, and nutrient poor. Chimpanzees and gorillas are known to infrequently ingest the bark and wood of several plant species (e.g. Huffman and Wrangham, 1994; Cousins and Huffman, 2002).

While the bark of many plant species is ingested by primates, little is actually known about the contribution of bark to the diet and general health. The literature on African ethnomedicine provides evidence for the bioactivity of some of these species. The bark of *Pycnanthus angolensis* is ingested by chimpanzees at Mahale and used in West Africa by people as a purgative, laxative, digestive tonic, emetic, and reliever of toothaches (Abbiw, 1990). Bark strips of *Grewia platyclada* are sometimes chewed for the relief of stomach aches by local residents at Mahale (Huffman, unpublished data; Mohamedi Seifu Kalunde, personal communication). In 1992, one juvenile male chimpanzee suffering from diarrhea was observed chewing, but not ingesting fibers, of the bark of this species at Mahale (Huffman, 1997). Mahale chimpanzees occasionally eat the bark of *Erythrina abyssinica*. Significant plasmodicidal and anti-schistosomal activities were demonstrated from the bark of this species collected at Mahale (Ohigashi, 1995) and it is known for its ethnomedicinal use as an antiseptic, for stomach aches, and for intestinal parasites (Abbiw, 1990).

These accounts of the ingestion of peculiar fruit, leaf, bark, and wood items serve to demonstrate the diversity of inferred pharmacological activity present in primate diets.

Shelly Masi (Muséum National d'Histoire Naturelle, Paris) and colleagues (2012) later explored the concept of medicinal foods in Kibale chimpanzees and Bai Hokou (Central African Republic) lowland gorillas, calling these items "unusual and bioactive foods (UBF)."

24.6 Tongwe ethnomedicine and animal self-medication

People have long looked to animals for sources of herbal medicine. Numerous independently reported records of the possible use of plants as medicine by wild animals exist and humans are still using some of these plants today (e.g. Janzen, 1978; Cousins and Huffman, 2002; Engel, 2002; Huffman, 2002).

An unexpected by-product from the research on chimpanzee self-medication at Mahale was the realization in 2000, partway through my study, that the local Tongwe people (Chapter 49) have long believed that sick animals, such as the porcupine, bushpig, warthog, elephant, and baboon self-medicate. In long discussions with Mohamedi Seifu Kalunde based on episodes passed on to him from his mother Joha Kasante, his grandfather Kalunde, his uncle Musalangi (and some stories passed down from his grandfather Kabingo), it was found they have benefited directly from this knowledge, attributing the acquisition of several medicinal treatments for parasites, stomach upset, and other ailments from watching the behavior of sick animals (Huffman, 2007). During the course of my collaboration with him, Mohamedi developed a new treatment for diarrhea based on his observations of chimpanzees' use of *Trema orientalis*. Chimpanzees typically fold and swallow the leaves, which results in the physical expulsion of parasites (see above), but they will also sometimes eat and chew them as food (Huffman, personal observations). Based on the taste and smell of the leaves, Mohamedi made a water concoction from the crushed leaves and tested its use to control diarrhea. Mohamedi's mother, adopted this treatment for use on her patients in the Buhingu area (Chapter 2), her village of residence, with good effects (Mohamedi Seifu Kalunde, personal communication).

Another example of the practical applications of this research comes from work on the secondary compounds found in *V. amygdalina*. Based on the findings of Koshimizu *et al.* (1993) and previous work by Kupchan *et al.* (1969) showing significant inhibition of tumor promotion and immunosuppressive activities of vernodalin, two independent laboratories, one at Jackson State University, MS (United States) and the other at the University of Malaya (Malaysia), have recently demonstrated significant anti-cancer properties of vernodalin, in particular against breast cancer (Izevbigie, 2005; Looi *et al.*, 2013). While it is has never been the primary goal of chimpanzee self-medication research at Mahale, it can be said that such activities may lead to the future discovery of important drugs for human use.

References

Abbiw, D. K. (1990). *Useful Plants of Ghana*. Kew, UK: Intermediate Technology Publications and Royal Botanic Gardens.

Brack, M. (1987). *Agents Transmissible from Simians to Man*. Berlin: Springer-Verlag.

Constable, C. P. and Towers, G. H. N. (1989). The complex nature of the mechanism of toxicity and antibiotic dithiacylohexadiene polylines (thiarubrines) from the Asteraceae. *Planta Medica*, **55**, 35–7.

Cousins, D. and Huffman, M. A. (2002). Medicinal properties in the diet of gorillas – An ethnopharmacological evaluation. *African Studies Monographs*, **23**, 65–89.

Engel, C. (2002). *Wild Health*. Boston: Houghton Mifflin.

Fruth, B., Ikombe, N. B., Matshimba, G. K., *et al.* (2014). New evidence for self-medication in bonobos: *Maniophyton fulvum* leaf- and stemstrip-swallowing from LuiKotale, Salonga National Park, DR Congo. *American Journal of Primatology*, **76**, 146–58.

Githen, T. S. (1949). *Drug Plants of Africa*. Pittsburg, PN: University of Pennsylvania Press.

Goodall, J. (2013). *Seeds of Hope: Wisdom and Wonder from the World of Plants*. New York: Grand Central Publishing.

Howe, H. F. and Westley, L. C. (1988). *Ecological Relationships of Plants and Animals*. New York: Oxford University Press.

Huffman, M. A. (1993). [An investigation of the use of medicinal plants by wild chimpanzees. Current status and future prospects.] *Primate Research*, **9**, 179–87. In Japanese with English summary.

Huffman, M. A. (1994). The CHIMPP Group: A multi-disciplinary investigation into the use of medicinal plants by chimpanzees. *Pan Africa News*, **1**(1), 3–5.

Huffman, M. A. (1997). Current evidence for self-medication in primates: A multidisciplinary perspective. *Yearbook of Physical Anthropology*, **40**, 171–200.

Huffman, M. A. (2001). Self-medicative behavior in the African great apes: An evolutionary perspective into the origins of human traditional medicine. *BioScience*, **51**, 651–61.

Huffman, M. A. (2002). Animal origins of herbal medicine. In *From the Sources of Knowledge to the Medicines of the Future*, ed. J. Fleurentin, J-M. Pelt, and G. Mazars. Paris: IRD Editions, pp. 31–42.

Huffman, M. A. (2007). Primate self-medication. In *Primates in Perspective*, ed. C. Campbell, A. Fuentes, K. MacKinnon, M. Panger, and S. Bearder. Oxford: University of Oxford Press, pp. 677–90.

Huffman, M. A. (2010). Self-medication: passive prevention and active treatment. In *Encyclopedia of Animal Behavior*, ed. M. D. Breed and J. Moore. Oxford: Academic Press, pp. 125–31.

Huffman, M. A. and Caton J. M. (2001). Self-induced increase of gut motility and the control of parasitic infections in wild chimpanzees. *International Journal of Primatology*, **22**, 329–46.

Huffman, M. A and Hirata, S. (2004). An experimental study of leaf swallowing in captive chimpanzees: insights into the origin of a self-medicative behavior and the role of social learning. *Primates*, **45**, 113–18.

Huffman, M. A. and MacIntosh, A. J. J. (2012). Plant food diet of the Arashiyama-Kyoto Japanese macaques and its potential medicinal value. In *The Monkeys of Stormy Mountain: 60 Years of Primatological Research on the Japanese Macaques of Arashiyama*, ed. J-B Leca, M. A. Huffman, and P. L. Vasey. Cambridge: Cambridge University Press, pp. 356–431.

Huffman, M. A. and Seifu, M. (1989). Observations on the illness and consumption of a possibly medicinal plant *Vernonia amygdalina* (Del.), by a wild chimpanzee in the Mahale Mountains National Park, Tanzania. *Primates*, **30**, 51–63.

Huffman, M. A. and Wrangham, R. W. (1994). The diversity of medicinal plant use by chimpanzees in the wild. In *Chimpanzee Cultures*, ed. R. W. Wrangham, W. C. McGrew, F. B. deWaal and P. G. Heltne. Cambridge MA: Harvard University Press, pp. 129–48.

Huffman, M. A., Nishida, T., and Uehara, S. (1990). Intestinal parasites and medicinal plant use in wild chimpanzees: possible behavioral adaptation for the control of parasites. *Mahale Mountains Chimpanzees Research Project. Ecological Report*, **72**.

Huffman, M. A., Gotoh, S., Izutsu, D., Koshimizu, K., and Kalunde, M. S. (1993). Further observations on the use of *Vernonia amygdalina* by a wild chimpanzee, its possible effect on parasite load, and its phytochemistry. *African Study Monographs*, **14**, 227–40.

Huffman, M. A., Koshimizu, K., and Ohigashi, H. (1996a). Ethnobotany and zoopharmacognosy of *Vernonia amygdalina*, a medicinal plant used by humans and chimpanzees. In *Compositae: Biology & Utilization* Vol. **2**, ed. P. D. S. Caligari and D. J. N. Hind. Kew: The Royal Botanical Gardens, pp. 351–60.

Huffman, M. A., Page, J. E., Sukhdeo, M. V. K., *et al.* (1996b). Leaf-swallowing by chimpanzees, a behavioral adaptation for the control of strongyle nematode infections. *International Journal of Primatology*, **72**, 475–503.

Huffman, M. A., Gotoh, S., Turner, L. A., Hamai, M., and Yoshida, K. (1997). Seasonal trends in intestinal nematode infection and medicinal plant use among chimpanzees in the Mahale Mountains, Tanzania. *Primates*, **38**, 111–25.

Huffman, M. A., Pebsworth, P., Bakuneeta, C., Gotoh, S., and Bardi, M. (2009). Macro-habitat comparison of host–parasite ecology in two populations of chimpanzees in the Budongo Forest, Uganda and the Mahale Mountains, Tanzania. In *Primate Parasite Ecology: The Dynamics of Host-parasite Relationships*, ed. M. A. Huffman and C. Chapman. Cambridge: Cambridge University Press, pp. 311–30.

Huffman, M. A., Spiezio, C., Sgaravatti, A., and Leca, J-B. (2010). Option biased learning involved in the acquisition and transmission of leaf swallowing behavior in chimpanzees (*Pan troglodytes*)? *Animal Cognition*, **13**, 871–80.

Izevbigie, E. B. (2005). Phytochemotherapy for cancer. US Patent application 756241. January 13, 2004

Janzen, D. H. (1978). Complications in interpreting the chemical defenses of trees against tropical arboreal plant-eating vertebrates. In *The Ecology of Arboreal Folivores*, ed. G. G. Montgomery. Washington DC: Smithsonian Institute Press, pp. 73–84.

Jisaka, M., Kawanaka, M., Sugiyama, H., *et al.* (1992a). Antischistosomal activities of sesquiterpene lactones and steroid glucosides from *V. amygdalina*, possibly used by wild chimpanzees against parasite-related disease. *Bioscience, Biotechnology and Biochemistry*, **56**, 845–6.

Jisaka, M., Ohigashi, H., Takagaki, T., *et al.* (1992b). Bitter steroid glucosides, Vernoniosides A1, A2 and A3 and related B1 from a possible medicinal plant *Vernonia amygdalina*, used by wild chimpanzees. *Tetrahedron*, **48**, 625–32.

Jisaka, M., Ohigashi, H., Takegawa, K., Huffman, M. A., and Koshimizu, K. (1993a). Antitumoral and antimicrobial activities of bitter sesquiterpene lactones of *Vernonia amygdalina*, a possible medicinal plant used by wild chimpanzee. *Bioscience, Biotechnology and Biochemistry*, **57**, 833–4.

Jisaka, M., Ohigashi, H., Takegawa, K., *et al.* (1993b). Steroid glucosides from *Vernonia amygdalina*, a possible chimpanzee medicinal plant. *Phytochemistry*, **34**, 409–13.

Johns, T. (1990). *With Bitter Herbs They Shall Eat It*. Tucson, AZ: The University of Arizona Press.

Kawabata, M. and Nishida, T. (1991). A preliminary note on the intestinal parasites of wild chimpanzees of the Mahale Mountains, Tanzania. *Primates*, **32**, 275–8.

Kokwaro, J. (1976). *Medicinal Plants of East Africa*. Nairobi: East African Literature Bureau.

Koshimizu, K., Ohigashi, H., Huffman, M. A., Nishida, T., and Takasaki, H. (1993). Physiological activities and the active constituents of potentially medicinal plants used by wild chimpanzees of the Mahale Mountains, Tanzania. *International Journal of Primatology*, **14**(2), 345–56.

Krief, S., Hladik, C. M., and Haxaire, C. (2005). Ethnomedicinal and bioactive properties of plants ingested by wild chimpanzees in Uganda. *Journal of Ethnopharmacology*, **101**, 1–15.

Krief, S., Huffman, M. A., Sevenet, T., *et al*. (2006). Bioactive properties of plant species ingested by chimpanzees (*Pan troglodytes schweinfurthii*) in the Kibale National Park, Uganda. *American Journal of Primatology*, **68**, 51–71.

Kupchan, S. M., Hemingway, R. J., Karim, A., and Werner, D. (1969). Tumor Inhibitors XLVII. Vernodalin and vernomygdin, two new cytotoxic sesquiterpene lactones from *Vernonia amygdalina* Del. *Journal of Organic Chemistry*, **34**, 3908–11.

Looi, C. C., Arya, A., Chea, F. K., *et al*. (2013). Induction of apoptosis in human breast cancer cells via caspase pathway by vernodalin isolated from *Centratherum anthelminticum* (L.) seeds. *PLoS ONE*, **8**, e56643.

MacIntosh, A. J. J. and Huffman, M. A. (2010). Towards understanding the role of diet in host-parasite interactions: the case for Japanese macaques. In *The Japanese Macaques*, ed. N. Nakagawa, M. Nakamichi, and H. Sugiura. Tokyo: Springer, pp. 323–44.

Masi, S., Gustafsson, E., Saint Jalme, M., *et al*. (2012). Unusual feeding behavior in wild great apes, a window to understand origins of self-medication in humans: Role of sociality and physiology on learning process. *Physiology & Behavior*, **105**, 337–49.

McLennan, M. R. and Huffman, M. A. (2012). High frequency of leaf swallowing and its relationship to parasite expulsion in "village" chimpanzees at Bulindi, Uganda. *American Journal of Primatology*, **74**, 642–50.

Messer, E. J. and Wrangham, R. W. (1996). *In vitro* testing of the biological activity of *Rubia cordifolia* leaves on primate *Strongyloides* species. *Primates*, **37**, 105–8.

Neuwinger, H. D. (1996). *African Ethnobotany: Chemistry, Pharmacology, Toxicology*. London: Chapman & Hill.

Newton, P. N. and Nishida, T. (1990). Possible buccal administration of herbal drugs by wild chimpanzees, *Pan troglodytes. Animal Behaviour*, **39**, 798–801.

Nishida, T. (2012). *Chimpanzees of the Lakeshore: Natural History and Culture at Mahale*. Cambridge: Cambridge University Press.

Ohigashi, H. (1995). Plants used medicinally by primates in the wild and their physiologically active constituents. Report to the Ministry of Science, Education and Culture for 1994 Grant-in-Aid for Scientific Research (No. 06303012).

Ohigashi, H., Jisaka, M., Takagaki, T., *et al*. (1991). Bitter principle and a related steroid glucoside from *Vernonia amygdalina*, a possible medicinal plant for wild chimpanzees. *Agricultural and Biological Chemistry*, **55**, 1201–3.

Ohigashi, H., Huffman, M. A., Izutsu, D., *et al*. (1994). Toward the chemical ecology of medicinal plant-use in chimpanzees: the case of *Vernonia amygdalina* Del., a plant used by wild chimpanzees possibly for parasite-related diseases. *Journal of Chemical Ecology*, **20**, 541–53.

Page, J. E., Balza, F. F., Nishida, T., and Towers, G. H. N. (1992). Biologically active diterpenes from *Aspilia mossambicensis*, a chimpanzee medicinal plant. *Phytochemistry*, **31**, 3437–9.

Page, J. E., Huffman, M. A., Smith, V., and Towers, G. H. N. (1997). Chemical basis for medicinal consumption of *Aspilia* leaves by chimpanzees: a re-analysis. *Journal of Chemical Ecology*, **23**, 2211–25.

Pebsworth, P., Krief, S., and Huffman, M. A. (2006). The role of diet in self-medication among chimpanzees in the Sonso and Kanyawara communities, Uganda. In *Primates of Western Uganda*, ed. N. E. Newton-Fisher, H. Notman, V. Reynolds, and J. D. Paterson. New York: Springer, pp. 105–33.

Rodriguez, E. and Wrangham, R. W. (1993). Zoopharmacognosy: the use of medicinal plants by animals. In *Recent Advances in Phytochemistry: Vol. 27. Phytochemical Potential of Tropic Plants*. ed. K. R. Downum, J. T. Romeo, and H. Stafford. New York: Plenum Press, pp. 89–105.

Rodriguez, E., Aregullin, M., Nishida, T., *et al*. (1985). Thiarubrine A, a bioactive constituent of *Aspilia* (Asteraceae) consumed by wild chimpanzees. *Experientia*, **41**, 419–20.

Sapolsky, R. M. (1998). *The Trouble with Testosterone: And Other Essays on The Biology of the Human Predicament*. New York: Simon & Schuster.

Su, H., Su, Y., and Huffman, M. A. (2013). Leaf-swallowing and parasite infection in the Chinese lesser civet (*Viverricula indica*) in northern Taiwan. *Zoological Studies*, **52**, 22.

Suárez-Rodríguez, M., López-Rull, I., and Macías Garcia, C. (2013). Incorporation of cigarette butts into nests reduces nest ectoparasite load in urban birds: new ingredients for an old recipe? *Biology Letters*, **9**, 20120931.

Takasaki, H. and Hunt, K. (1987). Further medicinal plant consumption in wild chimpanzees? *African Studies Monographs*, **8**, 125–8.

Toubiana, R. and Gaudemer, A. (1967). Structure du vernolide, nouvel ester sesquiterpique isole de *Vernonia colorata*. *Tetrahedron Letters*, **14**, 1333–6. In French.

Towers, G. H. N., Abramowski, Z., Finlayson, A. J., and Zucconi, A. (1985). Antibiotic properties of thiarubrine-A, a naturally occurring dithiacyclohedadiene polyine. *Planta Medica*, **3**, 225–9.

Villalba, J. J. and Landau, S. Y. (2012). Host behavior, environment and the ability to self-medicate. *Small Ruminant Research*, **103**, 50–9.

Watt, J. M. and Breyer-Brandwinjk, M. G. (1962). *The Medicinal and Poisonous Plants of Southern and East Africa*. Edinburgh: E. and S. Livingstone.

Wink, M., Hofer, A., Bilfinger, M., *et al*. (1993). Geese and dietary allelochemicals: food palatability and geophagy. *Chemoecology*, **4**, 93–107.

Wrangham, R. W. (1975). *The Behavioural Ecology of Chimpanzees in Gombe National Park, Tanzania*. Ph.D. thesis. Cambridge: University of Cambridge.

Wrangham, R. W. (1995). Relationship of chimpanzee leaf-swallowing to a tapeworm infection. *American Journal of Primatology*, **37**, 297–303.

Wrangham, R. W. and Nishida, T. (1983). *Aspilia* spp. leaves: a puzzle in the feeding behavior of wild chimpanzees. *Primates*, **24**, 276–82.

25 Diseases and deaths: variety and impact on social life

SHUNKICHI HANAMURA, TAKANORI KOORIYAMA, AND KAZUHIKO HOSAKA

25.1 Introduction

Wild chimpanzees can contract various diseases, some of which can be fatal (Goodall, 1986a: 92–7). The accumulated knowledge gained from several long-term field sites has revealed that habituated chimpanzees and bonobos (*Pan paniscus*) have experienced several fatal epidemics, such as polio, Ebola, and respiratory diseases (including flu-like diseases) (Table 25.1). Roughly half of the inferred causes of chimpanzee deaths are diseases at Gombe, Tanzania (58%: Williams *et al.*, 2008) and Mahale (48%: Nishida *et al.*, 2003).

Many kinds of diseases can be transmitted by humans (*Homo sapiens*) to nonhuman primates and vice versa because of their phylogenetic proximity (Wolfe *et al.*, 1998). At Taï, Côte d'Ivoire, and Mahale, tissue or fecal samples from dead or sick chimpanzees during respiratory disease epidemics have contained zoonotic or human-originated viruses (Table 25.1). Although chimpanzees can contract various diseases without human contact, currently, interactions between humans and chimpanzees are multifaceted, in that they include social (e.g. research, tourism, filming, and poaching), ecological (e.g. forest destruction), and pathogenic elements (this third interaction has been well documented, at least in captivity: Kooriyama *et al.*, 2013; see Chapter 47 regarding internal parasites). In order to reduce the risk of disease transmission by humans, particularly by those who are in close contact with chimpanzees, it is necessary to explore the infection routes and to establish effective observation guidelines (e.g. Boesch, 2008; Chapters 48, 50). However, our knowledge of the diseases that infect wild chimpanzees, their characteristics, and their impact on chimpanzee society is still limited.

In this chapter, therefore, we review the history and the variety of diseases that have infected the chimpanzees of the K and M groups in Mahale. We then examine the seasonality of emergence, other relevant factors, age-specific mortality, and the possible pathogens and infection routes of respiratory diseases, which are the most frequent cause of chimpanzee deaths at some long-term field sites (Table 25.1). Finally, we review chimpanzee attitudes to sick or dead individuals.

25.2 Diseases infecting chimpanzees at Mahale

25.2.1 Difficulty of confirming disease symptoms and deaths

Wild chimpanzees live in a unit-group, and the group members split into temporary parties of varying size, composition, and duration (Chapters 6, 9). Researchers usually observe only some of the group members in a day, and it is common for them not to see some individuals for several days, weeks, or even over a month. Sick individuals, particularly those that are seriously ill, are more difficult to find, because they tend to spend more time alone or only with their dependent offspring. Therefore, researchers cannot always notice sick individuals, or their symptoms. Deaths are also difficult to confirm, for several reasons. First, it is rare to find corpses in the forest. Second, even

Mahale Chimpanzees: 50 Years of Research, ed. M. Nakamura, K. Hosaka, N. Itoh and K. Zamma. Published by Cambridge University Press.

Table 25.1 **Reported disease epidemics in habituated unit-groups of wild chimpanzees and bonobos**

Study group	Year	Disease	No. of deaths	Detected or suggested possible pathogens: Virus[a]	Detected or suggested possible pathogens: Bacteria	Source[b]
Pan troglodytes schweinfurthii						
Gombe Kasekela	1966	Polio	6			1–4
	1968	Respiratory	4 or 5			1–4
	1975	Respiratory	1			1–3
	1978	Respiratory	1			1–3
	1987	Respiratory	9 or 10			2–4
	1997	Mange	3			2–4
	2000	Respiratory	2		*Streptococcus pneumoniae; S. pyogenes*	3, 4
Gombe Mitumba	1996	Respiratory	8, 9, or 11			2, 3
Gombe Kalande	2002	Respiratory	> 2			3
Mahale K	1968	Respiratory	0			5
	1977	Flu-like	< 2			5
Mahale M	1968	Respiratory	0?			5
	1985–7	AIDS-like	> 3			6
	1992	Flu-like	< 1			5
	1993	Flu-like	< 12			5
	2003	Respiratory	< 2			5
	2005	Respiratory	< 2	Paramyxovirus?; Rotavirus?		5
	2006	Flu-like	< 12	hMPV		5
	2007	dermatophytosis	0			7

Table 25.1 **(cont.)**

Study group	Year	Disease	No. of deaths	Detected or suggested possible pathogens		Source[b]
				Virus[a]	Bacteria	
Pan troglodytes verus						
Taï North	1986	Conjunctivitis	0			8
	1987	Monkey pox	1			8
	1990	Flu-like	0			8
	1991	Flu-like	0			8
	1992	Ebola	8			8, 9
	1994	Ebola	10 or 12	Ebola virus		8, 9
	1999	Respiratory	> 6	hRSV (STLV; Foamy; LCV1)	*S. pneumoniae; Klebsiella pneumoniae; Haemophilus influenzae*	10, 11
Taï South	2004 (twice)	Respiratory	8	hMPV	*S. pneumoniae; Pasteurella multocida; K. pneumoniae*	10, 11
	2006	Respiratory	1	hRSV	*S. pneumoniae*	10, 11
Taï East	2006	Respiratory	2	hRSV	*S. pneumoniae; P. multocida*	10, 11
Taï North, Middle, South	2001–2	Anthrax	8	(STLV; Foamy; LCV1)	*Bacillus anthracis*	11, 12
Bossou	1992	Respiratory	1			13
	2003	Flu-like	5			13
Pan paniscus						
Wamba E1	2003	Flu-like	0			14
	2008	Flu-like	< 2			14

[a] hMPV, human metapneumovirus; hRSV, human respiratory syncytial virus; STLV, simian T-lymphotropic virus; foamy, foamy virus; LCV1, lymphocryptovirus type 1 (the pathogenicity of the latter three viruses remains hidden from view).

[b] 1: Goodall, 1986a; 2: Wallis and Lee, 1999; 3: Lonsdorf *et al.*, 2006; 4: Williams *et al.*, 2008; 5: see Table 25.2; 6: see Section 25.2.3; 7: see Section 25.2.4; 8: Boesch and Boesch-Achermann, 2000; 9: Formenty *et al.*, 1999; 10: Köndgen *et al.*, 2008; 11: Boesch, 2008; 12: Leendertz *et al.*, 2004; 13: Humle, 2011; 14: Sakamaki *et al.*, 2009.

if corpses were found, visual identification of the individuals is often difficult, due to decay and damage caused by scavengers, although, in rare cases, researchers may still be able to use morphological (Nakai *et al.*, 2004) or genetic evidence (cf. Nakazawa *et al.*, 2013) for identification. Therefore, researchers often judge a chimpanzee death retrospectively.

In this chapter, we define *death* and *death from disease* as follows.

Death: Besides confirmed deaths, chimpanzees that had disappeared for a long time (usually more than 3 months) were presumed to have died, except in the case of adolescent females having reached sexual maturity who had disappeared without any sign of injury or disease, because they can be considered to emigrate out of their native unit-groups.

Death from disease: Chimpanzees were presumed to have died due to disease if they had showed any symptoms of disease when they were last seen. The estimated number of deaths from respiratory disease epidemics using this definition, however, is likely to be an underestimate (Hanamura *et al.*, 2008). Therefore, in order to obtain the highest possible estimate of deaths from such epidemics, chimpanzees that had disappeared or were found dead during the epidemics were presumed to have died due to the epidemics, even if they were not observed exhibiting any symptoms. In such cases, however, if there was any evidence that death was caused by senility, serious injury, or other causes (Nishida *et al.*, 2003), then death due to these causes was presumed.

25.2.2 Respiratory disease epidemics

Eight respiratory disease epidemics, two in the K group and six in the M group, have been reported in Mahale over the past 50 years (Table 25.2). The first occurred in the K group in August–September 1968. By the end of September, many chimpanzees in the adjacent M group had also begun to cough. Considering the fact that some females interacted with individuals of both the K and M groups at that time (Nishida, 1973: 256–8), the disease might have been transmitted from the K to the M group. A similar epidemic occurred in the K group in April–May 1977. During this epidemic, an adult female (Wasalamba) who had lost her infant carried the corpse, and another adult female (Wabunengwa) appeared without her infant. In addition to these two epidemics, Nishida (1973: 250–3; 1981: 268) described the symptoms of respiratory disease exhibited by two adult females in the K group: Wamikambi coughed and had a poor appetite (unknown date); Wantangwa showed coughing, nasal discharge, and emaciation (September 8–22, 1979). These reports suggest that respiratory disease was common, and may sometimes have caused infant death in the K group.

The other five reported respiratory disease epidemics occurred in the M group, which was the research focus after the K group ceased to exist (Chapter 10). These five epidemics were presumed to cause deaths of chimpanzees (range of the highest estimate: 1–12 individuals). The epidemics in 1993 and 2006 were severe outbreaks that inflicted serious damage to the M group, because they were thought to kill more individuals than the other three epidemics (Table 25.2). Juveniles, adolescents, and even adults died, as well as infants (Table 25.3). The recent increase in reported epidemics does not necessarily reflect the actual increase in epidemics; rather, it may reflect our recent concerns about disease. Without doubt, however, respiratory diseases should be counted among the most serious threats to Mahale chimpanzees.

25.2.3 Unknown disease with AIDS-like symptoms

An unknown disease, described as “AIDS-like,” was presumed to have killed at least three adult females in the M group between 1985 and 1987 (Nishida *et al.*, 1990, 2003). These females exhibited progressive emaciation and hair blanching, and disappeared with their infants a few months after the onset of the disease. During this period, four other individuals, including one adult and one adolescent male and two adolescent females, exhibited similar symptoms, and the two females stopped their sexual cycling. These four individuals took 1 or 2 years to recover from the disease.

In later years, the researchers studying the M-group chimpanzees witnessed at least four more cases of

Table 25.2 **Reported epidemics of respiratory disease in the K and M groups**

Date[a]	Group	Observed symptoms	No. of individuals showing symptoms	Estimated no. of deaths[b]	No. of bodies found	References
Aug 30–at least Sep 6, 1968	K	Cough; nasal discharge	6 (21% of the group members)	0	0	Nishida, 1973: 250–3; Nishida, 1981: Figure 2 (group size cited in Nishida *et al.*, 1985)
End of Sep 1968	M	Cough	many	0?	0	Nishida, 1973: 250–3
Late Apr–the end of May 1977	K	Cough; nasal discharge; inability to emit normal vocalizations; lethargy	many	min: N/A max: 2	1	Nishida, 1981: 183–4; Uehara and Nyundo, 1983; Nishida *et al.*, 1990
Late Aug–Oct 15, 1992	M	Cough; sneeze; nasal discharge; crouch on the ground	52 (63% of the group members)	min: N/A max: 1	0	Nishida and Nakamura, 1993; MMCRP, unpublished data
Sep 28–Oct 21, 1993	M	Cough; sneeze; nasal discharge; weight loss; crouch on the ground; lethargy; tachypnea; bronchial congestion; inability to emit normal vocalizations	22 (26% of group members)	min: 11[c] max: 12[d]	6	Hosaka, 1995; Hosaka *et al.*, 2000; Shimizu *et al.*, 2002; Nishida *et al.*, 2003; Hosaka, unpublished data
Sep 10–22, 2003	M	Cough; sneeze; nasal discharge; respiratory distress; lethargy	57 (98% of the group members)	min: 2[e] max: 2[e]	1	Kaur *et al.*, 2008[e]; MMCRP, unpublished data[e]

Jul 15–Sep 7, 2005	M	cough; sneeze; nasal discharge; respiratory distress; lethargy	33 (52% of the group members)	min: 2 max: 2	1	Kaur *et al.*, 2008; MMCRP, unpublished data
Jun 3–around the end of Jul 2006	M	cough; sneeze; nasal discharge; crouch on the ground; respiratory distress; lethargy; tachypnea; signs consistent with pneumonia, malaise, and anorexia	23[f] (35% of the group members)	min: 3 max: 12	5	Hanamura *et al.*, 2006, 2008; Kaur *et al.*, 2008; Lukasik-Braum and Spelman, 2008[g]

[a] The last dates for some epidemics were not clear, because some individuals relapsed or began to show similar symptoms 1–2 weeks after the outbreaks had ceased.

[b] "Min" and "max" represent estimated numbers of "death from diseases" and "highest estimate of death from epidemics," respectively. "?" indicates uncertain number due to imperfect individual identification at the time.

[c] Including two disappearances and two deaths of three old individuals and one juvenile, who began to face aftereffects, including severe emaciation, complicated by feebleness or lack of maternal care.

[d] Including the disappearance of one infant who looked healthy but struggled without any parental transport for 3 days after his mother's disappearance.

[e] Kaur *et al.* (2008) reported a total of four deaths during the epidemic in 2003; this was a transcription error.

[f] We adopted the number reported by Hanamura *et al.* (2008), although Lukasik-Braum and Spelman (2008) reported a larger number.

[g] Although they described two other epidemics in November 2006 and October 2007, we did not list them in this table because of lack of some relevant data.

Table 25.3 **Highest estimates of deaths for different age–sex classes in recent reported epidemics of respiratory disease**

Age–sex classes			Aug–Sep 1992 ($N = 81$)	Sep–Nov 1993 ($N = 87$)	Sep 2003[a] ($N = 58$)	Jul–Sep 2005[a] ($N = 63$)	Jun–Jul 2006[a] ($N = 65$)
Infant	Male	0–4 years	1/9 (11%)	3/9 (33%)	1/4 (25%)[b]	1/7 (14%)	1/6 (17%)
	Female	0–4 years	0/7 (0%)	1/9 (11%)	1/8 (13%)[b]	1/9 (11%)	5/9 (56%)
	Unknown	0–4 years	0/0 (–)	0/0 (–)	0/0 (–)	0/0 (–)	1/1 (100%)
Juvenile	Male	5–8 years	0/4 (0%)	1/3 (33%)	0/4 (0%)	0/2 (0%)	1/3 (33%)
	Female	5–8 years	0/7 (0%)	1/9 (11%)	0/7 (0%)	0/8 (0%)	0/8 (0%)
Adolescent	Male	9–15 years	0/9 (0%)	1/10 (10%)	0/5 (0%)	0/5 (0%)	0/5 (0%)
	Female	9–12 years	0/9 (0%)	0/7 (0%)	0/3 (0%)	0/2 (0%)	0/3 (0%)
Adult	Male	≥ 16 years	0/8 (0%)	1/9 (11%)	0/7 (0%)	0/8 (0%)	1/8 (0%)
	Female	≥ 13 years	0/28 (0%)	4/31 (13%)	0/20 (0%)	0/22 (0%)	3/22 (14%)

[a] This classification of age–sex classes differs from that in Kaur *et al.* (2008), as we followed the typical classification used at Mahale (see Chapter 22). Consequently, the corresponding mortalities differ from those in the previous study.

[b] Kaur *et al.* (2008) reported a total of four deaths during the epidemic in 2003; this was a transcription error.

unknown diseases that resembled an "AIDS-like" disease, in that emaciation was the most noticeable symptom, but differed in that hair blanching was not observed. The first case was of Shike (adult male): in January 1992, when he was found after being missing for a month, he was emaciated and walked very slowly, and could not follow the other individuals (Hosaka, unpublished data). This was the last time that researchers saw him. The second case was of Ai (natal adolescent female): in June or July 1999, she began to exhibit an awkward movement in both her legs, which were without external injury. After that, she did not grow normally, and weakened. In May 2002, when she was found after being missing for 4 months, she was emaciated and frequently stumbled over the ground. One month later, she disappeared (Shimada and Matumula, 2004). The third case was of Dogura (adult male): in January 2005, he was found emaciated after being missing for a month. He stayed on a fruit tree for several days and was finally found dead the next month (Nakamura, 2004; see also Chapter 45). Finally, in July 2007, Michio (adolescent male) was found similarly emaciated after being missing for 1 month, but he gradually recovered and grew to adulthood (Michio Nakamura, personal communication).

Keele *et al.* (2009) conducted a pathological study of SIVcpz infecting Gombe chimpanzees. SIVcpz is one of over 40 different SIVs, and is the immediate precursor of HIV-1 (Gao *et al.*, 1999). Contrary to the prevailing view, SIVcpz is pathogenic, and leads to increases in mortality and decreases in female fertility in infected chimpanzees (Keele *et al.*, 2009). We should be cautious in assuming that the unknown diseases we have described here are identical to SIVcpz. Indeed, SIVcpz was not detected in fecal samples from at least some of the M-group chimpanzees at Mahale (Michael A. Huffman, personal communication; see also Sharp *et al.*, 2005). However, it is necessary to conduct a follow-up study that investigates whether Mahale chimpanzees are infected with this virus, since it could have a substantial, negative impact on their health, reproduction, and lifespan.

25.2.4 Other diseases

Skin diseases

Between May 2001 and October 2002, 13 chimpanzees of various age–sex classes in the M group had a skin disease that blanched several parts of their faces or hands, and lasted for 1–35 days (Nishida *et al.*, 2007c). This disease may have been a fungal infection, such as dermatophytosis (Nishida *et al.*, 2007c).

Subcutaneous swellings

Some chimpanzees in the M group, of various age–sex classes, developed a subcutaneous tumor as big as a ping-pong ball in their lower abdomen (at least 10 individuals between 1999 and 2007; Nishida *et al.*, 2007b). It may have been caused by a cyst, possibly due to an *Oesophagostomum stephanostomum* infection (Michael A. Huffman, personal communication; see also Chapters 24, 47). The swellings may not have been fatal, since the researchers could not confirm the occurrence of any subsequent deaths. However, Gwekulo (adult female) had such an abdominal swelling for over 10 years, which once filled with pus probably due to a secondary infection (Nishida *et al.*, 2007b). Nishida *et al.* (1990) also reported that two adult females, Chausiku (when she was still in her natal K group) and Happy, exhibited a swollen throat like a goiter. Similarly, at Gombe, an adult female had a severe goiter (Goodall, 1986a: 96).

Congenital diseases or handicaps

A male in the M group (Kasulamemba), who lived to approximately 19 years of age, may have had a congenital heart disease, because he had often been observed breathing heavily, even after slight exertion (Nishida *et al.*, 1990). A male infant (a son of Tula, M group), who died a month after birth, had patchy alopecia throughout his body when he was first observed (Hosaka *et al.*, 2000). A female infant (daughter of Christina, M group), who lived to be nearly 2 years old, had several pathological symptoms: a spinal bulge with hair thinning on the back, projection on the abdomen, thin floating sixth finger on the left hand, flaccid face, and defective grasp force in both the legs (Matsumoto *et al.*, in prep.).

Diarrhea

Diarrhea can be a symptom of various diseases, although chimpanzees sometimes have diarrhea during inter-unit-group auditory encounters (Chapters 11, 39) and intra-unit-group conflicts. Many instances of diarrhea have been reported as indicators of intestinal parasitic infection, in a series of studies on the medicinal plants used by M-group chimpanzees (Chapters 24, 47).

Age-related impairments

There have been some reports of age-related symptoms or illnesses (see Chapter 23 for details).

The variety of diseases described above represents the tip of the iceberg. Nishida (2012) speculated that the mysterious mass disappearance of the M-group chimpanzees from 1994 to 1996 (Chapter 7) might have been caused by an epidemic disease.

25.3 Multifaceted factors affecting the emergence of respiratory disease epidemics

As indicated in the previous section, infectious respiratory disease is one of the most frequent and fatal diseases in Mahale chimpanzees (Chapter 7). In this section, we discuss the seasonality of its occurrence, its age-specific mortality, and relevant meteorological and social factors, including the impact of humans as a vector for the pathogen.

25.3.1 Dataset and analyses

We compiled monthly data of (1) the number of individuals observed coughing at least once; (2) the number of infant deaths; and (3) the number of non-infant deaths in the M group over 117 months between August 1999 and July 2009, based on Nishida *et al.* (2007a) and unpublished data by MMCRP (data were not available for September 2008 and March–April 2009 due to the researchers' absence). The diagnosis of respiratory disease was only based on coughing (because it was the only symptom with available data), which was observed in all the reported respiratory disease epidemics at Mahale (Table 25.2).

We then calculated the proportion of individuals that coughed (median: 2; range: 0–53) as a percentage of the total number of individuals observed (median, 56; range: 38–78), on a monthly basis. This percentage is hereafter termed the *monthly morbidity* (note that there still remained some bias caused by different frequencies of observation during the given month). Regarding the number of deaths (hereafter *monthly mortality*), we excluded three deaths that were presumed to have been caused by factors other than disease, or caused by non-respiratory disease, but there were other deaths the cause of which is still unknown.

Figure 25.1 shows the long-term monthly morbidity and age-specific monthly mortality. If months with over 20% monthly morbidity were regarded as "epidemic periods," as in previous studies in Gombe (Williams *et al.*, 2008; note that the authors calculated morbidity over a period of up to 3 months, and included symptoms other than coughing), there were 12 epidemic periods between 1999 and 2009 (white arrows in Figure 25.1). The epidemics occurred once or twice every year, except for 2004, and sometimes continued for 2 (five cases) or 3 (one case) months. Months in which all three reported epidemics during this survey period occurred (black horizontal bars in Figure 25.1; see also Table 25.2) corresponded with the epidemic periods in this analysis.

25.3.2 Seasonality and relevant factors

Ten of the 12 epidemic periods began between May and October, which roughly corresponds to the dry season, with large diurnal temperature swings, although two of those cases began in May and October, which corresponds to the end and the beginning of the wet season, respectively (for climate, see Chapter 12). Four of the five epidemics reported before this survey period, including those in the K group, also occurred during the dry season

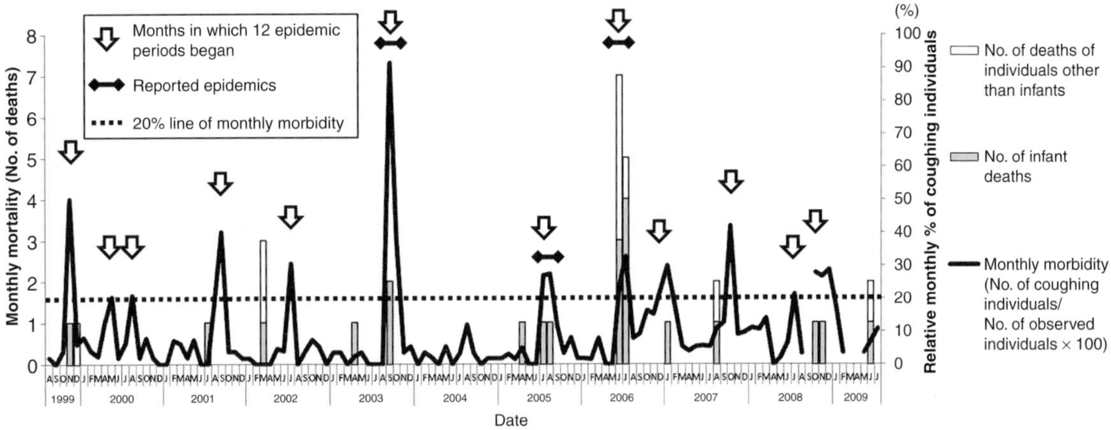

Figure 25.1 Monthly morbidity and age-specific mortality from August 1999 to July 2009. For detailed information about monthly morbidity and mortality, white arrows, and black horizontal bars, see Section 25.3.1.

(Table 25.2). These results suggest that there was seasonality in the emergence of respiratory disease epidemics in the M group and perhaps the K group, which was associated with dryness and/or large temperature swings. Similarly, at Gombe, the monthly prevalence of respiratory diseases in the study group has a strong negative correlation with rainfall (Lonsdorf *et al.*, 2011). The onset of human seasonal influenza is also related to low humidity (e.g. Shaman *et al.*, 2010).

However, the other two epidemic periods began in November and December, which is one of the two seasons that have the highest rainfall (see Chapter 12). Moreover, although there was no significant difference in mean monthly morbidity among months (Kruskal-Wallis test, $N = 12$, $H = 14.18$, $p = 0.22$), the months with higher mean monthly morbidities than the overall average (9.11%) were 5 months from July to November (range: 10.46–19.33%), and May and June had much lower mean monthly morbidities (range: 4.93–5.23%). Therefore, factors other than dryness could also have affected the emergence and spread of respiratory disease.

In fact, different types of pathogen have different relationships with meteorological factors, and the emergence of infectious diseases in humans and other animals is also affected by the contact rate between hosts, and that between hosts and vectors, if it is a vector-borne disease (Altizer *et al.*, 2006). Although there are few data on types of pathogen, following two social factors may be associated with an increase in morbidity.

In the M group, nomadic parties often begin to increase in size from July to September, and large nomadic parties are observed until about January (see Chapters 8, 9, 48). Since this season roughly corresponds with the months of higher morbidity, the degree of chimpanzee cohesiveness could also affect morbidity and the probability that epidemics would emerge (Hanamura *et al.*, 2008). In addition, July to October is the season when many tourists visit Mahale to observe the M-group chimpanzees (Chapter 48); therefore, during these months, the chimpanzees would have a higher probability of contracting human-borne pathogens and undergoing human-related stress. Fujita *et al.* (2009) conducted a multiple regression analysis of the occurrence of the chimpanzees' various disease symptoms and the number of researchers, the number of tourists, rainfall, and temperature, using detailed data collected from October 2006 to February 2009 at Mahale. The results of this analysis suggested that occurrences of coughing and diarrhea were associated with the number of tourists visiting the site.

As such, the factors that affect the emergence of respiratory disease epidemics should be considered as multifaceted.

25.3.3 High infant mortality

If we regard all the deaths during the 12 epidemic periods in Figure 25.1 as deaths due to the epidemics, there were six epidemics without deaths, five epidemics with one or two infant deaths, and one epidemic with seven infant and five non-infant deaths. This suggests that respiratory disease epidemics are associated with infant deaths, although some of these deaths may not have been caused directly by their own infections, but by a temporary reduction in care due to their mothers' infections (Hanamura *et al.*, 2008).

High infant mortality caused by respiratory diseases is also reported in Taï (Kuehl *et al.*, 2008). The epidemics that caused multiple infant deaths in Taï were followed by infant abundance the next year (caused by synchronization of the reproductive cycles of their mothers), which cyclically facilitated such epidemics when the infants began to play with each other (Kuehl *et al.*, 2008). Therefore, attention should be paid to such self-organized cycles of epidemic emergence related to infant deaths in future studies.

25.3.4 Possible human-borne pathogens and their infection routes

Kaur *et al.* (2008) analysed the fecal samples from M-group chimpanzees who had exhibited respiratory symptoms during the epidemics in 2005 and 2006. They found a paramyxovirus in some of these samples using electron microscopy, and genetically detected a virus that was closely related to human metapneumovirus (hMPV) in one sample collected in 2006. Because many kinds of paramyxovirus (including hMPV) are known to cause human respiratory disease (e.g. Chang and Dutch, 2012), they may have been the pathogens responsible for the two epidemics. Continuous research in Taï found that the chimpanzees were infected with hMPV and another paramyxovirus, human respiratory syncytial virus (hRSV), in conjunction with bacterial infections (Köndgen *et al.*, 2008; see Table 25.1). Since captive chimpanzees are highly susceptible to ubiquitous human viruses, many human pathogens could cause the next outbreak in wild chimpanzees (Kooriyama *et al.*, 2013).

There are several possible routes by which human-borne pathogens could have been transmitted to M-group chimpanzees. One possibility is that they may have been transported by local people from nearby villages. However, there was no relationship between the timing of the epidemics in the M group and the number of outpatients with the common cold at the local hospital (Nishida *et al.*, 2007a). Moreover, after the Mahale area became a national park in 1985, the distance between the nearest village and the home range of the M group is approximately 10 km, and few local people enter the national park.

Another possibility is that human-borne pathogens may have traveled with researchers and/or tourists from abroad or may have been transmitted by workers in the park. Indeed, the hMPV strain detected in one sample during the epidemic in 2006 was Japanese (Kaur *et al.*, 2008), demonstrating that the virus was introduced into the area from Japan. Moreover, just before, or around the same time, as the epidemics in 1977 (K group) and 2003 (M group) (see Table 25.2), a similar respiratory disease had infected some researchers, local research assistants, and park staff (Uehara and Nyundo, 1983; Zamma, 2005).

It is difficult to predict the timing, but there is always a risk of another human-borne disease outbreak occurring. Because separating habituated chimpanzees from humans is currently not realistic (as they are subject to ongoing research and tourism), the continuous health monitoring of chimpanzees is necessary, and viewing regulations (such as wearing masks and prohibiting humans that exhibit signs of respiratory disease) should be strictly observed (Chapters 48, 50).

25.4 Sociality in regard to the diseases and deaths of others

Our review of different diseases and the long-term survey of respiratory disease suggest that chimpanzees routinely interact with sick unit-group members, and are sometimes faced with the death of those individuals. In this section, chimpanzee attitudes toward sick and dead conspecifics are reviewed based on cases in Mahale and other field sites.

25.4.1 Attitudes to sick individuals

Goodall (1986a: 385–6) described two cases in Gombe that suggest that chimpanzees care for weakened individuals. One case was of an adult male (Humphrey) that attended an old, intimate male (Mr. McGregor) crippled by polio, even protecting him from a dominant male's attack, for as long as 14 days until he died. The other case was of an adolescent female (Honey Bee) that remained near her mother (Madam Bee), who had been fatally injured. She groomed her and kept flies away from her for the last 5 days of her life. In contrast, Goodall (1986b) reported that when most adult males and females met unit-group members that were paralysed in one arm (an adolescent male, Pepe) or both legs (Mr. McGregor) due to polio, they shunned or attacked those individuals, exhibited grins of fear, embraced each other for reassurance, watched the victims from a distance, and/or uttered a "huu" call of unease or puzzlement.

In Mahale, the chimpanzees have exhibited an ambivalent attitude toward sick individuals. An adolescent female, Ai, who was seriously weakened and tended to move alone, was attacked three times by non-kin adult females, but another non-kin adult female and an adolescent male intervened during two of the attacks (Shimada and Matumula, 2004). In weakened Shike's case, a non-kin adolescent male (Masudi) exhibited mixed reactions when Shike, who used to be a beta male before a 40-day absence, briefly rejoined the others (Hosaka, unpublished data). Masudi's first reaction was submissiveness, as when Shike was healthy. His second reaction was to harass Shike. When a middle-ranking male (Aji) charged Shike, Masudi also charged, from a distance. His final and most consistent reaction was keeping in close proximity. Masudi remained nearby, grooming and walking with Shike, after the other individuals had gone ahead and left Shike, who walked too slowly. The two remained together for more than 8.5 h until they made their beds. Shike benefited from this association by eating the *Saba comorensis* fruit that had been left over by his companion, because he could not climb high trees himself. However, there was no clear evidence of Masudi's compassion to care for Shike, and overall, his behavior could better be interpreted as curiosity.

Although chimpanzee attitudes to sick individuals vary from aggression, unease, ambivalence, to caring, they seem to result from unusual movement or lack of normal movement by the sick individuals. M-group chimpanzees paid no particular attention to the individuals having a subcutaneous swelling, who did not exhibit marked behavioral changes related to the affected body parts (Nishida *et al.*, 2007b).

25.4.2 Depression after losing close kin

Goodall (1986a: 101–4) reported that some infants and juveniles in Gombe became depressed (i.e. low activity and poor appetite) for a while after they had lost their mothers. Nishida (1981: 93) described a similar but opposite case, in that one M-group mother exhibited symptoms of depression after she had lost her infant. Shortly afterward, she was observed carrying another female's infant, which is a rare occurrence in parous females. During a respiratory disease epidemic in 2006, a sick mother frequently emitted "hoo" calls (Goodall, 1986a: 131), or a "lost" call, in the evening of the day when her infant, who was also sick, disappeared (Hanamura, personal observation; the infant was found dead 3 days later).

25.4.3 Attitudes to the dead bodies of conspecifics

Most reports of chimpanzee behavior toward dead conspecifics are related to the carrying of infant corpses by their mothers (Figure 25.2). This behavior is common to many primate species (Anderson, 2011) and the observed cases, particularly the prolonged ones, mainly occur during the dry season, or in species living in cold regions where corpse mummification is facilitated (Fashing *et al.*, 2011).

Figure 25.2 Wakusi carrying her mummified infant in 2006. ©Michio Nakamura.

In chimpanzees at Bossou, Guinea, all three cases that lasted for 19–68 days occurred during the dry season (Biro, 2011). In Mahale, of the 11 reported cases that lasted for 1–111 days, four and seven cases began in the dry season and during the transitional period, respectively (Table 25.4). However, contrary to expectations, two prolonged cases lasted until the wet season.

Some mothers in Mahale that were carrying their dead infants waved flies away from and/or groomed the corpses, without any sign of depression (Nishida, 1973: 251; Hosaka *et al.*, 2000), and one mother, Vera, even carried a stillborn infant that was her first child (Kooriyama, 2009). However, Goodall (1986a: 383) notes that the mothers at Gombe often handled their infant corpses in a rough manner. One mother at Mahale, Wakusi, sucked the bared bone from the neck of her mummified infant corpse several times (Hanamura, personal observation). Therefore, they seem to understand that their infants had changed into something different from what they were when they were alive or differed from the living infants of other mothers perhaps due to lack of any reaction.

Some individuals, other than mothers, at Mahale sometimes exhibited "exploratory" behavior, such as peering, smelling, and/or touching the dead infants, particularly if the corpse gave off a putrid smell (Nishida, 1973: 251; Hosaka *et al.*, 2000). Moreover, although there was one case in which an adult male snatched a dead infant from its mother and might have eaten it (Nishida, 1998), there were three other cases of individuals (other than the mother) carrying dead infants for up to a few hours without eating them (Table 25.4; see also Notman and Munn, 2003, for a similar case at Budongo, Uganda). In one case, an adolescent male (Bonobo) carried and groomed the corpse of an infant that had just died, and stared at it as if exploring the condition (Hosaka *et al.*, 2000). He continued these behaviors for over 2 h until an adult male snatched it away from him. In another case, a juvenile female (Serena) borrowed a mummified infant corpse from its mother, playfully carried it, and even played "airplane" with it (Hosaka *et al.*, 2000). These various behaviors may reflect differences in the performers' motivations, age-sex classes, and relationships with the infants, as well as the contexts of the deaths and the degree of decay of the corpses.

Compared to such attitudes to dead infants, those to dead or dying adults seem to be much stronger (Gombe: Teleki, 1973; Stewart *et al.*, 2012; Mahale:

Table 25.4 **Reported cases of dead infant carrying in the K and M groups**

			Performer		Dead infant			
Date[a] (Season[b])	Estimated duration	Group	Name	Identity (age[c] in yr)	Name	Age and sex	Inferred cause of death	References
Aug 6–Nov 5, 1968[d] (D to W)	> 90 d	K→M[e]	Wankungwe	Mother (32?)	WK66	2 yr ♂	Pyogenesis of bodily product or external injury on the back	Nishida, 1973: 250–3
Aug 1971 (D)	N/A	K	Wakasunga	Mother (17?)	WA70	1 yr ♂	External injury	Nishida, 1973: 250–3
May 1977 (T)	N/A	K	Wasalamba	Mother (31?)	Nina	2 mo ♂	Respiratory disease	Nishida, 1981: 183–4
Sep 1, 1992 (D)	1 d	M	Gwamwami	Mother (26)	Garbo	3 yr ♀	Two serious wounds	Nishida, 1998; Hosaka, unpublished data
May 13–19, 1993 (T)	7 d	M	Christina	Mother (18?)	XT93	< 3 wk ♂	Unknown	Hosaka *et al.*, 2000
May 15–Jun 9, 1993 (T to D)	26 d	M	Tula	Mother (12)	TL93	< 2 mo ♂	Feebleness, with patchy alopecia	Hosaka *et al.*, 2000
Oct 11, 1993–Jan 26, 1994[f] (T to W)	111–142 d	M	Gwamwami	Mother (27)	GM93	3.5–4.5 mo ♂	Respiratory disease	Hosaka *et al.*, 2000
Oct 21–7, 1993 (T)	7 d	M	Nkombo	Mother (23)	NK93	3.5 mo ♂	Respiratory disease	Hosaka *et al.*, 2000
Jul 1–22, 2006 (D)	22 d	M	Christina	Mother (31?)	XT05	1 yr ♀	Respiratory disease	Hanamura *et al.*, 2008; MMCRP, unpublished data

Table 25.4 (**cont.**)

			Performer		Dead infant			
Date[a] (Season[b])	Estimated duration	Group	Name	Identity (age[c] in yr)	Name	Age and sex	Inferred cause of death	References
Jul 15–Sep 15, 2006 (D)	90 d	M	Wakusi	Mother (45?)	WX05	1 yr ♀	Respiratory disease	Hanamura *et al.*, 2008; MMCRP, unpublished data
Oct 10, 2008[g] (T)	< 2 d	M	Vera	Mother (14?)	VR08	< 2 d ♂	Stillborn	Kooriyama, 2009, unpublished data
Oct 29, 1993 (T)	≥ 2 h	M	Bonobo[h]	Adol. ♂ (12)	Unknown	< 3 d ♂	Infanticide or stillborn	Hosaka *et al.*, 2000
Nov 13, 1993 (W)	< a night	M	Serena[h]	Juv. ♀ (6)	GM93	3.5–4.5 mo ♀	Respiratory disease	Hosaka *et al.*, 2000
Oct 10, 2008[g] (T)	< 30 min	M	Carmen	Adol. ♀ (11)	VR08	< 2 d ♂	Stillborn	Kooriyama, 2009, unpublished data

[a] From the date of the first observation to that of the last observation.

[b] D, dry season, June–September; W, wet season, November–April; T, transitional period, May and October (see Chapter 12).

[c] "?" indicates an estimated age.

[d] The researcher left the study site on this date.

[e] The mother moved also with the M-group members, but also interacted with the K-group members until the end of 1971 (Nishida, 1973).

[f] The mother had lost the dead infant on some unknown day between January 26 and February 26.

[g] The mother left the dead infant behind because of two adult males' displays toward her, and the two adult males and one adolescent female (Carmen) continued to keep the corpse, although the mother tried to retrieve it several times. Finally, Carmen picked it up and carried it for up to 30 min, before discarding it.

[h] See text for the details.

Hosaka *et al.*, 2000; Nishida, 2012: 252–4; Taï: Boesch and Boesch-Achermann, 2000: 248–51; Budongo: Reynolds, 2005: 45–8). That is, chimpanzees repeated "wraa" and "huu" calls that are thought to express fear or unease (Chapter 39), watched a dead body or a dying individual intently from a distance for a while, or performed charging displays toward it or toward nearby individuals. Such upset attitudes to dead or dying adult conspecifics may reflect the chimpanzees' deeper relationships with those individuals, and the more noticeable changes in adults compared with those in dead infants. However, after a period of time, they also exhibited calmer behaviors, such as resting nearby while looking at it, waving flies away from it with grins and "huu" calls (Hosaka *et al.*, 2000), and grooming it (Boesch and Boesch-Achermann, 2000: 248–50). In addition, there is one case that when one adult male found an adult corpse alone, he passed the corpse just looking at it (Hosaka *et al.*, 2000). Therefore, chimpanzees who find a corpse together may facilitate and alleviate each other's upset attitude to the corpse through their interactions.

Future studies should continue to collect and compile cases of chimpanzee behaviors when dealing with sick or dead unit-group members or their disappearances, because researchers have witnessed only a handful of cases even after decades of research. Knowledge of chimpanzee attitudes to such events may provide an insight into differences with humans, and the evolution of death awareness in humans. This knowledge is also important in understanding how the social lives of chimpanzees and humans are affected by such events.

References

Altizer, S., Dobson, A., Hosseini, P., *et al.* (2006). Seasonality and the dynamics of infectious diseases. *Ecological Letters*, **9**, 467–84.

Anderson, J. R. (2011). A primatological perspective on death. *American Journal of Primatology*, **73**, 410–14.

Biro, D. (2011). Chimpanzee mothers carry the mummified remains of their dead infants: three case reports from Bossou. In *The Chimpanzees of Bossou and Nimba*, ed. T. Matsuzawa, T. Humle, and Y. Sugiyama. Tokyo: Springer, pp. 241–50.

Boesch. C. (2008). Why do chimpanzees die in the forest? The challenges of understanding and controlling for wild ape health. *American Journal of Primatology*, **70**, 722–6.

Boesch, C. and Boesch-Achermann, H. (2000). *The Chimpanzees of Taï Forest: Behavioral Ecology and Evolution*. Oxford: Oxford University Press.

Chang, A. and Dutch, R. E. (2012). Paramyxovirus fusion and entry: multiple paths to a common end. *Viruses*, **4**, 613–36.

Fashing, P. J., Nguyen, N., Barry, T. S., *et al.* (2011). Death among geladas (*Theropithecus gelada*): a broader perspective on mummified infants and primate thanatology. *American Journal of Primatology*, **73**, 405–9.

Formenty, P., Boesch, C., Wyers, M., *et al.* (1999). Ebola virus outbreak among wild chimpanzees living in a rain forest of Côte d'Ivoire. *Journal of Infectious Disease*, **179** (supplement 1), S120–6.

Fujita, S., Zamma, K., Hanamura, S., *et al.* (2009). [Influence of ecotourism on health status of wild chimpanzees in the Mahale Mountains National Park.] *Primate Research*, **25** (supplement), S-11. In Japanese.

Gao, F., Bailes, E., Robertson, D. L., *et al.* (1999). Origin of HIV-1 in the chimpanzee *Pan troglodytes troglodytes*. *Nature*, **397**, 436–41.

Goodall, J. (1986a). *The Chimpanzees of Gombe: Patterns of Behavior*. Cambridge, MA: Bellknap.

Goodall, J. (1986b). Social rejection, exclusion, and shunning among the Gombe chimpanzees. *Ethology and Sociobiology*, **7**, 227–36.

Hanamura, S., Kiyono, M., Nakamura, M., *et al.* (2006). A new code of observation employed at Mahale: prevention against a flu-like disease. *Pan Africa News*, **13**, 13–16.

Hanamura, S., Kiyono, M., Lukasik-Braum, M., *et al.* (2008). Chimpanzee deaths at Mahale caused by a flu-like disease. *Primates*, **49**, 77–80.

Hosaka, K. (1995). Mahale: a single flu-like epidemics killed at least 11 chimps. *Pan Africa News*, **2**(2), 3–4.

Hosaka, K., Matsumoto-Oda, A., Huffman, M. A., and Kawanaka, K. (2000). [Reactions to dead bodies of conspecifics by wild chimpanzees in the Mahale Mountains, Tanzania.] *Primate Research*, **16**, 1–15. In Japanese with English summary.

Humle, T. (2011). The 2003 epidemic of a flu-like respiratory disease at Boosou. In *The Chimpanzees of Bossou and Nimba*, ed. T. Matsuzawa, T. Humle, and Y. Sugiyama. Tokyo: Springer, pp. 325–33.

Kaur, T., Singh, J., Tong S., *et al*. (2008). Descriptive epidemiology of fatal respiratory outbreaks and detection of a human-related metapneumovirus in wild chimpanzees (*Pan troglodytes*) at Mahale Mountains National Park, Western Tanzania. *American Journal of Primatology*, **70**, 755–65.

Keele, B. F., Jones, J. H., Terio, K. A., *et al*. (2009). Increased mortality and AIDS-like immunopathology in wild chimpanzees infected with SIVcpz. *Nature*, **460**, 515–19.

Köndgen, S., Kühl, H., N'Goran P. K., *et al*. (2008). Pandemic human viruses cause decline of endangered great apes. *Current Biology*, **18**, 260–4.

Kooriyama, T. (2009). The death of a newborn chimpanzee at Mahale: reactions of its mother and other individuals to the body. *Pan Africa News*, **16**, 19–21.

Kooriyama, T., Okamoto, M., Yoshida, T., *et al*. (2013). Epidemiological study of zoonoses derived from humans in captive chimpanzees. *Primates*, **54**, 89–98.

Kuehl, H. S., Elzner, C., Moebius, Y., Boesch, C., and Walsh, P. D. (2008). The price of play: self-organized infant mortality cycles in chimpanzees. *PLoS ONE*, **3**, e2440.

Leendertz, F. H., Ellerbrok, H., Boesch, C., *et al*. (2004). Anthrax kills wild chimpanzees in a tropical rainforest. *Nature*, **430**, 451–2.

Lonsdorf, E. V., Travis, D., Pusey, A. E., and Goodall, J. (2006). Using retrospective health data from the Gombe chimpanzee study to inform future monitoring efforts. *American Journal of Primatology*, **68**, 897–908.

Lonsdorf, E. V., Murray, C. M., Lonsdorf, E. V., *et al*. (2011). A retrospective analysis of factors correlated to chimpanzee (*Pan troglodytes schweinfurthii*) respiratory health at Gombe National Park, Tanzania. *EcoHealth*, **8**, 26–35.

Lukasik-Braum, M. and Spelman, L. (2008). Chimpanzee respiratory disease and visitation rules at Mahale and Gombe National Parks in Tanzania. *American Journal of Primatology*, **70**, 734–7.

Nakai, M., Itoh, N., Nakamura, M., Huffman, M. A., and Nishida, T. (2004). [Individual identification from skeletal remains: cases of wild chimpanzees of the Mahale Mountains National Park, Tanzania.] *Primate Research*, **20**, 1–9. In Japanese with English summary.

Nakamura, M. (2004). [In memoriam of Dogura.] *Mahale Chimpun*, **3**, 1–2. In Japanese.

Nakazawa, N., Hanamura, S., Inoue, E., Nakatsukasa, M., and Nakamura, M. (2013). A leopard ate a chimpanzee: first evidence from East Africa. *Journal of Human Evolution*, **65**, 334–7.

Nishida, T. (1973). [*The Children of the Mountain Spirits: Search for the Social Structure of Chimpanzees*.] Tokyo: Chikuma-shobō. In Japanese.

Nishida, T. (1981). [*The World of Wild Chimpanzees*.] Tokyo: Chūkō-shinsho. In Japanese.

Nishida, T. (1998). Deceptive tactic by an adult male chimpanzee to snatch a dead infant from its mother. *Pan Africa News*, **5**, 13–5.

Nishida T. (2012). *Chimpanzees of the Lakeshore: Natural History and Culture at Mahale*. Cambridge: Cambridge University Press.

Nishida, T. and Nakamura, M. (1993). Chimpanzee tool use to clear a blocked nasal passage. *Folia Primatologica*, **61**, 218–20.

Nishida, T., Hiraiwa-Hasegawa, M., Hasegawa, T., and Takahata, Y. (1985). Group extinction and female transfer in wild chimpanzees in the Mahale National Park, Tanzania. *Zeitschrift für Tierpsychologie*, **67**, 284–301.

Nishida, T., Takasaki, H., and Takahata, Y. (1990). Demography and reproductive profiles. In *The Chimpanzees of the Mahale Mountains: Sexual and Life History Strategies*, ed. T. Nishida. Tokyo: University of Tokyo Press, pp. 63–97.

Nishida, T., Corp, N., Hamai, M., *et al*. (2003). Demography, female life history, and reproductive profiles among the chimpanzees of Mahale. *American Journal of Primatology*, **59**, 99–121.

Nishida, T., Fujimoto, M., Fujita, S., *et al*. (2007a). [Prevalence of infectious disease among the chimpanzees of Mahale.] *Primate Research*, **23** (supplement), S-6. In Japanese.

Nishida, T., Fujita, S., Inaba, A., and Kooriyama T. (2007b). Note on a subcutaneous tumor found among wild chimpanzees. *Pan Africa News*, **14**, 31–2.

Nishida, T., Fujita, S., Matsusaka, T., Shimada, M., and Kitopeni, R. (2007c). Dermatophytosis of M group chimpanzees, Mahale Mountains, Tanzania. *Pan Africa News*, **14**, 5–6.

Notman, H. and Munn, J. (2003). A case of infant carrying by an adult male chimpanzee in the Budongo Forest. *Pan Africa News*, **10**, 7–9.

Reynolds, V. (2005). *The Chimpanzees of the Budongo Forest: Ecology, Behaviour, and Conservation*. Oxford: Oxford University Press.

Sakamaki, T., Mulavwa, M., and Furuichi, T. (2009). Flu-like epidemics in wild bonobos (*Pan paniscus*) at Wamba, the Luo Scientific Reserve, Democratic Republic of Congo. *Pan Africa News*, **16**, 1–4.

Shaman, J., Pitzer, V. E., Viboud, C., Grenfell, B. T., and Lipstich, M. (2010). Absolute humidity and the seasonal onset of influenza in the continental United States. *PLoS Biology*, **8**, e1000316.

Sharp, P. M., Shaw, G. M., and Hahn, B. H. (2005). Simian immunodeficiency virus infection of chimpanzees. *Journal of Virology*, **79**, 3891–902.

Shimada, M. and Matumula, M. (2004). Chimpanzees attitude toward a seriously weakened adolescent female at Mahale. *Pan Africa News*, **11**(1), 6–8.

Shimizu, D., Gunji, H., Hashimoto, H., *et al.* (2002). The four chimpanzee skulls collected in the Mahale Mountains, Tanzania. *Anthropological Science*, **110**, 251–66.

Stewart, F. A., Piel, A. K., and O'Malley R. C. (2012). Responses of chimpanzees to a recently dead community member at Gombe National Park, Tanzania. *American Journal of Primatology*, **74**, 1–7.

Teleki, G. (1973). Group response to accidental death of a chimpanzee in Gombe National Park, Tanzania. *Folia Primatologica*, **20**, 81–94.

Uehara, S. and Nyundo, R. (1983). One observed case of temporary adoption of an infant by unrelated nulliparous females among wild chimpanzees in the Mahale Mountains, Tanzania. *Primates*, **24**, 456–66.

Wallis, J. and Lee, D. R. (1999). Primate conservation: the prevention of disease transmission. *International Journal of Primatology*, **20**, 803–26.

Williams, J. M., Lonsdorf, E. V., Wilson, M. L., *et al.* (2008). Causes of death in the Kasekela chimpanzees of Gombe National Park, Tanzania. *American Journal of Primatology*, **70**, 766–77.

Wolfe, N. D., Escalante, A. A., Karesh, W. B., *et al.* (1998). Wild primate populations in emerging infectious disease research: the missing link? *Emerging Infectious Disease*, **4**, 149–58.

Zamma, K. (2005). [Reconsidering the relationships between humans and chimpanzees: a case of Mahale.] *Abstract of Presentation for SAGA8 Symposium*, p. 8. In Japanese.

26 Conspecific killings

MICHIO NAKAMURA AND NORIKO ITOH

26.1 Introduction

Conspecific killings or killing of individuals of one's own species were considered to be unique to humans at the time of Conrad Lorenz (Lorenz, 1963). Although aggression, in general, does occur within species of other animals, it was thought that there were some mechanisms in nonhuman animals that suppress aggression and prevent killings. Therefore, infanticide among hanuman langurs (*Semnopithecus entellus*) was a striking discovery (Sugiyama, 1965). The finding was at first rejected or at best considered a pathological behavior. Later, it was consistently placed in the sociobiological paradigm where infanticides by langurs were explained in terms of a normal male reproductive strategy (Hrdy, 1977). When it was found that infanticides also occurred in several other animal species as part of male strategies under certain conditions, conspecific killing was not regarded unique to our own species any more but quite widespread in the animal kingdom. Since infanticide has been confirmed in 35 primate species (Palombit, 2012), several authors now include it as an important factor when dealing with primate social evolution within socioecological theory (e.g. van Schaik, 2000).

Infanticide by chimpanzees (more accurately, cannibalism that may be inferred as infanticide) was first observed at Budongo Forest in Uganda (Suzuki, 1971), followed by direct observation of infanticide and cannibalism at Gombe, Tanzania (Bygott, 1972). Since then, there have been dozens of similar cases of infanticide and/or cannibalism at various chimpanzee study sites (reviewed in Manson and Wrangham, 1991; Wrangham, 1999; Wilson and Wrangham, 2003). Not only infants, but even adults are sometimes victims of chimpanzee conspecific killings. A famous example is the case at Gombe, sometimes referred to as the "war" (Goodall, 1986), where several adult males and females of the Kahama group were killed by the members of the Kasekela group. More recently, Ngogo chimpanzees at Kibale, Uganda, killed or fatally wounded 18 chimpanzees from other groups and consequently expanded their range (Mitani *et al.*, 2010).

Conspecific killings of chimpanzees may be shocking to many people because one cannot imagine such brutal acts by cute, clever, funny, and human-like chimpanzees they often see on TV and in zoos. It has also received wide academic attention and important discussions of such killings in chimpanzees address the possible origin of violence and warfare in our own species (e.g. Wrangham and Peterson, 1996). Because of the impact of the topic and perhaps because of the infrequency of the killings, some ambiguous cases are often accepted as circumstantial evidence and sometimes interpreted as killings.

In this chapter, we review both confirmed and suspected cases of conspecific killings at Mahale. We specifically focus on several cases to critically validate their interpretations.

26.2 Definition of conspecific killings

When researchers find a dead body of a chimpanzee after an unusual uproar among the chimpanzees, they may infer that conspecific killing has occurred.

Mahale Chimpanzees: 50 Years of Research, ed. M. Nakamura, K. Hosaka, N. Itoh and K. Zamma. Published by Cambridge University Press.

However, in such cases, there are other possible reasons for chimpanzee uproar: it may be because chimpanzees found a dead body or a dying chimpanzee or it may not be related to death. Therefore, in order to determine whether a conspecific killing *actually* occurred, we need to confirm the following three criteria:

26.2.1 Attacks prior to death

First, we have to confirm that attacks on a chimpanzee by other chimpanzees did occur immediately prior to his/her death. If cannibalism had been taking place at the time a researcher started the observation, we cannot determine whether conspecific killing actually occurred or cannibalism followed death caused by other factors.

26.2.2 Deaths of the victim

Conversely, even if severe attacks have been observed, it does not necessarily mean that they caused the death of the victim; for example, as Sakamaki *et al.* (2001) reported, attacks that seemed to have been premeditated to kill a baby did not result in the death of the victim. The victim, a female infant later named Rubicon, was not injured in the attacks, but she later lost several of the fingers on her right hand and left foot while still an infant. It was assumed that she was severely attacked afterward, although the presumed attack that led to the loss of her fingers was not observed. She survived and grew into a healthy adolescent female; she was last seen when she was 12 years and 5 months old, and we assumed that she emigrated to a different group at that age.

Furthermore, when a victim is an individual from an unhabituated group and we lose sight of the victim, sometimes we cannot confirm its death. In such cases, it is possible that the fleeing individual may have survived the attack.

26.2.3 Injuries to the dead body inflicted by other chimpanzees

A dead body is not evidence for conspecific killing, although researchers sometimes assume so after finding a half-decomposed or even skeletonized body. Because fresh injuries cannot be confirmed in such cases, we cannot ascertain whether death was caused by chimpanzee attacks. In order to infer conspecific killing, we need to confirm fresh injuries on a dead chimpanzee that suggest the chimpanzee attacks.

When all three of the above criteria (1–3) are confirmed, we consider death by conspecific killing as certain (C), and when two of the three criteria are confirmed, we consider death by conspecific killing as suspected (S). When only one or none of the criteria is confirmed, death by conspecific killing is not sufficiently convincing, although based on the circumstantial evidence, it cannot be ruled out. Here, we included the latter type of cases if it has already been inferred that conspecific killings might have occurred.

26.3 Conspecific killings at Mahale

At Mahale, from 1965 to 2013, 24 cases (29 victims) have been claimed or suggested to be due to conspecific killings (Table 26.1). According to the criteria described above, there were only five (0.10 cases/year) certain (hereafter, C) cases, and with the addition of suspected cases (hereafter, C+S cases), the number increased to 12 (0.24 cases/year). These figures are likely underestimated because of cases of actual killings that were not witnessed by observers. Among the 12 C+S cases, 8 were intragroup infanticides (4 C cases), 2 were intragroup killings of adults (1 C case), and 2 were intergroup infanticides (no C case). Notably, all the victims among these 12 C+S cases at Mahale were males. However, as stated above, Sakamaki *et al.* (2001) reported a case of an attempted killing of a female infant and several other cases of possible killings of female infants have been recorded (cases 11, 16, and 21 in Table 26.1).

Although, there have been cases of female-led infanticides at other chimpanzee study sites, such as Gombe (Goodall, 1977) and Budongo (Townsend *et al.*, 2007), there has not been a single case of a female committing a killing at Mahale. In other words, in all observed fatal attacks at Mahale, the attackers were always adult males.

Table 26.1 **Confirmed and suspected conspecific killings at Mahale**

Case no.	Date	Victim	Age	Sex	Group of attacker/victim	Evidence[a] Attack	Body	Wound	Certainty[b]	References
1	1970–83	Six males[c]	Adult	♂	M/K	+				Nishida *et al.*, 1985
2	1974/4/25	Shigeo[c]	3 yr	♂	K/M[d]		○	○	S	Nishida *et al.*, 1979
3	1976/1/15	Lukanda[c]	1.5 yr	♂	M/K			+[e]		Nishida *et al.*, 1979
4	1977/1/13	Humbe[c]	2.5 mo	♂	M/M	+	○	○	S	Norikoshi, 1982
5	1979/6/14	Wakasunga's	1.5 mo	♂	M/M[f]		○	○	S	Kawanaka, 1981
6	1981	Wakasunga's	6 days	♂	Unknown/M		○			Hiraiwa-Hasegawa *et al.*, 1984
7	1981	Wally	2 mo	♂	Unknown/M			+[e]		Hiraiwa-Hasegawa *et al.*, 1984
8	1983/7/5	Wantendele's	1 mo	♂	M/M	○	○	○	C	Takahata, 1985
9	1983/12/15	Chausiku's	3 mo	♂	M/M		○	○	S	Nishida and Kawanaka, 1985
10	1985/7/16	Tomato's	10 mo	♂	M/M	○	○	○	C	Hamai *et al.*, 1992
11	1987	Gwamwami's	10 mo	♀	Unknown/M	+				Hamai *et al.*, 1992
12	1988	Fanta's		♂	Unknown/M					Hamai *et al.*, 1992
13	1989/10/3	Mirinda's	6 mo	♂	M/M	○	○	○	C	Hamai *et al.*, 1992
14	1989	Juno's[c]		♂	Unknown/M					Hamai *et al.*, 1992
15	1990/7/25	Betty's	5 mo	♂	M/M	○	○	○	C	Hamai *et al.*, 1992
16	1992/9/1	Garbo	10 mo	♀	Unknown/M			○		Nishida, 1998
17	1993/10/29	Unknown	2 days	♂	M/M[d]		○	○	S	Hosaka *et al.*, 2000
18	1995/11/14	Ntologi	40 yr	♂	Unknown/M		○[g]	○	S	Nishida, 1996; Itoh, personal observation

19	1996	Jilba	21 yr	♂	M/M		○			Wrangham, 1999; Itoh, personal observation
20	1998/5/12	Rubicon[h]	1 day	♀	M/M	○				Sakamaki *et al.*, 2001
21	1999	Pinky's[c]	14 mo	♀	Unknown/M			+[d]		Sakamaki *et al.*, 2001
22	2000/12/12	Unknown	2.5 mo	♂	M/Unknown (Y?)	○		○	S	Kutsukake and Matsusaka, 2002
23	2003/6/15	Unknown	–	–	M/Unknown		○			MMCRP unpublished record
24	2011/10/2	Pimu	23 yr	♂	M/M	○	○	○	C	Kaburu *et al.*, 2013

[a] Attack: attack observed; Body: dead body seen; Wound: wound likely from observed chimpanzee attacks; ○: certain evidence; +: circumstantial evidence.
[b] C: certain; S: suspected.
[c] Inferred from missing.
[d] Inferred.
[e] The mother had wounds and was without the infant.
[f] She and her baby were also seen in the K group several times this year.
[g] Seriously weakened body was found and he died afterwards.
[h] She survived the attempted infanticide.

26.4 Infanticides

A characteristic feature of Mahale infanticides is that the majority occurred within a group. Among the 10 C+S infanticides, 8 were killings of M-group infants by M-group males. Among these eight cases, attacks and deaths were confirmed only in four cases, and in the other four cases, infanticide was inferred from observations of males holding and/or eating already dead infant bodies. Because details of most infanticide cases have already been published in English, here we show a case that was reported only in Japanese.

> On October 29, 1993, an adolescent male, Bonobo, was first observed carrying an unknown dead male infant. The umbilical cord was still attached to the infant. The infant was bleeding on the right side of the head and around the nose. Bonobo carried it for about 2 h but did not eat it (instead, he groomed it). Finally, the infant was snatched by Toshibo (an adult male), who ran away (Hosaka *et al.*, 2000, our translation).

Hosaka *et al.* (2000) suggested that the bleeding that was observed raises the possibility of infanticide. Because there have been two reported cases of infant snatching after stillbirth or infant's death from injury (Nishida, 1998; Kooriyama, 2009), we cannot reject the possibility that the infant died of other reasons and Bonobo somehow obtained the dead body afterwards (therefore, we consider this case as suspected).

Among 181 infants born to the M group in 33 years, from 1980 to 2012 (Chapter 7), 6 (3.3%) were victims of infanticides (C+S: cases 7, 8, 9, 12, 14, and 16 in Table 26.1). Strangely, all six cases occurred before 1993, and after 1994, there was no case of newborn infanticide in the M group. The percentage of victims per birth significantly differed between these two periods (1980–93: 6/99; 1994–2012: 0/82; Fisher's exact test, $p = 0.033$).

What could these biased occurrences of infanticide mean? Simple observation bias may not explain these numbers because observation conditions improved over time (better habituation and increased number of observation trails), and there was a higher number of researchers in the second period. All the victims in the 1980s were killed by the M-group males, which might be explained by some kind of social instability. The 1980s were atypical for the M group, because the group accepted many parous K-group females after the extinction of the K group (Chapter 10), thus increasing the group size to an unusual extent (see Chapter 7). In such a situation, increased numbers of individuals might have caused social instability and augmented social conflicts, which might have led to the increase in infanticides in the same period.

Another notable detail is that during 1980–93, the period of high level of infanticide at Mahale, Ntologi was the alpha male; his tenure as an alpha male was much longer than that of other alpha males (see Chapter 5). After Ntologi fell from the alpha status, infanticides did not occur under other alpha males. Thus, it is necessary to consider the possibility that the personality of the alpha male may influence the occurrence of infanticides.

26.5 Killings of adults

Among conspecific killings, killings of adults – often in the form of coalitionary killings – are especially notable in chimpanzees (Wrangham, 1999). The most famous case is that of Gombe where individuals of Kahama group, a splinter group from Kasekela group, were killed one after the other by the Kasekela males (Goodall, 1986). Because many incidents of these killings were actually observed, this case is direct evidence for chimpanzee coalitionary killings. Further, because the Kahama was a splinter group (which is unusual – no other chimpanzee study site has shown group fission), this case may be very special. Thus, in order to generalize the case, the extinction of the Mahale K group was often referred to. If the extinction of the K group was the result of coalitionary killings, then the Gombe case is more generalized than it is a special case accompanying the unusual event of group fission.

26.5.1 Were the K-group males killed by the M group?

Extinction of the K group at Mahale (Nishida *et al.*, 1985; Chapter 10) was one of the most significant incidents during the entire research history of Mahale. During the course of research, six adult males gradually disappeared for unknown reasons. These males were often considered to have been killed by the larger M group. Together with the shocking observations of the "war" at Gombe, the extinction of the K group was considered supporting evidence for lethal raiding of chimpanzees (Manson and Wrangham, 1991; Wrangham and Peterson, 1996). Certain circumstantial evidence supports this interpretation: growing tension between the groups and several incursions by the M group into the home range of the K group (Chapter 10).

Nevertheless, none of the deaths of these six males was confirmed, and they were considered dead because of their long absence. Furthermore, the M-group males were not observed attacking any of these six males prior to their disappearances. Actually, these males did not disappear at once, in a short period, but gradually over a rather long period. At first, there were six males from 1966 to 1969: one disappeared in 1970, two in 1975, one in 1978, one in 1979, and one in 1982.

We need to consider whether it is unusual to lose six males from a group in less than 20 years. Eventually, when we compared the data of the K and M groups for the same duration, we noted that the M group lost (either disappeared or died) more adult males than the K group did (Table 26.2). The number of males that disappeared for unknown reasons in both groups are about the same, indicating that the K group was not unusual in the number of male deaths (in reality, we can rarely confirm or know the causes of deaths). Therefore, we do not necessarily consider that the loss of six males from the K group is unusual, and this loss could be attributed to intergroup conflicts. It is noteworthy that at the same time, about the same number of adolescent males in the M group became adult males, thus maintaining the equilibrium in the total number of adult males. However, in the K group, only one adolescent male became an adult in 18 years, as shown in Table 26.2. When we look back at the demographic data of the K group, there were only one adolescent and two infant males in the group in 1968 (3 years from the start of the research), that is, before the number of individuals in the K group started decreasing (see Chapter 10). The youngest infant died the next year, and the K group, from the start, did not have enough younger males to cover the losses of adult males. Thus, the extinction of the K group might have been predictable from its demographic structure if we knew how many males could have potentially been lost during a certain period.

Researchers at the time faced a shocking and unbearable phenomenon that, at that time, the main study group was becoming extinct and they observed several occasions of incursions by the M group into the range of the K group. They might have also heard of the Gombe "war" that was going on in the same period. Under such circumstances, it is understandable that they strongly suspected that the decrease of the K group was because of the killings by the M group. However, we cannot rule out the possibility that the number of males in the K group *first* decreased for some other reasons (more commonly, diseases, injuries, or senility), and *then* as a result of this decrease the M group was able to make more frequent incursions. It is not known whether M group's incursions caused the decrease of the K-group males, or conversely, the decrease of the K group allowed the incursion of the M group.

26.5.2 Jilba's case

Wrangham (1999) included the name of Jilba in the list of intraspecific killings (Table 4 in Wrangham, 1999), although indicated with a question mark. Mahale researchers did not initially report this as a case of intraspecific killing (there is an informal comment by a researcher, discussed below) and, likewise, we consider that there is no strong evidence to support the case. We describe the reasons below.

Jilba (at the time a 21-year-old adult male) was found dead on January 9, 1996. The body had already decomposed and it was believed to be that of Jilba

Table 26.2 **Decrease and increase in number of adult males in Mahale K and M groups. Data for K group are from Nishida *et al.* (1985)**

		Cause of decrease								
Unit-group	Duration	Old age	Illness	Injury	Lethal aggression	Unknown	Total decrease	Increase[a]	Number of males at the beginning	Number of males at the end
K	1966–1983 (18 years)	−1	0	0	0	−6	−7	+1	6	0
M	1994–2011 (18 years)	0	−2	−1[b]	−1[c]	−7[d]	−11	+11	9	9

[a] Increase is achieved when adolescent males became adults during these periods.

[b] Ntologi's case, which is sometimes considered as a case of lethal aggression (see text).

[c] Pimu's case.

[d] Including Jilba's case, which is sometimes assumed to be a case of lethal aggression (see text).

based on the skull lacking two canines (Itoh, unpublished). From his entire skeleton, only the skull has been preserved. A specialist on bone traumas, Masashi Nakai, visited Mahale and investigated several bones, including Jilba's (see Chapter 45). However, he did not report any traumas that could identify the cause of death. Thus, on the basis of the remaining skull, it was not possible to suspect that he was killed by conspecifics. Further, according to the research assistant, who found the dead body, there was no circumstantial evidence to suspect intraspecific killing such as severe battles among males or unusual uproars.

Then why was Jilba assumed to be killed by conspecifics? This may be because he had been a victim of earlier repeated gang attacks (Nishida *et al.*, 1995); Jilba had been severely attacked by several chimpanzees, and the attacks could have been fatal. It is possible that some researchers recalled the previous gang attacks when Jilba's dead body was found.

However, the gang attacks had occurred in October 1991, more than 4 years before his dead body was found. Of course, we cannot completely rule out the possibility that males kept concealing their murderous intent toward him, while superficially appearing to be friendly for more than 4 years. However, we do not have any evidence for this. During Nakamura and Itoh's stay at Mahale from 1994 to 1996, Jilba was (or at least it appeared to us) an ordinal middle-ranking young male. It did not appear that the other males were waiting for an opportunity to kill Jilba or that he was afraid of other males.

A comment by a Mahale researcher in a photography book for the public is considered another possible reason why Jilba was believed to have been killed. In the text that was written mainly by the wife of the photographer, Michael Huffman mentions, "I have a feeling that Jilba might have been killed by members of his own group" (Hofer *et al.*, 2000: 76). Three reasons were pointed out as circumstantial evidence: (1) Jilba appeared healthy when he was seen *a few days* earlier; (2) Jilba had been gang attacked in 1991; and (3) Ntologi had died 2 months earlier (see below) and some other males had disappeared. The first account is not consistent with the MMCRP record because Jilba had not been seen for *more than a week* by researchers, including Huffman, and it is unlikely that the body decomposed in only a few days. The second account is not conclusive for reasons discussed above. The third evidence may or may not be a mere coincidence, as Huffman himself questioned. His comment ended with the following: "The scientist in me of course needs more concrete evidence, but as the ape within me speaks, the hair on the back of my neck rises and a cold shiver runs down my spine." (Hofer *et al.*, 2000: 77). As he clearly stated, someone might be able to *feel* that a killing occurred, but a scientific account requires concrete evidence to *conclude* so.

We can conclude that the dead body was Jilba's, but there was no information about the cause of his death. Apart from the reports of gang attacks that occurred more than 4 years earlier, we have no observations of similar attacks just before his death. Thus, we conclude that it was too imprudent to assume that gang attacks occurred again and caused Jilba's death.

26.5.3 Ntologi's case

Similar to Jilba's case, Wrangham (1999) included Ntologi in the list of reported intraspecific killings (Table 4 in Wrangham, 1999), but this time without a question mark (therefore, he assumed that the case was certain). In our Table 26.1, we listed Ntologi's case as suspected.

Ntologi (then a 40-year-old male) was found dying, and he eventually died the next day (Nishida, 1996). Although Nishida (1996) strongly believed that he was killed by conspecifics, there was no strong evidence to support his assumption. First, no one observed the initial attack. Although Ntologi had 10–20 injuries or cuts on his body surface, they were not fresh as they were not bleeding (probably 1 or 2 days old) (Itoh, personal observation). This was different from the case of Pimu (see below) who was apparently bleeding after the attack. Most injuries found on Ntologi were on his ears and limbs, and some were on the body trunk but were not very deep. At that time, there were daily quarrels among males, and all males, including Ntologi, had several injuries on their bodies. Itoh noticed that Ntologi was not in a condition worse than other males. Nevertheless, Ntologi's injuries did not include any obvious canine puncture marks by other chimpanzees, but some may have been the result of chimpanzee bites. Because veterinarians were not present at Mahale at the time, the cause of his death was not determined. According to Itoh, none of the scars was deep enough to reach bones, and later inspection of his bones did not show any traumas (such as broken bones).

Aji, an adult male who was present when the dying Ntologi was found, did not cause new injuries. Nishida (1996) described that Aji charged at Ntologi several times, but according to Itoh's observation, he also groomed Ntologi, stared at him, and rested beside him for a while. Thus, it is difficult to determine whether Aji intended to kill Ntologi (it would have been easy to do so if he had intended). It is also possible that Aji was rather confused, scared, and/or was trying to confirm that Ntologi was still alive.

Ntologi might have died as a result of the several injuries found on his body or of shock after being attacked by conspecifics, although the cause of his death remains unknown. Further, the observation of daily quarrels among males supports the interpretation that, when compared to Jilba, Ntologi was more likely killed by conspecifics. However, the lack of evidence of direct attacks prior to dying prevents us from concluding with "certainty" that he was killed by conspecifics. His old age at the time of his death indicates that he had likely lost his strength owing to senility or illness.

26.5.4 Pimu's case

Pimu was a 23-year-old alpha male from the M group when he died on October 2, 2011. Unfortunately, the incidents before his death were not observed by researchers but only by tour guides and tourists. According to interviews with them and video footage taken by a tour guide (Kaburu *et al.*, 2013), Pimu first attacked the beta male Primus for unknown reasons during a grooming session. Primus fought back and a severe fight ensued, which lasted for at least 35 s. Both were bleeding when they broke apart. About 10 min later, when Primus had already fled, four other adult males jointly charged at Pimu. In the beginning, he tried to fight back and escape, but soon he was motionless. After the males continuously charged at Pimu for about 2 h, Pimu finally died. This is the only certain case of conspecific killing of an adult at Mahale, because the attacks, injury by chimpanzees, and subsequent death have been confirmed. It is unknown whether the severe wound he received in the initial fight with Primus was already fatal or the successive charging by the four males (which authors assumed as "coalitional lethal violence") actually caused his death. Because the four males did not seem to have delivered the final blow by cutting Pimu's throat or by removing his scrotum (Kaburu *et al.*, 2013), which is more common in intergroup coalitional killings, we cannot conclude that the attacking males had clear intentions to kill.

26.6 Discussion

There is no doubt that chimpanzees sometimes kill conspecifics. However, we have to be careful not to overestimate the number of such killings. In particular, cases of adult killings, which are few in number, are sometimes discussed with rather weak evidence and inferences.

Among the cases that have been thought to be conspecific killings at Mahale, the incidents of the K-group males and Jilba are supported by weak evidence. In the former case, their death was not confirmed, and in the latter, we have no information on any attacks that could have possibly led to Jilba's death; Ntologi's case may be controversial.

In this chapter, we took a rather prudent approach to judge whether a given case is a conspecific killing. This is because we are concerned about certain expressions often found in secondary descriptions (especially in books for the general public). In the original reports of killings, usually scientific papers, authors carefully distinguish the observed killings from the suspected and inferred. However, secondary descriptions are written in a manner that would lead the readers to conclude that all cases were certain to be conspecific killings. For example, Boehm (1999), referring to Goodall (1986) and Nishida *et al.* (1985), misled the readers of his book that the Gombe and Mahale cases were almost identical type of killings by writing the following: "On record are two instances in which small communities [. . .] were lethally raided by large ones until the males of the smaller groups were virtually eliminated" (Boehm, 1999: 130).

Similarly, Furuichi (1999), in his popular Japanese book, introduced the "genocide" of Gombe (i.e. lethal raiding by Kasekela group) and continued, "similar intergroup conflict was observed between Mahale M and K groups. Here again, males of smaller K group were killed or disappeared . . ." (Furuichi, 1999: 62, our translation). He goes on "this (genocide of males and recruiting of females) is not a special incident because similar incidents happened at Gombe and Mahale (Furuichi, 1999: 63). This is a more cautious expression than that by Boehm but misleads the readers that the case at Mahale is similar to that at Gombe. He further generalizes that chimpanzees are "experts in eliminating other group members" (Furuichi, 1999: 63).

In these 10 and some years, several adult conspecific killings have been observed in Ngogo group (see summary by Mitani *et al.*, 2010). Although these killings have not eliminated the adjacent victim group, as a result of these lethal raids, the Ngogo group expanded its home range by reducing the size of the opponent group, as discussed by the authors. Because Ngogo chimpanzees, unlike Gombe and Mahale chimpanzees, have never experienced provisioning, assessments that artificial provisioning was the cause of unnatural intergroup aggression (e.g. Power, 1991) cannot be applied here (however, this does not completely eliminate the possibility that the cause of the Gombe "war" was provisioning). The Ngogo group is unusual in its size: it has more individuals than any chimpanzee unit-groups reported thus far and has more than double the number of males compared with the average unit-groups of chimpanzees. Such unusual demographic characteristics may have triggered coalitionary killings of other group members by the Ngogo chimpanzees. To date, certain cases of chimpanzee "wars" (series of coalitionary killings of other group members) have been reported for only two study groups, Gombe and Ngogo, both of which have special features: repeated contacts between splinter groups through limited feeding site in the former and extraordinarily large number of males compared with its home-range size in the latter. We are reluctant to conclude from these two cases that the chimpanzee, as a species, is *generally* prone to kill other group members and are *experts* in eliminating others.

Because coalitionary killing of conspecifics is a common feature in chimpanzees and humans, the leading academic opinion is that it has been significant in the evolutionary history of humans as well. We should not forget that such a "violent" view of chimpanzees is often emphasized when comparing them with the "peaceful" bonobos (e.g. de Waal, 2005). Because chimpanzees are close relatives to humans, such conspecific killings astonish us and affect the general public to a greater degree. Therefore, scientists should carefully scrutinize the rare cases of coalitionary killing. When one hypothesis is dominant in an academic domain, incomplete information may easily be interpreted as *for* the hypothesis rather than *against* it. Examples may include the cases of Jilba and Ntologi: the former was interpreted as a possible case of killing with almost no supporting evidence and the latter was interpreted as a certain case with only some uncertain evidence.

We can conclude that both chimpanzees and humans, under certain conditions, kill members of their own species, sometimes acting as a group. On the other hand, we are yet to conclude that both species are prone to kill others regularly for maximizing their reproductive success, let alone call them *experts* in killing. We need further careful investigations of limited observations in order to understand what types of situations and conditions result in such rare events of conspecific killings in these species.

References

Boehm, C. (1999). *Hierarchy in the Forest*. Cambridge MA: Harvard University Press.

Bygott, J. D. (1972). Cannibalism among wild chimpanzees. *Nature*, **238**, 410–11.

Furuichi, T. (1999). [*Evolution of Sex and Evolution of Man: From Observation of Bonobos*.] Tokyo: Asahi-sensho. In Japanese.

Goodall, J. (1977). Infant killing and cannibalism in free-living chimpanzees. *Folia Primatologica*, **28**, 259–82.

Goodall, J. (1986). *The Chimpanzees of Gombe: Patterns of Behavior*. Cambridge MA: Belknap.

Hamai, M., Nishida, T., Takasaki, T., and Turner, L. A. (1992). New records of within-group infanticide and cannibalisms in wild chimpanzees. *Primates*, **33**, 151–62.

Hiraiwa-Hasegawa, M., Hasegawa, T., and Nishida, T. (1984). Demographic study of a large-sized unit-group of chimpanzees in the Mahale Mountains, Tanzania: a preliminary report. *Primates*, **25**, 401–13.

Hofer, A., Huffman, M. A., and Ziesler, G. (2000). *Mahale: A Photographic Encounter with Chimpanzees*. New York: Sterling Publishing Co.

Hosaka, K., Matsumoto-Oda, A., Huffman, M. A., and Kawanaka, K. (2000). [Reactions to dead bodies of conspecifics by wild chimpanzees in the Mahale Mountains, Tanzania.] *Primate Research*, **16**, 1–15. In Japanese with English summary.

Hrdy, S. B. (1977). Infanticide as a primate reproductive strategy. *American Scientists*, **65**, 40–9.

Kaburu, S. S. K., Inoue, S., and Newton-Fisher, N. E. (2013). Death of the alpha: within-community lethal violence among chimpanzees of the Mahale Mountains National Park. *American Journal of Primatology*, **75**, 789–97.

Kawanaka, K. (1981). Infanticide and cannibalism in chimpanzees, with special reference to the newly observed case in the Mahale Mountains. *African Study Monographs*, **1**, 69–99.

Kooriyama, T. (2009). The death of a newborn chimpanzee at Mahale: Reactions of its mother and other individuals to the body. *Pan Africa News*, **16**, 19–21.

Kutsukake, N. and Matsusaka, T. (2002). Incident of intense aggression by chimpanzees against an infant from another group in Mahale Mountains National Park, Tanzania. *American Journal of Primatology*, **58**, 175–80.

Lorenz, K. (1963[1985]). *On Aggression*, trans. T. Hidaka and K. Kubo. Tokyo: Misuzu-shobō. Japanese translation.

Manson, J. H. and Wrangham, R. W. (1991). Intergroup aggression in chimpanzees and humans. *Current Anthropology*, **32**, 369–90.

Mitani, J. C., Watts, D. P., and Amsler, S. J. (2010). Lethal intergroup aggression leads to territorial expansion in wild chimpanzees. *Current Biology*, **20**, R507–8.

Nishida, T. (1996). The death of Ntologi, the unparalleled leader of M group. *Pan Africa News*, **3**(1), 3–4.

Nishida, T. (1998). Deceptive tactic by an adult male chimpanzee to snatch a dead infant from its mother. *Pan Africa News*, **5**, 13–15.

Nishida, T. and Kawanaka, K. (1985). Within-group cannibalism by adult male chimpanzees. *Primates*, **26**, 274–84.

Nishida, T., Uehara, S., and Nyundo, R. (1979). Predatory behavior among wild chimpanzees of the Mahale Mountains. *Primates*, **20**, 1–20.

Nishida, T., Hiraiwa-Hasegawa, M., Hasegawa, T., and Takahata, Y. (1985). Group extinction and female transfer in wild chimpanzees in the Mahale National Park, Tanzania. *Zeitschrift für Tierpsychologie*, **67**, 284–301.

Nishida, T., Nakamura, M., and Hamai, M. (1995). A within-group gang attack on young adult male chimpanzee: ostracism of an ill-mannered member? *Primates*, **36**, 207–11.

Norikoshi, K. (1982). One observed case of cannibalism among wild chimpanzees of the Mahale Mountains. *Primates*, **23**, 66–74.

Palombit, R. A. (2012). Infanticide: male strategies and female counterstrategies. In *The Evolution of Primate Societies*, ed. J. C. Mitani, J. Call, P. M. Kappeler, R. A. Palombit, and J. B. Silk. Chicago, IL: University of Chicago Press, pp. 432–68.

Power, M. (1991). *The Egalitarians: Humans and Chimpanzee*. Cambridge: Cambridge University Press.

Sakamaki, T., Itoh, N., and Nishida, T. (2001). An attempted within-group infanticide in wild chimpanzees. *Primates*, **42**, 359–66.

van Schaik, C. P. (2000). Infanticide by male primates: the sexual selection hypothesis revisited. In *Infanticide by Males and its Implications*, ed. C. P. van Schaik and C. H. Janson. Cambridge: Cambridge University Press, pp. 27–60.

Sugiyama, Y. (1965). On the social change of hanuman langurs (*Presbytis entellus*) in their natural condition. *Primates*, **6**, 381–418.

Suzuki, A. (1971). Carnivority and cannibalism observed among forest-living chimpanzees. *Journal of Anthropological Society of Nippon*, **79**, 30–48.

Takahata, Y. (1985). Adult male chimpanzees kill and eat a male newborn infant: newly observed intragroup infanticide and cannibalism in Mahale National Park, Tanzania. *Folia Primatologica*, **44**, 161–70.

Townsend, S. W., Slocombe, K. E., Emery Thompson, M., and Zuberbuehler, K. (2007). Female-led infanticide in wild chimpanzees. *Current Biology*, **17**, R355–6.

de Waal, F. (2005). *Our Inner Ape*. New York: Riverhead Books.

Wilson, M. L. and Wrangham, R. W. (2003). Intergroup relations in chimpanzees. *Annual Review of Anthropology*, **32**, 363–92.

Wrangham, R. W. (1999). Evolution of coalitionary killing. *Yearbook of Physical Anthropology*, **42**, 1–30.

Wrangham, R. and Peterson, D. (1996). *Demonic Males*. Boston, MA: Houghton Mifflin.

PART VI Social relations

27 Male–male relationships

KAZUHIKO HOSAKA AND MICHIO NAKAMURA

27.1 Introduction

Male–male relationships, being one of the most eminent social relationships within a chimpanzee unit-group, have been of great interest to many Mahale researchers, especially until the 1990s. For example, Nishida (1983) documented how three adult males in the K group tactically behaved in the course of struggle over alpha status. Kawanaka (1984, 1989, 1990, 1991, 1993, 2002) had continuously investigated social behaviors of M-group males, particularly focusing on how each male (from late adolescence to adulthood) assesses his position in a network of adult males and devotes his time to grooming others and keeping social proximity with others, so that he forms or maintains mature and stable relations with other adult males. Other researchers also documented details of M group's adult male–male relationships and coalitionary strategies, particularly focusing on high-ranking males (Takahata, 1990a; Nishida *et al.*, 1992; Uehara *et al.*, 1994; Hosaka, 1995; Nishida and Hosaka, 1996; Kaburu and Newton-Fisher, 2013).

In this chapter, we first summarize general patterns of social interactions, both affiliative and antagonistic, among adult males. Then, observations regarding social status and behavioral strategies to gain or maintain higher social status within the dominant–subordinate hierarchy will be described. Specific cases of ostracism observed at Mahale are then reviewed. Finally, the relationships of younger males are summarized in relation to their relevance to entering the adult male–male bonds.

27.2 Complex relationships intermingled with cooperation and conflict

27.2.1 Affiliative interactions

Adult males in a unit-group usually form strong affiliative bonds with each other. They often travel together, groom with each other for a long time, and share foods with each other (Chapter 20). Such strong bonding among males is often explained by a strategy to defend females and ranging areas of their unit-group against males of other unit-groups (see Chapter 11). However, an adult male in a unit-group does not equally make affiliative relationships with all the other males. He may flexibly form a coalition (defined as "a one-time cooperative action by at least two individuals or units against at least one other individual or unit": de Waal and Harcourt, 1992) or an alliance ("long-term cooperative relationships, that is, partnerships that form coalitions on a regular basis": de Waal and Harcourt, 1992) with a particular male or males.

From the early days of his long-term research at Mahale, Nishida (1979) has considered that adult males are the "core" of the unit-group. This was based on his own observations in which adult males most frequently appeared together in parties to the provisioning site (Nishida, 1968). Although, there are some critiques to this "male-core" view (Chapter 6), it seems correct that adult males associate most frequently with each other (Figure 27.1) compared to other combinations of age–sex classes, even without the influence of provisioning. Although Nishida (1979) noted that adult males were observed most

Mahale Chimpanzees: 50 Years of Research, ed. M. Nakamura, K. Hosaka, N. Itoh and K. Zamma. Published by Cambridge University Press.

Figure 27.1 Four M-group adult males, Kalunde, Primus, Fanana, and Bonobo (from front to back) traveling in a file. ©M. Nakamura.

frequently in mix-sexed parties in the K group, the average attendance rate in large nomadic parties of the M group (i.e. nomadic parties with more than half of the unit-group members) was slightly higher for adolescent males (85.8%) than adult males (82.9%) (MMCRP, unpublished data from 1994 to 2010).

Grooming is also more frequently observed among adult males than any other combinations of age–sex classes (Nishida, 1970). Prime adult males often groom reciprocally with each other (Kawanaka, 1989), whereas old males are often groomed unidirectionally particularly by the alpha male (Nishida and Hosaka, 1996; Chapter 23). Typical grooming among adult males is characterized by dyadic, intense, and long bouts. An adult male may often concentrate his grooming to the face of the partner (Figure 27.2), sometimes using lips (thus kissing the face) or even putting his index finger into the nostrils to eat mucus (Nishida and Hosaka, 1996). Such intense grooming between two males often gives an observer an impression that these males are very intimate with each other.

Vocal exchange (Chapter 39) is another important interaction among adult males (Mitani and Nishida, 1993). The authors showed that Mahale adult males traveled together more often before and after pant-hooting than expected, and males called more when alliance partners were nearby. By these results, they suggested that males pant-hoot to particular members, specifically to male allies, to maintain contact with them.

Sharing of meat is also important in regard to social relationships among adult males. With their frequent involvement in hunting and their ability to steal meat from females or adolescent males just

Figure 27.2 An alpha male, Pimu (right), grooming an adult male, Alofu (left), by putting his mouth on Alofu's forehead. ©M. Nakamura.

after prey capture, adult males are more often the possessor and the sharer of meat than others. Affiliative individuals and important alliance partners typically get more shares of meat (see Chapter 20 for details).

Another eminent affiliative social interaction among adult males is play, although play is more often observed among younger chimpanzees (Chapter 36). In nonhuman primates, play among adult males is rare, thus it may be one of the unique characteristics of the chimpanzee. Hayaki (1990a: 73–6) focused on the play among adult males and argued that chimpanzees can play even by concealing their potentially competitive relationships for a moment. Although quantitative analyses of adult male–male social play have not been conducted, it may be interesting to see how such childish interactions are integrated in the social relationships among adult males.

27.2.2 Competitive interactions

Intense competition among adult males is sometimes observed over access to estrous females. When a parous female is in estrous, many males follow her to look for the opportunity to mate. An alpha male does not usually permit other males, even his alliance partners, to mate with females during the periovulatory period (see the examples of Kalunde–Musa and Ntologi–Bakali in Chapter 23). This may be different from a quantitative analysis of data from an alpha male at Kanyawara, Uganda (Duffy *et al.*, 2007), suggesting that the alpha male rewards his allies by allowing them preferential access to mates.

Males' antagonistic interactions occur more often in the form of intimidation displays (Chapter 31) than direct aggressions (Chapter 32). However, there are cases that males get injured due to severe aggressions with each other (Chapter 26).

27.2.3 Is infanticide explained as male–male competition?

Chapter 26 reviews cases of infanticidal cannibalism that occurred at Mahale from the perspective of conspecific killing (see also Chapter 20 for the meat-eating aspect). Here we review some arguments over whether infanticide by males is explicable in terms of male–male competition or any other adaptive mechanism.

Hiraiwa-Hasegawa (1987) noted that, until the mid-1980s, the victims of within-group infanticides by males at Mahale were biased towards male infants, mostly born to newly immigrated females. Thus, she suggested that such within-group infanticides could be discussed in the context of male–male competition between different unit-groups (see similar discussions in Takahata, 1985; Nishida and Kawanaka, 1985). Since newly immigrated females sometimes take ambiguous ranging patterns, it is possible that they spend some ovulatory days with males of their natal unit-groups. In that case, it may well be adaptive for males to kill infants possibly sired by different unit-group males. However, it is also possible that the males killed infants of their own, because the females copulated with these new males during the time when the victimized infants had likely been conceived (Kawanaka, 1981; Takahata, 1985). Furthermore, there was no comparable observation from Gombe, another long-term study site in Tanzania, where only between-group infanticides by males and within-group infanticides by females were observed (Goodall, 1986). Thus, the adaptive significance of the infanticide in chimpanzees has been less obvious so far (Hiraiwa-Hasegawa and Hasegawa, 1994).

As for between-group infanticide, there was no evidence that infant-killers gain any immediate reproductive benefit such as copulation with mothers of killed infants or consequent transfer of the mothers (Hiraiwa-Hasegawa, 1987). Still, the adaptive significance has been argued since killing the male infants of a neighboring unit-group might serve to reduce the number of future adult males in that unit-group, thus raising the relative strength of their own unit-group in the long run (Hiraiwa-Hasegawa, 1987). However, due to the extreme rarity of between-group infanticide at Mahale (Chapter 26), it is difficult to claim that chimpanzees kill infants as such an adaptive mechanism.

Hiraiwa-Hasegawa and Hasegawa (1994) attempted to view both between-group and within-group infanticide by males as an accompanying mechanism of "local resource competition." In male-philopatric societies of chimpanzees (Chapter 6), local resource competition may occur among males (Hiraiwa-Hasegawa and Hasegawa, 1994). If such is the case, adult males may enhance local resource by optimizing the number of males in their unit-group. However, if the number of males exceeds some optimal size, within-group competition among males would be intense. In that case, it is plausible that males may limit recruitment of distantly related males. In other words, they may kill male infants if they are either of ambiguous paternity or of neighboring unit-groups, thus reducing the number of future competitors both within and between unit-groups (Hiraiwa-Hasegawa, 1987). Otherwise, they may split a too large unit-group into two as occurred at Gombe (Goodall, 1986). So far, due to the paucity of data concerning the intensity of competition over resources, the adaptive significance of the infanticide in chimpanzees has remained a puzzle (Hiraiwa-Hasegawa and Hasegawa, 1994). Additionally, infanticide itself has become so rare at Mahale recently (Chapter 26) that testing the above hypothesis would also be difficult.

27.3 Social status and social tactics of adult males

27.3.1 Pursuit of high rank?

Dominant–subordinate relationships (Chapter 33) among males and their dynamism seem to attract those who are familiar with the individuals. It is quite common for researchers and local assistants to casually discuss in their daily conversations a particular male's current position in the dominance hierarchy. In fact, many male chimpanzees seem to be pursuing to attain higher-ranking, the alpha status in particular. Although there can be dozens of biological benefits to being high-ranking, it seems to us that apart from such benefits, many male chimpanzees somehow have inclination to become high-ranking only in itself. More accurately, the tendency to pursue status does not necessarily mean that the dominant–subordinate relationships among males are unequivocal. According to the relationships among nine adult males in 1992 at Mahale, only two highest-ranking males and three lowest-ranking males were unequivocal in their ranks (Nishida and Hosaka, 1996). Pant-grunts (Chapter 33) are seldom heard among middle-ranking males, thus exact rankings of these males are often ambiguous (Hayaki *et al.*, 1989). Furthermore, adult males rarely display overt dominance to one another (Hayaki *et al.*, 1989). The authors argued that this is probably because the presence of other males affects their interactions. Some males seem to try to keep their dominant–subordinate relationship ambiguous because making it overt is not always advantageous. The authors argue that this may be a political way for males to coexist with one another in a unit-group (Hayaki *et al.*, 1989).

27.3.2 The alpha status

Special social status or just the highest ranking?

An alpha male is by definition, the most dominant individual in the dominant–subordinate hierarchy among males in a given period. There are some views that the alpha status is somehow special (e.g. Nishida, 1979) – the alpha status seems more than just a rank higher than the second rank. Nishida (1979) noted that a K-group male Kasonta, when he was an alpha, had been most prominent in frequencies of aggressive behaviors, grooming interactions, and occasions chosen as the target of following by other males. Usually, an alpha male among the dominant–subordinate hierarchy is very clear because all adult males pant-grunt to him (Hayaki *et al.*,1989; Nishida and Hosaka, 1996; Hayaki, 1998). This is clearly different from a beta male who is not always pant-grunted to by all the lower-ranking males (Hayaki, 1998). An alpha male is also exceptional in that he receives the majority of pant-grunts: Ntologi as an alpha male received 177 of 267 pant-grunts (66.3%) occurring among adult males (Nishida and Hosaka, 1996), 154 of 280 pant-grunts (55.0%) among adult and adolescent males (Hayaki, 1998), and 180 of 403 pant-grunts (44.7%) among males of all age classes (Hayaki *et al.*, 1989) in different periods of study. Such tendency is common for other alpha males as long as the status is stable. Fanana received 189 of 352 pant-grunts (53.7%) among adult and adolescent males in 2000–1 (Kutsukake, 2003) and Primus received 96 of 196 pant-grunts (49.0%) among adult and adolescent males in 2011 (Nakamura, unpublished data). Alpha males receive pant-grunts not only from other males but also a prominent amount of pant-grunts from females. For example, Fanana in 1999–2000 received 1350 of 2880 pant-grunts (46.9%) by females to adult males, four times more than the male who received the second most pant-grunts (Sakamaki, 2011).

The alpha status may also be special in that alpha males tend to possess and share meat with others (see below and Chapter 20).

Priority in mating

Alpha males had the highest mating frequency among males in the small-sized K group: e.g. Kasonta and Sobongo (Nishida, 1979, 1983). If alpha males did not have the highest overall mating frequency, they had the priority of making with ovulating females by showing possessive behavior to them (Figan: Tutin, 1979; Ntologi: Hasegawa and Hiraiwa-Hasegawa, 1983). Ntologi, as an alpha male, kept the highest frequency mating with ovulating females both soon after attaining the alpha position (Hasegawa and Hiraiwa-Hasegawa, 1990; Takahata, 1990b) and just before losing it (Nishida, 1997), as well as during the prime of his life. The results of paternity analyses at Mahale also support the priority of access in alpha males (Chapter 35).

Social tactics by alpha males

When rival males are grooming each other or in proximity, a male may direct charging to separate the rival males. De Waal (1982) termed it *separating intervention.* In the wild, alpha males frequently perform such interventions (Nishida and Hosaka, 1996). Among 25 interventions, 20 (80%) were made by the alpha male Ntologi and, among them, 17 (85%) were towards a group of males including the rivaling beta male Nsaba. Ntologi, especially in his second alpha male tenure, was thorough about this policy. In 1992, Ntologi and Nsaba never had grooming interactions (Hayaki, 1994). Hosaka, during his 1-year study period in 1993, observed only one instance where Ntologi groomed Nsaba. In the same year, Nsaba once tried to groom Ntologi, but in vain. As soon as Nsaba touched the back of Ntologi, Ntologi changed his posture so that Nsaba had to move backward.

Nishida *et al.* (1992) argued that an alpha male, Ntologi, used meat sharing as a coalition strategy to maintain his status. Ntologi showed a consistent tendency over 10 years to share meat with affiliative and/or old males while not sharing meat with rival beta males and young adult males who were rising in the dominance hierarchy.

Readers should bear in mind that the majority of published accounts about the alpha status at Mahale are about one alpha male, Ntologi, because he had stayed in power for an extraordinarily long time. Although quantitative data are yet to be published, at least some of the successive M-group alpha males (Kalunde, Nsaba, Fanana, Alofu, Pimu, and Primus: see Chapter 5) seem to employ different social strategies. Some maintained their statuses by suppressing others only by their physical power with their large bodies, while others made the best use of alliances with other males. Further comparative works are awaited in order to fully understand the variation of alpha males' strategies.

27.3.3 The beta status

Although not as evident as alpha males, beta males are also usually easy for the observers to recognize. The beta status is often marked with obvious rivalry with the alpha male, particularly when Ntologi was an alpha male. Ntologi clearly differentiated a beta male from other males, such as by not grooming him, not sharing meat with him, and directing separating interventions towards him. On the other hand, the current alpha male, Primus, would often groom the beta male, Alofu, probably because Alofu was the former alpha male and older than Primus and it was not likely that Alofu would challenge the alpha status again.

Hayaki (1994) particularly investigated social relationships of a beta male, Nsaba. Nsaba was frequently found in the proximity of the alpha male Ntologi but his daily proximity indices with other males fluctuated largely. The author suggested that Nsaba often tried to follow Ntologi when the latter allowed but he sometimes moved alone to prevent aggression from Ntologi. Nsaba's most frequent grooming partner was the gamma male and he also frequently groomed other males sometimes unidirectionally, suggesting that he paid considerable efforts to maintaining his social status by providing services to others.

Another notable beta male, Shike (see Chapter 5), assumed the status at the estimated age of 21 years in 1991, when Kalunde ostracized Ntologi. Shike might be distinguished from other beta males by high frequencies of some types of behaviors which are usually typical of alpha status. That is, he received pant-grunts from females and immatures and performed charging displays even more frequently than the alpha male Kalunde did (Hosaka, unpublished data). Shike performed charging displays usually out of sight from Kalunde. During focal follows, Kalunde was frequently observed to get nervous about sounds of commotions caused by Shike's charging displays. It seems that Shike began to show this tendency when he was a 16-year-old young adult in 1986 (Kawanaka, 1991). When he became one of the high-ranking males in 1989, he made pant-hoots even more frequently than the alpha male Ntologi (Mitani, 1994), whereas Shike dropped his frequency of pant-hoots in 1991; Kalunde pant-hooted most often among adult males (Hosaka, unpublished data).

27.3.4 Middle- and low-ranking males

At Mahale, the gamma position is not always clear as the alpha or the beta positions. For example, among the 196 pant-grunts among males in 2011 (Nakamura, unpublished data), the beta male received a total of 20 pant-grunts from all but one adult males who were ranked lower than him. On the other hand, a male who received the third-most frequent pant-grunts overall was pant-grunted to only three times by two adult males (other pant-grunts were given by adolescent males). He never received pant-grunts from 4–6th-ranking males. Thus, in such a case, the gamma position is unclear and he may be better categorized as middle ranking together with others.

Even middle- or low-ranking males intimidate other males who are ranked lower than themselves. When Alofu and Dogura were entering adulthood, they targeted the lowest-ranking adult male Masudi, who easily pant-grunted to them (Chapter 5). On the other hand, Fanana who displayed frequently against other adult males in 1991 only played with Masudi. Toshibo, low-ranking in 1991, made displays frequently around old males Bakali and Musa.

Older low-ranking males also intimidate younger males who are raising their ranks. In 1993, a low-ranking male Bembe frequently charged Fanana who had just reached adulthood (Chapter 5) and Musa frequently charged young Masudi (Chapter 31).

Being low-ranking does not necessary mean that such males are less successful in reproduction. Researchers often observe low-ranking males successfully mate with estrous females by ambushing (see an example of Masudi in Chapter 5). Females often easily accept such males quietly waiting in the bush to copulate, although there have been some observations of punishment-like attacks by higher-ranking males (Hosaka, personal observation). Some male chimpanzees may be employing an alternative strategy to reproduce offspring by staying low-ranking to avoid risky (sometimes fatal) aggressive competitions among high-ranking males. Future studies should examine whether there is such a strategy among low-ranking males.

27.3.5 Triadic relationships

When the alpha status becomes unstable and the beta male does not show submission, occasionally, the third male assumes a key role both for the alpha and the beta males. A famous and well-documented example is the triadic relationships among Kasonta, Sobongo, and Kamemanfu in the K group (Nishida, 1983; also see Chapter 5). Similar triadic relationships are also observed among Ntologi, Nsaba, and Kalunde in the M group, although there were six other adult males during the study period (Nishida and Hosaka, 1996).

Triadic relationships when Kalunde was an alpha were slightly different from these cases. He occasionally formed coalitions with the beta male Shike in order to suppress the younger and upcoming gamma male Nsaba. Kalunde, unlike Ntologi, also sometimes groomed and shared meat with the beta male Shike (Hosaka and Nishida, 2002).

27.4 Ostracism

Ostracism is "the general process of rejection and exclusion" from a social group, originally used for human groups but also applied to many other nonhuman species (Gruter and Roger, 1986). For the chimpanzee society in particular, Hosaka and Nishida (2002) used the term for when an adult male starts to travel alone and be outside the dominant–subordinate hierarchy, often after coalitionary attacks or a formation of alliances against him. At Mahale, some alpha males are reported to be ostracized (see below). At Gombe, the alpha male Goblin's case (Goodall, 1992) may be regarded as an example. After two coalitionary attacks against him, he was badly injured and became estranged from ranging with the large parties including males for 4 months. When he returned to the group, he was submissive to most of the males.

Uehara *et al.* (1994) reported that alpha males, Kasonta of the K group and Kajugi and Ntologi of M group, came to spend much of their time alone after being defeated by the challenging beta males. Among these, Kasonta and Ntologi returned to the group and regained the alpha status, whereas Kajugi could not return to the group and finally disappeared.

Ntologi's case of ostracism was outlined in several publications (Hamai, 1994; Nishida, 1994; Hosaka, 1995; Hosaka and Nishida, 2002). He had been an alpha

Figure 27.3 Grooming hand-clasp between Kalunde (right) and Jilba (left). Nishida (1994) noted that Jilba groomed the ground instead of grooming Kalunde reciprocally. ©Toshisada Nishida.

Figure 27.4 Ntologi (right) and Musa (left) in June 1993. Their alliance was unharmed despite Ntologi's long absence due to ostracism. ©K. Hosaka.

male since 1979, but due to the alliance between the beta male Kalunde and the fourth-ranking male Nsaba, he was unable to keep his position and was ostracized in March 1991. After 10 months of ostracized life, he finally regained the alpha status by reconciling with Kalunde (Hosaka and Nishida, 2002). It is possible that for many adult females and immature individuals, and even for some adult males, Ntologi kept being an alpha without participating in the group. For example, Jilba (Figure 27.3) who was also ostracized in a similar period, kept pant-grunting to Ntologi when they met, but he never did so to Kalunde (Nishida *et al.*, 1995), the alpha male of the M group during Ntologi's absence. Hosaka and Nishida (2002) argued that Ntologi chose to be ostracized avoiding pant-grunting at Kalunde, which would mean that he admitted that he was outranked by Kalunde (Figure 27.4).

Another notable example from the M group is of Fanana (Nishie, 2004a, b) who somehow moved away from other members, but sometimes met with a few group members, from 2003 to 2007 (see Chapter 5 for more detail). Whether he was literally "ostracized" by others is doubtful, because there were no apparent conflicts over the alpha status prior to his disappearance. When he happened to meet a large party of the M group including the next alpha male Alofu, Alofu did not show any signs of dominance over Fanana. Many females pant-grunted to Fanana as if he was still the alpha male. However, the next day, Fanana again disappeared (Nishie, 2004a). Thus, it is also possible that Fanana spontaneously chose the solitary life for more than 3 years.

We should note that not all ex-alpha males become solitary at Mahale. For example, Kalunde and Alofu stayed in the group and ranged together with other males as high-ranking males even after they fell from the alpha status (Alofu's case is described in Inaba, 2009). Nsaba disappeared after being an alpha male but was never observed traveling alone. Pimu ended his life as an alpha male through aggression by other males (Kaburu *et al.*, 2013).

Examples of non-alpha males being ostracized are relatively rare. The only examples known are that of Jilba at Mahale (Nishida *et al.*, 1995) and of Evered at Gombe who became more and more peripheral and avoided large groups (Riss and Goodall, 1977).

Then why do alpha males take such a risk of behavioral choice leading to ostracism? It seems so risky to become an alpha male when we think of the fact that so many alpha males are ostracized and/or die after falling from their status. From a behavioral ecological point of view, it is puzzling why male chimpanzees persist in trying to become high-ranking, particularly top ranking. There are hot debates that being an alpha may or may not be adaptive for the particular individual male. We should explore the answers to this question by accumulating more observations in the near future.

27.5 Younger males

Although we have basically focused on the relationships among adult males in the above sections, we should not forget that there are younger (infant, juvenile, and adolescent) males in a unit-group. Although they are not yet integrated in tight adult male social networks, they gradually become so until adulthood. A series of works by Hayaki who investigated social relationships of juvenile and adolescent males are important in this account.

When an older infant or juvenile male, who still follows his mother, meets with adult males, he tends to show a strong interest in them. He often rushes toward the males to pant-grunt even before his mother does so and may start to groom the males. In early adolescence, males start to leave their mothers to follow adult males. Typically, by late adolescence or early adulthood, males came to associate most frequently with adult males (Hayaki, 1988).

All adolescent males are unequivocally subordinate to all adult males, directing sometimes exaggerated pant-grunts with bobbing toward adult males, especially the alpha male (Hayaki, 1990b). On the other hand, they seldom direct pant-grunts to each other (Hayaki *et al.*, 1989). Adolescent males are often so afraid of adult males, that even the presence of adult males is enough for adolescent males to restrict themselves from copulating with females (Hayaki, 1985).

In a study on 5- to 20-year-old males in 1981–9, Hayaki (1992) showed that their time of being in proximity (10 m) to others decreases with age but grooming frequency increases. Although males increase the time allocated to grooming, adolescent or young adult males unidirectionally groom elders and receive little grooming from prime adult males (Kawanaka, 1989).

Adolescent males are sometimes observed to perform non-vocal intimidation display with piloerection when nobody is present (Hayaki, 1998). As they get older, they direct more and more displays toward adult females, and finally in late adolescence or early adulthood, they begin to direct displays toward adult males.

27.6 Discussion and future perspectives

As we stated in the introduction, male–male relationships in the Mahale M group have fascinated various researchers. Although the long-term observations of such relationships, initially based on animal sociological interests, have been accumulated, there have been relatively few reports of such details published in English. The long-term research at Mahale has elucidated some rare but intriguing social phenomena such as ostracism. Ostracisms of ex-alpha males are repeatedly observed at least at Mahale, but not well documented in other study sites. Further careful comparisons with other study sites is indispensable to knowing if such phenomena are universal to the chimpanzee. If ostracism of ex-alpha males is characteristic to Mahale, it may be interesting to investigate its causes and consequences to the society of the Mahale chimpanzees.

Male–male relationships and males' strategies to gain higher social status were relatively well documented when Ntologi was an alpha male and when he was ostracized. Because the personality of alpha males may strongly influence overall male–male relationships and each male's social strategy, we need to have a comparative view with other alpha males' eras. Finally, we should not forget that males are not only coping with other males in a chimpanzee society. As we will see in the next chapters (Chapters 28, 29), females are as important as males in understanding the complex and dynamic aspect of chimpanzee society and sociality.

References

Duffy, K. G., Wrangham, R. W., and Silk, J. B. (2007). Male chimpanzees exchange political support for mating opportunities. *Current Biology*, **17**, R586–7.

Goodall, J. (1986). *The Chimpanzees of Gombe: Patterns of Behavior*. Cambridge, MA: Belknap.

Goodall, J. (1992). Unusual violence in the overthrow of an alpha male chimpanzee at Gombe. In *Topics in Primatology, Vol. 1: Human Origins*, ed. T. Nishida, W. C. McGrew, P. Marler, M. Pickford, and F. B. M. de Waal. Tokyo: University of Tokyo Press, pp. 131–42.

Gruter, M. and Roger, D. M. (1986). Ostracism as a social and biological phenomenon: an introduction. *Ethology and Sociobiology*, **7**, 149–58.

Hamai, M. (1994). ["Attractive" females for male chimpanzees?] *Primate Research*, **10**, 33–40. In Japanese with English summary.

Hasegawa, T. and Hiraiwa-Hasegawa, M. (1983). Opportunistic and restrictive matings among wild chimpanzees in the Mahale Mountains, Tanzania. *Journal of Ethology*, **1**, 75–85.

Hasegawa, T. and Hiraiwa-Hasegawa, M. (1990). Sperm competition and mating behavior. In *The Chimpanzees of the Mahale Mountains: Sexual and Life History Strategies*, ed. T. Nishida. Tokyo: University of Tokyo Press, pp. 115–32.

Hayaki, H. (1985). Copulation of adolescent male chimpanzees, with special reference to the influence of adult males, in the Mahale National Park, Tanzania. *Folia Primatologica*, **44**, 148–60.

Hayaki, H. (1988). Association patterns of young chimpanzees in the Mahale Mountains National Park, Tanzania. *Primates*, **29**, 147–61.

Hayaki, H. (1990a). [*The Human within Chimpanzee*.] Tokyo: Shōka-bō. In Japanese.

Hayaki, H. (1990b). Social context of pant-grunting in young chimpanzees. In *The Chimpanzees of the Mahale Mountains: Sexual and Life History Strategies*, ed. T. Nishida. Tokyo: University of Tokyo Press, pp. 189–206.

Hayaki, H. (1992). [Grooming patterns of young males in a chimpanzee unit-group at the Mahale Mountains National Park, Tanzania.] *Bulletin of Faculty of Humanities and Sciences, Kobe Gakuin University*, **4**, 83–105. In Japanese.

Hayaki, H. (1994). [Social relations of the second ranking male chimpanzee at M group of the Mahale Mountains National Park, Tanzania.] *Primate Research*, **10**, 289–305. In Japanese with English summary.

Hayaki, H. (1998). [Agonistic interactions of young male chimpanzees at Mahale Mountains National Park.] *Bulletin of Faculty of Humanities and Sciences, Kobe Gakuin University*, **16**, 1–16. In Japanese.

Hayaki, H., Huffman, M. A., and Nishida, T. (1989). Dominance among male chimpanzees in the Mahale Mountains National Park, Tanzania: a preliminary study. *Primates*, **30**, 187–97.

Hiraiwa-Hasegawa, M. (1987). Infanticide in primates and a possible case of male-biased infanticide in chimpanzees. In *Animal Societies: Theories and Facts*, ed. Y. Ito, J. L. Brown, and J. Kikkawa. Tokyo: Japan Scientific Society Press, pp. 125–39.

Hiraiwa-Hasegawa, M. and Hasegawa, T. (1994). Infanticide in nonhuman primates: sexual selection and local resource. In *Infanticide and Parental Care*, ed. S. Parmigiani and F. S. vom Saal. Chur, Switzerland: Harwood Academic Publishers, pp. 137–53.

Hosaka, K. (1995). A rival yesterday is a friend today: a grand political drama in the forest. *Pan Africa News*, **2**(2), 10–11.

Hosaka, K. and Nishida, T. (2002). [Ostracism: Alpha male came back after a long social exclusion.] In [*The Mahale Chimpanzees: 37 Years of <Panthropology>*], ed. T. Nishida, S. Uehara, and K. Kawanaka. Kyoto: Kyoto University Press, pp. 439–70. In Japanese.

Inaba, A. (2009). Power takeover occurred in M group of the Mahale Mountains, Tanzania, in 2007. *Pan Africa News*, **16**, 13–15.

Kaburu, S. S. K. and Newton-Fisher, N. E. (2013). Social instability raises the stakes during social grooming among wild male chimpanzees. *Animal Behaviour*, **86**, 519–27.

Kaburu, S. S. K., Inoue, S., and Newton-Fisher, N. E. (2013). Death of the alpha: within-community lethal violence among chimpanzees of the Mahale Mountains National Park. *American Journal of Primatology*, **75**, 789–97.

Kawanaka, K. (1981). Infanticide and cannibalism in chimpanzees: with special reference to the newly observed case in the Mahale Mountains. *African Study Monographs*, **1**, 69–99.

Kawanaka, K. (1984). Association, ranging, and the social unit in chimpanzees of the Mahale Mountains, Tanzania. *International Journal of Primatology*, **5**, 411–34.

Kawanaka, K. (1989). Age differences in social interactions of young males in a chimpanzee unit-group at the Mahale Mountains National Park, Tanzania. *Primates*, **30**, 285–305.

Kawanaka, K. (1990). Alpha males' interactions and social skills. In *The Chimpanzees of the Mahale Mountains: Sexual and Life History Strategies*, ed. T. Nishida. Tokyo: University of Tokyo Press, pp. 171–87.

Kawanaka, K. (1991). [Males in a patrilineal group: social development of male chimpanzees.] In [*The Cultural History of Primates*], ed. T. Nishida, K. Izawa, and T. Kano. Tokyo: Heibon-sha, pp. 217–39. In Japanese.

Kawanaka, K. (1993). Age differences in spatial positioning of males in a chimpanzee unit-group at the Mahale Mountains National Park, Tanzania. *Primates*, **34**, 255–70.

Kawanaka, K. (2002). [Change of social relationships among high-ranking males.] In [*The Mahale Chimpanzees: 37 Years of <Panthropology>*], ed. T. Nishida, S. Uehara, and K. Kawanaka. Kyoto: Kyoto University Press, pp. 417–38. In Japanese.

Kutsukake, N. (2003). [*Influences of Relationship Quality on Social Stress and Conflict Resolution in Wild Chimpanzees*.] Doctoral Dissertation. Tokyo: The University of Tokyo. In Japanese.

Mitani, J. C. (1994). Ethological studies of chimpanzee vocal behavior. In *Chimpanzee Cultures*, ed. R. W. Wrangham, W. C. McGrew, F. B. M. de Waal, and P. Heltne. Cambridge, MA: Harvard University Press, pp. 195–210.

Mitani, J. C. and Nishida, T. (1993). Contexts and social correlates of long-distance calling by male chimpanzees. *Animal Behaviour*, **45**, 735–46.

Nishida, T. (1968). The social group of wild chimpanzees in the Mahali Mountains. *Primates*, **9**, 167–224.

Nishida, T. (1970). Social behavior and relationship among wild chimpanzees of the Mahali Mountains. *Primates*, **11**, 47–87.

Nishida, T. (1979). The social structure of chimpanzees of the Mahale Mountains. In *The Great Apes*, ed. D. A. Hamburg and E. R. McCown. Menlo Park, CA: Benjamin/Cummings, pp. 73–121.

Nishida, T. (1983). Alpha status and agonistic alliance in wild chimpanzees (*Pan troglodytes schweinfurthii*). *Primates*, **24**, 318–36.

Nishida, T. (1994). Review of recent findings on Mahale chimpanzees implications. In *Chimpanzee Cultures*, ed. R. W. Wrangham, W. C. McGrew, F. B. M. de Waal, and P. Heltne. Cambridge, MA: Harvard University Press, pp. 373–96.

Nishida, T. (1997). Sexual behavior of adult male chimpanzees of the Mahale Mountains national park, Tanzania. *Primates*, **38**, 379–98.

Nishida, T. and Hosaka, K. (1996). Coalition strategies among adult male chimpanzees of the Mahale Mountains, Tanzania. In *Great Ape Societies*, ed. W. C. McGrew, L. F. Marchant, and T. Nishida. Cambridge: Cambridge University Press, pp. 114–34.

Nishida, T. and Kawanaka, K. (1985). Within-group cannibalism by adult male chimpanzees. *Primates*, **26**, 274–84.

Nishida, T., Hasegawa, T., Hayaki, H., Takahata, Y., and Uehara, S. (1992). Meat-sharing as a coalition strategy by an alpha male chimpanzee? In *Topics in Primatology, Vol.1: Human Origins*, ed. T. Nishida, W. C. McGrew, P. Marler, M. Pickford, and F. B. M. de Waal. Tokyo: University of Tokyo Press, pp. 159–74.

Nishida, T., Hosaka, K., Nakamura, M., and Hamai, M. (1995). A within-group gang attack on a young adult male chimpanzee: ostracism of an ill-mannered member? *Primates*, **36**, 207–11.

Nishie, H. (2004a). [Disappearance of the alpha male Fanana and its consequences.] *Mahale Chimpun*, **3**, 5–6. In Japanese.

Nishie, H. (2004b). [The fight at Kasiha Valley: The fall of Fanana from the alpha status and Alofu's takeover.] *Mahale Chimpun*, **4**, 6–7. In Japanese.

Riss, D. and Goodall, J. (1977). The recent rise to the alpha-rank in a population of free-living chimpanzees. *Folia Primatologica*, **27**, 134–51.

Sakamaki, T. (2011). Submissive pant-grunt greeting of female chimpanzees in Mahale Mountains National Park, Tanzania. *African Study Monographs*, **32**, 25–41.

Takahata, Y. (1985). Adult male chimpanzees kill and eat a male newborn infant: newly observed intragroup infanticide and cannibalism in Mahale National Park, Tanzania. *Folia Primatologica*, **44**, 161–70.

Takahata, Y. (1990a). Social relationships among adult males. In *The Chimpanzees of the Mahale Mountains: Sexual and Life History Strategies*, ed. T. Nishida. Tokyo: University of Tokyo Press, pp. 149–70.

Takahata, Y. (1990b). Adult males' social relations with adult female. In *The Chimpanzees of the Mahale Mountains: Sexual and Life History Strategies*, ed. T. Nishida. Tokyo: University of Tokyo Press, pp. 133–48.

Tutin, C. E. G. (1979). Mating patterns and reproductive strategies in a community of wild chimpanzees (*Pan troglodytes schweinfurthii*). *Behavioral Ecology and Sociobiology*, **6**, 29–38.

Uehara, S., Hiraiwa-Hasegawa, M., Hosaka, K., and Hamai, M. (1994). The fate of defeated alpha male chimpanzees in relation to their social networks. *Primates*, **35**, 49–55.

de Waal, F. (1982). *Chimpanzee Politics: Power and Sex among Apes*. Baltimore, MD: The Johns Hopkins University Press.

de Waal, F. B. M. and Harcourt, A. H. (1992). Preface. In *Coalitions and Alliances in Humans and other Animals*, ed. A. H. Harcourt and F. B. M. de Waal. Oxford: Oxford University Press, pp. v–vi.

28 Female–female relationships

NORIKO ITOH AND MICHIO NAKAMURA

28.1 Introduction

Female chimpanzees have been considered "egocentric and individualistic" (Nishida, 1967, 1979), "less social," or "asocial" (see later). However, it is obvious that without considerable social skills, a female cannot raise her offspring for years (Chapter 22). Because an infant does not always behave as the mother pleases and sometimes may have conflicts with her, the mother must always monitor her offspring and negotiate with him/her. The mother must also predict potential dangers in relation to social life, such as being involved in fights with/between others. Goodall (e.g. 1986) described a number of episodes where mothers coped with various circumstances quickly and appropriately. Apart from the biological benefits of maternal investment to her offspring, females' sociality observed in such scenes should be taken more seriously.

Furthermore, if we admit that chimpanzees have a social unit and that females also belong to the society (Chapter 6), females' ability to create new social relationships, irrespective of the partners' age or sex, in the group she immigrates to should also be considered part of their sociality. It is notable that although females' "less sociable" nature may sometimes tacitly be assumed as a consequence of females' shyness toward other chimpanzees or humans, females are much bolder than males in some contexts: the first members of the M group who started to eat sugarcane at the feeding site in Mahale were an adolescent female accompanied by ex-K-group females who were already used to human observers and/or the feeding site (Nishida and Kawanaka, 1972). Similarly, when Itoh encountered an unknown unit-group (Itoh *et al.*, 1999), it was the females, including mothers, who were seemingly waiting in a tree for humans and left silently after verifying these unfamiliar creatures. Males, in contrast, promptly fled from the place emitting pant-hoots from the upper stream where the females headed afterward. We do not intend to assert here that females in general are bold, but without interests and adaptations to strangers and/or new situations, female immigration to other unit-groups may be unrealistic.

Compared with the number of studies of chimpanzee male–male social relationships, studies of female–female relationships are fewer (Lehmann and Boesch, 2008). This, regrettably, is also the case for Mahale. Thus, in this chapter, we avoid overgeneralization from only a few studies, but review what is known to date, including information from other sites, by taking into account the practical difficulty of studying female–female relationships.

28.2 Underlying complexity in understanding female–female relationships

A female chimpanzee's life is characterized not only by physical changes through development (Chapter 22), but also by dynamic and cyclic physical and behavioral changes due to estrus (Chapter 44), pregnancy, delivery, and child rearing (see also Chapter 7). Such dynamic changes may mean that the same natural and social environments differ for a female at different times. In addition to internal changes, because many

Mahale Chimpanzees: 50 Years of Research, ed. M. Nakamura, K. Hosaka, N. Itoh and K. Zamma. Published by Cambridge University Press.

females transfer to other unit-groups at sexual maturity (Chapters 6, 7), they must also cope with drastic changes in the natural and social environments. Boesch and Boesch-Achermann (2000: Chapter 3) reviewed female chimpanzees' life history from various perspectives by carefully considering the underlying complexity, including social situations and backgrounds that differ between females.

Relationships between two individual primates (e.g. dominant–subordinate relationship) are known to be influenced by third parties, for example, in Japanese macaques (*Macaca fuscata*; Kawamura, 1958). This influence is also found in chimpanzees, and many other factors such as sexual cycling and presence of offspring have also been mentioned as affecting relationships since the outset of research at Mahale (Nishida, 1979) and elsewhere (Goodall, 1986). In addition, under the fission–fusion social system in chimpanzees (Chapters 6, 9), association memberships always change. This fluidity poses the difficulty for researchers that they cannot take two behavioral samples in identical social situations.

Another, but related, issue concerns degrees of habituation. Although the Mahale research is long term, there were considerable interindividual differences in the degrees of habituation, at least for females, up to the late 1990s (see also Chapters 3, 7). Several females had not tolerated the presence of human observers from earlier on, and new females constantly immigrated into the M group. A similar issue of females' habituation has been reported from Taï (Boesch and Boesch-Achermann, 2000; Lehmann and Boesch, 2008). The degree of habituation and the presence of observers easily affect the actual social relationships and, in turn, bias the observation data taken in such a situation (see also Boesch and Boesch-Achermann, 2000: 106). For example, Itoh witnessed some females who had been considered peripheral and appeared only rarely stealthily following other members from bush to bush. When Itoh followed them, she almost never met other researchers following adult males. The researchers thought that she had given up following and gone back to camp early, but instead, she was staying with the focal female quietly in a dense bush. Itoh and the shy female could see other researchers and chimpanzees from inside the bush, but they could not be seen by them. Following such a shy female was hard because she would almost never use observation trails but moved from bush to bush and because the observer needed to keep an adequate distance from the target in order to not frighten her. Still, the female would feel nervous of the human presence and go into deeper bush, resulting in the follow being terminated. It is difficult to judge whether such a female initially had a solitary tendency or could not socialize with others because of avoiding humans. However, it is possible that the latter was the case (see also Chapter 7). The degree of habituation can even decrease: a female, Opal, who had tolerated humans well in 1997, had started to avoid them by 2005.

Demographic characteristics of a unit-group are also important factors that prevent easy generalization of sociality. Lehmann and Boesch (2008: 76) asserted that large biased sex ratios in most chimpanzee unit-groups lead to stronger dilution effects on the estimated "average" degree of sociality in the more frequent sex, female (see also Chapter 6). In addition to such biases, a unit-group sometimes faces drastic changes. In the Taï South group, the number of females decreased to about half in 1999–2006, and female gregariousness also decreased (Wittiger and Boesch, 2013). At Mahale, many systematic observations of the M group were made when the group greatly increased its number by receiving many females from the adjacent K group. Because such a mass immigration has not been observed since then, it is assumed that the situation at that time was extraordinary for the females (e.g. increased competition or disruption of equilibrium due to rapid influx of females; Chapters 6, 7). The M-group number also declined to about half in the mid-1990s (Chapters 6 and 7). At Gombe, Tanzania, one unit-group split into two branch groups through 1971–2 and one disappeared shortly thereafter (Bygott, 1974). The Gombe population may also have faced isolation caused by rapid deforestation around the small national park (Bygott, 1974). A similar isolation also occurred at Bossou, Guinea (Sugiyama, 2004). Such isolation can be an obstacle for females to transfer between groups. Eventually, only 6 out of 11 female residents had originated from other unit-groups at Gombe (Pusey *et al.*, 1997), with the Bossou group never having received new immigrant females from the onset of research (Sugiyama, 2004). Such situations are thought to be critical from a genetic point of view (Gombe: Pusey *et al.*, 1997; Bossou: Shimada *et al.*, 2009), but they may also influence female–female relationships in the groups. It may not be meaningful to assert which unit-group females are more social or asocial without considering the historical and demographic backgrounds of each unit-group; unit-groups with a small and stable membership are distinct from unit-groups with large membership and constant immigrations. Even within a unit-group, observed frequencies of interactions should be considered, taking into account the previous demographic changes as described earlier.

In short, several variables such as population dynamics, degree of habituation, interindividual changes of physical characteristics, and the fission–fusion nature of the society should be synthesized for evaluating the data of female–female relationships.

28.3 Features of female–female relationships

The image of females being less social derives from observations that females do not meet and interact with each other compared with males. Such an image seems to be nurtured especially in the studies at Gombe. In that most famous and longest-term study site, chimpanzees there often represent "the wild chimpanzee," especially for nonchimpanzee researchers. Indeed, many Gombe researchers have repeatedly emphasized females' lower sociability. In an early and influential study, Wrangham and Smuts (1980: 13) wrote "females are less gregarious and less sociable than males." In a later commentary paper in *Science*, Wrangham (1997: 774) summarized that "[m]any females meet each other rarely" and "[a]t Gombe, each mother's foraging is concentrated in her own core area within a larger community range defended by natal males." Similar accounts are repeated in more recent works: "females are rarely aggressive and lead relatively secluded lives" (Townsend *et al.*, 2007: R355), "East African chimpanzees (*P. t. schweinfurthii*) generally conform to a male-bonded model in which females are less social than males" (Murray *et al.*, 2006: 1020).

However, a growing body of studies from other study sites places these statements at one extreme of female sociality and female–female relationships, and it is now questioned whether Gombe females represent "the chimpanzee (or the eastern chimpanzee) females." Studies on Gombe females enlivened studies on "Subtle, Secret Female Chimpanzees" (Wrangham, 1997), as summarized later.

28.3.1 Female gregariousness

Gombe males and females show a clear contrast in the proportions of time spent alone: 27% and 65% , respectively (Wrangham and Smuts, 1980), and female solitariness was suggested to be even higher in a later study (Pusey *et al.*, 1997). Kanyawara (Uganda) females showed similar proportions, spending 23.1% of their time alone and 69.8% when in fruiting trees (5.3% and 20.5%, respectively, for males; Wrangham *et al.*, 1992).

In contrast, lone female and male parties were observed only 1.8% and 3.3%, respectively, of the time in Taï (Boesch, 1996; Boesch and Boesch-Achermann, 2000). Ngogo (Uganda) females spent time with other females for 64% of scans (Wakefield, 2008). Mahale M-group females ($N = 8$) spent 82.9%, on average, of their time with others, whereas males ($N = 5$) spent 78.9% of their time with others (these observations are restricted to focal data with more than five total observation hours in the dataset in Chapter 17). In the Ngogo group, which has an unusually large unit-group size exceeding 100, females are known to form associative cliques within the unit-group (Wakefield, 2008).

Ranging patterns may partly explain the low gregariousness observed in Gombe and Kanyawara, and the higher gregariousness in other sites. Gombe females stay in their small core area, distributed in different areas with some overlaps, and travel over wider areas only when in estrus (Wrangham, 1979). Kanyawara females and males exhibit large differences in space use (Chapman and Wrangham, 1993). Ranging patterns of Mahale females and males were not so clearly different in the 1980s (Hasegawa, 1990), and the process of female transfer and seasonality has been suggested to affect space use (Uehara, 1981; Hasegawa, 1990); when the availability of the staple food *Saba comorensis* is high, females travel together all over the unit-group range. See also Matsumoto-Oda (2002) for similar results for M-group cycling females in the 1990s. Repeated observations of huge nomadic party formation, which may be unique to Mahale (Figure 28.1), also support female ranging with other individuals irrespective of physiological conditions (Itoh and Nishida, 2007; Chapter 8). As these members range together for consecutive months over wide areas, M-group females live in and lead lives of different social conditions from those of Gombe females. Taï females also show little difference from males in ranging pattern, irrespective of unit-group size and composition (Boesch, 1991; Lehmann and Boesch, 2005). Bossou also showed little sex difference in ranging (Sugiyama, 1988). Thus, as Wittiger and Boesch (2013: 1108) suggest, "simple dichotomy of sociality between eastern [e.g. Gombe, Mahale, Kanyawara] and western [e.g. Bossou, Taï] female chimpanzees" can no longer be inferred (site names in brackets added). They also found chronological changes of female relationships within a unit-group that were influenced by not only ecological but also social conditions.

Considering the M-group data in detail, interindividual variation in proportion of time spent alone was also high: 3.8–35.7% for females and 0.6–50.6% for males. Females spent 59.9%, on average, of their time with other adult females, but this behavior also showed a high interindividual variation: 17.7–96.2%. The lowest value seems to be biased by data collected when a female was separately ranging with a male for consortship. The female with the second-lowest value (19.6%) was almost always accompanied by her 9-year-old son and 4-year-old daughter. The highest value came from a cycling female who gave birth in March 1998 during the study period. The second-highest value (86.9%) came from a female who also gave birth in February 1998. The remaining four females (56–74%) were one lactating female, one cycling female almost always accompanied by her 7-year-old adopted daughter, and two cycling females without offspring (note that the latter two were also fully adult females although they did not have offspring at the time). Thus, both males and females show a large variation in association patterns, suggesting that the existence of a "typical" female (or male) be reconsidered.

28.3.2 Frequency of interactions

Another reason for females having been considered less social is the infrequency of explicit interactions compared with males.

Grooming is one such interaction (Chapter 34). Again, a clear contrast between females and males is observed in Gombe. Considering the distribution of grooming among different age–sex classes, females with offspring devoted only 14.2% ± 18.3% of grooming activity to the same-sex adults and adolescents, whereas males devoted 59.2% ±19.6% of their grooming activity to same-sex adults and adolescents (as calculated by Wrangham *et al.*, 1992, from data in Goodall, 1986). Amazingly, Kibale (Uganda) females were not observed to groom each other during about 85 h of focal follows of females (Wrangham *et al.*, 1992). Focal data from Mahale females showed a similar tendency (11.4% ± 8.6% for residents and 29.0% ± 20.7% for immigrants, as calculated by Wrangham *et al.*, 1992, from data in Nishida, 1989). Taï females groomed with same-sex individuals significantly less than expected, whereas males groomed the same sex significantly more than expected (Lehmann and Boesch, 2008). These findings may not be so surprising when we consider that females, especially mothers, need to allocate time for caring (e.g. grooming and nursing) their offspring. In fact, the time Gombe mothers devoted to grooming with their offspring (59.2% ± 35.3%) was comparable with that males devoted to grooming the same sex (Wrangham *et al.*, 1992). An exception is at Bossou, where female–female grooming and male–male grooming frequencies showed the reverse pattern (Sugiyama, 1988).

Figure 28.1 Part of a large nomadic party. At least 25 chimpanzees, including several mothers and offspring, are confirmed in this picture. ©Michio Nakamura.

More interestingly, considering the number of same-sex grooming partners relative to available same-sex partners, there were no differences between the sexes in Gombe, Bossou, Mahale, Ngogo, Budongo (Uganda), and Taï (Lehmann and Boesch, 2008). As there were differences in length of grooming bouts between sexes (male > female, Taï; Lehmann and Boesch, 2008), it can be assumed that females engage in grooming to balance time spent on a grooming interaction and covering the number of partners within the unit-group members. Similarly, Nakamura (2010) showed that among seven focal females (21 possible combinations) at Mahale, grooming was observed in 19 dyads (90.5%) in 1996–7. Even in the shorter observation time in 2004, grooming was observed in 32 out of 45 possible female dyads (71.1%). The comparable figure for males was 73.3% (11 out of 15 male–male dyads) in the former period. This observation showed that M-group females groom – although each grooming may not last long – the majority of other females. He also showed that females with offspring had 20.3 grooming partners on average and, at most, one female had 40 partners (two-thirds of the unit-group members) within only a 2-month study period. Thus, Mahale females have quite wide grooming networks in terms of partners.

Another interesting point is that M-group females groomed longer than males in large grooming clusters with more than five individuals (Figure 28.2), although they groomed less in smaller clusters (see Chapter 34 for details). Thus, female sociality in terms of grooming may also vary according to the social conditions such as how many individuals and who are participating *in situ*.

28.4 Long-term relationships

28.4.1 Dominant–subordinate relationships

Because aggression is also infrequent among females (e.g. Gombe: Bygott, 1979; Mahale: Nishida, 1989; Kanyawara: Muller, 2002; Ngogo: Wakefield, 2008), early researchers admitted that it was difficult to assign clear-cut dominant–subordinate relationships among females (Bygott, 1974; Goodall, 1986).

However, after the seminal paper of Pusey *et al.* published in 1997, many studies on female–female dominant–subordinate relationships were published from several study sites (e.g. Fawcett, 2000; Wittig and Boesch, 2003; Murray *et al.*, 2006, 2007; Murray, 2007; Kahlenberg *et al.*, 2008; Wakefield, 2008). Female dominant–subordinate relationships also seem to have variations in their structure and function, if any, among unit-groups. For this reason, further spatiotemporally wider comparisons and microlevel observations of actual female interactions are needed, with a consideration of the historical and demographic backgrounds of the unit-groups, as we have mentioned earlier.

Figure 28.2 A cluster of females with offspring, some grooming, some resting, and some playing. In this scene, four lactating mothers, their six children, one immigrant nulliparous female, and one adolescent male are present. ©Michio Nakamura (from video footage).

In Mahale, an age-related dominant–subordinate relationship was suggested in some early studies (Nishida, 1979, 1989), but a systematic investigation has not been conducted. Nishida (1979) reported that the dominant–subordinate relationships among females were unstable and complicated because they may change according to estrous status (a young female in estrus may dominate adult females). Such changes may also be influenced by a female's association or lack of one with a high-ranking male.

Another factor influencing dominant–subordinate relationships was babysitting. For example, although Chausiku had been dominant over Wakasila since 1969, the relationship reversed during a part of 1973–4 as Chausiku became an eager babysitter of Wakasila's baby. The reversed relationship continued even after Chausiku's delivery (Nishida, 1979).

Shyness in the presence of other chimpanzees also seemed to influence dominant–subordinate relationships. Some females become shyer and more timid when they give birth, and such shy females often avoid even subordinate females without infants by running away or grimacing when they approach (Nishida, 1979). Thus, for Mahale females, domination over other females may not be very important as they easily let go of their dominance at the expense of "mere" babysitting or rearing their own baby. However, dominance seems to benefit females in other study sites, as in Gombe (Pusey *et al.*, 1997; Murray *et al.*, 2006, 2007) and Kanyawara (Emery Thompson *et al.*, 2007; Kahlenberg *et al.*, 2008).

Although we do not go into further details of dominant–subordinate relationships, we mention some of their implications from the viewpoint of sociality or social capability, especially with respect to the most severe dominant–subordinate relationships occurring in Gombe. As we have seen earlier, Gombe females are said to rarely meet each other without interacting much but maintain long-term dominant–subordinate relationships (e.g. Pusey *et al.*, 1997). This observation is striking when we consider males that devote substantial time and energy to maintaining their intra-sex dominant–subordinate relationships all the time (Chapter 27). Dominant–subordinate relationships may differ in their quality or meaning for different sexes, but it can be said that, at least at Gombe, females maintain long-term relationships without frequent interactions or even meeting each other. At the same time, observations of dominant–subordinate relationships from Mahale suggest their capability to change flexibly in reflection of the context where they are situated.

28.4.2 Friendship

It is known that some females form friendly relationships that last several years or more. Taï females are famous for such relationships (Boesch and Boesch-Achermann, 2000). Although publications from Mahale did not use the term "friendship," similar relationships may exist. Frequent associations between Gwekulo and Wantendele (Nishida and Hiraiwa-Hasegawa, 1985) and long-term associations between Gwekulo and Wakapala (Nishida, 1989), both ex-K-group pairs who had immigrated to the M group, are such examples (see Chapter 31 for another friend of Gwekulo, Wakasunga).

Wittiger and Boesch (2013) suggested that females slowly form new social bonds, so that quantitative assessment of each female–female relationship may need a much longer time than that for males. The longevity of chimpanzees (Chapter 7) and constant immigration of new females may also make the short-term assessment less fruitful. In Mahale, we have no consecutive data to trace back long-term association patterns of females. However, from some anecdotal observations, we have become aware of some friendly relationships.

In the M group, a new immigrant female and a resident female form such a relationship. The Juno–Cynthia and Fatuma–Qanat pairs are such examples. For the former pair, Juno frequently let Cynthia carry her daughter while traveling, and the daughter herself alternated riding on either Cynthia or Juno while traveling in 1995–6.

Opal, Gwekulo, and Pinky were also frequent associates. Opal and Gwekulo were noticed to be friends for a long time (at least from the late 1990s).

We do not have the impression that they engaged in explicit interactions very frequently. However, for example, in the dispersal season (see Chapters 8, 9) in 1998, after Gwekulo, her adopted daughter Pipi, and another friend Pinky and her two offspring ranged together for a few days, they suddenly started to emit loud calls and rushed. When they arrived, Opal and her three offspring were there. Gwekulo and Opal stood face to face pant-grunting at each other. The excitement of reunion did not last long, and soon, they all started to peacefully rest together for a while. Their relationship lasted until the death of Opal in 2006. The Gwekulo–Pinky relationship eventually resulted in the adoption of Pinky's 6-year-old daughter Puffy by Gwekulo, after the death of Pinky also in 2006 (Chapter 30).

Another notable pair is Wakusi and Calliope. At least since the 2000s in their forties, these two old females frequently associated with each other. On one occasion, Itoh observed Wakusi rapidly climbing a tree near Kasiha Valley when males started to get excited. Males ran around and others rushed in different directions as if to escape from the scene. Many passed under Wakusi's tree, crossing rivers northward or climbing up a mountain eastward, but no one seemed to notice her. Calliope also passed a few meters downstream from Wakusi's tree without noticing her. However, after a while, she came back from the north, quickly retracing the same route, and looked around, probably in search of Wakusi. But again, she failed to see Wakusi and crossed the river southward. As the follow ended, their reunion could not be confirmed on that day. We have no previous data to decide when their friendship began, as Calliope had always been shy in the 1990s and probably before.

28.5 Explicit interactions under specific conditions

Female–female relationships, however unstable or vague they may seem, may still manifest under some "vital" situations, a circumstance that possibly benefits females. In Taï, female friends not only frequently associate but also frequently share food and support in conflicts (Boesch and Boesch-Achermann, 2000). In Mahale, a systematic study has not been performed, but some cases are known.

Opal and Gwekulo protected Opal's granddaughter from an attempted infanticide. One day, when Opal's daughter Ruby appeared with a newborn, the baby and/or Ruby were attacked by two males. Opal supported Ruby, and Gwekulo also came in support of them, driving the two males away and leaving a bleeding wound on one of the male's foot (Sakamaki *et al.*, 2001; see also Chapter 29). Because all Ruby could do was protect her baby from the two males, and because both were high-ranking males including the alpha, Opal might have faced difficulty fending them off alone if Gwekulo had not arrived.

Such coalitional supports (or situations of such need) between females are rare but have been observed since the early days at Mahale. In 1981, when Wantendele, Gwekulo's former friend, and her son were attacked by two males, Gwekulo behaved as if to take the boy away from the scene, although she failed and eventually ran away (Nishida and Hiraiwa-Hasegawa, 1985). In two other cases observed in 1981 and 1983, females intervened in conflicts among other females (Nishida, 1989). Coalitional attacks between resident and ex-K-group females were also observed four times (Nishida, 1989). Some additional cases of the M-group female–female coalitions against males are also introduced in Chapter 29. There are also reports from other sites of female coalitions (or alliances) against males (Budongo: Newton-Fisher, 2006) and females (Taï: Boesch and Boesch-Achermann, 2000).

The above cases can be regarded as support from friends in "vital" situations, and the short-term effects are relatively clear. However, there is another type of support by friends: carrying of offspring while traveling, as Cynthia did for Juno, may also help the mother. In the Gwekulo–Pinky case, the relationship helped Pinky's daughter after she became an orphan (Chapter 30). Thus, friendly relationships between females may also benefit them in the longer term.

28.6 Implicit interactions among females

As we have seen earlier, social interactions among females in their average frequencies seem to be less than those among males. So how do females form relationships among themselves, especially affiliative relationships? Are there any interactions, apart from salient interactions, such as grooming or direct support that may be important for them to form friendly relationships?

Nishida and Kawanaka (1972) described some female–female interactions, even though some of them may be atypical because the observations were made at the feeding station. Presenting and mounting interactions were observed between females when females of the ex-K-group immigrated to the M group. We have not recently observed female–female mounting. Open-mouth kissing was also observed once between a dominant (F103, Wankungwe) and an adolescent (F110, Wasobongo) female (Nishida and Kawanaka, 1972). These are relatively explicit interactions and possibly related to dominant–subordinate interactions.

There are also some interactions that are quite difficult to assess. For example, when several females were feeding on the same tree, one descended and rested under the tree. After a while, another started to descend, and as she came close to the female on the ground, she emitted a soft grunt-like sound. The female on the ground made no apparent response, and the one emitting the sound left. After a few seconds, the resting female followed her. Such soft vocalizations are sometimes observed among females. They may be a kind of pant-grunting to confirm their dominant–subordinate relationships (Chapters 33, 39), but may not be because the contexts are often much calmer than those toward males, and the interaction does not seem unidirectional.

On another occasion, one female, by occasionally sniffing at the ground, succeeded in tracing another female who had been feeding nearby but had already moved away. Such preceding and following occurs often during a prolonged time span, so it may easily be overlooked by observers. This kind of behavior does not actively or directly involve the other individual. Repeating such subtle interactions may be essential to form, even slowly, tolerance among females. Such a slow formation of tolerance may resemble what we human observers attempt in the process of habituating chimpanzees.

Finally, we mention some symmetrical interactions that are also sometimes observed among M-group females: mutual touching of sexual skin; mutual pant-grunting; mutual gazing into the face; or mutual touching of hands, most of which usually occur in relatively calm situations. Such interactions are also not easily assigned to any particular function. We have only recently begun to notice such subtle and implicit interactions among females, probably because of improved conditions for observation of females. By quantitative and short-term evaluations of explicit interactions and face-to-face associations that seem to be more suitable for assessing male–male relationships, female chimpanzees may easily be labeled "less social" than males. We expect that further investigations over longer time spans will shed light on the sociality of female chimpanzees, which may be "different" from that of males or those recognized as "important" to date, but need not be "less."

Here, we have briefly seen the amazing variability in female–female relationships and sociality, in general, between unit-groups and between pairs within a unit-group. From some studies, we now know that these features are influenced by ecological and social conditions over different time spans and also change in time. We obviously need more detailed and longer studies at Mahale as well as more comparative and cooperative studies between various sites to gain a deeper understanding of the variability and underlying factors.

References

Boesch, C. (1991). The effect of leopard predation on grouping patterns in forest chimpanzees. *Behaviour*, **117**, 220–41.

Boesch, C. (1996). Social grouping in Taï chimpanzees. In *Great Ape Societies*, ed. W. McGrew, L. Marchant, and T. Nishida. Cambridge: Cambridge University Press, pp. 101–13.

Boesch, C. and Boesch-Achermann, H. (2000). *The Chimpanzees of the Taï Forest: Behavioural Ecology and Evolution*. New York: Oxford University Press.

Bygott, J. D. (1974). *Agonistic Behaviour and Dominance in Wild Chimpanzees*. Ph.D. thesis. Cambridge: University of Cambridge.

Bygott, J. D. (1979). Agonistic behavior, dominance, and social structure in wild chimpanzees of Gombe National Park. In *The Great Apes*, ed. D. A. Hamburg and E. R. McCown. Menlo Park, CA: Benjamin/Cummings, pp. 405–27.

Chapman, C. A. and Wrangham, R. W. (1993). Range use of the forest chimpanzees of Kibale: implications for the understanding of chimpanzee social organization. *American Journal of Primatology*, **31**, 263–73.

Emery Thompson, M., Kahlenberg, S. M., Gilby, I. C., and Wrangham, R. W. (2007). Core area quality is associated with variance in reproductive success among female chimpanzees at Kibale National Park. *Animal Behaviour*, **73**, 501–12.

Fawcett, K. A. (2000). *Female Relationships and Food Availability in a Forest Community of Chimpanzees*. PhD. thesis. Edinburgh: University of Edinburgh.

Goodall, J. (1986). *The Chimpanzees of Gombe: Patterns of Behavior*. Cambridge, MA: Belknap.

Hasegawa, T. (1990). Sex differences in ranging patterns. In *The Chimpanzees of the Mahale Mountains*, ed. T. Nishida. Tokyo: University of Tokyo Press, pp. 99–114.

Itoh, N. and Nishida, T. (2007). Chimpanzee grouping patterns and food availability in Mahale Mountains National Park, Tanzania. *Primates*, **48**, 87–96.

Itoh, N., Sakamaki, T., Hamisi, M., *et al.* (1999). A new record of invasion by an unknown unit-group into the center of M group territory. *Pan Africa News*, **6**, 8–10.

Kahlenberg, S. M., Emery Thompson, M., and Wrangham, R. W. (2008). Female competition over core areas in *Pan troglodytes schweinfurthii*, Kibale National Park, Uganda. *International Journal of Primatology*, **29**, 931–47.

Kawamura, S. (1958). [The matriarchal social order in the Minoo-B group.] *Primates*, **1**, 149–56. In Japanese, with English summary.

Lehmann, J. and Boesch, C. (2005). Bisexually bonded ranging in chimpanzees (*Pan troglodytes verus*). *Behavioral Ecology and Sociobiology*, **57**, 525–35.

Lehmann, J. and Boesch, C. (2008). Sexual differences in chimpanzee sociality. *International Journal of Primatology*, **29**, 65–81.

Matsumoto-Oda, A. (2002). Behavioral seasonality in Mahale chimpanzees. *Primates*, **43**, 103–17.

Muller, M. N. (2002). Agonistic relations among Kanyawara chimpanzees. In *Behavioural Diversity in Chimpanzees and Bonobos*, ed. C. Boesch, G. Hohmann, and L. F. Marchant. Cambridge: Cambridge University Press, pp. 112–24.

Murray, C. M. (2007). Method for assigning categorical rank in female *Pan troglodytes schweinfurthii* via the frequency of approaches. *International Journal of Primatology*, **28**, 853–64.

Murray, C. M., Eberly, L. E., and Pusey, A. E. (2006). Foraging strategies as a function of season and rank among wild female chimpanzees (*Pan troglodytes*). *Behavioral Ecology*, **17**, 1020–8.

Murray, C. M., Mane, S. V., and Pusey, A. E. (2007). Dominance rank influences female space use in wild chimpanzees, *Pan troglodytes*: towards an ideal despotic distribution. *Animal Behaviour*, **74**, 1795–804.

Nakamura, M. (2010). Ubiquity of culture and possible social inheritance of sociality among wild chimpanzees. In *The Mind of the Chimpanzee: Ecological and Experimental Perspectives*, ed. E. V. Lonsdorf, S. R. Ross, and T. Matsuzawa. Chicago, IL: University of Chicago Press, pp. 156–67.

Newton-Fisher, N. E. (2006). Female coalitions against male aggression in wild chimpanzees of the Budongo Forest. *International Journal of Primatology*, **27**, 1589–99.

Nishida, T. (1967). [The society of wild chimpanzees: success of provisioning at Kasoge.] *Shizen*, **22**(8), 31–41. In Japanese.

Nishida, T. (1979). The social structure of chimpanzees of the Mahale Mountains. In *The Great Apes*, ed. D. A. Hamburg and E. R. McCown. Menlo Park, CA: Benjamin/Cummings, pp. 73–121.

Nishida, T. (1989). Social interactions between resident and immigrant female chimpanzees. In *Understanding Chimpanzees*, ed. P. G. Heltne and L. A. Marquardt. Cambridge, MA: Harvard University Press, pp. 68–89.

Nishida, T. and Hiraiwa-Hasegawa, M. (1985). Responses to a stranger mother-son pair in the wild chimpanzee: a case report. *Primates*, **26**, 1–13.

Nishida, T. and Kawanaka, K. (1972). Inter-unit-group relationships among wild chimpanzees of the Mahali Mountains. *Kyoto University African Studies*, **7**, 131–69.

Pusey, A. E., Williams, J., and Goodall, J. (1997). The influence of dominance rank on the reproductive success of female chimpanzees. *Science*, **277**, 828–31.

Sakamaki, T., Itoh, N., and Nishida, T. (2001). An attempted within-group infanticide in wild chimpanzees. *Primates*, **42**(4), 359–66.

Shimada, M. K., Hayakawa, S., Fujita, S., Sugiyama, Y., and Saitou, N. (2009). Skewed matrilineal genetic composition in a small wild chimpanzee community. *Folia Primatologica*, **80**, 19–32.

Sugiyama, Y. (1988). Grooming interactions among adult chimpanzees at Bossou, Guinea, with special reference to social structure. *International Journal of Primatology*, **9**, 393–407.

Sugiyama, Y. (2004). Demographic parameters and life history of chimpanzees at Bossou, Guinea. *American Journal of Physical Anthropology*, **124**, 154–65.

Townsend, S. W., Slocombe, K. E., Emery Thompson, M., and Zuberbühler, K. (2007). Female-led infanticide in wild chimpanzees. *Current Biology*, **17**, R355–6.

Uehara, S. (1981). [The social unit of wild chimpanzees: a reconsideration based on the diachronic data accumulated at Kasoje in the Mahale Mountains, Tanzania.] *Journal of African Studies*, **20**, 15–32. In Japanese with English summary.

Wakefield, M. L. (2008). Grouping patterns and competition among female *Pan troglodytes schweinfurthii* at Ngogo, Kibale National Park, Uganda. *International Journal of Primatology*, **29**, 907–29.

Wittig, R. M. and Boesch, C. (2003). Food competition and linear dominance hierarchy among female chimpanzees of the Taï National Park. *International Journal of Primatology*, **24**, 847–67.

Wittiger, L. and Boesch, C. (2013). Female gregariousness in western chimpanzees (*Pan troglodytes verus*) is influenced by resource aggregation and the number of females in estrus. *Behavioral Ecology and Sociobiology*, **67**, 1097–111.

Wrangham, R. W. (1979). Sex differences in chimpanzee dispersion. In *The Great Apes*, ed. D. A. Hamburg and E. R. McCown. Menlo Park, CA: Benjamin/Cummings, pp. 481–9.

Wrangham, R. W. (1997). Subtle, secret female chimpanzees. *Science*, **277**, 774–5.

Wrangham, R. W. and Smuts, B. B. (1980). Sex differences in the behavioural ecology of chimpanzees in the Gombe National Park, Tanzania. *Journal of Reproduction and Fertility*, **Suppl. 28**, 13–31.

Wrangham, R. W., Clark, A. P., and Isabirye-Basuta, G. (1992). Female social relationships and social organization of Kibale Forest chimpanzees. In *Topics in Primatology Vol. 1: Human Origins*, ed. T. Nishida, W. C. McGrew, P. Marler, M. Pickford, and F. B. M. de Waal. Tokyo: University of Tokyo Press, pp. 81–98.

29 Male–female relationships: affiliative, interventional, and dominant–subordinate interactions

SHUNKICHI HANAMURA

29.1 Introduction

Chimpanzee males stay in their natal unit-group for their whole life. In contrast, most females transfer to different unit-groups on reaching sexual maturity and, after first giving birth, they stay in the unit-group for the rest of their life (Chapter 6). Accordingly, these males and females from different unit-groups form various relationships in the course of their long lives.

The relative strength of dyadic "affinity" or "bond" among the group members in various group-living animals has been assessed by frequencies of each dyad's associations and affiliative interactions (Whitehead, 2008). Such bond strength within- and between-sexes has often been studied in relation to the social structure of the group. In this chapter, I first briefly review such general association patterns of chimpanzee males and females at Mahale and other sites. Second, I introduce cases of affiliative relationships formed by particular male–female dyads of both kin and non-kin, and then consider the possible influence of such relationships on males' conflicts. Third, I introduce cases of females' interventions in male aggression toward others, focusing on the influence of length of relationships and age gap between males and females on their interactions. Fourth, I review studies on male–female dominant–subordinate relationships. Finally, these male–female relationships are reconsidered through the perspective of the females' immigration process.

I lay stress on the long-term nature and nonsexual aspect of male–female relationships (for sexual interactions, see Chapter 35). By doing so, I reconsider the general view of male–female relationships, that is often based only on their sexual aspects, and their significance to chimpanzees' social life. Hereafter, I generally refer to adult males as "males" and adult females as "females," and add a given age–sex class for others (for definitions of age–sex classes, see Chapter 22).

29.2 General association patterns of males and females

From the early studies at Mahale, Nishida considered that a unit-group of eastern chimpanzees was organized mainly by strong bonds among males, and that females were integrated into the group through weaker bonds with males, mainly when they were in estrus (Nishida, 1968, 1970, 1979; Nishida and Hiraiwa-Hasegawa, 1987). He concluded this based on the following two tendencies.

First, there are sex differences in association patterns, where females, compared to males, have fewer associations and affiliative interactions with other group members, and spend more time alone or with only their dependent offspring (although there are some contradicting views: see Chapters 6, 28). Second, females change their association patterns largely depending on their reproductive status (i.e. cycling or lactating) and/or the stage of their estrous cycle (i.e. estrus or not) as follows. Typically, females have more associations with males when they are cycling than when lactating and when they are in

Mahale Chimpanzees: 50 Years of Research, ed. M. Nakamura, K. Hosaka, N. Itoh and K. Zamma. Published by Cambridge University Press.

estrus than not (Takahata, 1990; Matsumoto-Oda, 1999; for female's estrous cycle, see Chapter 35).

These two points have also been reported in other eastern chimpanzee unit-groups (e.g. Gombe, Tanzania: Goodall, 1986; Ngogo, Uganda: Pepper *et al.*, 1999; Kanyawara, Uganda: Machanada *et al.*, 2013) and are often explained together. When the amount of food is relatively small, chimpanzees should have fewer associations and form smaller parties in order to avoid feeding competitions, and females would suffer more severely from those competitions than others due to increased costs of traveling with offspring (Wrangham, 2000). Potential infanticide or aggression by males may also cause females' low frequency of associations with males (Otali and Gilchrist, 2006). Accordingly, it is often assumed that females do not have a reason to associate frequently with others, especially males, except when they are in estrus and need to increase mating opportunities (Nishida and Hiraiwa-Hasegawa, 1987).

29.3 Affiliative relationships formed by particular male–female dyads

Although it has been assumed that a male and a female eastern chimpanzee generally associate more frequently when the female is in estrus, there are some exceptions. In this section, I focus on such exceptional male–female pairs that continue affiliative relationships beyond the female's estrous periods. Here, I define an affiliative relationship as the relationship between two individuals who have more associations (i.e. spending time in the same party or in proximity) and/or affiliative interactions (e.g. grooming, food sharing, etc.) than other pairs (cf. Silk, 2002). I do not consider paternal kinship, because chimpanzee males do not show parental behaviors toward their offspring and most females are from different unit-groups.

29.3.1 Affiliative relationships between kin male and female

It is unsurprising that a male and a female with maternal kinship often form an affiliative relationship. Because chimpanzee sons stay and mature in their natal unit-group, it is common that some mother–son pairs form such relationships for the duration of their life (K group: Nishida, 1979; M group: Hayaki, 1988; Takahata, 1990). In addition to frequent association and grooming, the mother may receive a share of meat from her mature son (Nishida *et al.*, 1992; Chapter 20). Early studies at Gombe also reported that some mother–son affiliative relationships continued for several years after the sons' weaning and that a mature son rushed to support his mother when she screamed due to aggression by others or, conversely, a mother rushed to the scene when her mature son had fights with another male (Pusey, 1983; Goodall, 1986: 376).

At some study sites, the core ranging areas of some mature sons overlapped with those of their mothers (Budongo, Uganda: Newton-Fisher, 2002; Gombe: Murray *et al.*, 2008; Ngogo: Langergraber *et al.*, 2013). This suggests that males often feed in the familiar places where they spent their childhood, with their mothers, and/or that they have continual associations with their mothers. In 2006, I observed that a mother in the Mahale M group, Fatuma (with her two dependent daughters), and her 18-year-old mature son, Pimu, often traveled together for many hours (maximum: 6 h in a day) in the area where Fatuma often ranged in the dispersing season. They sometimes exchanged pant-hoots (Chapter 39) when they parted along the way.

Brothers and sisters on the maternal side, especially those who are close in ages and hence spent time with their mother concurrently, often form affiliative relationships in childhood. Such relationships may continue until the sister emigrates from the natal unit-group or throughout life if the sister remains in the group (for various sibling relationships in the Mahale M group, see Nishida, 2012: 116–20).

Such male–female affiliative relationships formed through childhood may not be limited to maternal siblings. For example, at Gombe, a female who stayed in her natal unit-group had lifelong affiliative relationships with two non-kin males who had been her playmates in childhood (Goodall, 1986: 182, 197–8). This suggests that the basis of an affiliative relationship among maternal male–female siblings is not their biological kinships themselves, but is the actual experiences of spending time together during childhood. Similarly, the aforementioned mother–son affiliative relationships would also be nurtured through their actual parental-dependent interactions in the sons' childhood.

29.3.2 Affiliative relationships between non-kin male and female

For the Mahale M group, Matsumoto-Oda (2002) reported that some male and cycling female pairs continued their affiliative relationships beyond the females' estrous periods. The males were often high-ranking, and the pairs tended to mate more frequently than other pairs when females were in estrus. The author considered that the mating-associated affiliative relationships were shorter than those in other primate species, such as "particular proximate relationships" (PPRs) in Japanese macaques (*Macaca fuscata*: Kitamura, 1977; Takahata, 1982) and "friendships" in olive baboons (*Papio anubis*: Smuts, 1985; Strum, 1987), which last for several years.

In the M group, there is a male–female pair that is presumed to have continued a long-term affiliative relationship for around 20 years until the time of writing in 2014: a male, Kalunde and a female, Nkombo (Nishida, 2004, 2012: 255–6; Hosaka, 2012; Chapter 5). Although their frequent associations regardless of Nkombo's estrous cycle had been due mainly to her following Kalunde (Nishida, 2004), he also began to follow her after approximately 1996 (Hosaka, 2012). Since then they have come to associate and even play with each other more frequently than before, and local research assistants therefore call Nkombo "mke" ("wife" in Swahili) of Kalunde (Hosaka, 2012). When I focal followed Nkombo in some years between 2005 and 2014, I also often observed them traveling together and making night beds nearby on the same or adjacent trees (Figure 29.1). In addition, when Kalunde was high-ranking and confronted with another male, Fanana, in 1997 (Chapter 27), Nkombo supported Kalunde with two other females by barking and even chasing the opponent (Nishida, 2004, 2012: 255–6). Several

Figure 29.1 Kalunde (left) and Nkombo (right) in 2006. (a) Sitting and laying their hands on the same tree; (b) lying down beside each other. ©S. Hanamura.

similar cases of Nkombo's support for Kalunde's opposing other males were observed, at least from 1999 to 2000 (Sakamaki, 2011) and from 2005 to 2006 (Hanamura, personal observation).

This long-term affiliative relationship has possibly continued as an extension of a mating-associated affiliative relationship between a high-ranking male and a cycling female as mentioned above. The prolongation may be because Nkombo has been cycling all the time: since she gave birth to, but soon lost, her first child in 1993, she has not successfully given birth. "Consortship" (Chapter 35) can be another candidate to springboard such prolonged associations because it lasts up to 3 months in the M group (Nishida, 2012: 215); hence the association can continue well beyond an estrous period (12–13 days) of the female partner.

Figure 29.1 (cont.)

Intriguingly, it seems that Kalunde and Nkombo recently became less interested in mating with each other, even though each mated with others (Hosaka, 2012). Although more data are needed, I have also never witnessed their mating, at least during my focal follows of Nkombo since 2005. The PPRs in Japanese macaques are also marked by a characteristic that the pairs gradually cease mating (Takahata, 1982). In many animal species, including humans (*Homo sapiens*), a male and a female who spent childhood together tend to avoid mating and this tendency is considered to be a mechanism for inbreeding avoidance (for a review, see Wolf, 1993). The examples of PPRs, and perhaps the Kalunde–Nkombo relationship, suggest that similar mating avoidance among affiliative individuals may also arise after sexual maturity.

Long-term affiliative relationships between non-kin males and females have also been reported from other study sites. In a 12-year study of male–female association patterns at Kanyawara, Machanda *et al.* (2013) briefly mentioned in their discussion that some male–female pairs had strong associations comparable to male–male pairs, although they did not give further details. Langergraber *et al.* (2013) recently demonstrated that some frequently associated male–female pairs at Ngogo continued to associate after several years, when the females were not in estrus. Although the authors considered that the frequent associations may be mainly due to selective ranging by both sexes in the same area, they also mentioned that such associations may be more than passive aggregations. Moreover, they showed that the pairs with such strong associations had a greater probability of reproducing, implying that at least some mating had occurred between them. Therefore, despite some other possibilities, such strong male–female associations at Ngogo might have started as mating-associated affiliative relationships, as we have seen above for Mahale. We need to wait for further observations to see whether such frequently associated pairs at Ngogo will avoid mating with each other in future, as is the case for Kalunde and Nkombo.

29.3.3 Influence of male–female affinity on politics among males

Chimpanzee males sometimes compete over their dominance ranks and they are known to act "politically" on such occasions (Chapter 27). In the Arnhem captive colony in the Netherlands, male–

female affinity has a large impact on such politics among males because females often take sides based on their affinity to the males (de Waal, 1982, 1984). In the wild, such female involvements in males' politics are considered to be limited (Uehara *et al.*, 1994; Nishida and Hosaka, 1996).

Though rare, there have been some reports of some females' participation in males' fights at Mahale (Nishida and Hosaka, 1996; Sakamaki, 2011; Nishida, 2012: 243–4, 255–6). Some female support to males in the wild may also be based on affinity as suggested by the aforementioned case of Nkombo. We still need to accumulate such observations in the wild to determine the actual influence of male–female affinity on politics among males.

On the other hand, how females act toward an "ostracized ex-alpha male" presumed to spend most of his time alone, avoiding contact with other males, probably due to political conflicts with some males can only be observed in the wild (see Chapter 27). At Mahale, Uehara *et al.* (1994) suggested that in the two cases of ostracized ex-alpha males (Kasonta and Kajugi), several females continued to meet with them. From 2005 to 2006, I observed another ex-alpha male, Fanana, who is presumed to have been in a similar situation to these two males since 2003 (Nishie, 2004). Although he maintained contact with a few particular males, at least in my study period, he occasionally ranged with several females, most of whom were lactating, and their offspring for hours. On such days, he groomed with the females and even spent a night with them at the same sleeping site. Thus during these males' "lone" life, they could actually have associations with others, to some extent, because of such females' continuous affinity to them irrespective of males' conflicts.

29.4 Females' interventions in male aggression toward others

As mentioned in the previous section, a female occasionally intervenes in males' fights to *support* one of the males based on her affinity to the male. However, there have been reports that females intervened to *stop* males fighting and to *stop* a male aggression toward another female. In this section, I introduce such examples and explore possible male–female relationships underlying such interventions.

29.4.1 Females' interventions to stop male fights

In the Arnhem captive colony, de Waal (1982, 1996) described several cases of "mediations": a female mediated between two opposing males by prompting the males to groom one another or by disarming one male of his weapon, a stone or a tree branch. A recent study revealed that in several captive colonies, high-ranking females, though rare, sometimes show "impartial interventions" (i.e. threatening/attacking both sides or serving as a buffer between both sides without taking sides) in fights including those among males (von Rohr *et al.*, 2012).

There have been no reports on such interventions in the wild. However, a case observed at Mahale suggests the possibility that females in the wild also show similar interventions in male–male fights. Kaburu *et al.* (2013) reported that in 2011, the alpha male, Pimu, died after repeated aggression by some males (Chapter 26). In the first recorded fight between Pimu and the beta male Primus, a 15-year-old female, Vera, and a 25-year-old female, Effie, attempted to pull back Primus and Pimu, respectively, and Effie eventually bit Pimu. Despite the females' interventions, the two males continued the fight for a while and Primus finally ran away.

Because the two females attempted to separate the males by pulling each back in opposite directions, their behaviors may be interpreted as a joint intervention to stop the fight. The intervention, however, did not seem to be effective, and the authors stressed that Effie's bite to Pimu gave Primus an opportunity to attack Pimu.

Such interventions, including this possible case at Mahale, cannot be simply explained by females' affinity to either of the males. Instead, in captivity, it was suggested that impartial interventions need a certain length of relationship with the colony members because newly introduced females never performed such interventions despite being high-ranking (von Rohr *et al.*, 2012). In other words, some acquaintance with the group members may be necessary for such interventions to occur. Although dominance ranks among females are not clear at Mahale (Chapter 28), Vera and Effie had resided in the M group for 5 and 14 years, respectively, and hence it is plausible to think that they had cultivated continuing relationships with the target males by the time of their intervention. Such long relationships with the males may be one of the contributing factors in their intervention.

29.4.2 Females' interventions to stop male aggression toward other females

As mentioned above, females occasionally intervene in male aggressions toward other females, and such interventions have also been reported in the wild (Gombe: Boehm, 1994; Mahale: Sakamaki *et al.*, 2001; Budongo: Newton-Fisher, 2006).

In the Mahale case observed by Sakamaki *et al.* (2001) in 1998, a female, Opal (then estimated to be 27 years old), intervened in the attacks by two males (Fanana and Kalunde, then estimated to be 20 and 35 years old, respectively) toward her 12-year-old daughter, Ruby, who had just given birth to her first child. During the first series of attacks, Opal extended her hand to the males and interposed herself between Ruby and the males. Thanks partly to such Opal's behaviors, Ruby could flee. An hour-and-a-half later, while Ruby was still absent, Opal met the males again and continued to move with them. When they reunited with Ruby with some uproar, Fanana started to groom Ruby. When Kalunde approached them, Opal suddenly came close and pant-grunted to him, and accordingly Kalunde groomed Opal. After a while, however, the males resumed attacking Ruby and attempted to snatch Ruby's infant. During this second series of attacks, Opal interposed herself between Ruby and the males at least three times. Consequently, Opal was also attacked by the males, and began to fight back. Finally, Opal directly attacked the males together with an older non-kin female, Gwekulo (then estimated to be 36 years old), who had arrived at the scene in the middle of the episode after hearing their loud vocalizations. Following the counterattack by these females, the males moved away.

Males sometimes, though not frequently, attack lactating females severely, some of which may lead to infanticides (see Chapter 26). However, as this case shows, other females sometimes intervene in such male aggressions or potential aggressions to protect the victims. Although Opal's protection to Ruby in this case can easily be interpreted as kin-mediated affinity, if we consider the aspect of a female intervention in males' aggression, the following two points need attention. First, Opal's non-offensive but active interventions, such as extending a hand and interposing herself, may be a manner to *negotiate* with physically stronger individuals. Opal's close associations with the males during Ruby's temporal absence might also have contributed to change the context into a nonantagonistic one. However, the second point is that Opal did not hesitate to attack the males depending on the situation, and the males' retreat at the end seems to have been achieved by a coalitional attack on the males with Gwekulo. Such flexible interventions by Opal depending on the situation, and probably on the mood of males, may have been partly because she had been in the M group for 15 years at that point, and had known the males for a long period.

Another case (Hanamura, personal observation) shows a similar, but solo intervention by a more aged female. On January 6, 2006, the alpha male, Alofu (then 24 years old), started to display in front of several males and on the way slapped a 15-year-old, cycling female, Qanat, who had immigrated into this unit-group 3 years earlier. While Alofu held down and stamped on screaming Qanat many times, an old female, Calliope (then estimated to be 46 years old), who came to the site upon hearing the screams, walked straight up to Alofu with Qanat crouching under his foot and gave a soft bark while standing fullface to him. A moment after Calliope's barking, Alofu moved away from Qanat and disappeared into the bush. Then Calliope walked back the way she had come.

Males occasionally attack cycling females, some of which may be related to sexual coercion (see

Chapter 32). In this case, however, because Alofu began to attack Qanat in the course of his charging display toward other males, he might have used her as a "tool" for the display. It seemed to me that Calliope came to the site to stop this one-sided aggression. Qanat was non-kin of Calliope and the frequencies of her associations and affiliative interactions with Calliope were not especially high, at least in those days (Hanamura, unpublished data), suggesting that females intervene in male aggressions even if they do not have kinships and/or particular affiliative relationships with the victims.

Although the situations were not the same, compared with Opal's intervention, Calliope's was more offensive from the beginning. It is noteworthy that Alofu moved away just by Calliope's barking. Accordingly, this interaction looks to me as if she *reprehended* Alofu, rather than *negotiated* with him. Because when Alofu was born, Calliope was 22 years old and was already residing in the M group, they have been acquaintances since Alofu's infancy. This age gap and their long relationship may help explain Calliope's large social influence on Alofu.

A previous study at Budongo showed that most female interveners in male aggressions toward other females were high-ranking (Newton-Fisher, 2006). However, since long residency is linked to high status for females in the group (Reynolds, 2005: 132), it might also be possible to interpret females' social influence on males in terms of long acquaintances with the males, rather than dominance rank among females. Social influence of old females on males is also suggested in other social situations such as when some old females receive selective shares of meat from an alpha male (Nishida *et al.*, 1992). Because food sharing usually occurs by tolerating the begging individual (Chapter 20), such old females might have received his "respect" (Chapter 23).

The cases introduced in this section imply that females occasionally perform impartial or victim-protecting interventions in male aggressions, in various manners. Although more cases are needed to judge conclusively, the females' actions and the target males' responses may partially be affected by the length of relationships and/or the age gap between them. We should investigate such a possibility in addition to other factors such as females' dominance rank, personality, reproductive status, and contexts of aggressions.

The role of long-term relationships could also be considered in reverse, i.e. male impartial or victim-protecting interventions in female–female fights (e.g. captive colonies: de Waal, 1982; von Rohr *et al.*, 2012; Gombe: Goodall, 1986; Boehm, 1994; Mahale: Nishida, 1989; Kanyawara: Kahlenberg *et al.*, 2008). Although such males' interventions have been interpreted as the males' tactic for gaining respect, ostentatious shows of dominance, or roles for policing and reproductive strategy, there may be room to rethink how the male–female long-term relationships affect occurrences of such interventions.

29.5 Male–female dominant–subordinate relationships

Although males' dominance over females is often postulated because males are physically stronger than females due to sexual dimorphism, this is still an open question in primate studies (Smuts, 1987). In chimpanzees, de Waal (1982) discussed two types of dominant–subordinate relationships between sexes that are associated with one another. *Real* dominance is abstracted from the directions of threat/attack–retreat interactions, and *formal* dominance from the directions of pant-grunting, which can be regarded as "submissive greeting" because it is given by subordinates toward dominants in the former dominance (see Chapter 33). The author suggested that in captivity, the former directions can reverse depending on the situations but the latter directions never reverse.

Similarly at Mahale, females can threaten/attack and chase away even the alpha male in some situations as mentioned above, but all males receive pant-grunts from females and there are very few reversals (Hayaki, 1990; Takahata, 1990; Sakamaki, 2011). However, there are several aged females who never pant-grunt to the youngest males (Nishida,

1970), and females never or rarely pant-grunt to immature males (Hayaki, 1990).

Nishida (2003) reported that juvenile and adolescent males at Mahale start to perform various displays toward adolescent and adult females while receiving their counterattacks (see also Chapter 22). The author interpreted the males' behaviors as "harassments" to claim dominance over females that continue until they begin to receive pant-grunts from females. Such harassments of females by immature males are also reported at Gombe (Pusey, 1990) and this interpretation is supported by observations that adolescent males stop harassments after they begin to receive pant-grunts from females (captive colony: de Waal, 1982; Gombe: Goodall, 1986: 431). Thus, for pubertal males to become "adults" socially may not only require developing strong bonds with males (Chapter 27), but also receiving pant-grunts from females. However, because females can chase away even such "adult" males if the need arises, the females' pant-grunting would only *formally* be performed.

In fact, at Mahale, although females often pant-grunt to males when the males perform displays or attack them, many of the females pant-grunts to males are uttered spontaneously without such male aggression and the spontaneous pant-grunting can occasionally trigger the target male's aggression (Sakamaki, 2011). Besides, females do not always pant-grunt to males they meet, and if they do so, they tend to pant-grunt only to the highest-ranking males in the scene, and hence each female never or rarely pant-grunts to middle- or low-ranking males (Sakamaki, 2011; Chapter 33). Thus, it is difficult to regard females' pant-grunting to males only as a submissive response to actual or potential male aggressions.

In summary, males' dominance over females, though males sometimes claim it by performing displays, is not determined unilaterally by the physically stronger males. Such male–female dominant–subordinate relationships, if any, are supported by females choosing to hold a subordinate position (i.e. not performing threats, attacks, and counterattacks) and may be expressed merely for form's sake by females' pant-grunting.

29.6 Male–female relationships in relation to female immigration

Male–female relationships have also been studied from the viewpoint of the female immigration process. Several studies have suggested that resident females threatened or attacked newly immigrated females due to feeding competitions, and males protected immigrants from such aggressions to increase their mating opportunities (Gombe: Pusey, 1980; Mahale: Nishida, 1989; Kanyawara: Kahlenberg *et al.*, 2008). Thus, it has been emphasized that newly immigrated females follow males to receive protection from them (e.g. Nishida, 2012: 187–8).

However, the study at Mahale found these tendencies based mainly on the observations of females who had transferred from the disappearing K group to the adjacent M group (Nishida, 1989), and thus may not be typical. In fact, M-group males sometimes attack newly immigrated females and resident females sometimes protect them from such aggressions by males, as in the case of Calliope's intervention (see Section 29.4.2). Preliminary analysis on five typical newly immigrated females who transferred into the M group from 2002 to 2011 suggested that they received more aggressions from adolescent and adult males than from resident females, and occasionally were protected by resident females, as well as males in such agonistic interactions with other individuals (Hanamura, unpublished data).

On the other hand, given that almost all newly immigrated females are cycling females, including those studied by Nishida (1989), being able to mate with males would be important for their immigration (Chapter 35). However, their relationships with males are not limited to sexual ones, as reviewed in this chapter. Several years after immigration, females may develop long-lasting affiliative relationships with males beyond their estrous period. If a female gives birth to a son, she may continue a lifelong affiliative relationship with the son.

Moreover, given that some of the male aggression directed to newly immigrated females are "harassments" by juvenile or adolescent males, immigrants may have already begun to cultivate formal dominant–subordinate relationships with such males. Furthermore, females soon after immigration, associate, groom, and play with resident individuals of various age–sex classes (Nakamura and Itoh, 2005; Hayakawa *et al.*, 2011). In the course of such daily lives, they may also begin to deepen relationships with various individuals including males, under which conditions females may intervene in male aggression to others.

In conclusion, eastern chimpanzees' male–female bonds seem to be stronger and more diverse than previously thought. Male–male bonds, which have been thought of as central to unit-group organization, may be partly influenced by male–female affiliative relationships. In addition, both males and females occasionally intervene in each other's aggression toward other individuals who live together in the same unit-group, and their dominant–subordinate relationships, if any, result from male–female interactions. As such, the chimpanzee unit-group is organized through male–female mutual involvements.

References

Boehm, C. (1994). Pacifying interventions at Arnhem Zoo and Gombe. In *Chimpanzee Cultures*, ed. R. W. Wrangham, W. C. McGrew, F. B. M. de Waal, and P. G. Heltne. Cambridge, MA: Harvard University Press, pp. 221–6.

Goodall, J. (1986). *The Chimpanzees of Gombe: Patterns of Behavior*. Cambridge, MA: Belknap.

Hayakawa, T., Nakashima, M., and Nakamura, M. (2011). Immigration of a large number of adolescent female chimpanzees into the Mahale M group. *Pan African News*, **18**, 8–10.

Hayaki, H. (1988). Association patterns of young chimpanzees in the Mahale Mountains National Park, Tanzania. *Primates*, **29**, 147–61.

Hayaki, H. (1990). Social context of pant-grunting in young chimpanzees. In *The Chimpanzees of the Mahale Mountains: Sexual and Life History Strategies*, ed. T. Nishida. Tokyo: University of Tokyo Press, pp. 189–206.

Hosaka, K. (2012). [Introduction to Mahale chimps 19: Nkombo.] *Mahale Chimpun*, **19**, 6. In Japanese.

Kaburu, S. S. K., Inoue, S., and Newton-Fisher, N. E. (2013). Death of the alpha: within-community lethal violence among chimpanzees of the Mahale Mountains National Park. *American Journal of Primatology*, **75**, 789–97.

Kahlenberg, S. M., Emery Thompson, M., Muller, M. N., and Wrangham, R. W. (2008). Immigration costs for female chimpanzees and male protection as an immigrant counterstrategy to intrasexual aggression. *Animal Behaviour*, **76**, 1497–509.

Kitamura, K. (1977). [Particular proximity among individuals in Japanese macaques.] *Kikan Jinruigaku*, **8**(3), 3–39. In Japanese.

Langergraber, K. E., Mitani, J. C., Watts, D. P., and Vigilant, L. (2013). Male–female socio-spatial relationships and reproduction in wild chimpanzees. *Behavioral Ecology and Sociobiology*, **67**, 861–73.

Machanda, Z. P., Gillby, I. C., and Wrangham, R. W. (2013). Male–female association patterns among free-ranging chimpanzees (*Pan troglodytes schweinfurthii*). *International Journal of Primatology*, **34**, 917–38.

Matsumoto-Oda, A. (1999). Mahale chimpanzees: grouping patterns and cycling females. *American Journal of Primatology*, **47**, 197–207.

Matsumoto-Oda, A. (2002). Social relationships between cycling females and adult males in Mahale chimpanzees. In *Behavioural Diversity in Chimpanzees and Bonobos*, ed. C. Boesch, G. Hohmann, and L. F. Marchant. Cambridge: Cambridge University Press, pp. 168–80.

Murray, C. M., Gilby, I. C., Mane, S. V., and Pusey, A. E. (2008). Adult male chimpanzees inherit maternal ranging patterns. *Current Biology*, **18**, 20–4.

Nakamura, M. and Itoh, N. (2005). Notes on the behavior of a newly immigrated female chimpanzee to the Mahale M group. *Pan Africa News*, **12**, 20–2.

Newton-Fisher, N. E. (2002). Ranging patterns of male chimpanzees in the Budongo Forest, Uganda: range structure and individual differences. In *New Perspectives in Primate Evolution and Behaviour*, ed. C. S. Harcourt and B. R. Sherwood. Otley, West Yorkshire: Westbury, pp. 287–308.

Newton-Fisher, N. E. (2006). Female coalitions against male aggression in wild chimpanzees of the Budongo Forest. *International Journal of Primatology*, **27**, 1589–99.

Nishida, T. (1968). The social group of wild chimpanzees in the Mahali Mountains. *Primates*, **9**, 167–224.

Nishida, T. (1970). Social behavior and relationship among wild chimpanzees of the Mahali Mountains. *Primates*, **11**, 47–87.

Nishida, T. (1979). The social structure of chimpanzees of the Mahale Mountains. In *The Great Apes*, ed. D. A. Hamburg and E. R. McCown. Menlo Park, CA: Benjamin/Cummings, pp. 73–121.

Nishida, T. (1989). Social interactions between resident and immigrant female chimpanzees. In *Understanding Chimpanzees*, ed. P. G. Heltne and L. A. Marquardt. Cambridge, MA: Harvard University Press, pp. 68–89.

Nishida, T. (2003). Harassment of mature female chimpanzees by young males in the Mahale Mountains. *International Journal of Primatology*, **24**, 503–14.

Nishida, T. (2004). [Introduction to Mahale chimps 3: Kalunde.] *Mahale Chimpun*, **3**, 6. In Japanese.

Nishida, T. (2012). *Chimpanzees of the Lakeshore: Natural History and Culture at Mahale*. Cambridge: Cambridge University Press.

Nishida, T. and Hiraiwa-Hasegawa, M. (1987). Chimpanzees and bonobos: cooperative relationships among amales. In *Primate Societies*, ed. B. B. Smuts, D. L. Cheney, R. M. Seyfarth, R. W. Wrangham and T. T. Struhsaker. Chicago, IL: The University of Chicago Press, pp. 165–77.

Nishida, T. and Hosaka, K. (1996). Coalition strategies among adult male chimpanzees of the Mahale Mountains, Tanzania. In *Great Ape Societies*, ed. W.C. McGrew, L. F. Marchant, and T. Nishida. Cambridge: Cambridge University Press, pp. 114–34.

Nishida, T., Hasegawa, T., Hayaki, H., Takahata, Y., and Uehara, S. (1992). Meat-sharing as a coalition strategy by an alpha male chimpanzee? In *Topics in Primatology Vol. 1: Human Origins*, ed. T. Nishida, W. C. McGrew, P. Marler, M. Pichford, and F. B. M. de Waal. Tokyo: University of Tokyo Press, pp. 159–74.

Nishie, H. (2004). [The fight at Kasiha Valley: The fall of Fanana from the alpha status and Alofu's takeover.] *Mahale Chimpun*, **4**, 6–7. In Japanese.

Otali, E. and Gilchrist, J. S. (2006). Why chimpanzee (*Pan troglodytes schweinfurthii*) mothers are less gregarious than nonmothers and males: the infant safety hypothesis. *Behavioral Ecology and Sociobiology*, **59**, 561–70.

Pepper, J. W., Mitani, J. C., and Watts, D. P. (1999). General gregariousness and specific social preferences among wild chimpanzees. *International Journal of Primatology*, **20**, 613–32.

Pusey, A. E. (1980). Inbreeding avoidance in chimpanzees. *Animal Behaviour*, **28**, 543–52.

Pusey, A. E. (1983). Mother–offspring relationships in chimpanzees after weaning. *Animal Behaviour*, **31**, 363–77.

Pusey, A. E. (1990). Behavioural changes at adolescence in chimpanzees. *Behaviour*, **115**, 203–46.

Reynolds, V. (2005). *The Chimpanzees of the Budongo Forest: Ecology, Behaviour, and Conservation*. Oxford: Oxford University Press.

von Rohr, C. R., Koski, S. E., Burkart, J. M., *et al.* (2012). Impartial third-party interventions in captive chimpanzees: a reflection of community concern. *PLoS ONE*, **7**, e32494.

Sakamaki, T. (2011). Submissive pant-grunt greeting of female chimpanzees in Mahale Mountains National Park, Tanzania. *African Study Monographs*, **32**, 25–41.

Sakamaki, T., Itoh, N., and Nishida, T. (2001). An attempted within-group infanticide in wild chimpanzees. *Primates*, **42**, 359–66.

Silk, J. B. (2002). Using the 'F'-word in primatology. *Behaviour*, **139**, 421–46.

Smuts, B. B. (1985). *Sex and Friendship in Baboons*. New York: Aldine de Gruyter.

Smuts, B. B. (1987). Gender, aggression, and influence. In *Primate Societies*, ed. B. B. Smuts, D. L. Cheney, R. M. Seyfarth, R. W. Wrangham, and T. T. Struhsaker. Chicago, IL: The University of Chicago Press, pp. 400–12.

Strum, C. S. (1987). *Almost Human: A Journey into the World of Baboons*. New York: Random House.

Takahata, Y. (1982). Social relations between adult males and females of Japanese monkeys in the Arashiyama B troop. *Primates*, **23**, 1–23.

Takahata, Y. (1990). Adult males' social relations with adult females. In *The Chimpanzees of the Mahale Mountains: Sexual and Life History Strategies*, ed. T. Nishida. Tokyo: University of Tokyo Press, pp. 133–48.

Uehara, S., Hiraiwa-Hasegawa, M., Hosaka, K., and Hamai, M. (1994). The fate of defeated alpha male chimpanzees in relation to their social networks. *Primates*, **35**, 49–55.

de Waal, F. B. M. (1982). *Chimpanzee Politics: Power and Sex among Apes*. London: Jonathan Cape.

de Waal, F. B. M. (1984). Sex differences in the formation of coalitions among chimpanzees. *Ethology and Sociobiology*, **5**, 239–55.

de Waal, F. B. M. (1996). *Good Natured: Origins of Right and Wrong in Humans and Other Animals*. Cambridge, MA: Harvard University Press.

Whitehead, H. (2008). *Analyzing Animal Societies: Quantitative Methods for Vertebrate Social Analysis*. Chicago, IL: The University of Chicago Press.

Wolf, A. P. (1993). Westermarck redivivus. *Annual Review of Anthropology*, **22**, 157–75.

Wrangham, R. W. (2000). Why are male chimpanzees more gregarious than mothers? A scramble competition hypothesis. In *Primate Males: Causes and Consequences of Variation in Group Composition*, ed. P. M. Kappeler. Cambridge: Cambridge University Press, pp. 248–58.

30 Orphans and allomothering

MICHIO NAKAMURA AND KAZUHIKO HOSAKA

30.1 Introduction

Chimpanzee infants depend on their mother's milk for quite a long period after birth (see Chapter 22 for more details), and if infants lose their mothers before weaning, i.e. before the age of 4 or 5 years, they usually do not survive (Teleki *et al.*, 1976; Goodall, 1983). However, if they lose their mothers at about the time of weaning, they still have a chance of survival. Nevertheless, because chimpanzee youngsters follow and feed with mothers for several years after weaning, young surviving orphans need to be taken care of or at least need someone to follow. Several cases of young orphans that were adopted by nonrelated females were observed at Mahale. Such adoptions can be regarded as an extension to allomothering of infants who still have mothers. In this chapter, we first introduce the general pattern of allomothering behavior toward infants, including non-orphans, and then proceed to explain cases of orphaning and consequent details.

30.2 Allomothering behavior

Allomothering behavior (also called alloparental behavior, nonmaternal care, or allocare) is a mother-like behavior, such as handling, carrying, playing, and grooming, exhibited by non-mother individuals toward infants. Among anthropoid primates, the rate of allomothering in chimpanzees is relatively low (Ross and MacLarnon, 2000): the median is 2.6% and maximum 8% of the time spent for allomothering of a chimpanzee infant under age 2 years (Nishida, 1983). In other words, mothers usually care for their own infants most of the time. In their dichotomous categorization depending on frequencies of carrying by nonmaternal conspecifics, Mitani and Watts (1997) classified chimpanzees as a non-allocaretaking species. Perhaps due to the low frequency of allomothering, few studies have focused on chimpanzees' allomothering behaviors, with the exception of an early study by Nishida (1983), who provided a detailed description of alloparental behaviors in the Mahale K-group chimpanzees. Here, we introduce the general pattern of allomothering according to the study by Nishida (1983) as well as our own observations.

30.2.1 Allomothering by young females

Juvenile and adolescent females, i.e. nulliparous young females, are the ones that show earnest allomothering behavior (Nishida, 1983). These young females seem to be very much attracted to small infants. A young female often approaches an adult female with a small infant and starts to groom her until the mother becomes relaxed. The infant also seems to be interested in such individuals and often tries to touch her. Then the young female starts to play gently with the infant who is still on the belly or in the lap of its mother. If the infant does not resist and descends to the ground near its mother, especially when its mother is engaged in grooming with others, the young female may slowly pick it up and walk away with the infant on her belly or her back. Such attempts are often not successful when the infant is too young because the mother intervenes or retrieves

Mahale Chimpanzees: 50 Years of Research, ed. M. Nakamura, K. Hosaka, N. Itoh and K. Zamma. Published by Cambridge University Press.

the infant immediately. A mother often allows her daughter (older sister of the infant) to carry the baby from a relatively young age and becomes tolerant of allomothering by others when her infant gets older.

Allomothering behaviors by young females can be considered practice for child rearing (Nishida, 1983). When the female allomother herself is a juvenile, such attempts are mostly playful and the handling of infants is sometimes rough, but when she is in her late adolescence, the handling becomes very tender and behavior is more maternalistic than playful.

30.2.2 Allomothering by males

Males also sometimes take care of infants, and among them, older brothers are the most frequent allomothers (Figure 30.1). Nishida (1983) did not discuss allomothering by brothers because there were no brothers of the target infants at that time. Similar to the behavior exhibited by a sister, a brother also tries to play with, handle, groom, and sometimes carry his infant sibling, and the mother often permits such attempts by her son. In 2005, Orion, then a 14-year-old male, frequently carried his 1-year-old infant brother on his back. During one observation, Orion displayed bipedal posture and charging behavior at females, with his infant brother clinging to his back. Their mother Opal soon rushed to the place and retrieved the infant son, but the infant did not seem afraid.

Nonrelated adult males also play frequently with infants. Most commonly, the playing includes gentle touching or poking the infant with a finger during a grooming sequence with the infant's mother (Figure 30.2). If the mother does not care, the male may put the infant on his belly while lying on his back or sometimes carry the infant on his belly or his back, away from the mother. The mother usually retrieves the infant if the adult male takes her infant too far from her; thus, carrying of infants by adult males often does not last long.

When a mother Chausiku left her son Katabi (then 5 years old) in the K group and moved to the M group for 10 months, an adult male, Kamemanfu, took care of Katabi, mostly in the form of grooming (Hiraiwa-Hasegawa and Hasegawa, 1988). However, this may not be regarded as active allomothering by Kamemanfu, because it was Katabi who actively followed and made his sleeping bed close to Kamemanfu. This relationship is quite similar to the one between an orphan male and an adult male described below.

Figure 30.1 A 9-year-old male, Caesar (right), holding his 8-month-old infant sister, Cissy. Their mother, Cynthia (left), comes to pick up the baby. ©M. Nakamura.

Figure 30.2 An adult male, Alofu (right), gently touching a 5-month-old infant male. The infant's mother is grooming the back of a male who is at the front. ©M. Nakamura.

30.2.3 Allomothering by grandmothers

Although a case of a grandmother adopting an infant while the infant's mother (daughter of the grandmother) was still alive has been reported from Gombe, Tanzania (Wroblewski, 2008); no such case was observed at Mahale. This is mainly due to there being less opportunity for a grandmother and a mother to coexist in a group, because most of the natal females at Mahale emigrate to different groups upon maturity (see Chapters 6, 7). Among the relatively rare cases when a female stayed in the M group and gave birth is Ruby who still had a living mother Opal. When Ruby and her infant daughter Rubicon were attacked by adult males, Opal fought with the males (Sakamaki *et al.*, 2001; see Chapters 27, 28). When Rubicon reached the late infant stage, Ruby often left Rubicon behind and did not return to retrieve Rubicon even when the infant made whimper tantrums. In such cases, Opal frequently took care of her granddaughter. Opal would often approach the distressed Rubicon and groom her as if to assure her. Because Opal had a son, Oscar, who was about the same age as Rubicon, we often observed Oscar and Rubicon following Opal as if they were both Opal's children. Nishida (2006a) reported an interesting case when the 2-year-old Rubicon fell into a hole about 50-cm deep. Although both Ruby and Opal were present nearby, it was Opal who rescued Rubicon from the hole.

30.2.4 Gwekulo: a specialist of allomothering

When we discuss allomothering at Mahale, we cannot overlook Gwekulo. She was identified for the first time in 1972 as an adolescent female of about 10 years in the K group. In 1979, she transferred to the M group. She was in estrous cycle periodically and frequently copulated with males, but never gave birth; therefore, she was assumed to have been sterile for some reason.

Gwekulo's career as an allomother can be traced back to when she was still in the K group. When Chausiku caught common cold in 1977, Gwekulo took care of Chausiku's 2-year-old son Katabi for a week until his mother recovered (Uehara and Nyundo, 1983). She continued to be an earnest allomother after she moved to the M group: she took care of Wakasunga's son, Linta; Wakampompo's daughter, Blondy, and son, Brutus; Pulin's daughter, Pipi; Pinky's son, Primus, and daughter, Puffy; Christina's sons, Christmas and XT08 (Figure 30.3); and Fatuma's son, Figaro (Nishida, 2006b and our personal observations), by frequently carrying the infants. This caretaking of unrelated infants by Gwekulo seems to be based on strong relationships between Gwekulo and the mothers. Each time she formed a strong and long-lasting bond with a mother, she almost always ranged together with the mother and groomed with her for a long time. Consequently, she often took care of the mother's successive children. When such offspring was attacked or harassed by others, the mother and Gwekulo often retaliated jointly even when the attacker was an adult male. The strong bonds with offspring were also long lasting, even in the absence of the mothers. Gwekulo continued to take care of Pipi and Puffy after their mothers had died and the foster mother–daughter relationship continued until the foster daughters matured sexually (see below).

Among adult females, only those who do not have their own offspring, such as Gwekulo and Gwakakumo (see below), generally take care of unrelated infants. However, this does not mean all females without offspring do so: Nkombo, another adult female who did not have offspring for a long time, seldom allomothered except for playing with youngsters.

Figure 30.3 Eleven-month-old, XT08, on the back of Gwekulo. XT08's mother is feeding on the same tree. ©M. Nakamura.

30.2.5 Characteristics of mother and offspring affecting allomothering

Whether allomothering will take place does not depend only on the allomothers' characteristics. It strongly depends on the personality of the mother and offspring and the child-rearing experience of the mother. For example, Christina is one of the mothers who tolerate allomothering: in 2009, Christina carried her 11-month-old son less than half the observation time, whereas the allomothers Gwekulo, Xantip (Christina's daughter), Puffy (nonrelated adolescent female and foster-daughter of Gwekulo), and Flavia (nonrelated adolescent female) carried the infant for the rest of the time (Nakamura, 2010). Because a typical 11-month-old chimpanzee is almost never away from its mother (van Lawick-Goodall, 1967), Christina was considered to be very tolerant of allomothering. It is also important to note that Christina's infant son was also very active, did not mind being carried away from his mother, and did not whimper while being carried by allomothers.

30.3 Orphans

The mother, being older than her offspring by nature, is expected to die before her offspring. What is relevant to the offspring is its age when the mother dies. If a mother dies when the offspring is very young and completely dependent on her, its survival is a challenge. In Mahale, chimpanzee infants that are orphaned before the age of 3 years usually cannot survive. On the other hand, if the offspring has grown into a full adult, the mother's death does not directly affect the offspring's survival although, similar to humans, it may have a psychological effect. If the offspring's age at the time of its mother's death is between 3 years and a full adult, the offspring may survive but may face some disadvantages and need to compensate for the loss of its mother. Here, by *orphans* we mean those who have lost their mothers before becoming fully adults.

Table 30.1 summarizes the orphans in the M group who had lost their mothers from the ages of 3 to 13 years and survived for at least several months. There were 22 male and 17 female orphans by 2012. The manner in which the orphans interacted with others or others supported the orphans may differ with the orphans' age. They need to somehow compensate for the care otherwise obtained from their own mothers, such as being carried by or being allowed to range and feed together.

30.3.1 Infant orphans

At Mahale, infants who were orphaned before their third birthday never survived mothers' deaths. However, there are three cases of 3-year-old orphans that survived. Usually, 3-year-old infants are carried on the back of their mothers, thus all such young orphans were "adopted" by an individual who would almost always carry them and share their bed.

Maggie's (MG) case

Maggie's case has already been described by Nishida (1994, 2012). She was the daughter of Wantendele and lost her mother at the age of 3 years and 1 month. On August 17, 1990, nulliparous immigrant female Pinky was carrying Maggie on her back but she soon abandoned Maggie when a male directed a charging display toward her. After that, Maggie's 13-year-old brother Masudi was observed carrying her. By September 6, she was being carried by the 10-year-old Tula who herself was an orphan female, and Nishida (1994) described Tula "adopting" Maggie. Tula carried Maggie, groomed her, played with her, shared her bed with her (Nishida, 1994), and also shared some meat (Huffman and Kalunde, 1993). When Maggie was harassed by others, Tula rushed to her and hugged her. Tula provided almost every aspect of maternal care for Maggie except for lactation. Tula emigrated from the M group in 1994 when Maggie was 7 years old. Maggie successfully reached the age of 11 years and was assumed to have emigrated to a different group in 1998.

Pipi's (PP) case

A daughter of Pulin, Pipi was orphaned in October 1993 when she was 3 years and 9 months old. Although Nishida (2012) emphasized that Gwekulo

Table 30.1 **List of individuals who were orphaned at the age between 3 and 13 years. Those who died immediately after mothers' death are excluded**

Name	Sex	Year of birth	Mother[a]	Orphaned age in years	Age at death in years[b]	Life without mother in years[c]
Chopin	male	1985	Chausiku	5	8	3.3
Aji	male	1973	?	6	23	16.6
Darwin	male	1988	Dar	6	24+	17.9+
Kenji	male	1970	?	7	12	4.9
Hiro	male	1976	?	7	15	8.2
Mumu	male	1979	Momo	7	8	0.6
Nick	male	1982	Ndilo	7	14	6.5
Lukaja	male	1965	?	8	26	18.0
Mtwale	male	1975	Mtimpu	8	14	5.9
Charles	male	1985	?	8	14	6.4
Kalunde	male	1963	?	10	49+	39.0+
Fancy	male	1978	Fanta	10	13	2.4
Michio	male	1996	Miya	10	16+	6.2+
Hit	male	1972	?	11	16	4.6
Hawazi	male	1960	?	12	15	3.0
Nishida	male	1960	?	12	18	6.0
Kimaro	male	1971	Adija	12	14	2.3
Hanby	male	1980	Happy	12	22	9.4
Sinsi	male	1984	Fahari	12	13	0.6
Nsaba	male	1973	Sada	13	24	10.3
Toshibo	male	1976	Wasobongo	13	19	6.5
Masudi	male	1977	Wantendele	13	29	15.8
Maggie	female	1987	Wantendele	3	11e	7.6+
Chelsea	female	1990	?	3	7	3.4
Pipi	female	1990	Pulin	3	12e	8.7+
Shinako	female	1978	Adija	5	10e	5.8+
Ako	female	1981	Imelda	5	31+	26.5+
Fususi	female	1991	Fahari	5	5	0.5
Mitsue	female	2001	Miya	5	11+	6.2+
Tula	female	1980	Sada	6	14	8.1
Puffy	female	2000	Pinky	6	12+	6.1+
Wasiwasi	female	1978	Gwepala	7	10e	2.6+

Table 30.1 **(cont.)**

Name	Sex	Year of birth	Mother[a]	Orphaned age in years	Age at death in years[b]	Life without mother in years[c]
Abi	female	1982	Wasobongo	7	30+	23.2+
Jidda	female	1995	Juno	7	12e	5.1+
Lusia	female	1975	Wamkime	10	13e	2.8+
Mazuruka	female	1980	Tango	10	14e	3.7+
Serena	female	1987	Silafu	10	11e	0.9+
Rubicon	female	1998	Ruby	11	12e	1.4+

[a] Question mark indicates that the mother is unknown. The individual is assumed to be orphaned, because he/she was first identified without a mother. Birth years of these individuals are estimations.

[b] Years with plus indicate that they were alive at the time of data analysis (July 2012). Females with "e" indicate those that possibly emigrated to other unit-groups at that age.

[c] Years with plus indicate that they lived or may have lived longer, because they are still alive or they are estimated to have emigrated to other unit-groups.

adopted her and provided all maternal care except lactation, it was not clear whether Gwekulo helped Pipi during the most critical period. Hosaka was at Mahale when Pulin died, but he did not see Gwekulo carrying Pipi for 5 months after the mother's death because at that time Gwekulo was busy babysitting Brutus, a son of Wakampompo. After Pipi saw the researchers collecting the dead body of her mother, she walked by herself in the direction of a party of other chimpanzees (Hosaka *et al.*, 2000). When she met adult males, she approached them and pant-grunted at them. For a while, she was often seen with adult males, but during the dispersing season (Chapter 8), she was observed several times with Tula and Maggie, the other two previously mentioned orphans. A strong bond with Gwekulo was made after June 1994, and thereafter, Pipi almost always followed Gwekulo. They frequently groomed each other as real mother and daughter would do (Figure 30.4), and Pipi's first grooming hand-clasp partner, which is usually a mother, was Gwekulo (Nakamura and Nishida, 2013). Pipi and Gwekulo's intimate mother–daughter-like relationship continued until 2002, when Pipi emigrated from the M group at the age of 12 years.

Chelsea's (CS) case

Chelsea's mother was unknown. She was first identified in 1993 together with her older brother, Charles, and based on her body size she was estimated to be 3 years old. It was assumed that Charles and Chelsea's mother, probably a peripheral female not habituated to humans, had recently died, and the orphans appeared in the large party of other chimpanzees. An old female Gwakakumo, who had given birth several times but never successfully raised an offspring, was carrying Chelsea. We imagine that Gwakakumo might have had an intimate relationship with Chelsea's mother when she was alive. Gwakakumo and Chelsea looked like a real mother–daughter pair. In 1996 when Gwakakumo disappeared (assumed to have died; a dead body was later identified as hers; see Chapter 45), Chelsea was frequently observed to be with Charles but in 1997 Chelsea too disappeared (assumed to have died).

30.3.2 Older orphans

When offspring reach the age of 5 years, they are usually not carried by their mothers anymore, but they walk on their own and follow their mothers.

Figure 30.4 Gwekulo grooming 6-year-old Pipi. They groomed each other as a real mother–daughter pair would. ©M. Nakamura.

Therefore, there have been no cases of 5-year-old or older orphans being carried by others. These older orphans usually follow particular individuals instead of their mothers. A male orphan tends to follow a particular adult, often a low-ranking male, and a female orphan tends to follow a particular adult female. However, the partner may change over time and sometimes this is not very obvious.

Males

Male orphans who lost their mothers at an age of 5 to around 10 years tend to follow particular adult males; for example, Charles, Chelsea's older brother (see above) who was estimated to be 8 years old when orphaned, at first followed Gwakakumo, the foster mother of his infant sister and eventually intervened in the copulation of Gwakakumo (offspring often intervene in their mother's copulation). After a while, Charles started to walk with a low-ranking adult male, Bembe. In 1994, 47% of the time Charles was in the proximity of somebody else, it was with Bembe (Nakamura, unpublished data). Charles also spent the longest grooming time with Bembe, but it was always Charles who groomed, Bembe never groomed him back. Thus, the adult male may have allowed Charles to follow him, but the active maintenance of the relationship seemed to have been done from the side of the orphan.

Similarly, Aji (orphaned at the age of 6 years) actively followed Rashidi; Mtwale (orphaned at the age of 8 years) followed Lubulungu (Hayaki, 1987, 1988); Kalunde (orphaned around the age of 10 years) followed Kagimimi (Nishida, 2004); and Nick (orphaned at the age of 7 years) followed Aji; all of those followed by orphans were adult males at that time. On the other hand, Darwin (orphaned at 6 years) did not follow a particular partner but soon after he lost his mother, he joined a large party including the alpha and other adult males. A few years later, he was often observed following Bembe together with Charles but not to the extent of other orphaned males.

Male orphans who lost mothers after the age of 10 years do not usually follow a particular partner, but they tend to be attracted to and join a party of many adult males, just like adolescent males with living mothers usually do. For example, Hayaki (1987) wrote that mothers' deaths were not so influential to the 12-year-old Kimaro and 14-year-old Shike. Similarly, Michio and Hanby, who were orphaned at the ages 11 and 12 years, respectively, did not seem to differ from other adolescent males of the same age regarding whom they followed.

Females

Shinako, who lost her mother at 5 years of age, followed her older brother Kimaro for a while, but after Kimaro died, she often followed an adult female, Lunkukuma (Hayaki, 1987). Two orphaned females, Labule and Lusia, were frequently observed accompanying an old female, Wankuma, when she was focal followed (Huffman, 1990). Puffy, who lost her mother at 6 years of age, followed an adult female

Gwekulo. As mentioned above, Gwekulo had already been an allomother to Puffy when Puffy's mother was alive. Thus, the intimate relationship continued after the death of Puffy's mother. Unlike adult males who were followed by male orphans, Gwekulo evidently cared for Puffy when ranging together. For example, Gwekulo often waited under a tree while Puffy was still feeding on the tree, which is commonly observed within a mother–offspring pair. Puffy's first grooming hand-clasp partner was also Gwekulo (Nakamura and Nishida, 2013). As Puffy is now in her late adolescence and has begun her estrous cycle, she often moves away from Gwekulo nowadays.

Some orphan females do not have obvious partners to follow. For example, Mitsue, orphaned at 5, was observed to follow Michio, her brother; Darwin, an adult male; Linda, an adult female friend of her mother; and Wakusi, an old female, each of them for a short period of time. The reason for her frequent changes of partners to follow may be because these individuals did not care about Mitsue to the degree that Gwekulo did about Puffy. As a result, Mitsue appeared to follow anybody who was relatively easy to follow at that moment. Jidda (orphaned at 7 years of age) and Serena (orphaned at 10 years of age) did not follow any particular individual nor did they try to follow; instead, they seemed to move freely on their own.

30.3.3 Survival of male orphans after weaning

As mentioned above, wild chimpanzee infants cannot survive when they lose their mothers before they are weaned (Teleki *et al.*, 1976; Goodall, 1983). It is often assumed that a chimpanzee orphan can survive only if it is weaned (Gibbons, 2008). However, in theory, because a chimpanzee mother and her offspring are in the same social group, it may be beneficial for the mother to support her weaned offspring. If this is the case, orphans may be affected by the loss of their mothers. Therefore, we tested whether males who were orphaned after weaning died at a younger age than expected (Nakamura *et al.*, 2014). Figure 30.5 shows the summary of the results.

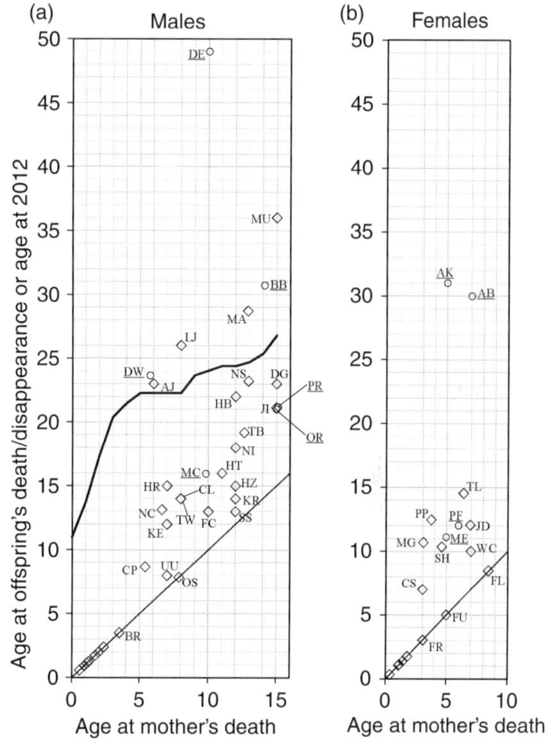

Figure 30.5 Age at death of orphan males (a) and orphan females (b) plotted against the age at their mothers' deaths. Two capital letters indicate abbreviated names of individual chimpanzees. Diamond symbols indicate those who had already died and circle symbols with underlined names indicate those who were still alive at the time of data analysis (in July 2012). The values on the vertical axis indicate age at death or disappearance for diamond symbols and age in July 2012 for circle symbols. Females who disappeared after reaching 9 years of age were assumed to have emigrated to different groups. A thick line in the males' graph shows the expected age of death at each age. A thin straight line indicates the offspring's age at mother's death. If the plot is on the same line, it means that the offspring died simultaneously with the mother.

Among the 37 orphaned males observed thus far, only seven lived longer than the expected age of death (orphaned age + life expectancy at that age; shown as a thick line in Figure 30.5). On the other hand, 27 males died before the expected age of death. This difference between the observed and expected ages of deaths was statistically significant (Wilcoxon matched-pairs signed-rank test, $T = 521$, $p < 0.00014$). The significance still held when we excluded eight males who were less than 5 years old (i.e. before weaning) at the time they lost their mothers and were thus more impacted by their mothers' deaths (T = 285, p = 0.0056). Differences between the observed and expected ages of deaths correlated positively with orphaned age (Kendall's rank correlation test, $\tau = 0.84$, $z = 6.93$, $p = 4.24 \times 10^{-12}$).

Direct comparisons between sexes are not possible, as we were unable to calculate the life expectancy of females because they emigrate to neighboring groups at the age of about 9 years, at the earliest, and we do not have subsequent data on survival of those who were believed to have emigrated to groups that are not habituated to humans. However, if we observe the survival/death ratio in females orphaned prior to their age of 9 years, only about half of female infants died immediately after their mothers' death (five lived, six died). This ratio differed significantly from the ratio of male infants who died under the same conditions (Fisher's exact test, $p = 0.044$). Furthermore, in general, the female orphans seemed to have a higher survival rate: 10 out of 14 females who were orphaned after 3 years of age survived up to the possible emigration age (9 years old). This ratio did not differ from that of females with living mothers, where 39 out of 56 females survived (Fisher's exact test, $p = 1.0$).

30.4 Discussion

30.4.1 Allomothering

Unlike some New World monkeys with allomothering system as a species character, chimpanzees' allomothering is not a fixed quality. Allomothering behavior in young females, which is the most common type of allomothering behavior in chimpanzees, may have some adaptive value for the young females since they can practice child rearing before they actually give birth (Nishida, 1983). If the allomother is trustworthy, the mother may also benefit from allomothering because she does not have to carry the infant on her back thus reducing her energy expense, and she can more easily consume food while entrusting her infant to the allomother. It may also be beneficial to the infant's social development. Such direct benefits of allomothering have not been investigated in detail and they warrant further research.

Gwekulo's allomothering, on the other hand, is not as straightforward, because it cannot be considered a practice for her own future offspring. Gwekulo is considered to be sterile, and her allomothering behavior may be a "wrong" activation of maternalistic behaviors that were to be directed at her own offspring if she had a child (Nishida, 1999). On the other hand, others view adoption as an altruistic behavior (Boesch *et al.*, 2010). Because adoption is a kind of allomothering, allomothering in general can also be regarded altruistic. We need additional information to assess which of the two postulations is correct.

On a different level, we may discuss the importance of being engaged with other chimpanzees. Gwekulo's allomothering might have reduced the mothers' costs and helped orphaned Pipi and Puffy to survive. However, we do not necessarily think that Gwekulo had such intentions. Perhaps her motivation to take care of various infants was just the need to hug small babies. In the case of Pipi and Puffy, it is possible that Gwekulo also needed somebody as a constant ranging partner. Hence, Gwekulo needed them too, even though they were still juveniles, although not as much as the orphans needed her.

The connection between males' allomothering and their own reproductive success is not straightforward. They do not have to practice child rearing, nor is there any direct evidence that allomothering works as agonistic buffering. It is likely that taking care of an infant may be useful for building a good relationship with the mother or the offspring in the long term; this possibility has not been investigated fully yet.

30.4.2 Orphans

In general, male orphans at Mahale were never subject to adoption but usually followed adult males, while some female orphans were adopted by nonrelated females and tended to form stronger bonds with the foster mothers. This tendency is different from other chimpanzee study sites. At Gombe, adoptions by older siblings (including brothers) were the most common (Teleki *et al.*, 1976), although there was a case of a 5-year-old female orphan, Skosha, who was finally adopted by her mother's friend and survived (Goodall, 1983). At Taï, Côte d'Ivoire, 18 out of 36 orphans were adopted by others (Boesch *et al.*, 2010). Of these, eight were adopted by females (including one by a sister and three by mothers' friends), and even lactation by the foster mother was observed in two cases. Taï adult males also sometimes adopted orphans (Boesch *et al.*, 2010). Lactation by a nonrelated female and adoption by males have not been observed at Mahale.

A common feature among the three study sites is that a friend of the mother adopts the orphan. The difference is that sibling adoption is more common at Gombe, while there has been no such case at Mahale: although Maggie, Chelsea, Puffy, and Mitsue had older brothers who took care of these orphans to some extent, they were not completely adopted by their brothers. For some reason, all the observed cases of sibling orphans at Mahale are the combination of an older brother and a younger sister; it is not known whether an older sister would adopt a younger sibling or what would happen between same-sex sibling pairs. The large difference observed between Taï and Mahale is that in the former, adult males sometimes adopted orphans, while all the obvious cases of adoptions at Mahale were by females. Boesch *et al.* (2010) explained the existence of male adoptions at Taï because of the high predation pressure by leopards. However, leopards are very frequent at Mahale, and there was a case of a chimpanzee being eaten by a leopard (Nakazawa *et al.*, 2013). Hence, such a clear-cut explanation may not be applicable.

The results of male orphans' survival (Nakamura *et al.*, 2014) suggest that in the male-philopatric societies of chimpanzees, the existence of living mothers is important for the survival of sons who remain in their natal group. In particular, it should be noted that the loss of a mother decreases the likelihood of survival not only in infants but also in juveniles and early adolescents in the long term. The juvenile and adolescent males no longer depend on mothers' milk and have ceased to receive other direct maternal care such as bed sharing or transportation. Hence, this indicates that they receive other types of presumably indirect and relatively infrequent benefits from living mothers.

References

Boesch, C., Bolé, C., Eckhardt, N., and Boesch, H. (2010). Altruism in forest chimpanzees: the case of adoption. *PLoS ONE*, **5**, e8901.

Gibbons, A. (2008). The birth of childhood. *Science*, **322**, 1040–3.

Goodall, J. (1983). Population dynamics during a 15 year period in one community of free-living chimpanzees in the Gombe National Park, Tanzania. *Zeitschrift für Tierpsychologie*, **61**, 1–60.

Hayaki, H. (1987). [On orphans of wild chimpanzees.] *Hoiku Nenpou*, **7**, 1–8.

Hayaki, H. (1988). Association partners of young chimpanzees in the Mahale Mountains National Park, Tanzania. *Primates*, **29**, 147–61.

Hiraiwa-Hasegawa, M. and Hasegawa, T. (1988). A case of offspring desertion by a female chimpanzee and the behavioral changes of the abandoned offspring. *Primates*, **29**, 319–30.

Hosaka, K., Matsumoto-Oda, A., Huffman, M. A., and Kawanaka, K. (2000). [Reactions to dead bodies of conspecifics by wild chimpanzees in the Mahale Mountains, Tanzania.] *Primate Research*, **16**, 1–15. In Japanese with English summary.

Huffman, M. A. (1990). Some socio-behavioral manifestations of old age. In *The Chimpanzees of the Mahale Mountains*, ed. T. Nishida. Tokyo: University of Tokyo Press, pp. 237–55.

Huffman, M. A. and Kalunde, M. A. (1993). Tool-assisted predation on a squirrel by a female chimpanzee in the Mahale Mountains, Tanzania. *Primates*, **34**, 93–8.

Mitani, J. C. and Watts, D. (1997). The evolution of non-maternal caretaking among anthropoid primates: do helpers help? *Behavioral Ecology and Sociobiology*, **40**, 213–20.

Nakamura, M. (2010). [What does it mean a society to be complex?: Question about relationship between society and individual.] *Primate Research*, **26**, 131–42. In Japanese with English summary.

Nakamura, M. and Nishida, T. (2013). Ontogeny of a social custom in wild chimpanzees: Age changes in grooming hand-clasp at Mahale. *American Journal of Primatology*, **75**, 186–96.

Nakamura, M., Hayaki, H., Hosaka, K., Itoh, N., and Zamma, K. (2014). Orphaned male chimpanzees die young even after weaning. *American Journal of Physical Anthropology*, **153**, 139–43.

Nakazawa, N., Hanamura, S., Inoue, E., Nakatsukasa, M., and Nakamura, M. (2013). A leopard ate a chimpanzee: The first evidence from East Africa. *Journal of Human Evolution*, **65**, 334–7.

Nishida, T. (1983). Alloparental behavior in wild chimpanzees of the Mahale Mountains, Tanzania. *Folia Primatologica*, **41**, 1–33.

Nishida, T. (1994). [*Thirty-six Stories of Wild Chimpanzees*.] Tokyo: Kinokuniya. In Japanese.

Nishida, T. (1999). [*Where does Human Nature Come from?: An Approach from Primatology*.] Kyoto: Kyoto University Press. In Japanese.

Nishida, T. (2004). [Introduction to Mahale chimps 3: Kalunde.] *Mahale Chimpun*, **3**, 6. In Japanese.

Nishida, T. (2006a). [Introduction to Mahale chimps 8: Opal.] *Mahale Chimpun*, **8**, 7. In Japanese.

Nishida, T. (2006b). [Introduction to Mahale chimps 7: Gwekulo.] *Mahale Chimpun*, **7**, 7. In Japanese.

Nishida, T. (2012). *Chimpanzees of the Lakeshore: Natural History and Culture at Mahale*. Cambridge: Cambridge University Press.

Ross, C. and MacLarnon A. (2000). The evolution of non-maternal care in anthropoid primates: A test of the hypotheses. *Folia Primatologica*, **7**, 93–113.

Sakamaki, T., Itoh, N., and Nishida, T. (2001). An attempted within-group infanticide in wild chimpanzees. *Primates*, **42**, 359–66.

Teleki, G., Hunt Jr., E. E., and Pfifferling, J. H. (1976). Demographic observations (1963–1973) on the chimpanzees of Gombe National Park, Tanzania. *Journal of Human Evolution*, **5**, 559–98.

Uehara, S. and Nyundo, R. (1983). One observed case of temporary adoption of an infant by unrelated nulliparous females among wild chimpanzees in the Mahale Mountains, Tanzania. *Primates*, **24**, 456–66.

van Lawick-Goodall, J. (1967). Mother – offspring relationships in free-ranging chimpanzees. In *Primate Ethology*, ed. D. Morris. London: Weidenfeld & Nicolson, pp. 287–346.

Wroblewski, E. E. (2008). An unusual incident of adoption in a wild chimpanzee (*Pan troglodytes*) population at Gombe National Park. *American Journal of Primatology*, **70**, 995–8.

PART VII Social behavior

31 Intimidation display

KAZUHIKO HOSAKA

31.1 Introduction

Intimidation displays are conspicuous and spectacular aspects of wild chimpanzees' social lives, and researchers have been intrigued by these behaviors since the pioneering days of chimpanzee research (Reynolds and Reynolds, 1965; van Lawick-Goodall, 1968; Sugiyama, 1969; Nishida, 1970). On the other hand, intimidation displays are usually prolonged and complicated behaviors composed of many types of actions, vocalizations, postures, and piloerection. Nevertheless, few studies have analysed the structure and function of intimidation displays as behavioral systems, except for a pilot study by Mori (1982, 1984). Instead, they have been studied merely as "threat display" with diverse forms and intensities in relation to agonistic interactions among adult males (e.g. Bygott, 1979; Nishida and Hosaka, 1996; Muller, 2002). Thus, intimidation displays are sometimes referred to as "agonistic displays" and broadly categorized as "aggression" (Chapter 32). Furthermore, since chimpanzees compose their intimidation displays by taking advantage of environmental elements, they develop novel patterns like metal house drumming at Mahale (Section 31.5.4), which has attracted the attention of researchers interested in chimpanzee culture (Chapter 38), and use of tools (Chapter 42).

This chapter aims to describe intimidation displays in terms of their natural history, without deviating deeply into other topics that will be addressed in subsequent chapters (Chapters 32, 33), while reviewing relevant studies about behavioral patterns, function, development, flexibility, and interspecies comparison.

31.2 Terminology

Van Lawick-Goodall (1968) defined "charging displays" as a type of behavior that involves all or some of the following behavioral patterns: charging, stamping, dragging, throwing, and drumming. Goodall (1986) distinguished between "charging displays" and "intimidation displays" (e.g. swaggering and branch shaking). However, the distinction was not always clear, because swaggering and branch shaking also occur in a series of charging displays. Mahale researchers have used both terms in various ways. Some authors (e.g. Kawanaka, 1989; Takahata, 1990) counted the frequency of "charging displays" by males (*sensu* van Lawick-Goodall, 1968). Other authors used only "intimidation displays" to denote both (e.g. Uehara *et al.*, 1994). Nishida *et al.* (2009) used both terms but with a wider definition for the latter, while Nishida (2012) treated both terms as synonymous. In studies of captive chimpanzees of Arnhem Colony, the Netherlands, van Hooff and de Waal (1981) called both of them "bluff."

According to the ethogram of the Mahale chimpanzees compiled by Nishida *et al.* (2010) (Chapter 37), the components of charging displays are <scratch dead leaves>, <drag branch>, <shake branch and sway woody vegetation>, <bristle>, <compress lips>, <slap>, <stamp>, <slap-stamp>, <flail>, <throw>, <drum>, <rake>, <swagger, bipedal>, and <pant-hoot>. One additional behavior <charge> should have been added.

In this chapter, "intimidation display" is defined broadly as a series of one or more of the behavioral

Mahale Chimpanzees: 50 Years of Research, ed. M. Nakamura, K. Hosaka, N. Itoh and K. Zamma. Published by Cambridge University Press.

components listed above used in intimidating contexts. For example, a single component such as <charge> that results in a brief attack or <pant-hoot> emitted as intimidation (see also Chapter 39) is grouped into "intimidation displays."

31.3 Behavioral patterns and functions of typical intimidation displays

31.3.1 "Intimidation displays" in aggressive contexts

"Charging displays" (intimidation displays with "charge") have long been studied to examine agonistic relations among chimpanzees (e.g. Bygott, 1979; Muller, 2002). "Charge" is the important component that indicates the target, direction, and level of aggression.

"Charge" is an action of brisk dashing, typically starting as quadrupedal running with piloerection, and often progressing into bipedal swagger with hunched shoulders. "Charge" can be classified as "directed" or "undirected" depending on whether the charger rushes toward a particular target or not.

In most directed charging displays, physical contact with the target is not intended by the charger. The charger usually allows the target to flee and does not further pursue him/her. However, "charge" is categorized as "attack" if it accompanies physical contact with the target at any point in the action, or as "chase" if the attacker pursues the fleeing target. Both "attack" and "chase" could be regarded as overt aggression; therefore, Muller (2002) called them *high-level aggression*, while calling charging displays without "attack" or "chase" *low-level aggression*.

There is no apparent target in undirected charging displays, which are often performed to impress audiences with the charger's physical strength and social status, much like rain dance and streambed display (see Section 31.5). However, on other occasions, the charger may have had a target in his mind (Case 31.1).

Both target and non-target individuals nearby may respond to the intimidation displays, typically by pant-grunting or other submissive behaviors toward the performer (Cases 31.1 and 31.2). There have been arguments about whether such submissive responses have an appeasement function (Chapters 32, 33).

Case 31.1 February 25, 1992

At 16:50 h, when the M-group chimpanzees were moving northward along the trail, I observed Nsaba (the beta male) sitting on the trail. Then, Ntologi (the alpha male) appeared 5 m south of Nsaba and stood with his hair bristling and his lips compressed. Nsaba moved slowly 1 m into the thicket and began to feed on the pith of *Renealmia engleri*. A 4-year-old infant female, Serena, along with Musa (old male) and Silafu (Serena's mother), reached Ntologi. As Serena was about to walk past Ntologi, he suddenly seized Serena's hand with his right hand, stood up hunching over, and beat Serena against the ground. Immediately, Silafu rushed over to rescue her daughter with a full grin and screaming loudly (Figure 31.1). Ntologi jumped on, slapped, and grabbed Silafu who swiftly secured Serena under her belly. Ntologi bit the back of Silafu's neck, held her down, and slapped her once more. Without interruption, Ntologi began to charge away northward, first dragging grass by the left hand, next dragging a dead branch by the right hand, and finally throwing another branch upward. He stopped 15 m ahead and sat 10 m north of Nsaba. The entire series of Ntologi's aggressive displays lasted for 16 s. Two adult females approached and pant-grunted

Figure 31.1 A charging display of Ntologi in February 1992. (a) Ntologi initiated the display by seizing Serena and beating her against the ground. The victim's mother, Silafu, rushed to rescue her daughter. (b) Ntologi jumped on Silafu to slap her. (c) Ntologi started to lunge off, leaving the victims. (d) Ntologi dragged a branch on the way. (e) Ntologi threw a branch just before finishing the display. ©K. Hosaka.

very loudly at Ntologi. Nsaba, who had sheltered in the bush, made no particular response.

This case describes one intimidation display bout composed of 13 components: <bristle>, <compress lips>, <hunch over>, <beat victim against the ground>, <jump on victim>, <slap and grab victim>, <bite victim>, <hold victim down>, <slap victim>, <charge>, <drag vegetation>, <drag branch>, and <throw branch>.

Overall, this case could be categorized as an undirected charging display. However, it is plausible that Nsaba was a virtual target of Ntologi's display. It could also be interpreted as one of the repeated interactions at that time in which Ntologi threatened his rival, Nsaba (Chapter 27). Nsaba appeared to be used to these threats and made the least effort to escape from Ntologi's display. Nsaba sheltered in the bush by moving a few meters and engaged in feeding, as if he had moved there for another reason. Ntologi's charging display could have been rather effective at demonstrating dominance over Nsaba in front of an audience.

In contrast, instead of emitting evert aggression at Nsaba, Ntologi attacked Serena and her mother. For these victims, this was a directed charging display starting as attacks against them. Because Ntologi had no reason to attack these victims, it could be interpreted that he used them as tools to make his display more spectacular and effective. Thus, this charging display could be interpreted as two faceted; the undirected charging display with a target in mind and the directed charging display with instrumental attacks against third parties.

31.3.2 Joint intimidation display

Occasionally, two or more chimpanzees synchronize or coordinate their charging displays to make them more effective. Goodall (1986) called this pattern "joint charging display" and noted that this performance typically represents the formation of a stable coalition among males (whether against a rival or not).

This view might be extended to non-charging intimidation displays (whether among males or females) and stable coalitions among females. Case 31.2 shows an example of two adult females in a "friendship" (see Chapter 28) performing a 5-min joint intimidation display session composed of charging and non-charging display bouts.

Case 31.2 December 25, 1991

At 08:21 h, when I conducted *ad libitum* sampling on the trail to the north of the Ntale Valley (Chapter 2), two adult females, Wakasunga (full sexual swelling) and Gwekulo (partial sexual swelling), appeared while performing a joint charging display in the absence of adult males nearby. When they reached a tree on which another adult female, Wakusi (lactating), sat, the two climbed up the woody vines hanging from the tree and swung back and forth. They then paused while hanging from the vines while Wakusi sat silently on a vine above the two.

Fourteen seconds later, Wakasunga resumed swinging with piloerection and initiated a loud pant-hoot. Gwekulo synchronized her swinging and chorused the call. On the ground, another adult female, Nkombo (partial sexual swelling), began to scream loudly.

At 08:22 h, as Wakasunga descended to the ground, Wakusi's son Alofu (9-year-old) greeted her and showed a brief grin while standing bipedally, pressing his nipples with his right hand and peering into Wakasunga's face. Two meters ahead, Wakasunga's son Linta (7-year-old) copulated with Nkombo, who was still screaming loudly. As Wakasunga and others started heading south, Gwekulo remained on the vine and emitted a loud pant-hoot. During the excitement caused by several reply calls combined with the screams of Nkombo and a 9-year-old female named Abi, Gwekulo performed a vigorous display while hanging from the woody vine with piloerection and swinging up and down for 10 s.

At 08:24 h, when Wakasunga and Gwekulo moved together on the trail southward, they performed a joint charging display, apparently triggered by a pant-grunt by Linta toward Gwekulo. They stood side by side, swaggered bipedally, and then charged in different directions. Wakasunga's charging drove off some females, including Nkombo and Abi, whose prolonged screams got louder. Gwekulo came back to join Wakasunga, dragging a branch with her right hand and then swaying vegetation. Nkombo and Abi ceased screaming.

At 08:26 h, when Wakasunga and Gwekulo were sitting side by side on a woody vine, Wakusi joined and climbed a woody vine above the two. Gwekulo and Wakasunga then emitted a pant-hoot in chorus and charged directly toward Wakusi, who fled to a higher place and screamed loudly. Wakasunga began to climb down the slope into a ravine. Gwekulo followed while emitting a strong pant-hoot.

This session of intimidation displays comprised five bouts separated by pause intervals of 14–103 s. Only the third bout was a solo display by Gwekulo composed of <bristle>, <pant-hoot>, and <swing up and down>. The other bouts were joint displays by Wakasunga and Gwekulo composed of <bristle>, <charge>, <swing back and forth>, <pant-hoot>, <drag branch>, and <sway vegetation>. Most of the charges were undirected except for the last charge toward Wakusi.

Such joint intimidation displays by adult females are relatively rare compared to those by adult males. However, females occasionally perform these intimidation displays if there are no adult males nearby. In 1991 and 1993, Wakasunga and Gwekulo had partnerships in terms of association, grooming, and alliance (unpublished data), and Wakasunga could be added to the list of Gwekulo's successive "friends" (Chapter 28). I witnessed several other cases of their joint intimidation displays. They even made a coalitionary attack against an adult male who was too sexually coercive (Chapter 5).

What motivated these two females to perform the described joint display was unknown, but two consequences should be pointed out. First, it caused great excitement full of screams by some females. Earlier that morning, a party, including many adult males, had moved ahead and climbed the mountain. The two estrous females might have intended to remain in lowland areas and attempted to convey their intention to the males by their uproar, although there was no evidence to corroborate it. Second, the two females seemed to have persistently threatened Wakusi, one of the high-ranking females at that time (Toshisada Nishida, personal communication). Wakusi did not scream during the first three bouts although she was nearby, but at last, she screamed loudly when she was charged. It was uncertain whether this session should be considered a dominance-related conflict, but it is possible that Wakusi would not have screamed had she not been directly charged by the coalition of Gwekulo and Wakasunga.

Interestingly, the females' joint intimidation display elicited submissive greeting behaviors from juvenile and early adolescent males who were not the target of the aggressions. Such submissive greetings could well have an appeasement function (Chapter 33).

31.3.3 "Intimidation displays" in non-aggressive contexts

Some components of intimidation displays, such as bipedal swagger and shaking branches, are also exhibited in sexual contexts as components of courtship displays (Chapter 35). Some components, such as dragging branches and stamping, are also exhibited in playful contexts to invite or facilitate play interactions (Chapter 36). These displays are not intimidation displays in the strict sense as defined in this chapter, but it is interesting to note that they share the same behavioral components as intimidation displays. Rather than drawing a demarcation line between intimidation displays and non-intimidation displays, it might be more productive to discuss how and why chimpanzees employ the same behaviors in different contexts.

31.3.4 Vocal and non-vocal displays

Bygott (1979) distinguished between vocal and non-vocal displays, based on his studies on chimpanzee agonistic behaviors at Gombe, Tanzania. The former is accompanied by a pant-hoot, while the latter is not. He suggested that vocal displays are rarely directed toward specific individuals – most often, such displays are directed toward distant individuals, which coincides with the function of the pant-hoot as a long-distance call (Chapter 39). On the other hand, non-vocal displays are frequently directed toward a specific individual within sight and are more likely to be followed by an attack than any other form of threat is. Muller (2002) tested this hypothesis at Kibale, Uganda, and provided supporting data.

However, de Waal (1982) described an interesting finding during his observation of rank reversals among Arnhem Colony males. He further discriminated between two types of vocal displays according to breathing patterns: ventilating and inflated displays. He noticed that the former type was evident in the challenging party in agonistic interactions over dominance. Once the period of instability was over and the rival had submitted, the challenger switched to the inflated display that reflects his self-confidence and assertiveness.

Therefore, the interpretation that vocal displays do not include intentions to attack particular individuals, as suggested for Gombe, may be applicable to ordinary situations, whereas a different pattern can be observed in unusual circumstances, such as in rank reversals. Nishida (1981), who observed the struggle for alpha status by the K-group males in 1976, mentioned that the challenging male Sobongo repeated pant-hoots near the alpha male, Kasonta. Until this observation, the author did not notice that pant-hoots may have the function of intimidating a nearby individual, in addition to the known function of long-distance communication. This notion is further supported by my own observation in December 1991, when the alpha status in the M group was unstable; the challenger, Nsaba, repeated intimidation displays accompanied by pant-hoots with a prolonged introduction phase towards the alpha male, Kalunde (Chapter 27). The derivative function of pant-hoot vocalizations (Chapter 39) may be an important topic for further research.

31.4 Intimidation displays by males in terms of development and dominance

31.4.1 From the juvenile period to adolescence

Pusey (1980) suggested that juvenile males at Gombe rarely performed charging displays with piloerection. She noted that even when they did so, most of them were brief and directed at the observer or baboons; however, a few were directed at other chimpanzees. Displays with piloerection seemed to increase in the transition period from early to late adolescence. In addition, she found that vocal displays with piloerection were only shown by males of 10 years or older. At Mahale, juvenile and adolescent males' "harassments" towards females (Nishida, 2003a) gradually advance to the typical form of intimidation display as the males grow older (Chapters 22, 29).

31.4.2 Who performed intimidations most frequently among adult males?

Takahata (1990) analysed social relationships among adult males at Mahale in 1981 and found no positive correlation between male dominance rank and the frequency of charging displays. This result was due to the fact that the 24-year-old male Lubulungu most frequently initiated and participated in charging displays (12 times), even more than the alpha male Ntologi (10 times).

According to Kawanaka (1989), who observed the M-group males in 1986, charging displays were performed most frequently by a 16-year-old young adult male, Shike (1.18 times/h), rather than the alpha male, Ntologi (0.61 times/h). Shike also performed the displays most frequently among males when he became a 21-year-old prime adult in 1991 (Hosaka, unpublished data). Thus, his frequent displays may be due to his personality rather than an age-related effect.

However, young adult males most commonly do not dare to perform displays in front of older males. More than 60% of the alpha male Ntologi's charging displays were performed when prime or old adult males were within 10 m, while other males except Shike never did so (Kawanaka, 1989). Shike, despite his high frequency of displays, did not perform displays in front of two high-ranking, middle-aged adult males, Ntologi (alpha) and Bakali, but in front of a middle-aged adult male, Lubulungu. Kawanaka (1989) interpreted this as Shike having already virtually acquired dominance over Lubulungu at that time by increasing the number of displays in front of him.

Figure 31.2 Cadmus's streambed display in September 2008. ©Michio Nakamura.

31.5 Some characteristic patterns of intimidation displays

31.5.1 Rain dance

Chimpanzees, especially adult males, often perform intimidation displays when it begins to rain heavily, while most other chimpanzees sit calmly on trees (Goodall, 1986). There are two patterns of rain dance at Mahale (Nishida *et al.*, 2010). One is a noisy, active display accompanied by pant-hoots. The other is performed without vocalizing at a more prolonged and slower tempo than are typical non-vocal displays. The latter type, in particular, looks as if the males are "dancing" in response to the rain, thus the term "rain dance."

The function of rain dance remains unclear. The performer gets wet to the skin as a result of this display, which seems less adaptive in terms of physical health. One might speculate that chimpanzees exploit natural phenomena to make their displays more spectacular, but it is difficult to verify this hypothesis through conventional ethological methods.

31.5.2 Streambed display

Mahale chimpanzees, mostly adult males, often engage in intimidation displays when they cross the rock-strewn streambed of some valleys, such as the Kasiha Valley (see Chapter 2 for geographical information). They typically throw large stones, either using one or both hands, into the water to make a large splash (see Chapters 41, 42 for behavioral details) and swing or shake woody vines hanging down over the streambed (Figure 31.2). Most commonly, these displays are presented as prolonged performances in front of audiences rather than for intimidating particular individuals.

31.5.3 Intimidation displays as a contest

Nishida *et al.* (2010) noted that there could be up to five participants in a series of intimidation displays. As one male starts displaying, more dominant or similar-ranking individuals join with chain reactions; typically, such a "contest" of displays ends with the alpha male's appearance.

I observed an unusually long such contest, lasting from 17:02 to 17:30 h, among adult males on August

25, 2009 at the Kasiha Valley. When Fanana and Kalunde arrived at the Kasiha Valley, Fanana started a bipedal directed charging display towards Bonobo, who also opposed bipedally. Soon it became a large and noisy series of directed and undirected joint displays including seven adult males. First, Orion screamed as a result of coalitional charges by Alofu, Kalunde, and Fanana. The coalitions and the targeted individual then charged one after another. For example, Kalunde, with Alofu behind, charged Cadmus. The contest of displays continued by using the full width of the riverbed and finally calmed down when the alpha male, Pimu, appeared at the scene with some females and charged Fanana. Two other adult males and an adult female were present on the sidelines and did not join the display contest.

31.5.4 Drumming display

Wild chimpanzees are known to generate long-distance sounds across their habitats via "buttress drumming," which involves hitting and/or kicking the buttress of a tree. Arcadi *et al.* (1998) suggested that there is a function of interindividual signature in such drumming, such as in pant-hoots, from comparative analyses at Taï, Côte d'Ivoire, and Kibale. There has been no comparable study at Mahale on this account.

In 1979, Mahale chimpanzees acquired the habit of "metal house drumming" (Figure 31.3; see also Chapter 42). When they pass the Kansyana Camp (see Chapter 2 for its location), they slap and/or kick the metal walls of two researchers' facilities (one of which was demolished) in a similar way to buttress drumming. In 1991 and 1993, I witnessed male chimpanzees drumming other metal houses in Kasiha and Sisindiko (see Chapter 2 for location), although it was uncertain when and how this behavior had spread. This behavior has not only persisted up to now, but also became one of the most spectacular intimidation displays of Mahale adult males. In fact, they usually quicken their pace when approaching the camp, seemingly motivated by this behavior and endeavoring to reach the camp as soon as possible. Nishida (2003b) pointed out that there are marked individual differences in the ways of banging the metal houses and suggested that the metal house drumming functions as a "signature," as suggested by Arcadi *et al.* (1998) for buttress drumming.

Figure 31.3 Alofu's metal house drumming display in October 2004. ©Michio Nakamura.

Interestingly, only socially mature males perform metal house drumming. Immature males do not drum the metal house even if they have started buttress drumming (Nishida, 2003b). Nishida and I observed that the 7-year-old male Chopin once hesitantly slapped the metal wall; this was the earliest reported age for the behavior (Nishida, 2003b).

On September 24, 1991, I observed the 14-year-old adolescent male Masudi hesitantly drumming the metal wall; he stood bipedally looking around, as if to confirm that nobody was nearby, and with a knuckle of his hand he drummed – or to be more accurate, pushed the metal house. I was only able to hear a faint hollow sound standing at a distance of 15 m.

On April 18, 1993, I observed that Masudi, a then 16-year-old low-ranking adult male, appeared at Kansyana Camp and pant-hooted without drumming the metal house. After this, he sniffed the ground, appearing slightly nervous about an old male Musa (low-ranking but dominant over Masudi) who had passed by the spot 3 min earlier. However, Masudi pant-hooted again after 1min and this time powerfully hit the metal wall with his right hand and foot. Musa then rushed back from the bush and charged Masudi. As Masudi fled screaming loudly, Musa chased him. When Masudi stopped, Musa briefly hunched over Masudi's back and drummed two metal houses in succession with his right foot and both feet, respectively. Masudi screamed loudly with a temper tantrum as Musa rushed away. After 5 min, when Musa pant-hooted with drumming again, Masudi gave a call in chorus.

31.5.5 Some innovations in intimidation displays

Sound effects

On October 23, 1991, when Shike (the beta male) passed Kansyana Camp, he abruptly charged three helmeted guineafowl that researchers kept for provisions. He did not attempt to capture the birds but made sharp sounds by stamping the ground. The birds leapt away and began to give noisy calls, which made Shike's charging display even more spectacular.

Light effects

The M-group adult males sometimes perform a "red sunset display," a type of charging display that seems to make use of the red sunset as a visual effect. It occurs only on a sandy trail between Kansyana Camp and Kasiha Beach. As the adult male chimpanzees walk with piloerection along the trail before sunset, they often deliberately raise a cloud of sandy dust, thereby silhouetting themselves against the blaze of red dust (Figure 31.4). I first witnessed this type of display when the alpha male, Pimu, performed it several times in 2008 and 2009. Video material recorded by ANC Production, Inc. showed that Pimu had begun this display in 2007 just after usurping alpha status.

31.6 Intimidation display against heterospecific animals

31.6.1 Against prey

When chimpanzees hunt red colobus monkeys, they often repeat charging displays on the ground. There are several explanations for this, for example, subduing aggressive prey monkeys by intimidation, expression of psychological uplift in relation to hunting, or agitation for group hunting (see Chapters 16, 20 for details).

31.6.2 Against humans

Mahale chimpanzees sometimes deliberately charge past human observers or lightly slap the humans' knees or hips. Such slaps are usually gentle and painless. Judging from the contexts and audiences, such behaviors may be motivated to display "boldness" to other chimpanzees.

In February 1992, the beta male Nsaba began to direct charging displays toward me and a TV crew of four or five staff members. Nsaba may have been motivated to show his "boldness" to his rival, Ntologi, who had just come back as the alpha male by overpowering Nsaba with the allied aid of the gamma male Kalunde (Chapter 27). Although Nsaba had at

Figure 31.4 Pimu's red sunlight display in September 2009. (a) Pimu (rear) was walking along the trail, raising a cloud of sandy dust, thereby silhouetting himself against the blaze of red dust. Darwin (middle) and Cadmus (front) had just finished pant-grunting at Pimu and began to walk in front of Pimu. (b) Then, Pimu walked past Darwin with piloerection and his shoulders hunched up and received a pant-grunt from Darwin. ©K. Hosaka.

least two opportunities to seriously harm me, he never did. In the first incident, when I was awkwardly walking on a narrow cliff path, Nsaba came out performing a charging display on the same path. While I was at a loss over where to shelter, he detoured as if to avoid me and did not dare to thrust me off the cliff. In the second incident, while I was walking cautiously on a natural log bridge, Nsaba came from behind to cross the same bridge, gently slapping the bridge with both hands alternately in a hunching position as if to speed me up. Nevertheless, he waited for me to get across the bridge. These anecdotes suggest that Nsaba modulated his charging displays to avoid harming humans.

On the other hand, Ntologi's displays were sometimes dangerous to humans. In 1993, when I was observing another adult male, he approached me silently and slowly from as close as 2 m behind (Kenji Kawanaka, personal communication). He then abruptly threw a large stone (15-cm diameter) toward me that hit below my right knee and caused acute pain; fortunately, I got away with only a bruise. After throwing the stone, Ntologi did not charge but stood still quadrupedally with his lips compressed, hair bristling, and his shoulders hunched up. On another occasion, he threw two stones at the head of a research assistant who barely escaped the danger. Once, I witnessed Ntologi approach a female researcher from behind while dragging a bough and use it to strike the backs of her knees, causing her to fall on her behind.

One reason why Ntologi was more dangerous than Nsaba might be that Ntologi actively used tools such as stones and boughs in his intimidation displays against humans. The other reason might be that he intended to show "ruthlessness" as the alpha male rather than "boldness," although such an argument has not been elaborated in conventional ethology.

From a conservation perspective, it is very important to prevent intimidation displays against humans. Such aggression against humans might not only reflect the various adult male tactics discussed above but also a form of stress caused by the close proximity of humans. In the early 1990s, the habituation of Mahale chimpanzees reached its maximum (Toshisada Nishida, personal communication), which enabled humans to observe chimpanzees very closely. However, when Pimu became the alpha in 2007 (Chapter 27), there were some incidents in which he punched several researchers, bruising them (Michio Nakamura, personal communication). Pimu showed his "ruthlessness" as the alpha male without tools, which might have been necessary for Ntologi. Alternatively, Pimu might have become uneasy when humans got too close, which probably triggered his charging display. Since researchers have practiced the self-imposed rule of keeping a safe distance from the chimpanzees to prevent transmission of human diseases (Chapters 48, 50), such chimpanzee aggression toward humans has decreased.

31.7 Comparison with bonobos

Chimpanzees can be noisy animals. While most of the time they are relatively quiet, they can get explosively excited in groups wherein they engage in prolonged exchanges of loud vocalizations and intimidation displays. Such a noisy situation has been called "carnival" (Reynolds and Reynolds, 1965), "booming" (Sugiyama, 1969), and "mass excitement" (Nishida, 1970).

Mori (1982) considered that charging displays play pivotal roles in composing such booming sessions. During a 6-month study in 1972–3, Mori (1982) observed the Mahale K-group chimpanzees at the provisioning site to analyse charging displays as intra- and interindividual sequences of various behavior elements. During a 4-month study in 1979–80, Mori (1984) conducted similar observations on bonobos (*Pan paniscus*) at the provisioning site of Wamba, Zaïre, for comparison with Mahale chimpanzees. One notable finding was that bonobos did not often show the "bipedal" posture in their undirected charging displays initiated by "drag branch," while this posture was an important component in the corresponding displays of chimpanzees. However, he found that bonobos showed the bipedal posture more frequently in initiating sexual behaviors and affiliative dominant–subordinate behaviors. There was also a sharp contrast between sexes and age groups, in that female and young bonobos often performed drag branch displays, whereas adult male chimpanzees were the predominant performers of the same behavior. Thus, Mori argued that the meaning of "bipedal" might have changed from male antagonism in chimpanzees to affiliation in bonobos.

31.8 Discussion and future perspectives

Intimidation displays can be viewed as an example of animal display behaviors. Their primary function consists of "threat" or "bluff." The main function of directed intimidation displays might simply be to dominate a specific individual, while the undirected form aims to show the strength of the displayer to all individuals within sight. However, such a statement is clearly an oversimplification. For instance, an alpha male often lunges toward adult females or immature individuals (e.g. Case 31.1) in the course of his intimidation display, but its function is a mere performance because his dominance over such individuals is obvious. On the other hand, the performer of undirected intimidation display may have the target in mind (e.g. Case 31.1) while pretending not to direct aggression toward a specific individual. Furthermore, undirected joint displays by adult males A and B can sometimes abruptly change into a directed charge from A to B (Hosaka, unpublished data), although Goodall (1986) insisted that joint charging display demonstrates the formation of a stable coalition among males. It is probable that chimpanzees themselves do not always foresee the outcomes of an intimidation display they initiate or join, especially if the display has some aspects of a contest. Thus, researchers should be cautious when sampling any type of aggressive behaviors from intimidation displays.

In general, animal "threat displays" are adaptive in conserving energy or avoiding risk of being wounded, making it unnecessary for the animal to have a real fight. The statement may be applicable to chimpanzee intimidation displays of short duration. However, the intimidation display of long duration may cost chimpanzees considerable energy loss, since it can persist for up to 20 min (as in the case of the prolonged streambed display in 31.5.3). Nishida (2012) suggested that the alpha male demonstrates his strength by prolonging his charging display as long as possible, although no study has analysed the relation between the stability of alpha status and the duration of charging displays. It is plausible that the chimpanzee intimidation display is polymorphic and multifunctional. For example, the alpha male may make a low-cost, momentary lunge to elicit a submissive response from another adult male, whereas he may perform a high-cost, prolonged intimidation display encompassing almost all behavioral components to show his strength.

Finally, I must stress that females occasionally perform intimidation displays, including joint displays. In future studies, researchers should investigate the intimidation display as a communication system as well as a behavior related to aggression or culture. Just as the pant-hoot has dual functions of "long-distance call" and "threat" (Chapter 39; see also Section 31.3.4), the intimidation display may have similar dual functions, although mainly the latter function has been studied so far. Females may play an important role in the former function, because they create noisy situations by emitting loud calls (pant-grunts, screams, barks, and pant-hoots in chorus) in response to the display by adult males. The joint display performed by Gwekulo and Wakasunga (Case 31.2) is a good instance of adult females actively communicating with distant parties by producing "booming" situations if there are no adult males nearby.

References

Arcadi, A. C., Robert, D., and Boesch, C. (1998). Buttress drumming by wild chimpanzees: temporal patterning, phrase integration into loud calls, and preliminary evidence for individual distinctiveness. *Primates*, **39**, 505–18.

Bygott, J. D. (1979). Agonistic behavior, dominance, and social structure in wild chimpanzees of the Gombe National Park. In *The Great Apes*, ed. D. A. Hamburg and E. R. McCown. Menlo Park, CA: Benjamin/Cummings, pp. 405–27.

Goodall, J. (1986). *The Chimpanzees of Gombe: Patterns of Behavior*. Cambridge, MA: Belknap.

van Hooff, J. A. R. A. M. and de Waal, F. B. M. (1981). Side-directed communication and agonistic interactions in chimpanzees. *Behaviour*, **77**, 164–98.

Kawanaka, K. (1989). Age differences in social interactions of young males in a chimpanzee unit-group at the Mahale Mountains National Park, Tanzania. *Primates*, **30**, 285–305.

van Lawick-Goodall, J. (1968). The behaviour of free-living chimpanzees in the Gombe Stream Reserve. *Animal Behaviour Monographs*, **1**, 161–311.

Mori, A. (1982). An ethological study on chimpanzees at the artificial feeding place in the Mahale Mountains, Tanzania: with special reference to the booming situation. *Primates*, **23**, 45–65.

Mori, A. (1984). An ethological study of pygmy chimpanzees in Wamba, Zaïre: a comparison with chimpanzees. *Primates*, **25**, 255–78.

Muller, M. N. (2002). Agonistic relations among Kanyawara chimpanzees. In *Behavioural Diversity in Chimpanzees and Bonobos*, ed. C. Boesch, G. Hohmann, and L. F. Marchant. Cambridge: Cambridge University Press, pp. 112–24.

Nishida, T. (1970). Social behavior and relationships among chimpanzees of Mahali Mountains. *Primates*, **11**, 47–87.

Nishida, T. (1981). [*The World of Wild Chimpanzees*.] Tokyo: Chūkō-shinsho. In Japanese.

Nishida, T. (2003a). Harassment of mature female chimpanzees by young males in the Mahale Mountains. *International Journal of Primatology*, **24**, 503–14.

Nishida, T. (2003b). Individuality and flexibility of cultural behavior patterns in chimpanzees. In *Animal Social Complexity: Intelligence, Culture, and Individualized Societies*, ed. F. B. M. de Waal and P. L. Tyack. Cambridge, MA: Harvard University Press, pp. 392–417.

Nishida, T. (2012). *Chimpanzees of the Lakeshore: Natural History and Culture at Mahale*. Cambridge: Cambridge University Press.

Nishida, T. and Hosaka, K. (1996). Coalition strategies among adult male chimpanzees of the Mahale Mountains, Tanzania. In *Great Ape Societies*, ed. W. C. McGrew, L. F. Marchant, and T. Nishida. Cambridge: Cambridge University Press, pp. 114–34.

Nishida, T., Matsusaka, T., and McGrew, W. C. (2009). Emergence, propagation or disappearance of novel behavioral patterns in the habituated chimpanzees of Mahale: a review. *Primates*, **50**, 23–36.

Nishida, T., Zamma, K., Matsusaka, T., Inaba, A., and McGrew, W. C. (2010). *Chimpanzee Behavior in the Wild: An Audio-Visual Encyclopedia*. Tokyo: Springer.

Pusey, A. E. (1980). Behavioural changes at adolescence in chimpanzees. *Behaviour*, **115**, 203–46.

Reynolds, V. and Reynolds, F. (1965). Chimpanzees of the Budongo Forest. In *Primate Behavior*, ed. I. DeVore. New York: Holt, Rinehart and Winston, pp. 368–424.

Sugiyama, Y. (1969). Social behavior of chimpanzees in the Budongo Forest, Uganda. *Primates*, **10**, 197–225.

Takahata, Y. (1990). Social relationships among adult males. In *The Chimpanzees of the Mahale Mountains: Sexual and Life History Strategies*, ed. T. Nishida. Tokyo: University of Tokyo Press, pp. 149–70.

Uehara, S., Hiraiwa-Hasegawa, M., Hosaka, K., and Hamai, M. (1994). The fate of defeated alpha male chimpanzees in relation to their social networks. *Primates*, **35**, 49–55.

de Waal, F. B. M. (1982). *Chimpanzees Politics: Power and Sex among Apes*. New York: Harper and Row.

32 Aggression and conflict management*

NOBUYUKI KUTSUKAKE AND KAZUHIKO HOSAKA

32.1 Introduction: aggression in chimpanzees

Most behavioral studies involve three stages: description, explanation, and synthesis (Altmann and Altmann, 2003). Studies of aggression, one of the distinctive characteristics of chimpanzees, provide a good example of this three-stage transition. Since the early days of chimpanzee research, numerous studies have investigated the frequency, effects, contexts, determinants, functions, and proximate underpinnings of aggression (Nishida, 1970, 1983, 1989, 2003; Goodall, 1986; Nishida *et al.*, 1999, 2010; recent reviews by Muller, 2002; Muller and Mitani, 2005). Researchers have also investigated social interactions that are temporally or functionally associated with aggression (see Section 32.5). By combining findings, researchers can examine both the ultimate and the proximate aspects of aggression from ecological and evolutionary perspectives. One aim of our chapter is to review the transitions between the three stages of studies regarding aggression and to elucidate what is known and what is not known regarding chimpanzee aggression. Because of space limitations, we do not aim to provide a comprehensive review of chimpanzee aggression (see Goodall, 1986; de Waal, 1996; Muller, 2002; Muller and Mitani, 2005; Wittig, 2010). We will focus on only the intragroup aggression observed on a daily basis in chimpanzee groups but will not discuss intergroup aggression (see Chapters 10, 11) or aggressive events involving gang attacks or killing within a group (Chapters 26, 27).

32.2 Description: patterns and contexts

There are different degrees of aggression, ranging from mild displays (e.g. threats without physical contact, agonistic displays, acts of physical aggression) to extreme behaviors that result in death (Nishida *et al.*, 2010). Although aggression is usually directed at a specific individual(s), this may not always be the case for agonistic displays (Chapter 31). The recipient of aggression may retaliate against an aggressor, and the situation may escalate into bidirectional aggression. Bidirectional aggression is commonly seen in chimpanzees because of their relatively less despotic dominance style, and this form of aggression is particularly likely to occur when a subordinate initiates the aggression (Wittig and Boesch, 2003a). Patterns of aggression are complicated when uninvolved bystanders participate in an ongoing aggressive event or when the original aggressive interaction results in other aggressive interactions (Goodall, 1986). Polyadic aggression can be classified into several categories according to the recipient of agonistic support (partial interventions involve supporting the winner or loser, whereas impartial interventions involve attempts to pacify or police the participants; de Waal, 1978, 1982; Boehm, 1994; Nishida and Hosaka, 1996; von Rohr *et al.*, 2012). One well-studied pattern of polyadic aggression involves alliance and coalition formation

* We thank Teresa Romero and Shunkichi Hanamura for critical and constructive comments.

Mahale Chimpanzees: 50 Years of Research, ed. M. Nakamura, K. Hosaka, N. Itoh and K. Zamma. Published by Cambridge University Press.

(Harcourt and de Waal, 1992), which is observed during dominance competitions among males (de Waal, 1982; Nishida, 1983; Nishida and Hosaka, 1996; Gilby *et al.*, 2013; see Chapter 27) or mate guarding (Watts, 1998). Side-directed behavior is often used to solicit support and/or seek assurance (de Waal and van Hooff, 1981; Hemelrijk *et al.*, 1991; Nishida and Hosaka, 1996). Social interactions following aggression (postconflict behavior) can be classified according to the type, initiator, and combination of the interactants. Following aggression, the opponents may affiliate (reconciliation; see Section 32.5) or renew the aggression (Wittig and Boesch, 2003b; Koski *et al.*, 2007b). A bystander sometimes affiliates with an aggressor (appeasement; Kutsukake and Castles, 2004; Koski and Sterck, 2009; Romero and de Waal, 2011; Romero *et al.*, 2011) or with a victim (consolation; de Waal and van Roosemalen, 1979; Wittig and Boesch, 2003b; Kutsukake and Castles, 2004; Fraser and Aureli, 2008; Fraser *et al.*, 2009; Romero and de Waal, 2010; Romero *et al.*, 2010; but see Arnold and Whiten, 2001). A recipient of aggression may redirect the aggression toward a bystander, or an aggressor may direct aggression toward uninvolved bystanders (Goodall, 1986). As a result of the successive acts of aggression, one bout may snowball into aggression involving numerous individuals (Goodall, 1986; Nishida and Hosaka, 1996).

From a proximate perspective, the occurrence of aggression is mediated by many factors, including individual intrinsic (genes, hormones, personality characteristics, etc.) and individual extrinsic (behavioral context, history of interactions, the value of resources, dominance stability, etc.) factors (Kutsukake, 2009). Previous studies have clarified several important individual intrinsic factors, including sex, with males being more aggressive than females (Muller, 2002). Dominance rank strongly affects rates of aggression, with an alpha male usually being more aggressive than other males (Muller, 2002). Age and social and reproductive stage (for females) also affect aggression. Juvenile or adolescent males challenge and outrank adult females because of the aggressiveness of the former (Nishida, 2003). Resident females frequently attack female immigrants, whereas males protect those newcomers (Nishida, 1989; Kahlenberg *et al.*, 2008a, b; see also Chapter 29 for contradictory cases observed at Mahale). Estrous females are more likely than non-estrous ones to be attacked by males (see below). Furthermore, comparisons among alpha males suggest individual differences in aggressiveness and social tactics (Foster *et al.*, 2009).

In terms of individual extrinsic factors, aggression has been classified into several categories according to context (reunion, sexual competition, plant-food competition, meat competition, social excitement, protection, and miscellaneous; Goodall, 1986; Muller, 2002). Research on different groups (the Mahale, Gombe, and Kanyawara groups in Kibale) found that the likelihood of initiating each kind of aggression was affected by the sex of an actor; that is, males were more likely to be aggressive in the context of reunions and sexual competitions, and females were more likely to be aggressive in situations involving food and protection (Muller, 2002; Muller and Mitani, 2005).

As Muller and Mitani (2005) indicated, basic information about rates of aggression among wild chimpanzees (based on relatively unbiased observational methods such as focal observations) is scarce (e.g. 0.15 interactions/h in males in Mahale; Hayaki *et al.*, 1989). Aggression rates fluctuate and are affected by many factors, such as the demographic characteristics of a group, dominance (in)stability, and ecological factors affecting food competition. For example, Miller *et al.* (2014) reported that the rate of aggression by females in Gombe was higher within than outside females' core areas. Thus, studies of aggression should consider the determinants of aggression, and simple comparisons of rates of aggression should be performed with care.

In this context, descriptions can reveal patterns and lead to the subsequent explanation stage, in which researchers examine how each factor affects aggression rates and how each aggressive act affects the behavior of the individuals involved.

32.3 Explanation: causes and consequences of aggression

From a functional and ultimate perspective, aggression usually results from interindividual conflict (e.g. interindividual conflicts of interests; Aureli and de Waal, 2000). Conflict arises for various reasons, such as competition for limited resources (e.g. food or mating opportunities), dominance competition, and other fitness-related purposes (e.g. protection of an infant or an important social partner). Because decision making during conflict and aggression is based on the costs and benefits to participants (Wittig and Boesch, 2003a), it has been assumed that evolutionary processes have led to particular adaptive patterns of aggression.

Wittig and Boesch (2003a) conducted a systematic investigation of the causes of aggression. Their study extended the relational model (de Waal, 1996), which holds that aggression, tolerance, and avoidance are social options during interindividual conflict, by considering the expected outcomes (costs and benefits) of initiating aggression at an individual level (see also Wittig and Boesch, 2003b, c; 2005). Wittig and Boesch (2003a) investigated the factors affecting the likelihood of initiating aggression as well as of the characteristics of aggression. The authors found that aggression initiated by dominants was targeted primarily at less important partners and that such aggression was long in duration and high in intensity. Aggression initiated by subordinates was usually short and less intense, and the social costs related to the aggression were reduced by reconciliation with a good social partner. Thus, Wittig and Boesch (2003a) showed that the characteristics and outcomes of aggression can be understood by carefully considering social context and associated outcomes.

Studies investigating the consequences of aggression have tested predictions derived from hypotheses about its adaptive value. A good example is aggression used as sexual coercion; Muller *et al.* (2007, 2011) studied male aggression toward estrous females in Kanyawara and showed that aggression increased both the probability of mating and the attractiveness of aggressive males (female choice) to females. In contrast, Stumpf and Boesch (2010) studied the consequences of a similar type of aggression in Taï Forest, Côte d'Ivoire, but did not find such effects. It is beyond the scope of this chapter to discuss the possible reason for the discrepancy between these studies, but it is important to note that studies devoted to testing hypotheses are powerful ways to assess adaptive explanations.

32.4 Unanswered questions in aggression studies

Careful testing of hypotheses about the causes and consequences of aggression derived from early descriptive studies has advanced our understanding of aggression. However, many questions remain unaddressed and unanswered. In theory, the causes of aggression can be understood by carefully disentangling individual intrinsic and extrinsic factors. However, this is a difficult task for practical reasons. For example, aggression occurs as a result of the history of social interactions, but researchers usually cannot observe the complete histories of wild groups. Because detailed analyses of the consequences of aggression have not been performed for all types of aggression, the functions served by aggression have remained untested under many conditions. Costs, benefits, and fitness consequences are notoriously difficult to quantify in any species of animal, and this problem is exacerbated in chimpanzees, who live long and are slow to reproduce.

One remaining challenge for researchers studying chimpanzee aggression is that individuals attack group members in situations in which there are no apparent resources to gain (at least from the perspective of human observers) or following non-agonistic social interactions (Goodall, 1986; Kawanaka, 1989; Hayaki, 1990). Early studies in Mahale have reported on this type of aggression in detail. Kawanaka (1989: 302) noted, "The alpha male (NT) carried out aggression towards targets of most age classes. These interactions mostly occurred following interactions accompanied by pant-grunts towards him by those who were attacked."

Hayaki (1990: 202) addressed the question "[W]hy would a subordinate chimpanzee pant-grunt at a dominant individual who may direct aggression toward him rather than be soothed." Some cases of aggression previously classified as "reunion" or "social excitement" may be included in this category because no obvious reasons for the occurrence of aggression can be discerned in these contexts. Aggression is costly for an actor in terms of energy consumption and the opportunity costs related to other activities (Muller and Wrangham, 2004). Therefore, it is reasonable to assume that the cost of aggression is balanced by some sort of benefit. Such types of aggression may function to compete for and/or reinforce the dominance position of an aggressor (Muller, 2002). However, the classification of this type of aggression is usually based on a process of elimination or the application of a post hoc explanation, which leads us to conclude that the reasons for such aggression among chimpanzees remain unclear.

Understanding this type of aggression may require integrating knowledge about aggression-related behaviors into a single framework, because aggression does not occur independently of other social interactions but is embedded in a cascade of social interactions. Below we will discuss how studies of conflict management and resolution can offer insights into the functions of these types of aggression.

32.5 Toward a synthesis: conflict management and resolution

Previous research regarding aggression per se and regarding chimpanzees in particular has provided ample data on aggression-related behaviors. Specifically, conflict management (i.e. a behavior that settles conflict but does not necessarily end it) and resolution (a behavior that eliminates the incompatibility between conflicting individuals; Aureli and de Waal, 2000) are important research areas related to the examination of aggression in chimpanzees. Aggression disrupts social relationships and group cohesion; in contrast, conflict management and resolution function to maintain group cohesion and to regulate social relationships. Therefore, these two social forces operate in opposite directions, and their integration into a single framework will increase our understanding of roles of aggression in chimpanzee societies. Here we briefly review two types of conflict management and resolution, submission and reconciliation, and discuss their implications for aggression.

32.5.1 Submission

Submission may function to confirm the status quo dominance relationship and prevent the escalation of aggression by a dominant (Hayaki, 1990; de Waal, 1996; Sakamaki, 2011; see Chapter 33). In chimpanzees, pant-grunt vocalization is a reliable signal of submission (Bygott, 1979; Noë and de Waal, 1980; see Chapter 33). Rival subordinate males do not emit pant-grunts when their relative dominance is unstable, and the absence of submissive signals may alter the behavior of a dominant individual (de Waal, 1982; Nishida, 1983; Nishida *et al.*, 1995). De Waal (1986, 1996) proposed that submission is a signal of the acceptance of formal dominance, and a dominant may offer a conditioned reassurance only when the subordinate emits that signal. Thus, preemptive submission may function as conflict management.

However, submission is sometimes followed by aggressive reactions from an aggressor (Kawanaka, 1989; Hayaki, 1990; Sakamaki, 2011; see Kutsukake and Clutton-Brock, 2008, for an example in other species), suggesting that preemptive submission does not always function to reduce aggression in chimpanzees. Thus, it remains unclear why individuals offer a submissive signal rather than fleeing from the aggressor (Matsumura and Hayden, 2006). If a subordinate attempts to minimize the risk of being attacked, avoidance would be expected to be the most effective behavior (Sommer *et al.*, 2002; Wittig and Boesch, 2003b; Kutsukake and Clutton-Brock, 2008). Avoidance must be particularly useful for chimpanzees living in a fission–fusion society (Chapters 6, 9), because individuals can leave a party and avoid further interactions with opponents (Wittig and Boesch, 2003b). In addition, pant-grunt vocalizations and the accompanying greeting behavior appear to have a certain cost; that is, an actor needs to rush to meet a recipient from a distance and spend a certain length of time enacting pant-grunt and other behaviors (Kawanaka, 1989).

Given that submission is costly, we would expect that submissive signals would serve an adaptive function for subordinates. One possible benefit is that submissive signals reduce, but do not completely eliminate, the risk of aggression by a dominant individual. Alternatively, it is possible that some cases of pant-grunt vocalizations labeled as submission by researchers are not actually submissive signals. Pant-grunts sometimes occur bidirectionally (Hayaki, 1990), and the utterance of pant-grunt vocalizations is flexibly moderated by the identities of nearby group members (Laporte and Zuberbühler, 2010). These studies suggest that such signals may express more than just subordination (reviewed in Laporte and Zuberbühler, 2010).

Understanding of the function of submission requires consideration of the perspectives of both subordinates and aggressors. The apparently purposeless aggression of dominants must also have adaptive significance that overrides the preemptive function (if any) of subordinates' spontaneous submission. It may be that the formation of a dominance hierarchy reduces conflict and unnecessary contests (Aureli and de Waal, 2000). However, this does not mean that aggression never occurs. In chimpanzees, the frequency of aggression by males increases and their submission decreases during periods in which dominance is unstable; however, the frequency of aggression is also constant when dominance is stable (Muller, 2002). This suggests that dominance relationships are maintained by the constant efforts of dominant individuals to signal their relative superiority or of subordinates to express their submissiveness; otherwise, dominance relationships may become unstable. In stable periods with clear dominance relationships among individuals, the physical cost of aggression that reinforces dominance may be low relative to that in periods in which dominance is unstable. In other words, aggression toward an individual who is clearly subordinate is less costly because it is less likely that the subordinate will retaliate against the dominant. In such situations, the extreme aggressiveness involved in attacking subordinates, even without an immediately apparent reason, may be adaptive in that it preserves rank-related benefits. Silk (2002) hypothesized that randomly timed aggression toward randomly chosen subordinates may be effective for reinforcing dominance because unpredictable aggression or even threats by dominants (Cant, 2011) can impose continuous physical and psychological costs (stress) on subordinates (see also Miller, 1997, for a similar argument). It would be interesting to investigate whether aggression is random and whether/how a social threat affects the stress level of wild chimpanzees (cf. Kutsukake, 2003, 2006, 2007).

This discussion suggests that the interplay between aggression and submission in chimpanzees is not straightforward. Previous analyses have not always separated preemptive from post-aggression signals (for exceptions, see Flack and de Waal, 2007, in macaques; Sakamaki, 2011) or measured the cost (intensity and duration) of pant-grunt vocalizations. It is critical to examine how different levels of submission are related to the probability of subsequent aggression.

32.5.2 Reconciliation

Affiliation between opponents soon after aggression is called reconciliation, which constitutes conflict resolution (de Waal and van Roosemalen, 1979; Aureli and de Waal 2000; Arnold *et al.*, 2010). Reconciliation was initially studied in a group of chimpanzees in Burgers' Zoo in Arnhem, the Netherlands (de Waal and van Roosemalen, 1979). Later, postconflict behavior was studied in various chimpanzee groups as well as in other species (reviewed in Aureli and de Waal, 2000; Arnold *et al.*, 2010). Reconciliation has been observed in all chimpanzee study sites in which systematic surveys have been conducted (see Wittig, 2010, for a recent review). Typically, reconciliation occurs immediately after aggression, usually within 1 or 2 min. The behaviors used for reconciliation include grooming, brief touching, embracing, other forms of physical contact, and kissing (de Waal and van Roosemalen, 1979).

The relationship quality, or valuable relationship, hypothesis states that individuals with beneficial relationships reconcile frequently (Aureli and de Waal, 2000; Watts, 2006; Arnold *et al.*, 2010). Although this hypothesis has been supported in wild chimpanzees, there have been a few exceptions. Studies in Budongo and Taï found that (a) baseline affiliation level was positively associated with the conciliatory tendency; and (b) the sex of the opponents predicted the conciliatory tendency, with females reconciling least frequently (Arnold and Whiten, 2001; Wittig and Boesch, 2005; Wittig, 2010). In the Ngogo group in Kibale, the occurrence of reconciliation among males was positively correlated with the frequencies of coalition formation and grooming (Watts, 2006). In contrast, a study in Mahale showed that females reconciled at a frequency similar to that of male–male and male–female dyads and that association patterns did not predict the occurrence of reconciliation (Kutsukake and Castles, 2004). Muller (2002) also reported that females reconciled at a rate similar to that of other dyads in Kanyawara. This intraspecific variation suggests the need for additional studies regarding how the dimensions of relationship quality affect the occurrence of reconciliation. Social relationships involve at least three dimensions: value, security, and compatibility (Aureli and de Waal, 2000; Fraser *et al.*, 2008a; Arnold *et al.*, 2010). However, the effects of security and compatibility on reconciliation have not been fully investigated in the wild (see Koski *et al.*, 2007a). Detailed classifications of relationship quality (see Fraser *et al.*, 2008a) and examinations of the effect of each dimension on the occurrence of reconciliation in wild groups are needed.

With regard to its function, reconciliation can restore a relationship that has been damaged by aggression perpetrated by opponents. In general, affiliation (e.g. grooming) has a calming effect on both the initiator and the recipient (e.g. Aureli *et al.*, 1999). Indeed, reconciliation reduces the probability of renewed aggression between opponents in many species (Arnold *et al.*, 2010). Although the same effect has not been reported in wild chimpanzees, aggression reduces tolerance between the opponents (Wittig and Boesch, 2005). This suggests that reconciliation functions to restore tolerance to normal levels in wild groups.

According to another hypothesis, the function of reconciliation is to reduce stress. This hypothesis holds that aggression increases the level of stress of participants and that reconciliation reduces stress to normal levels. Although this hypothesis has been supported in many species of captive primates and chimpanzees (Koski and Sterck, 2007; Fraser *et al.*, 2008b; but also see Koski *et al.*, 2007a), the results regarding wild chimpanzees are far from clear. Using self-directed behavior as a behavioral indicator of stress, Arnold and Whiten (2001) showed that the scratching rate decreased more after reconciled than non-reconciled aggression. Although this result appears to support the stress reduction hypothesis, this study also reported that the scratching rate did not increase more after aggression than after a control period. The scratching rate of Mahale chimpanzees also did not increase during a postconflict period (unpublished analysis using data in Kutsukake and Castles, 2004). Thus, it remains unclear whether aggression increases stress and whether reconciliation reduces stress in wild chimpanzees. Kutsukake (2003) found that the self-scratching rate was affected by the activity of the actor in that the rate was lower during foraging and moving than during resting. This result is not surprising, as these activities physically constrain the arms and hands, which are used for self-scratching. Thus, the absence of increases and reductions in the stress indices of wild chimpanzees may be a result of such constraints, because the activity performed soon after aggression may preclude self-scratching. It is necessary to use other behavioral indicators, such as social monitoring or vigilance (Kutsukake, 2006, 2007), to measure postconflict stress. In the ideal case, physiological indicators, such as cortisol level or heart rate, would be used, but those are practically difficult to measure in the field.

Thus, how do studies of reconciliation increase our understanding of aggression? Reconciliation is expected to occur when relationships have been damaged by aggression (Aureli and de Waal, 2000; Arnold *et al.*, 2010). If this is the case, it may be possible to investigate the effect of aggression on social relationships by examining the factors affecting the occurrence of reconciliation (conciliatory tendencies). Fraser *et al.* (2010) showed that the conciliatory tendency of chimpanzees in Chester Zoo was lower after aggression in the form of agonistic display than after other forms of aggression. The authors suggested that agonistic displays may not disturb social relationships because they are sometimes directed at inanimate objects. Food-related aggression is not followed by reconciliation in wild long-tailed macaques (Aureli, 1992), suggesting that food-related aggression may not disrupt social relationships. Thus, certain types of aggression may not damage the social relationship between opponents. Studies also suggest that a retrospective analysis of post-aggression behavior is useful for understanding how aggression affects the social relationship between the opponents.

In chimpanzees, either an aggressor or a victim can initiate affiliation with the opponent after the aggression: reconciliation is initiated by an aggressor in approximately 11–40% of cases (Arnold and Whiten, 2001; Kutsukake and Castles, 2004; Wittig and Boesch, 2005). The initiation of reconciliation by an aggressor seems contradictory because that individual has just attacked the recipient of the conciliatory behavior. Aggressor-initiated reconciliation would be expected to serve a different function from that served by recipient-initiated reconciliation, because the risk involved in reconciliation should be low for an aggressor. Nonetheless, most studies have not distinguished between these events (for an exception, see Wittig and Boesch, 2005), and the factors that predict who initiates a reconciliation remain unclear.

32.6 Toward a further synthesis

As discussed above, an examination of conflict management and resolution helps us to understand the nature of aggression and vice versa. It may also be possible to address unanswered questions by integrating different topics into a single framework. In conclusion, we will briefly discuss one such relevant question: how relationship uncertainty affects the pattern of and decision making regarding aggression and submission. It has been proposed that aggressiveness in chimpanzees may be increased by a fission–fusion compared with a cohesive grouping system (Muller, 2002). Individuals in fission–fusion societies sometimes lose either the opportunity to interact with or the social knowledge about other group members, which may destabilize intragroup social relationships. When the nature of relationships (e.g. dominance) is highly uncertain, it may be necessary for dominants to continuously reinforce a dominance relationship to reduce this uncertainty (Muller, 2002). It remains unclear how relationship certainty decays in the wild as a function of separation time and whether separation intervals predict the occurrence of aggression (see Okamoto *et al.*, 2001, for a discussion of animals in captivity). Relationship uncertainty may also be related to the evolution of exaggerated forms of aggression and submission. The costs of social signals contribute to ensuring their honesty (Zahavi, 1977). Individuals may need to emit costly signals when relationship uncertainty is high, which may explain the costly forms of both aggression and submission in chimpanzees (see above). It would be useful to study the evolution of these costly signals from a theoretical perspective and to investigate how the cost of signals, which can be measured empirically by quantifying their intensity and duration, is determined. Clarification of these issues will increase our understanding of sociality in chimpanzees as well as in other animals.

References

Altmann, S. A. and Altmann, J. (2003). The transformation of behaviour field studies. *Animal Behaviour*, **65**, 413–23.

Arnold, K. and Whiten, A. (2001). Post-conflict behavior of wild chimpanzees (*Pan troglodytes schweinfurthii*) in Budongo Forest, Uganda. *Behaviour*, **138**, 649–90.

Arnold, K., Fraser, O. N., and Aureli, F. (2010). Postconflict reconciliation. In *Primates in Perspective*, 2nd edn., ed. C. J. Campbell, A. Fuentes, K. C. MacKinnon, S. K. Bearder, and R. M. Stumpf. Oxford: Oxford University, pp. 608–25.

Aureli, F. (1992). Post-conflict behaviour among wild long-tailed macaques (*Macaca fascicularis*). *Behavioral Ecology and Sociobiology*, **31**, 329–37.

Aureli, F. and de Waal, F. B. M. (2000). *Natural Conflict Resolution*. Oakland, CA: University of California Press.

Aureli, F., Preston, S. D., and de Waal, F. B. M. (1999). Heart rate responses to social interactions in free-moving rhesus macaques (*Macaca mulatta*): a pilot study. *Journal of Comparative Psychology*, **113**, 59–65.

Boehm, C. (1994). Pacifying interventions at Arnhem Zoo and Gombe. In *Chimpanzee Cultures*, ed. R. W. Wrangham, W. C. McGrew, F. B. M. de Waal and P. G. Heltne. Cambridge, MA: Harvard University Press, pp. 221–6.

Bygott, J. D. (1979). Agonistic behavior, dominance, and social structure in wild chimpanzees of the Gombe National Park. In *The Great Apes*, ed. D. A. Hamburg and E. R. McCown. Menlo Park, CA: Benjamin/Cummings, pp. 405–27.

Cant, M. A. (2011). The role of threats in animal cooperation. *Proceedings of the Royal Society B: Biological Sciences*, **278**, 170–8.

Flack, J. C. and de Waal, F. (2007). Context modulates signal meaning in primate communication. *Proceedings of the National Academy of Sciences of the USA*, **104**, 1581–6.

Foster, M. W., Gilby, I. C., Murray, C. M., *et al.* (2009). Alpha male chimpanzee grooming patterns: implications for dominance "style." *American Journal of Primatology*, **71**, 136–44.

Fraser, O. N. and Aureli, F. (2008). Reconciliation, consolation and postconflict behavioral specificity in chimpanzees. *American Journal of Primatology*, **70**, 1114–23.

Fraser, O. N., Schino, G., and Aureli, F. (2008a). Components of relationship quality in chimpanzees. *Ethology*, **114**, 834–43.

Fraser, O. N., Stahl, D., and Aureli, F. (2008b). Stress reduction through consolation in chimpanzees. *Proceedings of the National Academy of Sciences of the USA*, **105**, 8557–62.

Fraser, O. N., Koski, S. E., Wittig, R. M., and Aureli, F. (2009). Why are bystanders friendly to recipients of aggression. *Communicative and Integrated Biology*, **2**, 285–91.

Fraser, O. N., Stahl, D., and Aureli, F. (2010). The function and determinants of reconciliation in *Pan troglodytes*. *International Journal of Primatology*, **31**, 39–57.

Gilby, I. C., Brent, L. J., Wroblewski, E. E., *et al.* (2013). Fitness benefits of coalitionary aggression in male chimpanzees. *Behavioral Ecology and Sociobiology*, **67**, 373–81.

Goodall, J. (1986). *The Chimpanzees of Gombe: Patterns of Behavior*. Cambridge, MA: Belknap.

Harcourt, A. H. and de Waal, F. B. M. (1992). *Coalitions and Alliances in Humans and Other Animals*. Oxford: Oxford University Press.

Hayaki, H. (1990). Social context of pant-grunting in young chimpanzees. In: *The Chimpanzees of the Mahale Mountains: Sexual and Life History Strategies*, ed. T. Nishida. Tokyo: University of Tokyo Press, pp. 189–206.

Hayaki, H., Huffman, M. A., and Nishida, T. (1989). Dominance among male chimpanzees in the Mahale Mountains National Park, Tanzania: a preliminary study. *Primates*, **30**, 187–97.

Hemelrijk, C. K., Klomberg, T. J., Nooitgedagt, J. H., and van Hooff, J. A. (1991). Side-directed behaviour and recruitment of support in captive chimpanzees. *Behaviour*, **118**, 89–102.

Kahlenberg, S. M., Emery Thompson, M., Muller, M. N., and Wrangham, R. W. (2008a). Immigration costs for female chimpanzees and male protection as an immigrant counterstrategy to intrasexual aggression. *Animal Behaviour*, **76**, 1497–509.

Kahlenberg, S. M., Emery Thompson, M., and Wrangham, R. W. (2008b). Female competition over core areas in *Pan troglodytes schweinfurthii*, Kibale National Park, Uganda. *International Journal of Primatology*, **29**, 931–47.

Kawanaka, K. (1989). Age differences in social interactions of young males in a chimpanzee unit-group at the Mahale Mountains National Park, Tanzania. *Primates*, **30**, 285–305.

Koski, S. E. and Sterck, E. H. M. (2007). Triadic post-conflict affiliation in captive chimpanzees; does consolation console? *Animal Behaviour* **73**, 133–42.

Koski, S. E. and Sterck, E. H. M. (2009). Post-conflict third-party affiliation in chimpanzees: what's in it for the third party? *American Journal of Primatology*, **71**, 409–18.

Koski, S. E., Koops, K., and Sterck, E. H. (2007a). Reconciliation, relationship quality, and postconflict anxiety: testing the integrated hypothesis in captive chimpanzees. *American Journal of Primatology*, **69**, 158–72.

Koski, S. E., de Vries, H., van den Tweel, S. W., and Sterck, E. H. (2007b). What to do after a fight? The determinants and inter-dependency of post-conflict interactions in chimpanzees. *Behaviour*, **144**, 529–56.

Kutsukake, N. (2003). Assessing relationship quality and social anxiety among wild chimpanzees using self-directed behavior. *Behaviour*, **140**, 1153–71.

Kutsukake, N. (2006). The context and quality of social relationships affect vigilance behavior in wild chimpanzees. *Ethology*, **112**, 581–91.

Kutsukake, N. (2007). Conspecific influences on vigilance behavior in wild chimpanzees. *International Journal of Primatology*, **28**, 907–18.

Kutsukake, N. (2009). Complexity, dynamics and diversity of sociality in group-living mammals. *Ecological Research*, **24**, 521–31.

Kutsukake, N. and Castles, D. L. (2004). Reconciliation and post-conflict third-party affiliation among wild chimpanzees at the Mahale Mountains, Tanzania. *Primates*, **45**, 157–65.

Kutsukake, N. and Clutton-Brock, T. H. (2008). Do meerkats engage in conflict management following aggression? Reconciliation, submission and avoidance. *Animal Behaviour*, **75**, 1441–53.

Laporte, M. N. and Zuberbühler, K. (2010). Vocal greeting behavior in wild chimpanzee females. *Animal Behaviour*, **80**, 467–73.

Matsumura, S. and Hayden, T. J. (2006). When should signals of submission be given?–A game theory model. *Journal of Theoretical Biology*, **240**, 425–33.

Miller, G. F. (1997). Protean primates: the evolution of adaptive unpredictability in competition and courtship. In *Machiavellian Intelligence II: Extensions and Evaluations*, ed. A. Whiten and R. W. Byrne. Cambridge: Cambridge University Press, pp. 312–40.

Miller, J. A., Pusey, A. E., Gilby, I. C., *et al.* (2014). Competing for space: female chimpanzees are more aggressive inside than outside their core areas. *Animal Behaviour*, **87**, 147–52.

Muller, M. N. (2002). Agonistic relations among Kanyawara chimpanzees. In *Behavioural Diversity in Chimpanzees and Bonobos*, ed. C. Boesch, G. Hohmann, and L. F. Marchant. Cambridge: Cambridge University Press, pp. 112–24.

Muller, M. N. and Mitani, J. C. (2005). Conflict and cooperation in wild chimpanzees. *Advances in the Study of Behavior*, **35**, 275–331.

Muller, M. N. and Wrangham, R. W. (2004). Dominance, cortisol and stress in wild chimpanzees (*Pan troglodytes schweinfurthii*). *Behavioral Ecology and Sociobiology*, **55**, 332–40.

Muller, M. N., Kahlenberg, S. M., Thompson, M. E., and Wrangham, R. W. (2007). Male coercion and the costs of promiscuous mating for female chimpanzees. *Proceedings of the Royal Society B: Biological Sciences*, **274**, 1009–14.

Muller, M. N., Thompson, M. E., Kahlenberg, S. M., and Wrangham, R. W. (2011). Sexual coercion by male chimpanzees shows that female choice may be more apparent than real. *Behavioral Ecology and Sociobiology*, **65**, 921–33.

Nishida, T. (1970). Social behavior and relationship among wild chimpanzees of the Mahali Mountains. *Primates*, **11**, 47–87.

Nishida, T. (1983). Alpha status and agonistic alliance in wild chimpanzees (*Pan troglodytes schweinfurthii*). *Primates*, **24**, 318–36.

Nishida, T. (1989). Social interactions between resident and immigrant female chimpanzees. In *Understanding Chimpanzees*, ed. P. G. Heltne and L. A. Marquardt. Cambridge MA: Harvard University Press, pp. 68–89.

Nishida, T. (2003). Harassment of mature female chimpanzees by young males in the Mahale

Mountains. *International Journal of Primatology*, **24**, 503–14.

Nishida, T. and Hosaka, K. (1996). Coalition strategies among adult male chimpanzees of the Mahale Mountains, Tanzania. In *Great Ape Societies*, ed. W. C. McGrew, L. F. Marchant, and T. Nishida. Cambridge: Cambridge University Press, pp. 114–34.

Nishida, T., Hosaka, K., Nakamura, M., and Hamai, M. (1995). A within-group gang attack on a young adult male chimpanzee: ostracism of an ill-mannered member? *Primates*, **36**, 207–11.

Nishida, T., Kano, T., Goodall, J., McGrew, W. C., and Nakamura, M. (1999). Ethogram and ethnography of Mahale chimpanzees. *Anthropological Science*, **107**, 141–88.

Nishida, T., Zamma, K., Matsusaka, T., Inaba, A., and McGrew, W. C. (2010). *Chimpanzee Behavior in the Wild: An Audio-Visual Encyclopedia*. Tokyo: Springer.

Noë, R. and de Waal, F. B. M. (1980). Types of dominance in a chimpanzee colony. *Folia Primatologica*, **34**, 90–110.

Okamoto, K., Agetsuma, N., and Kojima, S. (2001). Greeting behavior during party encounters in captive chimpanzees. *Primates*, **42**, 161–5.

von Rohr, C. R., Koski, S. E., Burkart, J. M., *et al.* (2012). Impartial third-party interventions in captive chimpanzees: a reflection of community concern. *PLoS ONE*, **7**, e32494.

Romero, T. and de Waal, F. B. M. (2010). Chimpanzee (*Pan troglodytes*) consolation: Third-party identity as a window on possible function. *Journal of Comparative Psychology*, **124**, 278–86.

Romero, T. and de Waal, F. B. M. (2011). Third-party postconflict affiliation of aggressors in chimpanzees. *American Journal of Primatology*, **73**, 397–404.

Romero, T., Castellanos, M. A., and de Waal, F. B. M. (2010). Consolation as possible expression of sympathetic concern among chimpanzees. *Proceedings of the National Academy of Sciences of the USA*, **107**, 12110–15.

Romero, T., Castellanos, M. A., and de Waal, F. B. M. (2011). Post-conflict affiliation by chimpanzees with aggressors: other-oriented versus selfish political strategy. *PLoS ONE*, **6**, e22173.

Sakamaki, T. (2011). Submissive pant-grunt greeting of female chimpanzees in Mahale Mountains National Park, Tanzania. *African Study Monographs*, **32**, 25–41.

Silk, J. B. (2002). Practice random acts of aggression and senseless acts of intimidation: the logic of status contests in social groups. *Evolutionary Anthropology*, **11**, 221–5.

Sommer, V., Denham, A., and Little, K. (2002). Postconflict behaviour of wild Indian langur monkeys: avoidance of opponents but rarely affinity. *Animal Behaviour*, **63**, 637–48.

Stumpf, R. M. and Boesch, C. (2010). Male aggression and sexual coercion in wild West African chimpanzees (*Pan troglodytes verus*). *Animal Behaviour*, **79**, 333–42.

de Waal, F. B. M. (1978). Exploitative and familiarity-dependent support strategies in a colony of semi-free-living chimpanzees. *Behaviour*, **66**, 268–312.

de Waal, F. B. M. (1982). *Chimpanzee Politics: Power and Sex among Apes*. Baltimore, MD: Johns Hopkins University Press.

de Waal, F. B. M. (1986). The integration of dominance and social bonding in primates. *Quarterly Review of Biology*, **61**, 459–79.

de Waal, F. B. M. (1996). Conflict as negotiation. In *Great Ape Societies*, ed. W. C. McGrew, L. F. Marchant, and T. Nishida, Cambridge: Cambridge University Press, pp. 159–72.

de Waal, F. B. M. and van Hooff, J. A. (1981). Side-directed communication and agonistic interactions in chimpanzees. *Behaviour*, **77**,164–98.

de Waal, F. B. M. and van Roosemalen, A. (1979). Reconciliation and consolation among chimpanzees. *Behavioral Ecology and Sociobiology*, **5**, 55–66.

Watts, D. P. (1998). Coalitionary mate guarding by male chimpanzees at Ngogo, Kibale National Park, Uganda. *Behavioral Ecology and Sociobiology*, **44**, 43–55.

Watts, D. P. (2006). Conflict resolution in chimpanzees and the valuable-relationships hypothesis. *International Journal of Primatology*, **27**, 1337–64.

Wittig, R. M. (2010). Function and cognitive underpinnings of post-conflict affiliation in wild chimpanzees. In *Chimpanzee Mind*, ed. E. Lonsdorf, S. Ross, and T. Matsuzawa. Chicago, IL: Chicago University Press, pp. 208–19.

Wittig, R. M. and Boesch, C. (2003a). 'Decision-making' in conflicts of wild chimpanzees: an extension of the relational model. *Behavioral Ecology and Sociobiology*, **54**, 491–504.

Wittig, R. M. and Boesch, C. (2003b). The choice of post-conflict interactions in wild chimpanzees (*Pan troglodytes*). *Behaviour*, **140**, 1527–59.

Wittig, R. M. and Boesch, C. (2003c). Food competition and linear dominance hierarchy among female chimpanzees of the Taï National Park. *International Journal of Primatology*, **24**, 847–67.

Wittig, R. M. and Boesch, C. (2005). How to repair relationships: reconciliation in wild chimpanzees. *Ethology*, **111**, 736–63.

Zahavi, A. (1977). The cost of honesty: further remarks on the handicap principle. *Journal of Theoretical Biology*, **67**, 603–5.

33 Greetings and dominance

TETSUYA SAKAMAKI AND HITOSHIGE HAYAKI

33.1 Introduction

Chimpanzees routinely exchange "greetings" as part of their social lives. A greeting may be defined as a component of social interactions after some period of separation (van Lawick-Goodall, 1968). Chimpanzee social unit is a multi-male and multi-female unit-group (Chapter 6). Group members rarely congregate in a single location, but instead associate in temporary subgroups or parties that vary in size, composition, and duration (Nishida, 1968; Goodall, 1986; Chapter 9). Because of this "fission–fusion" grouping behavior, meetings often occur between members of the same unit-group after a period of separation. In such reunions, some chimpanzees may show their physical strength through charging displays, shaking vines while standing upright, rolling rocks, and running with their hairs erect (Chapter 31). By contrast, others may offer "submissive greetings" (de Waal, 1982) to the dominant individuals by bending down and uttering appropriate vocalizations, such as pant-grunts or a series of soft or loud grunts. The exchange of these aggressive and submissive displays may involve many individuals in a state of increasing excitement (Nishida, 1970; Sakamaki, 2009). During these times of excitement, some individuals may seek body contact with others by reaching out, touching, patting, bringing their faces together, kissing, embracing, mounting, copulating, and so on. As the excitement fades, they may start to groom each other and sometimes form large grooming clusters (Bauer, 1979; Nakamura, 2003).

Chimpanzee greetings may have developed as a result of flexible fission–fusion grouping behavior (van Lawick-Goodall, 1968; Nishida, 1970; Hayaki, 1990). When individuals meet after a period of separation, they may need to re-establish mutual tolerance, confirm their past familiarity, and fill in the psychological blank of absence. In such situations, clear-cut behaviors like pant-grunting and charging displays may be used to quickly confirm recognition between individuals with established dominance relationships (van Lawick-Goodall, 1968; Hayaki *et al.*, 1989). Meetings may occur after a week, or even after a month or more of separation. Considering chimpanzee's xenophobia and hostility toward unfamiliar individuals (Bygott, 1979; Wrangham, 1999), a reunion after a relatively long period of separation might be a stressful and risky time, even among members of the same group. Ritualized submissive behaviors like pant-grunting may have developed as a response to this hostility and aggressiveness.

Greeting behaviors are related to the social structure and bonding of a unit-group. Repeated, unilateral performance of clearly submissive or aggressive displays may maintain social bonds and structure of chimpanzee society in terms of dominance relationships, although joint displays with two or more individuals are also possible. On the other hand, some interactions such as social grooming (Chapter 34) and social play (Chapter 36), in which individuals may alternate roles (e.g. groomer and groomee), or even perform the same roles simultaneously, may strengthen equal or equipotential social bonds. Both unidirectional and mutual/reciprocal interactions are important factors in chimpanzee social structure and bonding.

Mahale Chimpanzees: 50 Years of Research, ed. M. Nakamura, K. Hosaka, N. Itoh and K. Zamma. Published by Cambridge University Press.

One unique chimpanzee behavior is the pant-grunt vocalization, which may grade into subtle pants, louder pant-barks, and shrieking pant-screams; together, these vocalizations are reliable indicators of dominance relationships (Bygott, 1979; de Waal, 1982; Goodall, 1986; Hayaki *et al.*, 1989). Hereafter, we will use the general term "pant-grunt" to denote all related vocalizations, including pants, pant-grunts, pant-barks, and pant-screams. Dominance relationships measured by the target of pant-grunts ("formal dominance"; de Waal, 1982) predict aggressive behaviors such as physical attacks and charges, but not necessarily priority of access to attractive foods or outcomes of non-agonistic competitive interactions (Bygott, 1974; Noë *et al.*, 1980). This is an intriguing feature when considering the significance of formal dominance and pant-grunt greetings in chimpanzee society.

This chapter primarily focuses on dominance rank as reflected by submissive pant-grunt greetings (Figures 33.1–33.3). First, we review the social functions of pant-grunts. Some functions are interpreted in the context of behavioral sequences and social relationships, such as submission, appeasement, expression of respect, and tolerance assessment. Another feature of pant-grunts is their role as behavioral components linking aggressive behavior and loud vocalization, which can readily lead to mass excitement among many individuals. We also describe appropriate etiquette for reinforcing bonds within a male cluster. In Section 33.3, we discuss dominance ranks in terms of alpha status, linearity of the dominance hierarchy, reproductive benefits, and hormones. Finally, we highlight intriguing aspects of chimpanzee dominance with respect to the underlying equality that may exist in their sociality.

33.2 Functions of pant-grunt greetings

33.2.1 Submissive signals

Pant-grunts are clearly unidirectional, given only by lower-ranking individuals toward dominants (Bygott, 1979; Noë *et al.*, 1980; Goodall, 1986; Hayaki *et al.*, 1989). In this sense, the pant-grunt is a "submissive greeting" (de Waal, 1982) and is thus the standard index used to measure dominance relationships. The "greeting" here does not include behaviors that occur in reunions of separated parties. Formal dominance, manifested by the pant-grunt, does not vary with social context (e.g. the presence or absence of coalition partners), whereas the directions of

Figure 33.1 An example of pant-grunting: a female puts her face close to a male's. ©T. Sakamaki.

Figure 33.2 An example of pant-grunting: two females approach two seated males. The target of their pant-grunting is clearly one of the two males. ©T. Sakamaki.

Figure 33.3 An example of pant-grunting: a passing female pant-grunts to a recumbent male. ©T. Sakamaki.

threat–retreat interactions can be inverted depending on the situation (de Waal, 1982; Chapter 29). Chimpanzees regularly confirm dominance relationships in an almost ritualized manner through the use of pant-grunts or submissive greetings.

On the other hand, the social variables driving pant-grunts are not well understood. For example, when encountering a higher-ranking male, subordinates do not always perform pant-grunts. Furthermore, pant-grunts actually occur between pairs whose dominance relationships are already apparent due to differences in body size or to other agonistic interactions (Hayaki *et al.*, 1989; Hayaki, 1990; Takahata, 1990; Sakamaki, 2011). These observations suggest that the social function of pant-grunts is more complex than the mere expression of subordination or submission, as described below.

33.2.2 Appeasement signals

Previous studies have proposed other social functions for pant-grunts. One of them is that they may serve as an appeasement signal. Pant-grunts often occur in moments of tension or excitement, sometimes in response to an aggressive display. In such cases, pant-grunts may restrain or deflect the aggression of dominant individuals, suggesting that they might function as a kind of appeasement signal (van Lawick-Goodall, 1968; Nishida, 1970; Marler and Tenaza, 1977). Because of this, pant-grunting occurs not only during reunions but also when subordinates remain in the vicinity of dominants (Hayaki, 1990). This function corresponds to the general observation that dominant males that receive pant-grunts from others are more frequently aggressive (Bygott, 1979; Muller, 2002).

On the other hand, pant-grunts are sometimes directed toward individuals who show no aggressive intent. These spontaneous pant-grunts without prior provocation may occur when the performer anticipates the aggression of dominant individuals. In such cases, the pant-grunter may attempt to appease the dominant and deter their aggression before it occurs (Hayaki, 1990). Aggressive behaviors frequently occur during reunions (Bygott, 1979). Therefore, some individuals may perform pant-grunts when they encounter dominant individuals even if the dominants show no signs of aggression.

33.2.3 Expressions of support

Next, pant-grunts may function as a token of respect and support (de Waal, 1982; Goodall, 1986). De Waal (1982) reported a case of captive chimpanzees in which the alpha male, Yeroen, was deposed, yet for 2 years, he received more frequent pant-grunts from adult females and immature individuals than did the new alpha male. During this period, some females supported the former alpha male against the new alpha, but none supported the new against the old. In this case, pant-grunts may have been an expression of support for a preferred male. This type of expression could appear only during a stage of uncertainty in the alpha status; when male dominance rank, especially alpha male status, is stable, agonistic confrontations among adult males are rare, and the alpha male receives the most pant-grunts from other group members (Sakamaki, 2011). Goodall (1986: 425–6) reported a case in which, as the alpha male, Mike, grew old, another big adult male, Humphrey, did not try to overcome him but instead continued pant-grunting to Mike, apparently out of respect. Goodall regarded this as a good example of the dominance status being maintained by daily greeting habits.

33.2.4 Checking tolerance

According to a study on pant-grunting in young chimpanzees (Hayaki, 1990), vigorous and exaggerated pant-grunts seem to signify excitement and, furthermore, excite dominant males. Thus, young individuals may perform pant-grunts to make their presence known to the receiver and to assess how tolerant the individual is likely to be (Hayaki, 1990). Young individuals make bobbing gestures and sometimes display aggressive behaviors such as slapping the ground, shaking woody vines, and throwing branches while pant-grunting to adult males (Hiraiwa-Hasegawa, 1989; Hayaki, 1990; Sakamaki, 2011). Compared with pant-grunts by adult females toward adult males, immature individuals more often showed aggressive or bobbing gestures while pant-grunting toward adult males (Sakamaki, 2011). Moreover, although adult females also pant-grunted persistently toward adult males, they received less aggressive reactions from adult males than did immature individuals (Sakamaki, 2011). These observations suggest that it is more important for immature individuals than for adult females to make their presence known and to determine the tolerance of the pant-grunt receiver.

33.2.5 Components of noisy mass excitement

Chimpanzees are often involved in increasingly excited social interactions with many individuals, a phenomenon known as "snowballing" (Goodall, 1986). Pant-grunting is a component that links behavioral elements such as aggression and loud calling to the social interactions of mass excitement (van Hooff, 1973). Several studies have described such mass excitement, which sometimes lasts for more than 10 min, as an intriguing feature of chimpanzee societies (Reynolds and Reynolds, 1965; Nishida, 1970; Mori, 1982). De Waal (1992) called social excitement before feeding "celebrations," and pointed out that it immediately established peaceful coexistence for feeding.

Even when chimpanzees split into smaller parties, a relatively large core group usually remains (Nishida, 1968). These core parties often include the alpha and other high-ranking males, along with females and immature individuals. The core party may engage in noisy mass excitement consisting of pant-hoots, drumming, pant-grunts, waa-barks, and so on several times a day. This behavior might inform other, scattered individuals of the whereabouts of the alpha male, who generally receives the most pant-grunts (Nishida, 1968; Sakamaki, 2009). Pant-grunts, sometimes escalating into louder pant-barks and pant-screams, play a role in consolidating dominance and tolerance relationships in face-to-face interactions and in maintaining contact between scattered parties and individuals (Sakamaki, 2009).

33.2.6 Reinforcement of male clusters

Regular pant-grunting among males plays a role in reinforcing adult male clusters or coalition networks. A notable case from Mahale of a gang attack on a young adult male, Jilba, and the ostracism that ensued (Nishida *et al.*, 1995; Nishida, 2012; see Chapter 27 for further details) suggests that dominant males, especially the alpha, demand the proper pant-grunting etiquette from young males. Because males of similar dominance ranks rarely pant-grunt at each other, as discussed below (Section 33.3.2), they may maintain social bonds in other ways, such as by social grooming and close association, or they may instead avoid interaction altogether due to existing rivalries. Frequent and appropriately delivered pant-grunting is especially important for adolescent and young adult males, who are still establishing their positions in the adult male cluster (Hayaki, 1988; Hayaki *et al.*, 1989). This aspect is also related to assessing the tolerance of dominant males.

Greetings might also represent a confirmation of group membership or social identity (e.g. Nettle and Dunbar, 1997). However, chimpanzee greetings do not seem to conform to this idea, though chimpanzees do seem to make a clear distinction between members and non-members of their unit-groups (Nishida, 2012). Nevertheless, there is no evidence of group-specific greetings akin to the dialects of human language.

33.2.7 Multiple functions of pant-grunt greetings

Pant-grunting is a strongly directional behavior. As a general rule, where individual *A* consistently attacks individual *B*, *B* consistently directs pant-grunting at *A*. This is described as a dominance–subordination relationship, and pant-grunting is a submissive signal in this context. Moreover, although the directions of threat–retreat interactions can be inverted depending on the situation, such as in the presence or absence of coalition partners, the direction of pant-grunting does not vary (de Waal, 1982; Chapter 29). Therefore, pant-grunt greetings as ritualized shows of submission characterize chimpanzee dominance relationships.

In this section, we have reviewed the various social functions of pant-grunt greetings. Some functions are interpreted in the context of behavioral sequences, and others in the context of social relationships. These types of pant-grunting are not mutually exclusive, and some are closely related. For example, when a female chimpanzee pant-grunts to an adult male, we may infer that she is reconfirming her submissive relationship to him and, at the same time, appeasing him to restrain his aggression. Furthermore, if the rank order among males is unstable, we may infer support in her pant-grunting. Thus, several explanations of pant-grunting may be compatible. Then again, functions may be contradictory in other cases. For example, when a young male approaches an adult male while pant-grunting, he may be announcing himself to the receiver. However, this persistent and exaggerated pant-grunting is contrary to the appeasement signal, as it may lead to an aggressive response.

In states of mass excitement, aggressive charging displays may lead to submissive pant-grunting and vice versa, and many individuals make various loud calls including waa-barks and screams. Pant-grunting could be an important component in mass excitement, linking aggressive and submissive behaviors and resulting in a continuous series of loud calls. Mass excitement itself is a remarkable phenomenon that helps to maintain the social structure.

33.3 Dominance rank

33.3.1 Recipients of pant-grunting

The alpha male is conspicuous in the dominance order as characterized by submissive pant-grunt greetings. For example, at Mahale, the alpha male received 45% of all the pant-grunts among 10 adult, 9 adolescent, and 8 immature males in one study (Hayaki *et al.*, 1989). In other studies, the alpha male received 47% of all pant-grunts from 8 adult males and 20 adult females (Sakamaki, 2011) and 84% of those among 8 adult males alone (Sakamaki, 2005). Chimpanzees at other study sites also show the same tendency. For example, in the Kanyawara group of Kibale, Uganda, 83% of the pant-grunts among 11 adult males were directed to the alpha male (Muller and Wrangham, 2004a). When most pant-grunts are directed at the alpha male, his status is clearly stable, and he usually holds conspicuous social power. One study of the Sonso group at Budongo, Uganda, showed that when the dominance relationships of two top-ranking males were ambiguous, both received many pant-grunts from others (Newton-Fisher, 2004). This study took place during a time of transition from the former alpha male to the new one. Even after the alpha status has changed to the new male, however, the former alpha male may still receive the most pant-grunts out of habit and respect, as in the cases of Yeroen (de Waal, 1982) and Mike (Goodall, 1986) discussed above.

How does the alpha male receive the most pant-grunts? When two or more males coexist, the precise social order is related to the pant-grunts performed by others because the males do not have equal social status. For example, when females that have previously been associated with the alpha male encounter other adult males, the females rarely pant-grunt toward them, even though they pant-grunt to other adult males in other situations (Sakamaki, 2011). In other words, the presence of the alpha male inhibits females' pant-grunting to other dominant males (Laporte and Zuberbühler, 2010). Indeed, the choice to pant-grunt may be even more complex. For example, even if females encounter the alpha male while others are interacting with him in an excited state, some females may remain bystanders (Sakamaki, 2011). Therefore, whether a subordinate performs pant-grunts toward a dominant also depends on the current social situation, though the direction of pant-grunting does not vary with social context.

These examples involve females pant-grunting toward males, and females are formally subordinate to all adult males (Hayaki *et al.*, 1989; Hayaki, 1990; Takahata, 1990). We need to assume greater complexity in cases of male pant-grunting, as these situations involve both dominance relationships and political strategies (de Waal, 1982; Nishida, 1983; Chapter 27). Chimpanzees judge the current balance of social power by watching interactions between existing and newly arrived individuals.

33.3.2 Linear hierarchies in dominance ranking

It is not always possible to determine a precise linear hierarchy in the dominance ranks of a chimpanzee society. Bygott (1974) concluded that there were four basic levels of this hierarchy (the alpha, high, middle, and low ranks) among 15 males by analysing patterns of agonistic relationships. After his study, the focal group split into two separate groups, and the number of males per group decreased. At that time, there was a clear linear hierarchy in their dominance ranking (Riss and Goodall, 1977). The eight adult males of the Mahale M group, for example, were determined to have three levels (the alpha, three high-ranking, and four low-ranking males) based on 11 months of data beginning in October 1999, including 505 pant-grunts (Sakamaki, 2005). In the subsequent period from September 2000, dominance rank based on the direction of pant-grunts revealed a precise linear hierarchy (Kutsukake, 2003). During both periods, the status of the alpha male was stable. The reason for this difference in the number of dominance levels is not clear.

Adolescent males first begin to acquire dominance over females, and later over other males (Goodall, 1986; Hayaki *et al.*, 1989; Pusey, 1990). Adult males are usually dominant over adolescents, but adolescent males do not necessarily establish dominance relationships with one another (Hayaki *et al.*, 1989). They may try to keep their dominance relationships ambiguous by not showing submission or aggression against others of their age class. On the other hand, in the extraordinarily large Ngogo group at Kibale, 17 adolescent males were ranked relative to one another in a linear hierarchy (although some relationships remained unclear), possibly because of competition for access to resources (Sherrow, 2012).

Females are formally subordinate to all adult males at Mahale and Gombe, Tanzania (Bygott, 1979; Goodall, 1986; Hayaki *et al.*, 1989; Hayaki, 1990; Takahata, 1990), although in exceptional cases, interactions in the opposite direction can occur (Sakamaki, 2005). On the other hand, adult females sometimes appeared to have an influence on the dominance ranking of adult males (de Waal, 1982; Boesch and Boesch-Achermann, 2000; Wittig and Boesch, 2003). How female–female coalitions and the female gregariousness are related to dominance relationships between males and females is an intriguing topic for future study (Sakamaki *et al.*, 2001; Newton-Fisher, 2006; Chapters 28, 29).

33.3.3 Dominance ranks among females

Linear hierarchies also exist to some extent in the dominance rankings of females, but they are usually more ambiguous than those among males because agonistic interactions are rare (Mahale: Nishida, 1989; Gombe: Pusey *et al.*, 1997; Budongo: Newton-Fisher, 2006; Kanyawara: Emery Thompson *et al.*, 2007; see also Chapter 28). On the other hand, observations at Taï, Côte d'Ivoire, showed a clear linear hierarchy in female dominance rank as determined by the direction of submissive greetings (Wittig and Boesch, 2003). Possible reasons for this difference include food competition, predation risk, and observation time. However, it is still unclear whether the linear hierarchy in Taï females is unique among western chimpanzees.

Chimpanzees sometimes pant-grunt spontaneously without prior provocation by dominants. De Waal (1982, 1986) found that such pant-grunts occur more frequently between females than between males. In other words, the pant-grunts of males were more closely associated with intimidation, whereas female pant-grunts occurred more spontaneously. He argued that acceptance of dominance is more important than proving dominance among females.

33.3.4 Reproductive benefit of dominance to males

Here, we briefly review the direct reproductive benefits of male dominance (see also Chapter 27), although there are many other potential explanations that are indirectly related to reproductive benefit, such as the costs of seeking social status, preferential access to prime foods, and kin-selected benefits such as the effect of a mother's status on her son's rank. The priority of access model (Altmann, 1962) proposes that dominance rank and reproductive success should be positively correlated because dominant males enjoy preferential access to females in estrus. Previous studies in chimpanzees support this hypothesis with respect to mating success. In other words, dominant males usually monopolize attractive parous females in estrus during the most fertile period in the menstrual cycle (review: Muller and Mitani, 2005: 283–7).

Recent genetic studies have revealed the paternity of chimpanzees, supporting the priority of access model. At Mahale, two alpha males sired 45.5% of all offspring analysed (Inoue *et al.*, 2008; Chapter 35). Non-alpha males also sired offspring, but their reproductive success was relatively low. Studies of paternity at other sites reached similar conclusions (see Chapter 35).

33.3.5 Hormones related to male dominance

A few studies at Kibale reported hormonal dynamics related to male dominance ranking. Muller and Wrangham (2004a) showed that the testosterone level of 11 males was associated with aggression but not with sexual behavior. Another study showed that male dominance rank was positively correlated with cortisol level and that cortisol level was also positively correlated with the frequency of male aggression (Muller and Wrangham, 2004b). The relationship between cortisol and rank may be driven by energetic, rather than psychosocial, factors. More information on hormonal studies of chimpanzees is given in Chapter 44.

33.3.6 Reproductive benefit of dominance to females

Parental investment is heavily biased toward females. As sexual selection theory predicts, male chimpanzees are more frequently aggressive than females, which is related to competition over dominance rank, whereas most female aggression is related to competition over food or defense of offspring (Bygott, 1979; Goodall, 1986; Nishida, 1989; Muller, 2002; Wittig and Boesch, 2003). Although dominance rank among females is still ambiguous, two studies from Gombe suggest that competition among females is significant. First, rank has significant effects on females' reproduction through increased infant survivorship and decreased interbirth interval (Pusey *et al.*, 1997). Second, dominant females are heavier and show less variation in weight across seasons, suggesting better access to prime foods (Pusey *et al.*, 2005). In these studies, direct interactions with which to evaluate the dominance relationship among females were rare, and dominance was ranked from observations over 2 years. Therefore, the benefit of dominance to females might be related to scramble competition and may not be the result of direct aggression. A study at Kibale revealed that food quality in the core area was associated with variance in reproductive success among females although dominance rank was not discussed due to infrequent interactions (Emery Thompson *et al.*, 2007).

33.4 Equality and dominance

33.4.1 Equality and inequality principles

One of the most important objectives for early Japanese primatologists was to understand society and sociality in nonhuman primates and, ultimately, relate these to human society and sociality. On one hand, comparative studies were conducted on social structure in nonhuman primates; on the other, studies of behaviors and social interactions that contribute to maintaining society, social structure, and the coexistence of individuals were also popular. Building on the coexistence principles of primate and human society discussed by Itani (1986, 1988), the origins of equality in human society were explored based on knowledge of dominance ranking in nonhuman primates.

With Itani's ideas in mind, we can highlight some intriguing aspects of dominance rank in chimpanzee societies. His discussion described two contrasting principles: the equality principle and the inequality principle. The equality principle may be divided into *original equality* and *conditional equality*. First, original equality is assumed in the pair-based social units that can be found in various clades of primates. In this type of society, a particular male and female tolerate one another and form a stable social unit. The two are equal, and their sexual dimorphism remains minimal. In more primitive cases, elemental societies, such as those of solitary nocturnal prosimians, are assumed. In this type of society, individuals are anonymous and essentially avoid each other except at mating time. There are no stable social units in elemental societies.

By contrast, societies of larger groups with female philopatry, such as those of Japanese macaques, are based on the inequality principle. Under the inequality principle, individuals are differentiated within stable social units: strong–weak, superior–inferior, dominant–subordinate, and so on. Over time, stable dominance relationships emerge. This type of society is characterized by strict dominant–subordinate interactions. What is important is that coexistence according to the inequality principle is supported by the self-restraint of subordinates. For example, when a piece of prime food is set between a dominant and a subordinate, the dominant acquires the food as the result of self-restraint by the subordinate. The coexistence of individuals within a group is founded on both the inequality and equality principles. In a troop of Japanese macaques, for example, the inequality principle is apparently fundamental, while the equality principle moderates it.

By extension, the equality principle, or conditional equality, is founded on the negation of the inequality principle (Itani, 1986, 1988). It functions in a society based on the inequality principle, but where that principle has been suspended. The equality principle is typically observed in social play and food sharing. In social play, such as chasing and wrestling, a dominant and subordinate may play together and may even exchange roles. When an adult plays with a juvenile, "self-handicapping" may occur; that is, the adult restrains its strength to balance that of the juvenile (Hayaki, 1983, 1985). In these interactions, participants act as if the distinction between them, such as dominant–subordinate or strong–weak, does not exist. Hayaki (1985) argued that each behavioral component is not play in itself but that play is instead the context that enables such behavior to be interpreted as playful (see also Chapter 36). Itani (1988) termed this a "fictitious world." Yet individuals engaging in social play are indeed enjoying it as a real entity. In food sharing among chimpanzees (Chapter 20), subordinates may beg a dominant for prime foods like meat. In this case, the subordinate beggars are unrestrained, but the dominant possessor restrains himself when he allows them to take his food. In contrast to the inequality principle, food sharing among chimpanzees is supported by the self-restraint of the dominant.

33.4.2 The equality principle in chimpanzee dominance

Dominance ranking in chimpanzee societies is different from that in societies with female philopatry, such as Japanese macaques, which are based on the inequality principle and in which dominance relationships are strict. In chimpanzee society, dominance relationships are realized by endless displays of domination and self-restraint by dominants (Kuroda, 2010). Subordinates also endlessly perform submissive acts, but subordination does not always constrain their behavior. In dominant–subordinate and other social interactions, equality may serve as a common base. Its nature is related to the individuality of chimpanzees; in other words, each chimpanzee in a fission–fusion society can live independently in its ecological and social environment (Kuroda, 2010).

As part of our discussion of the equality principle, it may be useful to point out features of the formal dominance expressed by pant-grunt greetings. First, though dominants ceaselessly display their dominance, this is not enough to ensure their status: they must also receive submissive responses from others. Chimpanzees perform pant-grunts not only as responses to aggression but also spontaneously, and dominant status is realized through the submissive acts of subordinates. As de Waal (1982) argued, it is important not only to prove dominance but also to assure that it is accepted by others. Thus, both dominants and subordinates hold power in determining dominance relationships. This fact is the basis of chimpanzees' behavioral flexibility, which could not emerge in a society supported by strict dominant–subordinate relationships.

Second, formal dominance in chimpanzees is not connected directly with priority in accessing prime resources. For example, a female chimpanzee may approach and pant-grunt toward an adult male who is feeding, and then displace him and eat in his place. Likewise, subordinates sometimes beg for prime food from a dominant. These situations bear some similarity to social play; that is, formal dominance is temporarily suspended, resulting in a conditional equality. Similarly, exaggerated and persistent pant-grunt greetings seem to be performed merely for the sake of the greeting interaction itself – though this does not deny the reproductive or biological benefits of dominance rank. Nevertheless, this could assure behavioral flexibility, unconstrained by strict dominance relationships. Without such flexibility, chimpanzees could not share food. Flexibility may also aid chimpanzees in developing social recognition, sympathy, and self-objectification (Itani, 1988).

Third, a social structure characterized by formal dominance and manifested through pant-grunt greetings may allow chimpanzees to form extraordinarily large unit-groups, such as the Ngogo group at Kibale, which comprises over 150 individuals. In formal dominance, the alpha male has a special position; that is, he receives the most pant-grunts from almost every other individual. Observations at Mahale showed that chimpanzees ran to pant-grunt toward the alpha male on a daily basis whenever they met him (Sakamaki, 2009, 2011). On the other hand, they interacted differently with other group members, engaging in short-range interactions with many, and merely passing others by altogether. Moreover, low-ranking males did not interact directly with all the individuals in direct visual contact (Sakamaki, 2009). Chimpanzees interact directly with the alpha male but not with all others in visual range, yet still establish mutual tolerance and maintain unit-group membership. This may allow the formation of large unit-groups; if chimpanzees needed to interact intimately and regularly with every group member, the upper limit of group size would be smaller.

The frequency and type of interactions is another key to tolerance and coexistence among members in a society. Bonobos (*Pan paniscus*), who also live in a fission–fusion society, may not form unit-groups as large as those of chimpanzees because they lack equivalent pant-grunt repertoires. Because of this, there is no conspicuous center to their social interactions, such as the alpha male of chimpanzees or any other individual who becomes the center of the group's attention. Social structure, or the pattern of social interactions and the resulting social relationships, is closely related to the form of social interactions: unidirectional, symmetrical, or reciprocal. Bonobos have symmetrical repertoires of interactions, such as genito-genital rubbing, and rump–rump contact (Kano, 1980; Kuroda, 1980), and bonobo society is said to be more egalitarian (de Waal, 1995). Pant-grunt greetings and formal dominance, which represent conspicuous differences between chimpanzees and bonobos (Nishida and Hiraiwa-Hasegawa, 1987), elucidate aspects of equality and dominance in the evolution of human societies.

References

Altmann, S. (1962). A field study of the sociobiology of rhesus monkeys, *Macaca mulatta*. *Annals of the New York Academy of Science*, **102**, 338–435.

Bauer, H. R. (1979). Agonistic and grooming behavior in the reunion context of Gombe Stream chimpanzees. In *The Great Apes*, ed. D. A. Hamburg and E. R. McCown. Menlo Park, CA: Benjamin/Cummings, pp. 395–403.

Boesch, C. and Boesch-Achermann, H. (2000). *The Chimpanzees of the Taï Forest:*

Behavioural Ecology and Evolution. Oxford: Oxford University Press.

Bygott, J. D. (1974). *Agonistic Behavior and Dominance in Wild Chimpanzees*. PhD thesis. Cambridge: University of Cambridge.

Bygott, J. D. (1979). Agonistic behavior, dominance, and social structure in wild chimpanzees of the Gombe National Park. In *The Great Apes*, ed. D. A. Hamburg and E. R. McCown, Menlo Park, CA: Benjamin/Cummings, pp. 405–27.

Emery Thompson, M., Kahlenberg, S. M., Gilby, I. C., and Wrangham, R. W. (2007). Core area quality is associated with variance in reproductive success among female chimpanzees at Kibale National Park. *Animal Behaviour*, **73**, 501–12.

Goodall, J. (1986). *The Chimpanzees of Gombe: Patterns of Behavior*. Cambridge, MA: Belknap.

Hayaki, H. (1983). Social interactions of juvenile Japanese monkeys on Koshima Islet. *Primates*, **24**, 139–53.

Hayaki, H. (1985). Social play of juvenile and adolescent chimpanzees in the Mahale Mountains National Park, Tanzania. *Primates*, **26**, 343–60.

Hayaki, H. (1988). Association partners of young chimpanzees in the Mahale Mountains National Park, Tanzania. *Primates*, **29**, 147–61.

Hayaki, H. (1990). Social context of pant-grunting in young chimpanzees. In *The Chimpanzees of the Mahale Mountains: Sexual and Life History Strategies*, ed. T. Nishida. Tokyo: University of Tokyo Press. pp. 189–206.

Hayaki, H., Huffman, M. A., and Nishida, T. (1989). Dominance among male chimpanzees in the Mahale Mountains National Park, Tanzania: a preliminary study. *Primates*, **30**, 187–97.

Hiraiwa-Hasegawa, M. (1989). Sex differences in the behavioral development of chimpanzees at Mahale. In *Understanding Chimpanzees*, ed. P. G. Heltne and L. Marquardt, Cambridge, MA: Harvard University Press, pp. 104–15.

van Hooff, J. A. R. A. M. (1973). A structure analysis of the social behaviour of a semi-captive group of chimpanzees. In *Social Communication and Movement*, ed. M. von Cranach and I. Vine, London: Academic Press, pp. 75–162.

Inoue, E., Inoue-Murayama, M., Vigilant, L., Takenaka, O., and Nishida, T. (2008). Relatedness in wild chimpanzees; influence of paternity, male philopatry, and demographic factors. *American Journal of Physical Anthropology*, **137**, 256–62.

Itani, J. (1986). [Discourse on the origin of equality.] In [*Anthropology of Natural Societies*], ed. J. Itani and J. Tanaka, Kyoto: Academia Shuppan-kai, pp. 349–89. In Japanese.

Itani, J. (1988). The origin of human equality. In *Social Fabrics of the Mind*, ed. M. R. A. Chance. Hove, UK: Lawrence Erlbaum Associates Ltd, pp. 137–56.

Kano, T. (1980). Social behavior of wild pygmy chimpanzees (*Pan paniscus*) of Wamba: a preliminary report. *Journal of Human Evolution*, **9**, 243–60.

Kuroda, S. (1980). Social behavior of the pygmy chimpanzees. *Primates*, **21**, 181–97.

Kuroda, S. (2010). [Descent of equality in the "Discourse on the origin of equality."] *Ningen Bunka*, **27**, 3–6. In Japanese.

Kutsukake, N. (2003). [*Influences of Relationship Quality on Social Stress and Conflict Resolution in Wild Chimpanzees*.] Doctoral Dissertation. Tokyo: The University of Tokyo. In Japanese.

Laporte, M. N. C. and Zuberbühler, K. (2010). Vocal greeting behavior in wild chimpanzee females. *Animal Behaviour*, **80**, 467–73.

van Lawick-Goodall, J. (1968). The behavior of free-living chimpanzees in the Gombe Stream Reserve. *Animal Behaviour Monographs*, **1**, 161–311.

Marler, P. and Tenaza, R. (1977). Signaling behavior of apes with special reference to vocalization. In *How Animals Communicate*, ed. T. A. Sebeok. London: Indiana University Press, pp. 965–1033.

Mori, A. (1982). An ethological study on chimpanzees at the artificial feeding place in the Mahale Mountains, Tanzania, with special reference to the booming situation. *Primates*, **23**, 45–65.

Muller, M. N. (2002). Agonistic relations among Kanyawara chimpanzees. In *Behavioural Diversity in Chimpanzees and Bonobos*. ed. C. Boesch, G. Hohmann, and L. F. Marchant, Cambridge: Cambridge University Press, pp. 112–24.

Muller, M. N. and Mitani, J. C. (2005). Conflict and cooperation in wild chimpanzees. *Advances in the Study of Behavior*, **35**, 275–331.

Muller, M. N. and Wrangham, R. W. (2004a). Dominance, aggression and testosterone in wild chimpanzees: a test of the "challenge hypothesis." *Animal Behaviour*, **67**, 113–23.

Muller, M. N. and Wrangham, R. W. (2004b). Dominance, cortisol and stress in wild chimpanzees (*Pan troglodytes schweinfurthii*). *Behavioral Ecology and Sociobiology*, **55**, 332–40.

Nakamura, M. (2003). "Gathering" of social grooming among wild chimpanzees: implications for evolution of sociality. *Journal of Human Evolution*, **44**, 59–71.

Nettle, D. and Dunbar, R. I. M. (1997). Social markers and the evolution of reciprocal exchange. *Current Anthropology*, **38**, 93–9.

Newton-Fisher, N. E. (2004). Hierarchy and social status in Budongo chimpanzees. *Primates*, **45**, 81–7.

Newton-Fisher, N. E. (2006). Female coalitions against male aggression in wild chimpanzees of the Budongo Forest. *International Journal of Primatology*, **27**, 1589–99.

Nishida, T. (1968). The social group of wild chimpanzees in the Mahali Mountains. *Primates*, **9**, 167–224.

Nishida, T. (1970). Social behavior and relationship among wild chimpanzees of the Mahali Mountains. *Primates*, **11**, 47–87.

Nishida, T. (1983). Alpha status and agonistic alliance in wild chimpanzees (*Pan troglodytes schweinfurthii*). *Primates*, **24**, 318–36.

Nishida, T. (1989). Social interactions between resident and immigrant female chimpanzees. In *Understanding Chimpanzees*, ed. P. G. Heltne and L. A. Marquardt. Cambridge, MA: Harvard University Press, pp. 68–89.

Nishida, T. (2012). *Chimpanzees of the Lakeshore: Natural History and Culture at Mahale*. Cambridge: Cambridge University Press.

Nishida, T. and Hiraiwa-Hasegawa, M. (1987). Chimpanzees and bonobos: cooperative relationships among males. In *Primate Societies*, ed. B. B. Smuts, D. L. Cheney, R. M. Seyfarth, R. W. Wrangham, and T. T. Struhsaker. Chicago: University of Chicago Press, pp. 165–77.

Nishida, T., Hosaka, K., Nakamura, M., and Hamai, M. (1995). A within-group gang attack on a young adult male chimpanzee: ostracism of an ill-mannered member? *Primates*, **36**, 207–11.

Noë, R., de Waal, F. B. M., and van Hoof J. A. R. A. M. (1980). Types of dominance in a chimpanzee colony. *Folia Primatologica*, **34**, 90–110.

Pusey, A. E. (1990). Behavioural changes at adolescence in chimpanzees. *Behaviour*, **115**, 203–46.

Pusey, A., Williams, J., and Goodall, J. (1997). The influence of dominance rank on the reproductive success of female chimpanzees. *Science*, **277**, 828–31.

Pusey, A. E., Oehlert, G. W., Williams, J. M., and Goodall, J. (2005). Influence of ecological and social factors on body mass of wild chimpanzees. *International Journal of Primatology*, **26**, 3–31.

Reynolds, V. and Reynolds, F. (1965). Chimpanzees of the Budongo Forest, Uganda. In *Primate Behavior: Field Studies of Monkeys and Apes*, ed. I. DeVore. New York: Holt, Rinehart and Winston, pp. 368–424.

Riss, D. and Goodall, J. (1977). The recent rise to the alpha-rank in a population of free-living chimpanzees. *Folia Primatologica*, **27**, 134–51.

Sakamaki, T. (2005). [*Studies on Submissive Vocalization, Pant-Grunt, in Wild Chimpanzees of the Mahale Mountains National Park*]. Doctoral Dissertation. Kyoto: Kyoto University. In Japanese.

Sakamaki, T. (2009). Group unity of chimpanzees elucidated by comparison of sex differences in short-range interactions in Mahale Mountains National Park, Tanzania. *Primates*, **50**, 321–32.

Sakamaki, T. (2011). Submissive pant-grunt greeting of female chimpanzees in Mahale Mountains National Park, Tanzania. *African Study Monographs*, **32**, 25–41.

Sakamaki, T., Itoh, N., and Nishida, T. (2001). An attempted within-group infanticide in wild chimpanzees. *Primates*, **42**, 359–66.

Sherrow, H. M. (2012). Adolescent male chimpanzees at Ngogo, Kibale National Park, Uganda have decided dominance relationships. *Folia Primatologica*, **83**, 67–75.

Takahata, Y. (1990). Adult males' social relations with adult females. In *The Chimpanzees of the Mahale Mountains: Sexual and Life History Strategies*, ed. T. Nishida. Tokyo: University of Tokyo Press. pp. 133–48.

de Waal, F. B. M. (1982). *Chimpanzee Politics: Power and Sex among Apes*. Baltimore, MD: Johns Hopkins University Press.

de Waal, F. B. M. (1986). The integration of dominance and social bonding in primates. *Quarterly Review of Biology*, **61**, 459–79.

de Waal, F. B. M. (1992). Appeasement, celebration, and food sharing in the two *Pan* species. In *Topics in Primatology, Vol. 1. Human Origins*: ed. T. Nishida, W. C. McGrew, P. Marler, M. Pickford and F. B. M. de Waal. Tokyo: University of Tokyo Press, pp. 37–50.
de Waal, F. B. M. (1995). Bonobo sex and society. *Scientific American*, **272**, 58–64.
Wittig, R. M. and Boesch, C. (2003). Food competition and linear dominance hierarchy among female chimpanzees of the Taï National Park. *International Journal of Primatology*, **24**, 847–67.
Wrangham, R. W. (1999). Evolution of coalitionary killing. *Yearbook of Physical Anthropology*, **42**, 1– 30.

34 Grooming: its hygienic and social aspects*

KOICHIRO ZAMMA AND MICHIO NAKAMURA

34.1 What is grooming?

Grooming is found widely within Animalia, from insects to fishes to mammals, including primates (Sparks, 1967). While prosimians use lower incisors as a comb (Buettner-Janusch and Andrew, 1962; Rose *et al.*, 1981) in a similar way to rodents (Murray, 1961) and ungulates (McKenzie, 1990), Anthropoidea primates use hands, fingers, mouths, and teeth during grooming. They acquired fine eyes with retinal foveas (Woollard, 1927) that allow them to find small objects in the hairs of the body, and dexterous fingers (Bishop, 1962), which let them part the hairs and pick small objects from the body surface (van Lawick-Goodall, 1968; Zamma, 2011). In addition to the fingers, chimpanzees use their lips to part hair and their mouths and teeth to remove small objects from the body surface.

There are essentially two types of grooming: social grooming and self-grooming. Social grooming has been often studied as a reflection of social relationships between individuals, and self-grooming, wherein the subject cleans his/her own body, has been considered selfish behavior.

Most grooming is directed toward conspecifics, but chimpanzees have also been known to groom other species on rare occasions. Bakuneeta (1996) reported that chimpanzees and an unknown monkey groomed each other at the Kaniyo–Pabidi Ecotourism Site, Uganda. Chimpanzees also sometimes groom the dead bodies of animals, such as the red colobus, which they hunt (Nishida *et al.*, 2010). In Bossou, Guinea, a female chimpanzee groomed a dead hyrax (*Dendrohyrax dorsalis*) while she was carrying, flapping, beating, and toying with it (Hirata *et al.*, 2001). Mahale chimpanzees were observed to groom the dead body of a leopard, one of their potential predators (Nakazawa *et al.*, 2013), in the course of performing other behaviors such as inspecting, sniffing, slapping, and biting (Nishida *et al.*, 2010).

Chimpanzees also groom leaves. "Leaf grooming" is recognized as a cultural behavior of eastern chimpanzees (Figure 34.1; Whiten *et al.*, 1999; see Chapter 42). Typically, a chimpanzee picks a leaf from a shrub or herb after grooming conspecific animals, puts a small object on the leaf from the mouth, folds the leaf, and crushes the place where the object was enfolded (van Lawick-Goodall, 1968; Boesch, 1995, 1996; Zamma, 2002a; Assersohn *et al.*, 2004; Figure 34.1). The small object is typically an ectoparasite (Boesch, 1995, 1996), such as a chimpanzee louse (*Pediculus schaeffi*) or its egg (Zamma, 2002a, 2006). It was also observed that biting flies such as *Haematopota fasciatapex* were leaf-groomed by Mahale chimpanzees (Zamma, personal observation). Other chimpanzees are sometimes attracted by leaf grooming and peer into the hands of the performer. Indeed, chimpanzees may occasionally perform leaf grooming solely to attract others' attention. The mean frequency of leaf grooming among adult chimpanzees was 0.19 instances per hour (range = 0–0.53, $N = 11$), and there

* We thank I. Tanaka for advice on observing grooming behavior and G. Yamakoshi for advice on investigating louse eggs on hairs left on the beds of chimpanzees. We also thank M. Cole for advice on climbing trees, and N. Kuze for advice on the numbers of hairs in primates.

Mahale Chimpanzees: 50 Years of Research, ed. M. Nakamura, K. Hosaka, N. Itoh and K. Zamma. Published by Cambridge University Press.

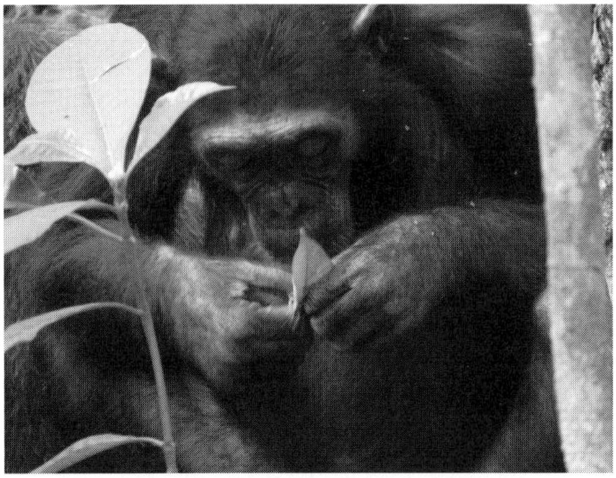

Figure 34.1 Leaf-grooming behavior performed by an adult female, Puffy, on November 14, 2013. ©K. Zamma.

was no sex difference in this behavior (female: $N = 5$, male: $N = 6$, $U = 13.5$, $p = 0.78$, Mann–Whitney U-test) in data collected by Zamma from 1999 to 2000 (see Zamma, 2011, for information on data collection).

Grooming an object or substrate has been observed in Mahale since 1991 (Nishida, 1994). This behavior is often performed by a groomee who is groomed by another: the groomee presses the tip of a finger gently on the surface of the ground or tree trunk, and moves the finger back and forth over the substrate. Although the groomee could receive grooming without doing anything, he/she grooms the ground intently. This may be considered a displacement behavior: the groomee might be in conflict between willingness to continue being groomed and a desire to reciprocate. The frequency of this behavior was 0.05/h (range = 0–0.19, $N = 11$), and there was no sex difference (female: $N = 5$, male: $N = 6$, $U = 10$, $p = 0.35$; see Zamma, 2011, for information on data collection). Hand movements in leaf grooming and grooming substrates differ from those of grooming a conspecific, but they are also traditionally called "grooming." This may be because these behaviors have often been observed in grooming contexts, and they have superficial similarities to grooming in that chimpanzees move their hands or fingers on an object while looking at it closely.

These are rather special aspects of "grooming," and grooming of a conspecific is observed more frequently. Chimpanzees may be very intent in grooming others, or they may groom another drowsily in a very relaxed mood. A chimpanzee may start grooming another without any solicitation from the latter, or the groomee may show obvious solicitation by raising his arm or showing his back to the prospective groomer.

Compared with other primates, chimpanzees seem to groom in more varied ways. One example is mutual grooming, wherein two chimpanzees sit face to face and groom each other simultaneously. Other primates, such as Japanese macaques (*Macaca fuscata*), rarely engage in such mutual grooming (Nakamura, 2003). This is probably because such monkeys usually groom using both hands, which does not provide space for the partner to groom back. Because chimpanzees often groom with one hand, as well as with both hands, they can groom each other simultaneously. The grooming hand-clasp is an odd manner of grooming, akin to a ritual: the two chimpanzees groom each other simultaneously using either of their hands, and their opposite hands are grasped or crossed over their heads (more interestingly, grasping hands are always the corresponding hands: either right–right or left–left) (McGrew and Tutin, 1978; McGrew *et al.*, 2001; Nakamura, 2002b; Nakamura and Uehara, 2004). Chimpanzees sometimes scratch others during social grooming, a behavior called "social scratching" (Nakamura *et al.*, 2000). The grooming hand-clasp and social scratching are not universal among chimpanzees; they are observed in only some populations, including Mahale, and are regarded as cultural behaviors (see Chapter 38).

Grooming has several functions, and many authors have reviewed the functional aspects of the behavior (e.g. Goodall, 1986). Here, we review grooming relationships among chimpanzees by considering the underlying hygienic factors.

34.2 Social relationships between grooming dyads

Grooming has been considered as an a-priori indicator of social bonding (e.g. Furuya, 1957). To measure the affinitive relationship between two individuals, the frequency and/or duration of social grooming is often examined, as well as other measures such as frequency of open-mouth kissing (Nishida and Hosaka, 1996) or attendance at the same party (Hayaki, 1988). One aspect of Mahale chimpanzee social relationships evaluated by grooming behaviors can be summarized as follows.

Infants, after starting self-grooming first at around 1-year-old, start to groom socially at around 2 years of age, and provide a substantial amount of grooming after they reach 4 years of age (Nishida, 1988). Grooming among adult males, who remain in their natal group, is observed more frequently than among adult females (Nishida, 1970, 1979), who have usually transferred from different unit-groups. Although grooming is usually limited to group members, because intergroup relationships of chimpanzees tend to be agonistic (Nishida, 1979; Goodall, 1986), a case has been observed where chimpanzees belonging to the M and K unit-groups groomed each other during the period when the K group was becoming extinct (Kawanaka, 1982).

When a male solicits a female for consortship just prior to her estrus, he tends to groom her in a unilateral way (Hayaki, 1997; Zamma, personal observation). When grooming is observed after copulation, males usually initiate it (Nishida, 1997). When a female gives birth to an infant, she spends a large proportion of her grooming time on her offspring (Nishida, 1988). An alloparent (see Chapter 30) also frequently grooms the infant whom he/she takes care of (Nishida, 1983b). The alloparent may also groom the infant's mother immediately before he/she tries to move the infant away from the mother (Nishida, 1983b). Grooming between a male and a non-estrous female is less frequent than male–male grooming, and grooming from a non-alpha male to a non-estrous female is observed more frequently than vice versa (Takahata, 1990).

Grooming among chimpanzees is also important in making and maintaining alliances (Chapter 27). Third-ranking males have been observed to manipulate the unstable dominance relationship between two higher-ranking males by changing their coalition partner from moment to moment (de Waal, 1982; Nishida, 1983a; Nishida and Hosaka, 1996). Nishida and Hosaka (1996) reported that the two highest-ranking males frequently groomed the faces of other males, which, they argued, was possibly because the groomer can read the signs of infidelity by gazing at the face of the groomee.

Alpha males at Mahale tend to exchange grooming with others reciprocally (Kawanaka, 1990). In contrast, grooming strategies of alpha males at Gombe, Tanzania, differed with their body mass: smaller alpha males tended to exchange grooming reciprocally with others, but larger alpha males tended to be groomed by others more than vice versa (Foster *et al.*, 2009). An alpha male in Mahale, Ntologi, was considered one of the largest in the M group; however, he exchanged grooming reciprocally with others (Kawanaka, 1990). Thus, balancing the amount of grooming given and received may be a characteristic of alpha males at Mahale.

Nevertheless, it was observed that an alpha male, Fanana, was groomed for a much longer time than he groomed others. Fanana held the alpha rank from 1997 to 2003 (Nishida, 2012), and was smaller than Ntologi. In the following section, we look at this case in more detail and discuss what was exchanged in this grooming.

34.3 Reciprocating grooming time or removal movements?

The observation of Fanana was conducted by Zamma from 1999 to 2000. Fanana spent 3.27 h grooming 14 individuals, but was groomed for 8.97 h by 23 individuals during 55.9 h of total observation time (see Zamma, 2011). To investigate the exchange of grooming in each dyad, we extracted dyads that represented more than 5% (> 0.61 h) of Fanana's total grooming time. We found asymmetry in grooming durations among these dyads ($T = 0$, $N = 6$, $p = 0.03$, Wilcoxon signed-rank test). Within all the dyads, Fanana was groomed longer than he groomed the partner (Figure 34.2).

Why did Fanana not exchange grooming time symmetrically? It may be because he could not do so even though he tried. Fanana was often observed to be groomed by multiple groomers at a time. One can be groomed by more than one groomer, but can

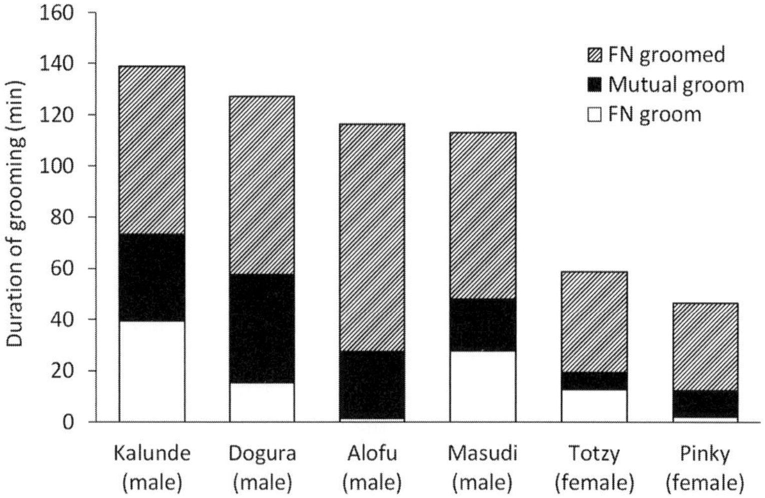

Figure 34.2 Reciprocity of grooming duration between Fanana (FN) and six M-group chimpanzees.

groom only one at a time. Fanana spent 5.8% of observation time in grooming behaviors, which was higher than the mean of 11 adult chimpanzees in the study period (4.6%). If he had reciprocated grooming with all partners by balancing the times of grooming given and received, he would have needed more than twice the time to groom others. He did not do this, probably because time was limited. So, what did he do instead?

Duration has typically been used to measure the quantity of grooming, perhaps because the duration of grooming was assumed to reflect the amount of the hygienic benefit. The hygienic function of social grooming among chimpanzees was supported by a study showing that the removal efficiency of social grooming is superior to that of self-grooming (Zamma, 2011). This study also suggested that chimpanzee grooming is not always uniform in its hygienic efficiency. As mentioned above, chimpanzees sometimes groom others intently, but sometimes do so slowly and lazily, suggesting that hygienic efficiency may depend on the groomer. Thus, the number of removal movements that directly reflect the hygienic function may be a more useful measure for evaluating the social relationship between members of a grooming dyad.

The number of removal movements was compared between Fanana and six grooming partners, and there was no significant difference between the number of removal movements performed by Fanana and the number performed by others ($T = 12$, $N = 6$, $p = 0.84$). He received 676 removal movements during 8.97 h of receiving grooming, and he performed 804 removal movements during 3.27 h of providing grooming. This result suggests that Fanana made a well-balanced exchange of the number of removal movements, although the grooming duration was unbalanced between Fanana and others.

This result indicates the flexibility of grooming among chimpanzees. Fanana might have a need to reciprocate in terms of the "quality" of grooming to maintain his alpha status, but other members, such as non-alpha males, adult females, and immature individuals, did not need to do the same. Both intense grooming by Fanana and slow grooming by other members serve the function of cleaning the hairs of grooming partners and maintaining intimate relationships among members of the group, but individuals seem to modify the intensity or efficiency of grooming depending on the situation.

34.4 Let's get together for grooming

The formation of large grooming clusters (de Waal, 1982; Goodall, 1986) is also a unique characteristic of chimpanzee grooming. Nakamura (2002a, 2003) conducted a detailed analysis on grooming clusters at Mahale. He defined a *grooming clique* as a group of chimpanzees directly connected by grooming at any given time, and a *grooming cluster* as a temporospatial set of cliques within a distance of 3 m and within 5 min (if a member of a clique creates another clique within 5 min, these cliques are considered to be in a same grooming cluster). Nakamura (2003) reported that Mahale chimpanzees formed 27 types of grooming cliques (later he observed four more: Nakamura, personal observation) consisting of 2–7 individuals (Figure 34.3).

The frequent occurrence of polyadic cliques, which is less common in other primates, is a unique characteristic of chimpanzees, and the ability to groom others while being groomed by others may also be a key factor in the formation of large grooming cliques (Nakamura, 2003). Nakamura (2002a) predicted that an individual, taking the roles of a groomer and a groomee at the same time, may perform lazy grooming because of the two different roles. To assess this prediction, we conducted a comparison of removal efficiencies of a groomer when he/she was being groomed and when not; we used data collected by Zamma from 1999 to 2000 (see Zamma, 2011). The results confirmed the prediction: the removal efficiency was lower when the groomer was being groomed by another than when this was not the case ($T = 7$, $N = 11$, $p = 0.02$; Figure 34.4). This result also may suggest that groomers are more relaxed in large grooming cliques with many partners.

Polyadic and mutual grooming may have evolved as an effective bonding mechanism. Although Dunbar (1996) hypothesized that human language evolved as a more efficient social-bonding mechanism than primate grooming, Nakamura (2000) found that chimpanzee grooming had about the same efficiency as human conversation in terms of the quantity of social interactions occurring at a given time.

Figure 34.3 A grooming clique comprising seven individuals (three adult females, four adult males). A juvenile is peering in (upper left). The center of this clique is Wakusi, the oldest female. This picture was taken on December 4, 2013. ©K. Zamma.

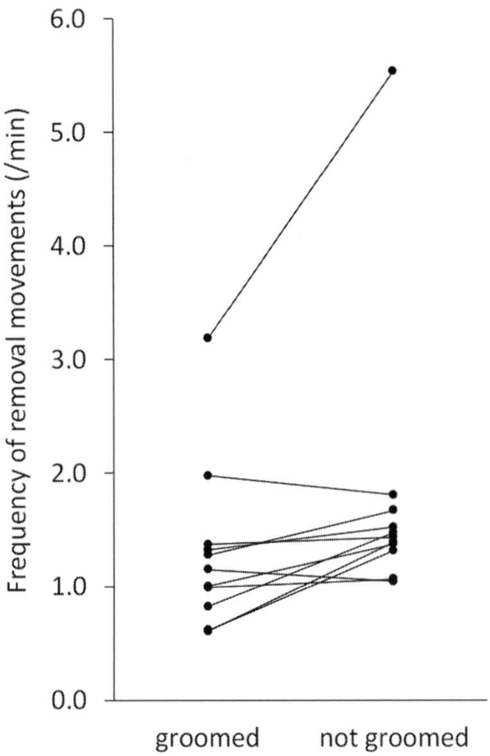

Figure 34.4 Comparison of frequencies of removal movements when a groomer was groomed and when not groomed.

Although a typical monkey can interact with only one partner at a time with their dyadic social grooming, a human can simultaneously interact with multiple individuals in conversation (Dunbar, 1993), and chimpanzees can also have multiple interactions at a time with polyadic and mutual grooming (Nakamura, 2000). This conclusion suggests that both humans and chimpanzees have a need to make social bonds with a large number of group members, and they have acquired the ability to satisfy the demand with different media, namely language and grooming, through evolutionary history.

The size of the grooming cluster sometimes reaches almost half of the unit-group members (Nakamura, 2003). Grooming clusters can be large when chimpanzees form large cliques, continue to groom with changing members, and/or groom near other grooming cliques. Although a grooming cluster is just an accumulation of groomer–groomee relationships, characteristics of grooming interactions in a large cluster are not the same as those with dyadic grooming. Although adult males groomed more in relatively smaller clusters (size 2–4), adult females groomed more in clusters of five or more individuals (Nakamura, 2003). Mothers were more likely to groom their infant offspring in small clusters (Nakamura, 2002a), but adult females, including mothers, tended to have more interactions with a variety of individuals in larger grooming clusters (Nakamura, 2002a, 2003). Such characteristics of grooming at Mahale enable mothers to have many grooming partners (mean 20.3, maximum 40) in only a 2-month period (Nakamura, 2010).

Unlike most primates, chimpanzees live in fission–fusion societies, and members of a unit-group do not always range together (Nishida, 1968). Chimpanzees may have limited time to interact directly with other members, so polyadic and mutual grooming may have an advantage in allowing for multiple social interactions. The relatively egalitarian sociality and the cognitive ability of chimpanzees may also be related to the formation of large grooming clusters (Nakamura, 2003). Chimpanzees may sometimes aspire to gather together, and the unique form of their grooming interactions may enable them to do so (Nakamura, 2002a, 2003).

34.5 Ectoparasites and epizoochorous seeds on chimpanzees

In this section, we focus on the objects removed by chimpanzees. It has been confirmed to remove lice and louse eggs (Zamma, 2002a, 2006; Figure 34.5), and ticks (*Ixodes pseudorasus*) infest them (Nishida, 1988). Epizoochorous seeds, which are likely to adhere to or be caught in the hair of chimpanzees, have also been observed to be removed. The ectoparasites are often too small to identify in the field. However, a sand flea, or jigger (*Tunga penetrans*), has a special feature that can be identified

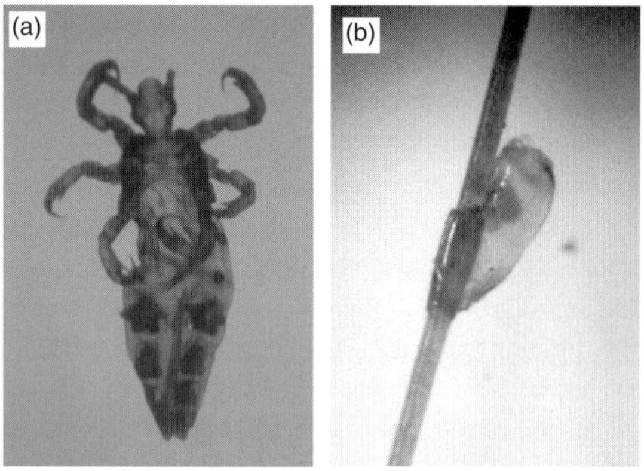

Figure 34.5 (a) An adult chimpanzee louse (*Pediculus schaeffi*) (Zamma, 2002a) and (b) a louse egg. The length of an adult louse is about 3 mm, and that of louse egg is about 0.9 mm. ©K. Zamma.

readily: the flea has a habit of burrowing between the toenail and skin, and its egg mass is white and has a globular shape of a few millimeters in diameter. Chimpanzees in Mahale usually remove it using fingers and sucking it out with the mouth (Nishida, 2002). A chimpanzee was also observed trying to remove an egg mass using the sharp edge of a twig (Nishida *et al.*, 2010; see Chapter 42).

It is hard to evaluate how many of these small ectoparasites, such as lice and ticks, and seeds are found on chimpanzees. Thus, the loads of ectoparasites and seeds have been evaluated in an indirect way.

Ticks have a habit of spending their life both on the host and on the ground. They lay eggs on the ground, and the larvae climb onto a host to feed on blood. When they suck enough blood, they drop off and shed their skins. They repeat climbing up on a host and dropping off to the ground until they become adults (Kettle, 1990); thus, the number of ticks on the ground can be used as an index of tick load on the host (Mooring, 1995; Saunders and Hausfater, 1988). To assess the amount of ticks in the habitat of the M-group chimpanzees, a dragging method (Mooring, 1995; Saunders and Hausfater, 1988) was used from August 1999 to July 2000 in the MMNP. A white flannel cloth (size = 1.0 × 0.6 m) was dragged by Zamma along a 50-m transect for 150 s between 08:00 h and 13:00 h, and the number of ticks that were attached to it was counted (Zamma, 2002b). Epizoochorous seeds also attached to the cloth, and the number was counted. Transects were selected from the paths used by chimpanzees within 15 days before each sampling day. Zamma walked eight transects each month, but four transects in August 1999, and analysed the mean number of ticks per 50-m transect per month.

In this study period, 512 ticks were collected in total. Numbers of tick in the habitat differed significantly by month ($\chi^2 = 42.8$, $df = 10$, $p < 0.01$, Kruskal–Wallis test), and more ticks appeared during the dry season (June to September) than during the wet season (October to May; Figure 34.6a, b). The number of ticks was negatively correlated with rainfall ($\tau = -0.64$, $N = 11$, $p = 0.006$, Kendall's correlation coefficient by ranks). A total of 9943 epizoochorous seeds were counted, and the numbers of seeds in the habitat differed significantly by month ($\chi^2 = 43.3$, $df = 10$, $p < 0.01$). Many were found from May to August, peaking in June (Figure 34.6c), but the occurrence was not correlated with rainfall ($\tau = -0.42$, $N = 11$, $p = 0.09$).

Ticks in Mahale showed a dramatic decrease in the rainy season, but this tendency differed from the seasonality of ticks reported in the Kyle Recreational Park, Zimbabwe; there, larger numbers of adult ticks were observed in the warm/wet season, and the numbers decreased in the hot/dry season (Mooring, 1995). The mean annual rainfall in the home range of the M-group chimpanzees was 1739 mm (Chapter 12), but it was 828 mm at Kyle (Mooring, 1995). Ticks may prefer moderately wet conditions such as the dry season of the Mahale forest or the wet season of the Kyle savanna–woodland.

In contrast to evaluation of ticks and epizoochorous seeds, the number of lice is difficult to assess because they never leave the body surface of hosts in their life cycle. An egg of a louse is "glued on" to the hair of its host, and the hatched louse sucks the blood of the host over stages from nymph to adult (Buxton, 1947). It seems that there is no method to investigate the louse load on chimpanzees in an indirect way, but there is one way to evaluate the

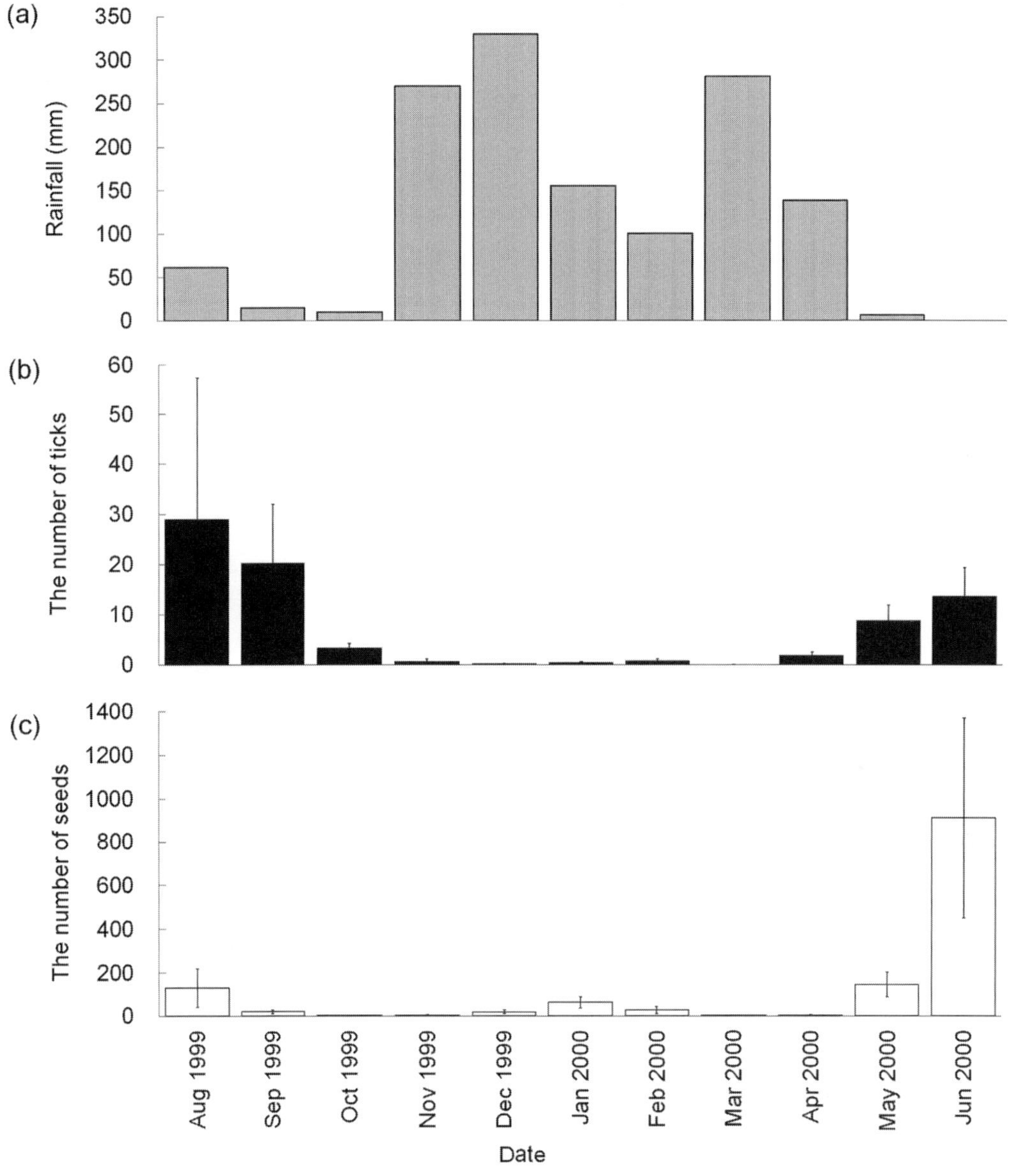

Figure 34.6 Seasonal changes in rainfall (a), the number of ticks (b), and the number of epizoochorous seeds (c) in chimpanzee habitat. Error bars indicate SE.

number of lice that have dropped off their host, namely to count the louse eggs attached to fallen hairs collected in chimpanzee beds, which are located in trees.

Zamma investigated louse egg infection from 1999 to 2011 at Mahale. In total, 140 beds were investigated, and 5550 chimpanzee hairs were obtained (Table 34.1). The mean number of hairs per bed was 39.6 (range = 1–295). A microscope was used to check for louse eggs attached to hairs, and nine well-formed eggshells were observed. Hatched eggs as well as unhatched ones were

Table 34.1 **Numbers of louse eggs on hairs from the beds of chimpanzees in Mahale**

Year	Month	Season[a]	No. of beds	No. of hairs	No. of louse eggs with good shapes	No. of louse eggs with only marks
1999	Aug	D	5	135	2	0
	Sep	D	3	44	0	0
	Oct	R	1	47	0	0
2000	Jun	D	41	1493	3	4
2003	Aug	D	11	356	0	3
	Sep	D	14	222	0	0
	Oct	R	6	100	1	2
	Nov	R	12	491	1	7
2004	May	R	2	73	0	0
	Jun	D	4	261	0	2
	Jul	D	2	109	0	0
2006	Oct	R	7	196	0	2
	Nov	R	6	883	0	1
	Dec	R	6	252	1	1
2007	Jan	R	3	186	0	0
2008	Nov	R	8	137	0	0
	Dec	R	4	367	1	0
2010	Dec	R	2	138	0	1
2011	Jan	R	3	60	0	0
Total			140	5550	9	23

[a] D, dry season; R, rainy season.

considered because hatched eggs are also targets of chimpanzee grooming (Zamma, 2006). The louse egg density was 1.62/1000 hairs. Monthly differences in louse egg infections could not be examined because of the small numbers of eggs; thus, infections were compared between dry and rainy seasons. The numbers of louse eggs per hair did not differ between seasons (dry season: $N = 7$, wet season: $N = 12$, $U = 40$, $p = 0.88$).

Some hairs had marks of louse eggs; in most cases, a portion of the egg shell was lost, but the adhesive cement remained on the hairs. The eggshells may have been broken by removal movements or by scratching by the chimpanzees, or they may have been lost due to weather condition such as rain, wind, or ultraviolet sun rays. The marks of louse eggs were observed on 23 hairs, but these were not analysed because the shape was readily distinguishable from the "well-formed" eggs that were confirmed as the targets of chimpanzee grooming.

34.6 Relationships between chimpanzees and ectoparasites

To consider the relationships between chimpanzee grooming and these small objects, we next looked at the seasonality of grooming. The data were collected from August 1999 to June 2000 in Mahale, and 11 focal animals of the M group (five adult females, six adult males) were observed (see Zamma, 2011, for details of the observation procedures). The focal animals were observed for 7 or 8 days through the year. Any removal movements directed to the body surface of the focal animal (i.e. those performed by others grooming the focal animal and those by the focal animal himself/herself in self-grooming) were compared with the abundance of objects likely attached on the body surface. In the study period, a total of 30.4 h of being groomed and 17.0 h of self-grooming were observed. Video analyses were available for 22.4 h and 12.4 h, respectively, and 1800 and 586 removal movements were counted, respectively (see also Zamma, 2011).

The removal efficiency (the number of removal movements per grooming duration) was about 1.0/min through the year, and there was no monthly difference in this efficiency ($\chi^2 = 9.68$, $df = 10$, $p = 0.47$, Kruskal–Wallis test; Figure 34.7). The monthly removal efficiency was not correlated with the monthly numbers of ticks ($\tau = -0.05$, $N = 11$, $p = 0.88$, Kendall's correlation coefficient) or epizoochorous seeds appearing in the chimpanzees habitat ($\tau = -0.2$, $N = 11$, $p = 0.45$). The removal efficiency also did not differ between the dry and wet seasons ($T = 13$, $N = 11$, $p = 0.08$, Wilcoxon signed-rank test).

These results suggest that ticks and seeds have no significant effects on grooming overall. Indeed, we occasionally observed a chimpanzee ignoring epizoochorous seeds on the body surface of the groomee, although he/she was able to remove them. The epizoochorous seeds may not be removed, as they are not harmful to chimpanzees. However, ticks are undesirable ectoparasites that suck blood, although their effects are minor. One possible reason for this is the low rate of tick infestations on chimpanzees. Chimpanzees walk on the ground, but they also spend much time in the trees feeding and sleeping. The effect of tick load on grooming was positive among ungulates (Mooring, 1995), but negative among baboons (Saunders and Hausfater, 1988). Chimpanzees are arboreal as well as terrestrial, so tick infestation may be limited.

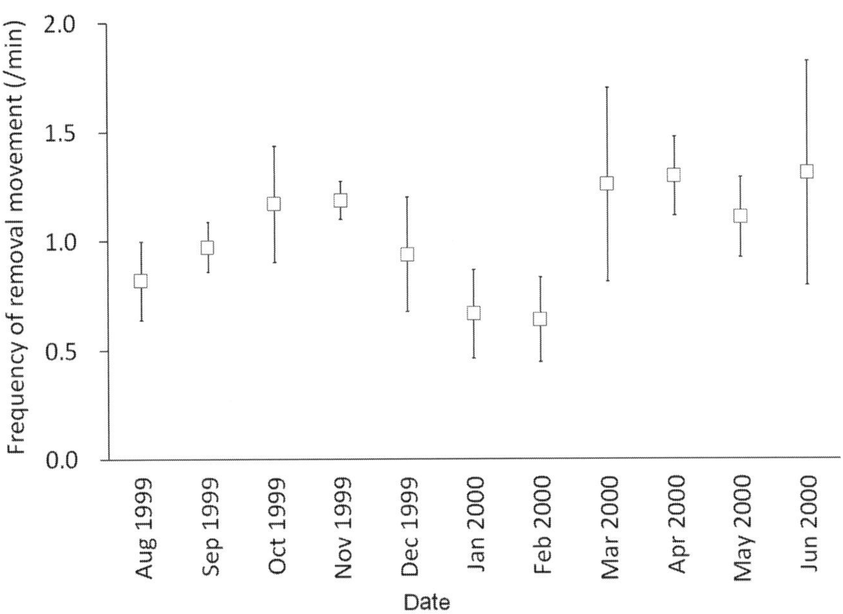

Figure 34.7 Monthly frequencies of removal movements among chimpanzees of Mahale. Error bars indicate SE.

The effect of ticks may be relatively lower than that of lice among chimpanzees. Ticks were almost absent during the wet season in Mahale, but removal movements were also observed in this period. Most of the removed objects may be lice or louse eggs, as louse infestation may be present all year round; the density of louse eggs did not differ by season. Chimpanzees may remove ticks, but the load of ticks may be much lower than that of lice; thus, no effect of ticks was found in considering removal efficiency.

The density of louse eggs on chimpanzees was 1.62 (per 1000 hairs), greater than that in Japanese macaques (0.14/1000 hairs) (calculated from data in Zamma, 2002c). This difference in louse egg infestation may be associated with the difference in hair density between these animals. Schultz (1931) investigated the hair density of a chimpanzee and estimated a total of 187 000 hairs on the three body parts examined (scalp, back, and chest). Using these data, the number of louse eggs on these three body parts was estimated to be 367. This value is comparable to the estimated number of louse eggs, 288, on the same three body parts of Japanese macaques, calculated from the data of Zamma (2002c). Chimpanzee lice suck blood, warm themselves on the body surface of their host, and can readily increase in this rich environment (e.g. the natural increase of human lice, *Pediculus humanus*, is 0.111: Evans and Smith, 1952). As lice are significant ectoparasites for Japanese macaques and humans, chimpanzees may also be greatly affected by louse infection.

34.7 Removing lice and creating a large cluster

Lice are commonly found on chimpanzees in the wild. Thus, grooming is required to remove these lice to decrease the "costs" of lice, such as itching. However, chimpanzees sometimes seem to be fascinated by removing small objects in grooming. Kuhn (1968) vividly described a series of actions of a captive male chimpanzee who found a louse as follows: "the moment he saw a tiny chimp-louse the ape was wide awake, quite excited, and he tried very hard to catch the little insect with his lips and fingers." Chimpanzees inspect ectoparasites placed on leaves (Budongo, Uganda: Assersohn *et al.*, 2004), and they squash them on their arm (Taï, Côte d'Ivoire: Boesch, 1996), on the palm (Bossou, Guinea: Nakamura and Nishida, 2006), or on a leaf (Gombe: van Lawick-Goodall, 1968, Boesch, 1995; Mahale: Zamma, 2002a). In Mahale, chimpanzees perform rhythmic "lip smacking" and "teeth clacking" sounds (Nishida *et al.*, 2010) when they find something such as a louse during grooming (Nishida *et al.*, 2004). This rhythmical beat of the sound gives the impression that they are expressing excitement.

Chimpanzees need to be groomed and appear to have fun grooming one another. Thus, grooming must be used as a tool for communication. When one chimpanzee meets another, both chimpanzees are potentially motivated to give and receive grooming because they have lice on their bodies. Thus, grooming behavior connects chimpanzees and can lead to the formation of large grooming clusters.

References

Assersohn, C., Whiten, A., Kiwede, Z. T., Tinka, J., and Karamagi, J. (2004). Use of leaves to inspect ectoparasites in wild chimpanzees: a third cultural variant? *Primates*, **45**, 255–8.

Bakuneeta, C. (1996). Chimpanzees grooming an unknown monkey. *Pan Africa News*, **3**(2), 10.

Bishop, A. (1962). Control of the hand in lower primates. *Annals of the New York Academy of Sciences*, **102**, 316–37.

Boesch, C. (1995). Innovation in wild chimpanzees (*Pan troglodytes*). *International Journal of Primatology*, **16**, 1–16.

Boesch, C. (1996). Three approaches for assessing chimpanzee culture. In *Reaching into Thought*, ed. A. E. Russon, K. A. Bard, and S. T. Parker. Cambridge: Cambridge University Press, pp. 404–29.

Buettner-Janusch, J. and Andrew, R. J. (1962). The use of the incisors by primates in grooming.

American Journal of Physical Anthropology, **20**,127–9.
Buxton, P. A. (1947). *The Louse: An Account of the Lice which Infest Man, Their Medical Importance and Control*, 2nd edn. London: Edward Arnold.
Dunbar, R. I. M. (1993). Coevolution of neocortical size, group size and language in humans. *Behavioral and Brain Sciences*, **16**, 681–735.
Dunbar, R. I. M. (1996). *Grooming, Gossip and Evolution of Language*. Cambridge, MA: Harvard University Press.
Evans, F. C. and Smith, F. E. (1952). The intrinsic rate of natural increase for the human louse, *Pediculus humanus. The American Naturalist*, **86**, 299–310.
Foster, M. W., Gilby, I. C., Murray, C.M., *et al*. (2009). Alpha male chimpanzee grooming patterns: implications for dominance "style." *American Journal of Primatology*, **71**, 136–44.
Furuya, Y. (1957). [Grooming behavior in the wild Japanese monkeys.] *Primates*, **1**, 47–68. In Japanese.
Goodall, J. (1986). *The Chimpanzees of Gombe: Patterns of Behavior*. Cambridge, MA: Belknap.
Hayaki, H. (1988). Association partners of young chimpanzees in the Mahale Mountains National Park, Tanzania. *Primates*, **29**, 147–61.
Hayaki, H. (1997). [A case of consortship in M group chimpanzees of Mahale Mountains National Park, Tanzania.] *Bulletin of Faculty of Humanities and Sciences of Kobe Gakuin University*, **14**, 15–30. In Japanese.
Hirata, S., Yamakoshi, G., Fujita, S., Ohashi, G., and Matsuzawa, T. (2001). Capturing and toying with hyraxes (*Dendrohyrax dorsalis*) by wild chimpanzees (*Pan troglodytes*) at Bossou, Guinea. *Primates*, **53**, 93–7.
Kawanaka, K. (1982). A case of inter-unit-group encounter in chimpanzees of the Mahale Mountains. *Primates*, **23**, 558–62.
Kawanaka, K. (1990). Alpha males' interactions and social skills. In *The Chimpanzees of the Mahale Mountains*, ed. T. Nishida. Tokyo: University of Tokyo Press, pp. 133–48.
Kettle, D. S. (1990). *Medical and Veterinary Entomology*. Oxford: CAB International.
Kuhn, H. J. (1968). Parasites and the phylogeny of the catarrhine primates. In *Taxonomy and Phylogeny of Old World Primates with References to the Origin of Man*, ed. B. Chiarelli. Torino: Rosenberg and Sellier, pp. 187–95.
van Lawick-Goodall, J. (1968). The behavior of free-living chimpanzees in the Gombe Stream Reserve. *Animal Behaviour Monographs*, **1**, 161–311.
McGrew, W. C. and Tutin, C. E. G. (1978). Evidence for a social custom in wild chimpanzees? *Man*, **13**, 234–51.
McGrew, W. C., Marchant, L. F., Scott, S. E., and Tutin, C. E. G. (2001). Intergroup differences in a social custom of wild chimpanzees: the grooming hand-clasp of the Mahale Mountains. *Current Anthropology*, **42**, 148–53.
McKenzie, A. A. (1990). The ruminant dental grooming apparatus. *Zoological Journal of the Linnean Society*, **99**, 117–28.
Mooring, M. S. (1995). The effect of tick challenge on grooming rate by impala. *Animal Behaviour*, **50**, 377–92.
Murray, M. D. (1961). The ecology of the louse *Polyplax serrate* (Burm.) on the mouse *Mus musculus* L. *Australian Journal of Zoology*, **9**, 1–13.
Nakamura, M. (2000). Is human conversation more efficient than chimpanzee grooming?: comparison of clique sizes. *Human Nature*, **11**, 281–97.
Nakamura, M. (2002a). Grooming-hand-clasp in Mahale M group chimpanzees: implication for culture in social behaviors. In *Behavioural Diversity in Chimpanzees and Bonobos*, ed. C. Boesch, G. Hohmann, and L. F. Marchant. Cambridge: Cambridge University Press, pp. 71–83.
Nakamura, M. (2002b). [*Grooming as gatherings*.] In [*The Mahale Chimpanzees: 37 Years of <Panthropology>*], ed. T. Nishida, K. Kawanaka, and S. Uehara. Kyoto: Kyoto University Press, pp. 345–67. In Japanese.
Nakamura, M. (2003). "Gatherings" of social grooming among wild chimpanzees: implications for evolution of sociality. *Journal of Human Evolution*, **44**, 59–71.
Nakamura, M. (2010). Ubiquity of culture and possible social inheritance of sociality among wild chimpanzees. In *The Mind of the Chimpanzee: Ecological and Experimental Perspectives*, ed. E. V. Lonsdorf, S. R. Ross, and T. Matsuzawa. Chicago: University of Chicago Press, pp. 156–67.
Nakamura, M. and Nishida, T. (2006). Subtle behavioral variation in wild chimpanzees, with special reference

to Imanishi's concept of *kaluchua*. *Primates*, **47**, 35–42.
Nakamura, M. and Uehara, S. (2004). Proximate factors of different types of grooming hand-clasp in Mahale chimpanzees: implications for chimpanzee social customs. *Current Anthropology*, **45**, 108–14.
Nakamura, M., McGrew, W. C., Marchant, L. F., and Nishida, T. (2000). Social scratch: another custom in wild chimpanzees? *Primates*, **41**, 237–48.
Nakazawa, N., Hanamura, S., Inoue, E., Nakatsukasa, M., and Nakamura, M. (2013). A leopard ate a chimpanzee: first evidence from East Africa. *Journal of Human Evolution*, **65**, 334–7.
Nishida, T. (1968). The social group of wild chimpanzees in the Mahali Mountains. *Primates*, **9**, 167–224.
Nishida, T. (1970). Social behavior and relationship among wild chimpanzees of Mahali Mountains. *Primates*, **11**, 47–87.
Nishida, T. (1979). The social structure of chimpanzees of the Mahale Mountains. In *The Great Apes*, ed. D. A. Hamburg and E. R. McCown. Menlo Park, CA: Benjamin/Cummings, pp. 73–121.
Nishida, T. (1983a). Alpha status and agonistic alliance in wild chimpanzees (*Pan troglodytes schweinfurthii*). *Primates*, **24**, 318–36.
Nishida, T. (1983b). Alloparental behavior in wild chimpanzees of the Mahale Mountains, Tanzania. *Folia Primatologica*, **41**, 1–33.
Nishida, T. (1988). Development of social grooming between mother and offspring in wild chimpanzees. *Folia Primatologica*, **50**, 109–23.
Nishida, T. (1994). Review of recent findings on Mahale chimpanzees: implications and future research directions. In *Chimpanzee Cultures*, ed. R. W. Wrangham, W. C. McGrew, and F. B. M. de Waal and P. G. Heltne. Cambridge, MA: Harvard University Press, pp. 373–96.
Nishida, T. (1997). Sexual behavior of adult male chimpanzees of the Mahale Mountains National Park, Tanzania. *Primates*, **38**, 379–98.
Nishida, T. (2002). A self-medicating attempt to remove the sand flea from a toe by a young chimpanzee. *Pan Africa News*, **9**, 5–6.
Nishida, T. (2012). *Chimpanzees of the Lakeshore: Natural History and Culture at Mahale*. Cambridge: Cambridge University Press.
Nishida, T. and Hosaka, K. (1996). Coalition strategies among adult male chimpanzees of the Mahale Mountains. In *Great Ape Societies*, ed. W. C. McGrew, L. F. Marchant, and T. Nishida. Cambridge, Cambridge University Press, pp. 114–34.
Nishida, T., Mitani, J.C., and Watts, D. P. (2004). Variable grooming behaviours in wild chimpanzees. *Folia Primatologica*, **75**, 31–6.
Nishida, T., Zamma, K., Matsusaka, T., Inaba, A., and McGrew, W. C. (2010). *Chimpanzee Behavior in the Wild*. Tokyo: Springer.
Rose, K. D., Walker, A., and Jacobs, L. L. (1981). Function of the mandibular tooth comb in living and extinct mammals. *Nature*, **289**, 583–5.
Saunders, C. D. and Hausfater, G. (1988). The functional significance of baboon grooming behavior. *Annals of the New York Academy of Sciences*, **525**, 430–2.
Schultz, A. H. (1931). The density of hair in primates. *Human Biology*, **3**, 303–21.
Sparks, J. (1967). Allogrooming in primates: a review. In *Primate Ethology*, ed. D. Morris. London: Weidenfeld & Nicolson, pp. 148–75.
Takahata, Y. (1990). Adult males' social relations with adult females. In *The Chimpanzees of the Mahale Mountains*, ed. T. Nishida. Tokyo: University of Tokyo Press, pp. 133–48.
de Waal, F. B. M. (1982). *Chimpanzee Politics: Power and Sex among Apes*. New York: Harper & Row.
Whiten, A., Goodall, J., McGrew, W. C., *et al.* (1999). Cultures in chimpanzees. *Nature*, **399**, 682–5.
Woollard, H. H. (1927). The differentiation of the retina in the primates. *Proceedings of the Zoological Society of London*, **97**, 1–18.
Zamma, K. (2002a). Leaf-grooming by a wild chimpanzee in Mahale. *Primates*, **43**, 87–90.
Zamma, K. (2002b). Seasonal change of the density of ticks and grooming among chimpanzees at Mahale Mountains National Park, Tanzania. *Anthropological Science*, **110**, 128.
Zamma, K. (2002c). Grooming site preferences determined by lice infection among Japanese macaques in Arashiyama. *Primates*, **43**, 41–9.
Zamma, K. (2006). A louse egg left on a leaf. *Pan Africa News*, **13**, 8–10.
Zamma, K. (2011). Frequency of removal movements during social versus self-grooming among wild chimpanzees. *Primates*, **52**, 323–8.

35 Sexual behavior and mating strategies

SHIHO FUJITA AND EIJI INOUE

35.1 Life history and sexual activity

As they grow, both male and female chimpanzees develop the sexual behaviors needed to participate in the complicated adult mating system. Males are able to copulate with intromission 15 months after birth, although they cannot ejaculate (Hasegawa and Hiraiwa-Hasegawa, 1983; Hasegawa, 1991). Such nonprocreative sexual activity in infant and juvenile males is thought to be a learning process in preparation for entering adulthood. At about 9 years of age, males gain the ability to ejaculate (Hasegawa, 1991). While the copulation rate of adolescent males is higher than that of adult males (Hayaki, 1985; Takahata *et al.*, 1996; Takahata, 2002), adolescent males typically copulate with females when they are unlikely to conceive (Hasegawa and Hiraiwa-Hasegawa, 1990; Takahata, 1990; Hasegawa, 1991). In adolescent males, copulation usually accompanies courtship displays to and grooming of a female partner (Hayaki, 1985). The rate of copulation with females at their fertile stage reaches a peak at males' prime (i.e. in their early twenties). After their late thirties, the copulation rate in males begins to decrease, perhaps reflecting low spermatogenesis at that age (Hasegawa and Hiraiwa-Hasegawa, 1990; Hasegawa, 1991).

Females start to show minor sexual swelling between 8 and 9 years of age (Figure 35.1), but they do not copulate during this period. They show maximal sexual swelling starting around 10 years of age, at which point they begin to copulate. Although some females experience first parturition within their natal group, most emigrate from their natal group before their first parturition at 11 (range 9.0–13.0) years of age (Nishida, 2002; see Chapter 7 for more detail). On average, females deliver their first offspring at the age of 13 years and continue to reproduce at 3.6–4.6-year intervals until their late thirties (Hiraiwa-Hasegawa *et al.*, 1984; Nishida *et al.*, 1990, 2003; Nishida, 2002; see Chapters 7, 44). Some females still show estrus over 40 years of age without delivering (Nishida, 2002; see Chapter 23).

As in males, the copulation rate in females is negatively correlated with age (Hasegawa and Hiraiwa-Hasegawa, 1990; Hasegawa, 1991). Adolescent and young adult females mate frequently, especially with immature males (under 12 years old) (Takasaki, 1985; Hasegawa and Hiraiwa-Hasegawa, 1990). Females' rate of copulation with middle- and high-ranking males, on the other hand, reaches a peak in their early twenties (Hasegawa and Hiraiwa-Hasegawa, 1990; Hasegawa, 1991).

The female reproductive period consists of three phases: cycling, pregnancy, and lactating sterile (Figure 35.2). An estrous cycle occurs not only during the cycling period but also in the early pregnancy period (Takahata, 2002). Females repeat an estrous cycle every 32–33 days during which they are receptive to copulation for 12–13 days when they show maximum genital swelling (Takahata, 1990; Matsumoto-Oda *et al.*, 2007) (Figure 35.3).

35.2 Sexual behavior

35.2.1 Copulation

Copulation in chimpanzees (Figure 35.4) is characterized by a single brief intromission and a high frequency of ejaculation. This may reflect high sexual

Mahale Chimpanzees: 50 Years of Research, ed. M. Nakamura, K. Hosaka, N. Itoh and K. Zamma. Published by Cambridge University Press.

Figure 35.1 An adolescent female (front) showing a small swelling. A swelling around puberty has an irregular cycle, and its maximal size is not as large as that of adult females. ©S. Fujita.

Figure 35.3 A female showing a swelling of the anogenital region during cycling phase. ©S. Fujita

Figure 35.2 A female's sexual swelling stops after parturition. Postpartum amenorrhea continues for 3–6 years if her infant survives until weaning. ©S. Fujita.

Figure 35.4 Copulation between an adult male and an adult female. ©Kazuhiko Hosaka.

activity in a multi-male/multi-female mating system in which sperm competition pressures are greatest (Dixson, 2009). Chimpanzees have the shortest duration of intromission among primate species at a mean of only 7–8 s (Hasegawa and Hiraiwa-Hasegawa, 1983, 1990; Hasegawa, 1991; Nishida, 1997). Females are largely responsible for this short copulation, because they leave their partners as soon as possible following ejaculation (Hasegawa, 1991, 1992). The duration of intromission tends to be longer in immature than in adult males (Hasegawa and Hiraiwa-Hasegawa, 1990). Among adult males, the mean duration of intromission was not correlated with either dominance rank (Nishida, 1997) or age (Hasegawa and Hiraiwa-Hasegawa, 1990; Nishida, 1997). However, decline of social status was reported to relate with the shortening of intromission. In the case of one adult male, Kalunde, as he declined in rank from the alpha to the third between 1991 and 1992, his mean duration of intromission declined from 7.1 to 3.7 s (Nishida, 1997). Physiological and/or psychological status might be a cause of the shortening of intromission.

At Mahale, the mean hourly number of copulations in adult males was 0.19–0.21 (Hasegawa and Hiraiwa-Hasegawa, 1990; Hasegawa, 1991; Takahata *et al.*, 1996; Nishida, 1997; Takahata, 2002). In other words, males copulate two times per daytime hour. On the other hand, females at the stage of maximal genital swelling copulate on average 0.79 and 1.09 times per hour with adult and adolescent males, respectively. While copulation rates with immature males do not change throughout the estrous cycle, those with adult males increase as ovulation approaches (Hasegawa, 1991). As in the most extreme case, for one adult female, Wakilufya, the rates of copulation were 2.61 times/h and 1.25 times/h with adult and immature males, respectively, for 3 days when she was most likely to conceive (Hasegawa, 1992).

35.2.2 Genital inspection, courtship, and sexual coercion

To gather information on reproductive status in females, adult males typically perform a "genital inspection." When a male encounters a female, he routinely inserts his index finger into her vagina and then sniffs the finger. Cycling females are inspected more frequently than are lactating and pregnant females. Among cycling females, the frequency of genital inspection increases dramatically for a few days before the start of swelling and gradually decreases as the day of ovulation approaches (Nishida, 1997). Genital inspection might have the function of gathering information on the onset of swelling (Wallis and Lemmon, 1986).

Both males and females exhibit courtship gestures to advertise the intention to copulate. Males perform courtship gestures more often than females regardless of whether it results in copulation or not (Nishida, 1997). In addition, males use a more diversified repertoire of courtship gestures such as sit hunching, leaf clipping, arm stretching, bipedal swaggering, branch shaking, shrub bending, thumping, and leading. With any pattern of courtship gestures by a male, a female is likely to perceive his intention as long as he presents his erect penis (Nishida, 1997, 2012). Adolescent males often exhibit leaf-clipping and shrub-bending displays (see Chapter 42), which are subtler signals compared with other types of displays and, thus, less likely to provoke adult males to interrupt (Nishida, 1980; Hayaki, 1985). On the other hand, an estrous female typically courts a male only by presenting her swollen buttocks in front of the male. Estrous females also use leaf clip and branch shake to court males, especially when males do not show interest (Nishida, 1997, 2012).

Even though males more often demonstrate the courtship gesture, copulation could not occur unless females cooperate with them. As males are larger than females, they theoretically could force females to mate. In fact, when females show reluctance to comply with males' solicitation, adult males sometimes display aggressive behavior by bluffing, hitting, or kicking them. Estrous females, however, stubbornly resist such violence and intervention and almost never succumb to it. In addition, females scream until dominant males approach the conflicting pair, expecting such solicitation to stop. Of 12 cases in which the two oldest males, Musa and Bakali, resorted to violence in response to females' refusal, none resulted in mating (Nishida, 1997).

35.2.3 Non-reproductive sexual behavior

Homosexual behaviors are not typically observed between mature males or between estrous females at Mahale (Nishida, 1997). However, juvenile and adolescent females at Mahale sometimes engage in non-reproductive sexual behavior with each other. They have been observed to embrace ventro-ventrally and rub each other's sexual skin together (Nishida, 1997; Zamma and Fujita, 2004; Shimada, 2014; see Chapter 36). This behavior is similar to the genital rubbing frequently observed among female bonobos (Kano, 1980).

Male masturbation is common in some primate species living in groups made up of multiple sexually mature males and females (Thomsen *et al.*, 2003). In Mahale chimpanzees, mature and immature males are occasionally observed fumbling with their erect penises, but no ejaculation by masturbation has been observed (Nishida, 1997).

35.3 Mating strategies and reproductive outcomes

35.3.1 Male mating strategies

Tutin (1979, 1980) classified the mating patterns of wild chimpanzees at Gombe, Tanzania, into three types: (a) opportunistic mating, when an estrous female may be mated with non-competitively by all the males in the group; (b) possessiveness, when a male establishes a special short-term relationship with an estrous female; and (c) consortship, when a male and a female leave the group together and maintain an exclusive mating relationship. The latter two mating patterns are accompanied by herding behavior (Nishida 1997). Although all types of mating are observed at Mahale, opportunistic mating accounts for more than 90% of observed mating (Hasegawa and Hiraiwa-Hasegawa, 1983; Takasaki, 1985; Matsumoto-Oda, 1999; Takahata, 2002). This figure was slightly higher at Mahale than at Gombe, where the rate is 73% (Tutin, 1979, 1980).

In opportunistic mating, females commonly mate with males one after another, and both males and females are less discriminating in their mating partners. Under this condition, males do not always compete directly for the chance to mate. However, when a female mates with two or more males, competition between the gametes of rival males can occur for access to her ovum. Such competition encourages the evolution of mechanisms to optimize sperm numbers in the ejaculate and to enhance fertility. Chimpanzee males have larger testes than expected given their body weight, which reflects the high competition in this species' post-copulation levels (Dixson, 2009).

Possessiveness mating is characteristic of the alpha males because only alpha males can mate with females without interference by other males (Hasegawa and Hiraiwa-Hasegawa, 1983; Takasaki, 1985; Nishida, 1997; Takahata, 2002). All physical intervention observed, though not very frequent at Mahale, was by males who were higher-ranking than the copulating male, and aggression in such intervention was directed mainly toward the copulating male rather than females (Hasegawa and Hiraiwa-Hasegawa, 1990; Figure 35.5). The intervention in the copulation by an adult male was, thus, probably intended to prevent lower-ranking males from mating rather than to copulate with the females. Although such direct intervention is rare, it is also possible that the presence of the dominant males itself prevents lower-ranking males from mating. The males who are rivals to the alpha male and low-ranking males tend to take females out of the group for consortship over a long period (Hasegawa, 1992). Possessiveness mating by an alpha male accounted for 80% (16/20) in Hasegawa and Hiraiwa-Hasegawa (1983) and 94.4% (17/18) of all possessiveness mating in Takasaki (1985). Moreover, an alpha male more frequently shows possessiveness toward females as they approach ovulation time (Hasegawa and Hiraiwa-Hasegawa, 1983; Takasaki, 1985). An alpha male, thus, has priority of access to receptive females.

Figure 35.5 A male with an injury on his back. Male–male aggression sometimes causes severe injuries in competing males. ©S. Fujita.

An episode of rank reversal observed in the K group supports the idea of the reproductive advantage of the alpha status. When rank reversal occurred between two adult males, Kasonta and Sobongo, the frequency of copulation in these competing males changed drastically. When Kasonta still held the alpha status, his copulation rate was 85.6% (83/97). However, once Sobongo outranked Kasonta, Sobongo's copulation rate rose to 87.0% (60/69). Sobongo displaced

Kasonta in reproductive status as well as in dominance rank (Nishida, 1983). In addition, a male could manipulate social relationships to improve his copulatory share. In the rank reversal described above, Kamemanfu, the third-ranking male, showed the highest copulatory rate (19/37, 51.4%) when the alpha status was unstable. During inter-male competition, Kamemanfu occasionally changed alliance between Kasonta and Sobongo, and apparently had the deciding vote. Under this condition, neither of the dominant males seemed to intervene in Kamemanfu's copulation because he might change allegiance from one to the other (Nishida, 1983).

The ratio of females to males, however, influences success in monopolization by an alpha male among primates (Altman, 1962). When there are more receptive females in a group, mating patterns can be more promiscuous, because it is difficult for an alpha male to monopolize all estrous females. Under this condition, subordinate males have more chance to copulate with them. Hasegawa and Hiraiwa-Hasegawa (1983) compared mating patterns between two groups, the M group and the K group, with a different number of members in each group. From 1980 to 1982, the M group consisted of 12 adult males and 40 adult females and the K group of 2 adult males and 3 adult females. The percentages of opportunistic mating by adult males were 88% and 48% in the M and K groups, respectively. In addition, the percentages of possessiveness mating by the alpha males were 8% and 52% in the M and K groups, respectively. These results indicate that, in the M group, more males could have the chance to copulate with receptive females. In addition to demographic factors, males' ability to discriminate exact time of ovulation can also influence the reproductive success of an alpha male, because female chimpanzees show maximum sexual swelling beyond their fertile period (Emery Thompson and Wrangham, 2008). However, this ability of male chimpanzees has not been investigated at Mahale.

While possessiveness mating is dominated by an alpha male, other males sometimes consort with females so that they are not interfered with by mating competitors. In two observed or suspected cases of consortship at Mahale (less than 1% of total mating), one was by the second-ranking male, and the other was by a young adult male (Hasegawa and Hiraiwa-Hasegawa, 1983). When a male does not want to be bothered by other males, he takes his partner away to travel with him to the edge of the range. The female has the choice of whether or not to follow him, and she typically resists the male's proposal for consortship. She agrees to travel with him only when she finally gives up because no one comes to help her or the male's aggression is too strong (Nishida, 2012). Although consortship was not frequently observed at Mahale, this does not necessarily mean that this type of mating seldom occurs. It is probably due to the difficulty of observing this behavior as the consorting pair tries to avoid other group members and moves to the edge of range (Hasegawa and Hiraiwa-Hasegawa, 1983), reducing the probability that researchers will witness them. A pair is suspected to be under consortship when a particular male and female are simultaneously absent from the group. Consortship lasts for between a few days (Hasegawa and Hiraiwa-Hasegawa, 1983) and, in the longest case, 3 months (Nishida, 2012). Consortship was also rare at Gombe; around 2% of the observed copulations occurred during consortships (Tutin, 1979). However, 3 of 14 females conceived during consortships (Constable *et al.*, 2001). This result implies the importance of this strategy, although there are no data to confirm whether this strategy is also effective in Mahale.

Some reports showed that chimpanzee males prefer older females. In copulation by high-ranking males or in possessiveness mating, where the males might demonstrate their preference for mating partners, males copulated more with older (16 years old and over) females than with younger ones (Hasegawa and Hiraiwa-Hasegawa, 1983; Takasaki, 1985). This tendency was similar to the behavior of Kanyawara chimpanzees at Kibale, Uganda (Muller *et al.*, 2006). Muller *et al.* (2006) revealed that female attractiveness measured in terms of males' interest or male–male competition increased with female age. In addition, males did not primarily avoid nulliparous females because males distinguished between old (over 30 years old) parous females and young (under 30 years old) parous females and preferred the former (Muller *et al.*, 2006). Further, at Gombe, when adult males had a choice between two estrous females, adult males copulated with older females (Tutin, 1979), and higher-ranking males fathered offspring more often with older (Wroblewski *et al.*, 2009) or multiparous females (Newton-Fisher *et al.*, 2010). These finding suggests that males prefer older females because of their greater experience of rearing infants.

35.3.2 Female mating strategies

The basic characteristic of mating in chimpanzees is promiscuity; each female mates with multiple males. As female estrous progresses, the frequency of mating with adult males and the number of adult male mating partners increase (Hasegawa, 1991; Matsumoto-Oda, 1999). This promiscuity might come from female proceptivity. At Mahale, it is the females rather than the males that take the initiative in copulation. In preceding social interactions related to copulation, females unilaterally approach the adult males and more often solicit them to copulate than the reverse (Takahata *et al.*, 1996).

While males can impregnate more females through promiscuous mating, females are less likely to gain the benefit, because they produce only one ovum in an estrous cycle. In addition, for females, mating is a potentially costly activity in terms of time, energy, risk of predation, contracting disease, and injury. Many hypotheses have been proposed to explain why females of some mammalian species mate with multiple males, such as paternity uncertainty to counter infanticide or increasing the genetic diversity of offspring (for review see Soltis, 2002; Wolff and Macdonald, 2004). Hasegawa (1991, 1992) pointed out two plausible explanations for female promiscuity in chimpanzees. First, female mating with each male could avoid conflict among males over an estrous female, leading to strengthened male coalitions. Females may also gain benefit from such coalitions, which can counter invasions of the males of neighboring unit-groups. Second, mating with multiple males promotes sperm competition, which results in insemination by the highest quality males. If the qualities of the sperm of victorious male are heritable and correlated in some way with male quality, such competition could enhance the inclusive fitness of a female. The Mahale females demonstrate their initiative in promiscuous mating and may, therefore, exercise cryptic mate choice by promoting sperm competition. The counterstrategy of infanticide could also be an explanation of female promiscuity. However, since within-group infanticide by adult males has been reported at Mahale (Kawanaka, 1981; Nishida and Kawanaka, 1985; Takahata, 1985; Hamai *et al.*, 1992; see Chapter 26), even though the female chimpanzees whose infants were killed had engaged in promiscuous mating, promiscuous mating may not guarantee safety from aggression by mated males.

Although females are promiscuous throughout the estrous cycle as a whole, they may adopt one or both of two mating strategies (many-male and best-male) depending on the stage. When the likelihood of conception is expected to be highest, mating patterns tend to become restrictive, and the number of repeated mating incidents with high-ranking males, especially with an alpha male, increases (Hasegawa and Hiraiwa-Hasegawa, 1983; Takasaki, 1985; Matsumoto-Oda, 1999; Nishida 2012). This is partly because the alpha male has priority of access to receptive females (Hasegawa and Hiraiwa-Hasegawa, 1983; Takasaki, 1985). Females also prefer to mate with high-ranking males when they are most likely to conceive (Matsumoto-Oda, 1999). The frequency of female grooming, which is thought to be female mate choice behavior, increased when the likelihood of conception was greatest (Matsumoto-Oda, 1999). Female primates sometimes choose large, powerful, or dominant males when these males provide access to food, paternal care, or protection from other infanticidal males (Paul, 2002). Also, at other study sites, females used two mating strategies and exhibited different strategies across the estrous period. In Taï, Côte d'Ivoire, females are more selective when they are most fertile and less selective when they are least likely to conceive (Stumpf and Boesch, 2005, 2006). On the other hand, although Kanyawara females had strong mate preference during both fertile and non-fertile periods, they shifted in mate preference between these periods (Pieta, 2008). Such non-reproductive mating during non-fertile periods may allow females to obscure paternity.

Muller *et al.* (2007, 2011) pointed out the importance of the influence of male aggression in studies of female choice. They reported that at Kibale each female copulated more frequently with males that were aggressive toward her. They concluded that aggression against females forced females to mate with the aggressor males, and mating with high-ranking males was caused not by female choice but by this sexual coercion. At Mahale, solicitations by females, though few, were more frequently observed when higher-ranking males were absent than when they were present (present:absent, 4:12 for female to adult male, 1:9 for female to adolescent male; Matsumoto-Oda, 1999). This suggests that the presence of higher-ranking males at least constrains female choice. In order to see the effectiveness of male coercion at Mahale, more detailed data are necessary to address the influence of male aggression on mating success.

Physical interventions by females in the copulation of another pair have rarely been observed at Mahale (Hasegawa and Hiraiwa-Hasegawa, 1990; Hasegawa, 1991; Nishida, 1997), and female competition for male mates is unlikely to occur. However, when mates or sperm have limited availability, such as when females' fertile periods are highly synchronized and few male mating partners are available, female competition for access to males can be increased. Under this condition, female reproductive success is expected to decrease due to the depletion of the available mating resources. Matsumoto-Oda *et al.* (2007) analysed the data regarding female estrous synchrony at Mahale for more than 10 years using a newly proposed index. The results revealed that females avoid synchronizing their estrous periods. In addition, Matsumoto-Oda *et al.* (2007) found a significant inverse correlation between female birthrate and the breeding sex ratio (the number of cycling females per adult male). It is possible, then, that females incite male–male competitions through estrous asynchrony, which leads to increases in female reproductive success.

For females that are newly immigrated to a unit-group, copulation with males may have a non-reproductive function. Most immigrant females (76%) were adolescent at Mahale and copulated more frequently with males than did adult females (Hasegawa and Hiraiwa-Hasegawa, 1983, 1990; Takasaki, 1985; Hasegawa, 1991). In addition, regardless of female age, copulation rate in immigrant females was higher than in resident females (Hasegawa and Hiraiwa-Hasegawa, 1983; Takasaki, 1985). Between 1979 and 1982, all estrous females in K group immigrated into the adjacent M group because of the extinction of the males in K group (Chapter 10), and copulation rate of these immigrated females increased compared to that before immigration (Hasegawa, 1991). Immigrant females tended to copulate opportunistically with young (13–19 years old) or middle- and low-ranking males and took the initiative in such copulations (Hasegawa and Hiraiwa-Hasegawa, 1983; Hasegawa, 1991). By mating with as many males as possible, immigrant females might be aiming to extract males' support in establishing their social status in a new group. Practically, when a female immigrates to a new group, she sometimes faces harassment from other female members. When the adult males see the resident females attacking immigrant females, the males tend to take the side of the immigrant females or interfere and stop the assault (Nishida, 2012, but see Chapter 29 for a different view). Immigrant females might get a "social passport" to become a member of a new group by mating with group males (Boesch and Boesch-Achermann, 2000).

35.3.3 Paternity

Both male and female mating strategies could affect reproductive outcomes. The "priority of access" model predicts that high-ranking males can monopolize females to ensure higher reproductive success (Altmann, 1962). In addition, female mate choice for high-ranking males can promote reproductive success biased toward these males. At Mahale, as the expected date of ovulation approached, the proportion of restrictive mating increased, especially with an alpha male (Hasegawa and Hiraiwa-Hasegawa, 1983; Takasaki, 1985; Matsumoto-Oda, 1999). On the other hand, copulations during opportunistic mating still formed a relatively high proportion (79.6–86.7%) at that time (Hasegawa and Hiraiwa-Hasegawa, 1983; Takasaki, 1985). Hasegawa and Hiraiwa-Hasegawa (1990) identified mating patterns during the conception cycle, and in 10 of the 18 conceptions (55.6%), opportunistic mating was observed at least once. This result indicates that males other than the alpha male also have a chance to sire offspring via sperm competition.

The results of a recent study in which paternity of 11 offspring born from 1999 to 2005 at Mahale were analysed to support the priority of access in alpha males (Inoue *et al.*, 2008). Two different alpha males fathered 45.5% of offspring; when Fanana was the alpha male, he sired 4 of 10 offspring, and when Alofu was the alpha male, he sired 1. The fact that the alpha males sired more offspring than the others was consistent with the other study sites (Tai: Boesch *et al.*, 2006; Gombe: Wroblewski *et al.*, 2009; Budongo: Newton-Fisher *et al.*, 2010). These results indicated that alpha males enjoy preferential access to estrous females. The proportion of offspring sired by alpha males decreased as the number in the group increased (Boesch *et al.*, 2006; Inoue, 2011). Although the number of males and females in estrus may skew the male reproductive data, it is difficult to distinguish which effect has a stronger influence because the number of males is associated with female mating synchrony.

On the other hand, males other than alpha males could also sire offspring, though the proportion was small in Mahale (Inoue *et al.*, 2008). Among 10–13 paternal candidates, four males other than the alphas could father at most one offspring per male. Therefore, sperm competition through opportunistic mating might be the most effective strategy for such males, which explains the advantage of adolescent and young adult males. At Mahale, all the non-alpha males who sired offspring were less than 20 years old. Similar results were obtained at Gombe (Wroblewski *et al.*, 2009). The study showed that males aged 15–19 years old were the most reproductively successful. Therefore, age, in addition to dominance rank, is probably an important factor that influences chimpanzee males' ability to reproduce.

35.3.4 Incest avoidance

Sex-biased dispersal is common in primates, and it seems to serve as a method of preventing incest (Pusey and Wolf, 1996). In chimpanzees, females typically transfer between groups (Nishida and Kawanaka, 1972). It is possible that inbreeding occurs in the case of a mature male whose mother is still alive. However, no copulation between a mother and her mature son has been observed at Mahale (Nishida, 1997). In the case of sexually immature males aged 4–5, it is not rare for them to mate with their mothers (Nishida, 1997). In contrast to Mahale, an alpha male at Gombe copulated with his mother and produced offspring (Constable *et al.*, 2001). Another possibility for inbreeding is between a female who stayed in her natal group (see Chapter 7 for such cases) and her father or mature brother. Although such cases are rare at Mahale, copulations have been observed between an alpha male and his younger sister (Takuya Matsumoto, personal communication). These results suggest that the rarity of inbreeding is probably due to the reduced opportunity for adult males to copulate with their relatives (see Chapter 29).

References

Altmann, S. A. (1962). A field study of the sociobiology of the rhesus monkeys (*Macaca mulatta*). *Annals of the New York Academy of Sciences*, **102**, 338–435.

Boesch, C. and Boesch-Achermann, H. (2000). *The Chimpanzees of the Taï Forest*. Oxford: Oxford University Press.

Boesch, C., Kohou, G., Néné, H., and Vigilant, L. (2006). Male competition and paternity in wild chimpanzees of the Taï Forest. *American Journal of Physical Anthropology*, **130**, 103–15.

Constable, J. L., Ashley, M. V., Goodall, J., and Pusey, A. E. (2001). Noninvasive paternity assignment in Gombe chimpanzees. *Molecular Ecology*, **10**, 1279–300.

Dixson, A. F. (2009). *Sexual Selection and the Origin of Human Mating Systems*. New York: Oxford University Press.

Emery Thompson, M. and Wrangham, R. W. (2008). Male mating interest varies with female fecundity in *Pan troglodytes schweinfurthii* of Kanyawara, Kibale National Park. *International Journal of Primatology*, **29**, 885–905.

Hamai, M., Nishida, T., Takasaki, H., and Turner, L. A. (1992). New records of within-group infanticide and cannibalism in wild chimpanzees. *Primates*, **33**, 151–62.

Hasegawa, T. (1991). [Sexual behavior of wild chimpanzees.] In [*The Cultural History of Primates*], ed. T. Nishida, K. Izawa, and T. Kano. Tokyo: Heibonsha, pp. 371–80. In Japanese.

Hasegawa, T. (1992). [The evolution of female promiscuity: chimpanzees and Japanese macaques.] In [*Aggression and Cooperation among Animal Societies*], ed. Y. Ito. Tokyo: Tōkaidaigaku-shuppankai, pp. 223–50. In Japanese.

Hasegawa, T. and Hiraiwa-Hasegawa, M. (1983). Opportunistic and restrictive matings among wild chimpanzees in the Mahale Mountains, Tanzania. *Journal of Ethology*, **1**, 75–85.

Hasegawa, T. and Hiraiwa-Hasegawa, M. (1990). Sperm competition and mating behavior. In *The Chimpanzees of the Mahale Mountains: Sexual and Life History Strategies*, ed. T. Nishida. Tokyo: University of Tokyo Press, pp. 115–32.

Hayaki, H. (1985). Copulation of adolescent male chimpanzees, with special reference to the influence of adult males, in the Mahale National Park, Tanzania. *Folia Primatologica*, **44**, 148–60.

Hiraiwa-Hasegawa, M., Hasegawa, T., and Nishida, T. (1984). Demographic study of a large-sized unit-group of chimpanzees in the Mahale Mountains, Tanzania: a preliminary report. *Primates*, **25**, 401–13.

Inoue, E. (2011). Male reproductive skew and paternal kin-biased behavior in primates. In *From Genes to Animal Behavior*, ed. M. Inoue-Murayama, S. Kawamura, and A. Weiss. Tokyo: Springer, pp. 67–81.

Inoue, E., Inoue-Murayama, M., Vigilant, L., Takenaka, O., and Nishida, T. (2008). Relatedness in wild chimpanzees: influence of paternity, male philopatry, and demographic factors. *American Journal of Physical Anthropology*, **137**, 256–62.

Kano, T. (1980). Social behavior of wild pygmy chimpanzees (*Pan paniscus*) of Wamba: a preliminary report. *Journal of Human Evolution*, **9**, 243–60.

Kawanaka, K. (1981). Infanticide and cannibalism in chimpanzees, with special reference to the newly observed case in the Mahale Mountains. *African Study Monographs*, **1**, 69–99.

Matsumoto-Oda, A. (1999). Female choice in the opportunistic mating of wild chimpanzees (*Pan troglodytes schweinfurthii*) at Mahale. *Behavioral Ecology and Sociobiology*, **46**, 258–66.

Matsumoto-Oda, A., Hamai, M., Hayaki, H., *et al.* (2007). Estrus cycle asynchrony in wild female chimpanzees, *Pan troglodytes schweinfurthii. Behavioral Ecology and Sociobiology*, **61**, 661–8.

Muller, M. N., Thompson, M. E., and Wrangham, R. W. (2006). Male chimpanzees prefer mating with old females. *Current Biology*, **16**, 2234–8.

Muller, M. N., Kahlenberg, S. M., Thompson, M. E., and Wrangham, R. W. (2007). Male coercion and the costs of promiscuous mating for female chimpanzees. *Proceedings of the Royal Society B: Biological Sciences*, **274**, 1009–14.

Muller, M. N., Thompson, M. E., Kahlenberg, S. M., and Wrangham, R. W. (2011). Sexual coercion by male chimpanzees shows that female choice may be more apparent than real. *Behavioral Ecology and Sociobiology*, **65**, 921–33.

Newton-Fisher, N. E., Thompson, M. E., Reynolds, V., Boesch, C., and Vigilant, L. (2010). Paternity and social rank in wild chimpanzees (*Pan troglodytes*) from Budongo Forest, Uganda. *American Journal of Physical Anthropology*, **142**, 417–28.

Nishida, T. (1980). The leaf-clipping display: a newly-discovered expressive gesture in wild chimpanzees. *Journal of Human Evolution*, **9**, 117–28.

Nishida, T. (1983). Alpha status and agonistic alliance in wild chimpanzees (*Pan troglodytes schweinfurthii*). *Primates*, **24**, 318–36.
Nishida, T. (1997). Sexual behavior of adult male chimpanzees of the Mahale Mountains National Park, Tanzania. *Primates*, **38**, 379–98.
Nishida, T. (2002). [Demography.] In [*The Mahale Chimpanzees: 37 Years of <Panthropology>*], ed. T. Nishida, S. Uehara, and K. Kawanaka. Kyoto: Kyoto University Press, pp. 171–202. In Japanese.
Nishida, T. (2012). *Chimpanzees of the Lakeshore: Natural History and Culture at Mahale*. Cambridge: Cambridge University Press.
Nishida, T. and Kawanaka, K. (1972). Inter-unit-group relationships among wild chimpanzees of the Mahali Mountains. *Kyoto University African Studies*, **7**, 131–69.
Nishida, T. and Kawanaka, K. (1985). Within-group cannibalism by adult male chimpanzees. *Primates* **26**, 274–84.
Nishida, T., Takasakai, H., and Takahata, Y. (1990). Demography and reproductive profiles. In *The Chimpanzees of the Mahale Mountains: Sexual and Life History Strategies*, ed. T. Nishida. Tokyo: University of Tokyo Press, pp. 63–97.
Nishida, T., Corp, N., Hamai, M., *et al.* (2003). Demography, female life history, and reproductive profiles among the chimpanzees of Mahale. *American Journal of Primatology*, **59**, 99–121.
Paul, A. (2002). Sexual selection and mate choice. *International Journal of Primatology*, **23**, 877–904.
Pieta, K. (2008). Female mate preferences among *Pan troglodytes schweinfurthii* of Kanyawara, Kibale National Park, Uganda. *International Journal of Primatology*, **29**, 845–64.
Pusey, A. and Wolf, M. (1996). Inbreeding avoidance in animals. *Trends in Ecology & Evolution*, **11**, 201–6.
Shimada, M. (2014). Homosexual interactions among young female wild chimpanzees: an example of social pretend play? *Pan Africa News*, **21**, 6–8.
Soltis, J. (2002). Do primate females gain nonprocreative benefits by mating with multiple males? Theoretical and empirical considerations. *Evolutionary Anthropology*, **11**, 187–97.
Stumpf, R. and Boesch, C. (2005). Does promiscuous mating preclude female choice? Female sexual strategies in chimpanzees (*Pan troglodytes verus*) of the Taï National Park, Côte d'Ivoire. *Behavioral Ecology and Sociobiology*, **57**, 511–24.
Stumpf, R. and Boesch, C. (2006). The efficacy of female choice in chimpanzees of the Taï Forest, Côte d'Ivoire. *Behavioral Ecology and Sociobiology*, **60**, 749–65.
Takahata, Y. (1985). Adult male chimpanzees kill and eat a male newborn infant: newly observed intragroup infanticide and cannibalism in Mahale National Park, Tanzania. *Folia Primatologica*, **44**, 161–70.
Takahata, Y. (1990). Adult males' social relations with adult females. In *The Chimpanzees of the Mahale Mountains: Sexual and Life History Strategies*, ed. T. Nishida. Tokyo: University of Tokyo Press, pp. 133–48.
Takahata, Y. (2002). [Comparison of "sex" – chimpanzee and bilya.] In [*The Mahale Chimpanzees: 37 Years of <Panthropology>*], ed. T. Nishida, S. Uehara, and K. Kawanaka. Kyoto: Kyoto University Press, pp. 393–416. In Japanese.
Takahata, Y., Ihobe, H., and Idani, G. (1996). Comparing copulations of chimpanzees and bonobos: do females exhibit proceptivity or receptivity? In *Great Ape Societies*, ed. W. C. McGrew, L. F. Marchant, and T. Nishida. Cambridge: Cambridge University Press, pp. 146–55.
Takasaki, H. (1985). Female life history and mating patterns among the M group chimpanzees of the Mahale National Park, Tanzania. *Primates*, **26**, 121–9.
Thomsen, R., Soltis, J., and Teltscher, C. (2003). Sperm competition and the function of male masturbation in non-human primates. In *Sexual Selection and Reproductive Competition in Primates: New Perspectives and Directions*, ed. C. B. Jones. Norman, OK: American Society of Primatologist, pp. 437–53.
Tutin, C. E. (1979). Mating patterns and reproductive strategies in a community of wild chimpanzees (*Pan troglodytes schweinfurthii*). *Behavioral Ecology and Sociobiology*, **6**, 29–38.
Tutin, C. E. G. (1980). Reproductive behaviour of wild chimpanzee in the Gombe National Park, Tanzania. *Journal of Reproduction and Fertility*, **Suppl. 28**, 43–57.

Wallis, J. and Lemmon, W. B. (1986). Social behavior and genital swelling in pregnant chimpanzees (*Pan troglodytes*). *American Journal of Primatology*, **10**, 171–83.

Wolff, J. and Macdonald, D. (2004). Promiscuous females protect their offspring. *Trends in Ecology & Evolution*, **19**, 127–34.

Wroblewski, E. E., Murray, C. M., Keele, B. F., *et al.* (2009). Male dominance rank and reproductive success in chimpanzees, *Pan troglodytes schweinfurthii. Animal Behaviour*, **77**, 873–85.

Zamma, K. and Fujita, S. (2004). Genito-genital rubbing among wild chimpanzees. *Pan Africa News*, **11**(2), 5–8.

36 Social play: history of the studies at Mahale and a new perspective

MASAKI SHIMADA, TAKAHISA MATSUSAKA, AND HITOSHIGE HAYAKI

36.1 Studies of chimpanzee social play: an overview

Since Johan Huizinga (1950) called humans *Homo ludens* (man the player), and stated that human play was an essential part of humanity, many authors from various research fields, such as philosophy (e.g. Fink, 1960; Henriot, 1969), cultural anthropology (e.g. Huizinga, 1950; Aoyagi, 1977), sociology (e.g. Caillois, 1961; Bateson, 1972), or developmental psychology (e.g. Piaget, 1945; Vygotsky, 1967), have been interested in play behavior in humans. The beginning of the study of play among nonhuman animals goes back to the nineteenth century when Karl Groos wrote a pioneering book *The Play of Animals* in 1898. However, many ethologists started to study play in nonhuman animals in the 1970s (Fagen, 1981; Burghardt, 2005; Shimada, 2011), after the grand theory of sociobiology was established to explain function as an ultimate factor of animal behavior in general, including animal play (Wilson, 1975).

In the tradition of ethology, researchers have divided animal play behavior into the following three categories (Fagen, 1981; Burghardt, 2005; Shimada, 2012): "locomotor play (solo play)" as solitary activities, such as running, leaping, swimming, or rolling; "object play" as solitary activities involving holding detached portable objects, including pushing, pulling, or breaking an object; and "social play" as interactions among two or more individuals, such as play fighting or play chasing. Previous researchers of chimpanzee behavior also have agreed to divide play into lone (solo) play and social play (van Lawick-Goodall, 1968; Goodall, 1986). For example, Hayaki (1990: 68–9) described the phenomena of typical social play between a juvenile and an adult male in Mahale as follows:

> I have observed 4- or 5-year-old juveniles thoroughly and loudly slap and clap the backs of adult males several times. In these cases, adult males expressed play-faces when they were slapped. Sometimes, they emitted vocal expressions – sounds like "gué-gué . . .," which is known as a play-pant, a characteristic vocal repertoire of chimpanzees, and often accompanying social play. It may seem strange, but adult males look pleased to be slapped by juveniles.

Although it is difficult or impossible to define the ethological concept of "play" specifically or generally (Wilson, 1975; Burghardt, 2005; Shimada, 2011), after the long-term fieldwork on wild chimpanzees in Mahale, Nishida *et al.* (2010: 136) revealed that the interactive elements of chimpanzee social play included <play bite>, <push head into ventral>, <chase>, <circle>, <drag>, <wrestle with fingers>, <hang wrestle>, <kick heel>, <kick>, <kick back>, <pinch>, <play walk>, <pull>, <push>, <step on>, <rub>, <slap>, <spar>, <tickle>, <pull object from opposing sides>, <wrestle> (Table 36.1). Social play among chimpanzees can be observed and described as a temporal sequence of multiple actions from the elements shown above (Hayaki, 1985; Shimada, 2013a). As previous authors have pointed out, many of these behavioral elements of social play, such as <chase>, <kick>, or <spar>, are common to both play and "aggressive" actions or interactions (Symons, 1978; Fagen, 1981; Burghardt, 2005; Shimada, 2010a).

Mahale Chimpanzees: 50 Years of Research, ed. M. Nakamura, K. Hosaka, N. Itoh and K. Zamma. Published by Cambridge University Press.

Table 36.1 **List of play behavior arranged by Nishida et al. (2010)**

Type A[a]	Type B[a]	Type C[a]	Type D[a]
bounce foot/feet up and down rhythmically	balance	chase	fend
circle orthograde	bite and pull hairs	flee	handicap self
circle quadrupedal	drag and circle	kick other	hang wrestle
drag by hand	drop	push/pull swing	join play
drag by mouth	grasp and heave		play socially
drag other by hand	hurl self		struggle free
drag other by mouth	kick back		travel and play
drop self	leap down and wait		
grab	mouth and shake		
grab and shake	nod and mouth		
grapple	pinch clitoris		
grasp hand	play face		
hand and stamp	play in bed		
hang upside-down by feet	play, rough and tumble		
hang upside-down by hands	play socially with object		
hang with legs pitterpat	play with sand		
headstand	press object on		
hold body part in mouth	pull each other		
kick heel	pull object from opposing sides		
leap on	sit on		
leap up	somersault		
lie on other	stamp		
mount, misdirected	tag		
mouth	throw at animate object		
nod with body part in mouth	tickle		
play bite	trample		
play face, full	wrestle		
play face, half	wrestle bipedal		
play-pant			
poke			
press down			
press neck with lower arm			
pull			

Table 36.1 **(cont.)**

Type A[a]	Type B[a]	Type C[a]	Type D[a]
pull with mouth			
push			
push head into ventral			
release infant to fall			
roll			
shake object up and down			
shake off			
shake other			
slap other			
somersault, backward			
somersault, forward			
somersault, side			
spar			
splash water			
stamp other			
stamp quadrupedal			
stand with head down, bottom up			
swing			
wrestle with fingers			

This table is modified from Table 2 in Shimada (2013a: 248).
For detailed descriptions or definitions of the behaviors in the list, see Nishida *et al.* (2010).
[a] Type A, simple anatomical terms; Type B, complex anatomical terms; Type C, simple functional terms; Type D, complex functional terms (Nishida *et al.*, 2010: 12).

Some of these play elements themselves have attracted special attention. For example, <wrestle with fingers> (finger wrestling: van Lawick-Goodall, 1968; handwrestle: van Hooff, 1973) is known to be a playful interactive pattern unique to chimpanzees and humans, unlike <wrestle>, which can be observed among juveniles in most mammal species (Burghardt, 2005). Van Lawick-Goodall (1968) termed this interaction "finger wrestling," and described it as "gentle fondling, holding squeezing, tickling of one hand (occasionally foot) of intended play partner. This is how a mature individual typically seeks to initiate play with another adult. It may lead to a vigorous bout of social play." Hayaki (1990: 73) also pointed out that chimpanzees normally started finger wrestling calmly; later on, they often emitted play-pants, and looked very excited. Interestingly, Kano (1998), following his long-term research, reported that this interactive pattern has not been observed in wild bonobos (*Pan paniscus*).

<Circle> (chase-and-turn-around: van Lawick-Goodall, 1968; Nishida, 2012) is another conspicuous interactive pattern, described as "two or more individuals may chase round and round tree, another individual, or without any pivot-point" (Nishida *et al.*, 2010). Hayaki (1990) emphasized the interactive feature of this pattern as follows: "the role of 'chaser' is equivalent to the role of 'chasee' among individuals engaging in the circle behavior. Participants are

simultaneously playing dual roles and doubling their pleasure in play, as they do not have to interrupt their interaction to reverse their roles."

As these examples show, many researchers seem to have been attracted to the study of chimpanzee social play, and have generated many descriptions (e.g. Hayaki, 1990; Nakamura, 2009; Nishida, 2012). However, so far there have been only a few systematical studies focusing on social play among wild chimpanzees in the long research history of Mahale; social play in Mahale chimpanzees remains a largely unexplored field of research. Actually, few expert primatologists have studied social play in wild chimpanzees as a main research topic, with the exception of studies by the authors of this chapter (Hayaki, 1985; Matsusaka, 2004; Shimada, 2013a, b, 2014) and Nishida (2004, 2012). In this chapter, we introduce the studies of social play that have been carried out in Mahale, and then we present ongoing social play studies. Studies on lone (solo) play will be introduced in Chapter 40.

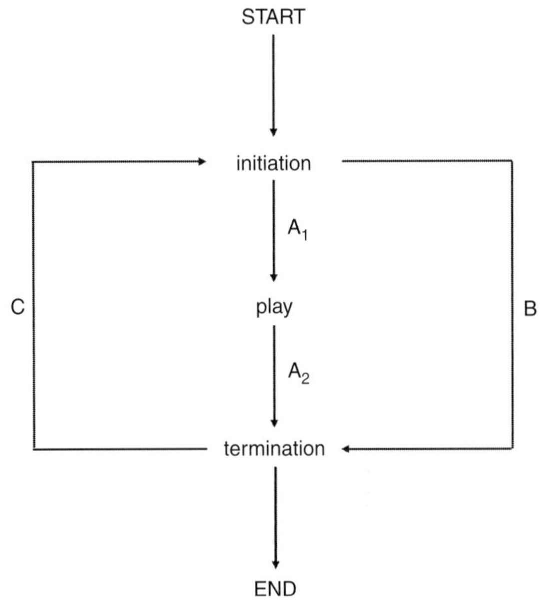

Figure 36.1 Social play in a simple scheme. A play bout is defined as an interaction from initiation to termination through A_1 and A_2, and a para-play bout through B. Para-play and play bouts often occur repeatedly (through C). Such a series of play and para-play bouts is called a play session (from START to END). Redrawn based on Fig. 1 in Hayaki (1985).

36.2 Studies of social play of Mahale chimpanzees

Although wild chimpanzee social play was first reported from the Gombe Stream, Tanzania, in the 1970s in the context of mother–infant interactions (van Lawick-Goodall, 1968; Plooij, 1979, 1984), the study of wild chimpanzee social play behavior as a main research topic began in Mahale. Hayaki started his pioneering research on the social play of M-group chimpanzees in Mahale in the 1980s. His study (Hayaki, 1985) has had a large influence on subsequent research on social play among chimpanzees.

Hayaki studied the social play of juvenile and adolescent chimpanzees in Mahale by analysing the processes of social play, and interindividual relationships in play. The methodological contrivance of his study describing social play interaction was to define not only play bouts, but also so-called "para-play" bouts. A play bout was defined as beginning when one individual directed social play elements to its partner, and as ending when the participants stopped their activities or one of them moved away. A para-play bout was defined as being initiated with an invitation, approach, or direct contact, but terminated without play interactions. Play and para-play bouts often occurred in series. Hayaki illustrated this as Figure 36.1. In this study, Hayaki took the view that social play was part of communication (Bateson, 1972).

Hayaki discussed how communication about play depended on the receiver's interpretation. That is, chimpanzees can interpret the sender's behavior as play, referring to (1) play signals accompanied by the behavior, (2) transformation of the behavior in timing, strength, or rhythm, and (3) situation of the occurrence. Depending on the receiver's interpretation, initiation attempts by the sender sometimes failed, because the receiver hesitated in playing with the other. Hayaki also stressed that the phenomenon of play "escalation" is an important aspect of social play. Although the stronger individual often reduced his/her activity during play, play tended to escalate in activity, and players may enjoy such escalation. Play seems to incorporate a mechanism that prevents it escalating into fighting. Hayaki found that play was sometimes influenced by individuals other than the players, and that the third party's movements often affected the players' interaction. The term play does not indicate a behavior itself, but the context of the behavior. That is, the players interpret their behaviors in their play context. In addition to demonstrating his idea of the cognitive or "communicative" aspect of social play, Hayaki also revealed the basic ethological data of social play among wild chimpanzees. For example, most chimpanzee social play was observed in dyadic bouts, and lasted no longer than a minute. His intensive observation data of social play among wild chimpanzees, and his perspective on the phenomena of social play, have had a large influence on subsequent researchers. These ethological features of social play are known to be shared with chimpanzees at other research sites, such as Gombe (van Lawick-Goodall, 1968; Goodall, 1986).

About 30 years after Hayaki's study, Shimada (2013a) asked why chimpanzees played in dyads. Shimada studied and described the overall structure and temporally changing configuration of social play participants among the chimpanzees of Mahale. In order to describe the complex features of chimpanzee social play, Shimada borrowed the idea of "clique" and "cluster" from Nakamura's pioneering study of grooming interactions among chimpanzees (Nakamura, 2003; see also Chapter 34). Here, microscopic clique levels were conceptualized as directly connected configurations through social play behavior, and macroscopic cluster levels were conceptualized as indirectly connected gatherings of members of adjacent multiple cliques at the same time and within the same space. Shimada found that most playing clusters started as dyads. Although the cumulative number of participants increased, each clique size remained at two for most of the 1-min observational units. Dyadic cliques were more stable and lasted longer than larger cliques, although 43 types of unit cliques were observed during the study period in total, and the maximum size of the observed unit cliques was six (Figure 36.2). Only five types were observed for more than 1% of the total duration of social play: Dy1.1 (1.7%), Dy2.1 (78.3%), Tr3.1 (1.9%), Tr4.1 (7.2%), and Tr6.1 (3.5%; see Figure 36.2 for clique types). Of all clusters' playing fields, 64.7% had maximum diameters of 3 m. In summary, chimpanzees played stably in dyadic pairs most of the time. As time passed, other chimpanzees often joined in the playing groups to form large polyadic clusters. Even when all chimpanzees in a cluster played socially at the same time, they normally did so in separate dyadic pairs, forming multiple dyadic cliques simultaneously in a small space. The overall features of social play among chimpanzees quantified by this study corresponded with those found for previous studies at Mahale (Hayaki, 1985) and other sites (van Lawick-Goodall, 1968; Goodall, 1986). Shimada suggested that social play dynamics in chimpanzees could be explained using a model derived from the sociological idea of "balance."

Balance theory was proposed by the socio-psychologist Fritz Heider (1946, 1958). He explained that, for example, people felt comfortable if they agreed with others whom they liked, while people felt uncomfortable if they disagreed with others whom they liked. Heider (1946) simplified the combination of such sentiments among entities as balanced or unbalanced. He proposed that two special properties of balance are thought to emerge in societies organized by people: "mutuality (or symmetry)," where individual *i* directs the behavior to individual *j* and *j* directs the behavior to *i*, and "transitivity," where individual *i* directs the behavior to individual *j*, *j* directs the behavior to individual *k*, and *i* directs the behavior to *k* (Heider, 1946). Shimada proposed that balanced cliques are those in which all the existing pairs form mutual dyads; during social play, they tend to avoid unbalanced, and maintain balanced relationships. As a result, larger cliques are difficult to maintain for long periods and tend to transition into dyadic mutual cliques. Thus, Heider's balance theory can potentially explain not only human social

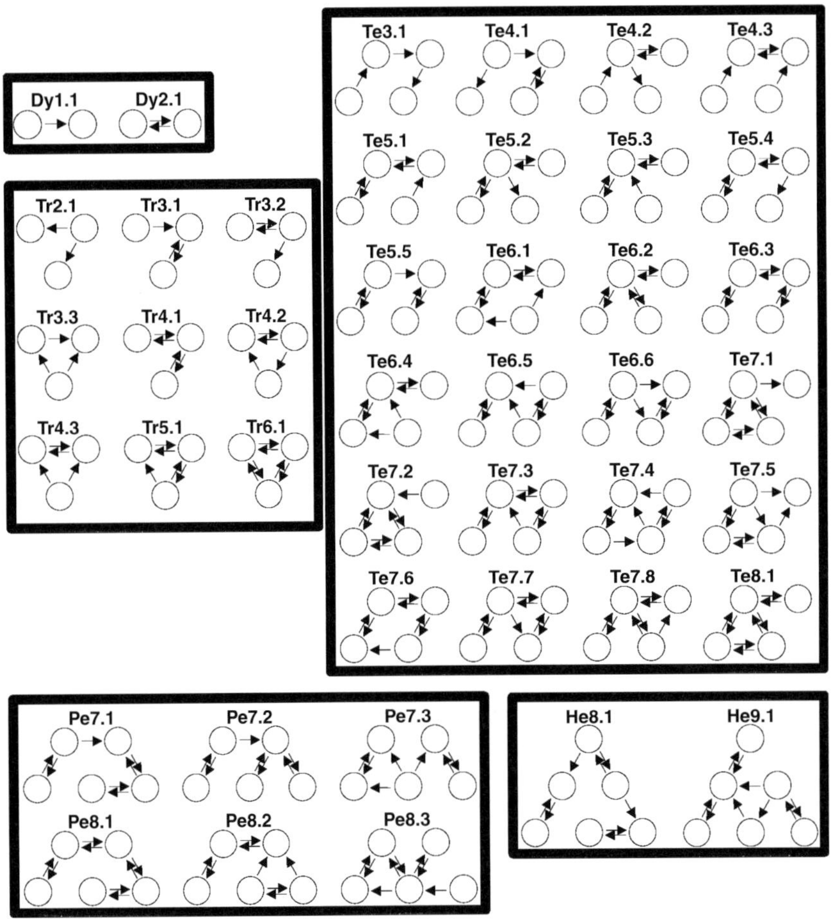

Figure 36.2 List of all the observed configurations of unit cliques. This figure is reprinted from Fig. 4 in Shimada (2013a).

phenomena, but also the proximate mechanism of the structure and temporal change of social play among wild chimpanzees. Shimada pointed out that in various human networks, both mutual and transitive relationships are balanced, whereas only mutual relationships among socially playing chimpanzees tend to be balanced.

Recently, Shimada and Sueur (2014) have studied that aspect as "social networks" of social play among chimpanzees in M group. Along with social grooming and food sharing, social play is considered an affiliative interaction among wild chimpanzees. However, infant, juvenile, and adolescent animals participate in social play more frequently than do adult animals, unlike for other affiliative interactions, such as social grooming or food sharing. Shimada has studied the social play of M-group chimpanzees at Mahale during two research periods in 2010 and 2011. Social network analysis (SNA; Wasserman and Faust, 1994; Croft *et al.*, 2008; Whitehead, 2008) revealed several features of their play network. In both periods, most members of the M group participated in social play at least once (Figure 36.3). The play network degree centralities, defined as the number of associations that individuals had (e.g. Wasserman and Faust, 1994), were significantly larger in infants, juveniles, and adolescents than in adults. There was a significant, positive correlation between participation in social play and the degree centrality of play networks. Partial play networks and partial

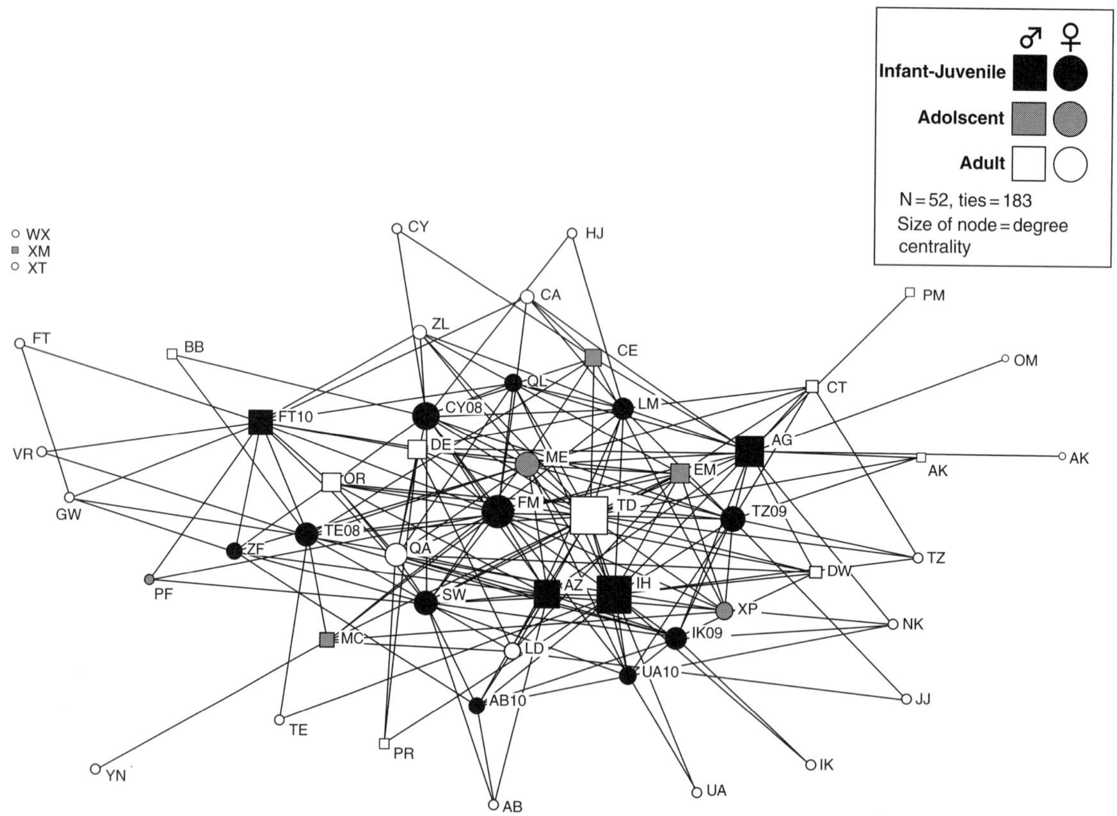

Figure 36.3 Play network of the M-group chimpanzees. The network was built using NetDraw2.123 and UCINET 6 software (Borgatti *et al.*, 2002). Nodes represent individuals. The two letters indicate the names of individuals. The graphs were drawn by a spring-embedding method (Croft *et al.*, 2008). The size of a node is directly related to the individual degree centrality. This figure is reprinted from Figure 1a in Shimada and Sueur (2014).

association networks, which were defined as consisting of individuals in the same age categories, were positively correlated for infants and juveniles, although they were not correlated for adolescents or adults. These results suggest that infants, juveniles, and adolescents who played frequently were more central, and that adults who played infrequently were more peripheral. In addition, the overall structure of the network of social play was stable over time. It was suggested that the frequency of participation in social play positively contributed to the formation of affiliative social relationships among chimpanzees. The social play network may allow for the development of social techniques that help the individual to acquire a central position in a society, and for the formation of affiliative relationships during the infant or juvenile period. Interestingly, however, this function seems to disappear when individuals enter adolescence. These findings may provide a new evolutionary viewpoint from which to consider the social networks of human children.

After long-term fieldwork at Mahale, Nishida (2004) noticed that although the play patterns of chimpanzees were multifold and complex, "group play" was completely lacking among them. In human group play, such as soccer, baseball, or the card game bridge, two groups of people compete or mock fight with each other for victory. Although many chimpanzees do engage in play fights and grapple with one another, we have never seen a team of two or more chimpanzees play fight against another team of a similar number. Nishida suggested that, according

Figure 36.4 Xantip (left) is mounting Enju (right) dorsoventrally during the first bout. This figure is reprinted from Figure 1 in Shimada (2014).

to the motor-training hypothesis (see Section 36.4) of social play, the function of group play in human beings is practice for warfare, or an organized battle between two groups (Nishida, 2012). Nishida pointed out that even in antagonistic confrontations between unit-groups of chimpanzees, only one side tends to be an organized multi-male party, with the counterpart usually being a lone individual who is victimized.

The main purposes of Hayaki (1985) and Shimada (2013a) were to reveal the communicative features or proximate factors of social play; that is, the mechanism (e.g. Davies *et al.*, 2012) by which multiple chimpanzees play socially, rather than the function or ultimate factors of social play. We will discuss why there has been little research to clarify the function of play in Section 36.4.

36.3 Other aspects of social play

Although some authors have not focused directly on social play as their main research topic, they have referred to social play among Mahale chimpanzees indirectly, in order to understand other types of behavior in relation to social play.

During social play, chimpanzees often emit play-pants. Matsusaka (2004) investigated when play panting was emitted in social play, and how play interactions were affected by its occurrence, to clarify the social functions of play panting in chimpanzees. Matsusaka found that chimpanzees emitted play panting when they were tickled or chased, but rarely did so when they tickled or chased others. This suggested that chimpanzee play panting did not have the function of a play signal, communicating that these "aggressive" actions were performed in play, not aggression. Matsusaka also found that chimpanzees emitted play panting more often when they received "aggressive" actions that supposedly elicited higher arousal, and a chimpanzee tended to continue to perform "aggressive" actions when the target emitted play panting. Thus, play panting activated social play interactions by encouraging the performer to continue tickling or chasing. Matsusaka suggested that chimpanzee play panting served as positive feedback to the play partner to encourage the continuation of somewhat fragile interactions. These types of interactions might involve the risk of excessive arousal, and the risk that the performer of "aggressive" actions might confuse "defensive" actions by the target with real efforts to escape the situation.

Since both social play and social grooming bring chimpanzees into close physical contact (van Lawick-Goodall, 1968), some authors have focused on the relationship between these two activities. Nakamura (2010) reported that chimpanzees often engaged in grooming and social play interaction by turns. Shimada (2013b) observed three cases of the simultaneous performance of social grooming and social play behaviors by individuals of M-group chimpanzees. While a chimpanzee was grooming the back of an adult bimanually, an infant or a juvenile approached the grooming individual. The individual then started playing with the infant/juvenile using only the right hand, while simultaneously grooming the back of the adult with the left hand. In one case, an old female, Calliope, continued the simultaneous performance for about 1 min. Such performances probably occur at low frequency because they are not often required. The similarity in both the neurobiological bases (Trezza *et al.*, 2010; Vanderschuren, 2010) and the functions of social grooming and social play behaviors (Nakamura, 2010), both of which include repetitive contact with the body of another individual, may facilitate their simultaneous performance. Furthermore, both social grooming and touching other individuals reduces tension (Terry, 1970; Schino *et al.*, 1988; Dunbar, 2010).

One study focused on homosexual behavior between adolescent female chimpanzees (see also Chapter 35 for information about sexual behavior among Mahale chimpanzees). It is known that wild female bonobos frequently engage in homosexual interactions, such as genito-genital rubbing (e.g. Furuichi, 1989; Idani, 1991), whereas wild female chimpanzees rarely engage in homosexual interactions (Anestis, 2004; Zamma and Fujita, 2004). Shimada (2014) intermittently observed two homosexual bouts between adolescent females, Xantip and Enju, which lasted for 38 s and 3 s, respectively (Figure 36.4). Enju's presenting, and Xantip's mounting that followed were homosexual behaviors in which their genital regions did not make contact, unlike genito-genital rubbing interactions between female bonobos. The interactions occurred in a rather calm situation, and the direct function of the interaction was unclear. The first interaction lasted substantially longer (38 s) than typical heterosexual copulations, which last for an average of 7.1 s (range 2–13 s) in the wild (Nishida, 1997), and 6.7 s (range 1–30 s) in captivity (Anestis, 2004). Even though homosexual interactions occurred very rarely, the observed properties of the interaction between Xantip and Enju suggest that it was a pattern of social play, since it fulfills the five criteria of play (Burghardt, 2005). During these interactions, Xantip and Enju took the roles of male and female, respectively, as they occur in normal heterosexual interactions. Shimada suggested that they might have synchronized their movements only to "pretend as if" they were interacting heterosexually. Pretense can be recognized if a subject intentionally performs an idea that he/she knows to be a fiction, and controls the action (Mitchell, 2007). Although no evidence of social pretend play among play partners in wild nonhuman animals has been confirmed thus far (Tomasello and Call, 1997; Bjorklund and Pellegrini, 2002), the present observation might be interpreted as a rare case of social pretend play with a division of roles.

36.4 Conclusion and future topics

We agree with the view that research of social play, as well as other types of play, should be one of the major targets of chimpanzee studies (Nishida, 2012), since we can expect that ethological studies of play behavior in nonhuman primates – our close evolutionary relatives – may develop our knowledge about the origin and evolution of humanity (Shimada, 2010b). However, we also have noticed the difficulties in studying animal play. For example, Shimada has discussed and summarized two main reasons why it is difficult for ethologists to study animal play behavior (Shimada, 2010b, 2011). The first reason is the difficulty in reaching a definition of play behavior with which all researchers are satisfied (Dugatkin, 2003: 513), even though several researchers have made efforts to do so since the sociobiologist Edward O. Wilson complained about the lack of an ethological definition of animal play behavior in his book, *Sociobiology* (Wilson, 1975: 364). Shimada (2010b: 376) pointed out that "we have no common view on what behavior or which behavioral category should be called 'play' thus far, although anyone can easily imagine and perceive the meaning of the word 'play', or 'play behavior.'" For example, Burghardt (2005) criticized the definitions of play put forward by Fagen (1981), Bekoff and Byers (1981), and Martin and Caro (1985), whereas Dugatkin (2003) criticized Bekoff and Byers' (1998) definition of play (Shimada, 2011). We show several examples of definitions of play proposed by previous researchers in Table 36.2. Since the five criteria of play proposed by Burghardt (2005) are actually useful to judge whether a particular animal behavior is play or not, some authors have applied his idea to their studies (e.g. Cordoni and Palagi, 2011; Shimada, 2014).

The second reason is the difficulty of applying the hypothesis verification style to studies, since play behavior seems to have no direct function (Shimada, 2010b, 2011). Shimada (2010b: 376) pointed that:

Table 36.2 **A glossary of definitions of play**

Author(s)	Page	Definition
Huizinga, 1950	28	Play is a voluntary activity or occupation executed within certain fixed limits of time and place, according to rules freely accepted but absolutely binding, having its aim in itself and accompanied by a feeling of tension, joy and the consciousness that it is "different" from "ordinary life."
Fagen, 1981	65	(Play is) a behavior that functions to develop, practice, or maintain physical or cognitive abilities and social relationships, including both tactic and strategies, by varying, repeating, and/or recombining already functional subsequences of behavior outside their primary context.
Bekoff and Byers, 1981	300	Play is all motor activity performed postnatally that appears to be purposeless, in which motor patterns from other contexts may often be used in modified forms and altered temporal sequencing.
Martin and Caro, 1985	65	Play is all motor activity performed postnatally that appears to an observer to have no obvious immediate benefits for the player, in which motor patterns resembling those used in serious functional contexts may be used in modified form. The motor acts constituting play have some or all of the following structural characteristics: exaggeration of movements, repetition of motor acts, and fragmentation or disordering of sequences of motor acts.
Fagen, 1993	182	Play is improvised performance, with variations, of skilled motor and communicative actions in a context separate from the environment in which behavior including these actions proximately increases reproductive success.
Burghardt, 2005	82	Play is repeated, incompletely functional behavior differing from more serious versions structurally, contextually, or ontogenetically, and initiated voluntarily when the animal is in a relaxed or low-stress setting.

many modern researchers studying animal behavior tend to design their research plans to reveal a function as an ultimate factor of the behavior they want to investigate. They build hypotheses about the function of the behavior, and then verify these hypotheses. Because many researchers agree to define play as behavior that seems to have no direct function, most researchers who adopt the so-called hypothesis verification approach inevitably hesitate or avoid studying play behavior.

Even though the function of play remains a mystery so far, many researchers from various fields believe that there must be some adaptive benefits to social play, because the costs of this behavior are obviously remarkable (e.g. Fagen, 1981; Burghardt, 2005). Pellis *et al.* (2010: 279–80) suggested three likely reasons for the difficulty with identifying the adaptive value of play. First, play is likely to be multifunctional. Second, the benefits of play may be either immediate or delayed. Third, play tends to be a behavior engaged in only after all other needs are met, so the opportunity to engage in play can vary with food availability and with other environmental stressors. As Martin and Caro (1985) have indicated, for many species, absence of play is not equal to developmental disadvantages. Thus, we cannot say that play is essential for proper development in a particular species, only that play may refine or facilitate development (Martin and Caro, 1985; Pellis *et al.*, 2010). Pellis *et al.* (2010) reviewed previous studies and suggested two possible and verifiable (but unexclusive) hypotheses about the benefits of play in general, including social play, which satisfy the three conditions mentioned above: the motor-training hypothesis (e.g. Groos, 1898; Byers and Walker, 1995), and the training for the unexpected hypothesis (e.g. Baldwin and Baldwin, 1977; Špinka *et al.*, 2001). The former hypothesis assumes that play behavior occurs frequently during the juvenile period to prepare the motor system for the adult period, while the latter hypothesis assumes that animals play to expose themselves to a varied range of actions, many of which lead to unpredictable outcomes (Pellis *et al.*, 2010: 281–2). It is possible that this training for the unexpected induces the benefit of forming a play network, which enables juvenile chimpanzees to learn social rules for living in a group (Shimada and Sueur, 2014; see also Section 36.2). Adult social play may be beneficial for assessing and manipulating social partners, as proposed by Palagi *et al.* (2004). Exploring the benefit or function of chimpanzee social play must be one of the most promising and exciting areas of this field of research.

In spite of these inevitable difficulties, as we have seen above, many authors have taken up the challenge of making social play a direct/indirect target of ethological studies of chimpanzees. Here, we introduce some additional possible research topics relating to social play among wild chimpanzees.

Juvenile Japanese macaques (*Macaca fuscata*) sometimes participate in "social object play (SOP)," in which they play socially with portable object(s). Shimada (2006, 2010b) found that SOP in a provisioned troop in Arashiyama had the following particular interactive features. First, multiple juveniles treated only one object as a target of play; second, the holder of the target object escaped from the non-holder(s), and finally, when holder-reversals happened, the new holder assumed the role of chasee. SOP consisting of these interactive features was a fixed play pattern among juveniles in Arashiyama. Shimada (2012) suggested that this play pattern should be added to the list of traditions of Japanese macaques (Perry and Manson, 2003).

Like Japanese macaques, chimpanzees often use objects during social play (Hayaki, 1985; Nishida *et al.*, 2010). Hayaki listed the behavioral patterns, such as "shake, bite or fold branch," that are social play invitations. Nishida *et al.* (2010) also pointed out that wild chimpanzees often use objects to invite a social play partner to start playing. However, there has been no further information about SOP among wild chimpanzees. Studies of the interactive patterns of SOP among chimpanzees, and the developmental process of object handling during social play, comparable with that of Japanese macaques, must be one of the most fruitful potential research areas for the future. This is particularly so since there have been no reports of traditions of social play patterns in wild chimpanzees so far, except for a report of "leaf-pile pulling," a play pattern found in Mahale (Nishida and Wallauer, 2003), which may be a possible example of traditional social play patterns among wild chimpanzees (see Chapters 38, 42).

Hayaki (1985: 356) also observed a special way of using an object during play, which he regarded as imaginary social play (see Chapter 40 for details). Doll-like manipulation of objects or carcasses has also been reported for other study sites (e.g. Wrangham and Peterson, 1996; Hirata *et al.*, 2001; see also Chapter 40). These suggested cases of imaginary social play by chimpanzees might offer interesting and valuable information about their cognitive capacity for primatologists, as well as researchers in other fields, such as psychology or philosophy. As we discussed in Section 36.3, there is still little evidence to demonstrate pretense in chimpanzees (Tomasello and Call, 1997; Bjorklund and Pellegrini, 2002).

We should emphasize, finally, that in order to understand social play among chimpanzees, the comparative view is important. We need to facilitate comparisons between different chimpanzee groups, between social play and other types of social interactions, and/or between different animal species including humans.

References

Anestis, S. F. (2004). Female genito-genital rubbing in a group of captive chimpanzees. *International Journal of Primatology*, **25**, 477–88.

Aoyagi, M. (1977). *[Cultural Anthropology of "Play".]* Tokyo: Kōdan-sha. In Japanese.

Baldwin, J. D. and Baldwin, J. (1977). The role of learning phenomena in the ontogeny of exploration and play. In *Primate Bio-Social Development: Biological, Social, and Ecological Determinants*, ed. S. Chevalier-Skolnikoff and F. E. Poirier. New York: Garland, pp. 343–406.

Bateson, G. G. (1972). A theory of play and fantasy. In *Steps to an Ecology of Mind*, ed. G. Bateson. New York: Ballantine Books, pp. 177–93.

Bekoff, M. and Byers, J. A. (1981). A critical reanalysis of the ontogeny and phylogeny of mammalian social and locomotor play: An ethological hornet's nest. In *Behavioral Development: The Bielefeld Interdisciplinary Project*,. ed. K. Immelman, G. W. Barlow, L. Petrinovich, and M. Main. Cambridge: Cambridge University Press, pp.296–337.

Bekoff, M. and Byers, J. A, eds. (1998). *Animal Play: Evolutionary, Comparative, and Ecological Perspectives*. Cambridge: Cambridge University Press.

Bjorklund, D. F. and Pellegrini, A. D. (2002). *The Origins of Human Nature: Evolutionary Developmental Psychology*. New York: American Psychological Association.

Borgatti, S. P., Everett, M. G., and Freeman, L. C. (2002). *UCINET for Windows: Software for Social Network Analysis*. Harvard, MA: Analytic Technologies.

Burghardt, G. M. (2005). *The Genesis of Animal Play: Testing the Limits*. Cambridge, MA: MIT Press.

Byers, J. A. and Walker, C. (1995). Refining the motor training hypothesis for the evolution of play. *American Naturalist*, **146**, 25–40.

Caillois, R. (1961). *Man, Play and Games*. New York: Free Press of Glencoe.

Cordoni, G. and Palagi, E. (2011). Ontogenetic trajectories of chimpanzee social play: Similarities with humans. *PLoS ONE*, **6**, e27344.

Croft, D. P., James, R., and Krause, J. (2008). *Exploring Animal Social Networks*. Princeton, NJ: Princeton University Press.

Davies, N. B., Krebs, J. R., and West, S. A. (2012). *An Introduction to Behavioural Ecology*. West Sussex, UK: John Wiley & Sons.

Dugatkin, L. A. (2003). *Principles of Animal Behavior*. New York: W.W. Norton & Company.

Dunbar, R. I. M. (2010). The social role of touch in humans and primates: Behavioural function and neurobiological mechanisms. *Neuroscience and Biobehavioral Reviews*, **34**, 260–8.

Fagen, R. (1981). *Animal Play Behavior*. Oxford: Oxford University Press.

Fagen, R. (1993). Primate juveniles and primate play. In *Juvenile Primates: Life History, Development, and Behavior*, ed. M. E. Pereira and L. A. Fairbanks. New York: Oxford University Press, pp. 182–96.

Fink, E. (1960). *Spiel als Weltsymbol*. Stuttgart: Kohlhammer W.

Furuichi, T. (1989). Social interactions and the life history of female *Pan paniscus* in Wamba, Zaire. *International Journal of Primatology*, **10**, 173–97.

Goodall, J. (1986). *The Chimpanzees of Gombe: Patterns of Behavior*. Cambridge, MA: Belknap.

Groos, K. (1898). *The Play of Animals*. New York: D. Appleton.

Hayaki, H. (1985). Social play of juvenile and adolescent chimpanzees in the Mahale Mountains National Park, Tanzania. *Primates*, **26**, 343–60.

Hayaki, H. (1990). [*The Human within Chimpanzee*.] Tokyo: Shōka-bō. In Japanese.

Heider, F. (1946). Attitudes and cognitive organization. *Journal of Psychology*, **21**, 107–12.

Heider, F. (1958). *The Psychology of Interpersonal Relations*. New York: Wiley.

Henriot, J. (1969). *Le Jeu*. Paris: Presses Universitaires de France.

Hirata, S., Yamakoshi, G., Fujita, S., Ohashi, G., and Matsuzawa, T. (2001). Capturing and toying with hyraxes (*Dendrohyrax dorsalis*) by wild chimpanzees (*Pan troglodytes*) at Bossou, Guinea. *American Journal of Primatology*, **53**, 93–7.

van Hooff, J. A. R. A. M. (1973). A structural analysis of the social behaviour of a semi-captive group of chimpanzees. In *Social Communication and Movement, ed.* M. von Caranach and I. Vine. London: Academic Press, pp. 75–162.

Huizinga, J. (1950). *Homo Ludens*. Boston: Beacon.

Idani, G. (1991). Social relationships between immigrant and resident Bonobo (*Pan paniscus*) females at Wamba. *Folia Primatologica*, **57**, 83–95.

Kano, T. (1998). A preliminary glossary of bonobo behaviors at Wamba. In *Comparative Study of the Behavior of the Genus Pan by Compiling Video Ethogram*, ed. T. Nishida. Kyoto: Nisshindo, pp. 39–81.

van Lawick-Goodall, J. (1968). The behaviour of free-living chimpanzees in the Gombe Stream Reserve. *Animal Behaviour Monographs*, **1**, 161–311.

Martin, P. and Caro, T. M. (1985). On the function of play and its role in behavioral development. *Advances in the Study of Behavior*, **15**, 59–103.

Matsusaka, T. (2004). When does play panting occur during social play in wild chimpanzees? *Primates*, **45**, 221–9.

Mitchell, R. W. (2007). Pretense in animals: The continuing relevance of children's pretense. In *Play and Development: Evolutionary, Sociocultural, and Functional Perspectives, ed.* A. Göncü and S. Gaskins. New York: Taylor & Francis, pp. 51–75.

Nakamura, M. (2003). 'Gatherings' of social grooming among wild chimpanzees: Implications for evolution of sociality. *Journal of Human Evolution*, **44**, 59–71.

Nakamura, M. (2009). [*Chimpanzees: What They Have Told Me without Language*.] Tokyo: Chūkō-shinsho. In Japanese.

Nakamura, M. (2010). [What does it mean a society to be complex?: Questions about relationship between society and individual.] *Primate Research*, **26**, 131–42. In Japanese with English summary.

Nishida, T. (1997). Sexual behavior of adult male chimpanzees of the Mahale Mountains National Park, Tanzania. *Primates*, **38**, 379–98.

Nishida, T. (2004). Lack of "group play" in wild chimpanzees. *Pan Africa News*, **11**(1), 2–3.

Nishida, T. (2012). *Chimpanzees of the Lakeshore: Natural History and Culture at Mahale*. Cambridge: Cambridge University Press.

Nishida, T. and Wallauer, W. (2003). Leaf-pile pulling: An unusual play pattern in wild chimpanzees. *American Journal of Primatology*, **60**, 167–73.

Nishida, T., Zamma, K., Matsusaka, T., Inaba, A., and McGrew, W. C. (2010). *Chimpanzee Behavior in the Wild: An Audio-Visual Encyclopedia*. Tokyo: Springer.

Palagi, E., Cordoni, G., and Borgognini Tarli, S. M. (2004). Immediate and delayed benefits of play behaviour: New evidence from chimpanzees (*Pan troglodytes*). *Ethology*, **110**, 949–62.

Pellis, S. M., Pellis, V. C., and Bell, H. C. (2010). The function of play in the development of the social brain. *American Journal of Play*, **2**, 278–98.

Perry, S. and Manson, J. (2003). Tradition in monkeys. *Evolutional Anthropology*, **12**, 71–81.

Piaget, J. (1945). *Play, Dreams, and Imitation in Childhood*. New York: Norton.

Plooij, F. X. (1979). How wild chimpanzee babies trigger the onset of mother-infant play and what the mother makes of it. In *Before Speech: The Beginning of Interpersonal Communication*, ed. M. Bullowa. Cambridge: Cambridge University Press, pp. 223–43.

Plooij, F. X. (1984). *The Behavioral Development of Free-Living Chimpanzee Babies and Infants. Monographs on Infancy*. Norwood, NJ: Ablex Publishing Corporation.

Schino, G., Scucchi, S., Maestripieri, D., and Turillazzi, P. G. (1988). Allogrooming as a tension-reduction mechanism: a behavioral approach. *American Journal of Primatology*, **16**, 43–50.

Shimada, M. (2006). Social object play among young Japanese macaques (*Macaca fuscata*) in Arashiyama, Japan. *Primates*, **47**, 342–9.

Shimada, M. (2010a). [Why do we call the interaction "play"?: From the image analysis of body contacts among juveniles of wild Japanese macaques.] In [*Boundary and Conjunction of Social Interaction: Studies in Nonhuman Primates, Humans and Conversation*], ed. D. Kimura, M. Nakamura, and K. Takanashi. Kyoto: Shōwa-dō, pp. 142–63. In Japanese.

Shimada, M. (2010b). Social object play among juvenile Japanese macaques. In *The Japanese Macaques, ed.* N. Nakagawa, M. Nakamichi, and H. Sugiura. Tokyo: Springer, pp. 375–85.

Shimada, M. (2011). [Logical justification of the ethological study of play: Ethology meets prototype theory.] *Primate Research*, **27**, 127–39. In Japanese with English summary.

Shimada, M. (2012). Social object play among juvenile Japanese macaques in provisioned Arashiyama troop, compared with that of non-provisioned Kinkazan troop. In *The Monkeys of Stormy Mountain: 60 Years of Primatological Research on the Japanese Macaques of Arashiyama,* ed. J-B. Leca,

M. A. Huffman, and P. L. Vasey. Cambridge: Cambridge University Press, pp. 258–302.

Shimada, M. (2013a). Dynamics of the temporal structures of playing clusters and cliques among wild chimpanzees in Mahale Mountains National Park. *Primates*, **54**, 245–57.

Shimada, M. (2013b). Wild chimpanzees can perform social grooming and social playing behavior simultaneously. *Primates*, **54**, 315–17.

Shimada, M. (2014). Homosexual interactions among young female wild chimpanzees: An example of social pretend play? *Pan Africa News*, **21**, 6–8.

Shimada, M. and Sueur, C. (2014). The importance of social play networks for wild chimpanzees at Mahale Mountains National Park, Tanzania. *American Journal of Primatology*, **76**, 1025–36.

Špinka, M., Newberry, R. C., and Bekoff, M. (2001). Mammalian play: can training for the unexpected be fun? *Quarterly Review of Biology*, **76**, 141–68.

Symons, D. (1978). *Play and Aggression: A Study of Rhesus Monkeys*. New York: Columbia University Press.

Terry, R. L. (1970). Primate grooming as a tension reduction mechanism. *The Journal of Psychology*, **76**, 129–36.

Tomasello, M. and Call, J. (1997). *Primate Cognition*. Oxford: Oxford University Press.

Trezza, V., Baarendse, P. J. J., and Vanderschuren, L. J. M. J. (2010). The pleasures of play: pharmacological insights into social reward mechanisms. *Trends in Pharmacological Sciences*, **31**, 463–9.

Vanderschuren, L. J. M. J. (2010). How the brain makes play fun. *American Journal of Play*, **2**, 315–37.

Vygotsky, L. S. (1967). Play and its role in the mental development of the child. *Soviet Psychology*, **5**, 6–18.

Wasserman, S. and Faust, K. (1994). *Social Network Analysis: Methods and Applications*. Cambridge: Cambridge University Press.

Whitehead, H. (2008). *Analyzing Animal Societies. Quantitative Methods for Vertebrate Social Analysis*. Chicago, IL: The University of Chicago Press.

Wilson, E. O. (1975). *Sociobiology: The New Synthesis*. Cambridge, MA: Belknap Press.

Wrangham, R. W. and Peterson, D. (1996). *Demonic Males: Apes and the Origins of Human Violence*. Boston: Houghton Mifflin.

Zamma, K. and Fujita, S. (2004). Genito-genital rubbing among the chimpanzees of Mahale and Bossou. *Pan Africa News*, **11**(2), 5–8.

37 Ethograms and the diversity of behaviors

KOICHIRO ZAMMA AND TAKAHISA MATSUSAKA

37.1 Benefits of chimpanzee ethograms

Some researchers have published ethograms or glossaries of chimpanzee behavior (e.g. Plooij, 1984; Goodall, 1989; Nishida *et al.*, 1999, 2010). One of the purposes of making a chimpanzee ethogram is to synthesize various terms used for a behavior or similar behaviors to discuss them without confusion. A chimpanzee ethogram is a good tool for people who want to learn about this species. Using the ethogram, beginners or experts can easily recognize and discuss the behavior of chimpanzees. One can also define a new term for behavior when the term is not found in the existing ethogram. The content of an ethogram in a study field becomes richer as the study field is developed (Plooij, 1984; Nishida *et al.*, 2010).

Ethograms are also useful for comparative studies. A comparison of the ethograms of *Pan* and *Homo* is useful for estimating the possible behavioral patterns of the last common ancestor of these genera, and comparison of the ethograms of chimpanzees and bonobos (*Pan paniscus*) clarifies the evolutionary trends in the species differences (Nishida *et al.*, 2010). By changing the level of the chimpanzee group that is compared, researchers can discuss various behavioral differences. Chimpanzee culture (Chapter 38) can be discussed by comparing the ethograms of chimpanzee populations; chimpanzee development (Chapter 22) can be discussed by comparing generations of a chimpanzee group; chimpanzee individuality can be discussed by comparing individual chimpanzees in a unit-group. Zookeepers can discuss behavioral differences of chimpanzees in zoos worldwide or between wild and captive animals, which can be used to enrich the lives of zoo animals. These comparative studies are activated by making ethograms for each level of group.

In addition to the many video clips of chimpanzee behaviors on websites, one video ethogram for wild chimpanzees has been published (Nishida *et al.*, 2010). Video ethograms are a good supplement to text descriptions. Although a video image of a behavior itself might not help someone to understand the meaning of the behavior, by watching video images with text explanations, those who have never seen chimpanzees can easily develop an accurate picture of the behavior, while a text definition of a behavior inevitably ignores some information on subtle movements. For example, knuckle walking is defined as "quadrupedal walk with knuckles on the ground or a large horizontal bough" (Nishida *et al.*, 2010), but the speed of movement or the order of hand and foot movements is not described. One can obtain such information only by watching video clips. Video ethograms are also useful for researchers interested in comparative studies of chimpanzees and they can detect subtle variation easily from such video ethograms.

This chapter describes the importance of updating a chimpanzee ethogram by adding novel behaviors. We will also describe some "subtle" behaviors of chimpanzees in Mahale and discuss the diversity of chimpanzee behaviors.

37.2 Making and updating an ethogram

Chimpanzees, like other animals, are mobile from birth to death. The movements involve spatiotemporal changes in various actions of various body parts, and

Mahale Chimpanzees: 50 Years of Research, ed. M. Nakamura, K. Hosaka, N. Itoh and K. Zamma. Published by Cambridge University Press.

it is impossible to describe the entire actions in words. Instead of describing all of the complex, continuous actions, researchers have divided the animal movements into a series of behaviors. A behavior is defined as "a natural unit," including a category of a characteristic action pattern distinguished from other behaviors (Martin and Bateson, 1986), and an ethogram is a catalog that describes the entire or partial repertoire of the behaviors of an animal species (Martin and Bateson, 1986).

A behavior defined in this way can be described only as a set of actions by ignoring small differences. Therefore, a behavior inevitably includes some subtle variation. For example, researchers can describe "slapping" behavior as an action in which the palm of the hand hits a substrate, but this definition does not consider variation in the speed of hitting or the range of lifting the hand. When researchers start a behavioral study, it is necessary to adequately define the behaviors in which they are interested (Martin and Bateson, 1986). Consequently, the researchers should consider whether a term used for a behavior in one study is used to refer to the same action patterns in other studies.

Behaviors described in this way tend to give the impression that chimpanzees perform them in a single sequence, but chimpanzees sometimes perform more than one behavior simultaneously. For example, chimpanzees sometimes urinate while walking, share food while eating it, or copulate with a female while kissing her weaning offspring who interferes with the copulation (Nishida *et al.*, 2010). These behaviors can be categorized as a new behavior (e.g. urination walking) or as a combination of two behaviors (e.g. urination and walking). Researchers can choose the term to describe the behavior in which they are interested.

A chimpanzee ethogram can be updated when a researcher discovers a new chimpanzee behavior, and the manner in which a new behavior is detected differs among study epochs. When chimpanzee studies began, many findings, such as tool use by wild chimpanzees (van Lawick-Goodall, 1968), were described as novel because very little was known about wild chimpanzees. These pioneer works provided thorough descriptions of the exciting and varied lives of chimpanzees and led to the discovery of many new behaviors.

When studies are continued at one site and results accumulate, new behaviors can be detected by comparing behaviors between study sites. As mentioned above, a behavior designated by a term always includes some variation. When a researcher regards such variation as an important difference, the researcher might coin a new term for the variation. This new term might be added as an independent behavior in the updated ethogram. For example, in Mahale, social scratch had been considered a variation of grooming behavior. It was described independently when it was found to be unique to Mahale and absent at other sites (Nakamura *et al.*, 2000).

When a long-term study of a chimpanzee group has been conducted, most of the chimpanzee behaviors have been defined and included in the ethogram. Then, researchers might be able to detect a novel behavior that had not been performed in the group before. In Mahale, as at other study sites, some behaviors have been confirmed to have been initiated in recent decades. As the details of these behaviors are mentioned in the chapters on tool use (Chapter 42) and play (Chapter 40), we address such examples below briefly.

In Mahale, a young adult male chimpanzee performed a novel behavior called "ground grooming" (Chapter 34) for the first time in 1991, and a decade later the behavior was performed by 10 chimpanzees belonging to the group (Nishida, 1994, 2003). Leaf sponge and leaf spoon might have been brought to M group by immigrant females or might have been innovations devised by a few chimpanzees in the group (Nishida, 1990, 2012; Matsusaka *et al.*, 2006). After 2000, the behavior spread to many group members (Matsusaka *et al.*, 2006; Chapter 42). Genito-genital rubbing among Mahale chimpanzees was first reported in 2004 (Zamma and Fujita, 2004), and the behavior was still observed in 2013 with some variation (Shimada, 2014; Chapter 40). Some novel behaviors spread to other members of the group, while others do not (Nishida, 2012). "Nasal probe" was initiated by an adult male (Nishida and Nakamura,

1993), but only two chimpanzees have been observed performing the behavior in the subsequent decades (Marchant and McGrew, 1999; Nishida *et al.*, 2009; Chapter 42). "Carrying guineafowl feathers" was initiated by an adult female (Nakamura, 2009; Chapter 42), but so far, no other chimpanzees have done this. In 2010, a young female started "eating oil palm pith" (Zamma *et al.*, 2011; Chapter 40), but because the female probably emigrated from M group and has been absent since 2012 (MMCRP, unpublished data), the behavior might have come to an end, at least in M group.

These novel behaviors tell us that chimpanzees do not behave within an existing behavioral set, but they occasionally innovate some behaviors and stop performing others. Consequently, an ethogram should be updated continuously. Recording the changes in chimpanzee behaviors in a continuously updated ethogram is useful as basic data for discussion of chimpanzee culture, development, and individuality.

37.3 Records of subtle behavioral variations

The chimpanzee ethogram records species- or group-specific behaviors; however, not all chimpanzees always behave in such ways. Some chimpanzees display idiosyncratic behaviors, and some behaviors are observed only once. Some of those behaviors were studied because they were thought to have a potential to expand in the group as cultural behavior (Nishida *et al.*, 2010). Other behaviors were thought to have developed at the individual level or to have little function, and are rarely the focus of research. While such behaviors are difficult to discuss in chapters on a specialized discipline, they are components of chimpanzee behavioral diversity. In this section, we describe such behaviors to show the behavioral diversity of chimpanzees.

37.3.1 Nipple fumbling

In 1994, Nishida (1994) reported "nipple press" (or nipple fumbling) as an idiosyncratic behavior performed by an adolescent male in Mahale, named Alofu. Alofu sometimes pressed his nipples with his fingers when he pant-grunted to adult males (Nishida, 1994; Marchant and McGrew, 1999), and occasionally he pressed both nipples with one hand while standing (Marchant and McGrew, 1999; Nishida *et al.*, 2010). Nishida *et al.* (2010) described Alofu's unique behavior as follows: "Alofu dexterously alternates between the right and left nipple when anxious and pant-grunting. When using the right hand, he simultaneously fumbles the right nipple with his thumb and the left nipple with his little finger, but when he is using the left hand, he fumbles the left nipple with his thumb and the right nipple with his little finger."

Since Alofu typically did this when he pant-grunted to high-rank males, this behavior was speculated to serve as a visual signal of pant-grunting (Marchant and McGrew, 1999). However, even when he was the alpha male from 2004 to 2006 (Nishida, 2012), he occasionally pressed his nipples. On July 27, 2004, Alofu consorted with an estrous female, Pinky, and when she pant-grunted, Alofu stood bipedally and pressed his nipples with his hand. On July 29, 2004, he pressed his nipples with his hand before he started to hunt red colobus. On October 27, 2006, while Alofu was consorting with the estrous female Christina, five adult males approached him while displaying and pant-hooting. Alofu also displayed to them, and at the end of the display exchange, Alofu touched his nipples with his hand and lay on the ground. This nipple press behavior was thought to be self-reassurance or self-stimulation activity (Nishida *et al.*, 2010), and Alofu continued to perform the behavior after he fell from alpha status in 2007 (Figure 37.1).

The nipple fumbling behavior has been also observed in other M-group chimpanzees, but very infrequently. Nishida (1997) reported that at least three adult females (Wakilufya, Fatuma, and Wakusi) touched their nipples with their fingers, and Marchant and McGrew (1999) observed three other adult females (Abi, Gwekulo, and Ikocha) perform the behavior. Nishida (2010) reported that he and Matsusaka observed nipple press or nipple fumbling behavior in three adult females (Abi, Fatuma, and

Figure 37.1 Alofu, an adult male, presses his nipples with his right thumb and forefinger on December 17, 2010 (after he fell from alpha status in M group). ©K. Zamma.

Figure 37.2 Charles, an adolescent male, pinches his left nipple with his right thumb and forefinger on September 11, 1999. ©K. Zamma.

Wakusi), an adult male (Fanana), four adolescent females (Puffy, Qanat, Carmen, and Rubicon), and two adolescent males (Pimu and Christmas). During 24 months of field studies from 1999 to 2008, Zamma observed at least seven chimpanzees besides Alofu perform the behavior: two adolescent males (Charles and Christmas), two adult females (Wakusi and Abi), an adolescent immigrant female (Qanat), and two juvenile females (Athena and Liz) (Zamma, personal observation).

Although the frequency of observation was low, we notice two characteristics of this behavior. First, the nipple press behavior is observed in various age–sex classes, including males, nulliparous females, and females with an infant offspring. The nipple is used to breastfeed infants, and lactating females or females who have lost a newborn have occasionally been observed to suck their own nipples (Matsumoto-Oda, 1997; Nishida *et al.*, 2010). Therefore, females with infants might be more interested in their nipples, but this might not be related to the nipple press behavior. Second, although the behavior is defined as manipulation of an individual's own nipples with a thumb or finger (Nishida *et al.*, 2010), it includes three variations: pressing the nipple with the thumb or finger (Christmas), pinching the nipple with the thumb and finger (Wakusi, Liz, Athena, and Charles, Figure 37.2), and rubbing the nipple with a thumb or fingers (Abi and Qanat, Figure 37.3a, b).

These characteristics suggest that the performers of nipple fumbling did not acquire the behavior through social learning, but on their own. Nishida (2010) noted that some relatives shared the behavior and discussed the possibility of genetic inheritance of the behavior: Puffy is Alofu's daughter; Alofu and Athena are Wakusi's offspring; and Fatuma is Pimu's mother. Nishida (2010) observed the behavioral similarity between Puffy and her father, Alofu, and their behavioral pattern was described as massaging the nipple with the thumb; that was not same as the behavioral pattern used by Athena who is Alofu's sister (see above). The nipple might be a sensitive organ in chimpanzees, and the sensitivity of the nipples might vary among individuals. The nipple press behavioral pattern might not be inherited, but nipple sensitivity can be inherited. An individual who has sensitive nipples might find that tension is relaxed by fumbling the nipple, and the others who have less sensitive nipples might not find utility in the behavior.

37.3.2 Lip flipping

Chimpanzees have dexterous lips, which they use when feeding on plants (Chapter 17) and small ants (Chapter 21). The lips are also used for capturing lice while grooming (Chapter 34) and for communicating via facial expressions, vocalizations, and kisses (Goodall, 1986). Lip flipping is such a chimpanzee

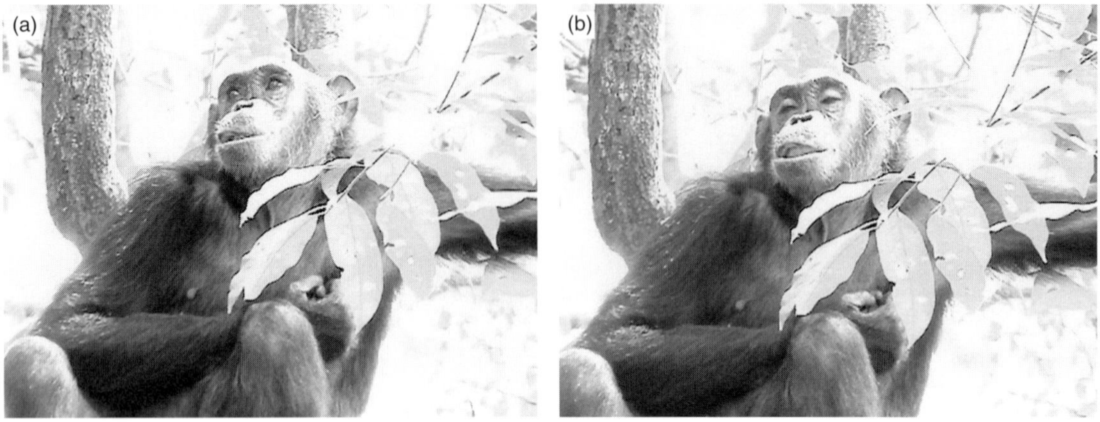

Figure 37.3 Qanat, an immigrant adolescent female, rubs her left nipple with her right thumb on October 21, 2003. The thumb moves from the top of her nipple (a) to the bottom (b). ©K. Zamma.

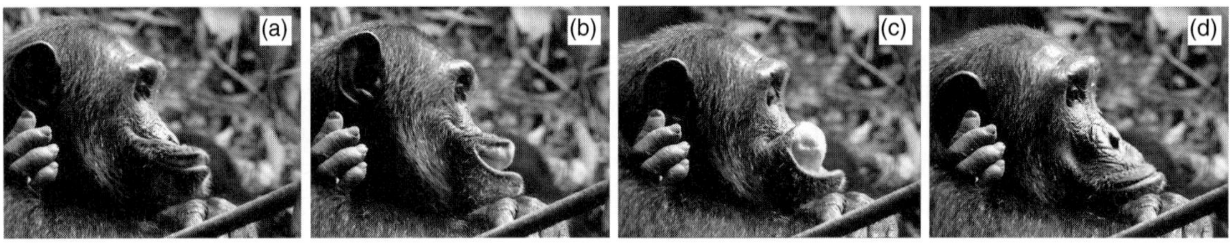

Figure 37.4 Masudi, an adult male, flips his upper lip on December 28, 1999. ©K. Zamma.

behavior and it is defined as the upper lip being rolled up and back towards the nose (Figure 37.4a–d; Goodall, 1989; Nishida *et al.*, 2010). The function of this behavior is unknown, but it was reported that chimpanzees perform the behavior when they are relaxed or bored (Goodall, 1989; Nishida *et al.*, 2010).

Zamma observed lip-flipping behavior in M-group chimpanzees 88 times in the 24 months of observations between 1999 and 2008, and 14 chimpanzees performed the behavior (Table 37.1). The observation times differed among the individuals, so the individual differences in frequencies cannot be discussed. However, the behavior was observed in almost all age-sex classes. Some chimpanzees repeated the behavior within short intervals; for example, Fanana, an adult male, flipped his lip three times during 2 min beginning at 11:44 h on May 8, 2000.

There might be slight variation in the action of the lips; chimpanzees flipped their lips in different situations, such as while resting, after feeding, and while communicating with others. Many lip-flipping behaviors were observed when chimpanzees were resting. For example, Masudi, an adult male, flipped his lip when he was lying on the ground alone at 10:43 h on October 18, 1999, and Acadia, a juvenile female, flipped her lip when she was resting at 15:01 h on June 17, 2004. As Nishida *et al.* (2010) described, the lip-flipping chimpanzees appeared bored.

Chimpanzees sometimes moved and flipped their lips after eating fruit or ants. On December 22, 1999, Bonobo, an adult male chimpanzee, ate *Saba comorensis* fruit and rubbed his upper and lower lips with each other to remove the white sap of the fruit from his lips. One minute later, he flipped his upper lip. On October 6, 1999, Abi, an adult female, flipped her lip just after eating *Crematogaster* ants. These observations suggest that chimpanzees sometimes flip their lips to reduce a feeling of strangeness of the lips caused by certain foods.

Lip-flipping behavior is occasionally observed simultaneously with a social behavior. On April 3,

Table 37.1 **Lip-flipping behavior observed during 24 months of field research in Mahale from 1999 to 2008**

Sex	Age class	Name	*N*
Female	Adult	Abi	1
		Miya	11
		Qanat[a]	1
		Sally	1
	Adolescent	Qanat[a]	3
	Juvenile	Acadia	1
		Rubicon	13
	Infant	Aqua	1
		Swala	1
Male	Adult	Bonobo	11
		Dogura	1
		Fanana	10
		Masudi	27
	Adolescent	Charles	2
	Juvenile	Michio	4

[a] Qanat was observed performing lip-flipping behavior when she was an adolescent and as an adult.

2000, Fanana exhibited lip flipping while shaking a branch to solicit copulation from an estrous female. On September 7, 1999, Bonobo flipped his lip as he looked at Masudi, who was eating fruit. On July 30, 2004, Rubicon, a juvenile female, approached her grandmother while lip flipping.

Chimpanzees performed lip flipping in multiple contexts; therefore, chimpanzees might not use the behavior for a single specific function. We imagine the following scenario leading to the acquisition of lip-flipping behavior. First, a chimpanzee discovers that the lip has the potential to be rolled up when he/she moved his/her lips after eating certain foods that caused lip discomfort. After making this discovery, he/she might do it to pass time. Then, some individuals started to use the behavior as a facial expression to express their own intention to reduce another's tension in the same manner as "grinning," for example. However, it would be difficult for another individual to understand the meaning of a novel facial expression; therefore, lip flipping as a facial expression failed to spread among the group members.

Another facial expression called "funny face" also failed to spread among M-group members. Alofu was the only individual who was reported to use funny face, which was defined as a facial expression that involved moving the lips and squeezing the eyes (Marchant and McGrew, 1999, Figure 37.5). He showed funny face to estrous females and adult males who were unwilling to follow him despite frequent solicitation; therefore, its function might be to reassure (Nishida *et al.*, 2010), although researchers might have interpreted the facial expression simply by reading other social behaviors, such as hand extension, that were performed along with the facial expression. Funny face might have little meaning by itself, but it might reinforce an

Figure 37.5 Funny face expressed by Alofu (right). ©K. Zamma.

intentional behavior, similar to the phrase "very much" in English.

37.3.3 Leaf for scratching a persistent itch

Chimpanzees sometimes perform subtle behaviors that have clear objectives. At 08:45 h on March 16, 2000, Opal, an adult female, was eating *Ficus trichopoda* fruit. She stopped eating at 09:07 h, and started resting and performed some self-grooming for a short time. At 09:17 h, she scratched her left ear with her left thumb, and at 09:22 h, she scratched it with her right thumb. She continued scratching for 2 min. Then, she scratched her head once before starting to scratch the left ear again. At 09:25 h, after scratching her leg, she picked a leaf from a branch with her right hand and rubbed it on the left ear. She threw the leaf away shortly afterwards and started scratching her ear with her hand again. At 09:29 h, after scratching her belly, back, chest, and arms, she picked another leaf and rubbed her left ear with the leaf twice using her right hand, looked at the leaf, and threw it away. Then, she started scratching her left ear with her hand and continued for 12 min, before leaving.

It was clear that Opal used the leaves to try to resolve the persistent itch of her left ear; however, the behavior has not been observed again, perhaps because the leaf did not bring about a desirable effect. Trial-and-error does not necessarily resolve a problem, but it might be a source of new behavioral patterns, and some of which, if they become fixed, would increase the behavioral diversity of chimpanzee.

37.4 Conclusion

This chapter focused on the behavioral diversity of chimpanzees and the importance of recording it in an ethogram. A new behavioral pattern can be a variation of a known behavior or a completely novel behavior formed by a chimpanzee. Such behaviors are important not only because they are the seeds of local culture (Nishida *et al.*, 2010) but also because they reflect the individuality and flexibility of chimpanzees. As Nishida *et al.* (2010: 203) mentioned, "we have currently reached only the starting point of understanding the behavioral diversity and flexibility of chimpanzees." Chimpanzee researchers occasionally witness unrepeated subtle behaviors. The accumulation of these findings will clarify the evolutionary trends in the behavioral diversity of chimpanzees.

A video image or a photograph is useful for recording a behavior in an ethogram because it includes the subtle variation in a behavior that is difficult to describe in a written definition. Even old photographs or video images can provide new findings. Nakamura and Uehara (2004) examined photographs of the chimpanzees of an extinct group and found a behavioral variation of the grooming hand-clasp between two unit-groups (see Chapter 38). "Subtle" differences in behaviors might be regarded as such based on the standards of researchers, but such seemingly subtle behaviors or variations might be significant for chimpanzees and useful for answering some research questions. A video ethogram is considered to be more helpful for comparative studies of chimpanzees because it includes "subtle" variation in behaviors.

Chimpanzee behaviors are always challenging (Nishida *et al.*, 2010), and researchers will continue to discover new behaviors. Updating the chimpanzee ethogram might contribute to understanding the finer details of their behaviors.

References

Goodall, J. (1986). *The Chimpanzees of Gombe: Patterns of Behavior*. Cambridge, MA: Belknap.

Goodall, J. (1989). *Glossary of chimpanzee behaviors*. Tucson: Jane Goodall Institute.

van Lawick-Goodall, J. (1968). The behaviour of free-living chimpanzees in the Gombe Stream Reserve. *Animal Behaviour Monographs*, **1**, 165–311.

Marchant, L. F. and McGrew, W. C. (1999). Innovative behavior at Mahale: new data on nasal probe and nipple press. *Pan Africa News*, **6**, 16–18.

Martin, P. and Bateson, P. (1986). *Measuring Behaviour: An Introductory Guide*. Cambridge: Cambridge University Press.

Matsumoto-Oda, A. (1997). Self-suckling behavior by a wild chimpanzee. *Folia Primatologica*, **68**, 342–3.

Matsusaka, T., Nishie, H., Shimada, M., *et al*. (2006). Tool-use for drinking water by immature chimpanzees of Mahale: prevalence of an unessential behavior. *Primates*, **47**, 113–22.

Nakamura, M. (2009). Aesthete in the forest? A female chimpanzee at Mahale collected and carried guineafowl feathers. *Pan Africa News*, **16**, 17–19.

Nakamura, M. and Uehara, S. (2004). Proximate factors of different types of grooming hand-clasp in Mahale chimpanzees: implications for chimpanzee social customs. *Current Anthropology*, **45**, 108–14.

Nakamura, M., McGrew, W. C., Marchant, L. F., and Nishida, T. (2000). Social scratch: another custom in wild chimpanzees? *Primates*, **41**, 237–48.

Nishida, T. (1990). A quarter century of research in the Mahale Mountains: an overview. In *The Chimpanzees of the Mahale Mountains: Sexual and Life History Strategies, ed*. T. Nishida. Tokyo: University of Tokyo Press, pp. 3–35.

Nishida, T. (1994). Review of recent findings on Mahale chimpanzees: implications and future research directions. In *Chimpanzee Cultures*, ed. R. W. Wrangham, W. C. McGrew, F. B. M. de Waal, and P. G. Heltne. Cambridge, MA: Harvard University Press, pp. 373–96.

Nishida, T. (1997). Sexual behavior of adult male chimpanzees of the Mahale Mountains National Park, Tanzania. *Primates*, **38**, 379–98.

Nishida, T. (2003). Individuality and flexibility of cultural behavior patterns in chimpanzees. In *Animal Social Complexity: Intelligence, Culture, and Individualized Societies*, ed. F. B. M. de Waal and P. L. Tyack. Cambridge, MA: Harvard University Press, pp. 392–413.

Nishida, T. (2010). Puffy inherits a habit from her father? *Pan Africa News*, **17**, 10–11.

Nishida, T. (2012). *Chimpanzees of the Lakeshore: Natural History and Culture at Mahale*. Cambridge: Cambridge University Press.

Nishida, T. and Nakamura, M. (1993). Chimpanzee tool use to clear a blocked nasal passage. *Folia Primatologica*, **61**, 218–20.

Nishida, T., Kano, T., Goodall, J., McGrew, W. C., and Nakamura, M. (1999). Ethogram and ethnography of Mahale chimpanzees. *Anthropological Science*, **107**, 141–88.

Nishida, T., Matsusaka, T., and McGrew, W. C. (2009). Emergence, propagation or disappearance of novel behavioral patterns in the habituated chimpanzees of Mahale: a review. *Primates*, **50**, 23–36.

Nishida, T., Zamma, K., Matsusaka, T., Inaba, A., and McGrew, W. C. (2010). *Chimpanzee Behavior in the Wild: An Audio-Visual Encyclopedia*. Tokyo: Springer.

Plooij, F. X. (1984). *The Behavioral Development of Free-Living Chimpanzee Babies and Infants*. Norwood, NJ: Ablex.

Shimada, M. (2014). Homosexual interactions among young female wild chimpanzees: an example of social pretend play? *Pan Africa News*, **21**, 6–8.

Zamma, K. and Fujita, S. (2004). Genito-genital rubbing among the chimpanzees of Mahale and Bossou. *Pan Africa News*, **11**(2), 5–8.

Zamma, K., Nakashima, M. and Ramadhani, A. (2011). Mahale chimpanzees start to eat oil palm. *Pan Africa News*, **18**, 6–8.

PART VIII Behavioral diversity

38 Culture

HITONARU NISHIE AND MICHIO NAKAMURA

38.1 Introduction

Two comprehensive ethograms of Mahale chimpanzees based on data accumulated over 50 years have been published (Nishida *et al.*, 1999, 2010; see Chapter 37). These descriptions of behavioral patterns in the Mahale chimpanzees allowed us to identify novel behaviors not previously included in ethograms, follow the process of their diffusion within the unit-group, and compare behavioral repertoires among the wild chimpanzee unit-groups/populations (see Table 38.1 for the list of putative cultural behaviors in the Mahale K and M groups). Accordingly, studies of wild chimpanzee cultures have focused on variations in behavioral patterns within/among unit-group(s) of chimpanzees that cannot be explained by genetic or ecological factors or on the mechanisms of diffusion and transmission of innovative behaviors within a unit-group (Nishida, 1987; McGrew, 1992, 2004; Wrangham *et al.*, 1994; Whiten *et al.*, 1999, 2001; de Waal, 2001).

Many studies have investigated tool use in wild chimpanzees in terms of culture (for a review, see McGrew, 1992). Such material culture studies have an advantage in that tool/object use can be readily recognized and analysed in unhabituated as well as habituated populations. Furthermore, tool use is associated with higher cognitive ability, which is tacitly regarded as a prerequisite for human culture (Nakamura, 2010a). However, as tool/object use is further discussed in Chapter 42, this chapter focuses on putative cultural behaviors in the social domain of the wild chimpanzees at Mahale. Here, we will review how studies of the Mahale population have contributed to the current understanding of wild chimpanzee culture.

38.2 Cultural behaviors in Mahale chimpanzees: overview of major findings

38.2.1 In the 1960s and 1970s

Although rarely referred to in the current discussions of chimpanzee culture, the first major findings in Mahale chimpanzees were identification of the "unit-group" as the basic social group and the "fission–fusion" characteristics of the daily ranging pattern (Nishida, 1968; see Chapters 6, 9). Prior to Nishida's work with the Mahale chimpanzees in the 1960s, Japanese primatologists had conducted empirical and theoretical studies of the ecology, social structure, and culture of Japanese macaques (*Macaca fuscata*) (Itani and Nishimura, 1973; see Chapter 3). Following previous studies in Japanese macaques, Nishida adopted methods such as individual identification and provisioning for habituation to human observers. He also set up a long-term study that allowed him to identify the unit-group as the basic social unit of wild chimpanzees (Nishida, 1968), and identify several cultural behaviors (summarized below). Moreover, under the mentorship of Kinji Imanishi (Imanishi, 1952), the founder of Japanese primatology, Nishida recognized that nonhuman animals with continuous group living had their own culture (Nishida, 1987, 2003). Imanishi's view of culture, which has been

Mahale Chimpanzees: 50 Years of Research, ed. M. Nakamura, K. Hosaka, N. Itoh and K. Zamma. Published by Cambridge University Press.

Table 38.1 **Cultural behaviors observed in the Mahale K and M groups**

Behavior	Mahale K	Mahale M	Source
Aimed throw	Present	Customary	K: Nishida and Hiraiwa, 1982 M: Nishida, 1994, 2003
Ant-fish	Customary	Customary	K: Nishida, 1973; Nishida and Hiraiwa, 1982 M: Nishida and Hiraiwa, 1982; Nishie, 2011
Bee probe	Present	–	K: Nishida and Hiraiwa, 1982
Drum metal wall	–	Customary	M: Nishida, 1994, 2003
Expel/stir	Habitual	Habitual	K: Nishida, 1973; Nishida and Hiraiwa, 1982 M: Huffman and Kalunde, 1993; Nakamura and Itoh, 2008
Fluid dip	Habitual	Habitual	K: Nishida and Hiraiwa, 1982 M: Matsusaka and Kutsukake, 2002; Matsusaka *et al.*, 2006
Ground groom	–	Present	M: Nishida, 1994, 2003
Hand-clasp grooming	Customary	Customary	K: McGrew and Tutin, 1978; McGrew *et al.*, 2001; Nakamura and Uehara, 2004 M: McGrew *et al.*, 2001; Nakamura, 2002; Nakamura and Uehara, 2004
Knuckle knock	Customary	Customary	K: Whiten *et al.*, 1999 M: Nishida, 1997 (*thump*)[a]; Nishida *et al.*, 1999 (*rap*)
Leaf clip	Customary	Customary	K: Nishida, 1980b M: Nishida, 1980b, 1987, 1997
Leaf groom	Customary	Customary	K: Nishida, 1980b M: Zamma, 2002, 2006
Leaf napkin	Present	Present	K: Nishida and Hiraiwa, 1982 M: Nishida, 2003; Corp *et al.*, 2009
Leaf strip	–	Present	M: Nishida *et al.*, 1999 (*pull-through*)
Nipple press	–	Present	M: Nishida, 1994; Marchant and McGrew, 1999
Rain dance	Customary	Customary	K: Whiten *et al.*, 1999 M: Nishida *et al.*, 1999 (*display, rain*)
Shrub bend	–	Customary	M: Nishida, 1987, 1994, 1997
Social scratch	Customary	Customary	K: Nishida, 1983 (*scratch or rub*) M: Nakamura *et al.*, 2000
Stem pull-through	–	Habitual	M: Nishida, 1994, 1997
Termite fish	Customary	Present	K: Nishida and Uehara, 1980; Uehara, 1982 M: Takahata, 1982

The putative cultural behaviors on the list and the status of each behavior are based on Whiten *et al.* (2001).
[a] The italic names in parentheses are synonyms used in respective references.

tacitly incorporated in Japanese primatology, was not limited to higher cognitive behaviors such as the sweet potato washing observed in Japanese macaques or tool use in chimpanzees, nor did it have connotations of the noble, intentional, sophisticated, and complex aspects of intellectual activities often regarded as prerequisites for human culture. Imanishi's notion extended to broader social phenomena among group-living animals including social structure, dominance rank, and the characters of individuals in various social contexts (Itani and Nishimura, 1973; Nishida, 1987, 2003; Nakamura and Nishida, 2006; Nakamura, 2010a). Accordingly, Mahale researchers recently proposed the application of broader and more holistic perspectives to chimpanzee cultural studies including social structure, ranging pattern, microscopic formation of social gatherings, and the manner of social interaction (Nakamura, 2003, 2010a; Nakamura and Nishida, 2006; Nishie, 2008, 2012; see below).

In the early to mid-1970s, tool-use behaviors such as *Camponotus* ant fishing, termite fishing, and leaf clipping, and medicinal use of leaves were discovered one after another (see reviews in Chapters 24, 42). Current cultural primatologists point to these early discoveries as examples of variation and plasticity of the chimpanzee behavioral repertoire and the extensive knowledge of their habitat, which appears to be learned and shared socially among members of the unit-group.

The grooming hand-clasp was first documented as a social custom among the K-group chimpanzees at Mahale in the mid-1970s (McGrew and Tutin, 1978). The authors, who had studied chimpanzees at Gombe, Tanzania, witnessed the behavior at Mahale and noted that it had never been observed at Gombe. They concluded that this difference between two chimpanzee unit-groups was derived from social customs or cultural variations of the respective groups. However, several influential cultural anthropologists were highly critical of this interpretation and rejected it, insisting that the concept of culture could not be applied to nonhuman species because they do not have the linguistic ability to teach or learn knowledge as tradition (Lévi-Strauss, 1967; Washburn and Benedict, 1979; Sperber, 1982).

38.2.2 In the 1980s and 1990s

By the 1980s, knowledge accumulated over 15 years enabled comparisons between Mahale and Gombe chimpanzees and the discovery of novel behavioral patterns at Mahale.

Nishida (1980a) compared the reaction of Mahale and Gombe chimpanzees to water and concluded that the Mahale chimpanzees were less hydrophobic than those in Gombe. Mahale, but not Gombe, chimpanzees were observed wading in streams, drinking water from the lake, and becoming wet while gathering aquatic plants from the swamp. (Additional observations have been reported sporadically at Mahale; Hosaka, 1995.) Nishida suggested that these differences were cultural rather than idiosyncratic.

Nishida *et al.* (1983) identified differences in the plant-food repertoires of the Mahale and Gombe chimpanzees. They argued that because the differences in plant-food species and feeding techniques could not be fully explained by ecological factors in the respective habitats, and the plant-feeding habit in each group was likely to be transmitted across generations through social learning, they represented cultural differences between two unit-groups.

Between 1981 and 1982, Mahale chimpanzees were observed to incorporate fruit species such as guava, mango, and lemon into their food repertoire (Takasaki, 1983; Takahata *et al.*, 1986). Mangos did not become an important dietary item for the M group because most were eaten by baboons before ripening, although chimpanzees have been observed to eat them sporadically. On the other hand, lemons and guavas are currently important foods for the M group (Nishida, 2003).

In 1987, the M-group chimpanzees were observed to consume leaves and juice from the pith of shoots for medicinal use on two occasions (see Chapter 24).

In 1997, the first case of predation on a yellow baboon by the M-group chimpanzees was observed, and several additional cases followed (see Chapter 20). Newly observed cases of tool use for probing tree holes to hunt mammalian prey were reported in the 1990s, and similar cases were reported in the following decade (see Chapter 42).

In the late 1990s, the "social scratch" (see below for behavioral description) behavior observed in the M group was reported as another example of a locale-specific social custom, as the behavior had not been observed at any of the other long-term study sites (Nakamura *et al.*, 2000). However, the social scratch was later observed at Gombe (Shimada, 2002) and Ngogo, Uganda (Nishida *et al.*, 2004), although the type of scratching differed among sites (see below).

In the late 1990s, the function of leaf grooming, in which chimpanzees seize and manipulate a leaf in a grooming context as if they "groom" the leaf, was found to be squashing ectoparasites (lice) (Zamma, 2002). Subsequent studies revealed local variations in the pattern of squashing ectoparasites among the Mahale (leaf grooming; Zamma, 2002, 2006), Gombe (leaf grooming; Boesch, 1995), Taï, Côte d'Ivoire (index hit; Boesch, 1995), and Bossou, Guinea (index to palm; Nakamura and Nishida, 2006) chimpanzees (see also Chapters 34, 42).

The first exhaustive ethogram of Mahale chimpanzees was published in 1999 (Nishida *et al.*, 1999). In the same year, a seminal article on chimpanzee culture was published, summarizing local variations in the behavioral repertoires of wild chimpanzees in the seven long-term study sites across Africa (Whiten *et al.*, 1999).

38.2.3 After the 2000s

The flourishing studies of chimpanzee culture in the 1990s (McGrew, 1992; Wrangham *et al.*, 1994; Nishida *et al.*, 1999; de Waal, 1999; Whiten *et al.*, 1999) have been followed by several recent investigations of local variation in the behavioral repertoires of wild chimpanzee unit-groups/populations.

Detailed comparisons of the behavioral patterns involved in the grooming hand-clasp were conducted between the K and M groups at Mahale (McGrew *et al.*, 2001; Nakamura and Uehara, 2004). Nakamura (2002) investigated local variations in the grooming hand clasp across wild chimpanzee populations that could not be explained by genetic or ecological conditions in the respective populations.

The use of tools for drinking water was rare among Mahale chimpanzees in the 1990s, but became prevalent among immature chimpanzees in the M group in the early 2000s (see Chapter 42). Moreover, leaf-pile pulling, a playful behavior, has been observed frequently among immature chimpanzees of the M group (see Chapter 42 for details). Leaf-pile pulling has never been documented in other chimpanzee populations, with the exception of a few cases at Gombe, and thus is regarded as an example of chimpanzee cultural variation (Nishida and Wallauer, 2003). Corp *et al.* (2009) described muzzle/hand rubbing, a tool/object use behavior that increased from 1998, reached a peak around 2000, and declined after 2002 (Nishida, 2003; see Chapter 42 for details).

The Y group, an unhabituated chimpanzee group at Mahale, has been observed and heard occasionally in the former range of the K group since the late 1990s (Chapter 11). A preliminary survey of the Y group revealed novel feeding habits that differed from those of the M group, including frequently feeding on *Camponotus brutus* ants and young stems of the *Cissus oliveri*, both of which were rarely, if ever, eaten by the K and M groups (Nishida and Uehara, 1983; Sakamaki *et al.*, 2007; Nishie, 2012). Although the Y-group chimpanzees are not habituated to human observers, we expect to document more behavioral diversity between the neighboring unit-groups in Mahale.

In the 2000s, Mahale researchers visited other long-term study sites as a "culture hunting" (Nishida, 2012) mission to compare behavioral patterns and repertoires among the long-term wild chimpanzee study sites. Nishida (2003, 2012; Nishida *et al.*, 2004) visited Ngogo in August 2001 and observed several behavioral differences between Ngogo and Mahale chimpanzees, such as the type of social scratch (see above and below), the sounds uttered during social grooming ("lip smacking" and "teeth clacking" at Mahale versus "sputtering" at Ngogo; Nishida *et al.*, 2004), and the frequency of pant-grunt among adult males (frequently observed from adult males to the alpha male at Mahale, but rarely observed among adult males at Ngogo; Nishida, 2012). Shimada (2002) visited Gombe and observed social scratch by three chimpanzees, which had not been previously observed in Gombe chimpanzees. In 2003, Nakamura visited Bossou and reported subtle behavioral variations compared with the Mahale chimpanzees, such as "mutual genital touch," "heel tap," "index to palm," and "sputter" (Nakamura and Nishida, 2006).

In 2010, a comprehensive ethogram of Mahale chimpanzees was published as an audio-visual encyclopedia based on video images recorded since 1999 (Nishida *et al.*, 2010). The ethogram will enable chimpanzee researchers who have never visited Mahale to view the behavioral patterns performed by Mahale chimpanzees on the video footage and to compare them with those of chimpanzees in other study sites. Thus, the audio-visual encyclopedia is expected to facilitate identification of local variations in the behavioral patterns/repertoires of wild chimpanzees, which is a possible starting point in characterizing specific chimpanzee cultures (Whiten *et al.*, 1999; McGrew, 2004).

38.3 Cultural behaviors in the social domain

In this section, we review in detail the social behaviors thought to signify culture and discuss the social aspects of culture in wild chimpanzees.

38.3.1 Grooming behaviors

Grooming hand-clasp

The grooming hand-clasp (McGrew and Tutin 1978; Nakamura, 2002) is a classic example of cultural variation outside tool use. For this behavior, two chimpanzees sit face to face, clasp their corresponding hands or wrists (either right or left) overhead to form an "A-frame" and groom each other's underarm with their free hands (Figure 38.1). This behavioral pattern

Figure 38.1 (a) Grooming hand clasp between an adult male and an adult female of the Mahale K group. ©Shigeo Uehara. (b) Grooming hand clasp between two adult males of the Mahale M group. ©M. Nakamura.

is customarily performed by chimpanzees at Mahale (M and K groups), Kibale (Kanyawara and Ngogo groups: Uganda), Kalinzu (Uganda), and Lopé (Gabon), whereas it has been observed in only a few chimpanzees at Taï and has never been reported at Gombe, Budongo (Uganda), or Bossou (Nakamura, 2002). The pattern has also been observed in captive chimpanzees at the Yerkes Primate Center in the United States (de Waal and Seres, 1997; Bonnie and de Waal, 2006).

McGrew and Tutin (1978) first documented this behavior describing it as a candidate for a social custom, or even "culture," if the definition of culture were sufficiently broad to include nonhuman species. However, apart from one negative response (Washburn and Benedict, 1979), their discovery was largely forgotten by anthropologists and primatologists because the time for "cultural primatology" had not yet arrived, at least in the West (McGrew, 2004). Cultural primatology came of age in the 1990s, following McGrew's review (1992) of tool use among wild chimpanzees (de Waal, 1999). The landmark *Nature* paper by Whiten *et al.* (1999) included social behaviors such as the grooming hand-clasp; however, most of the behavioral patterns they described were related to feeding or material objects.

McGrew *et al.* (2001) described two different patterns of the grooming hand-clasp: palm-to-palm and non-palm-to-palm hand-clasp. In the former pattern, groomers clasp each other's palm, whereas in the latter, their wrists are often flexed and one hand rests on the other. The authors noted that the palm-to-palm pattern dominated in the Mahale K group, but was not observed in the M group. However, re-examination of a larger dataset revealed that the palm-to-palm pattern was performed in the M group, albeit infrequently; however, all of the observed cases involved an immigrant female from the K group, Gwekulo (Nakamura and Uehara, 2004). Thus, the authors concluded that the palm-to-palm pattern was not transmitted successfully to the M group, but was performed only during interactions with Gwekulo, who continued to perform the customary pattern of the K group.

Nakamura (2002) reviewed the distributions of hand-clasp and branch-clasp grooming among the long-term study sites; the behaviors differ only in that chimpanzees grasp an overhead branch instead of clasping their hands in branch-clasp grooming. He found that in contrast to the hand-clasp, branch-clasp grooming has been observed in all the long-term wild chimpanzee study sites, and it therefore may be considered universal. Nakamura noted that hand-clasp grooming is a rare example of culture as a "pure" social behavior that does not use material objects. He argued that inter-population variations in the behavior could not be attributed to ecological or genetic differences and that the use of hand-clasp grooming was completely arbitrary because the same function, cleaning the underarm region, could be attained using the ubiquitous form of branch-clasp grooming.

Social scratch

The social scratch, in which an individual scratches the body, usually the back, of another individual during social grooming, is customarily performed at Mahale (Nishida *et al.*, 1999; Nakamura *et al.*, 2000). This behavior has been observed at Gombe (Shimada, 2002) and Ngogo (Nishida *et al.*, 2004), but not at Kanyawara, Bossou, or Taï (Nakamura *et al.*, 2000). This interaction is so straightforward – simply scratching another's body – that Nishida (2012) was amazed to find that chimpanzees at other sites did not engage in the behavior. Furthermore, Nishida *et al.* (2004) noted a difference in social scratch behavioral patterns between Mahale and Ngogo chimpanzees such that Ngogo chimpanzees scratched using their fingers to *poke* the body of their grooming partner while keeping their fingers straight, whereas most Mahale chimpanzees used flexed fingers to *stroke* the body of their partner, although an adult female in the Mahale M group, Ako, and her daughter, Acadia, displayed the poke-type behavior (Nishida *et al.*, 2004; Nakamura, 2010b).

Grooming "density" and development of a grooming network

Most studies of chimpanzee culture have focused on distinctive patterns of complex behaviors (implying complex cognitive abilities and intelligence) such as tool/object use; however, Nakamura (2010a) argued that there are more quantitative than qualitative local differences in social behavior among sites that can be described as cultural variations. Using published datasets, Nakamura compared the densities of male–male and female–female grooming networks, determined by plotting the number of grooming partners against the number of individuals of each sex, among the long-term wild chimpanzee study sites. He found that male grooming networks were very dense in general, although the density of Ngogo male networks appeared to be lower than that of the other sites, presumably owing to the large number of males at Ngogo. The female grooming networks were relatively sparse in Gombe, Kanyawara (Kibale), and Sonso (Budongo), whereas the grooming densities of Bossou and Mahale females were as high as those of the males. At Mahale, mothers had up to 40 grooming partners and averaged about 20.3 partners each, similar to the pattern of adult males (average: 19.3 partners each), whereas juvenile and older female infants had 10–20 grooming partners (average: 14.3 partners each).

Nakamura (2010a) argued that the differences among female grooming densities across sites could not be explained simply by differences in group size, subspecies, degree of habituation to human observers, observation methods, or local cultural variations, but was likely a result of a confluence of these factors. Nevertheless, it may be that female sociality and other social behaviors have cultural significance in their respective societies because culture and other factors are not mutually exclusive, but rather are interactive. Moreover, Nakamura suggested that grooming networks among mothers could affect those of their offspring; thus, it may be that the social breadth (sociality) of female chimpanzees is socially inherited as culture in respective societies of wild chimpanzees.

38.3.2 Feeding habits

Nishida *et al.* (1983) compared the types of food and food species consumed by chimpanzees at Mahale and Gombe. Among the 143 potential plant-food species, 85 (59.4%) were eaten at both sites. Of the 286 potential food types, 104 (36.4%) were eaten at both sites, whereas the remaining 182 (63.6%) types were ignored by either of the populations. Furthermore, the authors observed differences in feeding techniques; for example, Gombe chimpanzees banged the hard-shelled fruits of *Strychnos* spp. against a tree trunk or a rock to open them, whereas Mahale chimpanzees always bit them open.

Local differences in diet have been largely ignored in studies of wild chimpanzee culture because feeding habits, with the exception of tool/object use feeding behaviors, have been regarded as subsistence activities that are largely determined by ecological, not cultural, factors. Thus, feeding habits or local differences in diet have been studied in terms of ecology, not culture, because subsistence activities do not meet the classic criteria for culture in nonhuman animals (cf. McGrew and Tutin, 1978). However, as early as the 1980s, Nishida *et al.* (1983) maintained that if local differences in diet were transmitted from generation to generation through social learning, they may be reasonably referred to as cultural differences, despite their being influenced by or originating from various environmental factors, including ecological ones. This point resonated with recent animal culture debates, some of which criticized the dichotomy between culture and nature (ecology/genetics).

At Mahale, chimpanzees incorporated new food items into their diet between the 1980s and 1990s. Conversely, the consumption of other food species or items decreased. For example, Sally, a newly immigrated adolescent female, was observed eating algae in a stream three times in 1997 (Sakamaki, 1998; Chapter 11); however, the behavior has not been observed since (Nishida *et al.*, 2009). The impact of ecological and cultural factors on chronological changes in feeding habits requires further study.

38.3.3 Play behaviors

With the exception of leaf-pile pulling (see above and Chapters 36, 40, 42), play behavior among chimpanzees has not been studied in terms of culture;

thus, little is known about local variation in play. Perhaps play has been ignored because it is often viewed as "functionless"; however, play behaviors may provide examples of local variation, innovation, and behavioral diffusion (Chapter 40), and they should thus be considered in studies of chimpanzee culture, much like children's culture has recently received growing attention in studies of human behavior.

38.3.4 Grouping patterns

Although grouping patterns have not been considered in current studies of wild chimpanzee culture, here we review differences in grouping patterns or cohesiveness among study sites that may have been shaped by culture.

Itoh and Nishida (2007) investigated the correlation between party size (cohesiveness) and food (fruit) availability in the M-group chimpanzees at Mahale. They found that the size of the "nomadic party" was correlated with fruiting plant density and mean potential patch size, whereas the size of the "face-to-face party" was weakly or not correlated with density and potential patch size (see Chapter 9 for definitions of the two types of parties). Furthermore, the authors compared nomadic and face-to-face party sizes among unit-groups. The cohesiveness of the Mahale and Bossou unit-groups, estimated by relative nomadic party size, was greater than that of the Kanyawara unit-group. The maximum face-to-face party size relative to unit-group size was higher among the Mahale K and M, Bossou, and Budongo groups than among groups in other sites in some years. Thus, the Mahale K and M and the Bossou unit-groups were likely to be capable of higher cohesiveness when ranging, whereas cohesiveness when ranging tended to be lower at the unit-group level in other sites. Nakamura (Chapter 8) also noted similarities in the grouping pattern of Mahale and Bossou chimpanzees based on his observations at both sites.

Grouping pattern, or cohesiveness of the unit-group, has been studied primarily in relation to ecological factors such as food availability and seasonal fluctuation or patch size and distribution; however, few studies have considered grouping pattern as a cultural entity. The concept of culture is generally applied narrowly to locale-specific behavioral variation that cannot be explained by ecological or genetic factors (e.g. Nishida, 1987; Whiten *et al.*, 1999; de Waal, 2001). However, several researchers (Fragaszy, 2003; Laland and Janik, 2006, 2007) have criticized this "ethnographic" method (Wrangham, *et al.*, 1994) or "method of exclusion" (Krützen *et al.*, 2007) for seeking to isolate cultural variation by ruling out alternative explanations for behavioral differences among populations. These critics argued that culture is not a stand-alone phenomenon, but rather one that is influenced by several variables including ecological and genetic factors; thus, all factors are expected to be mutually interactive rather than mutually exclusive (Fragaszy, 2003; Laland and Janik, 2006, 2007). If we accept this argument, it follows that grouping pattern or cohesiveness, which has been traditionally considered a response to ecological conditions, is a cultural phenomenon in which animals learn and reproduce conventional or customary responses to specific environmental conditions. For instance, Itoh and Nishida (2007) reported that the Mahale M group showed high cohesiveness and maintained large nomadic parties because the liana species *Saba comorensis* was hyper-dispersed during the high fruiting season. Itoh (2013) subsequently questioned this conclusion: chimpanzees are *merely able to aggregate* when many fruits are available, but they *need not aggregate* because *Saba* fruits are widespread in their home range. She argued that Mahale chimpanzees detected the signs of *Saba* fruiting and used the information to aggregate in conventional ways, which could be described as their doing "*how all others usually do* in such a situation" (Itoh, 2013). In this respect, grouping patterns may be discussed in terms of culture if chimpanzees behave according to "others' ways of doing"; thus, this meets one of the basic definitions of culture proposed by McGrew (2004): "*the way we do things.*"

38.4 Implications and future directions for chimpanzee culture studies: taking "social" seriously

Culture is a difficult term to define, even in anthropology; however, most would agree that culture is "social" rather than "individual," whether referring to human or nonhuman animal culture. Here, we review the importance and implications of focusing on the social domain in studies of chimpanzee culture.

Culture is, by definition, social. Nonetheless, most studies of nonhuman animal culture have focused on the "behavior of the individual" as a unit of analysis, downplaying the social aspect as merely "the addition of individuals" or "the process of transmitting information between individuals." For example, Whiten *et al.* (1999, 2001) referred to local variations in behavioral repertoires as wild chimpanzee cultures and classified the behavioral patterns in relation to their local frequencies of occurrence, i.e. customary, habitual, present, and absent, based on *the number of individuals* who exhibited each behavioral pattern in the respective groups. Thus, their focus was firmly on the behavior of the individual, and "social" merely indicated "the addition of individuals." Similarly, cognitive psychologists have typically viewed culture as representative knowledge and information installed into an individual's mind (or brain) to generate certain behavioral patterns or skills and as the process of social learning through which representative knowledge and information is transmitted between individuals. Here, the social domain is trivialized as "the process of transmitting information or knowledge between individuals." These mainstream contemporary studies of nonhuman animal culture tacitly imply that "atomistic individuals" are the basic agents of culture and "the additions of individuals" and "the transmission of information or knowledge between individuals" is the basis for the social aspect of culture. This view of culture and society (social) has been criticized as a "metaphor of transmission" or "Xerox model" (Ingold, 1998), "individual agent-based information transfer model of culture" (Nakamura, 2010a), and "cognition-transmission model" (Nishie, 2008, 2012) in which animal culture is viewed in terms of an individual's cognitive ability to transmit and receive information and knowledge, and to exhibit certain skills based on that knowledge. These critics argue that we should deal with culture not as assemblies of representative knowledge in individuals minds or as behavioral patterns exhibited by individuals, but as the process of ongoing social practice through which knowledge, not *detachable* from one's body but *embodied* or *embedded* in it, far from being transmitted or "jumping" from one individual to another, undergoes continuous generation and regeneration in the course of one's life history within the respective environments (Ingold, 1998).

If we take this argument seriously, it is reasonable to broaden the concept of "social" to include other environmental factors such as "genetic" or "ecological." We typically use the concept of "social" not only as the opposite of "individual" but also as the opposite of "material" or "biological" which is the root of the culture/nature dichotomy. However, from the more developmental and genuinely ecological (Ingold, 1996) perspective on culture, it makes no sense to describe activities with conspecifics and inanimate components as "social" (or "culture") and "material" (or "nature": "ecological" and "genetic"), respectively, because conspecifics, as well as inanimate objects, are part of the total environmental situation in which one dwells, and material objects are as incorporated in the course of one's activities as are conspecifics (Ingold, 1996, 1998; Fragaszy, 2003; Nishie, 2008, 2012). Rather, we should view social, material (ecological and genetic), and all other factors as mutually interactive to generate the ongoing processes of activities. During such continual processes, knowledge, as a synthesis of embodied skills for actions and perceptions, is formed through ontogeny and persists through generations. This process-oriented view may encompass the holistic process of culture as a continual practice that is grounded in, not detached from, the *context* in which all of these factors contribute to the generation and development of embodied knowledge and skills.

This line of reasoning allows us to deal with a much broader range of phenomena in terms of culture, such as social behaviors (Section 38.3.1), feeding habits (Section 38.3.2), play behaviors (Section 38.3.3), and grouping patterns (Section 38.3.4), that have not necessarily been studied as indices of culture in wild chimpanzees. In this respect, we need not seek local variation in behavioral repertoires as possible candidates for culture, nor must we exclude local variation that is ecologically explicable. This approach resonates with contemporary works in anthropology that focus on the process of practice, in which embodied knowledge and skills are embedded (Ingold, 1998). Thus, it may help us to broaden our perspective and make fair comparisons between human and nonhuman cultures.

References

Boesch, C. (1995). Innovation in wild chimpanzees (*Pan troglodytes*). *International Journal of Primatology*, **16**, 1–16.

Bonnie, K. E. and de Waal, F. B. M. (2006). Affiliation promotes the transmission of a social custom: handclasp grooming among captive chimpanzees. *Primates*, **47**, 27–34.

Corp, N., Hayaki, H., Matsusaka, T., *et al.* (2009). Prevalence of muzzle-rubbing and hand-rubbing behavior in wild chimpanzees in Mahale Mountains National Park, Tanzania. *Primates*, **50**, 184–9.

Fragaszy, D. (2003). Making space for traditions. *Evolutionary Anthropology*, **12**, 61–70.

Hosaka, K. (1995). [Water play.] *Monkey*, **39**(2), 20–1. In Japanese.

Huffman, M. A. and Kalunde, M. S. (1993). Tool-assisted predation on a squirrel by a female chimpanzee in the Mahale Mountains, Tanzania. *Primates*, **34**, 93–8.

Imanishi, K. (1952). [Evolution of humanity.] In [*Man*], ed. K. Imanishi. Tokyo: Mainichi-shinbun-sha, pp. 36–94. In Japanese.

Ingold, T. (1996). Situating action VI: a comment on the distinction between the material and the social. *Ecological Psychology*, **8**, 183–7.

Ingold, T. (1998). Comments. *Current Anthropology*, **39**, 606–7.

Itani, J. and Nishimura, A. (1973). The study of infrahuman culture in Japan. In *Precultural Primate Behavior*, ed. E. W. Menzel. Basel: Karger, pp. 26–50.

Itoh, N. (2013). [Dual-rings of mode of coexistence and act choices: Chimpanzee group and emergence of institution.] In [*Practices, Conventions and Institutions: the Evolution of Human Sociality*], ed. K. Kawai. Kyoto: Kyoto University Press, pp. 143–66. In Japanese.

Itoh, N. and Nishida, T. (2007). Chimpanzee grouping patterns and food availability in Mahale Mountains National Park, Tanzania. *Primates*, **48**, 87–96.

Krützen, M., van Schaik, C., and Whiten, A. (2007). The animal cultures debate: response to Laland and Janik. *Trends in Ecology & Evolution*, **22**, 6.

Laland, K. N. and Janik, V. M. (2006). The animal cultures debate. *Trends in Ecology & Evolution*, **21**, 542–7.

Laland, K. N. and Janik, V. M. (2007). Response to Krützen *et al.*: further problems with the "method of exclusion". *Trends in Ecology & Evolution*, **22**, 7.

Lévi-Strauss, C. (1967). *Les Structures Élémentaires de la Parenté*. The Hague: Mouton & Co and Maison des Sciences de l'Homme.

Marchant, L. F. and McGrew, W. C. (1999). Innovative behavior at Mahale: new data on nasal probe and nipple press. *Pan Africa News*, **6**, 16–18.

Matsusaka, T. and Kutsukake, N. (2002). Use of leaf-sponge and leaf-spoon by juvenile chimpanzees at Mahale. *Pan Africa News*, **9**, 6–9.

Matsusaka, T., Nishie, H., Shimada, M., *et al.* (2006). Tool-use for drinking water by immature chimpanzees of Mahale: prevalence of an unessential behavior. *Primates*, **47**, 113–22.

McGrew, W. C. (1992). *Chimpanzee Material Culture: Implications for Human Evolution*. Cambridge: Cambridge University Press.

McGrew, W. C. (2004). *The Cultured Chimpanzee: Reflections on Cultural Primatology*. Cambridge: Cambridge University Press.

McGrew, W. C. and Tutin, C. E. G. (1978). Evidence for a social custom in wild chimpanzees? *Man*, **13**, 234–51.

McGrew, W. C., Marchant, L. F., Scott, S. E., and Tutin, C. E. G. (2001). Intergroup differences in a social custom of wild chimpanzees: The grooming hand-clasp of the Mahale Mountains. *Current Anthropology*, **42**, 148–53.

Nakamura, M. (2002). Grooming-hand-clasp in Mahale M group chimpanzees: implication for culture in social behaviors. In *Behavioural Diversity in Chimpanzees and Bonobos*, ed. C. Boesch, G. Hohmann, and L. F. Marchant. Cambridge: Cambridge University Press, pp. 71–83.

Nakamura, M. (2003). [Questions about chimpanzee culture studies.] *Ecosophia*, **12**, 55–61. In Japanese.

Nakamura, M. (2010a). Ubiquity of culture and possible social inheritance of sociality among wild chimpanzees. In *The Mind of the Chimpanzee*, ed. E. V. Lonsdorf, S. R. Ross, and T. Matsuzawa. Chicago, IL: The University of Chicago Press, pp. 156–67.

Nakamura, M. (2010b). Poke-type social scratching persists at Mahale. *Pan Africa News*, **17**, 15–17.

Nakamura, M. and Itoh, N. (2008). Hunting with tools by Mahale chimpanzees. *Pan Africa News*, **15**, 3–6.

Nakamura, M. and Nishida, T. (2006). Subtle behavioral variation in wild chimpanzees, with special reference to Imanishi's concept of *kaluchua*. *Primates*, **47**, 35–42.

Nakamura, M. and Uehara, S. (2004). Proximate factors of different types of grooming hand-clasp in Mahale chimpanzees: implications for chimpanzee social custom. *Current Anthropology*, **45**, 108–14.

Nakamura, M., McGrew, W. C., Marchant, L. F., and Nishida T. (2000). Social scratch: another custom in wild chimpanzees? *Primates*, **41**, 59–71.

Nishida, T. (1968). The social group of wild chimpanzees in the Mahali Mountains. *Primates*, **9**, 167–224.

Nishida, T. (1973). The ant-gathering behaviour by the use of tools among wild chimpanzees of the Mahali Mountains. *Journal of Human Evolution*, **2**, 357–70.

Nishida, T. (1980a). Local differences in responses to water among wild chimpanzees. *Folia Primatologica*, **33**, 189–209.

Nishida, T. (1980b). The leaf-clipping display: a newly-discovered expressive gesture in wild chimpanzees. *Journal of Human Evolution*, **9**, 117–28.

Nishida, T. (1983). Alloparental behavior in wild chimpanzees of the Mahale Mountains, Tanzania. *Folia Primatologica*, **41**, 1–33.

Nishida, T. (1987). Local traditions and cultural transmission. In *Primate Societies*, ed. B. B. Smuts, D. L. Cheney, R. M. Seyfarth, R. W. Wrangham, and T. T. Struhsaker. Chicago, IL: The University of Chicago Press, pp. 462–74.

Nishida, T. (1994). Review of recent findings on Mahale chimpanzees: implications and future research directions. In *Chimpanzee Cultures*, ed. R. W. Wrangham, W. C. McGrew, F. B. M. de Waal, and P. G. Heltne. Cambridge, MA: Harvard University Press, pp. 373–96.

Nishida, T. (1997). Sexual behavior of adult male chimpanzees of the Mahale Mountains National Park, Tanzania. *Primates*, **38**, 379–98.

Nishida, T. (2003). Individuality and flexibility of cultural behavior patterns in chimpanzees. In *Animal Social Complexity: Intelligence, Culture, and Individualized Societies*, ed. F. B. M. de Waal and P. Tyack. Cambridge, MA: Harvard University Press, pp. 392–413.

Nishida, T. (2012). *Chimpanzees of the Lakeshore: Natural History and Culture at Mahale*. Cambridge: Cambridge University Press.

Nishida, T. and Hiraiwa, M. (1982). Natural history of a tool-using behavior by wild chimpanzees in feeding upon wood-boring ants. *Journal of Human Evolution*, **11**, 73–99.

Nishida, T. and Uehara, S. (1980). Chimpanzees, tools, and termites: another example from Tanzania. *Current Anthropology*, **21**, 671–2.

Nishida, T. and Uehara, S. (1983). Natural diet of chimpanzees (*Pan troglodytes schweinfurthii*): long-term record from the Mahale Mountains, Tanzania. *African Study Monographs*, **3**, 109–30.

Nishida, T. and Wallauer, W. (2003). Leaf-pile pulling: an unusual play pattern in wild chimpanzees. *American Journal of Primatology*, **60**, 167–73.

Nishida, T., Wrangham, R. W., Goodall, J., and Uehara, S. (1983). Local differences in plant-feeding habits of chimpanzees between the Mahale Mountains and Gombe National Park, Tanzania. *Journal of Human Evolution*, **12**, 467–80.

Nishida, T., Kano, T., Goodall, J., McGrew, W. C., and Nakamura, M. (1999). Ethogram and ethnography of Mahale chimpanzees. *Anthropological Science*, **107**, 141–88.

Nishida. T., Mitani, J. C., and Watts, D. P. (2004). Variable grooming behaviors in wild chimpanzees. *Folia Primatologica*, **75**, 31–6.

Nishida, T., Matsusaka, T., and McGrew, W. C. (2009). Emergence, propagation or disappearance of novel behavioral patterns in the habituated chimpanzees of Mahale: A review. *Primates*, **50**, 23–36.

Nishida, T., Zamma, K., Matsusaka, T., Inaba, A., and McGrew, W. C. (2010). *Chimpanzee Behavior in the Wild*. Tokyo: Springer.

Nishie, H. (2008). [Chimpanzee culture and sociality: reconsidering "the metaphor of transmission of knowledge".] *Primate Research*, **24**, 73–90. In Japanese with English abstract.

Nishie, H. (2011). Natural history of *Camponotus* ant fishing by the M group chimpanzees at the Mahale Mountains National Park, Tanzania. *Primates*, **52**, 329–42.

Nishie, H. (2012). [*Chimpanzee Culture and Sociality.*] Doctoral Dissertation. Kyoto: Kyoto University. In Japanese.

Sakamaki, T. (1998). First record of algae-feeding by a female chimpanzee at Mahale. *Pan Africa News*, **5**, 1–3.

Sakamaki, T., Nakamura, M., and Nishida, T. (2007). Evidence of cultural differences in diet between two neighboring unit groups of chimpanzees in Mahale Mountains National Park, Tanzania. *Pan Africa News*, **14**, 3–5.

Shimada, M. (2002). Social scratch among chimpanzees in Gombe. *Pan Africa News*, **9**, 21–3.

Sperber, D. (1982). *Le Savoir des Anthropologues*. Paris: Hermann.

Takahata, Y. (1982). Termite fishing observed in M group chimpanzees. *Mahale Mountains Chimpanzee Research Project Ecological Report*, **18**.

Takahata, Y., Hiraiwa-Hasegawa, M., Takasaki, H., and Nyundo, R. (1986). Newly acquired feeding habits among the chimpanzees of the Mahale Mountains National Park, Tanzania. *Human Evolution*, **1**, 277–84.

Takasaki, H. (1983). Mahale chimpanzees taste mangoes: toward acquisition of a new food item? *Primates*, **24**, 273–75.

Uehara, S. (1982). Seasonal change in the techniques employed by wild chimpanzees in the Mahale Mountains, Tanzania, to feed on termites (*Pseudacanthotermes spiniger*). *Folia Primatologica*, **37**, 44–76.

de Waal, F. B. M. (1999). Cultural primatology comes of age. *Nature*, **399**, 635–6.

de Waal, F. B. M. (2001). *The Ape and the Sushi Master*. New York: Basic Books.

de Waal, F. B. M. and Seres, M. (1997). Propagation of handclasp grooming among captive chimpanzees. *American Journal of Primatology*, **43**, 339–46.

Washburn, S. and Benedict, B. (1979). Non-human primate culture. *Man*, **14**, 163–4.

Whiten, A., Goodall, J., McGrew, W. C., *et al.* (1999). Cultures in chimpanzees. *Nature*, **399**, 682–5.

Whiten, A., Goodall, J., McGrew, W. C., *et al.* (2001). Charting cultural variation in chimpanzees. *Behaviour*, **138**, 1481–516.

Wrangham, R. W., de Waal, F. B. M., and McGrew, W. C. (1994). The challenge of behavioral diversity. In *Chimpanzee Cultures*, ed. R. W. Wrangham, W. C. McGrew, F. B. M. de Waal, and P. G. Heltne. Cambridge, MA: Harvard University Press, pp.1–18.

Zamma, K. (2002). Leaf-grooming by a wild chimpanzee in Mahale. *Primates*, **43**, 87–90.

Zamma, K. (2006). A louse egg left on a leaf. *Pan Africa News*, **13**, 8–10.

39 Vocal communication

KAZUHIKO HOSAKA, TAKAHISA MATSUSAKA, AND SHUNKICHI HANAMURA

39.1 Introduction

This chapter outlines studies on wild chimpanzee vocalizations, especially those from Mahale. We discuss the comparisons between chimpanzees and humans, and elucidate some unanswered questions.

Comparisons with human language are common in studies on vocalizations of nonhuman animals. Three approaches have been adopted in such studies. The first is semantics, which explores meaningful signals in primate calls. For example, a famous series of studies on symbolic communication in vervet monkeys (Cheney and Seyfarth, 1990) showed that they use acoustically distinct alarm calls as referential signals to denote different predators. The second is syntax or grammar, which explores any system to generate new meanings by ordering or combining acoustic elements or signals. The third is communication, the process by which animals interact with each other by producing, altering, or suppressing calls in various social contexts. Many primatologists study how vocalizations affect or reflect the dynamics of social interactions and relationships.

Following findings of referential communication in birds and monkeys up to the 1980s (Cheney and Seyfarth, 1990), some scholars expected wild apes to have more human-like communication. For example, Diamond (1992) predicted that wild chimpanzee researchers would overcome logistical problems and find results akin, even superior, to those for vervets. Around that time, empirical studies on wild chimpanzee vocalizations sharply increased.

Classic studies on chimpanzee vocalizations were conducted by Marler (Marler and Hobbett, 1975; Marler, 1976; Marler and Tenaza, 1977), who collected tape-recordings at Gombe, Tanzania. Additionally, Goodall (1986) examined chimpanzee vocalizations based on her substantial field experience. Since these classic accounts largely concur with chimpanzee vocalizations elsewhere, chimpanzees' vocal repertoires seem simpler than their rich variations in tool use and gestures (Chapters 38, 42).

At Mahale, Mori (1983) first studied chimpanzee loud calls accompanying charging displays. In the 1980s, several researchers recorded chimpanzee vocalizations. In 1989–94, Mitani, who had studied wild gibbon and orangutan vocal behavior, joined the Mahale team to study the long-distance calls, or pant-hoots, of M-group chimpanzees. Concurrently, Arcadi (Clark before 1996) studied the same calls at Kibale Forest, Uganda. These two projects reactivated research on wild chimpanzee vocalizations.

In the early 1990s, due to a growing interest among researchers in social bonds of male chimpanzees in Mahale (Chapter 27), there was a reasonable expectation of understanding social interactions among adult males in terms of vocal exchange. Nishida (1994) attempted several simultaneous focal follows of multiple males to confirm whether and how males exchanged pant-hoots. Regrettably, these data remain unpublished.

See Chapter 33 for a review of studies on pant-grunting, another major vocalization type, with respect to social relationships among chimpanzees. Some specific vocalizations are reviewed below.

Mahale Chimpanzees: 50 Years of Research, ed. M. Nakamura, K. Hosaka, N. Itoh and K. Zamma. Published by Cambridge University Press.

39.2 Pant-hoot

39.2.1 Structure

Pant-hoot calls have four phases, introduction, buildup, climax, and letdown, with a total duration of 7–11 s (Hasegawa and Kajikawa, 2002). The introduction is a low-frequency tonal phase, prolonged as if a caller is saving energy. The buildup is a rhythmic phase of repeating inhalation and exhalation at faster rates. The climax comprises high-frequency, high-amplitude elements that resemble screams. Chimpanzees usually repeat one or two elements in a climax phase, but sometimes they emit more (e.g. Figure 39.1). Letdown resembles buildup in its acoustic structure, but is shorter and weaker and sometimes omitted.

Goodall (1986) proposed four distinct pant-hoots that are distinguishable according to different contexts in her studies at Gombe (for skeptical views, see Nishida, 1994, and Mitani, 1994). Clark and Wrangham (1993) found that pant-hoots with letdown phases occurred more frequently upon arrival at fruiting trees in Kibale. However, they were cautious about concluding that such pant-hoots broadcast information specific to the ecological context, because they were also heard at other times during feeding bouts. From Budongo Forest, Uganda, Notman and Rendall (2005) reported an association between pant-hoots with a "letdown" phase and specific contexts, such as travelling and arrival at a food source.

39.2.2 Flexibility and inherency

The repertoire of chimpanzee vocalizations is roughly similar in study sites across Africa. In captivity, a deaf female in the Arnhem Colony, the Netherlands, could vocalize sufficiently to function normally in a social context (de Waal, 1982). Although a thorough, intraspecific comparison of chimpanzee vocalizations has not been done to date, it seems reasonable that chimpanzees acquire their species-specific repertoire without large modifications during development.

Individual differences

Some fieldworkers can distinguish chimpanzees from only their pant-hoots (Nishida, 1994), although they avoid publishing those data. It seems likely that chimpanzees can distinguish the pant-hoots of most or all of their unit-group members and can aurally monitor movements of their group members.

Marler and Hobbett (1975) made the first quantitative analyses of pant-hoots of seven chimpanzees at Gombe and revealed that each animal had distinctive acoustic differences, probably sufficient to permit chimpanzees to identify individuals. Using a matching-to-sample task at the Primate Research Institute, Kyoto University, experimental psychologists showed that a female

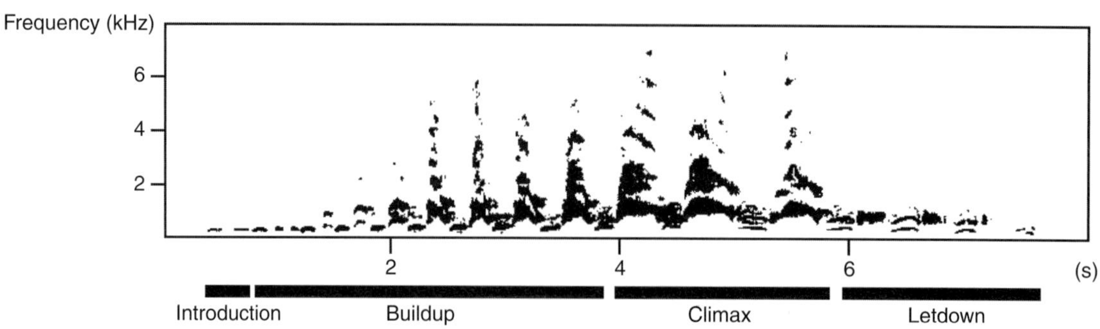

Figure 39.1 A spectrogram of a pant-hoot emitted by a 23-year-old male, Masudi, on January 16, 2000. The vertical axis shows the frequency (kHz), and the horizontal axis shows time (s). This pant-hoot was atypical in that it had a shorter introduction, louder buildups, more climaxes, and more prolonged letdowns.

captive chimpanzee could identify all her group members using only vocal sounds (pant-hoots, pant-grunts, and screams) (Kojima, 2003).

Mitani *et al.* (1996) investigated factors facilitating the evolution of individual vocal signatures by comparing pant-hoots and pant-grunts. Since pant-hoots are delivered beyond visual range, while pant-grunts are directed at visible receivers, they predicted that natural selection would favor the elaboration of individually distinctive acoustic features in the former. Acoustic analyses supported this prediction; pant-hoots were more consistent within individuals and variable between individuals than pant-grunts.

Call matching during choruses

Mitani and Brandt (1994) found that male chimpanzees modified their pant-hoots, especially when they chorused with other males, to the extent that within-individual acoustic variation exceeded that between individuals. Males who called together appeared to match each other's pant-hoot characteristics. Mitani and Gros-Louis (1998) found that male chimpanzees accommodated each other vocally through active alteration of their calls during choruses. As pant-hoot choruses are related to male alliances, call matching may help strengthen social bonds between males. Thus, social factors can affect the acoustic morphology of calls.

Dialects in chimpanzee calls?

Although variation in pant-hoots is minor, vocal plasticity may exist in the process of developing pant-hoots. Mitani *et al.* (1992) detected a slight variation in the acoustic structure of male pant-hoots between Mahale and Gombe. Mahale males had significantly shorter buildup phases at faster rates, while Mahale males uttered higher-frequency climaxes that spanned greater frequency ranges. The researchers interpreted such variation as "dialects," because other explanations, such as differences in body size or vegetation density, seemed unsatisfactory. Mitani *et al.* (1999) also found differences in acoustic features, especially the buildup phase, between pant-hoots in Mahale and Kibale. However, they could not rule out ecological factors and body size effects, so they avoided the term "dialect."

Sexual differences in acoustic structure

Based on spectrogram analyses, Marler and Hobbett (1975) concluded that the pant-hooting of adult females differs consistently from that of adult and adolescent males in the absence of a "climax" phase. However, according to our field observations at Mahale, adult females also utter pant-hoots, including a climax. Further studies are needed to ascertain whether such differences exist.

39.2.3 Functions

Long-distance call

Since pant-hoots are a kind of long-distance call, several hypotheses have been proposed and tested to explain the social and ecological factors for this call. The first is the "cohesion hypothesis"; chimpanzees call to maintain cohesion in widely scattered, large, nomadic, fruit-foraging parties (Reynolds and Reynolds, 1965). The second is the "recruiting hypothesis"; adult males call to recruit favorable partners, such as allies or mates, within earshot but not visible (Mitani and Nishida, 1993; Clark and Wrangham, 1994). The third is the "altruism hypothesis"; chimpanzees emit pant-hoots after reaching fruiting trees to help others to aggregate at good food sources (Wrangham, 1977).

Based on their observations at Mahale, Mitani and Nishida (1993) argued against the "cohesion hypothesis," emphasizing that there was no correlation between party size variables and the call rate of individual males. Instead, they argued in favor of their "recruiting hypothesis," saying that male chimpanzees call to particular audiences. Clark and Wrangham (1994) tested the "altruism hypothesis" at Kibale, with the negative result that arrival pant-hoots by males did not increase the arrival of others and that the probability of calling was not correlated with the amount of ripe fruits in the tree. Moreover, they showed that lower-ranking individuals that arrived at unoccupied trees without calling were not punished. They suggested that, instead of providing information about food, arrival pant-hoots mark the status of high-ranking males in the parties.

Intimidation in face-to-face interactions

Pant-hoots often accompany intimidation displays (see "vocal displays" in Chapter 31). Moreover, pant-hoots per se may function as intimidation in face-to-face interactions (Chapter 31). When the dominance relationship between the alpha male Kalunde and his rival Nsaba was unstable in December 1991, Nsaba emitted 47 pant-hoots for 2 h and 22 min during two episodes where Kalunde shared meat with his allies (Hosaka, unpublished data). Since adult males produce one or two pant-hoots per hour on average (Mitani and Nishida, 1993), this rate was extraordinarily high. Thirty-six (76.6%) of them had incomplete fading before grading into the buildup phase. Nsaba followed or accompanied the pant-hoots with intimidation displays toward Kalunde such as charging (six times), kicking (twice), and approaching with piloerection and his back hunched up (four times).

The intimidation function of pant-hoots may have derived from their original function as long-distance calls. High-ranking males pant-hoot more frequently than low-ranking males at both Mahale (Mitani and Nishida, 1993) and Kibale (Clark and Wrangham, 1994). In Kibale, low-ranking males did not pant-hoot when arriving at an unoccupied fruiting tree (Clark and Wrangham, 1994). Anecdotal evidence suggested that low-ranking males are punished if they call before high-ranking males arrive at the tree, possibly because arrival pant-hoots attract estrous females and their production by low-ranking males could be recognized as a challenge to high-ranking males (Clark and Wrangham, 1994). Thus, higher-ranking males may assume the "privilege" of emitting pant-hoots before others. Anecdotal evidence from Mahale suggests that lower-ranking males could be punished by higher-ranking males when the former initiates a pant-hoot (Hosaka, personal observation). For example, in March 1992, when middle-ranking male Jilba started the introduction phase of a pant-hoot, Kalunde, the party's highest-ranking male, immediately charged him.

Intimidation of neighboring unit-groups

At Mahale, pant-hoots appear to function as counter-threats to strangers' calls, so that chimpanzees avoid direct encounter with neighboring unit-groups (see Chapter 11). Despite their rarity, adult males occasionally become intensely excited on hearing a distant stranger's call, quickly showing piloerection and repeating shrill pant-hoot choruses while performing joint charging displays.

Pant-hoots emitted as counter-calls to strangers might be modulated in pitch or pace than those emitted in usual situations, although there are no systematic studies on this in wild settings at Mahale. Kajikawa and Hasegawa (2000) played back pant-hoots of Mahale males to captive chimpanzees of the Japan Monkey Centre, and revealed that chimpanzees modified pant-hoots in response, displaying higher rates of delivery, shorter duration of buildup, and lower minimum fundamental frequency of climax than spontaneous calls. Such modification might reflect the emotional states of the callers on hearing strangers' calls.

At Kibale, Wilson *et al.* (2001) conducted the first playback experiments in the wild, testing whether chimpanzees regulate their response to a single unfamiliar male's pant-hoot depending on the number of adult males in the listening party. Chimpanzees in parties with three or more males responded to the playback in a chorus of loud calls and approached the speaker, whereas chimpanzees in parties with fewer males usually stayed silent and approached the speaker less often.

Playback experiments were also conducted at Taï Forest, Côte d'Ivoire, to investigate how chimpanzees respond to pant-hoots of the same unit-group members, "neighbors" from adjacent unit-groups, and "strangers" from a distant unit-group (Herbinger *et al.*, 2009). The authors discriminated between neighbors and strangers because the latter are potentially more dangerous without experience of previous interactions. Chimpanzees responded differently to the three categories of calls, suggesting that they can associate pant-hoots with categories of individuals.

Sexual differences in vocal production

Marler's (1976) study in Gombe revealed that chimpanzees of each sex produce vocalizations of all types with equal frequency. However, Clark (1993) observed that adult female Kibale chimpanzees were generally quieter than adult males, but increased vocal production depending on social contexts; females were quiet when alone, emitted submissive vocalizations when with females only, and emitted submissive and nonsubmissive vocalizations most when in larger, mixed parties. Based on field observations at Mahale, we believe that adult females participate in vocal exchanges between parties, especially by pant-hooting (Section 39.5). Further studies on this are necessary.

39.3 Pant-grunt

39.3.1 Structure

Pant-grunts are vocalizations with low-pitched sounds on both inhalation and exhalation (Marler and Tenaza, 1977). Pant-grunts comprise an average of approximately five breathing cycles per second, compared to one per second in pant-hooting (Marler and Tenaza, 1977). Pant-grunts vary in morphology within an individual depending on social situations (Nishida, 1994) or arousal state. Some pant-grunts, called "pants," are very quiet, as they are emitted without accompanying vibration of vocal cords. Other pant-grunts might be called "pant-barks" or "pant-screams," if ordinary pant-grunts are escalated into louder ones and assume components overlapping barks or screams.

39.3.2 Development and functions

Pant-grunts seem intrinsically related to fission–fusion and dominance structures in chimpanzee societies (Chapter 33). Specifically, they are given unidirectionally from subordinate individuals toward dominant individuals. Alpha males are usually the most frequent receiver of pant-grunts (Hayaki, 1990; Laporte and Zuberbühler, 2010; Sakamaki, 2011), although sometimes another male receives more pant-grunts, e.g. Shike of Mahale M group in 1991 (Chapter 27).

Pant-grunts are usually uttered in greeting contexts, but not always. For instance, juvenile and adolescent males tend to pant-grunt at adult males frequently in non-greeting contexts, which may serve to test the tolerance of dominant males for coexistence (Hayaki, 1990). Such pant-grunts are closely related to the process of attaining social maturity. In contrast, female chimpanzees may perform pant-grunts to maintain existing relations with adult males (Sakamaki, 2011; also see Chapter 33).

Although pant-grunts are not categorized as long-distance calls and are less variable between individuals than pant-hoots (Section 39.2.2), some adult females or adolescent males utter these so loudly that other individuals walking separately seem to understand "*who* pant-grunted at *whom*." Thus, pant-grunts may allow chimpanzees to aurally recognize social interactions. Nishida (1994) claimed that the pant-grunts of some adult females could be identified by human observers, not to mention chimpanzees.

39.4 Comparison with other great apes

39.4.1 Loud calls

Mori (1983) first compared the vocalizations of three wild African great apes (chimpanzees at Mahale; bonobos [*Pan paniscus*] at Wamba, Zaïre; and eastern gorillas [*Gorilla beringei*] at Kahuzi-Biega, Zaïre). He proposed that the common ancestor of *Gorilla* and *Pan* might have had one function of the "hoot" as a territorial call accompanying charging displays in common, whereas only *Pan* might have developed another function for hoots as long-distance calls causing cohesion between temporary parties. He also hypothesized that the "wraah," as "threat or alarm to an enemy (different species)," is homologous across African apes. He further hypothesized that the ancestral vocal display was "hoot" and "wraah" calls accompanied by charging display as a male territorial behavior.

Mori (1983) combined gorillas' "wraah" and "roar" calls into one category, "wraah," based on their similarities mentioned by Fossey (1972). However, Marler (1976) considered that the gorilla "roar" has no chimpanzee equivalent and that the gorilla "wraah" corresponds to two chimpanzees' calls, "waa-bark" and "wraaa." Mori (1983) described that his chimpanzee "wraah" mostly means "waa-bark," which is shorter than "wraaa" and is heard daily in contexts such as reunions of temporary parties. The chimpanzee "wraaa" is heard typically in unusual contexts such as encounters with predators (see below). Thus, it may validly correspond to the gorilla "wraah" judging from the contextual similarities, although Mitani (1996) noted that the chimpanzee "wraaa" is a unique call with no homologous equivalent in other African apes.

Wraa

"Wraa" (the same as "wraaa" in Goodall, 1986) is probably an acoustic variant of the waa-bark (Marler and Tenaza, 1977). Both calls begin as similar shrill loud sounds, but the last tonal section of "wraa" is drawn out into a howl that may last 1 s or more (Marler and Tenaza, 1977). Waa-barks have an overall tendency for the pitch to rise and fall audibly during the course of the sound, but often with a heavy overlay of noise (Marler and Tenaza, 1977). Waa-barks are typically emitted by adolescent or young adult males singly and in succession only briefly when they hear some uproar caused by distant adult male charging displays.

Schel *et al.* (2013) provided spectrograms and audio files of "soft huu," "alarm huu," and "waa-bark"; their "soft huu" corresponds to Goodall's (1986) "huu" and their "waa-bark" corresponds to "wraaa" (Marler and Tenaza, 1977; Goodall, 1986). We agree with Schel *et al.* (2013) that the three calls grade into each other, because at Mahale, we sometimes heard all these calls together (Chapters 16, 25).

Wraas are rarely heard in within-group interactions, but are typically emitted in encounters with potential predators (Chapter 16). If chimpanzees encounter such animals alone, they tend to continue emitting wraas with some threatening gestures for more than 10 min until they appear exhausted. Similarly, chimpanzees may emit wraas in encounters with dangerous animals like buffaloes (Goodall, 1986) or when seeing humans if chimpanzees are unhabituated (Toshisada Nishida, personal communication).

Wraas deserve attention since many authors described them as chimpanzee "alarm calls" (e.g. van Lawick-Goodall, 1968; Tsukahara, 1993; Boesch and Boesch-Achermann, 2000; Sugiyama and Humle, 2011). If wraas were emitted only in the alarm situations stated above, then they could be considered alarm calls, because they might fit the definition of transferring information about dangers to others. Recently, at Budongo, field experiments assuming wraa and soft "huu" as alarm calls were conducted to examine whether chimpanzees can control vocal production intentionally depending on their social knowledge (Crockford *et al.*, 2012; Schel *et al.*, 2013).

However, wraas are also heard in other contexts than alarm situations. First, chimpanzees utter this call when they encounter dead bodies of conspecifics (Hosaka *et al.*, 2000; Stewart *et al.*, 2012; Chapter 25) and heterospecifics (Chapter 16). Some of these can be considered alarm situations, because chimpanzees may perceive that the predators that killed the animal are nearby (e.g. encounters with freshly killed bushbucks and aardvarks in Chapter 16). That may not be the case, though, because chimpanzees do emit wraas when they encounter putrid bodies of bushbucks or conspecifics (Hosaka, personal observation). Further, Gombe chimpanzees emitted wraas when they were fully aware that a chimpanzee had died falling from a tree (Teleki, 1973). At Mahale, chimpanzees emitted wraas, barks, and screams together with social interactions involving pant-barks just after experiencing a big earthquake (Fujimoto and Hanamura, 2008). Thus, chimpanzees may emit wraas in situations that arouse their fear or anxiety. Note that even the escape behavior of wraa receivers in alarm situations could be explained by emotional empathy and contextual information.

If alarm calls are defined broadly as calls without specific referents but that may produce danger-avoidance behaviors in receivers, then wraas could be considered "alarm calls." However, since many empirical studies have revealed that "referential alarm calls" are widespread in birds and mammals and typically evoke unambiguous responses (Macedonia and Evans, 1993), one should be cautious about terming wraas "alarm calls" to avoid confusion with "referential alarm calls"; a parsimonious view of this call is still as a rather emotional expression.

Scream

Screams are, with grunts, hoots, and barks, one of the four basic components of vocal repertoires of African apes (Mitani, 1996). Mitani and Gros-Louis (1995) revealed that screams from chimpanzees, emitted at relatively low frequencies, were distinguishable from the higher-pitched calls by bonobos. They suggested that these interspecific variations may be attributed to differences in laryngeal mechanisms and social dispersion between the two species and sexes; since higher-frequency sounds attenuate faster than lower-frequency sounds, the stronger fission–fusion tendency of chimpanzees may have favored low-pitched screams for efficient transmission.

It seems probable that individual differences in chimpanzee screams enable receivers to identify callers (see Section 39.2.2 about evidence from captive studies), as was demonstrated empirically in vervet monkeys (Cheney and Seyfarth, 1990). Goodall (1986) classified four kinds of chimpanzee screams according to intensity, context, and audible acoustic differences: the victim scream, tantrum scream, SOS scream, and copulation scream.

39.4.2 Short-distance calls

As Mitani (1996) noted, male chimpanzees emit grunts much less frequently than male gorillas, but emit pant-hoots much more frequently than male gorillas emit hoot series. Further, Harcourt *et al.* (1993) observed that gorillas appear to have a richer repertoire of "close" calls than chimpanzees. Seyfarth *et al.* (1994) found that double grunts by male gorillas were individually distinctive and suggested that gorillas exchange these grunts communicatively. Considering gorillas have more cohesive groups than chimpanzees, it seems reasonable that they evolved many close calls to regulate close-range social interactions.

Chimpanzees do have a repertoire of "close" calls, even though they may be uttered less frequently than those of gorillas. Hasegawa and Kajikawa (2002) noted that if primatologists explore any precursor of vocal language in chimpanzee vocalizations, the study of mild calls emitted in calm situations would be inevitable. Itani (1963) first emphasized the importance of examining non-affective short-distance calls (e.g. "muttering" in Japanese macaques) in one-to-one communication to explore the evolution of language from a primatological perspective. He even mentioned the possibility of finding various muttering-like sounds in chimpanzee vocalizations through the then-new chimpanzee research project at Tanganyika (Itani, 1963). Current primatology has not fully responded to this expectation. In this section, we deal with the prospective topics of play-pants and soft/extended grunts.

Play-pant

All extant ape species have panting or grunting sounds uttered during play (Marler and Tenaza, 1977). Probably, these sounds have the same evolutionary origin in all ape species (Davila-Ross *et al.*, 2009). Such sounds emitted by playing chimpanzees are often called "laughter" (e.g. Goodall, 1986) since they sound like human laughter. Nishida *et al.* (1999) adopted a more neutral term "play-pant."

Provine (1996) emphasized two differences in acoustic structures between human laughter and chimpanzee play-pants. First, chimpanzee play-pants lack the clear harmonic structure of human laughter. Second, while the discrete notes of human laughter ("ha") are produced by chopping a single exhalation, the notes of chimpanzee play-pants are produced during each exhalation and inhalation. However, Bachorowski *et al.* (2001) claimed that previous work on human laughter only examined typical voiced laughter, disregarding two kinds of unvoiced laughter. Taking this into account, the differences that Provine (1996) emphasized seem less valid.

Play-pants are uttered in response to tickling and some social play contexts (Chapter 40). Since play-pants accompany the play face, a typical facial expression when chimpanzees play, van Hooff (1972) proposed that the play-pant and play face in primates were the phylogenetic origins of human laughter. However, subsequent primate studies influenced by this hypothesis mainly investigated the facial expressions of grin face and play face (e.g. Preuschoft and van Hooff, 1997), and very few have focused on play-pants.

Matsusaka (2004) indicated that play-pants in Mahale chimpanzees function as a play signal to prevent the partner from quitting play due to misinterpretation of playful escape, and also to activate social play (for more information, see Chapter 36).

Chimpanzee "close" calls

At least three kinds of chimpanzee grunts, "soft grunt," "extended-grunt," and "nest-grunt" (Goodall, 1986), appear functionally comparable to gorilla close calls.

Soft-grunts are contact calls that may be exchanged between close chimpanzees, particularly kin individuals, during foraging or traveling together, and these grunts may function to regulate movement and cohesion (Goodall, 1986).

The extended-grunt is "a long, drawn-out sound that may be composed of two syllables – *the double grunt*, eh-mmmmm – is heard during rest sessions" (Goodall, 1986). This is a rare call whose significance even Goodall could not confidently explain. However, it seems that it is species-typical because it is also heard at Mahale and in captivity. Hopkins *et al.* (2007) claimed that chimpanzees of the Yerkes Primate Research Center emit the extended-grunt as attention-getting signals. At Mahale, human observers sometimes discover chimpanzees resting in a bush because this sound is uttered in response to the approaching humans (Nishida *et al.*, 1999), possibly as an expression of mild disturbance. In the early 1990s, Hosaka heard these grunts occasionally while observing adult male chimpanzees resting on the ground in a hollow space in a thicket tangled with vines. Extended grunts sound milder than cough-barks and are not accompanied by threatening gestures.

The bed-grunt (a.k.a. "nest-grunt") is "another distinctive double grunt, given when an individual begins to look around for a suitable bed site, during the making of the bed, or as he settles down for the night" (Goodall, 1986). At Mahale, some researchers agree that bed-grunts usually comprise more than two syllables, which are shorter and repeated faster than in extended grunts. More empirical studies are necessary to examine whether bed-grunts differ between Gombe and Mahale or vary within sites.

39.5 Discussion and future perspectives

The studies led by Mitani in the 1990s greatly influenced subsequent chimpanzee vocal communication research. For example, the discovery of geographical variations in pant-hoots (Mitani *et al.*, 1992, 1999) provided Crockford *et al.* (2004) an insight for their studies of intergroup variations at Taï. Further, the discoveries of call matching in pant-hoot chorusing (Brandt, 1994; Mitani and Mitani and Gros-Louis, 1998) and its function in male social bonds (Mitani and Nishida, 1993) were developed in studies from Kibale examining how this behavior affects adult male social bonds (Fedurek *et al.*, 2013a) and how acoustic structures were modified depending on audience reactions (Fedurek *et al.*, 2013b).

Interestingly, two investigators of chimpanzee vocalizations in the 1990s reached similar conclusions that may have disappointed the expectations of Diamond (1992). That is, they claimed that there is no evidence that chimpanzee vocal behavior is qualitatively different from that of other primates (Mitani, 1996) and that chimpanzee vocal exchanges show no special similarity to human conversations (Arcadi, 2000). Instead, Mitani (1996) stressed the value of close-range social interactions where more complex skills might be required, as in conversation-like vocal exchanges by gorillas (Harcourt *et al.*, 1993; Seyfarth *et al.*, 1994). Further, Mitani (1996) mentioned the need for analyses of dynamic process of vocal communication with more focus on receivers.

Arcadi (2000) did not find conversational attributes in vocal exchanges, since chimpanzees vocalized at low rates and tended not to respond to calls that they heard. However, Hanamura (2010) suggested that pant-hoot exchanges function as dynamic interactions between callers and receivers; receivers can either respond to a call within 10 s or keep silent, either of which may affect callers' subsequent actions.

In contrast to the skeptical views of Mitani and Arcadi, many recent studies on chimpanzee vocal behaviors have searched for precursors of human language, i.e. semantics and syntax, because such a paradigm is still productive. For example, empirical studies from Budongo found referential aspects in screams (Slocombe and Zuberbühler, 2007). Furthermore, Crockford and Boesch (2003) revealed that frequent signal combinations of bark types and drums at Taï increased specificity levels almost equivalent to the vervet monkey predator alarm calls. They claimed that chimpanzees use the system to generate novel meaning by combining context-specific signals, but this result also showed that the referential communication could barely match the levels vervets attain with single signals. These findings may plausibly lead to some new paradigm to explore the intrinsic mechanisms of chimpanzee social interactions that have been favored in their fission–fusion lives with auditory contact after divergence from the human lineage.

References

Arcadi, A. C. (2000). Vocal responsiveness in male wild chimpanzees: implications for the evolution of language. *Journal of Human Evolution*, **39**, 205–23.

Bachorowski, J.-A., Smoski, M. J., and Owren, M. J. (2001). The acoustic features of human laughter. *The Journal of the Acoustical Society of America*, **110**, 1581–97.

Boesch, C. and Boesch-Achermann, H. (2000). *The Chimpanzees of the Taï Forest: Behavioural Ecology and Evolution*. Oxford: Oxford University Press.

Cheney, D. L. and Seyfarth, R. M. (1990). *How Monkeys See the World: Inside the Mind of Another Species*. Chicago, IL: University of Chicago Press.

Clark, A. P. (1993). Rank differences in the production of vocalizations by wild chimpanzees as a function of social context. *American Journal of Primatology*, **31**, 159–79.

Clark, A. P. and Wrangham, R. W. (1993). Acoustic analysis of wild chimpanzee pant hoots: do Kibale Forest chimpanzees have an acoustically distinct food arrival pant hoot? *American Journal of Primatology*, **31**, 99–109.

Clark, A. P. and Wrangham, R. W. (1994). Chimpanzee arrival pant-hoots: do they signify food or status? *International Journal of Primatology*, **15**, 185–205.

Crockford, C. and Boesch, C. (2003). Context-specific calls in wild chimpanzees, *Pan troglodytes verus*: analysis of barks. *Animal Behaviour*, **66**, 115–25.

Crockford, C., Herbinger, I., Vigilant, L., and Boesch, C. (2004). Wild chimpanzees produce group-specific calls: a case for vocal learning? *Ethology*, **110**, 221–43.

Crockford, C., Wittig, R. M., Mundry, R., and Zuberbühler, K. (2012). Wild chimpanzees inform ignorant group members of danger. *Current Biology*, **22**, 142–6.

Davila-Ross, M., Owren, M., and Zimmermann E. (2009). Reconstructing the evolution of laughter in great apes and humans. *Current Biology*, **19**, 1106–11.

Diamond, J. (1992). *The Third Chimpanzee: The Evolution and Future of the Human Animal*. New York: HarperCollins.

Fedurek, P., Machanda, Z. P., Schel, A. M., and Slocombe, K. E. (2013a). Pant hoot chorusing and social bonds in male chimpanzees. *Animal Behaviour*, **86**, 189–96.

Fedurek, P., Schel, A. M., and Slocombe, K. E. (2013b). The acoustic structure of chimpanzee pant-hooting facilitates chorusing. *Behavioral Ecology and Sociobiology*, **67**, 1781–9.

Fossey, D. (1972). Vocalizations of the mountain gorilla (*Gorilla gorilla beringei*). *Animal Behaviour*, **20**, 36–53.

Fujimoto, M. and Hanamura, S. (2008). Responses of wild chimpanzees (*Pan troglodytes schweinfurthii*) toward seismic aftershocks in the Mahale Mountains National Park, Tanzania. *Primates*, **49**, 73–6.

Goodall, J. (1986). *The Chimpanzees of Gombe: Patterns of Behavior*. Cambridge, MA: Belknap.

Hanamura, S. (2010). [Long-distance calls by chimpanzees: how they connect actions and organize their social field beyond visual contact.] *Primate Research*, **26**, 159–76. In Japanese with English summary.

Harcourt, A. H., Stewart, K. J., and Hauser, M. D. (1993). Functions of wild gorilla "close" calls. I. Repertoire, context, and interspecific comparison. *Behaviour*, **124**, 89–122.

Hasegawa, T. and Kajikawa, S. (2002). [What do chimpanzees communicate by pant-hoot calls?] In [*The Mahale Chimpanzees: 37 Years of <Panthropology>*], ed. T. Nishida, S. Uehara, and K. Kawanaka. Kyoto: Kyoto University Press, pp. 303–19. In Japanese.

Hayaki, H. (1990). Social context of pant-grunting in young chimpanzees. In *The Chimpanzees of the Mahale Mountains: Sexual and Life History Strategies*, ed. T. Nishida. Tokyo: University of Tokyo Press, pp. 189–206.

Herbinger, I., Papworth, S., Boesch, C., and Zuberbühler, K. (2009). Vocal, gestural and locomotor responses of wild chimpanzees to familiar and unfamiliar intruders: a playback study. *Animal Behaviour*, **78**, 1389–96.

van Hooff, J. A. R. A. M. (1972). A comparative approach to the phylogeny of laughter and smiling. In *Non-Verbal Communication*, ed. R. A. Hinde. Cambridge: Cambridge University Press, pp. 209–23.

Hopkins, W. D., Taglialatela, J. P., and Leavens, D. A. (2007). Chimpanzees differentially produce novel vocalizations to capture the attention of a human. *Animal Behaviour*, **73**, 281–6.

Hosaka, K., Matsumoto-Oda, A., Huffman, M. A., and Kawanaka, K. (2000). [Reactions to dead bodies of conspecifics by wild chimpanzees in the Mahale Mountains, Tanzania.] *Primate Research*, **16**, 1–15. In Japanese with English summary.

Itani, J. (1963). Vocal communication of the wild Japanese monkey. *Primates*, **4**, 11–66.

Kajikawa, S. and Hasegawa, T. (2000). Acoustic variation of pant hoot calls by male chimpanzees: a playback experiment. *Journal of Ethology*, **18**, 133–9.

Kojima, S. (2003). *A Search for the Origins of Human Speech: Auditory and Vocal Functions of the Chimpanzee*. Kyoto: Kyoto University Press.

Laporte, M. N. C. and Zuberbühler, K. (2010). Vocal greeting behaviour in wild chimpanzee females. *Animal Behaviour*, **80**, 467–73.

van Lawick-Goodall, J. (1968). The behaviour of free-living chimpanzees in the Gombe Stream Reserve. *Animal Behaviour Monographs*, **1**, 161–312.

Macedonia, J. M. and Evans, C. S. (1993). Variation among mammalian alarm call systems and the problem of meaning in animal signals. *Ethology*, **93**, 177–97.

Marler, P. (1976). Social organization, communication and graded signals: The chimpanzee and the gorilla. In *Growing Points in Ethology*, ed. P. P. G. Bateson and R. A. Hinde. Cambridge: Cambridge University Press, pp. 239–80.

Marler, P. and Hobbett, L. (1975). Individuality in a long-range vocalization of wild chimpanzees. *Zeitschrift für Tierpsychologie*, **38**, 37–109.

Marler, P. and Tenaza, R. (1977). Signaling behavior of apes with special reference to vocalization. In *How Animals Communicate*, ed. T. A. Sebeok. Bloomington, IN: Indiana University Press, pp. 965–1033.

Matsusaka, T. (2004). When does play panting occur during social play in wild chimpanzees? *Primates*, **45**, 221–9.

Mitani, J. C. (1994). Ethological studies of chimpanzee vocal behavior. In *Chimpanzee Cultures*, ed. R. W. Wrangham, W. C. McGrew, F. B. M. de Waal, and P. Heltne. Cambridge, MA: Harvard University Press, pp. 195–210.

Mitani, J. C. (1996). Comparative studies of African ape vocal behavior. In *Great Ape Societies*, ed. W. C. McGrew, L. F. Marchant, and T. Nishida. Cambridge: Cambridge University Press, pp. 241–54.

Mitani, J. C. and Brandt, K. L. (1994). Social factors influence the acoustic variability in the long-distance calls of male chimpanzees. *Ethology*, **96**, 233–52.

Mitani, J. C. and Gros-Louis, J. (1995). Species and sex differences in the screams of chimpanzees and bonobos. *International Journal of Primatology*, **16**, 393–411.

Mitani, J. C. and Gros-Louis, J. (1998). Chorusing and call convergence in chimpanzees: tests of three hypotheses. *Behaviour*, **135**, 1041–64.

Mitani, J. C. and Nishida, T. (1993). Contexts and social correlates of long-distance calling by male chimpanzees. *Animal Behaviour*, **45**, 735–46.

Mitani, J. C., Hasegawa, T., Gros-Louis, J., Marler, P., and Byrne, R. W. (1992). Dialects in wild chimpanzees? *American Journal of Primatology*, **27**, 233–43.

Mitani, J. C., Gros-Louis, J., and Macedonia, J. M. (1996). Selection for acoustic individuality within the vocal repertoire of wild chimpanzees. *International Journal of Primatology*, **17**, 569–83.

Mitani, J. C., Hunley, K. L., and Murdoch, M. E. (1999). Geographic variation in the calls of wild chimpanzees: a reassessment. *American Journal of Primatology*, **47**, 133–51.

Mori, A. (1983). Comparison of the communicative vocalizations and behaviors of group ranging in eastern gorillas, chimpanzees and pygmy chimpanzees. *Primates*, **24**, 486–500.

Nishida, T. (1994). Review of recent findings on Mahale chimpanzees implications. In *Chimpanzee Cultures*, ed. R. W. Wrangham, W. C. McGrew, F. B. M. de Waal, and P. Heltne. Cambridge, MA: Harvard University Press, pp. 373–96.

Nishida, T., Kano, T., Goodall, J., McGrew, W. C., and Nakamura, M. (1999). Ethogram and ethnography of Mahale chimpanzees. *Anthropological Science*, **107**, 141–88.

Notman, H. and Rendall, D. (2005). Contextual variation in chimpanzee pant hoots and its implications for referential communication. *Animal Behaviour*, **70**, 177–90.

Preuschoft, S and van Hooff, J. A. R. A. M. (1997). The social function of "smile" and "laughter": variations across primate species and societies. In *Nonverbal Communication: Where Nature Meets Culture*, ed. U. Segerstråle and P. Molnár. Mahwah, NJ: Erlbaum, pp. 171–89.

Provine, R. R. (1996). Laughter. *American Scientist*, **84**, 38–45.

Reynolds, V. and Reynolds, F. (1965). Chimpanzees of the Budongo Forest. In *Primate Behavior: Field Studies of Monkeys and Apes*, ed. I. DeVore. New York: Holt, Rinehart, and Winston, pp. 368–424.

Sakamaki, T. (2011). Submissive pant-grunt greeting of female chimpanzees in Mahale Mountains National Park, Tanzania. *African Study Monographs*, **32**, 25–41.

Schel, A. M., Townsend, S. W., Machanda, Z. P., Zuberbühler, K., and Slocombe, K. E. (2013). Chimpanzee alarm call production meets key criteria for intentionality. *PLoS ONE*, **8**, e76674.

Seyfarth, R., Cheney, D., Harcourt, A., and Stewart, K. (1994). The acoustic features of gorilla double grunts and their relation to behavior. *American Journal of Primatology*, **33**, 31–50.

Slocombe, K. E. and Zuberbühler, K. (2007). Chimpanzees modify recruitment screams as a function of audience composition. *Proceedings of the National Academy of Sciences of the USA*, **104**, 17228–33.

Stewart, F. A., Piel, A. K., and O'Malley, R. C. (2012). Responses of chimpanzees to a recently dead community member at Gombe National Park, Tanzania. *American Journal of Primatology*, **74**, 1–7.

Sugiyama, Y. and Humle, T. (2011). A wild chimpanzee uses a stick to disable a snare at Bossou, Guinea. *Pan Africa News*, **18**, 3–4.

Teleki, G. (1973) Group response to the accidental death of a chimpanzee in Gombe National Park, Tanzania. *Folia Primatologica*, **20**, 81–94.

Tsukahara, T. (1993). Lions eat chimpanzees: the first evidence of predation by lions on wild chimpanzees. *American Journal of Primatology*, **29**, 1–11.

de Waal, F. (1982). *Chimpanzees Politics: Power and Sex among Apes*. New York: Harper and Row.

Wilson, M. L., Hauser, M. D., and Wrangham, R. W. (2001). Does participation in intergroup conflict depend on numerical assessment, range location, or rank for wild chimpanzees? *Animal Behaviour*, **61**, 1203–16.

Wrangham, R. W. (1977). Feeding behaviour of chimpanzees in Gombe National Park, Tanzania. In *Primate Ecology: Studies of Feeding and Ranging Behavior in Lemurs, Monkeys and Apes*, ed. T. H. Clutton-Brock. London: Academic Press, pp. 503–38.

40 Diversity of play

TAKAHISA MATSUSAKA, MASAKI SHIMADA, AND MICHIO NAKAMURA

40.1 Introduction

Humans are the world's most playful creatures, and they often engage in play of a great variety. Children in particular spend a good part of the day in various forms of play. Although only immature individuals play in most animal species, human adults often enjoy play activities such as sports, gambling, music, and dance. Playful attitudes can also be seen in various contexts other than the domain of so-called play (e.g. leisure and entertainment), such as hunting, fishing, conversation, sex, art, and science.

Chimpanzees also play often and have a large repertoire of play. Solo play is done in a great variety of ways, such as locomotor play involving various motor movements and object play using various physical items. Social play, such as wrestling and chasing, also includes various and flexible interactions. However, few studies have extensively investigated the play of wild chimpanzees, and the variation of the play has rarely been described systematically. Therefore, in this chapter, we summarize the diversity of play found in the wild chimpanzees of Mahale.

We have collected descriptions of various types of play behaviors from our own observations as well as from published reports on Mahale chimpanzees. We have especially referred to a thorough compilation of the behavioral diversity of Mahale chimpanzees (Nishida *et al.*, 2010). Then we classified the variety of play behaviors according to the types of interaction observed between a player and the environment (including playmates).

40.1.1 Definition of play

It is often said that making a strict definition of play is difficult in the study of animal play. Play is generally recognized by the features of its actions, such as exaggeration, repetition, relaxed mood, and flexibility. Another point often used to discern play is that it has no obvious immediate benefit. In chimpanzees, play face and play panting can be indicators of play, although they are not always useful because play behaviors do not always accompany them. In this chapter, we compile typical play behaviors using these conventional criteria. However, we also include other play-like behaviors such as exploratory behaviors, since they are difficult to discern from play and seem to share similar functions for the development of young chimpanzees.

40.1.2 Brief overview of the play of Mahale chimpanzees

Infants and juveniles play most frequently, at any time of the day, even during travels or when adults are feeding. Adolescent and adult chimpanzees also occasionally play, mainly during resting times. In this chapter, we use "young chimpanzees" to generally refer to non-adult chimpanzees, but we specify the age or age class (Chapter 22) whenever applicable. The play of chimpanzees can be divided into solo play, social play, and intermediate forms of solo and social play. Solo play is performed by a single individual, including various patterns of locomotor play, object play, imaginary play, and other exploratory play. Social play is performed by two or more chimpanzees,

Mahale Chimpanzees: 50 Years of Research, ed. M. Nakamura, K. Hosaka, N. Itoh and K. Zamma. Published by Cambridge University Press.

including wrestling, chasing, and tickling. We categorize some play patterns as "intermediate of solo and social play" when they are observed in two or more chimpanzees but lack obvious mutual play interaction. Solo play is usually seen in infants and juveniles, while social play occasionally involves adolescent and adult chimpanzees.

40.2 Diversity of solo play

Young chimpanzees of Mahale engage in a great variety of solo play, which can be classified into four categories: (1) locomotor play, (2) object play, (3) imaginary play, and (4) other exploratory play. Some patterns of play can be classified into two of these categories, but this classification is useful for understanding the diversity of solo play. Patterns of solo play are often elicited by various features encountered in the environment, such as slopes, holes, water, and objects (e.g. leaves, branches, animal carcasses).

40.2.1 Locomotor play

Locomotor play (or locomotor-rotational play) consists of vigorous motor acts that are typically redundant and exaggerated. It is often seen when a party of chimpanzees travels on the ground or rests on the ground or in trees.

A small infant, aged younger than 1 year, often repeats climbing and descending movements on short shrubs. He/she also walks awkwardly on the ground while staggering or stumbling. These simple locomotor behaviors in small infants can be regarded as play, since they repeat the movements with no obvious destination or objective. They are probably adaptive behaviors, since infants can train their motor skills. Older infants and juveniles engage in a greater variety of locomotor play as described below.

Locomotor play in trees

Various patterns of arboreal locomotor play are observed on various forms of physical substrate. An infant often repeats the sequence of climbing shrubs or trees and diving to lower vegetation (or the ground). He/she also often hangs from a horizontal branch by one hand and spins the body. A young chimpanzee occasionally hangs upside-down from a branch, or swings forward and rotates upward like the human "forward upward circling" movement. He/she sometimes hangs and walks upside-down under a branch with hands and feet like a sloth. He/she also often clings to a woody vine and shakes it or swings on it.

Locomotor play on ground

Young chimpanzees often engage in locomotor play while traveling on the ground, especially in open areas such as observation trails. They perform various types of acrobatics such as turning somersaults (forward, sideways, or backward) and "pirouette" by spinning around the vertical axis while advancing forward in a straight line (Nishida and Inaba, 2009). Infants often show an irregular pattern of playful walking, "play walk," in which they walk quadrupedally with the head tilted or shook while staggering. Another type of irregular walking, which has rarely been observed, is an infant walking quadrupedally not on the knuckles but on the backs of the hands. When a party of chimpanzees rests on the ground, an infant sometimes runs around and around a stem. An infant also occasionally repeats hopping on a spot.

As for play in general, some patterns of locomotor play are stimulated by specific features of the environment, such as slopes, dry leaves, and sand, which we describe separately in the following sections.

Locomotor play on a slope

A slope often stimulates chimpanzees to engage in locomotor play. When a chimpanzee party descends on a gentle slope, infants or juveniles may turn somersaults, or roll down like a tube with the arms straightened over head. An infant also occasionally slides repeatedly down a large slanted rock.

Locomotor play with dry leaves

When a party travels in a procession on ground covered with dry leaves, chimpanzees may perform "leaf-pile pulling," in which they turn around and walk backwards quadrupedally while raking dry leaves with the hands (Nishida and Wallauer, 2003; see Chapters 38, 42). Leaf-pile "pushing" play has also been observed but is less common than pulling. A young chimpanzee sometimes collects many dry leaves on his/her belly and somersaults backward, scattering the leaves. A juvenile female, Athena, once lay supine and slid downward with head first on a slope covered with dry leaves, while kicking the ground with her feet and rotating her arms like human "backstroke" swimming.

Locomotor play on sandy soil

Sandy soil is rare within the M group's range, but it is sometimes available when national park staff bring in some sand for construction of trail beacons, etc. When young chimpanzees encounter sandy ground, they may roll, somersault, or slide supine on it. As a result, they occasionally get covered all over with sand.

40.2.2 Object play

Chimpanzees at Mahale use various objects in play. They occasionally engage in object play with attached objects, such as manipulation of tree bark and slapping on tree buttresses, but more often they play with detached objects. The great variety of objects used in play includes branches, fruits, leaves, flowers, pieces of bark, stones, sand, the remains of animals (e.g. carcass, pieces of skin), and human artifacts.

Various behaviors are observed during object play, according to the shape, size, weight, and other features of the object. These behaviors can be classified into four categories: (a) holding (e.g. picking up, lifting, and holding in mouth, hand, groin pocket, or neck pocket), (b) transportation (e.g. dragging with hand or mouth and carrying in hand, foot, mouth, groin pocket, and on back or nape), (c) manipulation (e.g. touching, handling, inspecting, rotating, slapping, mouthing), and (d) orientation (an object is oriented to another object, substrate, or individual as in tool use, e.g. hitting, inserting, digging, throwing at). An infant may pick up a dry leaf from the ground and try to hold it between the chest and chin. Infants also carry a twig or a bunch of fruits in the mouth or groin pocket and walk around in a tree. This pattern of play can also be regarded as locomotor play, which may be useful for practicing various types of locomotion with objects.

Object play with sticks

Sticks (branches, twigs, etc.) are often used in object play. Chimpanzees may carry a branch and brandish or throw it. Sticks are also used as tools in various ways, such as touching or hitting something and inserting into holes. A 5-year-old female, Imani, once repeatedly inserted sticks inside the rind of a fruit of *Saba comorensis* to obtain its flesh, although she could eat it only by using her finger. In a dry season, two juvenile females were observed to use a stick to dig a trench in the wet ground of a stream to drink water from it (Nishida *et al.*, 2009). Sticks are also used by some adolescent and young adult chimpanzees in attempts to hunt for small animals such as squirrels and hyraxes (see Chapter 42). Some of these cases may not be genuine play, since they seem to have a substantial benefit, but they suggest the great curiosity and flexible tool-using ability of chimpanzees.

Object play with stones and fruits

Infants and juveniles occasionally pick up a stone, fruit, or hard rind of a fruit, and carry it. They may raise and drop it, put it on their back, or hold it in the groin pocket. They sometimes throw a stone or fruit, like adult males do in charging display. Throwing is performed in both underarm and overarm movements. When an infant or a juvenile throws underarm, the object occasionally flies not forward but upward or even backward. A stone or a fruit can be used like a ball. A juvenile male, Michio, once lay supine on the ground, extended his legs upward, put a large fruit of *Myrianthus* on his feet, and rotated the fruit like a ball with his hands and feet (Nishida *et al.*, 2010).

Object play with dry leaves and sand

Infants and juveniles occasionally grab dry leaves and repeat the sequence of lifting and dropping them. They also rake dry leaves, scatter the pile, and then sometimes throw them. Similar play behaviors are observed for sand. Infants also grasp and lift up sand and then let it fall little by little from the upraised hand. Chimpanzees seem to enjoy producing rustling sounds in these kinds of play. There is no report of "drawing" on sandy soil, though some lines occasionally appear as a result of playing with sand.

Play with tree buttresses and metal walls

When a party travels near a tree buttress, infants and juveniles occasionally approach and slap or kick it several times. They appear to enjoy producing sounds, though they are not as loud as the banging sounds made by adult males in a charging display. In contrast to a tree buttress, a metal wall is rarely slapped or kicked during play, though it is often beaten in adult charging display (see Chapters 31, 42). Juvenile or adolescent males sometimes only touch the wall softly, as if they were afraid to make loud sounds with it (Nishida, 2012: 221).

Making bed play

An infant occasionally engages in "making bed" play in a tree in daytime. He/she bends a few twigs and piles them up in a place to make a bed, sit or lie on it, and then often destroy it by hopping on it. This sequence is sometimes repeated several times. Infants sleep in a mother's bed at night but have to make a bed by themselves after weaning, and this type of play is probably useful for practicing how to make beds (see Chapter 43). A juvenile or adolescent chimpanzee also makes a bed during play, but it is usually not for amusement in itself but for creating a platform on which he/she plays socially with another infant.

Object play with the remains of animals

Chimpanzees occasionally encounter animal remains such as carcasses, and they may pick them up to inspect and carry them. A 10-year-old female, Liz, was once seen carrying the carcass of a mongoose. After hunting, a remaining part of the prey, such as a piece of skin or a tail, is sometimes carried by young chimpanzees: they do not eat it but brandish or groom it. A similar case was once also seen in an adult female, Gwekulo, who carried the arm of a red colobus for more than 7 h while traveling with another adult female carrying the mummified dead body of her infant (see also Chapter 5; Hosaka *et al.*, 2000). An adolescent female, Mimi, was once seen to carry a dead leopard cub, and she showed alloparental behaviors toward it on a day bed, such as grooming and tickling (Hiraiwa-Hasegawa *et al.*, 1986). Nakamura (2009) reported that an old female, Wakusi, collected and carried several guineafowl feathers. Wakusi did not show any attempt to manipulate the feathers in a playful manner but just kept holding them in her hand, lips, or groin pocket, perhaps because she was fascinated by them (see also Chapter 42). Hosaka *et al.* (2000) reported a case where an adolescent male, Bonobo, carried the dead body of an unknown infant chimpanzee. He carried it for about 2 h and did not eat it but groomed it (see Chapter 26).

An encounter with the carcass of a large animal is so rare that it elicits more intense exploratory behaviors. Nishida (2012) described the response of chimpanzees to a freshly dead body of a leopard. Many chimpanzees gathered to the spot and touched, sniffed, slapped, and even groomed the body. Most of these actions were assumed to be aimed at collecting information on the leopard as a predator (see also Chapter 16).

Object play with human artifacts

Although chimpanzee contact with human artifacts should be avoided as much as possible, and we are currently taking maximum care to avoid such contact in order to reduce the risk of disease transmission from humans (see Chapter 25), there had been several cases where young chimpanzees tried to touch or even steal human belongings. When they do this, they often inspect the items and play with them. Equipment used for research activities, such as wooden boards indicating the location and plastic tags on trees for phenology censuses, have also been occasionally carried by chimpanzees. An 11-year-old

Figure 40.1 A juvenile male, Michio, drums on a clay pot. ©T. Matsusaka.

male, Cadmus, once found a clay pot discarded long before by the former villagers, and he lifted it and put it on his head like a helmet (Matsusaka, 2012). Similarly, a 6-year-old male, Michio, found a discarded clay water pot and drummed on it repeatedly (Figure 40.1; Matsusaka, 2012).

Object orientation to one's own body

An object is occasionally oriented to the body of the player. Young chimpanzees occasionally drape a bunch of fruit, leaves, or animal skin over their shoulders, nape, or head. McGrew and Marchant (1998) reported that a young adult female, Ako, once draped around her neck a piece of skin of a red colobus that was formed into a circle with a single overhand knot (see also Chapter 42). A juvenile female, Imani, once wore a cylindrical bark on her lower arm, like a gauntlet (Nishida *et al.*, 2010). A 5-year-old female, Athena, once lay on her back and rubbed her belly with a smooth stone, probably enjoying the touch. Sticks are also occasionally applied to one's own body. Nishida (2002) reported that an 8-year-old female, Ivana, once applied a probe to her toe in an attempt to remove a sand flea, but she failed (see also Nishida *et al.*, 2009; Chapter 42). A two-year-old female, Mitsue, once inserted a tiny stick into the nostril, without inducing sneeze as in the cases reported by Nishida and Nakamura (1993) (see Chapter 42). These may be attempt at self-medicating but also could be regarded as flexible and playful use of sticks. Using a stick to tickle one's own body is reported to be habitual at Gombe, Tanzania (Whiten *et al.*, 1999) but absent at Mahale.

40.2.3 Imaginary play

Although observed only rarely, some reports describe imaginary play (pretend play or fantasy play) that apparently involves an imaginary playmate (see also Chapter 36). Hayaki (1985) reported that an 8-year-old male, Aji, played solo with branches of shrubs as if he were wrestling with another chimpanzee; he entwined himself in the branches and slapped, bit, and mounted them while emitting play panting. An adult female once lay supine with all limbs raised and emitted play panting as if playing "airplane" with the infant who had died a few months ago (Nishida *et al.*, 1999). A juvenile female, Serena, similarly once played "airplane" with a mummified dead infant of another adult female (see Chapter 25, Table 25.4; Hosaka *et al.*, 2000). Object play with an animal carcass, described above, can also be regarded as imaginary play like human "doll play," in which alloparental behaviors such as hugging and grooming are observed. Similar solo play with an object or carcass is also observed at Bossou, Guinea (Matsuzawa, 1997; Hirata *et al.*, 2001). Nakamura (2012) reported another type of solo imaginary play observed at Bossou: a 10-year-old male held and bit his own leg while emitting play panting as if he struggled to catch and bite the leg of a playmate.

40.2.4 Other exploratory play

Chimpanzees sometimes pay attention to and inspect a part of the environment, such as holes, water, and other animal species. They also sometimes inspect and manipulate their own body.

Play at a hole

Holes occasionally attract the interest of chimpanzees, especially of the young. When they find a hole in the ground that has been dug by other animals such as an aardvark, an infant or a juvenile may look inside or even enter it. Chimpanzees also sometimes investigate tree hollows or the crannies on a rock; they may look inside, put a finger or arm into it, insert a stick, and then sniff the finger or stick. Some of these activities may possibly be feeding-related behaviors to search for insects or small animals, but in most cases, nothing is extracted from the hole.

Play with water

Adult chimpanzees generally avoid water contact, and they usually use stepping-stones or overhead vegetation to cross a large river. However, since the early 1990s and, in particular, the early 2000s, young chimpanzees have begun to play with water (Nishida *et al.*, 2009; Matsusaka *et al.*, in prep.). At gentle streams or water puddles, they may splash the water surface with the hand, foot, or lower lip, stir water with the hand or foot, turn over a stone underwater, or pick up a stone and throw it. A 7-year-old male, Chopin, once entered a stream up to his knees and even dunked his face into the water (Hosaka, 1995). At a water puddle, a 6-year-old male, Christmas, once repeatedly immersed his hand and then dripped water from the back of the hand.

They have also begun to play with their self-image reflected in the "water mirror" (Nishida *et al.*, 2009, 2010). Juvenile and adolescent chimpanzees occasionally crouch by a stream and look at the water surface intently from a close distance. Some chimpanzees also made body movements, such as shaking hands or head, or rocking the body, while looking at their reflection in the water.

Similarly, chimpanzees have been handling the water collected in tree hollows since the early 2000s. Many infants and juveniles have begun to drink the water of tree hollows using various objects: They use a crumpled leaf like a sponge, a flat leaf like a spoon, and a stick, often one after the other (Matsusaka *et al.*, 2006; Figure 40.2). Various tool-using behaviors are also observed at streams, where they can easily and efficiently drink water directly with the mouth. They are assumed to be stimulated by curiosity about water, holes, and the function of various materials. Thus, it is likely that these tool-using behaviors are performed

Figure 40.2 A juvenile female, Jidda, uses a leaf to drink water from a tree hollow. ©T. Matsusaka.

as object play rather than subsistence tool-using behavior out of necessity (see also Chapter 42).

Exploratory behaviors for food items

Small infants occasionally put a fruit or a leaf into the mouth without ingesting it. They occasionally put even items that are not eaten by adult chimpanzees into the mouth. Through these exploratory feeding behaviors, along with dental and physical development and the social influence of the mother's feeding behaviors, they may learn little by little the food repertoire of the group (Chapter 22). Exploratory behaviors could also be the origin of new feeding habits. Zamma *et al.* (2011) reported that a juvenile female, Mitsue, started to eat a pith of oil palm, which has never been eaten by adult chimpanzees at Mahale. They conjectured that her high propensity for playing with carried objects (in this case, an oil palm frond cut by a human) might be a trigger for this innovative feeding behavior.

Play with one's own body

Chimpanzees occasionally perform exploratory play behaviors oriented to their own body. We have already described the examples of play with an object oriented to one's own body, and imaginary play using one's own leg (see above). Another example is solo sexual play, in which chimpanzees engage in genital stimulation (see also Chapter 35). A male occasionally touches or pulls his penis, leading to penile erection but never culminating in ejaculation. A female also fumbles with her clitoris using a hand, but only rarely. They may also rub genitals against a substrate such as ground or branches, or rub a branch or stone against their own genitals. Several males show a peculiar style in that they rub their penises with their own heel or sole (Nishida *et al.*, 2009).

Since about 1999, several young chimpanzees have begun to slap their own belly to produce drum-like sounds while hanging from a tree (Nishida *et al.*, 2009). This is first performed as solo play, but in later years, it is also used in play solicitation or intimidation displays to others. Several chimpanzees occasionally flip their lips, making the upper lips rolled up. This playful facial movement may reflect boredom (Nishida *et al.*, 2010; Chapter 37).

Play with another animal

At Gombe, young chimpanzees commonly engage in social play with olive baboons (*Papio anubis*) (Goodall, 1986). This kind of mutual play interaction with other nonhuman animals has never been observed at Mahale. A small animal can be a toy of chimpanzees. A 6-year-old female, Imani, once played with a moth (Nakamura, 2013): She rolled with the moth flapping in her mouth, while emitting play panting. Another type of interaction, playful teasing, is occasionally observed at Mahale, in which a chimpanzee threatens other animals, such as warthogs or baboons, by chasing or throwing a stone or branch. Zamma (2002) reported that an adult female, Nkombo, teased a squirrel by hitting, biting, and dragging it as well as liberating it to allow the squirrel to counterattack her. She swung her arm in presentation to the squirrel as a decoy to attract its attention, while showing a play face. In other cases, chimpanzees seem to be motivated more by investigating other animals, rather than by seeking amusement (see also Chapter 16). Matsusaka (2007) reported threat-like behaviors toward a porcupine: chimpanzees cautiously placed themselves some distance from the porcupine while shaking branches or clubbing the ground near it. They probably tried to elicit some reaction from the porcupine through threatening behavior as a way to learn about the animal. Zamma (2011) reported that a group of chimpanzees took the trouble of cautiously approaching to watch a python, despite its status as a potential predator.

Play with humans

Until researchers had begun to notice the seriousness of possible disease transmission to chimpanzees from humans (Chapter 25, 50), they tended to get closer to chimpanzees than they do now. In such situations, some chimpanzees occasionally tried to interact with human observers. Some young chimpanzees supposedly solicited a human to play with them by touching the trousers and going away while pirouetting, or hitting and running away while showing a play face. Several juvenile and adolescent chimpanzees occasionally threatened or teased a human by throwing a branch or clubbing the ground with a branch. An adolescent male, Darwin, repeatedly put his head between an observer's thighs and pushed from behind (Nishida *et al.*, 2009). A juvenile female, Imani, occasionally passed through the legs of a standing human. As a general rule, researchers at Mahale should not play with chimpanzees but try to ignore these actions and immediately move away from them in such cases. However, in some cases we actually could not help laughing or screaming briefly. These responses may have stimulated some young chimpanzees at that time to repeatedly amuse themselves with humans.

40.3 Intermediate forms of solo and social play

In this section, we describe play behaviors observed in two or more chimpanzees but lacking obvious mutual play interaction.

40.3.1 Teasing play

Teasing play is a social interaction in which one chimpanzee provokes another to upset or anger. This cannot be called "social play" since the target is not playful or amused at all, while the teaser may enjoy the response of the target. Nishida (2012: 115) reported that a 10-year-old female, Tula, once tried to "fish" for a 4-year-old male, Linta, with a colobus tail. She suspended the tail from a tree, and whenever Linta jumped for it from the ground, she withdrew it, so that he could not reach it. Nakamura (1995) reported a similar interaction, in which a juvenile female in a tree made fun of an adolescent male on the ground. Juvenile and adolescent males often pester older females by threatening behavior such as shaking and throwing branches. These behaviors (called "harassment") resemble teasing play but are usually performed more aggressively and thought to be a process of dominating females (Nishida, 2003; see Chapters 22, 29, 42).

40.3.2 Solo play using mother's body

A small infant younger than 1 year old often repeats climbing and descending on the mother's body. Older infants occasionally play on the mother's body during travels. They may hang below the mother's belly with only one or both hands, while touching dry leaves on the ground with their feet. When riding the mother's back, an infant may lie supine or stand on his/her head (Figure 40.3). An older infant often leaps down from the mother's back, runs ahead and climbs a shrub or fallen log, waits for the mother to catch up, and then jumps on her back again.

40.3.3 Solo play under social influence

Chimpanzees are sometimes attracted to the object of others' attention, and they approach to gaze at what others do, such as eating, trying to remove an ectoparasite, inspecting wounds, or playing with an object. Observers occasionally perform similar behavior subsequently. Playful tool use for drinking water (mentioned above) is assumed to have propagated among young chimpanzees at Mahale through such observations based on curiosity (Matsusaka *et al.*, 2006). A similar process is also suggested for other play behaviors, such as leaf-pile

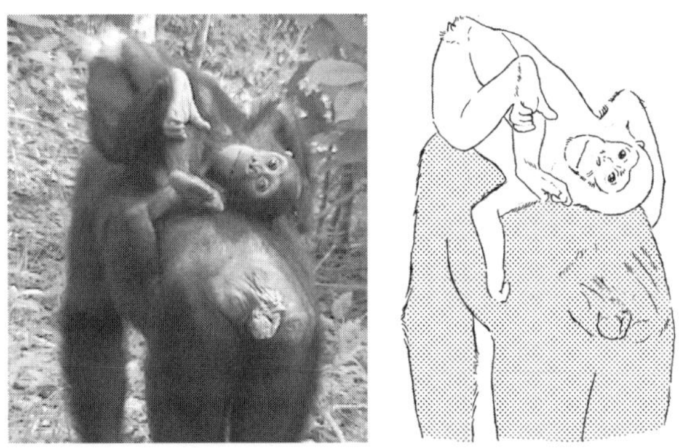

Figure 40.3 An infant female, Aqua, stands on her head while riding on her mother's back. ©T. Matsusaka.

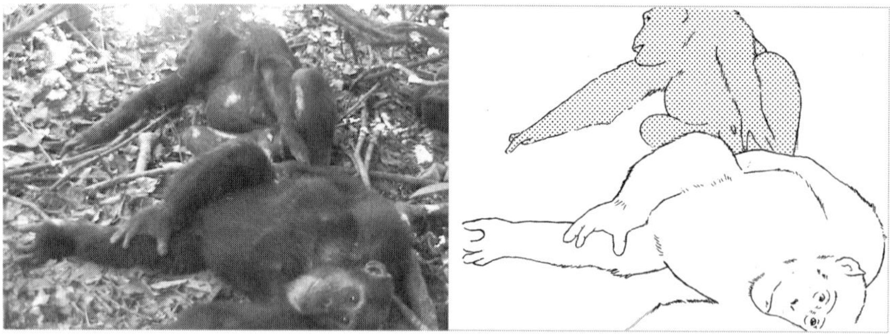

Figure 40.4 Kalunde tickles Masudi's foot with his right hand, and Masudi shows play face. ©T. Matsusaka.

pulling (Nishida and Wallauer, 2003) and play with water at streams (Nishida *et al.*, 2009).

40.3.4 Parallel play

Two or more infants around 1 year old occasionally gather in one place and play together, but each plays solo instead of interacting with one other. They may walk about on branches, or repeat the sequence of climbing and leaping down one after another. Older infants come to mutually respond to the actions of the playmate, and then they begin to engage in social play such as wrestling and chasing.

40.4 Diversity of social play

In this section, we concisely cover the diversity of social play (see Chapter 36 for more detailed description on social play).

40.4.1 Diversity of patterns of social play

Various behavioral patterns are observed during the social play of chimpanzees at Mahale (Hayaki, 1985; Nishida *et al.*, 2010). Rough-and-tumble play, involving chasing, wrestling, and tickling, is the most common pattern of social play. Rough-and-tumble play is accompanied by various aggressive actions (chase, grab, press down, mouth, tickle, slap, drag, etc.) as well as defensive actions (run away, struggle to flee, squirm, fend against tickle, etc.). It is also often accompanied by play face and play panting (Matsusaka, 2004). Some patterns of solo play are also observed during rough-and-tumble play, such as somersaults and hanging upside-down. Variations of rough-and-tumble play include round-chasing (or "circle," where players run in a circle: Hayaki, 1985), finger-wrestling (gentle tickling play between adults; Figure 40.4), and hang-wrestling (arboreal wrestling usually seen in infants hanging from a branch). Mother or allomothers show another specific type of play with an infant, "airplane (or balance)," in which they lie on the back and lift the infant by his/her limbs.

Chimpanzees occasionally play socially with an object, such as a branch, fruit, or animal skin. Players seldom throw the object at the playmate, but they may pull it from opposing sides. Chimpanzees also press a branch against a playmate while wrestling. Chimpanzees rarely play socially with water, but two adolescent chimpanzees once grabbed hands and pulled each other at a stream, one standing in the water while the other stood on a stepping-stone.

Another type of social play is sexual play (see also Chapters 35, 36). Infant or juvenile males occasionally engage in copulation with estrous females. This involves intromission but no ejaculation, and it often turns into wrestling play. Genital contact is sometimes also seen. Rubbing one's genitals against another's genitals has occasionally been seen in females (Zamma and Fujita, 2004; Nishida *et al.*, 2010). During social play, a player occasionally pinches the clitoris of an infant or juvenile female as tickling.

40.4.2 Social interactions accompanying social play

Social play is also a good opportunity to experience several kinds of social interactions with various partners.

Solicitation of play

Various gestures, postures, and facial expressions are used for solicitation of social play, such as slapping ground, shaking branches, hitting and running away, extending a hand, leaf clipping, head standing, and showing play face (Nishida *et al.*, 2010). Some of these gestures are also used in other contexts such as courtship. Play solicitation attempts often fail (Hayaki, 1985), but infants and juveniles may learn how to use gestures appropriately through these experiences.

Alloparental care and interaction with mother

Alloparental care for an infant is often regarded as a form of play (play mothering), or at least often involves social play such as tickling or playing airplane (see also Chapter 30). Nulliparous adolescent females are the most earnest caretakers, and their behaviors are assumed to be useful for practicing maternal behaviors (Nishida, 1983). Chimpanzees also perform tactical grooming in the course of allomothering; they often groom the mother before they take the infant in order to get her permission.

Conflicts in social play

Chimpanzees often experience conflicts in social play. Social play occasionally turns into a quarrel between the players. Competition over play partners also occurs in triadic play (Hayaki, 1985). Young chimpanzees may learn how to deal with conflicts with various individuals by experiencing a lot of social play.

40.5 Discussion

40.5.1 Diversity of solo play and its significance

Young chimpanzees often engage in solo play stimulated by objects or features in the environment, such as branches, vines, dry leaves, slopes, holes, water, and so on. When they encounter them, they may pay attention to them and try to examine various ways to manipulate or interact with them. If they find a novel way of manipulation or interaction, it would probably further stimulate their curiosity to repeat it as well as modify it, and thus various types of play behaviors emerge.

Through these activities, young chimpanzees are assumed to learn various ways to interact with the environment. Locomotor play might function as training not only for physical development but also for developing the locomotor skills needed to avoid predators and to feed or travel on precarious substrates. Object play can contribute at an early stage to the development of skills in holding and transporting objects, such as foods and tools. It is likely that object play also provides an important opportunity to learn tool-using techniques: young chimpanzees experience various ways to manipulate objects and may learn the function of materials as tools, as well as the causal and physical relationships between objects. Other types of exploratory play may be important for collection of information on something unfamiliar in the environment.

40.5.2 Diversity of social play and its significance

Social play includes several types of interaction, such as wrestling, chasing, and tickling. These are performed between group members of various age–sex combinations. Social play might be important for promoting the ability to fight against or escape from conspecifics or predators, for adjusting one's behaviors flexibly according to the body size and behavior of the playmate, and for getting information about playmates, such as physical ability and personality. Play mothering and sexual play may serve as practice of the maternal and sexual behaviors performed later in adulthood. Social play could also be a good source of opportunities to experience other kinds of social interaction, such as conflicts between playmates and conciliation of the reluctant mother of the target infant of allomothering. These experiences are expected to be beneficial to learning about how to deal with conflicts and how to negotiate with other individuals having a different interest or demand.

40.5.3 Conclusion and remaining issues

In summary, various types of play have developed from the richness of the natural and social environment. Chimpanzees interact with the natural environment in solo play, and with other chimpanzees in social play. The great variety of play is likely important for the physical, behavioral, cognitive, emotional, and social development of young chimpanzees. Further studies on wild chimpanzee play behaviors are needed to clarify the developmental process of play and its significance to the development of chimpanzees in general.

Further studies on the variety of play behaviors in wild chimpanzees are also needed to clarify the local variations of play among study sites. Humans exhibit a great cultural variation in the domain of play, but such a phenomenon has not yet been verified in the case of chimpanzees. A few patterns of play have been proposed as cultural play behaviors of a certain population, such as leaf-pile pulling at Mahale (Nishida and Wallauer, 2003; Chapters 38, 42) and using sticks for self-tickling at Gombe (Whiten *et al.*, 1999). However, it is plausible that there are more variations in play, especially when we consider that reports of such variation have not been made at most study sites and that innovation and propagation are believed to occur occasionally (Matsusaka *et al.*, 2006; Nishida *et al.*, 2009).

References

Goodall, J. (1986). *The Chimpanzees of Gombe: Patterns of Behavior*. Cambridge, MA: Belknap.

Hayaki, H. (1985). Social play of juvenile and adolescent chimpanzees in the Mahale Mountains National Park, Tanzania. *Primates*, **26**, 343–60.

Hiraiwa-Hasegawa, M., Byrne, R. W., Takasaki, H., and Byrne, J. M. E. (1986). Aggression toward large carnivores by wild chimpanzees of Mahale Mountains National Park, Tanzania. *Folia Primatologica*, **47**, 8–13.

Hirata, S., Yamakoshi, G., Fujita, S., Ohashi, G., and Matsuzawa, T. (2001). Capturing and toying with hyraxes (*Dendrohyrax dorsalis*) by wild chimpanzees (*Pan troglodytes*) at Bossou, Guinea. *American Journal of Primatology*, **53**, 93–7.

Hosaka, K. (1995). [Water play.] *Monkey* **39**(2), 20–1. In Japanese.

Hosaka, K., Matsumoto-Oda, A., Huffman, M. A., and Kawanaka, K. (2000). [Reactions to dead bodies of conspecifics by wild chimpanzees in the Mahale Mountains, Tanzania.] *Primate Research*, **16**, 1–15. In Japanese with English summary.

Matsusaka, T. (2004). When does play panting occur during social play in wild chimpanzees? *Primates*, **45**, 221–9.

Matsusaka, T. (2007). Exploratory-threat behaviors in wild chimpanzees encountering a porcupine. *Pan Africa News*, **14**, 29–31.

Matsusaka, T. (2012). Playful drumming by immature wild chimpanzees at Mahale: do they enjoy making sounds? *Pan Africa News*, **19**, 23–5.

Matsusaka, T., Nishie, H., Shimada, M., *et al.* (2006). Tool-use for drinking water by immature chimpanzees of Mahale: prevalence of an unessential behavior. *Primates*, **47**, 113–22.

Matsuzawa, T. (1997). The death of an infant chimpanzee at Bossou, Guinea. *Pan Africa News*, **4**, 4–6.

McGrew, W. C. and Marchant, L. F. (1998). Chimpanzee wears a knotted skin "necklace." *Pan Africa News*, **5**, 8–9.

Nakamura, M. (1995). ["Teasing" play.] *Monkey*, **39**, 18–19. In Japanese.

Nakamura, M. (2009). Aesthete in the forest? A female chimpanzee at Mahale collected and carried guineafowl feathers. *Pan Africa News*, **16**, 17–19.

Nakamura, M. (2012). Playing with his leg: a case of imaginary social play by an adolescent male chimpanzee at Bossou? *Pan Africa News*, **19**, 1–3.

Nakamura, M. (2013). A juvenile chimpanzee played with a live moth. *Pan Africa News*, **20**, 22–4.

Nishida, T. (1983). Alloparental behavior in wild chimpanzees of the Mahale Mountains, Tanzania. *Folia Primatologica*, **41**, 1–33.

Nishida, T. (2002). A self-medicating attempt to remove the sand flea from a toe by a young chimpanzee. *Pan Africa News*, **9**, 5–6.

Nishida, T. (2003). Harassment of mature female chimpanzees by young males in the Mahale Mountains. *International Journal of Primatology*, **24**, 503–14.

Nishida, T. (2012). *Chimpanzees of the Lakeshore: Natural History and Culture at Mahale*. Cambridge: Cambridge University Press.

Nishida, T. and Inaba, A. (2009). Pirouettes: the rotational play of wild chimpanzees. *Primates*, **50**, 333–41.

Nishida, T. and Nakamura, M. (1993). Chimpanzee tool use to clear a blocked nasal passage. *Folia Primatologica*, **61**, 218–20.

Nishida, T. and Wallauer, W. (2003). Leaf-pile pulling: an unusual play pattern in wild chimpanzees. *American Journal of Primatology*, **60**, 167–73.

Nishida, T., Kano, T., Goodall, J., McGrew, W. C., and Nakamura, M. (1999). Ethogram and ethnography of Mahale chimpanzees. *Anthropological Science*, **107**, 141–88.

Nishida, T., Matsusaka, T., and McGrew, W. C. (2009). Emergence, propagation or disappearance of novel behavioral patterns in the habituated chimpanzees of Mahale: a review. *Primates*, **50**, 23–36.

Nishida, T., Zamma, K., Matsusaka, T., Inaba, A., and McGrew, W. C. (2010). *Chimpanzee Behavior in the Wild: An Audio-Visual Encyclopedia*. Tokyo: Springer.

Whiten, A., Goodall, J., McGrew, W. C., *et al.* (1999). Cultures in chimpanzees. *Nature*, **399**, 682–5.

Zamma, K. (2002). A chimpanzee trifling with a squirrel: Pleasure derived from teasing? *Pan Africa News*, **9**, 9–11.

Zamma, K. (2011). Responses of chimpanzees to a python. *Pan Africa News*, **18**, 13–15.

Zamma, K. and Fujita, S. (2004). Genito-genital rubbing among the chimpanzees of Mahale and Bossou. *Pan Africa News*, **11**(2), 5–8.

Zamma, K., Nakashima, M. and Ramadhani, A. (2011). Mahale chimpanzees start to eat oil palm. *Pan Africa News*, **18**, 6–8.

41 Laterality of hand function

LINDA F. MARCHANT

41.1 Introduction

This chapter has three goals. First, to situate and review studies of chimpanzee laterality of hand function in Mahale. Second, to more broadly address conceptual and methodological issues in laterality studies that bear directly on research design and how this in turn impacts the usefulness of results obtained. Finally, to briefly suggest directions for future laterality research not only in chimpanzees but in other apes. The second goal is more fully explored in Marchant and McGrew (2013).

This chapter will not attempt to incorporate the burgeoning results from captive studies of chimpanzee laterality of hand use for reasons of space and lack of ecological validity. That is not to say that captive studies have less value – they have many benefits including regular access to subjects, and with that, closer scrutiny of ontogeny, as well as the ability to design specific tasks that may elicit stronger lateralized responses. Ideally researchers benefit from both perspectives, and in the rare instance, conduct ethological and experimental laterality research with the same chimpanzee subjects. The most salient example of this approach is that of Matsuzawa and colleagues working at Bossou, Guinea (for example, see Matsuzawa *et al.*, 2001; Humle and Matsuzawa, 2009).

While authors use a multiplicity of terms to characterize primate laterality, often the default term has been "handedness." This choice has led to more confusion than clarity. To add more precision, this chapter relies on a framework proposed by Marchant and McGrew (1994). (See Marchant and McGrew, 2013, for an extended discussion of this framework and how it applies to laterality studies in chimpanzees.) To be brief, the number of individuals who are observed (subject *N*) and the number of behavioral patterns (often labeled "tasks") provides a framework to interpret laterality results. If a single subject on a single task shows a statistically significant result in direction of lateralization it is labeled "hand preference." If multiple subjects on a single task show a statistically significant result in direction of lateralization it is labeled "task specialization." If a single subject on multiple tasks shows a statistically significant result in direction of lateralization it is labeled "manual specialization." Finally, if multiple subjects on multiple tasks show a statistically significant result in direction of lateralization it is labeled "handedness" (see Table 1 in Marchant and McGrew, 2013).

"Handedness" as defined here, is the pattern observed in modern *Homo sapiens sapiens* with approximately 90% of individuals right-handed. This strict definition (*sensu stricto*) has been argued by some (see Marchant *et al.*, 1995; Marchant and McGrew, 2013) to be a human universal, i.e. a species-typical trait, and that this trait is derived in our taxon. Importantly, even this universality in humans is most clearly expressed in contexts of tool use and especially precision tool use. (See Marchant *et al.*, 1995 for an ethological study of hand use in three traditional societies: G/wi San of Botswana, Himba of Namibia, and Yanomanö of Venezuela.) Although this chapter will focus solely on chimpanzee behavior, this manual asymmetry in *H. sapiens* is linked to human cerebral asymmetries, hemispheric specialization, and the evolution of language abilities and other lateralized functions. Given the phylogenetic proximity of *Pan*

Mahale Chimpanzees: 50 Years of Research, ed. M. Nakamura, K. Hosaka, N. Itoh and K. Zamma. Published by Cambridge University Press.

Table 41.1 **Laterality of hand-use studies at Mahale**

Citation	Behavioral pattern(s)	Result
Nishida, 1973	Ant fishing	Ambilateral?
Nishida and Hiraiwa, 1982	Ant fishing	Ambilateral?
Norikoshi, 1994, 1998	Ant fishing	No individual data
Nakamura *et al.*, 2000	Social scratching	Ambilateral
McGrew and Marchant, 2001	35 patterns (excluding tool use)	Ambilateral
Corp and Byrne, 2004	Bimanual process *Saba comorensis* and *Citrus limon*	Hand preference; sex difference: females right biased; males left biased
Marchant and McGrew, 2007	Ant fishing	Ambilateral
Nishida *et al.*, 2012	Throwing	Ambilateral

and *Homo* (chimpanzees and humans are sometimes referred to as sister taxa), it is not surprising that the chimpanzee has been a critical focus for comparative laterality research (Bradley and Vigilant, 2002).

The study of laterality in wild chimpanzees is well into its fifth decade. Efforts to characterize, interpret, and compare patterns of chimpanzee hand use (laterality) at long-term research sites have grown accordingly (e.g. Nishida, 1973; Boesch, 1991; Sugiyama *et al.*, 1993; Lonsdorf and Hopkins, 2005; McGrew *et al.*, 2013). Over these nearly 50 years the focus of publications has ranged widely from natural history descriptions of specific tool-use patterns (e.g. Nishida and Hiraiwa, 1982) to providing exhaustive ethological descriptions of patterns of laterality for all individuals at a study site (see Marchant and McGrew, 1996; McGrew and Marchant, 2001).

41.2 Laterality studies at Mahale

41.2.1 Summary of studies

The chimpanzees of Mahale have provided insight into multiple aspects of laterality in the chimpanzee. This study site, founded in 1965 by Toshisada Nishida, has frequently led the way in our understanding of the biology and behavior of wild chimpanzees (Nishida, 2012). So, it is not surprising that Japanese researchers working at Mahale were among the first field workers to characterize patterns of manual behavior (see Table 41.1 for a list of publications and behavioral patterns described). Five of the publications (Nishida, 1973; Nishida and Hiraiwa, 1982; Norikoshi, 1994, 1998; Marchant and McGrew, 2007; Nishida *et al.*, 2012) report on tool-use patterns, ant fishing and throwing. (The two publications by Norikoshi on ant fishing present much the same information with one in Japanese and one in English). The other three publications scrutinize a single pattern, social scratching (Nakamura *et al.*, 2000); report on 35 behavioral measures from a manual ethogram (McGrew and Marchant, 2001); and describe bimanual food processing for two fruits, a wild fruit, *Saba comorensis*, and an introduced domesticated fruit, the lemon (*Citrus limon*) (Corp and Byrne, 2004).

41.2.2 Ant-fishing studies

A review of Table 41.1 indicates the prominence of ant fishing as a research focus at Mahale that began as early as Nishida's (1973) publication. Nishida described this pattern of arboreal extractive foraging in which the chimpanzee makes a flexible probe of vegetation and inserts it into an entry hole of a nest of wood-boring ants (*Camponotus* spp.) (see Chapter 42). Three of the later publications (Norikoshi, 1994, 1998; Marchant and McGrew, 2007) focused on laterality as the primary purpose of the report. The two earlier publications (Nishida, 1973; Nishida and Hiraiwa, 1982) include discussions of laterality but these papers are principally superb examples of meticulous natural history with exhaustive discussion of ant and tree species, habitat description, categories of raw material for tools, tool-using behavior, and the 1973 paper has some of the first photographic images of ant fishing. Also of note, these earlier papers are based on field observations of the K group with only a single individual from M group contributing data to the 1982 paper.

In the 1973 paper, Nishida provides some qualitative impressions of laterality in hand use, or rather the lack of it. He provides quantitative data for one subject, Wakilufya, and gives a precise report of one session (observation no. 27: see Fig. 4 and Table 3 in Nishida, 1973). Overall he concludes that for this tool-using behavior the chimpanzees do "not always require the exclusive use of one hand" (Nishida, 1973: 362). Nishida is prescient in the discussion section of this paper. He anticipates a number of issues that are salient in today's laterality research including arboreal versus terrestrial tool use, the constraints and potential danger of falling in arboreal tool use to access prey insect taxa, origins of tool use, and the division of labor between the two hands in hominin tool manufacture.

In the paper that followed, Nishida and Hiraiwa (1982) supplied individual data for ant fishing for 20 individuals (19 from K group and one from M group; see Table 5 of their paper). Frustratingly, only five subjects have six or more bouts to allow for the binomial test. The authors note that most of the chimpanzees (17 of 20) use "different hands in different bouts, or both hands in the same bout" (Nishida and Hiraiwa, 1982: 85). They infer that chimpanzees "are ambidextrous and can fish for ants with either hand." Anticipating a later paper by Marchant and McGrew (2007) the authors also observe that it is the nest entrance of the ants that often dictates which hand the chimpanzee will use in order to ensure stability while doing this extractive foraging.

Two more papers, both by Norikoshi (1994, 1998), also targeted hand laterality in ant fishing. The author had 56 subjects and 201 bouts of ant fishing. However, the data were pooled for analysis so it was not possible to test for individual laterality. Like Nishida (1973) and Nishida and Hiraiwa (1982), Norikoshi (1998) thought that the nature of the arboreal task and the position of ants' holes constrained the chimpanzees' postures and access to the prey.

Marchant and McGrew's (2007) paper on ant fishing sought to test this hypothesis about arboreality and positional constraints. They also sought to see whether ant fishing would depart from a well-documented pattern for tool use in wild chimpanzees. Specifically that for tool use most individual chimpanzees are statistically significantly lateralized to one side or the other (see Table 41.2 reprinted from Marchant and McGrew, 2007). Twenty-three members of M group were observed to ant fish; data were collected in field notebooks and video records. Fifteen of the 23 subjects had six or more bouts of ant fishing and so individual statistical testing was possible. The result was surprising but also supported the ideas of Nishida and Norikoshi. That is, unlike other patterns of tool use where the majority of individuals are lateralized, sometimes exclusively, for this group of 15 chimpanzees, the majority (eight) were ambilateral. Of the remaining seven, one fished significantly more with the left, four significantly more with the right, and two only with the right (see Table 41.2).

This is a paradoxical result compared with other reported laterality studies of chimpanzee tool use. Yet the result accords with the earlier hypotheses about the siting of ant nests, the need to secure a safe posture to ant fish, and the constraints on the body, especially the upper limbs, to perform the behavior in an arboreal context. Marchant and McGrew (2007) examined this issue further by using the video records to examine the rate of changes of hand as a function of whether the "other" hand was providing major

Table 41.2 **Laterality of hand use in object manipulation of wild chimpanzees, by individual, comparing tool use and non-tool use (modified from Table 1 of Marchant and McGrew, 2007 with permission. ©Japan Monkey Centre and Springer, Japan)**

Behavioral pattern	Arboreal/ terrestrial	Always LH[a]	Signific. LH[b]	Ambi-lateral[c]	Signific. RH[d]	Always RH[e]	N[f]	Study site
Non-tool use								
Reach/pick up (Boesch, 1991)	T	1	2	**13**	3	1	20	Taï
Reach/pick up (Sugiyama *et al.*, 1993)	T	0	1	**18**	0	0	19	Bossou
Pluck food (Marchant and McGrew, 1996)	A/T	0	0	**32**	0	0	32	Gombe
Pluck food (McGrew and Marchant, 2001)	A/T	0	1	**39**	0	0	40	Mahale
Eat (McGrew and Marchant, 2001)	A/T	0	0	**40**	1	0	41	Mahale
Peel *Saba* (Corp and Byrne, 2004)	A/T	**13**	2	5	6	12	38	Mahale
Tool use								
Wadge dip (Boesch, 1991)	T	3	1	3	0	**9**	16	Taï
Nut crack (Boesch, 1991)	T	8	**10**	5	8	**10**	41	Taï
Nut crack (Matsuzawa *et al.*, 2001)	T	6	1	2	1	**10**	20	Bossou
Strychnos pound (McGrew *et al.*, 1999)	T	4	1	2	2	**5**	14	Gombe
Termite fish (McGrew and Marchant, 1996)	T	**12**	4	9	5	6	36	Gombe
Termite fish (Lonsdorf and Hopkins, 2005)	T	2	**10**	1	3	1	17	Gombe
Ant fish (Marchant and McGrew, 2007)	A	0	1	**8**	4	2	15	Mahale

On each row, the figure in bold is the mode.

[a] All bouts done with left hand (LH), $N \geq 6$.

[b] Statistically significant left-hand bias, binomial test.

[c] No statistically significant difference between left and right hand.

[d] Statistically significant right-hand bias.

[e] All bouts done with right hand (RH).

[f] Number of individuals for whom enough data were obtained for binomial testing.

support, or minor support or no support (see Table 2 in Marchant and McGrew, 2007). Sessions with change of hands had a statistically significant correlation with major support.

41.2.3 Other laterality studies

McGrew and Marchant's (2001) study of manual laterality of the Mahale chimpanzees was a replication and extension of a similar study of the chimpanzees in Gombe National Park, Tanzania, approximately 150 km north of Mahale (Marchant and McGrew, 1996). The 27 adult (age ≥ 14 years) and 17 subadult (≤ 12 years) chimpanzees of M group served as subjects. The ethogram consisted of 35 patterns of manual behavior (excluding tool use). The data were collected in bouts, not events, to avoid inflated sample size which may result in false-positive, statistically significant results. A bout was the first performance of a behavioral pattern after an intervening bout of another behavioral pattern. In the pooled data, a total of 14 696 bouts was recorded; 6575 bouts with the left hand (51% of unimanual bouts), 6277 bouts with the right hand (49% of unimanual bouts), and 1844 bimanual bouts. The 15 most frequent patterns were scratch, eat, groom, pluck, self-groom, nose wipe, pull, cradle, arm wrestle, hold, chin rest, pick-up, social scratch, reach, and arm cross; these 15 patterns accounted for 94% of all bouts recorded. An analysis of manual specialization (within subject and across all patterns of hand use) revealed that of the 44 subjects with $p < 0.05$, two-tailed, 8 were left biased, 2 were right biased, and 34 were unlateralized. For the 15 most frequent patterns, and for individuals who had enough bouts for a binomial test, of a possible 660 outcomes (35 patterns × 44 subjects), 318 cases had too few data, 320 cases were ambipreferent, and only 22 cases (15 L vs. 7 R) indicated an individual being lateralized for a pattern. Additional analyses for sex and age did not reveal sex or age differences in laterality of hand use in the Mahale chimpanzees. The authors concluded "Wild chimpanzees at Mahale are notably unlateralized for their upper limb movements, whether at the individual or population level" (McGrew and Marchant, 2001: 349).

Nakamura and colleagues (2000) examined a pattern of manual behavior called "social scratch" (see also Chapters 34, 38). Nishida (1983) had earlier described the pattern as "scratch and rub" when he observed it in alloparental care in K-group chimpanzees assuming it was commonplace. In this behavior the individual chimpanzee rakes a "hand back and forth across the body of another." In their publication the authors purposely chose a title that was reminiscent of an earlier title that used the word "custom" (McGrew and Tutin, 1978). The bulk of the publication is devoted to describing the behavioral pattern and considering four hypotheses to account for the behavior. However, the laterality data come from overlapping field seasons of three of the authors. Nineteen subjects had adequate observations for a binomial test. Of the 19, two individuals were significantly left preferent, three were right preferent, and the remaining 14 were ambidextrous.

The Mahale publication that presents strong evidence for laterality of hand preference, unlike the other studies in Table 41.1, is that of Corp and Byrne (2004). Thirty-nine M-group chimpanzees were observed as they processed two fruits, *Saba comorensis* and *Citrus limon*. Data for individuals consisted of both bouts (minimum of six) and fruits (minimum of six) and importantly, this task required bimanual processing with manual role differentiation (also see Chapter 17). The hand labeled "*preferred*", dominant, was the one that made precision actions, delicate manipulations, in contrast to the non-preferred hand that grasped the fruit in an initial cup grip (Marzke and Wullstein, 1995). In the results the authors noted "Overall, there were no population trends toward right- or left-handedness " (Corp and Byrne, 2004: 64), but unlike previous studies they demonstrated a robust hand preference for individuals as well as a sex difference. Indeed, the majority of chimpanzees were exclusively left or right preferent (*Saba* 25/36; *Citrus* 26/30). Of the 39 subjects, 19 were right biased, 18 were left biased, and only 2 unlateralized. In addition, the sexes differed in the direction and strength of hand preference; females were right biased; while males were more left biased and stronger in their preference.

The most recent laterality publication from Mahale is that of Nishida and colleagues (2012) report on a form of tool use – specifically, throwing. Data were collected by Nishida from 1999 to 2004 and were available in the dataset of McGrew and Marchant (2001). Sixteen subjects (15 male and 1 female) were recorded for a total of 556 throws with each throw considered an independent data point. Sixty-three throws were two handed, typically occurring in male charging displays. Of the remaining 493 one-handed throws, 226 (46%) were with the left hand and 267 (54%) were with the right hand. Of the 16 subjects, 12 were unlateralized, and 4 subjects showed statistically significant laterality on a binomial test, two-tailed, $p < 0.05$; three were biased to the right versus one to the left. This result contrasts sharply with the only captive data available. Hopkins *et al.* (2005) reported for a group of 89 captive chimpanzees, 50 were right biased, 23 were left biased, and 16 were unbiased.

Both the captive study and the Mahale study required six or more observations of an individual throwing for the binomial test to be used and like Mahale it appears observations were collected ad lib, although two of the three social contexts include chimpanzees throwing at unfamiliar humans near their cage and throwing to capture the attention of an inattentive human or conspecific. It is unclear whether there are parallels to these situations for wild chimpanzees. Interestingly, of the 89 subjects in Hopkins *et al.* (2005), 55 were female and 34 were males. In their analysis they note that sex was not a significant influence in handedness. However, this contrasts strongly with the Mahale results with only a single female subject. With respect to rearing history (mother reared, nursery reared, wild caught) wild-caught individuals have absolutely and relatively the fewest numbers of throwers ($N = 4$) versus non-throwers ($N = 48$) compared with the other two rearing conditions. As they argue in their discussion, rearing history is confounded with age since most wild-caught individuals are also the oldest and that overall the mean age of throwers was significantly younger than the mean age of non-throwers.

Nishida *et al.* (2012) is the only publication on manual laterality in throwing for *any* wild nonhuman primate species. As such, it would be useful for other researchers, especially those at long-term chimpanzee sites, to examine their records to see if comparable datasets are available for comparison. This is important not only to compare other wild chimpanzee groups to Mahale, but to see if the wild versus captive distinction is supported.

Several other laterality publications from Mahale merit attention. Interest in cradling and nipple preference led Nishida to examine this phenomenon in Mahale chimpanzees (Nishida, 1993). Of the 34 pairs of mother–infants who were observed, 13 of 15 male infants suckled more often from the left than the right nipple. Six of 19 female infants suckled more often from the left than the right nipple. Of those infants who showed a statistically significant laterality for nipple preference, all eight, six males and two females, had a left nipple preference. Nishida suggested this preference might be enabled by maternal cradling with the left arm but did not have adequate data to test this. When McGrew and Marchant's (2001) data on cradling are examined only seven females have adequate numbers of cradling bouts to test statistically (McGrew and Marchant, 2001: Table 5b). Of the seven subjects, six were ambipreferent and one was significantly left preferent.

Finally, in a Japanese publication (Nishida, 1994; Michio Nakamura, personal communication), Nishida observed strong individual hand preference when the chimpanzees bimanually processed the fruit of *Myrianthus arboreus* (Figure 41.1). He collected data on 13 individuals and stated that 12 subjects had a strong preference for either the right or left, but that the numbers were not skewed in either direction, i.e. about the same number of left and right preferent. Although data are not available for sex, this may be similar to what emerged in the Corp and Byrne (2004) study when it is necessary to differentiate roles in a bimanual task.

In sum, the laterality studies conducted at Mahale have contributed significantly to our understanding of the complexity of chimpanzee hand use in nature. In addition these studies added substantively to the natural history descriptions of wild chimpanzee

Figure 41.1 Bimanual processing of a *Myrianthus* fruit. ©Michio Nakamura.

behavior as well as distinctive cultural behavioral patterns shown by Mahale chimpanzees.

Early on, pioneering Japanese primatologists documented manual behavior in a tool-use behavioral pattern, ant fishing (see Table 41.1, Nishida, 1973; Nishida and Hiraiwa, 1982, Norikoshi, 1994, 1998; later joined by Marchant and McGrew, 2007). This set of publications not only set a standard for careful ethological description, they also examined one of a few arboreal patterns of tool use by wild chimpanzees. The results suggested the differing selection pressures that may prevail in arboreal contexts and also suggested how a shift to increased terrestrial tool use may have enabled increased levels of exclusive or increased bias for apes to express hand preference in tool use contents. The other tool-use pattern, throwing, as reported in Nishida *et al.* (2012) is still the only publication on this behavior in wild nonhuman primates.

Nakamura *et al.* (2000) added social scratch to the growing ethnography of chimpanzee behavior that is variously labeled culture, tradition, or custom. Mahale's K and M groups had already been well documented in the cross-site comparison of long-term field sites (Whiten *et al.*, 1999) and social scratching was yet another pattern added to Mahale's cultural repertoire.

McGrew and Marchant's (2001) report of ambilaterality in manual behavior (excluding tool use) for M-group chimpanzees was a persuasive replication of a similar study done at Gombe (Marchant and McGrew, 1996). Neither study revealed evidence of strong individual or group-level bias in direction of laterality.

Corp and Byrne's (2004) study of bimanual processing of two fruit species by Mahale chimpanzees revealed a fascinating sex difference in hand preference, with 37 subjects showing clear hand preference but with females right biased and males left biased and yet no evidence of a group-level bias. They also rightly brought attention to the role differentiation of the two hands, i.e. complex manual processing. This result highlights the notion that tool use per se is not a necessary requirement to evoke strong individual lateralized responses.

41.3 Further thoughts

Increasingly the goal in studying laterality in wild chimpanzees is to probe aspects of the apes' behavior that may provide insights and explicit comparisons with the human condition. This is a useful goal. To help further that discussion, several conceptual issues need to be more clearly defined and ideally researchers will make explicit the level of comparison they are seeking to assert.

In an effort to demonstrate a purported continuity of the *Homo–Pan* condition some authors have chosen to assert "population" or "population-level" laterality in publications (e.g. see Lonsdorf and Hopkins, 2005; Bogart *et al.*, 2012). However, it quickly becomes apparent that "population" means different things to different authors. The term is not used in the genetic or population biology sense; that is the authors are not using a biogeographic designation. In these publications the term is parsed to imply species typicality with the implication that the result is the equivalent of handedness, *sensu stricto*. But usually the reported result is applied to a single "task" or behavior–task specialization in Marchant and McGrew's (1994) framework.

Interestingly, in Bogart *et al.*'s (2012) publication the authors conflate several laterality terms ("hand preference" and "population-level left-hand preference") while combining data for termite fishing from two subspecies, *Pan troglodytes verus* in Fongoli, Senegal, and *Pan troglodytes schweinfurthii* in Gombe. Given the small sample size at Fongoli the authors chose to use results from two Gombe studies (McGrew and Marchant, 1992, 1996) and state in the abstract "When the data were combined across all studies, wild chimpanzees showed a population-level left-hand preference for termite fishing." Leaving aside the open questions of combining subspecies data and whether the termite-fishing behavior at the two sites is similar with respect to tools used (e.g. raw materials, tool dimensions, etc.), when the data are examined for the Fongoli subjects ($N = 27$), 8 of the 16 subjects whose hand preference is determined to be left have five or fewer bouts of termite fishing (range 1–5) in making that assignment. To be fair, in their statistical analyses the authors chose to use events rather than bouts.

In seeking to characterize chimpanzee, or other ape lateralized hand use, it is worth considering the strength of the claim that is made. When arguing for continuity with extant humans the strength of the claim is often generalized beyond the study subjects. The most conservative claim is a group-level bias or lack thereof. For example, when describing results on a bimanual tube task used with captive bonobos (*Pan paniscus*) in three European zoos and the Lola Ya Bonobo Sanctuary in the DRC, Chapelain *et al.* (2011: 491) noted strong individual hand preference with most individuals being lateralized but "no group-level bias even when (combining) data from the sanctuary and the zoos to enlarge the sample to 77."

In contrast, as noted previously, population-level handedness is sometimes used within and across chimpanzee subjects from different wild communities (or unit-groups), from either the same or different chimpanzee subspecies in considerably different habitat types. It is also useful to consider the *degree* of lateral bias that is reported. The default comparison with *Homo sapiens* is the ubiquitous population-level, species-typical report of 90% right-handedness. Earlier in this chapter this "invariant" percentage was tempered by reports from traditional societies. Nonetheless, it is still correct to characterize the taxon as right-handed.

Unlike the condition of *Homo sapiens*, the *direction* of lateral bias for several behavioral patterns in wild chimpanzees has been presented as "population-level handedness" but with "task-specific" differences in the direction of laterality for patterns of tool use (see Lonsdorf and Hopkins, 2005). In that publication termite fishing is described as population-level (left) handedness for wild chimpanzees; in contrast, wild chimpanzees are described as showing population-level (right) handedness for nut cracking. This is not problematic from a descriptive point of view yet it is clearly different from the behavioral phenotype of laterality of hand use in *Homo sapiens*.

As field researchers continue to tackle the topic of laterality of hand function in wild chimpanzees it is essential that they appreciate the distinction between handedness *sensu lato* (in the broad sense) and handedness *sensu stricto* (in the strict sense). The chimpanzee is a useful model for studying "handedness" *sensu lato* as it presents an intermediate phylogenetic state of manual lateralization and is arguably the taxon that resembles most closely an ape form after the last common ancestor (LCA; see Moore, 1996) but *before* "handedness" *sensu stricto* emerged in the hominin lineage.

Far more scrutiny of task complexity in other hominoid taxa may reveal selective pressures that shaped our own evolutionary journey, in our taxon, to species-typical right-handedness, *sensu stricto.* For example, Byrne and Byrne's (1991, 1993) study of food processing in mountain gorillas (*Gorilla beringei*) at Karisoke revealed complex and multi-step unimanual and bimanual sequences to process nettles (*Laportea alatipes*) and gallium (*Galium ruwenzoriense*). In these multi-staged processes "each stage has a laterality bias and each stage is sequentially dependent on the last " (Byrne and Byrne, 1993). Morino's recent (2011) paper on wild siamangs (*Symphalangus syndactylus*) reported individual and population "left-hand preference for a complex manual task" – collecting and drinking water ("task specialization" in Marchant and McGrew's 1994 framework).

An ongoing set of concerns centers on statistical considerations and whether the investigator makes explicit the null hypothesis of ambilaterality and uses two-tailed statistical tests. Following on that, there is concern for type I errors, rejecting the null hypothesis, $H_{0,}$ of ambilaterality, where in fact statistically significant right handedness (or left handedness) may have arisen due to chance. Equally of concern are type II errors where one accepts the null hypothesis of ambilaterality when H_0 is false. For both type I and type II errors, the number of subjects in the sample, and the nature of data collection – bouts versus events – may play an important role. Data collection that depends on the use of "bouts" – the first performance of a pattern, after an intervening bout of another behavioral pattern – has sparked considerable discussion in laterality studies (see McGrew and Marchant, 1997, for an extended discussion of bouts versus events, independence of data points, and inflation of datasets; see Hopkins, 1999 for a contrasting point of view).

A related statistical issue emerges when funnel plots – "scatter plots of percent right-hand use vs. sample size" are applied to the results of laterality studies (Palmer, 2002). Applying funnel plots to a publication claiming "population-level right handedness in 140 individual captive chimpanzees" (Hopkins, 1994) revealed a higher percentage of right-hand use for individuals with fewer observations. Indeed, the proportion of ambilateral chimpanzees "increased with increasing numbers of observations per individual" (Palmer, 2002).

As more results from wild chimpanzee study sites emerge, it may be useful to look at the nature of the data reported and to consider whether it is valid to combine results within or across study sites to increase sample size of subjects and datasets. A number of issues pertain to such comparisons – is the nature of the task/behavioral pattern the same, i.e. is it ethologically similar enough to be combined? Corp and Byrne's (2004) study of Mahale chimpanzees processing *Saba* and lemons presented individual results for each subject for the two fruits and then, given the ethological similarity of opening and processing both species of fruit, the authors correctly combined their results.

More problematic is Lonsdorf and Hopkins (2005) report. It appears that individual subject laterality results for termite fishing (McGrew and Marchant, 1992, 1996) and ant fishing (Nishida and Hiraiwa, 1982) were combined with more recent termite-fishing data from Gombe (Lonsdorf, 2005) in the data analyses despite the behavioral patterns being distinct as well as one being terrestrial and the other arboreal. (See legend for Fig. 1 in Lonsdorf and Hopkins, 2005, for histogram for "previous wild data".)

41.4 Future directions

It would be useful if laterality researchers defined terms more carefully and were aware of the wide variation in usage by different researchers. It seems unlikely that there will be universal adoption of any specific perspective, for example, Marchant and McGrew's (1994) laterality framework outlined earlier in this chapter. However, if researchers could appreciate the apparent differences between "handedness" *sensu stricto* and "handedness" *sensu lato* this would advance discussion and research design. More explicit description of procedures and rationales would also greatly benefit communication between researchers.

For long-term field sites such as Mahale it is a reasonable goal to collect laterality data using cross-sectional and longitudinal studies. The ontogeny of laterality is an area ripe for investigation, even with a long-lived taxon like the chimpanzee.

More field study of complex spontaneous manual behavior, especially of other great ape species, would also enlarge the discussion of the functions of lateralization. The work of Byrne and Byrne (1991, 1993) on mountain gorilla complex food processing is an exemplary demonstration of such an approach. Byrne and Byrne documented the more challenging behavioral patterns that enlist both hands (bimanual) and differentiate complementary roles for the two hands. It is therefore not so surprising that Corp and Byrne's (2004) focus on bimanual food processing of *Saba* and lemon by the Mahale chimpanzees revealed strong individual hand preference not unlike that seen more routinely for tool-use patterns in wild chimpanzees (Boesch, 1991; Sugiyama *et al.*, 1993; McGrew *et al.*, 1999).

A recent publication by Hobaiter and Byrne (2013) is a welcome contribution to the growing literature on gesture in wild apes and may promote new avenues of laterality research. In this instance the subjects are chimpanzees of the Sonso community in the Budongo Forest Reserve, Uganda. Hobaiter and Byrne report "Despite strong evidence for ambilateral hand use in gestural communication in the wild Sonso chimpanzees, it is difficult to completely dismiss the pattern of small but consistently right-hand biases we have found." They note that no subject used a single hand exclusively and interestingly, with increasing amounts of data, the estimates of strength of hand preference decreased (Palmer, 2002). What is most intriguing is that the most striking right-hand bias is in active object manipulation. This suggests fertile ground for other field primatologists to pursue and also reinforces that there is much to be learned about the relationship between manual behavior, objects, tool use, and role differentiation of the forelimbs in bimanual tasks.

References

Boesch, C. (1991). Handedness in wild chimpanzees. *International Journal of Primatology*, **12**, 541–58.

Bogart, S. L., Pruetz, J. D., Ormiston, L. K., *et al.* (2012). Termite fishing laterality in the Fongoli savanna chimpanzees (*Pan troglodytes verus*): further evidence of a left hand preference. *American Journal of Physical Anthropology*, **149**, 591–8.

Bradley, B. J. and Vigilant, L. (2002). The evolutionary genetics and molecular ecology of chimpanzees and bonobos. In *Behavioural Diversity in Chimpanzees and Bonobos, ed.* C. Boesch, G. Hohmann, and L.F. Marchant. Cambridge: Cambridge University Press, pp. 259–76.

Byrne, R. W. and Byrne, J. M. (1991). Hand preference in the skilled gathering tasks of mountain gorillas (*Gorilla g. beringei*). *Cortex*, **27**, 521–46.

Byrne, R. W. and Byrne, J. M. (1993). Complex leaf-gathering skills of mountain gorillas (*Gorilla g. beringei*): variability and standardization. *American Journal of Primatology*, **31**, 241–61.

Chapelain, A. S., Hogervorst, E. Mbonzo, P., and Hopkins, W. D. (2011). Hand preference for bimanual coordination in 77 bonobos (*Pan paniscus*): replication and extension. *International Journal of Primatology*, **32**, 491–510.

Corp, N. and Byrne, R. W. (2004). Sex differences in chimpanzee handedness. *American Journal of Physical Anthropology*, **123**, 62–8.

Hobaiter, C. and Byrne, R. W. (2013). Laterality in the gestural communication of wild chimpanzees. In *The Evolution of Human Handedness*, ed. W. C. McGrew, L. F. Marchant, and W. Schiefenhövel. *Annals of the New York Academy of Sciences*, **1288**, 9–16.

Hopkins, W.D. (1994) Hand preferences for bimanual feeding in 140 captive chimpanzees (*Pan troglodytes*): rearing and ontogenetic determinants. *Journal of Comparative Psychology*, **109**, 291–7.

Hopkins, W. D. (1999). On the other hand: statistical issues in the assessment and interpretation of hand preference data in non-human primates. *International Journal of Primatology*, **20**, 851–66.

Hopkins, W. D., Russell, J. L., Cantalupo, C., Freeman, H., and Shapiro, S. J. (2005). Factors influencing the prevalence and handedness for throwing in captive chimpanzees (*Pan troglodytes*). *Journal of Comparative Psychology*, **119**, 363–70.

Humle, T. and Matsuzawa, T. (2009). Laterality in hand use across four tool-use behaviors among the wild chimpanzees of Bossou, Guinea, West Africa. *American Journal of Primatology*, **71**, 40–8.

Lonsdorf, E. V. (2005). Sex differences in the development of termite-fishing skills in wild chimpanzees (*Pan troglodytes schweinfurthii*) of Gombe National Park, Tanzania. *Animal Behaviour*, **70**, 673–83.

Lonsdorf, E. and Hopkins, W. D. (2005). Wild chimpanzees show population-level handedness for tool use. *Proceedings of the National Academy of Sciences of USA*, **102**, 12634–8.

Marchant, L. F. and McGrew, W. C. (1994). An ethological perspective on human handedness. In *The Ethological Roots of Culture*, ed. R. A. Gardner, A. B. Chiarelli, B. T. Gardner, and F. X. Plooij. Dordrecht, the Netherlands: Kluwer, pp. 375–84.

Marchant, L. F. and McGrew, W. C. (1996). Laterality of limb function in wild chimpanzees of Gombe National Park: comprehensive study of spontaneous activities. *Journal of Human Evolution*, **30**, 427–43.

Marchant, L. F. and McGrew, W. C. (2007). Ant fishing by wild chimpanzees is not lateralised. *Primates*, **48**, 22–6.

Marchant, L. F. and McGrew, W. C. (2013). Handedness is more than laterality: lessons from chimpanzees. In *The Evolution of Human Handedness*, ed. W. C. McGrew, L. F. Marchant, and W. Schiefenhövel. *Annals of the New York Academy of Sciences*, **1288**, 1–8.

Marchant, L. F., McGrew, W. C., and Eibl-Eibesfeldt, I. (1995). Is human handedness universal? Ethological analyses from three traditional cultures. *Ethology*, **101**, 239–58.

Marzke, M. W. and Wullstein, K. L. (1995). Chimpanzee and human grips: a new classification with a focus on evolutionary morphology. *International Journal of Primatology*, **17**, 117–39.

Matsuzawa, T., Biro, D., Humle, T., *et al.* (2001). Emergence of culture in wild chimpanzees: education by master-apprenticeship. In *Primate Origins of Human Cognition and Behavior*, ed. T. Matsuzawa. Tokyo: Springer, pp. 557–74.

McGrew, W. C. and Marchant, L. F. (1992). Chimpanzees, tools, and termites: hand preference or handedness? *Current Anthropology*, **33**, 114–19.

McGrew, W. C. and Marchant, L. F. (1996). On which side of the apes: ethological study of laterality of hand use. In *Great Ape Societies*, ed. W. C. McGrew, L. F. Marchant, and T. Nishida. Cambridge: Cambridge University Press, pp. 255–72.

McGrew, W. C. and Marchant, L. F. (1997). On the other hand: current issues and meta-analysis of the behavioral laterality of hand function in non-human primates. *Yearbook of Physical Anthropology*, **40**, 201–32.

McGrew, W. C. and Marchant, L. F. (2001). Ethological study of manual laterality in the chimpanzees of the Mahale Mountains, Tanzania. *Behaviour*, **138**, 329–58.

McGrew, W. C. and Tutin, C. E. G. (1978). Evidence for a social custom in wild chimpanzees? *Man*, **13**, 234–51.

McGrew, W. C., Marchant, L. F., Wrangham, R. W., and Klein, H. (1999). Manual laterality in anvil use: wild chimpanzees cracking *Strychnos* fruits, *Laterality*, **4**, 79–97.

McGrew, W. C., Marchant, L. F., Payne, C. L. R., Webster, T. H., and Hunt, K. D. (2013). Well digging by Semliki chimpanzees: new data and possible significance of hydrology. *Pan Africa News*, **20**, 5–8.

Moore, J. (1996). Savanna chimpanzees, referential models and the last common ancestor. In *Great Ape Societies*, ed. W. C. McGrew, L. F. Marchant, and T. Nishida. Cambridge: Cambridge University Press, pp. 275–92.

Morino, L. (2011). Left-hand preference for a complex manual task in a population of wild siamangs (*Symphalangus syndactylus*). *International Journal of Primatology*, **32**, 793–800.

Nakamura, M., McGrew, W. C., Marchant, L. F., and Nishida, T. (2000). Social scratch: another custom in wild chimpanzees? *Primates*, **41**, 237–48.

Nishida, T. (1973). The ant-gathering behaviour by the use of tools among wild chimpanzees of the Mahali Mountains. *Journal of Human Evolution*, **2**, 357–70.

Nishida, T. (1983). Alloparental behavior in wild chimpanzees of the Mahale Mountains, Tanzania. *Folia Primatologica*, **41**, 1–33.

Nishida, T. (1993). Left nipple suckling preference in wild chimpanzees. *Ethology and Sociobiology*, **14**, 45–52.

Nishida, T. (1994). [*Thirty-six Stories of Wild Chimpanzees*.] Tokyo: Kinokuniya. In Japanese.

Nishida, T. (2012). *Chimpanzees of the Lakeshore: Natural History and Culture at Mahale*. Cambridge: Cambridge University Press.

Nishida, T. and Hiraiwa, M. (1982). Natural history of a tool-using behavior by wild chimpanzees in feeding upon wood-boring ants. *Journal of Human Evolution*, **11**, 73–99.

Nishida T., McGrew, W. C., and Marchant, L.F. (2012). Wild chimpanzees at Mahale are not manually lateralized for throwing. *Pan Africa News*, **19**, 21–3.

Norikoshi, K. (1994). [Hand preference of wild chimpanzees observed in their tool behavior.] *Primate Research*, **10**, 315–19. In Japanese with English summary.

Norikoshi, K. (1998). Hand preference of wild chimpanzees observed in their tool behaviour. In: *Resource Use Patterns and Social Structure among Chimpanzees*, ed. T. Nishida. Kyoto: Nisshindo, pp. 97–104.

Palmer, A. R. (2002). Chimpanzee right-handedness reconsidered: evaluating the evidence with funnel plots. *American Journal of Physical Anthropology*, **118**, 191–9.

Sugiyama, Y., Fushima, T., Sakaru, O., and Matsuzawa, T. (1993). Hand preference and tool use in wild chimpanzees. *Primates*, **34**, 151–9.

Whiten, A., Goodall, J., McGrew, W. C., *et al.* (1999). Cultures in chimpanzees. *Nature*, **399**, 682–5.

42 Use of tools and other objects

HITONARU NISHIE

42.1 Introduction

Since Goodall (1963) first discovered termite-fishing behavior among wild chimpanzees at Gombe in Tanzania, it has become well known that not only humans but also wild chimpanzees customarily make and use tools in their natural habitat. Goodall's discovery of tool use among wild chimpanzees had a great impact on the controversy over where and how to draw the boundary between humans and animals, because the making and use of tools had long been considered one of the critical characteristics of human beings (Goodall, 1990). Since the discovery at Gombe, many more examples of tool making and tool use by chimpanzees at Mahale and other long-term study sites have been discovered, and we now know that these behaviors are widespread and diverse among chimpanzee groups and populations, implying that these wild chimpanzee societies have their own cultures (McGrew, 1992; see also Chapter 38).

In this chapter, I review observations of tool making, tool use, and object use among wild chimpanzees at Mahale, and compare them with the behaviors of other groups and populations of wild chimpanzees. Hereafter, I use the word "tool" according to the definition by Shumaker *et al.* (2011):

> The external employment of an unattached or manipulable attached environmental object to alter more efficiently the form, position, or condition of another object, another organism, or the user itself, when the user holds and directly manipulates the tool during or prior to use and is responsible for the proper and effective orientation of the tool.

In their definition, "manipulable attached object" and "unattached object" were included as tools to refer to many examples of possible tool use by various animal species. In addition, they also introduced the concept of "object use" to refer to non-manipulable objects, such as large stones, tree buttresses, and stout branches used as anvils for smashing nuts, which has also been called "substrate use." Thus, this chapter broadly summarizes the use of external objects including both "tools" and "objects or substrates" by Mahale chimpanzees.

42.2 Tool use for feeding

42.2.1 Tool use for feeding on carpenter ants (*Camponotus* spp.)

The first discovery of tool use by Mahale chimpanzees was "ant fishing," which is to feed on arboreal carpenter ants by using probes made of plant matter as "fishing rods" such as peeled bark, vine, branch, or midrib of leaves (Nishida, 1973; Figure 42.1). On November 2, 1971, when Nishida (1981) was observing K-group chimpanzees in the rain, Chausiku, an adult female chimpanzee, climbed 15 m up a *Brachystegia* tree, grasped a stick, inserted it into a hollow of a branch, and then withdrew it and put it in her mouth. Nishida and his assistant picked up and collected the sticks discarded by Chausiku as well as specimens of large ants that had fallen from the tree. Thus, they ascertained that Chausiku used tools as fishing probes to harvest ants (Nishida, 1981).

Mahale Chimpanzees: 50 Years of Research, ed. M. Nakamura, K. Hosaka, N. Itoh and K. Zamma. Published by Cambridge University Press.

Figure 42.1 An adult female chimpanzee of the M group, Miya, using a probe to harvest *Camponotus* ants. ©Michio Nakamura.

Ant-fishing behavior among Mahale chimpanzees has been studied, mainly in the K group by Nishida (1973) and Nishida and Hiraiwa (1982); and in the M group by Nishie (2011) (see also Chapter 41 for information on a lack of laterality of hand preference in ant fishing). Ant-fishing behavior is composed of a sequence of four actions: creating and modifying a fishing probe made of plant matter such as peeled bark, vine, branch, or the midrib of a leaf; inserting the probe into a tree hole where wood-boring carpenter ants are nesting; withdrawing the probe with soldier ants attached; and removing them with the lips and tongue (Nishida *et al.*, 1999). Nishida and Hiraiwa (1982) described four species of *Camponotus* ants as the target species of ant fishing in the K group: *C. vividus, C. brutus, C. maculatus,* and C. sp. (*Myromotrema*). Nishie (2011) identified only two in the M group, C. sp. (*chrysurus*-complex) and *C. brutus* (Table 42.1; see also Chapter 21). Nishie (2011) implied that this difference in target species between the groups might be at least partly due to different ecological conditions (see also Chapter 21). *Camponotus* predation using tools has been reported in other wild chimpanzee populations. Gombe chimpanzees use tools to prey on at least three species of *Camponotus* ants (*Camponotus vividus, C. chrysurus*, and *C. brutus*) (O'Malley *et al.*, 2012). At Bossou in Guinea, Yamamoto *et al.* (2008) observed two sessions of ant fishing for *C. brutus* by a juvenile male chimpanzee. *Camponotus* ant fishing has also been observed among chimpanzees at Lopé in Gabon (*C. brutus*: Tutin and Fernandez, 1992; Tutin *et al.*, 1995) and Gashaka in Nigeria (*C. chrysurus*: Fowler and Sommer, 2007). At Assirik in Senegal, *Camponotus* ant fishing was inferred from the presence of remaining tools and ants (McGrew, 1983, 1992). We recently launched a comparative study of *Camponotus* ant fishing between Mahale and Gombe chimpanzees (O'Malley and Nishie, 2013). Preliminary analyses revealed some differences; for example, the chimpanzees at Gombe (Kasekela) fish for *Camponotus* ants at a lower frequency, for shorter durations, in fewer tree species, and at fewer locations than at Mahale. We also found that *Camponotus* ant fishing is mostly an arboreal activity at Mahale, whereas this is not the case at Gombe.

Chimpanzees prepare tools in advance of arriving at the tree where they fish for *Camponotus* ants (Nishida, 1973; Nishida and Hiraiwa, 1982). When a tool becomes dull or bent in the process of repeated fishing, they discard it and make another one from vegetation near the nest hole of the ants. Nishida and Hiraiwa (1982) collected 206 tools used for *Camponotus* ant fishing by the K-group chimpanzees, and classified them into six types as follows (see "Ant" column of K group in the "Tool material" row in Table 42.1): bark tools (B type), modified branches (MBr type), unmodified branches (UBr type), unmodified vines or grasses (UVG type), bark of branches (BBr type), and midribs of leaves (MR type). Tool materials were most frequently derived from vines (71.4%) and secondly from trees (19.9%). The most common plant species used to make tools was *Uvaria angolensis*, which accounted for 48.1% of tools. The top 10 species listed in Table 42.1 accounted for 80.1% of all tools. The mean length of ant-fishing

Table 42.1 **Tool use for eating insects by Mahale chimpanzees**

	B group		K group		M group	
Prey type	Termite	Ant	Termite	Ant	Termite	Ant
Species fished for	(2)*, (6) *Macrotermes herus*	(3)** *Camponotus* sp.	(2), (3) *Pseudacanthotermes spiniger*	(1), (4) *Camponotus vividus* *C. brutus* *C. maculatus* *C.* sp (*Myromotrema*)	(5) *Pseudacanthotermes* sp.	(7) *Camponotus* sp. (*chrysurus-complex*) *C. brutus*
Type of tool material	(6) Bark: $N = 218$ Sedge: $N = 43$ Vine: $N = 21$ Twig: $N = 6$ Leaf: $N = 1$	–	(3) B: $N = 13$ MBr: $N = 3$ UVG: several	(4) B: $N = 118$ MBr: $N = 32$ UVG: $N = 28$ BBr: $N = 12$ UBr: $N = 8$ MR: $N = 8$	(5) B: $N = 2$	–
Species of tool material	(6) [Woody plants] *Acalypha chirindica* *Artabotrys monteirode* *Bauhinia petersiana* *Carpolobia alba* *Cyphostemma* sp. [Sedge] *Cyperus pseudoleplocladus*	–	(3) [B type] *Uvaria angolensis* *Artabotrys monteiroae* *Grewia platyclada* [MBr type] *Paullinia pinnata* [UVG type] *Olyra latifolia*	(4) *Uvaria angolensis* *Brachystegia bussei* *Tinospora caffra* *Combretum molle* *Grewia forbesii* *Landolphia* sp. *Glycine* sp. *Olyra latifolia* *Setaria candula* *Canthium rubrocostatum*	(5) *Uvaria angolensis* *Artabotrys monteiroae* *Monanthotaxis* sp.	–

Tool length	(3) [*N* = 97] Mean: 54.6 cm Range: 21.7–125.6 cm (6) [*N* = 287] Mean: 37.7 cm Median: 35.9 cm Range: 12–84 cm	–	(3) [Nov. 24, 1977; *N* = 16] (Total) Median: 51.5 cm Range: 29.5–97.2 cm (Effective) Median: 30 cm Range: 20–45 cm [20 Nov. 1978; *N* = 9] Median: 45.2 cm Range: 32.0–157.1 cm	(1) [*N* = 37] (Total) Mean: 33.2 cm Range: 10.0–80.0 cm (Effective; *N* = 19) Mean: 14.0 cm Range: 7.3–20.1 cm (3) [*N* = 28] Mean: 21.4 cm Range: 10.5–58.8 cm	(5) [*N* = 2] 18.2 cm ca. 40 cm	–

* Numbers in parentheses represent the references: (1) Nishida, 1973; (2) Nishida and Uehara, 1980; (3) Uehara, 1982; (4) Nishida and Hiraiwa, 1982; (5) Takahata, 1982; (6) McGrew and Collins, 1985; (7) Nishie, 2011.

** *Campnotus* ant-fishing in B group was inferred from circumstantial evidence (Used probes and *Camponotus* ants remained in fecal samples were found in B group) (Uehara, 1982).

tools was 33.2 cm ($N = 37$; range: 10.0–80.0 cm) collected and measured by Nishida (1973), and was 21.4 cm ($N = 28$; range: 10.5–58.8 cm) by Uehara (1982). There are no comparable data from other study sites where *Camponotus* ant fishing has been observed, and the characteristics of tools used for *Camponotus* ant fishing among the M-group chimpanzees remain to be studied (Nishie, 2011).

The frequency of *Camponotus* ant fishing was 0.37 sessions per 10 observation hours for K-group chimpanzees (Nishida and Hiraiwa, 1982), and 0.64 sessions per 10 h for M-group chimpanzees (Nishie, 2011). The mean ant-fishing bout length was 33.2 min by the K group (Nishida and Hiraiwa, 1982) and 24.75 min by the M group (Nishie, 2011). However, Nishie (2011) implied that these differences in the frequency and bout length of *Camponotus* ant fishing are likely due to differences in observational conditions or analysis methods between the two studies, rather than actual differences in behavior between the two groups. In both the K and M groups, females fish for *Camponotus* ants more frequently and in longer bouts than males (Nishida and Hiraiwa, 1982; Nishie, 2011). These consistencies in ant fishing between the K and M groups indicate that, except for some differences caused by local ecological conditions, chimpanzees of the K and M groups have similar customs of *Camponotus* ant fishing with regard to the target ant species and the frequency and duration of ant fishing (Nishie, 2011).

42.2.2 Tool use for feeding on termites

When Nishida discovered *Camponotus* ant fishing among the Mahale chimpanzees in 1971, he hypothesized that they were also likely to use tools to feed on termites, as the Gombe chimpanzees did (Nishida, 1973). However, K-group chimpanzees, which were the main study subjects at that time, were sometimes observed feeding on termites using their fingers but never with tools (Nishida, 1973).

In 1975, however, some circumstantial evidence of feeding on termites with tools among B-group chimpanzees, which were distributed north of the K group, was discovered (Nishida and Uehara, 1980). On November 14, 1975, at the beginning of the wet season, a field assistant encountered many chimpanzees gathered together on the ground in the woodland of the B group's home range. After a while, he approached them and found many grass and vine stems protruding from a termite mound after they had fled. On February 6, 1976, when Nishida approached chimpanzees of the B group feeding in the gallery forest, he found a termite mound that was partially destroyed with four fishing probes projecting from the mound. Nishida also collected 133 fishing probes on termite mounds in the B group's home range between November and December 1979. These probes were made of bark (*Uvaria angolensis, Bauhinia petersiana, Grewia platyclada, G. mollis, Acalypha* sp.), vines (*Rhynchosia luteola, Typlophora* sp.), twigs (*Rothmania manganjae, Milletia angustidentata*), and grass, bamboo, or sedge (*Olyra latifolia, Hyparrhenia rufa, Cyperus diffusus*). In 1977, Uehara observed K-group chimpanzees using probes to feed on termites (Figure 42.2). Although this is the only direct observation of termite fishing with tools by K-group chimpanzees, there has been other sporadic indirect evidence such as remnants of probes on termite mounds (Uehara, 1982). Among the M-group chimpanzees, except for the two cases of termite fishing with tools observed in November 1981 by Takahata (1982), no other evidence of tool use for feeding on termites has been observed (Uehara, 1999).

The intergroup differences in termite fishing among the B, K, and M groups at Mahale described above can be explained by differences in the ecological conditions of the home ranges of the different groups (Nishida and Uehara, 1980; Uehara, 1982; McGrew and Collins, 1985; Collins and McGrew, 1987). The home ranges of the B, K, and M groups are distributed adjacently from north to south (see Chapter 4). Rainfall decreases toward the north, resulting in drier environments and more woodlands in the B group's range, while the home ranges of the K and M groups are more forested. Because of these differences in rainfall and vegetation, *Macrotermes* termites, which can be caught with fishing tools, are more abundant in the home range of the B group than in the K and M groups. *Pseudacanthotermes* termites, which can be more easily harvested without tools by destroying

Figure 42.2 An infant male chimpanzee of the K group, Katabi, using a probe to harvest termites (*Pseudacanthotermes spiniger*). ©Shigeo Uehara.

their mounds, are more common in the home range of the K group (Collins and McGrew, 1987). Thus, termite fishing is seldom observed in the K and M groups at Mahale, because few if any *Macrotermes* termites are present in their home ranges (McGrew and Collins, 1985; Collins and McGrew, 1987; Uehara, 1999).

Tools used for termite fishing by Mahale chimpanzees in the B and K groups were collected and analysed in detail (summarized in Table 42.1). In the B group, most tools for termite fishing were made of bark, sedge, or vines, with a mean length of 54.6 cm (range: 21.7–125.6 cm) measured by Uehara (1982) and 37.7 cm (range: 12–84 cm) by McGrew and Collins (1985). The length of termite-fishing tools measured by McGrew and Collins (1985) was as long as that of *Camponotus* ant-fishing tools measured by Nishida (1973). However, Uehara (1982) pointed out that their effective length, defined as the portion of the probe actually inserted into a hole, was about twice as long as that of *Camponotus* ant-fishing tools, and that greater effective length and higher elasticity of materials are necessary to fish for termites than for ants. Therefore, fishing tools made of the midribs of leaves (MR type), bark of branches (BBr type), or unmodified branches (UBr type) would be of less use for termite fishing because they tend to be short (MR type) and/or non-elastic (BBr and UBr types) (Uehara, 1982).

Although termite fishing is a relatively minor tool-use behavior in Mahale chimpanzees, it has been observed in many other populations of wild chimpanzees across Africa: at Gombe (McGrew *et al.*, 1979; Goodall, 1986; Lonsdorf, 2005), Kasakati (Tanzania: Suzuki, 1966), Ndakan and Bai Hokou (Central Africa: Fay and Carroll 1994), Ndoki (Congo: Suzuki *et al.*, 1995), Lossi (Congo: Bermejo and Illera, 1999), Goualougo (Congo: Sanz *et al.*, 2004), Campo (Cameroon: Sugiyama, 1985; Muroyama, 1991), Dja (Cameroon: Deblauwe *et al.*, 2006), Belinga (Gabon: McGrew and Rogers, 1983), Okorobikó (Equatorial Guinea: Jones and Sabater Pi, 1969; Sabater Pi, 1974), Bossou (Sugiyama and Koman, 1979; Humle, 1999), Mt. Assirik (McBeath and McGrew, 1982; Bermejo *et al.*, 1989), and Fongoli (Senegal: McGrew *et al.*, 2005; Bogart and Pruetz, 2008).

42.2.3 Hunting with tools

Huffman and Kalunde (1993) reported the first case of tool-assisted hunting of mammalian prey by a chimpanzee at Mahale. On December 16, 1991, they observed a 12-year-old female, Tula, use a modified branch to probe a tree hole, rouse and grab a squirrel from the hole, and then eat it, sharing a small piece with a 4.5-year-old female, Maggie. The tool used for rousing the squirrel was 73 cm long, 9 cm in circumference, and weighed 155 g.

Nakamura and Itoh (2008) observed additional cases of hunting with tools in 1995 and 2004. On October 4, 1995, Maggie, the chimpanzee who was previously observed begging Tula for squirrel meat and now 8 years old, was found poking a stick (about 50 cm long and 3 cm in diameter) into a tree hollow. After a while, she took out an immobilized (either dying or already dead) squirrel from the tree hollow, and ate it.

On October 5, 2004, Darwin (16-year-old male) and Cadmus (13-year-old male) inserted and poked tools (vine and stem: about 60 cm long and 1–3 cm in diameter) into a cave at the bottom of a large rock, and wielded a tool (vine: 80 cm long and 0.5 cm in diameter) like a whip at the entrance of the cave. After a while, Orion (a 13-year-old male) came and also wielded the whip-like tool that had been used by Darwin and Cadmus. The cave was thought to be the nest of yellow-spotted hyraxes, although the existence of the hyraxes could not be confirmed.

The tools used to hunt mammalian prey in these (two successful and one attempted) cases were longer and thicker than those used for ants and termites, because they were used to poke into and rummage within tree hollows or a cave. Hunting with tools by wild chimpanzees is habitually observed at Fongoli, where chimpanzees hunt lesser bushbabies by using sticks as "spears" (Pruetz and Bertolani, 2007). Nakamura and Itoh (2008) implied that Mahale chimpanzees use tools to hunt mammalian prey more frequently than observed.

42.3 Courtship behaviors with tools

42.3.1 Leaf clipping

Mahale chimpanzees use a leaf as a kind of communicatory tool in social interactions. This is termed leaf-clipping behavior (Figure 42.3). The sequence of this behavior is that a chimpanzee picks off a leaf, grasps the petiole between the thumb and the index finger, and repeatedly pulls it between its lips or teeth while removing the leaf blade, producing a conspicuous sound that attracts attention from a prospective sex partner (Nishida, 1980b; Nishida *et al.*, 1999; Chapter 35). Nishida (1980b) analysed 41 cases of leaf clipping among K-group chimpanzees, and classified them as follows: 23 cases (56.1%) in sexual contexts such as soliciting copulation or "herding" behavior, 13 cases (31.7%) in frustration or play, and 5

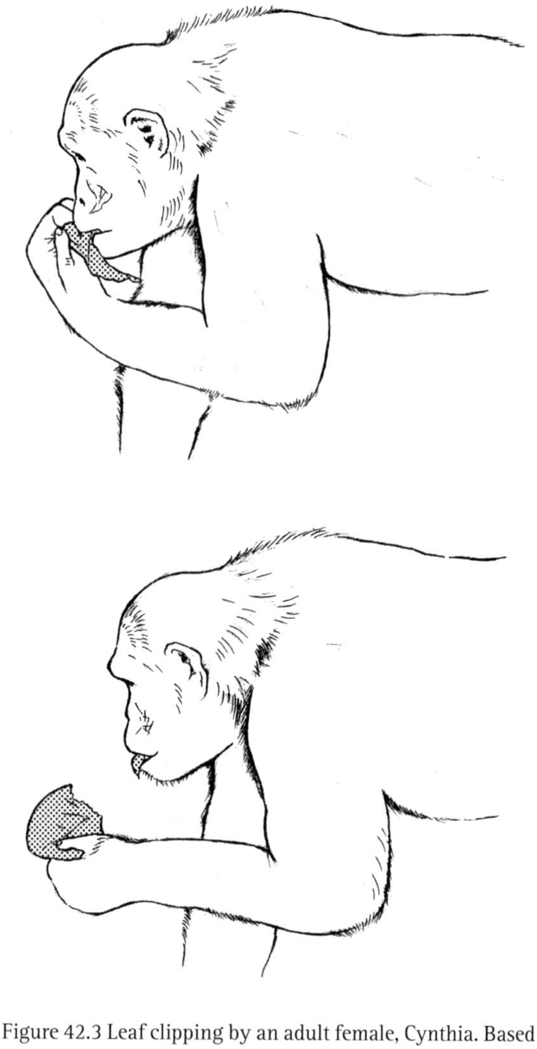

Figure 42.3 Leaf clipping by an adult female, Cynthia. Based on image ©Michio Nakamura.

cases (12.2%) in demanding food from the human observer. Nishida (1997) studied the sexual behaviors of adult male chimpanzees of the M group, and indicated that leaf clipping was observed in 26.4% (19 cases) of the 72 total cases of courtship displays observed. Thus, Nishida *et al.* (1999) concluded that leaf clipping was one of the most common courtship displays in Mahale, although he also noted that it sometimes occurred in other contexts, such as frustration or play (Nishida, 1980b, 1987).

Recently, we re-examined the social context of leaf clipping among M-group chimpanzees (Nishie *et al.*, 2014). We analysed 536 cases of leaf clipping observed between 1994 and 2012, and reconfirmed that leaf clipping frequently occurred in sexual contexts, but there were also several cases of leaf clipping in nonsexual contexts, such as a male leaf clipping toward another male or a non-estrous female leaf clipping toward her offspring. Thus, we presumed that these multiple uses of leaf clipping imply that the behavior itself is not a fixed signal of courtship as indicated by Nishida (1980b), but that it is widely used as a gesture to get the attention of a recipient(s) for subsequent interactions.

42.3.2 Shrub bend

In the late 1980s, another style of courtship display was discovered among M-group chimpanzees: shrub bend. Typically, a male sitting on the ground or in a tree faces an estrous female and makes a crude day bed or cushion by bending 2–4 shrubs to the ground, which he then sits on and repeatedly stamps with one foot (Nishida, 1987, 1997; Nishida *et al.*, 1999). Nishida (1997) showed that shrub bend accounted for only 4.2% (three cases) of the 72 courtship displays observed. Shrub bend has not been observed in the K group. This behavior has been reported as habitual in Bossou and customary in Budongo, Uganda (Whiten *et al.*, 2001).

42.3.3 Stem pull-through

Nishida (1997) described stem pull-through as a courtship display performed by only two adult males of the M group, Musa and Ntologi. Typically, they pulled the leafy branch of a shrub or a clump of grass stems through their hand by a rapid movement of the forearm, and then released the stem, producing a conspicuous sound while seated and watching an estrous female. This behavior has been reported as customary in Bossou, habitual in Kibale (Uganda), and present in Gombe (Whiten *et al.*, 2001).

42.4 Intimidation displays with tools

42.4.1 Throwing objects

Throwing objects such as stones, rocks, branches, sticks, handfuls of grass or sand, and so on, has been documented in many populations of wild chimpanzees as intimidation displays (Nishida *et al.*, 1999). In particular, a pattern of rock throwing displayed by Mahale chimpanzees is unique. When M-group chimpanzees cross a streambed, adult males, and more rarely adolescent males, typically throw rocks into the stream, producing a loud splash that intimidates others (Nishida, 1994, 2003b; Nishida *et al.*, 1999; see also Chapter 31 for descriptions of "streambed display"). This pattern of rock throwing into water ("throw splash": Nishida *et al.*, 1999) has been observed only at Mahale, although it is not known when this type of display began (Nishida, 2003b). Older skillful males often stand bipedally, choose a large rock weighing 10 kg or more, lift it with both hands, and make an aimed throw into the water.

42.4.2 Drumming objects

Mahale chimpanzees drum objects such as buttress roots, metal walls or wooden doors of the researchers' and tourists' camp, and metal barrels as intimidation displays (Nishida, 1994, 2003b; Nishida *et al.*, 1999). Drumming natural objects such as buttress roots or trunks is common among many populations (Whiten *et al.*, 2001), whereas drumming artificial objects such as metal walls or barrels is an idiosyncratic pattern of

Mahale chimpanzees, although a Gombe male, Mike, learned to incorporate empty kerosene cans into his intimidation displays (Goodall, 1971). At Mahale, chimpanzees initially ignored and simply passed by the metal houses for some years after they were built in 1976; however, by August 1979, many adult males began to use the metal walls of the houses as drumming objects (Nishida, 1994, 2003a). Drumming natural objects is performed not only by adult males but also by adolescent, juvenile, and even 4-year-old males as well as by adult, adolescent, and juvenile females. However, drumming metal walls is performed exclusively by adult males. Adult males vary in their patterns of drumming on metal walls. Some regularly stand on their feet and slap quickly with both hands, while others stand bipedally and slap slowly with one hand, and others stand on their feet and slap and kick simultaneously. This variation in drumming behavior among males partly derives from individual preferences in the loudness and rhythm of drumming; this in turn could represent individualistic "signature patterns" that are potentially recognizable to the others in the same unit-group (Nishida, 2003a; see also Chapter 31).

42.4.3 Other intimidation displays with tools/objects

Nishida (2003b) reported that immature male chimpanzees sometimes threaten older individuals, adult females in particular, by clubbing, flailing, or throwing branches. In all, 25 behaviors were confirmed to be "harassment" mostly by adolescent and juvenile males toward adult females, 14 of which included tool or object use such as shaking, throwing, dragging, or flailing branches; clubbing or stamping the ground; or throwing dead leaves. Shaking branches was the most frequently observed harassment behavior; however, interestingly, shaking branches has also been observed in other broad contexts, including courtship (solicitation for copulation), continuation of consortship (herding technique), enticement to travel together, and solicitation for play (Nishida *et al.*, 1999).

42.5 Hygienic behaviors with tools

42.5.1 Muzzle rubbing and hand rubbing with objects

Leaf napkins, which are tools used for wiping off feces, semen, blood, and other excreted matter, have been customarily observed in Gombe, Kibale, and Budongo, but only rarely at Mahale (Goodall, 1986; Nishida, 1994, 2003a; Nishida *et al.*, 1999; Whiten *et al.*, 2001). However, in 1998, four chimpanzees began to regularly wipe lemon juice from their mouths with leaves and branches without detaching them from the lemon trees (Nishida, 2003a; Corp *et al.*, 2009). In 1999, 18 and 12 chimpanzees were observed wiping their faces or hands, respectively. This behavior, termed "muzzle rubbing" (Nishida *et al.*, 1999; Corp *et al.*, 2009), differed from leaf napkin behavior because Mahale chimpanzees rub their bodies (muzzles and hands) with undetached plant materials, not with detached leaves or twigs. Muzzle rubbing at Mahale increased from 1998, reached a peak around 2000, and declined after 2002. Corp *et al.* (2009) implied that the muzzle and hand rubbing at Mahale became prevalent via the influence of the alpha male, Fanana, who was one of the first individuals to exhibit muzzle rubbing while eating lemons, but was never observed after 2002, despite more than 300 h of focal observations in 2002–5.

42.5.2 Leaf grooming

The chimpanzees of Gombe (van Lawick-Goodall, 1968; Boesch, 1995), Mahale (Nishida, 1980b), Kibale, and Budongo (Whiten *et al.*, 2001; Assersohn *et al.*, 2004) have been observed to seize a leaf and manipulate it in a grooming context as if they were grooming the leaf (see also Chapter 34). However, the function of this behavior was a puzzle for decades (Nishida *et al.*, 1999; Zamma, 2002).

Zamma (2002, 2006) identified a function of leaf grooming at Mahale by collecting and analysing leaves discarded by chimpanzees after leaf grooming. Zamma (2002) observed a case of leaf grooming in which an adult male, Masudi, groomed with a leaf and a louse was found on the discarded leaf. In this case, Masudi picked a leaf and transferred a louse from his lower lip to it, folded it and crushed the edge where the louse was placed inside, opened it and touched the crushed point with his lips. Zamma (2002) indicated that at least one function of leaf grooming is "squashing ectoparasites." In addition, Zamma (2006) found a hatched egg of a louse on a leaf discarded after leaf grooming, and suggested that chimpanzees have the skill to remove louse eggs as well as adult lice and ticks by grooming, and squash them by leaf grooming.

42.5.3 Probe use for removing sand fleas from a toe

On September 25, 2001, Nishida (2002) observed an 8-year-old female of the M group, Ivana, broke a small twig from a shrub, held its leafy part, and pushed the sharp edge under the nail of a toe on her left foot with her right hand. After trying to insert the probe several times over 5 min, she threw the probe away and put her toe into her mouth, apparently trying to suck something out. This use of a probe inserted under the nail with visual and manual inspection strongly suggested that Ivana was trying to remove a sand flea (or fleas), because such behavior is performed by humans in the same circumstances (Nishida, 2002).

42.5.4 Nasal probe

Nishida and Nakamura (1993) reported a case in which an adult male, Kalunde, used a "toothpick-like tool" to induce sneezing that cleared the nasal passage during a flu-like illness in 1992. Kalunde showed this behavior for more than 10 years (which was also observed by Kenji Kawanaka in 1996, by Toshisada Nishida in 1997, and by Koichiro Zamma in 2000; Nishida *et al.*, 2009). Marchant and McGrew (1999) observed a similar case in which a 9-year-old female, Maggie, held a small twig in her left hand to probe her left nostril on October 8, 1996. She removed mucus by probing her nostril and ate it, putting the twig crosswise in her mouth. Soon after, she repeated the procedure and succeeded in inducing a sneeze that yielded more mucus, which she ate with some leaves.

42.6 Play with objects

42.6.1 Tool use for drinking water

Tool use for drinking water is customary among the chimpanzees of Bossou, Taï, Gombe, Kibale, and Budongo (Whiten *et al.*, 2001). In contrast, it was observed only rarely at Mahale (Whiten *et al.*, 2001; Matsusaka *et al.*, 2006) until the 1990s. Subsequently, 42 cases of drinking water with tools were observed between 1999 and 2004 among M-group chimpanzees, all of which were performed by immature individuals (2.3–10.8 years old: Matsusaka *et al.*, 2006). In these cases, water was obtained from tree holes in 20 cases and from streams in 22 cases. Chimpanzees successively used several objects such as leaves (leaf sponge, leaf spoon, uncrumpled leaf) and sticks (twig, vine, grass stem, peeled vine, midrib of leaf) during one episode of tool use for drinking water. In two cases, the use of a "tool set" was observed: the chimpanzees used a stick to pick a leaf sponge out of a tree hole. In contrast to other study sites, Mahale chimpanzees rarely modified leaf sponges before soaking them in water, but usually soaked an uncrumpled leaf in water and chewed the drenched leaf, consequently making the leaf like a sponge. The increase in this behavior in the 2000s among only immature chimpanzees at Mahale was likely the result of social learning because it occurred within a very short period and adults paid little attention to the behavior while immature individuals often observed the tool use of others with great interest. This suggests that tool use for drinking water at Mahale seems to be a playful activity, i.e. not done out of necessity (see also Chapter 40).

42.6.2 Leaf-pile pulling

Leaf-pile pulling is a playful behavior customarily performed by Mahale chimpanzees (Nishida and Wallauer, 2003). Typically, when a party of chimpanzees moves in a procession down a slope, a chimpanzee turns around and walks backward while raking many dry leaves on the ground with both hands. This activity accumulates many dry leaves while producing a lot of sound. A performer usually faces another individual that is immediately following him or her in the procession.

Since leaf-pile pulling is performed by one individual as a locomotory movement, includes manipulation of objects (dead leaves), and is often displayed to an individual following the performer, it comprises all three components of primate play (locomotory, object, and social play; Nishida and Wallauer, 2003), although few detailed studies of play have been conducted among primates (see also Chapter 40).

According to Nishida and Wallauer (2003), leaf-pile pulling was first recorded in 1989 at Mahale, and 73 cases were observed during 339 observation hours between 1999 and 2002. The performers were 11 infant males, 11 infant females, 22 juvenile males, 19 juvenile females, 6 adolescent males, 1 adolescent male, and 3 adult males. Thus, leaf-pile pulling was performed mostly by infants and juveniles (63/73 = 86.3%). The contexts in which leaf-pile pulling was displayed were travel (39/73 = 53.4%), play (13/73 = 17.8%), and rest (8/73 = 11.0%).

Leaf-pile pulling has not been reported at Taï, Bossou, Kibale, Budongo, or Kalinzu (Uganda). At Gombe, this behavior was observed in five different chimpanzees on 10–15 occasions between 1996 and 2002. At Mahale, leaf-pile pulling is a standardized play pattern that is displayed by most immature chimpanzees of the M group and can be called customary (Whiten *et al.*, 2001). Because this behavior has been performed for more than two decades, it might be socially learned and thus be a tradition of the M-group chimpanzees.

42.7 Carrying objects

42.7.1 Knotted skin "necklace"

McGrew and Marchant (1998) reported a case of a knot tied in a piece of the skin of a red colobus monkey worn by a chimpanzee at Mahale. On November 7, 1996, chimpanzees of the M group killed and ate a red colobus monkey, and the next morning they still carried "leftovers." On the morning of November 8, a juvenile female, Ai, was playing with and grooming a strip of skin. It was stolen by a juvenile male, Primus, while playing. After a while, an adolescent female, Ako, was seen wearing what appeared to be the skin draped around her neck. The knot created a "necklace" of 68 cm circumference, with a "tail" of 31 cm length. McGrew and Marchant (1998) noted that this was the first case in which a draped adornment was made by nonhuman animals.

42.7.2 Carrying guineafowl feathers

Nakamura (2009) reported a case in which an old female, Wakusi, collected some guineafowl feathers and carried them for more than 30 min without displaying any emotional expressions. It is unlikely that Wakusi assumed the feathers represented bird meat because she never sniffed or licked them. The feathers did not seem to be "toys" for Wakusi either, because she did not show any attempt to manipulate them in a playful manner, but just held them in her hand, lips, or groin pocket. Nakamura (2009) noted that Wakusi did not seem to collect and carry the feathers for a particular functional reason, and that it is possible she did it with a kind of aesthetic sensibility that might be shared with humans.

42.8 Conclusion

Mahale chimpanzees have a smaller repertoire of tool and object use than chimpanzees at other long-term study sites. In particular, the repertoire of tool use for feeding is not rich at Mahale. For example, nut-cracking behavior using stone or wood hammers and anvils to obtain kernels inside hard shells of nuts, which has been widely observed among wild chimpanzee populations in West Africa (Sugiyama and Koman, 1979; Boesch and Boesch, 1990), and *Dorylus* ant-dipping behavior, which is popular among many populations across Africa (McGrew, 1974; Sugiyama *et al.*, 1988; Boesch and Boesch, 1990), have never been observed at Mahale. Furthermore, as noted above (42.2), although termite-fishing behavior has been reported from many populations of wild chimpanzees across Africa, at Mahale it has been observed customarily only in the B group, and seldom in the K and M groups, indicating it is a relatively minor tool-using behavior at Mahale. In sum, Mahale chimpanzees do not seem to be eager to forage for potential food resources that are hard to obtain without tools.

On the other hand, the repertoire of tool and object use in social, hygienic, and play behaviors of Mahale chimpanzees is rather rich. This may imply that Mahale chimpanzees spend more time on non-subsistence activities such as social grooming and play compared with chimpanzees at other sites. Although a detailed comparison of food availability across chimpanzee study sites has not yet been conducted, there might be relatively more food resources at Mahale, making Mahale chimpanzees less eager to forage for potential food resources that require tools to be accessed. The relationship between tool use for food extraction and ecological factors, such as food availability or time allocated to foraging, remains to be studied in detail.

It is also notable that many "rare cases" of tool and object use, which have not become customary in a group, have been reported from Mahale. This chapter reviewed only some of them: termite fishing with tools by M-group chimpanzees, hunting with tools, probe use for removing sand fleas from toes, nasal probing, wearing of a skin "necklace," and carrying guineafowl feathers. There are many other reported cases of idiosyncratic behavioral patterns of tool and object use that this chapter does not describe in detail. For instance, Nishida (1994) reported a case of "washing" a colobus coat in a riverbed by dunking and shaking it underwater and by stamping on the skin with one foot on a rock by an adult male. Nishida *et al.* (2009) reported a case of digging with a stick in a streambed by a juvenile female to obtain water. Only long-term, painstaking observation enables researchers to encounter and detect such rare behaviors; therefore, the many rare or idiosyncratic cases of tool or object use reported from Mahale are the fruit of 50 years of long-term study (Nakamura and Nishida, 2012).

Although this chapter failed to include them all, some other tool- or object-use behaviors have been reported in Mahale chimpanzees as well, such as bed making (see Chapter 43), ground grooming (see Chapter 34), and back rubbing against a tree trunk after rain (Nishida, 1980a). Further studies on tool- or object-use behaviors of Mahale chimpanzees are necessary to elucidate changes in these behaviors over time, the relationship between tool or object use and ecological factors, and the mode of tool or object use in social interactions.

References

Assersohn, C., Whiten, A., Kiwede, Z. T., Tinka, J., and Karamagi, J. (2004). Use of leaves to inspect ectoparasites in wild chimpanzees: a third cultural variant? *Primates*, **45**, 255–8.

Bermejo, M. and Illera, G. (1999). Tool-set for termite-fishing and honey extraction by wild chimpanzees in the Lossi Forest, Congo. *Primates*, **40**, 619–27.

Bermejo, M., Illera, G., and Sabater Pi, J. (1989). New observations on the tool-behavior of chimpanzees from Mt. Assirik (Senegal, West Africa). *Primates*, **30**, 65–73.

Boesch, C. (1995). Innovation in wild chimpanzees (*Pan troglodytes*). *International Journal of Primatology*, **16**, 1–16.

Boesch, C. and Boesch, H. (1990). Tool use and tool making in wild chimpanzees. *Folia Primatologica*, **54**, 86–99.

Bogart, S. L. and Pruetz, J. D. (2008). Ecological context of savanna chimpanzee (*Pan troglodytes verus*) termite fishing at Fongoli, Senegal. *American Journal of Primatology*, **70**, 605–12.

Collins, D. A. and McGrew, W. C. (1987). Termite fauna related to differences in tool-use between groups of chimpanzees (*Pan troglodytes*). *Primates*, **28**, 457–71.

Corp, N., Hayaki, H., Matsusaka, T., *et al.* (2009). Prevalence of muzzle-rubbing and hand-rubbing behavior in wild chimpanzees in Mahale Mountains National Park, Tanzania. *Primates*, **50**, 184–9.

Deblauwe, I., Guislain, P., Dupain, J., and van Elasacker, L. (2006). Use of tool-set by *Pan troglodytes troglodytes* to obtain termites (*Macrotermes*) in the periphery of the Dja Biosphere Reserve, southeast Cameroon. *American Journal of Primatology*, **68**, 1191–6.

Fay, J. M. and Carroll, R. W. (1994). Chimpanzee tool use for honey and termite extraction in Central Africa. *American Journal of Primatology*, **34**, 309–17.

Fowler, A. and Sommer, V. (2007). Subsistence technology of Nigerian chimpanzees. *International Journal of Primatology*, **28**, 997–1023.

Goodall, J. (1963). Feeding behaviour of wild chimpanzees: a preliminary report. *Symposia of Zoological Society of London*, **10**, 39–47.

Goodall, J. (1971). *In the Shadow of Man*. London: William Collins.

Goodall, J. (1986). *The Chimpanzees of Gombe: Patterns of Behavior*. Cambridge, MA: Belknap.

Goodall, J. (1990). *Through a Window: Thirty Years with the Chimpanzees of Gombe*. London: Weidenfeld & Nicolson Ltd.

Huffman, M. A. and Kalunde, M. S. (1993). Tool-assisted predation on a squirrel by a female chimpanzee in the Mahale Mountains, Tanzania. *Primates*, **34**, 93–8.

Humle, T. (1999). New record of fishing for termites (*Macrotermes*) by the chimpanzees of Bossou (*Pan troglodytes verus*), Guinea. *Pan Africa News*, **6**, 3–5.

Jones, C. and Sabater Pi, J. (1969). Sticks used by chimpanzees in Rio Muni, West Africa. *Nature*, **223**, 100–1.

van Lawick-Goodall, J. (1968). The behaviour of free-living chimpanzees in the Gombe Stream Reserve. *Animal Behaviour Monographs*, **1**, 161–311.

Lonsdorf, E. V. (2005). Sex differences in the development of termite-fishing skills in the wild chimpanzees, *Pan troglodytes schweinfurthii*, of Gombe National Park, Tanzania. *Animal Behaviour*, **70**, 673–83.

Marchant, L. F. and McGrew, W. C. (1999). Innovative behavior at Mahale: new data on nasal probe and nipple press. *Pan Africa News*, **6**, 16–18.

Matsusaka, T., Nishie, H., Shimada, M., *et al.* (2006). Tool-use for drinking water by immature chimpanzees of Mahale: prevalence of an unessential behavior. *Primates*, **47**, 113–22.

McBeath, N. M. and McGrew, W. C. (1982). Tools used by wild chimpanzees to obtain termites at Mt Assirik, Senegal: the influence of habitat. *Journal of Human Evolution*, **11**, 65–72.

McGrew, W. C. (1974). Tool use by wild chimpanzees in feeding upon driver ants. *Journal of Human Evolution*, **3**, 501–8.

McGrew, W. C. (1983). Animal foods in the diets of wild chimpanzees (*Pan troglodytes*): why cross-cultural variation? *Journal of Ethology*, **1**, 46–61.

McGrew, W. C. (1992). *Chimpanzee Material Culture: Implications for Human Evolution*. Cambridge: Cambridge University Press.

McGrew, W. C. and Collins, D. A. (1985). Tool use by wild chimpanzees (*Pan troglodytes*) to obtain termites (*Macrotermes herus*) in the Mahale Mountains, Tanzania. *American Journal of Primatology*, **9**, 47–62.

McGrew, W. C. and Marchant, L. F. (1998). Chimpanzee wears a knotted skin "necklace." *Pan Africa News*, **5**, 8–9.

McGrew, W. C. and Rogers, M. E. (1983). Chimpanzees, tools, and termites: new record from Gabon. *American Journal of Primatology*, **5**, 171–4.

McGrew, W. C., Tutin, C. E. G., and Baldwin, P. J. (1979). Chimpanzees, tools, and termites: cross-cultural comparisons of Senegal, Tanzania, and Rio Muni. *Man*, **14**, 185–214.

McGrew, W. C., Pruetz, J. D., and Fulton, S. J. (2005). Chimpanzees use tools to harvest social insects at Fongoli, Senegal. *Folia Primatologica*, **76**, 222–6.

Muroyama, Y. (1991). Chimpanzees' choices of prey between two sympatric species of *Macrotermes* in the Campo Animal Reserve, Cameroon. *Human Evolution*, **6**, 143–51.

Nakamura, M. (2009). Aesthete in the forest? A female chimpanzee at Mahale collected and carried guineafowl feathers. *Pan Africa News*, **16**, 17–19.

Nakamura, M. and Itoh, N. (2008). Hunting with tools by Mahale chimpanzees. *Pan Africa News*, **15**, 3–6.

Nakamura, M. and Nishida, T. (2012). Long-term field studies of chimpanzees at Mahale Mountains National Park, Tanzania. In *Long-Term Field Studies of Primates*, ed. P. M. Kappeler and D. P. Watts. Heidelberg, Germany: Springer-Verlag, pp. 339–56.

Nishida, T. (1973). The ant-gathering behaviour by the use of tools among wild chimpanzees of the Mahali Mountains. *Journal of Human Evolution*, **2**, 357–70.

Nishida, T. (1980a). Local differences in responses to water among wild chimpanzees. *Folia Primatologica*, **33**, 189–209.

Nishida, T. (1980b). The leaf-clipping display: a newly-discovered expressive gesture in wild chimpanzees. *Journal of Human Evolution*, **9**, 117–28.

Nishida, T. (1981). [*The World of Wild Chimpanzees*.] Tokyo: Chūkō-shinsho. In Japanese.

Nishida, T. (1987). Local traditions and cultural transmission. In *Primate Societies*, *ed.* B. B. Smuts, D. L. Cheney, R. M. Seyfarth, R. W. Wrangham, and T. T. Struhsaker. Chicago, IL: University of Chicago Press, pp. 462–74.

Nishida, T. (1994). Review of recent findings on Mahale chimpanzees: implications and future research directions. In *Chimpanzee Cultures*, ed. R. W. Wrangham, W. C. McGrew, F. B. M. de Waal, and P. G. Heltne. Cambridge, MA: Harvard University Press, pp. 373–96.

Nishida, T. (1997). Sexual behavior of adult male chimpanzees of the Mahale Mountains National Park, Tanzania. *Primates*, **38**, 379–98.

Nishida, T. (2002). A self-medicating attempt to remove the sand flea from a toe by a young chimpanzee. *Pan Africa News*, **9**, 5–6.

Nishida, T. (2003a). Harassment of mature female chimpanzees by young males in the Mahale Mountains. *International Journal of Primatology*, **24**, 503–14.

Nishida, T. (2003b). Individuality and flexibility of cultural behavior patterns in chimpanzees. In *Animal Social Complexity and Intelligence*, ed. F. B. M. de Waal and P. Tyack. Cambridge, MA: Harvard University Press, pp. 392–413.

Nishida, T. and Hiraiwa, M. (1982). Natural history of a tool-using behavior by wild chimpanzees in feeding upon wood-boring ants. *Journal of Human Evolution*, **11**, 73–99.

Nishida, T., and Nakamura, M. (1993). Chimpanzee tool use to clear a blocked nasal passage. *Folia Primatologica*, **61**, 218–20.

Nishida, T. and Uehara, S. (1980). Chimpanzees, tools, and termites: another examples from Tanzania. *Current Anthropology*, **21**, 671–2.

Nishida, T. and Wallauer, W. (2003). Leaf-pile pulling: an unusual play pattern in wild chimpanzees. *American Journal of Primatology*, **60**, 167–73.

Nishida, T., Kano, T., Goodall, J., McGrew, W. C., and Nakamura, M. (1999). Ethogram and ethnography of Mahale chimpanzees. *Anthropological Science*, **107**, 141–88.

Nishida, T., Matsusaka, T., and McGrew, W. C. (2009). Emergence, propagation or disappearance of novel behavioral patterns in the habituated chimpanzees of Mahale: A review. *Primates*, **50**, 23–36.

Nishie, H. (2011). Natural history of *Camponotus* ant fishing by the M group chimpanzees at the Mahale Mountains National Park, Tanzania. *Primates*, **52**, 329–42.

Nishie, H., Inoue, E., Matsusaka, T., Nakamura, M., and Zamma, K. (2014). Leaf-clipping is not a token of "courtship display": from the descriptions of leaf-clipping behavior by the M group chimpanzees at Mahale, Tanzania. *Abstracts of the 25th Congress of the International Primatological Society*, 176.

O'Malley, R. C. and Nishie, H. (2013). Patterns of ant fishing for carpenter ants (*Camponotus* spp.) by Gombe and Mahale chimpanzees. *Program of the 82nd Annual Meeting of the American Association of Physical Anthropologists*, pp. 209.

O'Malley, R. C., Wallauer, W., Murray, C. M., and Goodall, J. (2012). The appearance and spread of ant fishing among the Kasekela chimpanzees of Gombe: a possible case of intercommunity cultural transmission. *Current Anthropology*, **53**, 650–63.

Pruetz, J. D. and Bertolani, P. (2007). Savanna chimpanzees, *Pan troglodytes verus*, hunt with tools. *Current Biology*, **17**, 412–17.

Sabater Pi, J. (1974). An elementary industry of the chimpanzees in the Okorobikó Mountains, Rio Muni (Republic of Equatorial Guinea), West Africa. *Primates*, **15**, 351–64.

Sanz, C., Morgan, D., and Gulick, S. (2004). New insights into chimpanzees, tools, and termites from the Congo Basin. *The American Naturalist*, **164**, 567–81.

Shumaker, R. W., Walkup, K. R., and Beck B. B. (2011). *Animal Tool Behavior: The Use and Manufacture of Tools by Animals*. Baltimore, MD: The Johns Hopkins University Press.

Sugiyama, Y. (1985). The brush-stick of chimpanzees found in south-west Cameroon and their cultural characteristics. *Primates*, **26**, 361–74.

Sugiyama, Y. and Koman, J. (1979). Tool-using and -making behavior in wild chimpanzees at Bossou, Guinea. *Primates*, **20**, 513–24.

Sugiyama, Y., Koman, J., and Sow, M. B. (1988). Ant-catching wands of wild chimpanzees at Bossou, Guinea. *Folia Primatologica*, **51**, 56–60.

Suzuki, A. (1966). On the insect-eating habits among wild chimpanzees living in the savanna woodland of western Tanzania. *Primates*, **7**, 481–7.

Suzuki, S., Kuroda, S., and Nishihara, T. (1995). Tool-set for termite-fishing by chimpanzees in the Ndoki Forest, Congo. *Behaviour*, **132**, 219–35.

Takahata, Y. (1982). Termite fishing observed in M group chimpanzees. *Mahale Mountains Chimpanzee Research Project Ecological Report*, **18**.

Tutin, C. E. G. and Fernandez, M. (1992). Insect-eating by sympatric lowland gorillas (*Gorilla g. gorilla*) and chimpanzees (*Pan t. troglodytes*) in the Lopé Reserve, Gabon. *American Journal of Primatology*, **28**, 29–40.

Tutin, C. E. G., Ham, R., and Wrogemann, D. (1995). Tool-use by chimpanzees (*Pan t. troglodytes*) in the Lopé Reserve, Gabon. *Primates*, **36**, 181–92.

Uehara, S. (1982). Seasonal change in the techniques employed by wild chimpanzees in the Mahale Mountains, Tanzania, to feed on termites (*Pseudacanthotermes spiniger*). *Folia Primatologica*, **37**, 44–76.

Uehara, S. (1999). Why don't the chimpanzees of M group at Mahale fish for termites? *Pan Africa News*, **6**, 22–4.

Whiten, A., Goodall, J., McGrew, W. C., *et al*. (2001). Charting cultural variation in chimpanzees. *Behaviour*, **138**, 1481–516.

Yamamoto, S., Yamakoshi, G., Humle, T., and Matsuzawa, T. (2008). Invention and modification of a new tool use behavior: ant fishing in trees by a wild chimpanzee (*Pan troglodytes*) at Bossou, Guinea. *American Journal of Primatology*, **70**, 699–702.

Zamma, K. (2002). Leaf grooming by a wild chimpanzee in Mahale. *Primates*, **43**, 87–90.

Zamma, K. (2006). A louse egg left on a leaf. *Pan Africa News*, **13**, 8–10.

43 Bed making and nocturnal behavior

KOICHIRO ZAMMA AND HIROSHI IHOBE

43.1 A chimpanzee in the dark

Chimpanzees are diurnal apes. Various behaviors – such as feeding, moving, and grooming – are frequently observed in the daytime, and chimpanzees are almost inactive at night. However, this does not mean that they do nothing. Sleeping, and its associated brain activity (Hobson, 2005), is one of their main nocturnal behaviors. They also make beds for sleeping and sometimes move about (Izawa and Itani, 1966; van Lawick-Goodall, 1968; Pruetz and Bertolani, 2009; Nishida, 2012). Nocturnal chimpanzee behavior has been little studied; consequently, little about it is known (Nishida, 1994). In this chapter, we explore the significance of nocturnal activities in chimpanzees.

43.2 Beds

All great apes make dish-shaped beds, with raised edges and a depressed center, from elliptical bundles of branches or grasses (Nissen, 1931; Goodall, 1962; Schaller, 1963; MacKinnon, 1974; Kano, 1983; Stewart *et al.*, 2007; van Casteren *et al.*, 2012; Figure 43.1). Most of these are built in trees, but ground beds are also seen, except in orangutans (e.g. Kano, 1983; Iwata and Ando, 2007; Tagg *et al.*, 2013). Beds are typically built at dusk or night, and occasionally in the daytime (e.g. Schaller, 1963; van Lawick-Goodall, 1968).

Beds of great apes are similar to birds' nests in shape, but differ in their constituent materials and functions. They are generally made of living branches and twigs, whereas birds' nests are made of various materials, such as dead twigs, leaves, grasses, mud, and so on, depending on the species (Hansell, 2000). The major function of birds' nests is to hold eggs during the breeding season (Hansell, 2000), whereas great apes make new beds each night for sleeping (Goodall, 1962; Schaller, 1963). The word "nest" has been used to describe the platforms made by great apes, but we will use "bed" in this chapter because great apes' platforms function similarly to human beds (Videan, 2006), and because bed making is thought to be a congenetic behavior initiated by a common hominid ancestor (e.g. Baldwin *et al.*, 1981).

Most catarrhine primates sleep on branches or the ground without building a sleeping structure (e.g. Lumsden, 1951; Hall, 1965; Tsuchiya, 1978; Hausfater and Meade, 1982; Ogawa and Takahashi, 2003); hence, bed-making behavior is thought to have evolved in the hominid lineage millions of years ago (Baldwin *et al.*, 1981). Various hypotheses have been proposed to explain the evolution of bed making. One is that it evolved to support the large body weights of resting great apes (Baldwin *et al.*, 1981); that is, sleeping on a branch without the aid of a platform would require considerable energy to maintain balance. In this view, beds could provide a comfortable, more energy efficient, night's sleep. Alternately, Fruth and Hohmann (1996) speculated that beds in trees originated as platforms that aimed to compete with monkeys for food. In this view, ape ancestors bent and broke branches to reach fruits, and the resulting stack of branches allowed them to sit stably without grasping other branches for support (Fruth and Hohmann, 1996).

Some additional benefits of beds have been proposed. For example, chimpanzee beds in trees may

Mahale Chimpanzees: 50 Years of Research, ed. M. Nakamura, K. Hosaka, N. Itoh and K. Zamma. Published by Cambridge University Press.

Figure 43.1 A bed in a tree, January 8, 2007. ©K. Zamma.

help their occupants avoid predation (Goodall, 1962; Pruetz *et al.*, 2008; Stewart and Pruetz, 2013), stay warm (Goodall, 1962; Stewart, 2011; but see also Samson and Hunt, 2012), rest comfortably (Stewart *et al.*, 2007), and avoid biting insects such as mosquitoes (Stewart, 2011). Some have proposed that peaceful and secure beds improve sleep quality, which might have promoted the evolution of hominid cognitive ability (Baldwin *et al.*, 1981; Fruth and Hohmann, 1996). Chimpanzees spend 1–5 min making their beds (van Lawick-Goodall, 1968). These few minutes of effort provide various benefits, potentially promoting the behavior among great apes.

43.2.1 Sleeping sites and bed lifespan

Beds are one of the indicators used to determine chimpanzee presence in a habitat, and their locations have been discussed in terms of sleeping site preferences (e.g. Kano, 1972). These are thought to be affected by several factors, including vegetation, feeding ecology, predation pressure, and so on. For example, Kano (1972), who surveyed bed distribution in Western Tanzania, concluded that chimpanzees preferred to sleep in forests or *Brachystegia bussei* woodlands (see Chapter 4). Other studies have shown that chimpanzees sleep more frequently near fruit trees (Furuichi and Hashimoto, 2004), in certain food tree species (Basabose and Yamagiwa, 2002; Stanford and O'Malley, 2008), or in trees on hillsides (Hernandez-Aguilar, 2009; Ogawa *et al.*, 2013).

Several beds are often closely clumped, and the benefit of sleeping clusters has been also discussed. Ogawa *et al.* (2007), who conducted a long-term study in Ugalla, Tanzania, suggested that chimpanzees form large sleeping clusters to lower predation risk and/or to meet chimpanzees that range in separate parties during the daytime.

By taking bed lifespan (aging, longevity, or decay) into account, it is possible to estimate the population density of great apes over large areas of their habitat (e.g. Kano, 1972; Ghiglieri, 1984; Tutin and Fernandez, 1984; Plumptre and Reynolds, 1996; Chapter 4). It is important to determine bed age as accurately as possible, as a doubling of bed age would halve the resulting population size estimate, as shown by the formula in Chapter 4.

Bed lifespan can be measured using two methods. One considers the period between the day when a new bed is built and the day when the bed loses its last leaf ("bed lifespan by leaves" in Table 43.1), or the period between the day when a new bed is built and the day when its structure is completely decayed ("bed lifespan by structure" in Table 43.1). There are seasonal differences in bed lifespans by structure, but tendencies differ among and even within study sites. For example, at Ugalla, beds made in forests had longer lifespans in the rainy season than in the dry season, whereas those made in woodlands were the opposite (Stewart *et al.*, 2011). In the forest of Mahale, the lifespan of beds made in the rainy season was shorter than that in the dry season (Ihobe, 2005; Zamma and Makelele, 2012). Thus, seasonality seems to alter bed lifespans differently in different habitats, as Kouakou *et al.* (2009) suggested. On the other hand, bed lifespan by leaves shows similar seasonality across sites: longer in the rainy season at both Budongo, Uganda, and Mahale (Table 43.1). As Plumptre and Reynolds (1996) suggested, dried leaves may decay more easily than wet.

Bed lifespans may vary among the vegetation types in which beds are built (Table 43.1). In Ugalla and Mahale, beds built in woodlands last longer than those in forest (Ihobe, 2005; Stewart *et al.*, 2011; Zamma and Makelele, 2012). The tree species used

Table 43.1 **Relative lifespans of chimpanzee beds among study sites**

					Bed lifespan by structure (days)		Bed lifespan by leaves (days)		
Study site	Country	Rainfall (mm)	Vegetation	Season	mean (range)	*N*	mean (range)	*N*	Source
Ugalla	Tanzania	980 [a]	Woodland	Dry, rainy	260	10			Ogawa *et al.*, 2007
Ugalla	Tanzania	980 [a]	Forest	Rainy	118.9 (14–350)	25			Stewart *et al.*, 2011
Ugalla	Tanzania	980 [a]	Forest	Dry	83.3 (14–224)	26			Stewart *et al.*, 2011
Ugalla	Tanzania	980 [a]	Woodland	Rainy	139.2 (7–427+)	91			Stewart *et al.*, 2011
Ugalla	Tanzania	980 [a]	Woodland	Dry	185.5 (14–357+)	49			Stewart *et al.*, 2011
Budongo	Uganda	1600 [b]	Forest [b]	Dry			37.2 (10–154)	48	Plumptre and Reynolds, 1996
Budongo	Uganda	1600 [b]	Forest [b]	Rainy			54.6 (14–161)	48	Plumptre and Reynolds, 1996
Kibale	Uganda	1664	Forest/ Woodland	–	110.8 (15–202)	28			Ghiglieri, 1984
Taï	Côte d'Ivoire	*c.* 1800	Forest	Dry, rainy	73.3 (7–290)	26			Marchesi *et al.*, 1995
Taï	Côte d'Ivoire	*c.* 1800	Forest	Dry, rainy	62.9 (SD = 51.79)	119			Kouakou *et al.*, 2009
Mahale (Kasoje)	Tanzania	1739 [c]	Forest	Dry	131 (30–189)	41	49 (18–110)	41	Ihobe, 2005
Mahale (Kasoje)	Tanzania	1739 [c]	Forest	Rainy	87 (21–231)	9	76 (14–158)	9	Zamma and Makelele, 2012
Mahale (Miyako)	Tanzania	1774 [d]	Woodland	Rainy	164 (13–233+)	7	126 (13–157)	7	Zamma and Makelele, 2012
Belinga	Gabon	*c.* 2000	Forest	–	113.6 (35–151)	49			Tutin and Fernandez, 1984
Campo and Ma'an	Cameroon	2129 [e]	Forest	Dry, rainy	130	40			Matthews and Matthews, 2004

[a] Ogawa *et al.*, 2007; [b] Reynolds, 2005; [c] Chapter 12; [d] Takasaki *et al.*, 1990; [e] Muroyama, 1991.

may also influence bed lifespans differently in forest than in woodland (Zamma and Makelele, 2012).

The within-site ranges of bed lifespans appear to vary considerably among sites (Table 43.1). For example, lifespans in the woodland of Ugalla varied widely during the rainy season, from 1 week to more than 1 year (Stewart *et al.*, 2011). This suggests that the lifespan is affected by various factors, such as the weather on the day the bed was abandoned, bed robustness, season, or vegetation type. This variance must be considered when bed lifespan is used to estimate population size.

43.2.2 Characteristics of bed trees

In Mahale, use of at least 28 tree species (18 families) in beds has been recorded (Table 43.2). As in previous studies (Basabose and Yamagiwa, 2002; Stanford and O'Malley, 2008), many of these species yield chimpanzee foods (20 species), and 87.6% of beds were made directly on them. However, this does not necessarily mean that the tree provided food at the time the beds were made.

The preference for food tree species in bed making may be due to familiarity from climbing these trees for feeding. Oil palm (*Elaeis guineensis*) is a common bed tree in some study sites where oil palm products are commonly eaten (e.g. Goodall, 1962; Yamakoshi and Sugiyama, 1995; Sousa *et al.*, 2011). The chimpanzees of Mahale have not been observed to eat any oil palm products (McGrew, 1992) with the exception of pith (Zamma *et al.*, 2011), nor even to climb on the species. Therefore, they do not make beds on the trees, though, in rare cases, they may use oil palm leaves as a secondary bedding material (Zamma, personal observation).

Table 43.2 **Tree species used for the beds of Mahale chimpanzees**

			Number of beds		
Area	Family	Scientific name	Total	Ihobe (2005)	Zamma and Makelele (2012)
Kasoje	Anacardiaceae	*Pseudospondias microcarpa* (#1) [a]	13		13
	Annonaceae	*Xylopia longipetala* (#2) [a]	10	7	3
	Apocynaceae	*Tabernaemontana pachysiphon* [a]	1	1	
	Bignoniaceae	*Stereospermum kunthianum* [a]	2	1	1
	Boraginaceae	*Cordia millenii* [a]	18		18
	Campanulaceae	*Lobelia stricklandiae*	1	1	
	Chrysobalanaceae	*Parinari curatellifolia* [a]	3	3	
	Combretaceae	*Combretum molle* (#3) [a]	2		2
	Dracaenaceae	*Dracaena usambarensis* (#4)	2	1	1
	Euphorbiaceae	*Bridelia atroviridis* [a]	2		2
		Bridelia micrantha (#5) [a]	3		3
		Croton sylvaticus [a]	5		5
	Leguminosae-Caesalpinioideae	*Senna spectabilis* (#6)	6		6
		#6 (dead) [b]	2		2

Table 43.2 **(cont.)**

Area	Family	Scientific name	Number of beds		
			Total	Ihobe (2005)	Zamma and Makelele (2012)
	Loganiaceae	*Anthocleista schweinfurthii* [a] & #2 [a, c]	1		1
	Moraceae	*Ficus sur* [a]	2	1	1
		Ficus vallis-choudae [a]	1	1	
		Milicia excelsa [a]	1	1	
	Myristicaceae	*Pycnanthus angolensis* (#7) [a]	40	21	19
	Myrtaceae	*Syzygium guineense* [a]	5		5
	Sapindaceae	*Aporrhiza paniculata*	1		1
		Lecaniodiscus fraxinifolius [a]	3		3
	Sapotaceae	*Synsepalum brevipes* [a]	3	1	2
	Sterculiaceae	*Sterculia tragacantha* [a]	1	1	
	?	?	3	3	
		#1 [a, c] & #2 [a]	2		2
		#1 [a, c] & #3 [a]	1		1
		#1 [a, c] & #4	1		1
		#1 [a, c] & #7[a]	1		1
		#2 [a, c] & #5 [a]	1		1
		#2 [a, c] & #7 [a]	1		1
Miyako	Euphorbiaceae	*Croton megalocarpus*	1		1
	Sapindaceae	*Blighia unijugata*	1		1
	Sapotaceae	*Mimusops* sp. [a]	4		4
	?	?	1		1
Total			145	43	102

[a] Fruits, leaves, and/or pith of these plants are eaten by Mahale chimpanzees (see Chapter 17).

[b] Two beds were constructed at the base of dead trees, but were built from living vines including *Saba comorensis* (Apocynaceae), *Keetia venosa* (Rubiaceae), and *Tetracera potatoria* (Dilleniaceae). Ants (*Crematogaster* spp.) living inside the dead branches of this tree species are eaten by Mahale chimpanzees (see Chapter 21).

[c] Two tree species were used in bed construction.

It is possible that Mahale chimpanzees have specific preferences in sleeping trees, as nearly one-quarter of beds were made on *Pycnanthus angolensis* (Table 43.2). On the other hand, it is also possible that chimpanzees are less discriminate, as 10 of 28 species were used only once during the study periods. Chimpanzees, except for infants, make beds every night, which means that they are skilled at the activity. Chimpanzees may sometimes select tree species that are especially suitable for making beds, but their skills and experience allow them to make comfortable beds using the branches of any familiar tree species.

Zamma investigated the characteristics of bed tree leaves and twigs from 2006 to 2007. Branch samples were taken from bed trees and investigated for their length, number of twigs, number of leaves, leaf size, and the strength of their leaf-stem connections. Chimpanzees were indiscriminate with respect to both twig (0.5–151.5 per meter of branch) and leaf number (9.6–567.4 per meter branch) (Table 43.3). There was a positive correlation between the number of both twigs and leaves per meter ($\tau = 0.70$, $p < 0.01$, Kendall's rank correlation coefficient). On the other hand, the size of leaves (in square meters) correlated negatively with the number of leaves per meter of branch ($\tau = -0.71$, $p < 0.01$). In other words, a branch with more twigs has more leaves of smaller size.

We predicted that branches with greater total leaf area would be preferred for making leafy beds. We estimated the total area of leaves attached to a 1-m branch (Table 43.3), and *P. angolensis*, the most common species of bed tree (Table 43.2), had the largest area (1.01 m^2), followed by *Pseudospondias microcarpa* (0.99 m^2) and *Xylopia longipetala* (0.89 m^2), which are also major bed species. This result suggests that the total area of leaves per unit length of branch is a factor in chimpanzee bed tree selection. By using such tree species, chimpanzees can make leafier beds.

We investigated the strength of the leaf-stem connection to assess its relationship with bed lifespan by leaves. When leaves are easily removed, the lifespan of beds made from a given tree species is expected to be shorter. The strength of connection was measured for 10 leaves by pulling a leaf via a spring scale until it separated from the stem; we then calculated the mean value. The strength of leaf-stem connection ranged from 0.15 to 3.78 kg (Table 43.3) and was correlated with the estimated leaf area ($\tau = 0.70$, $p < 0.01$); that is, larger leaves are connected more tightly to the stem. Comparison of these strength data with bed lifespans by leaves (data from Zamma and Makelele, 2012) showed a negative correlation that was only marginally significant (Figure 43.2, $\tau = -0.62$, $p = 0.069$). Although smaller leaves are more easily detached, the lifespans of beds in such species were probably not markedly shorter because leaves were sandwiched and retained between branches with abundant twigs.

We also investigated temporal changes in the strength of leaf-stem connections. Strength was measured (a) when the branch was sampled, (b) after 15 days, (c) after 30 days, and (d) after 60 days. There were three types of trends in the strength changes (Figure 43.3). Type 1, "drastic decline," occurred in seven species (species no. 2, 4, 6, 7, 8, 9, and 15; see Table 43.3 for the full names of all species). The strength of the leaf-stem connection declined to less than half its original value over the first 15 days in these species. Type 2, "slow monotonic decline," occurred in eight species (no. 1, 5, 13, 14, 17, 22, 24, and 25). The connection strength of these species decreased slowly throughout the study period. The remaining 10 species were categorized as Type 3, "others," and showed little change in leaf connection strength. *Psydrax parviflora* (No. 19) was anomalous in showing a monotonic increase in leaf connection strength during the study period. It is difficult to elucidate the mechanisms of the temporal change in leaf-stem connection strength, but they may differ from the process of normal leaf abscission. A branch in a bed does not necessarily drop its leaves immediately, probably because the branch is separated from the tree. The initial strength of the leaf-stem connection, and its tendency to change temporally, may be the key factors affecting the lifespans of chimpanzee beds.

43.2.3 Bed making

Chimpanzees use an average of 7.3 main branches (range 4–10) when making their beds, with an average diameter of 20 mm (range 10–48) (Izawa and Itani, 1966). Branches and twigs are usually taken from the tree on which the bed is made, or from other trees within arm's reach (Goodall, 1962). However, at Mahale, Nishida (1981) observed an adult female, Wakilufya, break a branch on the ground, carry it into a tree, and construct a bed with it and other branches. He assumed that she wanted to make her bed near that of the alpha male, Kasonta, with whom she was in consort.

Table 43.3 **Characteristics of the twigs and leaves of bed tree species**

				Leaf size					
Family	Scientific name	Number of twigs per 1-m branch	Number of leaves per 1-m branch (N)	Major vein length (mm) (L)[a]	Width (mm) (W)[a]	Area of a leaf (mm^2) (A)[b]	Total leaf area attached to a 1-m branch (m^2)[c]	Strength of picking leaf (kg)[d]	No. in Figure 43.3
Anacardiaceae	*Pseudospondias microcarpa*	21.4	141.7	156.8	56.9	7003.7	0.99	1.99	1
Annonaceae	*Annona senegalensis* [e]	8.7	37.7	122.8	79.1	7625.1	0.29	1.39	2
	Xylopia longipetala	151.5	506.1	76.6	29.2	1755.8	0.89	0.81	3
Bignoniaceae	*Stereospermum kunthianum*	9.2	53.8	102.1	41.4	3318.1	0.18	0.99	4
Boraginaceae	*Cordia millenii*	4.4	52.7	124.8	89.0	8719.2	0.46	2.17	5
Combretaceae	*Combretum molle*	13.8	46.3	170.2	79.0	10555.0	0.49	0.94	6
Dracaenaceae	*Dracaena usambarensis*	1.2	92.6	212.9	23.7	3960.9	0.37	0.89	7
Euphorbiaceae	*Bridelia atroviridis*	16.9	175.9	83.5	35.0	2294.2	0.40	0.75	8
	Bridelia micrantha	34.3	127.1	112.8	55.3	4896.7	0.62	1.43	9
	Croton sylvaticus	28.6	185.7	91.4	50.3	3609.0	0.67	0.84	10
	Croton megalocarpus	38.6	185.7	104.4	40.6	3327.3	0.62	1.17	11
Leguminosae-Caesalpinioideae	*Brachystegia bussei* [e]	79.4	406.3	54.7	20.0	858.8	0.35	0.39	12
	Brachystegia spiciformis [e]	74.7	424.1	42.7	21.9	734.1	0.31	0.41	13

Table 43.3 (cont.)

Family	Scientific name	Number of twigs per 1-m branch	Number of leaves per 1-m branch (*N*)	Leaf size: Major vein length (mm) (*L*)[a]	Leaf size: Width (mm) (*W*)[a]	Leaf size: Area of a leaf (mm^2) (*A*)[b]	Total leaf area attached to a 1-m branch (m^2)[c]	Strength of picking leaf (kg)[d]	No. in Figure 43.3
	Senna spectabilis	13.1	300.0	76.2	17.4	1040.8	0.31	0.23	14
Loganiaceae	*Anthocleista schweinfurthii*	0.5	9.6	244.2	114.9	22026.0	0.21	3.78	15
Moraceae	*Ficus sur*	37.2	124.4	116.2	66.4	6056.8	0.75	1.61	16
Myristicaceae	*Pycnanthus angolensis*	9.2	84.2	227.3	67.4	12026.2	1.01	3.25	17
Myrtaceae	*Syzygium guineense*	16.2	118.9	118.7	48.2	4491.3	0.53	1.62	18
Rubiaceae	*Psydrax parviflora* [e]	20.8	109.1	94.7	37.7	2802.6	0.31	1.62	19
Sapindaceae	*Aporrhiza paniculata*	4.9	18.5	187.7	86.7	12774.8	0.24	2.28	20
	Blighia unijugata	104.2	202.8	95.3	36.9	2760.5	0.56	0.90	21
	Lecaniodiscus fraxinifolius	54.8	226.0	93.4	35.9	2632.2	0.59	0.69	22
Sapotaceae	*Synsepalum brevipes*	6.3	67.2	222.9	71.0	12423.3	0.83	1.50	23
	Mimusops sp.	22.5	125.0	113.6	41.0	3656.2	0.46	1.38	24
?	?	133.7	567.4	44.0	18.6	642.4	0.36	0.15	25

[a] Mean length and width of ten leaves.

[b] Leaf surface area calculated as an ellipse; i.e. $A = L/2 \times W/2 \times \pi$.

[c] Calculated by multiplying the area of a leaf by the number of leaves per 1-m branch; i.e. $A \times (1/1000)^2 \times N$.

[d] The connection between the petiole and stem. A spring scale was used to measure the force required to separate a leaf from the stem by pulling. Measurements were conducted on ten leaves; the mean values are shown.

[e] These species are not listed in Table 43.2, but are confirmed bed trees in Mahale.

A chimpanzee generally makes only one (Goodall, 1962), and rarely multiple, beds in a single night. Nishida (1981) observed that Kasonta made three beds in an evening when he was concerned about the movement of the beta male and an estrous female.

Beds are occasionally re-used (e.g. Goodall, 1962; van Lawick-Goodall, 1968; Plumptre and Reynolds, 1997), but the frequency of this varied among study sites (Budongo: 9.6–11.4%, Plumptre and Reynolds, 1997; Brownlow *et al.*, 2001; Mahale: 4.7%, Ihobe, 2005; Ugalla: 6%, Stewart *et al.*, 2011). A bed can be re-used by the same individual, and sometimes by a different animal (van Lawick-Goodall, 1968). There are various reasons why a chimpanzee might re-use an old bed. On November 24, 2006, Zamma observed an adult female with a severe cough sleep on an old bed supplemented by a few new branches. She may have done so because she lacked the energy to make a new one. On the other hand, Nishida (1981) described a rare case in which a bed was re-used twice: first by an adult male, Kamemanfu, on February 25, 1976, and again by Kasonta 3 days later. Although Nishida did not discuss why they re-used the bed, it appears that physical condition is not the only reason for such behavior, as the individuals in question appeared healthy at the time.

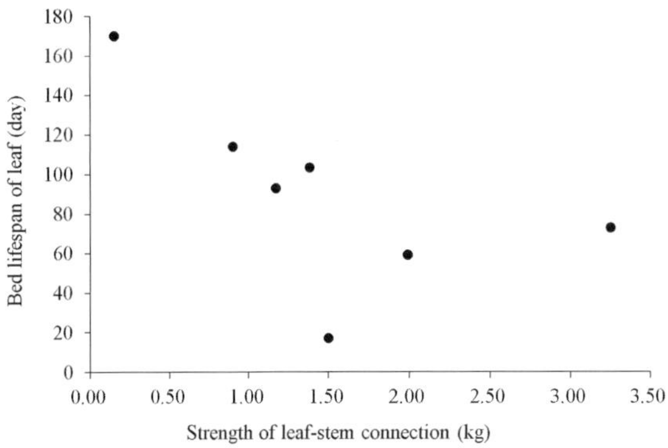

Figure 43.2 Correlation between the strength of the leaf-stem connection and bed lifespan by leaves.

43.2.4 Development of bed construction techniques

Chimpanzees construct their beds in only a few minutes (Goodall, 1962), interweaving branches and twigs using their hands and feet to produce complex three-dimensional structures (Goodall, 1962; Izawa

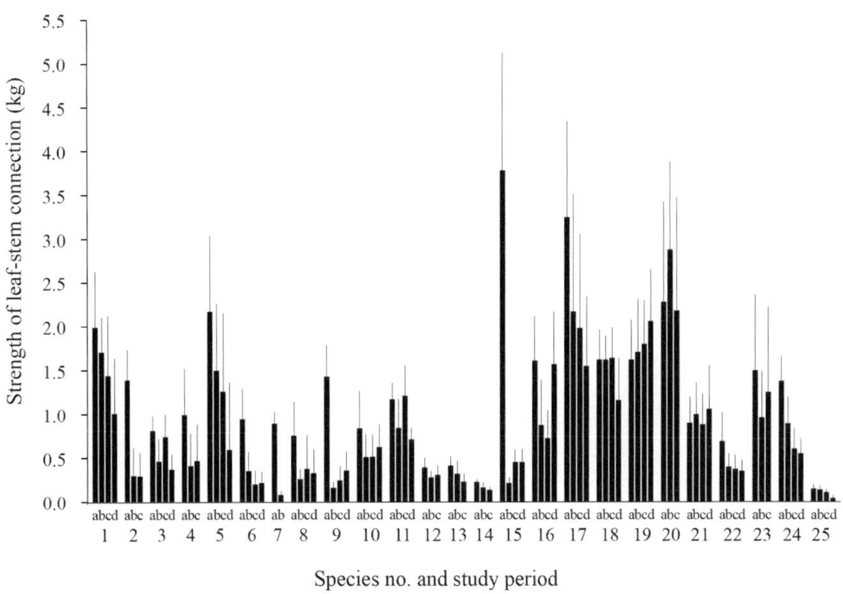

Figure 43.3 Temporal change in the strength of the leaf-stem connection. The number shown on the x-axis indicates the tree species number from Table 43.3. (a) The day the branch was sampled, (b) after 15 days, (c) after 30 days, and (d) after 60 days. The black column indicates mean force (kg) and the vertical bar represents standard deviation.

and Itani, 1966). This manipulation of branches and twigs may represent the foundation of tool use (Fruth and Hohmann, 1996).

Chimpanzees learn bed making from their mother (Goodall, 1962; Videan, 2006). Infants initially sleep with the mother (Goodall, 1962; Hiraiwa-Hasegawa, 1989), and start to sleep in their own bed by making it near their mothers' when they are around 5 years old (Hiraiwa-Hasegawa, 1989). As adults, males build beds lower than do females (Fruth and Hohmann, 1996; Brownlow *et al.*, 2001; Reynolds, 2005).

We investigated the development of bed construction techniques and bed heights during the rainy season at Mahale (Zamma, in prep.). The bed builders were identified by observing them before dusk or after dawn.

As previous studies have pointed out (Goodall, 1962; Izawa and Itani, 1966), Mahale chimpanzees in all age–sex classes construct the lower layer of their beds with more bent and curved branches, and the higher layers with more broken branches. The proportion of processed branches, however, differed among age classes (Figure 43.4). Around 80% of branches used for the upper layer of juveniles' beds were broken, whereas they accounted for less than 60% of those in adults' beds. Juvenile chimpanzees that may be vulnerable to cold may need cushion-like beds with more leaves of broken branches (Zamma, in prep.). For the lower layer, juvenile and adolescent chimpanzees used more curved than bent branches, whereas the opposite was true for adults (Figure 43.4). Both curved and bent branches serve to strengthen the bed structure. Immature individuals probably used curved branches more frequently because they lack the strength to bend branches. It has already been suggested that they have learned how to accumulate

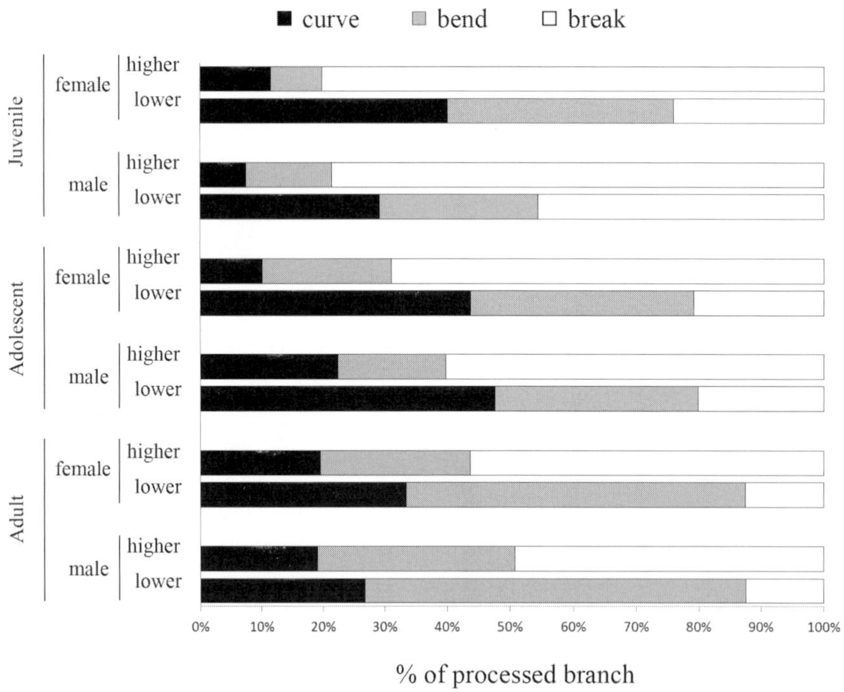

Figure 43.4 Proportion of processed branches used in the higher and lower bed layers by each age–sex class. Processed branches were classified as follows. Curve: branch changed to form an arch without damaging its inflection. Bend: branch changed to form an angle, damaging its inflection. Break: branch separated into two or more pieces and incorporated into the structure. For age–sex classes see Chapter 22. A bed is constructed through the accumulation of layers of processed branches; the upper 50% of all layers was categorized as the higher layer, and the remainder as the lower layer.

processed branches (Goodall, 1962; Videan, 2006), but it seems they can also adjust the proportion of differently processed branches depending on their needs and powers (Zamma, in prep.).

Bed height among females was mostly unchanged from juveniles to adults but, among males, decreased sharply in early adolescence. Thus, sexual differences in bed height started at the immature stage (Zamma, in prep.). Early adolescent males, whose weight is half that of adults (Pusey, 1990), begin to follow adult males instead of their mothers (Hayaki, 1988, Pusey, 1990). Thus, they may learn bed height and other behavior from adult males (Zamma, in prep.).

43.2.5 Day beds

Chimpanzees make beds in the daytime as well as at night. In Mahale, day beds were observed more often when chimpanzees rested in the rain than in fine or cloudy weather (Zamma, unpublished data). This may be because they are unwilling to rest on the wet, muddy ground.

Day bed building begins at around 1 year old, which is earlier than for night beds (Hiraiwa-Hasegawa, 1989). Day beds made by such young infants often fall apart due to shoddy construction (Hiraiwa-Hasegawa, 1989). The frequency of day bed construction increases by the time infants are 3 years of age (4.4 times/10 observation hours), and then decreases as they grow older (Hiraiwa-Hasegawa, 1989). Immature chimpanzees play socially or alone on day beds, and occasionally make beds as part of play (Nishida *et al.*, 2010; see Chapter 40).

Adult female chimpanzees make day beds more frequently than males, possibly due to the tendency of adult males to lie on the ground when they rest during daytime (Hiraiwa-Hasegawa, 1989).

43.3 Nocturnal behaviors

Nocturnal behaviors refer to activities performed at night, though most are not unique to the night. For example, sleeping is one of the main behaviors at night, but chimpanzees also nap in the daytime. Chimpanzees shift the frequency of activities from day to night, but do not change their activities completely.

Those who have not stayed in a forest at night may have the idea that it is totally dark. However, this is not always true. One cannot see a hand in front of one's eyes at new moon, but one can both see and walk about in the forest on moonlit nights. Unlike sunshine, the illuminance of moonlight changes with the phase of the moon. Therefore, chimpanzees that sleep in the forest canopy are surely affected by moonlight, and this may help to explain the similarity of activities between night and day.

43.3.1 Sleep

As with humans and other mammals, chimpanzees experience rapid eye movement (REM) sleep and four stages of non-REM sleep (Freemon *et al.*, 1969, 1970, 1971). Investigation of captive chimpanzees revealed that seven to nine periods of REM sleep occurred per night, with an average interval of about 85 min (Freemon *et al.*, 1971). Body movement was more frequent during late REM than during early REM (Freemon *et al.*, 1970). One male chimpanzee adopted a prone position during REM sleep early in the night, but the adoption of this position decreased later in the REM period. On the other hand, the chimpanzee was more likely to adopt a lateral position during the non-REM period (Freemon *et al.*, 1970).

Wild chimpanzees, as discussed, generally sleep in beds in trees. They sleep solitarily and use one bed each, except for mother–infant pairs (Goodall, 1962). They usually start to make their beds around sunset, and utter bed grunts when they lie in the finished product (Goodall, 1986). Few signs of Mahale chimpanzee activity or vocalization are heard between 21:00 and 22:00 h (Zamma, 2014), the time corresponding to the second or third stage of non-REM sleep (Freemon *et al.*, 1971), and thus the chimpanzees apparently sleep deeply at that time. They may also sleep deeply around 03:00 and 04:00 h, judging from the frequency of vocalizations and activity (Zamma, 2014).

Chimpanzees generally sleep in the supine, prone, or lateral position (Goodall, 1962). We observed the lateral position most frequently in video recordings of nocturnal behavior at Mahale (Zamma, in prep.). From 2010 to 2013, Zamma recorded chimpanzees in their night beds from the higher vantage point of nearby hillsides, but obtained only two clear video recordings, on December 28, 2010 and January 11, 2011. Frequent body movements were observed when chimpanzees took a supine position, indicating that wild chimpanzees may experience shallow sleep when supine. On the other hand, they may experience deep sleep when in the lateral position, as they displayed infrequent body movements.

Observations of three sleeping chimpanzees showed that their heads were aligned along the major axis of their ellipse-shaped beds, and that this orientation was almost unchanged during the night (Zamma, in prep.). These results suggest that chimpanzees design the shape of their beds to accommodate their favored sleeping position.

REM sleep was also observed during the daytime nap (Freemon *et al.*, 1971). Thus, to compare daytime rest with nighttime sleep, Zamma (in prep.) also investigated the resting posture of Mahale chimpanzees. A lying posture was observed in about 40% of resting cases, while sitting postures accounted for more than half. During daytime resting, chimpanzees adopted supine postures more often than during nighttime sleep. This does not necessarily mean they were in REM sleep during their daytime naps, because Zamma did not always check the status of their eyes. Alternatively, the supine position may aid in hearing sounds or vocalizations during the daytime.

43.3.2 Nocturnal activities

Chimpanzees occasionally remake their beds, or even construct new ones, during the night. The sound of bed making was heard after dark (from 20:00 to 21:00 h) and midnight (01:00 h) on new moons at Mahale (Zamma, 2014). Chimpanzees also vocalize in the night. Izawa and Itani (1966) observed chimpanzee screams and/or pant-hoots in Kasakati, Tanzania, at midnight, which they described as "night greetings." Chimpanzees also utter loud calls and pant-hoots in response to carnivore vocalizations (Tutin *et al.*, 1981). The activities of nocturnal predators and arboreal animals are one of the causes of nocturnal chimpanzee behaviors (Tutin *et al.*, 1981; Stewart, 2011), and vocalizations are detected more frequently on moonlit nights (van Lawick-Goodall, 1968). Ghiglieri (1984) recorded pant-hoots and buttress drumming among chimpanzees of Kibale Forest, Uganda, in the night. On moonless nights, vocalizations such as pant-hoots, bed grunts, and soft calls were heard frequently between 23:00 and 02:00 h (Zamma, 2014).

Chimpanzees excrete from the bed at night. Zamma (2014) heard the sound of defecation and urination around 01:00 h on moonless nights, and proposed that night defecation is one reason that chimpanzees awake around midnight. Goodall (1962) observed that chimpanzees in Gombe, Tanzania, defecated and urinated over the edge of the bed in the night, and further reported that no excrement was found in the bed. In Mahale, an adult male chimpanzee was videotaped taking a prone position, projecting his posterior out from the bed, and defecating (Figure 43.5). When Zamma investigated beds for

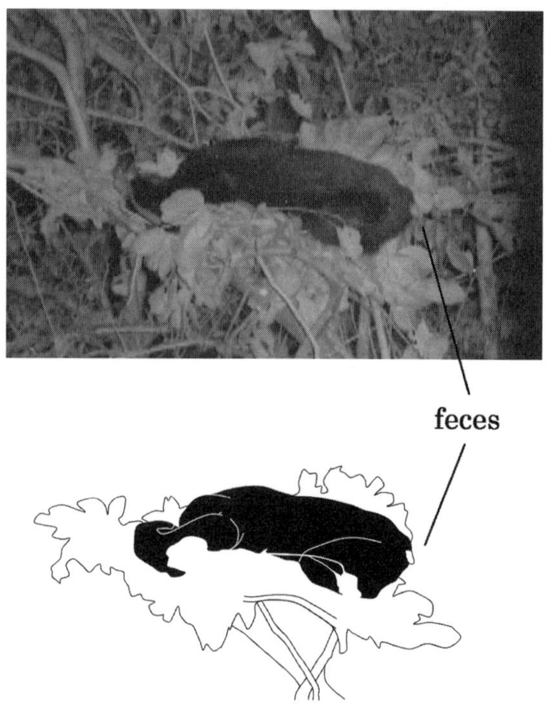

Figure 43.5 An adult male, Carter, projects his posterior out past the edge of the bed and defecates during the night.

their construction (see above), feces were not found at the beds' center, but a few fecal remains were found at the edge of 6 of 32 beds. Cockroaches were found in 8 of the 32 beds, but most of them ($N = 7$) were found in the beds without feces; thus, the insects may feed on decaying branches rather than feces.

When it rains in the night, chimpanzees sit up and hunch over in their beds (Goodall, 1962). Chimpanzees do not usually make roofs or shelters from the rain (Goodall, 1962), but Izawa and Itani (1966) observed one example of a rain protection device in Kasakati: at 17:41 h, an old adult male broke off a thick leafy branch and covered himself with it on the bed. In Mahale, a male chimpanzee adopted a sitting posture when it started raining in the night, but as the rain worsened and thunder began, he started displaying and descended from the tree (Zamma, in prep.). Rain displays are common during daytime rains (Goodall, 1989; Nishida *et al.*, 2010), but are also seen on stormy nights.

There are some reports of nocturnal traveling in chimpanzees. Izawa and Itani (1966) heard vocalizations in Kasakati as the chimpanzees moved in the moonlit night just before daybreak. Darkness is incomplete on moonlit nights, and moonlight is sufficient for chimpanzees to walk (e.g. Izawa and Itani, 1966; van Lawick-Goodall, 1968). Pruetz and Bertolani (2009) observed that chimpanzees in Fongoli, Senegal, moved up to 1 km on moonlit nights. Fongoli is hot and dry, and thus the authors proposed that the cooler nighttime temperatures were more suitable for traveling. In Gombe, chimpanzees have been observed coming to a feeding area on both dark and moonlit nights (van Lawick-Goodall, 1968). Nishida (1994) observed an estrous female chimpanzee from Mahale visit and copulate with adult males, one after another, at dawn. He also observed some adult chimpanzees with fresh wounds in the early morning, and speculated that aggression often occurs at night (Nishida, 1994). Moonless nights may represent a good opportunity for low-ranking males to covertly copulate with females; however, this has yet to be confirmed.

43.4 The other half of the day

This chapter has discussed the various behaviors of chimpanzees at night. The main nocturnal behavior of chimpanzees is sleeping in their beds. They choose their bed tree from among certain preferred species, consider its height, and make their bed. They prefer lateral over supine or prone positions for sleeping. Given the relative frequencies of body movements, sleep quality may be different among sleep positions, and may influence chimpanzee cognitive ability (Fruth and Hohmann, 1996).

On the other hand, chimpanzees frequently engage in other nocturnal activities when light or temperature conditions allow (Izawa and Itani, 1966; van Lawick-Goodall, 1968; Pruetz and Bertolani, 2009), when they are awakened by nocturnal animals (Tutin *et al.*, 1981; Stewart, 2011), or for excretion (Zamma, 2014). These observations suggest that nocturnal behaviors are not uncommon among wild chimpanzees.

Studies of chimpanzees have been conducted for 50 years in Mahale, most of which were during the daytime. In other words, a total of 25 years of nighttime activities has been missed. Further studies of nocturnal behaviors will compensate for our lack of understanding of the other half of chimpanzees' lives.

References

Baldwin, P. J., Sabater Pí, J., McGrew, W. C., and Tutin, C. E. G. (1981). Comparisons of nests made by different populations of chimpanzees (*Pan troglodytes*). *Primates*, **22**, 474–86.

Basabose, A. K. and Yamagiwa, J. (2002). Factors affecting nesting site choice in chimpanzees at Tshibati, Kahuzi-Biega National Park: influence of sympatric gorillas. *International Journal of Primatology*, **23**, 263–82.

Brownlow, A. R., Plumptre, A. J., Reynolds, V., and Ward, R. (2001). Sources of variation in the nesting behavior of chimpanzees (*Pan troglodytes schweinfurthii*) in the Budongo Forest, Uganda. *American Journal of Primatology*, **55**, 49–55.

van Casteren, A., Sellers, W. I., Thorpe, S. K. S., *et al.* (2012). Nest-building orangutans demonstrate engineering know-how to produce safe, comfortable beds. *Proceedings of the National Academy of Sciences of the USA*,**109**, 6873–7.

Freemon, F. R., McNew, J. J., and Adey, W. R. (1969). Sleep of unrestrained chimpanzee: cortical and subcortical recordings. *Experimental Neurology*, **25**, 129–37.

Freemon, F. R., McNew, J. J., and Adey, W. R. (1970). Sleep of unrestrained chimpanzee: differences between first and last rapid eye movement periods. *Folia Primatologica*, **13**, 144–9.

Freemon, F. R., McNew, J. J., and Adey, W. R. (1971). Chimpanzee sleep stages. *Electroencephalography and Clinical Neurophysiology*, **31**, 485–9.

Fruth, B. and Hohmann, G. (1996). Nest building behavior in the great apes: the great leap forward? In *Great Ape Societies*, ed. W. C. McGrew, L. Marchant, and T. Nishida. Cambridge: Cambridge University Press, pp. 225–40.

Furuichi, T. and Hashimoto, C. (2004). Botanical and topographical factors influencing nesting-site selection by chimpanzees in Kalinzu Forest, Uganda. *International Journal of Primatology*, **25**, 755–65.

Ghiglieri, M. P. (1984). *The Chimpanzees of Kibale Forest: A Field Study of Ecology and Social Structure*. New York: Columbia University Press.

Goodall, J. (1962). Nest building behavior in the free ranging chimpanzee. *Annals of the New York Academy of Sciences*, **102**, 455–67.

Goodall, J. (1986). *The Chimpanzees of Gombe: Patterns of Behavior*. Cambridge, MA: Belknap.

Goodall, J. (1989). *The Glossary of Chimpanzee Behaviors*. Tucson, AZ: Jane Goodall Institute.

Hall, K. R. L. (1965). Behaviour and ecology of the wild patas monkey, *Erythrocebus patas*, in Uganda. *Journal of Zoology*, **148**, 15–87.

Hansell, M. H. (2000). *Bird Nests and Construction Behaviour*. Cambridge: Cambridge University Press.

Hausfater, G. and Meade, B. J. (1982). Alternation of sleeping groves by yellow baboons (*Papio cynocephalus*) as a strategy for parasite avoidance. *Primates*, **23**, 287–97.

Hayaki, H. (1988). Association partners of young chimpanzees in the Mahale Mountains National Park, Tanzania. *Primates*, **29**, 147–61.

Hernandez-Aguilar, R. A. (2009). Chimpanzee nest distribution and site reuse in a dry habitat: implications for early hominin ranging. *Journal of Human Evolution*, **57**, 350–64.

Hiraiwa-Hasegawa, M. (1989). Sex differences in the behavioral development of chimpanzees at Mahale. In *Understanding Chimpanzees,* ed. P. G. Heltne and L. A. Marquardt. Cambridge, MA: Harvard University Press, pp. 104–15.

Hobson, J. A. (2005). Sleep is of the brain, by the brain and for the brain. *Nature*, **437**, 1254–6.

Ihobe, H. (2005). Life span of chimpanzee beds at the Mahale Mountains National Park, Tanzania. *Pan Africa News*, **12**, 10–12.

Iwata, Y. and Ando, C. (2007). Bed and bed-site reuse by western lowland gorillas (*Gorilla g. gorilla*) in Moukalaba-Doudou National Park, Gabon. *Primates*, **48**, 77–80.

Izawa, K. and Itani, J. (1966). Chimpanzees in Kasakati Basin, Tanganyika. 1. Ecological study in the rainy season 1963–1964. *Kyoto University African Studies*, **1**, 73–156.

Kano, T. (1972). Distribution and adaptation of the chimpanzee on the eastern shore of Lake Tanganyika. *Kyoto University African Studies*, **7**, 37–129.

Kano, T. (1983). An ecological study of the pygmy chimpanzees (*Pan paniscus*) of Yalosidi, Republic of Zaire. *International Journal of Primatology*, **4**, 1–31.

Kouakou, C. Y., Boesch, C., and Kuehl, H. (2009). Estimating chimpanzee population size with nest counts: validating methods in Taï National Park. *American Journal of Primatology*, **71**, 447–57.

van Lawick-Goodall, J. (1968). The behaviour of free-living chimpanzees in the Gombe Stream Reserve. *Animal Behaviour Monographs*, **1**, 165–311.

Lumsden, W. H. R. (1951). The night-resting habits of monkeys in a small area on the edge of the Semliki Forest, Uganda: a study in relation to the epidemiology of Sylvan yellow fever. *The Journal of Animal Ecology*, **20**, 11–30.

MacKinnon, J. (1974). The behaviour and ecology of wild orang-utans (*Pongo pygmaeus*). *Animal Behaviour*, **22**, 3–74.

Marchesi, P., Marchesi, N., Fruth, B., and Boesch, C. (1995). Census and distribution of chimpanzees in Côte d'Ivoire. *Primates*, **36**, 591–607.

Matthews, A. and Matthews A. (2004). Survey of gorillas (*Gorilla gorilla gorilla*) and chimpanzees (*Pan troglodytes troglodytes*) in Southwestern Cameroon. *Primates*, **45**, 15–24.

McGrew, W. C. (1992). *Chimpanzee Material Culture*. Cambridge: Cambridge University Press.

Muroyama, Y. (1991). Chimpanzees' choices of prey between two sympatric species of *Macrotermes* in the Campo Animal Reserve, Cameroon. *Human Evolution*, **6**, 143–51.

Nishida, T. (1981). [The World of Wild Chimpanzees.] Tokyo: Chūkō-shinsho. In Japanese.

Nishida, T. (1994). Review of recent findings in Mahale chimpanzees: implications and future research directions. In *Chimpanzee Cultures*, ed. R. W. Wrangham, W. C. McGrew, F. B. M. de Waal, and P. G. Heltne. Cambridge, MA: Harvard University Press.

Nishida, T. (2012). *Chimpanzees of the Lakeshore: Natural History and Culture at Mahale*. Cambridge: Cambridge University Press.

Nishida, T., Zamma, K., Matsusaka, T., Inaba, A., and McGrew, W. C. (2010). *Chimpanzee Behavior in the Wild*. Tokyo: Springer.

Nissen, H. W. (1931). A field study of the chimpanzee: observations of chimpanzee behavior and environment in western French Guinea. *Comparative Psychology Monographs*, **8**, 1–122.

Ogawa, H. and Takahashi, H. (2003). Triadic positions of Tibetan macaques huddling at a sleeping site. *International Journal of Primatology*, **24**, 591–606.

Ogawa, H., Idani, G., Moore, J., Pintea. L., and Hernandez-Aguilar, A. (2007). Sleeping parties and nest distribution of chimpanzees in the savanna woodland, Ugalla, Tanzania. *International Journal of Primatology*, **28**, 1397–412.

Ogawa, H., Yoshikawa, M., and Idani, G. (2013). The population and habitat preferences of chimpanzees in non-protected areas of Tanzania. *Pan Africa News*, **20**, 1–5.

Plumptre, A. J. and Reynolds, V. (1996). Censusing chimpanzees in the Budongo Forest, Uganda. *International Journal of Primatology*, **17**, 85–99.

Plumptre, A. J. and Reynolds, V. (1997). Nesting behavior of chimpanzees: implications for census. *International Journal of Primatology*, **18**, 475–85.

Pruetz, J. D. and Bertolani, P. (2009). Chimpanzee (*Pan troglodytes verus*) behavioral responses to stresses associated with living in a savanna-mosaic environment: implications for hominin adaptations to open habitats. *PaleoAnthropology*, **2009**, 252–62.

Pruetz, J. D., Fulton, S. J., Marchant, L. F., *et al.* (2008). Arboreal nesting as anti-predator adaptation by savanna chimpanzees (*Pan troglodytes verus*) in southeastern Senegal. *American Journal of Primatology*, **70**, 393–401.

Pusey, A. E. (1990) Behavioural changes at adolescent in chimpanzees. *Behaviour*, **115**, 203–46.

Reynolds, V. (2005). *The Chimpanzees of the Budongo Forest: Ecology, Behaviour, and Conservation*. New York: Oxford University Press.

Samson, D. R. and Hunt, K. D. (2012). A thermodynamic comparison of arboreal and terrestrial sleeping sites for dry-habitat chimpanzees (*Pan troglodytes schweinfurthii*) at the Toro-Semliki Wildlife Reserve, Uganda. *American Journal of Primatology*, **74**, 811–18.

Schaller, G. (1963). *The Mountain Gorilla: Ecology and Behavior*. Chicago: The University of Chicago Press.

Sousa, J., Barata, A. V., Sousa, C., Casanova, C. C. N., and Vicente, L. (2011). Chimpanzee oil-palm use in southern Cantanhez National Park, Guinea-Bissau. *American Journal of Primatology*, **73**, 485–97.

Stanford, C. B. and O'Malley, R. C. (2008). Sleeping tree choice by Bwindi chimpanzees. *American Journal of Primatology*, **70**, 642–9.

Stewart, F. A. (2011). Why sleep in a nest? Empirical testing of the function of simple shelters made by wild chimpanzees. *American Journal of Physical Anthropology*, **146**, 313–18.

Stewart, F. A. and Pruetz, J. D. (2013). Do chimpanzee nests serve an anti-predatory function? *American Journal of Primatology*, **75**, 593–604.

Stewart, F. A., Pruetz, J. D., and Hansell, M. H. (2007). Do chimpanzees build comfortable nests? *American Journal of Primatology*, **69**, 930–9.

Stewart, F. A., Piel, A. K., and McGrew, W. C. (2011). Living archaeology: artifacts of specific nest site fidelity in wild chimpanzees. *Journal of Human Evolution*, **61**, 388–95.

Tagg, N., Willie, J., Petre, C.-A., and Haggis, O. (2013). Ground night nesting in chimpanzees: new insights from central chimpanzees (*Pan troglodytes troglodytes*) in South-East Cameroon. *Folia Primatologica*, **84**, 362–83.

Takasaki, H., Nishida, T., Uehara, S., *et al.* (1990). Summary of meteorological data at Mahale research camps, 1973–1988. In *The Chimpanzees of the Mahale Mountains: Sexual and Life History Strategies*, ed. T. Nishida. Tokyo: University of Tokyo Press, pp. 291–300.

Tsuchiya, S. (1978). [Nocturnal behavior of Japanese monkeys.] [*Monkey*], **22**(6), 6–13. In Japanese.

Tutin, C. E. G. and Fernandez, M. (1984). Nationwide census of gorilla (*Gorilla g. gorilla*) and chimpanzee (*Pan t. troglodytes*) populations in Gabon. *American Journal of Primatology*, **6**, 313–36.

Tutin, C. E. G., McGrew, W. C., and Baldwin, P. J. (1981). Responses of wild chimpanzees to potential predators. In *Primate Behavior and Sociobiology*, ed. A. B. Chiarelli and R. S. Corruccini. Berlin: Springer-Verlag, pp. 136–41.

Videan, E. N. (2006). Bed-building in captive chimpanzees (*Pan troglodytes*): the importance of early rearing. *American Journal of Primatology*, **68**, 745–51.

Yamakoshi, G. and Sugiyama, Y. (1995). Pestle-pounding behavior of wild chimpanzees at Bossou, Guinea: a newly observed tool-using behavior. *Primates*, **36**, 489–500.

Zamma, K. (2014).What makes wild chimpanzees wake up at night? *Primates*, **55**, 51–7.

Zamma, K. and Makelele, M. (2012). Comparison of the longevity of chimpanzee beds between two areas in the Mahale Mountains National Park, Tanzania. *Pan Africa News*, **19**, 25–8.

Zamma, K., Nakashima, M., and Ramadhani, A. (2011). Mahale chimpanzees starts to eat oil palm. *Pan Africa News*, **18**, 6–8.

PART IX From field to lab

44 Field endocrinology

SHIHO FUJITA

44.1 Introduction

Hormones are involved in maintaining physiological mechanisms in systems, including reproduction, growth, and metabolism. Furthermore, hormones also affect behavior by influencing the interaction components of behavior, such as the sensory system, central nervous system, and effector organs (Nelson, 2000). The hormonal profile, therefore, has been used to monitor the physiological state of animals or to investigate hormone–behavior interactions. The recent development of noninvasive techniques to measure hormone levels from samples have allowed us to apply endocrine data to behavioral and ecological study, especially on animals in their natural environment (Whitten *et al.*, 1998; Strier and Ziegler, 2005).

In this chapter, sampling methods will be discussed with consideration of the precautions that should be made in the field, and findings from field endocrinology, mainly in wild chimpanzees, will be described.

44.2 Sample collection and preservation under field conditions

The hormones to be investigated should be selected based on the purpose of the study. Hormones can be classified into four different types based on their chemical structure: peptides (including proteins), steroids, amino acid derivatives, and fatty acid derivatives. In behavioral studies, peptides and steroids are the main focus of interest. Each type of hormone has different physiological properties. When field researchers plan to measure hormone levels in excreted substances such as urine and feces, they should consider the pathway of hormone excretion. Steroids are mainly conjugated as glucuronides or sulfates in the liver, and then excreted via urine and feces. The excreted substance contains both unconjugated (free) and conjugated steroids; the ratio of these depends on the study species (Whitten *et al.*, 1998; Ziegler and Wittwer, 2005). On the other hand, peptides are excreted from the bloodstream to urine but cannot be detected in feces.

Sex steroids such as estrogens, progestins, and androgens have been widely investigated in field studies. Estradiol-17β (E_2), a major estrogen steroid, and progesterone (P_4), a major progestin steroid, are produced by the ovaries, and are secreted from females during folliculogenesis (follicular phase) or during corpus luteum formation (luteal phase), respectively. Progesterone is produced by the placenta, as well as the ovaries, and is increased during early pregnancy. Therefore, E_2, P_4, and their metabolites have been used to monitor the female reproductive state in a wide range of primates (Hodges and Heistermann, 2011). Profiles of these steroids provide information about the ovulation time (e.g. chimpanzees: Deschner *et al.*, 2003; Emery Thompson, 2005) and conception (e.g. baboons: Stavisky *et al.*, 1995; Beehner *et al.*, 2006; macaques: Fujita *et al.*, 2001; Möhle *et al.*, 2005). On the other hand, testosterone, a major androgen steroid, is the main sex steroid in males and has been used to assess gonadal activities in relation to variations in seasons, age, and/or dominance rank (e.g. gorilla: Robbins and Czekala, 1997; bonobo: Sannen *et al.*, 2004; chimpanzee: Seraphin *et al.*, 2008).

Mahale Chimpanzees: 50 Years of Research, ed. M. Nakamura, K. Hosaka, N. Itoh and K. Zamma. Published by Cambridge University Press.

Sex steroids have also been used in behavioral studies. While these steroids affect the sex drive in both males and females (Dixson, 1998), they can also influence other social behaviors. Studies in captive primates revealed that E_2 and P_4 affect maternal behaviors (e.g. macaque: Maestripieri, 1998; Bardi *et al.*, 2001; gorilla: Bahr *et al.*, 2001; baboon: Bardi *et al.*, 2004). Testosterone is widely studied in association with aggressive and competitive behavior in males (see later) and also in females (e.g. lemur: von Engelhard *et al.*, 2000; Drea, 2007; baboon: Beehner *et al.*, 2005). Adrenal glucocorticoids, which are responsible for energy regulation through glucose metabolism, are secreted in response to stress and have been studied in behavioral and ecological investigations because of their association with social rank and social interactions (see later).

Although peptide hormones have been less commonly studied in the field, luteinizing hormone (LH) and chorionic gonadotropin (CG) are useful to monitor female reproductive states. A surge in LH occurs immediately before ovulation and is useful to identify the day of ovulation. CG is produced in the placenta, and the presence of CG in urine is indicative of implantation. In my experience, urinary test strips designed for human use to detect LH and CG can also be used in chimpanzees. Although other peptide hormones have been investigated in relation to social behaviors in mammals (Nelson, 2000), this field remains relatively unexplored in primates, especially in the wild. For example, prolactin mediates maternal behavior. Oxytocin is important for the formation of an exclusive bond between mothers and offspring in several species. Vasopressin appears to be important for pair bonding in both males and females.

Urine samples can be collected under field conditions by aspiration using a pipette or absorption onto filter paper from vegetation, or setting a plastic sheet to collect urine drops under the night bed. During the wet season, contamination by rain should be avoided during collection. A total of 0.5–1.0 ml may be sufficient to measure hormone levels, although it depends on the assay system used and the number of hormones to be assayed. Screw-cap vials are preferable to avoid leakage. Urine samples should be frozen and maintained at ≤−20°C. To avoid repeated freeze–thaw cycles, these samples should remain frozen throughout all stages of transport by using dry ice or ice packs. Although electric, solar, and kerosene freezers are commercially available, they are not always available in the field. Other common methods for storage include the use of preservatives such as ethanol (the final solution should be within 30–50%) or sodium azide (0.1%), at room temperature. In most African countries with primate habitat, 90% ethanol can be purchased from local pharmacies.

Fecal samples are easier to collect than urine under field conditions. As soon as an individual defecates, feces are scooped into an appropriately labeled plastic bag, and if possible, kept cool by placing the bag into a cooler bottle with an ice pack until returning to the camp. Because steroid breakdown due to bacterial metabolism can be a potential problem in fecal samples, immediate preservation after collection should be considered. A mass (*ca.* 1.0 g) of feces is sufficient to extract and measure hormones. Where a freezer is available at the camp, fecal samples should be stored at ≤−20°C. An alternative method to freezing is to use alcohol as a preservative. Feces can be stored by placing them in tubes with >80% ethanol at room temperature. Drying is also an effective preservation technique. The sample is packaged in foil, pressed flat, and oven-dried at 55–85°C for 2–3 h (Whitten *et al.*, 1998). Another method of drying is to place the fecal sample into a plastic tube and keep it in an airtight container filled with silica gel at room temperature. In my experience, it takes a few days for the fecal samples to dry, depending on the initial moisture content. After fecal samples are transported to the laboratory, they are completely dried in a vacuum oven at 80°C for 2 h. Field extraction is an alternative method to preserve and transport fecal samples. After extracting and filtering fecal samples using an alcohol-based solution, the extract is loaded onto a C-18 solid phase extraction cartridge (Whitten *et al.*, 1998; Ziegler and Wittwer, 2005). This cartridge requires little storage space and helps prevent leaking.

Because animal urine and feces are potential sources of infection, samples should be handled with care at all stages of collection, storage, transport, and subsequent analysis. Thus, appropriate precautions, such as wearing gloves, should be taken to reduce the risk of infection. Preservation by using alcohol can cause leakage especially during air transportation. Capped tubes should be sealed with tape to prevent them from becoming loose, and tube labels should be covered to prevent marker pen writing from being washed off by any alcohol leaks.

Sampling frequency is determined by the study endpoint and the type of information required. To detect ovarian cycles, i.e. to determine whether there is follicular development or formation of corpus luteum, minimum sample collection of every 2–3 days or twice a week is essential. To determine the exact day of ovulation, daily samples around the time of ovulation would be required. To follow the course of pregnancy, samples once or twice a week should be sufficient. In contrast, to investigate interindividual differences, seasonal change, or variation between populations, information on overall physiological condition requires less frequent sampling. A minimum of six samples per condition should be collected, and 10 samples per condition are required to compensate for intra-individual (sample-to-sample) variation (Hodges and Heistermann, 2011).

When the timing of events is determined by the hormonal profile detectable in the excreted samples, lag times of excretion should be taken into consideration. Peak levels of urinary steroids are usually excreted within 4–8 h, and are completely excreted within 24 h. In contrast, the time lag of steroids appearing in fecal samples is longer, requiring 0.3–3 days, and is more variable between and within species. Large-bodied species generally have longer lag time, which reflects a longer duration of gut passage. The time of collection should also be controlled. Circadian rhythms of secretion are known for some hormones such as cortisol and testosterone (Muller and Wrangham, 2004a, b). These patterns are more obvious for steroids in urine compared with fecal samples. Whenever possible, samples should be collected at the same time each day.

Hormones and their metabolites can be measured using a number of methods. Enzyme immunoassays (EIAs) are widely used because of their high sensitivity and ease of handling: EIAs avoid the problems associated with the use and disposal of radiolabeled hormones, which are used in radioimmunoassays (RIAs). Immunoassays use antibodies specific to each hormone or metabolite. Excreted hormone samples, however, contain numerous metabolites of the parent hormone, and the relative abundance of each metabolite varies between species (Ziegler and Wittwer, 2005). Accurate measurements are often difficult and might be less useful in cases where the antibody detects only metabolites of low abundance. In such cases, group-specific assays, which use antibodies with cross-reactivity to several metabolites of related structure, can be useful. Alternatively, high-performance liquid chromatography (HPLC) can be used to separate and identify the major steroid and steroid metabolites present in extracts. This method also helps to test the accuracy and cross-reactivity of antibodies in the chosen assay system, and to develop a new assay system when information about the main steroid metabolites is unavailable for the target species. Recently, other assay systems, such as liquid chromatography-tandem mass spectrometry (LC-MS/MS) have been developed to measure the levels of excreted steroids in fecal samples in some species including primates (Surbeck *et al.*, 2012; Weltring *et al.*, 2012).

Detailed information about the methods used in field endocrinology is provided by Whitten *et al.* (1998), and by Hodges and Heistermann (2011).

44.3 Monitoring female reproductive cycles

Estrogen and progestin are mainly secreted from ovaries, and data regarding the levels of these steroids enables the presence and timing of ovarian cycles, and the occurrence of pregnancies to be determined. Hormonal data are useful in species in which external signs of ovulation, such as sexual swellings, are not obvious. Even in species where behavioral and physical cues of ovulation are conspicuous, therefore, hormonal data are important because these signals are occasionally dissociated from physiological states (Dixson, 1998). Using hormonal data, the temporal relationships between sexual activity and physiological events can be investigated, enabling us to precisely assess mating strategies. Chimpanzees are sexually active during extended periods beyond ovulation (see Chapter 35), and hormonal profiles are expected to provide a more detailed analysis of mating strategies by discriminating between the fertile and non-fertile phases (Deschner *et al.*, 2003, 2004; Emery Thompson, 2005; Emery Thompson and Wrangham, 2008).

In the field studies of chimpanzees, sexual swelling has been traditionally used to monitor ovarian cycles. At Mahale, Hasegawa and Hiraiwa-Hasegawa (1983) reported that the median cycle length was 31.5 days ($N = 44$) with a range from 27 to 39 days, and the median length of maximum tumescence was 12.5 days with a range from 8 to 19 days. Matsumoto-Oda *et al.* (2007) reported that the mean cycle length was 33.4 (SD: 6.3) days ($N = 108$), and the mean length of maximum tumescence was 12.5 (SD: 4.0) days ($N = 194$). These figures are close to those obtained in eastern chimpanzees at Kibale, Uganda, and Gombe, Tanzania (mean cycle length: 34.7 days, mean length of maximum tumescence: 12.7 days) while they are relatively shorter than in western chimpanzees at Taï, Côte d'Ivoire (mean cycle length: 40.3 days, mean length of maximum tumescence: 10.9 days) (Deschner *et al.*, 2003). Although the mechanisms underlying sexual swelling remain unclear, water balance in swollen tissues seems to shift due to E_2 secretion (Machatschke *et al.*, 2006).

The probability of ovulation in relation to sexual swelling has been investigated using hormonal data from female chimpanzees in the wild. These studies have shown that ovulation occurs within the last 6–7 days of maximal swelling, and most frequently on day-1 to day-3 relative to the last day of maximum tumescence (day 0), although there is variation between females and cycles (Deschner *et al.*, 2004; Emery Thompson, 2005). The size of swelling, measured using video captures, showed that a 15–20% increase in swelling occurred during the period of traditionally defined maximum tumescence, and the periovulatory phase was associated with the largest swelling size. Practically, males are more likely to initiate soliciting sexual behavior and compete with other males when females are most likely to conceive (Emery Thompson and Wrangham, 2008). These results indicate that males can detect the fertility of a female using multiple cues including those based on morphology, olfaction, and behavior.

Figure 44.1 shows the profile of urinary E_1C, an E_2 metabolite, and PdG, a P_4 metabolite, during the ovarian cycle in a female at Mahale. LH surge, which was detected using a urinary test strip, indicated the time of ovulation. E_1C gradually increased from 5 to 6 days before the LH surge, and PdG increased after the LH surge. This profile is typical of an ovarian cycle. In this case, ovulation occurred during the last day of sexual detumescence.

44.4 Ovarian steroids and reproductive constraints

The abundance and fluctuation of food availability in a habitat can lead to energetic stress in females, which may indirectly influence their ovarian function as a consequence of their nutritional states (Bronson, 1989). Modulation of ovarian function is thought to occur along a graded continuum from a fully fecund ovarian cycle, through follicular and luteal suppression, anovulation, and oligomenorrhea, to amenorrhea (Ellison *et al.*, 1989, 1993). Therefore, profiles of E_2 and P_4, which are secreted from ovaries, provide a meaningful way to monitor ovarian function (Ellison *et al.*, 1986; Bentley *et al.*, 1997).

Long-term studies on wild chimpanzee populations across different study sites have revealed that wide variability of female reproductive parameters, such as age at menarche, age at first parturition, and interbirth intervals, exist within the species (Table 44.1; see also Chapter 7). This suggests that the reproductive potential of wild chimpanzees is constrained by environmental factors. Although genetic differences are a possible explanation for such variation, subspecies differences are not found in the literature. For example, the parameters of Taï females (*Pan troglodytes verus*) are closer to those of Mahale and Gombe females (*P. t. schweinfurthii*) than the same subspecies at Bossou, Guinea (Table 44.1).

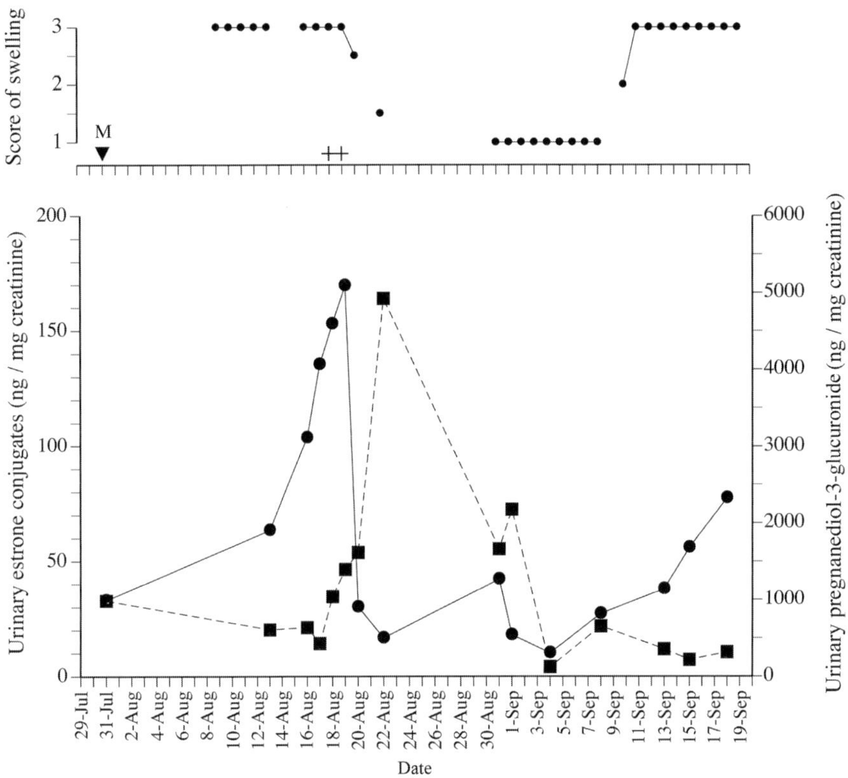

Figure 44.1 Upper: changes in scores denoting the degree of sexual swelling and occurrence of menstruation (M) and LH surge (+) in a female. Swelling was scored from 1 to 3 at intervals of 0.5: (1) no swelling, (2) partial swelling, and (3) maximum swelling. A gap in the scores is due to a lack of information available for that observation day. Menstruation and LH surge were verified by a urinary test strip. It is thought that the second menstruation occurred in early September but this was not tested because of lack of sampling. Below: profile of urinary estrone conjugates (black circle) and pregnanediol-3-glucuronide (black square).

In order to determine how the reproductive potential was affected by different habitat conditions, data were collected between July and October 2001 at Mahale and between December 2001 and March 2002 at Bossou. Reproductive parameters suggest that Mahale females reached sexual maturity later, recovered from postpartum sexual swelling later, and bred more slowly than did Bossou females (Table 44.1). Interbirth intervals consisted of the period of postpartum amenorrhea, waiting time to conception, and gestation length. The period of postpartum amenorrhea was significantly longer in Mahale females than in Bossou females, while waiting time to conception did not significantly differ between the two sites. Assuming that gestation length is similar between individuals compared with the period of postpartum amenorrhea and waiting time to conception, earlier recovery of postpartum swelling leads to a shortened interbirth interval for Bossou females. Comparison of ovarian hormones revealed that fecal estrogen levels detected during the follicular phase were significant lower in Mahale females ($N = 5$) than in Bossou ($N = 4$) (E_2: mean ± SD, 50.2 ± 51.0 ng/g vs. 119.7 ± 90.6 ng/g, $z = -6.35$, $p < 0.0001$ by Mann–Whitney's U-test; E_1C: 13.3 ± 7.4 ng/g vs. 28.2 ±15.1 ng/g, $z = -6.46$, $p < 0.0001$). In addition, fecal progestin levels were also significantly lower during the luteal phase (P_4: 42.6 ± 68.1 ng/g vs. 193.1 ± 205.6 ng/g, $z = -4.64$, $p < 0.0001$; PdG: 578.0 ± 359.2 ng/g vs. 1494.0 ±1355.4 ng/g, $z = -4.09$, $p <$ 0.0001) (Fujita, 2003). Although the nutritional status of females was not directly investigated, these results suggest that the reproductive potentials of Mahale females are more strongly constrained than in Bossou females.

Table 44.1 **Comparisons of reproductive parameters (mean, and range in parentheses) in wild chimpanzees**

Site	Age at first swelling (yrs)	Adolescent subfecundity (yrs)	Age at first parturition (yrs)	Interbirth interval (yrs)	Postpartum amennorhea (yrs)	Waiting time to conception (mo)
Mahale[a]	10.2 (9.5–11.3)	2.5 (1.1–5.4)	13.8 (12.0–16.8)	5.8 (4.3–7.3)	4.6 (2.8–6.2)	8.5 (0.4–25.4)
Gombe[b]	10.8 (8.5–13.5)	2.4 (0.6–4.9)	13.3 (11.1–17.2)	5.2 (3.3–7.8)	3.9 (2.4–5.7)	4.7 (0.5–13.4)
Kibale[c]	11.1	1.6 (0.3–2.5)	15.4 (14–18)	6.2 (2.3–10.0)	4.4 (1.7–6.5)	–
Bossou[d]	7.6 (7.1–8.3)	2.1 (2.0–2.2)	10.6 (9.5–13.0)	5.3 (4–11)	3.9 (2.8–4.9)	13.8 (2.0–25.4)
Taï[e]	–	2.6 (1.4–5.7)	14.3 (12.5–18.5)	5.8 (4.0–10.0)	2.0 (0.2–5.1)	26.9

Chimpanzees at Mahale, Gombe, and Kibale are *Pan troglodytes schweinfurthii* while those at Bossou and Taï are *P. t. verus*.

Adolescent fecundity and postpartum amenorrhea, when these data are not available, were calculated by subtracting the gestation length of 228 days from the period from the first maximal swelling to the first birth and the periods from the resumption of maximal swelling to the subsequent birth, respectively.

[a] Nishida, 2002; Nishida *et al.*, 2003.

[b] Wallis, 1997.

[c] Knott, 2001.

[d] Sugiyama, 2004; Sugiyama and Fujita 2011.

[e] Boesch and Boesch-Achermann, 2000.

Limited data show that the levels of ovarian hormone vary between populations or subgroups in primates. For example, women in subsistence horticulture in the Ituri Forest of Zaïre showed a lower level of salivary P_4 and a lower frequency of ovulation compared with urban women in Boston, USA (Ellison *et al.*, 1989). In chimpanzees (Emery Thompson, 2005) and baboons (Wasser, 1996), comparisons of ovarian steroid levels between conceptive and non-conceptive cycles revealed that conception is associated with higher levels of estrogens during the follicular phase, and/or higher levels of estrogens and progestins during the luteal phase compared with non-conceptive cycles. These results suggest that females with low levels of hormonal activity are not likely to conceive. In addition, in baboons, the levels of P_4 during the luteal phase of the conceptive ovarian cycle were higher during the dry season and among low-ranking females, which indicates that the threshold of P_4 concentration required for successful implantation is higher in harsh conditions where the probability of successful reproduction is low (Wasser, 1996). The profile of ovarian steroid levels can be used as a tool to assess the effect of ecological and social environments on reproductive control, especially in slow reproducing species, such as chimpanzees, for which data on the reproductive rate require a long time to be collected.

44.5 Social dominance and hormone levels

Since female mammals invest more in their offspring compared with males, in terms of gestation, lactation, and infant transport, access to environmental resources such as food should be the primary constraint on their reproductive success. In contrast, male reproductive success is primarily constrained by access to mates. Thus, male gonadal function might be more responsive to social factors than to the nutritional states. Testosterone has been associated with male rank and aggression. The "challenge hypothesis" argues that testosterone does not influence all aspects of aggression but rather specifically regulates mate guarding and other aspects of reproductive aggression (Wingfield *et al.*, 1990). As predicted by this hypothesis, basal levels of testosterone during the breeding season are associated with mating systems across bird species. Monogamous birds maintain high testosterone levels during territory formation and breeding, and react strongly to challenges from conspecifics by increasing testosterone production. Polygynous birds, on the other hand, show high levels of testosterone throughout the breeding season and do not show an endocrine response to reproductive challenge because their testosterone levels are already close to the physiological maximum (Wingfield *et al.*, 1990). Since testosterone has inhibitory effects on parental behavior, monogamous birds decrease testosterone production while providing paternal care, which provides a trade-off with the benefits of mate defense. Even in species that do not exhibit paternal behavior, high levels of testosterone might be costly because of the direct and indirect energy costs and immunosuppression. Thus, males will exhibit high testosterone levels only when it is of sufficient benefit to do so.

The challenge hypothesis predicts that males will increase testosterone during periods of high reproductive competition such as mating seasons or in the presence of estrous females. This hypothesis has been tested for social mammals such as carnivores and primates. A number of studies utilizing primates have generated evidence to support this hypothesis (Anestis, 2010). For example, in primates that breed seasonally, a significant increase in testosterone levels is observed during the copulatory period and is associated with aggression between males (e.g. lemur: Cavigelli and Pereira, 2000; capuchin: Lynch *et al.*, 2002; macaque: Barrett *et al.*, 2002). In contrast, in species in which intense sperm competition rather than contest aggression over access to mates is important, significant changes in testosterone levels are not observed at the onset of the breeding season (muriquis: Strier *et al.*, 1999).

The challenge hypothesis has been tested in the Kanyawara and Ngogo chimpanzee groups at Kibale. Chimpanzee males generally mate with females restrictively when they show maximum sexual swelling, and most of the copulations occur within the group and on an opportunistic basis. During the periovulatory period, i.e. when the females are more fertile, however, male–male competition is more intense and high-ranking males can often access females exclusively (see Chapter 35). Thus, the challenge hypothesis predicts that male testosterone levels increase in the presence of fertile females, i.e. when reproductive aggression is more intense. Data obtained at Kanyawara and Ngogo were consistent with this prediction. Testosterone levels in male chimpanzees increased only in the presence of parous (attractive) swelling females but did not respond in the presence of nulliparous swelling females (Muller and Wrangham, 2004a). In addition, when one popular female came into estrus after her infant son died, both testosterone levels and rates of aggression significantly increased (Muller and Wrangham, 2001). In relation to the dominance rank, high-ranking males were more aggressive than low-ranking males (Muller and Wrangham, 2004a), and produced higher levels of testosterone (Muehlenbein *et al.*, 2004; Muller and Wrangham, 2004a). Sapolsky (1983) found that an increased testosterone level was associated with increased dominance status only during periods of instability in the male dominance hierarchy in a group of wild olive baboons. In male chimpanzees, however, testosterone levels remained positively associated with dominance rank even when the dominance hierarchy was stable (Muller and Wrangham, 2004a). This discrepancy may be accounted for by the fission–fusion nature of chimpanzee society. In such society, dominant males are consistently at risk from coalitions formed by subordinate males, and thus, the elevated levels of testosterone may be in response to the difficulty of predicting challenges from rivals (Muller and Wrangham, 2004a).

Stress response in relation to social dominance has also been assessed across a wide range of species in birds and mammals, including primates (Creel, 2001; Abbott *et al.*, 2003). Glucocorticoids are widely studied to assess stress response. The most conspicuous physiological effects of glucocorticoids involve regulation of metabolism by increasing the availability of glucose in the bloodstream. Although acute rises in cortisol represent an essential, adaptive response to short-term stressors, chronically high levels of glucocorticoid secretion can have adverse effects including gastric ulcers, muscle wasting, and immunosuppression. Stressors are physiological, e.g. energy expenditure, and psychological, e.g. social interaction. Early studies in captive animals argued that chronic stress is imposed on subordinates, and that their reproduction is suppressed by glucocorticoids. Conversely, recent studies in the wild revealed that dominant individuals have elevated glucocorticoids more often than do subordinates because of the cost of dominance (Creel, 2001). Among primate species there is pronounced variation in the relationship between social status and glucocorticoids (Abbott *et al.*, 2003). In addition, in some species, a previous negative correlation between cortisol and rank can disappear when the dominance hierarchy is unstable. Sapolsky (1992) explained that in a stable hierarchy, high-ranking males feel secure with predictability and control, while in an unstable hierarchy, their position might be threatened, leading to a heightened stress response.

The relationship between cortisol, the primary glucocorticoid in primates, and male dominance rank in wild chimpanzees was investigated in the Kanyawara group (Muller and Wrangham, 2004b). Male dominance rank was found to be positively correlated with cortisol levels, although the dominance hierarchy was stable throughout the study period. Furthermore, cortisol levels were positively correlated with the rates of male aggression. These results suggest that high-ranking males experience higher levels of metabolic stress due to aggression rather than psychological ones such as the unpredictability of social hierarchy.

44.6 Application of field endocrinology to conservation

Field endocrinology can be used to address many issues in conservation biology. Human disturbance in wild animals, particularly by tourism, can lead the animals to experience a stress response, and monitoring the glucocorticoid levels can be used to assess the stress levels of the animals (Walker *et al.*, 2005; Busch and Hayward, 2009). There have been a few studies showing the human impact of tourism on free-living primates. In Barbary macaques, male glucocorticoid levels were positively related to the rate of aggressive behavior of tourists toward the monkeys (Maréchal *et al.*, 2011). The habituation process could also cause the target animals to experience stress. In western lowland gorillas, the group undergoing a habituation program had the highest glucocorticoid levels (Shutt *et al.*, 2014). In orangutans, fecal corticoid metabolite levels were lower in the habituated females than their unhabituated counterparts (Muehlenbein *et al.*, 2012). This knowledge of human impacts in wild animals is required for their conservation, and monitoring their stress levels should help to establish an appropriate habituation program during tourism and at the start of a research program in a new field site.

44.7 Conclusion

Inter- and intra-species variation in reproductive characteristics is well recognized but the reason for such variation remains unclear, and is a question that should be addressed in field endocrinology. In addition, individual variation in endocrine physiology is one important factor influencing variation of behavior among animals, and hormonal data can provide the explanation for this variation. Field endocrinology enables these issues to be addressed.

References

Abbott, D. H., Keverne, E. B., Bercovitch, F. B., *et al.* (2003). Are subordinates always stressed? A comparative analysis of rank differences in cortisol levels among primates. *Hormones and Behavior*, **43**, 67–82.

Anestis, S. F. (2010). Hormones and social behavior in primates. *Evolutionary Anthropology*, **19**, 66–78.

Bahr, N. I., Martin, R. D., and Pryce, C. R. (2001). Peripartum sex steroid profiles and endocrine correlates of postpartum maternal behavior in captive gorillas (*Gorilla gorilla gorilla*). *Hormones and Behavior*, **40**, 533–41.

Bardi, M., Shimizu, K., Fujita, S., Borgognini-Tari, S., and Huffman, M. A. (2001). Hormonal correlates of maternal style in captive macaques (*Macaca fuscata, M. mulatta*). *International Journal of Primatology*, **22**, 647–62.

Bardi, M., French, J. A., Ramirez, S. M., and Brent, L. (2004). The role of the endocrine system in baboon maternal behavior. *Biological Psychiatry*, **55**, 724–32.

Barrett, G. M., Shimizu, K., Bardi, M., Asaba, S., and Mori, A. (2002). Endocrine correlates of rank, reproduction, and female-directed aggression in male Japanese macaques (*Macaca fuscata*). *Hormones and Behavior*, **42**, 85–96.

Beehner, J. C., Phillips-Conroy, J. E., and Whitten, P. L. (2005). Female testosterone, dominance rank, and aggression in an Ethiopian population of hybrid baboons. *American Journal of Primatology*, **67**, 101–19.

Beehner, J. C., Nguyen, N., Wango, E. O., Alberts, S. C., and Altmann, J. (2006). The endocrinology of pregnancy and fetal loss in wild baboons. *Hormones and Behavior*, **49**, 688–99.

Bentley, G. R., Vitzthum, V. J., Caceres, E., *et al.* (1997). Reproduction and ecology in Provincia Aroma, Bolivia: fecundity of women with low levels of salivary progesterone. *American Journal of Physical Anthropology*, **Supplement 26**, 110.

Boesch, C. and Boesch-Achermann, H. (2000). *The Chimpanzees of the Taï Forest*. Oxford: Oxford University Press.

Bronson, F. H. (1989). *Mammalian Reproductive Biology*. Chicago: The University of Chicago Press.

Busch, D. S. and Hayward, L. S. (2009). Stress in a conservation context: a discussion of glucocorticoid actions and how levels change with conservation-relevant variables. *Biological Conservation*, **142**, 2844–53.

Cavigelli, S. and Pereira, M. (2000). Mating season aggression and fecal testosterone levels in male ring-tailed lemurs (*Lemur catta*). *Hormones and Behavior*, **37**, 246–55.

Creel, S. (2001). Social dominance and stress hormones. *Trends in Ecology & Evolution*, **16**, 491–7.

Deschner, T., Heistermann, M., Hodges, K., and Boesh, C. (2003). Timing and probability of ovulation in relation to sex skin swelling in wild West African chimpanzees, *Pan troglodytes verus. Animal Behaviour*, **66**, 551–60.

Deschner, T., Heistermann, M., Hodges, K., and Boesch, C. (2004). Female sexual swelling size, timing of ovulation, and male behavior in wild West African chimpanzees. *Hormones and Behavior*, **46**, 204–15.

Dixson, A. F. (1998). *Primate Sexuality: Comparative Studies of the Prosimians, Monkeys, Apes, and Human Beings*. Oxford: Oxford University Press.

Drea, C. (2007). Sex and seasonal differences in aggression and steroid secretion in *Lemur catta*: are socially dominant females hormonally 'masculinized'? *Hormones and Behavior*, **51**, 555–67.

Ellison, P. T., Peacock, N. R., and Lager, C. (1986). Salivary progesterone and luteal function in two low-fertility populations of northeast Zaire. *Human Biology*, **58**, 487–3.

Ellison, P. T., Peacock, N. R., and Lager, C. (1989). Ecology and ovarian function among lese women of the Ituri Forest, Zaire. *American Journal of Physical Anthropology*, **78**, 519–26.

Ellison, P. T., Panter-Brick, C., Lipson, S. F., and O'Rourke, M. T. (1993). The ecological context of human ovarian function. *Human Reproduction*, **8**, 2248–58.

Emery Thompson, M. (2005). Reproductive endocrinology of wild female chimpanzees (*Pan troglodytes schweinfurthii*): methodological considerations and the role of hormones in sex and conception. *American Journal of Primatology*, **67**, 137–58.

Emery Thompson, M. and Wrangham, R. W. (2008). Male mating interest varies with female fecundity in *Pan troglodytes schweinfurthii* of Kanyawara, Kibale National Park. *International Journal of Primatology*, **29**, 885–905.

von Engelhard, N., Kappeler, P. M., and Heistermann, M. (2000). Androgen levels and female social dominance in *Lemur catta*. *Proceedings of the Royal Society of London. Series B: Biological Sciences*, **267**, 1533–9.

Fujita, S. (2003). [*Reproductive Biology in Wild Female Primates: Variability in Hormonal Profiles, Behavior and Reproductive Parameters*]. Doctoral Dissertation. Kyoto: Kyoto University. In Japanese.

Fujita, S., Mitsunaga, F., Sugiura, H., and Shimizu, K. (2001). Measurement of urinary and fecal steroid metabolites during the ovarian cycle in captive and wild Japanese macaques, *Macaca fuscata*. *American Journal of Primatology*, **53**, 167–76.

Hasegawa, T. and Hiraiwa-Hasegawa, M. (1983). Opportunistic and restrictive matings among wild chimpanzees in the Mahale Mountains, Tanzania. *Journal of Ethology*, **1**, 75–85.

Hodges, J. K. and Heistermann, M. (2011). Field endocrinology: monitoring hormonal changes in free-ranging primates. In *Field and Laboratory Methods in Primatology: A Practical Guide*, 2nd edn., ed. J. M. Setchell and D. J. Curtis. Cambridge: Cambridge University Press. pp. 353–70.

Knott, C. (2001). Female reproductive ecology of the apes: implication for human evolution. In *Reproductive Ecology and Human Evolution*, ed. P. T. Ellison. New York: Aldine de Gruyter, pp. 429–1463.

Lynch, J. W., Ziegler, T. E., and Strier, K. B. (2002). Individual and seasonal variation in fecal testosterone and cortisol levels of wild male tufted capuchin monkeys, *Cebus apella nigritus*. *Hormones and Behavior*, **41**, 275–87.

Machatschke, I. H., Dittami, J., and Wallner, B. (2006). Morphometric and hormonal changes during the chimpanzee menstrual cycle. *Journal of Medical Primatology*, **35**, 331–40.

Maestripieri, D. (1998). Parenting styles of abusive mothers in group-living rhesus macaques. *Animal Behaviour*, **55**, 1–11.

Maréchal, L., Semple, S., Majolo, B., *et al*. (2011). Impacts of tourism on anxiety and physiological stress levels in wild male Barbary macaques. *Biological Conservation*, **144**, 2188–93.

Matsumoto-Oda, A., Hamai, M., Hayaki, H., *et al*. (2007). Estrus cycle asynchrony in wild female chimpanzees, *Pan troglodytes schweinfurthii*. *Behavioral Ecology and Sociobiology*, **61**, 661–8.

Möhle, U., Heistermann, M., Dittami, J., *et al*. (2005). Patterns of anogenital swelling size and their endocrine correlates during ovulatory cycles and early pregnancy in free-ranging Barbary macaques (*Macaca sylvanus*) of Gibraltar. *American Journal of Primatology*, **66**, 351–68.

Muehlenbein, M. P., Watts, D. P., and Whitten, P. L. (2004). Dominance rank and fecal testosterone levels in adult male chimpanzees (*Pan troglodytes schweinfurthii*) at Ngogo, Kibale National Park, Uganda. *American Journal of Primatology*, **64**, 71–82.

Muehlenbein, M. P., Ancrenaz, M., Sakong, R., *et al*. (2012). Ape conservation physiology: fecal glucocorticoid responses in wild *Pongo pygmaeus morio* following human visitation. *PLoS ONE*, **7**, e33357.

Muller, M. N. and Wrangham, R. W. (2001). The reproductive ecology of male hominoids. In *Reproductive Ecology and Human Evolution*, ed. P. T. Ellison. New York: Aldine de Gruyter, pp. 397–427.

Muller, M. N. and Wrangham, R. W. (2004a). Dominance, aggression and testosterone in wild chimpanzees: a test of the 'challenge hypothesis'. *Animal Behaviour*, **67**, 113–23.

Muller, M. N. and Wrangham, R. W. (2004b). Dominance, cortisol and stress in wild chimpanzees (*Pan troglodytes schweinfurthii*). *Behavioral Ecology and Sociobiology*, **55**, 332–40.

Nelson, R. J. (2000). *An Introduction to Behavioral Endocrinology*. Sunderland, MA: Sinauer Associates.

Nishida, T. (2002). [Demography.] In [*The Mahale Chimpanzees: 37 Years of <Panthropology>*], ed. T. Nishida, S. Uehara, and K. Kawanaka. Kyoto: Kyoto University Press, pp. 171–202. In Japanese.

Nishida, T., Corp, N., Hamai, M., *et al*. (2003). Demography, female life history, and reproductive profiles among the chimpanzees of Mahale. *American Journal of Primatology*, **59**, 99–121.

Robbins, M. M. and Czekala, N. M. (1997). A preliminary investigation of urinary testosterone and cortisol levels in wild male mountain gorillas. *American Journal of Primatology*, **43**, 51–64.

Sannen, A., Van Elsacker, L., Heistermann, M., and Eens, M. (2004). Urinary testosterone-metabolite levels and dominance rank in male and female bonobos (*Pan paniscus*). *Primates*, **45**, 89–96.

Sapolsky, R. M. (1983). Endocrine aspects of social instability in the olive baboon (*Papio anubis*). *American Journal of Primatology*, **5**, 365–79.

Sapolsky, R. M. (1992). Cortisol concentrations and the social significance of rank instability among wild baboons. *Psychoneuroendocrinology*, **17**, 701–9.

Seraphin, S. B., Whitten, P. L., and Reynolds, V. (2008). The influence of age on fecal steroid hormone levels in male Budongo Forest chimpanzees (*Pan troglodytes schweinfurthii*). *American Journal of Primatology*, **70**, 661–9.

Shutt, K., Heistermann, M., Kasim, A., *et al.* (2014). Effects of habituation, research and ecotourism on faecal glucocorticoid metabolites in wild western lowland gorillas: implications for conservation management. *Biological Conservation*, **172**, 72–9.

Stavisky, R., Russell, E., Stallings, J., *et al.* (1995). Fecal steroid analysis of ovarian cycles in free-ranging baboons. *American Journal of Primatology*, **36**, 285–97.

Strier, K. B. and Ziegler, T. E. (2005). Advances in field-based studies of primate behavioral endocrinology. *American Journal of Primatology*, **67**, 1–4.

Strier, K. B., Ziegler, T. E., and Wittwer, D. J. (1999). Seasonal and social correlates of fecal testosterone and cortisol levels in wild male muriquis (*Brachyteles arachnoides*). *Hormones and Behavior*, **35**, 125–34.

Sugiyama, Y. (2004). Demographic parameters and life history of chimpanzees at Bossou, Guinea. *American Journal of Physical Anthropology*, **124**, 154–65.

Sugiyama, Y. and Fujita, S. (2011). The demography and reproductive parameters of Bossou females. In *The Chimpanzees of Bossou and Nimba*, ed. T. Matsuzawa, T. Humle, and Y. Sugiyama. Tokyo: Springer, pp. 23–34.

Surbeck, M., Deschner, T., Schubert, G., Weltring, A., and Hohmann, G. (2012). Mate competition, testosterone and intersexual relationships in bonobos, *Pan paniscus. Animal Behaviour*, **83**, 659–69.

Walker, B. G., Boersma, P. D., and Wingfield, J. C. (2005). Field endocrinology and conservation biology. *Integrative and Comparative Biology*, **45**, 12–18.

Wallis, J. (1997). A survey of reproductive parameters in the free-ranging chimpanzees of Gombe National Park. *Journal of Reproduction and Fertility*, **109**, 297–307.

Wasser, S. K. (1996). Reproductive control in wild baboons measured by fecal steroids. *Biology of Reproduction*, **55**, 393–9.

Weltring, A., Schaebs, F. S., Perry, S. E., and Deschner, T. (2012). Simultaneous measurement of endogenous steroid hormones and their metabolites with LC–MS/MS in faeces of a New World primate species, *Cebus capucinus. Physiology and Behavior*, **105**, 510–21.

Whitten, P. L., Brockman, D. K., and Stavisky, R. C. (1998). Recent advances in noninvasive techniques to monitor hormone-behavior interactions. *Yearbook of Physical Anthropology*, **41**, 1–23.

Wingfield, J., Hegner, R., Dufty, A. J., and Ball, G. F. (1990). The "challenge hypothesis": theoretical implications for patterns of testosterone secretion, mating systems, and breeding strategies. *American Naturalist*, **136**, 829–46.

Ziegler, T. E. and Wittwer, D. J. (2005). Fecal steroid research in the field and laboratory: improved methods for storage, transport, processing, and analysis. *American Journal of Primatology*, **67**, 159–74.

45 Skeletal and dental morphology

DAISUKE SHIMIZU

45.1 Introduction

Ever since Toshisada Nishida started studying chimpanzees in Mahale in 1965 (Chapter 3), several articles have been published on the skeletal remains of chimpanzees collected from here (see below for details). The skeletal remains of an infant female (Amina), adult females (Pulin, Betty, and Wansombo), and an adult male (Musa) were taken to Japan with the permission of the Government of Tanzania, and stored in the Japan Monkey Centre, where Nishida was based. The other skeletal remains were stored at the Kansyana Camp (see Chapter 2) in Mahale.

Long-term research in the field means that there will be every possibility of identifying such skeletal remains. This is a great advantage in the treatment of the skeletal remains of wild animals, because of the information obtained from their age, even though sometimes the ages have to be estimated (see Chapter 7). Age is very important for conducting growth or aging study of the animals. However, wild-shot specimens stored in museums or institutes rarely have any age-related information, only rough age-class estimations from tooth eruptions, skeletal growths, or impressions from the corpse. Therefore, captive animals in zoos or institutes have been used for this purpose (e.g. Gavan, 1971; Hamada *et al.*, 1996; Kuykendall, 1996). Captive animals, however, grow more rapidly and age more slowly than their wild counterparts (Tutin, 1996; Boesch and Boesch-Achermann, 2000; Zihlman *et al.*, 2004). Conditions are usually better in captivity than in the wild, such as sufficient food, less need for exercise, no predation threat, and treatment of disease or injury by veterinarians. Psychologically, however, they may be highly stressed due to living in a much smaller space, people observing them, and having to take various tests.

In Mahale, several morphological studies in the field have been conducted that have measured body weight, positional behavior, and the size of male testes. The results of such studies can be integrated with the information obtained from the skeletons of the animals, which results in a better understanding of the condition of skeletal materials. Therefore, such studies will be briefly reviewed, in addition to skeletal studies.

45.2 Descriptions of reported skeletons

Here, I will summarize the information on seven individual skeletons, the details of which have been published. Although some had already been scavenged and were not complete, fortunately most of them were found intact.

Amina, a female infant, was born in March between 6 and 28, 1980, died in December between 1 and 6, 1981, and was 18–19 months old. The cause of death was not known, although she behaved normally when she was last observed. The body was found intact, and the soft tissues had decomposed in the field. Although almost every part of the skeleton was collected after the maceration of the body, some parts were missing, i.e. the right upper first deciduous incisor, some distal phalanges, and some epiphyses (Yasui and Takahata, 1983).

Betty, a young adult female, joined the M group in February 1986, at an estimated age of 13 years. She gave birth to three offspring, but two did not survive for long. Her body was found on October 15, 1993, so her age at

Mahale Chimpanzees: 50 Years of Research, ed. M. Nakamura, K. Hosaka, N. Itoh and K. Zamma. Published by Cambridge University Press.

death was estimated to be 20 years. The body was already putrefied when found; therefore, it was not possible to identify her. She was identified as Betty because she was the only missing individual at the time. Because of putrefaction and animals scavenging on the corpse, several skeletal parts of Betty were missing, i.e. the mandible, some cervical and thoracic vertebrae, and most of the ribs (Shimizu *et al.*, 2002).

Pulin, another young adult female, joined the M group in February 1982, at an estimated age of 12 years. She was 23 years old and primiparous when her body was found on October 10, 1993. The body had also been scavenged; however, the skeleton was almost intact, except for the fact that the left pterygoid hamulus was broken (Shimizu *et al.*, 2002). An epidemic viral infection occurred in the M group during late September to early October 1993 (see Chapter 25). The symptoms included coughing, bronchial congestion, and weight loss. The infection was suspected to have killed 11 chimpanzees, including Betty and Pulin, during this period.

Wansombo, an old female, had already been a member of the M group in 1968 when her age was estimated to be 16 years. She gave birth to five offspring, and weaned two males and two females. Her last birth was in July 1988 at the age of *c.* 39, therefore she had already ceased sexual cycling when she was observed dying on January 20, 1994. She may have died of old age, because she was then *c.* 42 years old. All her skeletal parts were collected without any damage to them (Shimizu *et al.*, 2002).

Musa, an old male, had also already been a member of the M group in 1968, and was estimated to be 11 years old. His body was found on December 25, 1993 at an estimated age of 36, and had been partly scavenged; some skeletal parts were broken. According to Michael A. Huffman's investigation (see Chapter 24), Musa was invasively infected with parasitic nodular worms. This infection had caused bleeding through, and the blockage of, his urogenital tract during his last few days, which may have been the proximate cause of death (Shimizu *et al.*, 2002).

Another body of a chimpanzee was found on January 26, 1996. Since the body was putrefied, age class and sex were not identifiable at the time. The skeleton was almost intact, was later identified as a female by the size of the canines, and was estimated to be in her mid-thirties by comparison with the above-mentioned three adult females' dental wear pattern data (Shimizu *et al.*, 2002). Her right humerus and the fourth proximal phalanx of her left hand were much shorter than those on the opposite side, which was caused by arthritis during growth. The nasal notch of the right maxilla was deformed by myositis ossificans traumatica, caused by trauma. The body was identified as the adult female Gwakakumo by comparing these osteological characteristics and her live physical characteristics (Nakai *et al.*, 2004). Gwakakumo joined the M group in 1973, at an estimated age of 12 years, and died at the age of 35 years.

Another unidentified body was found on November 27, 1996, without any apparent damage caused by scavengers. It was judged to be a female based on canine size and the genital organs. Because the skin color was much paler than that of typical adult females, she was assumed to be adolescent at the time. However, because all of the sutures of her long bones were fused, and the degree of dental wear and the fusion of her cranial sutures, her age was later estimated to be approximately 20 years. Although there was no useful osteopathological evidence for identification, only a female called Amisa was missing from the M group at the time the body was found. Therefore, the body was identified as Amisa (Nakai *et al.*, 2004), and her age was estimated to be 19 years.

There still are skeletal remains at the Kansyana Camp that are yet to be reported. Nakai *et al.* (2004) mentioned four other individuals, apart from the two individuals reported by them, which were three identified adults (Ntologi, a male that was *c.* 40 years old; Wakilufya, a female that was *c.* 37 years old; and Jilba, a male that was 21 years old) and one unidentified infant. Five more bodies were later found, skeletonized, and stored at the camp. Three of them were identified when they were found, namely Dogura (a 23-year-old male found in 2004), Rhea (a 2-year-old female found in 2006; Hanamura *et al.*, 2008), and Pimu (a 23-year-old male found in 2011; Chapter 26). The other two, an adult female and a juvenile of unknown gender, were not identifiable with any confidence (Hanamura *et al.*, 2008).

The body of Dogura was found on February 10, 2004. His body had already been putrefied and scavenged, presumably by a bushpig. When he was observed alive at the end of January, he had apparently lost weight and was unable to move from a fig tree. Therefore, he was assumed to have died from a disease that caused emaciation (Chapter 25).

Rhea died of a flu-like epizootic disease that spread through the M group for 2 months, starting in June 2006 (Chapter 25). Twenty-three individuals (out of the 65 members of the group) caught the disease, and at least three (possibly 12) were killed. Rhea had exhibited typical symptoms of the disease (coughing, sniveling, and clear nasal discharge) for several days before her body was found. Two unidentified individuals were found dead at almost the same time as Rhea. One was deemed to be an adult female, based on the remaining part of the inguinal region and the size of the canines. When she was found, two females were missing (26-year-old Miya and 35-year-old Opal). Therefore, further morphological investigations should tell us exactly who she was. Whoever she was, she probably died of the same disease as Rhea.

Part of a neurocranium and a left scapula from the other unidentified individual were found on July 3, 2006; they had been completely skeletonized (Hanamura *et al.*, 2008). Hanamura *et al.* (2008) assumed that the bones belonged to a juvenile, because of the small size of the scapula. Oscar was the only missing juvenile when the bones were found. He, however, was observed alive on June 5, 2006, and it is not certain that all the soft tissues had completely decomposed within a month, so it is possible that the animal died much earlier. The bones need to be morphologically investigated in order to be identified.

Pimu was a victim of within-group coalitional violence (Chapter 26). His body was collected and his external condition was examined. He had suffered several severe wounds, and quite a few lacerations, and may have died from traumatic shock. After an examination of its external condition, the body was decomposed and skeletonized. The ilium, and a rib, had minor cracks.

45.3 Skeletal and dental development and growth

Developmental studies of chimpanzee skeletons or dentitions have often been carried out using captive individuals of known age. However, captive conditions are completely different from those in the wild, which can affect individual development, such as the timing of teeth eruption or sexual maturity. There have only been a few studies conducted on skeletal development and growth in wild chimpanzees of known age (Yasui and Takahata, 1983; Matsuzawa *et al.*, 1990; Zihlman *et al.*, 1990, 2007). Zihlman *et al.* (2007) reported the timing of dental eruptions in 22 immature wild chimpanzees of known chronological age from Taï, Côte d'Ivoire, and showed that captive chimpanzees may mature 3 years earlier than their wild counterparts. Differences in skeletal and dental development and growth between subspecies are yet to be elucidated, because the number of wild specimens of known age is still low.

The skeletons of three immature individuals have been found to date in Mahale. Yasui and Takahata's (1983) report of Amina was the first detailed record of dental development and skeletal growth in a wild chimpanzee infant of known age. They reported the condition of sutures, epiphyses, and dental eruptions, and took basic measurements of the skeletons. Almost all the skull sutures were unfused. The internasal suture and the maxilla-premaxillary sutures had fused. The ventral arch of the atlas and the dens of the axis had not yet fused with the other parts. The other cervical vertebrae had already fused, but only the dorsal parts. Although the vertebral bodies and arches of all the thoracic vertebrae had not yet fused, they had fused to each other at all the lumbar vertebrae. Only the ischiopubic ramus suture was fused in the pelvis. All of Amina's deciduous teeth, except for the lower canines, had fully erupted. The crowns of the third and fourth premolar had started to calcify, and the crown formation of the first molar was almost complete in the mandible. Amina's metric and non-metric characteristics fitted well into the infant

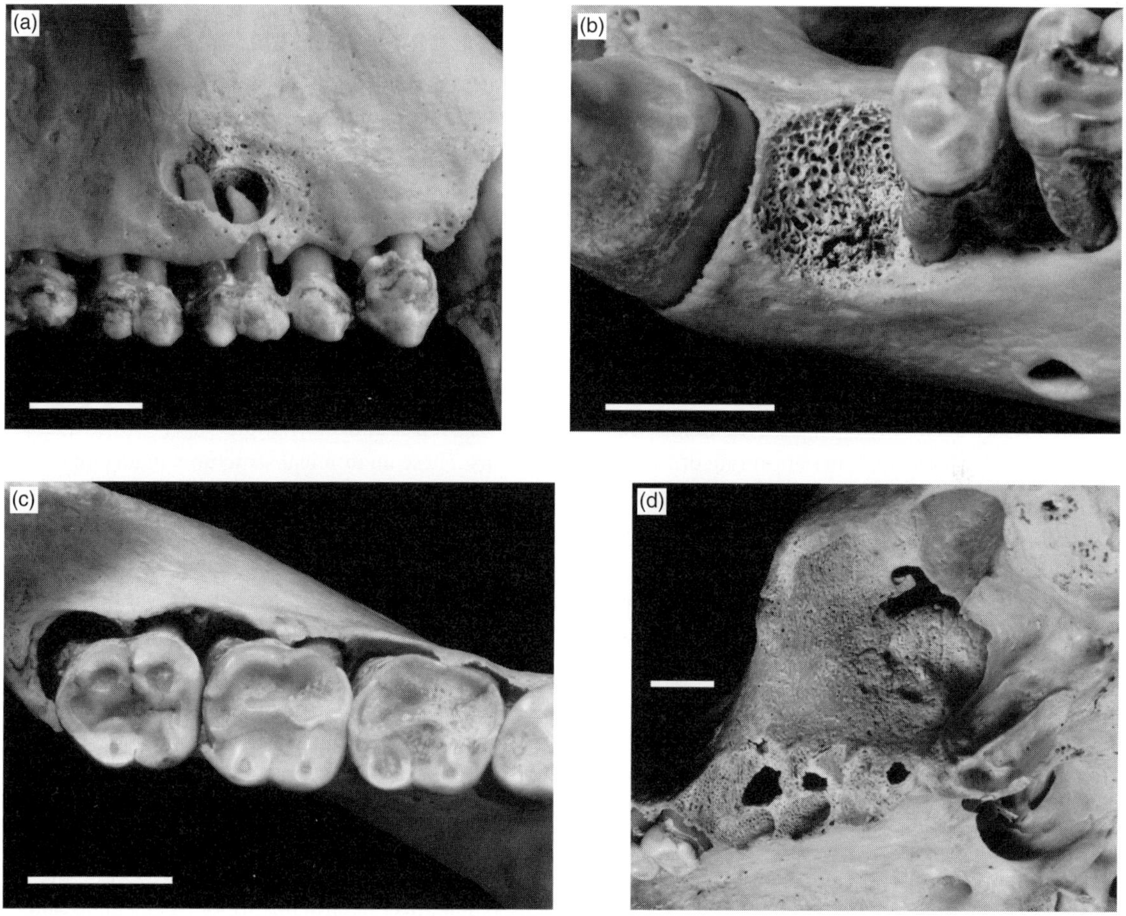

Figure 45.1 Pathological traits of Mahale chimpanzees. ©D. Shimizu. **(a)** Lateral view of Musa's maxilla, showing an abscess around the root of the first molar. **(b)** Oblique view of Musa's mandible, showing a severe loss of alveolar bone around the third premolar to the first molar. **(c)** Occlusal view of Wansombo's mandible, showing periodontal pockets on the left side alveolar ridge from the first to the third molar. **(d)** Oblique view of Wansombo's maxilla, showing slight porotic hyperostosis around the temporal fossa.

category of wild chimpanzees (e.g. Fenart and Deblock, 1973; Zihlman *et al.*, 2007).

45.4 Pathology and aging

Occasionally, the skeletal remains of wild animals exhibit pathological traits (Figure 45.1). Dental problems are the most common pathological traits that remain on skeletons in chimpanzees (Kilgore, 1989). Shimizu *et al.* (2002) found that all four Mahale individuals that they investigated had some kind of dental problem. Musa's upper first molars and left canine were broken before his death. Consequently, the pulp cavities of these teeth were exposed and had decayed. All four individuals also had periodontal disease. Two old individuals (Wansombo and Musa) exhibited significant alveolar reductions, so much so that Musa had lost his lower left third premolar while he was alive, and the socket of the premolar was partially resorbed. Musa, Wansombo, and Pulin also had abscess formations. This high frequency of periodontal disease and abscess formation is commonly observed among wild eastern chimpanzees in Gombe, Tanzania (Kilgore, 1989), Kibale, Uganda (Carter *et al.*, 2008), and Kalinzu, Uganda (Shimizu *et al.*, in prep.). According to Shimizu *et al.* (in prep.), approximately 60% of museum specimens of *Pan troglodytes schweinfurthii* had periodontal disease.

Periodontal disease is a progressive and age-related disease, and older individuals tend to suffer most from it. The disease is caused by dental plaque, which is composed of bacteria (70%) and its extracellular polysaccharide (30%). This is mainly transferred by direct or indirect contact between mouths. During food sharing (Chapter 20), bacteria may stick to food that had already been in contact with the mouth of another individual. The high infection rate of this disease that is observed among wild chimpanzees could be exacerbated by their social system, which involves food sharing between many individuals (Shimizu *et al.*, in prep.). Once an individual contracts this disease, the infection easily spreads throughout the group by direct or indirect mouth contact (e.g. food sharing, grooming, kissing, and touching the mouth). Furthermore, the disease may spread to a new group via female transfer. An infinite reiteration of this would result in the disease spreading to the entire species.

The morphology and mechanical properties of long bones do not differ between wild and captive primates (e.g. Burr *et al.*, 1989; Morimoto *et al.*, 2011). The bone mineral density of Gombe and Mahale adult chimpanzees is also within the normal range seen in captive chimpanzees (Sumner *et al.*, 1989; Gunji *et al.*, 2003), yet an old Mahale female, Wansombo, had a much lower bone mineral density than the average for adult female chimpanzees, and she was categorized as having osteoporosis (Gunji *et al.*, 2003). This may have been caused by a combination of factors (pregnancy, lactation, menopause, and physical and physiological aging). Wansombo's bones were, however, still free from any fractures. In Gombe, fractures in two old female chimpanzees were reported (Jurmain, 1989). Although one female (43 years old) had several fractures, all of them happened when she was young, and they had completely healed. The other female (over 40 years old) also had several fractures. One of them exhibited visible callus formation, indicating healing was still active when she died. Therefore, it seems that a low bone mineral density might not always lead to bone fractures in old wild chimpanzees. In humans, bone fractures caused by a decreased bone mineral density often occur. A larger force in the craniocaudal direction is concentrated on human vertebral bodies than on those of quadrupedal animals (Shi *et al.*, 1995). Chimpanzees are knuckle walkers (quadrupedal). Chimpanzee forelimbs can support their body weight, which reduces the effects of body weight on the femoral neck or vertebrae. Chimpanzees also have fewer and shorter lumbar vertebrae than humans. These anatomical structures reduce the momentum of upper body rotation (Shi *et al.*, 1995), and in chimpanzees, these may also reduce stress on the lumbar vertebrae. A study of age-related behavior in captive chimpanzees indicated that old individuals moved less than young chimpanzees (Baker, 2000). Huffman (1990) reported that in old chimpanzees in Mahale there was a gradual decrease in activity, particularly in traveling and grooming (see also Chapter 23). Such decreases in activity might result in a low fracture frequency during old age. Further studies are needed to confirm this.

Although degenerative joint disease is very rare in wild chimpanzees (Jurmain, 2000), Wansombo and Musa exhibited the symptoms of the disease (Gunji *et al.*, 1998; Shimizu *et al.*, 2002) (Figure 45.2). Slight hyperostosis was observed over the whole of Wansombo's cranium, and evidence of arthritis was found in both temporomandibular joints. Osteophytosis was subsequently observed on both of the mandibular fossae, and both of the mandibular heads had resorbed. Osteophytes were observed on the anterior medial side of both occipital condyles. The osteophyte on the right side was well developed, and a joint facet had formed with the dens of the axis. Wansombo also had osteoarthritis of the knee joints, and spondylosis deformans in the cervical vertebrae. Musa also had spondylosis deformans in the cervical vertebrae. Furthermore, Musa exhibited ossification of the anterior longitudinal ligament in the lumbar vertebrae, and some pitting and osteophytes around the joint surfaces.

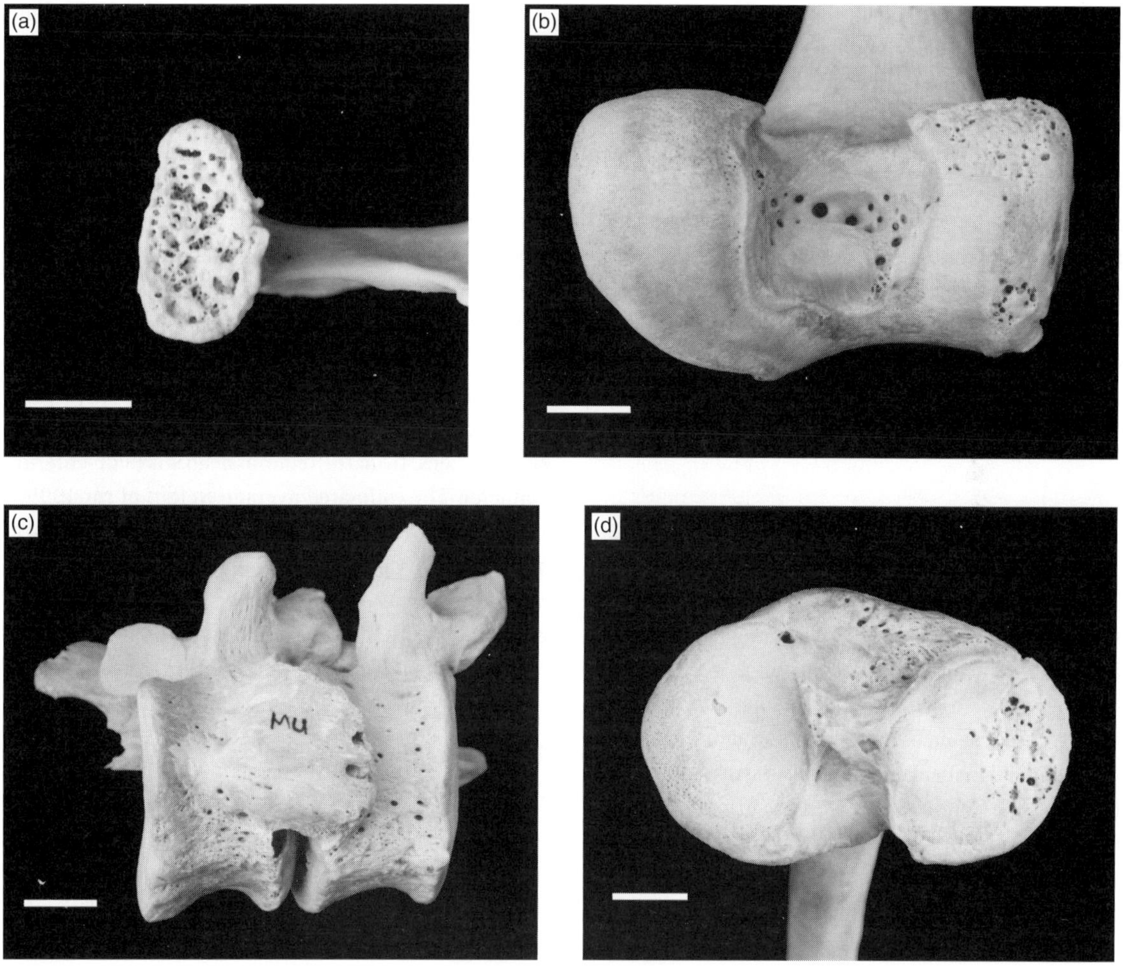

Figure 45.2 Bone degeneration in Mahale chimpanzees. ©D. Shimizu. **(a)** Upper view of Wansombo's right mandibular condyle showing advanced arthritis. **(b)** Femoral condyle of Wansombo with arthritis. **(c)** The third and fourth lumbar vertebrae of Musa, showing calcification of the anterior longitudinal ligament. **(d)** Tibial condyle of Wansombo with arthritis.

Trauma was the most common pathology seen in the postcranial skeletons of Gombe chimpanzees (Jurmain, 1989, 1997). In Mahale, however, trauma was not observed, except in Gwakakumo. There was a distortion and inflammatory mark on her nasal aperture, to the right maxilla, which appeared to be a bite wound. The left transverse process of her third lumbar vertebra had been fractured, and the fracture surface did not show any sign of healing. This meant that the fracture happened around the time of her death. Her left 6–12 ribs had also been fractured, and showed no vital reaction. These fractures indicated that she had experienced a strong force on the left side of her back around the 10th rib, where the spleen and left kidney are situated. Therefore, her cause of death might have been hemorrhagic shock caused by damage to her spleen and kidney. There were no high trees or branches in the bush where her corpse was found, and neither was there any evidence of scavenging. Therefore, she probably survived for a while after a fall from a tree, and moved into the bush by herself.

Gwakakumo exhibited other pathological characteristics (Nakai *et al.*, 2004). Her right humerus was 7 cm shorter than her left one, and her left fourth metacarpal and proximal phalanx were also short and deformed. These were caused by a developmental disorder. She had arthritis during her growth period, and the head of the humerus, the distal epiphysis of the metacarpal, and the proximal epiphysis of the proximal phalanx were fused during the early stages of growth. Her name, Gwakakumo, originated from the local words "gwa" (daughter) and "kakumo" (finger), and her deformed left fourth finger certainly gained the attention of field researchers when she was first identified. This meant that her left fourth finger was already deformed when she joined the M group.

45.5 Field studies related to chimpanzee morphology

In this section, I will briefly describe field studies that have investigated chimpanzee morphology. Relatively few such studies have been conducted, probably because measuring living wild animals is not easy. Because capture may cause stress and social disturbance to chimpanzees, it is not usually permitted to capture them to conduct morphological measurements. However, some studies have overcome such difficulties and have yielded useful information.

45.5.1 Body weights of wild chimpanzees

Weight is obviously an important and basic parameter for every aspect of studies on animals. At Mahale, Uehara and Nishida (1987) measured the weights of 14 chimpanzees without capturing them. At Miyako and Kansyana Camps, they attached a rope to a spring balance, waited until a chimpanzee climbed the rope to get food provided at the top, and read the scale with binoculars. This method is far less invasive than capturing, but still we cannot currently conduct it because provisioning has completely ceased at Mahale. The weights of six adult males at Mahale averaged 42.0 kg (30.3–52.0 kg), and those of eight adult females averaged 35.2 kg (30.0–45.5 kg). Body weight tended to decrease at the end of the rainy season (Uehara and Nishida, 1987). Nishida (2012) assumed that some large males that were not measured, such as Ntologi and Sankaku, probably weighed close to 60 kg each. Ntologi weighed only 38.5 kg after his death (Nishida, 1996). Similar measurements with a spring balance were conducted at Gombe, where the median weight of males was 39 kg and that of females was 31.3 kg (Pusey *et al.*, 2005).

Some estimates of body weights from appendicular skeletons have been published. At Kibale, the estimated average weight of two males was 50 kg, and one female was estimated to weigh 40.5 kg (Peterhans *et al.*, 1993). Carlson *et al.* (2006) also estimated the body weights of six Taï chimpanzees and four Mahale chimpanzees, from the femoral head superior–inferior diameter. The estimated average weight of three Mahale females was 31.3 kg, which was less than their measured average of 35.2 kg. In addition, an adult male (Musa) was estimated to weigh 35.6 kg, which was lighter than his measured weight of 37.0–44.0 kg (Uehara and Nishida, 1987).

45.5.2 Positional behavior of wild chimpanzees

Positional behavior (i.e. posture and locomotion) in the natural environment can also provide important information in understanding the function of chimpanzee anatomical specializations. In this regard, Hunt (1991a, b, 1992a, b, 1994a, b) conducted intensive field observations of 26 adult males and females at Mahale from 1986 to 1987. He distinguished 65 different positional modes (divided into 20 categories), and compared their frequencies with those of similar-body-sized baboons. Overall, chimpanzees exhibited more variety in their positional behaviors than did baboons. Regarding suspensory adaptation, brachiation (i.e. hand-over-hand suspensory locomotion) was observed on only a few occasions, and unimanual arm-hanging was the most commonly associated with a smaller diameter of supporting branch. As 95.7% of the unimanual arm-hanging was observed during feeding bouts, Hunt (1992a) assumed that the suspensory behavior was an adaptation to harvest fruit from the terminal branches. In comparison with other apes, unimanual arm-hanging is more often observed in chimpanzees than in orangutans (Hunt, 1991b).

Such characteristics of positional behavior were further discussed by Hunt (1991a) in relation to the morphological characteristics of chimpanzees. He argued that most osteoligamentous specializations of the upper body, such as an abductible humerus, a broad thorax, a corn-shaped torso, and a narrow scapula, are part of a coadapted functional complex that reduces muscle action and structural fatigue during arm-hanging. In addition, most muscular specializations are adaptations to vertical climbing. He also investigated the effect of body weight on some positional modes. No significant difference was found between the largest and the smallest males in branch-use diameter and the amount of arm-hanging. However, social rank consistently predicted branch diameter choice: high-ranking males used larger supports (Hunt, 1992b). For vertical climbing, large males used smaller weight-bearing structures; he argued, therefore, that large males minimized climbing and optimized arm-hanging by carefully selecting the appropriate support diameters, perhaps because vertical climbing is disproportionally expensive for large animals (Hunt, 1994a).

Circumstances in which bipedalism occurred were also investigated. Bipedalism was observed most commonly when the chimpanzees fed on small fruit trees in the open forest. In addition to considerations regarding australopithecine anatomy, Hunt (1994b) argued that hominid bipedalism may have evolved, in conjunction with arm-hanging, as a terrestrial feeding posture rather than as an adaptation to walking.

Later, Carlson *et al.* (2006) combined the structural limb data that they had obtained from skeletal remains and behavioral data collected by Hunt, and compared these data with skeletal materials collected from other long-term study sites, such as Gombe and Taï (Carlson *et al.*, 2008). Although interindividual and inter-unit-group differences in the shape of the diaphysis were observed, the authors could not explain how locomotor activity and the habitat characteristics could have contributed to variation in diaphyseal shape. Matsumura *et al.* (2010) have also pointed out the unique distribution of cortical bones on the femoral neck of Mahale chimpanzees (Musa, Wansombo, Pulin, and Betty). The chimpanzees had thin cortical bone on the superior part and thick cortical bone on the inferior part, which is often observed in humans but not in chimpanzees. Matsumura *et al.* (2010) suggested that this unique character was related to arboreal locomotor activity.

45.5.3 Asymmetry of male testes

Nakai and Zamma (2004) observed that the testes of mature male chimpanzees were asymmetric in the crotch, and 9 out of 12 males had a right-dangling testis. Although the sample size was small, this is contrary to the usual case in male humans, in which the majority has a left-dangling testis. Such outer morphological characteristics observable in the field have not gained enough attention, except for the purposes of individual identification. However, because morphology and behavior are tightly linked, future studies should gather such information in relation to behavior.

45.6 Note to collect chimpanzee skeletons

As described above, the skeletons of wild chimpanzees can provide a great deal of information. Here I will summarize a few important points when collecting skeletons. When a researcher finds a corpse of a chimpanzee in the field, he/she should never directly touch the corpse, which may have been infected with a virus or bacteria. In order to avoid any risk of infection, investigators should protect themselves with at least the following: medical disposable gloves, a mask, goggles, and a medical coat or raincoat. These items must be burned after use. If the corpse is still fresh, ask a veterinarian to perform an autopsy, which provides valuable information, such as the cause of death and the condition of the individual when alive. Then, measurements of anthropometric traits (see Table 45.1) and weight should be taken. For these

Table 45.1 **Criteria for anthropometric traits**

Anthropometric trait	Criteria
Sitting height (Figure 45.3a)[a]	The horizontal distance between the vertex and the most caudal point of the hip around the ischial tuberosity. The vertex is the most cranial point from the eye-ear plane on the sagittal plane. The corpse must be completely extended and lying in a dorsal position (i.e. facing upward)
Head and body length (Figure 45.3b)[a]	The horizontal distance between the tip of the nose and the most caudal point of the hip around the ischial tuberosity. The corpse must be completely extended and lying in a dorsal position
Anterior body length (Figure 45.3b)[a]	The horizontal distance between the suprasternale and the symphysion. The suprasternale is the intersection point at the jugular notch and the sagittal plane. The symphysion is the intersection point at the superior margin of the pubic symphysis and the sagittal plane. The corpse must be completely extended and lying in a dorsal position
Hand length (Figure 45.4a)[b]	The length between the dactylion and the mid-point between the stylion and stylion ulnare. The dactylion is the tip of the third finger. The stylion is the tip of the styloid process of the radius. The stylion ulnare is the tip of the styloid process of the ulna
Hand breadth (Figure 45.4a)[b]	The length between the metacarpale radiale and the metacarpale ulnare. The metacarpale radiale is the most lateral point of the head of the second metacarpal. The metacarpale ulnare is the most interior point of the head of the fifth metacarpal
Foot length (Figure 45.4. b)[b]	The length between the acropodion and the pternion. The acropodion is the tip of the longest toe in the direction of the axis of the foot. The acropodion is not necessarily a third toe. The pternion is the most posterior point of the heel
Foot breadth (Figure 45.4b)	The length between the metatarsale fibulare and the metatarsale tibiale. The metatarsale fibulare is the most lateral point of the head of the fifth metatarsal. The metatarsale tibiale is the most interior point of the head of the second metatarsal

[a] Anthropometer/tape measure.
[b] Standard slide caliper.

measurements, an anthropometer with a 2 m range, a standard slide caliper with a 30 cm range, and a weighing scale are required. If an anthropometer is not available, use a tape measure instead. The criteria for the measurements are presented in Table 45.1 and Figures 45.3 and 45.4. These measurements cannot be taken from the skeleton, only from the corpse. Once rigor mortis (postmortem stiffening of the muscles) sets in, some of the measurements cannot be taken, depending on the corpse's posture. Rigor mortis in humans becomes noticeable approximately 3 to 4 h after death, and the whole body becomes rigid after about 12 h (Gunn, 2009), before gradually decreasing after 36 h. The speed and duration of rigor mortis are strongly influenced by the environmental temperature. Therefore, to take measurements, you may leave the corpse until the rigor mortis subsides or terminate the rigor mortis artificially. It can be terminated by pulling forcibly on the affected parts.

Once the autopsy is over, or is difficult to perform (e.g. no veterinarians are available or the corpse is

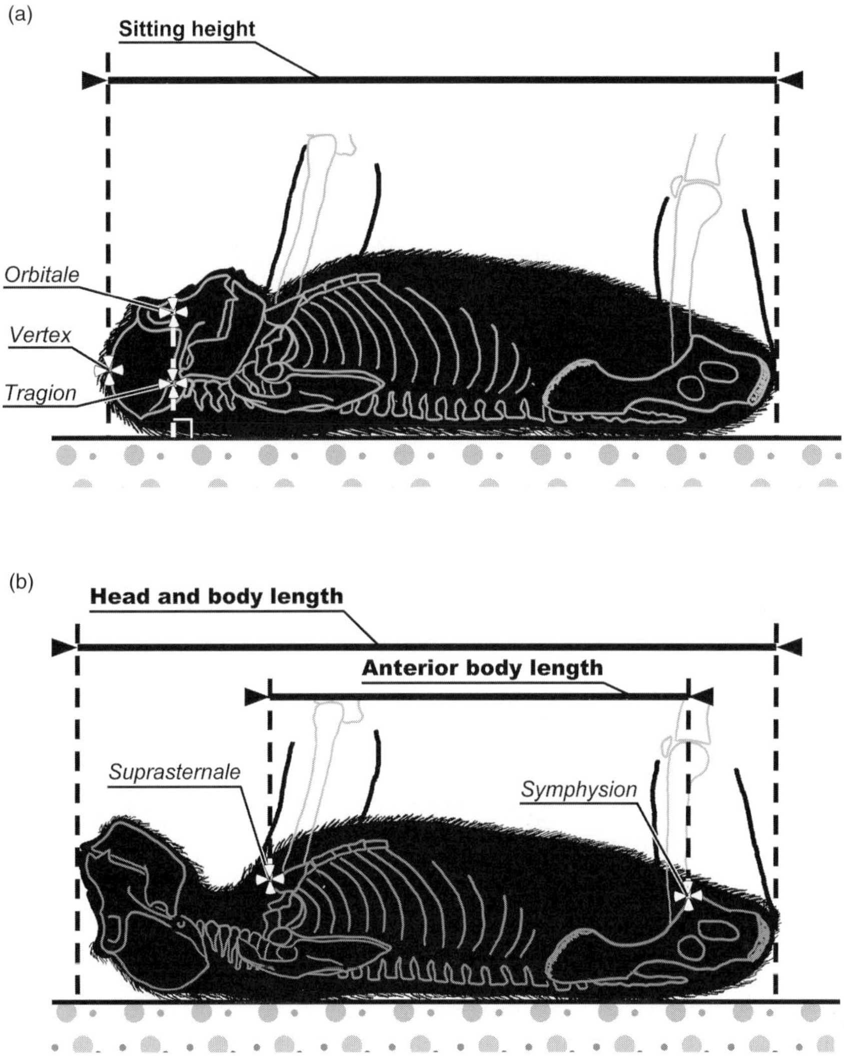

Figure 45.3 Criteria for the measurement of the trunk. Black arrows indicate measurement points and white stars indicate reference points. **(a)** Sitting height. The eye–ear plane is defined as a plane through the left and right tragions and left orbitale (white dashed line). **(b)** Head and body length and anterior body length.

already putrefied), decomposition of the soft tissue should be allowed to occur, so that the skeleton can be removed. The easiest way to do this is to put the corpse in a plastic bag and bury it in a deep pit, so it is not scavenged. This way, no flies can access the corpse, a putrid odor is avoided, and no parts of the skeleton should be lost. If this is too difficult to carry out, the plastic bag may be hung on a tree bough, but consequently there will be problems with maggots and a foul odor. The time taken for complete decomposition depends on the environment and the season. At Mahale, it takes roughly 3–6 months (rainy and dry seasons, respectively). After decomposition of the soft tissue, the bones should be boiled for 1 or 2 h over a low flame in a large pot, in order to kill bacteria and viruses and remove the remaining soft tissue and fat. Protection should be used up to this point to

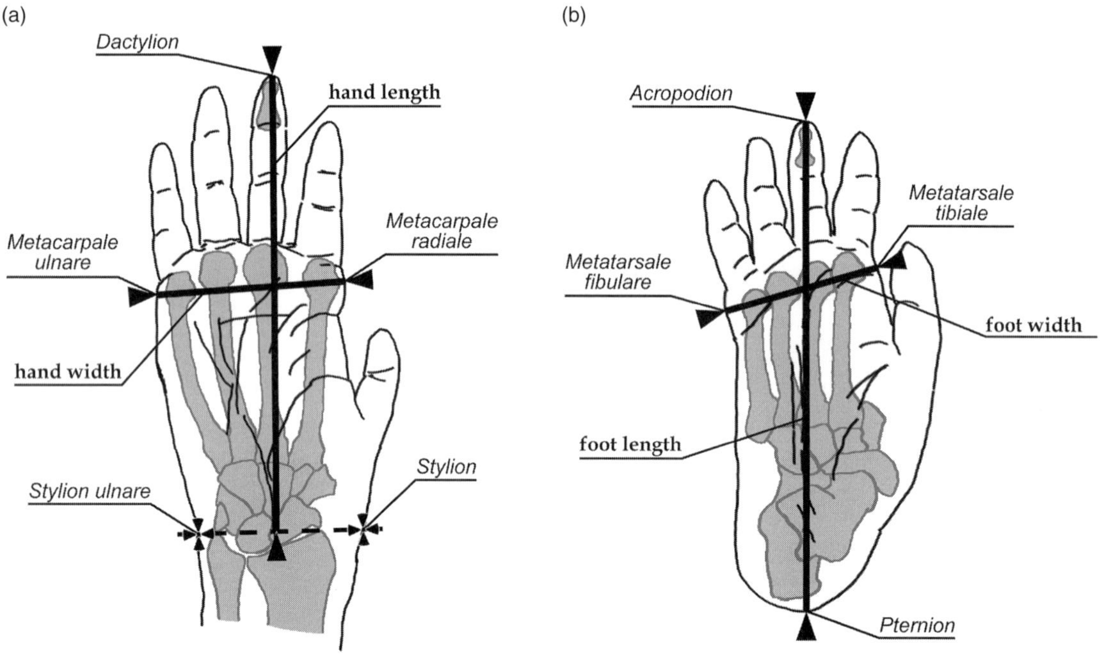

Figure 45.4 Criteria for the measurement of the hands and feet. Black arrows indicate measurement points and black stars indicate reference points. (a) Palmar view of the right hand, showing the hand length and the hand width. (b) Plantar view of the right foot, showing the foot length and the foot width.

avoid potential infection. After boiling, the bones can be handled without any protection. The bones should then be dried in the shade, packed in several zippered plastic bags with desiccating agent and insect repellent, and stored in a (preferably airtight) container. To avoid contamination with other individuals and the loss of any parts, a list of bones should be placed in the container, and a protocol number should be given to each set of bones.

In conclusion, wild chimpanzee skeletons collected from long-term study sites provide much-needed morphological information. Therefore, chimpanzee skeletons in Mahale should be collected at every opportunity.

References

Baker, K. C. (2000). Advanced age influences chimpanzee behavior in small social groups. *Zoo Biology*, **19**, 111–19.

Boesch, C. and Boesch-Achermann, H. (2000). *The Chimpanzees of the Taï Forest: Behavioural Ecology and Evolution*. Oxford: Oxford University Press.

Burr, D. B., Ruff, C. B., and Johnson, C. (1989). Structural adaptations of the femur and humerus to arboreal and terrestrial environments in three species of macaque. *American Journal of Physical Anthropology*, **79**, 357–67.

Carlson, K. J., Doran-Sheehy, D. M., Hunt, K. D., *et al.* (2006). Locomotor behavior and long bone morphology in individual free-ranging chimpanzees. *Journal of Human Evolution*, **50**, 394–404.

Carlson, K. J., Sumner, D. R., Morbeck, M. E., *et al.* (2008). Role of nonbehavioral factors in adjusting long bone diaphyseal structure in free-ranging *Pan troglodytes*. *International Journal of Primatology*, **29**, 1401–20.

Carter, M. L., Pontzer, H., Wrangham, R. W., and Peterhans, J. K. (2008). Skeletal pathology in *Pan troglodytes schweinfurthii* in Kibale National Park, Uganda. *American Journal of Physical Anthropology*, **135**, 389–403.

Fenart, R. and Deblock, R. (1973). *Pan paniscus* et *Pan troglodytes* craniométrie. Etude comparative et ontogénique selon les méthodes classiques et vestibulaire. *Muse Royal de L'Afrique Centrale, Tervuren, Belgique Annales, Série In-8°, Sciences Zoologiques*, **204**, 1–593.

Gavan, J. A. (1971). Longitudinal, postnatal growth in chimpanzee. In *The Chimpanzee, Vol. 4, Behavior, Growth, and Pathology of Chimpanzees*, ed. G. H. Bourne. Basel: Karger, pp. 46–102.

Gunji, H., Hosaka, K., Huffman, M. A., *et al.* (1998). Note on some chimpanzee skeletons from the Mahale Mountains National Park. *Mahale Mountains Chimpanzee Research Project Ecological Report*, **109**.

Gunji, H., Hosaka, K., Huffman, M. A., *et al.* (2003). Extraordinarily low bone mineral density in an old female chimpanzee (*Pan troglodytes schweinfurthii*) from the Mahale Mountains National Park. *Primates*, **44**, 145–9.

Gunn, A. (2009). *Essential Forensic Biology*, 2nd edn. Chichester, UK: John Wiley & Sons Ltd.

Hamada, Y., Udono, T., Teramoto, M., and Sugawara, T. (1996). The growth pattern of chimpanzees: somatic growth and reproductive maturation in *Pan troglodytes. Primates*, **37**, 279–95.

Hanamura, S., Kiyono, M., Lukasik-Braum, M., *et al.* (2008). Chimpanzee deaths at Mahale caused by a flu-like disease. *Primates*, **49**, 77–80.

Huffman, M. A. (1990). Some socio-behavioral manifestations of old age. In *The Chimpanzees of the Mahale Mountains*, ed. T. Nishida. Tokyo: University of Tokyo Press, pp. 237–55.

Hunt, K. D. (1991a). Mechanical implications of chimpanzee positional behavior. *American Journal of Physical Anthropology*, **86**, 521–36.

Hunt, K. D. (1991b). Positional behavior in the Hominoidea. *International Journal of Primatology*, **12**, 95–118.

Hunt, K. D. (1992a). Positional behavior of *Pan troglodytes* in the Mahale Mountains and Gombe Stream National Parks, Tanzania. *American Journal of Physical Anthropology*, **87**, 83–105.

Hunt, K. D. (1992b). Social rank and body size as determinants of positional behavior in *Pan troglodytes*. *Primates*, **33**, 347–57.

Hunt, K. D. (1994a). Body size effects on vertical climbing among chimpanzees. *International Journal of Primatology*, **15**, 855–65.

Hunt, K. D. (1994b). The evolution of human bipedality: ecology and functional morphology. *Journal of Human Evolution*, **26**, 183–202.

Jurmain, R. (1989). Trauma, degenerative disease, and other pathologies among the Gombe chimpanzees. *American Journal of Physical Anthropology*, **80**, 229–37.

Jurmain, R. (1997). Skeletal evidence of trauma in African apes, with special reference to the Gombe chimpanzees. *Primates*, **38**, 1–14.

Jurmain, R. (2000). Degenerative joint disease in African great apes: an evolutionary perspective. *Journal of Human Evolution*, **39**, 185–203.

Kilgore, L. (1989). Dental pathologies in ten free-ranging chimpanzees from Gombe National Park, Tanzania. *American Journal of Physical Anthropology*, **80**, 219–27.

Kuykendall, K. (1996). Dental development in chimpanzees (*Pan troglodytes*): the timing of tooth calcification stages. *American Journal of Physical Anthropology*, **99**, 135–57.

Matsumura, A., Gunji, H., Takahashi, Y., Nishida, T., and Okada, M. (2010). Cross-sectional morphology of the femoral neck of wild chimpanzees. *International Journal of Primatology*, **31**, 219–38.

Matsuzawa, T., Sakura, O., Kimura, T., Hamada, Y., and Sugiyama, Y. (1990). Case report on the death of a wild chimpanzee (*Pan troglodytes verus*). *Primates*, **31**, 635–41.

Morimoto, N., Ponce de León, M. S., and Zollikofer, C. P. E. (2011). Exploring femoral diaphyseal shape variation in wild and captive chimpanzees by means of morphometric mapping: a test of Wolff's low. *The Anatomical Record*, **294**, 589–609.

Nakai, M. and Zamma, K. (2004). Position and movement of the testes of wild chimpanzees at Mahale. *Pan Africa News*, **11**(1), 4–6.

Nakai, M., Itoh, N., Nakamura, M., Huffman, M. A., and Nishida, T. (2004). [Individual identification from skeletal remains: cases of wild chimpanzees of the Mahale Mountains National Park, Tanzania.] *Primate Research*, **20**, 1–9. In Japanese with English abstract.

Nishida, T. (1996). The death of Ntologi, the unparalleled leader of M group. *Pan Africa News*, **3**(1), 3–4.
Nishida, T. (2012). *Chimpanzees of the Lakeshore: Natural History and Culture at Mahale*. Cambridge: Cambridge University Press.
Peterhans, J. C. K., Wrangham, R. W., Carter, M. L., and Hauser, M. D. (1993). A contribution to tropical rain forest taphonomy: retrieval and documentation of chimpanzee remains from Kibale Forest, Uganda. *Journal of Human Evolution*, **25**, 485–514.
Pusey, A. E., Oehlert, G. W., Williams, J. M., and Goodall, J. (2005). Influence of ecological and social factors on body mass of wild chimpanzees. *International Journal of Primatology*, **26**, 3–31.
Shi, C., Nishizawa, S., Adachi, K., and Endo, B. (1995). [A comparative morphological analysis of vertebral bodies in humans and some other mammals.] *Anthropological Science*, **103**, 467–84. In Japanese.
Shimizu, D., Gunji, H., Hashimoto, H., *et al*. (2002). The four chimpanzee skulls collected in the Mahale Mountains, Tanzania. *Anthropological Science*, **110**, 251–66.
Sumner, D. R., Morbeck, M. E., and Lobick, J. J. (1989). Apparent age-related bone loss among adult female Gombe chimpanzees. *American Journal of Physical Anthropology*, **79**, 225–34.
Tutin, C. E. G. (1996). Reproductive success story: variability among chimpanzees and comparisons with gorillas. In *Chimpanzee Cultures*, ed. R. W. Wrangham, W. C. McGrew, F. B. M. de Waal, and P. G. Heltne. Cambridge, MA: Harvard University Press, pp. 181–93.
Uehara, S. and Nishida, T. (1987). Body weights of wild chimpanzees (*Pan troglodytes schweinfurthii*) of the Mahale Mountains National Park, Tanzania. *American Journal of Physical Anthropology*, **72**, 315–21.
Yasui, K. and Takahata, Y. (1983). Skeletal observation of a wild chimpanzee infant (*Pan troglodytes schweinfurthii*) from the Mahale Mountains, Tanzania. *African Study Monographs*, **4**, 129–38.
Zihlman, A. L., Morbeck, M. E., and Goodall, J. (1990). Skeletal biology and individual life history of Gombe chimpanzees. *Journal of Zoology*, **221**, 37–61.
Zihlman, A., Bolter, D., and Boesch, C. (2004). Wild chimpanzee dentition and its implications for assessing life history in immature hominin fossils. *Proceedings of the National Academy of Science of the* United States of America, **101**, 10541–3.
Zihlman, A. L., Bolter, D. R., and Boesch, C. (2007). Skeletal and dental growth and development in chimpanzees of the Taï National Park, Côte D'Ivoire. *Journal of Zoology*, **273**, 63–73.

46 Genetic studies*

EIJI INOUE

46.1 Introduction

Genetic analyses can reveal the history, dispersal pattern, and diversity of a population. For example, genetic analyses revealed the subspecies of chimpanzees. Chimpanzees are currently divided into four subspecies: *Pan troglodytes verus, P. t. ellioti* (formerly, *P. t. vellerosus*), *P. t. troglodytes, P. t. schweinfurthii* (Gagneux *et al.*, 1999; Oates *et al.*, 2008). *P. t. schweinfurthii* (eastern chimpanzee), which ranges in Tanzania, was reported to be genetically close to *P. t. troglodytes* (Becquet *et al.*, 2007; Stone *et al.*, 2010; Gonder *et al.*, 2011).

Several genetic studies have been conducted in Mahale and neighboring habitats. First, I summarize the results of noninvasive samples that are used for genetic analyses of chimpanzees. Then, studies investigating maternally inherited mitochondrial DNA, the paternally inherited marker, Y-chromosomal microsatellite (Y-STR), and an autosomal microsatellite marker are summarized. Furthermore, I include new genetic data from Mahale chimpanzee groups other than M group, and reanalyse the phylogenetic relationships of that population, including the published data. After describing the genetic characteristics of Mahale and neighboring habitats, these are compared between Mahale and other populations to determine the genetic structure of eastern chimpanzees. Genetic analyses also revealed the paternity in Mahale (Inoue *et al.*, 2008a). These data are summarized in Chapter 35.

* I am grateful to Drs. O. Takenaka, M. Inoue-Murayama, and L. Vigilant, and members of their laboratories for their assistance conducting the genetic analyses; and to Drs. Yamagiwa and Nakagawa for their advice.

46.2 Noninvasive DNA samples

Noninvasive genetic techniques are necessary for DNA analyses of wild mammals, in particular endangered mammals, because capturing wild animals may cause injury and disturbance. Most genetic research on wild chimpanzees has been conducted using fecal samples (Constable *et al.*, 2001; Vigilant *et al.*, 2001), hair samples, either shed in the beds (or nests) or from self-scratching (Goldberg and Wrangham, 1997; Gagneux *et al.*, 1999). Urine and food wadges (discarded food remains) can also provide enough DNA to study some genetic markers (Hashimoto *et al.*, 1996; Hayakawa and Takenaka, 1999). A small number of studies have assessed which samples are useful for the analysis of wild chimpanzee DNA. Below is a summary of such studies, which were conducted in Mahale.

Inoue *et al.* (2007b) studied samples from infant chimpanzees. Although DNA samples from infant chimpanzees are necessary to analyse paternity, it was difficult to collect feces, which have commonly been used in genetic analyses of infant chimpanzees (Boesch *et al.*, 2006). Inoue *et al.* (2007b) showed that urine and saliva samples were relatively easy to collect compared with fecal samples from small infant chimpanzees. They also showed that the quantity of

Mahale Chimpanzees: 50 Years of Research, ed. M. Nakamura, K. Hosaka, N. Itoh and K. Zamma. Published by Cambridge University Press.

Table 46.1 **Quantity of DNA extracted from noninvasive samples**

	DNA amount (ng)				
	No. analysed samples	Average	SD	Percentage of good samples (more than 2.5 ng[a])	Preservation method
Feces	11	24.9	47.0	64	Desiccated with silica gel
Urine	40	37.2[b]	84.3	50	70% ethanol
Saliva	6	22.9	26.6	100	70% ethanol
Drops of blood from wound	4	127.5	98.6	100	70% ethanol
Hair by self-scratching	2	128.5	102.2	100	Kept in a plastic bag
Food wadge	1	81.5	–	100	70% ethanol

[a] DNA extracts containing more than 2.5 ng were regarded as usable samples because a DNA template more than 25 pg per PCR can produce reliable genotypes (Morin *et al.*, 2001), and PCR amplification was normally conducted using 1/100 of total DNA extracts.

[b] The maximum value was extraordinary high (6390) compared with the other sample. This value was the average of the samples except for the maximum.

DNA extracted from saliva and urine is sufficient for further genetic analyses, such as paternity testing, and concluded that such samples are useful, in particular for infant chimpanzees.

Inoue *et al.* (2007a) compared the quantity of DNA isolated between noninvasive samples. Feces, urine, food wadges, saliva remaining after chimpanzees licked branches, shed hair samples, which remained after chimpanzees scratched, and drops of blood from wounds of chimpanzees were collected in Mahale for DNA analyses. The results are summarized in Table 46.1. Fecal samples have been commonly used for DNA analyses, but the quantity of DNA is low. As expected, drops of blood from wounds and from hair following self-scratching have been found to yield high quality DNA. However, the opportunity to obtain these samples is rare. Although improved sampling methods can increase the amount of DNA extracted from feces (Nsubuga *et al.*, 2004), the amount of DNA from fecal samples may not be sufficient to permit analysis of detailed genomic information. Collecting several kinds of noninvasive samples will provide us with an opportunity for conducting several genetic topics efficiently.

46.3 Mitochondrial variation

46.3.1 Gene flow of eastern chimpanzees

Female chimpanzees typically emigrate upon reaching maturity (Nishida and Kawanaka, 1972; Pusey, 1979; Boesch and Boesch-Achermann, 2000; Chapters 6 and 7). As expected from these findings, mitochondrial haplotypes were found to be shared by chimpanzees from different habitats, and they did not sort into location-specific clusters (Morin *et al.*, 1994; Goldberg and Ruvolo, 1997a; Inoue *et al.*, 2013). Even if mitochondrial haplotypes were shared by animals across several habitats, mitochondrial DNA analyses were able to reveal the history of the chimpanzee population. Analysis of mitochondrial divergence suggests that eastern chimpanzees expanded recently (between 20 000 and 61 000 years ago) from eastern DRC (Goldberg and Ruvolo, 1997b; Gagneux *et al.*, 1999).

Eastern chimpanzees are distributed in the DRC, Uganda, Rwanda, Burundi, and Tanzania (Inskipp, 2005). Tanzania is located at the southern and eastern end of the range of chimpanzee distribution and the

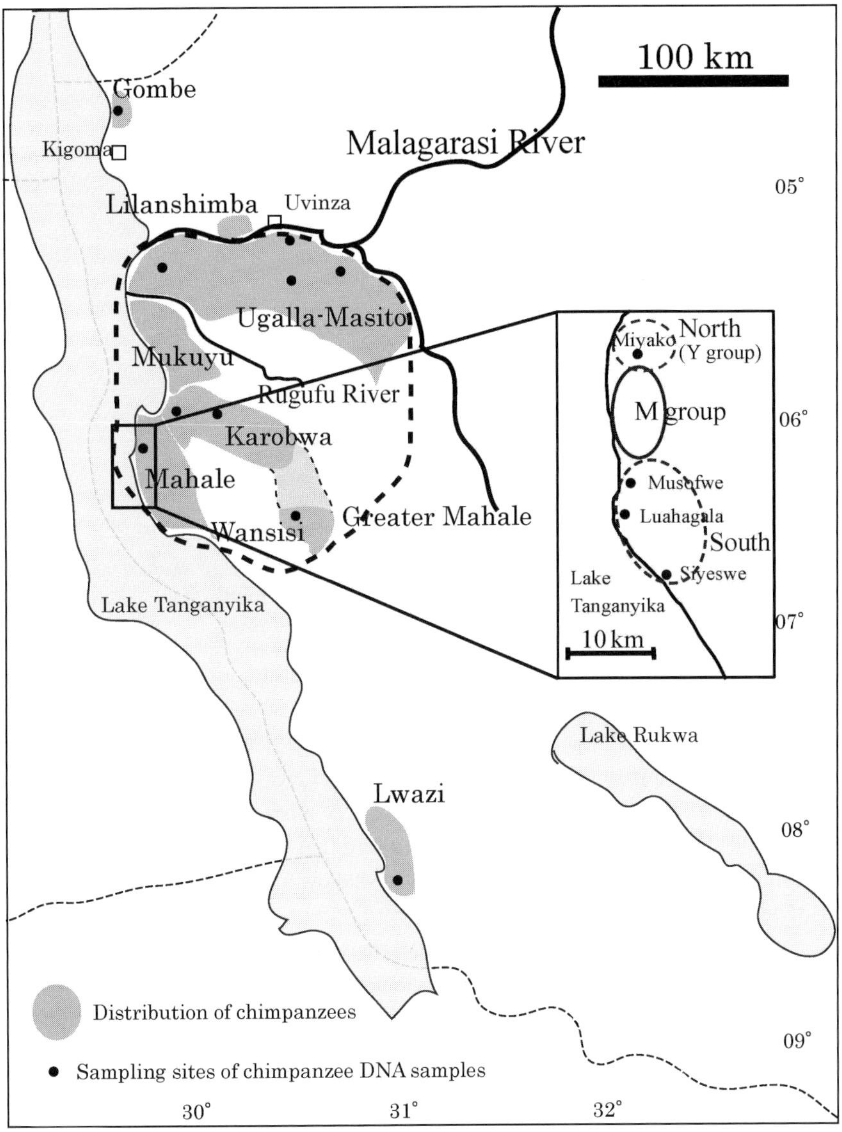

Figure 46.1 Distribution of chimpanzees and sampling sites in Tanzania. See Inoue *et al.* (2013) and references therein.

estimated number of chimpanzees was only 1500–2500 in this region (Inskipp, 2005). Yoshikawa *et al.* (2008) summarized the current distribution of chimpanzees in Tanzania and estimated that there were four populations from Gombe, Lilanshimba, Ugalla–Masito–Mukuyu–Mahale–Karobwa–Wansisi, and Lwazi. The number of chimpanzees outside the national parks (Gombe and Mahale) was estimated at less than 700 (Yoshikawa *et al.*, 2008). These four populations are now isolated from each other.

Inoue *et al.* (2013) reported the detailed genetic structure of these habitats except for Lilanshimba and Mukuyu. Their study analyzed the mitochondrial sequences from 138 chimpanzee samples collected across six sites (Lwazi, Wansisi, Mahale, Karobwa, Ugalla–Masito, and Gombe, Figure 46.1). The authors analysed the level of genetic differentiation (F_{ST} values) between habitats. The levels between Greater Mahale (Wansisi, Mahale, Karobwa, and Ugalla–Masito) were low, and statistical analysis showed that

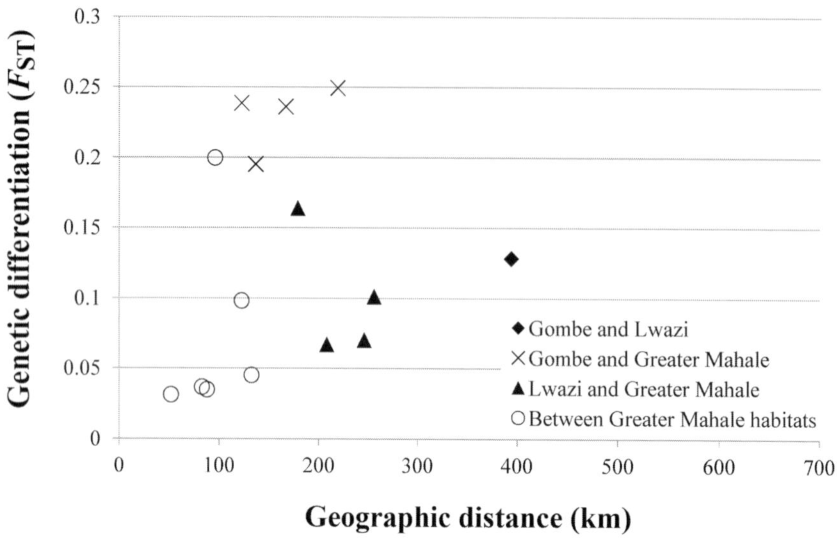

Figure 46.2 Relationship between the genetic and geographic distances (modified from Figure 2 in Inoue *et al.*, 2013). Genetic distances between Gombe and the other habitats were recalculated including the data in Gombe (Keele *et al.*, 2009; see text).

Karobwa and other habitats were not considered to be genetically different, although other population pairs within Greater Mahale were considered to be genetically different. Considering that Karobwa is located in the center of the Greater Mahale habitat, these results suggest that Greater Mahale habitats may be regarded as one continuous population.

Furthermore, Inoue *et al.* (2013) compared the geographic and genetic distance among these habitats and the animals within them. Although the southernmost habitat, Lwazi, was about 150 km away from the nearest habitat (Figure 46.1), the genetic distance between chimpanzees in Lwazi and Greater Mahale habitats was not high, considering the geographic distance (Figure 46.2). In contrast, animals from the northernmost habitat, Gombe, were genetically far from the other habitats, considering the geographic distance (Figure 46.2). The genetic distance between Gombe and Greater Mahale habitats was greater than that between Lwazi and Greater Mahale habitats, although Greater Mahale habitats were geographically closer to Gombe than to Lwazi. These results suggested that Gombe chimpanzees were genetically distant from the chimpanzees in other Tanzanian habitats. From these results, Inoue *et al.* (2013) suggested that the Malagarasi River, which runs between Gombe and the southern habitats, might act to limit gene flow.

Recently, Piel *et al.* (2013) showed contradictory results: the Malagarasi River did not act as an absolute barrier for chimpanzee movement. Their study found that there was a ford that could possibly be used to cross the river in shallow areas, along with the discovery of recent chimpanzee beds, and fresh knuckle prints leading right up to the riverbed (Piel *et al.*, 2013). Furthermore, they showed that some haplotypes were shared between chimpanzees in Gombe and Ugalla habitats and proposed that the finding by Inoue *et al.* (2013) was due to the small sample size (only six samples) from Gombe chimpanzees.

To understand the reasons for the differences between these two studies, I reanalysed the genetic structure using the data published by Inoue *et al.* (2013) and that published from the Gombe data (Keele *et al.*, 2009). The haplotype composition is summarized in Table 46.2. As Piel *et al.* (2013) noted, some haplotypes were shared between animals from the Gombe and the other Tanzanian habitats (Table 46.2). However, the genetic distance between chimpanzees from Gombe and the other Tanzanian habitats was not different from that described by Inoue *et al.* (2013), and was still relatively high considering the geographical distance

Table 46.2 **Haplotype constitutions in Western Tanzania (modified from Table 1 in Inoue et al., 2013)**

	Mitochondrial haplotype[a]																										
Habitat	A	B	C	D	E	F	G	H	I	J	K	L	M	N	O	P	Q	R	S	T	V	GM 3	GM 4	GM 5	GM 10	GM 12	GM 13	Total
Lwazi	1	6	2		1	4																						14
Wansisi	1	3		2			4																					10
Mahale	16	22	11	2																								51
Karobwa	3	1		3			1	1	1																			10
Ugalla–Masito		22		5			3			1	1	3	1	2	1	1	1	1										42
Gombe	5		3															4	16	19	14	7	2	10	6	8	5	99
Total	26	54	16	12	1	4	8	1	1	1	1	3	1	2	1	1	1	5	16	19	14	7	2	10	6	8	5	226

[a] I added the information of Gombe by Keele *et al.* (2009) to Table 1 in Inoue *et al.* (2013). Keele *et al.* (2009) reported 13 haplotypes (GM1–13) and some haplotypes were the same as the haplotype (A–V) in Inoue *et al.* (2013). Corresponding haplotypes are shown below. Analysed sequence length in Keele *et al.* (2009) was longer than that in Inoue *et al.* (2013). Consequently, GM1 and GM2 in Keele *et al.* (2009) were the same as T in Inoue *et al.* (2013).

GM1 = T, GM2 = T, GM6 = V, GM7 = C, GM8 = R, GM9 = A, GM11 = S

(Figure 46.2). This was because the haplotypes that were shared with other Tanzanian habitats were not the major haplotypes at Gombe (12/99, Table 46.2). Therefore, the criticism of Inoue *et al.* (2013) by Piel *et al.* (2013) regarding the sample size was unfounded, and gene flow between Gombe and the other southern habitats was relatively limited.

The results of Inoue *et al.* (2013) and Piel *et al.* (2013) may not be inconsistent. There is one possible explanation for these contradictory results. It is possible that there were no chimpanzees around Lilanshimba until recently when some chimpanzees settled there from Ugalla–Masito. There is a gap in chimpanzee distribution between Gombe and Lilanshimba (Figure 46.1). To my knowledge, there is no evidence to explain the existence of such a gap. There is limited knowledge about the chimpanzee population dynamics in this area and, thus, we cannot make a firm conclusion about which genetic barrier has caused the genetic difference between Gombe and Greater Mahale habitats. However, the Malagarasi River may be one possible genetic barrier. Even though chimpanzees can currently cross the river (Piel *et al.*, 2013), it remains possible that the river previously acted as a genetic barrier, particularly if the place where chimpanzee could cross the river had been limited in time and place. Otherwise, other geographic features between Gombe and the Malagarasi River might have limited chimpanzee movements. Historical data and/or other mammal data are needed to determine the existence of such genetic barriers.

46.3.2 Mitochondrial variation within Mahale

There are several chimpanzee unit-groups around the Mahale M group (Chapter 4; Nishida *et al.*, 1990). Most of the M-group females emigrated from such groups and female immigration into the M group was common (Nishida *et al.*, 2003). This suggests that mitochondrial haplotypes may be shared between neighboring groups. I reanalysed mitochondrial variation at Mahale using the samples from other chimpanzee groups. I used the published data on the mitochondrial D-loop of the M group at Mahale (Inoue *et al.*, 2013) and included samples from the north and south of the region (Figure 46.1). Three food wadge samples were collected in the northern area (Miyako) and were considered to be from members of the Y group (Sakamaki and Nakamura, 2007; Chapter 11). Three hair samples were collected in the southern area (one sample at Musofwe, one at Luahagala, and one at Siyeswe, in Figure 46.1). Although the sampled areas were in or around the range of the N or L groups (Nishida *et al.*, 1990; Chapter 4), it was difficult to determine the identity of each group. Accordingly, I regarded those samples as those from the southern area. I extracted DNA from these samples and sequenced the mitochondrial D-loop region following the method described by Inoue *et al.* (2013).

Four haplotypes were in the M group (Table 46.3; Inoue *et al.*, 2013). Two haplotypes were found in the Y group, which were shared with M-group members (Table 46.3). Three haplotypes were found in the southern areas, and one of them was not found in M-group chimpanzees (Table 46.3). The haplotype was not found in the Western Tanzania population and was close to the haplotype T in the previous report (Inoue *et al.*, 2013). Although the number of analysed samples was not enough to make a firm conclusion, these results suggest that the transfer of female chimpanzees between the Y and M groups occurred more frequently than transfer between the southern and M groups. Additional analyses in the neighboring groups may help to clarify the frequency of female transfer between groups.

46.4 Y-STR analyses

46.4.1 Backgrounds

In primate groups in which females are commonly philopatric, (e.g. macaques), the maternally inherited marker, mitochondrial DNA, has been studied to investigate the intergroup variation and phylogeography (Kawamoto *et al.*, 2007). In contrast, in chimpanzees, it is the females that typically emigrate upon maturity (Nishida and Kawanaka, 1972; Pusey, 1979; Boesch and Boesch-Achermann, 2000). The paternally inherited marker, Y-STR, can

Table 46.3 **Mitochondrial haplotypes of chimpanzees at Mahale**

	Number[a]			Sequence variation																
Haplotype	M	Y	South	71	84	86	91	126	149	162	176	177	181	187	192	194	198	214	230	247
A	16		1	C	C	G	C	C	A	C	T	G	C	A	A	A	T	C	C	C
B	22	2						T		T										
C	11	1	1		T	A	T		T	T	C			G	G		C	T		
D	2																		T	
New			1	T	T	A				T	C	A	T			C		T		T

[a] In the M group, the number shows the number of individuals. In the south and north group, the number shows the number of analysed samples. In the latter cases, the samples may include those from the same individuals.

provide information on intergroup variation and the history of the local regions. As expected, Y-STR haplotypes differentiated the populations of chimpanzees and bonobos (Eriksson *et al.*, 2006; Langergraber *et al.*, 2007). I combined the data from a previous report at Mahale (Inoue *et al.*, 2008b) with some new data, and reanalysed the Y-STR variation.

46.4.2 Y-STR analyses in Mahale

Data and methods

Inoue *et al.* (2008b) analysed 11 Y-STR loci from male chimpanzees in the M group and in the northern area (Y group) of Mahale (Figure 46.1). In this study, Y-STR haplotypes of 16 male samples from the M group and 5 male samples from the Y group were determined. In addition to these published data, I used the data regarding the Y-STR haplotypes of male samples obtained in the southern area to the M group's range. Three hair samples were collected from chimpanzee beds in the southern area (two from Luahagala and one from Musofwe, Figure 46.1). I extracted DNA from those samples and genotyped 11 Y-STR loci following the method described by Inoue *et al.* (2008b).

Y-STR variation within the M group

Out of the 11 studied Y-STR loci, two were polymorphic and four haplotypes were found in the M group (Table 46.4). Among 16 males in the M group, 11 shared the same haplotype. One of the males which had haplotype D was the alpha male (Fanana) at that time. The other two males with haplotype D were the infants of this alpha male (Inoue *et al.*, 2008a). This result suggests that the frequency of a haplotype within a group may increase when the haplotype is shared by the alpha male. In the M group, haplotype D was found only rarely among adult males (1/7), whereas half of the males younger than 10 years old at the time of sampling (2/4) had this haplotype (Inoue *et al.*, 2008b). This finding was consistent with the fact that the alpha male was the father of around half of the offspring in this group (Inoue *et al.*, 2008a). This suggests that recent male reproductive skew affected Y-STR variation within the group.

Y-STR variation between groups

Males in neighboring groups shared Y-STR haplotypes in Mahale (Table 46.4, Inoue *et al.*, 2008b). One of the M-group males, Masudi, was born in the neighboring K group, and transferred into the M group with his mother when he was 5 years old (Nishida *et al.*, 1985; Takahata and Takahata, 1989). This male had haplotype A, which was common in the M group (Table 46.4). The K group had ranged in the northern habitat of the M group (around Miyako in Figure 46.1) and become extinct (Nishida *et al.*, 1985; Chapter 10), and currently, Y group use the area that the K group used (Sakamaki and Nakamura, 2007). Furthermore, haplotype A was identified in five samples from the Y group (Table 46.4). Males in the M, K, and Y groups shared the same haplotype, which was the most common haplotype in the M group. This is suggestive of past occurrences of group fission in Mahale, because a group fission is known to have occurred at Gombe (Goodall, 1983). Alternatively, the transfer of males between groups, as observed by Masudi, might have occurred several times in the past.

On the other hand, males in the southern area did not share Y-STR haplotypes with males belonging to the M group (Table 46.4). Two haplotypes were identified in three samples taken from the south area. The haplotypes found in the southern area were different from those found in the M group and were phylogenetically close to each other (see also Figure 46.4). These data suggest that the males in the M, K, and Y groups were genetically closer than males from the M group and the southern area.

It is difficult to evaluate the genetic difference between the southern and the M groups. Data regarding the mutation rate can help us to estimate the frequency at which new haplotypes appear. The mutation rate of Y-STR in human was $2.2–2.8 \times 10^{-3}$ mutations per locus (Kayser and Sajantila, 2001; Kurihara *et al.*, 2004). The rate of mismatch between the haplotypes of biological fathers and son using 11 loci is expected to be $2.4–3.1 \times 10^{-2}$. This means that one Y-STR mismatch can occur in 32–42 father–son pairs. Chimpanzees belonging to the M group at Mahale had a birth rate of 5–10% and the group comprised around 80 individuals on average (Nishida

Table 46.4 **Y-STR haplotypes at Mahale (modified from Table 1 in Inoue et al., 2008b)**

	Number[a]			Y-STR marker[b]										
Haplotype	M	Y	South	DYS 439	DYS 469	DYS 510	DYS 517	DYS 520	DYS 533	DYS 562	DYS 588	DYS 612	DYS 630	DYS 632
A	11	5		11	20	9	11	6	10	9	12	14	17	6
B	1			11	20	9	11	6	11	9	12	14	17	6
C	1			11	20	9	11	6	10	9	12	15	17	6
D	3			11	20	9	11	6	9	9	12	15	17	6
E			1	11	20	9	11	6	10	9	13	12	17	6
F			2	11	20	9	11	6	10	9	12	12	17	6

[a] In the M group, the number shows the number of individuals. In the south and north group, the number shows the number of analyzed samples. In the latter cases, the samples may include those from the same individuals.

[b] The number indicate the estimated number of repeated unit.

et al., 2003). Using these data, we can surmise that 2–4 male babies were born per year in the group. Assuming that the mutation rate in chimpanzees is similar to human, a new Y-STR haplotype may appear approximately once per 15 years per group. However, because a new haplotype can spread by genetic drift, or by the reproductive success of a male with the new haplotype, it is not easy to reconstruct the exact history of these groups.

Y-STR variation analysis revealed that M-group males were genetically closer to males in the north than those in the south. This was consistent with the result of mitochondria: one mitochondrial haplotype was found only in the south area. Gene flow between the M group and southern area might have been limited in comparison.

46.5 Genetic structure within group

Female-biased dispersal patterns in chimpanzees may affect the kin structure within the group. The relatedness of male chimpanzees has been assumed to be higher than that of females (Morin *et al.*, 1994), which may be related to male–male cooperation in this species. However, the relatedness between males was rarely significantly higher than that among females in Taï, Côte d'Ivoire, in Gombe, Tanzania, and in Budongo, Uganda (Vigilant *et al.*, 2001; Lukas *et al.*, 2005). Using mathematical predictions, Lukas *et al.* (2005) showed that the relatedness among the philopatric sex decreased as the number of the individuals in the group increased.

In Mahale, Inoue *et al.* (2008a) analysed the genetic structure within M group. The average relatedness among males (N = 12 individuals, R = 0.038) was significantly higher than that among females (N = 19 individuals, R = −0.056). According to the previous prediction (Lukas *et al.*, 2005), a significant difference between sexes is expected only in a small group. However, the group size did not explain the significant difference in the M group because the number of males in this group was higher than that in the groups at Taï and comparable to that of the group at Gombe. Inoue *et al.* (2008a) showed that relatedness of Kalunde, the oldest male, to other males was high, and the average relatedness among males except for Kalunde (N = 11 individuals, R = 0.008) was not significantly higher than the relatedness among females. Five mature males were conceived while Kalunde was the high-ranking male, and he was not genetically excluded as the father of two of these five males. Therefore, Kalunde might have paternal relatives such as sons or nephews, which lead to the significant difference in relatedness between sexes.

Inoue *et al.* (2008a) also analysed the effect of immigration. There were four natal females who gave birth while in the M group (Inoue *et al.*, 2008a). The relatedness among natal female–male pairs (N = 48 dyads, R = 0.056) was significantly higher than that among immigrant female–male pairs (N = 144 dyads, R = −0.015). This suggests that the risk of inbreeding was higher among natal females than among immigrant females as expected. This result also suggested that female transfer might act to maintain low average relatedness within the group.

46.6 Comparison between Mahale and other populations

46.6.1 Haplotype sharing

The study of four chimpanzee unit-groups in Uganda showed that the genetic patterns of mitochondrial DNA and Y-STR reflected female-biased dispersal (Langergraber *et al.*, 2007). Using these data in Uganda (Langergraber *et al.*, 2007) and the data on Mahale (Inoue *et al.*, 2008b, 2013; this study), I constructed a neighbor-joining network tree of Y-STR and mitochondrial D-loop with NETWORK 4.6.1.0 (Fluxus Technology). A network tree of mitochondrial D-loop (Figure 46.3) was different from that of Y-STR (Figure 46.4). As expected from the female-biased dispersal patterns in chimpanzees (Nishida and Kawanaka, 1972; Pusey, 1979; Boesch and Boesch-Achermann, 2000), Y-STR haplotypes at Mahale were different from the haplotypes in Uganda (Figure 46.4), and mitochondrial haplotypes did not sort into location-specific clusters (Figure 46.3). Mitochondrial

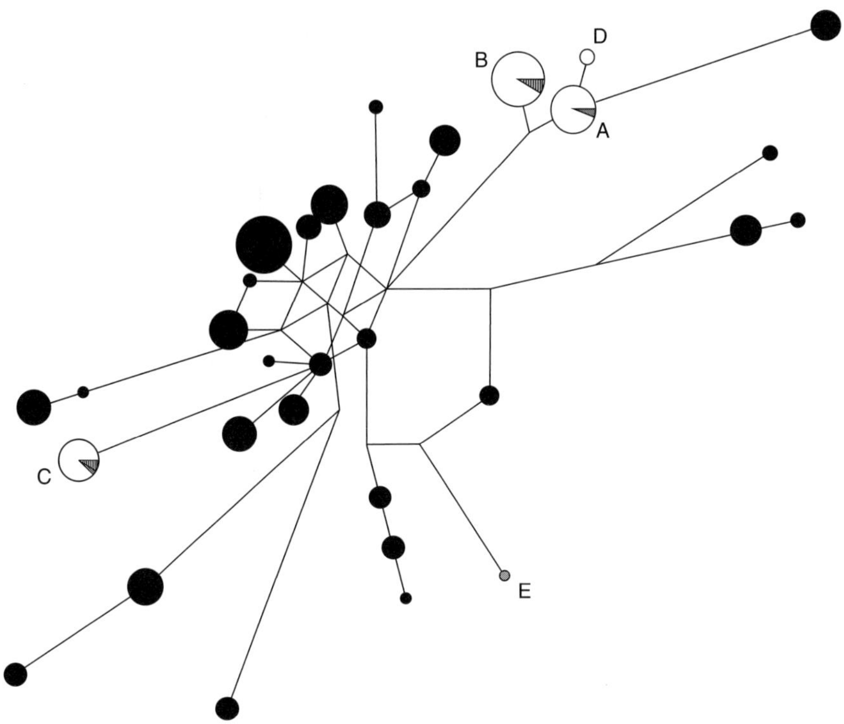

Figure 46.3 Median-joining networks of mitochondrial D-loop in the chimpanzee population: Mahale M (white), Mahale Y (vertical), Mahale south (gray), Uganda (black; Langergraber *et al.*, 2007). Circle size is proportional to haplotype frequency.

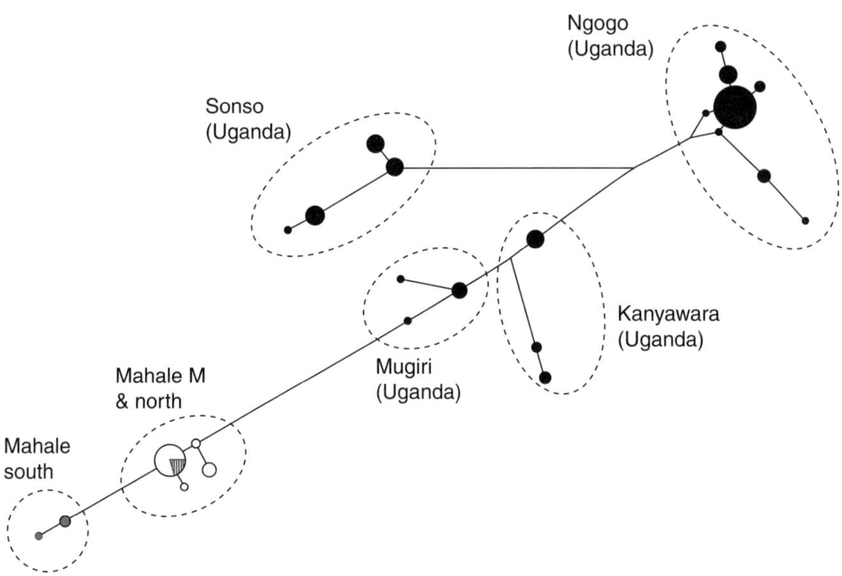

Figure 46.4 Median-joining networks of Y-STR in the chimpanzee population: Mahale M (white), Mahale Y (vertical), Mahale south (gray), Uganda (black; Langergraber *et al.*, 2007). Circle size is proportional to haplotype frequency.

haplotypes were not shared between Mahale and Uganda, and some haplotypes found in Uganda, however, were close to the haplotypes found in Mahale. As this result shows, the comparison between mitochondrial DNA and Y-STR patterns is useful to describe the sex-biased dispersal pattern.

Recent studies comparing the variations between mitochondrial and Y-chromosomal haplotypes clarified sex-biased dispersal patterns in gorillas (Douadi *et al.*, 2007) and orangutans (Nater *et al.*, 2011; Nietlisbach *et al.*, 2012). In both apes, mitochondrial haplotypes were relatively restricted to the specific habitats compared with Y-chromosomal haplotypes (Douadi *et al.*, 2007; Nater *et al.*, 2011; Nietlisbach *et al.*, 2012). These genetic patterns were quite different from those of chimpanzees. Therefore, extremely female-biased dispersal pattern is a specific character of *Pan* species among great apes.

46.6.2 Genetic diversity

Genetic diversity in the peripheral population of the species distribution is expected to be lower than that in the center of the distribution (Hoffmann and Blows, 1994). Garner *et al.* (2004) showed that genetic diversity was higher in the place closer to the postglacial refugium. The Western Tanzanian chimpanzee population is the southernmost population of eastern chimpanzees, which have recently expanded from the eastern DRC (Goldberg and Ruvolo, 1997a). These facts imply that the genetic diversity in the Tanzanian population is lower than that in other populations of eastern chimpanzees. Mitochondrial analyses showed that the genetic diversity in the Tanzanian population was in fact lower than that in the Uganda population (Inoue *et al.*, 2013).

Here, I calculated the haplotype diversity (H; Nei, 1987) of Y-STR in Mahale using Arlequin software (Excoffier *et al.*, 2005), in order to compare Y-STR diversity in Mahale with that in Uganda (Langergraber *et al.*, 2007). The haplotype diversity of Y-STR in Mahale M group ($H = 0.52$) was slightly lower than that in four Ugandan groups ($H = 0.63$, $SD = 0.08$, Langergraber *et al.*, 2007). Previous reports showed that the haplotype diversity of the mitochondrial D-loop in the Mahale M group ($H = 0.68$, Inoue *et al.*, 2013) was lower than that in the four Ugandan groups ($H = 0.87$, $SD = 0.04$, Langergraber *et al.*, 2007). The diversity of both Y-STR and mtDNA was lower in Mahale, but the difference in Y-STR was smaller than that in mtDNA considering the SD values of Ugandan groups. Little difference in Y-STR diversity between two populations probably reflects female-biased dispersal with chimpanzees. Male-mediated gene flow between groups rarely occurs in chimpanzees, and therefore Y-STR diversity may not be affected by the distance from the postglacial refugium, but may be affected by genetic drift or male reproductive skew like the Mahale M group (see Section 46.4.2).

References

Becquet, C., Patterson, N., Stone, A., Przeworski, M., and Reich, D. (2007). Genetic structure of chimpanzee populations. *PLoS Genetics*, **3**(4), e66.

Boesch, C. and Boesch-Achermann, H. (2000). *The Chimpanzees of the Taï Forest: Behavioural Ecology and Evolution*. Oxford: Oxford University Press.

Boesch, C., Kohou, G., Nene, H., and Vigilant, L. (2006). Male competition and paternity in wild chimpanzees of Taï Forest. *American Journal of Physical Anthropology*, **130**, 103–15.

Constable, J., Ashley, M., Goodall, J., and Pusey, A. (2001). Noninvasive paternity assignment in Gombe chimpanzees. *Molecular Ecology*, **10**, 1279–300.

Douadi, M. I., Gatti, S., Levrero, F., *et al.* (2007). Sex-biased dispersal in western lowland gorillas (*Gorilla gorilla gorilla*). *Molecular Ecology*, **16**, 2247–59.

Eriksson, J., Siedel, H., Lukas, D., *et al.* (2006). Y-chromosome analysis confirms highly sex-biased dispersal and suggests a low male effective population size in bonobos (*Pan paniscus*). *Molecular Ecology*, **15**, 939–49.

Excoffier, L., Laval, G., and Schneider, S. (2005). Arlequin ver. 3.0: an integrated software package for population genetics data analysis. *Evolutionary Bioinformatics Online*, **1**, 47–50.

Gagneux, P., Wills, C., Gerloff, U., *et al.* (1999). Mitochondrial sequences show diverse evolutionary histories of African hominoids. *Evolution*, **96**, 5077–82.

Garner, T. W., Pearman, P. B., and Angelone, S. (2004). Genetic diversity across a vertebrate species' range: a test of the central–peripheral hypothesis. *Molecular Ecology*, **13**, 1047–53.

Goldberg, T. L. and Ruvolo, M. (1997a). The geographic apportionment of mitochondrial genetic diversity in East African chimpanzees, *Pan troglodytes schweinfurthii. Molecular Biology and Evolution*, **14**, 976–84.

Goldberg, T. L. and Ruvolo, M. (1997b). Molecular phylogenetics and historical biogeography of East African chimpanzees. *Biological Journal of the Linnean Society*, **61**, 301–24.

Goldberg, T. L. and Wrangham, R.W. (1997). Genetic correlates of social behaviour in wild chimpanzees: evidence from mitochondrial DNA. *Animal Behaviour*, **54**, 559–70.

Gonder, M. K., Locatelli, S., Ghobrial, L., *et al.* (2011). Evidence from Cameroon reveals differences in the genetic structure and histories of chimpanzee populations. *Proceedings of the National Academy of Sciences of the USA*, **108**, 4766–71.

Goodall, J. (1983). Population dynamics during a 15 year period in one community of free-living chimpanzees in the Gombe National Park, Tanzania. *Zeitschrift für Tierpsychologie*, **61**, 1–60.

Hashimoto, C., Furuichi, T., and Takenaka, O. (1996). Matrilineal kin relationship and social behavior of wild bonobos (*Pan paniscus*): Sequencing the D-loop region of mitochondrial DNA. *Primates*, **37**, 305–18.

Hayakawa, S. and Takenaka, O. (1999). Urine as another potential source for template DNA in polymerase chain reaction (PCR). *American Journal of Primatology*, **48**, 299–304.

Hoffmann, A. A. and Blows, M. W. (1994). Species borders: ecological and evolutionary perspectives. *Trends in Ecology & Evolution*, **9**, 223–7.

Inoue, E., Inoue-Murayama, M., Nishida, T., Vigilant, L., and Takenaka, O. (2007a). [Paternity analyses in a wild chimpanzee group.] *DNA Polymorphism*, **15**, 54–8. In Japanese.

Inoue, E., Inoue-Murayama, M., Takenaka, O., and Nishida, T. (2007b). Wild chimpanzee infant urine and saliva sampled noninvasively usable for DNA analyses. *Primates*, **48**, 156–9.

Inoue, E., Inoue-Murayama, M., Vigilant, L., Takenaka, O., and Nishida, T. (2008a). Relatedness in wild chimpanzees: influence of paternity, male philopatry, and demographic factors. *American Journal of Physical Anthropology*, **137**, 256–62.

Inoue, E., Inoue-Murayama, M., Vigilant, L., Takenaka, O., and Nishida, T. (2008b). [Y-STR polymorphism in a wild chimpanzee group.] *DNA Polymorphism*, **16**, 21–4. In Japanese.

Inoue, E., Tashiro, Y., Ogawa, H., *et al.* (2013). Gene flow and genetic diversity of chimpanzees in Tanzanian habitats. *Primate Conservation*, **26**, 67–74.

Inskipp, T. (2005). Chimpanzee (*Pan troglodytes*). In: *World Atlas of Great Apes and their Conservation*, ed. J. Caldecott and L. Miles. Berkeley, CA: University of California Press, pp.53–81.

Kawamoto, Y., Shotake, T., Nozawa, K., *et al.* (2007). Postglacial population expansion of Japanese macaques (*Macaca fuscata*) inferred from mitochondrial DNA phylogeography. *Primates*, **48**, 27–40.

Kayser, M. and Sajantila, A. (2001). Mutation at Y-STR loci: implications for paternity testing and forensic analysis. *Forensic Science International*, **118**, 116–21.

Keele, B. F., Jones, J. H., Terio, K. A., *et al.* (2009). Increased mortality and AIDS-like immunopathology in wild chimpanzees infected with SIVcpz. *Nature*, **460**, 515–19.

Kurihara, R., Yamamoto, T., Uchihi, R., *et al.* (2004). Mutations in 14 Y-STR loci among Japanese father–son haplotypes. *International Journal of Legal Medicine*, **118**, 125–31.

Langergraber, K. E., Mitani, J. C., and Vigilant, L. (2007). The limited impact of kinship on cooperation in wild chimpanzees. *Proceedings of the National Academy of Sciences of the USA*, **104**, 7786–90.

Lukas, D., Reynolds, V., Boesch, C., and Vigilant, L. (2005). To what extent does living in a group mean living with kin? *Molecular Ecology*, **14**, 2181–96.

Morin, P. A., Moore, J. J., Chakraborty, R., *et al.* (1994). Kin selection, social structure, gene flow, and the evolution of chimpanzees. *Science*, **265**, 1193–201.

Morin, P. A., Chambers, K. E., Boesch, C., and Vigilant, L. (2001). Quantitative polymerase chain reaction analysis of DNA from noninvasive samples for accurate microsatellite genotyping of wild chimpanzees (*Pan troglodytes verus*). *Molecular Ecology*, **10**, 1835–44.

Nater, A., Nietlisbach, P., Arora, N., *et al.* (2011). Sex-biased dispersal and volcanic activities shaped phylogeographic patterns of extant orangutans (genus: *Pongo*). *Molecular Biology and Evolution*, **28**, 2275–88.

Nei, M. (1987). *Molecular Evolutionary Genetics*. New York: Columbia University Press.

Nietlisbach, P., Arora, N., Nater, A., *et al.* (2012). Heavily male-biased long-distance dispersal of orang-utans (genus: *Pongo*), as revealed by Y-chromosomal and mitochondrial genetic markers. *Molecular Ecology*, **21**, 3173–86.

Nishida, T. and Kawanaka, K. (1972). Inter-unit-group relationships among wild chimpanzees of the Mahali Mountains. *Kyoto University African Studies*, **7**, 131–69.

Nishida, T., Hiraiwa-Hasegawa, M., and Hasegawa, T. (1985). Group extinction and female transfer in wild chimpanzees in the Mahale National Park, Tanzania. *Zeitschrift für Tierpsychologie*, **67**, 284–301.

Nishida, T., Takasaki, H., and Takahata, Y. (1990). Demography and reproductive profiles. In *The Chimpanzees of the Mahale Mountains*, ed. T. Nishida. Tokyo: University of Tokyo Press. pp. 63–97.

Nishida, T., Corp, N., Hamai, M., *et al.* (2003). Demography, female life history, and reproductive profiles among the chimpanzees of Mahale. *American Journal of Primatology*, **59**, 99–121.

Nsubuga, A. M., Robbins, M. M., Roeder, A. D., *et al.* (2004). Factors affecting the amount of genomic DNA extracted from ape faeces and the identification of an improved sample storage method. *Molecular Ecology*, **13**, 2089–94.

Oates, J. F., Tutin, C. E. G., Humle, T., *et al.* (2008). *Pan troglodytes*. The *IUCN Red List of Threatened Species. Version 2013.1*. Available at: www.iucnredlist.org. Accessed August 30, 2013.

Piel, A. K., Stewart, F. A., Pintea, L., *et al.* (2013). The Malagarasi River does not form an absolute barrier to chimpanzee movement in Western Tanzania. *PLoS ONE*, **8**(3), e58965.

Pusey, A. E. (1979). Intercommunity transfer of chimpanzees in Gombe National Park. In *The Great Apes*, ed. D. Hamburg and E. McCown. Menlo Park, CA: Benjamin/Cummings. pp. 465–79.

Sakamaki, T. and Nakamura, M. (2007). Preliminary survey of unhabituated chimpanzees in the Mahale Mountains National Park, Tanzania: Behavioral diversity across neighboring unit-groups and intergroup relationships. In *Formation of a Strategic Base for Biodiversity Studies, The 21st Century COE Program of Kyoto University*, pp. 278–80.

Stone, A. C., Battistuzzi, F. U., Kubatko L. S., *et al.* (2010). More reliable estimates of divergence times in *Pan* using complete mtDNA sequences and accounting for population structure. *Philosophical Transactions of the Royal Society B: Biological Sciences*, **365**, 3277–88.

Takahata, H. and Takahata, Y. (1989). Inter-group transfer of an immature male of the common chimpanzee and his social interactions in the non-natal group. *African Study Monographs*, **9**, 209–20.

Vigilant, L., Hofreiter, M., Siedel, H., and Boesch, C. (2001). Paternity and relatedness in wild chimpanzee communities. *Proceedings of the National Academy of Sciences of the USA*, **98**, 12890–5.

Yoshikawa, M., Ogawa, H., Sakamaki, T., and Idani, G. (2008). Population density of chimpanzees in Tanzania. *Pan Africa News*, **15**, 17–20.

47 Internal parasites

TAKANORI KOORIYAMA AND HIDEO HASEGAWA

47.1 Intestinal parasites at Mahale

Parasitological studies of chimpanzees in Mahale have been conducted since the late 1970s, mainly on K and M groups in the Kasoje area. Kawabata and Nishida (1991) first reported on the parasite fauna, and recorded protozoans, cestodes, trematodes, and nematodes from chimpanzee feces. Parasite fauna and prevalence varied according to the chimpanzee study site (Ashford *et al.*, 2000). The prevalence depends on the habitat environment, climate, food resources, group size, home range, and behavior. Among chimpanzee habitats in Tanzania, Mahale and Gombe consist of semi-deciduous forest. Both are located on the shores of Lake Tanganyika, but more than 100 km apart. The habitat at Ugalla is woodland, and is located inland from the lake, between two national parks. These three areas are separated mostly by human habitations. The estimated chimpanzee population size at Ugalla is smaller than that at Mahale, and fewer than 700 chimpanzees have been estimated to reside outside the national parks (Yoshikawa *et al.*, 2008).

Parasite species recorded in chimpanzee fecal samples in Mahale are listed in Table 47.1. The prevalence of parasites among chimpanzees differs by parasite species: chimpanzees show higher prevalences of *Oesophagostomum* spp., *Strongyloides* sp., *Probstmayria gombensis*, *Troglodytella abrassarti*, and *Troglocorys cava*, but lower prevalences of *Trichuris* spp., *Bertiella* sp., Dicrocoeliidae gen sp., and *Entamoeba* sp., as determined by microscopic analysis of fecal specimens. Spirurida nematode infection was not observed in chimpanzees despite infection of other primate species at Mahale (Kooriyama *et al.*, 2012).

Modes of infection are also important (see Table 47.2). These are divided into two types, with or without an intermediate host, which affects parasite prevalence in combination with other factors. The parasite species and their characteristics are summarized below.

47.2 Parasite examination

Several studies used microscopy to identify parasite species morphologically. However, the preparation methods for microscopic examination varied: direct smear methods (Kawabata and Nishida, 1991), McMasters floatation and formalin–ether concentration method, combined with EPG calculation (Huffman *et al.*, 1997), saturated NaCl floatation method (Nigi *et al.*, 1998), formalin–ether concentration and Harada-Mori culture methods (Hasegawa *et al.*, 2010), and formalin–ether concentration method (Kooriyama *et al.*, 2012). For the examination of protozoans, basic iodine staining or polyvinyl alcohol fixation techniques for feces smeared on slide glass is recommended. Microscopic fecal examinations can yield lower prevalences of parasite eggs, such as *Bertiella* sp., because such cestodes normally do not liberate eggs in the host intestine but pass their senile proglottids in feces.

Morphological analysis is the basic technique, but species identification from eggs is difficult or impossible for some species. Recent trends in parasitology are to use molecular biological techniques; such methods should also be used for parasitological investigation of the Mahale chimpanzees.

Mahale Chimpanzees: 50 Years of Research, ed. M. Nakamura, K. Hosaka, N. Itoh and K. Zamma. Published by Cambridge University Press.

Table 47.1 **Prevalence of intestinal parasites of chimpanzees in Mahale Mountains National Park, Tanzania**

Parasite type																Remarks
Group		K	M	M		M		M	M	M	M		M		M	
Study period		1975–76		1989–90		1991–2		1993–4	1997	2005	2007		2008		2009	
		Sep–Feb		Aug–Feb		Aug–Jan		Nov–Feb	Aug–Oct	Jun–Aug	Jul–Oct		Apr–Oct		Oct	
No. of fecal samples investigated		72	81	161		156		86	46	52	201		51		2	
No. of chimpanzees providing samples		30	90		49 (48)[b]		49	17	34	48		43		29		
Examination methods[a]		D		FE, M		FE, M		FE, M	S	D, F	FE		FE		FE	
Parasite species																
Protozoa																
Amoebida	*Entamoeba* spp.										11	14	10	14		
	Entamoeba coli			21	8	14	12				34	23	37	24	50	
	Endolimax nana			7	2	6	2									
	Iodamoeba buetschlii			6	2	2	1									
Diplomonadida																
	Giardia lamblia			2	1											
Ciliata																
	Troglodytella abrassarti			24	12					92	34	67	51	62	100	
	Troglocorys cava										26	54	39	45	50	
Acanthocephala																
Oligancanthorhynchidae	*Prosthenorchis* sp.	1														
Trematoda																
Dicrocoeliidae	*Dicrocoelium dendriticum* (syn. *D. lanceoatum*)			2	1	0	0									

Cestoda																	
Anoplocephalidae	*Bertiella studeri*	1										2	5	2	3		
Nematoda																	
Trichuroidea	*Trichuris* spp.	2								9		3	12	4	7		
	Trichuris trichiura			16	7	27	24	36	41								
Strongyloidea	*Oesophagostomum* spp.	55								91							
	Oesophagostomum stephanostomum			41 (42)[b]	18	59	37	52	71			57	77	76	59	50	Genetically proved (Gasser *et al.*, 1999)
		64	46														
Rhabditoidea	*Strongyloides* spp.	22								65		47	72	53	48	50	
		24	21														
	Strongyloides fuelleborni			38 (35)[b]	14	31	13	47	65								Genetically proved (Hasegawa *et al.*, 2010)
	Strongyloides stercoralis																
Cosmoceroidea																	
	Probstmayria gombensis											12	42	14	21		
Unidentified nematoda		1		12	5	31	11	10	41	29							
References		Kawabata and Nishida, 1991		Ashford *et al.*, 2000						Nigi *et al.*, 1998	Kaur *et al.*, 2010[c]	Kooriyama *et al.*, 2012					

[a] D, Direct; FE, formalin–ether; M, McMaster; S, saturated NaCl; F, formalin.

[b] The values have some minor deviation among these sources, Ashford *et al.* (2000) modified the data on Huffman *et al.* (1993, 1996, 1997).

[c] *Troglodytella abrassarti* has been surveyed.

Table 47.2 **List, characteristics, and mode of infection of chimpanzee parasites found in Mahale Mountains National Park, Tanzania. Sources of the parasite species: Huffman et al., 1990, 1997; Kawabata and Nishida, 1991; Nigi et al., 1998; Gasser et al., 1999; Kooriyama et al., 2012**

Group	Species	Characteristics of forms appearing in the feces	Mode of infection
Protozoa			
Amoebida	*Entamoeba coli*	Cysts: 10–35 μm, spherical, but may be elongated. Mature cysts possess 8, 16, or more nuclei. Compact or diffused karyosomes. Peripheral chromatin coarse and granular, and irregularly arranged along the nuclear membrane. Diffused glycogen.	Ingestion of cyst
	Entamoeba spp.	Cysts: 8–15 μm, smaller and fewer than four nuclei. These species were not discriminated from other *Entamoeba* species but *Entamoeba coli* (Kooriyama *et al.*, 2012).	
	Endolimax nana	Cysts: 5–10 μm, spherical to ellipsoidal in shape. Mature cysts possess four nuclei with large karyosomes and no peripheral chromatin. Diffused glycogen in cytoplasm without chromatoid bodies.	Ingestion of cyst
	Iodamoeba buetschlii	Cysts: 5–20 μm, large glycogen vacuole, a large karyosome.	Ingestion of cyst
Diplomonadida	*Giardia lamblia*	Cysts: 8–19 μm, oval or ellipsoidal, 4 nuclei. Trophozoite: 10–20 μm, pear shaped or leaf shaped, 2 nuclei, 4 lateral, 2 ventral, 2 caudal flagella.	Ingestion of cyst
Ciliata	*Troglodytella abrassarti*	Trophozoite: 108–195 × 66–110 μm, circular corn shape and helical cilia from the bottom to the apex; nucleus L-shaped.	Ingestion of trophozoite
	Troglocorys cava	Trophozoite: 32–42 × 25–35 μm, a spherical body with a frontal lobe, a long vestibulum, a cytoproct located at the posterior dorsal side of the body, an ovoid macronucleus, a contractile vacuole near the cytoproct, and a large concavity on the left surface of the body.	Ingestion of trophozoite
Acanthocephala			
Oligancanthorhynchidae	*Prosthenorchis* sp.	Eggs: 60–80 × 42–50 μm, ellipsoid with thick 3-layered shell, containing an embryo (acanthor) with visible spines at one end.	Ingestion of intermediate host
Trematoda			

Dicrocoeliidae	*Dicrocoelium dendriticum* (syn. *D. lanceolatum*)	Eggs: 35–45 × 20–30 μm, ellipsoidal, operculated, and characteristically thick-shelled with yellow-brown coloration.	Ingestion of intermediate host
Cestoda			
Anoplocephalidae	*Bertiella studeri*	Eggs: 43–48 × 43–48 μm, spherical with thick wall, with pyriform apparatus containing onchosphere.	Ingestion of intermediate host (oribatid mites)
Nematoda			
Trichuroidea	*Trichuris trichiura*/ *Trichuris* spp.	Eggs: 45–71 × 23–40 μm, barrel-shaped, thick, brown shell with plugs at each pole. Demonstrations of *Trichuris* species were based on their egg morphology, but identifications depended on sources.	Ingestion of larvated egg
Strongyloidea	*Oesophagostomum stephanostomum*/ *Oesophagostomum* spp.	Eggs: 57–93 × 37–60 μm, elliptical, thin-shelled, and containing blastomeres. Demonstrations of *Oesophagostomum* species were based on their egg morphology, but their shapes or sizes are indistinguishable from the other strongyles. The identifications depended on sources.	Ingestion of larva
Rhabditoidea	*Strongyloides fuelleborni*	Eggs: 40–63 × 20–40 μm, elliptical, thin-shelled, and larvated.	Skin penetration by infective larva
	Strongyloides stercoralis	Rhabditiform larva: 200–300 × 16–20 μm, buccal cavity very short, esophagus of rhabditoid form.	Skin penetration by infective larva
	Strongyloides spp.	Hatched larva of *S. fuelleborni* is indistinguishable from *S. stercoralis*, thus these two species were referred to collectively depending on sources.	
Cosmoceroidea	*Probstmayria gombensis*	Six globose lips at the mouth, large excretory vesicle; female didelphic, viviparous, with one well-developed larva in uterus along with some eggs in progressive stages of development in the oviduct; male with gubernaculum of concave triangular shape, four pairs of small, inconspicuous, postanal papillae; cloaca lined with thick transparent cuticle.	Ingestion of larva

47.3 Protozoa

47.3.1 Amoebida

Entamoeba coli, Endolimax nana, and *Iodamoeba buetschlii* have been reported from Mahale (Huffman *et al.*, 1997; Kooriyama *et al.*, 2012). *Entamoeba* species can be identified by the number of nuclei as well as their size and shape by iodine or trichrome staining at the cyst stage: *Entamoeba coli* forms multi-nucleated (eight nuclei) cysts (Figure 47.1a), *E. nana* is a small amoeba with four nuclei, each with a large

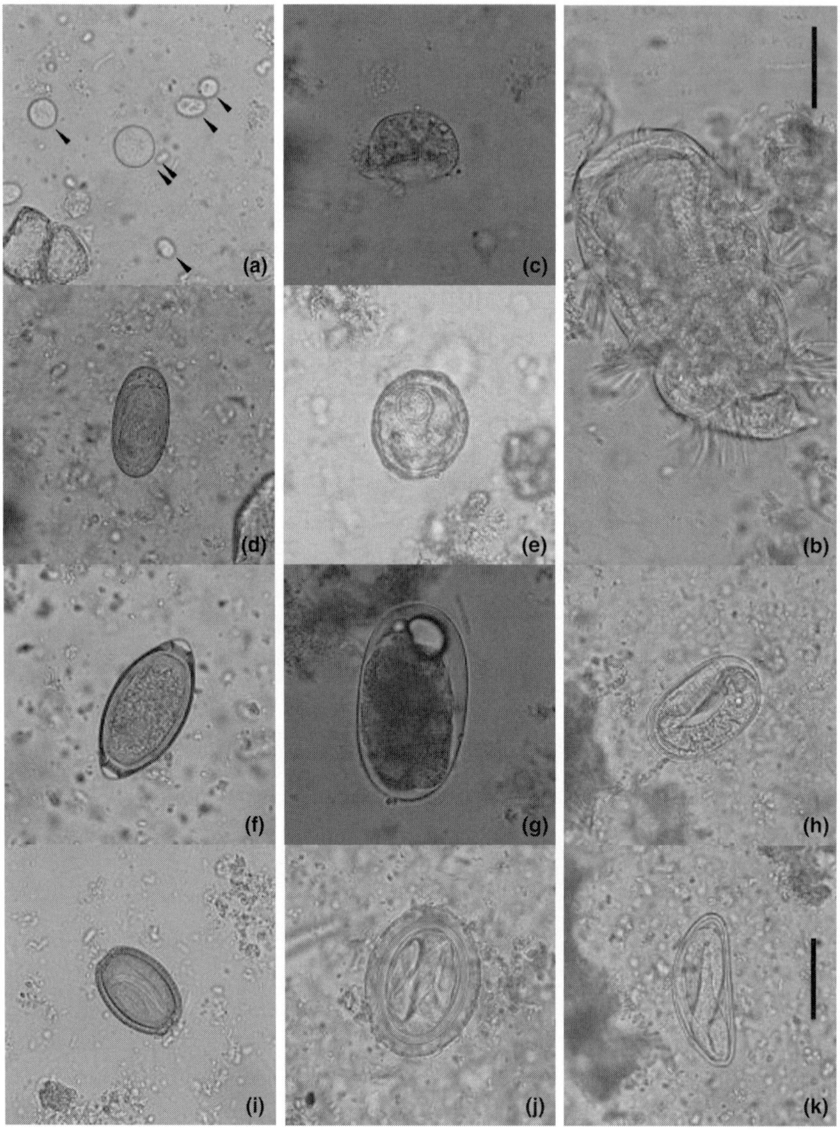

Figure 47.1 Protozoans and eggs of parasitic helminths in five primates in Mahale. (a) *Entamoeba coli* (double arrow heads) and *Entamoeba* spp. (single arrow head) of a red colobus, (b) *Troglodytella abrassarti* trophozoite of a chimpanzee, (c) *Troglocorys cava* of a chimpanzee, (d) Dicrocoeliidae gen sp. of a yellow baboon, (e) *Bertiella studeri* of a chimpanzee, (f) *Trichuris* sp. of a red colobus, (g) *Oesophagostomum stephanostomum* of a chimpanzee, (h) *Strongyloides fuelleborni* of a chimpanzee, (i) *Chitwoodspirura* sp. of a red-tailed monkey, (j) Spirurina fam. gen. sp. or spp. in a yellow baboon, and (k) *Enterobius* sp. of a red colobus (scale bar = 30 µm). Photographs adapted from Kooriyama *et al.* (2012).

karyosome, and *Iodamoeba buetschlii* has a nucleus with a large karyosome and large glycogen vacuole. Infection of *Entamoeba* spp. occurs by ingestion of cysts, followed by proliferation of the trophozoite in the host intestine. These three non-pathogenic *Entamoeba* species have been identified in Mahale, but the pathogenic *E. histolytica* has not. Molecular biological analyses are necessary to differentiate between *E. histolytica* and *E. dispar* (non-pathogenic) because these two species are indistinguishable morphologically.

47.3.2 Ciliata

In Mahale, the entodiniomorphid ciliates *Troglodytella abrassarti* and *Troglocorys cava* have been found in chimpanzees (Huffman *et al.*, 1990, 1997; Kaur *et al.*, 2010; Kooriyama *et al.*, 2012). *Neobalantidium coli* (syn. *Balantidium coli*; Pomajbíková *et al.*, 2013) has not been detected in chimpanzees, even though sympatric yellow baboons were infected with *N. coli* (Kooriyama *et al.*, 2012). This is similar to the situation in Mt. Assirik and Fongoli, Senegal (McGrew *et al.*, 1989; Howells *et al.*, 2011). In contrast, *N. coli* infections among chimpanzees and among sympatric olive baboons have been reported in Gombe (Gillespie *et al.*, 2010). *N. coli* infection is common in captive apes and can be pathogenic under captive conditions (Kim *et al.*, 1978). However, the risks to nonhuman primates have not been determined, although the pathogenicity of *N. coli* in humans is well known (Gillespie *et al.*, 2010).

Troglodytella abrassarti is commonly found in chimpanzees and has a circular corn shape and helical cilia from the bottom to the apex (Figure 47.1b). *T. abrassarti* contributes to cellulose digestion in the chimpanzee colon (Profousová *et al.*, 2011). Its appearance in feces may indicate that the ciliate proliferated in the host intestine under "good" enteric conditions. Vallo *et al.* (2012) surveyed the molecular diversity of the small subunit ribosomal DNA (SSU rDNA) region of *T. abrassarti* collected from four chimpanzee subspecies (*Pan troglodytes verus, P. t. ellioti, P. t. troglodytes*, and *P. t. schweinfurthii*), and bonobos (*P. paniscus*). The *T. abrassarti* of *P. t. schweinfurthii* belonged to type II together with those of *P. t. ellioti, P. t. troglodytes*, and bonobos; only *T. abrassarti* of *P. t. verus* belonged to Type I. *T. abrassarti* is a mutualistic ciliate of great apes; indeed, these may be coevolved species. A previous study on mitochondrial DNA showed that the western chimpanzee, *P. t. verus*, is exclusive, as well as geographically separated, from the three other chimpanzee subspecies.

An entodiniomorphid ciliate, *Troglocorys cava*, was detected in chimpanzees at Mahale (Figure 47.1c). This ciliate was demonstrated in wild chimpanzees from East to West African study fields, including Ugalla, Tanzania, but it has not been detected in captive chimpanzees or chimpanzees reintroduced after captivity (Pomajbíková *et al.*, 2012).

47.3.3 Other protozoa

Giardia lamblia has been found in Mahale (Huffman *et al.*, 1997), but not as a major infection. *Giardia* is potentially pathogenic to chimpanzees and zoonotic between human and nonhuman primates (Graczyk *et al.*, 2002). The transmission routes of *Giardia* in Mahale have not been determined. However, human settlements were present in Mahale before it became a national park (Chapter 2). Thus, past sharing of the living environment of the chimpanzees with local people and their livestock might have resulted in *Giardia* transmission (Ashford *et al.*, 2000; Bakuza and Nkwengulila, 2009). *Chilomastix* sp., *Entamoeba chattoni, E. hartmanni*, and *Blastocystis* sp. have not been found in chimpanzees of Mahale, but have been reported at other chimpanzee study sites in Africa (Ashford *et al.*, 2000).

47.4 Trematoda

47.4.1 Dicrocoeliidae

Dicrocoelium dendriticum (syn. *D. lanceolatum*) infection has been reported in Mahale, but it is not a major infection (Huffman *et al.*, 1997). Dicrocoeliid eggs are ellipsoidal, operculated, and characteristically thick-shelled with a yellow–brown coloration. Dicrocoeliidae gen. sp. was highly prevalent among vervet monkeys and yellow baboons in Mahale (Figure 47.1d). Dicrocoeliids generally require intermediate insect hosts, such as ants and orthopteran insects. Orthopteran insects live on the ground, mostly on grasses, and are prey of these monkeys. Although chimpanzees often eat ants (Chapter 21), the frequently eaten species (*Camponotus* spp.) have not been demonstrated to be dicrocoeliid vectors. Interestingly, dicrocoeliid infection is not common in chimpanzees but is common in bonobos (Hasegawa *et al.*, 1983).

47.5 Cestoda

47.5.1 Anoplocephalidae

Bertiella eggs and senile proglottids have been found in the feces of chimpanzees in Mahale albeit at a low prevalence (Kawabata and Nishida, 1991; Huffman *et al.*, 2009; Kooriyama *et al.*, 2012). *Bertiella* eggs are spherical with thick walls, and have a characteristic pyriform apparatus containing the onchosphere (Figure 47.1e). The presence of eggs in feces is rare, but senile proglottids filled with eggs are often found. Eggs containing pyriform apparatus are ingested by oribatid mites that live in the soil or lower branches of fruit trees, and so mite-contaminated foods (such as fruits) are possible sources of infection (Stunkard *et al.*, 1964; Bhagwant, 2004). Fallen fruits may have a higher risk of infection than fruits on trees.

47.6 Acanthocephala

47.6.1 Oligacanthorhynchidae

Kawabata and Nishida (1991) demonstrated *Prosthenorchis* sp. eggs in the feces of chimpanzees of Mahale with a low prevalence, 0.7%. *Prosthenorchis* sp. is an acanthocephalan and a concern for New World primates, especially tamarins, squirrel monkeys, and marmosets, because the adult worms are large and often induce intestinal obstruction (Yamashita, 1963). Other reports from Mahale did not demonstrate *Prosthenorchis* sp.

47.7 Nematoda

47.7.1 Trichuroidea

Trichuris eggs were found in chimpanzees in all previous reports from Mahale. *Trichuris* sp. eggs have a thick, brown shell with a plug at each pole (Figure 47.1f). Transmission occurs through the ingestion of soil or food contaminated with infective larvated eggs. The eggs become infective in 15–30 days (Bezjian *et al.*, 2008). A recent report showed *Trichuris* infections in red colobus monkeys, red-tailed monkeys, vervet monkeys, and yellow baboons in Mahale, and the prevalence among chimpanzees was low compared with that in other primates (Kooriyama *et al.*, 2012). This low prevalence is similar to the results of studies at Gombe and Fongoli (Murray *et al.*, 2000; Howells *et al.*, 2011).

47.7.2 Strongyloidea

Oesophagostomum spp. infections were identified morphologically by eggs and/or adult worms in all reports from Mahale. *Oesophagostomum* eggs are elliptical, thin-shelled, and contain blastomeres (Figure 47.1g). *Oesophagostomum* infection is primarily acquired by oral ingestion or skin penetration by infective larvae. Adult worms were sometimes collected from relatively soft feces of chimpanzees, and can assist its identification as *O. stephanostomum* morphologically (Kooriyama *et al.*, 2012). Specifically, the worms have a funnel-shaped buccal cavity and corona radiata with numerous elements (Figure 47.2a). Huffman *et al.* (1996) examined adult worms using scanning electron microscopy and identified them to be *O. stephanostomum*, morphologically. Using molecular biological techniques, *O. stephanostomum* was previously identified in chimpanzees of Mahale

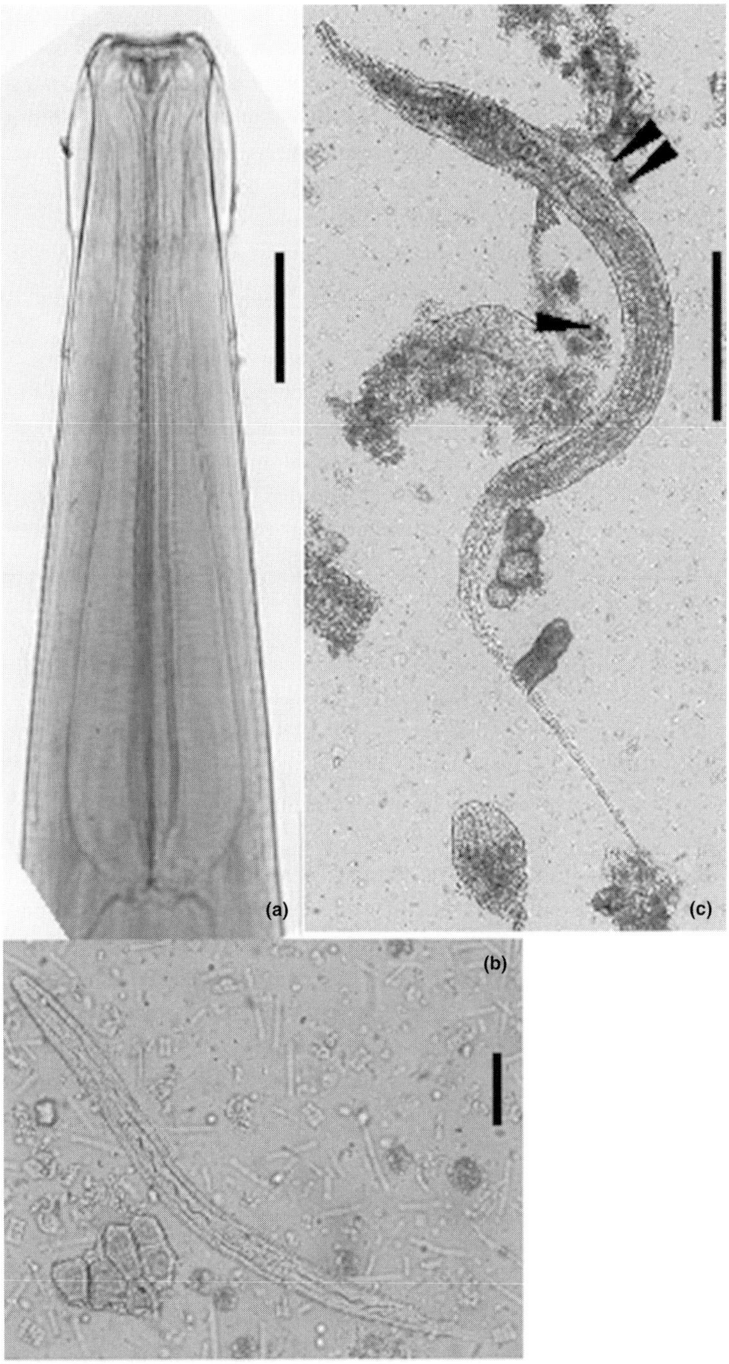

Figure 47.2 Nematodes appearing in feces of primates in Mahale. (a) *Oesophagostomum stephanostomum* of a chimpanzee; (b) *Strongyloides* sp., rhabditoid larva, of a chimpanzee; (c) *Probstmayria gombensis* of a chimpanzee, larva (arrow head) and egg (double arrow heads) are seen in its body (scale 200 μm for a and c; 30 μm for b). Photographs were adapted from Kooriyama *et al.* (2012) and high-resolution image of "photo a" is courtesy of Dr. T. Iwaki.

(Gasser *et al.*, 1999). *Oesophagostomum* spp. are common in African monkeys and apes, and were observed among many primates of Mahale (chimpanzees, red colobus monkeys, red-tailed monkeys, vervet monkeys, and yellow baboons) (Kooriyama *et al.*, 2012). Murray *et al.* (2000) compared the prevalence of *Oesophagostomum* infection coprologically between chimpanzees (terrestrial/arboreal) and baboons (mostly terrestrial/partly arboreal) in the same lakeshore forest at Gombe. The results showed that prevalence in chimpanzee was higher than in baboons, similar to the results from Mahale (Kooriyama *et al.*, 2012). Terrestriality and arboreality may also be important in sustaining *Oesophagostomum* infection, combined with other factors, such as group size, home range, and climate. *Oesophagostomum* is known as the nodular worm in ruminants, pigs, nonhuman primates, and humans. Its infection produces oesophagostomiasis by producing multinodular lesions (granulomas) in the gut wall. Wild chimpanzees rarely show signs of oesophagostomatitis even when infected with *Oesophagostomum* at a high rate. However, granulomas of *Oesophagostomum* spp. were demonstrated in wild chimpanzees aged over 1 year in Gombe at necropsy (Terio *et al.*, 2011). Thus, chimpanzees in Mahale may also suffer from granulomas in their gut wall (Kooriyama *et al.*, 2012).

47.7.3 Rhabditoidea

Strongyloides infections in chimpanzees were reported in all previous studies from Mahale. *Strongyloides* eggs are elliptical, thin-shelled, and larvated (Figure 47.1h). Two *Strongyloides* species, *S. fuelleborni* and *S. stercoralis*, are common. They are distinguishable by the forms that appear in the feces. In *S. fuelleborni* infection, only eggs are found in the feces. In contrast, *S. stercoralis* eggs hatch in the intestinal mucosa, then rhabditiform larvae are found in the feces (Figure 47.2b). Specific identification is difficult when samples are old and/or in cases of mixed infections. *Strongyloides* infection routes are (1) infection by filariform larvae, which penetrate the skin and (2) autoinfection: filariform larvae penetrate the host intestinal mucosa or perianal skin. Thus, terrestrial primates may show a higher *Strongyloides* infection rate than arboreal primates. Indeed, more-terrestrial chimpanzees, vervet monkeys, and yellow baboons showed higher prevalence than less-terrestrial red colobus monkeys at Mahale. Vertical transmission by way of breast milk to infants may also occur (Kooriyama *et al.*, 2012). Recently, molecular analyses focused on specific and intraspecific variations related to geographical dispersal of *Strongyloides* spp. have been developed (Hasegawa *et al.*, 2009). Furthermore, human strongyloidiasis, which originated from chimpanzee *S. fuelleborni* infection, has been reported (Hasegawa *et al.*, 2010). Thus, such cross-infection of *Strongyloides* may occur among primates in Mahale.

47.7.4 Spirurida

Spirurid nematodes were reported in chimpanzees in Gombe and Ugalla (*Physaloptera* sp., *Abbreviata caucasica*, Spirurida fam. gen. sp., *Protospirura muricola*), but not in Mahale (Bakuza and Nkwengulila, 2009; Barbora, 2010). Furthermore, spirurid nematodes (*Chitwoodspirura* sp., *Streptopharagus* sp., Spirurina fam. gen. sp.) were found in red-tailed monkeys, vervet monkeys, and yellow baboons in Mahale (Figure 47.1i, j) (Kooriyama *et al.*, 2012). Spirurid nematode infection requires ingestion by an intermediate host. Chimpanzees in Mahale may not eat the intermediate host insects, unlike other primates.

47.7.5 Oxyuroidea

Enterobius eggs are flattened on one side, thick-shelled, and larvated. Because adult female pinworms lay eggs on perianal skin, it is rare to find their eggs in a normal fecal examination. *Enterobius* infection in chimpanzees was reported from Rubondo Island, Tanzania (Petrzelková *et al.*, 2010). No *Enterobius* eggs have been found in chimpanzees of Mahale or the neighboring Gombe or Ugalla areas. *Enterobius* eggs were found in one red colobus fecal sample at Mahale (Figure 47.1k).

47.7.6 Cosmoceroidea

Probstmayria gombensis was recently recorded from Mahale chimpanzees (Kooriyama *et al.*, 2012). *P. gombensis* is viviparous and transmitted by the fecal-oral route. Characteristic morphology is six globose lips at the mouth, a long esophagus with a posterior bulb, and a large excretory vesicle (Figure 47.2c). Females are didelphic and viviparous, and one well-developed larva is often seen in the uterus along with one or two eggs in progressive stages of development in the oviduct. *P. gombensis* infection in chimpanzees has previously been reported from Gombe (File, 1976; Murray *et al.*, 2000) but has not been detected at Ugalla (Barbora, 2010).

47.7.7 Other nematodes

Strongyles, *Necator* sp. and *Ascaris* have not been found in Mahale, but were demonstrated in Gombe (Murray *et al.*, 2000). *Primasubulura* sp. was found in red-tailed monkeys, vervet monkeys, and yellow baboons in Mahale, but not in chimpanzees.

47.8 Seasonality

The prevalence of parasite infection differs between the rainy (wet) season and the dry season. Kawabata and Nishida (1991) and Huffman *et al.* (1997) noted that the prevalence of nematodes was higher in the rainy season than the dry season, but that it depended on the parasite species. The reason for such seasonal differences in prevalence may be a moisture-mediated increase in the number of hatching eggs in the soil, which also enhances the survival rate of the larva (better than on dried ground). Therefore, the infection rate of nematode larvae increases in the rainy season. However, heavy rain may wash away larvae and/or eggs. Nematodes, such as *Trichuris* spp., that require a long time (15–30 days) to become infective may have less opportunity to infect chimpanzees (Brown, 1927).

47.9 Chimpanzee ecology and parasite infection

47.9.1 Behavior (social interaction)

Chimpanzees are social animals that frequently interact using affiliative behaviors such as grooming each other with their hands and/or mouth (Chapter 34), sharing chewed food between mother and child or sometimes between nonrelated individuals (Chapter 20), and open-mouth kissing. These behaviors increase the probability of parasite transmission. The degree of such social interactions may be affected by kinship, dominance status, gender, and the reproductive status of the performer.

47.9.2 Status in the dominance hierarchy

An alpha male in a multi-male chimpanzee group makes considerable efforts to maintain his status in the dominance hierarchy (Chapters 27, 33). The testosterone levels of the alpha are highest among the males, which is thought to provide the passion and strength to affiliate or maintain his high dominance rank (Muehlenbein *et al.*, 2004; see also Chapter 44). However, a high testosterone level imposes several costs, including increased metabolic rates and immunosuppression. Meanwhile, maintaining the alpha status is stressful; such stress increases the cortisol level (Muller and Wrangham, 2004). High cortisol levels usually suppress immunity. The prevalence and infection rate of internal parasites are higher in the alpha male than in the other males (Muehlenbein and Watts, 2010), possibly because the alpha male is under stress owing to his efforts to maintain his status in the group. Thus, dominance rank may be directly associated with helminth parasite richness.

47.9.3 Parasite infection cycles with diet and drinking water

Chimpanzees eat a variety of fruits, leaves, piths, gum (Chapter 17), animal prey (vertebrates and insects: Chapter 20 and Chapter 21, respectively), and soil.

Chimpanzees drink water from streams, ponds, puddles, and hollows of trees, some of which are old and muddy. Consuming plants and water may lead to direct intake of infective parasite eggs or larvae. Prey insects and mites on fruits are intermediate hosts of Acanthocephala, Cestoda, and Trematoda. Thus it is possible that chimpanzees eat leaves or piths contaminated with ants or orthopteran insects that are hosts for the parasites, such as *Dicrocoelium* spp.

47.9.4 Fission–fusion, home range, and female immigration

Chimpanzees repeat fission and fusion within their group home range (Chapters 6, 8, 9), and the degree of gregariousness is affected by the available food resources and existence of estrous females. Infection rates may increase when chimpanzees form large parties because of the high population density therein. Infection rates may also be affected by the duration for which chimpanzees reside in an area and the frequency with which they visit the same area again. Parasite eggs may contaminate chimpanzee feces, which are often dropped on the trails on which they walk, and may also be spread on the leaves and branches that they grab. Then, another chimpanzee may acquire the infection when he/she later walks on the same ground, climbs on the same tree, or sleeps in a bed on the same tree, if the parasites are in their infective periods.

Female chimpanzees emigrate to and immigrate from neighboring groups at an adolescent stage, usually on reaching sexual maturity (Chapters 6 and 7). These exchanges of individuals contribute to the rate of encountering novel parasite hosts. Mahale and Gombe share similar environments, but the variety of parasites differs. However, anthropogenic environmental gaps prevent female emigration and immigration between these two populations, and thus there is no opportunity for parasite exchange.

47.9.5 Pathogenic or zoonotic parasites

Oesophagostomum spp. and *Strongyloides* spp. are pathogenic to the host. *O. bifurcum* and *O. stephanostomum* infections have been reported in several African countries and these nematodes cause oesophagostomiasis not only in chimpanzees but also in humans (Thomas, 1910). These *Oesophagostomum* spp. infections induced granulomatous nodules around the small intestine and nodules are formed in the large intestine (Polderman *et al.*, 1999). *Strongyloides* sp. is also a pathogenic nematode and zoonotic parasite. *Strongyloides fuelleborni* has been found in Old World monkeys, apes, and humans (Yamashita, 1963), and *S. stercoralis* has been found in humans and apes. These *Strongyloides* spp. cause strongyloidiasis, and cross-infections have been reported in humans and apes in Mahale (Hasegawa *et. al.*, 2010). Some cases of bertiellosis were reported as simian *Bertiella studeri* infections after ingesting the intermediate host. Most of the cases were asymptomatic in human children and involved some contact or association with nonhuman primates.

47.10 Parameters of host animal health

Host animal immunity is activated by a parasite infection and then controls successive infections. Pathogenic parasites induce diseases in their host, but the severity depends on the host animal's health condition. In individuals with immunodeficiency viruses or with malnutrition, pathogenic parasites tend to cause more serious disease.

Good intestinal digestion generally improves the host's health condition. The entodiniomorphid ciliate, *Troglodytella abrassarti*, is a cellulolytic organism in the chimpanzee intestine. These ciliates play a role in cellulose digestion, assist nutrient absorption (Howells *et al.*, 2011), and contribute to polysaccharide hydrolytic activities in the chimpanzee colon (Howells *et al.*, 2011; Profousová *et al.*, 2011). Thus, the presence of *T. abrassarti* may be an indicator of the health and/or the nutritional condition of the host chimpanzee.

47.11 Conclusion

Parasite fauna varies among habitats because each chimpanzee population lives in accordance with their environment. Past studies have revealed various parasites infecting the chimpanzees of the K and M groups in Mahale. However, to gain a better understandings of the parasite fauna, additional surveillance using molecular biological techniques that can identify species more precisely should be performed. Furthermore, whole-year surveillance should be performed in the future to assess seasonal changes in parasite prevalence. Each individual should be sampled for parasite infection 3–4 times on non-consecutive days to provide more accurate diagnoses, as pointed out by Muehlenbein (2005). Recently, the number of tourists has increased; more than 1000 visitors come to Mahale per year to watch the chimpanzees (Chapter 48). Since the establishment of the National Park, tourism seems to have been a major transporter of infectious diseases into the park from outside. Thus, continuous parasite examination is also required to monitor the chimpanzees' health conditions, which will contribute to the early discovery of zoonotic and/or pathogenic parasite invasions in Mahale.

References

Ashford, R. W., Reid. G. D. F., and Wrangham, R. W. (2000). Intestinal parasites of the chimpanzee *Pan troglodytes*, in Kibale Forest, Uganda. *Annals of Tropical Medicine and Parasitology*, **94**, 173–9.

Bakuza, S. J. and Nkwengulila, G. (2009). Variation over time in parasite prevalence among free-ranging chimpanzees at Gombe National Park, Tanzania. *International Journal of Primatology*, **30**, 43–53.

Barbora, K. (2010). *Ecology and Parasite Fauna of Chimpanzees of Dry Habitats with Emphasis to Chimpanzees of Ugalla–Tongwe Forest Reserve, Tanzania*. Bachelor's thesis. Brno, Czech Republic: Faculty of Science Department of Botany and Zoology, Masaryk University.

Bezjian, M., Gillespie, T. R., Chapman, C. A., and Greiner, E. C. (2008). Coprologic evidence of gastrointestinal helminths of forest baboons, *Papio anubis*, in Kibale National Park, Uganda. *Journal of Wildlife Diseases*, **44**, 878–87.

Bhagwant, S. (2004). Human *Bertiella studeri* (family Anoplocephalidae) infection of probable Southeast Asian origin in Mauritian children and an adult. *American Journal of Tropical Medicine and Hygiene*, **70**, 225–8.

Brown, H. W. (1927). Studies on the rate of development and viability of the eggs of *Ascaris lumbricoides* and *Trichuris trichiura* under field conditions. *The Journal of Parasitology*, **14**, 1–15.

File, S. K. (1976). *Probstmayria gombensis* sp. n. (Nematoda: Atractidae) from the chimpanzee. *Journal of Parasitology*, **62**, 256–8.

Gasser, R. B., Woods, W. G., Huffman, M. A., *et al.* (1999). Molecular separation of *Oesophagostomum stephanostomum* and *Oesophagostomum bifurcum* (Nematoda: Strongyloidea) from non-human primates. *International Journal of Parasitology*, **29**, 1087–91.

Gillespie, T. R., Lonsdorf, E. V., Canfield, E. P., *et al.* (2010). Demographic and ecological effects on patterns of parasitism in eastern chimpanzees (*Pan troglodytes schweinfurthii*) in Gombe National Park, Tanzania. *American Journal of Physical Anthropology*, **143**, 534–44.

Graczyk, T. K., Bosco-Nizeyi, J., Ssebide, B., *et al.* (2002). Anthropozoonotic *Giardia duodenalis* genotype (assemblage) A infections in habitats of free-ranging human-habituated gorillas, Uganda. *Journal of Parasitology*, **88**, 905–9.

Hasegawa, H., Kano, T., and Mulavwa, M. (1983). A parasitological survey on the feces of pygmy chimpanzees, *Pan paniscus*, at Wamba, Zaïre. *Primates*, **24**, 419–23.

Hasegawa, H., Huffman, M. A., and Chapman, C. A. (2009). Useful diagnostic references and images of protozoans, helminths, and nematodes commonly found in wild primates. In *Primate Parasite Ecology*,

ed. M. A. Huffman and C. A. Chapman. Cambridge: Cambridge University Press, pp. 507–13.
Hasegawa, H., Sato, H., Fujita, S., *et al.* (2010). Molecular identification of the causative agent of human strongyloidiasis acquired in Tanzania: Dispersal and diversity of *Strongyloides* spp. and their hosts. *Parasitology International*, **59**, 407–13.
Howells, M. E., Pruetz, J., and Gillespie, T. R. (2011). Patterns of gastro-intestinal parasites and commensals as an index of population and ecosystem health: the case of sympatric western chimpanzees (*Pan troglodytes verus*) and guinea baboons (*Papio hamadryas papio*) at Fongoli, Senegal. *American Journal of Primatology*, **73**, 173–9.
Huffman, M. A., Nishida, T., and Uehara, S. (1990). Intestinal parasites and medical plant use in wild chimpanzees: possible behavioral adaptation for the control of parasites. *Mahale Mountains Chimpanzee Research Project Ecological Report*, **72**, 1–14.
Huffman, M. A., Gotoh, S., Izutsu, D., *et al.* (1993). Further observations on the use of the medicinal plant, *Vernonia amygdalina* (Del), by a wild chimpanzee, its possible affect on parasite load, and its phytochemistry. *African Study Monograph*, **14**, 227–40.
Huffman, M. A., Page, J. E., Sukhdeo, M. V. K., *et al.* (1996). Leaf-swallowing by chimpanzees, a behavioral adaptation for the control of strongyle nematode infections. *International Journal of Primatology*, **72**, 475–503.
Huffman, M. A., Gotoh, S., Turner, L. A., *et al.* (1997). Seasonal trends in intestinal nematode infection and medical plant use among chimpanzees in the Mahale Mountains, Tanzania. *Primates*, **38**, 111–25.
Huffman, M. A., Pebsworth, P., Bakuneeta, C., *et al.* (2009). Chimpanzee–parasite ecology at Budongo Forest (Uganda) and the Mahale Mountains (Tanzania): influence of climatic differences on self-medicative behavior. In *Primate Parasite Ecology*, ed. M. A. Huffman and C. A. Chapman. Cambridge: Cambridge University Press, pp. 331–50.
Kaur, T., Singh, J., and Lindsay, D. S. (2010). Prevalence of *Troglodytella abrassarti* Brumpt and Joyeux, 1912 in wild chimpanzees (*Pan troglodytes schweinfurthii*) at Mahale Mountains National Park in Western Tanzania. *Journal of Parasitology*, **96**, 209–10.
Kawabata, M. and Nishida, T. (1991). A preliminary note on the intestinal parasites of wild chimpanzees in the Mahale Mountains, Tanzania. *Primates*, **32**, 275–8.
Kim, J. C., Abee, C. R., and Wolf, R. H. (1978). Balantidiosis in a chimpanzee (*Pan troglodytes*). *Laboratory Animals*, **12**, 231–3.
Kooriyama, T., Hasegawa, H., Shimozuru, M., *et al.* (2012). Parasitology of five primates in Mahale Mountains National Park, Tanzania. *Primates*, **53**, 365–75.
McGrew, W. C., Tutin, C. E. G., Collins, D. A., and File, S. K. (1989). Intestinal parasites of sympatric *Pan troglodytes* and *Papio* spp. at two sites: Gombe (Tanzania) and Mt. Assirik (Senegal). *American Journal of Primatology*, **14**, 147–55.
Muehlenbein, M. P. (2005). Parasitological analyses of the male chimpanzees (*Pan troglodytes schweinfurthii*) at Ngogo, Kibale National Park, Uganda. *American Journal of Primatology*, **65**, 167–79.
Muehlenbein, M. P. and Watts, D. P. (2010). The costs of dominance: testosterone, cortisol and intestinal parasites in wild male chimpanzees. *BioPsychoSocial Medicine*, **4**, 21.
Muller, M. N. and Wrangham, R. W. (2004). Dominance, aggression and testosterone in wild chimpanzees: a test of the 'challenge hypothesis'. *Animal Behaviour*, **67**, 113–23.
Muehlenbein, M. P., Watts, D. P., and Whitten, P. L. (2004). Dominance rank and fecal testosterone levels in adult male chimpanzees (*Pan troglodytes schweinfurthii*) at Ngogo, Kibale National Park, Uganda. *American Journal of Primatology*, **64**, 71–82.
Murray, S., Stem, C., Boudreau, B., and Goodall, J. (2000). Intestinal parasites of baboons (*Papio cynocephalus anubis*) and chimpanzees (*Pan troglodytes*) in Gombe National Park. *Journal of Zoo and Wildlife Medicine*, **31**, 176–8.
Nigi, H., Nishida, T., Itoh, N., and Sakamaki, T. (1998). Helminthic parasites infection in wild chimpanzees in the Mahale Mountains National Park, Tanzania, in dry season. In *Resource Use Patterns and Social Structure among Chimpanzees*, ed. T. Nishida. Kyoto: Nisshindo Printer, pp. 57–62.
Petrzelková, K. J., Hasegawa, H., Appleton, C. C., *et al.* (2010). Gastrointestinal parasites of the chimpanzee

population introduced onto Rubondo Island National Park, Tanzania. *American Journal of Primatology*, **72**, 307–16.

Polderman, A. M., Anemana, S. D., and Asigri, V. (1999). Human oesophagostomiasis: a regional public health problem in Africa. *Parasitology Today*, **15**, 129–30.

Pomajbíková, K., Petrželková, K. J., Petrášová, J., *et al.* (2012). Distribution of the entodiniomorphid ciliate *Troglocorys cava* Tokiwa, Modrý, Ito, Pomajbíková, Petrželková, & Imai, 2010, (Entodiniomorphida: Blepharocorythidae) in wild and captive chimpanzees. *Journal of Eukaryotic Microbiology*, **59**, 97–9.

Pomajbikova, K., Obornik, M., Horak, A., *et al.* (2013). Novel insights into the genetic diversity of *Balantidium* and *Balantidium*-like cyst-forming ciliates. *PLoS Neglected Tropical Disease*, **7**, e2140.

Profousová, I., Mihaliková, K., Laho, T., *et al.* (2011). The ciliate, *Troglodytella abrassarti*, contributes to polysaccharide hydrolytic activities in the chimpanzee colon. *Folia Microbiologica (Praha)*, **56**, 339–43.

Stunkard, H. W., Koivastik, T., and Healy, G. R. (1964). Infection of a child in Minnesota by *Bertiella studeri* (Cestoda: Anoplocephalidae). *American Journal of Tropical Medicine and Hygiene*, **13**, 402–9.

Terio, K. A., Kinsel, M. J., Raphael, J., *et al.* (2011). Pathologic lesions in chimpanzees (*Pan troglodytes schweinfurthii*) from Gombe National Park, Tanzania, 2004–2010. *Journal of Zoo and Wildlife Medicine*, **42**, 597–607.

Thomas, H. W. (1910). The pathological report of a case of oesophagostomiasis in man. *Annals of Tropical Medicine and Hygiene*, **4**, 57–88.

Vallo, P., Petrželková, K. J., Profousová, I., *et al.* (2012). Molecular diversity of entodiniomorphid ciliate *Troglodytella abrassarti* and its coevolution with chimpanzees. *American Journal of Physical Anthropology*, **148**, 525–33.

Yamashita, J. (1963). Ecological relationships between parasites and primates. I. Helminth parasites and primates. *Primates*, **3**, 1–96.

Yoshikawa, M., Ogawa, H., Sakamaki, T., and Idani, G. (2008). Population density of chimpanzees in Tanzania. *Pan Africa News*, **15**, 17–20.

PART X People and chimpanzees

48 Current status of tourism

MICHIO NAKAMURA

48.1 Introduction

At Mahale, tourism has focused on chimpanzee viewing. Tourism is economically important because it brings revenue to the MMNP (Chapter 2), which is essential for park management and effective conservation practices. It is also important for environmental education, providing an opportunity for visitors to consider the necessity of conserving chimpanzees and other wildlife by observing chimpanzees in their natural environment. However, tourism poses some threats to chimpanzees, because tourists can unintentionally introduce diseases and their presence can be stressful to the chimpanzees. In this chapter, I will briefly introduce the history of tourism at Mahale and will review the current status of tourism in more detail.

48.2 Characteristics and brief history of tourism at Mahale

Mahale has several unique characteristics that may attract tourists from all over the world. First, the landscape is spectacular and diversified. It consists of forests, wooded savannas, high mountains, valleys, waterfalls, and the lake. Second, the park has great biodiversity with flora and fauna representative of western Africa, such as chimpanzees, brush-tailed porcupines, and crested guineafowl, in addition to those from eastern and southern Africa (see Chapter 15). Third, the park is extremely isolated, and this remoteness is highly valued by tourists from large cities, particularly from the West. Fourth, the park was designed as a "foot-walking national park," and no large permanent guesthouses can be built. Unlike most African safaris in which tourists ride in a safari car to see savanna animals, at Mahale, people can enjoy walking in the forest. Finally, the park has a long-term ongoing chimpanzee research program. Visitors are provided with the latest research information and can learn about individual life histories and hear interesting stories of the group activities.

Soon after Mahale was designated a national park in 1985, a private tour company introduced small-scale chimpanzee tourism. Up to six visitors at a time were accepted into a tented camp and were allowed to spend up to 1 h observing the chimpanzees. In the mid-1990s, the MMNP began to accept tourists in its own tourist camp. Two more companies joined the tourist business at Mahale in the 2000s, increasing the number of tourists in the park from 200 per year in the early 1990s to more than 1000 in 2005 (Nishida and Nakamura, 2008). Most tourists stay 3–4 days (sometimes a week); therefore, the total number of man-days is much higher. Although we do not have statistics on the nationality of tourists, most are from North America and Europe, with a small number from Asia. Apart from visitors from South Africa, there are very few native African tourists.

At the introduction of tourism at Mahale, the MMNP employed some ex-research assistants, who were used to locating chimpanzees, as park guides. Groups of tourists were only accompanied by a park guide, and together they looked for chimpanzees. Commonly, researchers and tourists were waiting for the calls of chimpanzees in the same locations suitable for listening to chimpanzee calls (see Chapter 2). Inevitably,

Mahale Chimpanzees: 50 Years of Research, ed. M. Nakamura, K. Hosaka, N. Itoh and K. Zamma. Published by Cambridge University Press.

sometimes the tourists failed to see chimpanzees, even if they went into the forest. Because the guides initially employed by the MMNP spoke minimal English, they could not provide sufficient guidance to tourists in English although they had good knowledge of the chimpanzees and other local animals and plants.

In the late 1990s, the tour camp hired an ex-research assistant as a private tracker, or scout, who would go into the forest in the early morning to locate chimpanzees. The tourists no longer had to walk around the forest to locate chimpanzees but could wait at the campsite. When chimpanzees were located, the tracker immediately communicated with the camp by a transceiver, and an MMNP guide led the tourists directly to the location. The system worked well, at least for tourism, and the number of private trackers at the camp was increased. Currently, when the camp has guests, two or three trackers are sent out early in the morning to locate chimpanzees. These trackers try to locate different parties of chimpanzees to offer the tourists the best option (e.g. a larger party with an alpha male or with infants is preferred by tourists over a lone, low-ranking male). Because the chimpanzees are usually on the move, the trackers continue to follow the party, sometimes for more than 4 h, until the tourists arrive.

In the early 2000s, in addition to an official MMNP guide, a camp manager, a native speaker of English also fluent in Swahili, almost always accompanied groups of tourists as a private guide. This was a bonus for the tourists because they could learn more about the chimpanzees through English translation by the manager. Later, when the manager changed, the tour company employed Tanzanian guides fluent in English to fill the role of the former manager. These privately employed camp guides, who were usually not local but from large cities in Tanzania, had little knowledge about the chimpanzees although they had relatively high levels of education. After they arrived in Mahale, they were trained to give a general presentation about chimpanzees.

Another two tour companies took up the system and also employed their own trackers and private guides. Because there are now three tourist camps at Mahale, there are often 3–5 trackers from different camps following a party of chimpanzees, in addition to researchers and research assistants. This increases the number of people visiting chimpanzees and sometimes causes overcrowding around small numbers of chimpanzees (see later).

In addition to an official park guide, each tourist group is led by a camp guide (and/or a camp manager). Although most of the park guides now also speak English and some are locally hired, they do little guide work with tourists. In fact, some know little of the identities and behaviors of chimpanzees and instead, camp guides fulfill that role. The regulations require that a park guide accompany the tourist groups, and the tourists are required to pay fees for this service. A tour group usually stays with the chimpanzees for 1 h, the time permitted by the park regulations, and the park guide keeps track of the time. After tourists arrive at a site, the camp trackers wait nearby, because if the chimpanzees move to an area where they cannot be observed, the trackers must follow them to locate an alternative place where the chimpanzees will be visible to the tourists. In addition, the camp guides, who have limited knowledge of the chimpanzees' identities, often need to ask the trackers in Swahili about the identities of chimpanzees, which they translate into English for the tourists.

48.3 Regulations of tourism at Mahale

According to the park regulations (Tanzania National Parks, 2006), the maximum chimpanzee viewing group size is seven, six tourists and one park guide. This figure is similar to gorilla tourism with group sizes set to six in Uganda and eight in Rwanda and DRC (Homsy, 1999). In Mahale, in addition to these seven, one tracker and one tour camp provided guide may accompany the tracking excursion. However, once the chimpanzees are sighted, only the six tourists and the official park guide may approach the chimpanzees. The trackers and the tour camp provided guide must remain 200 m away from the chimpanzees at all times.

The maximum number of groups permitted to observe chimpanzees per day is three, and the timing has to be arranged so that only one group visits the chimpanzees at any one time. Therefore, the park

Figure 48.1 The status of tourism at Mahale in 2005. It is obvious that tourists get too close to chimpanzees. Some well-habituated chimpanzees, particularly adult males, do not care about people, at least superficially, but shy chimpanzees might not dare to get close to tourist groups. ©M. Nakamura.

regulations allow 18 tourists or 21 people in total to visit the chimpanzees per day. This figure is different from gorilla tourism with only one group permitted to visit per day (Homsy, 1999). In addition to the tourists and guides, three researchers and three research assistants are permitted to follow the M-group chimpanzees.

There are regulations to stipulate the distance of observations (10 m for tourists and 7.5 m for researchers) and the duration of observations (1 h for tourists). Other regulations include restriction of the use of flashlights, eating in the presence of or near chimpanzees, defecating in the forest, and leaving belongings unattended. The regulations do not stipulate a minimum or maximum chimpanzee party size for visitation.

Observation rules had been proposed by researchers well before the park specifically included them in their 2006 regulations, but the rules were not made clear to tourists, and some of the rules were not observed at all. For example, visitors used to get as close to the chimpanzees as they wished (Figure 48.1). However, in 2006, following a severe outbreak of respiratory disease that killed several chimpanzees (Chapter 25; Hanamura *et al.*, 2008), park staff, tour camp managers, and researchers resolved to adhere to the rules more strictly. In addition to the MMNP regulations, researchers proposed additional precautions (Hanamura *et al.*, 2006), e.g. everyone should wear a facemask when in the vicinity of chimpanzees. The MWCS (Chapter 2) provided facemasks to the tour camps soon after the outbreak, and now everyone is willing to wear masks. Tourists also maintain more distance from chimpanzees, and therefore, the tourism situation is better than before the outbreak (Figure 48.2).

48.4 Studies on tourism at Mahale

There have been a few systematic studies on the status of tourism at Mahale. Although the MMNP and the Frankfurt Zoological Society requested a preliminary assessment of tourism at Mahale (Walpole, 2004), the assessment was inadequate because it was conducted

Figure 48.2 The status of tourism at Mahale in 2009. Tourists now maintain an acceptable distance from chimpanzees, and they also wear facemasks. ©M. Nakamura.

during a relatively low tourist season and only lasted for 1 week. Furthermore, because the chimpanzees were in the higher part of their range at the time, the assessor did not actually observe any chimpanzees during the study week. This meant that the assessor could not investigate how tourists observed chimpanzees.

Other reports from Mahale include annual statistics provided by the MMNP on the number of tourists (Nishida and Mwinuka, 2005) and reports suggesting an overabundance of tourists (Nakamura and Nishida, 2004; Nishida and Nakamura, 2008) although these impressions of overabundance were not accompanied by empirical data. In 2009, Nishida and I presented the first, although preliminary, quantitative data on tourists from 2006 to 2008 (Nakamura and Nishida, 2009).

48.5 Current status of chimpanzee tourism at Mahale

Here I analyse the data on tourists at Mahale collected over 325 days (observation days) from July 2009 to February 2013 (hereafter referred to as 2009–13 data). The basic methods of data collection were identical to those used in our previous study (Nakamura and Nishida, 2009), but in the present analysis, I use data obtained by a research assistant. When possible, I compare this new data with the previously published data (Nakamura and Nishida, 2009) (hereafter referred to as 2006–8 data).

Among the 325 days when the assistant was in the field to observe chimpanzees, he encountered tourists on 165 days (tourist days). Because several groups of tourists can visit in a day, he encountered 344 groups in these 165 days. The accumulated number of tourists was 1461 (the same tourist visiting on two different days was counted as two), and the cumulative number of people, including guides and trackers was 2572. As discussed later, the numbers of tourists differ considerably from month to month. When the numbers are corrected with actual observations of tourist numbers by month and summed for a year, the estimation becomes 1595.9 per year for tourists and 2813.4 per year for total people. However, these figures are an underestimation because tourists who were not encountered by the assistant were not included.

48.5.1 Number of people in a tourist group

Table 48.1 shows the number of people in tourist groups visiting chimpanzees according to the 2009–13 data. Remember, the park regulations stipulate an upper limit of six tourists in a group (Tanzania National Parks, 2006). Twenty-four groups among the 344 observed groups (7.0%) exceeded the upper limit, and this is significantly more than the 1.3% (3/233 groups) in the 2006–8 data (Fisher's exact test, $p = 0.001$). In the 2006–8 data, all oversized groups had seven tourists (only one extra person). However, in the 2009–13 data, some groups contained 8, 9, or 13 tourists, suggesting that the regulations on the number of tourists in a group are not being strictly followed.

Table 48.1 **Number of people in tourist groups visiting chimpanzees. Park regulations permit six tourists and one guide per group**

Number of people in a group	Tourists only (number of groups)	Including guides and trackers (number of groups)
1	11	0
2	86	5
3	20	18
4	62	15
5	62	38
6	79	38
7	13	53
8	8	52
9	2	48
10	0	46
11	0	25
12	0	2
13	1	4
Total	344	344

The median number of tourists in a group was four (range: 1–13), which did not differ significantly from the 2006–8 data (Wilcoxon rank sum test, $W = 43\,106.5$, $p = 0.11$). However, the median number of total people (including trackers and guides) in a group was eight (range: 2–13), which had increased significantly compared with the 2006–8 data (Wilcoxon rank sum test, $W = 32\,066$, $p = 0.0035$).

48.5.2 Number of tourist groups in a day

Table 48.2 shows the number of tourist groups in a day. On 12.7% of days in 2009–13, the regulation was violated, i.e. four or more groups visited chimpanzees in a day. This had decreased from 23.1% in the 2006–8 data. The maximum group number was six in 2006–8 but was five in 2009–13. However, the median number of groups did not differ significantly between the two study periods (Wilcoxon rank sum test, $W = 8730.5$, $p = 0.059$).

48.5.3 Number of people visiting in a day

In the 2006–8 data, the median number of tourists per day was 8 (range 1–27, $N = 107$ days), and the median number of total people per day was 13 (range 2–39, $N = 102$ days). In 2009–13, the median values were 7 (1–28, $N = 165$) (no significant difference compared with 2006–8 data: $W = 9670$, $p = 0.65$) and 13 (2–45, $N = 165$) (significantly higher compared with 2006–8 data: $W = 8133$, $p = 0.0074$), respectively. The overall number of people visiting chimpanzees appears to be increasing. The maximum number of people visiting in a day was 39 in the former study and 45 in the current study.

Table 48.2 **Number of tourist groups per day (N = 165 observation days)**

Number of groups	Observed days	Percentage to the total
1	72	43.6%
2	33	20.0%
3	39	23.6%
4	16	9.7%
5	5	3.0%
Total	165	

48.5.4 Seasonality of tourism

In the earlier dataset, we could not analyse the seasonality of tourism because the observation days were limited to particular months. The current dataset covers a range of months. Therefore, here I discuss the seasonal variation of tourism.

Figure 48.3 shows an estimation of tourist numbers per month. During August, the peak tourism month, an estimated 303.8 tourists and 504.7 people in total visited chimpanzees. There were no tourists at all in April, probably because the largest tour company closes its camp in April.

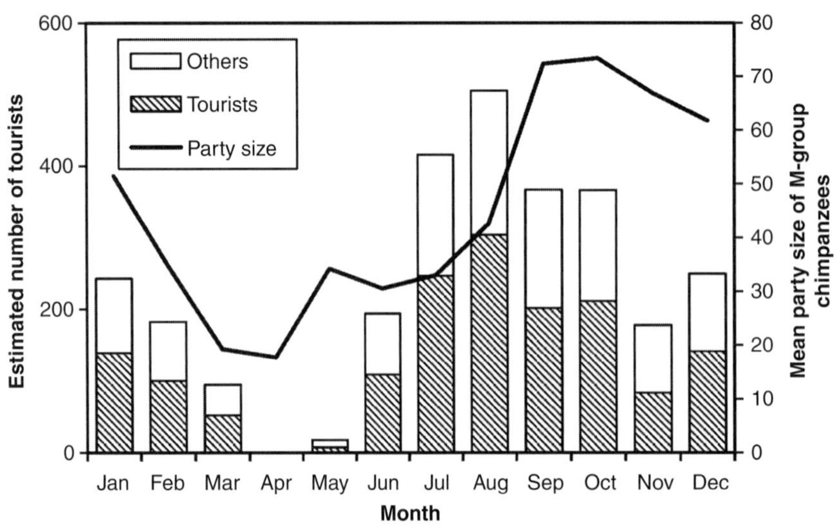

Figure 48.3 The estimated number of tourists and tour operators and the mean party size of the M-group chimpanzees in different months. The estimated tourist numbers were calculated from observational data between July 2009 and February 2013, and the mean party size was calculated from the MMCRP records from 1994 to 2008. "Others" include official park guides, camp guides, and camp trackers.

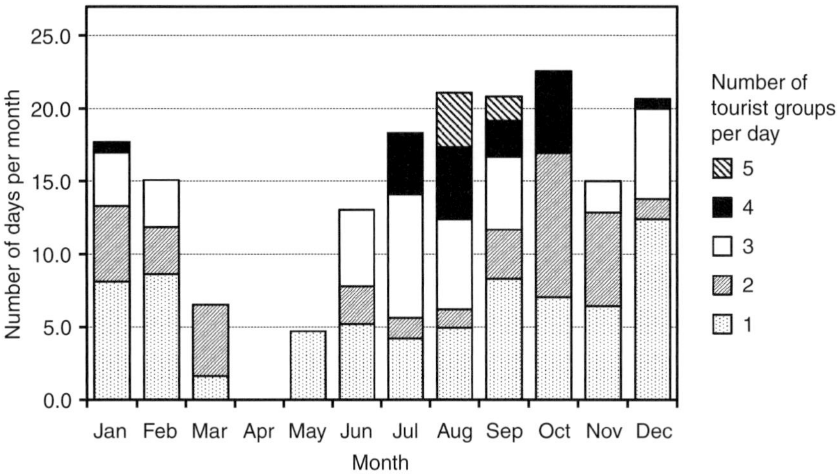

Figure 48.4 The estimated number of days with different numbers of tourist groups/day in different months. For example, in January, tourists are estimated to have visited chimpanzees on 13.3 of 31 days. Only one tourist group visited on 8.1 days, two groups on 5.2 days, and so on. Note that there are more days in July–October when four or more groups visited per day, violating the regulations.

The monthly estimation of tourist numbers was not significantly correlated with the average number of chimpanzees observed per day in each month (a rough representation of the observability of chimpanzees) (Kendall's rank correlation test, $\tau = 0.42$, $p = 0.06$). Tourist numbers were highest during summer vacation time (July and August) although the number of chimpanzees observed is not so large during that period, whereas the number of tourists decreased in November–January when chimpanzees can be observed more easily. This may be because tourists avoid the rainy season that usually starts in mid-October (Chapter 12).

Figure 48.4 shows the proportion of days with different numbers of tourist groups in each month. As expected, when there were more tourists (i.e. July–October), there were more violations of the group number regulations (i.e. four or more groups). These violations may not be intentional. In the high tourist season, each tour company often has two or three tourist groups at a time. Because there is limited communication between tourist camps and the MMNP management to allocate numbers of groups to camps, the violations occur as a result of each camp taking two or more tourist groups in a day during the high season.

48.5.5 Characteristics of tourist camps

Table 48.3 summarizes the number of people sent out by each of four tourist camps (including one run by the MMNP) during the observation days. More than half of the tourists were sent by the company that operates Camp A. This large company operates safari tours in several places in Tanzania and, with its ability to attract more customers, regularly brings tourists on shared charter flights from large cities such as Arusha and Dar es Salaam. A tour combined with a visit to Katavi National Park, another park that is not easy to access by car, is very popular. Camp A is also notable for the number of camp guides and trackers. It always has a camp guide, in addition to an official park guide, and always sends two or more trackers for a tourist group. On average, it sends 0.61 guides/trackers to one tourist, whereas other camps only send 0.4. Because Camp A constantly receives guests, the company can provide more people and services. Although increased services are good for tourists, they are not necessarily good for conservation because it means more people make contact with chimpanzees. In fact, the recent increases in numbers of people in a group and in a day were mainly a result of such an increase in services.

Table 48.3 **Number of people sent out by each tour camp in 325 days (165 tourist days) between 2009 and 2013**

Camp	Tourists	MMNP guides	Camp guides	Trackers	Total number of people
A	905	199	199	356	1659
B	388	96	84	93	661
C	112	42	0	6	160
D	56	12	12	12	92
Total	1461	349	295	467	2572

48.1 Case study: July 2013

Here I present an example of tourism on July 24, 2013, when a large number of tourists visited a small group of chimpanzees.

08:02 h	My tracker and I find a party of six chimpanzees, two adult females, Tarnie and Zola, and their children, on a tree near Kasiha Valley, R-2 (see Chapter 2 for the location). I start my focal follow of Tarnie. About the same time, a tracker from Camp B also finds the group and starts communicating with a transceiver.
08:11 h	Four tourists from Camp B (hereafter T1), together with a park guide, a camp guide, and another tracker, arrive. The chimpanzees are still high on the tree.
08:22 h	As Zola starts to descend from the tree, the party of chimpanzees begins to move. I follow the party and the T1 also follows us.
08:30 h	Zola and Tarnie climb different trees about 30 m away from each other. T1 follows and observes Zola on the tree.
08:39 h	The Camp B tracker calls another group of tourists with the transceiver.
08:53 h	Another group from Camp B (T2), with two tourists, a park guide, a camp guide, and a tracker, arrives. A tracker from Camp A also arrives. T1 leaves the location and makes way for T2, but T1 waits nearby (about 50 m away).
08:56 h	Tarnie starts to move on the ground and passes in front of T1.
09:00 h	Zola also comes and they all climb a tree and start to feed on fruits. Soon T2 arrives at the tree and starts to observe the chimpanzees.
09:20 h	T1 is still waiting within visual range.
09:29 h	A tourist in T2 coughs several times.
09:35 h	The Camp A tracker who has been nearby calls for another group of tourists with his transceiver.
09:39 h	T2 leaves the location and T1 returns.
10:24 h	T1 leaves and T2 returns.

10:43 h	T2 leaves. The tracker from Camp A remains, sometimes communicating with the transceiver.
11:25 h	A group from Camp A (T3), with five tourists, a park guide, and a camp guide, arrives.
11:51 h	T3 leaves the location.
11:55 h	Another group from Camp A (T4), with five tourists and a camp guide, arrives. The guide is not wearing a facemask on arrival, so I tell him to put one on.
11:57 h	The chimpanzees start to descend from the tree and move into the bush. As we follow them, the camp tracker follows us.
12:13 h	As Zola starts to climb a tree the tracker calls the tourist group with transceiver.
12:20 h	T4 arrives again.
12:28 h	Tarnie starts to move on the ground, and on the way she meets T3 who are waiting nearby (about 30 m away from T4). Zola also starts to move.
12:41 h	As Tarnie sits on the ground T4 catches up with her. Zola is in the nearby bush.
12:47 h	Tarnie starts to go into the bush and moves further away. We follow them and the camp tracker also follows.
12:51 h	The camp tracker goes back. It seems that the tourists have stopped following chimpanzees.

On this day, a total of 16 tourists visited only six chimpanzees and the groups alternately stayed in the vicinity of the chimpanzees for more than 4 h.

48.6 Discussion

Wildlife tourism is an important source of foreign currency for Tanzania. Furthermore, at Mahale, the revenue from tourism is essential for the MMNP to operate wildlife conservation and to promote public relationships with adjacent communities. Tourism also provides employment opportunities for local people. However, in addition to the problems associated with safari tours targeting larger savanna-dwelling mammals, there are several concerns relating to tourism targeting great apes. In particular, extreme caution is needed because great apes are more susceptible to human diseases than other mammals.

48.6.1 Numbers of visitors

The estimated number of tourists who visited chimpanzees in the recent observation period was more than 1500 per year. This estimation was based on the *observed* number of tourists encountered during chimpanzee research. Therefore, the number of tourists actually visiting chimpanzees is likely to be more than 2000. This is much higher than the 500 per year limit that Nishida thought optimal when tourism started at Mahale in 1989 (Nishida, 1989). From the researchers' perspective, the current number of tourists is too high, particularly in peak tourist months. This not only affects the chimpanzees themselves but also has impacts on research activities.

The increase in contact between people and chimpanzees may increase the risk of chimpanzee exposure to novel pathogens because tourists come from all over the world, and being surrounded by numbers of unfamiliar people can also cause stress to the chimpanzees. When a group includes more than four or five tourists, it is often difficult to observe the chimpanzees without disturbing their natural behaviors. Because most tourists wish to take their own photos, they tend to make a line transverse to the direction the chimpanzees are traveling, forming a wall of people. As the number of tourists in a group increases, not all members of the group can access the best position for observation, and this may decrease tourist satisfaction. To resolve this issue, the numbers of tourists should be strictly controlled using a booking system.

Today, many tourist camps provide attractions other than chimpanzee viewing, such as fishing, cruising, and walking to naturally made water pools in a valley. However, as I mentioned at the beginning of this chapter, there are many more charming and interesting locations in Mahale. Efforts are being made to further diversify tourism and to increase revenue while minimizing the impact on chimpanzees.

48.6.2 Seasonality

When we looked at the number of tourists visiting chimpanzees, we found that many come during summer vacation time when chimpanzee observability is limited. As a result, there is overcrowding of people observing a relatively small number of chimpanzees. Conversely, there are not so many tourists in the best season to observe chimpanzees, when they form large nomadic parties, namely from November to January. This is probably because tourists try to avoid the rainy season. To improve this mismatch, the introduction of seasonal fees should be considered, with entrance fees set higher in the dry season and set lower in the rainy season (Nishida and Mwinuka, 2005).

48.6.3 Problematic behaviors of tourists, guides, and camp trackers

Our previous report (Nakamura and Nishida, 2009) listed several problematic behaviors of tourists and others: approaching too close to chimpanzees, tourists without park guides, coughing, disturbing chimpanzees, two or more tourist groups in one place, going into the bush, disturbing researchers, using flashlights, staying longer than permitted, not wearing facemasks, leaving unattended belongings, and defecating in the forest.

These behaviors are still observed, and no improvements have been made. Before their visit, tourists should be told why the distance rule must be adhered to and must be prohibited from tipping guides to get closer to the apes. Finally, here I repeat the five recommendations stated previously (Nakamura and Nishida, 2009), which are still applicable and important 5 years later.

1. The number of groups per day must be strictly limited to three. This can be achieved by introducing a unified booking system controlled by the MMNP. For the time being, the MMNP should mediate constant communication among tourist camps. If the MMNP needs more annual revenue, higher entrance fees or the introduction of "the chimpanzee fee" (Nishida and Mwinuka, 2005) is an appropriate option.
2. Only one group of tourists should be allowed to visit the chimpanzees at any one time. This can best be achieved when the first recommendation (i.e. the limit on the number of groups allowed per day) is strictly enforced. The MMNP should coordinate and arrange the timing of observations by different tourist groups.
3. The total number of people in a group should be reduced. Capacity building of park guides is crucial. This would eliminate the need for camp guides and would benefit both tourism companies and conservation.
4. The presence of tourists must not disturb the behaviors of chimpanzees. When chimpanzees approach them, tourists should quietly step back to the specified distance before they take

photographs. Tourists must be reminded that chimpanzees have the ability to severely injure humans. Tourists should receive comprehensive instruction before they are allowed to observe chimpanzees.

5. Increased hygiene measures should be considered, such as presenting updated vaccination certificates for potentially dangerous diseases and washing shoes to prevent the introduction and dissemination of pathogens.

References

Hanamura, S., Kiyono, M., Nakamura, M., *et al.* (2006). A new code of observation employed at Mahale: prevention against a flu-like disease. *Pan Africa News*, **13**, 1–16.

Hanamura, S., Kiyono, M., Lukasik-Braum, M., *et al.* (2008). Chimpanzee deaths at Mahale caused by a flu-like disease. *Primates*, **49**, 77–80.

Homsy, J. (1999). Ape tourism and human diseases: how close should we get? A critical review of the rules and regulations governing park management and tourism for the wild mountain gorilla, *Gorilla gorilla beringei*. Report for the International Gorilla Conservation Programme.

Nakamura, M. and Nishida, T. (2004). [Present status of conversation on chimpanzees and their habitat in Mahale.] *Primate Research*, **20**, 57–61. In Japanese.

Nakamura, M. and Nishida, T. (2009). Chimpanzee tourism in relation to the viewing regulations at the Mahale Mountains National Park, Tanzania. *Primate Conservation*, **24**, 85–90.

Nishida, T. (1989). [A journey to the Mahale National Park: its past and present.] *Seien*, **484**, 24–9. In Japanese.

Nishida, T. and Mwinuka, C. (2005). Introduction of seasonal park fee system to Mahale Mountains National Park: a proposal. *Pan Africa News*, **12**, 17–19.

Nishida, T. and Nakamura, M. (2008). Long-term research and conservation in the Mahale Mountains, Tanzania. In *Science and Conservation in African Forests: The Benefits of Long-term Research*, ed. R. W. Wrangham and E. Ross. Cambridge: Cambridge University Press, pp. 173–83.

Tanzania National Parks (2006). *Mahale Mountains National Park: General Management Plan 2006–2016*. Arusha, Tanzania: TANAPA.

Walpole, M. (2004). *An Assessment of Tourism in Mahale Mountains National Park, Tanzania*. A report compiled in prep. for the General Management Plan. Arusha, Tanzania: TANAPA and FZS.

49 Culture and subsistence ecology of the Tongwe, and their significance in chimpanzee research*

MICHIO NAKAMURA

49.1 Introduction

There are two major reasons why I present the culture and livelihood of the Tongwe people (*Batongwe* in their own language) living in Western Tanzania, within this book commemorating the 50th year of chimpanzee research in Mahale. First, the Tongwe have always been indispensable to the research in Mahale. Most of the research assistants there have been Tongwe, and their frank characters and rich knowledge of nature have always been a fundamental support to the chimpanzee studies. Second, in the early days of the research, a number of human-ecological studies were conducted on the Tongwe society parallel to the chimpanzee studies, by the same Kyoto University research team (see Chapter 3 for the historical background of the research team). When the research began, the Tongwe were largely dependent on the nature in and around Mahale. Because the research team was motivated by anthropological rather than biological interests, the Tongwe way of life in the wilderness was very useful to the research team.

In his book, *Chimpanzees in the Wilderness* (1977a), Jun'ichiro Itani showed that studies of the Tongwe society and those of chimpanzees were inseparable. While this book mainly aimed to publish extensive research findings on wild chimpanzees in Tanzania, Itani includes much description of the livelihood and culture of the Tongwe. For instance, he discusses their genealogy, hunting practices, belief in spirits, perception of animals, and traditional curing rituals. Toshisada Nishida, the founder of Mahale chimpanzee research, spent a long time with the Tongwe people, and described many aspects of their life and customs in his first book, *The Children of the Mountain Spirits* (1973). It was perhaps inevitable that these early researchers, in order to develop a holistic understanding of the environment of the chimpanzees, would develop an interest in the lives and practices of the Tongwe people. Thus, Nishida writes in the book's postscript:

> During the course of following chimpanzees, I could not help being concerned about Mahale as a whole. I have become eager to know and wonder about everything that I see there, such as the variety of animals, birds, fishes, butterflies, beetles, and the forest itself with its trees and herbs. And I have also become fascinated by the Tongwe who live there, and have helped me so much through their relationship with nature, worldview, history, and society. (Nishida, 1973, my translation)

The title of this book, which seems slightly fanciful for a book about chimpanzees, refers to the traditional chief, Sobongo, who took care of young Nishida at Kasoje. He thought of chimpanzees as "children of mountain spirits" and prayed to the spirits of Tongwe on Nishada's behalf when it was difficult to locate the chimpanzees. For the early researchers, then, the study

* This chapter was originally intended to be co-authored by the late Professor Makoto Kakeya who, among all the researchers, knew the most about the Tongwe. However, this was not possible because of his untimely death. I would like to express my utmost gratitude to him for his contribution to knowledge of the Tongwe. I also thank Dr. Juichi Itani for his helpful comments on the manuscript.

Mahale Chimpanzees: 50 Years of Research, ed. M. Nakamura, K. Hosaka, N. Itoh and K. Zamma. Published by Cambridge University Press.

of chimpanzees at Mahale was inextricably connected with the Tongwe, and this provided the motivation to learn about their life and culture.

During 1971–2, Makoto Kakeya, a human ecologist and a student of Itani, visited Mahale and stayed in several settlements where he began intensive anthropological research on Tongwe subsistence and culture, particularly focusing on traditional curing rituals (Kakeya, 1974, 1976). However, the traditional way of livelihood of the Tongwe had largely diminished by 1974, because of the governmental policy to move people living in scattered groups to large communal villages called *kijii cha ujamaa*. Thus, the descriptions given below are mostly based on the writing of Itani, Nishida, and Kakeya, who witnessed the traditional Tongwe way of life from the 1960s to early 1970s. It has been necessary to review these older papers for English readers, because some are written only in Japanese. In addition to this review, I will describe the current situation of the Tongwe, with whom I have built an intimate relationship through my own research on chimpanzees.

49.2 Overview of the Tongwe ethnic group

Itani describes the Tongwe as follows:

> Their bodies are not as slender as the Ha people living to the north, nor are they as small as Bembe or Tabwa people in Zaïre on the other side of the lake. They are generally lean but tall, and have muscular builds. They are not assimilated into civilization but prefer to live alone in the wilderness. They are gracious but sometimes arrogant, not as hard-working as the Ha or Bembe, and many have reflective as well as "big" personalities. (Itani, 1977b, my translation)

The land of the Tongwe (*Utongwe*) covers about 20 000 km^2 of the Kigoma and Mpanda regions in Western Tanzania (Kakeya, 1974). The northern, eastern, and southern boundaries are the Malagarasi River, the Mpanda–Uvinza road, and the Mpanda–Karema road, respectively. Wild chimpanzees inhabit most of the area (Kano, 1972; Yoshikawa *et al.*, 2008). According to the 1967 national census of Tanzania, the Tongwe population was 21 120, 69% of which (14 595) lived in Tongwe land (Kakeya, 1974). Thus the population density there was quite low, i.e. less than 0.8 heads/km^2.

Kano (1972) who has ranged widely through Tongwe land in order to estimate the distribution of chimpanzees in Western Tanzania, also recorded the distribution of Tongwe settlements. Although there were some relatively large settlements along the shore of Lake Tanganyika, most traditional inland Tongwe settlements usually included only 5–10 houses. Thus it was characteristic of the Tongwe to have many small Tongwe settlements spread throughout the broad woodland (Figures 49.1, 49.2). As mentioned above, however, their traditional way of living in such small settlements with a strong dependency on nature, came to an end with the governmental policy to relocate people to a small number of larger communal villages. Most Tongwe, therefore, live in one of such communal villages now.

For everyday use, the Tongwe speak, or used to speak, their own ethnic language: Tongwe or *Sitongwe* (one of the Bantu languages) (see Kakeya and Nishida, 1976 for glossary). However, they also speak Swahili fluently; being an official language of Tanzania it is used as a common language for communication among ethnic groups. The Tongwe are usually Muslims or Christians, but have combined these monotheistic religions with their beliefs in animism and myriad spirits (Nishida, 1973).

Broadly speaking, the Tongwe refer to themselves as "Tongwe" or "Bende." The languages of these two groups differ only to a slight degree in terms of dialect. Thus, both Itani (1977c) and Kakeya (1974) consider Tongwe and Bende as "Tongwe." Abe (2006), in her recent linguistic study of the Bende language, also acknowledges two dialects. However, she notes that, although previously these two dialects have been highly uniform, people are becoming less aware of the fact that they originate from the same ethnic group.

The Tongwe have a patrilineal, polygamous society, which includes more than 15 clans called *mulahilo*. There is no single, paramount chief who rules over the whole Tongwe people. Each clan has multiple lineages

Figure 49.1 A Tongwe settlement in the wilderness. ©Toshisada Nishida, 1969.

Figure 49.2 A traditional Tongwe house. ©Toshisada Nishida, 1968.

(*ibufi*), and each lineage has its own, traditional chief (*mwami*: see below for more details), and a sub-chief (*mtwale*), who recognize the consanguineous relationships between the people of their own clan. Each *mulahilo* has its own oral tradition, each of which says that they originated from completely different ethnic groups than the other clans. This suggests that several ethnic groups of different origins migrated to the area and converged under the common Tongwe language (Kakeya, 1991).

In the 1970s, there were several settlements in Kasoje, the central stage for the Mahale chimpanzee research (see Chapter 2), of more than 100 people, most of whom belonged to the *Mujonga* clan (Kakeya, 1974).

49.3 Subsistence of the Tongwe

The Tongwe mainly cultivated cassava in the semi-permanent lakeshore fields; however, in the mid- to high-altitudinal areas, they depended more on maize, using slash-and-burn cultivation (Kakeya, 1974). They usually cleared riverine forests and planted crops in the open land, but they never cultivated broad Miombo woodland and savanna areas.

The Tongwe in the 1970s, cultivated 42 species of crops, hunted 43 species of wild animals, caught 33 species of fish (only staple fish), gathered 82 species of wild plants, and 9 species of mushrooms (although only 32 species of plants and mushrooms were actively sought) (Kakeya, 1974, 1976; Nishida, 1974). Thus, although they were farmers, their daily subsistence largely depended on the hunting of wild animals and the gathering of plants. Dependence on the meat of wild animals was especially high in the mountainous areas. For example, within the inland settlements of Ilumbi and Sitete, 58.6% and 47.4%, of subsidiary foods were obtained from wild animals or plants, respectively. In a medium-altitude settlement and a lakeshore settlement that were investigated, the dependency was more on river fish (65.5%) and lake fish (95.3%), respectively. Wild animals and plants were not only used for food, but also for traditional medicines and rituals (see below). Insectivory was not common in the 1970s: people only used the larvae of long-horned beetles, alate termites, the larvae of bees, and a type of stingless bee. In the lakeshore settlement where fish is easily caught, insects were seldom eaten. Beekeeping for honey-collecting in the Miombo woodlands was also an important subsistence activity (Takeda, 1976). Making bamboo and winnowing baskets, straw mats, and clay pots (Figure 49.3), sometimes brought monetary income.

Kakeya (1986) uses two key terms to explain the Tongwe approach to subsistence: "minimal efforts" and the "leveling of foods." In terms of the former, he notes that "they have an inclination to rely on food that is most easily available in the nearby environment, with minimal subsistence effort" and in terms of the latter he explains that "when foods are scarce in one settlement, people visit settlements nearby or those inhabited by their kin to request for support. Those with enough stocks cannot refuse such requests. There is a mechanism for mutual assistance with food among the settlements." He argued that "the motivation behind the practice of the leveling of foods

Figure 49.3 A Tongwe woman making a clay pot. ©Toshisada Nishida, 1968.

is fear of ancestral ghosts or myriad spirits, the envy of others, and grudge curses" (Kakeya, 1991). The next section summarizes the Tongwe's beliefs about ancestral ghosts, myriad spirits, and curses.

49.4 Society and spiritual world

The Tongwe have a unique manner of greeting: a form of hand clapping called *kusyesia* (Itani, 1977c; Kakeya, 1986). Typically, the younger one bends forward in front of the elder, and claps hands 10 or more times. This *kusyesia* is specific to Tongwe society, and it is thought to differentiate them from other ethnic groups. There are several types of *kusyesia* that differ in form and rules, such as those between friends, between acquaintances of different ages and class, or to a chief of one's own lineage (Itani, 1977c).

Itani (1977c) identifies three important social systems in Tongwe society: *buwami*, *bufumo*, and *bujege*. He explains that each deals, respectively, with ancestral ghosts, people, and nature. Kakeya (1982a) is more specific, stating that they correspond, respectively, to the chief accession rite, the rite to become a traditional doctor, and the rite to expel the evil spirit of an elephant.

The *buwami* is an institution for chiefdom, and has a backbone of consanguinity. The person who administers it is called an *mwami* (Figure 49.4) and is basically the head of the patrilineal consanguineous lineage. An *mtwale* (sub-chief) becomes an *mwami* through a *buwami* rite, and is then given official authority. However, an *mwami* is not a political leader but someone who is responsible for enshrining their ancestral ghosts. The Tongwe, who lived scattered in the broad woodland environment, regarded this system as governing their society with a public authority underpinned by supernatural spiritual powers. Kakeya (1991) gives further information about this *mwami* system and the *buwami* rite.

Bufumo bridges the physical and spiritual worlds, and its administrator is an *mfumo*. In more common usage, *Mfumo* is the title of a traditional doctor who diagnoses and heals illnesses caused by an *mlosi* (sorcerer) or an *mkuli* (evil spirit of a dead person). An *mfumo* also has a good knowledge of traditional medicines called *dawa*. *Dawa* are divided into *mti* (plant medicines; Kakeya, 1982b) and *sikomelo* (animal medicines; Kakeya, 1978). Kakeya, himself, became an *mfumo* by formally undertaking a *bufumo* rite. Through participatory observations, he was able to describe the daily activities of the traditional doctor, such as collection of medicines, divinations, and curing rituals (Kakeya, 1982b).

Figure 49.4 A lion-skin crown symbolizes an *mwami*, a traditional chief. ©Toshisada Nishida, 1969.

As mentioned above, the fear of curses is fundamental to the Tongwe society (Kakeya, 1986). Nishida (1973) asserted, in the early days of research, that when some Tongwe assistants died from unknown causes, people attributed it to curses or evil spirits.

Bujege is a rite which aims to appease the evil spirit (*mkuli*) of elephants, the administrator of which is called an *mjege*. A hunter who shoots an elephant becomes an *mjege* through a formal rite. The elephant is a symbolic representation of the force of nature, thus an *mjege* who resists such forces is also the representative of a true hunter. Birth of an *mjege* is honorable to the family group, but when an *mjege* dies, one of the family members should succeed the role to avoid misfortune visiting the family. Kakeya (1982a) describes the details of a *bujege* rite held in 1980.

Symbolic artifacts confirm the Tongwe's belief in ancestral ghosts and spirits (Itani *et al.*, 1973, includes many photographs). In the traditional settlements, numerous leopard or lion skulls are displayed in dead branches (called *nsongole*; Figure 49.5) to denote the location of the chief, *mwami* (Itani, 1984). Other such artifacts are: *mgabo* (two half-buried rocks wrapped in white cloths which enshrine the ancestral ghosts), *mahiga* (a rock fireplace close to an *mgabo*, used to cook meals for the ancestors), *msimu* (small house-shaped artifacts which enshrine the ghost of sub-chiefs; Figure 49.6), *kasindye* (two Y-shaped branches with a third, straight branch over them, placed partially into the ground in order to appease the ghosts of breech babies), and *mahasa* (two *kasindye*-shaped structures with many branches over them, for appeasing the ghosts of twins). As such, the Tongwe "have been living with myriad spirits in mountains and valleys and ancestral ghosts" (Itani, 1977c).

49.5 Tongwe people as research assistants

In 1961, Itani, who pioneered the ape research, began developing an intimate relationship with the Tongwe. He hired them as guides and porters to assist with the chimpanzee research, and after 1965 he ranged widely in Tongwe land and visited many settlements (Itani, 1984). When Nishida began his research at Mahale, he stayed in a room in the house of the oldest son of the *mwami* of the Kasiha settlement, Saidi Sobongo (Nishida, 1973). Itani and Nishida's research approach was quite different from that of their predecessor, Kinji Imanishi, who brought a lot of research equipment from Japan at considerable expense and built a cutting-edge base camp with the help of a construction engineer from Japan (Nishida, 2012).

Figure 49.5 Leopard skulls showing the location of an *mwami*, a traditional chief. ©Toshisada Nishida, 1968.

Figure 49.6 Symbolic artifacts enshrining the ancestral ghosts of the Tongwe. ©Toshisada Nishida, 1968.

After Itani took over leadership of the project, he and his co-researchers continued the research while living and integrating with the villagers.

49.5.1 Knowledge of plants

Identifying plant species (Chapters 13, 17) is an essential aspect of chimpanzee research, and Mahale researchers owe a lot to the Tongwe people in terms of this. While the Tongwe do not, generally, have access to higher education, they do have a wide knowledge of local nature. Tongwe terminology includes the names of many plants, the majority of which correspond to the scientific nomenclature of the same plant species. A list of Tongwe–Latin names of plants, compiled by Nishida and Uehara (1981), has been useful for ecological studies of chimpanzees and other animals at Mahale (see Appendix I for an updated version). Many Tongwe names for plants are much shorter than scientific names and are thus easier to remember, even for Japanese researchers. When a researcher cannot identify plants that chimpanzees eat, if a local assistant (or perhaps a village elder) knows its Tongwe name, the scientific name may be found by referring to the list of Tongwe–Latin names of plants (Nishida and Uehara, 1981). Knowledge and understanding of the way in which the Tongwe identify and classify plants, also makes a valuable contribution to ethnotaxonomy. In my experience, both researchers and local assistants in other chimpanzee study sites, refer to plants by their scientific names. Both researchers and local assistants in Mahale, however, refer to plants by their Tongwe names, and this policy helps to facilitate their involvement as research assistants.

In some cases, local knowledge of plants has led to new discoveries. The most evident example may be the discovery of medicinal plants consumed by chimpanzees. It should be noted that Tongwe traditional, indigenous knowledge of medicines, together with behavioral observations and analyses of chemical compounds by researchers, has significantly contributed to the finding (please see Chapter 24 for details).

49.5.2 Knowledge of animals

The Tongwe's knowledge of wild animals has also greatly assisted wildlife research at Mahale. The Tongwe, when they were still heavily dependent on nature, possessed detailed knowledge about local animals. Itani (1977b) collected more than 300 local names of animals during his surveys on chimpanzees, and listed 275 of them with their corresponding Latin names. He listed 62 invertebrates, 58 fish (excluding unidentified species), 21 amphibians and reptiles, 73 birds, and 61 mammals. Among the 61 mammalian species, 57 of the Tongwe names correspond exactly with their Latin names. Exceptions are rodents, bats, and insectivores, which are known by collective names (Itani, 1977c). Of the avian species, 45 correspond with their Latin names, and others are known by collective names. Tongwe's higher classification comprises: *libwasi* (plants), *limujye* (bugs: including mollusks, annelids, arthropods, amphibians, and reptiles), *isembe* (fish), *inyoni* (birds), and *inyuwele* (mammals and some large reptiles) (Itani, 1977b). Kakeya (1978) later added 11 more local names of animals in his systematic analysis of animal medicine (*sikomelo).*

Itani explains the Tongwe's attitude to animals as follows:

> Many of them are extremely good nature observers and their views of nature are more straightforward and objective than one would think. Their overall attitude towards animals in general is impersonal and practical, thus they almost never show compassion to animal life [...] But they look very precisely at the animals' behaviors and habits. (Itani, 1977b, my translation)

The Tongwe's detailed knowledge of animals may be related to their way of life, as in addition to farming, they also hunt wild animals using various methods. Itani (1977c) provides detailed descriptions of 29 hunting methods, including the use of guns, spears, bows, and arrows, and various kinds of traps and snares, which they use to target different animals. They are not considered to be compassionate towards animals, but their use of various hunting methods helps to develop their knowledge of animals. Their knowledge of animal habits is also passed on through their names for plants, which commonly refer to their association with animals. For instance, their name for the plant *mulyansuja* means "the foods of *nsuja* [bushbucks.]"

49.5.3 Indigenous knowledge about chimpanzees

Itani also collected rich narratives, poetry, and folktales about a variety of animals. These show us the way in which the Tongwe regard each animal species. The following section focuses on their attitude towards chimpanzees.

In this region, the chimpanzee is often called *sokomtu* in local Swahili (the standard Swahili is *sokwe* or *sokwe mtu*). A research assistant gave the following reason for the fact that this name includes *mtu*, which means "man":

> *Sokomtu* females have nipples on their breasts just as human women do, and hold their babies in their arms. Unlike other animals, their waist is straight and they can stand on two legs. Also, *sokomtu*'s skin is as black as that of African people. They look like *mtu* (humans), so we call them *sokomtu*. (Itani, 1964, my translation)

Itani (1977b) asked five Tongwe people, to which higher classification of animals (see above) the chimpanzees belonged. Interestingly, two of them thought that chimpanzees are included in the *inyuwele* category (nonhuman mammals), while one answered that they were originally humans, and the other three answered that they are between the categories of *inyuwele* and humans. Thus, Itani concluded that the *inyuwele* category is connected to humans through chimpanzees.

Mahale and the surrounding Tongwe land are suitable for chimpanzee research, partly because the indigenous Tongwe does not eat chimpanzees. This, the Tongwe say, is "because chimpanzees and four cercopithecoid monkey species are close to humans, especially because they have hands similar to us" (Itani, 1977c). However, the Tongwe people did not always avoid hunting chimpanzees, as they once used them in traditional medicines and for making drum skins.

There is a rationale behind the way in which the Tongwe use each animal species in traditional medicine (*sikomelo*) (Kakeya, 1978). In terms of chimpanzees, Kakeya (1978) reports that they use their heads as anti-sorcery medicine (*sindiko*). The logic behind this is "chimpanzees change their nest every day, even if a sorcerer tries to put curse on a house, he will change his mind and go away the next day." The muscle of the right hand of a chimpanzee is used as a medicine when praying to have strong arms, because "chimpanzees have very powerful arms." Furthermore, because "chimpanzees form a group and efficiently find out sufficient foods," they are used as a medicine to pray for a good catch of fish. These descriptions show us the extent of their knowledge of the behaviors and habits of chimpanzees.

Descriptions of chimpanzee's sociality can also be found in Tongwe poetry:

> *Mansoko* [chimpanzees], they are together in a group.
> *Manguje* [baboons], they unite in a separate tribe.
> Thus, never treat children badly (we should also live in harmony). (Itani, 1977c, my translation)

49.6 Tongwe people today

As mentioned above, the communal-living policy of the mid-1970s forced the Tongwe way of life, with its strong dependency on nature, to change. Further, when Mahale was designated as a national park in 1985, a large part of Tongwe land was included in the protected area, which means that people can no longer visit the area without permission from the park authority. Consequently, for example, Mt. Nkungwe (the name of which represents the strongest mountain spirit of Tongwe; Nishida, 1973) has become a mountaineering site, frequently visited by foreign tourists, but for most Tongwe people it can only be viewed in the far distance from their communal villages. Former traditional settlements, such as Mahembe and Ilumbi (both located within the park), have become almost desolate wilderness, with few signs of former residents (Hanamura, 2011, 2012). Ilumbi was once the site of a settlement of mountain-dwellers, the chief of which was regarded by the people living along the lakeshore as being of ancient and honorable origin (Kakeya, 1991). Today, the Tongwe, except for a few research assistants, are never able to visit the places that were once important spiritual locations. The consequences of rapid economic growth in Tanzania, such as the proliferation of large paved roads and mobile phone usage, are also contributing greatly to change in the traditional way of life that Itani, Nishida, and Kakeya witnessed half a century ago.

Such changes have affected the Tongwe's usage of wild plants and animals. The fish of Lake Tanganyika are still an important source of protein intake, but wild mammals are seldom eaten today (hunting wild mammals is prohibited and regarded as poaching), and the use of wild plants or mushrooms is also limited for those living in a communal village (a young local research assistant told me that he has only eaten wild mushrooms once in his life). Prohibition or limitation of the use of wildlife is important for conservation, but it should not be forgotten that it also leads to the disappearance of indigenous knowledge among the younger generations who have fewer opportunities to establish contact with nature (see Nakamura, 2014). In 2012, I visited the primary school at Katumbi village (see Chapter 2 for the location) and showed pupils photographs of 10 animal species commonly seen in Mahale in order to examine whether they knew their names. On an average, the students could correctly identify only 3.95 species. Even when they recognized a species, most of them gave the Swahili, rather than the Tongwe, name for it. Red colobus and red-tailed monkeys are abundant in the Mahale Park (see Chapter 15), but are rare or absent in other areas of Tanzania. Thus, they have only Tongwe names, but, nevertheless, only two and three students, respectively, knew what they were called. In a science textbook widely used in primary education in Tanzania, such species are never mentioned. For example, a third grade science textbook written in Swahili (Kalugula and Kilama, 2005), designates names such as *mbega, nyani*, and *kima*. *Mbega* is a Swahili term for the Abyssinian black-and-white colobus (*Colobus guereza*) of Northern Tanzania (the name is also used to refer to the Angolan black-and-white colobus of Mahale); *nyani* is a common term for baboons (*Papio* spp.: although sometimes, mistakenly, used to denote apes), which are widespread in Tanzania; and *kima* is a general term for monkeys (typically *Cercopithecus* monkeys). If such generic education continues, traditional knowledge of indigenous animals may soon be extinct.

The younger generation's use of the Tongwe language, itself, is diminishing. This is partly because of the prevalence of primary education in Swahili, as well as an increase in inter-ethnic marriages. Such tendencies seem to be more common in the communal villages around Lake Tanganyika. But, from a different perspective, such changes are essential and important for the development of the Tongwe, and not a target of accusation by a foreigner. However, it seems unfortunate that younger people have less access to the diverse plants and animals in their local environment; their traditional knowledge and awe and respect for nature that their grandparents once had is reducing. Probably with similar concerns, on hearing the news of the deaths of chiefs whom he had met through his long-term research, Itani wrote:

> Such sad news of great chiefs' deaths echoed in our ears as if a curfew for the culture of the Tongwe, who had been living in the wilderness. But, was it really just an auditory hallucination? (Itani, 1977c, my translation)

However, not all Tongwe culture is extinct. When researchers visit Tongwe villages today, they still see people performing *kusyesia* (the hand-clapping salutation) to the elders; slightly further inland, artifacts for ancestral ghosts can still be seen (Hanamura, 2013). There are some Tongwe people still living remotely in the mountains outside the park. In 2006, a student and I visited a settlement called Kakungu during the course of an extensive survey. The house where we stayed for some nights was about 20-min walk from the nearest house, and half a day's walk from the nearest communal village with its schools and dispensary, etc. An extended family was living there almost self-sufficiently, and the master of the house happened to be an *mfumo* (traditional doctor). The doctor showed me various types of *dawa* (traditional medicines): powdered or oily materials from plants and animals, which were kept in small glass bottles or plastic cases. When my student colleague contracted eczema of an unknown cause, the doctor mashed up a medicinal plant and applied it to the eczema, which was almost cured by the next morning. The children of the house brought us several plants to show us that they knew their names and usages. Forty years after Itani began to worry about the extinction of Tongwe culture, traditional knowledge was still alive in the wilderness.

In summary, chimpanzee research at Mahale has been conducted from a local perspective, which includes the Tongwe people's knowledge and ideas about nature and the world. The majority of Tanzanian research assistants who support our research in Mahale today are the children or grandchildren of the first generation of Tongwe assistants. These younger trackers are in their twenties or thirties and were born after Mahale became a national park. Thus they have spent most of their life in the communal villages outside the park. However, during the course of assisting our chimpanzee research, they quickly learn the names of plants and animals, and when out on safari for survey, they learn the old ways of living in the wilderness from the older Tongwe assistants. It should not be forgotten that the people and culture of the Tongwe have made a significant contribution to Mahale chimpanzee research, and support is needed to ensure that their wonderful knowledge is passed down from one generation to the next, before it disappears.

References

Abe, Y. (2006). *Bantu Vocabulary Series 13: A Bende Vocabulary*. Tokyo: Research Institute for Language and Cultures of Asia and Africa.

Hanamura, S. (2011). [Following elephants in Ilumbi, part 2: Through Nkungwe to Mahembe.] *Mahale Chimpun*, **17**, 5–6. In Japanese.

Hanamura, S. (2012). [Following elephants in Ilumbi, part 4: Wisdom of *mjege* and a curse of an elephant.] *Mahale Chimpun*, **20**, 3–4. In Japanese.

Hanamura, S. (2013). [Following elephants in Ilumbi, part 5: Guarded by spirits of Nkungwe and Likapuapua.] *Mahale Chimpun*, **21**, 2–3. In Japanese.

Itani, J. (1964). [*Animals of Africa*.] Tokyo: Kawade-shobō. In Japanese.

Itani, J. (1977a[1993]). *In Search of Chimpanzees in the Wilderness*. Tokyo: Heibon-sha Library. In Japanese.

Itani, J. (1977b[2008]). [Ethnozoology of the Batongwe.] Reprinted in [*Works of Jun'ichiro Itani, Vol. IV: Heartbeat of Ecological Anthropology*], Tokyo: Heibon-sha, pp. 148–246. In Japanese.

Itani, J. (1977c[1993]). [*The Savanna of Chimpanzees: In Search of Logic of the Wild*.] Tokyo: Heibon-sha Library. In Japanese.

Itani, J. (1984[2008]). [The natural villages of the Batongwe.] Reprinted in [*Works of Jun'ichiro Itani, Vol. IV: Heartbeat of Ecological Anthropology*.] Tokyo: Heibon-sha, pp. 7–32. In Japanese.

Itani, J., Nishida, T., and Kakeya, M. (1973). [*The Eastern Hinterland of Lake Tanganyika: Its Nature and People*.] Tokyo: Chikuma-shobō. In Japanese.

Kakeya, M. (1974). [Livelihood maintenance system of Tongwe: Living environment, subsistence and diet.] *Kikan Jinruigaku*, **5**(3), 3–90. In Japanese.

Kakeya, M. (1976). Subsistence ecology of the Tongwe, Tanzania. *Kyoto University African Studies*, **10**, 143–212.

Kakeya, M. (1978). [*Sikomelo*: Its materials and logic: animal medicine of the Tongwe.] *Journal of African Studies*, **17**, 1–33. In Japanese.

Kakeya, M. (1982a). [Exorcism of evil spirit of elephants: Mentality of Tongwe living in the wilderness.] *Kikan Minzokugaku*, **19**, 30–41.

Kakeya, M. (1982b). Curing ritual of the Tongwe traditional doctor: Its process and logic. *African Study Monographs*, **Suppl. 1**, 105–39.

Kakeya, M. (1986). [Livelihood of traditional horticultural people with special reference to Tongwe.] In [*Anthropology of Natural Societies*], ed. J. Itani and J. Tanaka. Tokyo: Academia Shuppan, pp. 217–48. In Japanese.

Kakeya, M. (1991) [Between equality and inequality: Tongwe's mwami institution.] In [*Natural History of Man*], ed. J. Tanaka and M. Kakeya. Heibon-sha, pp. 59–88. In Japanese.

Kakeya, M. and Nishida, T. (1976). *A Glossary of Sitongwe*. Unpublished report.

Kalugula, C. and Kilama, H. (2005). [*Primary School Science: 3rd Grade*.] Oxford: Oxford University Press. In Swahili.

Kano, T. (1972). Distribution and adaptation of the chimpanzee on the eastern shore of Lake Tanganyika. *Kyoto University African Studies*, **7**, 37–129.

Nakamura, M. (2014). [Change of the knowledge on animals at the Tongwe land: To what extent do people near Mahale know the wild animals?] *News Letter of the Society of Ecological Anthropology*, **20**, 2–4. In Japanese.

Nishida, T. (1973). [*The Children of the Mountain Spirits: Search for the Social Structure of Chimpanzees*.] Tokyo: Chikuma-shobō. In Japanese.

Nishida, T. (1974). [Ecology of wild chimpanzees.] In [*Human Ecology*], ed. R. Ohtsuka, J. Tanaka, and T. Nishida. Tokyo: Kyōritsu-shuppan, pp. 15–60. In Japanese.

Nishida, T. (2012). *Chimpanzees of the Lakeshore: Natural History and Culture at Mahale*. Cambridge: Cambridge University Press.

Nishida, T. and Uehara, S. (1981). Kitongwe name of plants: a preliminary listing. *African Study Monographs*, **1**, 109–31.

Takeda, J. (1976). An ecological study of the honey-collecting activities of the Tongwe, western Tanzania, East Africa. *Kyoto University African Studies*, **10**, 13–247.

Yoshikawa, M., Ogawa, H., Sakamaki, T., and Idani, G. (2008). Population density of chimpanzees in Tanzania. *Pan Africa News*, **15**, 17–20.

50 Conservation and the future

KAZUHIKO HOSAKA AND MICHIO NAKAMURA

50.1 Conservation risks for wild chimpanzees

50.1.1 General continent-wide threats to chimpanzees

Chimpanzees are facing extinction throughout their range ("endangered" in the IUCN red list; Oates *et al.*, 2008). Kobayashi (2002) participated in establishment of a management plan at Mahale and summarized five main problems for the protection of chimpanzees: (1) habitat destruction; (2) consumption in the human diet; (3) capture and export in the pet trade; (4) use in medical experiments; and (5) disease transmitted via contact with humans.

The first and the largest problem for chimpanzees is rapid destruction of their habitat all over Africa. Destruction of tropical forests, the major habitat for chimpanzees, is mainly caused by human activity. Recent economic growth in Africa, such as road construction, and clearing for agricultural cultivation, timber, and firewood (Figure 50.1) have dramatically reduced the extent of forest. Global warming is also likely to be an indirect factor in the reduction of tropical forests.

The bushmeat trade is another significant problem for great apes, including chimpanzees, in many parts of Africa (e.g. Kano and Asato, 1994; Rose, 1996); this problem is not as severe at Mahale because local Tanzanians do not generally consume chimpanzees. However, caution must be paid because poaching can cause problems, such as injury by wire snares intended to capture small mammals, even in locations in which chimpanzees are not directly hunted (e.g. Quiatt *et al.*, 2002).

Capture for the pet trade is a possible cause of the loss of wild chimpanzees in Tanzania. For example, Kobayashi (2002) reported a case of attempted illegal international trade of juvenile chimpanzees at Lake Tanganyika in 1987. A similar case of attempted commercial trafficking occurred in 1994, when six infant chimpanzees were confiscated in Kigoma, a town along the shore of Lake Tanganyika (Massawe, 1995).

The use of chimpanzees in medical experiments was a problem even before the establishment of the Convention on International Trade in Endangered Species of Wild Fauna and Flora (CITES); vigilance should be maintained towards the illegal capture and international trade of wild chimpanzees for medical use as long as this is known to occur. We should also bear in mind that many African apes exported to the United States, Europe, and Japan before CITES are still living in laboratories, zoos, and sanctuaries.

Transmission of disease from humans to chimpanzees has been one of the most severe problems at most chimpanzee study sites, including Mahale, for the past two decades. This is partly a result of increased visitors from outside of Africa (mostly from Europe and the United States) participating in benignly named "ecotourism" (see Chapter 25).

50.1.2 Conservation problems specific to Mahale

Since Mahale is a national park, anthropogenic disturbances such as poaching or illegal logging

Mahale Chimpanzees: 50 Years of Research, ed. M. Nakamura, K. Hosaka, N. Itoh and K. Zamma. Published by Cambridge University Press.

Figure 50.1 A common example of deforestation caused by cutting of trees for firewood and charcoal. According to local residents, it has become difficult to obtain enough firewood near the shore of Lake Tanganyika because of the population influx. This firewood was collected adjacent to the national park and will be brought to more populated villages for sale. ©M. Nakamura.

within the park are limited (Nakamura and Nishida, 2004). This is especially true in the Kasoje area (see Chapter 2), which has been a center for research and tourism; the daily presence of park staff, researchers, and tourists in this area acts as a deterrent to such destructive activities. However, there are several issues of concern to wildlife conservation at Mahale.

Itoh *et al.* (2012) revealed trends in climatic change in Mahale that were thought to be related to global warming (see Chapter 12). Climate change may pose a threat to Mahale wildlife, including chimpanzees, because it can influence vegetation (Chapter 13) and plant phenology (Chapter 14), which in turn influences the availability of plant foods (e.g. leaves, flowers, fruits).

Despite efforts made by park management, patrolling has been insufficient outside the Kasoje area, and bushfires have occurred during the dry season in most years (Figure 50.2) (Nakamura and Fukuda, 1999; Nakamura *et al.*, 2005). Animal (e.g. elephant and monkey) carcasses are occasionally observed in burned fields. The areas damaged by bushfire are mainly Miombo woodlands; because the dominant plants in these environments are usually fire resistant, the woodlands recover during the rainy season. Such periodical renewal of vegetation by burning can be beneficial for many herbivorous animals, but it might present a serious threat to forest-dwelling animals including chimpanzees. The causes of bushfires are not confirmed, but it is assumed that they are set by poachers or by slash-and-burn farmers who enter the national park without permits.

New threats have emerged in relation to dynamic changes occurring in modern Tanzania. Rapid development, such as construction of paved roads, is taking place around Mahale. Together with deforestation in the adjacent Ugalla Area, severe destruction of vegetation is occurring as a result of the recent immigration of agropastoralists with large herds of cattle (Gen'ichi Idani, personal communication).

Political issues are the source of additional problems. Political uncertainty and civil war is a large problem for wildlife research and conservation projects in Africa. Although the Mahale project has faced few such disturbances thanks to exceptional

Figure 50.2 A burned field in the Mahale Mountains National Park. Large areas in the southern and the eastern parts of the park often burn during the dry season. ©M. Nakamura.

political stability in Tanzania, it has been affected by sporadic political unrest in adjacent countries such as Burundi, Rwanda, and the DRC. For instance, many refugees from Zaïre (the current DRC) have settled in areas close to the national park, and a large refugee camp was established in the wilderness area between Gombe and Mahale. The large population influx into this remote area has had significant effects on the environment. Refugees cut trees for firewood, and some with different food cultures have hunted primates including chimpanzees. Such problems are most obvious in the chimpanzee habitat near Lilansimba where a refugee camp was located (Ogawa *et al.*, 2006).

We must mention another major concern for the health and welfare of chimpanzees at Mahale. To date, research and tourism have focused on the only habituated group, the M group, which could result in disease transmission and continual stress on chimpanzees. The current status and problems with tourism at Mahale are reviewed in Chapter 48, and past epidemics of infectious respiratory diseases at Mahale are summarized in Chapter 25.

50.2 Establishment of the Mahale Mountains National Park

50.2.1 JICA project

An important contribution of long-term study by a team of Japanese researchers at Mahale was the effort to establish Mahale as a national park (see Chapter 3). From the early stages of their research, the Japanese scientists were aware of the need to protect Mahale. In 1967 (the research began in 1965), Nishida requested that the acting director of the game department make Mahale a game reserve (Nishida and Nakamura, 2008). The JICA (see Chapter 2 for abbreviations of the organizations that appear hereafter) project was started in 1975 as the official source of overseas aid by the Japanese government to prepare for the establishment of MMNP. Considering that most international aid projects funded by JICA in the 1970s to 1980s were oriented toward economic development (e.g. road construction or agricultural support), this conservation-oriented project was unusual and forward-thinking at that time.

Table 50.1 **Names of JICA experts worked at Mahale (based on Nishida, 2002)**

Year	Names of experts
1975	Toshisada Nishida, Kohshi Norikoshi
1976	Nishida, Norikoshi, Shigeo Uehara
1977	Norikoshi, Uehara
1978	Uehara, Kenji Kawanaka
1979	Kawanaka, Jun'ichiro Itani, Kisaburo Tokuda, Toshikazu Hasegawa
1980	T. Hasegawa, Mariko Hasegawa
1981	T. Hasegawa, M. Hasegawa
1982	T. Hasegawa, M. Hasegawa, Yukio Takahata
1983	Takahata
1984	Takahata
1985	Ken'ichi Masui
1986	Masui
1987	Masui
1988	Satoshi Kobayashi
1989	Kobayashi
1994	Fumio Fukuda
1995	Fukuda
1996	Fukuda

Many Japanese researchers were dispatched as experts for the JICA project (Table 50.1) to collect basic information on Mahale flora and fauna, and on chimpanzees in particular (Nishida, 1990). This work resulted in a final report that proposed the establishment of a national park at Mahale (Anonymous, 1980). On June 14, 1985, President Julius Nyerere designated Mahale as the 11th national park of Tanzania (Nishida, 1985). This was the first case in which a national park was established by Japanese international governmental aid (Kobayashi, 2002).

50.2.2 Effects of establishing the national park on wildlife conservation

Several positive effects resulted from the designation of Mahale as a protected area. Bushfires were prohibited with the start of the JICA project in 1975, and residency and farming were prohibited in the park by 1985. Notable forest recovery has occurred; formerly cultivated fields are now almost completely forested. Except for remnant patches of oil palms, mangos, and guavas, it is difficult to identify the locations of the once-cultivated areas (see Chapter 13).

Changes in the mammalian fauna have occurred as vegetation has recovered. For example, the frequency of observation of bushbucks (which prefer disturbed habitat) has decreased, while observation of blue duiker and some primate species that prefer more forested habitat has increased (Chapter 15). Some animals appear to have become less afraid of humans, probably because of the lack of hunting pressure.

50.2.3 Initial stage of park management

The governing structure of the MMNP was not completed when the park was established in 1985. At that time, tourists visiting Mahale were taken in by

Serengeti Wildlife Research Institute (SWRI, the current TAWIRI) officials or by the Japanese researchers, until the first park warden, Mr. A. H. Seki, arrived in 1989 (Toshisada Nishida, personal communication). Although it was the mid-1990s before the MMNP resolved its staffing and equipment deficiencies, the situation has improved greatly.

50.3 Foundation of the Mahale Wildlife Conservation Society

50.3.1 History and mission

MWCS is a Tanzanian NGO established in 1994. Hosea Y. Kayumbo (Professor at University of Dar es Salaam), Toshisada Nishida, and Toshimichi Nemoto initially served as the chairman, co-chairman, and secretary-general of the society, respectively. Since 2011, Hosaka has served as the co-chairman (representative of the Japanese branch). Members of the MWCS consist mostly of researchers who have conducted their studies at Mahale, and of other individuals who are interested in the activities of the MWCS/MMCRP. The mission of the MWCS is to promote wildlife research and conservation inside and around Mahale. MWCS pursues community-based conservation, a strategy that involves the local community through environmental education while providing improvements to school and healthcare infrastructure. MWCS has published an English language newsletter, *Pan Africa News* (see below), since 1994, and a Japanese newsletter, *Mahale Chimpun*, since 2003.

50.3.2 *Pan Africa News*: its role and contribution to conservation and panthropology

One of the well-known roles of the MWCS is to publish *Pan Africa News* (*PAN*), a biannual newsletter that covers not only chimpanzee research and conservation at Mahale, but also papers about wild *Pan* species (chimpanzees and bonobos) from other African study sites. The first Chief Editor, Toshisada Nishida, launched *PAN* in response to an idea generated at the symposium "Understanding chimpanzees," which took place in Chicago in November 1986 (Toshisada Nishida, personal communication). Jane Goodall and other chimpanzee experts felt that they needed a newsletter in which fieldworkers studying wild chimpanzees at various sites could share information and experiences. At that time, the Gombe and Mahale projects had accumulated research experience spanning more than two decades, and many other projects exploring the two *Pan* species had emerged across equatorial Africa. However, researchers were not always familiar with ongoing studies and the conservation status of chimpanzees at sites other than their own; information published in journals with wider readership was limited. *PAN* was intended to serve as a forum in which fieldworkers interested in the *Pan* species could communicate informally to facilitate their research and conservation efforts.

PAN is published in June and December, and an online version is also available (http://mahale.main.jp/PAN/). In 1997, a new editorial system was adopted to manage the submission and peer-review process to meet the needs of fieldworkers who wished to publish small findings, given that other scientific journals tend not to publish papers based on a single or rare observation. By accumulating rare events, *PAN* has provided an information reservoir that can be useful for understanding the behavioral diversity of *Pan* species (Chapters 37, 38) and the species- or genus-specific characteristics of these animals (e.g. Wilson *et al.*, 2014).

50.4 Achievements of MWCS and other organizations over these two decades

In this section, conservation-related activities at Mahale by MWCS and other governmental or NGOs since the establishment of MWCS in 1994 will be summarized.

50.4.1 Educational and medical support to the local community

Today, MWCS mainly conducts activities that support education and medical needs of the people who originally lived in the Mahale area. The Tongwe, the indigenous people of Mahale, accepted the government's decision to make their land into a national park and moved out of the area without protest or complaint (see also Chapter 49). MWCS's support to the local people has an aspect of compensation for those who lost their land and also offers a strategy for implementing community-based conservation based on the belief that few activities can be effective in the absence of improving educational and medical infrastructure. In 1998, MWCS started an environmental education project with financial support from the Japan Environment Corporation, and donated a visitor center, which was built next to the park headquarters in Bilenge (Nishida, 2000). The society also prepared the first information leaflet in Swahili, to raise the awareness of local residents. In 2001, MWCS built a primary school (Figure 50.3) in Katumbi, one of the communal villages to which original inhabitants of Mahale moved with financial aid from the Grant Assistance for Grass-Roots Projects from the Embassy of Japan in Tanzania (Nishida, 2004a). Classroom furniture was donated to the school by MWCS, but educational staff and furnishings have remained insufficient for the >800 schoolchildren that presently attend (Hosaka, 2008). Since 2006, MWCS has collaborated with GRASP-Japan, which granted the fund for conservation activities in the Greater Mahale region (Nishida, 2004b). Japanese researchers have visited the primary school and donated supplies such as textbooks and stationery (Zamma, 2011; Shimada, 2011; Hanamura, 2012), as well as giving lectures and videos. We have also helped the local government to develop a fund for construction of a clinic in the village (the Katumbi Clinic) and donated medical supplies (Inoue, 2011). After 2010, MWCS provided a scholarship for a young, talented Tongwe man to study and obtain a diploma in wildlife management at the College of African Wildlife Management, Mweka (Nyundo, 2011, 2012, 2013; Hosaka, 2013). Although such academic accomplishments are essential for

Figure 50.3 The Katumbi Primary School, built with aid from the Embassy of Japan. ©M. Nakamura.

obtaining work at national parks or in conservation-related organizations in Tanzania, very few Tongwe youth have studied there.

In addition to MWCS, the Watoto Foundation (founded in 1997 by one of the pioneer researchers, Kenji Kawanaka and his wife, Hatsuko Kawanaka) and the Tongwe Trust (supported by Roland Purcell, the first tour operator at Mahale) have provided scholarships and support to the Katumbi Clinic and the nursery class of the Katumbi Primary School.

50.4.2 Eradication of invasive alien tree species

One of the ecological problems in the Kasoje area in Mahale after the 1990s was the spread of an invasive alien plant species, *Senna spectabilis* (Figure 50.4) (for details, see Nishida, 1996; Turner, 1996; Lukosi, 1997; Wakibara, 1998; Wakibara and Mnaya, 2002). From the late 1990s, with financial support from MWCS and the ANC Production, Inc., MMNP started a project to eliminate *S. spectabilis* by cutting or peeling the bark of parent trees and by burning fallen seeds. Because complete removal of seeds and saplings was impossible, many trees reestablished after a few years. Therefore, continuous efforts were required to repeatedly remove the trees. By 2014, after more than 15 years, *S. spectabilis* patches appear to have decreased considerably. A test plot, in which James Wakibara exhaustively removed *S. spectabilis* and planted seedlings germinated from chimpanzee feces, has undergone succession and recovered to a dense thicket that includes chimpanzee food plants.

50.4.3 Actions against the outbreak of epidemic diseases

The M group experienced two serious outbreaks of infectious respiratory diseases of suspected human origin, in 1993 and 2006 (Chapter 25). The 1993 outbreak occurred before the national park was properly structured and when former inhabitants of Kasoje freely visited Kasiha Village (the only residential area inside the M group's range at that time) to stay with family members working in the park. Although the number of tourists remained small (approximately 1–3 per week), some individuals who acquired respiratory conditions en route to Mahale

Figure 50.4 *Senna spectabilis* in the 1990s in the Kasoje area. Such extensive patches of *S. spectabilis* are no longer observed at Kasoje. ©M. Nakamura.

tracked chimpanzees in the park. In response to advice offered by Hosaka after the outbreak, the park administration took urgent measures to improve hygiene conditions in Kasiha (Hosaka, 1995) and in later years came to more strictly regulate the entry of former inhabitants into the park. Such actions were vital to preventing the spread of human diseases, such as cholera, amoebic dysentery, and typhoid, from adjacent villages to chimpanzees. Other important measures, such as maintaining distance to animals, were specified in the park regulations, but no official park guides were trained at that time to direct English-speaking tourists to observe park regulations. Such language problems have gradually been resolved.

After the 2006 outbreak, researchers, tour operators, tour guides, and MMNP wardens convened a meeting and resolved to improve observance of park regulations for chimpanzee viewing. In addition to existing rules, members of MWCS requested everyone (including researchers, tourists, and guides) who visited chimpanzees to wear facemasks (Hanamura *et al.*, 2006). MWCS provided masks for tourists and a MWCS member visited the tourist camps and asked for the cooperation of camp managers in directing guests to wear masks in the forest. Wearing of masks has recently taken root as a general rule and some tourist camps now prepare masks themselves.

50.4.4 Participation in the decision-making process for park management

A workshop that aimed to discuss problems of conservation and management in and around Mahale and to finalize a general 10-year management plan for the national park was held at MMNP headquarters in February 2004. The workshop was organized by Frankfurt Zoological Society (FZS) and sponsored by the EU and the executive committee was composed of FZS members and MMNP wardens. Stakeholders from organizations such as TAWIRI, TANAPA, MMWRC, and the Jane Goodall Institute, as well as tour operators and influential persons from villages around Mahale, were invited to the workshop. Nakamura, then conducting research at Kasoje, was invited to represent MWCS/MMCRP. Although the executive committee probably intended to invite all stakeholders, Tongwe research assistants, who have contributed substantially to Mahale research and who have the best knowledge about Mahale, were not invited. The workshop was conducted in English and Swahili (approximate ratio of 8:2).

During the workshop, Mahale's exceptional resource values, purposes, current problems, and long-term strategies (for the next 10 years) were discussed and ranked. Because these rankings were decided by majority vote, some minority opinions tended to be discarded such that only moderate opinions remained. For example, in a statement about the park's value, Nakamura proposed to include the name "Tongwe," who had made major contributions to the protection of chimpanzee habitat well before establishment of the national park (see Chapter 49) and who were required to move from their homeland. However, this proposal was not adopted by majority vote.

The significance of this workshop was unquestionable, as this was one of the first occasions to bring diverse stakeholders, including tour operators and village chiefs, together. However, some important decisions made during this workshop have not yet been executed. For example, introduction of an advanced booking system for chimpanzee tracking permits was proposed to mitigate the effects on the M group of overcrowding by excess tourism. This initiative obtained the most votes from stakeholders. Regrettably, it has not been realized as of 2014, 8 years after completion of the General Management Plan (GMP) in August 2006, and the number of tourists is not yet under good control (Chapter 48).

50.4.5 Collaboration with television and film production

Mahale researchers have been collaborating with television and film producers. From 1989, The ANC Production, Inc., led by Miho Nakamura (a former student of Nishida) made several television documentaries about chimpanzees by following them for several months, methods similar to those used by researchers. Some programs have been translated into English and have obtained high international repute. The East Entertainment, Inc. led by Masayasu Mori has filmed sporadically at Mahale since 1982, and after 1994, when Mitsue Matsuya joined as a director, the television crew visited Mahale almost every year for several years. The chimpanzee stories produced by East Entertainment, Inc. often appeared on *Dōbutsu Kisōtengai* [*Amazing Animals*], a popular prime-time program that was broadcasted until 2009. Other television crews, such as the NHK (Nippon Hōsō Kyōkai or Japan Broadcasting Corporation) and the BBC (British Broadcasting Corporation), also filmed chimpanzees at Mahale. Currently, no television crews are filming on a regular basis, probably because of increased filming fees.

Such television programs have helped the general public in Japan (and worldwide) to become familiar with chimpanzees, and have made some individual chimpanzees at Mahale, such as old Kalunde (see Chapters 5, 23, 27), famous among Japanese people. The researchers often assisted the television crew by providing information on-site or by supervising the edited programs. Although chimpanzees are often described in an unduly anthropomorphic way in such television programs, such appearances are beneficial in that they raise public awareness about chimpanzees and the need to promote their conservation.

50.5 Future perspectives

50.5.1 Reconstruction of community-based conservation

As we have outlined in this chapter, several conservation activities have been conducted at Mahale, and most of these efforts have been fruitful. However, these activities were mostly led by researchers or by park management and involved addressing specific problems largely on an ad hoc basis. In order for effective conservation measures to continue long term, a framework is needed in which local people can think and act proactively. Internationally recognized values of biodiversity, and efforts made by researchers or conservationists, do not always have the same significance to local people; local communities should be engaged and given foremost consideration in the conservation framework.

The Tongwe used to have good knowledge of animals and plants, as represented by their indigenous taxonomy (Chapter 49). Many Mahale animals and plants are endemic and often do not have common names in Swahili, the official language in Tanzania. The Tongwe, through their traditional knowledge of nature, have also contributed much as research assistants to the Japanese team. Regrettably, there have been few opportunities for Tongwe youth from communal villages to enter the national park and observe animals and plants, which may accelerate a loss of traditional knowledge about nature. This situation is worsened by the loss of the Tongwe language, which is nearly extinct as a result of frequent inter-ethnic marriages and infrequent use of the language in multi-ethnic communal villages.

There is a serious concern about the decline of natural and cultural diversity in places that were originally rich in both but that are facing rapid modernization and globalization. It is necessary to reaffirm the importance of biodiversity, which is valuable from an international/scientific perspective but also from a local/indigenous perspective. Then, the establishment of a locally maintainable education system that can pass traditional knowledge on through the generations should be explored. To realize this, it might also be necessary to create a framework by which the community can realize economic benefits – for example, by using traditional knowledge for tourism – under the mutual agreement of stakeholders. We are currently running such a project, called "Localizing biodiversity: introduction of an environmental education system to support spontaneous conservation efforts by the local community in Western Tanzania" (project leader: Michio Nakamura), with financial support from the Research Grant Program 2013 of the Toyota Foundation.

50.5.2 Merits of the long-term study

The long-term presence of researchers in the field site undoubtedly helps to protect chimpanzees (e.g. Pusey *et al.*, 2007; Wrangham and Ross, 2008; Tranquilli *et al.*, 2012). Researchers' daily walks in the forest to search for or follow chimpanzees help to deter poaching and illegal logging. It is also essential to accumulate long-term demographic and health data for the preparation of an effective conservation plan.

In the future, closer collaboration with researchers studying chimpanzees around Mahale might need to be facilitated. Considering that the Mahale chimpanzee population is connected with populations outside the park, habitat loss outside Mahale may cause the in-park population to deteriorate. At the same time, we need to establish a research team that includes non-primatologists. Relatively little research on Mahale mammalian fauna has been performed (see Chapter 15), and of this, most has been conducted by primatologists. Recently, some students have started studies of non-primate mammals at Mahale. For example, a preliminary survey by Nobuko Nakazawa confirmed for the first time that a chimpanzee was eaten by a leopard (Nakazawa *et al.*, 2013). Collaboration with non-primatologists might improve understanding of Mahale's complex ecosystem, which is important for its conservation.

50.5.3 Environmental education

Chimpanzee research at Mahale is little known to the public in Japan and elsewhere, despite its long history and academic achievements. This is made clear when we compare its worldwide recognition with that of Gombe. Better dissemination of the achievements of long-term research, academic findings, and conservation activities is needed. For this purpose, we may pursue efficient use of mass media and consistent education of schoolchildren and the general public.

50.5.4 Collaboration with other organizations

Finally, we point out the necessity of collaboration with other organizations. Chimpanzee conservation is no longer an issue for just one location; it is necessary to confront multifaceted problems in this rapidly globalizing world. It is obvious that the MWCS, which is basically an assemblage of researchers, cannot cope with these problems alone. We need to collaborate with researchers from other study sites, international organizations such as the United Nations Environment Programme (UNEP), conservation organizations such as the Wildlife Conservation Society (WCS), local government, TANAPA, TAWIRI, tour companies, and FZS, with whom we share the ultimate purpose of conserving chimpanzees. It might also be essential to train researchers who can promote practical work in conservation biology and to coordinate efforts with other governmental or conservation organizations.

References

Anonymous (1980). *Mahale: Study for the Proposed Mahale Mountains National Park. Final Report, May 1980.* Tokyo: Japan International Cooperation Agency.

Hanamura, S. (2012). [Visiting Katumbi primary school.] *Mahale Chimpun*, **19**, 4. In Japanese.

Hanamura, S., Kiyono, M., Nakamura, M., *et al.* (2006). A new code of observation employed at Mahale: prevention against a flu-like disease. *Pan Africa News*, **13**, 1–16.

Hosaka, K. (1995). Mahale: a single flu epidemic killed at least 11 chimps. *Pan Africa News*, **2**(2), 3–4.

Hosaka, K. (2008). [Visiting Katumbi primary school.] *Mahale Chimpun*, **12**, 3. In Japanese.

Hosaka, K. (2013). The first MWCS scholarship student awarded diploma. *Pan Africa News*, **20**, 24.

Inoue, S. (2011). [Completion of a clinic at Katumbi.] *Mahale Chimpun*, **17**, 3–4. In Japanese.

Itoh, N., Nakamura, M., Ihobe, H., *et al.* (2012). Long-term changes in the social and natural environments surrounding the chimpanzees of the Mahale Mountains National Park. In *The Ecological Impact of Long-Term Changes in Africa's Rift Valley*, ed. A. J. Plumptre. New York: Nova Science, pp. 249–77.

Kano, T. and Asato, R. (1994). Hunting pressure on chimpanzees and gorillas in the Motaba river area, northeastern Congo. *African Study Monographs*, **15**, 143–62.

Kobayashi, S. (2002). [Tanzania's nature conservation policy and park management at Mahale.] In [*The Mahale Chimpanzees: 37 Years of <Panthropology>*], ed. T. Nishida, S. Uehara, and K. Kawanaka. Kyoto: Kyoto University Press, pp. 153–68. In Japanese.

Lukosi, N. (1997). A brief note on possible control of *Senna spectabilis*, an invasive exotic tree at Mahale. *Pan Africa News*, **4**, 18.

Massawe, E. (1995). Commercial trafficking in chimpanzees: Six young chimpanzees were confiscated in Kigoma, Tanzania. *Pan Africa News*, **2**(2), 8–9.

Nakamura, M. and Fukuda, F. (1999). Chimpanzees to the east of the Mahale Mountains. *Pan Africa News*, **6**, 5–7.

Nakamura, M. and Nishida, T. (2004). [Present status of conservation on chimpanzees and their habitat in Mahale.] *Primate Research*, **20**, 57–61. In Japanese.

Nakamura, M., Nishie, H., and Mwinuka, C. (2005). Survey of the southern part of the Mahale Mountains. *Pan Africa News*, **12**, 5–8.

Nakazawa, N., Hanamura, S., Inoue, E., Nakatsukasa, M., and Nakamura, M. (2013). A leopard ate a chimpanzee: the first evidence from East Africa. *Journal of Human Evolution*, **65**, 334–7.

Nishida, T. (1985). The Mahale Mountains declared a national park. *IPPL Newsletter*, **12**, 9.

Nishida, T. (1990). A quarter century of research in the Mahale Mountains: an overview. In *The Chimpanzees of the Mahale Mountains: Sexual and Life History Strategies*, ed. T. Nishida. Tokyo: University of Tokyo Press, pp. 3–35.

Nishida, T. (1996). Eradication of the invasive, exotic tree *Senna spectabilis* in the Mahale Mountains. *Pan Africa News*, **3**(2), 6–7.

Nishida, T. (2000). New visitor centre in Mahale Park. *Pan Africa News*, **7**, 4–5.

Nishida, T. (2002). [A brief history of Mahale research.] In [*The Mahale Chimpanzees: 37 Years of <Panthropology>*], ed. T. Nishida, S. Uehara, and K. Kawanaka. Kyoto: Kyoto University Press, pp. 5–28. In Japanese.

Nishida, T. (2004a). [Completion of Katumbi primary school.] *Mahale Chimpun*, **4**, 6. In Japanese.

Nishida, T. (2004b). [GRASP and foundation of GRASP-Japan.] *Mahale Chimpun*, **4**, 5. In Japanese.

Nishida, T. and Nakamura, M. (2008). Long-term research and conservation in the Mahale Mountains, Tanzania. In *Science and Conservation in African Forests*, ed. R. W. Wrangham and E. Ross. Cambridge: Cambridge University Press, pp. 173–83.

Nyundo, B. R. (2011). [First year at MWECA.] *Mahale Chimpun*, **18**, 9–10. In Japanese.

Nyundo, B. R. (2012). [Finishing two years at MWECA.] *Mahale Chimpun*, **20**, 8. In Japanese.

Nyundo, B. R. (2013). My teaching experience at my hometown. *Pan Africa News*, **20**, 24–5.

Oates, J. F., Tutin, C. E. G., Humle, T., *et al.* (2008). *Pan troglodytes. The IUCN Red List of Threatened Species. Version 2010.4.* Available at: www.iucnredlist.org. Accessed March 15, 2014.

Ogawa, H., Moore, J., and Kamenya, S. (2006). Chimpanzees in the Ntakata and Kakungu areas, Tanzania. *Primate Conservation*, **21**, 97–101.

Pusey, A. E., Pintea, L., Wilson, M. L., Kamenya, S., and Goodall, J. (2007). The contribution of long-term research at Gombe National Park to chimpanzee conservation. *Conservation Biology*, **21**, 623–34.

Quiatt, D., Reynolds, V., and Stokes, E. J. (2002). Snare injuries to chimpanzees (*Pan troglodytes*) at 10 study sites in east and West Africa. *African Journal of Ecology*, **40**, 303–5.

Rose, A. L. (1996). The African great ape bushmeat crisis. *Pan Africa News*, **3**(2), 1–6.

Shimada, M. (2011). [Donation to villages around Mahale: Activity at Kalya.] *Mahale Chimpun*, **18**, 3. In Japanese.

Tranquilli, S., Abedi-Lartey, M., Amsini, F., *et al.* (2012). Lack of conservation effort rapidly increases African great ape extinction risk. *Conservation Letters*, **5**, 48–55.

Turner, L. A. (1996). Invasive plant in chimpanzee habitat at Mahale. *Pan Africa News*, **3**(1), 5.
Wakibara, J. V. (1998). Observations on the pilot control of *Senna spectabilis*, an invasive exotic tree in the Mahale Mountains National Park, Western Tanzania. *Pan Africa News*, **5**, 4–6.
Wakibara, J. V. and Mnaya, B. J. (2002). Possible control of *Senna spectabilis* (Caesalpiniaceae), an invasive tree in Mahale Mountains National Park, Tanzania. *Oryx*, **36**, 357–63.
Wilson, M. L., Boesch, C., Fruth, B., *et al.* (2014). Lethal aggression in *Pan* is better explained by adaptive strategies than human impacts. *Nature*, **513**, 414–17.
Wrangham, R. W. and Ross, E., eds. (2008). *Science and Conservation in African Forests*. Cambridge: Cambridge University Press.
Zamma, K. (2011). [Donation to villages around Mahale: Activity at Katumbi.] *Mahale Chimpun*, **18**, 2. In Japanese.

APPENDIX I

Floral list

COMPILED BY NORIKO ITOH

This list is based on Itoh (*in prep.*) with synonyms excluded. Please note that confidence level on identification, name, and existence at Mahale is variable. For those who are in search of names before correction and those who are looking for corresponding local or Latin names from the plant observed in the field, see Itoh (*in prep.*) where detailed information are available. Here, only the identification status for those records not fully identified by specialists at herbariums and universities are provided (**by obs.**: identified by observation in the field; **iD ign.**: name published without description; **IiDα**: incomplete identification by a specialist (probably insufficient specimen); **IiDβ**: incomplete identification obtained by obs. or OSSED; **OSSED**: names obtained from Oxford's *Standard Swahili-English Dictionary*; **UiD**: not identified). Species names and family names are shown in bold font style and small capital letters, respectively, to avoid character corruption. Unique referential number (MLID) for names in Itoh (*in prep.*) appear in brackets.

Abrus precatorius L. LEGUMINOSAE-PAPILIONOIDEAE [548]
Abutilon mauritianum (Jacq.) Medik. MALVACEAE [1102]
Acacia amythethophylla A.Rich. LEGUMINOSAE-MIMOSOIDEAE [1286]
Acacia drepanolobium Harms ex Sjostedt LEGUMINOSAE-MIMOSOIDEAE [463] (by obs.)
Acacia hockii De Wild. LEGUMINOSAE-MIMOSOIDEAE [464]
Acacia polyacantha Willd. subsp. **campylacantha** (A.Rich.) Brenan LEGUMINOSAE-MIMOSOIDEAE [466.]
Acacia sieberiana DC. LEGUMINOSAE-MIMOSOIDEAE [469]
Acacia sp. LEGUMINOSAE-MIMOSOIDEAE [844] (IiDα)
Acalypha brachiata Krauss EUPHORBIACEAE [953]
Acalypha chirindica S.Moore EUPHORBIACEAE [293]
Acalypha ornata Hochst. ex A.Rich. EUPHORBIACEAE [294]
Acanthospermum hispidum DC. COMPOSITAE [395]
Acanthus sp. ACANTHACEAE [1] (IiDα)
Achyranthes aspera L. AMARANTHACEAE [27]
Acridocarpus sp. MALPIGHIACEAE [1208] (IiDα)
Adenia reticulata (De Wild. & Th.Dur.) Engl. PASSIFLORACEAE [602] (IiDα).
Adenia rumicifolia Engl. & Harms PASSIFLORACEAE [603]
Adiantum lunulatum Burm. f. ADIANTACEAE [1425]

Adiantum poiretii Wikstr. ADIANTACEAE [1298]
Aeschynomene burttii Baker f. LEGUMINOSAE-PAPILIONOIDEAE [549]
Aeschynomene cristata Vatke LEGUMINOSAE-PAPILIONOIDEAE [1426]
Aeschynomene leptophylla Harms LEGUMINOSAE-PAPILIONOIDEAE [551]
Aeschynomene sp. LEGUMINOSAE-PAPILIONOIDEAE [885] (IiDα)
Aframomum alboviolaceum (Ridl.) K.Schum. ZINGIBERACEAE [784]
Aframomum mala (K.Schum. ex Engl.) K.Schum. ZINGIBERACEAE [785]
Aframomum sp. ZINGIBERACEAE [845] (iD ign)
Afzelia africana Pres. LEGUMINOSAE-CAESALPINIOIDEAE [121]
Afzelia quanzensis Welw. LEGUMINOSAE-CAESALPINIOIDEAE [122]
Agathisanthemum globosum (Hochst. ex A.Rich.) Klotzsch RUBIACEAE [1307] (IiDα)
Agelaea pentagyna (Lam.) Baill. CONNARACEAE [1293]
Ageratum conyzoides (L.) L. COMPOSITAE [791]
Albizia adianthifolia (Schumach.) W.F.Wight LEGUMINOSAE-MIMOSOIDEAE [470]
Albizia antunesiana Harms LEGUMINOSAE-MIMOSOIDEAE [471]
Albizia glaberrima (Schum. & Thonn.) Benth. LEGUMINOSAE-MIMOSOIDEAE [1427]
Albizia gummifera (J.F.Gmel.) C.A.Sm. LEGUMINOSAE-MIMOSOIDEAE [473]
Albizia versicolor Oliver LEGUMINOSAE-MIMOSOIDEAE [475]
Albuca abyssinica Jacq. HYACINTHACEAE [1002]
Alchornea cordifolia (Schumach. & Thonn.) Müll.Arg. EUPHORBIACEAE [1076]
Allophylus africanus P.Beauv. s.l. SAPINDACEAE [694] (IiDα)
Allophylus africanus P.Beauv. var. **griseotomentosus** (Gilg) Verdc. SAPINDACEAE [1308]
Allophylus congolanus Gilg SAPINDACEAE [695]
Allophylus richardsiae Exell SAPINDACEAE [696]
Aloe sp. ALOACEAE [403] (IiDα)
Alternanthera sessilis (L.) R.Br.ex Roth AMARANTHACEAE [28]
Amaranthus dubius Mart. ex Thell. AMARANTHACEAE [29]
Amaranthus hybridus L. AMARANTHACEAE [30]
Ampelocissus abyssinica (Hochst. ex A.Rich.) Planch. VITACEAE [1302]
Ampelocissus africana (Lour.) Merr. VITACEAE [826]
Ampelocissus obtusata (Welw. ex Baker) Planch. subsp. **kirkiana** (Planch.) Wild & R.B. Drumm. VITACEAE [1103]
Anchomanes difformis (Blume) Engl. ARACEAE [1309] (IiDα)
Anchomanes sp. ARACEAE [71] (IiDα)
Ancylobothrys amoena Hua APOCYNACEAE [51]
Andropogon schirensis Hotchst. GRAMINEAE [341]
Aneilema beniniense (P.Beauv.) Kunth COMMELINACEAE [182]
Aneilema nyasense C.B.Clarke COMMELINACEAE [183]

Aneilema spekei C.B.Clarke COMMELINACEAE [185]
Anisopappus chinensis (L.) Hook. & Arn. subsp. **africanus** (Hook.f.) S.Ortiz & Paiva COMPOSITAE [1310]
Anisophyllea boehmii Engl. ANISOPHYLLEACEAE [628]
Annona senegalensis Pers. ANNONACEAE [39]
Anthericum sp. ANTHERICACEAE [404] (IiDα)
Anthocleista schweinfurthii Gilg LOGANIACEAE [411]
Anthonotha macrophylla P.Beauv. LEGUMINOSAE-CAESALPINIOIDEAE [1363] (IiDα)
Anthonotha noldeae (Rossberg) Exell & Hillcoat LEGUMINOSAE-CAESALPINIOIDEAE [123]
Antiaris sp. MORACEAE [485] (IiDα)
Antidesma venosum E.Mey. ex Tul. EUPHORBIACEAE [299]
Aphanocalyx richardsiae (J.Leonard) Wieringa LEGUMINOSAE-CAESALPINIOIDEAE [1343]
Apodytes dimidiata E.Mey. ex Arn. ICACINACEAE [389]
Aporrhiza paniculata Radlk. SAPINDACEAE [699]
Arachis hypogaea L. LEGUMINOSAE-PAPILIONOIDEAE [1377] (by obs.)
Argomuellera macrophylla Pax EUPHORBIACEAE [1000]
Argyrolobium tomentosum (Andrews) Druce LEGUMINOSAE-PAPILIONOIDEAE [1107]
Aristea alata Baker IRIDACEAE [1005]
Aristolochia albida Duch. ARISTOLOCHIACEAE [1281]
Artabotrys monteiroae Oliv. ANNONACEAE [40]
Arthropteris orientalis (Gmel.) Posth. DAVALLIACEAE [1105]
Asparagus africanus Lam. ASPARAGACEAE [405]
Asparagus buchananii Baker ASPARAGACEAE [1303] (IiDα)
Asparagus flagellaris (Kunth) Baker ASPARAGACEAE [406]
Asparagus sp. ASPARAGACEAE [408] (IiDα)
Aspilia africana (Pers.) C.D.Adams subsp. **magnifica** (Chiov.) Wild COMPOSITAE [1312]
Aspilia mossambicensis (Oliv.) Wild COMPOSITAE [197]
Aspilia pluriseta Schweinf. ex Schweinf. COMPOSITAE [1108]
Aspilia rudis Oliv. & Hiern COMPOSITAE [1429]
Asystasia gangetica (L.) T.Anderson ACANTHACEAE [2]
Athrixia rosmarinifolia (Sch.Bip. ex Walp.) Oliv. & Hiern COMPOSITAE [1109]
Azanza garckeana (F.Hoffm.) Exell & Hillc. MALVACEAE [421]
Baissea major (Stapf) Hiern APOCYNACEAE [52]
Baphia capparidifolia Baker subsp. **multiflora** (Harms) Brummitt LEGUMINOSAE-PAPILIONOIDEAE [553]
Bauhinia petersiana Bolle LEGUMINOSAE-CAESALPINIOIDEAE [125]
Bauhinia thonningii Schum. LEGUMINOSAE-CAESALPINIOIDEAE [126]
Begonia princeae Gilg BEGONIACEAE [102]

Begonia wakefieldii Gilg ex Engl. BEGONIACEAE [103]
Beilschmiedia ugandensis Rendle LAURACEAE [1110]
Beilschmiedia sp. LAURACEAE [851] (IiDα)
Berlinianche aethiopica (Welw.) Vattimo RAFFLESIACEAE [1282] (IiDα)
Bersama abyssinica Fresen. MELIANTHACEAE [454]
Bersama abyssinica Fresen. subsp. **paullinioides** (Planch.) Verdc. MELIANTHACEAE [1313]
Bidens cinerea Sherff COMPOSITAE [1299]
Bidens grantii (Oliv.) Sherff COMPOSITAE [200]
Bidens kilimandscharica (O.Hoffm.) Sherff var. **typica** Sherff COMPOSITAE [1112]
Bidens pilosa L. COMPOSITAE [202]
Bidens steppia (Steetz) Sherff COMPOSITAE [203]
Biophytum helenae Buscal. & Muschl. OXALIDACEAE [540]
Biophytum nyikense Exell OXALIDACEAE [541]
Biophytum umbraculum Welw. OXALIDACEAE [1314]
Bixa orellana L. BIXACEAE [112]
Blepharis buchneri Lindau ACANTHACEAE [3]
Blighia unijugata Baker SAPINDACEAE [701]
Blumea sp. COMPOSITAE [205] (IiDα)
Boehmeria macrophylla Hornem. URTICACEAE [1376]
Boerhavia diffusa L. NYCTAGINACEAE [523]
Boophone disticha (L.f.) Herb. AMARYLLIDACEAE [32]
Borassus sp. PALMAE [543] (by obs.)
Boscia salicifolia Oliv. CAPPARACEAE [1114]
Brachiaria brizantha (A.Rich.) Stapf GRAMINEAE [344]
Brachycorythis kalbreyeri Rchb.f. ORCHIDACEAE [1006]
Brachycorythis lastii Rolfe ORCHIDACEAE [926]
Brachystegia × longifolia Benth. LEGUMINOSAE-CAESALPINIOIDEAE [131]
Brachystegia allenii Burtt Davy & Hutch. LEGUMINOSAE-CAESALPINIOIDEAE [127]
Brachystegia boehmii Taub. LEGUMINOSAE-CAESALPINIOIDEAE [128]
Brachystegia bussei Harms LEGUMINOSAE-CAESALPINIOIDEAE [129]
Brachystegia floribunda Benth. LEGUMINOSAE-CAESALPINIOIDEAE [130]
Brachystegia manga De Wild. LEGUMINOSAE-CAESALPINIOIDEAE [132]
Brachystegia microphylla Harms LEGUMINOSAE-CAESALPINIOIDEAE [1115]
Brachystegia spiciformis Benth. LEGUMINOSAE-CAESALPINIOIDEAE [133]
Brachystegia stipulata De Wild. LEGUMINOSAE-CAESALPINIOIDEAE [134]
Brachystegia utilis Burtt Davy & Hutch. LEGUMINOSAE-CAESALPINIOIDEAE [1116]
Brassica juncea (L.) Czern. CRUCIFERAE [1316]
Bridelia atroviridis Müll.Arg. EUPHORBIACEAE [300]

Bridelia brideliifolia (Pax) Fedde EUPHORBIACEAE [301]

Bridelia micrantha (Hochst.) Baill. EUPHORBIACEAE [302]

Brillantaisia owariensis P.Beauv. ACANTHACEAE [1284]

Bromus leptoclados Nees GRAMINEAE [1007]

Buchnera speciosa Skan SCROPHULARIACEAE [1117]

Buchnera usuiensis Oliv. SCROPHULARIACEAE [1118]

Bulbostylis sp. CYPERACEAE [839] (IiDα)

Cajanus cajan (L.) Millsp. LEGUMINOSAE-PAPILIONOIDEAE [554]

Canarium schweinfurtii Engl. BURSERACEAE [117]

Canna indica L. CANNACEAE [154] (by obs.)

Canthium oligocarpum Hiern subsp. **captum** (Bullock) Bridson RUBIACEAE [1009]

Carapa grandiflora Sprague MELIACEAE [444]

Cardiospermum sp. SAPINDACEAE [1501] (IiDβ)

Carduus sp. COMPOSITAE [206] (IiDα)

Carica papaya L. CARICACEAE [1382] (by obs.)

Carpolobia alba G.Don POLYGALACEAE [612]

Cassia sp. LEGUMINOSAE-CAESALPINIOIDEAE [153] (IiDα)

Cassine aethiopica Thunb. CELASTRACEAE [1122]

Cassipourea malosana (Baker) Alston RHIZOPHORACEAE [1123]

Cassytha filiformis L. LAURACEAE [401]

Catha sp. CELASTRACEAE [856] (IiDα)

Catunaregam sp. RUBIACEAE [951] (IiDα)

Cayratia gracilis (Guill. & Perr.) Suess. VITACEAE [775]

Ceiba sp. BOMBACACEAE [113] (IiDα)

Celosia trigyna L. AMARANTHACEAE [31]

Celtis africana Burm.f. ULMACEAE [748]

Centaurea praecox Oliv. & Hiern COMPOSITAE [1124]

Cephalosphaera usambarensis (Warb.) Warb. MYRISTICACEAE [1125]

Ceropegia or **Cynanchum** sp. ASCLEPIADACEAE [80] (IiDα)

Ceropegia sp. ASCLEPIADACEAE [81] (IiDα)

Chaetachme aristata Planch. ULMACEAE [749]

Chamaecrista kirkii (Oliv.) Standl. LEGUMINOSAE-CAESALPINIOIDEAE [1318]

Chassalia cristata (Hiern) Bremek. RUBIACEAE [644]

Chionanthus africanus (Knobl.) Stearn OLEACEAE [1338]

Chionanthus niloticus (Oliv.) Stearn OLEACEAE [532]

Chironia sp. GENTIANACEAE [1127] (IiDα)

Chlorophytum stolzii (K.Krause) Weim. ANTHERICACEAE [890]

Chrysophyllum gorungosanum Engl. SAPOTACEAE [716]

Chytranthus sp. SAPINDACEAE [711] (IiDα)
Cissampelos mucronata A.Rich. MENISPERMACEAE [456]
Cissus integrifolia (Baker) Planch. VITACEAE [776]
Cissus oliveri (Engl.) Gilg ex Engl. VITACEAE [777]
Cissus petiolata Hook f. VITACEAE [778]
Cissus petiolata Hook.f. VITACEAE [1369] (IiDα)
Cissus rubiginosa (Welw. ex Baker) Planch. VITACEAE [779]
Cissus sp. VITACEAE [1410] (IiDβ)
Citrus limon (L.) Osbeck RUTACEAE [690] (by obs.)
Citrus sp. RUTACEAE [1292] (IiDβ)
Clausena anisata (Willd.) Hook.f. ex Benth. RUTACEAE [691]
Cleistanthus sp. EUPHORBIACEAE [859] (IiDα)
Clematis brachiata Thunb. RANUNCULACEAE [624]
Clematis villosa DC. RANUNCULACEAE [1280]
Clerodendrum myricoides (Hochst.) R.Br. ex Vatke VERBENACEAE [762]
Clerodendrum or **Premna** sp. VERBENACEAE [761] (IiDα)
Clerodendrum schweinfurthii Gürke VERBENACEAE [763]
Clerodendrum umbellatum Poir. VERBENACEAE [1381] (iD ign.)
Clerodendrum sp. VERBENACEAE [764] (IiDα)
Coccinia barteri (Hook.f.) Keay or **C. adoensis** (A.Rich.) Cogn. CUCURBITACEAE [1365] (IiDα)
Cola sp. STERCULIACEAE [861] (IiDα)
Colocasia esculenta (L.) Schott ARACEAE [72]
Combretum adenogonium Steud. ex A.Rich. COMBRETACEAE [168]
Combretum apiculatum Sond. COMBRETACEAE [169]
Combretum collinum Fresen. COMBRETACEAE [170]
Combretum molle R.Br. ex G.Don COMBRETACEAE [172]
Combretum paniculatum Vent. COMBRETACEAE [173]
Combretum zeyheri Sond. COMBRETACEAE [174]
Combretum sp. COMBRETACEAE [175] (IiDα)
Commelina benghalensis L. COMMELINACEAE [187]
Commelina diffusa Burm.f. COMMELINACEAE [188] (IiDα)
Commelina imberbis Ehrenb. ex Hassk. COMMELINACEAE [189]
Commelina petersii Hassk. COMMELINACEAE [190]
Commiphora eminii Engl. subsp. **zimmermannii** (Engl.) J.B.Gillett BURSERACEAE [118]
Commiphora sp. BURSERACEAE [120] (IiDα)
Conyza sp. COMPOSITAE [207] (IiDα)
Corchorus olitorius L. TILIACEAE [741]
Cordia africana Lam. BORAGINACEAE [114]

Cordia millenii Baker BORAGINACEAE [115]
Costus afer Ker Gawl. COSTACEAE [899]
Costus spectabilis (Fenzl) K.Schum. COSTACEAE [787]
Craibia affinis (De Wild.) De Wild. LEGUMINOSAE-PAPILIONOIDEAE [555]
Craibia grandiflora (Micheli) Baker f. LEGUMINOSAE-PAPILIONOIDEAE [556]
Crassocephalum crepidioides (Benth.) S.Moore COMPOSITAE [209]
Crassocephalum rubens (Juss. ex Jacq.) S.Moore COMPOSITAE [211]
Crassocephalum vitellinum (Benth.) S.Moore COMPOSITAE [212]
Craterispermum laurinum (Poir.) Benth. RUBIACEAE [1011]
Cremaspora triflora (Thonn.) K.Schum. RUBIACEAE [1012] (IiDα)
Crocosmia aurea (Pappe ex Hook.) Planch. IRIDACEAE [390]
Crossopteryx febrifuga (Afzel. ex G.Don) Benth. RUBIACEAE [645]
Crotalaria goetzei Harms LEGUMINOSAE-PAPILIONOIDEAE [557]
Crotalaria lachnophora A.Rich. LEGUMINOSAE-PAPILIONOIDEAE [558]
Crotalaria ochroleuca G.Don LEGUMINOSAE-PAPILIONOIDEAE [559]
Crotalaria pallida Aiton LEGUMINOSAE-PAPILIONOIDEAE [1436]
Crotalaria pallida Aiton var. **obovata** (G.Don) Polhill LEGUMINOSAE-PAPILIONOIDEAE [560]
Crotalaria sp. LEGUMINOSAE-PAPILIONOIDEAE [563] (IiDα)
Crotalaria sp. LEGUMINOSAE-PAPILIONOIDEAE [1366] (IiDα)
Crotalaria subcapitata De Wild. LEGUMINOSAE-PAPILIONOIDEAE [562]
Croton macrostachyus Hochst. ex Delile EUPHORBIACEAE [304]
Croton megalocarpus Hutch. EUPHORBIACEAE [305]
Croton sylvaticus Hochst. EUPHORBIACEAE [956]
Cucumis dipsaceus Ehrenb. ex Spach. CUCURBITACEAE [1323] (IiDα)
Cucumis spp. CUCURBITACEAE [1520] (IiDβ)
Cucurbita maxima Duchesne CUCURBITACEAE [260]
Culcasia falcifolia Engl. ARACEAE [73]
Culcasia scandens P.Beauv. ARACEAE [1131]
Curcuma longa L. ZINGIBERACEAE [933]
Cussonia arborea Hochst. ex A. Rich. ARALIACEAE [1014]
Cussonia zimmermannii Harms ARALIACEAE [75]
Cyanastrum johnstonii Baker CYANASTRACEAE [739]
Cyanotis speciosa (L.f.) Hassk. COMMELINACEAE [1015]
Cyathea dregei Kunze CYATHEACEAE [268]
Cyathula prostrata (L.) Blume AMARANTHACEAE [792]
Cyclosorus sp. THELYPTERIDACEAE [96] (IiDα)
Cymbopogon densiflorus (Steud.) Stapf GRAMINEAE [345]
Cynanchum sp. ASCLEPIADACEAE [82] (IiDα)

Cynodon aethiopicus Clayton & Harlan GRAMINEAE [346]

Cynometra alexandri C.H.Wright LEGUMINOSAE-CAESALPINIOIDEAE [1132]

Cynometra sp. LEGUMINOSAE-CAESALPINIOIDEAE [1472] (IiDα)

Cynometra sp. LEGUMINOSAE-CAESALPINIOIDEAE [1473] (IiDα)

Cyperus articulatus L. CYPERACEAE [269]

Cyperus cyperoides (L.) Kuntze CYPERACEAE [1432]

Cyperus diffusus Vahl CYPERACEAE [270]

Cyperus papyrus L. CYPERACEAE [271] (OSSED)

Cyperus sp. CYPERACEAE [272] (IiDα)

Cyperus sp. CYPERACEAE [1448] (IiDβ)

Cyperus sp. CYPERACEAE [1516] (IiDα)

Cyphostemma sp. VITACEAE [864] (IiDα)

Dalbergia boehmii Taub. LEGUMINOSAE-PAPILIONOIDEAE [564]

Dalbergia lactea Vatke LEGUMINOSAE-PAPILIONOIDEAE [565]

Dalbergia malangensis Sousa LEGUMINOSAE-PAPILIONOIDEAE [566]

Dalbergia nitidula Baker LEGUMINOSAE-PAPILIONOIDEAE [567]

Deinbollia fulvotomentella Baker f. SAPINDACEAE [703]

Delphinium dasycaulon Fresen. RANUNCULACEAE [1134]

Desmodium adscendens (Sw.) DC. LEGUMINOSAE-PAPILIONOIDEAE [568]

Desmodium barbatum (L.) Benth. var. **dimorphum** (Baker) B.G.Schub. LEGUMINOSAE-PAPILIONOIDEAE [570]

Desmodium gangeticum (L.) DC. LEGUMINOSAE-PAPILIONOIDEAE [1135]

Desmodium velutinum (Willd.) DC. LEGUMINOSAE-PAPILIONOIDEAE [571]

Desmodium sp. LEGUMINOSAE-PAPILIONOIDEAE [572] (IiDα)

Dichaetanthera verdcourtii A.& R.Fernandes MELASTOMATACEAE [439]

Dichapetalum sp. DICHAPETALACEAE [1213] (IiDα)

Dichrostachys cinerea (L.) Wight & Arn. subsp. **nyassana** (Taub.) LEGUMINOSAE-MIMOSOIDEAE [1324]

Dictyophleba lucida (K.Schum.) Pierre APOCYNACEAE [53]

Digitaria diagonalis (Nees) Stapf GRAMINEAE [347]

Dioscorea alata L. DIOSCOREACEAE [281]

Dioscorea cochleariapiculata De Wild. DIOSCOREACEAE [903]

Dioscorea dumetorum (Kunth) Pax DIOSCOREACEAE [283]

Dioscorea hirtiflora Benth. DIOSCOREACEAE [284]

Dioscorea praehensilis Benth. DIOSCOREACEAE [904]

Dioscorea preussii Pax DIOSCOREACEAE [69]

Dioscorea schimperiana Hochst. ex Kunth DIOSCOREACEAE [286]

Dioscorea sp. DIOSCOREACEAE [287] (IiDα)

Dioscorea sp. DIOSCOREACEAE [1474] (IiDα)

Dioscoreophyllum volkensii Engl. MENISPERMACEAE [457]

Diospyros kirkii Hiern EBENACEAE [290]

Diospyros mespiliformis Hochst. ex. A.DC. EBENACEAE [291] (IiDα)

Diplorhynchus condylocarpon (Müll.Arg.) Pichon APOCYNACEAE [54]

Dissotis arborescens A. Fern. & R. Fern. MELASTOMATACEAE [441]

Dissotis pachytricha Gilg. ex R.E. Fr. MELASTOMATACEAE [1137]

Dissotis rotundifolia (Sm.) Triana MELASTOMATACEAE [440]

Dodonaea viscosa (L.) Jacq. SAPINDACEAE [704]

Dolichopentas decora (S.Moore) Kårehed & B.Bremer RUBIACEAE [1269]

Dolichos kilimandscharicus Taub. LEGUMINOSAE-PAPILIONOIDEAE [573]

Dombeya acutangula Cav. STERCULIACEAE [1326]

Dombeya kirkii Mast. STERCULIACEAE [728] (by obs.)

Dombeya rotundifolia (Hochst.) Planch. STERCULIACEAE [729]

Dombeya sp. STERCULIACEAE [731] (IiDα)

Dombeya sp. STERCULIACEAE [1482] (IiDβ)

Dovyalis sp. FLACOURTIACEAE [867] (IiDα)

Dracaena afromontana Mildbr. DRACAENACEAE [1019]

Dracaena laxissima Engl. DRACAENACEAE [21]

Dracaena mannii Baker DRACAENACEAE [22]

Dregea schimperi (Dence.) Bullock ASCLEPIADACEAE [83] (IiDα)

Drynaria volkensii Hieron. POLYPODIACEAE [619]

Dryopteris inaequalis (Schltdl.) Kuntze DRYOPTERIDACEAE [97]

Drypetes gerrardii Hutch. EUPHORBIACEAE [307]

Drypetes natalensis (Harv.) Hutch. EUPHORBIACEAE [308]

Dyschoriste trichocalyx (Oliv.) Lindau ACANTHACEAE [5]

Dysphania ambrosioides (L.) Mosyakin & Clemants CHENOPODIACEAE [1320]

Echinops amplexicaulis Oliv. COMPOSITAE [213]

Ectadiopsis oblongifolia (Meisn.) Schult. ASCLEPIADACEAE [84]

Elaeis guineensis Jacq. PALMAE [544] (OSSED)

Elephantopus scaber L. COMPOSITAE [214]

Eleusine indica (L.) Gaertn. GRAMINEAE [1421]

Englerophytum magalismontanum (Sond.) T.D.Penn. SAPOTACEAE [979]

Englerophytum natalense (Sond.) T.D.Penn. SAPOTACEAE [980]

Ensete ventricosum (Welw.) Cheesman MUSACEAE [925]

Entada abyssinica A.Rich. LEGUMINOSAE-MIMOSOIDEAE [479]

Entada sp. LEGUMINOSAE-MIMOSOIDEAE [481] (IiDα)

Entandrophragma utile (Dawe & Sprague) Sprague MELIACEAE [445]

Eragrostis aspera (Jacq.) Nees GRAMINEAE [349]
Eragrostis racemosa (Thunb.) Steud. GRAMINEAE [1021]
Eremospatha haullevilleana De Wild. PALMAE [545]
Erica benguelensis (Welw. ex Engl.) E.G.H.Oliv ERICACEAE [1272]
Erica silvatica (Welw. Ex Engl.) Beentje ERICACEAE [1315]
Erigeron aegyptiacus L. COMPOSITAE [1439]
Erigeron floribundus (H.B. & K.) Sch.Bip. COMPOSITAE [215]
Erythrina abyssinica DC. LEGUMINOSAE-PAPILIONOIDEAE [574]
Erythrina excelsa Baker LEGUMINOSAE-PAPILIONOIDEAE [575]
Erythrococca kirkii (Müll.Arg.) Prain EUPHORBIACEAE [1022] (IiDα)
Erythrococca trichogyne (Müll.Arg.) Prain EUPHORBIACEAE [1023]
Erythrophleum suaveolens (Guill. & Perr.) Brenan LEGUMINOSAE-CAESALPINIOIDEAE [137]
Eulophia mechowii (Rchb.f.) T.Durand & Schinz ORCHIDACEAE [1027]
Eulophia odontoglossa Rchb.f. ORCHIDACEAE [1025]
Euphorbia tirucalli L. EUPHORBIACEAE [309]
Euphorbia sp. EUPHORBIACEAE [310] (IiDα)
Euphorbia sp. EUPHORBIACEAE [539] (IiDα)
Faidherbia albida (Delile) A.Chev. LEGUMINOSAE-MIMOSOIDEAE [1278]
Faurea rochetiana (A.Rich.) Chiov. ex Pic.Serm. PROTEACEAE [1328]
Faurea saligna Harv. PROTEACEAE [1142]
Ficalhoa laurifolia Hiern THEACEAE [740]
Ficus artocarpoides Warb. MORACEAE [488]
Ficus asperifolia Miq. MORACEAE [489]
Ficus conraui Warb. MORACEAE [995]
Ficus cyathistipula Warb. MORACEAE [492]
Ficus exasperata Vahl MORACEAE [493]
Ficus glumosa Delile MORACEAE [494]
Ficus ingens (Miq.) Miq. MORACEAE [496]
Ficus platyphylla Delile MORACEAE [1028]
Ficus scassellatii Pamp. MORACEAE [506] (IiDα)
Ficus sur Forssk. MORACEAE [500]
Ficus sycomorus L. MORACEAE [501]
Ficus thonningii Blume MORACEAE [502]
Ficus trichopoda Baker MORACEAE [994]
Ficus vallis-choudae Delile MORACEAE [504]
Ficus sp. MORACEAE [487] (IiDα)
Ficus sp. MORACEAE [1419] (IiDβ)
Ficus sp. MORACEAE [1392] (iD ign.)

Flacourtia indica (Burm.f.) Merr. FLACOURTIACEAE [335]
Funtumia africana (Benth.) Stapf APOCYNACEAE [55]
Galium ossirwaense K.Krause RUBIACEAE [1030] (IiDα)
Garcinia huillensis Welw. GUTTIFERAE [383]
Garcinia livingstonei T.Anderson GUTTIFERAE [1145]
Garcinia smeathmanii (Planch. & Friena) Oliv. GUTTIFERAE [1144]
Garcinia sp. GUTTIFERAE [384] (IiDα)
Gardenia imperialis K.Schum. RUBIACEAE [646]
Gardenia ternifolia Schumach. & Thonn. subsp. **jovis-tonantis** (Welw.) Verdc. RUBIACEAE [942]
Geophila repens (L.) I.M.Johnst. RUBIACEAE [793]
Geranium aculeolatum Oliv. GERANIACEAE [1146]
Girardinia diversifolia (Link) Friis URTICACEAE [757]
Gladiolus atropurpureus Baker IRIDACEAE [1036]
Gladiolus dalenii Van Geel IRIDACEAE [1434]
Gladiolus erectiflorus Baker IRIDACEAE [1330] (IiDα)
Gladiolus sp. IRIDACEAE [393] (IiDα)
Gloriosa lindenii (Baker) J.C.Manning & Vinn. COLCHICACEAE [1339]
Glycine sp. LEGUMINOSAE-PAPILIONOIDEAE [1475] (IiDα)
Glycine sp. LEGUMINOSAE-PAPILIONOIDEAE [580] (IiDα)
Gomphia densiflora (De Wild. & T.Durand) Verdc. OCHNACEAE [1306]
Gomphia reticulata P.Beauv. OCHNACEAE [1384]
Gonatopus boivinii (Decne.) Engl. ARACEAE [1038]
Gongronema sp. ASCLEPIADACEAE [85] (IiDα)
Gossypium sp. MALVACEAE [1147] (IiDα)
Grewia flavescens Juss. TILIACEAE [742]
Grewia forbesii Harv. ex Mast. TILIACEAE [743]
Grewia mollis Juss. TILIACEAE [744]
Guizotia scabra (Vis.) Choiv. COMPOSITAE [1148]
Gymnanthemum coloratum (Willd.) H.Rob. & B.Kahn COMPOSITAE [1431]
Gymnanthemum myrianthum (Hook.f.) H.Rob. COMPOSITAE [1428]
Gymnanthemum thomsonianum (Oliv. & Hiern) H.Rob. COMPOSITAE [1424]
Gymnema sylvestre (Retz.) R.Br. ex Sm. ASCLEPIADACEAE [86]
Gymnosporia senegalensis (Lam.) Loes. CELASTRACEAE [1342]
Gynandropsis gynandra (L.) Briq. CAPPARACEAE [155]
Gynura scandens O.Hoffm. COMPOSITAE [216]
Haemanthus sp. AMARYLLIDACEAE [1039] (IiDα)
Hallea rubrostipulata (K.Schum.) J.-F.Leroy RUBIACEAE [944]

Haplocoelum foliolosum (Hiern) Bullock SAPINDACEAE [1149]
Harungana madagascariensis Lam. ex Poir GUTTIFERAE [385]
Haumaniastrum villosum (Benth.) A.J.Paton LABIATAE [970]
Hebenstretia dentata L. SCROPHULARIACEAE [1150]
Helichrysum foetidum (L.) Cass. COMPOSITAE [1151]
Helichrysum kirkii Oliv. & Hiern COMPOSITAE [217]
Helichrysum patulifolium Baker COMPOSITAE [218]
Helichrysum stenopterum DC. COMPOSITAE [219]
Helichrysum wittei Hutch. & B.L.Burtt COMPOSITAE [220]
Heteromorpha trifoliata (H.L.Wendl.) Eckl. & Zeyh. UMBELLIFERAE [751]
Hewittia malabarica (L.) Suresh CONVOLVULACEAE [240]
Hewittia scandens (J. König ex Milne) Mabb. CONVOLVULACEAE [241]
Hexalobus sp. ANNONACEAE [50] (IiDα)
Hibiscus aponeurus Sprague & Hutch. MALVACEAE [422]
Hibiscus calyphyllus Cav. MALVACEAE [423]
Hibiscus cannabinus L. MALVACEAE [424]
Hibiscus (aff.) **cannabinus** L. MALVACEAE [1373] (IiDα)
Hibiscus debeerstii De Wild. & Th.Dur. MALVACEAE [425]
Hibiscus diversifolius Jacq. MALVACEAE [426]
Hibiscus mechowii Garcke MALVACEAE [1153]
Hibiscus meyeri Harv. MALVACEAE [1331] (IiDα)
Hibiscus micranthus L.f. MALVACEAE [1277]
Hibiscus nyikensis Sprague MALVACEAE [427] (IiDα)
Hibiscus rostellatus Guill. & Perr. MALVACEAE [429]
Hibiscus shirensis Sprague & Hutch. MALVACEAE [1155]
Hibiscus surattensis L. MALVACEAE [430]
Hibiscus vitifolius L. MALVACEAE [1156]
Hibiscus sp. MALVACEAE [434] (IiDα)
Hibiscus sp. MALVACEAE [1462] (IiDα)
Hoslundia opposita Vahl LABIATAE [1041]
Hydrostachys polymorpha Klotzsch HYDROSTACHYACEAE [1157]
Hymenocardia acida Tul. EUPHORBIACEAE [311]
Hymenodictyon floribundum (Hochst. & Steud.) B.L.Rob. RUBIACEAE [648]
Hyparrhenia cymbaria (L.) Stapf GRAMINEAE [350]
Hyparrhenia newtonii (Hack.) Stapf GRAMINEAE [1433]
Hyparrhenia nyassae (Rendle) Stapf GRAMINEAE [352]
Hyparrhenia pilgeriana C.E.Hubb. GRAMINEAE [353]
Hyparrhenia rufa (Nees) Stapf GRAMINEAE [354]

Hyparrhenia variabilis Stapf	GRAMINEAE	[355]		
Hypericophyllum compositarum Steetz	COMPOSITAE	[1158]		
Hypericum quartinianum A.Rich.	GUTTIFERAE	[1159]		
Hypericum revolutum Vahl	GUTTIFERAE	[1160]		
Hyperthelia dissoluta (Nees ex Steud.) Clayton	GRAMINEAE	[1042]		
Hypoestes aristata (Vahl) Sol. ex Roem. & Schult.	ACANTHACEAE	[1161]	(IiDα)	
Hypoestes sp.	ACANTHACEAE	[1463]	(IiDα)	
Hypoestes sp.	ACANTHACEAE	[6]	(IiDα)	
Ilex mitis (L.) Radlk.	AQUIFOLIACEAE	[1162]		
Illigera pentaphylla Welw.	HERNANDIACEAE	[387]		
Impatiens assurgens Baker	BALSAMINACEAE	[1163]	(IiDα)	
Impatiens hamata Warb.	BALSAMINACEAE	[101]		
Imperata cylindrica (L.) Raeusch.	GRAMINEAE	[914]		
Indigofera emarginella A.Rich.	LEGUMINOSAE-PAPILIONOIDEAE	[581]		
Indigofera podocarpa Baker f. & Martin	LEGUMINOSAE-PAPILIONOIDEAE	[582]		
Inula glomerata Oliv. & Hiern	COMPOSITAE	[1166]		
Inula shirensis Oliv.	COMPOSITAE	[1167]		
Inula sp.	COMPOSITAE	[221]	(IiDα)	
Ipomoea alba L.	CONVOLVULACEAE	[1168]		
Ipomoea batatas (L.) Lam.	CONVOLVULACEAE	[243]		
Ipomoea cairica (L.) Sweet	CONVOLVULACEAE	[244]		
Ipomoea eriocarpa R. Br.	CONVOLVULACEAE	[246]		
Ipomoea muricata (L.) Jacq.	CONVOLVULACEAE	[248]		
Ipomoea obscura (L.) Ker Gawl.	CONVOLVULACEAE	[249]		
Ipomoea ochracea (Lindl.) G.Don.	CONVOLVULACEAE	[250]		
Ipomoea pileata Roxb.	CONVOLVULACEAE	[1332]		
Ipomoea rubens Choisy	CONVOLVULACEAE	[251]		
Ipomoea turbinata Lag.	CONVOLVULACEAE	[1294]		
Ipomoea wightii (Wall.) Choisy	CONVOLVULACEAE	[253]		
Ipomoea wightii (Wall.) Choisy var. **kilimandschari** (Dammer) Verdc.	CONVOLVULACEAE	[254]		
Ipomoea sp.	CONVOLVULACEAE	[255]	(IiDα)	
Isoberlinia angolensis (Benth.) Hoyle & Brenan var. **niembaënsis** (De Wild.) Brenan	LEGUMINOSAE-CAESALPINIOIDEAE	[139]		
Jasminum dichotomum Vahl	OLEACEAE	[533]		
Jasminum sp.	OLEACEAE	[1169]	(IiDα)	
Jatropha curcas L.	EUPHORBIACEAE	[312]		
Julbernardia globiflora (Benth.) Troupin	LEGUMINOSAE-CAESALPINIOIDEAE	[140]		
Julbernardia magnistipulata (Harms) Troupin	LEGUMINOSAE-CAESALPINIOIDEAE	[141]		

Julbernardia paniculata (Benth.) Troupin LEGUMINOSAE-CAESALPINIOIDEAE [142]
Julbernardia seretii (De Wild.) Troupin LEGUMINOSAE-CAESALPINIOIDEAE [143]
Julbernardia unijugata J.Léonard LEGUMINOSAE-CAESALPINIOIDEAE [145]
Justicia nyassana Lindau ACANTHACEAE [7]
Justicia striata (Klotzsch) Bullock, s.l. ACANTHACEAE [8] (IiDα)
Justicia striolata Mildbr. ACANTHACEAE [1171]
Keetia gueinzii (Sond.) Bridson RUBIACEAE [649]
Keetia venosa (Oliv.) Bridson RUBIACEAE [650]
Khaya grandifoliola C.DC. MELIACEAE [446]
Khaya nyasica Stapf ex Baker f. MELIACEAE [999] (IiDα)
Khaya sp. MELIACEAE [447] (IiDα)
Kigelia africana (Lam.) Benth. BIGNONIACEAE [1333]
Kyllinga alba Nees CYPERACEAE [1305] (IiDα)
Lagenaria siceraria (Molina) Standl. CUCURBITACEAE [261]
Lagenaria sp. CUCURBITACEAE [262] (IiDα)
Laggera alata (D.Don) Sch.Bip. ex Oliv. COMPOSITAE [1290]
Landolphia buchananii (Hallier f.) Stapf APOCYNACEAE [1047]
Landolphia kirkii Dyer APOCYNACEAE [56]
Landolphia owariensis P.Beauv. APOCYNACEAE [57]
Landolphia sp. APOCYNACEAE [59] (IiDα)
Landolphia sp. APOCYNACEAE [1525] (IiDα)
Lannea schimperi (Hochst. ex A.Rich.) Engl. ANACARDIACEAE [36]
Lannea schimperi (Hochst. ex A.Rich.) Engl. ANACARDIACEAE [35] (IiDα)
Lannea schweinfurthii (Engl.) Engl. var. **stuhlmannii** (Engl.) Kokwaro ANACARDIACEAE [1334] (IiDα)
Laportea aestuans (L.) Chew URTICACEAE [1300]
Lecaniodiscus fraxinifolius Baker SAPINDACEAE [706]
Leea guineensis G. Don LEEACEAE [781]
Leersia hexandra Sw. GRAMINEAE [357]
Lefebvrea sp. UMBELLIFERAE [1174] (IiDα)
Leonotis nepetifolia (L.) R.Br. LABIATAE [398]
Lepistemon owariense (P. Beauv.) Hallier f. CONVOLVULACEAE [256]
Linum thunbergii Eckl. & Zeyh. LINACEAE [1179]
Lippia plicata Baker VERBENACEAE [766]
Lobelia stricklandiae Gilliland CAMPANULACEAE [1291]
Loranthus sp. LORANTHACEAE [419] (IiDα)
Loudetia arundinacea (A.Rich.) Hochst. ex Steud. GRAMINEAE [358]
Loudetia simplex (Nees) C.E.Hubb. GRAMINEAE [359]

Luffa cylindrica (L.) M.Roem. CUCURBITACEAE [263]

Macaranga kilimandscharica Pax EUPHORBIACEAE [313]

Macrotyloma axillare (E.Mey.) Verdc. LEGUMINOSAE-PAPILIONOIDEAE [1180]

Maerua sp. CAPPARACEAE [156] (IiDα)

Maesa lanceolata Forskk. MYRSINACEAE [514]

Magnistipula butayei De Wild. subsp. **bangweolensis** (R.E.Fr.) F.White CHRYSOBALANACEAE [1340]

Mallotus sp. EUPHORBIACEAE [1067] (IiDα)

Mangifera indica L. ANACARDIACEAE [815]

Manihot dichotoma Ule EUPHORBIACEAE [314]

Manihot esculenta Crantz EUPHORBIACEAE [1527] (OSSED)

Marantochloa leucantha (K.Schum.) Milne-Redh. MARANTACEAE [438]

Margaritaria discoidea (Baill.) G.L.Webster EUPHORBIACEAE [316]

Markhamia lutea (Benth.) K.Schum. BIGNONIACEAE [106]

Markhamia obtusifolia (Baker) Sprague BIGNONIACEAE [107] (ID ign.)

Markhamia zanzibarica (Bojer ex DC.) K.Schum. BIGNONIACEAE [1341]

Maytenus undata (Thunb.) Blakelock CELASTRACEAE [1399]

Melanthera pungens Oliv. & Hiern var. **albinervia** (O.Hoffm.) Beentje COMPOSITAE [1288]

Melanthera scandens (Schum. & Thonn.) Roberty subsp. **madagascarensis** (Baker) Wild COMPOSITAE [223]

Melastomastrum capitatum (Vahl) A. Fern. & R. Fern. MELASTOMATACEAE [442]

Melinis minutiflora P.Beauv. GRAMINEAE [360]

Melinis repens (Willd.) Zizka GRAMINEAE [1084]

Memecylon sp. MELASTOMATACEAE [443] (IiDα)

Merremia pterygocaulos (Choisy) Hall.f. CONVOLVULACEAE [257]

Mesogyne sp. MORACEAE [1215] (IiDα)

Microglossa pyrifolia (Lam.) Kuntze COMPOSITAE [225]

Mikania sp. COMPOSITAE [226] (IiDα)

Milicia excelsa (Welw.) C.C.Berg MORACEAE [508]

Millettia angustidentata De Wild. LEGUMINOSAE-PAPILIONOIDEAE [583]

Millettia sp. LEGUMINOSAE-PAPILIONOIDEAE [601] (IiDα)

Mimulopsis sp. ACANTHACEAE [1476] (IiDα)

Mimulopsis sp. ACANTHACEAE [16] (IiDα)

Mimusops bagshawei S.Moore SAPOTACEAE [718]

Mimusops penduliflora Engl. SAPOTACEAE [719]

Mimusops sp. SAPOTACEAE [720] (IiDα)

Momordica cissoides Planch. ex Benth. CUCURBITACEAE [264]

Momordica pterocarpa A. Rich. CUCURBITACEAE [1184]

Monanthotaxis buchananii (Engl.) Verdc. ANNONACEAE [41]
Monanthotaxis parvifolia (Oliv.) Verdc. ANNONACEAE [1301] (IiDα)
Monanthotaxis poggei Engl. & Diels ANNONACEAE [42]
Mondia whitei (Hook.f) Skeels ASCLEPIADACEAE [87]
Monotes elegans Gilg DIPTEROCARPACEAE [288]
Monotes rufotomentosus Gilg DIPTEROCARPACEAE [1185]
Moraea schimperi (Hochst.) Pic.Serm. IRIDACEAE [394]
Moraea ventricosa Baker IRIDACEAE [1050]
Morinda lucida Benth. RUBIACEAE [652]
Morus alba L. MORACEAE [1385]
Mucuna coriacea Baker LEGUMINOSAE-PAPILIONOIDEAE [584]
Mucuna gigantea (Willd.) DC. subsp. **quadrialata** (Baker) Verdc. LEGUMINOSAE-PAPILIONOIDEAE [585]
Mucuna pruriens (L.) DC. LEGUMINOSAE-PAPILIONOIDEAE [586]
Mukia maderaspatana (L.) M.Roem. CUCURBITACEAE [265]
Multidentia crassa (Hiern) Bridson & Verdc. RUBIACEAE [945]
Murdannia simplex (Vahl) Brenan COMMELINACEAE [195]
Musa sp. MUSACEAE [829] (iD ign.)
Mussaenda arcuata Poir. RUBIACEAE [654]
Mussaenda erythrophylla Schumach. & Thonn. RUBIACEAE [655]
Myrianthus arboreus P.Beauv. CECROPIACEAE [509]
Myrianthus holstii Engl. CECROPIACEAE [510]
Myrica salicifolia Hochst. ex A. Rich. MYRICACEAE [1186]
Nelsonia canescens (Lam.) Spreng. ACANTHACEAE [10]
Neoboutonia sp. EUPHORBIACEAE [317] (IiDα)
Neonotonia wightii (Wight & Arn.) J.A.Lackey LEGUMINOSAE-PAPILIONOIDEAE [1187]
Neonotonia wightii (Wight & Arn.) J.A.Lackey subsp. **pseudojavanica** (Taub.) J.A.Lackey LEGUMINOSAE-PAPILIONOIDEAE [1297]
Neorautanenia mitis (A.Rich.) Verdc. LEGUMINOSAE-PAPILIONOIDEAE [587]
Newtonia buchananii (Baker) G.C.C.Gilbert & Boutiqu LEGUMINOSAE-MIMOSOIDEAE [483]
Neyraudia arundinacea (L.) Henrard GRAMINEAE [361]
Nidorella spartioides (O.Hoffm.) Cronquist COMPOSITAE [1188]
Nuxia congesta R.Br. ex Fresen. BUDDLEJACEAE [414]
Ochna cyanophylla N. Robson OCHNACEAE [525]
Ochna holstii Engl. OCHNACEAE [1358] (IiDα)
Ochna leptoclada Oliv. OCHNACEAE [528]
Ochna mossambicensis Klotzsch OCHNACEAE [526]
Ocimum basilicum L. LABIATAE [399]

Olyra latifolia L. GRAMINEAE [362]
Oncinotis tenuiloba Stapf APOCYNACEAE [952]
Oncoba bukobensis (Gilg) S. Hul & Breteler FLACOURTIACEAE [1337]
Oncoba spinosa Forssk. FLACOURTIACEAE [337]
Ophiuros exaltatus (L.) Kuntze GRAMINEAE [918]
Oplismenus hirtellus (L.) P.Beauv. GRAMINEAE [795]
Oreobambos buchwaldii K.Schum. GRAMINEAE [363]
Oreosyce africana Hook f. CUCURBITACEAE [266]
Otiophora caerulea (Hiern) Bullock RUBIACEAE [1056]
Oxyanthus pyriformis (Hochst.) Skeels RUBIACEAE [946]
Oxyanthus speciosus DC. subsp. mollis (Hutch.) Bridson RUBIACEAE [658]
Oxyanthus sp. RUBIACEAE [659] (IiDα)
Oxytenanthera abyssinica (A.Rich.) Munro GRAMINEAE [364]
Ozoroa insignis Delile subsp. reticulata (Baker f.) J.B.Gillett ANACARDIACEAE [835]
Ozoroa mucronata (Bernh. ex Krauss) R. Fern. & A. Fern. ANACARDIACEAE [1271]
Ozoroa obovata (Oliv.) R.& A.Fernandes ANACARDIACEAE [834]
Pachycarpus eximius (Schltr.) Bullock ASCLEPIADACEAE [1191]
Pachystela excelsa Louis SAPOTACEAE [1192]
Pancovia sp. SAPINDACEAE [710] (IiDα)
Pandanus sp. PANDANACEAE [547] (IiDα)
Panicum maximum Jacq. GRAMINEAE [365]
Panicum repens L. GRAMINEAE [366]
Panicum sp. GRAMINEAE [368] (IiDα)
Panicum trichocladum K.Schum. GRAMINEAE [367]
Parinari curatellifolia Planch. ex Benth. CHRYSOBALANACEAE [629]
Parinari excelsa Sabine CHRYSOBALANACEAE [1344]
Parkia filicoidea Oliv. LEGUMINOSAE-MIMOSOIDEAE [484]
Paspalum scrobiculatum L. GRAMINEAE [796]
Paullinia pinnata L. SAPINDACEAE [707]
Pavetta crassipes K.Schum. RUBIACEAE [660]
Pavetta hymenophylla Bremek. RUBIACEAE [1361] (IiDα)
Pavetta molundensis K.Krause RUBIACEAE [661]
Pavetta tarennoides S.Moore RUBIACEAE [1380]
Pavetta sp. RUBIACEAE [663] (IiDα)
Pavonia urens Cav. var. tomentosa (Ulbr.) Brenan MALVACEAE [1194]
Pavonia sp. MALVACEAE [435] (IiDα)
Peddiea fischeri Engl. THYMELAEACEAE [1345]
Pennisetum macrourum Trin. GRAMINEAE [1064]

Pennisetum polystachion (L.) Schult. subsp. **atrichum** (Stapf & Hubbard) J.N. Brunken GRAMINEAE [1063]

Pennisetum purpureum Schumach. GRAMINEAE [369]

Pennisetum unisetum (Nees) Benth. GRAMINEAE [913]

Pentas bussei K.Krause RUBIACEAE [1066]

Pergularia daemia (Forssk.) Chiov. ASCLEPIADACEAE [88]

Pericopsis angolensis (Baker) Meeuwen LEGUMINOSAE-PAPILIONOIDEAE [588]

Pericopsis tinctorius Welw. LEGUMINOSAE-PAPILIONOIDEAE [1379] (iD ign.)

Periploca linearifolia Quart.-Dill. & A.Rich. ASCLEPIADACEAE [89]

Persicaria decipiens (R.Br.) K.L.Wilson POLYGONACEAE [1437]

Phaulopsis imbricata (Forssk.) Sweet ACANTHACEAE [797]

Phoenix reclinata Jacq. PALMAE [546] (OSSED)

Phragmanthera usuiensis (Oliv.) M.G.Gilbert LORANTHACEAE [1435]

Phragmites mauritianus Kunth GRAMINEAE [370]

Phyllanthus muellerianus (Kuntze) Exell EUPHORBIACEAE [319]

Phyllanthus ovalifolius Forssk. EUPHORBIACEAE [958]

Phyllanthus reticulatus Poir. EUPHORBIACEAE [320]

Phyllanthus sp. EUPHORBIACEAE [321] (IiDα)

Phyllobotryon paradoxum (Baill.) Hul FLACOURTIACEAE [1346]

Physalis angulata L. SOLANACEAE [724]

Phytolacca dodecandra L'Hér. PHYTOLACCACEAE [608]

Pimpinella engleriana (Engl.) H. Wolff UMBELLIFERAE [1199]

Pimpinella volkensii Engl. UMBELLIFERAE [753] (IiDα)

Piper capense L.f. PIPERACEAE [609]

Piper umbellatum L. PIPERACEAE [610]

Pistia stratiotes L. ARACEAE [74] (by obs.)

Pittosporum viridiflorum Sims PITTOSPORACEAE [1422]

Platostoma rotundifolium (Briq.) A.J.Paton LABIATAE [1032]

Platycoryne buchananiana (Kraenzl.) Rolfe ORCHIDACEAE [1068]

Plectranthus esculentus N.E.Br. LABIATAE [1070] (IiDα)

Pleiocarpa sp. APOCYNACEAE [1071] (IiDα)

Pleiotaxis dewevrei O.Hoffm. ex T.Durand & De Wild. COMPOSITAE [227]

Pleopeltis mildbraedii (Hieron) Pic. Serm. POLYPODIACEAE [1201]

Podocarpus sp. PODOCARPACEAE [738] (IiDα)

Polyalthia sp. ANNONACEAE [1202] (IiDα)

Polygala bakeriana Chodat POLYGALACEAE [1203]

Polygala exelliana Troupin POLYGALACEAE [613]

Polygala gomesiana Welw. ex Oliv. POLYGALACEAE [1204]

Polygala lancifolia A. St.-Hil. & Moq. POLYGALACEAE [1296]

Polygala ukirensis Gürke POLYGALACEAE [1205]

Polyscias fulva (Hiern.) Harms ARALIACEAE [76]

Polysphaeria lanceolata Hiern var. **pedata** Brenan RUBIACEAE [1072]

Polystachya sp. ORCHIDACEAE [1073] (IiDα)

Pouteria adolfi-friedericii (Engl.) A.Meeuse subsp. **australis** (J.H.Hemsl.) L.Gaut. SAPOTACEAE [978]

Premna angolensis Gürke VERBENACEAE [1219] (IiDα)

Premna resinosa (Hochst.) Schauer subsp. **holstii** (Gürke) Verdc. VERBENACEAE [987]

Premna sp. VERBENACEAE [768] (IiDα)

Protea angolensis Welw. PROTEACEAE [1222] (IiDα)

Protea gaguedi J.F.Gmel. PROTEACEAE [621]

Protea petiolaris (Hiern) Baker & C.H.Wright PROTEACEAE [622]

Protea rupestris R.E.Fr. PROTEACEAE [623]

Prunus africana (Hook. f.) Kalkman ROSACEAE [1350]

Pseudarthria hookeri Wight & Arn. LEGUMINOSAE-PAPILIONOIDEAE [589]

Pseudolachnostylis maprouneifolia Pax EUPHORBIACEAE [322]

Pseudolachnostylis maprouneifolia Pax var. **dekindtii** (Pax) Radcl.-Sm. EUPHORBIACEAE [1077]

Pseudospondias microcarpa (A.Rich.) Engl. ANACARDIACEAE [37]

Psidium guajava L. MYRTACEAE [517]

Psophocarpus scandens (Endl.) Verdc. LEGUMINOSAE-PAPILIONOIDEAE [590]

Psorospermum febrifugum Spach GUTTIFERAE [1268]

Psychotria capensis (Eckl.) Vatke subsp. **riparia** (K.Schum.& K.Krause) Verdc. RUBIACEAE [665]

Psychotria eminiana (Kuntze) E.M.A.Petit RUBIACEAE [666]

Psychotria eminiana (Kuntze) E.M.A.Petit var. **heteroclada** E.M.A.Petit RUBIACEAE [667]

Psychotria eminiana (Kuntze) E.M.A.Petit var. **tenuifolia** Verdc. RUBIACEAE [1362] (IiDα)

Psychotria fractinervata E.M.A.Petit RUBIACEAE [669]

Psychotria heterosticta E.M.A.Petit RUBIACEAE [948]

Psychotria mahonii C.H.Wright RUBIACEAE [949]

Psychotria peduncularis (Salisb.) Steyerm. RUBIACEAE [824] (iD ign)

Psychotria peduncularis (Salisb.) Steyerm. var. **nyassana** (K.Krause) Verdc. RUBIACEAE [672]

Psychotria peduncularis (Salisb.) Steyerm. var. **suaveolens** (Schweinf. ex Hiern) Verdc. RUBIACEAE [673]

Psychotria sp. RUBIACEAE [675] (IiDα)

Psychotria sp. RUBIACEAE [1531] (IiDα)

Psydrax parviflora (Afzel.) Bridson RUBIACEAE [1420]

Psydrax parviflora (Afzel.) Bridson subsp. **rubrocostata** (Robyns) Bridson RUBIACEAE [941]
Pteridium aquilinum (L.) Kuhn. DENNSTAEDTIACEAE [1270]
Pteridium centrali-africanum (Hieron.) Alston DENNSTAEDTIACEAE [1279]
Pteris catoptera Kunze PTERIDACEAE [20]
Pterocarpus angolensis DC. LEGUMINOSAE-PAPILIONOIDEAE [591] (OSSED)
Pterocarpus tinctorius Welw. LEGUMINOSAE-PAPILIONOIDEAE [1488]
Pterygota macrocarpa K.Schum. STERCULIACEAE [732]
Pycnanthus angolensis (Welw.) Warb. MYRISTICACEAE [513]
Pycnostachys meyeri Gürke ex Engl. LABIATAE [1080]
Pyrostria bibracteata (Baker) Cavaco RUBIACEAE [940]
Ramalina sp. USNEACEAE [758] (IiDα)
Rapanea melanophloeos (L.) Mez. MYRSINACEAE [515]
Rapanea pulchra Gilg & G.Schellenb. MYRSINACEAE [516]
Rauvolfia caffra Sond. APOCYNACEAE [61]
Rawsonia lucida Harv. FLACOURTIACEAE [1227] (IiDα)
Renealmia engleri K.Schum. ZINGIBERACEAE [790]
Rhamnus prinoides L'Hér. RHAMNACEAE [1229]
Rhoicissus tridentata (L.f.) Wild & R.B.Drumm. VITACEAE [782]
Rhoicissus sp. VITACEAE [783] (IiDα)
Rhynchosia luteola (Hiern) K.Schum. LEGUMINOSAE-PAPILIONOIDEAE [593]
Rhynchosia sp. LEGUMINOSAE-PAPILIONOIDEAE [1466] (IiDα)
Rhynchosia sp. LEGUMINOSAE-PAPILIONOIDEAE [595] (IiDα)
Ricinodendron heudelotii (Baill.) Heckel var. **tomentellum** (Hutch. & E.A.Bruce) Radcl.-Sm. EUPHORBIACEAE [1275]
Ricinus communis L. EUPHORBIACEAE [324]
Rinorea ilicifolia (Welw. ex Oliv.) Kuntze VIOLACEAE [1230]
Riocreuxia profusa N.E.Br. ASCLEPIADACEAE [90]
Rothmannia engleriana (K.Schum.) Keay RUBIACEAE [677]
Rothmannia longiflora Salisb. RUBIACEAE [678]
Rothmannia manganjae (Hiern) Keay RUBIACEAE [679]
Rourea thomsonii (Baker) Jongkind CONNARACEAE [238]
Rubia cordifolia L. subsp. **conotricha** (Grand.) Verdc. RUBIACEAE [799]
Rubus pinnatus Willd. ROSACEAE [633]
Rumex abyssinicus Jacq. POLYGONACEAE [617]
Rumex usambarensis (Dammer) Dammer POLYGONACEAE [618]
Rytigynia acuminatissima (K.Schum.) Robyns RUBIACEAE [680]
Rytigynia neglecta (Hiern) Robyns RUBIACEAE [1086]
Rytigynia uhligii (K.Schum. & K.Krause) Verdc. RUBIACEAE [950]

Rytigynia sp. RUBIACEAE [683] (IiDα)
Saba comorensis (Bojer ex A.DC.) Pichon APOCYNACEAE [62]
Sabicea orientalis Wernham RUBIACEAE [1087]
Saccharum officinarum L. GRAMINEAE [831]
Salacia cerasifera Welw. ex Oliv. CELASTRACEAE [1378]
Salacia congolensis De Wild. & Th.Dur. CELASTRACEAE [1370] (IiDα)
Salacia erecta (G.Don) Walp. CELASTRACEAE [1351] (IiDα)
Salacia madagascariensis (Lam.) DC. CELASTRACEAE [162]
Salacia sp. CELASTRACEAE [165] (IiDα)
Sansevieria sp. DRACAENACEAE [26] (IiDα)
Sapindus saponaria L. SAPINDACEAE [708]
Sapium ellipticum (Hochst.) Pax EUPHORBIACEAE [325]
Satyrium carsonii Rolfe ORCHIDACEAE [1088]
Schefflera abyssinica (Hochst. ex A.Rich.) Harms ARALIACEAE [78]
Schefflera goetzenii Harms ARALIACEAE [1386]
Schrebera alata (Hochst.) Welw. OLEACEAE [535]
Schrebera trichoclada Welw. OLEACEAE [536]
Scleria racemosa Poir. CYPERACEAE [278]
Sclerocarya birrea (A.Rich.) Hochst. subsp. **caffra** (Sond.) Kokwaro ANACARDIACEAE [1289]
Scolopia sp. FLACOURTIACEAE [339] (IiDα)
Securidaca longipedunculata Fresen. POLYGALACEAE [615]
Senecio ruwenzoriensis S.Moore COMPOSITAE [1232]
Senna siamea (Lam.) H.S.Irwin & Barneby LEGUMINOSAE-CAESALPINIOIDEAE [150]
Senna singueana (Delile) Lock LEGUMINOSAE-CAESALPINIOIDEAE [1319] (IiDα)
Senna spectabilis (DC.) Irwin & Barneby LEGUMINOSAE-CAESALPINIOIDEAE [151]
Sesamum angolense Welw. PEDALIACEAE [605]
Sesamum indicum L. PEDALIACEAE [606]
Sesamum sp. PEDALIACEAE [607] (IiDα)
Sesbania sesban (L.) Merr. var. **nubica** Chiov. LEGUMINOSAE-PAPILIONOIDEAE [596]
Setaria megaphylla (Steud.) T.Durand & Schinz GRAMINEAE [1364] (IiDα)
Setaria petiolata Stapf & C.E.Hubb. GRAMINEAE [373]
Setaria poiretiana (Schult.) Kunth GRAMINEAE [919]
Setaria sphacelata (Schumach.) Stapf & C.E.Hubb. ex Moss GRAMINEAE [1090]
Sida rhombifolia L. MALVACEAE [436]
Siphonochilus aethiopicus (Schweinf.) B.L.Burtt ZINGIBERACEAE [937]
Smilax anceps Willd. SMILACACEAE [931]
Solanecio angulatus (Vahl) C.Jeffrey COMPOSITAE [1322]
Solanecio mannii (Hook.f.) C.Jeffrey COMPOSITAE [727]

Solanum americanum Mill. SOLANACEAE [1354]
Solanum indicum L. SOLANACEAE [725]
Solanum terminale Forssk. SOLANACEAE [1423]
Solanum sp. SOLANACEAE [726] (IiDα)
Sonchus luxurians (R.E.Fr.) C.Jeffrey COMPOSITAE [229]
Sonchus schweinfurthii Oliv. & Hiern COMPOSITAE [230]
Sopubia trifida Buch.-Ham. ex D. Don SCROPHULARIACEAE [1235]
Sorghum arundinaceum (Desv.) Stapf GRAMINEAE [921]
Sorghum bicolor (L.) Moench GRAMINEAE [376]
Sorindeia sp. ANACARDIACEAE [1216] (IiDα)
Spathodea campanulata P.Beauv. BIGNONIACEAE [108]
Spermacoce dibrachiata Oliv. RUBIACEAE [939]
Sphenostylis stenocarpa (A.Rich.) Harms LEGUMINOSAE-PAPILIONOIDEAE [597]
Sporobolus fimbriatus (Trin.) Nees GRAMINEAE [1304] (IiDα)
Steganotaenia araliacea Hochst. UMBELLIFERAE [754]
Stephania abyssinica (Quart.-Dill. & A.Rich.) Walp. MENISPERMACEAE [459]
Sterculia quinqueloba (Garcke) K.Schum. STERCULIACEAE [733]
Sterculia subviolacea K.Schum. STERCULIACEAE [1236]
Sterculia tragacantha Lindl. STERCULIACEAE [735]
Stereospermum kunthianum Cham. BIGNONIACEAE [110]
Strombosia scheffleri Engl. OLACACEAE [530]
Strychnos cocculoides Baker LOGANIACEAE [415]
Strychnos innocua Delile LOGANIACEAE [416]
Strychnos scheffleri Gilg LOGANIACEAE [1237] (IiDα)
Strychnos sp. LOGANIACEAE [418] (IiDα)
Strychnos sp. LOGANIACEAE [1467] (IiDα)
Strychnos spinosa Lam. LOGANIACEAE [417]
Swartzia madagascariensis Desv. LEGUMINOSAE-PAPILIONOIDEAE [1217] (IiDα)
Synadenium sp. EUPHORBIACEAE [1092] (IiDα)
Synsepalum brevipes (Baker) T.D.Penn. SAPOTACEAE [722]
Syzygium cordatum Hochst. ex Krauss MYRTACEAE [518]
Syzygium cumini (L.) Skeels MYRTACEAE [519]
Syzygium guineense (Willd.) DC. MYRTACEAE [520]
Tabernaemontana pachysiphon Stapf APOCYNACEAE [65]
Tacazzea apiculata Oliv. ASCLEPIADACEAE [91]
Tacca leontopetaloides (L.) Kuntze TACCACEAE [737]
Tamarindus indica L. LEGUMINOSAE-CAESALPINIOIDEAE [152]

Tarenna pavettoides (Harv.) Sim. RUBIACEAE [684]
Telosma africana (N.E.Br.) N.E.Br. ASCLEPIADACEAE [92]
Temnocalyx sp. RUBIACEAE [686] (IiDα)
Tephrosia interrupta Engl. subsp. **mildbraedii** (Harms) J.B.Gillett LEGUMINOSAE-PAPILIONOIDEAE [598]
Teramnus labialis (L.f.) Spreng. LEGUMINOSAE-PAPILIONOIDEAE [334]
Terminalia kaiseriana F.Hoffm. COMBRETACEAE [176]
Terminalia kilimandscharica Engl. COMBRETACEAE [1238]
Terminalia mollis M.A.Lawson COMBRETACEAE [177]
Terminalia sericea Burch. ex DC. COMBRETACEAE [178]
Terminalia stenostachya Engl. & Diels COMBRETACEAE [179]
Tetracera alnifolia Willd. DILLENIACEAE [1239] (IiDα)
Tetracera potatoria Afzel. ex G.Don DILLENIACEAE [280]
Tetradenia riparia (Hochst.) Codd LABIATAE [974]
Themeda triandra Forssk. GRAMINEAE [379]
Thonningia sanguinea Vahl BALANOPHORACEAE [100]
Thunbergia alata Boj. ex Sims ACANTHACEAE [11]
Thunbergia lancifolia T.Anderson ACANTHACEAE [1242]
Thunbergia lancifolia T.Anderson var. **ciliata** De Wild. ACANTHACEAE [1240]
Thunbergia petersiana Lindau ACANTHACEAE [1372] (IiDα)
Thunbergia sp. ACANTHACEAE [15] (IiDα)
Tinnea apiculata Robyns & Lebrun LABIATAE [400]
Tinospora caffra (Miers) Troupin MENISPERMACEAE [461]
Toddalia asiatica (L.) Lam. RUTACEAE [693]
Tragia brevipes Pax EUPHORBIACEAE [326]
Trema orientalis (L.) Blume ULMACEAE [750]
Tricalysia sp. RUBIACEAE [687] (IiDα)
Trichilia dregeana Sond. MELIACEAE [448]
Trichilia emetica Vahl MELIACEAE [449]
Trichilia prieuriana A.Juss. MELIACEAE [450]
Trichilia sp. MELIACEAE [451] (IiDα)
Trichodesma physaloides (Fenzl) A. DC. BORAGINACEAE [116]
Trichoscypha sp. ANACARDIACEAE [843] (IiDα)
Triumfetta cordifolia A.Rich. TILIACEAE [746]
Triumfetta rhomboidea Jacq. TILIACEAE [1244]
Triumfetta sp. TILIACESE [747] (IiDα)
Tylophora sylvatica Decne. ASCLEPIADACEAE [93] (IiDα)

Tylosema sp. LEGUMINOSAE-CAESALPINIOIDEAE [1522] (IiDβ)
Typha sp. TYPHACEAE [1447] (IiDβ)
Uapaca kirkiana Müll.Arg. EUPHORBIACEAE [328]
Uapaca nitida Müll.Arg. EUPHORBIACEAE [329]
Uapaca sansibarica Pax EUPHORBIACEAE [330]
Urelytrum sp. GRAMINEAE [1246] (IiDα)
Urena lobata L. MALVACEAE [801]
Usnea goniodes Stirton USNEACEAE [759]
Usnea sp. USNEACEAE [760] (IiDα)
Uvaria acuminata Oliv. ANNONACEAE [1247]
Uvaria angolensis Welw. ex Oliv. ANNONACEAE [45]
Uvaria leptocladon Oliv. ANNONACEAE [1248]
Uvaria welwitschii (Hiern) Engl. & Diels ANNONACEAE [46]
Vangueria infausta Burch. RUBIACEAE [1438]
Vepris nobilis (Delile) Mziray RUTACEAE [1355]
Vernonia amygdalina Del. COMPOSITAE [231]
Vernonia campanea S.Moore COMPOSITAE [1250]
Vernonia nestor S.Moore COMPOSITAE [1251]
Vernonia sp. COMPOSITAE [235] (IiDα)
Vernonia sp. COMPOSITAE [1468] (IiDα)
Vigna heterophylla A.Rich. LEGUMINOSAE-PAPILIONOIDEAE [1295]
Vigna racemosa (G.Don) Hutch. & Dalziel LEGUMINOSAE-PAPILIONOIDEAE [1252]
Vigna schimperi Baker LEGUMINOSAE-PAPILIONOIDEAE [1253] (IiDα)
Vigna sp. LEGUMINOSAE-PAPILIONOIDEAE [600] (IiDα)
Vigna sp. LEGUMINOSAE-PAPILIONOIDEAE [1471] (IiDα)
Vitex doniana Sweet VERBENACEAE [769]
Vitex ferruginea Schumach. & Thonn. VERBENACEAE [770]
Vitex fischeri Gürke VERBENACEAE [771]
Vitex madiensis Oliv. VERBENACEAE [772]
Vitex sp. VERBENACEAE [773] (IiDα)
Vitex sp. VERBENACEAE [1530] (IiDα)
Voacanga africana Stapf ex Scott-Elliot APOCYNACEAE [67]
Voacanga sp. APOCYNACEAE [688] (IiDα)
Vossia cuspidata (Roxb.) Griff. GRAMINEAE [380]
Wahlenbergia napiformis (A.DC.) Thulin CAMPANULACEAE [1336]
Whitfieldia sp. ACANTHACEAE [1254] (IiDα)
Whitfieldia stuhlmannii (Lindau) C.B.Cl. ACANTHACEAE [1218] (IiDα)

Wissadula amplissima (L.) R.E.Fr. MALVACEAE [1357]
Wissadula periplocifolia (L.) Thwaites MALVACEAE [437]
Ximenia americana L. OLACACEAE [531]
Xylopia longipetala De Wild. & T.Durand ANNONACEAE [1287]
Xymalos monospora (Harv.) Baill. MONIMIACEAE [1258]
Ypsilopus sp. ORCHIDACEAE [1097] (IiDα)
Yushania alpina (K.Schum.) W.C.Lin GRAMINEAE [912]
Zanha golungensis Hiern SAPINDACEAE [709]
Zea mays L. GRAMINEAE [827]
Zehneria sp. CUCURBITACEAE [267] (IiDα)
Ziziphus abyssinica Hochst. ex A.Rich. RHAMNACEAE [626]
Ziziphus mucronata Willd. RHAMNACEAE [627]
Species not identified yet but possibly belonging to different species listed above.
Sp. Bhumbumsi [1481] (UiD)
Sp. Bhusume ANNONACEAE [805] (IiDα)
Sp. Bhusume EUPHORBIACEAE [804] (IiDα)
Sp. Bhutabhwelunfu no.2 [1524] (UiD)
Sp. Ikalambegha B [1496] (UiD)
Sp. Ilale [1529] (UiD).
Sp. Ipa [1497] (UiD)
Sp. Isalike SAPINDACEAE [1449] (IiDα)
Sp. Kabhwalongo [1526] (UiD).
Sp. Kakambakamba [1415] (UiD)
Sp. Lunsamba type A [1512] (UiD)
Sp. Likakwenkwe CUCURBITACEAE [1519] (IiDβ)
Sp. Lughughu LEGUMINOSAE-MIMOSOIDEAE [1521] (IiDβ)
Sp. Mubhalebhale [1485] (UiD)
Sp. Mulyasungula ?COMPOSITAE (IiDβ) [1502] (UiD)
Sp. Munsoka [1513] (UiD)
Sp. Munyegha LEGUMINOSAE-PAPILIONOIDEAE [1503] (IiDβ)
Sp. Musyasya MALVACEAE [1523] (IiDβ)
Sp. Ndabhalabha [1487] (UiD)
Sp. Ng'unga [1418] (UiD)
Sp. Nr1121 [1411] (UiD)
Sp. Nr1271 [1412] (UiD)
Sp. Nr6202 [1409] (UiD)
Sp. Sinyiasya [1505] (UiD)

Sp. Siponda no.1 [1407] (UiD)
Sp. Siponda no.2 [1408] (UiD)
Species possibly belonging to the species listed above.
Acalypha or **Hibiscus** sp./spp. [1458] (IiDβ)
Aframomum sp. (poss.) ZINGIBERACEAE [1444] (IiDβ)
Blepharis buchneri [3] (prob.) ACANTHACEAE [1461] (IiDβ)
Cissus sp. VITACEAE [1454] (IiDβ)
Dioscorea sp. DIOSCOREACEAE [1443] (IiDβ)
Ficus sp. MORACEAE [1406] (IiDβ)
Ficus sp. MORACEAE [1452] (IiDβ)
Ficus sp./spp. MORACEAE [1451] (IiDβ)
Ficus sp./spp. MORACEAE [1491] (IiDβ)
Ficus spp. MORACEAE [1446] (IiDβ)
Gymnanthemum sp. COMPOSITAE [1511] (IiDβ)
Hibiscus sp./spp. MALVACEAE [1453] (IiDβ)
Indigofera emarginella [581] or **podocarpa** [582] LEGUMINOSAE-PAPILIONOIDEAE [1514] (IiDβ)
Ipomoea or **Hewittia** sp./spp. CONVOLVULACEAE [1477] (IiDβ)
Ipomoea sp. CONVOLVULACEAE [1459] (IiDβ)
Keetia sp. RUBIACEAE [1417] (IiDβ)
Neonotonia sp./spp. LEGUMINOSAE-PAPILIONOIDEAE [1479] (IiDβ)
Piper sp./spp. PIPERACEAE [1460] (IiDβ)
Pteridium sp. DENNSTAEDTIACEAE [1455] (IiDβ)
Rhynchosia or **Vigna** sp. LEGUMINOSAE-PAPILIONOIDEAE [1395] (iD ign.)
Rumex sp. POLYGONACEAE [1490] (IiDβ)
Solanecio angulatus [1322] or **Gynura scandens** [216] COMPOSITAE [1394] (iD ign.)
Sp. Ikonjogholo APOCYNACEAE [1414] (IiDβ)
Sp. Ilyungulyungu CF [1478] (IiDβ)
Sp. Itesa COMMELINACEAE [1445] (IiDβ)
Sp. Luhosyo [1515] (UiD)
Sp. Lujongololo type undet. ANNONACEAE [1416] (IiDβ)
Sp. Lumpululu [1456] (IiDβ)

Reference

Itoh, N. (in prep.) *Flora of Mahale and Tongwe Taxonomy*.

APPENDIX II

Dietary list

COMPILED BY NORIKO ITOH, KOICHIRO ZAMMA, TAKUYA MATSUMOTO, HITONARU NISHIE, AND MICHIO NAKAMURA

With the increase in study period and the number of researchers, new chimpanzee food records are now available. In addition, records in Nishida and Uehara (1983) (*Natural Diet of Chimpanzees*, hereafter, NDC) were obtained when the extinct K group was the main target for observation. Thus, feeding data, irrespective of methodology, collected by the authors, who studied the M group on multiple occasions were added to NDC. The observation period is 1994–2014 (no data in 1999). We excluded items when we could not specify species or parts eaten. Months of feeding on fruit items (but not other items) are supplemented by MMCRP (Chapter 2) records from January 2008 to December 2013 (see Chapter 17 for the dataset). Food species were identified by local names *in situ*, and roughly examined to determine whether the observed plant matches with identified species for the local name in *Flora of Mahale and Tongwe Taxonomy* (Itoh in prep., hereafter, FMTT), which was compiled based on the previous lists by Nishida and Uehara (1981, 1983) and some other publications with synonymy updated. It is recommended that users examine FMTT for species record details as a proportion of records were ambiguous and other identification may be required to identify plants observed in the field. Beside these, within the new list, there was one species (**Ficus thonningii**) applied to two local names (*Mujimo* and *Kajimonsole*). Our observation details for *Mujimo* were too poor to specify species and this plant was sampled only once in the past for identification at the herbarium (see Nishida and Uehara 1981 or FMTT). Thus, we could not verify whether these local names are assigned to different species but identified by mistake (in the field and/or at the herbarium because of insufficient information) or were synonymously used. One further possibility is that there is inconsistency between scientific and local taxonomies, i.e. **Ficus thonningii** is locally discriminated into two types. For hope on future verification on *Mujimo*, we did not combine feeding records on *Mujimo* with *Kajimonsole*.

In the list, 443 items are included. Within these, 6.1% may overlap with another item or items in the list because these species were difficult to specify in the field. Thus, the remaining 407 items belonging to 224 species (excluding nine items known to be consumed by unit-groups other than M) were actually consumed by the M-group chimpanzees. Of these items, 36.1% were common between NDC and this study, 21.9% were newly confirmed, and 42.0% were only confirmed in NDC but not in this study. We do not think all of these 42.0% items are no longer consumed by the M group, but there possibly are many items we did not observe because of the short observation period (19 years by five researchers, see Chapter 17 for relation between dietary variation and observation period). However, as bark- and wood-feeding behavior are relatively infrequent in the M group compared with the K group (Chapter 17), some proportion of such items may not be included in the food repertoire or may be only rarely eaten by the M-group chimpanzees. Items such as sap and resin parts may be underestimated as such food species are more difficult to identify than the species whose young shoot systems are eaten. For such young shoot systems, leaves, piths, petioles, and shoots had the largest species diversity (184 species), followed by fruits and seeds (127 species). In contrast, there were only 36 species whose flowers were consumed by chimpanzees. Besides these, an unidentified algae was eaten by a female (Chapter 11).

The food list is composed of probable plant names, unique IDs for each item, parts eaten, unit-group names other than the M group, eaten months, and record source. A detailed explanation follows.

Plant name

Plant names are expressed by species name with authority (infra-taxa if identified), followed by species ID in FMTT and Appendix I (in brackets), and local name (for family names, see Appendix I). Species names are shown in bold if full identification is available in FMTT. Those species identified by direct observation in the field or from the *Standard Swahili-English Dictionary* are marked as "=by obs." and "=OSSED," respectively. Species with incomplete identifications are shown in italics with their identification status just after species names as follows: "IiDα" = specimen that could not be fully identified at herbarium; "IiDβ" = plant identified by direct observation in the field; "UiD" = not identified yet; "iD ign." = names given in NDC but the verification unknown.

FDID: Unique ID for each food item.

Part (parts eaten):

Fl: Flowers including blossoms. For those species where leaves are consumed together with tiny flowers or blossoms, we may have underestimated.

Fr: Fruits, seeds, including pods, and grass seeds (seeds were a separate category in NDC).

Le: Leaves and buds. Leaves may be eaten together with meat or other items to make a wadge. This type of feeding was excluded from food items.

Pe: Petioles. Applied only when leaf petioles are selectively consumed but leaves are ignored.

P: Piths. Applied only when live inner unlignified tissue extracted from stem or leaf petiole is selectively consumed with leaves ignored.

Sh: Shoots. Young stems of vines and lianas may sometimes be eaten without extracting pith. Leaves and/or petioles may also be eaten together.

R: Resins. Applied to solid or soft semisolid substance secreted from live trunk or branch of tree and only when it is chewed.

Sa: Saps. Any licked substance including liquid from live or dead trunk or branch of tree.

B: Barks. Applied only when inner bark or cambium is eaten.

W: Wood. Applied to live or dry dead wood and only when it is chewed.

Ro: Roots.

*: Observed only once.

More details of the parts eaten are provided just below "part," if needed.

Gr: Group names if observed for unit-groups other than the M group.

Months eaten: – = not observed to be eaten, ○ = recorded only in NDC, ● = common in NDC and this study, ♦ = only in this study. Data supplemented by MMCRP records only will appear as † and ‡ if common with NDC.

Source: NDC = Nishida and Uehara 1983; figures = number of observation in feeding records collected by MN, KZ, TM, NI, HN (combination of observed year and observer counted as 1); + = additional observation records other than feeding records (NI).
Other abbreviations used in the list:
NA: not available; nom.: name; obs.: observation; prob.: probably

Plant name	FDID	Part	Gr	Month(s) eaten J F M A M J J A S O N D	Source
Acacia hockii De Wild. [464]: Kasemele	1571	Fr		-------○○---	NDC
		Seeds eaten.			
Acacia sieberiana DC. [469]: Mughunga	1572	Fr		------○○○---	NDC
		Seeds eaten.			
Acalypha chirindica S.Moore [293]: Lusieti type B	1538	Le		○-○---------	NDC
Acalypha ornata Hochst. ex A.Rich. [294]: Lusieti type A	1539	Le		○○-------○○○	NDC
Acalypha or *Hibiscus* sp./spp. =IiDβ [1458]: Lusieti type unc. CF Prob. sp./spp. identified under the local supra-nom. Lusieti in FMTT	1870	Le		♦♦--♦♦♦----♦	7
Adenia rumicifolia Engl. & Harms [603]: Ilemo	1503	Fl*		---○--------	NDC
	1504	Le		○○○----○--○○	NDC
Aeschynomene sp. = IiDα [885]: Isila Sp. identification not certain	1922	W		♦♦♦♦--♦♦♦---	1
Aframomum alboviolaceum (Ridl.) K.Schum. [784]: Itungulu lubhanda	1773	P		○○○○○○--○○○○	NDC
		Piths of young stems eaten.			
Aframomum mala (K.Schum. ex Engl.) K.Schum. [785]: Itungulu isigho	1774	Fr		○○●○○○○○○○○○○	NDC, 1
	1775	P		○○●○○○○○○○○○○	NDC, 1
		Piths of young stems eaten.			
Aframomum sp. = iD ign. [845]: Itungulu lubhanda B Sp. could not be specified for our record, thus may be A. **alboviolaceum** [784]	1776	Fr		--●○○○○○----	NDC, 1
	1777	P		NA	NDC
		Piths of young stems eaten.			
Aframomum sp. (poss.) =IiDβ [1444]: Itungulu type undet. CF Include sp./spp. identified under the local supra-nom. Itungulu in FMTT	1832	Fr		-♦♦♦♦♦♦♦††-†	10
	1833	P		♦♦♦♦♦♦♦♦♦♦♦♦	34
		Piths of young stems eaten.			
Afzelia africana Pres. [121]: Mukola Could be A. **quanzensis** [122] instead	1557	Fr*		--○---------	NDC
		Seeds eaten.			
	1934	Fl		------♦-----	1

Plant name	FDID	Part	Gr	Month(s) eaten J F M A M J J A S O N D	Source
Albizia glaberrima (Schum. & Thonn.) Benth. [1427]: Kafunampasa type A	1835	Fr*		---------♦--	1
		Seeds eaten.			
	1573	Le		○○---------○	NDC
	1574	R		♦-♦----♦♦♦●○	NDC, 12
	1923	Sa*		---------♦--	1
	1575	B*		-----○------	NDC
Allophylus congolanus Gilg [695]: Lulamilo	1861	Fr*		----♦-------	1
	1663	Le		-○○---------	NDC
Ampelocissus abyssinica (Hochst. ex A.Rich.) Planch. [1302]: Lukosyo	1643	Fr		♦●●●●♦--♦---	NDC, 9
	1644	Le		○○---------○	NDC
	1645	P		●●♦♦♦♦♦●●●♦♦	NDC, 30
		Piths of young stems eaten.			
Ampelocissus africana (Lour.) Merr. [826]: Lisisusalukosyo type B	1642	Fr*		---○--------	NDC
Ancylobothrys amoena Hua [51]: Ikonjogholo type B	1686	Fr		○-----------	NDC
Annona senegalensis Pers. [39]: Bhufila	1481	Fr		○○-------○○○	NDC
	1482	Le		○○○------○○○	NDC
	1483	B		○-------○---	NDC
	1856	W		--♦-------♦-	2
Anthocleista schweinfurthii Gilg [411]: Mughongoghongo	1682	Fr		----------○-	NDC
	1932	W*		--------♦---	1
Antidesma venosum E.Mey. ex Tul. [299]: Mutimpu	1540	Fr		♦‡♦‡○♦♦-----	NDC, 4+
Aporrhiza paniculata Radlk. [699]: Muhoko type A	1897	Le*		----♦-------	1
		Petioles and leaf bases ignored.			
	1896	W*		♦-----------	1
Artabotrys monteiroae Oliv. [40]: Lujongololo bhusie	1484	Fr		-----♦●-----	NDC, 1
Aspilia mossambicensis (Oliv.) Wild [197]: Ighoghola Not sure for the sp. for our record (A. rudis [1429] included?)	1723	Le		●○●○○○○○○○○●	NDC, 5
Asystasia gangetica (L.) T.Anderson [2]: Katikamonga	1757	Fl*		---○--------	NDC
	1758	Le		○○○○--♦○♦-○-	NDC, 2

Plant name	FDID	Part	Gr	Month(s) eaten J F M A M J J A S O N D	Source
Azanza garckeana (F.Hoffm.) Exell & Hillc. [421]: Mutobho	1530	Fl		○○○---------	NDC
	1531	Fr		--○○○●●●●●●-	NDC, 16
	1879	Pe*		----------♦-	1
	1880	W*		-----------♦	1
Baphia capparidifolia Baker subsp. **multiflora** (Harms) Brummitt [553]: Ijubhila	1578	Fl		-----♦○●●●●○	NDC, 11
	1579	Fr		---------♦○-	NDC, 1
		Since the individual soon quit eating in our additional record, chimpanzees may occasionally try to eat.			
	1580	Le		●●●●●●●●●●●●	NDC, 30
	1815	Pe*		------♦-----	1
Bauhinia petersiana Bolle [125]: Iketya	1558	Fl		○--○--------	NDC
	1559	Fr		---○--○-♦---	NDC, 1
		Seeds eaten.			
	1560	Le		○○○○---○-○○○	NDC
Bauhinia thonningii Schum. [126]: Musakansaka	1570	Fr		---○-♦●●●♦--	NDC, 12
		Pods eaten together with seeds.			
Berlinianche aethiopica (Welw.) Vattimo =IiDα [1282]: NA	1499	Fr		------○-----	NDC
Bidens grantii (Oliv.) Sherff [200]: Lisanda type A	1724	Fl		----○-------	NDC
	1725	Le		○---○-------	NDC
Blepharis buchneri Lindau [3]: Isomang'ombe type A	1759	Le		○○○○○○○----○	NDC
	1760	P		○-----------	NDC
Blepharis buchneri [3] (prob.) =IiDβ [1461]: Isomang'ombe type unc. CF	1828	Le		-♦--♦♦♦----♦	3
Brachiaria brizantha (A.Rich.) Stapf [344]: Mbute	1795	Fr		---○--------	NDC

Plant name	FDID	Part	Gr	Month(s) eaten J F M A M J J A S O N D	Source
Brachystegia bussei Harms [129]: Mukoma	1561	Fr		–––––○○○○–––	NDC
		Commonly seeds eaten.			
	1563	Le		○––––○–––○○○	NDC
	1565	R		––––––○○–––○	NDC
	1566	Sa		–○––––––––––	NDC
	1564	B		○○○○○–––––○○	NDC
Brachystegia spiciformis Benth. [133]: Mutulu	1567	Fr		–○–––♦●‡–––‡	NDC, 3
		Seeds eaten.			
Bridelia atroviridis Müll.Arg. [300]: Kamilaninga	1541	Fr		–––––○○–––––	NDC
Bridelia micrantha (Hochst.) Baill. [302]: Munyonyi type A	1898	Fr		♦––––♦––––––	2
Cajanus cajan (L.) Millsp. [554]: Kabhalama	1581	Fr		––○–––––––––	NDC
		Seeds eaten.			
Canarium schweinfurtii Engl. [117]: Sighonfi type A	1662	Fr*		–––––––––––○	NDC
Celtis africana Burm.f. [748]: Kahefu	1610	Le		●♦♦–♦♦––♦♦♦♦	NDC, 9
Ceropegia sp. =IiDα [81]: Lumpululu type A	1706	Le		○––––––––––○	NDC
Ceropegia or *Cynanchum* sp. =IiDα [80]: Lumpululu type A	1703	Fl		○––––––––––○	NDC
	1704	Fr		–––––○––○–––	NDC
	1705	Le		○○○○○○○–○○○○	NDC
Chassalia cristata (Hiern) Bremek. [644]: Lisisusanandi	1713	Fr		––––––––––○–	NDC
	1714	Le		–––––○○–––––	NDC
Cissus oliveri (Engl.) Gilg ex Engl. [777]: Lighaghaja type A	1646	Le*		○–––––––––––	NDC
	1647	W		–––––––––––○	NDC
Cissus petiolata Hook.f. =IiDα [1369]: Mukombelonda type unc.	1649	Fr		–––○––––––––	NDC
	1650	P		––––––––○–––	NDC
Cissus rubiginosa (Welw. ex Baker) Planch. [779]: Mukombelonda type A	1648	Fr		–○○○––––––––	NDC
Cissus sp. = IiDβ [1454]: Lighaghaja type undet. CF Prob. C. oliveri [777] or C. petiolata [778]	1928	Le*		–––––––––––♦	1
Citrus limon (L.) Osbeck =by obs. [690]: Limauwe	1658	Fr		♦♦♦♦♦♦♦♦●●●♦	NDC, 25
Citrus sp. =IiDβ [1292]: Ichunghwa	1812	Fr		–♦♦♦♦♦♦–––––	4
Clematis villosa DC. [1280]: Mpisye ngosi	1494	Fr*		––––––○–––––	NDC

Plant name	FDID	Part	Gr	Month(s) eaten J F M A M J J A S O N D	Source
Clerodendrum schweinfurthii Gürke [763]: Linkumbwe type A	1764	Fl		---○-♦♦♦♦♦--	NDC, 7
	1846	Fr		------♦♦---♦	3
	1765	Le		-♦♦--♦♦●○●○-	NDC, 6
		Galls may be consumed together.			
Clerodendrum sp. =IiDα [764]: Linkumbwe type B Identified by obs. and not sure whether it exactly matches with sp. [764] in FMTT since its identification is also incomplete	1915	Le*		----------♦-	1
Coccinia barteri (Hook.f.) Keay or *C. adoensis* (A.Rich.) Cogn. =IiDα [1365]: NA	1505	Le		------○-----	NDC
Combretum molle R.Br. ex G.Don [172]: Mulama	1510	Le		○○-------○○○	NDC
	1511	R		------♦♦●♦♦♦	NDC, 12
	1890	Sa		------♦-♦---	2
	1891	W*		----♦-------	1
Cordia africana Lam. [114]: Mukibhu ghwakakamba	1899	Fl*		-------♦----	1
	1736	Fr		-----●●●‡-†♦	NDC, 11
Cordia millenii Baker [115]: Mukibhu ghwesimbwa	1737	Fl		-------♦●---	NDC, 4
	1738	Fr		♦●●●●●●●†○●●	NDC, 16
	1739	Le		---○-----○-○	NDC
	1740	P		---○-------♦	NDC, 1
		A pith of petiole was eaten in our record.			
	1741	R		-○○○--------	NDC
	1900	Sa*		--------♦---	1
		Licked after removing the outer bark.			
Costus afer Ker Gawl. [899]: Omoji	1778	Fr	B	------○-----	NDC
	1779	P		♦●●●●●●●●●●●	NDC, 23
		Piths of young stems eaten.			

Plant name	FDID	Part	Gr	Month(s) eaten J F M A M J J A S O N D	Source
Crassocephalum crepidioides (Benth.) S.Moore [209]: Litiku type B	1727	Fl*		--○---------	NDC
	1728	Le*		--○---------	NDC
Crassocephalum vitellinum (Benth.) S.Moore [212]: Litiku type A	1729	Le*		-○----------	NDC
Crotalaria sp. =IiDα [563]: Lijeghele Sp. identifications are not confident for our records	1582	Le		-○♦○○●♦-♦---	NDC, 2
Crotalaria sp. =IiDα [1366]: Lisisusalijeghele	1583	Le		-○○○--------	NDC
Croton sylvaticus Hochst. [956]: Ibhonobhono	1811	Fl		----------♦-	1
	1542	Fr		---○--------	NDC
	1543	Le		---○---○○○○-	NDC
	1921	Sa		--------♦---	1
Culcasia falcifolia Engl. [73]: Nandi	1882	Pe*		--------♦---	1
Culcasia scandens P.Beauv. [1131]: NA	1784	Le*		----○-------	NDC
Cyperus papyrus L. =OSSED [271]: Ibhongo	1793	P		--○---------	NDC
Cyperus sp. =IiDα [272]: Lulagho	1794	P		NA	NDC
Cyphostemma sp. =IiDα [864]: NA	1651	Fr	B	----------○-	NDC
	1652	P	B	----------○-	NDC
Dalbergia boehmii Taub. [564]: Msiji	1584	Fl		-------○----	NDC
Dalbergia malangensis Sousa [566]: Itambuka	1829	Fl		------♦-----	2
	1585	Le		-----○●♦---○	NDC, 5
Deinbollia fulvotomentella Baker f. [703]: Mpongolwela	1664	Fr		●---------♦♦	NDC, 12
Dioscorea praehensilis Benth. [904]: Sihama sha mulinga	1785	Fr		--○○--------	NDC
	1786	Le		--○---------	NDC
Dioscorea schimperiana Hochst. ex Kunth [286]: Lisunsu	1787	Fr		--○○--------	NDC
	1788	Le		○○○------○●●	NDC, 2
	1789	P		---------○♦♦	NDC, 7
		Piths of young stems eaten.			

Plant name	FDID	Part	Gr	Month(s) eaten J F M A M J J A S O N D	Source
Dioscorea sp. =IiDβ [1443]: Sihama type undet. CF Prob. **D. hirtiflora** [284] or **D. praehensilis** [904]	1902	Fl*		--♦---------	1
	1903	Fr		♦♦♦♦----†--♦	7
		Sometimes, only seeds are eaten.			
	1904	Le		--♦---------	2
	1905	P		-------♦♦-♦-	4
		Piths of young stems eaten.			
Dioscoreophyllum volkensii Engl. [457]: Sifu	1495	Le		-○------○---	NDC
Diospyros kirkii Hiern [290]: Ikonghwa	1678	Fr	B	------○-----	NDC
Diplorhynchus condylocarpon (Müll.Arg.) Pichon [54]: Musongati	1687	Fl		---------○●♦	NDC, 2
	1688	Fr		○●●●●○●●●♦--	NDC, 15
		Seeds are also digested.			
	1689	Le		○○------○○●○	NDC, 1
	1690	P*		----○-------	NDC
Dombeya rotundifolia (Hochst.) Planch. [729]: Musubhu	1519	Fl		-------○----	NDC
	1520	Le		----------○○	NDC
Dombeya sp. =IiDβ [1482]: Katobhotobho	1918	Le*		------♦-----	1
	1845	P*		-----------♦	1
		Pith of young stem was eaten.			
Dracaena mannii Baker [22]: Bhulonje	1790	Le		○-----○-----	NDC
	1791	B*		○-----------	NDC
	1809	W*		♦-----------	1
Dregea schimperi (Dence.) Bullock =IiDα [83]: Lumpululu type A	1707	Le*		----------○-	NDC
Dyschoriste trichocalyx (Oliv.) Lindau [5]: Lusaka	1761	Le		-------●○○○○	NDC, 4
Elaeis guineensis Jacq. =OSSED [544]: Sighasi	1883	P		♦-----------	1
		Only once by two infants.			
Englerophytum magalismontanum (Sond.) T.D.Penn. [979]: Mutwentwe type unc.	1680	Fr		----------○-	NDC
Ensete ventricosum (Welw.) Cheesman [925]: Kabhunditoke	1769	P		-♦♦-♦----●--	NDC, 4
	1936	Le*		------- ♦----	1

Plant name	FDID	Part	Gr	Month(s) eaten J F M A M J J A S O N D	Source
Erythrina abyssinica DC. [574]: Muko	1586	Fl		-------○●○--	NDC, 1
	1587	Fr		--------○○--	NDC
	1588	Le		○○○○----●○●●	NDC, 4
	1589	B*		------○-----	NDC
Faidherbia albida (Delile) A.Chev. [1278]: Ikese	1816	R*		--------♦---	1
Ficus artocarpoides Warb. [488]: Mululu	1876	Fr		♦-♦--♦♦♦♦♦-♦	6
	1913	Le*		------♦-----	1
Ficus asperifolia Miq. [489]: Kankolonkombe	1629	Fr		●●●●●●●●●●●●	NDC, 32
	1630	Le		●●●●●●●●●●●●	NDC, 35
	1855	Pe		---------♦♦-	2
Ficus cyathistipula Warb. [492]: Kajimonsole type D NDC cite this sp. as syn. of F. kirkii [505] As detailed information NA, currently assigned to **F. cyathistipula** only	1619	Fr		--○--♦------	NDC, 1
Ficus exasperata Vahl [493]: Lwagho	1620	Fr		●●♦-♦†●●●●●●	NDC, 26
	1621	Le		●●●●●●●●●●●●	NDC, 32
	1872	Pe*		-----------♦	1
	1871	P*		----------♦-	1
		A pith of young stem was eaten.			
	1873	Sa*		---------♦--	1
Ficus glumosa Delile [494]: Ikuku type A Two spp. (**F. glumosa**, **F. sonderi** [498]) in NDC are lumped here (syn.)	1622	Fr		-○----○○-○--	NDC
Ficus ingens (Miq.) Miq. [496]: Kajimonsole type B	1623	Fr		---------○---	NDC
	1624	Le		-----------○	NDC
Ficus sur Forssk. [500]: Ikubhila	1614	Fr		●●●●●●●●●●●●	NDC, 33
	1615	Le		-○○----♦-○--	NDC, 1
	1818	Sa		♦--------♦-♦	3
	1616	W		-------○○--♦	NDC, 1

Plant name	FDID	Part	Gr	Month(s) eaten J F M A M J J A S O N D	Source
Ficus sycomorus L. [501]: Ikuku type B	1914	Fr		----♦♦--♦-♦	5
Ficus thonningii Blume [502]					
As Kajimonsole type A	1628	Fr		---○--○○○---	NDC
As Mujimo	1886	Fr		--♦♦†♦♦♦♦---	7
	1887	Le		--♦--♦♦♦♦---	5
Ficus trichopoda Baker [994]: Muhololo	1617	Fr		♦♦●○♦♦●♦-‡○♦	NDC, 8
	1618	Le		---○○-♦-●-○●	NDC, 5
	1885	Sh*		-----------♦	1
		Young stems and petioles eaten.			
Ficus vallis-choudae Delile [504]: Ihambwa	1631	Fr		●●●●●●●●●●●●	NDC, 29
	1632	Le		-○○○○○-○-○-○	NDC
	1813	P		-----♦-♦♦--♦	5
		Usually piths of petioles eaten. Piths of young stems just below the petioles are also eaten.			
	1814	Sa		---------♦♦-	2
	1633	W		-○♦○---♦----	NDC, 2
Ficus sp./spp. =IiDβ [1451]: Ikuku type undet. Prob. F. **sycomorus** [495] and/or F. **glumosa** [494]	1819	Fr		†♦♦♦-♦♦---♦-	7
	1820	Le*		--------♦---	1
	1821	Sa*		----♦-------	1
	1822	W		--♦---♦-----	2
Ficus sp. =IiDα [487]: Kajimonsole type E	1625	Fr		------○-○---	NDC
	1626	Le		○-----○----○	NDC
Ficus sp. =iD ign. [1392]: Kajimonsole type F	1627	Fr		-------○-○--	NDC
Ficus spp. =IiDβ [1446]: Kajimonsole type undet. CF A Prob. belongs to any *Ficus* sp./spp. identified under the local nom. Kajimonsole in FMTT	1838	Fr		-♦♦♦♦♦♦♦-†--	10
	1839	Le		-♦----♦♦♦♦♦♦	6
		Galls eaten in Nov.			
Ficus sp. =IiDβ [1452]: Kajimonsole type undet. CF B	1837	Le		-----♦♦-----	1
Ficus sp./spp. =IiDβ [1491]: Mutolighana Could be F. **artocarpoides** [488] or any other *Ficus* sp./spp.	1892	Fr		†♦♦♦♦♦♦♦♦♦-♦	10

Plant name	FDID	Part	Gr	Month(s) eaten J F M A M J J A S O N D	Source
Flacourtia indica (Burm.f.) Merr. [335]: Bhusungunimba	1502	Fr		○○●--○○-----	NDC, 1
Garcinia huillensis Welw. [383]: Kasolyo	1512	Fr		●-----●●●●●●	NDC, 18
	1514	Sa		--------♦♦-●	NDC, 4
	1841	W*		-----------♦	1
Grewia flavescens Juss. [742]: Bhunkukuma	1516	Fl		--○---------	NDC
	1517	Fr		--○○‡●●‡‡---	NDC, 3
	1518	Le		---○--○-♦---	NDC, 1
	1869	P*		-♦----------	1
	1868	B*		--♦---------	1
Grewia mollis Juss. [744]: Mukole	1515	Fr		--○○○○------	NDC
Gymnanthemum myrianthum (Hook.f.) H.Rob. [1428]: Isonswesilu	1734	Fl		------○○----	NDC
	1735	P		○-●----♦-○--	NDC, 3
Gymnanthemum sp. =IiDβ [1511]: Mujonsojonso Resembles **G. coloratum** [1431]	1917	Pe*		-----------♦	1
	1916	P*		-----------♦	1
		A pith of young stem was eaten.			
Haplocoelum foliolosum (Hiern) Bullock [1149]: NA	1665	Le		------------	NDC
Harungana madagascariensis Lam. ex Poir [385]: Mutunu	1513	Fr		---○●●●●●---	NDC, 8
Hewittia malabarica (L.) Suresh [240]: Lisisusamughubhi type B	1742	Le		○○---○○-○---	NDC
Hibiscus aponeurus Sprague & Hutch. [422]: Lusieti type C	1532	Le		-○○○--------	NDC
Hibiscus cannabinus L. [424]: Kakonda kaghosi A	1533	Le		------○-----	NDC
Hibiscus nyikensis Sprague =IiDα [427]: Kafungu	1534	Le		○-○○○○-♦----	NDC, 1
Hibiscus rostellatus Guill. & Perr. [429]: Lyakubhabha type A	1535	Le		----○○○-----	NDC
Hibiscus surattensis L. [430]: Kakonda kakema	1537	Le		-○--♦---○---	NDC, 1
Hibiscus sp. =IiDα [1462]: Lyakubhabha type nud. CF	1536	Le		------○○----	NDC

Plant name	FDID	Part	Gr	Month(s) eaten J F M A M J J A S O N D	Source
Hibiscus sp./spp. =IiDβ [1453]: Kakonda type unc. CF H. sp./spp. including **H. diversifolius** [426], **rostellatus** [429], **surattensis** [430] and/or **vitifolius** [1156]	1850	Le		♦♦---♦♦----♦	4
Hyparrhenia variabilis Stapf [355]: Bhusali type A	1796	P*		○-----------	NDC
Illigera pentaphylla Welw. [387]: Ilyungulyungu type C	1488	Le*		----------○-	NDC
Indigofera podocarpa Baker f. & Martin [582]: Katwala kaghosi	1593	Fl		--○○--------	NDC
	1594	Le*		---○--------	NDC
Indigofera emarginella [581] or *podocarpa* [582] =IiDβ: Katwala type undet. CF Could not specify in the field	1926	Fl*		--♦---------	1
	1927	Le*		--♦---------	1
Ipomoea cairica (L.) Sweet [244]: Lunsengasenga	1743	Le		○♦-○○-○●○♦--	NDC, 3
Ipomoea muricata (L.) Jacq. [248]: Iteleka	1744	Le		-----○------	NDC
Ipomoea ochracea (Lindl.) G.Don. [250]: Lisisusamughubhi type C	1745	Le		-----○------	NDC
Ipomoea rubens Choisy [251]: Mughubhi type A	1746	Le		○------○--○-	NDC
Ipomoea sp. =IiDα [255]: Mutuntumba	1747	Fl		-----○------	NDC
	1748	Fr		-----○------	NDC
	1749	Le		○--○○○------	NDC
Ipomoea sp. =IiDβ [1459]: Mughubhi type unc. CF Prob. I. **rubens** [251]	1906	Le		♦---♦--♦♦--♦	6
	1909	Sh		♦♦♦♦♦♦♦♦♦♦-♦	17
		Stems and/or petioles eaten. Hard fibers are spitted out after wadging. Leaves may be consumed together.			
Ipomoea or *Hewittia* sp./spp. =IiDβ [1477]: Kasonsokemba type unc. CF	1842	Fl*		-----♦------	1
	1843	Le		-♦♦♦♦♦-♦-♦♦♦	9
Jasminum sp. =IiDα [1169]: NA	1685	Le		----------○○	NDC
Julbernardia seretii (De Wild.) Troupin [143]: Mwako	1568	Fr		----------○-	NDC
		Seeds eaten.			
	1569	Le		○----------○	NDC
Keetia venosa (Oliv.) Bridson [650]: Luntafwanengwa bhusie	1710	Fr		●○○○--†●●●○●	NDC, 11
	1711	Le		○○-------♦--	NDC, 1
		Not sure of the sp. eaten in this study.			

Plant name	FDID	Part	Gr	Month(s) eaten J F M A M J J A S O N D	Source
Landolphia owariensis P.Beauv. [57]: Mpila	1691	Fr		●●●----♦○●●●	NDC, 17
	1692	Le		○----●--♦-○○	NDC, 2
	1694	P		♦●●●●●●●●●●●	NDC, 23
		Pith of young stem eaten.			
	1693	Sa*		---○--------	NDC
Lannea schimperi (Hochst. ex A.Rich.) Engl. [36]: Kabhumbu	1671	Fr		--------○○○-	NDC
	1672	Le		○--------○○○	NDC
	1673	R		--○------♦--	NDC, 1
Lecaniodiscus fraxinifolius Baker [706]: Kafulujeghe	1666	Fr		---------♦○○	NDC, 1
	1834	R*		-------♦----	1
Leea guineensis G. Don [781]: Lingogha	1653	Fr		♦♦---○○○○---	NDC, 2
	1654	Le		○---○-------	NDC
	1655	P		----○○-----♦	NDC, 1
		In our record, a whole young stem was eaten.			
Lepistemon owariense (P. Beauv.) Hallier f. [256]: Lisisusamughubhi type A	1750	Le*		-○----------	NDC
Lippia plicata Baker [766]: Lumumbu	1766	Le*		--○---------	NDC
Maesa lanceolata Forsskk. [514]: Katunula	1935	Fr*		-----♦------	1
		Only once as a feeding record but local assistants know this as food source.			
Mangifera indica L. [815]: Muwembe	1677	Fr		-------♦♦●--	NDC, 5+
	1881	Pe*		--♦---------	1
Marantochloa leucantha (K.Schum.) Milne-Redh. [438]: Kalubhwabhakulu	1781	Fr*		------○-----	NDC
	1852	Le		--------♦---	1
	1782	P		♦♦♦♦-●●●●●●●	NDC, 30
		Piths of stems eaten.			

Plant name	FDID	Part	Gr	Month(s) eaten J F M A M J J A S O N D	Source
Margaritaria discoidea (Baill.) G.L.Webster [316]: Lulyolwankanga	1544	Fr		○○○○--------	NDC
	1545	Le		○----------○	NDC
Markhamia lutea (Benth.) K.Schum. [106]: Kabhulampako type A	1919	Fr*		---†--------	
	1752	Le		○○-----○--○○	NDC
	1753	B*		-----------○	NDC
Merremia pterygocaulos (Choisy) Hall.f. [257]: NA	1751	Le*		-----○------	NDC
Milicia excelsa (Welw.) C.C.Berg [508]: Mukamba	1612	Fl		-----♦♦●♦---	NDC, 9
		Galls may be consumed together.			
	1613	Le		♦----♦♦●♦♦--	NDC, 11
		Galls are often selectively eaten.			
Mimusops bagshawei S.Moore [718]: Mulonje type C	1895	Fr		---†♦♦♦-----	2
Mimusops penduliflora Engl. [719]: Mulonje type B	1681	Fr		-----○○-----	NDC
Monanthotaxis poggei Engl. & Diels [42]: Lujongololo mabhosa masogha	1858	P*		----♦-------	1
		Piths of young stems eaten.			
Monotes elegans Gilg [288]: Mukokoti	1507	Fl*		○-----------	NDC
	1508	R*		-------○----	NDC
Morus alba L. [1385]: Mulangi	1875	Fr		--------♦♦--	2+
Mucuna gigantea (Willd.) DC. subsp. **quadrialata** (Baker) Verdc. [585]: Ighombyase	1595	Le		○-----------	NDC
Mukia maderaspatana (L.) M.Roem. [265]: Kakobhakansimba	1506	Le		○---○-------	NDC
Multidentia crassa (Hiern) Bridson & Verdc. [945]: Bhungogholo type A	1709	Fr		----†●●●○---	NDC, 4
Musa sp. =iD ign. [829]: Ikonde	1770	Fr		○○○○○○○○○○○○	NDC
	1771	Le		--○----○○---	NDC
	1772	P		○○○○○○○○○○○○	NDC
Mussaenda arcuata Poir. [654]: Mumpindu	1877	Fl		-♦----------	2
	1715	Fr		---------○--	NDC
	1716	Le		-♦●○♦-○-----	NDC, 3
	1717	P		--○---------	NDC

Plant name	FDID	Part	Gr	Month(s) eaten J F M A M J J A S O N D	Source
Myrianthus arboreus P.Beauv. [509]: Isakama type A	1825	Fr		†◆◆---◆◆◆◆◆◆	20
	1827	Le		◆◆----------	2
	1826	P		◆◆◆◆◆◆◆◆◆◆◆◆	20
		Petioles or piths of petioles eaten. Sometimes, only base of petioles may be eaten.			
Myrianthus holstii Engl. [510]: Isakama type unc. We assume this was also **M. arboreus** [509]	1634	Fr		--------○○○-	NDC
	1635	Le		○○○○○○○○○--○	NDC
	1636	P		---○○------○	NDC
	1637	B*		○-----------	NDC
Neonotonia wightii (Wight & Arn.) J.A.Lackey [1187]: Kamtutula type A	1590	Fl		---○○-------	NDC
	1591	Fr*		------○-----	NDC
	1592	Le		○○○○○○○----○	NDC
Neonotonia sp./spp. =IiDβ [1479]: Kamtutula type undet. CF Prob. belongs to sp./spp. identified under the local supranom. Kamtutula in FMTT	1854	Le		◆◆◆-◆◆◆◆◆◆◆◆	20
Neorautanenia mitis (A.Rich.) Verdc. [587]: Nkolo	1596	Fr		○-○---------	NDC
Olyra latifolia L. [362]: Lintentwa	1797	Fr		---○○-○○----	NDC
	1798	P		----○----○--	NDC
Oncinotis tenuiloba Stapf [952]: Ikonjogholo type A	1695	Fr		---------○○○	NDC
Oncoba spinosa Forssk. [337]: Kaposo	1840	Fr		--------◆◆--	2
Oxytenanthera abyssinica (A.Rich.) Munro [364]: Lulonje	1799	P	B	---------○○-	NDC
Panicum maximum Jacq. [365]: Katengalujinga	1800	Fr		---○○--○----	NDC
Parinari curatellifolia Planch. ex Benth. [629]: Mubhula type A Could be **P. excelsa** [1344]	1552	Fr		◆◆●○○◆●●---◆	NDC, 8
	1553	Le		○○○------○○○	NDC
	1555	Sa		-----------●	NDC, 1
	1554	B*		-----------○	NDC
Parkia filicoidea Oliv. [484]: Iseha	1576	Fl*		---------○--	NDC
	1577	Fr		--------‡●●●	NDC, 15
		Seeds and pulps eaten.			

Plant name	FDID	Part	Gr	Month(s) eaten J F M A M J J A S O N D	Source
Paullinia pinnata L. [707]: Mpatwe	1667	Fr		---------○○--	NDC
	1668	Le		-----○○○----	NDC
Pavetta sp. =IiDα [663]: Lintinfu	1718	Fr		-----○○--○--	NDC
Pennisetum purpureum Schumach. [369]: Iswe	1801	P		●●●●●●●●●●●●	NDC, 29
		Piths of young stems eaten.			
Phragmites mauritianus Kunth [370]: Ibhano	1802	P		●●●♦●♦●●●●●●	NDC, 19
		Piths of young shoots eaten.			
Piper umbellatum L. [610]: Ilendelyakenyinamwami type A	1500	Fr		-----○○-----	NDC
	1501	P		○---○-------	NDC
Piper sp./spp. =IiDβ [1460]: Ilendelyakenyinamwami type undet. CF **P. capense** [609] and/or **P. umbellatum** [610]	1823	P		-♦♦-----♦♦♦♦	7
		Whole leaf petioles or piths of them are eaten. Piths of young stems also eaten.			
Premna sp. =IiDα [768]: Lulindi type A Not confident of the sp. identification for our record	1767	Le		--○○--♦-----	NDC, 1
Pseudarthria hookeri Wight & Arn. [589]: Katyebhelele	1597	Le		○-----------	NDC
Pseudospondias microcarpa (A.Rich.) Engl. [37]: Bhuhono	1674	Fr		●●--†♦♦♦♦♦♦●	NDC, 25
	1675	Le		○○○○♦--○--○○	NDC, 1
	1676	B*		---○--------	NDC
	1807	W*		-----------♦	1
	1808	Ro		---------♦--	1+
Psidium guajava L. [517]: Mpela	1874	Fr		♦♦♦♦♦♦♦†♦♦♦♦	16
Psophocarpus scandens (Endl.) Verdc. [590]: Mulyampongo	1598	Le		○----♦○-○---	NDC, 1
Psorospermum febrifugum Spach [1268]: Kangululungululu	1925	Fl*		--♦---------	1
Psychotria peduncularis (Salisb.) Steyerm. =iD ign. [824]: Lulyolwakape	1719	Fr		--●●●●●-----	NDC, 11
Psydrax parviflora (Afzel.) Bridson [1420]: Kamwibhi **C. rubrocostatum** [638] (now **P. parviflora** subsp. **rubrocostata** [941]) was also suspected in NDC together with **C. vulgare** [1010] (now **P. parviflora**), but based on obs., we specified to the latter	1712	Fr		---●●●●○----	NDC, 9

Plant name	FDID	Part	Gr	Month(s) eaten	Source
				J F M A M J J A S O N D	
Pteridium aquilinum (L.) Kuhn. [1270]: Linsilu type A	1480	Le		○-○---○○-○-○	NDC
	1487	P		----○--○----	NDC
Pteridium sp. =IiDβ [1455]: Linsilu type undet. CF Prob. **P. aquilinum** [1270]	1848	Le		♦-♦♦-♦♦---♦♦	9
Pterocarpus tinctorius Welw. [1488]: Mwenje	1599	Fl		-●●●♦-------	NDC, 4
	1600	Fr		--♦○●●●●●♦--	NDC, 13
		Seeds are extracted to eat when ripe.			
	1602	Le		●●●●●●●●●●●●	NDC, 26
	1894	W*		---------♦--	1
Pterygota macrocarpa K.Schum. [732]: Mukungu	1521	Le		---♦---♦○---	NDC, 2
Pycnanthus angolensis (Welw.) Warb. [513]: Lulumasya	1489	Fl		--●---------	NDC, 1
	1490	Fr		●●---○●●●●●●	NDC, 24
	1491	Le		---------○-○	NDC
	1862	R*		-----------♦	1
	1863	Sa		♦---♦♦--♦♦♦♦	13
	1492	B		--○-♦○----○○	NDC, 1
	1493	W		♦-○○♦○-●●♦●●	NDC, 11
Renealmia engleri K.Schum. [790]: Itungulu type A Prob. some of our records for *Aframomum* sp. [845] should be included here	1780	P		--○○○○○○----	NDC
		Piths of young stems eaten.			
Rhoicissus sp. =IiDα [783]: Lisisusalukosyo type A	1656	Le		------○-----	NDC
	1657	P		------○-----	NDC
Rhynchosia sp. =IiDα [595]: Ikalamalungwe	1603	Le*		-○----------	NDC
Rhynchosia or *Vigna* sp. =iD ign. [1395]: Lwighalila type B	1806	Le		○-○-------○○	NDC
Ricinodendron heudelotii (Baill.) Heckel var. **tomentellum** (Hutch. & E.A.Bruce) Radcl.-Sm. [1275]: Sitobhagha type A	1546	Fl	B	----------○-	NDC
	1547	Le	B	---------○--	NDC
Riocreuxia profusa N.E.Br. [90]: Mukobhoso type C	1708	Le		○-----○----○	NDC

Plant name	FDID	Part	Gr	Month(s) eaten J F M A M J J A S O N D	Source
Rothmannia manganjae (Hiern) Keay [679]: Nyundo nkema	1720	Le		○-○---------	NDC
Rubus pinnatus Willd. [633]: Bhutandula	1556	Fr		-----○○-----	NDC
Rumex sp. =IiDβ [1490]: Kamujiola Prob. **R. abyssinicus** [617] or **R. usambarensis** [618]	1912	Le*		-------♦----	1
Rytigynia sp. =IiDα [683]: Bhukuja	1721	Fr*		○-----------	NDC
	1722	Le		○-----------	NDC
Saba comorensis (Bojer ex A.DC.) Pichon [62]: Ilombo	1696	Fr		●●●♦♦♦●●●●●●	NDC, 35
		NDC reported seed use as coprophagy, which we have not observed in the M group. Seeds may be eaten with pulp occasionally or by extracting when unripen.			
	1698	Le		○♦--○○-●-○-○	NDC, 2
		May not be a usual food item for the M group, but frequently consumed when wadging meat (not included as food item in this list).			
	1699	P		-●●○○○●●♦●-♦	NDC, 7
		Piths of young stems eaten.			
Saccharum officinarum L. [831]: Isumbu	1803	P		○○○○○○○○○○○○	NDC
Salacia sp. =IiDα [165]: Ighandamakungu	1638	Fr		---------○--	NDC
	1639	Le*		○-----------	NDC
Scolopia sp. =IiDα [339]: Musilantumbalo	1878	Fr	Y	------------	MN
Sesbania sesban (L.) Merr. var. **nubica** Chiov. [596]: Musilie	1606	Sa		--○---------	NDC
	1605	W		○-------○---	NDC
	1604	B		○-○○♦-○-○--○	NDC, 1
Smilax anceps Willd. [931]: Linselele	1929	Fr*		------♦-----	1
	1783	Le		●●○○●♦♦♦●●●●	NDC, 23
		Young stems may also be eaten together.			
	1847	Sh		-♦♦-♦---♦-♦♦	6
		Whole young stems eaten.			

Plant name	FDID	Part	Gr	Month(s) eaten J F M A M J J A S O N D	Source
Solanecio angulatus [1322] or **Gynura scandens** [216] =iD ign. [1394]: Kahingo	1726	Le		○○●○○–––––––	NDC, 1
Sonchus schweinfurthii Oliv. & Hiern [230]: Ipempu	1730	Le		○○––––––––––	NDC
Spathodea campanulata P.Beauv. [108]: Lipulu	1754	Fr		–––––––○––♦–	NDC, 1
		Seeds eaten.			
Stephania abyssinica (Quart.-Dill. & A.Rich.) Walp. [459]: Lyumile	1496	Le*		––––––––––○–	NDC
Sterculia quinqueloba (Garcke) K.Schum. [733]: Mukungumwelu	1522	Fl		––––––♦●––––	NDC, 2
	1523	Fr		–––––♦♦●‡○○–	NDC, 5
		Carpels and seeds are eaten together when unripe but ignored when ripe.			
	1524	Le		––––––○○○○––	NDC
	1525	B		––○–––––––––	NDC
Sterculia tragacantha Lindl. [735]: Kakubhabholo	1526	Fl		––––––○○––––	NDC
	1527	Fr		––○○○○○○––––	NDC
	1528	Le		●●●○●○●●●●●●	NDC, 18
	1924	Pe		–♦–––––––––	1+
	1851	R		–––––––♦––––	1
	1529	B		○–○–––––––––	NDC
Stereospermum kunthianum Cham. [110]: Mutelele	1755	Le		○––––––––♦○–	NDC, 1
	1756	Sa*		––––––○–––––	NDC
Strychnos innocua Delile [416]: Kankundu	1683	Fr		––○○––––○○○○	NDC
	1684	Le		○–––––––––––	NDC
Synsepalum brevipes (Baker) T.D.Penn. [722]: Mulyansekesi	1679	Fr		–○●●●●●––††–	NDC, 8
Syzygium guineense (Willd.) DC. [520]: Musabhasabha type A	1509	Fr		––––––––†●–♦	NDC, 4
Tabernaemontana pachysiphon Stapf [65]: Ikolyoko no.2	1700	Fr		–––––♦●♦●♦––	NDC, 20
	1701	Le*		–––––––––––○	NDC
Tacca leontopetaloides (L.) Kuntze [737]: Kabhugha	1792	Fr		––○–––––––––	NDC

Plant name	FDID	Part	Gr	Month(s) eaten J F M A M J J A S O N D	Source
Tarenna pavettoides (Harv.) Sim. [684]: Mukombelonda type B	1901	Fr		---♦-♦♦♦----	7
Terminalia kaiseriana F.Hoffm. [176]: Kaselenje	1844	Fr*		--------♦---	1
Thunbergia alata Boj. ex Sims [11]: Nkulumonga	1762	Le		○○○○○-------	NDC
Tinospora caffra (Miers) Troupin [461]: Musomombo	1920	Fr		-†-†--------	
	1497	Le		●●●●●●●○-♦○●	NDC, 14
	1498	P		--○---------	NDC
Toddalia asiatica (L.) Lam. [693]: Katapansima	1660	Fr		----○●●○○○--	NDC, 4+
Trema orientalis (L.) Blume [750]: Muhehu	1611	Le		♦♦♦♦♦--♦--○○	NDC, 8
	1884	W*		-♦----------	1
Trichilia prieuriana A.Juss. [450]: Kamoko	1661	Le		--○○○○○-----	NDC
Trichilia sp. =IiDα [451]: Kamoko Although NDC specifies Kamoko as **T. prieuriana**, further validation needed for our records	1853	Le		---♦♦♦------	4
Uapaca kirkiana Müll.Arg. [328]: Ikusu	1548	Fr		-------○○○○-	NDC
Uapaca nitida Müll.Arg. [329]: Lulobhe	1549	Fl		--○---------	NDC
	1550	Fr		---○--♦●●●●♦	NDC, 11
Uapaca sansibarica Pax [330]: Kakusufinyia	1551	Fr		-------○○○--	NDC
Uvaria angolensis Welw. ex Oliv. [45]: Lujongololo bhosa	1485	Fr		-○○●●♦†♦♦♦†-	NDC, 8
		Dried fruits eaten at latter half of the year.			
	1486	Le		○○○○------○○	NDC
Vepris nobilis (Delile) Mziray [1355]: Mupemba	1659	Fr		♦---♦-------	NDC, 2
		NDC cited this item as food for the B group only (on month(s) eaten, NA).			

Plant name	FDID	Part	Gr	Month(s) eaten J F M A M J J A S O N D	Source
Vernonia amygdalina Del. [231]: Mujonso	1731	Le		○○○○––●○○–○○	NDC, 1
	1888	Pe		–––♦––––♦–––	2
	1732	P		●●●○●––●●♦●●	NDC, 13
	1733	B		––––––○○––––	NDC
	1933	Ro		––––––♦–––––	1
Vigna heterophylla A.Rich. [1295]: Kabhwase	1607	Fl		–––––○○–––––	NDC
	1608	Fr		–––––○○–––––	NDC
		Seeds eaten.			
	1609	Le		––○––○○–––––	NDC
Vitex doniana Sweet [769]: Bhufulu	1768	Fr		–––––○––––––	NDC
Voacanga africana Stapf ex Scott-Elliot [67]: Ikolyoko no.1	1702	Fr		––––○○○○○○––	NDC
Vossia cuspidata (Roxb.) Griff. [380]: Kasweghenda type A	1804	P		○○○○○○○○○○○○	NDC
Whitfieldia sp. =IiDα [1254]: NA	1763	Le*	B	–––––––––○––	NDC
Ximenia americana L. [531]: Bhusantu	1640	Fr		–––––––––○○–	NDC
Xylopia longipetala De Wild. & T.Durand [1287]: Kahwibhili	1836	Le*		–––––––––––♦	1
		Young leaves were selected.			
Zanha golungensis Hiern [709]: Sitalya	1669	Fr		––––––––––●♦	NDC, 3
	1670	Le		–––––––––○–––	NDC
Zea mays L. [827]: Sisaka	1805	P		○○–––––––––○	NDC
Ziziphus mucronata Willd. [627]: Kaghobhole Could be **Z. abyssinica** [626]	1641	Fr		––––––●‡––––	NDC, 1
Sp. Ikonjogholo =IiDβ [1414]: Ikonjogholo type undet. Sp. assigned to local supra-nom. Ikonjogholo in FMTT	1817	Fr*		–––––♦––––––	1
Sp. Ilyungulyungu CF =IiDβ [1478]: Ilyungulyungu type undet. CF Prob. **Illigera pentaphylla** [387]	1824	Le		♦––♦♦♦–♦–––♦	4
Sp. Itesa =IiDβ [1445]: Itesa type undet. CF Should belong to sp./spp. identified under the local supra-nom. Itesa in FMTT	1830	Le		–♦♦––––♦♦♦–♦	6
Sp. Luhosyo =UiD [1515]: Luhosyo Possibly *Conyza* sp. [207] or Sp. Luhosyo ASTERACEAE [98]	1930	Le		–♦––––––––––	1

Plant name	FDID	Part	Gr	Month(s) eaten J F M A M J J A S O N D	Source
Sp. Lujongololo type undet. =IiDβ [1416]: Lujongololo type undet. Should belong to sp./spp. identified under the local supra-nom. Lujongololo in FMTT	1931	Fr		--♦♦♦-------	1
	1860	Le		-------♦-♦♦-	4
Sp. Lumpululu =IiDβ [1456]: Lumpululu type undet. CF Any sp./spp. identified under local supra-nom. Lumpululu in FMTT could be applicable	1864	Fl*		---------♦--	1
		Prob. underestimated since chimpanzees may eat flowers together with leaves.			
	1865	Fr		--------♦♦♦♦	1
	1866	Le		♦♦♦♦♦♦♦♦♦♦♦♦	35
		Whole stems may be eaten together when young.			
	1867	P		♦-----------	2
		Piths of petiole eaten.			

References

Nishida, T. and Uehara, S. (1981). Kitongwe name of plants: A preliminary listing. *African Study Monographs*, **1**, 109–31.

Nishida, T. and Uehara, S. (1983). Natural diet of chimpanzees (*Pan troglodytes schweinfurthii*): Long-term record from the Mahale Mountains, Tanzania. *African Study Monographs*, **3**, 109–30.

APPENDIX III

Mammal list

COMPILED BY HIROSHI IHOBE

ORDER Family	Latin name	English name	Sources
PRIMATES			
Galagonidae	*Otolemur crassicaudatus*	Greater galago[a]	2
	Galago moholi	South African galago[a]	7
	Galagoides sp.	Dwarf galago	7
Cercopithecidae	*Procolobus rufomitratus*	Eastern red colobus[a]	1
	Colobus angolensis	Angolan black-and-white colobus[b]	2
	Papio cynocephalus	Yellow baboon	1
	Chlorocebus pygerythrus	Vervet monkey[a]	1
	Cercopithecus ascanius	Red-tailed monkey	1
	Cercopithecus mitis	Blue monkey	1
Hominidae	*Pan troglodytes*	Chimpanzee	1
CHIROPTERA			
Pteropodidae	*Epomophorus* sp.	Epauletted fruit bat	6
	Epomops franqueti	Singing fruit bat	7
Emballonuridae	*Taphozous perforatus*	Tomb bat	6
Vespertilionidae	*Pipistrellus* sp.	Pipistrelle bat	6
	Scotoecus albofuscus	Evening bat	7
SORICOMORPHA			
Soricidae	*Sylvisorex megalura*	Climbing shrew	6
	Crocidura sp.	White-toothed shrew[a]	6, 7
MACROSCELIDEA			
Macroscelididae	*Rhynchocyon cirnei*	Checkered elephant shrew	2
RODENTIA			
Sciuridae	*Heliosciurus rufobrachium*	Red-legged sun squirrel	5
	Protoxerus stangeri	African giant squirrel	2
Gliridae	*Graphiurus* sp.	African dormouse	6
Bathyergidae	*Heliophobius* sp.	Silky blesmol[a, c]	2
Hystricidae	*Hystrix africaeaustralis*	Cape crested porcupine[a]	8
	Atherurus africanus	Brush-tailed porcupine[a]	2

ORDER Family	Latin name	English name	Sources
Thryonomyidae	*Thryonomys swinderianus*	Greater cane rat	2, 3
Nesomyidae	*Cricetomys gambianus*	Forest pouched rat[a]	6
	Dendromus sp.	Climbing mouse	6
Muridae	*Otomys* sp.	Groove-toothed rat	7
	Lophuromys sp.	Brush-furred rat	6
	Praomys sp.	Soft-furred rat[a]	6
	Hylomyscus sp.	Montane wood mouse	6
	Mus sp.	Common mouse[a]	6
	Grammomys sp.	Thicket rat	6
	Oenomys hypoxanthus	Rusty-nosed rat	6
	Lemniscomys rosalia	Single-striped grass mouse	6
	Lemniscomys striatus	Common striped grass mouse	6
CARNIVORA			
Canidae	*Canis adustus*	Side-striped jackal[d]	2
	Lycaon pictus	Wild dog[d]	5
Mustelidae	*Mellivora capensis*	Ratel	2
	Aonyx capensis	African clawless otter[a]	2
	Lutra maculicollis	Spotted-necked otter	2
Herpestidae	*Herpestes sanguineus*	Slender mongoose	7
	Helogale parvula	Common dwarf mongoose	6
	Mungos mungo	Banded mongoose	3
	Ichneumia albicauda	White-tailed mongoose	2
	Bdeogale crassicauda	Bushy-tailed mongoose	3
Hyaenidae	*Crocuta crocuta*	Spotted hyena[d]	2
Viverridae	*Genetta* sp.	Genet[a]	7
	Civettictis civetta	African civet	2
Felidae	*Felis sylvestris*	African wild cat	3
	Felis serval	Serval cat[d]	2
	Panthera pardus	Leopard	2
	Panthera leo	Lion[d]	5
PHOLIDOTA			
Manidae	*Smutsia gigantea*	Giant pangolin[a]	7
TUBULIDENTATA			
Orycteropodidae	*Orycteropus afer*	Aardvark	2
HYRACOIDEA			
Procaviidae	*Heterohyrax brucei*	Yellow-spotted hyrax	3
	Dendrohyrax arboreus	Southern tree hyrax[c]	3

ORDER Family	Latin name	English name	Sources
PROBOSCIDEA			
Elephantidae	*Loxodonta africana*	African elephant[c]	2
PERISSODACTYLA			
Equidae	*Equus quagga*	Common zebra[a, c]	3
CETARTIODACTYLA			
Hippopotamidae	*Hippopotamus amphibius*	Hippopotamus	2
Suidae	*Potamochoerus larvatus*	Bushpig[a]	2
	Phacochoerus africanus	Common warthog[a]	2
Bovidae	*Syncerus caffer*	African buffalo[d]	2
	Tragelaphus scriptus	Bushbuck	2
	Sylvicarpa grimmia	Bush duiker[c]	2
	Cephalophus monticola	Blue duiker	4
	Raphicerus sharpei	Sharpe's grysbok[c]	4
	Oreotragus oreotragus	Klipspringer[c]	2
	Kobus ellipsiprymnus	Waterbuck[c]	9
	Hippotragus equinus	Roan antelope[c]	2

This table is modified from Table 1.1 in Nishida (1990) with some updated information. Scientific and English names are according to Kingdon (2001), Kingodon *et al.* (2013), and IUCN (2014).

[a] Identification, scientific names, or English names were changed from Table 1.1 in Nishida (1990) and see Chapter 15 for the changes in detail.

[b] Limited to the higher altitude near Mt. Nkungwe.

[c] Not found in the Kasoje area.

[d] Rarely observed in the Kasoje area but there have been some sightings.

Sources: 1: Kano, 1971; 2: Nishida *et al.*, 1979; 3: Anonymous, 1980; 4: Hasegawa and Hiraiwa-Hasegawa, 1980; 5: Nishida, 1985; 6: Stanley, 2004, 7; Moyer, 2006, 8; Kingdon *et al.*, 2013; 9: Koichiro Zamma, personal communication.

References

Anonymous (1980). *Mahale: Study for the Proposed Mahale Mountains National Park. Final Report, May 1980*. Tokyo: Japan International Cooperation Agency.

Hasegawa, T. and Hiraiwa-Hasegawa, M. (1980). Precipitation, atmospheric temperature, relative humidity and some other observations in Kasoje Chimpanzee Research Station, January–December 1979. *Mahale Mountains Chimpanzee Research Project, Metrological Report*, **5**.

IUCN (2014). *The IUCN Red List of Threatened Species*. Available at: http://www.iucnredlist.org/.

Kano, T. (1971). Distribution of the primates on the eastern shore of Lake Tanganyika. *Primates*, **12**, 281–304.

Kingdon, J. (2001). *The Kingdon Field Guide to African Mammals*. London: Academic Press.

Kingdon, J., Happold, D., Butynski, T., *et al.* (2013). *Mammals of Africa Vol. I–VI*. London: Bloomsbury Natural History.

Moyer, D. C. (2006). *Biodiversity of Mahale Mountains National Park, Tanzania*. New York: Wildlife Conservation Society.
Nishida, T. (1985). The Mahale Mountains declared a national park. *IPPL Newsletter*, **12**(3), 9.
Nishida, T. (1990). A quarter century of research in the Mahale Mountains: An overview. In *The Chimpanzees of the Mahale Mountains: Sexual and Life History Strategies*, ed. T. Nishida. Tokyo: University of Tokyo Press, pp. 3–35.
Nishida, T., Uehara, S., and Nyundo, R. (1979). Predatory behavior among wild chimpanzees of the Mahale Mountains. *Primates*, **20**, 1–20.
Stanley, W. T. (2004). *The Small Mammals of Mahale Mountains National Park*. Chicago, IL: The Field Museum.

APPENDIX IV

Meteorological data collected at Kansyana, 1983–2013

COMPILED BY NORIKO ITOH

In Mahale, data on precipitation, minimum and maximum temperatures, morning temperature, minimum and maximum humidity, and morning humidity have been and continue to be recorded daily since October 1973 at the weather station near Kansyana Camp. Records up to 1988 have been summarized with respect to monthly mean maximum and minimum temperatures and precipitation (Takasaki *et al.*, 1990).

Records from January 1, 1983 to December 31, 2013 are summarized in this appendix (detailed descriptions on climatological trends are provided in Chapter 12). Records before 1983 were excluded as the original data were not available (missing). Data was collected by researchers, or field assistants in the absence of a researcher, and the families of researchers also contributed (Takasaki *et al.*, 1990). From the climatic variables observed, I have summarized the records on precipitation and minimum and maximum temperatures. Due to the small sample size of the humidity records, which was limited to the earlier and recent years, they have been excluded from the summary. As morning temperature records do not differ significantly from the minimum temperature records, they were not analysed (see Chapter 12).

The maximum and minimum temperature records were screened prior to analysis, and some doubtful records were removed (e.g. apparent measurement errors as judged on the basis of a comparison with the data on morning temperature). Thus, 95% and 93% of the available records, respectively, are summarized here. Basically, all precipitation records were used as they appeared in the data sheets because their reliability could not be judged. Precipitation data up to 2008 were recorded from rain gauge amounts read manually every morning (Takasaki *et al.*, 1990; Itoh *et al.*, 2012). On October 25, 2011, a data logger was installed (Amenbo RF-3, T&D Corporation, Japan). As this logger omits rainfall of less than 0.5 mm, all manual data were corrected by rounding off the amounts down to 0.5 mm. On at least two occasions, this logger was not working during the recording period. In these cases, manual data records were used instead. The total number of observation days is 11 142 (manual: $N = 10\,420$; Amenbo: $N = 722$). Although a temperature data logger was also installed on June 17, 2010 (Ondotori RTR-53A and RTR-57 U, T&D Corporation, Japan), I basically used values obtained from a mercury thermometer for monthly mean estimation (see Chapter 12). The total observation days for maximum and minimum temperatures are 10 704 and 10 486, respectively. Of these, 0.9% of the maximum temperature data and 1.6% of data on the minimum temperature data are derived from an Ondotori recorder. The maximum and minimum temperatures recorded by the Ondotori were relatively lower (mean −0.3°C for those collected at 15-min intervals and −0.4°C for the 60-min intervals for both minimum and maximum temperatures, but the range was quite large for the 60-min intervals, which were not used in this dataset to any great extent).

Table IV-1 **Monthly rainfall (mm) at Kansyana, 1983–2013**

	Jan	Feb	Mar	Apr	May	Jun	Jul	Aug	Sep	Oct	Nov	Dec	mean/ day
1983	165.0	89.0	550.0	365.5	31.5	0.0	0.0	54.0	2.0	214.0	273.5	276.5	5.6
	(30)	(28)	(31)	(30)	(30)	(30)	(31)	(31)	(30)	(31)	(30)	(31)	
1984	215.5	195.5	279.0	211.5	21.5	0.0	0.0	0.0	1.5	87.0	402.0	238.0	4.5
	(31)	(29)	(31)	(30)	(31)	(30)	(31)	(31)	(30)	(31)	(30)	(31)	
1985	246.5	319.0	340.5	255.0	56.5	0.0	0.0	0.0	13.0	3.5	254.0	350.0	5.0
	(31)	(28)	(31)	(30)	(31)	(30)	(31)	(31)	(30)	(31)	(30)	(31)	
1986	183.0	338.0	337.5	260.0	133.0	2.5	0.0	*0.0*	68.5	129.0	313.5	318.0	6.3
	(31)	(28)	(31)	(30)	(24)	(29)	(31)	(6)	(30)	(31)	(30)	(31)	
1987	392.0	138.5	405.0	351.5	95.5	0.0	0.0	0.0	85.5	171.5	189.5	248.0	5.7
	(31)	(28)	(31)	(30)	(31)	(30)	(31)	(31)	(30)	(31)	(30)	(31)	
1988	211.5	143.0	194.0	263.5	0.0	2.0	0.0	4.5	4.5	102.0	243.0	309.0	4.0
	(31)	(29)	(31)	(30)	(31)	(30)	(31)	(31)	(30)	(31)	(30)	(31)	
1989	72.5	172.0	373.5	269.0	80.0	0.0	0.0	4.5	17.5	64.5	355.5	244.0	4.5
	(31)	(28)	(31)	(30)	(31)	(30)	(31)	(31)	(30)	(30)	(30)	(31)	
1990	276.5	321.0	321.5	162.5	141.0	0.0	0.0	0.0	35.5	72.0	201.0	521.5	5.7
	(31)	(28)	(31)	(30)	(31)	(22)	(31)	(31)	(30)	(31)	(30)	(31)	
1991	150.5	251.5	334.5	292.0	68.0	0.0	2.5	10.0	25.0	268.0	256.0	164.0	5.0
	(31)	(28)	(31)	(30)	(31)	(30)	(31)	(31)	(29)	(30)	(30)	(31)	
1992	396.0	192.5	NA	206.5	117.5	0.0	0.0	0.0	6.0	172.5	197.5	216.5	4.5
	(31)	(29)	(0)	(30)	(31)	(30)	(31)	(31)	(30)	(31)	(30)	(31)	
1993	243.0	275.0	93.5	162.5	121.5	0.0	0.0	13.0	11.0	10.5	130.5	222.5	3.5
	(31)	(28)	(31)	(30)	(31)	(29)	(31)	(31)	(30)	(30)	(30)	(31)	
1994	NA	220.0	405.5	237.5	54.0	0.0	2.0	7.5	0.5	158.0	204.5	234.5	4.6
	(0)	(28)	(31)	(30)	(31)	(30)	(31)	(31)	(30)	(31)	(30)	(31)	
1995	273.5	206.5	215.0	182.5	63.0	2.5	0.0	0.0	27.5	182.5	161.0	148.5	4.0
	(31)	(28)	(31)	(30)	(31)	(30)	(31)	(31)	(30)	(31)	(30)	(31)	
1996	228.0	385.0	431.0	429.5	31.5	0.0	0.0	0.0	21.0	175.5	144.5	237.0	5.7
	(31)	(28)	(31)	(30)	(31)	(30)	(31)	(31)	(30)	(31)	(30)	(31)	
1997	239.0	41.0	292.0	202.5	202.0	6.5	0.0	0.0	0.0	260.5	406.5	469.0	5.8
	(31)	(28)	(31)	(30)	(31)	(30)	(31)	(31)	(30)	(31)	(30)	(31)	
1998	518.5	88.0	361.5	277.5	162.0	0.0	0.0	23.0	1.0	104.0	162.5	315.0	5.5
	(31)	(28)	(31)	(30)	(31)	(30)	(31)	(31)	(30)	(31)	(30)	(31)	
1999	240.0	99.5	470.0	196.0	4.0	0.0	0.0	60.0	13.5	8.5	264.5	324.5	4.6
	(31)	(28)	(31)	(30)	(31)	(30)	(31)	(31)	(30)	(31)	(30)	(31)	

Table IV-1 **(cont.)**

	Jan	Feb	Mar	Apr	May	Jun	Jul	Aug	Sep	Oct	Nov	Dec	mean/ day
2000	151.0	95.5	275.5	134.5	6.0	0.0	0.0	5.5	1.0	6.5	334.5	361.0	3.7
	(31)	(29)	(31)	(30)	(31)	(30)	(31)	(31)	(30)	(31)	(30)	(31)	
2001	265.0	200.0	367.0	189.0	59.5	35.0	30.0	0.0	60.5	NA	NA	134.5	4.4
	(31)	(28)	(31)	(30)	(31)	(30)	(31)	(31)	(30)	(0)	(0)	(31)	
2002	498.5	241.0	256.0	217.5	27.0	0.0	0.0	0.5	5.0	124.0	371.5	217.5	5.4
	(31)	(28)	(31)	(30)	(31)	(30)	(31)	(31)	(30)	(31)	(30)	(31)	
2003	191.0	146.0	291.0	176.5	92.0	0.0	3.0	0.0	3.5	77.5	220.0	264.0	4.0
	(31)	(28)	(31)	(30)	(31)	(30)	(31)	(31)	(30)	(31)	(30)	(31)	
2004	350.5	234.0	161.5	238.0	0.0	0.0	0.0	0.0	43.5	116.0	141.0	294.5	4.3
	(31)	(29)	(31)	(30)	(31)	(30)	(31)	(31)	(30)	(31)	(30)	(31)	
2005	315.5	181.0	285.0	170.5	92.0	0.5	0.0	9.0	6.0	62.5	310.0	200.5	4.5
	(31)	(28)	(31)	(30)	(31)	(30)	(31)	(31)	(30)	(31)	(30)	(31)	
2006	274.0	228.0	174.0	324.5	101.5	0.0	0.0	18.0	6.5	32.0	402.0	622.0	6.0
	(31)	(28)	(31)	(30)	(31)	(30)	(31)	(31)	(30)	(31)	(30)	(31)	
2007	189.0	113.0	99.5	192.0	11.0	25.0	0.0	2.0	6.5	101.5	320.5	337.0	3.8
	(31)	(28)	(31)	(30)	(31)	(30)	(31)	(31)	(30)	(31)	(30)	(31)	
2008	244.5	299.0	180.5	308.5	45.0	1.0	0.0	0.5	18.5	126.0	214.5	212.5	4.5
	(31)	(29)	(31)	(30)	(31)	(30)	(31)	(31)	(30)	(31)	(30)	(31)	
2009	234.5	215.0	306.0	155.5	112.5	28.5	0.0	0.0	15.0	49.5	340.5	261.5	4.7
	(31)	(28)	(30)	(30)	(31)	(30)	(31)	(31)	(30)	(31)	(30)	(31)	
2010	151.5	118.5	363.0	162.5	49.0	4.5	0.0	0.0	0.0	126.0	182.5	216.0	3.8
	(31)	(28)	(31)	(30)	(31)	(30)	(31)	(31)	(30)	(31)	(30)	(31)	
2011	402.5	162.0	181.0	127.5	16.5	0.5	0.0	35.5	14.0	113.5	439.5	386.5	5.1
	(31)	(28)	(31)	(30)	(31)	(30)	(31)	(31)	(30)	(31)	(30)	(31)	
2012	183.5	192.5	230.5	131.5	122.0	0.5	3.5	60.5	7.0	44.5	384.0	423.5	4.9
	(31)	(29)	(31)	(30)	(31)	(30)	(31)	(31)	(30)	(31)	(30)	(31)	
2013	129.0	133.5	246.0	260.0	4.5	0.0	0.0	0.0	29.0	115.5	259.5	505.0	4.6
	(31)	(28)	(31)	(30)	(31)	(30)	(31)	(31)	(30)	(31)	(30)	(31)	
mean/ day	8.2	6.9	9.5	7.6	2.2	0.1	0.0	0.3	0.6	3.5	9.0	9.6	4.8

Figures in parentheses = *N*

Table IV-2 **Monthly mean daily maximum temperature (degree C) at Kansyana, 1983–2013**

	Jan	Feb	Mar	Apr	May	Jun	Jul	Aug	Sep	Oct	Nov	Dec	mean
1983	28.3	29.9	28.9	28.4	29.3	29.5	30.0	29.3	32.1	29.4	29.0	27.5	**29.3**
	(30)	(27)	(31)	(29)	(28)	(30)	(31)	(31)	(29)	(31)	(30)	(31)	
1984	27.8	28.0	28.3	28.8	30.3	29.3	29.4	31.1	31.6	30.2	27.0	26.9	**29.1**
	(31)	(29)	(31)	(30)	(27)	(20)	(28)	(28)	(30)	(30)	(30)	(31)	
1985	27.1	27.3	28.1	28.9	31.1	31.7	30.7	32.1	32.7	*32.8*	NA	NA	**30.3**
	(30)	(27)	(31)	(29)	(30)	(29)	(29)	(30)	(30)	(5)	(0)	(0)	
1986	27.8	28.5	28.2	30.1	30.6	30.9	30.7	*30.7*	31.6	29.5	27.1	26.4	**29.3**
	(29)	(27)	(31)	(30)	(26)	(29)	(31)	(7)	(30)	(28)	(26)	(27)	
1987	27.0	27.9	27.9	29.0	31.5	32.8	32.7	33.0	33.0	NA	28.5	27.9	**30.1**
	(31)	(18)	(31)	(27)	(28)	(28)	(23)	(27)	(27)	(0)	(28)	(31)	
1988	27.3	28.8	30.4	31.0	33.5	33.6	32.6	32.3	33.5	32.6	27.7	26.1	**30.8**
	(30)	(29)	(31)	(28)	(29)	(25)	(31)	(31)	(30)	(31)	(29)	(27)	
1989	28.2	28.2	28.5	30.3	30.5	31.9	31.2	31.9	31.9	30.7	28.7	27.3	**29.9**
	(31)	(28)	(30)	(30)	(31)	(29)	(30)	(31)	(29)	(30)	(29)	(31)	
1990	27.6	28.4	28.8	30.5	30.0	NA	NA	30.7	31.2	32.1	28.8	27.6	**29.6**
	(31)	(28)	(31)	(30)	(31)	(0)	(0)	(25)	(30)	(31)	(30)	(31)	
1991	30.3	30.8	29.9	29.3	29.0	30.9	29.5	30.3	31.2	28.3	28.0	28.6	**29.7**
	(31)	(28)	(31)	(29)	(31)	(30)	(31)	(31)	(29)	(30)	(30)	(29)	
1992	27.8	NA	NA	28.0	28.4	29.4	28.6	29.5	31.2	30.3	29.5	27.4	**29.0**
	(31)	(0)	(0)	(30)	(31)	(30)	(31)	(31)	(30)	(31)	(30)	(31)	
1993	27.7	28.3	28.8	28.5	28.5	29.7	28.7	30.1	31.7	32.6	29.7	29.2	**29.4**
	(18)	(28)	(31)	(30)	(31)	(29)	(31)	(30)	(30)	(28)	(29)	(31)	
1994	NA	29.1	28.8	28.5	29.3	30.0	28.6	30.7	31.8	29.8	27.8	27.0	**29.2**
	(0)	(28)	(25)	(27)	(31)	(30)	(31)	(31)	(30)	(31)	(30)	(31)	
1995	29.2	27.7	27.6	27.8	29.8	28.9	29.1	30.0	30.8	29.9	30.3	28.2	**29.1**
	(31)	(28)	(31)	(30)	(31)	(30)	(31)	(31)	(30)	(28)	(29)	(31)	
1996	27.8	26.7	27.5	27.9	29.4	*27.9*	28.1	29.2	30.7	30.1	27.9	27.4	**28.4**
	(31)	(23)	(31)	(30)	(15)	(4)	(30)	(26)	(30)	(31)	(30)	(31)	
1997	27.5	28.7	27.1	26.1	27.3	27.9	28.1	29.2	31.4	29.0	26.2	25.4	**27.8**
	(31)	(28)	(31)	(29)	(31)	(30)	(31)	(30)	(29)	(31)	(30)	(31)	
1998	26.1	28.1	27.1	27.6	27.6	27.5	27.5	28.5	30.2	30.8	29.1	26.3	**28.0**
	(31)	(28)	(31)	(30)	(31)	(30)	(31)	(31)	(30)	(31)	(30)	(31)	
1999	28.5	28.9	26.7	27.0	27.8	27.4	27.5	28.1	28.9	30.8	28.4	27.0	**28.1**
	(31)	(28)	(31)	(30)	(31)	(30)	(31)	(31)	(30)	(31)	(22)	(24)	
2000	27.6	28.0	27.8	28.1	28.9	28.0	27.6	29.7	32.4	32.6	29.1	26.5	**28.9**

Table IV-2 **(cont.)**

	Jan	Feb	Mar	Apr	May	Jun	Jul	Aug	Sep	Oct	Nov	Dec	mean
	(31)	(29)	(31)	(30)	(31)	(30)	(31)	(31)	(30)	(31)	(30)	(31)	
2001	26.3	27.2	27.1	27.2	27.6	26.6	26.7	28.4	29.3	NA	NA	26.2	**27.3**
	(31)	(28)	(31)	(30)	(31)	(30)	(31)	(30)	(30)	(0)	(0)	(30)	
2002	25.6	27.5	26.7	28.0	28.0	29.2	28.4	29.5	32.0	32.0	26.5	28.2	**28.5**
	(30)	(28)	(31)	(30)	(30)	(30)	(31)	(31)	(29)	(31)	(30)	(31)	
2003	29.2	29.0	28.5	28.7	28.5	27.3	27.3	29.1	32.0	31.7	29.6	28.8	**29.1**
	(31)	(28)	(31)	(30)	(31)	(30)	(31)	(31)	(30)	(30)	(29)	(31)	
2004	28.9	29.1	29.0	27.4	29.3	28.4	30.1	31.4	31.7	33.8	28.4	29.0	**29.7**
	(31)	(29)	(31)	(30)	(31)	(30)	(31)	(31)	(30)	(30)	(30)	(31)	
2005	30.2	31.3	30.6	30.6	29.4	32.6	31.8	32.1	33.3	33.6	31.1	28.6	**31.3**
	(31)	(28)	(31)	(30)	(31)	(30)	(31)	(30)	(30)	(31)	(30)	(31)	
2006	29.7	31.3	30.9	29.0	28.6	28.9	28.4	30.0	32.5	32.6	26.0	28.4	**29.7**
	(30)	(28)	(31)	(30)	(31)	(30)	(31)	(31)	(30)	(31)	(30)	(31)	
2007	28.9	29.4	28.7	27.9	27.6	26.2	27.2	28.8	31.1	29.8	26.6	25.9	**28.2**
	(31)	(28)	(31)	(30)	(31)	(30)	(31)	(31)	(30)	(31)	(30)	(31)	
2008	26.0	26.4	26.3	26.2	27.5	26.5	26.6	28.0	29.4	29.0	26.3	26.8	**27.1**
	(31)	(29)	(31)	(30)	(29)	(30)	(31)	(29)	(30)	(31)	(30)	(31)	
2009	27.3	28.3	28.4	26.6	26.1	26.7	26.3	28.0	29.8	30.9	28.3	27.3	**27.8**
	(31)	(27)	(31)	(30)	(31)	(30)	(31)	(31)	(30)	(29)	(30)	(20)	
2010	28.2	28.2	26.6	27.2	27.5	27.1	27.2	28.7	31.5	31.5	29.0	28.0	**28.4**
	(31)	(28)	(31)	(30)	(31)	(30)	(31)	(31)	(30)	(31)	(30)	(31)	
2011	27.0	27.8	26.5	27.5	27.0	27.2	27.6	28.3	29.6	28.5	26.4	25.8	**27.4**
	(31)	(28)	(29)	(29)	(30)	(30)	(31)	(31)	(29)	(31)	(30)	(29)	
2012	27.0	29.6	27.6	31.1	27.5	26.5	26.4	27.4	29.6	30.7	26.4	26.4	**28.0**
	(28)	(29)	(31)	(30)	(31)	(30)	(31)	(31)	(30)	(31)	(30)	(31)	
2013	27.6	29.7	27.6	27.1	26.7	26.2	27.2	29.9	32.0	32.5	29.5	27.8	**28.6**
	(31)	(28)	(31)	(30)	(31)	(30)	(31)	(31)	(30)	(31)	(30)	(30)	
Mean	**27.8**	**28.6**	**28.2**	**28.4**	**28.8**	**28.9**	**28.7**	**29.9**	**31.3**	**31.0**	**28.2**	**27.3**	

Figures in parentheses = *N*. Figures in italics are values with small sample size (<15 days).

Table IV-3 **Monthly mean daily minimum temperature (degree C) at Kansyana, 1983–2013**

	Jan	Feb	Mar	Apr	May	Jun	Jul	Aug	Sep	Oct	Nov	Dec	mean
1983	19.3	19.9	19.8	19.9	19.5	17.5	16.8	17.5	17.6	18.7	18.5	18.4	18.6
	(30)	(27)	(31)	(30)	(28)	(30)	(31)	(30)	(29)	(31)	(30)	(31)	
1984	18.4	18.7	18.7	18.8	18.0	16.3	16.3	16.3	17.2	18.6	18.6	18.6	17.9
	(31)	(29)	(31)	(30)	(27)	(20)	(28)	(28)	(30)	(30)	(29)	(31)	
1985	18.8	18.7	19.0	19.1	18.3	16.2	15.1	16.0	17.8	*19.2*	NA	NA	17.8
	(30)	(27)	(29)	(29)	(30)	(27)	(26)	(17)	(30)	(5)	(0)	(0)	
1986	18.9	18.3	18.5	19.3	18.6	*15.6*	15.0	*15.5*	17.4	19.8	19.3	19.0	17.9
	(29)	(27)	(30)	(30)	(25)	(14)	(17)	(5)	(30)	(18)	(20)	(26)	
1987	18.7	19.0	19.3	19.4	19.4	16.5	15.9	17.3	19.0	NA	19.3	19.1	18.4
	(31)	(18)	(28)	(29)	(29)	(28)	(23)	(29)	(27)	(0)	(28)	(29)	
1988	19.0	19.2	19.3	19.6	18.1	17.4	16.5	17.4	17.8	19.1	18.8	18.5	18.4
	(30)	(28)	(31)	(28)	(28)	(25)	(31)	(31)	(30)	(31)	(29)	(31)	
1989	18.9	18.7	18.7	18.9	18.5	16.8	15.9	16.5	17.6	19.2	18.9	18.9	18.1
	(31)	(26)	(29)	(30)	(31)	(29)	(30)	(31)	(30)	(29)	(29)	(31)	
1990	18.8	18.9	19.1	19.4	18.8	NA	NA	17.1	18.1	18.7	18.9	18.6	18.6
	(30)	(28)	(30)	(30)	(31)	(0)	(0)	(23)	(30)	(31)	(30)	(30)	
1991	19.5	18.9	19.2	19.0	19.4	16.9	16.0	16.7	17.4	18.2	18.4	18.9	18.2
	(22)	(28)	(31)	(29)	(31)	(30)	(31)	(31)	(29)	(30)	(20)	(24)	
1992	18.7	NA	NA	19.8	18.9	17.5	15.9	16.0	17.6	18.8	19.0	18.8	18.1
	(30)	(0)	(0)	(30)	(31)	(30)	(31)	(31)	(30)	(31)	(30)	(31)	
1993	19.0	18.7	18.4	19.5	18.9	17.2	16.0	16.1	16.5	19.1	19.7	19.4	18.2
	(17)	(28)	(29)	(28)	(28)	(29)	(29)	(30)	(30)	(28)	(29)	(30)	
1994	NA	19.3	19.2	19.5	19.3	16.8	16.4	17.1	18.9	19.6	19.3	19.3	18.6
	(0)	(28)	(25)	(27)	(31)	(30)	(31)	(31)	(30)	(31)	(30)	(31)	
1995	19.5	19.2	19.4	19.5	18.8	18.3	16.9	17.3	17.8	19.2	19.4	18.9	18.7
	(31)	(28)	(31)	(30)	(29)	(29)	(31)	(31)	(30)	(28)	(30)	(31)	
1996	18.9	19.1	19.3	19.3	19.3	*17.3*	16.0	16.1	18.2	19.2	19.5	19.1	18.5
	(27)	(18)	(29)	(29)	(15)	(4)	(30)	(26)	(30)	(31)	(30)	(31)	
1997	19.2	19.5	19.6	19.2	18.6	17.9	16.9	17.0	18.3	19.8	19.8	19.4	18.8
	(31)	(28)	(31)	(30)	(31)	(27)	(26)	(30)	(26)	(31)	(29)	(28)	
1998	19.8	20.2	20.4	20.5	19.1	17.3	16.7	17.9	19.2	19.7	19.5	18.8	19.1
	(31)	(27)	(30)	(30)	(31)	(30)	(30)	(30)	(29)	(30)	(30)	(31)	
1999	18.7	19.4	19.3	19.6	19.0	17.1	16.7	18.0	18.6	19.2	19.0	19.2	18.6
	(16)	(28)	(31)	(30)	(31)	(30)	(31)	(31)	(30)	(31)	(22)	(24)	
2000	18.8	18.6	18.9	19.2	18.3	17.2	15.9	17.7	18.7	19.7	19.3	19.5	18.5
	(30)	(28)	(31)	(29)	(31)	(30)	(31)	(31)	(28)	(29)	(28)	(31)	
2001	19.1	19.2	19.2	19.7	19.4	17.5	17.2	17.5	18.4	NA	NA	19.9	18.7
	(31)	(28)	(31)	(30)	(31)	(30)	(31)	(29)	(29)	(0)	(0)	(31)	

Table IV-3 **(cont.)**

	Jan	Feb	Mar	Apr	May	Jun	Jul	Aug	Sep	Oct	Nov	Dec	mean
2002	19.3	19.8	20.0	20.1	18.3	16.6	17.1	17.3	18.8	19.9	18.8	19.3	**18.8**
	(31)	(28)	(31)	(30)	(30)	(30)	(31)	(31)	(29)	(31)	(30)	(31)	
2003	19.7	19.5	19.6	19.6	19.1	18.0	16.3	17.0	18.8	19.4	19.6	19.5	**18.8**
	(31)	(28)	(31)	(30)	(31)	(30)	(31)	(31)	(30)	(30)	(30)	(31)	
2004	19.7	19.6	20.1	19.8	17.7	16.4	15.6	17.1	18.2	19.2	19.6	19.6	**18.6**
	(31)	(29)	(31)	(30)	(31)	(30)	(31)	(31)	(30)	(31)	(30)	(31)	
2005	19.3	20.0	20.4	20.4	19.0	17.7	16.1	17.4	18.4	20.4	18.8	18.7	**18.9**
	(31)	(28)	(31)	(29)	(31)	(28)	(31)	(30)	(28)	(24)	(28)	(27)	
2006	18.7	19.3	18.7	18.8	18.1	16.1	15.8	16.7	17.8	19.9	19.4	19.3	**18.2**
	(30)	(28)	(31)	(28)	(30)	(30)	(30)	(31)	(30)	(30)	(30)	(29)	
2007	19.4	19.6	19.3	19.9	19.4	17.3	17.2	17.2	18.7	19.3	19.2	18.9	**18.8**
	(30)	(27)	(31)	(29)	(30)	(29)	(30)	(31)	(29)	(29)	(30)	(31)	
2008	19.0	18.8	18.6	18.9	18.1	16.2	15.4	16.9	17.7	18.8	18.8	18.6	**18.0**
	(31)	(29)	(31)	(30)	(28)	(30)	(31)	(29)	(30)	(31)	(30)	(31)	
2009	18.3	18.4	18.1	17.6	17.5	17.3	15.7	16.9	17.6	19.4	19.6	18.7	**17.9**
	(31)	(27)	(31)	(30)	(31)	(28)	(31)	(31)	(28)	(28)	(29)	(20)	
2010	19.3	19.6	19.3	19.9	19.6	17.7	16.4	16.0	18.1	18.8	18.5	18.5	**18.5**
	(31)	(25)	(31)	(28)	(31)	(30)	(31)	(31)	(30)	(31)	(30)	(31)	
2011	18.6	18.5	18.3	18.9	18.6	17.1	16.1	17.0	18.4	19.0	18.8	18.9	**18.2**
	(31)	(27)	(28)	(30)	(29)	(30)	(30)	(31)	(29)	(30)	(30)	(29)	
2012	19.5	18.6	18.8	19.0	18.5	16.9	15.8	16.6	18.7	19.7	19.3	19.2	**18.4**
	(31)	(29)	(31)	(30)	(31)	(30)	(31)	(31)	(30)	(31)	(30)	(31)	
2013	19.5	19.3	19.5	19.4	18.6	16.1	15.4	17.0	19.3	19.5	19.3	18.9	**18.5**
	(31)	(28)	(30)	(30)	(31)	(30)	(31)	(30)	(30)	(31)	(29)	(30)	
mean	**19.1**	**19.2**	**19.2**	**19.4**	**18.7**	**17.0**	**16.2**	**16.9**	**18.1**	**19.3**	**19.1**	**19.0**	

Figures in parentheses = *N*. Figures in italics are values with small sample size (<15 days).

References

Itoh, N., Nakamura, M., Ihobe, H., *et al.* (2012). Long-term changes in the social and natural environments surrounding the chimpanzees of the Mahale Mountains National Park. In *The Ecological Impact of Long-Term Changes in Africa's Rift Valley*, ed. A. J. Plumptre. New York: Nova Science, pp. 249–77.

Takasaki, H., Nishida, T., Uehara, S., *et al.* (1990). Summary of meteorological data at Mahale research camps, 1973–1988. *In The Chimpanzees of the Mahale Mountains: Sexual and Life History Strategies*, ed. T. Nishida. Tokyo: University of Tokyo Press, pp. 291–300.

APPENDIX V

List of researchers who visited Mahale

COMPILED BY MICHIO NAKAMURA

Names in parentheses are visiting researchers. Individuals' first visits to Mahale are shown in bold face. Researchers who stayed for different years in one visit are shown for both years. As well as the visitors in this list, two Tanzanian professors, Hosea Y. Kayumbo and Kapepwa Tambila, have also visited Mahale. Tanzanian researchers from MMWRC, M-GWRC, TAWIRI, MMNP, and TANAPA are not shown in this appendix.

Year	Names of researchers
1965	**Kosei Izawa, Toshisada Nishida**
1966	T. Nishida, **Jun'ichiro Itani**
1967	T. Nishida, **(Takayoshi Kano)**, (Izawa)
1968	T. Nishida
1969	T. Nishida, Itani, **Kenji Kawanaka**
1970	Kawanaka
1971	T. Nishida, Itani, **Makoto Kakeya, (Richard W. Wrangham), (M. Hunkey)**
1972	Itani, Kawanaka, Kakeya, **Akio Mori, (Reizo Harako), (Jun Takeda)**
1973	T. Nishida, **Shigeo Uehara**
1974	Itani, **William C. McGrew, Caroline E. G. Tutin**
1975	T. Nishida, **Kohshi Norikoshi, Masato Kawabata, Hidemi Ishida, (Jiro Tanaka)**
1976	T. Nishida, Uehara, Norikoshi
1977	T. Nishida, Uehara, Norikoshi, **(Yukimaru Sugiyama), (Vernon Reynolds), (C Peters), (E O'Brien)**
1978	Kawanaka, Uehara
1979	T. Nishida, Itani, Kawanaka, **Mariko Hiraiwa-Hasegawa, Yukio Takahata, Masami Kimura, Toshikazu Hasegawa**
1980	Itani, Kakeya, Hiraiwa-Hasegawa, Hasegawa
1981	T. Nishida, Hiraiwa-Hasegawa, Takahata, Hasegawa, **Hitoshige Hayaki, Hiroya Kawanabe, Tetsuya Narita, Kenji Takamura**
1982	McGrew, Hiraiwa-Hasegawa, Takahata, Hasegawa, **D. Anthony Collins**
1983	T. Nishida, Kawanaka, Uehara, Takahata, Hayaki, **Hiroyuki Takasaki, Michio Hori, Tetsuo Kuwawura**
1984	Hiraiwa-Hasegawa, Takahata, Takasaki, **Richard W. Byrne, Jennifer M. Byrne, (J. David Bygott), (Jeannette P. Hanby)**
1985	T. Nishida, Hayaki, **Michael A. Huffman, Ken'ichi Masui, Juichi Itani**
1986	Kawanaka, Masui, **Kevin Hunt, (Itaru Ohta), (Satoshi Kobayashi)**

Year	Names of researchers
1987	T. Nishida, Uehara, Takasaki, Huffman, Masui, Hunt, R. Olomi, Takahiro Tsukahara
1988	McGrew, Hasegawa, Hayaki, Collins, Takasaki, Satoshi Kobayashi
1989	T. Nishida, Huffman, Tsukahara, Kobayashi, Miya Hamai, Linda A. Turner, John C. Mitani, Masaki Nishida
1990	Kawanaka, Hamai, Mitani, (Kaori Kawai)
1991	T. Nishida, Takasaki, Huffman, Hamai, Kozo Yoshida, Kazuhiko Hosaka, Akiko Matsumoto-Oda, Hiroko Yoshida, Koichi Koshimizu, Hajime Ohigashi, Mikio Kaji, (Barbara Smuts), (Craig B. Stanford)
1992	T. Nishida, Uehara, Norikoshi, Hayaki, Mitani, Hosaka, Matsumoto-Oda, Yoshida (C. Uhlenbrock)
1993	Kawanaka, Huffman, Hosaka, Matsumoto-Oda, (M. Nishida), (Kakeya), (Juichi Itani) Sarah Partan
1994	T. Nishida, Huffman, Turner, Mitani, Hosaka, Matsumoto-Oda, Fumio Fukuda, Michio Nakamura, (Hideshi Ogawa)
1995	T. Nishida, Norikoshi, Huffman, Koshimizu, Ohigashi, Fukuda, Hiroshi Ihobe, Noriko Itoh, Hitoshi Sasaki
1996	Kawanaka, Uehara, McGrew, Huffman, Fukuda, Nakamura, Itoh, Sasaki, Linda F. Marchant, Richard Magoti, (Gen Yamakoshi), (N. Koyama)
1997	T. Nishida, Nakamura, Ihobe, Itoh, Hideo Nigi, Tetsuya Sakamaki, E. B. Knox
1998	Kawanaka, Norikoshi, Hayaki, Matsumoto-Oda, Itoh, Sakamaki, Nadia Corp
1999	T. Nishida, Huffman, Koshimizu, Ohigashi, Sakamaki, Corp, Koichiro Zamma, Christophe Boesch
2000	T. Nishida, Uehara, Hosaka, Nakamura, Sakamaki, Zamma, James V. Wakibara, Nobuyuki Kutsukake, Takahisa Matsusaka, (Juichi Yamagiwa)
2001	T. Nishida, Kawanaka, Uehara, Hosaka, Nakamura, Wakibara, Kutsukake, Matsusaka, Shiho Fujita, Masaki Shimada, (Michael L. Wilson)
2002	T. Nishida, Uehara, Nakamura, Ihobe, Matsusaka, Shimada, Hitonaru Nishie, Masashi Nakai, (Izawa)
2003	T. Nishida, Huffman, Nakamura, Zamma, Matsusaka, Nishie, Eiji Inoue, (Frans B. M. de Waal)
2004	T. Nishida, Huffman, Nakamura, Sasaki, Zamma, Fujita, Nishie, E. Inoue
2005	T. Nishida, Hosaka, Nakamura, Itoh, Sakamaki, E. Inoue, Mariko Fujimoto, James Swenson, Shunkichi Hanamura
2006	Nakamura, Itoh, Sakamaki, Zamma, Fujimoto, Hanamura, Mieko Kiyono
2007	T. Nishida, Nakamura, Itoh, Sakamaki, Zamma, Matsusaka, Kiyono, Takanori Kooriyama, Agumi Inaba, Taranjit Kaur, Jatinder Singh
2008	Hosaka, Nakamura, Zamma, Matsusaka, Shimada, Kooriyama, Kaur, Singh
2009	T Nishida, Hosaka, Nakamura, Matsusaka, Shimada, Kooriyama, Eriko Iida
2010	Nakamura, Zamma, Shimada, Nishie, Kooriyama, Iida, Takuya Matsumoto, Mai Nakashima, Takashi Hayakawa, Sana Inoue, Stefano S. K. Kaburu
2011	Nakamura, Ihobe, Itoh, Zamma, Shimada, Matsumoto, S. Inoue, Kaburu
2012	Nakamura, Itoh, Nishie, Hanamura, Matsumoto, Nobuko Nakazawa
2013	Nakamura, Zamma, Shimada, Hanamura, Matsumoto, Nakazawa
2014	Nakamura, Zamma, Shimada, Nishie, Hanamura, Kooriyama, Nakazawa, (Alejandra Pascual-Garrido), Mia Otani

APPENDIX VI

List of abbreviations of Mahale chimpanzees' names

COMPILED BY MICHIO NAKAMURA

Abbreviation	Name	Sex	Group[a]	Note[b]
AA	Amisa	♀	?→M	(≠Asha=Muminsa=MM) (≠Kasola)
AB	Abi	♀	M	
AC	Acadia	♀	M→?	
AD	Adija	♀	?→M	[=Waluhanga=LH]
AG	Agano	♂	M	
AH	Asahi	♀	M	
AI	Ai	♀	M	
AJ	Aji	♂	M	
AK	Ako	♀	M	[=Akko]
AL	Alofu	♂	M	(≠Maliamu)
AM	Amina	♀	M	
AN	Kantamba	♀	M→?	
AO	Akiko	♀	?→M	(≠Alofu=AL)
AP	Apple	♀	?→M	
AQ	Aqua	♀	M→?	
AR	Argo	♂	M	
AS	Ashura	♀	K→M→?	[=Ashula]
AT	Athena	♀	M→?	
AY	Ayu	♀	M	
AZ	Azam	♂	M	
BA	Bakali	♂	M	
BB	Bonobo	♂	M	
BD	Badiri	♀	?→M	
BE	Bembe	♂	M	
BG	Bongo	♂	M	
BH	Brahim	♂	M	(≠Wabuhansa=Mpoko=MP)
BI	Binti	♀	M	

Abbreviation	Name	Sex	Group[a]	Note[b]
BL	Blondy	♀	M	
BO	Wansombo	♀	?→M	[=Wakanto=Bushy]
BR	Brutus	♂	M	
BT	Betty	♀	?→M	
BU	Bunde	♀	M	
CA	Calliope	♀	?→M	[=Caliope=Choko]
CB	Crab	♀	M→?	[=Poppo]
CC	Cocos	♀	M	
CD	Cadmus	♂	M	
CE	Caesar	♂	M	
CH	Chausiku	♀	K→M	
CI	Cissy	♀	M	
CL	Charles	♂	M	
CM	Mapinduzi	♀	M	[=Mapinduji]
CP	Chopin	♂	M	
CR	Carmen	♀	M→?	
CS	Chelsey	♀	M	
CT	Carter	♂	M	
CU	Chupi	♂	M	
CY	Cynthia	♀	?→M	
DA	Dar	♀	?→M	[=Wasoje]
DB	Derby	♀	M→?	[=Silvy]
DC	Deco	♀	M	
DE	Kalunde	♂	M	
DG	Dogura	♂	M	
DL	WD's child	♀	M	
DO	Dodoma	♀	?→M	
DR	Dora	♀	M	
DV	Devota	♀	?→M	
DW	Darwin	♂	M	
EA	Eagnes	♀	?→M	[=Agnes]
EB	Ebana	♀	M→?	[=Banana]
EC	Echo	♂	M	
ED	Edna	♀	M	
EF	Effie	♀	?→M	
EJ	Enju	♀	?→M	
EL	Elsa	♀	M	

Abbreviation	Name	Sex	Group[a]	Note[b]
EM	Emory	♂	M	
FA	Fanta	♀	?→M	[=Lenga]
FB	Fable	♀	M	[=Lemon]
FC	Fancy	♂	M	[=Lime]
FD				[=Faida=Lwela=LW]
FG	Figaro	♂	M	
FH	Fahari	♀	?→M	
FJ	Fuji	♀	?→M	
FL	Flora	♀	M	
FM	Fimbi	♀	M	
FN	Fanana	♂	M	
FR	Fairy	♀	M	
FT	Fatuma	♀	?→M	
FU	Fususi	♀	M	
FV	Flavia	♀	M→?	
FY	FA's child	♀	M	
GA	Ganmo	♀	?→M	[=Gwamatiwi=Gwamodoki]
GB	Geebee	♀	M	[=Gun](≠Garbo=GR)
GI	Gwekipala			(≠Gwamatwi=Ganmo=GA)
GK	Gwakakumo	♀	?→M	[=Wakakumo]
GM	Gwamwami	♀	?→K→M	
GN	Genie	♀	?→M	
GP	Gwepala	♀	?→M	[=Mpela]
GR	Garbo	♀	M	
GW	Gwekulo	♀	?→K→M	
HA	Happy	♀	?→M	
HB	Hanby	♂	M	
HI	Hideko	♀	?→M	
HJ	Hadija	♀	?→M	
HM	Hamadi			(≠Humbe=HU)
HO	Wakaluhola	♀	?→M	
HR	Hiro	♂	M	
HT	Hit	♂	M	
HU	Humbe	♂	M	
HW				[=Hawazi=HZ]
HZ	Hawazi	♂	M	

Abbreviation	Name	Sex	Group[a]	Note[b]
IA	Imelda	♀	?→M	
IB	Ibo	♀	M→?	
IC	Icon	♀	M→?	
ID				[=Indila=Ako=AK]
IH	Ichiro	♂	M	
II	Mimi	♀	M→?	
IK	Ikocha	♀	?→M	
IM	Imani	♀	M→?	
IR	Iris	♀	M	
IS	Isisu	♀	M	
IV	Ivana	♀	M→?	
IW	Iwan	♂	M	
JA	Wanja	♀	?→M	
JB	Jambo	♀	M	[=Kaswele]
JC	JN's child	♂	M	
JD	Jidda	♀	M→?	
JG	Kajugi	♂	M	
JH	John	♂	M	
JI	Jilba	♂	M	
JJ	Juju	♀	?→M	
JK	JN's child	♀	M	
JL				[=Jilba=JI]
JN	Juno	♀	?→M	
KA	Kasagula	♂	K	
KB	Katabi	♂	K	
KD	Kidole	♂	M	
KE	Kenji	♂	M	[=Kenta]
KG	Kaguba	♂	K	[=Kagba]
KH	Kahaso	♂	K	
KI	Kagimimi	♂	M	
KJ	Kajabala	♂	K	
KK	Keiko	♀	?→M	(≠Kenji=KE)
KL	Kalindimya	♂	M	
KM	Kamemanfu	♂	K	
KN	Kasanga	♂	K	
KO	Kabogo	♂	K	

Abbreviation	Name	Sex	Group[a]	Note[b]
KP	Koopy	♀	?→M	
KR	Kimaro	♂	M	
KS	Kasonta	♂	K	
KU				[=Sankaku=SK]
KZ	Kasangazi	♂	M	
LA	Labule	♀	?→M	
LB	Lubaju	♂	K	
LC	Lucy	♀	M	
LD	Linda	♀	?→M	
LH				[=Waluhanga=Adija=AD]
LI	Limongo	♂	K	
LJ	Lukaja	♂	M	
LK	Lukanda	♂	K	
LL	Lulemyo	♂	K→M	
LM	Lilim	♀	M	
LN	Lunkuma	♀	?→M	[=Lunkukuma]
LS	Lusia	♀	M→?	
LT	Linta	♂	M	
LU	Lubulungu	♂	M	
LW	Lwela	♀	M	[=Faida=FD]
LY	Lynso	♀	?→M	
LZ	Liz	♀	M→?	
MA	Masudi	♂	K→M	
MB	Simba			
MC	Michio	♂	M	
MD	Madina	♀	K→M	
ME	Mitsue	♀	M→?	
MG	Maggie	♀	M→?	[=Maggy] (≠Malange)
MI	Milongo	♀	K	
MJ	Miya	♀	?→M	
MK	Mimikire	♂	M	
ML	Milembe	♀	K→?	
MM	Muminsa	♀	M→?	[=Munminsa=Asha]
MN	Milinda	♀	?→M	
MO	Mao	♀	?→M	
MP	Mpoko	♀	?→M	[=Wabuhansa] (≠Mtimpu=TI)

Abbreviation	Name	Sex	Group[a]	Note[b]
MR	Mariko	♀	M→?	
MS	Masisa	♂	K	
MT	Mtintimya	♂	M	
MU	Musa	♂	M	
MW	Mwese	♀	K→?	
MY	Moyo	♀	M	[=Mpeko]
MZ	Mazuruka	♀	M→?	
NA	Nanasi	♀	?→M	
NB	Nobiko	♀	?→M	
NC	Nick	♂	M	
ND	Ndilo	♀	K→M	
NE	Nean			
NG	Wanaguma	♀	?→M	[=Mwajuma]
NI	Nishida	♂	M	
NK	Nkombo	♀	K→M	[=Nkonbo]
NN	Nina	♀	K	
NP	Nymph	♀	?→M	
NS	Nsaba	♂	M	
NT	Ntologi	♂	M	
NV	Nova	♀	?→M	
NY	Nyundo	♂	K	
OL	Omali	♂	M	
OM	Omo	♀	?→M	
OO	Momo	♀	?→M	[=Adijamodoki]
OP	Opal	♀	?→M	
OR	Orion	♂	M	
OS	Oscar	♂	M	
PA	Wantunpa	♀	M→?	[=Wamtumpa]
PE	Penny	♀	M	
PF	Puffy	♀	M	
PI	Pinky	♀	?→M	
PK				[=Pinky=PI]
PL	Plum	♀	?→M	
PM	Pimu	♂	M	[=Pim]
PN	Penelope	♀	M→?	[=Penerope]
PP	Pipi	♀	M→?	[=Pippi]

Abbreviation	Name	Sex	Group[a]	Note[b]
PR	Primus	♂	M	
PS	Pansy	♀	M	
PT	Patty	♀	?→M	
PU	Pulin	♀	?→M	
QA	Qanat	♀	?→M	
QL	Quilt	♀	M	
RA	Rashidi	♂	M	
RB	Ruby	♀	M	
RC	Rubicon	♀	M→?	
RH	Rhea	♀	M	
RJ	Rajua	♀	?→M	
RY	Ryo	♂	M	
SA	Sada	♀	K→M	
SB	Sobongo	♂	K	
SE	Serena	♀	M→?	(≠Semeni=SN)
SF	Sofia	♀	K→M	
SG	Shigeo	♂	M	
SH	Shinako	♀	M→?	
SI	Sibumba	♀	K→?	
SJ	Shinji	♂	M	
SK	Sankaku	♂	M	
SL	Silafu	♀	?→M	[=Wantinfu]
SM	Kasulamemba	♂	M	
SN	Semeni	♀	M	
SO	Wasobongo	♀	?→M	
SP	Sparta	♀	?→M	
SR	Sarah	♀	M	
SS	Sinsi	♂	M	
SU	Shike	♂	M	
SV	Silvy	♀	M	
SW	Swala	♀	M→?	
SY	Sally	♀	?→M	
TA	Tausi	♀	K→M	
TB	Toshibo	♂	M	
TD	Teddy	♂	M	
TE	Tarnie	♀	?→M	(≠Tende=Bembe=BE)

Abbreviation	Name	Sex	Group[a]	Note[b]
TG	Tango	♀	?→M	[=Waluhanga?]
TI	Mtimpu	♀	?→M	[=Wantinfu]
TL	Tula	♀	M→B	[=Tulalemwa]
TM	Tomato	♀	?→M	
TN	Tony	♂	M	
TO	Teto	♀	M	
TR	Tamara	♀	M	
TT	Tatu	♀	K→?	
TU				[=Tula=TL]
TW	Mtwale	♂	M	[=Mutwale=Muluta?]
TY	Katyentye	♀	M	[=Katientie=Kat]
TZ	Totzy	♀	M	
UA	Ua	♀	?→M	
UK	Mukoli	♂	M	
UM	Wankuma	♀	?→M	[=Gwamwego]
UP	Upepo	♀	M	
UU	Mumu	♂	M	
VD	Vandal	♂	M	
VL	Viola	♀	?→M	
VM	Vemu	♂	M	
VR	Vera	♀	?→M	
WA	Wakasunga	♀	?→K→M	
WB	Wabunengwa	♀	?→K→M	
WC	Wasiwasi	♀	M→?	
WD	Wantendele	♀	?→K→M	
WE	Wamkime	♀	?→M	[=Wamwelu]
WG	Wangulu	♀	?→K	[=Wobake]
WH	Wahanse	♀	K→?	
WI	Wakasila	♀	?→K→?	
WK	Wankungwe	♀	(?,K→)M	
WL	Wakilufya	♀	?→K→M	[=Wakiluhya]
WM	Wamikambi	♀	?→K	
WN	Wangala	♀	?→K→M	
WO	Wakampompo	♀	?→K→M	
WP	Wakapala	♀	?→K→M	
WS	Wasalamba	♀	?→K→M	

Abbreviation	Name	Sex	Group[a]	Note[b]
WT	Wantangwa	♀	?→K	
WW	Wabwema	♀	?→K	
WX	Wakusi	♀	?→M	
XM	Christmas	♂	M	[=Xmas]
XP	Xantip	♀	M	
XT	Christina	♀	?→M	[=Xtina]
YA	Yam	♀	?→M	
YB	Yuba	♀	M	
YK	Yukio	♂	M	
YN	Yuna	♀	?→M	
YR	Yuri	♀	?→M→?	
ZD	Zaida	♀	M	
ZF	Zorufa	♀	M	
ZH	Zuhura	♀	M→?	
ZL	Zola	♀	?→M	
ZM	Kazumi	♀	?→M	
ZN	Zainabu	♀	M→?	
ZO	Zoe	♀	?→M→?	
ZP	Zipu	♀	?→M	
ZW	Zawadi	♀	?→M→?	
ZZ	Zuzu	♀	?→M	

[a] An arrow shows confirmed or strongly suspected transfer between groups. For example, "K→M" means the individual was born in the K group and transferred to the M group. A question mark indicate an unknown group.

[b] A name in parentheses with ≠ indicates another individual with the same abbreviation (sometimes mistakenly used). For example, the abbreviation AA is basically used for Amisa but it was once used for Asha who was also called Muminsa (MM). A name in brackets with = indicates the same individual's different name or name in different spelling. For example, Adija (AD) was once called Waluhanga (LH), and Calliope (CA) was sometimes spelled as Caliope.

APPENDIX VII

Research and conservation funds for Mahale

COMPILED BY MICHIO NAKAMURA

Fund name and number	Year	Representative
Kakenhi (Grants-in-aid for scientific research by MEXT and JSPS)		
#X40160-94005	1965–1966	J. Itani
#X411601005	1966–1967	J. Itani
#X421601006	1967–1968	J. Itani
#X441601010	1969–1970	J. Itani
#X44060-0409	1969–1970	J. Itani
#X45060-0413	1971–1972	J. Itani
#X471018	1972–1973	J. Itani
#X481047	1973–1974	J. Itani
#X481016	1973–1974	J. Itani
#X491058	1974–1975	J. Itani
#X491036	1974–1975	T. Kano
#X521081	1977–1978	T. Kano
#X00160-404120	1979–1980	T. Nishida
#X00160-504319	1980–1981	T. Nishida
#56041018	1981–1982	T. Nishida
#57043014	1982–1983	T. Nishida
#58041025	1983–1984	T. Nishida
#59043022	1984–1985	T. Nishida
#60041020	1985–1986	T. Nishida
#62041021	1987–1988	T. Nishida
#62043067	1987–1988	T. Kano
#63043017	1988–1989	T. Nishida
#01041052	1989–1991	T. Nishida
#02041049	1990–1993	T. Kano
#03041048	1991–1993	K. Koshimizu
#03041046	1991–1994	T. Nishida

Fund name and number	Year	Representative
#04660143	1992–1994	H. Ohigashi
#06044122	1994–1995	H. Ohigashi
#06303012	1994–1996	H. Ohigashi
#062450	1994–1997	K. Hosaka
#07041139	1995–1996	H. Ohigashi
#07041138	1995–1998	T. Nishida
#08306005	1996–1998	H. Ohigashi
#08454278	1996–1998	T. Nishida
#10041184	1998–1999	K. Kawanaka
#10CE2005	1998–2003	O. Takenaka
#10440255	1998–2000	T. Nishida
#98J09099	1999–2000	K. Hosaka
#99J04368	1999–2001	N. Itoh
#99J04502	1999–2001	S. Fujita
#12740482	2000–2002	M. Nakamura
#00J09611	2000–2002	N. Kutsukake
#12640694	2000–2003	H. Ihobe
#12375003	2000–2004	T. Nishida
#12640700	2000–2004	Y. Takahata
#14604019	2002–2003	T. Nishida
#02J00985	2002–2004	M. Nakai
#15770159	2003–2005	K. Hosaka
#03J02628	2003–2005	S. Fujita
#15004835	2003–2005	K. Zamma
#16255007	2004–2007	T. Nishida
#16770186	2004–2007	M. Nakamura
#19107007	2007–2012	J. Yamagiwa
#19255008	2007–2011	T. Nishida
#19520715	2007–2009	H. Terashima
#07J02953	2007–2009	M. Shimada
#10360215	2007–2011	K. Hosaka
#20770199	2008–2011	T. Matsusaka
#21770262	2009–2012	M. Nakamura
#21820064	2009–2011	M. Shimada
#21681031	2009–2013	N. Itoh
#21255006	2009–2013	H. Ihobe

Fund name and number	Year	Representative
#24720399	2012–2014	M. Shimada
#24770234	2012–2014	T. Kooriyama
#12J00004	2012–2014	S. Hanamura
#12J04270	2012–2014	T. Hayakawa
#24255010	2012–2017	J. Yamagiwa
#20252413	2013–	H. Ihobe
#25711025	2013–2017	N. Kutsukake
#14J00963	2014–	H. Nishie
#14J00562	2014–2015	T. Matsumoto
#14J01073	2014–	N. Nakazawa
Wenner–Gren Foundation	1968	J. Ikeda
	1971	J. Itani
JSPS	1971	T. Nishida
JICA	1975–1989	See Table 50.1
	1994–1996	See Table 50.1
Sapporo University	1986–1997	S. Uehara
Primate Society of Japan	1993	M. A. Huffman
Faculty of Agriculture, Kyoto University	1993–1996, 1999	M. A. Huffman
The Wellcome Trust	1995–1997	M. A. Huffman
Mahale Wildlife Conservation Society	1994–present	T. Nishida, K. Hosaka
Philip and Elaina Hampton Fund (Miami University)	1996	M. F. Marchant, W. C. McGrew
Leaky Foundation	1997	M. A. Huffman
	1999	T. Nishida
Inamori Foundation	1997	H. Ihobe
	2009–2011	M. Nakamura
Japan Environment Cooperation	1997–2002	T. Nishida, S. Uehara
Japan Great Ape Conservation Fund	2000–present	T. Nishida, M. Nakamura
Sugiyama Jogakuen University	2000–2003, 2006, 2008, 2013	H. Ihobe
Sasakawa Scientific Research Grant from The Japan Science Society	2004–2005	M. Nakamura
	2008	E. Inoue
Inui Tasuke Memorial Fund	2004–2005	M. Nakamura
HOPE project	2004–2013	Young researchers at Primate Research Institute and Wildlife Research Center
Kobe Gakuin University	2005–2006	H. Terashima

Fund name and number	Year	Representative
Biodiversity Research of the 21st Century COE (A14)	2005–2006	N. Sato
Global Environment Research Fund (F0bl)	2006–2008	T. Nishida
GRASP–Japan	2006–2014	M. Nakamura
Global COE (F01)	2008–2012	H. Kida
The John D. and Catherine T. MacArthur Foundation (to Wildlife Conservation Society)	2010–2012	N. Itoh
The Toyota Foundation (D13-R-0577)	2013–2014	M. Nakamura
Asia-Africa Science Platform Program	2013–2014	G. Idani
Embassy of Japan in Tanzania	1992, 2001, 2010	MMWRC, MWCS, Kigoma Region, respectively

INDEX OF CHIMPANZEES' NAMES

SUBJECT INDEX